INTERNATIONAL BUILDING CODE®

CODE AND COMMENTARY

VOLUME II

2006

INTERNATIONAL
CODE COUNCIL®

First Printing: September 2006
Second Printing: March 2007
Third Printing: July 2007

ISBN -13: 978-1-58001-480-9
ISBN -10: 1-58001-480-1

PRINTED IN THE U.S.A.

PREFACE

The principal purpose of the Commentary is to provide a basic volume of knowledge and facts relating to building construction as it pertains to the regulations set forth in the 2006 *International Building Code*. The person who is serious about effectively designing, constructing and regulating buildings and structures will find the Commentary to be a reliable data source and reference to almost all components of the built environment

As a follow-up to the *International Building Code*, we offer a companion document, the *International Building Code Commentary—Volume II*. Volume II covers Chapters 16 through 35 of the 2006 *International Building Code*. The basic appeal of the Commentary is thus: it provides in a small package and at reasonable cost thorough coverage of many issues likely to be dealt with when using the *International Building Code* — and then supplements that coverage with historical and technical background. Reference lists, information sources and bibliographies are also included.

Throughout all of this, strenuous effort has been made to keep the vast quantity of material accessible and its method of presentation useful. With a comprehensive yet concise summary of each section, the Commentary provides a convenient reference for regulations applicable to the construction of buildings and structures. In the chapters that follow, discussions focus on the full meaning and implications of the code text. Guidelines suggest the most effective method of application, and the consequences of not adhering to the code text. Illustrations are provided to aid understanding; they do not necessarily illustrate the only methods of achieving code compliance.

The format of the Commentary includes the full text of each section, table and figure in the code, followed immediately by the commentary applicable to that text. At the time of printing, the Commentary reflects the most up-to-date text of the 2006 *International Building Code*. As stated in the preface to the *International Building Code,* the content of sections in the code which begin with a letter designation (i.e., Section 3401) are maintained by another code development committee. Each section's narrative includes a statement of its objective and intent, and usually includes a discussion about why the requirement commands the conditions set forth. Code text and commentary text are easily distinguished from each other. All code text is shown as it appears in the *International Building Code*, and all commentary is indented below the code text and begins with the symbol ❖.

Readers should note that the Commentary is to be used in conjunction with the *International Building Code* and not as a substitute for the code. **The Commentary is advisory only;** the code official alone possesses the authority and responsibility for interpreting the code.

Comments and recommendations are encouraged, for through your input, we can improve future editions. Please direct your comments to the Codes and Standards Development Department at the Chicago District Office.

The International Code Council would like to extend its thanks to the following individuals for their contributions to the technical control of this commentary:

Chris Marion   Gregory Cahanin
Jeff Tubbs   Rebecca Quinn
Joann Surmar   Vickie Lovell
James Milke   John Valiulis
Richard Walke   Marcelo Hirschler
Dave Adams   Edward Keith
Zeno Martin   Phillip Samblanet
Jason Thompson

TABLE OF CONTENTS

Chapter 16:
Structural Design

General Comments

This chapter contains the commentary for the following structural topics: definitions of structural terms, construction document requirements, load combinations, dead loads, live loads, snow loads, wind loads, soil lateral loads, rain loads, flood loads and earthquake loads.This chapter provides minimum design requirements so that all buildings and structures are proportioned to resist the loads and forces that are likely to be encountered. The loads specified herein have been established through research and service performance of buildings and structures. The application of these loads and adherence to the serviceability criteria will enhance the protection of life and property. The earthquake loads, wind loads and snow loads in this chapter are based on the 2005 edition of ASCE 7. The earthquake criteria and ASCE 7 load requirements are based on the National Earthquake Hazards Reduction Program's (NEHRP) *Recommended Provisions for the Development of Seismic Regulations for New Buildings* (FEMA 450). The NEHRP provisions were prepared by the Building Seismic Safety Council (BSSC) for the Federal Emergency Management Agency (FEMA).

Purpose

The purpose of this chapter is to prescribe minimum structural loading requirements for use in the design and construction of buildings and structures with the intent to minimize hazard to life and improve the occupancy capability of essential facilities after a design level event or occurrence.

SECTION 1601
GENERAL

1601.1 Scope. The provisions of this chapter shall govern the structural design of buildings, structures and portions thereof regulated by this code.

❖ While a significant portion of Chapter 16 is dedicated to the determination of minimum design loads, it also includes other important criteria that impact the design of structures, such as the permitted design methodologies as well as the combinations of design loads used to establish the required minimum strength of structural members. Unless stated otherwise, the criteria found in this chapter are applicable to all buildings and structures. See Chapter 34 for application of these requirements to alterations, additions or repairs to existing structures.

SECTION 1602
DEFINITIONS AND NOTATIONS

1602.1 Definitions. The following words and terms shall, for the purposes of this chapter, have the meanings shown herein.

❖ Certain terms are used in the code to describe something very specific to the related subject matter while other terms may have multiple meanings that can vary based on the subject matter and context. To avoid misunderstandings, it is preferable to reach a consensus on the meaning of these terms. Definitions are intended to facilitate the understanding of code provisions and to minimize potential confusion.

This section contains definitions of terms associated with structural design and minimum load criteria. Note that these terms are also listed in Chapter 2 with a cross reference to this section. Definitions that pertain primarily to structural design loads are included within this chapter to provide convenient access to them without having to refer back to Chapter 2. The use and application of all defined terms, including those defined herein, as well as undefined terms are set forth in Section 201.

ALLOWABLE STRESS DESIGN. A method of proportioning structural members, such that elastically computed stresses produced in the members by nominal loads do not exceed specified allowable stresses (also called "working stress design").

❖ This definition describes the allowable stress design method, which is one of the design approaches recognized under the code in Section 1604.1. In this approach, the computed stresses determined from the unfactored, or "nominal loads" (see definition below) cannot exceed allowable stresses, which provide a factor of safety. The material chapters of the code specify either the allowable stresses for a given material or in some cases provide allowable capacities of specific assemblies, such as the wood structural panel shear wall capacities specified in Chapter 23.

BALCONY, EXTERIOR. An exterior floor projecting from and supported by a structure without additional independent supports.

❖ This definition is provided to differentiate live loads between a balcony and a deck, based on the likelihood of different live loads being applied.

DEAD LOADS. The weight of materials of construction incorporated into the building, including but not limited to walls, floors, roofs, ceilings, stairways, built-in partitions, finishes, cladding and other similarly incorporated architectural and structural items, and the weight of fixed service equipment, such as cranes, plumbing stacks and risers, electrical feeders, heating, ventilating and air-conditioning systems and fire sprinkler systems.

❖ The definition of "Dead loads" identifies the type of items that must be accounted for. Also see Section 1607.11.2.3 for inclusion of items on landscaped roofs. This definition is necessary to distinguish dead loads from other loads, and for use in load combinations, as specified by Section 1605. The nominal dead load is to be determined in accordance with Section 1606.

The weights of service equipment, such as plumbing stacks and risers; heating, ventilating and air-conditioning (HVAC) equipment; elevators and elevator machinery; fire protection systems and similar fixed equipment are to be included in the dead load. For the most part tracking the weights of each utility system is not practical and the structural design is therefore based on a dead load allowance for these items. At times the actual weight of equipment to be installed is unknown during the design phase of a building because the supplier of the equipment has yet to be determined. The structural design must often proceed based on an estimated equipment dead load. For additional comments on dead load estimates, see the commentary to Section 1606.2.

DECK. An exterior floor supported on at least two opposing sides by an adjacent structure, and/or posts, piers or other independent supports.

❖ This definition is provided to differentiate live loads between a balcony and a deck, based on the different live loads specified in Table 1607.1.

DESIGN STRENGTH. The product of the nominal strength and a resistance factor (or strength reduction factor).

❖ This definition is needed to apply the strength design requirements in the code. The design strength is the nominal strength multiplied by a resistance or strength reduction factor that is less than one. The design strength and corresponding strength reduction factors are specified in the applicable material chapter of the code or a standard that is referenced therein.

DIAPHRAGM. A horizontal or sloped system acting to transmit lateral forces to the vertical-resisting elements. When the term "diaphragm" is used, it shall include horizontal bracing systems.

❖ Floor and roof diaphragms act to transfer the lateral forces due to wind or seismic loads to the vertical-resisting elements, such as shear walls or moment frames supporting them at their perimeter or intermittent locations.

Diaphragm, blocked. In light-frame construction, a diaphragm in which all sheathing edges not occurring on a framing member are supported on and fastened to blocking.

❖ Blocked diaphragms are horizontal or nearly horizontal assemblies designed to resist high shear forces in light-frame construction. Diaphragm sheathing may be applied with its long dimension either perpendicular or parallel to the main framing members. When the edge of the sheathing is not supported by the main framing member, it is considered to be unblocked. Blocking is accomplished by installing a framing member parallel to the otherwise unsupported edge of the sheathing. In some cases, the code may require blocking, while in other cases, the capacity of a diaphragm assembly will vary based on whether it is blocked or unblocked.

Diaphragm boundary. In light-frame construction, a location where shear is transferred into or out of the diaphragm sheathing. Transfer is either to a boundary element or to another force-resisting element.

❖ Diaphragm boundary is typically the connection between the floor or roof sheathing and the band board surrounding the diaphragm.

Diaphragm chord. A diaphragm boundary element perpendicular to the applied load that is assumed to take axial stresses due to the diaphragm moment.

❖ A diaphragm acts as a deep horizontal beam. The chords of the beam are the elements at the boundary of the diaphragm that are perpendicular to the direction of the applied load.

Diaphragm flexible. A diaphragm is flexible for the purpose of distribution of story shear and torsional moment where so indicated in Section 12.3.1 of ASCE 7, as modified in Section 1613.6.1.

❖ Because typical structural engineering design practice for flexible diaphragms differs from that of rigid diaphragms it is necessary to distinguish a rigid diaphragm from a flexible diaphragm. A flexible diaphragm deforms more than a rigid diaphragm when subjected to the same load. Where the diaphragm is determined to be flexible, the effect of diaphragm rigidity on the distribution of lateral forces is negligible and it need not be considered in the structural analysis. Load distribution from a flexible diaphragm to vertical supporting elements is usually based on the tributary area associated with those elements.

This definition is needed to apply the earthquake and wind load provisions that are referenced in this chapter. There is a direct reference to the ASCE 7 earthquake provision that addresses the classification of diaphragms as either flexible or rigid. That provision prescribes diaphragm construction that can be assumed to be flexible without the need of a substantiating calculation. Similarly, it prescribes diaphragm construction that can be assumed to be rigid. For diaphragms that do not meet either of these prescriptive classification criteria, the determination of whether a diaphragm is flexible is in accordance with the criteria of Section 12.3.1.3 of ASCE 7 that are illustrated Figure 1602.1. Note that the reference to "average story drift of adjoining vertical elements" is actually the average (relative) deflection of those elements for the story being evaluated.

Diaphragm, rigid. A diaphragm is rigid for the purpose of distribution of story shear and torsional moment when the lateral deformation of the diaphragm is less than or equal to two times the average story drift.

❖ This definition is needed to apply the earthquake and wind load provisions in this chapter. A rigid diaphragm deforms less than a flexible diaphragm when subjected to the same load. Load distribution from a rigid diaphragm is based on the stiffness of the vertical resisting elements.

DURATION OF LOAD. The period of continuous application of a given load, or the aggregate of periods of intermittent applications of the same load.

❖ An understanding of duration of load is necessary since the allowable design stresses for wood members and fasteners can be modified by a duration of load factor that is a function of the length of time that the wood member must resist an applied load. Note that it can be applied in the basic allowable stress load combinations of Section 1605.3.1 even though traditional allowable stress increases are not permitted.

ESSENTIAL FACILITIES. Buildings and other structures that are intended to remain operational in the event of extreme environmental loading from flood, wind, snow or earthquakes.

❖ This definition is needed to facilitate identification of Occupancy Category IV buildings under Section 1604.5. This is critical in determining the design earthquake, flood, snow and wind load criteria that apply to the facility.

FABRIC PARTITIONS. A partition consisting of a finished surface made of fabric, without a continuous rigid backing, that is directly attached to a framing system in which the vertical framing members are spaced greater than 4 feet (1219 mm) on center.

❖ This definition identifies which assemblies are considered fabric partitions for the purpose of applying the specific load criteria that is given in Section 1607.13.1. They are typically for uses, such as office furniture panel systems, for open floor plans in offices and similar occupancies.

FACTORED LOAD. The product of a nominal load and a load factor.

❖ This definition explains the term "factored loads" so that the loads to be applied are clear. Factored loads are determined for strength design (for load and resistance factor design) by multiplying nominal load (see definition) by a load factor, as in the load combinations of Section 1605.2. While this definition is applicable to most loads, the exception would be the earthquake load effect, E, determined in Section 1613, which is considered a strength level load without applying a load factor. This is based on the approach taken in the 1997 NEHRP provisions and is evidenced by a load factor of 1.0 being applied to E in the strength load combinations given by Equations 16-5 and 16-7.

GUARD. See Section 1002.1.

❖ This is simply a cross reference to the section where the complete definition is located. It is listed here since guard loads are addressed in this chapter.

IMPACT LOAD. The load resulting from moving machinery, elevators, craneways, vehicles and other similar forces and kinetic loads, pressure and possible surcharge from fixed or moving loads.

❖ This definition identifies the scope of the type of loading addressed in Section 1607.8. The effect of impact

Figure 1602.1
FLEXIBLE DIAPHRAGM DETERMINATION

loads on a structure can be significantly higher than the weight of the impacting elements because of their movement or vibration.

LIMIT STATE. A condition beyond which a structure or member becomes unfit for service and is judged to be no longer useful for its intended function (serviceability limit state) or to be unsafe (strength limit state).

❖ This definition is needed for a clear understanding of the load and resistance factor design methodology.

LIVE LOADS. Those loads produced by the use and occupancy of the building or other structure and do not include construction or environmental loads such as wind load, snow load, rain load, earthquake load, flood load or dead load.

❖ This definition identifies the scope of the type of loading included in Section 1607. Generally, live loads are not environmental loads or dead loads, but are transient in nature and will vary in magnitude over the life of a structure.

LIVE LOADS (ROOF). Those loads produced (1) during maintenance by workers, equipment and materials; and (2) during the life of the structure by movable objects such as planters and by people.

❖ This definition is needed for the proper application of the load combinations in this chapter. This definition clarifies that roof loads, such as snow loads, are not live loads.

LOAD AND RESISTANCE FACTOR DESIGN (LRFD). A method of proportioning structural members and their connections using load and resistance factors such that no applicable limit state is reached when the structure is subjected to appropriate load combinations. The term "LRFD" is used in the design of steel and wood structures.

❖ This definition describes the load and resistance factor design method, which is one of the design approaches recognized under the code in Section 1604.1. It is needed for the proper application of the steel design requirements in Chapter 22 and the wood design requirements in Chapter 23.

LOAD EFFECTS. Forces and deformations produced in structural members by the applied loads.

❖ This definition is needed to properly apply the structural load requirements in this chapter. "Load effects" is a collective term used to refer to the internal member forces and member deformations that result from the applied loads.

LOAD FACTOR. A factor that accounts for deviations of the actual load from the nominal load, for uncertainties in the analysis that transforms the load into a load effect, and for the probability that more than one extreme load will occur simultaneously.

❖ This definition clarifies the application of the load and resistance factor design load combinations in this chapter.

LOADS. Forces or other actions that result from the weight of building materials, occupants and their possessions, environ-

mental effects, differential movement and restrained dimensional changes. Permanent loads are those loads in which variations over time are rare or of small magnitude, such as dead loads. All other loads are variable loads (see also "Nominal loads").

❖ This definition is needed for the proper application of the structural load requirements in this chapter. It also includes a definition of "Permanent loads" that serves as the distinction between variable loads and permanent loads. This is important in Section 1605.1, which requires all load combinations to be investigated with one of more of the variable loads taken as zero.

The distinction made between variable loads and permanent loads is fairly straightforward. Dead load is specifically mentioned because it is an ideal example of a load that does not vary over time. But dead load is not necessarily the only load that could be considered permanent. An additional complication is the definition of certain loads that may include both permanent and variable components. An indicator of this is provided in the load combinations in Section 1605. For instance, in the allowable stress design basic load combinations given by Equations 16-11 and 16-13, the variable loads are reduced by a factor of 0.75. Based on the application of the reduction factor in these load combinations, live load, L, self-straining forces, T, roof live load, L_r, snow load, S, rain load, R, wind load, W, and earthquake load, E, are explicitly treated as variable loads. Since they are not reduced in these load combinations, pressure of a well-defined fluid, F, and lateral earth pressure, H, could be considered, at least in part, permanent loads.

NOMINAL LOADS. The magnitudes of the loads specified in this chapter (dead, live, soil, wind, snow, rain, flood and earthquake).

❖ This definition explains the term "nominal loads" so that the loads to be applied are clear. For the most part, the loads determined in this chapter are considered nominal loads, which are used directly in allowable stress design or multiplied by a load factor for use in strength or load and resistance factor design. The exception would be the earthquake load effect, E, determined in Section 1613, which is considered a strength level load without applying a load factor. This is based on the approach taken in the 1997 NEHRP provisions. In other words, the nominal (unfactored) earthquake load is more appropriately $0.7E$ when using the basic allowable stress design load combinations (Equations 16-12, 16-13 and 16-15) or $E/1.4$ when using the alternative basic allowable stress design load combinations (Equations 16-20 and 16-21).

OCCUPANCY CATEGORY. A category used to determine structural requirements based on occupancy.

❖ Occupancy category is a classification given to all buildings in Section 1604.5 based on the nature of the occupancy. This classification serves as a threshold for the earthquake, flood, snow and wind load requirements. Also see the definition of "Essential facilities" and the commentary for Section 1604.5.

OTHER STRUCTURES. Structures, other than buildings, for which loads are specified in this chapter.

❖ This definition is needed for application of the minimum load requirements to structures other than buildings that are included in this chapter. Note that buildings are defined in Chapter 2 of the code as "structures used to support or shelter an occupancy."

PANEL (PART OF A STRUCTURE). The section of a floor, wall or roof comprised between the supporting frame of two adjacent rows of columns and girders or column bands of floor or roof construction.

❖ This definition is needed to apply the structural load requirements in this chapter.

RESISTANCE FACTOR. A factor that accounts for deviations of the actual strength from the nominal strength and the manner and consequences of failure (also called "strength reduction factor").

❖ This definition is needed to apply the material strength adjustments by way of the resistance factor that is specified in the structural material chapters and the referenced structural standards.

STRENGTH, NOMINAL. The capacity of a structure or member to resist the effects of loads, as determined by computations using specified material strengths and dimensions and equations derived from accepted principles of structural mechanics or by field tests or laboratory tests of scaled models, allowing for modeling effects and differences between laboratory and field conditions.

❖ This definition is needed to apply the structural analysis and structural material design requirements in this chapter and the structural materials requirements in Chapters 19 through 23.

STRENGTH, REQUIRED. Strength of a member, cross section or connection required to resist factored loads or related internal moments and forces in such combinations as stipulated by these provisions.

❖ This definition is needed to apply the structural analysis requirements in this chapter and the structural material design requirements in Chapters 19 through 23.

STRENGTH DESIGN. A method of proportioning structural members such that the computed forces produced in the members by factored loads do not exceed the member design strength [also called "load and resistance factor design" (LRFD)]. The term "strength design" is used in the design of concrete and masonry structural elements.

❖ This definition describes the strength design method, which is one of the design approaches recognized under the code in Section 1604.1. It is needed to apply the structural analysis requirements in this chapter and the structural material design requirements in Chapters 19 through 23.

VEHICLE BARRIER SYSTEM. A system of building components near open sides of a garage floor or ramp or building walls that act as restraints for vehicles.

❖ This definition identifies vehicle barrier systems for the purposes of applying the minimum design loads specified in Section 1607.7.3.

NOTATIONS.

D = Dead load.

E = Combined effect of horizontal and vertical earthquake induced forces as defined in Section 12.4.2 of ASCE 7.

E_m = Maximum seismic load effect of horizontal and vertical seismic forces as set forth in Section 12.4.3 of ASCE 7.

F = Load due to fluids with well-defined pressures and maximum heights.

F_a = Flood load.

H = Load due to lateral earth pressures, ground water pressure or pressure of bulk materials.

L = Live load, except roof live load, including any permitted live load reduction.

L_r = Roof live load including any permitted live load reduction.

R = Rain load.

S = Snow load.

T = Self-straining force arising from contraction or expansion resulting from temperature change, shrinkage, moisture change, creep in component materials, movement due to differential settlement or combinations thereof.

W = Load due to wind pressure.

❖ These notations are used to refer to specific nominal loads that are determined in this chapter for use in the load combinations in Section 1605:

• D is the nominal dead load determined in Section 1606. Also see the definition of "Dead load."

• Earthquake load effect, E, in Section 12.4.2 of ASCE 7 includes the effects of the horizontal load, E_h, as well as a vertical component, E_v. E_h is the product of the redundancy factor, ρ, and Q_E, the effects of horizontal earthquake forces. E_v accounts for vertical acceleration due to earthquake ground motion, taken as $0.2S_{DS}D$. Note that its magnitude is not intended to represent a total vertical response, since that is not likely to coincide with the maximum horizontal response. It is essentially a portion of the dead load, D, that is added in "additive" load combinations or subtracted in "counteractive" load combinations. The term S_{DS}, design spectral response acceleration at short periods, is explained in the commentary to Section 1613.5.4.

For example, when this expression is used in the alternative allowable stress design load combinations of Section 1605.3.2 that include earthquake load effects the resulting combinations are as follows:

Equation 16-20
$$D + L + S + E/1.4 = (1+0.143S_{DS})D + L + S + \rho Q_E/1.4$$

Equation 16-21
$$0.9D + E/1.4 = (0.9-0.143S_{DS})D + \rho Q_E/1.4$$

Earthquake design criteria is provided in Section 1613, which, in turn, references the relevant ASCE 7 provisions for computation of the earthquake load effects. While these loads are necessary for establishing the required strength, the computed forces approximate the expected deformations under the design earthquake ground motions and are not applied to a structure in an actual earthquake.

- The maximum earthquake load effect, E_m, from Section 12.4.3 of ASCE 7 is used in the special seismic load combinations where "system overstrength" is a design consideration. See the commentary to Section 1605.4 for more information.

- F refers to the nominal load due of fluids having "well defined pressures and maximum heights." Unlike most other nominal loads, there is no code section governing the determination of fluid loads. Also note that F includes a vertical component (fluid weight), as well as a horizontal component (lateral pressure).

- F_a is used to refer to the flood load that is determined under Chapter 5 of ASCE 7. Note that Section 1612 does not directly refer to ASCE 7 for flood loads, but instead requires buildings in flood hazard areas to be designed and constructed in accordance with ASCE 24, *Flood Resistant Design and Construction*. This, in turn, refers to ASCE 7 for determination of flood loads. Note that F_a is not included under other loads listed for the alternative allowable stress design combination in Section 1605.3.2.1.

- H is used to refer to the nominal load resulting from lateral soil pressure, lateral pressure of ground water or the lateral pressure of bulk materials. Section 1610 specifies minimum requirements for lateral soil loads. Note that there are not specific provisions for the determination of load resulting from the lateral pressure of bulk materials.

- L in the nominal live load determined in accordance with Section 1607. Also see the definition of "Live load."

- L_r in the nominal roof live load determined in accordance with Section 1607.11.

- R is the nominal rain load determined in accordance with Section 1611.

- S is the nominal snow load determined in accordance with Section 1608.

- T is used to refer to self-straining forces resulting from contraction or expansion due to temperature change, shrinkage, moisture change or creep as well as movement due to differential settlement. A thermal gradient at an exterior wall is an example of a structural element where these self-straining forces can affect the design. Unlike for most other nominal loads, there is no code section governing the determination of self-straining forces.

- W is the nominal wind load determined in accordance with Section 1609.

SECTION 1603
CONSTRUCTION DOCUMENTS

1603.1 General. Construction documents shall show the size, section and relative locations of structural members with floor levels, column centers and offsets dimensioned. The design loads and other information pertinent to the structural design required by Sections 1603.1.1 through 1603.1.8 shall be indicated on the construction documents.

Exception: Construction documents for buildings constructed in accordance with the conventional light-frame construction provisions of Section 2308 shall indicate the following structural design information:

1. Floor and roof live loads.

2. Ground snow load, P_g.

3. Basic wind speed (3-second gust), miles per hour (mph) (km/hr) and wind exposure.

4. Seismic design category and site class.

5. Flood design data, if located in flood hazard areas established in Section 1612.3.

❖ The term "Construction documents" is defined in Chapter 2. It is commonly used to refer to calculations, drawings and specifications but it includes other data that is required to indicate compliance with the code as described in Section 106. The purpose of this section is to specifically require the design professional to provide the building official with the appropriate structural details, criteria and design load data for verifying compliance with the provisions of this chapter. Note that additional structural information may also be required to be incorporated by Chapters 17 through 23.

The construction documents are required to contain sufficient detail for the building official to perform plan review and field inspection, as well as for construction activity. Dimensions indicated on architectural drawings are not required to be duplicated on the structural drawings and vice versa. The design loads, to be indicated by the design professional on the construction documents, are to be consistent with the loads used in the structural calculations. Note that the loads are not required to be on the construction drawings but must be included within the construction documents in a manner such that the design loads are clear. The building official is to compare the loads on the construction documents with the applicable minimum required loads as specified by this chapter. The inclusion of the load design information is an indication that the structure has been designed for the loads required by the code. It should be emphasized that these requirements for construction documents are applicable regardless of the involvement of a registered design professional, which is regulated by the applicable state's licensing laws. The exception provides a less extensive list of structural data to be indicated for buildings constructed in accordance with the conventional wood-frame provisions of Section 2308. This is appropriate in view of the prescriptive nature of these requirements.

1603.1.1 Floor live load. The uniformly distributed, concentrated and impact floor live load used in the design shall be indicated for floor areas. Use of live load reduction in accordance with Section 1607.9 shall be indicated for each type of live load used in the design.

❖ The purpose of the requirement in this section is to provide information for the building official to facilitate the plan review process. The floor live loads, which are indicated on the construction documents by the design professional, are required to meet or exceed the loads in Section 1607. Any live load reductions taken are also to be indicated.

1603.1.2 Roof live load. The roof live load used in the design shall be indicated for roof areas (Section 1607.11).

❖ This section provides information allowing the building official to facilitate the plan review process. The roof live loads, indicated on the construction documents by the design professional, are required to meet or exceed the loads in Section 1607.11.

1603.1.3 Roof snow load. The ground snow load, P_g, shall be indicated. In areas where the ground snow load, P_g, exceeds 10 pounds per square foot (psf) (0.479 kN/m²), the following additional information shall also be provided, regardless of whether snow loads govern the design of the roof:

1. Flat-roof snow load, P_f.

2. Snow exposure factor, C_e.

3. Snow load importance factor, I.

4. Thermal factor, C_t.

❖ The roof snow load design basis, indicated on the construction documents (design drawings or specifications) by the design professional, provides information allowing the building official to facilitate the plan review process. The flat-roof snow load, snow exposure factor, snow load importance factor, I, and roof thermal factor are not to be less than the minimum requirements established by Section 1608.

1603.1.4 Wind design data. The following information related to wind loads shall be shown, regardless of whether

wind loads govern the design of the lateral-force-resisting system of the building:

1. Basic wind speed (3-second gust), miles per hour (km/hr).

2. Wind importance factor, I, and occupancy category.

3. Wind exposure, if more than one wind exposure is utilized, the wind exposure and applicable wind direction shall be indicated.

4. The applicable internal pressure coefficient.

5. Components and cladding. The design wind pressures in terms of psf (kN/m²) to be used for the design of exterior component and cladding materials not specifically designed by the registered design professional.

❖ The wind load design basis, indicated on the construction documents (design drawings or specifications) by the design professional, provides information allowing the building official to facilitate the plan review process. All five of the indicated items are to be on the submitted construction documents. Each of the indicated items is an important parameter in the determination of the wind resistance that is required in the building framework. The building official should verify that the information is on the construction documents during the plan review process. The correctness of the listed items is the responsibility of the owner or the owner's design professional.

1603.1.5 Earthquake design data. The following information related to seismic loads shall be shown, regardless of whether seismic loads govern the design of the lateral-force-resisting system of the building:

1. Seismic importance factor, I, and occupancy category.

2. Mapped spectral response accelerations, S_S and S_1.

3. Site class.

4. Spectral response coefficients, S_{DS} and S_{D1}.

5. Seismic design category.

6. Basic seismic-force-resisting system(s).

7. Design base shear.

8. Seismic response coefficient(s), C_S.

9. Response modification factor(s), R.

10. Analysis procedure used.

❖ The earthquake load design basis, indicated on the construction documents by the design professional, provides information that allows the building official to facilitate the plan review process. All buildings, except those indicated in the exceptions to Section 1613.1, are to be designed for earthquake effects. The earthquake design data for a specific building are required to meet or exceed the minimum requirements established by Section 1613.

1603.1.6 Flood design data. For buildings located in whole or in part in flood hazard areas as established in Section 1612.3, the documentation pertaining to design, if required in Section 1612.5, shall be included and the following information, referenced to the datum on the community's Flood Insurance Rate Map (FIRM), shall be shown, regardless of whether flood loads govern the design of the building:

1. In flood hazard areas not subject to high-velocity wave action, the elevation of the proposed lowest floor, including the basement.

2. In flood hazard areas not subject to high-velocity wave action, the elevation to which any nonresidential building will be dry floodproofed.

3. In flood hazard areas subject to high-velocity wave action, the proposed elevation of the bottom of the lowest horizontal structural member of the lowest floor, including the basement.

❖ The flood hazard elevation information to be shown on the construction documents by the design professional provides information allowing the building official to facilitate the plan review process. By providing the design documentation required in Section 1612.5 and by citing the specified flood information, the design professional is indicating that the building was designed in accordance with the flood hazard requirements in Section 1612. If any portion of a building is in a flood hazard area then the building must meet the corresponding flood requirements.

Depending on the nature of the designated flood hazard area, certain elevation requirements are to be met. In flood hazard areas not subject to high-velocity wave action (commonly called A/AE Zones), the lowest floor of all buildings and structures, or the elevation to which a nonresidential building is dry floodproofed, is to be located at or above the design flood elevation (DFE). In flood hazard areas subject to high-velocity wave action (commonly called V/VE Zones), the bottom of the lowest horizontal member of buildings and structures is to be located at or above the DFE. One of the most important requirements is to provide resistance to flood damage for buildings and structures located in designated flood hazard areas. It is also the main factor used in determining flood insurance premium rates for buildings and structures. Constructing a building or structure with its lowest floor (or extent of dry floodproofing) below the DFE will result in significantly higher flood insurance premiums for the building owner. It is important to note that, while not a requirement of the code, the National Flood Insurance Program (NFIP) requires dry floodproofing to extend to 1 foot (305 mm) above the DFE in order for the building or structure to receive a favorable flood insurance premium rate.

1603.1.7 Special loads. Special loads that are applicable to the design of the building, structure or portions thereof shall be indicated along with the specified section of this code that addresses the special loading condition.

❖ Indication of special loads on the construction documents by the design professional provides information allowing the building official to facilitate the plan review process. The design professional is expected to identify any special loads that the occupancy will impose on the structure. These could include the operating weight of specialty equipment, for instance. There are also instances outside of Chapter 16 where the code specifies loading criteria that the structural design must address. For example, in Section 415.7.3 liquefied petroleum gas distribution facilities that are separated from an adjacent use must have walls or floor assemblies that are designed for a static pressure of 100 pounds per square foot (psf) (4788 Pa).

1603.1.8 Systems and components requiring special inspections for seismic resistance. Construction documents or specifications shall be prepared for those systems and components requiring special inspection for seismic resistance as specified in Section 1707.1 by the registered design professional responsible for their design and shall be submitted for approval in accordance with Section 106.1. Reference to seismic standards in lieu of detailed drawings is acceptable.

❖ This section provides for construction documents for the systems and components specified in Section 1707.1. Generally, the systems and components are the seismic-force-resisting systems, certain mechanical and electrical equipment and systems and architectural components for buildings where the seismic performance category is high. Construction documents are needed for these important items to verify that they comply with code requirements.

1603.2 Restrictions on loading. It shall be unlawful to place, or cause or permit to be placed, on any floor or roof of a building, structure or portion thereof, a load greater than is permitted by these requirements.

❖ The loads that this section is referring to are the various structural loads specified in this chapter. For example, Table 1607.1 includes the minimum live loads for building design. Unless the building is designed for higher loads than specified in Table 1607.1, those values are not to be exceeded. Note that the loads in Table 1607.1 are minimum live loads. A building is permitted to be designed for higher loads, in which case the higher loads would be the limit of the actual applied loads.

1603.3 Live loads posted. Where the live loads for which each floor or portion thereof of a commercial or industrial building is or has been designed to exceed 50 psf (2.40 kN/m²), such design live loads shall be conspicuously posted by the owner in that part of each story in which they apply, using durable signs. It shall be unlawful to remove or deface such notices.

❖ This section requires that live loads be posted for most occupancies, since many of the live loads specified in Table 1607.1 exceed 50 psf (2.40 kN/m²). Where part of the floor is designed for 50 psf (2.40 kN/m²) or less and part for more than 50 psf (2.40 kN/m²), the live loads are required to be posted for those portions more than 50 psf (2.40 kN/m²). The code requires that the posting be done in the part where it applies. For example, an assembly area such as a restaurant would need to have the live load posted in the dining room.

This live load posting gives the building department easy access to the information for field verification. It also serves as a notice of the loading restriction that is stated in Section 1603.2.

1603.4 Occupancy permits for changed loads. Occupancy permits for buildings hereafter erected shall not be issued until the floor load signs, required by Section 1603.3, have been installed.

❖ The design live load signs required by Section 1603.3 need to be in place prior to the occupancy of the building for reference purposes. They serve as a record of the structural design loads for future reference, particularly when a change in occupancy is contemplated.

SECTION 1604
GENERAL DESIGN REQUIREMENTS

1604.1 General. Building, structures and parts thereof shall be designed and constructed in accordance with strength design, load and resistance factor design, allowable stress design, empirical design or conventional construction methods, as permitted by the applicable material chapters.

❖ This section identifies the various design methods that are permitted by the code and referenced design standards. The details for these design methods are either explicitly specified in the code or are located in the design standards that are referenced in the structural material design chapters for concrete, aluminum, masonry, steel and wood. For example, empirical design of masonry is addressed in Section 2109. The design of masonry using allowable stress design is required to be in accordance with ACI 530/ASCE 5/TMS 402 with the modifications in Section 2107.

1604.2 Strength. Buildings and other structures, and parts thereof, shall be designed and constructed to support safely the factored loads in load combinations defined in this code without exceeding the appropriate strength limit states for the materials of construction. Alternatively, buildings and other structures, and parts thereof, shall be designed and constructed to support safely the nominal loads in load combinations defined in this code without exceeding the appropriate specified allowable stresses for the materials of construction.

Loads and forces for occupancies or uses not covered in this chapter shall be subject to the approval of the building official.

❖ This section describes the strength and allowable stress design methods in the code. It also gives the building official approval authority for structural loads in buildings used for occupancies that are not specifically addressed in this chapter.

1604.3 Serviceability. Structural systems and members thereof shall be designed to have adequate stiffness to limit deflections and lateral drift. See Section 12.12.1 of ASCE 7 for drift limits applicable to earthquake loading.

❖ The stated objective in this section is to limit member deflections and system drifts by providing adequate stiffness. Note that the definition of "Limit state" describes a serviceability limit state as a condition be-

yond which a structure or member is no longer useful for its intended purpose. This would be in contrast to a strength limit state beyond which the structure or member is considered unsafe. The code does not directly address a serviceability limit state. Instead the code limits certain member deflections under specific nominal loads in addition to the story drift limits under earthquake loads.

The deflection limits for structural members must be in accordance with the requirements in this and subsequent sections. This section also provides a reference to the story drift limitations in the earthquake provisions. Generally, deflection limits are needed for the comfort of the building occupants and so that the structural member's deflection does not cause damage to supported construction. Excessive deflection can also contribute to excessive vibration, which is discomforting to the occupants of the building.

1604.3.1 Deflections. The deflections of structural members shall not exceed the more restrictive of the limitations of Sections 1604.3.2 through 1604.3.5 or that permitted by Table 1604.3.

❖ The deflection of structural members is limited by the code as well as certain material standards for damage control of supported construction and human comfort. Generally, the public equates visible deflection, or even detectable vibration, with a potentially unsafe condition (which in many cases is not true). The intent of this section is that deflection is not to exceed either the limitations in the applicable material design standard or the applicable specified requirements.

1604.3.2 Reinforced concrete. The deflection of reinforced concrete structural members shall not exceed that permitted by ACI 318.

❖ The deflection limitations in ACI 318 are not to be exceeded for reinforced (and prestressed) concrete (see ACI 318 for detailed deflection requirements).

1604.3.3 Steel. The deflection of steel structural members shall not exceed that permitted by AISC 360, AISI-NAS, AISI-General, AISI-Truss, ASCE 3, ASCE 8, SJI JG-1.1, SJI K-1.1 or SJI LH/DLH-1.1, as applicable.

❖ The design standard to be met depends on the type of steel structural member. For example, steel joists are to meet the deflection limitations in the Steel Joist Institute's (SJI) standard, which is applicable to the type of joist, and rolled steel members are to meet the deflection criteria in AISC 360.

TABLE 1604.3. See page 16-10.

❖ The deflection limits in this table apply when they are more restrictive than those in the structural design standards that are indicated in Sections 1604.3.2 through 1604.3.5. Note that the deflection limits for exterior walls and interior finishes vary with flexible or brittle finishes. A flexible finish is intended to be one that has been designed to accommodate the higher deflection indicated and remain serviceable. A brittle finish is any finish that has not been designed to accommodate the deflection allowed for a flexible finish. The deflection limit for a roof member supporting a plaster ceiling is intended to apply only for a plaster ceiling. The limit for a roof member supporting a gypsum board ceiling is that listed in the table for "supporting a nonplaster ceiling."

The answers to the following frequently asked questions provide further guidance on applying the deflection limits of the code.

TABLE 1604.3
DEFLECTION LIMITS[a, b, c, h, i]

CONSTRUCTION	L	S or W[f]	D + L[d,g]
Roof members:[e]			
Supporting plaster ceiling	$l/360$	$l/360$	$l/240$
Supporting nonplaster ceiling	$l/240$	$l/240$	$l/180$
Not supporting ceiling	$l/180$	$l/180$	$l/120$
Floor members	$l/360$	—	$l/240$
Exterior walls and interior partitions:			
With brittle finishes	—	$l/240$	—
With flexible finishes	—	$l/120$	—
Farm buildings	—	—	$l/180$
Greenhouses	—	—	$l/120$

For SI: 1 foot = 304.8 mm.

a. For structural roofing and siding made of formed metal sheets, the total load deflection shall not exceed $l/60$. For secondary roof structural members supporting formed metal roofing, the live load deflection shall not exceed $l/150$. For secondary wall members supporting formed metal siding, the design wind load deflection shall not exceed $l/90$. For roofs, this exception only applies when the metal sheets have no roof covering.

b. Interior partitions not exceeding 6 feet in height and flexible, folding and portable partitions are not governed by the provisions of this section. The deflection criterion for interior partitions is based on the horizontal load defined in Section 1607.13.

c. See Section 2403 for glass supports.

d. For wood structural members having a moisture content of less than 16 percent at time of installation and used under dry conditions, the deflection resulting from $L + 0.5D$ is permitted to be substituted for the deflection resulting from $L + D$.

e. The above deflections do not ensure against ponding. Roofs that do not have sufficient slope or camber to assure adequate drainage shall be investigated for ponding. See Section 1611 for rain and ponding requirements and Section 1503.4 for roof drainage requirements.

f. The wind load is permitted to be taken as 0.7 times the "component and cladding" loads for the purpose of determining deflection limits herein.

g. For steel structural members, the dead load shall be taken as zero.

h. For aluminum structural members or aluminum panels used in skylights and sloped glazing framing, roofs or walls of sunroom additions or patio covers, not supporting edge of glass or aluminum sandwich panels, the total load deflection shall not exceed $l/60$. For aluminum sandwich panels used in roofs or walls of sunroom additions or patio covers, the total load deflection shall not exceed $l/120$.

i. For cantilever members, l shall be taken as twice the length of the cantilever.

Frequently Asked Questions—Table 1604.3

Q1. For purposes of checking the deflection limits, of Table 1604.3, should the calculated wind load be used directly or the combined loads from Section 1605.3.1 or 1605.3.2?

A1. In computing deflections to verify compliance with Table 1604.3 limits, the loads shown in the column headings of Table 1604.3 are the only loads

that must be applied to the member. The load used to check the deflection in this case is W, the nominal wind load in accordance with Section 1609. It is not necessary to use the load combinations of Section 1605.3 for verifying that the deflection limits have been met.

Q2. While checking the member deflection of wall studs for a facility having a wind importance factor, I =1.15, could the wind importance factor be neglected?

A2. No. The nominal wind load, W, determined in accordance with Section 1609 includes the importance factor.

Q3. The allowable deflections for members due to wind load refer to Note f. This note allows the use of 0.7 times the "component and cladding" wind loads for determining the deflection value to check against these limits. Do these deflection limits only apply to secondary framing members, or must primary framing members (specifically rafters) meet these limits too?

A3. Where the components and cladding wind load is applicable to a roof framing member, the deflection may be verified using 0.7 times the component and cladding wind load. See definitions of "Main wind force resisting system (MWFRS)" and "Components and cladding (C&C)."

Q4. In Table 1604.3, Note g states "dead load shall be taken as zero for structural steel members." Would this apply to the precomposition check of composite beam deflection limits under wet weight of concrete?

A4. No. The serviceability requirements of Section 1604.3 apply to the finished construction. The loading condition described would be a construction consideration, which is not directly regulated by the serviceability criteria.

1604.3.4 Masonry. The deflection of masonry structural members shall not exceed that permitted by ACI 530/ASCE 5/TMS 402.

❖ The deflection of masonry beams and lintels is limited by ACI 530/ASCE 5/TMS 402 (see ACI 530/ASCE 5/TMS 402 for deflection requirement details).

1604.3.5 Aluminum. The deflection of aluminum structural members shall not exceed that permitted by AA ADM1.

❖ The deflection of aluminum structural members is limited to that in AA-94 for aluminum or that permitted by Table 1604.3 of the code.

1604.3.6 Limits. Deflection of structural members over span, l, shall not exceed that permitted by Table 1604.3.

❖ The limits specified in Table 1604.3 apply to the indicated members for any structural material. As indicated in Section 1604.3.1, the deflection limits in the structural material standards apply when they are more restrictive than indicated in the table.

The deflection limits that are applicable to a flat concrete roof member that does not support any nonstructural elements likely to be damaged by large deflections are summarized in Table 1604.3.6. As can be seen, the code deflection criteria is more stringent and would govern the design of this type of member.

TABLE 1604.3.6
CONCRETE ROOF MEMBER DEFLECTION CRITERIA

Construction	LOADS		
	L	*S* or *W*	*D + L*
IBC Table 1604.3			
Roof members	360	360	240
Concrete per ACI 318 See Table 9.5(b)			
Flat roofs	180	Note a	Note a

a. No deflection limit specified in ACI 318

1604.4 Analysis. Load effects on structural members and their connections shall be determined by methods of structural analysis that take into account equilibrium, general stability, geometric compatibility and both short- and long-term material properties.

Members that tend to accumulate residual deformations under repeated service loads shall have included in their analysis the added eccentricities expected to occur during their service life.

Any system or method of construction to be used shall be based on a rational analysis in accordance with well-established principles of mechanics. Such analysis shall result in a system that provides a complete load path capable of transferring loads from their point of origin to the load-resisting elements.

The total lateral force shall be distributed to the various vertical elements of the lateral-force-resisting system in proportion to their rigidities, considering the rigidity of the horizontal bracing system or diaphragm. Rigid elements assumed not to be a part of the lateral-force-resisting system are permitted to be incorporated into buildings provided their effect on the action of the system is considered and provided for in the design. Except where diaphragms are flexible, or are permitted to be analyzed as flexible, provisions shall be made for the increased forces induced on resisting elements of the structural system resulting from torsion due to eccentricity between the center of application of the lateral forces and the center of rigidity of the lateral-force-resisting system.

Every structure shall be designed to resist the overturning effects caused by the lateral forces specified in this chapter. See Section 1609 for wind loads, Section 1610 for lateral soil loads and Section 1613 for earthquake loads.

❖ This section includes the general requirements for structural analysis. The principles stated in this section are those commonly found in structural engineering textbooks. The requirement that structural analysis be capable of demonstrating a complete load path is essential to the adequate resistance of the structural system to wind loads or earthquake effects. The load path is to be capable of transferring all of the loads from their point of application onto the structure to the foundation. It is also important that nonstructural rigid elements be properly accounted for in the design. For example, a partial-height rigid masonry wall placed between steel columns in a steel frame will resist the horizontal shear load and cause bending in the column unless a flexible joint is provided between the wall and the column.

The definition of "Diaphragm" in Section 1602.1 includes definitions that distinguish a rigid diaphragm from a flexible diaphragm. The distribution of forces in buildings with flexible diaphragms differs from those having rigid diaphragms. Where the diaphragm is determined to be flexible, the effect of diaphragm rigidity on the distribution of lateral forces is considered to be negligible and can therefore be neglected in the structural analysis. Otherwise, for structures having rigid diaphragms, this section requires the engineer to distribute lateral forces to the vertical supporting elements in proportion to their rigidities, and to include the effect of the increased forces induced on the vertical supporting elements resulting from torsion due to eccentricity between the center of mass and the center of rigidity.

1604.5 Occupancy category. Buildings shall be assigned an occupancy category in accordance with Table 1604.5.

❖ This section requires classification of the occupancy category of any building in accordance with the nature of occupancy as described in Table 1604.5. The occupancy category serves as a threshold for a variety of code provisions related to earthquake, flood, snow and wind loads. Particularly noteworthy are the importance factors that are used in the calculation of design earthquake, snow and wind loads. The value of the importance factor generally increases with the importance of the facility. Structures assigned greater importance factors must be designed for larger forces. The result is a more robust structure that would be less likely to sustain damage under the same conditions than a structure with a lower importance factor. The intent is to enhance a structure's performance based upon its use or the need to remain in operation during and after a design event.

The impact of a higher occupancy category classification is not limited to increasing the design loads. Compared to Occupancy Category I, II or III, for instance, an Occupancy Category IV classification can lead to a higher seismic design category classification that can, in turn, require more stringent seismic detailing and limitations on the seismic-force-resisting system. This can also affect the seismic design requirements for architectural, mechanical and electrical components and systems.

TABLE 1604.5. See page 16-12.

❖ The occupancy category determined in this table generally increases with the importance of the facility, which generally relates to the availability of the facility after an emergency, and the consequence of a structural failure on human life. The categories range from Occupancy Category I, which represents the lowest hazard to life, through Occupancy Category IV, which encompasses essential facilities.

Occupancy Category IV: These are buildings that are considered to be essential in that their continuous use is needed, particularly in response to disasters. Fire, rescue and police stations and emergency vehicle garages must remain operational during and after major events such as earthquakes, floods or hurricanes. The phrase "designated as essential facilities" refers to designation by the building official that certain facilities are required for emergency response or disaster recovery. This provides jurisdictions the latitude to identify specific facilities that should be considered essential in responding to various types of emergencies. These could include structures that would not otherwise be included in this occupancy category. This designation would only be made with consideration of broader public policy as well as emergency preparedness planning within the jurisdiction in question. The reasons for including facilities such as fire stations, police stations emergency response operations center, etc., should be self-evident. Some items warranting additional discussion are as follows:

- *Hospitals and other health care facilities having surgery or emergency treatment facilities.* Hospitals are classified as Group I-2 in Section 308.3 and their important role in emergency response should be apparent. This item appears to group all health care facilities with emergency treatment or surgery facilities together with hospitals. Seemingly, this could include ambulatory surgery centers or outpatient units that typically do not have significant emergency capabilities and they could be classified as Group B occupancies, since they don't treat patients on a 24-hour basis. The existence of these facilities may

TABLE 1604.5
OCCUPANCY CATEGORY OF BUILDINGS AND OTHER STRUCTURES

OCCUPANCY CATEGORY	NATURE OF OCCUPANCY
I	Buildings and other structures that represent a low hazard to human life in the event of failure, including but not limited to: • Agricultural facilities. • Certain temporary facilities. • Minor storage facilities.
II	Buildings and other structures except those listed in Occupancy Categories I, III and IV
III	Buildings and other structures that represent a substantial hazard to human life in the event of failure, including but not limited to: • Covered structures whose primary occupancy is public assembly with an occupant load greater than 300. • Buildings and other structures with elementary school, secondary school or day care facilities with an occupant load greater than 250. • Buildings and other structures with an occupant load greater than 500 for colleges or adult education facilities. • Health care facilities with an occupant load of 50 or more resident patients, but not having surgery or emergency treatment facilities. • Jails and detention facilities. • Any other occupancy with an occupant load greater than 5,000. • Power-generating stations, water treatment for potable water, waste water treatment facilities and other public utility facilities not included in Occupancy Category IV. • Buildings and other structures not included in Occupancy Category IV containing sufficient quantities of toxic or explosive substances to be dangerous to the public if released.
IV	Buildings and other structures designated as essential facilities, including but not limited to: • Hospitals and other health care facilities having surgery or emergency treatment facilities. • Fire, rescue and police stations and emergency vehicle garages. • Designated earthquake, hurricane or other emergency shelters. • Designated emergency preparedness, communication and operation centers and other facilities required for emergency response. • Power-generating stations and other public utility facilities required as emergency backup facilities for Occupancy Category IV structures. • Structures containing highly toxic materials as defined by Section 307 where the quantity of the material exceeds the maximum allowable quantities of Table 307.1(2). • Aviation control towers, air traffic control centers and emergency aircraft hangars. • Buildings and other structures having critical national defense functions. • Water treatment facilities required to maintain water pressure for fire suppression.

not be apparent to emergency responders that need to direct the injured to a medical facility. Therefore, it may be more appropriate that the classification of such health care facilities should apply only where they are, in fact, "designated" as being essential in the emergency preparedness planning in that jurisdiction.

- *Designated emergency shelters and designated emergency response facilities.* These items repeat the term "designated" which is referring to designation by the building official that the facilities have been identified as necessary for sheltering evacuees or responding to emergencies (see discussion of "designated" above). For example, an elementary school having an occupant load of 275 would typically be considered an Occupancy Category III facility. If that school is designated as an emergency shelter, then the school will be considered an Occupancy Category IV building.

- *Facilities supplying emergency backup power for Occupancy Category IV.* A power-generating station or other utility (such as a natural gas facility) is to be classified as Occupancy Category IV only if the facility serves an emergency backup function for an Occupancy Category IV building such as a fire station or police station. Otherwise, the power-generating station or utility should be classified as Occupancy Category III.

- *Structures with highly toxic materials in excess of the quantities allowed in Table 307.1(2).* This applies only to "highly toxic materials" (see definition in Section 307.2), which are covered in the second row of Table 307.1(2). That table lists the maximum allowable quantities per control area of materials posing a health hazard. Since the use of control areas is permitted by Table 307.1(2), it seems reasonable to recognize the control area for the purpose of applying this occupancy category. In other words, this would apply only to occupancies classified as Group H-4 based on the quantities of highly toxic material. Recognizing control areas means the occupancy classification could be changed from a high hazard Group H-4 to the appropriate classification by adding either fire-resistance-rated walls or floor-ceiling assemblies in order to divide a building into a number of smaller "control areas." Also note that a Group H-4 occupancy classification could be based on exceeding the quantities permitted for toxics or corrosives, but the presence of those materials would not affect the classification of the facility.

Occupancy Category III: Occupancy Category III buildings include those occupancies that have relatively large numbers of occupants because of the overall size of the building. They also include uses that pose an elevated life safety hazard to the occupants, such as public assembly, schools or colleges.

In addition, Occupancy Category III includes uses where the occupant's ability to respond to an emergency is either restricted, such as in jails, or otherwise impaired, such as in nursing homes housing patients that require skilled nursing care. A discussion of some of the specific table listings follows:

- *Covered structure with a primary occupancy that is public assembly with an occupant load greater than 300.* Public assembly occupancies meeting this criterion will typically be classified as Group A in Chapter 3. These occupancies represent an elevated life safety risk due to higher concentrations of large numbers of people. In the 2006 edition this criterion has been changed to read "covered structures whose primary occupancy is public assembly with an occupant load greater than 300" to provide better consistency with prior practice under legacy codes, as well as the NEHRP provisions. The revised wording requires agreement on the determination that a building's "primary occupancy" is in fact public assembly. This could be as simple as verifying that the portion of the building housing the public assembly occupancy is more than 50 percent of the total building area. Also, the previous wording referred to the occupant load of 300 in "one area." Since the reference to one area has been dropped, it would now appear that the aggregate occupant load of all the public assembly uses would be used in making this determination.

- *Health care facilities with at least 50 resident patients and not providing emergency treatment.* This category applies to health care facilities with at least 50 resident patients. The term "resident patient" is not defined or used elsewhere in the code, but would seem to refer to locations where those patients receive around-the-clock (24 hour) care as opposed to ambulatory surgery centers or outpatient units. For the most part, this table entry covers facilities that would be classified as Group I-1 or I-2, which typically means these patients have difficulty responding to an emergency or are incapable of self-preservation. If such facilities also have significant emergency treatment capability, they would more appropriately require classification as Occupancy Category IV.

- *Buildings having an occupant load greater than 5,000.* Uses that pose elevated life safety concerns such as public assembly uses, schools and health care facilities are covered elsewhere and have much lower threshold based on the number of occupants. This table entry covers buildings that are large enough to have more than 5,000 occupants, providing added protection for the occupants of larger structures whatever the use happens to be. For example, an office building having an occupant load that is based on one occupant per 100 square feet (9.3 m^2) of gross floor area in accordance with Table

1004.1.2 would be classified as Occupancy Category III if the area exceeds 500,000 square feet (46 450 m²).

- *Power-generating stations, potable water treatment facilities, waste water treatment and other public utilities not included in Occupancy Category IV.* A failure and subsequent shutdown of these types of facilities would not pose an immediate threat to life safety. These infrastructure items are considered Occupancy Category III because of the impact an extended disruption in service can have on the public.

- *Buildings not included in Occupancy Category IV containing sufficient quantities of toxic or explosive substances that would be dangerous to the public if released.* Buildings included under Occupancy Category IV would be those containing quantities of highly toxic materials that exceed the permitted quantity per control area. This item addresses buildings with explosives or toxic substances, both of which are defined in Section 307.2. The wording refers to "sufficient quantities...to be dangerous to the public if released" without explaining what quantities are sufficient to pose such a danger. More importantly, it does not make any specific reference to the tabulated allowable quantities of either explosives or toxics in Section 307. In other words, the issue of whether the building needs to be classified as a high hazard, Group H occupancy is subject to interpretation, leaving the door open, perhaps, to consider occupancies not classified as high hazard (Group H) to be categorized as Occupancy Classification III. This possibility may be justifiable, considering that the use of control areas addresses fire safety issues that are reflected in the occupancy classification of the building. Table 1604.5 addresses the hazard posed by a structural failure. Whether or not the building has control areas allowing a different occupancy classification, could be irrelevant. It may be more logical that the occupancy category is determined by the total quantities of the hazardous material housed in the structure under consideration.

Occupancy Category II: Occupancy Category II buildings represent a lesser hazard to life because of fewer building occupants and smaller building size compared to those that are considered Occupancy Category III. Since Occupancy Categories III and IV represent buildings with higher risk or essential facilities, on a relative scale Occupancy Category II can be thought of as a "standard occupancy" building as evidenced by importance factors for earthquake, snow and wind that are all equal to 1.

Occupancy Category I: Occupancy Category I buildings exhibit the lowest hazard to life since they have little or no human occupants or, for those that are temporary, the exposure to the hazards of earthquakes,

floods, snow and wind would be considerably less than that of a permanent structure. Note that this category includes "minor storage facilities," but the code does not provide an explanation of which storage facilities could be considered minor.

1604.5.1 Multiple occupancies. Where a structure is occupied by two or more occupancies not included in the same occupancy category, the structure shall be assigned the classification of the highest occupancy category corresponding to the various occupancies. Where structures have two or more portions that are structurally separated, each portion shall be separately classified. Where a separated portion of a structure provides required access to, required egress from or shares life safety components with another portion having a higher occupancy category, both portions shall be assigned to the higher occupancy category.

❖ Buildings are frequently occupied by a mixture of uses or occupancies. A single-use building is probably the exception rather than the rule. Where multiple occupancies are proposed in a building, the occupancy category of each of those occupancies must be considered. In some cases, the proposed occupancies in a building will fall into more than one occupancy category and the requirements for multiple occupancies stated in Section 1604.5.1 must be satisfied. These requirements were previously part of the earthquake load provisions. In the 2006 edition, they have been relocated so that they now apply regardless of the type of load being considered. The code identifies two design options. The entire structure can be designed as a single unit based on the requirements for the most stringent occupancy category that will occupy the building. Alternatively, the engineer can separate portions of the structure containing distinct occupancy categories and design each portion accordingly.

This section also provides direction regarding access to and egress from adjacent structures that fall into different occupancy categories by making the requirements for the more stringent occupancy category applicable to both structures. This requirement is the result of lessons learned from events such as earthquakes in which essential functions have been rendered unusable because of a failure in an adjacent structure.

1604.6 In-situ load tests. The building official is authorized to require an engineering analysis or a load test, or both, of any construction whenever there is reason to question the safety of the construction for the intended occupancy. Engineering analysis and load tests shall be conducted in accordance with Section 1713.

❖ The building official has the option of requiring either a structural analysis, an on-site in-situ load test or both, in accordance with Section 1713, on an existing structure, building or portion thereof if there is reasonable doubt as to structural integrity. The building official should document his or her reasons for the testing requirement. Whenever possible, the concern should be addressed by structural analysis since load testing a structure is very expensive. One example would be an analysis by a third-party engineering firm acceptable to both the building official and the owner. The structural integrity may be examined for items such as visible signs of excessive settlement or lateral deflection, such as cracks in concrete foundation walls or excessive vibration when the assembly is loaded. The procedure must simulate the actual load conditions to which the structure is subjected during normal use.

1604.7 Preconstruction load tests. Materials and methods of construction that are not capable of being designed by approved engineering analysis or that do not comply with the applicable material design standards listed in Chapter 35, or alternative test procedures in accordance with Section 1711, shall be load tested in accordance with Section 1714.

❖ The alternative test procedure described in Section 1711 and the preconstruction load test procedure described in Section 1714 are intended to apply to materials or an assembly of structural materials that do not have an accepted analysis technique; thus, they are approved for use by way of the alternative test procedure. The preconstruction test procedure in Section 1714 includes the determination of the allowable superimposed design load (see Section 1714 for details).

1604.8 Anchorage.

1604.8.1 General. Anchorage of the roof to walls and columns, and of walls and columns to foundations, shall be provided to resist the uplift and sliding forces that result from the application of the prescribed loads.

❖ This section states a specific load path requirement for anchorage that is already required in more general terms by Section 1604.4. Required anchorage resistance to uplift and sliding is determined in accordance with the appropriate load combinations in Section 1605, which account for the effects of earthquakes, fluid pressures, lateral earth pressure or wind (also see Section 1604.9).

An example of the condition described in this section is roof uplift as a result of high winds. Table 2308.10.1 requires roof uplift connectors for wood-frame construction designed and installed in accordance with the prescriptive requirements for conventional light-frame construction. Note that net roof uplift occurs at all wind speeds indicated in the table; thus, uplift connectors are required for roof wood framing for all of the indicated wind speed locations.

1604.8.2 Concrete and masonry walls. Concrete and masonry walls shall be anchored to floors, roofs and other structural elements that provide lateral support for the wall. Such anchorage shall provide a positive direct connection capable of resisting the horizontal forces specified in this chapter but not less than a minimum strength design horizontal force of 280 plf (4.10 kN/m) of wall, substituted for "E" in the load combinations of Section 1605.2 or 1605.3. Walls shall be designed to resist bending between anchors where the anchor spacing exceeds 4 feet (1219 mm). Required anchors in masonry walls of hollow units or cavity walls shall be embed-

ded in a reinforced grouted structural element of the wall. See Sections 1609 for wind design requirements and see Section 1613 for earthquake design requirements.

❖ Because the response of building elements to earthquake ground motion is a function of their mass, heavy elements such as concrete or masonry walls experience significant lateral loading during a seismic event. A major problem during a severe earthquake has been concrete or masonry walls pulling away from their supporting floors or roofs. The result is a loss of vertical load-carrying elements designed to transmit the lateral forces from the diaphragms (floors and roofs) down to the foundations. A significant number of such anchorage failures were observed after the 1971 San Fernando earthquake and again in 1994 after the Northridge earthquake. Clearly, such failures are hazardous to life and property.

This section prescribes a minimum strength design load for anchorages between concrete or masonry walls and the horizontal diaphragms that provide lateral support. The connection requirements apply to both bearing walls and nonstructural walls. The wall anchorage must be a positive connection that does not rely on friction for load transfer. The reference to "E" is the earthquake load effect included in the load combinations in Section 1605. Other earthquake detailing provisions may specify higher anchorage design forces based on the seismic design category of the building.

1604.8.3 Decks. Where supported by attachment to an exterior wall, decks shall be positively anchored to the primary structure and designed for both vertical and lateral loads as applicable. Such attachment shall not be accomplished by the use of toenails or nails subject to withdrawal. Where positive connection to the primary building structure cannot be verified during inspection, decks shall be self-supporting. For decks with cantilevered framing members, connections to exterior walls or other framing members shall be designed and constructed to resist uplift resulting from the full live load specified in Table 1607.1 acting on the cantilevered portion of the deck.

❖ This requirement for the positive anchorage of decks is a result of failures that have occurred primarily on nonengineered residential decks. Toenail connections are very weak since many of the field connections split the wood framing member. Nails that are installed in line with applied tension forces pull out easily. Wood deck framing could be attached to a supporting ledger board by joist hangers that provide nails loaded in shear for the vertical and horizontal loads. Uplift occurs at the point of connection between the wood framing and the exterior wall for cantilevered deck framing. This section highlights the need for resistance to uplift by a positive connection at the exterior wall.

1604.9 Counteracting structural actions. Structural members, systems, components and cladding shall be designed to resist forces due to earthquake and wind, with consideration of overturning, sliding, and uplift. Continuous load paths shall be provided for transmitting these forces to the foundation. Where sliding is used to isolate the elements, the effects of friction between sliding elements shall be included as a force.

❖ This section requires that all elements of the structure be designed to resist the required earthquake and wind effects. The required level of resistance will be based on the appropriate load combinations of Section 1605. The term "counteracting" used in the section title refers to so-called counteracting load combinations as opposed to those in which the load effects are additive. The effects of overturning, uplift and sliding are most pronounced in these counteractive load combinations. These would consist of the following depending on the design method chosen:

- Load and resistance factor design or strength design, Section 1605.2.1:

   Equation 16-60.9D + 1.6W + 1.6H

   Equation 16-70.9D + 1.0E + 1.6H

- Basic allowable stress design load combinations, Section 1605.3.1:

   Equation 16-140.6D + W + H

   Equation 16-150.6D + 0.7E + H

- Alternative allowable stress design load combinations, Section 1605.3.2:

   Equation 16-210.9D + E/1.4

   (Note the adjustment to the vertical component of earthquake load effect that is permitted for evaluating foundations).

   Equation 16-17($^2/_3$)D + 1.3W

   (Equation 16-17 with $L = 0$ in accordance with Section 1605.1 requirement to investigate each load combination with one or more variable loads equal to zero, two-thirds of the dead load in accordance with requirement in Section 1605.3.2 and ω set equal to 1.3 for 3-second gust wind speed).

As can be seen, for the critical counteracting load combinations, resistance to overturning, uplift and sliding is largely provided by dead load. Note that Section 1605.3.2 requires the designer to consider only the dead load likely to be in place during a design wind event when using the alternative allowable stress load combinations. Regardless of the design methodology utilized, the designer should be cautious in estimating dead loads that resist overturning, sliding and uplift. In order to remain conservative, estimated dead load should be no more than the actual dead load in these cases. See design dead load in Section 1606.2. To keep a structure from sliding horizontally, the dead load must generate enough friction at the base of the structure to resist the horizontal base shear due to earthquake and wind. Otherwise, adequate anchorage must be provided to resist the base shear.

1604.10 Wind and seismic detailing. Lateral-force-resisting systems shall meet seismic detailing requirements and limita-

tions prescribed in this code and ASCE 7, excluding Chapter 14 and Appendix 11A, even when wind code prescribed load effects are greater than seismic load effects.

❖ This section is needed to clarify that the seismic detailing requirements of the code always apply even where the wind load effects are higher than the seismic load effects. This is required because the calculated earthquake load is dependent upon an assumed level of ductility to be provided in the selected seismic-force-resisting system. If the designer does not follow through by providing the detailing required to achieve the ductility of the selected system, the assumption is invalid.

For example, consider the case where the earthquake load provisions apply in accordance with Section 1613.1, the seismic-force-resisting system is an ordinary concentrically braced steel frame and, based on the seismic design category, Section 2205 requires that the AISC 341 seismic provisions are to be met. For this case, the requirements in Section 14 of AISC 341 must be satisfied in the design of the steel frame, even if the wind load effects exceed the seismic load effects.

SECTION 1605
LOAD COMBINATIONS

1605.1 General. Buildings and other structures and portions thereof shall be designed to resist the load combinations specified in Section 1605.2 or 1605.3 and Chapters 18 through 23, and the special seismic load combinations of Section 1605.4 where required by Section 12.3.3.3 or 12.10.2.1 of ASCE 7. Applicable loads shall be considered, including both earthquake and wind, in accordance with the specified load combinations. Each load combination shall also be investigated with one or more of the variable loads set to zero.

❖ Generally, there are two types of load combinations specified in the code: those to be used with a strength design or load and resistance factor design and those to be used with allowable stress design. Where additional load combinations are specified in the structural material chapters, they apply also. Note that this section also requires that engineers consider additional load cases where variable loads are not acting concurrently with other variable loads. It is necessary to explore such possibilities, since they can at times result in the most critical load effect for some parts of a structure. According to the definition of "Loads," loads that are not considered permanent are variable loads. In addition, the ASCE 7 sections that require use of the code's special seismic load combinations to account for system overstrength are listed (see commentary, Section 1605.4).

1605.2 Load combinations using strength design or load and resistance factor design.

1605.2.1 Basic load combinations. Where strength design or load and resistance factor design is used, structures and portions thereof shall resist the most critical effects from the following combinations of factored loads:

$$1.4(D + F) \qquad \text{(Equation 16-1)}$$

$$1.2(D + F + T) + 1.6(L + H) + 0.5(L_r \text{ or } S \text{ or } R) \qquad \text{(Equation 16-2)}$$

$$1.2D + 1.6(L_r \text{ or } S \text{ or } R) + (f_1 L \text{ or } 0.8W) \qquad \text{(Equation 16-3)}$$

$$1.2D + 1.6W + f_1 L + 0.5(L_r \text{ or } S \text{ or } R) \qquad \text{(Equation 16-4)}$$

$$1.2D + 1.0E + f_1 L + f_2 S \qquad \text{(Equation 16-5)}$$

$$0.9D + 1.6W + 1.6H \qquad \text{(Equation 16-6)}$$

$$0.9D + 1.0E + 1.6H \qquad \text{(Equation 16-7)}$$

f_1 = 1 for floors in places of public assembly, for live loads in excess of 100 pounds per square foot (4.79 kN/m²), and for parking garage live load, and

= 0.5 for other live loads.

f_2 = 0.7 for roof configurations (such as saw tooth) that do not shed snow off the structure, and

= 0.2 for other roof configurations.

Exception: Where other factored load combinations are specifically required by the provisions of this code, such combinations shall take precedence.

❖ This section lists the load combinations for strength design or load and resistance factor design methods. See the definitions in Section 1602 for an explanation of the load notations. The basis for these load combinations is Section 2.3.2 of ASCE 7. These combinations of factored loads are the agreed upon strength limit states that establish the required strength to be provided in the structural component being designed. In spite of the precise appearance of the load and resistance factor design load combinations one should keep in mind their probabilistic nature. The goal is to allow a wide variety of structures to be designed economically with an acceptably low probability that the strength of the structure will be exceeded. Doing so necessitates combining loads in scenarios that are likely to occur. Dead load is a permanent load and it appears in every combination. The load combinations are constructed by adding the dead load to one of the variable loads at its maximum value, which is typically indicated by the load factor of 1.6. In addition, other variable loads are included with load factors that are less than 1.0. Those so-called "companion loads" represent arbitrary point-in-time values for those loads. The exception is the maximum earthquake load that has a load factor of 1.0 (see the commentary to the

definitions "Factored load" and "Nominal load" in Section 1602.1).

1605.2.2 Other loads. Where F_a is to be considered in the design, the load combinations of Section 2.3.3 of ASCE 7 shall be used.

❖ Under a cooperative agreement with FEMA, ASCE completed an extensive analysis of flood loads and flood load combinations. The results of this study were first included in the 1998 edition of ASCE 7. For all buildings and structures located in designated flood hazard areas, this section specifies that flood load and flood load combinations using the strength design method are to be determined in accordance with Section 2.3.3 of ASCE 7, which states:

Load Combinations Including Flood Load. When a structure is located in a flood zone (Section 5.3.1), the following load combinations shall be considered:

1. In V Zones or Coastal A Zones, 1.6W in combinations (4) and (6) shall be replaced by 1.6W + 2.0F_a.

2. In noncoastal A Zones, 1.6W in combinations (4) and (6) shall be replaced with .8W + 1.0F_a.

1605.3 Load combinations using allowable stress design.

1605.3.1 Basic load combinations. Where allowable stress design (working stress design), as permitted by this code, is used, structures and portions thereof shall resist the most critical effects resulting from the following combinations of loads:

$$D + F \qquad \text{(Equation 16-8)}$$

$$D + H + F + L + T \qquad \text{(Equation 16-9)}$$

$$D + H + F + (L_r \text{ or } S \text{ or } R) \qquad \text{(Equation 16-10)}$$

$$D + H + F + 0.75(L + T) + 0.75(L_r \text{ or } S \text{ or } R) \qquad \text{(Equation 16-11)}$$

$$D + H + F + (W \text{ or } 0.7E) \qquad \text{(Equation 16-12)}$$

$$D + H + F + 0.75(W \text{ or } 0.7E) + 0.75L + 0.75(L_r \text{ or } S \text{ or } R) \qquad \text{(Equation 16-13)}$$

$$0.6D + W + H \qquad \text{(Equation 16-14)}$$

$$0.6D + 0.7E + H \qquad \text{(Equation 16-15)}$$

Exceptions:

1. Crane hook loads need not be combined with roof live load or with more than three-fourths of the snow load or one-half of the wind load.

2. Flat roof snow loads of 30 psf (1.44 kN/m²) or less need not be combined with seismic loads. Where flat roof snow loads exceed 30 psf (1.44 kN/m²), 20 percent shall be combined with seismic loads.

❖ See Section 1602 for an explanation of the notations. These basic load combinations for allowable stress design are based on Section 2.4.1 of ASCE 7. Note that a 0.75 factor is applied where these combinations include more than one variable load. This reduces the combined effect of these variable loads in recognition of the lower probability that two or more variable loads will reach their maximum values simultaneously.

Previous editions of the model codes specified that the overturning moment and sliding due to wind load could not exceed two-thirds of the dead load stabilizing moment; however, it was not typically applied to all elements in the building. In the code this limitation on dead load is accomplished through the load combinations. The applicable combination is $0.6D + W + H$. This load combination limits the dead load resisting wind loads to 60 percent ($^2/_3 = 0.67$; round down) but it applies to all elements. In this form, it is required that this safety factor on dead load applies to all actions where dead load is resisting wind loads.

The load combination, $0.6D + 0.7E + H$, applies throughout the structure and provides for overall stability of the structure similar to the load combination for wind. In determining seismic load effect, E, the definition of effective seismic weight requires inclusion of a portion of roof snow under certain conditions. Frequently the question of whether it is logical to include a component of snow load in determining E but not to use that same weight of snow in the load case to resist the overturning or uplift due to E is asked. This apparent inconsistency has been built into the seismic requirements for some time. It introduces a bit more conservatism into the earthquake loading where heavier snow loads are concerned.

Exception 1 for crane hook loads has been a long-standing allowance in the alternative basic load combinations of Section 1605.3.2 under the legacy model codes and is now permitted in this code. It allows special consideration in combining crane loads with wind, snow and roof live loads due to a lower probability that these maximum loads occur simultaneously. As noted above, these basic load combinations are based on ASCE 7 allowable stress load combinations, but there is no corresponding exception for crane loads under ASCE 7. In the development of the 2000 edition of this code, this exception was added to the basic load combinations in an attempt to provide "parity" with the alternative basic load combinations. It is worth noting that these two sets of load combinations come from two different sources as described above and, from an overall perspective, they never have provided parity. Crane loads are considered live loads (see Section 1607.12). Exception 1 has no effect on dead load, other floor live loads, rain loads or earthquake loads; thus, where crane live loads are to be considered, the exception eliminates the roof live load, L_r, altogether and reduces the snow load to $.75S$ and wind load to $.5W$. This exception does not negate the need to combine live loads other than the crane live loads in the usual manner with wind and snow loads. In other words, the load combinations in Equations 16-11 and 16-13 must be investigated without the crane live load and then modified versions of these

combinations that include crane live load (use L_c to denote crane live load) would be considered to make use of the exception as follows:

Equation 16-11
$$D + H + F + 0.75L + 0.75(L_r \text{ or } S \text{ or } R)$$

Equation 16-11c
$$D + H + F + 0.75[L + L_c] + 0.75(.75S \text{ or } R)$$

For Equation 16-13 with earthquake load, E:

Equation 16-13
$$D + H + F + 0.75(0.7E) + 0.75L + 0.75(L_r \text{ or } S \text{ or } R)$$

Equation 16-13c
$$D + H + F + 0.75(0.7E) + 0.75[L + L_c] + 0.75(.75S \text{ or } R)$$

For Equation 16-13 with wind load, W:

Equation 16-13
$$D + H + F + 0.75W + 0.75L + 0.75(L_r \text{ or } S \text{ or } R)$$

Equation 16-13c
$$D + H + F + 0.75(.5W) + 0.75[L + L_c] + 0.75(.75S \text{ or } R)$$

Exception 2 states that "flat roof snow loads of 30 psf or less need not be combined with seismic loads..." This is an exception to the requirement to combine the effects of snow load and earthquake load as would otherwise be required by Equation 16-13.

1605.3.1.1 Stress increases. Increases in allowable stresses specified in the appropriate material chapter or the referenced standards shall not be used with the load combinations of Section 1605.3.1, except that a duration of load increase shall be permitted in accordance with Chapter 23.

❖ Some code sections and some referenced standards permit increases in the allowable stress where load combinations include either wind or earthquake loads. Such increases are expressly prohibited with these load combinations. The exception is the duration of load increases that are recognized in Chapter 23.

1605.3.1.2 Other loads. Where F_a is to be considered in design, the load combinations of Section 2.4.2 of ASCE 7 shall be used.

❖ The study discussed in the commentary to Section 1605.2.2 determines flood load and flood load combination factors for use with the allowable stress method. This section specifies that flood load and flood load combinations using the allowable stress design method are to be determined in accordance with Section 2.4.2 of ASCE 7, which states:

Load Combinations including Flood Load. When a structure is located in a flood zone, the following load combinations shall be considered:

1. In V Zones or Coastal A Zones (Section 5.3.1), $1.5F_a$ shall be added to other load com-

binations (5), (6) and (7), and E shall be set equal to zero in (5) and (6).

2. In noncoastal A Zones, $0.75F_a$ shall be added to the combinations (5), (6) and (7), and E shall be set at zero in (5) and (6).

1605.3.2 Alternative basic load combinations. In lieu of the basic load combinations specified in Section 1605.3.1, structures and portions thereof shall be permitted to be designed for the most critical effects resulting from the following combinations. When using these alternative basic load combinations that include wind or seismic loads, allowable stresses are permitted to be increased or load combinations reduced where permitted by the material chapter of this code or the referenced standards. For load combinations that include the counteracting effects of dead and wind loads, only two-thirds of the minimum dead load likely to be in place during a design wind event shall be used. Where wind loads are calculated in accordance with Chapter 6 of ASCE 7, the coefficient ω in the following equations shall be taken as 1.3. For other wind loads, ω shall be taken as 1. When using these alternative load combinations to evaluate sliding, overturning and soil bearing at the soil-structure interface, the reduction of foundation overturning from Section 12.13.4 in ASCE 7 shall not be used. When using these alternative basic load combinations for proportioning foundations for loadings, which include seismic loads, the vertical seismic load effect, E_v, in Equation 12.4-4 of ASCE 7 is permitted to be taken equal to zero.

$$D + L + (L_r \text{ or } S \text{ or } R) \qquad \text{(Equation 16-16)}$$

$$D + L + (\omega W) \qquad \text{(Equation 16-17)}$$

$$D + L + \omega W + S/2 \qquad \text{(Equation 16-18)}$$

$$D + L + S + \omega W/2 \qquad \text{(Equation 16-19)}$$

$$D + L + S + E/1.4 \qquad \text{(Equation 16-20)}$$

$$0.9D + E/1.4 \qquad \text{(Equation 16-21)}$$

Exceptions:

1. Crane hook loads need not be combined with roof live loads or with more than three-fourths of the snow load or one-half of the wind load.

2. Flat roof snow loads of 30 psf (1.44 kN/m²) or less need not be combined with seismic loads. Where flat roof snow loads exceed 30 psf (1.44 kN/m²), 20 percent shall be combined with seismic loads.

❖ These are alternative load combinations to those in Section 1605.3.1 for use with allowable stress design. They are based on the allowable stress load combinations in the *Uniform Building Code®* (UBC).

Note that the dead load is limited to two-thirds of the dead load likely to be in place during a design wind event. Section 1604.9 requires that all elements of a building be anchored to resist overturning, uplift and sliding. Often, a considerable portion of this resistance is provided by the dead load of a building, including the weight of foundations and any soil directly above them. This section requires the designer to give consideration to the dead load used to resist wind loads by

stating that only the minimum dead load likely to be in place during a design wind event is permitted to be used. This, however, does not imply that certain parts or elements of a building are not designed to remain in place. This criteria simply cautions the designer against using dead loads that may not be installed, or in place, such as can occur where a design includes an allowance for a planned future expansion.

Exception 1 for crane hook loads has been a long-standing allowance under the legacy model codes that were also the source of these alternative basic load combinations. It allows special consideration in combining crane live loads with wind, snow and roof live loads due to a lower probability that these maximum loads occur simultaneously.

Exception 2 states that "flat roof snow loads of 30 psf (1.44 N/m²) or less need not be combined with seismic loads..." This is an exception to the requirement to combine the effects of snow load and earthquake load as would otherwise be required by Equation 16-20.

Also see the commentary to Section 1602.1 for discussion of earthquake load effect, E, as well as other loads required by these load combinations.

1605.3.2.1 Other loads. Where F, H or T are to be considered in the design, each applicable load shall be added to the combinations specified in Section 1605.3.2.

❖ See the definitions in Section 1602 for an explanation of the notation used in this section. As indicated, the applicable loads are to be added to each of the load combinations in Section 1605.3.2.

1605.4 Special seismic load combinations. For both allowable stress design and strength design methods where specifically required by Section 1605.1 or by Chapters 18 through 23, elements and components shall be designed to resist the forces calculated using Equation 16-22 when the effects of the seismic ground motion are additive to gravity forces and those calculated using Equation 16-23 when the effects of the seismic ground motion counteract gravity forces.

$$1.2D + f_1 L + E_m \qquad \text{(Equation 16-22)}$$

$$0.9D + E_m \qquad \text{(Equation 16-23)}$$

where:

E_m = The maximum effect of horizontal and vertical forces as set forth in Section 12.4.3 of ASCE 7.

f_1 = 1 for floors in places of public assembly, for live loads in excess of 100 psf (4.79 kN/m²) and for parking garage live load, or

= 0.5 for other live loads.

❖ The purpose of these strength load combinations is to account for the maximum earthquake load effect, E_m, which considers "system overstrength." This system characteristic is accounted for by multiplying the effects of the lateral earthquake load by the overstrength factor, Ω_0, for the seismic-force-resisting system that is utilized. It is representative of the upper bound system strength for purposes of designing nonyielding elements for the maximum expected load. Under the design earthquake ground motions, the forces generated in the seismic-force-resisting system can be much greater than the prescribed seismic design forces. If not accounted for, the system overstrength effect can cause failures of structural elements that are subjected to these forces. Because system overstrength is unavoidable, design for the maximum earthquake force that can be developed is warranted. The intent is to provide key elements with sufficient overstrength so that inelastic (ductile) response/behavior appropriately occurs within the vertical resisting elements. The ASCE 7 earthquake load provisions that require consideration of the special seismic load combinations are stated in Section 1605.1 as follows:

- Section 12.3.3.3 of ASCE 7 applies to structural elements that support discontinuous frames or shear wall systems where the discontinuity is severe enough to be deemed a structural irregularity. See Figure 1605.4(1) for examples of building configurations that include discontinuous vertical systems. See the commentary to Section C12.3.3.3 of ASCE 7.

- Section 12.10.2.1 of ASCE 7 applies to collector elements in structures assigned to Seismic Design Category C or higher. The term "collector" describes an element used to transfer the reaction from a diaphragm to the supporting element(s) of the lateral-force-resisting system where the length of the vertical element is less than the diaphragm dimension at that location. Figure 1605.4(2) illustrates this relationship. Failures of collectors can result in separation of

a portion of a building from its lateral-force-resisting system. Where a collector also supports gravity loads such a failure would result in the loss of a portion of the gravity load-carrying system. The intent is to produce collector elements that have sufficient overstrength so that any inelastic behavior appropriately occurs within the seismic-force-resisting system and rather than the collector elements.

Note that the special seismic load combinations are not general load cases, but are only to be applied where specified in the indicated earthquake load provisions (see above) or in the structural materials requirements. See Section 1808.2.23.2.3 or 1908.1.12 for an example of the latter.

Maximum earthquake load effect, E_m, in Section 12.4.3 of ASCE 7 includes effects of the horizontal load, E_{mh}, as well as a vertical component, E_v. E_{mh} is the product of the overstrength factor, Ω_0, and Q_E, the effects of horizontal earthquake forces. E_v is the same vertical component used for the earthquake load effect, E (see commentary, Section 1602.1). Substituting for E_m, the special seismic load combinations can be rewritten as follows:

Equation 16-22
$$1.2D + f_1L + E_m = (1.2+0.2S_{DS})D + f_1L + \Omega_0Q_E$$

Equation 16-23
$$0.9D + E_m = (0.9-0.2S_{DS})D + \Omega_0Q_E$$

The code requires the use of these strength combinations regardless of whether the structure is de-

HORIZONTAL IRREGULARITY TYPE 4... OUT-OF-PLANE OFFSETS OF THE LATERAL-FORCE-RESISTING ELEMENTS

VERTICAL IRREGULARITY TYPE 4... AN IN-PLANE OFFSET OF THE LATERAL-FORCE-RESISTING ELEMENTS

Figure 1605.4(1)
EXAMPLE OF DISCONTINUOUS SHEAR WALLS

signed using strength or allowable stress. The ASCE 7 standard does not use special seismic load combinations. Instead, the maximum earthquake load effect of Section 12.4.3 of ASCE 7 is intended to be used directly in the allowable stress or strength load combinations of that document. Where allowable stress design is used, Section 12.4.3.3 of ASCE 7 permits the allowable stresses to be increased by a factor of 1.2. This stress increase in ASCE 7 was derived from a former provision in both this code and UBC (see 2000 IBC Section 1617.1.2 or 2003 IBC Section 1617.1.1.2). That provision, which applied to allowable stress design, allowed the design strength under the special seismic load combinations to be determined using a stress increase of 1.7. Where a design is in accordance with ASCE 7 allowable stress load combinations, the factor of 0.7 converts the earthquake load effect from a strength load to a nominal load (similar to Equation 16-15). If that is done, then the 1.7 stress increase should be reduced by the same factor [1.7 x 0.7 = 1.19 (rounded to 1.2)]. Since the intent of the 1.2 stress increase in ASCE 7 is to provide a result that is comparable to what was previously permitted under the UBC and the code, it would be appropriate to consider the former 1.7 stress increase for determining a design strength when using these special seismic load combinations for structures designed using the allowable stress method.

1605.5 Heliports and helistops. Heliport and helistop landing areas shall be designed for the following loads, combined in accordance with Section 1605:

1. Dead load, D, plus the gross weight of the helicopter, D_h, plus snow load, S.

2. Dead load, D, plus two single concentrated impact loads, L, approximately 8 feet (2438 mm) apart applied anywhere on the landing area (representing the helicopter's two main landing gear, whether skid type or wheeled type), having a magnitude of 0.75 times the gross weight of the helicopter. Both loads acting together total one-and one half times the gross weight of the helicopter.

3. Dead load, D, plus a uniform live load, L, of 100 psf (4.79 kN/m²).

> **Exception:** Landing areas designed for helicopters with gross weights not exceeding 3,000 pounds (13.34 kN) in accordance with Items 1 and 2 shall be permitted to be designed using a 40 psf (1.92 kN/m²) uniform live load in Item 3, provided the landing area is identified with a 3,000 pound (13.34 kN) weight limitation. This 40 psf (1.92 kN/m²) uniform live load shall not be reduced. The landing area weight limitation shall be indicated by the numeral "3" (kips) located in the bottom right corner of the landing area as viewed from the primary approach path. The landing area weight limitation shall be a minimum of 5 feet (1524 mm) in height.

❖ Detailed requirements for helistops and heliports, including definitions, can be found in Section 412.5 of the code. For example Section 412.5.3 requires a minimum landing area size of 20 feet by 20 feet (6096 mm by 6096 mm) for helicopters weighing less than 3,500 pounds (15.57 kN). In addition, Section 1107 of the *International Fire Code®* (IFC®) regulates helistops and heliports. Section 1107.8 of the IFC requires Federal Aviation Administration (FAA) approval of these facilities.

These structural design requirements serve two purposes. First, they establish the minimum dead load and live load criteria that are specific to the design of the landing area and the supporting structural elements. Second, they specify how the dead and live loads are to be utilized in the load combinations of Section 1605.2 or 1605.3.

The first load case accounts for the gross weight of the helicopter by considering the helicopter weight as

PLAN VIEW: DIAPHRAGM AND SHEAR WALLS

COLLECTOR BEAM COLLECTS FORCE FROM DIAPHRAGM
AND DISTRIBUTES IT TO THE SHEAR WALL

Figure 1605.4(2)
COLLECTORS

a dead load on the landing area to be combined with other dead loads, as well as the design snow load. This simulates a condition where the helicopter is "parked" on the landing area for an extended period of time. The second load case considers other dead load along with concentrated live loads that are based on the weight of the helicopter plus an impact load from landing. This intends to account for the dynamic loads that are generated by a helicopter landing. The third load case includes all other dead loads combined with a uniform live load of 100 psf (4.79 kN/m²) on the heliport or helistop landing area.

The majority of helicopters used in general aviation have gross weights less than 3,000 pounds (13 kN). With weights comparable to those of passenger vehicles and considering the size of the landing area, the equivalent uniform load on these landing areas is actually lower than the minimum uniform live load required for a passenger vehicle parking garage. Thus, a reduced design live load is permitted for design of these landing areas. The exception to the use of a 100 psf (4.79 kN/m²) live load allows the use of a 40 psf (1.92 kN/m²) live load for helicopter landing areas. In consideration of the fact that well over half of all helicopters common to general aviation have gross weights not exceeding 3,000 pounds (13 kN), a maximum gross weight limitation of 3,000 pounds (13.34 kN) is incorporated in the exception. Since marking the weight limitation of the landing area is standard practice as well as an FAA recommendation, indicating the weight limitation on the landing area was included as a requirement under the exception.

SECTION 1606
DEAD LOADS

1606.1 General. Dead loads are those loads defined in Section 1602.1. Dead loads shall be considered permanent loads.

❖ The nominal dead load, D, is determined in accordance with Section 1606. Similar to the general section on live loads (see Section 1607.1), this section provides a reminder that the term "dead loads" is defined, establishing exactly what should be considered a dead load. It also states that dead loads are considered permanent. This affects how dead loads are classified (permanent versus variable), which is a necessary distinction to make when applying the provisions for load combinations.

1606.2 Design dead load. For purposes of design, the actual weights of materials of construction and fixed service equipment shall be used. In the absence of definite information, values used shall be subject to the approval of the building official.

❖ When determining the design dead loads, the actual weights of all materials, equipment, construction, etc., are to be used for the structural design. Often the exact type of materials and equipment to be installed has not yet been determined by the design team at the time the structural design is initiated. Where the actual weights are not known, it is common to use an esti-

mate of the anticipated materials and equipment. While the actual dead load itself does not vary, the estimate of the dead load that is used in structural computations can vary. Such estimates of dead loads are typically greater than the actual dead loads so that the computations are conservative and a redesign of the structure will not be necessary when the actual weights become known. This section clarifies that the building official must approve these estimated dead loads.

Overestimating the actual dead load, so that the design computations are conservative, is acceptable when considering load combinations in Section 1605 that are additive. But the same cannot be said when using counteracting load combinations (see commentary, Section 1604.9). In determining the anchorage required to resist the overturning or uplift effects of wind or seismic loads, resistance is typically provided by the dead load. For example, uplift forces due to wind cause tension to develop in hold-down connections. If the dead load used to resist the wind uplift is overestimated, the result may be an unconservative design. Note that Section 1605.3.2 restricts the dead load used to counteract the effects of overturning and uplift to be the minimum dead load likely to be in place during a design wind event.

As a design guideline, see the unit weights of common construction materials and assemblies in Tables C3-1 and C3-2 of ASCE 7. The unit dead loads listed in the tables for assembled elements are usually given in units of pounds per square foot of surface area (i.e., floor areas, wall areas, ceiling areas, etc.). Unit dead loads for materials used in construction are given in terms of density (pounds per cubic foot). The unit weights given in the tables are generally single values, even though a range of weights may actually exist. The average unit weights given are generally suitable for design purposes; however, where there is reason to believe that the actual weights of assembled elements or construction materials may substantially exceed the tabular values, then the situation should be investigated and the highest values used.

SECTION 1607
LIVE LOADS

1607.1 General. Live loads are those loads defined in Section 1602.1.

❖ Nominal live loads are determined in accordance with Section 1607. The live load requirements for the design of buildings and structures are based on the type of occupancy. Live loads are transient loads that vary with time. Generally, the design live load is that which is believed to be near the maximum transient load for a given occupancy.

TABLE 1607.1. See page 16-24.

❖ The design values of live loads for both uniform and concentrated loads are shown in the table as a function of occupancy. The values given are conservative and include both the sustained and variable portions of the live load. Section 1607.3 directs the designer to utilize the greater live loads produced by the intended occupancy, but not less than the minimum uniformly distributed live loads listed in Table 1607.1. It should also be noted that the "occupancy" category listed is not necessarily group specific. For example, an office building may be classified as Group B, but still contain incidental storage areas. Depending on the type of storage, the areas may warrant storage live loads of either 125 or 250 psf (5.98 or 11.9 kN/m²) to be applied to the space in question.

Residential attic live load: Table 1607.1 specifies minimum uniform live load in residential attics for three distinct conditions: uninhabitable attics without storage; uninhabitable attics with limited storage; and habitable attics and sleeping areas. Table 1607.1(1) summarizes the uniform live loads that apply to uninhabitable attics based on the criteria contained in Notes i, j and k to Table 1607.1. The process for determining which load is applicable is summarized in the flow chart in Figure 1607.1.

One noteworthy distinction made by Note k is that any uninhabitable attic that is served by a fixed stair must be designed using the load of 30 psf (1.44 kN/m²) that is applicable to habitable attics (see Note k). Since the 20 psf (0.96 kN/m²) load is described as "limited storage," the implication is that the presence of a fixed stair is conducive to larger quantities of storage and warrants a greater design live load. This recognizes

that the fixed stair is likely to serve an attic with greater headroom available, providing more storage capacity. For an uninhabitable attic that is accessed by means other than a fixed stair (e.g. framed opening or pull-down stairs), it is necessary to determine whether the limited storage load applies in accordance with the criteria contained in Notes i and j.

Historically a minimum load of 10 psf (0.48 kN/m²) has been viewed as appropriate where occasional access to the attic is anticipated for maintenance purposes, but significant storage is restricted by physical constraints such as low clearance or the configuration of truss webs. It provides a minimum degree of structural integrity, allowing for occasional access to an attic space for maintenance purposes. Allowing the application of this load to be independent of other live loads is deemed appropriate, since it would be rare for this load and other maximum live loads to occur at once.

Table 1607.1(1)
SUMMARY OF MINIMUM LIVE LOAD REQUIRED IN UNINHABITABLE ATTICS

Description	Uniform Load
Without storage	10 psf
With limited storage	20 psf
Served by fixed stair	30 psf

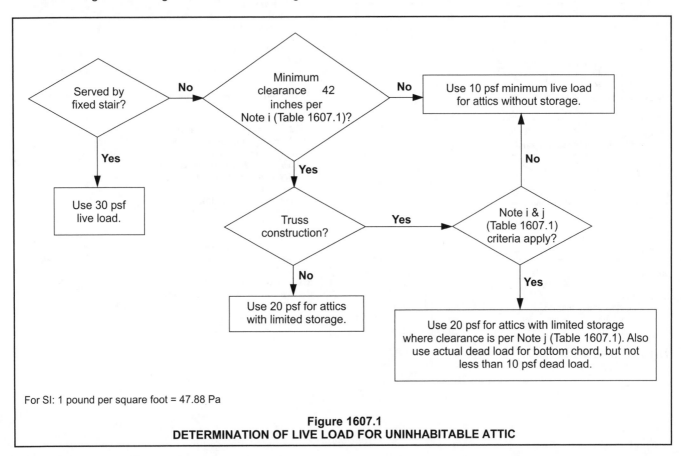

For SI: 1 pound per square foot = 47.88 Pa

Figure 1607.1
DETERMINATION OF LIVE LOAD FOR UNINHABITABLE ATTIC

TABLE 1607.1
MINIMUM UNIFORMLY DISTRIBUTED LIVE LOADS AND MINIMUM CONCENTRATED LIVE LOADS[g]

OCCUPANCY OR USE	UNIFORM (psf)	CONCENTRATED (lbs.)	OCCUPANCY OR USE	UNIFORM (psf)	CONCENTRATED (lbs.)
1. Apartments (see residential)	—	—	23. Libraries Corridors above first floor Reading rooms Stack rooms	 80 60 150[b]	 1,000 1,000 1,000
2. Access floor systems Office use Computer use	 50 100	 2,000 2,000	24. Manufacturing Heavy Light	 250 125	 3,000 2,000
3. Armories and drill rooms	150	—	25. Marquees	75	—
4. Assembly areas and theaters Fixed seats (fastened to floor) Follow spot, projections and control rooms Lobbies Movable seats Stages and platforms	 60 50 100 100 125	 — 	26. Office buildings Corridors above first floor File and computer rooms shall be designed for heavier loads based on anticipated occupancy Lobbies and first-floor corridors Offices	 80 — 100 50	 2,000 — 2,000 2,000
5. Balconies On one- and two-family residences only, and not exceeding 100 sq ft	100 60	 —	27. Penal institutions Cell blocks Corridors	 40 100	 —
6. Bowling alleys	75	—	28. Residential One- and two-family dwellings Uninhabitable attics without storage[i] Uninhabitable attics with limited storage[i, j, k] Habitable attics and sleeping areas All other areas except balconies and decks Hotels and multiple-family dwellings Private rooms and corridors serving them Public rooms and corridors serving them	 10 20 30 40 40 100	 —
7. Catwalks	40	300			
8. Dance halls and ballrooms	100	—			
9. Decks	Same as occupancy served[h]	—			
10. Dining rooms and restaurants	100	—			
11. Dwellings (see residential)	—	—			
12. Cornices	60	—			
13. Corridors, except as otherwise indicated	100	—	29. Reviewing stands, grandstands and bleachers	Note c	
14. Elevator machine room grating (on area of 4 in²)	—	300	30. Roofs All roof surfaces subject to maintenance workers Awnings and canopies Fabric construction supported by a lightweight rigid skeleton structure All other construction Ordinary flat, pitched, and curved roofs Primary roof members, exposed to a work floor Single panel point of lower chord of roof trusses or any point along primary structural members supporting roofs: Over manufacturing, storage warehouses, and repair garages All other occupancies Roofs used for other special purposes Roofs used for promenade purposes Roofs used for roof gardens or assembly purposes	 5 nonreduceable 20 20 Note 1 60 100	 300 2,000 300 Note 1
15. Finish light floor plate construction (on area of 1 in²)	—	200			
16. Fire escapes On single-family dwellings only	100 40	 —			
17. Garages (passenger vehicles only) Trucks and buses	40 Note a See Section 1607.6				
18. Grandstands (see stadium and arena bleachers)	—	—			
19. Gymnasiums, main floors and balconies	100	—			
20. Handrails, guards and grab bars	See Section 1607.7				
21. Hospitals Corridors above first floor Operating rooms, laboratories Patient rooms	 80 60 40	 1,000 1,000 1,000			
22. Hotels (see residential)	—	—			

(continued)

TABLE 1607.1—continued
MINIMUM UNIFORMLY DISTRIBUTED LIVE LOADS AND
MINIMUM CONCENTRATED LIVE LOADS[g]

OCCUPANCY OR USE	UNIFORM (psf)	CONCENTRATED (lbs.)
31. Schools		
Classrooms	40	1,000
Corridors above first floor	80	1,000
First-floor corridors	100	1,000
32. Scuttles, skylight ribs and accessible ceilings	—	200
33. Sidewalks, vehicular driveways and yards, subject to trucking	250[d]	8,000[e]
34. Skating rinks	100	—
35. Stadiums and arenas		
Bleachers	100[c]	—
Fixed seats (fastened to floor)	60[c]	
36. Stairs and exits		
One- and two-family dwellings	40	Note f
All other	100	
37. Storage warehouses (shall be designed for heavier loads if required for anticipated storage)		
Heavy	250	
Light	125	
38. Stores		
Retail		
First floor	100	1,000
Upper floors	75	1,000
Wholesale, all floors	125	1,000
39. Vehicle barriers	See Section 1607.7.3	
40. Walkways and elevated platforms (other than exitways)	60	—
41. Yards and terraces, pedestrians	100	—

For SI: 1 inch = 25.4 mm, 1 square inch = 645.16 mm^2,
1 square foot = 0.0929 m^2,
1 pound per square foot = 0.0479 kN/m^2, 1 pound = 0.004448 kN,
1 pound per cubic foot = 16 kg/m^3

a. Floors in garages or portions of buildings used for the storage of motor vehicles shall be designed for the uniformly distributed live loads of Table 1607.1 or the following concentrated loads: (1) for garages restricted to vehicles accommodating not more than nine passengers, 3,000 pounds acting on an area of 4.5 inches by 4.5 inches; (2) for mechanical parking structures without slab or deck which are used for storing passenger vehicles only, 2,250 pounds per wheel.

b. The loading applies to stack room floors that support nonmobile, double-faced library bookstacks, subject to the following limitations:
   1. The nominal bookstack unit height shall not exceed 90 inches;
   2. The nominal shelf depth shall not exceed 12 inches for each face; and
   3. Parallel rows of double-faced bookstacks shall be separated by aisles not less than 36 inches wide.

c. Design in accordance with the ICC *Standard on Bleachers, Folding and Telescopic Seating and Grandstands.*

d. Other uniform loads in accordance with an approved method which contains provisions for truck loadings shall also be considered where appropriate.

e. The concentrated wheel load shall be applied on an area of 20 square inches.

f. Minimum concentrated load on stair treads (on area of 4 square inches) is 300 pounds.

g. Where snow loads occur that are in excess of the design conditions, the structure shall be designed to support the loads due to the increased loads caused by drift buildup or a greater snow design determined by the building official (see Section 1608). For special-purpose roofs, see Section 1607.11.2.2.

h. See Section 1604.8.3 for decks attached to exterior walls.

i. Attics without storage are those where the maximum clear height between the joist and rafter is less than 42 inches, or where there are not two or more adjacent trusses with the same web configuration capable of containing a rectangle 42 inches high by 2 feet wide, or greater, located within the plane of the truss. For attics without storage, this live load need not be assumed to act concurrently with any other live load requirements.

j. For attics with limited storage and constructed with trusses, this live load need only be applied to those portions of the bottom chord where there are two or more adjacent trusses with the same web configuration capable of containing a rectangle 42 inches high by 2 feet wide or greater, located within the plane of the truss. The rectangle shall fit between the top of the bottom chord and the bottom of any other truss member, provided that each of the following criteria is met:
   i. The attic area is accessible by a pull-down stairway or framed opening in accordance with Section 1209.2, and
   ii. The truss shall have a bottom chord pitch less than 2:12.
   iii. Bottom chords of trusses shall be designed for the greater of actual imposed dead load or 10 psf, uniformly distributed over the entire span.

k. Attic spaces served by a fixed stair shall be designed to support the minimum live load specified for habitable attics and sleeping rooms.

l. Roofs used for other special purposes shall be designed for appropriate loads as approved by the building official.

1607.2 Loads not specified. For occupancies or uses not designated in Table 1607.1, the live load shall be determined in accordance with a method approved by the building official.

❖ Whenever an occupancy or use of a structure cannot be identified with the listing shown in Table 1607.1, then the live load values used for design are required to be determined by the design professional and subject to the approval of the building official. Aside from the obvious intent of this requirement, however, which is to prescribe a minimum design load value, some caution needs to be exercised by the design professional in determining the appropriate design live load value. For example, the table shows that heavy storage areas must be designed for a uniform live load of 250 psf (11.9 kN/m^2). This is a minimum value. Storage warehouses or storage areas within manufacturing facilities containing items such as automobile parts, electrical goods, coiled steel, plumbing supplies and bulk building materials generally have live loads ranging between 300 and 400 psf (14.4 and 19 kN/m^2). Similarly, storage facilities containing dry goods, paints, oil, groceries or liquor often have loadings that range between 200 to 300 psf (9.6 to 14.4 kN/m^2). Another example is a heavy manufacturing facility that makes generators for the electric power industry. Some of the production areas in this type of facility require structural floors that support loads of 1,000 psf (47.9 kN/m^2) or more, which is about seven times the live load specified in Table 1607.1.

1607.3 Uniform live loads. The live loads used in the design of buildings and other structures shall be the maximum loads expected by the intended use or occupancy but shall in no case be less than the minimum uniformly distributed unit loads required by Table 1607.1.

❖ Studies have shown that building live loads consist of a sustained portion based on the day-to-day use of the facilities, and a variable portion created by unusual events such as remodeling, temporary storage of materials, the extraordinary assemblage of people for an occasional business meeting or social function (i.e., holiday party) and similar events. The sustained portion of the live load will likely vary during the life of a building because of tenant changes, rearrangement of office space and furnishings, changes in the nature of the occupancy (i.e., number of people or type of business), traffic patterns and so on. In light of this variability of loadings that are apt to be imposed on a building, the code provisions simplify the design procedure by expressing the applicable load as either a uniformly distributed live load or a concentrated live load on the floor area. It should be pointed out that this section does not require the concurrent application of uniform live load and concentrated live load. In other words, this section requires that either the uniform load or the concentrated load be applied, so long as the type of load that produces the greater stress in the structural element under consideration is utilized.

1607.4 Concentrated loads. Floors and other similar surfaces shall be designed to support the uniformly distributed live loads prescribed in Section 1607.3 or the concentrated load, in pounds (kilonewtons), given in Table 1607.1, whichever produces the greater load effects. Unless otherwise specified, the indicated concentration shall be assumed to be uniformly distributed over an area 2.5 feet by 2.5 feet [6.25 square feet (0.58 m^2)] and shall be located so as to produce the maximum load effects in the structural members.

❖ A building or portion thereof is subjected to concentrated floor loads commensurate with the use of the facility. For example, in Group B, a law office may have stacks of books and files that impose large concentrations of loads on the supporting structural elements. An industrial facility may have a tank full of liquid material on a mezzanine that feeds a machine on the floor below. The structural floor of a stockroom may support heavy bins containing metal parts and so on. The exact locations or nature of such concentrated loadings is not usually known at the time of the design of the building. Furthermore, new sources of concentrated loadings will be added during the life of the structure, while some or all of the existing sources will be relocated; therefore, because of the uncertainties of the sources of concentrated loads as well as their weights and locations, the code provides typical loads to be used in the design of structural floors consistent with the type of use of the facility. The minimum concentrated loads to be used for design are contained in Table 1607.1. Concentrated loads are not required to be applied simultaneously, with the uniform live loads also specified in Table 1607.1. Concentrated loads are to be applied as an independent load condition at the location on the floor that produces the greatest stress in the structural members being designed. The single concentrated load is to be placed at any location on the floor. For example, in an office use area, the floor system is to be designed for either a 2,000-pound (8897 N) concentrated load (unless the anticipated actual concentrated load is higher) applied at any loca-

tion in the office area, or the 50 psf (2.40 kN/m^2) live load specified in Table 1607.1, whichever results in the greater stress in the supporting structural member.

1607.5 Partition loads. In office buildings and in other buildings where partition locations are subject to change, provisions for partition weight shall be made, whether or not partitions are shown on the construction documents, unless the specified live load exceeds 80 psf (3.83 kN/m^2). The partition load shall not be less than a uniformly distributed live load of 15 psf (0.74 kN/m^2).

❖ Provision for the weight of partitions must be made in the structural design. The weight of any built-in partitions should be considered a dead load in accordance with the definition in Section 1602. Buildings where partitions are readily relocated must include a live load of 15 psf (0.74 kN/m^2) if the uniform floor live load is 80 psf (3.83 kN/m^2) or less. This partition allowance is included under live loads because of its variable nature.

1607.6 Truck and bus garages. Minimum live loads for garages having trucks or buses shall be as specified in Table 1607.6, but shall not be less than 50 psf (2.40 kN/m^2), unless other loads are specifically justified and approved by the building official. Actual loads shall be used where they are greater than the loads specified in the table.

❖ The uniform load specified in this section is to be applied to the garage floor in accordance with Section 1607.6.1 (also see Table 1607.6 for the specified concentrated loads that are to be included in the design). The uniform and concentrated loads are to be applied as separate load cases and not at the same time.

TABLE 1607.6
UNIFORM AND CONCENTRATED LOADS

LOADING CLASS[a]	UNIFORM LOAD (pounds/linear foot of lane)	CONCENTRATED LOAD (pounds)[b]	
		For moment design	For shear design
H20-44 and HS20-44	640	18,000	26,000
H15-44 and HS15-44	480	13,500	19,500

For SI: 1 pound per linear foot = 0.01459 kN/m, 1 pound = 0.004448 kN, 1 ton = 8.90 kN.

a. An H loading class designates a two-axle truck with a semitrailer. An HS loading class designates a tractor truck with a semitrailer. The numbers following the letter classification indicate the gross weight in tons of the standard truck and the year the loadings were instituted.

b. See Section 1607.6.1 for the loading of multiple spans.

❖ These are the uniform and concentrated loads that have been established by the American Association of State Highway Transportation Officials (AASHTO) by truck live loads (see Section 1607.6.1 for directions regarding the application of these loads). The uniform load in the table is in pounds per linear foot (plf) of lane. The load is to be divided by 10 feet (3048 mm) to get the area loading of 64 or 48 psf (3.1 or 2.3 kN/m^2), depending on the truck loading used.

1607.6.1 Truck and bus garage live load application. The concentrated load and uniform load shall be uniformly distributed over a 10-foot (3048 mm) width on a line normal to the centerline of the lane placed within a 12-foot-wide (3658 mm) lane. The loads shall be placed within their individual lanes so as to produce the maximum stress in each structural member. Single spans shall be designed for the uniform load in Table 1607.6 and one simultaneous concentrated load positioned to produce the maximum effect. Multiple spans shall be designed for the uniform load in Table 1607.6 on the spans and two simultaneous concentrated loads in two spans positioned to produce the maximum negative moment effect. Multiple span design loads, for other effects, shall be the same as for single spans.

❖ This section specifies the application of live loads so as to produce the maximum stresses. This provision is the truck live loading that has been established by AASHTO.

1607.7 Loads on handrails, guards, grab bars and vehicle barriers. Handrails, guards, grab bars as designed in ICC A117.1 and vehicle barriers shall be designed and constructed to the structural loading conditions set forth in this section.

❖ The requirements of this section are intended to provide an adequate degree of structural strength and stability to handrails, guards, grab bars and vehicle barriers.

1607.7.1 Handrails and guards. Handrail assemblies and guards shall be designed to resist a load of 50 plf (0.73 kN/m) applied in any direction at the top and to transfer this load through the supports to the structure. Glass handrail assemblies and guards shall also comply with Section 2407.

Exceptions:

1. For one- and two-family dwellings, only the single concentrated load required by Section 1607.7.1.1 shall be applied.

2. In Group I-3, F, H and S occupancies, for areas that are not accessible to the general public and that have an occupant load less than 50, the minimum load shall be 20 pounds per foot (0.29 kN/m).

❖ The loading in this section is the maximum anticipated load from use of the handrail by a crowd of people on the stairway. The exceptions provide lower loads for circumstances where the public does not use the handrail. These loads, depicted in Figure 1607.7.1, are permitted to be applied independent of other loads.

1607.7.1.1 Concentrated load. Handrail assemblies and guards shall be able to resist a single concentrated load of 200 pounds (0.89 kN), applied in any direction at any point along the top, and have attachment devices and supporting structure to transfer this loading to appropriate structural elements of the building. This load need not be assumed to act concurrently with the loads specified in the preceding paragraph.

❖ The concentrated loading in this section is not to be applied with any other design load; it is a separate load case. The load simulates the maximum anticipated load from a person grabbing or falling into the handrail or guard.

1607.7.1.2 Components. Intermediate rails (all those except the handrail), balusters and panel fillers shall be designed to withstand a horizontally applied normal load of 50 pounds (0.22 kN) on an area equal to 1 square foot (0.093m²), including openings and space between rails. Reactions due to this loading are not required to be superimposed with those of Section 1607.7.1 or 1607.7.1.1.

❖ This is a localized design load for the guard members and is not to be applied with any other loads. It is to be applied horizontally at a 90-degree (1.57 rad) angle with the guard members. The number of balusters that would resist this load are those within the 1 square foot (0.093 m²) area in the plane of the guard as shown in Figure 1607.7.1.2.

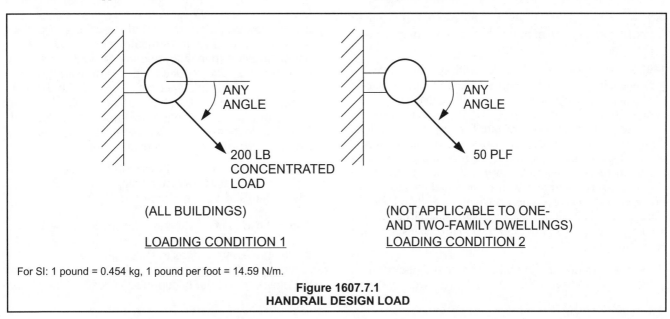

ANY ANGLE

200 LB CONCENTRATED LOAD

(ALL BUILDINGS)

LOADING CONDITION 1

ANY ANGLE

50 PLF

(NOT APPLICABLE TO ONE- AND TWO-FAMILY DWELLINGS)

LOADING CONDITION 2

For SI: 1 pound = 0.454 kg, 1 pound per foot = 14.59 N/m.

Figure 1607.7.1
HANDRAIL DESIGN LOAD

50 LB CONCENTRATED LOAD ON
1 FT x 1 FT SQUARE AT ANY POINT

For SI: 1 pound per square foot = 47.88 Pa.

Figure 1607.7.1.2
COMPONENT DESIGN LOAD

1607.7.1.3 Stress increase. Where handrails and guards are designed in accordance with the provisions for allowable stress design (working stress design) exclusively for the loads specified in Section 1607.7.1, the allowable stress for the members and their attachments is permitted to be increased by one-third.

❖ This section clarifies that the allowable stress design stresses are permitted to be increased by one-third when the handrail or guard is subjected to any one of the design loads specified in Sections 1607.7.1 through 1607.7.1.2. This is appropriate since the design loads are the maximum anticipated loads.

1607.7.2 Grab bars, shower seats and dressing room bench seats. Grab bars, shower seats and dressing room bench seat systems shall be designed to resist a single concentrated load of 250 pounds (1.11 kN) applied in any direction at any point.

❖ These live loads provide for the normal anticipated loads from the use of the grab bars, shower seats and dressing room bench seats.

1607.7.3 Vehicle barriers. Vehicle barrier systems for passenger cars shall be designed to resist a single load of 6,000 pounds (26.70 kN) applied horizontally in any direction to the barrier system and shall have anchorage or attachment capable of transmitting this load to the structure. For design of the system, the load shall be assumed to act at a minimum height of 1 foot, 6 inches (457 mm) above the floor or ramp surface on an area not to exceed 1 square foot (305 mm²), and is not required to be assumed to act concurrently with any handraolil or guard loadings specified in the preceding paragraphs of Section 1607.7.1. Garages accommodating trucks and buses shall be designed in accordance with an approved method that contains provision for traffic railings.

❖ Vehicle barriers provide a passive restraint system in locations where vehicles could fall to a lower level (see definition of "Vehicle barrier" in Section 1602). The loading depicted in Figure 1607.7.3 is required for the design of passenger car and light truck vehicle barriers. The 6,000-pound (26.70 kN) load considers impact. The loads are intended for use with a one-third increase in allowable stresses when using the allowable stress design method. For bus and heavy truck vehicle barrier design criteria, the state department of transportation should be contacted.

1607.8 Impact loads. The live loads specified in Section 1607.3 include allowance for impact conditions. Provisions shall be made in the structural design for uses and loads that involve unusual vibration and impact forces.

❖ In cases where "unordinary" live loads are likely to occur in a building that imposes impact or vibratory forces on structural elements (i.e., elevators, machinery, craneways, etc.), additional stresses and deflections are produced in the structural system. Where unusual vibration (dynamic) and impact loads are likely to occur, the code requires that the structural design take these effects into account. Typically, the dynamic effects are approximated through the application of an equivalent static load equal to the dynamic load effects. In most cases, an equivalent static load is sufficient. A dynamic analysis is usually not required.

1607.8.1 Elevators. Elevator loads shall be increased by 100 percent for impact and the structural supports shall be designed within the limits of deflection prescribed by ASME A17.1.

❖ The static load of an elevator must be increased to account for the effect of elevator motion. For example, when an elevator comes to a stop, the load on the elevator supports is significantly higher than the weight of the elevator and the occupants. This effect increases with the acceleration and deceleration rate of the elevator.

1607.8.2 Machinery. For the purpose of design, the weight of machinery and moving loads shall be increased as follows to allow for impact: (1) elevator machinery, 100 percent; (2) light machinery, shaft- or motor-driven, 20 percent; (3) reciprocating machinery or power-driven units, 50 percent; (4) hangers for floors or balconies, 33 percent. Percentages shall be increased where specified by the manufacturer.

❖ The specified increases for machinery loads include the vibration of the equipment, which increases the effective load. The load increase for reciprocating machinery versus rotating shaft-driven machinery is to account for the higher vibration. The increase for floor or balcony hangers recognizes that these structural systems are subject to vibration as a result of the way they are supported.

1607.9 Reduction in live loads. Except for roof uniform live loads, all other minimum uniformly distributed live loads, L_o, in Table 1607.1 are permitted to be reduced in accordance with Section 1607.9.1 or 1607.9.2.

❖ Small portions of a floor are more likely to be subjected to the full uniform load than larger floor areas. Unloaded or lightly loaded areas tend to reduce the total load on the structural members supporting those floors. The specified uniformly distributed live loads from Table 1607.1 are permitted to be reduced, with some notable exceptions or limitations, in recognition that the larger the tributary area of a structural member, the lower the likelihood that the full live load will be realized. The basis for the live load reduction in Sections 1607.9.1 through 1607.9.1.4 is ASCE 7. An alternative of live load reduction, retained from legacy model codes, is provided in Section 1607.9.2.

1607.9.1 General. Subject to the limitations of Sections 1607.9.1.1 through 1607.9.1.4, members for which a value of $K_{LL}A_T$ is 400 square feet (37.16 m²) or more are permitted to be designed for a reduced live load in accordance with the following equation:

$$L = L_o \left(0.25 + \frac{15}{\sqrt{K_{LL} A_T}} \right) \qquad \textbf{(Equation 16-24)}$$

For SI: $L = L_o \left(0.25 + \dfrac{457}{\sqrt{K_{LL} A_T}} \right)$

where:

L = Reduced design live load per square foot (meter) of area supported by the member.

L_o = Unreduced design live load per square foot (meter) of area supported by the member (see Table 1607.1).

K_{LL} = Live load element factor (see Table 1607.9.1).

A_T = Tributary area, in square feet (square meters).

L shall not be less than $0.50L_o$ for members supporting one floor and L shall not be less than $0.40L_o$ for members supporting two or more floors.

❖ This section provides a method of reducing uniform floor live loads that is based on the provisions of ASCE 7. The concept is that where the design live load is governed by the minimum live loads in Table 1607.1, the actual load on a large area of the floor is very likely to be less than the nominal live load in the table. Thus, the allowable reduction increases with the tributary area of the floor that is supported by a structural member; therefore, a girder that supports a large tributary area would be allowed to be designed for somewhat less live load per square foot than a floor beam that

6,000 POUNDS APPLIED AT 18 IN. ABOVE FLOOR ON 1 SQUARE FOOT AREA

VEHICLE BARRIER SYSTEM

18 IN.

> 12 INCHES
ADJACENT FLOOR LEVEL

For SI: 1 inch = 25.4 mm, 1 square foot = 0.0929 m², 1 pound = 4.448 N.

Figure 1607.7.3
PASSENGER CAR VEHICLE BARRIER REQUIREMENTS

supports a smaller total floor area. The following example demonstrates live load reduction method (refer to Figure 1607.9.1):

$$L = 50\left(.25 + \frac{15}{\sqrt{1500}}\right) = 32psf > .5L_0$$

TABLE 1607.9.1
LIVE LOAD ELEMENT FACTOR, K_{LL}

ELEMENT	K_{LL}
Interior columns	4
Exterior columns without cantilever slabs	4
Edge columns with cantilever slabs	3
Corner columns with cantilever slabs	2
Edge beams without cantilever slabs	2
Interior beams	2

All other members not identified above including:
  Edge beams with cantilever slabs
  Cantilever beams
  Two-way slabs
  Members without provisions for continuous shear
    transfer normal to their span
	1

❖ The purpose of this table is to provide tributary area adjustment factors, K_{LL}, for determining live load reductions in Section 1607.9.1. The factor converts the tributary area of the structural member to an "influence area." This "influence area" of a structural member is considered to be the adjacent floor area from which it derives any of its load. These adjustments to the tributary area range from 1 through 4 based on the type of structural element being designed and are meant to reflect the element's ability to share load with adjacent elements.

1607.9.1.1 Heavy live loads. Live loads that exceed 100 psf (4.79 kN/m²) shall not be reduced.

Exceptions:

1. The live loads for members supporting two or more floors are permitted to be reduced by a maximum of 20 percent, but the live load shall not be less than L as calculated in Section 1607.9.1.

2. For uses other than storage, where approved, additional live load reductions shall be permitted where shown by the registered design professional that a rational approach has been used and that such reductions are warranted.

❖ The purpose of this section is to prohibit live load reductions where the live loads exceed 100 psf (4.79 kN/m²). Such live loads are typically intended for storage or related purposes. It is more likely that the full live load will be realized at a given floor level for such occupancies. Thus, reduced live loads are not allowed for these conditions except as described in this section.

In Exception 1, the loads on structural members, such as columns and bearing walls that support two or more floors, are allowed to be reduced by 20 percent. Surveys have indicated that it is rare for the total live load on any story to exceed 80 percent of the tabulated uniform live loads. Conservatively, the full load should apply to beams and girders, but a member supporting multiple floors is allowed some live load reduction.

Recognizing that there are circumstances under which live loads exceed 100 psf (4.79 kN/m²) in occupancies other than storage, Exception 2 allows the

OFFICE BUILDING
L_o = 50 psf
D = 45 psf

DETERMINE REDUCED LIVE LOAD IN ACCORDANCE WITH SECTION 1607.9.1 FOR BEAM G1

2 @ 25 = 50

2 @ 30 = 60

G1

A_t = 25 x 30 = 750 SQ. FT.

For SI: 1 foot = 304.8 mm, 1 square foot = 0.0929 m², 1 pound per square foot = 47.88 Pa

Figure 1607.9.1
LIVE LOAD REDUCTION EXAMPLE

registered design professional to present a "rational" live load reduction proposal to be applied for members with larger tributary areas (e.g. girders, columns, foundations, etc.). Examples would be mechanical rooms, electrical rooms, process mezzanines in industrial buildings, etc. These types of areas may have very high localized uniform loads under the equipment footprints, for instance, but the live loads to members having larger tributary areas are much less, on average.

1607.9.1.2 Passenger vehicle garages. The live loads shall not be reduced in passenger vehicle garages except the live loads for members supporting two or more floors are permitted to be reduced by a maximum of 20 percent, but the live load shall not be less than L as calculated in Section 1607.9.1.

❖ This section limits the live load reduction for passenger vehicle garages to only those members that support more than two floors. Thus, floor framing members that support only a part of one floor do not warrant a reduction to the live load of 40 psf (1.92 kN/m²) that is specified in Table 1607.1. The rationale for allowing some live load reduction for members supporting multiple floors is similar to that given under Section 1607.9.1.1, Exception 1.

1607.9.1.3 Special occupancies. Live loads of 100 psf (4.79 kN/m²) or less shall not be reduced in public assembly occupancies.

❖ Due to the nature of public assembly occupancies, it is very likely that those areas will be fully occupied at times and, subsequently, there is a high probability that the entire floor area will actually be subjected to the full uniform live load. This section does not allow the live load to be reduced in a public assembly occupancy where the uniform live load is 100 psf (4.79 kN/m²) or less. This restriction is very similar to the prohibition on live load reductions for Group A occupancies under the alternative live load reduction method (see Section 1607.9.2, Item 1).

1607.9.1.4 Special structural elements. Live loads shall not be reduced for one-way slabs except as permitted in Section 1607.9.1.1. Live loads of 100 psf (4.79 kN/m²) or less shall not be reduced for roof members except as specified in Section 1607.11.2.

❖ One-way slabs are restricted from live load reductions based on the belief that this construction lacks redundancy. The live load reductions can be applied only where the slab supports more than one floor. The explanation for allowing some live load reduction for members supporting multiple floors is similar to that given under Section 1607.9.1.1, Exception 1. Roof load reductions are allowed based on the provisions of Section 1607.11.2.

1607.9.2 Alternate floor live load reduction. As an alternative to Section 1607.9.1, floor live loads are permitted to be reduced in accordance with the following provisions. Such reductions shall apply to slab systems, beams, girders, columns, piers, walls and foundations.

1. A reduction shall not be permitted in Group A occupancies.

2. A reduction shall not be permitted where the live load exceeds 100 psf (4.79 kN/m²) except that the design live load for members supporting two or more floors is permitted to be reduced by 20 percent.

3. A reduction shall not be permitted in passenger vehicle parking garages except that the live loads for members supporting two or more floors are permitted to be reduced by a maximum of 20 percent.

4. For live loads not exceeding 100 psf (4.79 kN/m²), the design live load for any structural member supporting 150 square feet (13.94 m²) or more is permitted to be reduced in accordance with the following equation:

$$R = 0.08 (A - 150) \qquad \textbf{(Equation 16-25)}$$

For SI: $R = 0.861 (A - 13.94)$

Such reduction shall not exceed the smallest of:

1. 40 percent for horizontal members;

2. 60 percent for vertical members; or

3. R as determined by the following equation.

$$R = 23.1 (1 + D/L_o) \qquad \textbf{(Equation 16-26)}$$

where:

A = Area of floor supported by the member, square feet (m²).

D = Dead load per square foot (m²) of area supported.

L_o = Unreduced live load per square foot (m²) of area supported.

R = Reduction in percent.

❖ This section includes an alternative floor live load method that is permitted to be used instead of the method indicated in Sections 1607.9 through 1607.9.1.4. The basis for this section is the 1997 UBC.

Where reductions are permitted, they are allowed at a rate of 0.08 percent per square foot of area in excess of 150 square feet (14 m²). This value cannot exceed 60 percent for vertical members, such as columns or bearing walls, or 40 percent for horizontal members such as beams or girders. Additionally, the reduction cannot be more than the value determined by Equation 16-26.

Example of an alternative floor live load reduction:

For beam G1 given in Figure 1607.9.1, determine the reduced floor live load in accordance with Section 1607.9.2.

Solution: A_T = 750 square feet (69 m²) > 150 square feet (14 m²); therefore, a reduction is permitted.

Equation 16-25
$$R = (0.08\%)(750 - 150) = 48\% > 40\%$$

Equation 16-26
$$R = (23.1\%)\left(1 + \frac{45}{50}\right) = 43.7\% > 40\%$$

Use the maximum 40 percent reduction allowed for horizontal members.

use $L = 50(1-0.4) = 30$ psf (144 kN/m^2)

1607.10 Distribution of floor loads. Where uniform floor live loads are involved in the design of structural members arranged so as to create continuity, the minimum applied loads shall be the full dead loads on all spans in combination with the floor live loads on spans selected to produce the greatest effect at each location under consideration. It shall be permitted to reduce floor live loads in accordance with Section 1607.9.

❖ For continuous floor members loaded such that the live loads of a building are distributed in some bays and not in others, some of the structural elements will be subjected to greater stresses because of partial loading conditions as compared to full loading on all spans. This code section requires the engineer to consider partial loadings that produce the greatest design forces for any location in the design of continuous floor elements.

For example, Figure 1607.10 shows a continuous multispan girder with partial loading. The Type I loading condition shows that only the alternate spans have uniform live loads, which produces:

- Maximum positive moments at the centers of the loaded spans (A-B, C-D, E-F) and
- Maximum negative moments at the centers of the unloaded spans (B-C, D-E, F-G).

The Type II live load distribution shows two loaded adjacent spans with alternate spans loaded beyond these, which produces:

- Maximum negative moment at Support D and
- Maximum girder shears.

To obtain the maximum total stresses imposed on the girder, the dead load moments and shears must be added to those produced by the partial live loadings.

1607.11 Roof loads. The structural supports of roofs and marquees shall be designed to resist wind and, where applicable, snow and earthquake loads, in addition to the dead load of construction and the appropriate live loads as prescribed in this section, or as set forth in Table 1607.1. The live loads acting on a sloping surface shall be assumed to act vertically on the horizontal projection of that surface.

❖ In addition to dead and live loads, the roof's structural system is to be designed and constructed to resist environmental loads caused by wind, snow and earthquakes. According to the definition of roof live loads in Section 1602.1, these are typically considered to be an allowance for maintenance of equipment and the roof system itself. Other live loads must be considered as appropriate, such as "special purpose roofs" (see Section 1607.11.2.2) or wherever an occupancy-related live load would be applicable.

1607.11.1 Distribution of roof loads. Where uniform roof live loads are reduced to less than 20 psf (0.96 kN/m^2) in accordance with Section 1607.11.2.1 and are involved in the design of structural members arranged so as to create continuity, the minimum applied loads shall be the full dead loads on all spans in combination with the roof live loads on adjacent spans or on alternate spans, whichever produces the greatest effect. See

Figure 1607.10
PARTIAL LOADING

Section 1607.11.2 for minimum roof live loads and Section 7.5 of ASCE 7 for partial snow loading.

❖ For continuous roof construction, where live loads are reduced to less tha 20 psf (0.96 kN/m²) as permitted in Section 1607.11.2.1, partial loadings must be included in the design of structural elements to determine the governing loading situation. For example, Figure 1607.10 shows a continuous multispan girder with partial loading. The Type I loading condition shows that only the alternate spans have uniform live loads, which produces:

• Maximum positive moments at the centers of the loaded spans (A-B, C-D, E-F) and

• Maximum negative moments at the centers of the unloaded spans (B-C, D-E, F-G).

The Type II live load distribution shows two loaded adjacent spans with alternate spans loaded beyond these, which produces:

• Maximum negative moment at Support D and

• Maximum girder shears.

To obtain the maximum total stresses imposed on the girder, the dead load moments and shears must be added to those produced by the partial live loadings.

1607.11.2 Reduction in roof live loads. The minimum uniformly distributed roof live loads, L_o, in Table 1607.1 are permitted to be reduced according to the following provisions.

❖ The minimum roof live loads typically reflect loads that occur during roof maintenance, construction or repair. In addition to the standard roof live load of 20 psf (0.96 kN/m²), Table 1607.1 includes roof live loads for special purpose roofs and fabric awnings. While this section seems to refer to reducing any of the tabulated uniformly distributed live loads, the actual reduction method provided in Section 1607.11.2.1 is limited to the 20 psf (0.96 kN/m²) live load. This is made evident by the limits on Equation 16-27 of $12 \leq L_r \leq 20$.

1607.11.2.1 Flat, pitched and curved roofs. Ordinary flat, pitched and curved roofs are permitted to be designed for a reduced roof live load as specified in the following equation or other controlling combinations of loads in Section 1605, whichever produces the greater load. In structures where special scaffolding is used as a work surface for workers and materials during maintenance and repair operations, a lower roof load than specified in the following equation shall not be used unless approved by the building official. Greenhouses shall be designed for a minimum roof live load of 12 psf (0.58 kN/m²).

$$L_r = L_o R_1 R_2 \qquad \text{(Equation 16-27)}$$

where: $12 \leq L_r \leq 20$

For SI: $L_r = L_o R_1 R_2$

where: $0.58 \leq L_r \leq 0.96$

L_r = Reduced live load per square foot (m²) of horizontal projection in pounds per square foot (kN/m²).

The reduction factors R_1 and R_2 shall be determined as follows:

$R_1 = 1$ for $A_t \leq 200$ square feet
(18.58 m²) **(Equation 16-28)**

$R_1 = 1.2 - 0.001A_t$ for 200 square
feet $< A_t < 600$ square feet **(Equation 16-29)**

For SI: $1.2 - 0.011A_t$ for 18.58 square meters $< A_t < 55.74$ square meters

$R_1 = 0.6$ for $A_t > 600$ square feet
(55.74 m²) **(Equation 16-30)**

where:

A_t = Tributary area (span length multiplied by effective width) in square feet (m²) supported by any structural member, and

$R_2 = 1$ for $F \leq 4$ **(Equation 16-31)**

$R_2 = 1.2 - 0.05 F$ for $4 < F < 12$ **(Equation 16-32)**

$R_2 = 0.6$ for $F > 12$ **(Equation 16-33)**

where:

F = For a sloped roof, the number of inches of rise per foot (for SI: $F = 0.12 \times$ slope, with slope expressed as a percentage), or for an arch or dome, the rise-to-span ratio multiplied by 32.

❖ This section provides a formula for the determination of the live load for design of flat, pitched or curved roofs. The live load from Table 1607.1 that applies is 20 psf (0.96 kN/m²). Reduced roof live loads are based on the roof slope and the tributary area of the member being considered. The portion of the live load reduction based on tributary area does not apply to roof members that support small tributary areas of less than 200 square feet (18.58 m²). The load can be reduced as the tributary area increases but never to less than 12 psf (0.58 kN/m²). For roof slopes between 4:12 and 12:12 live load reductions based on slope apply. Figure 1607.11.2.1 shows the roof live load, L_r, determined by Equation 16-27 for increments of roof slope. Since the relationship of variables is linear, intermediate values can be interpolated from the figure.

This section also provides for a lower roof live load for a greenhouse, since it is not likely that loads from maintenance or repair will exceed the specified 12 psf (0.58 kN/m²).

1607.11.2.2 Special-purpose roofs. Roofs used for promenade purposes, roof gardens, assembly purposes or other special purposes shall be designed for a minimum live load as required in Table 1607.1. Such roof live loads are permitted to be reduced in accordance with 1607.9.

❖ Roofs that are to be occupied during social events incidental to the principal use of the facility are to be designed for a minimum uniform live load of 60 psf (2.87 kN/m²). The promenade deck of a residential penthouse located on the main roof of an apartment building is an example of this type of use. Where roofs are designed to be used as roof gardens or to support large gatherings of people as a function accompanying the educational or assembly uses of a facility, the roofs are required to be designed to a minimum live load of 100 psf (4.79 kN/m²). The minimum live loads specified in Table 1607.1 are not required to be added to the design load requirements for special-purpose roofs specified in this section.

1607.11.2.3 Landscaped roofs. Where roofs are to be landscaped, the uniform design live load in the landscaped area shall be 20 psf (0.958 kN/m²). The weight of the landscaping materials shall be considered as dead load and shall be computed on the basis of saturation of the soil.

❖ Those areas of a special-purpose roof that are to be landscaped are required to be designed for a minimum uniform live load of 20 psf (0.96 kN/m²) to accommodate the occasional loads associated with the maintenance of plantings. The weight of landscaping materials is to be considered as dead loads in the design of the roof structure, which is to be combined with the live load (see Section 1605 for load combinations to be included).

1607.11.2.4 Awnings and canopies. Awnings and canopies shall be designed for uniform live loads as required in Table 1607.1 as well as for snow loads and wind loads as specified in Sections 1608 and 1609.

❖ Awning and canopy structures covered with fabric materials are to be designed to sustain a live load of 5 psf (0.24 kN/m²) as well as the specified snow and wind loads of Sections 1608 and 1609. The live load, snow load and wind load are to be combined according to Section 1605.

1607.12 Crane loads. The crane live load shall be the rated capacity of the crane. Design loads for the runway beams, including connections and support brackets, of moving bridge cranes and monorail cranes shall include the maximum wheel loads of the crane and the vertical impact, lateral and longitudinal forces induced by the moving crane.

❖ This section provides a general description of the crane loads that are required to be included in the design. The supporting structure for the crane is to be designed for a combination of the maximum wheel load, vertical impact and horizontal load as a simultaneous load combination. The typical arrangement for a top-running bridge crane is shown in Figure 1607.12.

1607.12.1 Maximum wheel load. The maximum wheel loads shall be the wheel loads produced by the weight of the bridge, as applicable, plus the sum of the rated capacity and the weight of the trolley with the trolley positioned on its runway at the location where the resulting load effect is maximum.

❖ The maximum vertical wheel load occurs when the trolley is moved as close as possible to the supporting beams under consideration. This results in the greatest portion of the crane weight, the design weight lifted load and the wheel vertical impact load on the supporting beams.

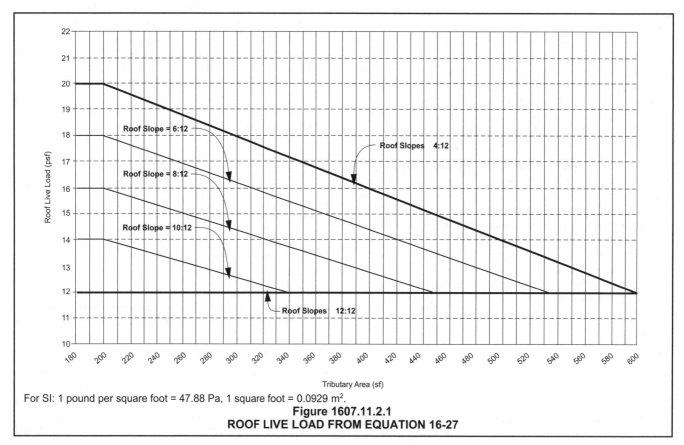

For SI: 1 pound per square foot = 47.88 Pa, 1 square foot = 0.0929 m².

Figure 1607.11.2.1
ROOF LIVE LOAD FROM EQUATION 16-27

1607.12.2 Vertical impact force. The maximum wheel loads of the crane shall be increased by the percentages shown below to determine the induced vertical impact or vibration force:

Monorail cranes (powered) · · · · · · · · · · · 25 percent

Cab-operated or remotely operated
bridge cranes (powered) · · · · · · · · · · · · · 25 percent

Pendant-operated bridge cranes (powered) · · · · 10 percent

Bridge cranes or monorail cranes with
hand-geared bridge, trolley and hoist · · · · · · · · 0 percent

❖ A vertical impact force is necessary to account for the impact from the starting and stopping movement of the suspended weight from the crane. Vertical impact is also created by the movement of the crane along the rails.

1607.12.3 Lateral force. The lateral force on crane runway beams with electrically powered trolleys shall be calculated as 20 percent of the sum of the rated capacity of the crane and the weight of the hoist and trolley. The lateral force shall be assumed to act horizontally at the traction surface of a runway beam, in either direction perpendicular to the beam, and shall be distributed according to the lateral stiffness of the runway beam and supporting structure.

❖ This section is necessary to define the design lateral force on the crane supports. Lateral force at the right angle to the crane rail is caused by the lateral movement of the lifted load and from the frame action of the crane.

1607.12.4 Longitudinal force. The longitudinal force on crane runway beams, except for bridge cranes with hand-geared bridges, shall be calculated as 10 percent of the maximum wheel loads of the crane. The longitudinal force shall be assumed to act horizontally at the traction surface of a runway beam, in either direction parallel to the beam.

❖ This section is needed to define the longitudinal force on the crane supports, which is caused from the longitudinal motion of the crane with the lifted load.

1607.13 Interior walls and partitions. Interior walls and partitions that exceed 6 feet (1829 mm) in height, including their finish materials, shall have adequate strength to resist the loads to which they are subjected but not less than a horizontal load of 5 psf (0.240 kN/m²).

Exception: Fabric partitions complying with Section 1607.13.1 shall not be required to resist the minimum horizontal load of 5 psf (0.24 kN/m²).

❖ The minimum lateral live load is intended to provide sufficient strength and durability of the wall framing and of the finished construction to provide a minimum level of resistance to nominal impact loads that commonly occur in the use of a facility, such as impacts from moving furniture or equipment, as well as to resist HVAC pressurization.

1607.13.1 Fabric partitions. Fabric partitions that exceed 6 feet (1829 mm) in height, including their finish materials, shall have adequate strength to resist the following load conditions:

1. A horizontal distributed load of 5 psf (0.24 kN/m²) applied to the partition framing. The total area used to determine the distributed load shall be the area of the fabric face between the framing members to which the fabric is attached. The total distributed load shall be uniformly applied to such framing members in proportion to the length of each member.

2. A concentrated load of 40 pounds (0.176 kN) applied to an 8-inch diameter (203 mm) area [50.3 square inches

**Figure 1607.12
TOP-RUNNING BRIDGE CRANE**

(32 452 mm²)] of the fabric face at a height of 54 inches (1372 mm) above the floor.

❖ This section provides criteria for fabric partitions (see definition in Section 1602.1) as an alternative to Section 1607.13. The construction of these partitions is unique, which makes it difficult to meet the full requirements of Section 1607.13. Condition 1 requires the partition framing to be capable of resisting a minimum lateral load. Condition 2 approximates the load of a person leaning against the fabric using their hand as the point of contact. This criterion is based on test standards that are used to evaluate the tip-over resistance of office furniture panel systems that are often used to provide open plan offices.

SECTION 1608
SNOW LOADS

1608.1 General. Design snow loads shall be determined in accordance with Chapter 7 of ASCE 7, but the design roof load shall not be less than that determined by Section 1607.

❖ The determination of the nominal snow load, S, must be in accordance with this section. The intent of this section is that the code requirements are based on the technical requirements in Chapter 7 of ASCE 7. The snow load provisions in ASCE 7 are based on over 40 years of ground snow load data, and include consideration of thermal resistance of the roof structure, a rain-on-snow surcharge, partial loading on continuous beam systems and ponding instability from melting snow or rain on snow. The variables that affect the determination of roof snow loads are:

- Ground snow load (p_g) – see Section 1608.2.
- Importance factor (I) – a factor determined in Section 7.3.3 of ASCE 7, ranging from 0.8 to 1.2, based on the occupancy category assigned in accordance with Section 1604.5.
- Exposure factor (C_e) and Thermal factor (C_t) – see Sections 7.3.1 and 7.3.2 of ASCE 7.

In addition to the above, the roof slope affects the snow load determination as well as other design considerations. Roofs with slopes equal to or less than 5 degrees (0.09 rad) are designed for a flat roof snow load, p_f, determined in accordance with Section 7.3 of ASCE 7, using the above criteria. Other considerations for low roof slopes include ponding instability and rain-on-snow surcharge loading. Where the roof slope is greater than 5 degrees (0.09 rad), the sloped roof snow load, p_s, in accordance with Section 7.4 of ASCE 7 applies. The sloped roof snow load essentially modifies the flat roof snow load by the slope factor (C_s), which varies from 1.0 at low a slope (i.e. no effect) to zero at a slope of 70 degrees (1.36 rad).

The flat roof snow load or sloped roof snow load applied uniformly to the entire roof is referred to as the balanced snow load condition. This loading is always a design consideration. Depending on factors such as the type of roof structural system, the geometry of the roof, etc., the following additional snow loadings may require evaluation:

- Partial loading (see Section 7.5 of ASCE 7) is a pattern consisting of balanced snow load and one-half of balanced snow load arranged to produce the maximum effects on the structural member being considered.
- Unbalanced snow loads (see Section 7.6 of ASCE 7) reflect an uneven loading pattern such as can occur on a sawtooth roof [see Figure 1608.1(1)].
- Drifting (see Sections 7.7 and 7.8 of ASCE 7) is a concern where adjacent roof surfaces are at different elevations [see Figures 1608.1(2) and (3)] or at projections above the roof level such as at equipment or parapets [see Figure 1608.1(4)].
- Sliding snow [see Section 7.9 of ASCE 7 and Figure 1608.1(5)].

Figure 1608.1(1)
BALANCED AND UNBALANCED LOADS FOR A SAWTOOTH ROOF

Figure 1608.1(2)
DRIFTING SNOW ON LOW ROOFS AND DECKS

For SI: 1 foot = 304.8 mm.

Figure 1608.1(3)
DRIFTING SNOW ONTO ADJACENT LOW STRUCTURE

1608.2 Ground snow loads. The ground snow loads to be used in determining the design snow loads for roofs shall be determined in accordance with ASCE 7 or Figure 1608.2 for the contiguous United States and Table 1608.2 for Alaska. Site-specific case studies shall be made in areas designated "CS" in Figure 1608.2. Ground snow loads for sites at elevations above the limits indicated in Figure 1608.2 and for all sites within the CS areas shall be approved. Ground snow load determination for such sites shall be based on an extreme value statistical analysis of data available in the vicinity of the site using a value with a 2-percent annual probability of being exceeded (50-year mean recurrence interval). Snow loads are zero for Hawaii, except in mountainous regions as approved by the building official.

❖ The ground snow loads on the maps in Figure 1608.2 of the code are generally based on over 40 years of snow depth records. The snow loads on the maps are those that have a 2-percent annual probability of being exceeded (a 50-year mean recurrence interval). The maps were generated from data through the winter of 1991-92, and from data through the winter of 1993-94 where the snows were heavy. The mapped snow loads are not increased much from a single snowy winter, since most reporting stations have more than 20 years of snow data. The map values indicate the ground snow load in pounds per square foot. In mountainous areas, the map also indicates the highest elevation that is appropriate for the use of the associated snow load. Where the elevation limit is exceeded, a site-specific case study is necessary to establish the appropriate ground snow load. In some areas the ground snow load is too variable to allow mapping. These areas are noted as "CS," which indicates a site-specific case study in necessary. Assistance in the determination of an appropriate ground snow load for these areas may be obtained from the U.S. Department of Army Cold Regions Research and Engineering Laboratory (CRREL) in Hanover, New Hampshire.

Figure 1608.1(4)
SNOW DRIFTING AT ROOF PROJECTIONS

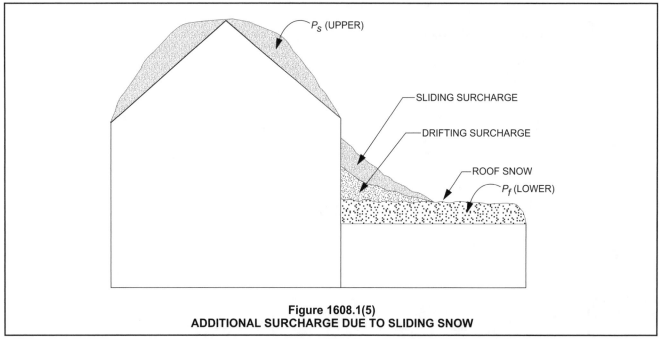

Figure 1608.1(5)
ADDITIONAL SURCHARGE DUE TO SLIDING SNOW

TABLE 1608.2. See below.

❖ Since Alaska in not shown on Figure 1608.2, this table provides the ground snow loads, P_g, for Alaskan locations. These values are needed to determine the appropriate roof snow loads for the indicated locations. The roof snow load is to be determined according to Section 7.3 of ASCE 7 for flat roofs and Section 7.4 of ASCE 7 for sloped roofs.

FIGURE 1608.2. See page 16-40.

❖ See the commentary to Section 1608.2 for an overview of the snow load map. See the commentary for Section 7.2 of ASCE 7 for a complete description of methodology used in developing the contour lines shown on the figure.

SECTION 1609
WIND LOADS

FIGURE 1609. See page 16-42.

❖ Figure 1609 is a duplication of Figure 6-1 of ASCE 7 and provides basic wind speeds based on 3-second-gusts at 33 feet (10 058 mm) above ground for Exposure Category C (see Section 1609.4). The reference to the 50-year mean recurrence interval (MRI) used in previous maps has been removed, reflecting the fact that the MRI is greater than 50 years along the hurricane coastline. However, nonhurricane wind speeds are based on a 50-year MRI as Section 1609.3 specifically requires when estimating basic wind speeds from local climatic data.

Earlier editions of ASCE 7 and the legacy model codes incorporated fastest-mile wind speed maps. "Fastest mile" is defined as the average speed of one mile of air that passes a specific reference point. It is important to recognize the differences in averaging times between fastest-mile and the 3-second gust maps. The averaging time for a 90-mph fastest-mile wind speed is ($t = 3,600/V$) 40 seconds. Obviously, due to greater averaging time, for a given location the fastest-mile wind speed will be less than the 3-second-gust wind speed. The fact that the wind speed values are higher does not necessarily indicate higher wind loads. Buildings and structures resist wind loads, not wind speeds. Wind speed, although a significant contributor, is only one of several variables and factors that affect wind forces. Wind loads are affected by atmosphere and aerodynamics. Other elements that affect actual wind forces as wind flows across a bluff body include shape factors (C_p), gust effect factors (G) and the velocity pressure that is a function of wind speed, exposure and topography, among others.

The change from the fastest-mile wind speed map to a 3-second-gust map was necessary for the following reasons. First, weather stations across the United States no longer collect fastest-mile wind speed data. Additionally, the perception of the general public will be more favorable where the code wind speeds are higher, although the design wind pressures were not changed significantly. The map includes a more complete analysis of hurricane wind speeds than previous maps, since more data was available for sites away from the coast. In western states, the 85- to 90-mph contour boundary follows along the Washington, Oregon and California eastern state lines. This is because inland wind data was such that there was no statistical basis to place them elsewhere. The design wind speeds do not include the effects of tornadoes. The probability of a tornadic event at any particular location is so low that tornadoes do not appear in the 50-year statistical storm data used to formulate the inland portion of the map. See the commentary to Section 6.5.4 in ASCE 7 for more information on basic wind speed.

TABLE 1608.2
GROUND SNOW LOADS, p_g, FOR ALASKAN LOCATIONS

LOCATION	POUNDS PER SQUARE FOOT	LOCATION	POUNDS PER SQUARE FOOT	LOCATION	POUNDS PER SQUARE FOOT
Adak	30	Galena	60	Petersburg	150
Anchorage	50	Gulkana	70	St. Paul Islands	40
Angoon	70	Homer	40	Seward	50
Barrow	25	Juneau	60	Shemya	25
Barter Island	35	Kenai	70	Sitka	50
Bethel	40	Kodiak	30	Talkeetna	120
Big Delta	50	Kotzebue	60	Unalakleet	50
Cold Bay	25	McGrath	70	Valdez	160
Cordova	100	Nenana	80	Whittier	300
Fairbanks	60	Nome	70	Wrangell	60
Fort Yukon	60	Palmer	50	Yakutat	150

For SI: 1 pound per square foot = 0.0479 kN/m².

FIGURE 1608.2
GROUND SNOW LOADS, p_g, FOR THE UNITED STATES (psf)

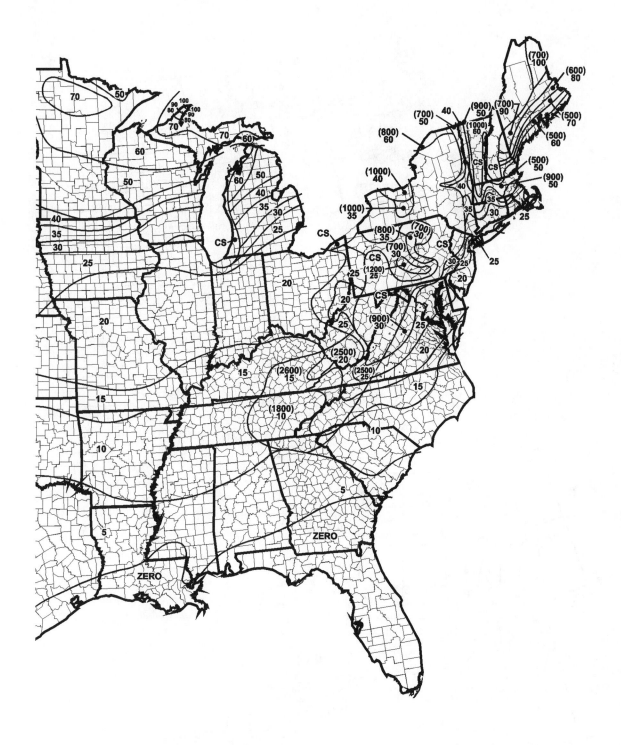

FIGURE 1608.2–continued
GROUND SNOW LOADS, p_g, FOR THE UNITED STATES (psf)

FIGURE 1609
BASIC WIND SPEED (3-SECOND GUST)

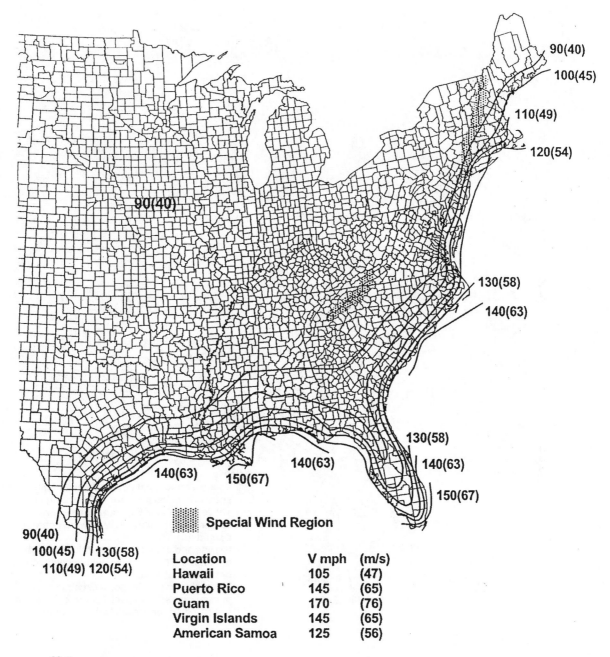

Location	V mph	(m/s)
Hawaii	105	(47)
Puerto Rico	145	(65)
Guam	170	(76)
Virgin Islands	145	(65)
American Samoa	125	(56)

Notes:

1. Values are nominal design 3-second gust wind speeds in miles per hour (m/s) at 33 ft (10 m) above ground for Exposure C category.
2. Linear interpolation between wind contours is permitted.
3. Islands and coastal areas outside the last contour shall use the last wind speed contour of the coastal area.
4. Mountainous terrain, gorges, ocean promontories, and special wind regions shall be examined for unusual wind conditions.

FIGURE 1609—continued
BASIC WIND SPEED (3-SECOND GUST)

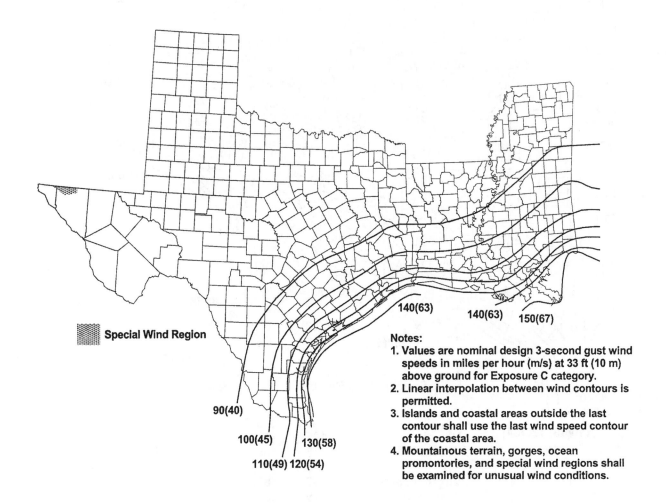

Special Wind Region

140(63)   140(63)   150(67)

90(40)

100(45)   130(58)

110(49) 120(54)

Notes:
1. Values are nominal design 3-second gust wind speeds in miles per hour (m/s) at 33 ft (10 m) above ground for Exposure C category.
2. Linear interpolation between wind contours is permitted.
3. Islands and coastal areas outside the last contour shall use the last wind speed contour of the coastal area.
4. Mountainous terrain, gorges, ocean promontories, and special wind regions shall be examined for unusual wind conditions.

FIGURE 1609–continued
BASIC WIND SPEED (3-SECOND GUST)
WESTERN GULF OF MEXICO HURRICANE COASTLINE

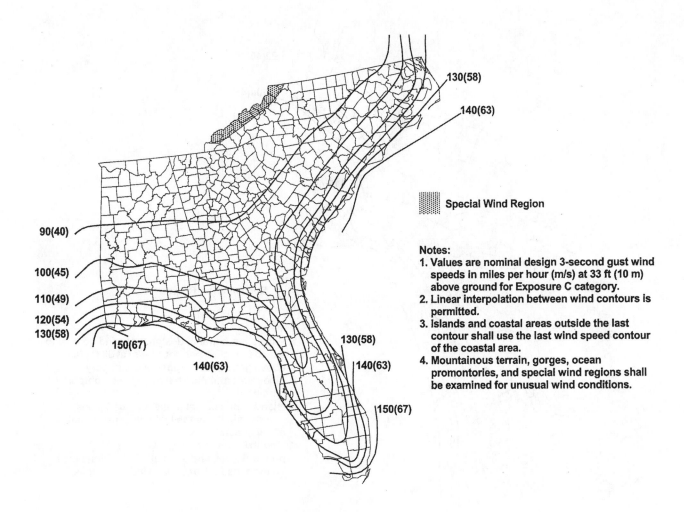

90(40)

100(45)

110(49)

120(54)
130(58)

150(67)

140(63)

130(58)

140(63)

150(67)

130(58)

140(63)

Special Wind Region

Notes:
1. Values are nominal design 3-second gust wind speeds in miles per hour (m/s) at 33 ft (10 m) above ground for Exposure C category.
2. Linear interpolation between wind contours is permitted.
3. Islands and coastal areas outside the last contour shall use the last wind speed contour of the coastal area.
4. Mountainous terrain, gorges, ocean promontories, and special wind regions shall be examined for unusual wind conditions.

FIGURE 1609–continued
BASIC WIND SPEED (3-SECOND GUST)
EASTERN GULF OF MEXICO AND SOUTHEASTERN U.S. HURRICANE COASTLINE

90(40)

100(45)

110(49)

120(54)

Special Wind Region

Notes:
1. Values are nominal design 3-second gust wind speeds in miles per hour (m/s) at 33 ft (10 m) above ground for Exposure C category.
2. Linear interpolation between wind contours is permitted.
3. Islands and coastal areas outside the last contour shall use the last wind speed contour of the coastal area.
4. Mountainous terrain, gorges, ocean promontories, and special wind regions shall be examined for unusual wind conditions.

FIGURE 1609–continued
BASIC WIND SPEED (3-SECOND GUST)
MID AND NORTHERN ATLANTIC HURRICANE COASTLINE

1609.1 Applications. Buildings, structures and parts thereof shall be designed to withstand the minimum wind loads prescribed herein. Decreases in wind loads shall not be made for the effect of shielding by other structures.

❖ The determination of the nominal wind load, *W*, must be in accordance with Section 1609. The intent of this section is to provide minimum criteria for the design and construction of buildings and other structures to resist wind loads. These regulations serve to reduce the potential for damage to property caused by windstorms and to provide an acceptable level of protection to building occupants. The objective also includes the prevention of damage to adjacent properties because of the possible detachment of major building components (e.g., walls, roofs, etc.), structural collapse or flying debris and for the safety of people in the immediate vicinity.

The criteria for wind design given in this section of the code generally reflect the wind load provisions of ASCE 7. For a better understanding of the wind load provisions it is important to have a fundamental knowledge of the effects of high-velocity wind forces on buildings and other structures. Wind/structure interactions can be characterized as follows:

When wind encounters a stationary object such as a building, the airflow changes direction and produces several effects on the building that are illustrated in Figure 1609.1(1). Exterior walls and other vertical surfaces facing the wind (windward side) and perpendicular to its path are subjected to inward (positive) pressures; however, wind does not stop on contact with a facing surface, but flows around and over the building surfaces. This airflow does not instantaneously change directions at surface discontinuities, such as corners of walls or eaves, or over ridges and roof corners. Instead it separates from the downwind surfaces due to high turbulence and localized pressures, resulting in outward (negative) pressures. This phenomena produces suction or outward pressures (negative) on the sidewalls, leeward wall and, depending on geometry, the roof.

Figure 1609.1(1) shows a flat roof and the resulting negative pressures caused by external wind; however, pressures may differ on sloping roofs in the direction perpendicular to the ridge. Roof surfaces (on the windward side) with shallow slopes are generally subjected to outward (negative) pressures—the same as flat roofs. Moderately sloping roofs [about 30 degrees (0.5235 rad)] may be subjected to either inward (positive) pressures, outward (negative) pressures or both—negative pressure in the lower part of the roof and positive pressure in the area of the ridge; however, the code does not require consideration of this scenario. High-sloping roofs (windward side) respond similar to walls and sustain positive wind pressures. Sloped roof sections on the leeward side are sub-

Figure 1609.1(1)
WIND-PRESSURES CAUSED BY EXTERNAL WIND FLOW

jected to negative pressures, regardless of the degree of slope. For sloped roofs where the wind direction is parallel to the ridge, the wind pressures act similarly to a flat roof where the roof is subject to outward (negative) pressures.

Openings in the building envelope can impact the magnitude of wind pressure on a structure by increasing the internal pressures that, in turn, will affect the net pressures on the structure. Figure 1609.1(2) illustrates the effects of openings in a building's exterior walls. An opening in the leeward wall causes negative pressure on the interior, increasing the total load on the windward wall. Openings on the windward wall cause positive internal pressure against all the walls from the interior of the building. As the figure shows,

the result is an increase in the total load on the leeward wall. In extreme cases, failures resulting from this type of wind flow appear as if the building has exploded.

The definition of "Openings" is worth noting, since in the nonstructural code provisions the term "openings" typically refers to doors, ducts or windows. The definition of "Openings" that is relevant to wind load is in Section 6.2 of ASCE 7. They are all described as "apertures or holes in the building envelope that allow air to flow through the building envelope and that are designed as 'open' during design wind loads." Such openings are then considered in classifying the building as enclosed, partially enclosed or open. The wind load considerations will differ based on that classification.

WIND

OPENING ON WINDWARD WALL

WIND

OPENING ON LEEWARD WALL

Figure 1609.1(2)
OPENING IN EXTERIOR WALL OF BUILDING

1609.1.1 Determination of wind loads. Wind loads on every building or structure shall be determined in accordance with Chapter 6 of ASCE 7. The type of opening protection required, the basic wind speed and the exposure category for a site is permitted to be determined in accordance with Section 1609 or ASCE 7. Wind shall be assumed to come from any horizontal direction and wind pressures shall be assumed to act normal to the surface considered.

Exceptions:

1. Subject to the limitations of Section 1609.1.1.1, the provisions of SBCCI SSTD 10 shall be permitted for applicable Group R-2 and R-3 buildings.

2. Subject to the limitations of Section 1609.1.1.1, residential structures using the provisions of the AF&PA WFCM.

3. Designs using NAAMM FP 1001.

4. Designs using TIA/EIA-222 for antenna-supporting structures and antennas.

❖ The intent of Section 1609 is to require that buildings and structures be designed and constructed to resist the wind loads quantified in Chapter 6 of ASCE 7. There are three methods for determining wind loads on a building in ASCE 7:

1. Method 1 is the simplified procedure found in Section 6.4 of ASCE 7. This approach is a projected area method – two applications are provided below.

2. Method 2 in Section 6.5 of ASCE 7 is the analytical procedure. Technically, there are two analytical procedures: rigid buildings of all heights and low-rise buildings [with mean roof heights less than or equal to 60 feet (18 288 mm)]. The simplified procedure is based on the low-rise buildings method.

3. Method 3 is the wind tunnel procedure found in Section 6.6 of ASCE 7.

There are four exceptions to using the provisions of ASCE 7 for the determination of wind loads.

Exception 1 provides for the use of the SBCCI SSTD 10 for Group R-2 and R-3 buildings where they are located within Exposure B or C as defined in Section 1609.4 and not sited on the upper half of an isolated hill, escarpment or ridge with the characteristics described in Section 1609.1.1.1. SBCCI SSTD 10 has prescriptive construction requirements and required load capacity tables that replace the requirement for structural analysis, which is intended to provide improved design construction details to achieve greater structural performance for single- and multiple-family dwellings in a high-wind event. SBCCI SSTD 10 applies to one- and two-story residential buildings of conventional wood-frame, masonry and concrete wall construction. The standard provides information for buildings sited in three fastest-mile wind climates: 90,

100 and 110 mph. See Table 1609.3.1 for conversions between the 3-second-gust wind speed and the fastest-mile wind speed. See SCBBI SSTD 10 for other detailed application limitations.

Exception 2 provides for the use of the AF&PA *Wood Frame Construction Manual for One- and Two-Family Dwellings* where the building is sited within Exposure B or C as defined in Section 1609.4 and not on the upper half of an isolated hill, escarpment or ridge with the characteristics described in Section 1609.1.1.1. The AF&PA *Wood Frame Construction Manual for One- and Two-Family Dwellings* has prescriptive construction requirements and required load-resistance tables that replace the requirement for structural analysis. The tabulated engineered and prescriptive design provisions apply to one- and two-family wood-frame dwellings where the fastest-mile basic wind speed is between 90 and 120 mph. See Table 1609.3.1 for conversion between the 3-second-gust and fastest-mile wind speeds (see the AF&PA *Wood Frame Construction Manual for One- and Two-Family Dwellings*, Chapter 1, for other detailed application limitations).

Exceptions 3 and 4 simply refer to national standards dealing specifically with the design of flagpoles and telecommunication towers.

Examples of calculation of wind pressures: The following examples illustrate the application of the wind criteria in Section 1609 and the ASCE 7 simplified procedure.

Example 1 [see Figure 1609.1.1(1)]

Given:

Enclosed, simple diaphragm building

V = 130 mph

Exposure Category = C

Roof slope = 7:12

Building width, W = 48 feet

Building length, L = 50 feet

Wall height = 18 feet.

K_{zt} and $I = 1.0$

Find: Horizontal main windforce-resisting system (MWFRS) wind loads for end zones (A and B), using the simplified wind procedure.

First, the mean roof height, H, has to be determined. From the building geometry, H is calculated as follows:

$H = 18 \text{ feet} + \frac{1}{2}(7/12)(48 \text{ feet}) = 25 \text{ feet}$

Roof slope = 7:12 = 30 degrees

Horizontal MWFRS wind loads for end zones A and B are found using Figure 6-2 of ASCE 7 for H = 30 feet and Exposure B. For V = 130 mph and roof angle of 30 degrees, the applicable end zone horizontal loads are as follows:

Transverse direction

Zone A – wall $\qquad p_{s30}$ = 30.1 psf

Zone B – roof $\qquad p_{s30}$ = 20.6 psf

Longitudinal direction (Zone A) $\qquad p_{s30}$ = 30.1 psf

These pressures have to be modified for mean roof height, exposure category, topographic factor and importance factor using Equation 6-1 of ASCE 7. From Figure 6-2 of ASCE 7, the height and exposure adjustment factor, λ, for H = 25 feet (7620 mm) and Exposure C is 1.35.

Therefore, the design horizontal wind loads are as follows:

p_s = (30.1 psf) x 1.35 x 1.0 = 40.6 psf (Zone A)

p_s = (20.6 psf) x 1.35 x 1.0 = 27.8 psf (Zone B)

These horizontal pressures on the MWFRS are to be applied as shown in Figures 1609.1.1(2) and 1609.1.1(3).

For SI: 1 foot = 304.8 mm.

Figure 1609.1.1(1)
EXAMPLE 1 BUILDING

For SI: 1 pound per square foot = 47.88 Pa.

Figure 1609.1.1(2)
HORIZONTAL MWFRS LOADS—TRANSVERSE DIRECTION

For SI: 1 pound per square foot = 47.88 Pa.

Figure 1609.1.1(3)
HORIZONTAL MWFRS LOADS—LONGITUDINAL DIRECTION

Example 2

Given:

Basic wind speed, V = 120 mph

Building mean roof height, H = 45 feet

Exposure Category = D

I and K_{zt} = 1.0

Find: Design component and cladding wind pressure for a 20-square-foot (1.9 m²) window located in an edge strip of wall (Zone 5).

Obtain the component and cladding design wind pressures for a building with H = 30 feet and Exposure B, from Figure 6-3 of ASCE 7. The window is located in a wall, Zone 5, with an effective wind area of 20 square feet (1.9 m²) and V = 120 mph. From Figure 6-3 of ASCE 7, the design pressures are:

p_{net30} = + 24.7 psf, -32.4 psf

These pressures have to be modified for mean roof height, exposure category and importance factor using Equation 6-2 of ASCE 7. From Figure 6-3 of ASCE 7, the height and exposure adjustment factor, λ, for H = 45 feet (13 716 mm) and Exposure D is 1.78.

Therefore, the design wind pressures are calculated as:

p_{net} = (+24.7 psf) x 1.78 x 1.0 = + 44 psf

p_{net} = (-32.4 psf) x 1.78 x 1.0 = -57.7 psf

1609.1.1.1 Applicability. The provisions of SSTD 10 are applicable only to buildings located within Exposure B or C as defined in Section 1609.4. The provisions of SBCCI SSTD 10 and the AF&PA WFCM shall not apply to buildings sited on the upper half of an isolated hill, ridge or escarpment meeting the following conditions:

1. The hill, ridge or escarpment is 60 feet (18 288 mm) or higher if located in Exposure B or 30 feet (9144 mm) or higher if located in Exposure C;

2. The maximum average slope of the hill exceeds 10 percent; and

3. The hill, ridge or escarpment is unobstructed upwind by other such topographic features for a distance from the high point of 50 times the height of the hill or 1 mile (1.61 km), whichever is greater.

❖ This section places limitations on the use of SBCCI SSTD 10 and the AF&PA *Wood Frame Construction Manual for One- and Two-Family Dwellings*. SBCCI SSTD 10 is based on the provisions of Section 1606.2 of the *Standard Building Code*, which is based on fastest-mile wind speed criteria. Neither of these standards accounts for the effect of isolated hills, ridges or escarpments. As illustrated in Figure 1609.1.1.1, the wind speed increases when a mass of air passes over those terrain features. This phenomenon is referred to as the "wind speed-up" effect and requires the use of ASCE 7.

1609.1.2 Protection of openings. In wind-borne debris regions, glazing in buildings shall be impact-resistant or protected with an impact-resistant covering meeting the requirements of an approved impact-resisting standard or ASTM E 1996 and ASTM E 1886 referenced therein as follows:

1. Glazed openings located within 30 feet (9144 mm) of grade shall meet the requirements of the Large Missile Test of ASTM E 1996.

2. Glazed openings located more than 30 feet (9144 mm) above grade shall meet the provisions of the Small Missile Test of ASTM E 1996.

Exceptions:

1. Wood structural panels with a minimum thickness of $^7/_{16}$ inch (11.1 mm) and maximum panel span of 8 feet (2438 mm) shall be permitted for opening protection in one- and two-story buildings. Panels shall be precut so that they shall be attached to the framing surrounding the opening containing the product with the glazed opening. Panels shall be secured with the attachment hardware provided. Attachments shall be designed to resist the components and cladding loads determined in accordance with the provisions of ASCE 7. Attachment in accordance with Table 1609.1.2 is permitted for buildings with a mean roof height of 33 feet (10 058 mm) or less where wind speeds do not exceed 130 mph (57.2 m/s).

2. Glazing in Occupancy Category I buildings as defined in Section 1604.5, including greenhouses that are occupied for growing plants on a production or research basis, without public access shall be permitted to be unprotected.

3. Glazing in Occupancy Category II, III or IV buildings located over 60 feet (18 288 mm) above the ground and over 30 feet (9144 mm) above aggregate surface roofs located within 1,500 feet (458 m) of the building shall be permitted to be unprotected.

❖ The purpose of this section is to address risks associated with wind-borne debris in high-wind areas. See the definitions of "Wind-borne debris region" and "Hurricane-prone regions" in Section 1609.2 for an expla-

nation of where these provisions apply. This section requires protection of glazed openings in buildings in wind-borne debris regions. This can be provided in the form of a protective assembly that is impact tested or by using impact-resistant glazing.

During a hurricane, buildings are impacted from wind-borne debris due to high-velocity winds. This debris can impact the glazing, causing breakage and creating an opening within the building envelope. The presence of openings in the building envelope can have a significant effect on the magnitude of the total wind pressure required to be resisted by each structural element of a building. Depending on the location of these openings with respect to wind direction and the amount of background porosity, external and internal pressures may act in the same direction to produce higher forces on some walls and the roof. An example of this is shown in Figure 1609.1.2. In this scenario, as the wind flows over the building, pressures are developed on the external surface as shown. Introduction of an opening in the windward wall causes the wind to rush into the building, exerting internal pressures (positive) against all interior surfaces. This type of opening has the net effect of producing potentially high internal pressures that will act in the same direction as the external pressures on the roof, side and leeward walls. Considering the high probability of wind-borne debris during a hurricane and the effect of an unintended opening in the building envelope, the code requires glazing in designated regions to be protected.

Where wind-borne protection is provided, the section specifies two types of tests to demonstrate adequate resistance: the large missile test to simulate large debris up to 30 feet (9144 mm) above grade and

Figure 1609.1.1.1
WIND SPEED-UP EFFECT

the small missile test to simulate smaller debris up to 60 feet (18 288 mm) above grade, both of which are common during very high winds. An example of small debris is gravel from the surrounding area that becomes air borne.

Impact-resistant coverings or glazing must meet the test requirements of an approved impact standard or ASTM E 1886 and ASTM E 1996. Other impact standards the building official may consider are SBCCI SSTD 12 and Miami-Dade County Protocols PA 201, 202 and 203. These standards specify similar-type testing with a large missile test (2 by 4), small missile test (2 gram balls) and cyclic pressure loading test. ASTM E 1886 and ASTM E 1996 work together with the standard test method (E 1886) and the test specification (E 1996), including scoping, technical requirements and pass/fail criteria.

Exception 1 permits the use of $^7/_{16}$-inch (11.1 mm) wood structural panels with maximum spans of 8 feet (2438 mm) as an impact-resistant covering for one- and two-story buildings. Panel attachments have to be designed to resist the component and cladding pressure from ASCE 7 or attached in accordance with Table 1609.1.2. This protective system has been tested and meets the requirements of SBCCI SSTD 12. The intent is that precut panel coverings and attachment hardware are provided on site.

Exception 2 exempts low-hazard (Occupancy Category I) buildings from this requirement for protecting openings against wind-borne debris. Exception 3 exempts openings in buildings that are classified in Occupancy Category II, III or IV where the opening locations meet the stated conditions.

TABLE 1609.1.2
WIND-BORNE DEBRIS PROTECTION FASTENING SCHEDULE FOR WOOD STRUCTURAL PANELS[a,b,c,d]

FASTENER TYPE	FASTENER SPACING (inches)		
	Panel Span ≤ 4 feet	4 feet < Panel Span ≤ 6 feet	6 feet < Panel Span ≤ 8 feet
No. 6 screws	16	12	9
No. 8 screws	16	16	12

For SI: 1 inch = 25.4 mm, 1 foot = 304.8 mm, 1 pound = 4.4 N, 1 mile per hour = 0.44 m/s.

a. This table is based on a maximum wind speed (3-second gust) of 130 mph and mean roof height of 33 feet or less.

b. Fasteners shall be installed at opposing ends of the wood structural panel. Fasteners shall be located a minimum of 1 inch from the edge of the panel.

c. Fasteners shall be long enough to penetrate through the exterior wall covering a minimum of 1.75 inches into wood wall framing; a minimum of 1.25 inches into concrete block or concrete; or into steel framing by at least three threads. Fasteners shall be located a minimum of 2.5 inches from the edge of concrete block or concrete.

d. Where screws are attached to masonry or masonry/stucco, they shall be attached utilizing vibration-resistant anchors having a minimum withdrawal capacity of 490 pounds.

❖ This table provides the connections for the wood structural panel impact-resistant covering that is described in the exception to Section 1609.1.2. The table lists the spacing of the screws around the perimeter of the panel for the indicated panel spans. Note that Table 1609.1.2 is only applicable to buildings with a mean roof height of 33 feet (10 058 mm) or less and located where the basic wind speed is 130 mph or less.

1609.1.2.1 Louvers. Louvers protecting intake and exhaust ventilation ducts not assumed to be open that are located within 30 feet (9144 mm) of grade shall meet requirements of an

Figure 1609.1.2
EFFECTS OF OPENINGS IN THE BUILDING ENVELOPE

approved impact-resisting standard or the Large Missile Test of ASTM E 1996.

❖ This provision provides direction for impact testing of louvers that cover intake or exhaust duct openings. Louvers often have blades or slats affixed to and covering an opening in the exterior envelope, making them similar to certain types of porous shutters used to protect glazed openings. The scope of ASTM E 1996 covers impact testing of exterior impact protection systems to be used in wind-borne debris regions. The ASTM standard has been referenced for these products in the absence of a standard that is specific to louvers. This test requirement does not apply where a duct penetration has been assumed to be an opening in the building envelope.

Section 1609.1.2.1 is a new requirement for the impact testing of louvers that cover the openings for intake and exhaust ducts in the coastal areas in hurricane-prone regions. When a louver in an exterior wall is either damaged or eliminated by wind-borne debris during a high wind event, the air leakage rated damper inside the ventilation duct may also be exposed to damage.

Prior to approval of this code change, a number of jurisdictions in hurricane prone regions were required testing to the Dade County standard TAS 203 or TAS 201, or the large missile impact test only of ASTM E 1996 to evaluate the impact resistance of louvers. However, the requirements for testing and the enforcement of such a requirement were inconsistent.

The scope of ASTM E 1996 covers impact testing of exterior building features, such as windows, glazed curtain walls, doors and storm shutters in buildings located in geographic regions that are prone to hurricanes, simulating impact by both large and small missiles. For glazed openings and nonporous shutters that protect the fenestration assembly, the impact testing is followed by a cyclic loading test. There is no specific provision in the standard for testing louvers that cover ventilation openings, either for impact or cycling loading.

The intent of this code section is not to require that louvers comply with ASTM E 1996 in its entirety; only the large missile impact test, excluding the cycling loading, or an approved impact standard that may be developed in the future that pertains specifically to louvers.

As with most engineering questions, some judgement by the test sponsor and the laboratory engineers is necessary, based on experience and technical knowledge to set the correct criteria for testing and evaluation, and such decisions must be documented in the test report for consideration by the authority having jurisdiction.

1609.2 Definitions. The following words and terms shall, for the purposes of Section 1609, have the meanings shown herein.

❖ Some terms are used in the code to describe something very specific to the related subject matter while other terms may have multiple meanings that can vary based on the subject matter and context. To avoid misunderstandings, it is preferable to reach a consensus on the meaning of these terms. Definitions are intended to facilitate the understanding of code provisions and to minimize potential confusion.

This section contains definitions of terms associated with wind load criteria. Note that these terms are also listed in Chapter 2 with a cross reference to this section. Definitions that pertain primarily to wind criteria are included within this section to provide convenient access to them without having to refer back to Chapter 2. The use and application of all defined terms, including those defined herein, as well as undefined terms are set forth in Section 201.

HURRICANE-PRONE REGIONS. Areas vulnerable to hurricanes defined as:

1. The U. S. Atlantic Ocean and Gulf of Mexico coasts where the basic wind speed is greater than 90 mph (40 m/s) and

2. Hawaii, Puerto Rico, Guam, Virgin Islands and American Samoa.

❖ This definition specifies the area wherein hurricane-force winds are expected to reach. These areas are also distinguished from nonhurricane wind speed regions by the inclusion of a hurricane importance factor implied in the wind speed contours of Figure 1609.

WIND-BORNE DEBRIS REGION. Portions of hurricane-prone regions that are within 1 mile (1.61 km) of the coastal mean high water line where the basic wind speed is 110 mph (48 m/s) or greater; or portions of hurricane-prone regions where the basic wind speed is 120 mph (53 m/s) or greater; or Hawaii.

❖ This definition identifies those areas that require consideration of the impact of wind-borne debris on the building envelope. See the commentary to Sections 1609.1 and 1609.1.2 for information on the effects of openings in the building envelope.

1609.3 Basic wind speed. The basic wind speed, in mph, for the determination of the wind loads shall be determined by Figure 1609. Basic wind speed for the special wind regions indicated, near mountainous terrain and near gorges shall be in accordance with local jurisdiction requirements. Basic wind speeds determined by the local jurisdiction shall be in accordance with Section 6.5.4 of ASCE 7.

In nonhurricane-prone regions, when the basic wind speed is estimated from regional climatic data, the basic wind speed shall be not less than the wind speed associated with an annual probability of 0.02 (50-year mean recurrence interval), and the estimate shall be adjusted for equivalence to a 3-second gust wind speed at 33 feet (10 m) above ground in Exposure Category C. The data analysis shall be performed in accordance with Section 6.5.4.2 of ASCE 7.

❖ This section establishes the basic wind speed that is to be used for design. The basic wind speed map identifies special wind regions where the speeds vary substantially within a very short distance due to topographic effects, such as mountains and valleys. This section specifies appropriate recurrence interval criteria to be used for estimating the basic wind speeds from regional climatic data in other than hurricane-prone regions. See Section 6.5.4 of ASCE 7 for issues to be addressed in the data analysis and the associated commentary for further details (also see commentary under Figure 1609).

1609.3.1 Wind speed conversion. When required, the 3-second gust basic wind speeds of Figure 1609 shall be converted to fastest-mile wind speeds, V_{fm}, using Table 1609.3.1 or Equation 16-34.

$$V_{fm} = \frac{(V_{3S} - 10.5)}{1.05}$$ **(Equation 16-34)**

where:

V_{3S} = 3-second gust basic wind speed from Figure 1609.

❖ Some referenced standards contain criteria or applications that are based on the fastest-mile wind speed data. For example, SBCCI SSTD 10 specifies prescriptive procedures that use the fastest-mile wind speed designation versus the 3-second-gust wind speed that is the basis of Figure 1609. Equation 16-34 or Table 1609.3.1 provide a means by which these standards can be used with new wind speeds until their requirements have been updated. The conversions were calculated using the gust factor curve of Figure C6-4 in the ASCE 7 commentary. It is important to note that the conversions provide equivalent wind speeds based on averaging times, and do not necessarily provide equivalent wind pressures.

TABLE 1609.3.1. See below.

❖ This table is used where any referenced standards are based on the fastest-mile wind speed. The table converts the 3-second-gust wind speed of a geographical location to the fastest-mile wind speed for use within the standards cited in Section 1609.1.1.

1609.4 Exposure category. For each wind direction considered, an exposure category that adequately reflects the characteristics of ground surface irregularities shall be determined for the site at which the building or structure is to be constructed. Account shall be taken of variations in ground surface roughness that arise from natural topography and vegetation as well as from constructed features.

❖ The concept of exposure categories provides a means of accounting for the relative roughness of the bound-

ary layer. The earth's surface exerts a horizontal drag force on wind due to ground obstructions that retard the flow of air close to the ground. The reduction in the flow of air is a function of height above ground and terrain roughness. Wind speeds increase with height above ground, and the relationship between height above ground and wind speed is exponential. The rate of increase in wind speeds with height is a function of the terrain features. The rougher the terrain (such as large city centers), the shallower the slope of the wind speed profile. The smoother the terrain (open water), the steeper the slope of the wind speed profile.

The definitions of exposure and roughness categories are used to account for this roughness in the boundary layer and are intended to provide an adequate assessment of surface roughness for most situations. Exposure B is considered the roughest boundary layer condition. Exposure D is considered the smoothest boundary layer condition. Accordingly, calculated wind loads are less for Exposure B, which has more surface obstructions, as compared to Exposure D, with all other variables the same. See ASCE 7 commentary for guidance on performing a more detailed analysis of surface roughness.

1609.4.1 Wind directions and sectors. For each selected wind direction at which the wind loads are to be evaluated, the exposure of the building or structure shall be determined for the two upwind sectors extending 45 degrees (0.79 rad) either side of the selected wind direction. The exposures in these two sectors shall be determined in accordance with Sections 1609.4.2 and 1609.4.3 and the exposure resulting in the highest wind loads shall be used to represent winds from that direction.

❖ An exposure category must be determined for each direction in which wind loading is being considered. This section describes the method for doing so, and requires use of the more restrictive exposure category where the two sectors in a given wind direction would be classified in different exposure categories.

1609.4.2 Surface roughness categories. A ground surface roughness within each 45-degree (0.79 rad) sector shall be determined for a distance upwind of the site as defined in Section 1609.4.3 from the categories defined below, for the purpose of assigning an exposure category as defined in Section 1609.4.3.

Surface Roughness B. Urban and suburban areas, wooded areas or other terrain with numerous closely spaced obstructions having the size of single-family dwellings or larger.

TABLE 1609.3.1
EQUIVALENT BASIC WIND SPEEDS[a,b,c]

V_{3S}	85	90	100	105	110	120	125	130	140	145	150	160	170
V_{fm}	71	76	85	90	95	104	109	114	123	128	133	142	152

For SI: 1 mile per hour = 0.44 m/s.
a. Linear interpolation is permitted.
b. V_{3S} is the 3-second gust wind speed (mph).
c. V_{fm} is the fastest mile wind speed (mph).

Surface Roughness C. Open terrain with scattered obstructions having heights generally less than 30 feet (9144 mm). This category includes flat open country, grasslands, and all water surfaces in hurricane-prone regions.

Surface Roughness D. Flat, unobstructed areas and water surfaces outside hurricane-prone regions. This category includes smooth mud flats, salt flats and unbroken ice.

❖ This section defines three surface roughness categories that are used in evaluating each sector and subsequently in determining an exposure category. These surface roughnesses were previously included in the corresponding exposure category definition. The required upwind distance that must be considered varies according to the exposure category definition in Section 1609.4.3.

A significant philosophical change in determination of exposure categories occurred with the inclusion of shorelines in hurricane-prone regions in the definition of Exposure Category C (Surface Roughness C). Exposure Category D (Surface Roughness D) had been used for wind flowing over open water until further research determined that wave action at the water's surface in a hurricane, due to the intensity of the turbulence, produced substantial surface obstructions and friction that reduces the wind profile values to be more in line with Surface Roughness C as opposed to Surface Roughness D. Surface Roughness D would still apply to inland waterways and shorelines that are not in the hurricane-prone regions, such as coastal California, Oregon, Washington and Alaska.

1609.4.3 Exposure categories. An exposure category shall be determined in accordance with the following:

Exposure B. Exposure B shall apply where the ground surface roughness condition, as defined by Surface Roughness B, prevails in the upwind direction for a distance of at least 2,600 feet (792 m) or 20 times the height of the building, whichever is greater.

> **Exception:** For buildings whose mean roof height is less than or equal to 30 feet (9144 mm), the upwind distance is permitted to be reduced to 1,500 feet (457 m).

Exposure C. Exposure C shall apply for all cases where Exposures B or D do not apply.

Exposure D. Exposure D shall apply where the ground surface roughness, as defined by Surface Roughness D, prevails in the upwind direction for a distance of at least 5,000 feet (1524 m) or 20 times the height of the building, whichever is greater. Exposure D shall extend inland from the shoreline for a distance of 600 feet (183 m) or 20 times the height of the building, whichever is greater.

❖ This section defines the three exposure categories that are used to determine various wind load parameters. Exposure categories are used along with basic wind speed elsewhere in the code as a threshold for wind requirements, such as establishing the need for structural observation in Section 1709.

Exposure B is the most common type of exposure category in the country. A recent study by the National

Association of Home Builders (NAHB) indicated that perhaps up to 80 percent of all buildings were located in Exposure B.

1609.5 Roof systems.

1609.5.1 Roof deck. The roof deck shall be designed to withstand the wind pressures determined in accordance with ASCE 7.

❖ This section specifies the wind load criteria for the roof deck. The roof deck is a structural component of the building and must resist the applicable wind pressures of ASCE 7. This section is referenced by Section 1609.5.2 as the criteria for the wind design for roof coverings that are relatively impermeable.

1609.5.2 Roof coverings. Roof coverings shall comply with Section 1609.5.1.

> **Exception:** Rigid tile roof coverings that are air permeable and installed over a roof deck complying with Section 1609.5.1 are permitted to be designed in accordance with Section 1609.5.3.

Asphalt shingles installed over a roof deck complying with Section 1609.5.1 shall be tested to determine the resistance of the sealant to uplift forces using ASTM D 6381.

Asphalt shingles installed over a roof deck complying with Section 1609.5.1 are permitted to be designed using UL 2390 to determine appropriate uplift and force coefficients applied to the shingle.

❖ This section establishes the wind design criteria for roof coverings. The exception references the use of Section 1609.5.3 for air-permeable rigid tile roof coverings. If the roof deck is relatively impermeable, wind pressures will act through it to the building frame system. The roof covering may or may not be subjected to the same wind pressures as the roof deck. If the roof covering is also relatively impermeable and fastened to the roof deck, the two components will react to and resist the same wind pressures. If the roof covering is not impermeable, the wind pressures will be able to develop on both the top of, and underneath, the roof covering. This "venting action" will negate some wind pressure on the roof covering.

Inclusion of UL 2390 provides a method to evaluate the wind uplift coefficient and to determine the force due to wind on asphalt shingles. It has been found that the sealant resistance to uplift is the critical factor in keeping asphalt shingles in place during windstorms. ASTM D 6381 provides a standardized test to determine the resistance of the sealant material to the applied uplift forces. The use of ASTM D 6381 in conjunction with UL 2390 provides the mechanism to determine the expected performance of asphalt shingles.

1609.5.3 Rigid tile. Wind loads on rigid tile roof coverings shall be determined in accordance with the following equation:

$$M_a = q_h C_L b L L_a [1.0 - GC_p] \qquad \textbf{(Equation 16-35)}$$

For SI: $M_a = \dfrac{q_h C_L bLL_a \left[1.0 - GC_p\right]}{1,000}$

where:

b = Exposed width, feet (mm) of the roof tile.

C_L = Lift coefficient. The lift coefficient for concrete and clay tile shall be 0.2 or shall be determined by test in accordance with Section 1715.2.

GC_p = Roof pressure coefficient for each applicable roof zone determined from Chapter 6 of ASCE 7. Roof coefficients shall not be adjusted for internal pressure.

L = Length, feet (mm) of the roof tile.

L_a = Moment arm, feet (mm) from the axis of rotation to the point of uplift on the roof tile. The point of uplift shall be taken at $0.76L$ from the head of the tile and the middle of the exposed width. For roof tiles with nails or screws (with or without a tail clip), the axis of rotation shall be taken as the head of the tile for direct deck application or as the top edge of the batten for battened applications. For roof tiles fastened only by a nail or screw along the side of the tile, the axis of rotation shall be determined by testing. For roof tiles installed with battens and fastened only by a clip near the tail of the tile, the moment arm shall be determined about the top edge of the batten with consideration given for the point of rotation of the tiles based on straight bond or broken bond and the tile profile.

M_a = Aerodynamic uplift moment, feet-pounds (N-mm) acting to raise the tail of the tile.

q_h = Wind velocity pressure, psf (kN/m²) determined from Section 6.5.10 of ASCE 7.

Concrete and clay roof tiles complying with the following limitations shall be designed to withstand the aerodynamic uplift moment as determined by this section.

1. The roof tiles shall be either loose laid on battens, mechanically fastened, mortar set or adhesive set.

2. The roof tiles shall be installed on solid sheathing which has been designed as components and cladding.

3. An underlayment shall be installed in accordance with Chapter 15.

4. The tile shall be single lapped interlocking with a minimum head lap of not less than 2 inches (51 mm).

5. The length of the tile shall be between 1.0 and 1.75 feet (305 mm and 533 mm).

6. The exposed width of the tile shall be between 0.67 and 1.25 feet (204 mm and 381 mm).

7. The maximum thickness of the tail of the tile shall not exceed 1.3 inches (33 mm).

8. Roof tiles using mortar set or adhesive set systems shall have at least two-thirds of the tile's area free of mortar or adhesive contact.

❖ This section includes the wind design method for clay or concrete rigid tile roofs. The method consists of the

calculation of the aerodynamic uplift moment from the wind that acts to raise the end of the tile. This section includes the characteristics and the type of installation of the concrete or clay roof tile that are required for the use of the design method.

In certain types of installations, the roof covering is not exposed to the same wind loads as the roof deck. Concrete and clay roof tiles are typical of this type of installation and are not subject to wind loads that would be obtained from current wind loading criteria. This is due to the gaps at tile joints allowing some equalization of pressure between the inner and outer face of the tiles, leading to reduced loads. A procedure has been developed through research for determining the uplift moment on loose-laid and mechanically fastened roof tiles when laid over sheathing with an underlayment. The procedure is based on practical measurements on real tiles to determine the effect of air being able to penetrate the roof covering.

SECTION 1610
SOIL LATERAL LOADS

1610.1 General. Basement, foundation and retaining walls shall be designed to resist lateral soil loads. Soil loads specified in Table 1610.1 shall be used as the minimum design lateral soil loads unless specified otherwise in a soil investigation report approved by the building official. Basement walls and other walls in which horizontal movement is restricted at the top shall be designed for at-rest pressure. Retaining walls free to move and rotate at the top are permitted to be designed for active pressure. Design lateral pressure from surcharge loads shall be added to the lateral earth pressure load. Design lateral pressure shall be increased if soils with expansion potential are present at the site.

Exception: Basement walls extending not more than 8 feet (2438 mm) below grade and supporting flexible floor systems shall be permitted to be designed for active pressure.

❖ Nominal loads attributable to lateral earth pressures are determined in accordance with this section. This section provides lateral loads for various soil types. This section requires that foundation and retaining walls be designed to be capable of resisting the lateral soil loads specified by Table 1610.1 where a specific soil investigation has not been performed. Consideration must be given to additional lateral soil pressures due to surcharge loads that result from sloping backfill, driveways or parking spaces that are close to a foundation wall as well as the foundation of an adjacent structure.

TABLE 1610.1. See page 16-58.

❖ The table lists at-rest and active soil pressures for a number of different types of moist soils. The basis of the soil classification into the various types listed is ASTM D 2487. Soils identified by Note b in Table 1610.1 have unpredictable characteristics. These are called expansive soils. Because of their ability to absorb water, they shrink and swell to a higher degree than other soils. As expansive soils swell, they are capable of exerting large forces on soil-retaining structures; thus, these types of soils are not to be used as backfill.

SECTION 1611
RAIN LOADS

1611.1 Design rain loads. Each portion of a roof shall be designed to sustain the load of rainwater that will accumulate on it if the primary drainage system for that portion is blocked plus the uniform load caused by water that rises above the inlet of the secondary drainage system at its design flow.

$$R = 5.2 (d_s + d_h) \qquad \text{(Equation 16-36)}$$

For SI: $R = 0.0098 (d_s + d_h)$

where:

d_h = Additional depth of water on the undeflected roof above the inlet of secondary drainage system at its design flow (i.e., the hydraulic head), in inches (mm).

d_s = Depth of water on the undeflected roof up to the inlet of secondary drainage system when the primary drainage system is blocked (i.e., the static head), in inches (mm).

R = Rain load on the undeflected roof, in psf (kN/m²). When the phrase "undeflected roof" is used, deflec-

tions from loads (including dead loads) shall not be considered when determining the amount of rain on the roof.

❖ The nominal rain load, *R*, is determined in accordance with this section. It represents the weight of accumulated rainwater, assuming a blockage of the primary roof drainage system. The design of the roof drainage systems must be in accordance with Chapter 11 of the *International Plumbing Code®* (IPC®). The primary roof drainage system can include roof drains, leaders, conductors and horizontal storm drains within the structure. Drainage system design is based on a specified design rainfall intensity as well as the roof area it drains. The criteria for sizing the components of the drainage system are provided in Section 1106 of the IPC. Where the building is configured such that water will not collect on the roof there is no requirement for a secondary drainage system [see Figure 1611.1(1)]. Likewise, there would be no rain load required in the design of the roof.

It is not uncommon to find that roof drains have become blocked by debris, leading to ponding of rainwater where the roof construction is conducive to retaining water. While the objective of providing roof drainage is typically to prevent the accumulation of water, the code also recognizes controlled drainage

TABLE 1610.1
SOIL LATERAL LOAD

DESCRIPTION OF BACKFILL MATERIAL[c]	UNIFIED SOIL CLASSIFICATION	DESIGN LATERAL SOIL LOAD[a] (pound per square foot per foot of depth)	
		Active pressure	At-rest pressure
Well-graded, clean gravels; gravel-sand mixes	GW	30	60
Poorly graded clean gravels; gravel-sand mixes	GP	30	60
Silty gravels, poorly graded gravel-sand mixes	GM	40	60
Clayey gravels, poorly graded gravel-and-clay mixes	GC	45	60
Well-graded, clean sands; gravelly sand mixes	SW	30	60
Poorly graded clean sands; sand-gravel mixes	SP	30	60
Silty sands, poorly graded sand-silt mixes	SM	45	60
Sand-silt clay mix with plastic fines	SM-SC	45	100
Clayey sands, poorly graded sand-clay mixes	SC	60	100
Inorganic silts and clayey silts	ML	45	100
Mixture of inorganic silt and clay	ML-CL	60	100
Inorganic clays of low to medium plasticity	CL	60	100
Organic silts and silt clays, low plasticity	OL	Note b	Note b
Inorganic clayey silts, elastic silts	MH	Note b	Note b
Inorganic clays of high plasticity	CH	Note b	Note b
Organic clays and silty clays	OH	Note b	Note b

For SI: 1 pound per square foot per foot of depth = 0.157 kPa/m, 1 foot = 304.8 mm.

a. Design lateral soil loads are given for moist conditions for the specified soils at their optimum densities. Actual field conditions shall govern. Submerged or saturated soil pressures shall include the weight of the buoyant soil plus the hydrostatic loads.

b. Unsuitable as backfill material.

c. The definition and classification of soil materials shall be in accordance with ASTM D 2487.

systems that are engineered to retain rainwater (see Section 1611.3). The important point is that wherever the potential exists for the accumulation of rainwater on a roof, whether it is intentional or otherwise, the roof must be designed for this load. Furthermore, Section 1101.7 of the IPC requires the maximum depth of water to be determined, assuming all primary roof drainage to be blocked. The water will rise above the primary roof drain until it reaches the elevation of the roof edge, scuppers or another serviceable drain. At the design rainfall intensity, this depth will be based on the flow rate of the secondary drainage system. This depth, referred to as the hydraulic head, can be determined from Figure 1611.1(2) for various types of drains and flow rates. Its use is illustrated in the example below.

Section 1107 of the IPC specifically requires a secondary roof drainage system where the building construction extends above the roof at the perimeter. This applies to parapet walls, stepped buildings or any other construction that would allow rainwater to pond on the roof. The sizing of a secondary drainage system is identical to the process used for the primary system. Instead of using a "piped" secondary system, designers may prefer to install scuppers to allow rainwater to overflow the roof. Examples of both types of secondary systems are shown in Figure 1611.1(3). Also note that the IPC requires a secondary system to be completely separate and to discharge above grade. Since the secondary system serves as an emergency backup, requiring it to discharge above grade provides a means of signaling that there is a blockage of the pri-

Figure 1611.1(1)
SECONDARY ROOF DRAINAGE NOT REQUIRED

DRAINAGE SYSTEM	FLOW RATE (gpm)									
	Depth of water above drain inlet (hydraulic head) (inches)									
	1	2	2.5	3	3.5	4	4.5	5	7	8
4-inch-diameter drain	80	170	180							
6-inch-diameter drain	100	190	270	380	540					
8-inch-diameter drain	125	230	340	560	850	1,100	1,170			
6-inch-wide, open-top scupper	18	50	*	90	*	140	*	194	321	393
24-inch-wide, open-top scupper	72	200	*	360	*	560	*	776	1,284	1,572
6-inch-wide, 4-inch-high, closed-top scupper	18	50	*	90	*	140	*	177	231	253
24-inch-wide, 4-inch-high, closed-top scupper	72	200	*	360	*	560	*	708	924	1,012
6-inch-wide, 6-inch-high, closed-top scupper	18	50	*	90	*	140	*	194	303	343
24-inch-wide, 6-inch-high, closed-top scupper	72	200	*	360	*	560	*	776	1,212	1,372

For SI: 1 inch = 25.4 mm, 1 gallon per minute = 3.785 L/m.
Source: Factory Mutual Engineering Corp. Loss Prevention Data 1-54.

Figure 1611.1(2)
FLOW RATE, IN GALLONS PER MINUTE, OF VARIOUS ROOF DRAINS AT
VARIOUS WATER DEPTHS AT DRAIN INLETS (INCHES)

mary drainage system.

Some roof failures have been attributed to the increased loads from ponding water. This section requires the roof to be capable of resisting the maximum water depth that can occur if the primary means of roof drainage becomes blocked. Blockages are typically caused by debris at the inlet to the primary roof drains but they can occur anywhere along the primary piping system, such as an under-slab pipe collapse. Computation of rain load, R, is in accordance with equation 16-36. The coefficient of that equation is merely the conversion of the unit weight of water to an equivalent unit load per inch of water depth as Figure 1611.1(4) illustrates. Two variables are considered to determine rain load: the depth of the water on the undeflected roof, as measured from the low point elevation to the inlet elevation of the secondary drain, and the additional depth of water at the secondary drainage flow, respectively referred to as static head and hydraulic head. The sum of these depths is the design depth for computing rain load, R, as indicated in equation 16-36. An example of the computation of rain load is provided in the following example [also see Figure 1611.1(5)].

EXAMPLE
Rain Load on Roof With Overflow Scuppers

Given:

Primary roof drain and overflow scupper shown in Figure 1611.1(5).

Static head, d_s = 7 inches

Tributary area, A = 5,400 square feet

Rainfall rate, i = 2.5 inches/hour = 0.208333 feet/hour

Determine:

Hydraulic head, d_h
Rain load, R

Calculate required flow rate, Q, at scupper in gallons per minute (gpm).

Q = $A \times i$ = 5400 square feet $\times$ 0.2083 feet/hour

= 1125 cubic feet/hour

= 18.75 cubic feet/minute

= 140.25 gpm

Look up hydraulic head using Figure 1611.1(2)
For 6 inches wide $\times$ 6 inches high scupper find 140 gpm.

d_h = 4 inches

Determine rain load, R

Total head = $(d_s + d_h)$ = 11 inches

R = 5.2 $(d_s + d_h)$

R = 57.2 psf

Figure 1611.1(3)
SEPARATE PRIMARY AND SECONDARY ROOF DRAINS

For SI: 1 inch = 25.4 mm, 1 foot = 304.8 mm,
1 cubic foot = 0.02832 m², 1 pound = 0.454 kg, 1 pound per square foot = 47.88 Pa.

Figure 1611.1(4)
EQUATION 16-36 COEFFICIENT

For SI: 1 inch = 25.4 mm.

Figure 1611.1(5)
RAIN LOAD EXAMPLE

1611.2 Ponding instability. For roofs with a slope less than $^1/_4$ inch per foot [1.19 degrees (0.0208 rad)], the design calculations shall include verification of adequate stiffness to preclude progressive deflection in accordance with Section 8.4 of ASCE 7.

❖ In roofs lacking sufficient framing stiffness, a condition known as "ponding instability" can occur where increasingly larger deflections caused by the continued accumulation of rain water are large enough to overload the structure and result in a roof collapse. This must be countered by providing adequate stiffness in order to prevent increasingly larger deflections due to the buildup of rain water. Another means to minimize the accumulation of rain water is to camber the roof framing. This section requires a check for ponding instability if the roof slope is less than $^1/_4$ inch per foot (1 : 48) [1.19 degrees (0.0208 rad)]. A ponding instability check is to be made assuming the primary roof drains are blocked. The determination of ponding instability is typically done by an iterative structural analysis where the incremental deflection is determined and the resulting increased rain load from the deflection is added to the original rain load.

1611.3 Controlled drainage. Roofs equipped with hardware to control the rate of drainage shall be equipped with a secondary drainage system at a higher elevation that limits accumulation of water on the roof above that elevation. Such roofs shall be designed to sustain the load of rainwater that will accumulate on them to the elevation of the secondary drainage system plus the uniform load caused by water that rises above the inlet of the secondary drainage system at its design flow determined from Section 1611.1. Such roofs shall also be checked for ponding instability in accordance with Section 1611.2.

❖ Controlled drainage is the limitation of the drainage flow rate to a rate that is less than the rainfall rate such that the depth of the water intentionally builds up on the roof during a design rainfall. Controlled flow roof drain systems must be designed in accordance with Section 1110 of the IPC. A secondary roof drain system is needed to limit the buildup of water to a specific depth for roof design. The depth of water on the roof is also to include the depth of the water above the inlet of the secondary drain when the design flow rate is reached. Consideration of the effect of the accumulated rainwater is identical to assuming a blockage in the primary drainage system as discussed under Section 1611.1.

SECTION 1612
FLOOD LOADS

1612.1 General. Within flood hazard areas as established in Section 1612.3, all new construction of buildings, structures and portions of buildings and structures, including substantial improvement and restoration of substantial damage to buildings and structures, shall be designed and constructed to resist the effects of flood hazards and flood loads. For buildings that are located in more than one flood hazard area, the provisions associated with the most restrictive flood hazard area shall apply.

❖ This section addresses requirements for all buildings and structures in designated flood hazard areas. These areas are commonly referred to as "floodplains" and are shown on a community's Flood Insurance Rate Map (FIRM) prepared by FEMA, or another adopted flood hazard map. Users should be aware that floods often affect areas outside the flood hazard area boundaries shown on FIRMs and can exceed the base flood elevation.

Through the adoption of the code, communities meet a significant portion of the floodplain management regulation requirements necessary to participate in the National Flood Insurance Program (NFIP). To participate in the NFIP, a jurisdiction must adopt, in addition to the flood-resistant requirements found in the code, Appendix G (Flood Resistant Construction) or a floodplain management ordinance that contains, at a minimum, the provisions contained in Appendix G. In either case, floodplain management requirements must be applied to all developments in designated flood hazard areas. Appendix G, Section 201.2, defines "Development" as "any man-made change to improved or unimproved real estate, including but not limited to buildings or other structures; temporary or permanent storage of materials; mining; dredging; filling; grading; paving; excavations; operations and other land-disturbing activities."

The NFIP was established to reduce flood losses; to better indemnify individuals from flood losses; and to reduce federal expenditures for disaster assistance. A community that has been determined to have flood hazard areas participates in the NFIP to protect health, safety and property and so that its citizens can purchase federally backed flood insurance. FEMA administers the NFIP, which includes monitoring community compliance with the floodplain management requirements of the NFIP.

Some communities choose to apply flood-resistant construction provisions to flood hazard areas that are more extensive than those shown on the community's FIRM. New buildings and structures, and substantial improvements to existing buildings and structures in flood hazard areas, are to be designed and constructed to resist flood forces to minimize damage. Flood forces include flotation, lateral (hydrostatic) pressures, moving water (hydrodynamic) pressures, wave impact and debris impact. Flood-related hazards may include erosion and scour.

It is important to note that many states and communities have elected to regulate floodplain development to a higher standard than the minimum required to participate in the NFIP. Communities considering using the code and other *International Codes*® to meet the floodplain management requirements of the NFIP are advised to consult with their state NFIP coordinator or the appropriate FEMA regional office.

If located in designated flood hazard areas, buildings and structures that are damaged by any cause are to be examined to determine if the damage constitutes substantial damage, in which the cost of repairing/restoring the building or structure to its predamaged condition equals or exceeds 50 percent of its market value before the damage occurred. All substantial improvements and repairs of buildings and structures that are substantially damaged are to meet the flood-resistant provisions of the code.

Some buildings are located such that they are in more than one flood zone or such that only a portion of the building is in a flood zone. Where this occurs, the entire building or structure is to be designed and constructed according to the requirements of the more restrictive flood zone. For example, if a building is partially in a flood hazard area subject to high-velocity wave action, then the entire building must meet the requirements for that area. Similarly, if a building is partially in a flood hazard area and partially out of the mapped floodplain, then the entire building must be flood resistant.

To provide additional guidance on how communities may use the code and other *International Codes* to participate in the NFIP, FEMA, the International Code Council® (ICC®), the American Society of Civil Engineers (ASCE), the Association of State Floodplain Managers (ASFPM) and others have developed a publication, *Reducing Flood Losses Through the International Code Series: Meeting the Requirements of National Flood Insurance Program.*

1612.2 Definitions. The following words and terms shall, for the purposes of this section, have the meanings shown herein.

❖ Some terms are used in the code to describe something very specific to the related subject matter while other terms may have multiple meanings that can vary based on the subject matter and context. To avoid misunderstandings, it is preferable to reach a consensus on the meaning of these terms. Definitions are intended to facilitate the understanding of code provisions and to minimize potential confusion.

This section contains definitions of terms associated with flood design criteria. Note that these terms are also listed in Chapter 2 with a cross reference to this section. Definitions that pertain primarily to flood design criteria are included within this section to provide convenient access to them without having to refer back to Chapter 2. The use and application of all defined terms, including those defined herein, as well as undefined terms are set forth in Section 201.

BASE FLOOD. The flood having a 1-percent chance of being equaled or exceeded in any given year.

❖ This term is used to define the land area along a body of water that is subject to flooding and within which flood-resistant design and construction requirements are applied. Typically, an authority such as FEMA or a state or local jurisdiction prepares flood hazard maps.

In doing so, the base flood is derived by applying hydrologic models or by examining regional or local flood history to determine the 1-percent annual chance event, also referred to as the "100-year flood." The 1-percent annual chance event has been the standard used to regulate flood hazard areas for more than 30 years.

BASE FLOOD ELEVATION. The elevation of the base flood, including wave height, relative to the National Geodetic Vertical Datum (NGVD), North American Vertical Datum (NAVD) or other datum specified on the Flood Insurance Rate Map (FIRM).

❖ The base flood elevation is the height to which floodwaters will rise during passage or occurrence of the base flood. FEMA uses commonly accepted computer models that estimate hydrologic and hydraulic conditions to determine the 1-percent annual chance (base) flood. Along rivers and streams, statistical methods and computer models may have been used to estimate runoff and to develop flood elevations. The models take into consideration watershed characteristics and the shape and nature of the floodplain, including natural ground contours and the presence of buildings, bridges and culverts. Along coastal areas, base flood elevations may be developed using models that take into account offshore bathymetry, historical storms and typical wind patterns. In many coastal areas, the base flood elevation includes wave heights.

BASEMENT. The portion of a building having its floor subgrade (below ground level) on all sides.

❖ This term, which is used only for buildings in flood hazard areas that are subject to the provisions of Section 1612, differs from the definition in Section 202. The general definition includes as basements those portions of buildings that are partially below grade. For the purpose of resisting flood damage, lack of drainage and a corresponding buildup of hydrostatic pressure on foundation walls are of concern when an area is below grade on all sides. In terms of federal flood insurance, buildings with basements (below grade on all sides) are subject to higher premium rates.

DESIGN FLOOD. The flood associated with the greater of the following two areas:

1. Area with a flood plain subject to a 1-percent or greater chance of flooding in any year; or

2. Area designated as a flood hazard area on a community's flood hazard map, or otherwise legally designated.

❖ The design flood is either the base flood (as defined above) or another flood based on other criteria. A state or local jurisdiction may choose to prepare and adopt flood hazard maps that show flood hazard areas that are not on maps prepared by FEMA. These may be areas that FEMA did not study or areas that were studied with different criteria. For example, as a general rule, FEMA is concerned primarily with flooding sources that have a drainage area of 1 square mile (2.590 km^2) or more. For another example, some communities elect to prepare flood hazard maps based on the assumption that the upland watershed is built out to existing zoning, often called "ultimate development.'

DESIGN FLOOD ELEVATION. The elevation of the "design flood," including wave height, relative to the datum specified on the community's legally designated flood hazard map. In areas designated as Zone AO, the design flood elevation shall be the elevation of the highest existing grade of the building's perimeter plus the depth number (in feet) specified on the flood hazard map. In areas designated as Zone AO where a depth number is not specified on the map, the depth number shall be taken as being equal to 2 feet (610 mm).

❖ The design flood elevation is the height to which floodwaters will rise during passage or occurrence of the design flood. The datum specified on the flood hazard map is important because it may differ from that used locally for other purposes.

Some flood hazard maps have areas denoted as AO zones, identifying areas subject to sheet flow flooding, shallow flooding or ponding. Specific instructions for determining the design flood elevation of these zones are provided, because the maps may specify a flood depth above grade rather than a height above datum.

FEMA refers to some areas subject to shallow flooding as AH zones. AH zones have specified base flood elevations and they are treated the same as the other A zones with base flood elevations.

DRY FLOODPROOFING. A combination of design modifications that results in a building or structure, including the attendant utility and sanitary facilities, being water tight with walls substantially impermeable to the passage of water and with structural components having the capacity to resist loads as identified in ASCE 7.

❖ All residential buildings and most nonresidential buildings in flood hazard areas are elevated so that the lowest floors are at or above the design flood elevation. Under certain circumstances, nonresidential buildings may be designed and constructed with enclosed areas below the design flood elevation, provided they incorporate dry floodproofing modifications. In addition to keeping water out of enclosed areas, dry floodproofing modifications are to be designed to withstand hydrostatic and hydrodynamic loads.

EXISTING CONSTRUCTION. Any buildings and structures for which the "start of construction" commenced before the effective date of the community's first flood plain management code, ordinance or standard. "Existing construction" is also referred to as "existing structures."

❖ For the NFIP, buildings and structures are considered to be "existing" if construction started before the local jurisdiction adopted its first regulation governing development in flood hazard areas. This distinction is important because new construction, for which the start of construction is after the date of the first regulation, is to be in full compliance with the regulation. The distinction between existing construction and new construction also affects the rates used by the NFIP for flood insurance premiums. When work is proposed for existing construction or structures, the costs of that work are to be examined, and, if the proposed work is a "substantial improvement" as defined later in this section, then the entire existing structure is to be brought into compliance with the flood-resistant provisions of the code.

EXISTING STRUCTURE. See "Existing construction."

❖ See the commentary for the definition of "Existing construction."

FLOOD or FLOODING. A general and temporary condition of partial or complete inundation of normally dry land from:

1. The overflow of inland or tidal waters.

2. The unusual and rapid accumulation or runoff of surface waters from any source.

❖ The term "flood" is broadly defined because the condition may occur near all types and sizes of bodies of water. Along streams and riverine areas, flooding results from the accumulation of rainfall runoff that drains from upland watersheds. Along coasts and the shorelines of large lakes, flooding is caused by wind-driven surges and waves that push water onshore, often augmented by tidal influences.

FLOOD DAMAGE-RESISTANT MATERIALS. Any construction material capable of withstanding direct and prolonged contact with floodwaters without sustaining any damage that requires more than cosmetic repair.

❖ A building or structure located within a flood hazard area will have certain structural and nonstructural elements below the design flood elevation, even when it is elevated to be compliant with the requirements of Section 1612. To minimize damage and facilitate cleanup, the materials composing those elements are to be resistant to damage by floodwater. In flood hazard areas subject to high-velocity wave action (commonly called "V Zones"), walls that are designed to break away during the design flood may be wetted during floods of lesser magnitude. When flood damage-resistant materials are used, the owners need only perform routine cleanup. In riverine and inland coastal flood hazard areas (commonly called "A Zones"), certain enclosures may be allowed if designed and constructed with openings that allow floodwater to enter and exit without causing structural damage. To minimize costs and facilitate cleanup, materials used for those enclosures are to resist flood damage. Prolonged contact is generally considered to be partial or total inundation for 72 hours in riverine areas (noncoastal) and 12 hours for areas subject to coastal flood conditions.

The National Evaluation Service, Inc., the predecessor to the ICC Evaluation Service, Inc. (ICC ES), developed a protocol providing guidance for testing flood damage-resistant materials, titled *Evaluation Plan for Determination of Flood Resistance of Building Elements* (April 2000).

FLOOD HAZARD AREA. The greater of the following two areas:

1. The area within a flood plain subject to a 1-percent or greater chance of flooding in any year.

2. The area designated as a flood hazard area on a community's flood hazard map, or otherwise legally designated.

❖ FEMA prepares FIRMs that delineate the land area that is subject to inundation by the 1-percent annual chance flood. Some states and local jurisdictions develop and adopt maps of flood hazard areas that are more extensive than the areas shown on FEMA's maps. For the purpose of the code, the flood hazard area within which the requirements are to be applied is the greater of the two delineated areas.

FLOOD HAZARD AREA SUBJECT TO HIGH VELOCITY WAVE ACTION. Area within the flood hazard area that is subject to high velocity wave action, and shown on a Flood Insurance Rate Map (FIRM) or other flood hazard map as Zone V, VO, VE or V1-30.

❖ Some coastal and lakeshore lines experience flooding that includes wind-driven waves. Flood hazard areas that are anticipated to have high-velocity wave action, which changes anticipated flood loads, are shown on FEMA's flood hazard area maps and commonly are referred to as "V Zones." In these areas wave heights are predicted to be greater than or equal to 3 feet (914 mm) or the wave runup elevations are predicted to reach 3 feet (914 mm) or more above grade.

FLOOD INSURANCE RATE MAP (FIRM). An official map of a community on which the Federal Emergency Management Agency (FEMA) has delineated both the special flood hazard areas and the risk premium zones applicable to the community.

❖ The FIRM shows the flood hazard areas along bodies of water where some level of assessment or study by FEMA has determined that there is a risk of flooding. In addition to showing the real extent of flood hazards, base flood elevations are shown where studies using detailed methods were conducted and floodways may be delineated. Users are cautioned that FIRMs may not show all bodies of water and may not show all areas prone to flooding, including areas with local drainage problems.

FLOOD INSURANCE STUDY. The official report provided by the Federal Emergency Management Agency containing the Flood Insurance Rate Map (FIRM), the Flood Boundary and Floodway Map (FBFM), the water surface elevation of the base flood and supporting technical data.

❖ A flood insurance study is prepared for each local jurisdiction for which FEMA has evaluated flood hazards. The report summarizes the information that describes and supports the flood hazard area maps, including narratives of records of past floods, descriptions of the methodologies used and (if applicable) floodway data tables with additional information for waterways for which floodways have been designated.

FLOODWAY. The channel of the river, creek or other watercourse and the adjacent land areas that must be reserved in order to discharge the base flood without cumulatively increasing the water surface elevation more than a designated height.

❖ FEMA delineates floodways along most rivers and streams that are studied using detailed methods. The floodway is the area that is to be kept clear of encroachments, such as fill and buildings, in order to pass the base flood without increasing the water surface more than a designated height. The designated height is found by referencing the floodway data table in the flood insurance study. In general, velocities are higher and water depths are greater in floodways than in adjacent fringe areas.

LOWEST FLOOR. The floor of the lowest enclosed area, including basement, but excluding any unfinished or flood-resistant enclosure, usable solely for vehicle parking, building access or limited storage provided that such enclosure is not built so as to render the structure in violation of this section.

❖ The lowest floor is the most important reference point when designing and constructing a building or structure in a flood hazard area. The term is specifically defined to include basements, which are any areas that are below grade on all sides. For elevated buildings, compliance with the flood-resistant design and construction provisions of the code is determined by an elevation survey that documents that the lowest floor, including the basement, is at or above the design flood elevation. It is important to note that certain enclosures may be built below otherwise elevated buildings provided they meet both the design criteria and use limitations (see Section 1612.5.1). If compliant with those limitations, then enclosures are not considered the lowest floor. This distinction also is important because the premium rate used to determine the cost of a federal flood insurance policy is very dependent on the elevation of the lowest floor.

SPECIAL FLOOD HAZARD AREA. The land area subject to flood hazards and shown on a Flood Insurance Rate Map or other flood hazard map as Zone A, AE, A1-30, A99, AR, AO, AH, V, VO, VE or V1-30.

❖ This term is used in the definition of "Flood insurance study" and in Section 1612.3, in which the flood hazard area is established. The special flood hazard area is the flood hazard area shown on maps prepared by FEMA.

START OF CONSTRUCTION. The date of permit issuance for new construction and substantial improvements to existing structures, provided the actual start of construction, repair, reconstruction, rehabilitation, addition, placement or other improvement is within 180 days after the date of issuance. The actual start of construction means the first placement of permanent construction of a building (including a manufactured home) on a site, such as the pouring of a slab or footings, installation of pilings or construction of columns.

Permanent construction does not include land preparation (such as clearing, excavation, grading or filling), the installation of streets or walkways, excavation for a basement, footings, piers or foundations, the erection of temporary forms or the installation of accessory buildings such as garages or sheds not occupied as dwelling units or not part of the main building. For a substantial improvement, the actual "start of construction" means the first alteration of any wall, ceiling, floor or other structural part of a building, whether or not that alteration affects the external dimensions of the building.

❖ This term is used in the definition of "Existing construction" and is applicable to improvements of existing buildings in order to capture all improvement work.

SUBSTANTIAL DAMAGE. Damage of any origin sustained by a structure whereby the cost of restoring the structure to its before-damaged condition would equal or exceed 50 percent of the market value of the structure before the damage occurred.

❖ This term is used in the definition of "Substantial improvement." Substantial damage is a special case of substantial improvement, and if the cost of restoring damage equals or exceeds 50 percent of the market value of the structure before the damage occurred, then compliance of the existing building is required. It is notable that a substantial damage determination is to be made regardless of what causes damage. Buildings have sustained substantial damage due to flood, fire, wind, earthquake, deterioration and other causes.

SUBSTANTIAL IMPROVEMENT. Any repair, reconstruction, rehabilitation, addition or improvement of a building or structure, the cost of which equals or exceeds 50 percent of the market value of the structure before the improvement or repair is started. If the structure has sustained substantial damage, any repairs are considered substantial improvement regardless of the actual repair work performed. The term does not, however, include either:

1. Any project for improvement of a building required to correct existing health, sanitary or safety code violations identified by the building official and that are the minimum necessary to assure safe living conditions.

2. Any alteration of a historic structure provided that the alteration will not preclude the structure's continued designation as a historic structure.

❖ One of the long-range objectives of the NFIP is to reduce the exposure of older buildings that were built in flood hazard areas before local jurisdictions adopted flood hazard area maps and regulations. Section 105.3 directs the applicant to state the valuation of the proposed work as part of the information submitted to obtain a permit. To make a determination as to whether a proposed repair, reconstruction, rehabilitation, addition or improvement constitutes substantial improvement or damage, the cost of the proposed work is to be compared to the market value of the building or structure before the work is started. In order to determine market value, the building official may require the applicant to provide such information, as allowed under Section 105.3. The 50-percent threshold

is a compromise that in theory is midway between no regulation of existing buildings and requiring full compliance for any alteration to an existing building. For additional guidance, refer to FEMA 213, *Answers to Questions About Substantially Damaged Buildings,* and FEMA 311, *Guidance on Estimating Substantial Damage Using the NFIP Residential Substantial Damage Estimator.*

1612.3 Establishment of flood hazard areas. To establish flood hazard areas, the governing body shall adopt a flood hazard map and supporting data. The flood hazard map shall include, at a minimum, areas of special flood hazard as identified by the Federal Emergency Management Agency in an engineering report entitled "The Flood Insurance Study for [INSERT NAME OF JURISDICTION]," dated [INSERT DATE OF ISSUANCE], as amended or revised with the accompanying Flood Insurance Rate Map (FIRM) and Flood Boundary and Floodway Map (FBFM) and related supporting data along with any revisions thereto. The adopted flood hazard map and supporting data are hereby adopted by reference and declared to be part of this section.

❖ Flood maps and studies are prepared by FEMA and are to be used as a community's official map, unless the community chooses to adopt a map that shows more extensive flood hazard areas. From time to time, FEMA's floodplain maps and studies may be revised and republished. When maps are revised and flood hazard areas are changed, FEMA involves the community and provides a formal opportunity to review the documents. Once the revisions are finalized, FEMA requires adoption of the new maps by the community. Communities may be able to minimize having to adopt each revision by referencing the date of the original map and study and all future revisions. This is a method by which subsequent revisions to flood maps and studies may be adopted administratively without requiring legislative action on the part of the community. Communities will need to determine whether this adoption by reference approach is allowed under their state's enabling authority and due process requirements. If not allowed, communities are to follow their state's requirements, which normally require public notices, hearings and specific adoption of revised maps by the community's legislative body.

1612.4 Design and construction. The design and construction of buildings and structures located in flood hazard areas, including flood hazard areas subject to high velocity wave action, shall be in accordance with ASCE 24.

❖ FEMA uses multiple designations for flood hazard areas shown on each FIRM, including AO, AH, A1-30, AE, A99, AR, AR/A1-30, AR/AE, AR/AO, AR/AH, AR/A, VO or V1-30, VE and V. Along many open coasts and lakeshores where wind-driven waves are predicted, the flood hazard area is commonly referred to as the "V Zone." For the purpose of the code and consistency with ASCE 7 and ASCE 24, these areas are called "flood hazard areas subject to high velocity wave action." Flood hazard areas that are inland of areas subject to high-velocity wave action, and flood hazard areas along rivers and streams, are commonly referred to as "A Zones." Due to waves, the flood loads in areas subject to high-velocity wave action differ from those in other flood hazard areas. ASCE 24 outlines in detail the specific requirements that are to be applied to buildings and structures in all flood hazard areas.

1612.5 Flood hazard documentation. The following documentation shall be prepared and sealed by a registered design professional and submitted to the building official:

1. For construction in flood hazard areas not subject to high-velocity wave action:

   1.1. The elevation of the lowest floor, including the basement, as required by the lowest floor elevation inspection in Section 109.3.3.

   1.2. For fully enclosed areas below the design flood elevation where provisions to allow for the automatic entry and exit of floodwaters do not meet the minimum requirements in Section 2.6.2.1 of ASCE 24, construction documents shall include a statement that the design will provide for equalization of hydrostatic flood forces in accordance with Section 2.6.2.2 of ASCE 24.

   1.3. For dry floodproofed nonresidential buildings, construction documents shall include a statement that the dry floodproofing is designed in accordance with ASCE 24.

2. For construction in flood hazard areas subject to high-velocity wave action:

   2.1. The elevation of the bottom of the lowest horizontal structural member as required by the lowest floor elevation inspection in Section 109.3.3.

   2.2. Construction documents shall include a statement that the building is designed in accordance with ASCE 24, including that the pile or column foundation and building or structure to be attached thereto is designed to be anchored to resist flotation, collapse and lateral movement due to the effects of wind and flood loads acting simultaneously on all building components, and other load requirements of Chapter 16.

   2.3. For breakaway walls designed to resist a nominal load of less than 10 psf (0.48 kN/m²) or more than 20 psf (0.96 kN/m²), construction documents shall include a statement that the breakaway wall is designed in accordance with ASCE 24.

❖ The NFIP requires that certain documentation be submitted in order to demonstrate compliance with provisions of the code that cannot be easily verified during a site inspection. The most common, described in Item 1.1, is the documentation of the lowest floor elevation. It provides evidence that the lowest floor is at or above the required minimum elevation. Elevation is one of the most important aspects of flood-resistant construction and is a significant factor used to determine flood insurance premium rates. FEMA Form 81-31, *Eleva-*

tion Certificate, which includes illustrations and instructions, is recommended [see Figure 1612.5.1(1)]. Building owners need elevation certificates to obtain federal flood insurance, and insurance agents use the certificates to compute the proper flood insurance premium rates. The criteria for the minimum number and size of flood openings are set by the NFIP to allow free inflow and outflow of floodwaters under all types of flood conditions. The statement described in Item 1.2 is required in the construction documents if the flood openings do not meet the minimum number and size criteria set forth in ASCE 24. For further guidance, refer to FEMA FIA-TB #1, *Openings in Foundation Walls for Buildings Located in Special Flood Hazard Areas.*

The statement described in Item 1.3 is to be included in the construction documents for nonresidential buildings that are designed to be dry floodproofed. It is important to note that dry floodproofing is allowed only for nonresidential buildings and structures that are located in flood hazard areas not subject to high-velocity wave action. The registered design professional who seals the construction documents is indicating that, based upon development or review of the structural design, specifications and plans for construction, the design and methods of construction are in accordance with accepted standards of practice for meeting the following provisions: (a) the structure, together with attendant utilities and sanitary facilities, is water tight to the floodproofed design elevation indicated with walls that are substantially impermeable to the passage of water and (b) all structural components are capable of resisting hydrostatic and hydrodynamic flood forces, including the effects of bouyancy and anticipated debris impact forces. The use of FEMA Form 81-65, *Floodproofing Certification,* is recommended [see Figure 1612.5.1(2)]. This certificate is used by insurance agents to determine federal flood insurance premium rates for dry floodproofed nonresidential buildings. For further guidance, refer to FEMA FIA-TB #3, *Nonresidential Floodproofing—Requirements and Certification for Buildings Located in Special Flood Hazard Areas.*

Certain documentation must be submitted in order to demonstrate compliance with provisions of the code that cannot be verified readily during a site inspection. The most common, in Item 2.1, provides evidence that the bottom of the lowest horizontal members of buildings constructed in flood hazard areas subject to high-velocity wave action (V Zones) are elevated to or above the minimum required height.

Buildings located in flood hazard areas subject to high-velocity wave action and winds are expected to experience significant flood and wind loads simultaneously. FEMA and coastal communities report significant damage to buildings that are not built to current code. The statement described in Item 2.2 is included in the construction documents to indicate that the design meets the flood load provisions of ASCE 24 and other loads required by this chapter.

The documentation described in Item 2.3 is used

only for specific situations in which properly elevated buildings in flood hazard areas subject to high-velocity wave action have enclosures beneath them, and then only if the walls of the enclosures are not designed to fail under a nominal load of between 10 psf (0.48 kN/m²) and 20 psf (0.96 kN/m²). Such enclosures are allowed only if their uses are restricted to building access, parking and limited storage. In these cases, the walls of the enclosures are to be designed to fail—or break away—under certain loads without transferring significant loads or causing damage to the building. Because breakaway walls will fail under flood conditions, building materials can become water-borne debris that may damage adjacent buildings. Refer to FEMA FIA-TB #5, *Free of Obstruction Requirements for Buildings Located in Coastal High Hazard Areas,* and FEMA FIA-TB #9, *Design and Construction Guidance for Breakaway Walls Below Elevated Buildings.*

U.S. DEPARTMENT OF HOMELAND SECURITY
Federal Emergency Management Agency
National Flood Insurance Program

ELEVATION CERTIFICATE

Important: Read the instructions on pages 1-8.

OMB No. 1660-0008
Expires February 28, 2009

SECTION A - PROPERTY INFORMATION

For Insurance Company Use:

A1. Building Owner's Name — Policy Number

A2. Building Street Address (including Apt., Unit, Suite, and/or Bldg. No.) or P.O. Route and Box No. — Company NAIC Number

City — State — ZIP Code

A3. Property Description (Lot and Block Numbers, Tax Parcel Number, Legal Description, etc.)

A4. Building Use (e.g., Residential, Non-Residential, Addition, Accessory, etc.) _____
A5. Latitude/Longitude: Lat. _____ Long. _____   Horizontal Datum: ☐ NAD 1927 ☐ NAD 1983
A6. Attach at least 2 photographs of the building if the Certificate is being used to obtain flood insurance.
A7. Building Diagram Number_____
A8. For a building with a crawl space or enclosure(s), provide:
  a) Square footage of crawl space or enclosure(s) _____ sq ft
  b) No. of permanent flood openings in the crawl space or enclosure(s) walls within 1.0 foot above adjacent grade _____
  c) Total net area of flood openings in A8.b _____ sq in
A9. For a building with an attached garage, provide:
  a) Square footage of attached garage _____ sq ft
  b) No. of permanent flood openings in the attached garage walls within 1.0 foot above adjacent grade _____
  c) Total net area of flood openings in A9.b _____ sq in

SECTION B - FLOOD INSURANCE RATE MAP (FIRM) INFORMATION

B1. NFIP Community Name & Community Number	B2. County Name	B3. State

B4. Map/Panel Number	B5. Suffix	B6. FIRM Index Date	B7. FIRM Panel Effective/Revised Date	B8. Flood Zone(s)	B9. Base Flood Elevation(s) (Zone AO, use base flood depth)

B10. Indicate the source of the Base Flood Elevation (BFE) data or base flood depth entered in Item B9.
☐ FIS Profile   ☐ FIRM   ☐ Community Determined   ☐ Other (Describe) _____
B11. Indicate elevation datum used for BFE in Item B9: ☐ NGVD 1929   ☐ NAVD 1988   ☐ Other (Describe) _____
B12. Is the building located in a Coastal Barrier Resources System (CBRS) area or Otherwise Protected Area (OPA)?   ☐ Yes   ☐ No
Designation Date_____   ☐ CBRS   ☐ OPA

SECTION C - BUILDING ELEVATION INFORMATION (SURVEY REQUIRED)

C1. Building elevations are based on: ☐ Construction Drawings*   ☐ Building Under Construction*   ☐ Finished Construction
*A new Elevation Certificate will be required when construction of the building is complete.
C2. Elevations – Zones A1-A30, AE, AH, A (with BFE), VE, V1-V30, V (with BFE), AR, AR/A, AR/AE, AR/A1-A30, AR/AH, AR/AO. Complete Items C2.a-g below according to the building diagram specified in Item A7.
Benchmark Utilized _____   Vertical Datum_____
Conversion/Comments _____

Check the measurement used.

a) Top of bottom floor (including basement, crawl space, or enclosure floor) _____.___ ☐ feet ☐ meters (Puerto Rico only)
b) Top of the next higher floor _____.___ ☐ feet ☐ meters (Puerto Rico only)
c) Bottom of the lowest horizontal structural member (V Zones only) _____.___ ☐ feet ☐ meters (Puerto Rico only)
d) Attached garage (top of slab) _____.___ ☐ feet ☐ meters (Puerto Rico only)
e) Lowest elevation of machinery or equipment servicing the building (Describe type of equipment in Comments) _____.___ ☐ feet ☐ meters (Puerto Rico only)
f) Lowest adjacent (finished) grade (LAG) _____.___ ☐ feet ☐ meters (Puerto Rico only)
g) Highest adjacent (finished) grade (HAG) _____.___ ☐ feet ☐ meters (Puerto Rico only)

SECTION D - SURVEYOR, ENGINEER, OR ARCHITECT CERTIFICATION

This certification is to be signed and sealed by a land surveyor, engineer, or architect authorized by law to certify elevation information. *I certify that the information on this Certificate represents my best efforts to interpret the data available. I understand that any false statement may be punishable by fine or imprisonment under 18 U.S. Code, Section 1001.*

☐ Check here if comments are provided on back of form.

PLACE SEAL HERE

Certifier's Name — License Number

Title — Company Name

Address — City — State — ZIP Code

Signature — Date — Telephone

FEMA Form 81-31, February 2006   See reverse side for continuation.   Replaces all previous editions

Figure 1612.5.1(1)
ELEVATION CERTIFICATE

(continued)

IMPORTANT: In these spaces, copy the corresponding information from Section A.			For Insurance Company Use:
Building Street Address (including Apt., Unit, Suite, and/or Bldg. No.) or P.O. Route and Box No.			Policy Number
City	State	ZIP Code	Company NAIC Number

SECTION D - SURVEYOR, ENGINEER, OR ARCHITECT CERTIFICATION (CONTINUED)

Copy both sides of this Elevation Certificate for (1) community official, (2) insurance agent/company, and (3) building owner.

Comments

Signature   Date

☐ Check here if attachments

SECTION E - BUILDING ELEVATION INFORMATION (SURVEY NOT REQUIRED) FOR ZONE AO AND ZONE A (WITHOUT BFE)

For Zones AO and A (without BFE), complete Items E1-E5. If the Certificate is intended to support a LOMA or LOMR-F request, complete Sections A, B, and C. For Items E1-E4, use natural grade, if available. Check the measurement used. In Puerto Rico only, enter meters.

E1. Provide elevation information for the following and check the appropriate boxes to show whether the elevation is above or below the highest adjacent grade (HAG) and the lowest adjacent grade (LAG).
    a) Top of bottom floor (including basement, crawl space, or enclosure) is _____. ____ ☐ feet ☐ meters ☐ above or ☐ below the HAG.
    b) Top of bottom floor (including basement, crawl space, or enclosure) is _____. ____ ☐ feet ☐ meters ☐ above or ☐ below the LAG.
E2. For Building Diagrams 6-8 with permanent flood openings provided in Section A Items 8 and/or 9 (see page 8 of Instructions), the next higher floor (elevation C2.b in the diagrams) of the building is _____. ____ ☐ feet ☐ meters ☐ above or ☐ below the HAG.
E3. Attached garage (top of slab) is _____. ____ ☐ feet ☐ meters ☐ above or ☐ below the HAG.
E4. Top of platform of machinery and/or equipment servicing the building is _____. ____ ☐ feet ☐ meters ☐ above or ☐ below the HAG.
E5. Zone AO only: If no flood depth number is available, is the top of the bottom floor elevated in accordance with the community's floodplain management ordinance? ☐ Yes ☐ No ☐ Unknown. The local official must certify this information in Section G.

SECTION F - PROPERTY OWNER (OR OWNER'S REPRESENTATIVE) CERTIFICATION

The property owner or owner's authorized representative who completes Sections A, B, and E for Zone A (without a FEMA-issued or community-issued BFE) or Zone AO must sign here. *The statements in Sections A, B, and E are correct to the best of my knowledge.*

Property Owner's or Owner's Authorized Representative's Name

Address	City	State	ZIP Code

Signature	Date	Telephone

Comments

☐ Check here if attachments

SECTION G - COMMUNITY INFORMATION (OPTIONAL)

The local official who is authorized by law or ordinance to administer the community's floodplain management ordinance can complete Sections A, B, C (or E), and G of this Elevation Certificate. Complete the applicable item(s) and sign below. Check the measurement used in Items G8. and G9.

G1. ☐ The information in Section C was taken from other documentation that has been signed and sealed by a licensed surveyor, engineer, or architect who is authorized by law to certify elevation information. (Indicate the source and date of the elevation data in the Comments area below.)

G2. ☐ A community official completed Section E for a building located in Zone A (without a FEMA-issued or community-issued BFE) or Zone AO.

G3. ☐ The following information (Items G4.-G9.) is provided for community floodplain management purposes.

G4. Permit Number	G5. Date Permit Issued	G6. Date Certificate Of Compliance/Occupancy Issued

G7. This permit has been issued for: ☐ New Construction ☐ Substantial Improvement
G8. Elevation of as-built lowest floor (including basement) of the building: _____. _____ ☐ feet ☐ meters (PR) Datum_____
G9. BFE or (in Zone AO) depth of flooding at the building site: _____. _____ ☐ feet ☐ meters (PR) Datum_____

Local Official's Name	Title
Community Name	Telephone
Signature	Date

Comments

☐ Check here if attachments

FEMA Form 81-31, February 2006   Replaces all previous editions

Figure 1612.5.1(1)—continued
ELEVATION CERTIFICATE

U.S. DEPARTMENT OF HOMELAND SECURITY
FEDERAL EMERGENCY MANAGEMENT AGENCY
National Flood Insurance Program

O.M.B. NO. 1660-0008
Expires February 28, 2009

FLOODPROOFING CERTIFICATE
FOR NON-RESIDENTIAL STRUCTURES

The floodproofing of non-residential buildings may be permitted as an alternative to elevating to or above the Base Flood Elevation; however, a floodproofing design certification is required. This form is to be used for that certification. Floodproofing of a residential building does not alter a community's floodplain management elevation requirements or affect the insurance rating unless the community has been issued an exception by FEMA to allow floodproofed residential basements. The permitting of a floodproofed residential basement requires a separate certification specifying that the design complies with the local floodplain management ordinance.

FOR INSURANCE COMPANY USE

BUILDING OWNER'S NAME

POLICY NUMBER

STREET ADDRESS (Including Apt., Unit, Suite, and/or Bldg. Number) OR P.O. ROUTE AND BOX NUMBER

COMPANY NAIC NUMBER

OTHER DESCRIPTION (Lot and Block Numbers, etc.)

CITY   STATE   ZIP CODE

SECTION I FLOOD INSURANCE RATE MAP (FIRM) INFORMATION

Provide the following from the proper FIRM:

COMMUNITY NUMBER	PANEL NUMBER	SUFFIX	DATE OF FIRM INDEX	FIRM ZONE	BASE FLOOD ELEVATION (In AO Zones, Use Depth)

SECTION II FLOODPROOFING INFORMATION (By a Registered Professional Engineer or Architect)

Floodproofing Design Elevation Information:

Building is floodproofed to an elevation of .... feet NGVD. (Elevation datum used must be the same as that on the FIRM.)

Height of floodproofing on the building above the lowest adjacent grade is .... feet.

(NOTE: for insurance rating purposes, the building's floodproofed design elevation must be at least one foot above the Base Flood Elevation to receive rating credit. If the building is floodproofed only to the Base Flood Elevation, then the building's insurance rating will result in a higher premium.)

SECTION III CERTIFICATION (By Registered Professional Engineer or Architect)

Non-Residential Floodproofed Construction Certification:

I certify that, based upon development and/or review of structural design, specifications, and plans for construction, the design and methods of construction are in accordance with accepted standards of practice for meeting the following provisions:

The structure, together with attendant utilities and sanitary facilities, is watertight to the floodproofed design elevation indicated above, with walls that are substantially impermeable to the passage of water.

All structural components are capable of resisting hydrostatic and hydrodynamic flood forces, including the effects of buoyancy, and anticipated debris impact forces.

I certify that the information on this certificate represents my best efforts to interpret the data available. I understand that any false statement may be punishable by fine or imprisonment under 18 U.S. Code, Section 1001.

CERTIFIER'S NAME   LICENSE NUMBER (or Affix Seal)

TITLE   COMPANY NAME

ADDRESS   CITY   STATE   ZIP CODE

SIGNATURE   DATE   PHONE

Copies should be made of this Certificate for: 1) community official, 2) Insurance agent/company, and 3) building owner.

FEMA Form 81-65, Feb 06   Replaces all previous editions   F-056 (2/06)

Figure 1612.5.1(2)
FLOOD PROOFING CERTIFICATE

SECTION 1613
EARTHQUAKE LOADS

1613.1 Scope. Every structure, and portion thereof, including nonstructural components that are permanently attached to structures and their supports and attachments, shall be designed and constructed to resist the effects of earthquake motions in accordance with ASCE 7, excluding Chapter 14 and Appendix 11A. The seismic design category for a structure is permitted to be determined in accordance with Section 1613 or ASCE 7.

Exceptions:

1. Detached one- and two-family dwellings, assigned to Seismic Design Category A, B or C, or located where the mapped short-period spectral response acceleration, S_S, is less than 0.4 g.

2. The seismic-force-resisting system of wood-frame buildings that conform to the provisions of Section 2308 are not required to be analyzed as specified in this section.

3. Agricultural storage structures intended only for incidental human occupancy.

4. Structures that require special consideration of their response characteristics and environment that are not addressed by this code or ASCE 7 and for which other regulations provide seismic criteria, such as vehicular bridges, electrical transmission towers, hydraulic structures, buried utility lines and their appurtenances and nuclear reactors.

❖ These provisions provide the requirements essential to determining a building's seismic design category. The balance of the earthquake load provisions are contained in the ASCE 7 load standard. The chapters that are noted as excluded from the ASCE 7 referenced standard are those that can create conflicts with Chapters 17 through 23. The balance of ASCE 7 earthquake load provisions that are referenced are as follows:

Chapter Subject

11 Seismic Design Criteria

12 Seismic Design Requirements for Building Structures

13 Seismic Design requirements for Nonstructural Components

15 Seismic Design Requirements for Nonbuilding Structures

16 Seismic Response History Procedures

17 Seismic Design Requirements for Seismically Isolated Structures

18 Seismic Design Requirements for Structures With Damping Systems

19 Soil Structure Interaction for Seismic Design

20 Site Classification Procedure for Seismic Design

21 Site-Specific Ground Motion Procedures for Seismic Design

22 Seismic Ground Motion and Long Period Transition Maps

23 Seismic Design Reference Documents

These seismic requirements are based, for the most part, on the NEHRP *Recommended Provisions for Seismic Regulations for New Buildings and Other Structures.* A new set of NEHRP recommended provisions has been prepared by the BSSC every three years since the first edition in 1985. The code uses the NEHRP recommended provisions as the technical basis for seismic design requirements because of the nationwide input into the development of these design criteria. The NEHRP recommended provisions present up-to-date criteria for the design and construction of buildings subject to earthquake ground motions that are applicable anywhere in the nation. The requirements are intended to minimize the hazard to life for all buildings, increase the expected performance of higher occupancy buildings as compared to ordinary buildings and improve the capability of essential facilities to function during and after an earthquake. These minimum criteria are considered to be prudent and economically justified for the protection of life safety in buildings subject to earthquakes. Achieving the intended performance, however, depends on a number of factors, including the type of structural framing type, configuration, construction materials and as-built details of construction.

There are four exceptions to seismic design requirements included in this section. The following discussion addresses each of the exceptions.

Exception 1 exempts detached one- and two-family dwellings under two conditions. The first applies to structures assigned to Seismic Design Category A, B or C (meaning that both $S_{DS} < 0.5g$ *and* $S_{D1} < 0.2g$). The second is for structures having a value for mapped short-period spectral response acceleration, S_S, less than 0.4g. The latter condition is derived from the ASCE 7 seismic provisions. Since it is based solely on the mapped value for short periods, it may allow structures to qualify more directly for this exception than will the first condition. In other words, the value of S_S should be checked first. If the structure qualifies based on S_S, then it is not necessary to check the seismic design category.

Exception 2 exempts conventional light-frame wood construction from the seismic requirements in this chapter. It should be noted that the limitations for conventional light-frame wood construction are included in Section 2308.2. There is no limitation on the use of the structure, except that Occupancy Category IV structures do not qualify under the conventional light-frame wood construction provision nor do irregular portions of Seismic Design Category D and E structures. The conventional light-frame wood construction provisions are deemed to provide equivalent seismic resistance as compared to construction designed in accordance with the requirements of this chapter based on the history of such conventional construction.

In Exception 3, agricultural buildings are exempt because they present a minimal life safety hazard due to the low probability of human occupancy.

Exception 4 corresponds to an ASCE 7 exception regarding structures that are covered under other regulations, making it unnecessary to comply with the earthquake load requirements of the code.

1613.2 Definitions. The following words and terms shall, for the purposes of this section, have the meanings shown herein.

❖ Some terms are used in the code to describe something very specific to the related subject matter while other terms may have multiple meanings that can vary based on the subject matter and context. To avoid misunderstandings, it is preferable to reach a consensus on the meaning of these terms. Definitions are intended to facilitate the understanding of code provisions and to minimize potential confusion.

This section contains definitions of terms associated with earthquake design criteria. Note that these terms are also listed in Chapter 2 with a cross reference to this section. Definitions that pertain primarily to earthquake design criteria are included within this section to provide convenient access to them without having to refer back to Chapter 2. The use and application of all defined terms, including those defined herein, as well as undefined terms are set forth in Section 201.

DESIGN EARTHQUAKE GROUND MOTION. The earthquake ground motion that buildings and structures are specifically proportioned to resist in Section 1613.

❖ A structure must be designed for the level of ground motion corresponding to the design earthquake. This is determined in Section 1613.5.4 as two-thirds of the maximum earthquake ground motions with adjustments to account for the site's soil profile.

MAXIMUM CONSIDERED EARTHQUAKE GROUND MOTION. The most severe earthquake effects considered by this code.

❖ The term "maximum considered earthquake" refers to the mapped earthquake ground motions that are provided in Figures 1613.5(1) through (14) for a Site Class B (rock) soil profile. It reflects the maximum level of earthquake ground shaking that is considered reasonable for design of new structures. These maximum considered earthquake ground motions are intended to provide an approximately uniform margin against collapse in structures that are designed in accordance with these provisions.

MECHANICAL SYSTEMS. For the purposes of determining seismic loads in ASCE 7, mechanical systems shall include plumbing systems as specified therein.

❖ This definition clarifies the application of ASCE 7 earthquake load provisions for mechanical components and systems that may include items that are covered under plumbing systems elsewhere in the *International Codes*.

ORTHOGONAL. To be in two horizontal directions, at 90 degrees (1.57 rad) to each other.

❖ This definition applies to the alternative seismic design category determination permitted under Section 1613.5.6.1. It makes it clear that the directions (axes) that are referred to are in the horizontal plane.

SEISMIC DESIGN CATEGORY. A classification assigned to a structure based on its occupancy category and the severity of the design earthquake ground motion at the site.

❖ The seismic design category serves as a trigger mechanism for many seismic requirements, including the following:

1. Permissible seismic-force-resisting systems.

2. Limitations on height.

3. Consideration of structural irregularities.

4. The type(s) of lateral force analysis that may be used.

5. The need for additional special inspections.

The first step in the design of a structure for earthquake forces is the determination of its seismic design category. Each building is assigned the more restrictive seismic design category classification depending upon the site soil profile, the mapped spectral response accelerations at the site and the nature of the uses that occupy the building, using both Tables 1613.5.6(1) and 1613.5.6(2). The following outlines the procedure used to determine a structure's seismic design category.

Step 1: Determine the mapped maximum considered earthquake spectral response acceleration at short periods, S_S, and at 1-second period, S_1, for the site location from Figures 1613.5(1) through 1613.5(14).

Step 2: Determine the (soil) site class in accordance with Table 1613.5.2.

Step 3: Determine the site coefficients F_a and F_v from Tables 1613.5.3(1) and 1613.5.3(2), respectively.

Step 4: Determine the 5-percent damped design spectral response acceleration at short periods, S_{DS}, and at 1-second period, S_{D1}, as follows:

$$S_{DS} = (2/3)(F_a)(S_S)$$

$$S_{D1} = (2/3)(F_V)(S_1)$$

Step 5: Determine the seismic design category as prescribed by Tables 1613.5.6(1) and 1613.5.6(2). The highest of the seismic design categories from the two tables is the category assigned to the building, unless Section 1613.5.6.1 is applicable. For example, if the seismic design category from Table 1613.5.6(1) is D and from Table 1613.5.6(2) it is C, then the building would be assigned to Seismic Design Category D.

SEISMIC-FORCE-RESISTING SYSTEM. That part of the structural system that has been considered in the design to provide the required resistance to the prescribed seismic forces.

❖ This definition is needed for the application of the seismic design provisions of the code (see Sections 1705 through 1708 for required special inspections of seismic-force-resisting systems). Unless classified as Seismic Design Category A, the type of seismic-force-resisting system must be selected to conform to Section 12.2 of ASCE 7. Providing a level of detailing that is appropriate for the selected system is critical to complying with the code's earthquake provisions.

SITE CLASS. A classification assigned to a site based on the types of soils present and their engineering properties as defined in Section 1613.5.2.

❖ This definition is necessary because the earthquake load on a structure is greatly affected by the soil at the site. There are six separate site classes that are defined in Table 1613.5.2.

SITE COEFFICIENTS. The values of F_a and F_v indicated in Tables 1613.5.3(1) and 1613.5.3(2), respectively.

❖ The site coefficients F_a and F_v are used to adjust the mapped ground motions to account for the characteristics of the site's soil profile when determining the soil adjusted earthquake spectral response accelerations in Section 1613.5.3. These coefficients are functions of the soil (site class) as well as the seismicity at the site of the structure.

1613.3 Existing buildings. Additions, alterations, modification, or change of occupancy of existing buildings shall be in accordance with Sections 3403.2.3 and 3406.4.

❖ In general, additions and alterations are governed by Section 3403. The referenced section provides more specific guidance with respect to the impact of an addition or alteration on the seismic resistance of an existing building. In general, a change in occupancy is governed by Section 3406. The referenced section provides more specific guidance with respect to the impact the occupancy change has on an existing building.

1613.4 Special inspections. Where required by Section 1705.3, the statement of special inspections shall include the special inspections required by Section 1705.3.1.

❖ This section provides a cross reference to Chapter 17, as Section 1705 contains special inspection provisions for earthquake-resistant construction. Assessment of past earthquake damage indicates some problems have been the result of poor quality construction. The code requires more extensive special inspections than some legacy codes in an attempt to improve the quality of construction in structures having a higher seismic risk.

1613.5 Seismic ground motion values. Seismic ground motion values shall be determined in accordance with this section.

❖ The design earthquake ground motion levels determined in this section may result in damage, both structural and nonstructural, from the high stresses that occur because of the dynamic nature of seismic events. For most structures, damage from the design earthquake ground motion would be repairable but might be so costly as to make it economically undesirable. For essential facilities, it is expected that damage from the design earthquake ground motion would not be so severe as to prevent continued occupancy and function of the facility. For ground motions greater than the design levels, the intent is that there be a low likelihood of structural collapse.

1613.5.1 Mapped acceleration parameters. The parameters S_s and S_1 shall be determined from the 0.2 and 1-second spectral response accelerations shown on Figures 1613.5(1) through

1613.5(14). Where S_1 is less than or equal to 0.04 and S_s is less than or equal to 0.15, the structure is permitted to be assigned to Seismic Design Category A.

❖ The mapped maximum considered earthquake spectral response accelerations at 0.2-second period (S_S) and 1-second period (S_1) for a particular site are to be determined from Figures 1613.5(1) through (14). Where a site is between contours, as would usually be the case, straight-line interpolation or the value of the higher contour may be used. Areas that are considered to have a low seismic risk based solely on the mapped ground motions are placed directly into Seismic Design Category A.

The mapped maximum considered earthquake spectral response accelerations for a site may also be obtained using a seismic parameter program that has been developed by the United States Geological Survey (USGS) in cooperation with the BSSC and FEMA. This program is available on CD or by accessing the USGS earthquake hazards website. The data are interpolated for a specific latitude-longitude or zip code, which the user enters. Output for an entry uses the built-in database to interpolate for the specific site. Caution should be used when using a zip code. In regions with highly variable mapped ground motions, the design parameters within a zip code may vary considerably from the value at the centroid of the zip code area. The code/NEHRP maximum considered earthquake (MCE) output for a site are the two spectral values required for design. The user may request that site coefficients be included in the two spectral values. The user may also use the program to calculate an MCE response spectrum, with or without site coefficients. Site coefficients can be calculated and included in calculations by simply selecting the site class; the program then calculates the site coefficient.

FIGURES 1613.5(1) through 1613.5(14). See page 16-76 to 16-92.

❖ These code figures provide the 5-percent damped maximum considered earthquake spectral response accelerations at 0.2-second period (SS) as well as at 1-second period (S1) for Site Class B soil profiles.

1613.5.2 Site class definitions. Based on the site soil properties, the site shall be classified as either Site Class A, B, C, D, E or F in accordance with Table 1613.5.2. When the soil properties are not known in sufficient detail to determine the site class, Site Class D shall be used unless the building official or geotechnical data determines that Site Class E or F soil is likely to be present at the site.

❖ Each site is to be classified as one of six site classes (A through F), as defined in Table 1613.5.2, based on one of three soil properties measured over the top 100 feet

(30 480 mm) of the site. If the top 100 feet (30 480 mm) are not homogeneous, as would typically be the case, Section 1613.5.5 indicates how to determine average properties. Site Class A is hard rock typically found in the eastern United States. Site Class B is softer rock, typical of the western parts of the country. Site Class C, D or E indicates progressively softer soils. From an earthquake-resistance perspective, rock is the best material for most structures to be founded on. Site Class F indicates soil so poor that site-specific geotechnical investigation and dynamic site response analysis are needed to determine appropriate site coefficients.

Site Class F is defined as any profile containing soils having one or more of the characteristics listed in the last row of Table 1613.5.2.

Any profile with more than 10 feet (30 480 mm) of soil (between the rock surface and the underside of the spread footing or mat foundation) having the characteristics listed in the next-to-the-last row of Table 1613.5.2, related to plasticity index (determined in accordance with ASTM D 4318), moisture content (determined in accordance with ASTM D 2216) and undrained shear strength (determined in accordance with ASTM D 2166 or D 2850) must be classified as Site Class E.

The three soil properties forming the basis of site classification are: shear wave velocity, standard penetration resistance or blow count (determined in accordance with ASTM D 1586) and undrained shear strength (determined in accordance with ASTM D 2166 or 2850). Site Class A and B designations must be based on shear wave velocity measurements or estimates.

Where soil property measurements to a depth of 100 feet (30 480 mm) are not feasible, the registered design professional preparing the soils report may estimate appropriate soil properties based on known geologic conditions.

When soil properties are not known in sufficient detail to determine the site class, the default site class is D, unless the building official determines that Site Class E or F soil may exist at the site.

Section 1613.5.5.1 provides a step-by-step procedure for classifying the soil at a site.

TABLE 1613.5.2. See page 16-75.

❖ This table is used to classify each site as one of six site classes based on one of three measured properties of soil: shear wave velocity, standard penetration resistance and undrained shear strength, except in the case of Site Class F. Plasticity index and moisture content are also used to classify a site as Site Class E or F.

TABLE 1613.5.2
SITE CLASS DEFINITIONS

SITE CLASS	SOIL PROFILE NAME	AVERAGE PROPERTIES IN TOP 100 feet, SEE SECTION 1613.5.5		
		Soil shear wave velocity, $\bar{v}_s$, (ft/s)	Standard penetration resistance, $\bar{N}$	Soil undrained shear strength, $\bar{s}_u$, (psf)
A	Hard rock	$\bar{v}_s > 5{,}000$	N/A	N/A
B	Rock	$2{,}500 < \bar{v}_s \leq 5{,}000$	N/A	N/A
C	Very dense soil and soft rock	$1{,}200 < \bar{v}_s \leq 2{,}500$	$\bar{N} > 50$	$\bar{s}_u \geq 2{,}000$
D	Stiff soil profile	$600 \leq \bar{v}_s \leq 1{,}200$	$15 \leq \bar{N} \leq 50$	$1{,}000 \leq \bar{s}_u \leq 2{,}000$
E	Soft soil profile	$\bar{v}_s < 600$	$\bar{N} < 15$	$\bar{s}_u < 1{,}000$
E	—	Any profile with more than 10 feet of soil having the following characteristics: 1. Plasticity index $PI > 20$, 2. Moisture content $w \geq 40\%$, and 3. Undrained shear strength $\bar{s}_u < 500$ psf		
F	—	Any profile containing soils having one or more of the following characteristics: 1. Soils vulnerable to potential failure or collapse under seismic loading such as liquefiable soils, quick and highly sensitive clays, collapsible weakly cemented soils. 2. Peats and/or highly organic clays ($H > 10$ feet of peat and/or highly organic clay where H = thickness of soil) 3. Very high plasticity clays ($H > 25$ feet with plasticity index $PI > 75$) 4. Very thick soft/medium stiff clays ($H > 120$ feet)		

For SI: 1 foot = 304.8 mm, 1 square foot = 0.0929 m^2, 1 pound per square foot = 0.0479 kPa. N/A = Not applicable

FIGURE 1613.5(1)
MAXIMUM CONSIDERED EARTHQUAKE GROUND MOTION FOR THE CONTERMINOUS UNITED STATES OF
0.2 SEC SPECTRAL RESPONSE ACCELERATION (5% OF CRITICAL DAMPING), SITE CLASS B

FIGURE 1613.5(1)—continued
MAXIMUM CONSIDERED EARTHQUAKE GROUND MOTION FOR THE CONTERMINOUS UNITED STATES OF
0.2 SEC SPECTRAL RESPONSE ACCELERATION (5% OF CRITICAL DAMPING), SITE CLASS B

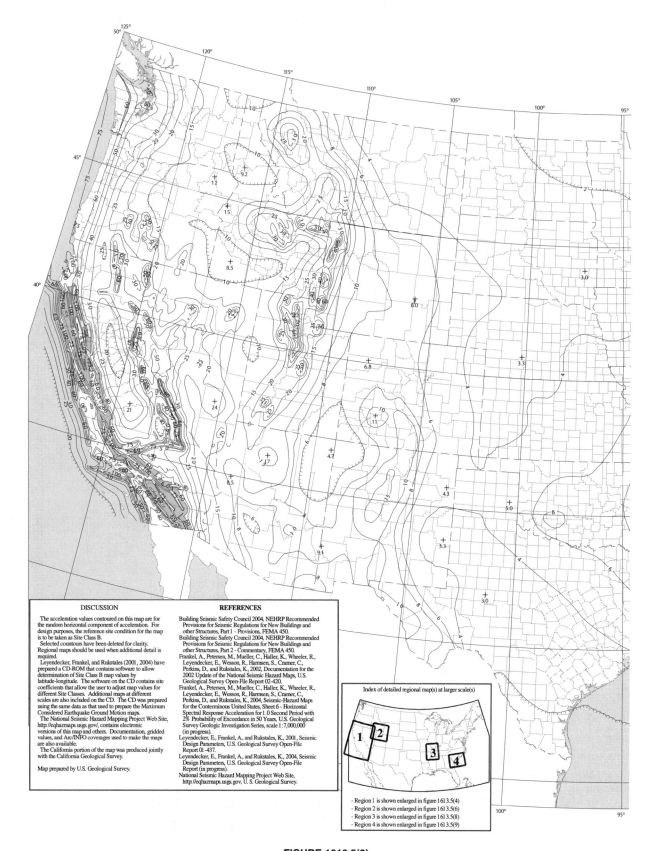

FIGURE 1613.5(2)
**MAXIMUM CONSIDERED EARTHQUAKE GROUND MOTION FOR THE CONTERMINOUS UNITED STATES
OF 1.0 SEC SPECTRAL RESPONSE ACCELERATION (5% OF CRITICAL DAMPING), SITE CLASS B**

FIGURE 1613.5(2)—continued
MAXIMUM CONSIDERED EARTHQUAKE GROUND MOTION FOR THE CONTERMINOUS UNITED STATES
OF 1.0 SEC SPECTRAL RESPONSE ACCELERATION (5% OF CRITICAL DAMPING), SITE CLASS B

FIGURE 1613.5(3)
MAXIMUM CONSIDERED EARTHQUAKE GROUND MOTION FOR REGION 1 OF
0.2 SEC SPECTRAL RESPONSE ACCELERATION (5% OF CRITICAL DAMPING), SITE CLASS B

FIGURE 1613.5(3)—continued
MAXIMUM CONSIDERED EARTHQUAKE GROUND MOTION FOR REGION 1 OF
0.2 SEC SPECTRAL RESPONSE ACCELERATION (5% OF CRITICAL DAMPING), SITE CLASS B

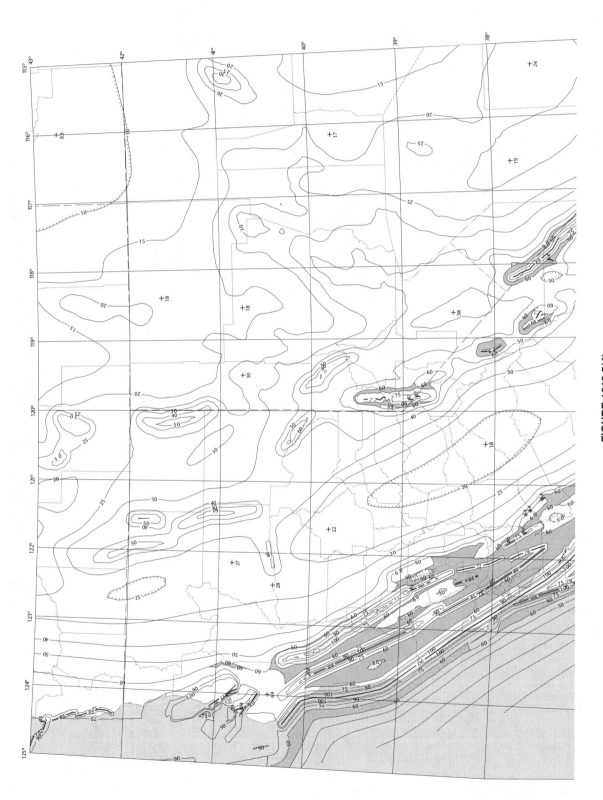

FIGURE 1613.5(4)
MAXIMUM CONSIDERED EARTHQUAKE GROUND MOTION FOR REGION 1 OF
1.0 SEC SPECTRAL RESPONSE ACCELERATION (5% OF CRITICAL DAMPING), SITE CLASS B

FIGURE 1613.5(4)—continued
MAXIMUM CONSIDERED EARTHQUAKE GROUND MOTION FOR REGION 1 OF
1.0 SEC SPECTRAL RESPONSE ACCELERATION (5% OF CRITICAL DAMPING), SITE CLASS B

FIGURE 1613.5.5(5)
MAXIMUM CONSIDERED EARTHQUAKE GROUND MOTION FOR REGION 2 OF
0.2 SEC SPECTRAL RESPONSE ACCELERATION (5% OF CRITICAL DAMPING), SITE CLASS B

FIGURE 1613.5(6)
MAXIMUM CONSIDERED EARTHQUAKE GROUND MOTION FOR REGION 2 OF
1.0 SEC SPECTRAL RESPONSE ACCELERATION (5% OF CRITICAL DAMPING), SITE CLASS B

Index map showing location of study area

FIGURE 1613.5(7)
MAXIMUM CONSIDERED EARTHQUAKE GROUND MOTION FOR REGION 3 OF
0.2 SEC SPECTRAL RESPONSE ACCELERATION (5% PERCENT OF CRITICAL DAMPING), SITE CLASS B

Index map showing location of study area

FIGURE 1613.5(8)
MAXIMUM CONSIDERED EARTHQUAKE GROUND MOTION FOR REGION 3 OF
1.0 SEC SPECTRAL RESPONSE ACCELERATION (5% OF CRITICAL DAMPING), SITE CLASS B

**FIGURE 1613.5(9)
MAXIMUM CONSIDERED EARTHQUAKE GROUND MOTION FOR REGION 4 OF
0.2 AND 1.0 SEC SPECTRAL RESPONSE ACCELERATION (5% OF CRITICAL DAMPING), SITE CLASS B**

0.2 SEC SPECTRAL RESPONSE ACCELERATION (5% OF CRITICAL DAMPING)

1.0 SEC SPECTRAL RESPONSE ACCELERATION (5% OF CRITICAL DAMPING)

Explanation

+ Point value of spectral response
6.2 acceleration expressed as a percent
     of gravity

—— 10 —— Contours of spectral response
          acceleration expressed as a percent
          of gravity. Hachures point in
          direction of decreasing values.

DISCUSSION

The acceleration values contoured on this map are for the random horizontal component of acceleration. For design purposes, the reference site condition for the map is to be taken as Site Class B.

The two areas shown as zone boundaries are the projection to the earth's surface of horizontal rupture planes at 9 km depth. Spectral accelerations are constant within the boundaries of the zones. The number on the boundary and inside the zone is the median spectral response acceleration times 1.5.

Leyendecker, Frankel, and Rukstales (2001, 2004) have prepared a CD-ROM that contains software to allow determination of Site Class B map values by latitude-longitude. The software on the CD contains site coefficients that allow the user to adjust map values for different Site Classes. Additional maps at different scales are also included on the CD. The CD was prepared using the same data as that used to prepare the Maximum Considered Earthquake Ground Motion maps.

The National Seismic Hazard Mapping Project Web Site, http://eqhazmaps.usgs.gov, contains electronic versions of this map and others. Documentation, gridded values, and Arc/INFO coverages used to make the maps are also available.

Map prepared by U.S. Geological Survey.

REFERENCES

Building Seismic Safety Council 2004, NEHRP Recommended Provisions for Seismic Regulations for New Buildings and other Structures, Part 1 - Provisions, FEMA 450.
Building Seismic Safety Council 2004, NEHRP Recommended Provisions for Seismic Regulations for New Buildings and other Structures, Part 2 - Commentary, FEMA 450.
Klein, F., Frankel, A., Mueller, C., Wesson, R. and Okubo, P., 2001, Seismic hazard in Hawaii: high rate of large earthquakes and probabilistic ground-motion maps, Bull. Seism. Soc. Am., v. 91, pp. 479-498.
Klein, F., Frankel, A., Mueller, C., Wesson, R. and Okubo, P., 1998, Seismic-Hazard Maps for Hawaii, Sheet 2 - 2% Probability of Exceedance in 50 Years for Peak Horizontal Acceleration and Horizontal Spectral Response Acceleration for 0.2, 0.3, and 1.0 Second Periods U.S. Geological Survey Geologic Investigation Series I-2724, scale 1:2,000,000.
Leyendecker, E., Frankel, A., and Rukstales, K., 2001, Seismic Design Parameters, U.S. Geological Survey Open-File Report 01-437.
Leyendecker, E., Frankel, A., and Rukstales, K., 2004, Seismic Design Parameters, U.S. Geological Survey Open-File Report (in progress).
National Seismic Hazard Mapping Project Web Site, http://eqhazmaps.usgs.gov, U. S. Geological Survey.

FIGURE 1613.5(10)
MAXIMUM CONSIDERED EARTHQUAKE GROUND MOTION FOR HAWAII OF
0.2 AND 1.0 SEC SPECTRAL RESPONSE ACCELERATION (5% OF CRITICAL DAMPING), SITE CLASS B

FIGURE 1613.5(11)
MAXIMUM CONSIDERED EARTHQUAKE GROUND MOTION FOR ALASKA OF
0.2 SEC SPECTRAL RESPONSE ACCELERATION (5% OF CRITICAL DAMPING), SITE CLASS B

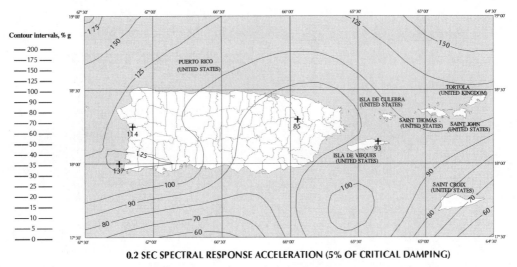

0.2 SEC SPECTRAL RESPONSE ACCELERATION (5% OF CRITICAL DAMPING)

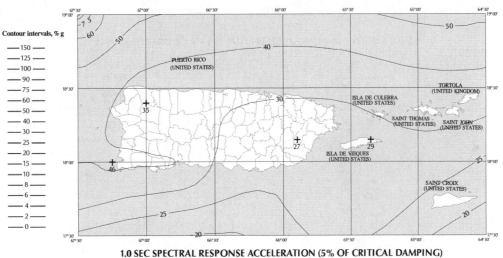

1.0 SEC SPECTRAL RESPONSE ACCELERATION (5% OF CRITICAL DAMPING)

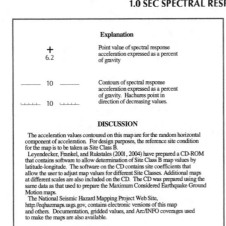

Explanation

+ 6.2   Point value of spectral response acceleration expressed as a percent of gravity

— 10 —   Contours of spectral response acceleration expressed as a percent of gravity. Hachures point in direction of decreasing values.

DISCUSSION

The acceleration values contoured on this map are for the random horizontal component of acceleration. For design purposes, the reference site condition for the map is to be taken as Site Class B.

Leyendecker, Frankel, and Rukstales (2001, 2004) have prepared a CD-ROM that contains software to allow determination of Site Class B map values by latitude-longitude. The software on the CD contains coefficients that allow the user to adjust map values for different Site Classes. Additional maps at different scales are also included on the CD. The CD was prepared using the same data as that used to prepare the Maximum Considered Earthquake Ground Motion maps.

The National Seismic Hazard Mapping Project Web Site, http://eqhazmaps.usgs.gov, contains electronic versions of this map and others. Documentation, gridded values, and Arc/INFO coverages used to make the maps are also available.

Map prepared by U.S. Geological Survey.

REFERENCES

Building Seismic Safety Council 2004, NEHRP Recommended Provisions for Seismic Regulations for New Buildings and other Structures, Part 1 - Provisions, FEMA 450.
Building Seismic Safety Council 2004, NEHRP Recommended Provisions for Seismic Regulations for New Buildings and other Structures, Part 2 - Commentary, FEMA 450.
Leyendecker, E., Frankel, A., and Rukstales, K., 2001, Seismic Design Parameters, U.S. Geological Survey Open-File Report 01-437.
Leyendecker, E., Frankel, A., and Rukstales, K., 2004, Seismic Design Parameters, U.S. Geological Survey Open-File Report (in progress).
Mueller, C., Frankel, A., Petersen, M., and Leyendecker, E., 2003, Documentation for 2003 USGS Seismic Hazard Maps for Puerto Rico and the U.S. Virgin Islands, U.S. Geological Survey Open-File Report 03-379.
Mueller, C., Frankel, A., Petersen, M., and Leyendecker, E., 2004, Seismic-Hazard Maps for Puerto Rico and the U.S. Virgin Island, Sheet 2 - 2% Probability of Exceedance in 50 Years for Peak Horizontal Acceleration and Horizontal Spectral Response Acceleration for 0.2, 0.3, and 1.0 Second Periods U.S. Geological Survey Geologic Investigation Series (in progress).
National Seismic Hazard Mapping Project Web Site, http://eqhazmaps.usgs.gov, U.S. Geological Survey.

FIGURE 1613.5(13)
MAXIMUM CONSIDERED EARTHQUAKE GROUND MOTION FOR PUERTO RICO, CULEBRA, VIEQUES, ST. THOMAS, ST. JOHN AND ST. CROIX OF 0.2 AND 1.0 SEC SPECTRAL RESPONSE ACCELERATION (5% OF CRITICAL DAMPING), SITE CLASS B

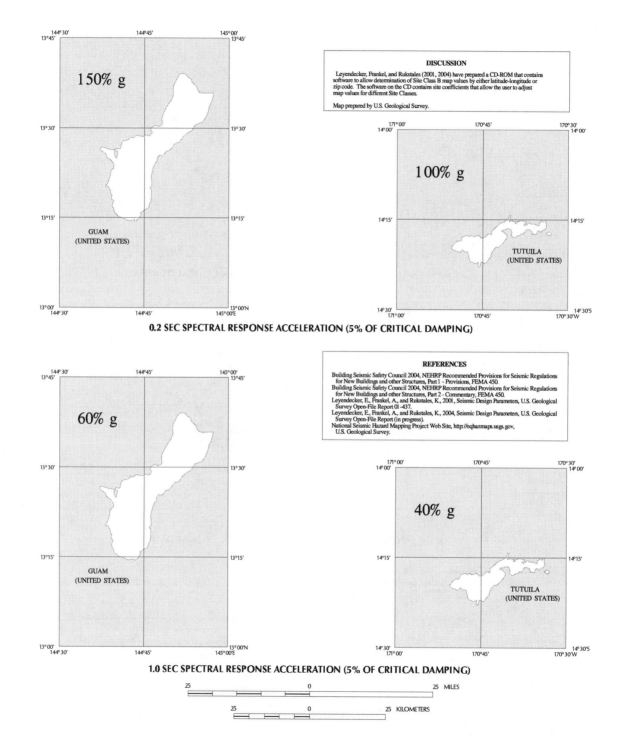

DISCUSSION

Leyendecker, Frankel, and Rukstales (2001, 2004) have prepared a CD-ROM that contains software to allow determination of Site Class B map values by either latitude-longitude or zip code. The software on the CD contains site coefficients that allow the user to adjust map values for different Site Classes.

Map prepared by U.S. Geological Survey.

0.2 SEC SPECTRAL RESPONSE ACCELERATION (5% OF CRITICAL DAMPING)

REFERENCES

Building Seismic Safety Council 2004, NEHRP Recommended Provisions for Seismic Regulations for New Buildings and other Structures, Part 1 - Provisions, FEMA 450.
Building Seismic Safety Council 2004, NEHRP Recommended Provisions for Seismic Regulations for New Buildings and other Structures, Part 2 - Commentary, FEMA 450.
Leyendecker, E., Frankel, A., and Rukstales, K., 2001, Seismic Design Parameters, U.S. Geological Survey Open-File Report 01 -437.
Leyendecker, E., Frankel, A., and Rukstales, K., 2004, Seismic Design Parameters, U.S. Geological Survey Open-File Report (in progress).
National Seismic Hazard Mapping Project Web Site, http://eqhazmaps.usgs.gov, U.S. Geological Survey.

1.0 SEC SPECTRAL RESPONSE ACCELERATION (5% OF CRITICAL DAMPING)

FIGURE 1613.5(14)
MAXIMUM CONSIDERED EARTHQUAKE GROUND MOTION FOR GUAM AND TUTUILLA OF
0.2 AND 1.0 SEC SPECTRAL RESPONSE ACCELERATION (5% OF CRITICAL DAMPING), SITE CLASS B

1613.5.3 Site coefficients and adjusted maximum considered earthquake spectral response acceleration parameters. The maximum considered earthquake spectral response acceleration for short periods, S_{MS}, and at 1-second period, S_{M1}, adjusted for site class effects shall be determined by Equations 16-37 and 16-38, respectively:

$$S_{MS} = F_a S_s \qquad \textbf{(Equation 16-37)}$$

$$S_{M1} = F_v S_1 \qquad \textbf{(Equation 16-38)}$$

where:

F_a = Site coefficient defined in Table 1613.5.3(1).

F_v = Site coefficient defined in Table 1613.5.3(2).

S_S = The mapped spectral accelerations for short periods as determined in Section 1613.5.1.

S_1 = The mapped spectral accelerations for a 1-second period as determined in Section 1613.5.1.

❖ Table 1613.5.3(1) defines an acceleration-related or short-period site coefficient, F_a, as a function of site class and the mapped spectral response acceleration at 0.2-second period, S_s. Table 1613.5.3(2) similarly defines a velocity-related or long-period site coefficient, F_v, as a function of site class and the mapped spectral response acceleration at 1-second period, S_1. The acceleration-related or short-period site coefficient F_a times S_s is S_{MS}, the 5-percent damped soil-modified maximum considered earthquake spectral response acceleration at short periods. The velocity-related or long-period site coefficient F_v times S_1 is S_{M1}, the 5-percent damped soil-modified maximum considered earthquake spectral response acceleration at 1-second period. Such modification by site coefficients is necessary because the mapped quantities are for Site Class B soils. Softer soils (Site Classes C through E) would typically amplify, and stiffer soils (Site Class A) would deamplify ground motion referenced to Site Class B.

As one would expect, both the short-period site coefficient, F_a, and the long-period site coefficient, F_v, are equal to unity for the benchmark Site Class B, irrespective of seismicity of the site. For Site Class A, both coefficients are smaller than unity, indicating reduction of benchmark Site Class B ground motion caused by the stiffer soils. For Site Classes C through E, both site coefficients, with the exception of F_a for Site Class E where $S_s \geq 1.00$, are larger than 1, indicating amplification of benchmark Site Class B ground motion on softer soils. For the same site class, each site coefficient is typically larger in areas of low seismicity than in areas of high seismicity. The basis of this lies in observations that low-magnitude subsurface rock motion is amplified to a larger extent by overlying softer soils than is high-magnitude rock motion. The site coefficients typically become larger for progressively softer soils. The only exception is provided by the short-period site coefficient in areas of high seismicity ($S_s > 0.75$), which remain unchanged or even decrease as the site class changes from D to E. The basis for this also lies in observations that very soft soils are not capable of amplifying the short-period components of subsurface rock motion; deamplification, in fact, takes place when the subsurface rock motion is high in magnitude.

TABLE 1613.5.3(1). See below.

❖ F_a, the acceleration-related or short-period site coefficient, is defined in this table as a function of site class and the seismicity at the site in the form of the mapped spectral response acceleration at 0.2-second period, S_s.

TABLE 1613.5.3(2). See page 16-94.

❖ F_v, the velocity-related or long-period site coefficient, is defined in this table as a function of site class and the seismicity at the site in the form of the mapped spectral response acceleration at 1-second period, S_1.

TABLE 1613.5.3(1)
VALUES OF SITE COEFFICIENT F_a [a]

SITE CLASS	MAPPED SPECTRAL RESPONSE ACCELERATION AT SHORT PERIOD				
	$S_s \leq 0.25$	$S_s = 0.50$	$S_s = 0.75$	$S_s = 1.00$	$S_s \geq 1.25$
A	0.8	0.8	0.8	0.8	0.8
B	1.0	1.0	1.0	1.0	1.0
C	1.2	1.2	1.1	1.0	1.0
D	1.6	1.4	1.2	1.1	1.0
E	2.5	1.7	1.2	0.9	0.9
F	Note b	Note b	Note b	Note b	Note b

a. Use straight-line interpolation for intermediate values of mapped spectral response acceleration at short period, S_s.

b. Values shall be determined in accordance with Section 11.4.7 of ASCE 7.

TABLE 1613.5.3(2)
VALUES OF SITE COEFFICIENT F_v [a]

SITE CLASS	MAPPED SPECTRAL RESPONSE ACCELERATION AT 1-SECOND PERIOD				
	$S_1 \leq 0.1$	$S_1 = 0.2$	$S_1 = 0.3$	$S_1 = 0.4$	$S_1 \geq 0.5$
A	0.8	0.8	0.8	0.8	0.8
B	1.0	1.0	1.0	1.0	1.0
C	1.7	1.6	1.5	1.4	1.3
D	2.4	2.0	1.8	1.6	1.5
E	3.5	3.2	2.8	2.4	2.4
F	Note b	Note b	Note b	Note b	Note b

a. Use straight-line interpolation for intermediate values of mapped spectral response acceleration at 1-second period, S_1.
b. Values shall be determined in accordance with Section 11.4.7 of ASCE 7.

1613.5.4 Design spectral response acceleration parameters.
Five-percent damped design spectral response acceleration at short periods, S_{DS}, and at 1-second period, S_{D1}, shall be determined from Equations 16-39 and 16-40, respectively:

$$S_{DS} = \frac{2}{3} S_{MS} \qquad \textbf{(Equation 16-39)}$$

$$S_{D1} = \frac{2}{3} S_{M1} \qquad \textbf{(Equation 16-40)}$$

where:

S_{MS} = The maximum considered earthquake spectral response accelerations for short period as determined in Section 1613.5.3.

S_{M1} = The maximum considered earthquake spectral response accelerations for 1-second period as determined in Section 1613.5.3.

❖ The design spectral response acceleration at 0.2-second period, S_{DS}, is two-thirds of the S_{MS} value calculated in accordance with Section 1613.5.3. The design spectral response acceleration at 1-second period, S_{D1}, is two-thirds of the S_{M1} value calculated in accordance with Section 1613.5.3. Two-thirds is the reciprocal of 1.5; thus, the design ground motion is $1/1.5$ times the soil-modified maximum considered earthquake ground motion. This is in recognition of the inherent margin contained in the NEHRP provisions that would make collapse unlikely under 1.5 times the design-level ground motion. The idea is to avoid collapse when a structure is subjected to the soil-modified maximum considered earthquake ground motion.

1613.5.5 Site classification for seismic design. Site classification for Site Class C, D or E shall be determined from Table 1613.5.5.

The notations presented below apply to the upper 100 feet (30 480 mm) of the site profile. Profiles containing distinctly different soil and/or rock layers shall be subdivided into those layers designated by a number that ranges from 1 to n at the bottom where there is a total of n distinct layers in the upper 100

feet (30 480 mm). The symbol i then refers to any one of the layers between 1 and n.

where:

v_{si} = The shear wave velocity in feet per second (m/s).

d_i = The thickness of any layer between 0 and 100 feet (30 480 mm).

where:

$$\bar{v}_s = \frac{\sum_{i=1}^{n} d_i}{\sum_{i=1}^{n} \dfrac{d_i}{v_{si}}} \qquad \textbf{(Equation 16-41)}$$

$$\sum_{i=1}^{n} d_i = 100 \text{ feet (30 480 mm)}$$

N_i is the Standard Penetration Resistance (ASTM D 1586) not to exceed 100 blows/foot (305 mm) as directly measured in the field without corrections. When refusal is met for a rock layer, N_i shall be taken as 100 blows/foot (305 mm).

$$\bar{N} = \frac{\sum_{i=1}^{n} d_i}{\sum_{i=1}^{n} \dfrac{d_i}{N_i}} \qquad \textbf{(Equation 16-42)}$$

where N_i and d_i in Equation 16-42 are for cohesionless soil, cohesive soil and rock layers.

$$\bar{N}_{ch} = \frac{d_s}{\sum_{i=1}^{m} \dfrac{d_i}{N_i}} \qquad \textbf{(Equation 16-43)}$$

where:

$$\sum_{i=1}^{m} d_i = d_s$$

Use d_i and N_i for cohesionless soil layers only in Equation 16-43.

d_s = The total thickness of cohesionless soil layers in the top 100 feet (30 480 mm).

m = The number of cohesionless soil layers in the top 100 feet (30 480 mm).

s_{ui} = The undrained shear strength in psf (kPa), not to exceed 5,000 psf (240 kPa), ASTM D 2166 or D 2850.

$$\bar{s}_u = \frac{d_c}{\sum\limits_{i=1}^{k} \dfrac{d_i}{s_{ui}}}$$

(Equation 16-44)

where:

$$\sum\limits_{i=1}^{k} d_i = d_c$$

d_c = The total thickness of cohesive soil layers in the top 100 feet (30 480 mm).

k = The number of cohesive soil layers in the top 100 feet (30 480 mm).

PI = The plasticity index, ASTM D 4318.

w = The moisture content in percent, ASTM D 2216.

Where a site does not qualify under the criteria for Site Class F and there is a total thickness of soft clay greater than 10 feet (3048 mm) where a soft clay layer is defined by: $s_u < 500$ psf (24 kPa), $w \geq 40$ percent, and $PI > 20$, it shall be classified as Site Class E.

The shear wave velocity for rock, Site Class B, shall be either measured on site or estimated by a geotechnical engineer or engineering geologist/seismologist for competent rock with moderate fracturing and weathering. Softer and more highly fractured and weathered rock shall either be measured on site for shear wave velocity or classified as Site Class C.

The hard rock category, Site Class A, shall be supported by shear wave velocity measurements either on site or on profiles of the same rock type in the same formation with an equal or greater degree of weathering and fracturing. Where hard rock conditions are known to be continuous to a depth of 100 feet (30 480 mm), surficial shear wave velocity measurements are permitted to be extrapolated to assess v_s.

The rock categories, Site Classes A and B, shall not be used if there is more than 10 feet (3048 mm) of soil between the rock surface and the bottom of the spread footing or mat foundation.

❖ Where different soil layers make up the upper 100 feet (30 480 mm) of soil profile at a site, this section provides formulas that average soil properties over the upper 100 feet (30 480 mm). There are specific con-

siderations given for the determination of Site Class A or B using shear wave velocity. The rock categories, Site Class A or B, must not be used if there is more than 10 feet (3048 mm) of soil between the rock surface and the underside of the spread footing or mat foundation.

TABLE 1613.5.5. See below.

❖ This table repeats the criteria of Table 1613.5.2 that apply specifically to the assignment of Site Class C, D or E by either of the methods specified in Item 3 of Section 1615.1.5.1.

1613.5.5.1 Steps for classifying a site.

1. Check for the four categories of Site Class F requiring site-specific evaluation. If the site corresponds to any of these categories, classify the site as Site Class F and conduct a site-specific evaluation.

2. Check for the existence of a total thickness of soft clay > 10 feet (3048 mm) where a soft clay layer is defined by: $\bar{s}_u < 500$ psf (24 kPa), $w \geq 40$ percent and $PI > 20$. If these criteria are satisfied, classify the site as Site Class E.

3. Categorize the site using one of the following three methods with $\bar{v}_s$, $\bar{N}$, and $\bar{s}_u$ and computed in all cases as specified.

   3.1. $\bar{v}_s$ for the top 100 feet (30 480 mm) ($\bar{v}_s$ method).

   3.2. $\bar{N}_{ch}$ for the top 100 feet (30 480 mm) ($\bar{N}$ method).

   3.3. $\bar{N}$ for cohesionless soil layers ($PI < 20$) in the top 100 feet (30 480 mm) and average, $\bar{s}_u$ for cohesive soil layers ($PI > 20$) in the top 100 feet (30 480 mm) ($\bar{s}_u$ method).

❖ This section outlines the specific steps leading to the classification of a site for other than Site Class A or B, since those are assigned solely on the basis of shear wave velocity as discussed in Section 1613.5.5.

The first step is to rule out the possibility that the soil meets the criteria of Table 1613.5.2 for Site Class F soil. The second step is to rule out the possibility that the soil meets the criteria of Table 1613.5.2 for Site Class E soil.

If none of the above conditions are satisfied, the third step is to classify the soil profile as either Site Class C, D or E on the basis of: (a) the average shear

TABLE 1613.5.5
SITE CLASSIFICATION[a]

SITE CLASS	$\bar{v}_s$	$\bar{N}$ or $\bar{N}_{ch}$	$\bar{s}_u$
E	< 600 ft/s	< 15	< 1,000 psf
D	600 to 1,200 ft/s	15 to 50	1,000 to 2,000 psf
C	1,200 to 2,500 ft/s	> 50	> 2,000

For SI: 1 foot per second = 304.8 mm per second, 1 pound per square foot = 0.0479 kN/m².

a. If the $\bar{s}_u$ method is used and the $\bar{N}_{ch}$ and $\bar{s}_u$ criteria differ, select the category with the softer soils (for example, use Site Class E instead of D).

wave velocity for the top 100 feet (30 480 mm); (b) the average standard penetration resistance for the top 100 feet (30 480 mm); or (c) the average standard penetration resistance of the cohesionless soil layers and the average undrained shear strength of the cohesive soil layers in the top 100 feet (30 480 mm). It should be noted that Table 1613.5.5 is actually an excerpt of Table 1613.5.2, except for the footnote added to Table 1613.5.5.

1613.5.6 Determination of seismic design category. Occupancy Category I, II or III structures located where the mapped spectral response acceleration parameter at 1-second period, S_1, is greater than or equal to 0.75 shall be assigned to Seismic Design Category E. Occupancy Category IV structures located where the mapped spectral response acceleration parameter at 1-second period, S_1, is greater than or equal to 0.75 shall be assigned to Seismic Design Category F. All other structures shall be assigned to a seismic design category based on their occupancy category and the design spectral response acceleration coefficients, S_{DS} and S_{D1}, determined in accordance with Section 1613.5.4 or the site-specific procedures of ASCE 7. Each building and structure shall be assigned to the more severe seismic design category in accordance with Table 1613.5.6(1) or 1613.5.6(2), irrespective of the fundamental period of vibration of the structure, T.

❖ The seismic design category classification provides a relative scale of earthquake risk to structures. The seismic design category considers not only the seismicity of the site in terms of the mapped spectral response accelerations, but also the site soil profile and the nature of the structure's occupancy category (see the step-by-step description of this process under the definition of "Seismic design category"). It is important to note that there are two tables that are used in order to establish the most restrictive seismic design category classification.

There are two conditions that allow the seismic design category to be determined based upon mapped spectral accelerations, making it unnecessary to go through the process described above. The first is Section 1613.5.1, which identifies areas that have a low seismic risk and are classified as Seismic Design Category A without the need to go through the usual steps. The second instance is an area that is close to a major active fault where S_1 is greater than or equal to 0.75g. These areas are deemed to be areas of considerable seismic risk, and the seismic design category is classified as E or F depending on the structure's occupancy category.

The seismic design category classification is a key criterion in using and understanding the seismic requirements because the analysis method, general design, structural material detailing and the structure's component and system design requirements are determined, at least in part, by the seismic design category. Some of the special inspection requirements in Section 1707 and structural observation requirements in Section 1709 are dependent on the seismic design category classification as well.

TABLE 1613.5.6(1)
SEISMIC DESIGN CATEGORY BASED ON
SHORT-PERIOD RESPONSE ACCELERATIONS

VALUE OF S_{DS}	OCCUPANCY CATEGORY		
	I or II	III	IV
$S_{DS} < 0.167g$	A	A	A
$0.167g \leq S_{DS} < 0.33g$	B	B	C
$0.33g \leq S_{DS} < 0.50g$	C	C	D
$0.50g \leq S_{DS}$	D	D	D

❖ This table defines seismic design category as a function of occupancy category and the short-period (0.2 second) spectral response acceleration at the site of a structure. As the value of S_{DS} increases, the structure is assigned a higher seismic design category and the earthquake design requirements become more stringent.

TABLE 1613.5.6(2)
SEISMIC DESIGN CATEGORY BASED ON
1-SECOND PERIOD RESPONSE ACCELERATION

VALUE OF S_{D1}	OCCUPANCY CATEGORY		
	I or II	III	IV
$S_{D1} < 0.067g$	A	A	A
$0.067g \leq S_{D1} < 0.133g$	B	B	C
$0.133g \leq S_{D1} < 0.20g$	C	C	D
$0.20g \leq S_{D1}$	D	D	D

❖ This table defines seismic design category as a function of occupancy category and the long-period (1-second) spectral response acceleration at the site of a structure. As the value of S_{D1} increases, the structure is assigned a higher seismic design category and the earthquake design requirements become more stringent.

1613.5.6.1 Alternative seismic design category determination. Where S_1 is less than 0.75, the seismic design category is permitted to be determined from Table 1613.5.6(1) alone when all of the following apply:

1. In each of the two orthogonal directions, the approximate fundamental period of the structure, T_a, in each of the two orthogonal directions determined in accordance with Section 12.8.2.1 of ASCE 7, is less than $0.8\ T_s$ determined in accordance with Section 11.4.5 of ASCE 7.

2. In each of the two orthogonal directions, the fundamental period of the structure used to calculate the story drift is less than T_s.

3. Equation 12.8-2 of ASCE 7 is used to determine the seismic response coefficient, C_s.

4. The diaphragms are rigid as defined in Section 12.3.1 in ASCE 7 or for diaphragms that are flexible, the distance

between vertical elements of the seismic-force-resisting system does not exceed 40 feet (12 192 mm).

❖ This section permits the seismic design category of structures meeting the four listed conditions to be based solely on the short-period design spectral coefficient, S_{DS}. This is similar to the approach taken in mapping seismic design categories for use in the *International Residential Code®* (IRC®) as well as for the simplified earthquake analysis procedure.

1613.5.6.2 Simplified design procedure. Where the alternate simplified design procedure of ASCE 7 is used, the seismic design category shall be determined in accordance with ASCE 7.

❖ A simplified earthquake procedure was first introduced under the 1997 UBC and continued on with minor changes through the first two editions of this code. The 2005 edition of ASCE 7 introduces the first stand-alone version of a simplified earthquake analysis for simple bearing wall or building frame systems that comply with the 12 limitations listed in Section 12.14.1.1 of ASCE 7. For structures that qualify to use this method of analysis, the simplified procedure permits the determination of seismic design category to be based solely on the short-period design spectral coefficient, S_{DS}.

1613.6 Alternatives to ASCE 7. The provisions of Section 1613.6 shall be permitted as alternatives to the relevant provisions of ASCE 7.

❖ This section is intended to make the modifications of the ASCE 7 earthquake load provisions optional rather than mandatory. The first item regarding flexible diaphragms may not be optional since the definition in Section 1602 specifically references this modification.

1613.6.1 Assumption of flexible diaphragm. Add the following text at the end of Section 12.3.1.1 of ASCE 7:

Diaphragms constructed of wood structural panels or untopped steel decking shall also be permitted to be idealized as flexible, provided all of the following conditions are met:

1. Toppings of concrete or similar materials are not placed over wood structural panel diaphragms except for nonstructural toppings no greater than $1^1/_2$ inches (38 mm) thick.

2. Each line of vertical elements of the lateral-force-resisting system complies with the allowable story drift of Table 12.12-1.

3. Vertical elements of the lateral-force-resisting system are light-framed walls sheathed with wood structural panels rated for shear resistance or steel sheets.

4. Portions of wood structural panel diaphragms that cantilever beyond the vertical elements of the lateral-force-resisting system are designed in accordance with Section 2305.2.5 of the *International Building Code*.

❖ This section adds conditions under which a diaphragm in light-frame shear wall buildings can be assumed to

be flexible under the code. The ASCE 7 provisions allow flexible diaphragm analysis for light-framed one- and two-family residential buildings. This modification provides additional conditions under which a flexible diaphragm assumption is permitted based solely on the building's construction.

The distinction currently made between rigid and flexible diaphragms is based on engineering judgment rather than by a rigorous analysis. These requirements came in response to failures that occurred during the last two major California earthquakes.

The Consortium of Universities for Research in Earthquake Engineering (CUREE)-Caltech Woodframe Project has shown that for regular, light-framed, wood diaphragm buildings, treating the diaphragms as flexible gives a better match with full size tests. Designs utilizing tributary area would require greater stiffness for wall lines composed of narrow panels than designs utilizing a rigid diaphragm-analysis, which would tend to transfer force away from these light wall lines, allowing them to be even lighter and exacerbating an undesirable condition. Permitting more light-framed wood structures to be analyzed using flexible diaphragms is accomplished with the following conditions:

• Item 1 requires the diaphragms to be flexible by ruling out toppings that are too stiff. Untopped decks have been shown by CUREE and by analysis to be sufficiently flexible to be accurately modeled by flexible diaphragm analysis. CUREE testing has also shown that a thin, lightweight, nonstructural cellular concrete or gypsum topping does not appreciably change the stiffness of a wood structural panel diaphragm.

• Item 2, requiring separate shear wall lines to meet the earthquake drift limits, provides vertical elements of the lateral-force-resisting system that are substantial enough to resist lateral load on a tributary area basis without torsional force redistribution.

• Item 3 limits this allowance to those seismic-force-resisting systems that traditionally have been designed assuming flexible diaphragm behavior.

• Since it is possible in some building configurations for portions of the horizontal diaphragm to cantilever without violating the plan irregularity limits, Item 4 requires a cantilevered portion to be short enough that the overall structure acts as a flexible diaphragm, and also that the cantilever portion be analyzed rigidly and torsional forces be dealt with properly.

1613.6.2 Additional seismic-force-resisting systems for seismically isolated structures. Add the following exception to the end of Section 17.5.4.2 of ASCE 7:

Exception: For isolated structures designed in accordance with this standard, the Structural System Limitations and the Building Height Limitations in Table 12.2-1 for ordinary steel concentrically braced frames (OCBFs) as defined in Chapter 11 and ordinary moment frames (OMFs) as defined in Chapter 11 are permitted to be taken as 160 feet (48 768 mm) for structures assigned to Seismic Design Category D, E or F, provided that the following conditions are satisfied:

1. The value of R_1 as defined in Chapter 17 is taken as 1.

2. For OMFs and OCBFs, design is in accordance with AISC 341.

❖ The ASCE 7 provisions include requirements for structures that utilize seismic base isolation. It is worth noting that use of seismic isolation is never required by the code. This method is an optional method of design for earthquake loading that has been recognized under legacy model codes such as the UBC. The modification adds an exception to the ASCE 7 requirements for structural systems above the isolation system. It corrects an oversight that unnecessarily changes prior practice by restricting the use of certain seismic-force-resisting systems that have been used successfully. As modified, it would allow ordinary concentric braced frames and ordinary moment frame systems with heights up to 160 feet (48 768 mm) when the stated conditions are met.

Bibliography

The following resource materials are referenced in this chapter or are relevant to the subject matter addressed in this chapter.

ACI 318-05, *Building Code Requirements for Structural Concrete.* Farmington Hills, MI: American Concrete Institute, 2005.

ACI 530/ASCE 5/TMS 402-05, *Building Code Requirements for Masonry Structures.* Farmington Hills, MI: American Concrete Institute; New York: American Society of Civil Engineers; Boulder, CO: The Masonry Council, 2005.

ADM 1-00, *Aluminum Design Manual: Part 1-A Aluminum Structures, Allowable Stress Design; and Part 1-B — Aluminum Structures, Load and Resistance Factor Design of Buildings and Similar Type Structures.* Washington, D.C.: Aluminum Association, 2000.

AF&PA WFCM-01, *Wood Frame Construction Manual for One- and Two-Family Dwellings.* Washington, D.C.: American Forest and Paper Association, 2001.

AISC 341-05, *Seismic Provisions for Structural Steel Buildings, including Supplement No. 1.* Chicago: American Institute of Steel Construction, Inc., 2005.

AISC 360-05, *Specification for Structural Steel Buildings.* Chicago: American Institute of Steel Construction, Inc., 2005.

AISI NAS-01, *North American Specification for Design of Cold-Formed Steel Structural Members, including 2004 Supplement.* Washington, D.C.: American Iron and Steel Institute, 2001.

Algermissen, S.T. *An Introduction to the Seismicity of the United States.* Oakland, CA: Earthquake Engineering Research Institute, 1983.

Ambrose, James and Dimitry Vergun. *Design for Lateral Forces.* New York: John A. Wiley & Sons, 1987.

ASCE 3-91, *Standard Practice for the Construction and Inspection of Composite Slabs.* Reston, VA: American Society of Civil Engineers, 1991.

ASCE 7-05, *Minimum Design Loads for Buildings and Other Structures, including Supplement No. 1.* New York: American Society of Civil Engineers, 2005.

ASCE 7-88, *Minimum Design Loads for Buildings and Other Structures.* Reston, VA: American Society of Civil Engineers, 1988.

ASCE 7-95, *Minimum Design Loads for Buildings and Other Structures.* New York: American Society of Civil Engineers, 1995.

ASCE 8-02, *Standard Specification for Design of Cold-formed Stainless Steel Structural Members.* Reston, VA: American Society of Civil Engineers, 2002.

ASCE 24-05, *Flood Resistance Design and Construction Standard.* Reston, VA: American Society of Civil Engineers, 2005.

ASME A17.1-00, *Safety Code for Elevators and Escalators.* New York: American Society of Mechanical Engineers, 2000.

ASME B31.4-95, *Boilers and Pressure Vessels Code.* New York: American Society of Mechanical Engineers, 1995.

ASTM D 2166-00, *Test Method for Unconfined Compressive Strength of Cohesive Soil.* West Conshohocken, PA: ASTM International, 2000.

ASTM D 2216-98, *Test Method for Laboratory Determination of Water (Moisture) Content of Soil and Rock by Mass.* West Conshohocken, PA: ASTM International, 1998.

ASTM D 2487-00, *Standard Classification of Soils for Engineering Purposes Unified Soil Classification System.* West Conshohocken, PA: ASTM International, 2000.

ASTM D 2850-03a, *Test Method for Unconsolidated, Undrained Triaxial Compression Test on Cohesive Soil.* West Conshohocken, PA: ASTM International, 2003.

ASTM D 4318-00, *Test Methods for Liquid Limit, Plastic Limt and Plasticity Index of Soils.* West Conshohocken, PA: ASTM International, 2000.

ASTM D 6381-03, *Standard Test Method for Measurement of Asphalt Shingle Uplift Resistance.* West Conshohocken, PA: ASTM International, 2003.

ASTM E 1886-02, *Test Method for Performance of Exterior Windows, Curtain Walls, Doors and Storm Shut-*

ters Impacted by Missiles and Exposed to Cyclic Pressure Differentials. West Conshohocken, PA: ASTM International, 2002.

ASTM E 1996-02, *Specification for Performance of Exterior Windows, Curtain Walls, Doors and Storm Shutters Impacted by Windborne Debris in Hurricanes*. West Conshohocken, PA: ASTM International, 2002.

Becker, Roy, Farzad Naeim and Edward J. Teal. *Seismic Design Practice for Steel Buildings*. Los Angeles: The Steel Committee of California, Revised 1988.

Bolt, Bruce A. *Earthquakes*. New York: W.H. Freeman and Company, 1988.

Chen, Wai-Fah and Charles Scawthorn. *Earthquake Engineering Handbook*. Boca Raton, FL: CRC Press LLC, 2003.

Choppra, Anil K. *Dynamics of Structures: A Primer*. Oakland, CA: Earthquake Engineering Research Institute, 1981.

Cook, N.J. *The Designer's Guide to Wind Loading of Building Structures, Part I: Building Research Establishment*. Cambridge: The University Press, 1985.

Diaphragms. Tacoma, WA: American Plywood Association, 1989.

Dowrick, David J. *Earthquake-Resistant Design for Engineers and Architects*, 2nd ed. New York: John Wiley & Sons, 1987.

Fanella, David A., S. K. Ghosh and Xuemei Liang. *Seismic and Wind Design of Concrete Buildings (2003 IBC, ASCE 7-02, ACI 318-02)*. Falls Church, VA: International Code Council, 2005.

FEMA 55, *Coastal Construction Manual*. Washington, D.C.: Federal Emergency Management Agency, 1999.

FEMA 213, *Answers to Questions About Substantially Damaged Buildings*. Washington, D.C.: Federal Emergency Management Agency, 1991.

FEMA 259, *Engineering Principles and Practices for Retrofitting Flood-Prone Residential Buildings*. Washington, D.C.: Federal Emergency Management Agency, 1995.

FEMA 302, *NEHRP Recommended Provisions for Seismic Regulations for New Buildings and Other Structures, Part 1*. Washington, D.C.: Prepared by the Building Seismic Safety Council for the Federal Emergency Management Agency, 1998.

FEMA 303, *NEHRP Recommended Provisions for Seismic Regulations for New Buildings and Other Structures Commentary, Part 2 Commentary*. Washington, D.C.: Prepared by the Building Seismic Safety Council for the Federal Emergency Management Agency, 1998.

FEMA 311, *Guidance on Estimating Substantial Damage Using the NFIP Residential Substantial Damage Estimator*. Washington, D.C.: Federal Emergency Management Agency, 1998.

FEMA 348, *Protecting Building Utilities From Flood Damage: Principles and Practices for the Design and Construction of Flood-Resistant Building Utility Systems*. Washington, D.C.: Federal Emergency Management Agency, 1999.

FEMA 450-1, *NEHRP Recommended Provisions for Seismic Regulations for New Buildings and Other Structures, Part 1*. Washington, D.C.: Prepared by the Building Seismic Safety Council for the Federal Emergency Management Agency, 2004.

FEMA 450-2, *NEHRP Recommended Provisions for Seismic Regulations for New Buildings and Other Structures, Part 2 Commentary*. Washington, DC: Prepared by the Building Seismic Safety Council for the Federal Emergency Management Agency, 2004.

FEMA (1996), *Floodproofing Certificate* (FEMA Form 81-65) [Online]. Available: http://www.fema.gov/nfip/ff81-65.pdf [1999, December 30].

FEMA (1999), *Elevation Certificate* (FEMA Form 81-31). [Online]. Available: http://www.fema.gov/library/elvcert.pdf [1999, December 30].

FEMA FIA-TB #1: *Openings in Foundation Walls for Buildings Located in Special Flood Hazard Areas*, Washington D.C.: Federal Emergency Management Agency, 1993.

FEMA FIA-TB #2: *Flood-Resistant Material Requirements for Buildings Located in Special Flood Hazard Areas*. Washington, D.C.: Federal Emergency Management Agency, 1993.

FEMA FIA-TB #3: *Nonresidential Floodproofing—Requirements and Certification for Buildings Located in Special Flood Hazard Areas*. Washington, D.C.: Federal Emergency Management Agency, 1993.

FEMA FIA-TB #4: *Elevator Installation for Buildings Located in Special Flood Hazard Areas*. Washington, D.C.: Federal Emergency Management Agency, 1993.

FEMA FIA-TB #5: *Free of Obstruction Requirements for Buildings Located in Coastal High Hazard Areas*. Washington, D.C.: Federal Emergency Management Agency, 1993.

FEMA FIA-TB #6: *Below-Grade Parking Requirements for Buildings Located in Special Flood Hazard Areas*. Washington, D.C.: Federal Emergency Management Agency, 1993.

FEMA FIA-TB #7: *Wet Floodproofing Requirements for Structures Located in Special Flood Hazard Areas*. Washington, D.C.: Federal Emergency Management Agency, 1993.

FEMA FIA-TB #8: *Corrosion Protection for Metal Connectors In Coastal Areas for Structures Located in*

Special Flood Hazard Areas. Washington, D.C.: Federal Emergency Management Agency, 1996.

FEMA FIA-TB #9: *Design and Construction Guidance for Breakaway Walls Below Elevated Buildings*. Washington, D.C.: Federal Emergency Management Agency.

FEMA (various dates). *NFIP Technical Bulletin Series*. Washington, D.C.: National Flood Insurance Program. [Online]. Available: http://www.fema.gov/mit/techbul.htm.

Fuller, Myron. *The New Madrid Earthquake*. Marion, IL: U.S. Department of the Interior/Geological Survey, 1912 (reprinted 1988).

Ghosh, S. K. *Seismic Design Using Structural Dynamics (2000 IBC)*. Falls Church, VA: International Code Council, 2003.

Glen, Berg V. *Seismic Design Codes and Procedures*. Oakland, CA: Earthquake Engineering Research Institute, 1983.

Green, Norman. *Earthquake-Resistant Building Design and Construction*. New York: Elsevier Science Publishing Company, 1987.

Guide for the Design and Construction of Mill Buildings (AISE Technical Report #13). Washington, D.C.: Association of Iron and Steel Engineers, August 1, 1979.

Hall, W.J. and N.M. Newmark. *Earthquake Spectra and Design*. Oakland, CA: Earthquake Engineering Research Institute, 1982.

IPC-06, *International Plumbing Code*. Falls Church, VA: International Code Council, 2006.

IRC-06, *International Residential Code*. Falls Church, VA: International Code Council, 2006.

Isyumor, N. and T. Tschanz. *Building Motion in Wind*. New York: American Society of Civil Engineers, 1986.

Liu, Henry. *Wind Engineering: A Handbook for Structural Engineers*. Englewood Cliffs, NJ: Prentice Hall, Inc., 1991.

Low Rise Building Systems Manual. Cleveland, OH: Metal Building Manufacturers Association, 1986.

Lubbock, Richard Marshall, Kishor C. Mehta and Dale Perry. *Guide to the Use of the Wind Load Provisions of ASCE 7-88 (formerly ANSI A58.1)*. New York: American Society of Civil Engineers, 1991.

Mehta, Kishor C. and Joseph E. Minor. "Wind Damage Observations and Implications." *Journal of the Structural Division*. New York: American Society of Civil Engineers, Vol. 105, November 1979.

NAAMM 1001-97, *Guide Specification for Design of Metal Flag Poles.* Chicago: National Association of Architectural Metal Manufacturers, 1997.

Naeim, Farzad. *The Seismic Design Handbook*. Norwell, MA: Kluwer Academic Publishers, 2001.

NCMA-TEK No. 109A, *Concrete Masonry Design for Seismic Forces*. Herndon, VA: National Concrete Masonry Association, Revised, 1989.

Pakiser, Louis C. *Earthquakes*. Washington, D.C.: U.S. Department of the Interior/Geological Survey, 1988.

Proceedings of the 20th Joint Meeting of the U.S.-Japan Cooperative Program in Natural Resources, Wind and Seismic Effects. National Institute of Standards and Technology, NIST SP 760, January 1989.

Proceedings of the First International Conference on Snow Engineering. Santa Barbara, CA: U.S. Army Corps of Engineers — Cold Regions Research & Engineering Laboratory, Special Report 89-6, July 1988.

Recommendations for Direct-Hung Acoustical Tile and Lay-in Panel Ceilings, Seismic Zones 0-2. St. Charles, IL: Ceiling and Interior Systems Construction Association, 1991.

Recommendations for Direct-Hung Acoustical Tile and Lay-in Panel Ceilings, Seismic Zones 3-4. St. Charles, IL: Ceiling and Interior Systems Construction Association, 1991.

Recommended Lateral Force Requirements and Commentary (SEAOC Blue Book). Sacramento, CA: Structural Engineers Association of California, 1999.

Recommended Lateral Force Requirements and Tentative Commentary. San Francisco: Structural Engineers Association of California, 1988.

Reducing Earthquake Hazards: Lessons Learned from Earthquakes. Publication No. 86-02. Oakland, CA: Earthquake Engineering Research Institute, 1986.

Reducing Flood Losses Through the International Code Series: Meeting the Requirements of the National Flood Insurance Program. Falls Church, VA: International Code Council, May 2000.

RMI-02, *Specifications for the Design, Testing, and Utilization of Industrial Steel Storage Racks*. Charlotte, NC: Rack Manufacturers Institute, 2002.

SEAW's Handbook of a Rapid-Solutions Methodology™ for Wind Design. Redwood City, CA: Structural Engineers Association of Washington, Applied Technology Council, 2004.

SEAW/ATC 60, *SEAW Commentary on Wind Code Provisions, Volume 1*. Redwood City, CA: Applied Technology Council, 2004.

SEAW/ATC 60, *SEAW Commentary on Wind Code Provisions, Volume 2—Example Problems*. Redwood City, CA: Applied Technology Council, 2004.

SBCCI SSTD 10-99, *Standard for Hurricane Resistant Residential Construction*. Birmingham, AL: Southern Building Code Congress International, 1999.

SJI JG-1.1-05, *Standard Specification for Joist Girders*. Myrtle Beach, SC: Steel Joist Institute, 2005.

SJI K-1.1-05, *Standard Specification for Open Web Steel Joists, K Series*. Myrtle Beach, SC: Steel Joist Institute, 2005.

SJI LH/DLH-1.1-05, *Standard Specification for Longspan Steel Joists, LH Series and Deep Longspan Steel Joists, DLH Series*. Myrtle Beach, SC: Steel Joist Institute, 2005.

SMACNA-HVAC-95, *HVAC Duct Construction Standards, Metal and Flexible*. Chantilly, VA: Sheet Metal & Air Conditioning Contractors' National Association, Inc., 1995.

SMACNA-Seismic-98, *Seismic Restraint Manual Guidelines for Mechanical Systems, 1991, including Appendix B, 1998*. Chantilly, VA: Sheet Metal & Air Conditioning Contractors' National Association, Inc., 1998.

Specifications for Electric Overhead Traveling Cranes (CMAA Specification #70). New York: Crane Manufacturers Association of America, Inc., 1988.

Structural Engineering Loads. Design Manual 2.02. Naval Facilities Engineering Command, September 1986.

Taly, Narendra. *Loads and Load Paths in Buildings: Principles of Structural Design*. Falls Church, VA: International Code Council, 2003.

Taly, Narendra. *Loads and Load Paths in Buildings: Principles of Structural Design — Problems and Solutions*. Falls Church, VA: International Code Council, 2005.

Taranath, Bungale. *Wind and Earthquake Resistant Buildings Structural Analysis and Design*. New York: Marcel Dekker, 2005.

The Loma Prieta Earthquake of October 17, 1989. Washington, D.C.: U.S. Department of the Interior/Geological Survey, 1989.

The Severity of an Earthquake. Washington, D.C.: U.S Department of the Interior/Geological Survey, 1985.

UBC-97, *Uniform Building Code*. Whittier, CA: International Conference of Building Officials, 1997.

UL 2390-04, *Test Method for Measuring the Wind Uplift Coefficients for Asphalt Shingles*. Northbrook, IL: Underwriters Laboratories, Inc., 2004.

Williams, Alan. *Seismic and Wind Forces Structural Design Examples, Second Edition*. Falls Church, VA: International Code Council, 2005.

Wind Loading and Wind-Induced Structural Response. Committee on Wind Effects. New York: American Society of Civil Engineers, 1987.

Chapter 17:
Structural Tests and Special Inspections

General Comments

In this chapter, the code sets minimum quality standards for the acceptance of materials used in building construction. It also establishes requirements for special inspections, structural observations and load testing.

Section 1701 contains the scope statement and the general statement for new and used materials.

The definitions of terms primarily related to this chapter are in Section 1702.

Section 1703 addresses the approval process.

Special inspections that are required are specified in Section 1704.

Section 1705 includes the requirements for the statement of special inspecting.

Requirements for a written statement by the contractor are provided in Section 1706.

Section 1707 contains the special inspection requirements based on the seismic design category of the structure.

Structural testing for seismic resistance is addressed in Section 1708.

Section 1709 establishes when structural observation by a registered design professional is required for high-seismic and high-wind areas.

The general requirements for determining the design strengths of materials are in Section 1710.

Section 1711 provides for an alternative test procedure in the absence of approved standards.

Provisions for a test load are addressed in Section 1712.

Section 1713 includes requirements for field load testing of a structure.

Preconstruction load testing of materials and methods of construction that are not capable of being designed by an approved analysis is covered by Section 1714.

Section 1715 includes specific material and test standards.

Chapter 17 provides information regarding the evaluation, inspection and approval process for any material or system proposed for use as a component of a structure. These are general requirements that expand on the requirements of Chapter 1 relating to the roles and responsibilities of the building official regarding approval of building components. Additionally, the chapter includes general requirements relating to the roles and responsibilities of the owner, contractor, special inspectors and architects or engineers.

Purpose

This chapter provides procedures and criteria for: testing materials or assemblies, labeling materials, systems and assemblies and special inspections of structural assemblies.

SECTION 1701
GENERAL

1701.1 Scope. The provisions of this chapter shall govern the quality, workmanship and requirements for materials covered. Materials of construction and tests shall conform to the applicable standards listed in this code.

❖ This chapter gives provisions for quality, workmanship, testing and labeling of all materials covered within. In general, all construction materials and tests must conform to the applicable standards listed in the code. This chapter provides requirements for materials and tests when there are no applicable standards; specific tests and standards are referenced in other chapters of the code. Additionally, this chapter provides basic requirements for labeling construction materials and assemblies, and for special inspections of structural systems and components.

1701.2 New materials. New building materials, equipment, appliances, systems or methods of construction not provided for in this code, and any material of questioned suitability proposed for use in the construction of a building or structure, shall be subjected to the tests prescribed in this chapter and in the approved rules to determine character, quality and limitations of use.

❖ Testing is required to be performed on materials that are not specifically provided for in the code. For example, the manufacturer of a sandwich panel consisting of aluminum skins and a foam plastic core wishes to use this panel as an exterior weather covering. The material does not conform to any of the standards referenced in Chapter 14, so an appropriate test protocol must be developed. The same provision for acceptance of alternative materials is already given in Section 104.11. That section provides a strong, definitive statement for performance requirements for alternative materials, requiring the proposed alternative to be equivalent to that prescribed by the code in quality, strength, effectiveness, durability and safety. Section 1701.2 simply reasserts that alternative materials (new materials) may be used, as long as the performance characteristics and quality can be established.

1701.3 Used materials. The use of second-hand materials that meet the minimum requirements of this code for new materials shall be permitted.

❖ Materials and assemblies may be reused, provided that they meet the requirements of the code for new materials (see Section 104.9.1 of the code regarding reuse of materials and equipment). Caution should be exercised in approving a used material for reuse. The applicable material standards must be consulted to determine if certain reuses are prohibited and to determine the characteristics of the used material that must be carefully checked before reuse is approved.

One example is a high-strength structural steel bolt. Reuse of the bolt is restricted by RCSC, *Specification for Structural Joints Using ASTM A315 or A490 Bolts*. Even a piece of structural steel, such as a wide flange, would need to be carefully checked to determine that dimensional tolerances for a new piece of structural steel are met (see ASTM A 6 and A 36).

SECTION 1702
DEFINITIONS

1702.1 General. The following words and terms shall, for the purposes of this chapter and as used elsewhere in this code, have the meanings shown herein.

❖ Some terms used in this code describe something very specific to the related subject matter while other terms may have multiple meanings that can vary based on the subject matter and context. To avoid misunderstandings, it is preferable to reach a consensus on the meaning of these terms. Definitions are intended to facilitate the understanding of code provisions and to minimize potential confusion.

This section contains definitions of terms associated with special inspections. Note that these terms are also listed in Chapter 2 with a cross reference to this section. Definitions that pertain primarily to special inspections are included within this chapter to provide convenient access to them without having to refer back to Chapter 2. The use and application of all defined terms, including those defined herein, as well as undefined terms are set forth in Section 201.

APPROVED AGENCY. An established and recognized agency regularly engaged in conducting tests or furnishing inspection services, when such agency has been approved.

❖ In order to identify the basic criteria or to understand what agencies are being referred to in Section 1703, there is a need to define the term "approved agency." The word "approved" means "acceptable to the building official" (see the definition of "Approved" in Section

202). The basis for approval of an agency for a particular activity by the building official may include, but is not necessarily limited to, the capacity and capability of the agency to perform the work in accordance with Section 1704 and other applicable sections. This is typically done through review of the résumés and references of the agency and its personnel.

APPROVED FABRICATOR. An established and qualified person, firm or corporation approved by the building official pursuant to Chapter 17 of this code.

❖ An approved fabricator is one who has received approval to perform work without a code-required special inspection. The approval is based upon review of the fabricator's written procedural and quality-control manuals and periodic auditing of fabrication practices by an approved special inspection agency.

CERTIFICATE OF COMPLIANCE. A certificate stating that materials and products meet specified standards or that work was done in compliance with approved construction documents.

❖ A certificate of compliance is a document issued by a supplier of materials and products that certifies they meet the specified requirements.

DESIGNATED SEISMIC SYSTEM. Those architectural, electrical and mechanical systems and their components that require design in accordance with Chapter 13 of ASCE 7 and for which the component importance factor, I_p, is greater than 1 in accordance with Section 13.1.3 of ASCE 7.

❖ Designated seismic systems are those architectural, electrical and mechanical components and systems in the referenced ASCE 7 provisions that are assigned a component importance factor of 1.5. The importance factor indicates that there is a higher level of importance placed on the system so that it will remain operational during and after an earthquake. This definition is important in correctly applying the provisions for special inspection and testing for designated seismic systems.

FABRICATED ITEM. Structural, load-bearing or lateral load-resisting assemblies consisting of materials assembled prior to installation in a building or structure or subjected to operations such as heat treatment, thermal cutting, cold working or reforming after manufacture and prior to installation in a building or structure. Materials produced in accordance with standard specifications referenced by this code, such as rolled structural steel shapes, steel-reinforcing bars, masonry units and wood structural panels shall not be considered "fabricated items."

❖ The term "fabricated item" can easily be misinterpreted to encompass a number of items for which the code does not intend special inspections; therefore, the term is defined to clarify the intent of the code (see Section 1704).

INSPECTION CERTIFICATE. An identification applied on a product by an approved agency containing the name of the manufacturer, the function and performance characteristics, and the name and identification of an approved agency that indicates that the product or material has been inspected and evaluated by an approved agency (see Section 1703.5 and "Label," "Manufacturer's designation" and "Mark").

❖ An inspection certificate is an identification applied to a product indicating that the individual product has been inspected by a third-party agency. The identification should include three items:

1. The name of the manufacturer;

2. The function and performance characteristics of the product; and

3. The name of the approved agency completing the inspection and evaluation for the product.

Note that the requirements for a label (see Section 1702 for the definition of "Label") differ from those of an inspection certificate. A label is issued by a third-party inspection agency that performs periodic inspections, while an inspection certificate is issued for the product at the time of inspection. The issuance of inspection certificates is an ongoing procedure.

LABEL. An identification applied on a product by the manufacturer that contains the name of the manufacturer, the function and performance characteristics of the product or material, and the name and identification of an approved agency and that indicates that the representative sample of the product or material has been tested and evaluated by an approved agency (see Section 1703.5 and "Inspection certificate," "Manufacturer's designation" and "Mark").

❖ A label provides verification of testing and inspection of materials, products or assemblies (see commentary, Section 1703.5).

MAIN WIND-FORCE-RESISTING SYSTEM. An assemblage of structural elements assigned to provide support and stability for the overall structure. The system generally receives wind loading from more than one surface.

❖ This definition identifies the structural components that are to comply with special inspection requirements. The main windforce-resisting system is the global structural system responsible for resisting wind loads on the building or structure. These include moment-resisting frames, braced frames, diaphragms and shear walls.

MANUFACTURER'S DESIGNATION. An identification applied on a product by the manufacturer indicating that a product or material complies with a specified standard or set of rules (see also "Inspection certificate," "Label" and "Mark").

❖ This represents terminology for a manufacturer's self-certification that a product complies with a given standard (see commentary, Section 1703.4).

MARK. An identification applied on a product by the manufacturer indicating the name of the manufacturer and the function of a product or material (see also "Inspection certificate," "Label" and "Manufacturer's designation").

❖ A mark represents the manufacturer's identification placed on a product, stating who made the product and describing its function. There is, however, no cer-

tification of compliance to any particular standard and no third-party quality control (see commentary, Section 1703.4).

SPECIAL INSPECTION. Inspection as herein required of the materials, installation, fabrication, erection or placement of components and connections requiring special expertise to ensure compliance with approved construction documents and referenced standards (see Section 1704).

❖ This category of inspection is intended to include those material connections or installations that require a special level of knowledge and attention. For example, special inspections are required for the installation of high-strength bolts, welded connections, concrete reinforcement, prestressed concrete, fabrication of laminated wood structural elements and pile installations to comply with the contract documents and the standards under which they are assembled.

SPECIAL INSPECTION, CONTINUOUS. The full-time observation of work requiring special inspection by an approved special inspector who is present in the area where the work is being performed.

❖ Continuous special inspection is full-time monitoring of specific tasks by a special inspector. These inspections must be carried out continuously over the duration of the particular task(s). "Full-time" should not be construed as requiring the special inspector to be present when these tasks are not being performed.

SPECIAL INSPECTION, PERIODIC. The part-time or intermittent observation of work requiring special inspection by an approved special inspector who is present in the area where the work has been or is being performed and at the completion of the work.

❖ Periodic special inspection allows intermittent monitoring of the specified tasks designated in the statement of special inspections as requiring periodic inspection. To avoid misunderstandings, the frequency or timing of these inspections should be established before a permit is issued.

SPRAYED FIRE-RESISTANT MATERIALS. Cementitious or fibrous materials that are spray applied to provide fire-resistant protection of the substrates.

❖ The cementitious or fibrous material is pneumatically projected onto a surface such that the density, thickness and cohesion/adhesion of the material will provide fire resistance to the surface.

STRUCTURAL OBSERVATION. The visual observation of the structural system by a registered design professional for general conformance to the approved construction documents at significant construction stages and at completion of the structural system. Structural observation does not include or waive the responsibility for the inspection required by Section 109, 1704 or other sections of this code.

❖ The registered design professional in responsible charge is required to visit the site to visually determine general conformance to the approved construction documents. Structural observation is in addition to and does not substitute for any required inspections by Section 109, 1704 or other sections of the code.

SECTION 1703
APPROVALS

1703.1 Approved agency. An approved agency shall provide all information as necessary for the building official to determine that the agency meets the applicable requirements.

❖ This section specifies the information that an approved agency must provide to the building official to enable him or her to determine if the agency meets the applicable requirements.

1703.1.1 Independent. An approved agency shall be objective and competent. The agency shall also disclose possible conflicts of interest so that objectivity can be confirmed.

❖ As part of the basis for a building official's approval of a particular inspection agency, the agency must demonstrate its objectivity and competence. The judgement of objectivity is linked to the financial and fiduciary independence of the agency. The competence of the agency is judged by its experience and organization, and the experience of its personnel.

   For example, suppose that ACME Agency is the inspection agency employed by Builder's, Inc. for factory-built fireplaces. During investigation of the agency, it is discovered that ACME and Builder's are subsidiaries of the same parent company, Conglomerate, Inc. The inspection agency and the manufacturer clearly have a relationship that is undesirable from the standpoint of independence.

1703.1.2 Equipment. An approved agency shall have adequate equipment to perform required tests. The equipment shall be periodically calibrated.

❖ As part of judging the ability of a testing or inspection agency, the building official should determine that the agency has the proper equipment to perform the required tests or inspections.

1703.1.3 Personnel. An approved agency shall employ experienced personnel educated in conducting, supervising and evaluating tests and/or inspections.

❖ The competence of an inspection or testing agency is also based on the experience and background of its personnel. For example, if 10 engineering graduates form an agency, the building official should question whether or not this newly formed agency is sufficiently experienced to perform the tests.

   If the services being provided by the inspection agency or test agency come within the purview of the professional registration laws of the state in which the building is being constructed, the building official must request evidence that the personnel are qualified to perform the work in accordance with this professional registration law as well.

1703.2 Written approval. Any material, appliance, equipment, system or method of construction meeting the requirements of this code shall be approved in writing after satisfactory completion of the required tests and submission of required test reports.

❖ In order to have a documented record of the approval and the basis for it, including any conditions or limitations, materials and systems must be approved in writing by the building official. The code also requires the approval to be granted within a reasonable period of time, after all documentation has been satisfactorily developed and submitted, so as to avoid any unnecessary delay in completion of construction.

1703.3 Approved record. For any material, appliance, equipment, system or method of construction that has been approved, a record of such approval, including the conditions and limitations of the approval, shall be kept on file in the building official's office and shall be open to public inspection at appropriate times.

❖ Written approvals must be kept on file by the building official and be available and open to the public. This provides reasonable access to the records on approvals of materials and systems should there be any subsequent investigation or further evaluation.

1703.4 Performance. Specific information consisting of test reports conducted by an approved testing agency in accordance with standards referenced in Chapter 35, or other such information as necessary, shall be provided for the building official to determine that the material meets the applicable code requirements.

❖ When conformance to the code is predicated on the performance and quality of materials, the building official must require the submittal of testing reports from an approved agency. In the absence of such reports, the building official must accept specific information and details that prove compliance with the intent of the applicable code requirements.

1703.4.1 Research and investigation. Sufficient technical data shall be submitted to the building official to substantiate the proposed use of any material or assembly. If it is determined that the evidence submitted is satisfactory proof of performance for the use intended, the building official shall approve the use of the material or assembly subject to the requirements of this code. The costs, reports and investigations required under these provisions shall be paid by the permit applicant.

❖ This section is usually used in conjunction with Section 104.11 when analysis of any construction material, such as new and innovative materials, is required to determine code compliance. The analysis is based entirely upon technical data. All costs of testing and investigations must be paid by the applicant.

1703.4.2 Research reports. Supporting data, where necessary to assist in the approval of materials or assemblies not specifically provided for in this code, shall consist of valid research reports from approved sources.

❖ Evaluation reports prepared by approved agencies, such as those published by organizations affiliated with model code groups, may be accepted as part of the information needed by the building official to form the basis for approval. Such reports can supplement the building department resources by eliminating the need for the building official to conduct a detailed analysis on each new product, material or system. It is critically important that such material be truly objective and credible and not consist merely of the manufacturer's brochures or similar proprietary information. It is also important to note that when the building official is utilizing research reports in evaluating compliance with the code, such as those issued by organizations affiliated with model code groups, he or she is not mandated to approve these research reports just because the code is the legally adopted building code in the jurisdiction. These reports are not code text; they are advisory only and intended for technical reference.

1703.5 Labeling. Where materials or assemblies are required by this code to be labeled, such materials and assemblies shall be labeled by an approved agency in accordance with Section 1703. Products and materials required to be labeled shall be labeled in accordance with the procedures set forth in Sections 1703.5.1 through 1703.5.3.

❖ This section provides requirements for third-party inspection of a manufacturer of a material or assembly when the code says that the material or assembly must be labeled. The materials or assemblies required to be labeled are given in other chapters of the code and the *International Mechanical Code®* (IMC®), *International Fire Code®* (IFC®) and *International Plumbing Code®* (IPC®). Labeling provides a readily available source of information that is useful for field inspection of installed products. The label identifies the product or material and provides other information that can be investigated further if there is any question as to its suitability for the specific installation.

Some examples are gas appliances, fire doors, prefabricated construction (when the building official does not inspect it), electrical appliances, glass, factory-built fireplaces, plywood and other wood members when used structurally, lumber and foam plastics.

1703.5.1 Testing. An approved agency shall test a representative sample of the product or material being labeled to the relevant standard or standards. The approved agency shall maintain a record of the tests performed. The record shall provide sufficient detail to verify compliance with the test standard.

❖ As a basis for the allowed use of an agency's label, the agency is required to perform testing on the material or product in accordance with the standard referenced by the code. For example, Section 903.1 of the IMC requires that factory-built fireplaces be tested in accordance with the referenced standard UL 127 and states that factory-built fireplaces are required to be listed and labeled by an approved agency.

1703.5.2 Inspection and identification. The approved agency shall periodically perform an inspection, which shall be in-plant if necessary, of the product or material that is to be

labeled. The inspection shall verify that the labeled product or material is representative of the product or material tested.

❖ The approved agency whose label is to be applied to a product must perform periodic inspections. The primary objective of these inspections is to determine that the manufacturer is, indeed, making the same product that was tested. For example, using the factory-built fireplace discussed in the commentary to Section 1703.5.1, if the fire chamber wall in the test was $^3/_8$-inch-thick (9.5 mm) steel, the inspection agency must check to see that this thickness is being used. If the manufacturer has decided to use $^1/_4$-inch (6.4 mm) steel, then the inspection agency would be forced to withdraw the use of its label and listing.

1703.5.3 Label information. The label shall contain the manufacturer's or distributor's identification, model number, serial number or definitive information describing the product or material's performance characteristics and approved agency's identification.

❖ This section lists the information that is required on a label (see Figure 1703.5.3). The purpose of this is so that there will be sufficient information for the inspector to determine that the installed product is the same as that which was approved during plan review.

ACME
MFG., INC.

ACME DOOR SERIES A
1-HOUR FIRE DOOR
TESTED IN ACCORDANCE WITH ASTM E 152

SERIAL NO. J99999

FIRELAB INSPECTORS, INC.

**Figure 1703.5.3
TYPICAL LABEL INFORMATION**

1703.6 Heretofore approved materials. The use of any material already fabricated or of any construction already erected, which conformed to requirements or approvals heretofore in effect, shall be permitted to continue, if not detrimental to life, health or safety to the public.

❖ If a material or system had been approved before the current code took effect, it can continue to be used as long as it can be shown that the material or system is not detrimental to the health or safety of the building occupants or the public. In other words, the code is not retroactive.

1703.7 Evaluation and follow-up inspection services. Where structural components or other items regulated by this code are not visible for inspection after completion of a prefabricated assembly, the permit applicant shall submit a report of each prefabricated assembly. The report shall indicate the complete details of the assembly, including a description of the assembly and its components, the basis upon which the assembly is being evaluated, test results and similar information and other data as necessary for the building official to determine conformance to this code. Such a report shall be approved by the building official.

❖ As an alternative to physical inspection by the building official in the plant or location where prefabricated components are manufactured, such as modular homes, trusses, etc., the building official has the option of accepting an evaluation report from an approved agency detailing such inspections.

1703.7.1 Follow-up inspection. The permit applicant shall provide for special inspections of fabricated items in accordance with Section 1704.2.

❖ The owner is required to provide special inspections of fabricated assemblies at the fabrication plant in accordance with Section 1704.2.

1703.7.2 Test and inspection records. Copies of necessary test and inspection records shall be filed with the building official.

❖ All testing and inspection records related to a fabricated assembly must be filed with the building official so as to maintain a complete and legal record of the assembly and erection of the building.

SECTION 1704
SPECIAL INSPECTIONS

1704.1 General. Where application is made for construction as described in this section, the owner or the registered design professional in responsible charge acting as the owner's agent shall employ one or more special inspectors to provide inspections during construction on the types of work listed under Section 1704. The special inspector shall be a qualified person who shall demonstrate competence, to the satisfaction of the building official, for inspection of the particular type of construction or operation requiring special inspection. These inspections are in addition to the inspections specified in Section 109.

Exceptions:

1. Special inspections are not required for work of a minor nature or as warranted by conditions in the jurisdiction as approved by the building official.

2. Special inspections are not required for building components unless the design involves the practice of professional engineering or architecture as defined by applicable state statutes and regulations governing the professional registration and certification of engineers or architects.

3. Unless otherwise required by the building official, special inspections are not required for occupancies

in Group R-3 as applicable in Section 101.2 and occupancies in Group U that are accessory to a residential occupancy including, but not limited to, those listed in Section 312.1.

❖ Special inspections provide a means of quality assurance. Structural properties of the concrete or steel that is used in most structures are not usually discernable by a mere visual inspection. Typically, construction materials must be tested and their installation must be monitored in order to provide a finished structure that performs in accordance with the construction documents. Trained specialists that provide these inspections give the building official and the engineer an indication that the required structural performance will be achieved. The permit applicant is responsible for hiring the special inspector and must incur all associated costs. According to Section 105.1, the permit applicant may be the owner or authorized agent in connection with the project (see Section 105.1 for further details). Special inspections do not replace inspections required of the jurisdiction. Rather, they are intended as an enhancement to those inspections. The code requires the special inspector to demonstrate competence to the satisfaction of the building official, but does not attempt to specify the inspector's qualifications. There are many certification and training programs for various facets of special inspections that may provide guidance to the building official in making this judgment.

Exceptions to the requirement for special inspections are minor work and work not required to be designed or sealed by a registered design professional, as regulated by the jurisdiction in which the project is located. Group R-3 occupancies or Group U occupancies that are accessory to a Group R-3 occupancy are typically not required to be designed by a registered design professional and Exception 2 exempts these. Since this is not true in all cases, Group R-3 and accessory Group U occupancies are specifically exempted by Exception 3.

Exception 1 refers to "conditions in the jurisdiction" as a possible exception. The primary "condition" envisioned is one in which the jurisdiction has the resources and skills to perform these inspection tasks. This exception should not be interpreted as one that can be invoked by the permit applicant. A local jurisdiction should not be obligated to invoke this exception. The purpose of this exception is merely to allow jurisdictions to perform these inspections where they desire to do so rather than requiring a special inspector.

Exception 2 eliminates the special inspection requirement for projects where a design professional is not required. The type of projects that do not require a design professional varies from state to state, but typically these are limited to small structures or those that can be prescriptively designed.

1704.1.1 Statement of special inspections. The permit applicant shall submit a statement of special inspections prepared by the registered design professional in responsible charge in accordance with Section 106.1 as a condition for permit issuance. This statement shall be in accordance with Section 1705.

Exceptions:

1. A statement of special inspections is not required for structures designed and constructed in accordance with the conventional construction provisions of Section 2308.

2. The statement of special inspections is permitted to be prepared by a qualified person approved by the building official for construction not designed by a registered design professional.

❖ The applicant must submit for approval a statement of special inspections in addition to other construction documents before issuance of the building permit. Section 1705.2 identifies the materials, components and work to be covered. The statement of special inspections is required to be prepared by the registered design professional responsible for the building or structure. This is because the special inspections statement relates directly to the construction documents, which are the responsibility of the registered design professional.

Exception 1 addresses where the statement of special inspections document is not needed, but does not exclude the special inspections themselves (see Section 1704.1). Exception 2 states conditions under which the statement of special inspections may be prepared by someone other than a registered design professional. This may be somewhat redundant, since Exception 2 in Section 1704.1 is likely to exclude this construction from special inspections.

1704.1.2 Report requirement. Special inspectors shall keep records of inspections. The special inspector shall furnish inspection reports to the building official, and to the registered design professional in responsible charge. Reports shall indicate that work inspected was done in conformance to approved construction documents. Discrepancies shall be brought to the immediate attention of the contractor for correction. If the discrepancies are not corrected, the discrepancies shall be brought to the attention of the building official and to the registered design professional in responsible charge prior to the completion of that phase of the work. A final report documenting required special inspections and correction of any discrepancies noted in the inspections shall be submitted at a point in time agreed upon by the permit applicant and the building official prior to the start of work.

❖ Records of each inspection must be submitted to the building official so as to compile a complete legal record of the project. These records must include all inspections made, violations and discrepancies. Before a certificate of occupancy is issued, a final report must be submitted indicating that all special inspections have been made and all discrepancies have been resolved or removed in order to show compliance with the applicable code requirements. It is the responsibil-

ity of the special inspector to document and submit inspection records to the building official and to the registered design professional in responsible charge of the project.

1704.2 Inspection of fabricators. Where fabrication of structural load-bearing members and assemblies is being performed on the premises of a fabricator's shop, special inspection of the fabricated items shall be required by this section and as required elsewhere in this code.

❖ Inspection of in-plant fabrications and the requirements for special in-plant inspections are addressed herein. This section should be used in conjunction with Section 1703.7 relating to evaluation and follow-up inspections.

1704.2.1 Fabrication and implementation procedures. The special inspector shall verify that the fabricator maintains detailed fabrication and quality control procedures that provide a basis for inspection control of the workmanship and the fabricator's ability to conform to approved construction documents and referenced standards. The special inspector shall review the procedures for completeness and adequacy relative to the code requirements for the fabricator's scope of work.

> **Exception:** Special inspections as required by Section 1704.2 shall not be required where the fabricator is approved in accordance with Section 1704.2.2.

❖ The special inspector is required to verify not only that the fabricator complies with the design details and in-house quality control procedures at the plant, but also its ability to construct/fabricate to the approved drawings, standards and specifications. An example of this would be an inspection of proper placement and rolling of truss-plate connectors at a wood-truss manufacturing plant. Improper procedures could result in the connectors "popping out" or "peeling back" after the truss is concealed and loaded, thus causing structural failure. Special inspections are not required if an approved independent agency conducts in-house inspections.

1704.2.2 Fabricator approval. Special inspections required by this code are not required where the work is done on the premises of a fabricator registered and approved to perform such work without special inspection. Approval shall be based upon review of the fabricator's written procedural and quality control manuals and periodic auditing of fabrication practices by an approved special inspection agency. At completion of fabrication, the approved fabricator shall submit a certificate of compliance to the building official stating that the work was performed in accordance with the approved construction documents.

❖ "Approved fabricator" is defined in Section 1702 and this section provides the basis for such approval by the building official. If the fabricator is approved by the building official, then its internal quality control procedures are deemed to be sufficient without the need for a special inspection. A certificate of compliance provides the building official with evidence on each project that the work has been performed in accordance with the code and the construction documents.

1704.3 Steel construction. The special inspections for steel elements of buildings and structures shall be as required by Section 1704.3 and Table 1704.3.

Exceptions:

1. Special inspection of the steel fabrication process shall not be required where the fabricator does not perform any welding, thermal cutting or heating operation of any kind as part of the fabrication process. In such cases, the fabricator shall be required to submit a detailed procedure for material control that demonstrates the fabricator's ability to maintain suitable records and procedures such that, at any time during the fabrication process, the material specification, grade and mill test reports for the main stress-carrying elements are capable of being determined.

2. The special inspector need not be continuously present during welding of the following items, provided the materials, welding procedures and qualifications of welders are verified prior to the start of the work; periodic inspections are made of the work in progress; and a visual inspection of all welds is made prior to completion or prior to shipment of shop welding.

   2.1. Single-pass fillet welds not exceeding $^{5}/_{16}$ inch (7.9 mm) in size.

   2.2. Floor and roof deck welding.

   2.3. Welded studs when used for structural diaphragm.

   2.4. Welded sheet steel for cold-formed steel framing members such as studs and joists.

   2.5. Welding of stairs and railing systems.

❖ The requirements to be followed by the special inspector for the erection and fabrication of structural steel elements of building construction are listed in Table 1704.3. This table includes minimum special inspection requirements for welding of steel construction based on the item and the type of weld. Under certain conditions listed in Exception 2, some of these welding inspections may be periodic; otherwise, they should be continuous (see definitions in Section 1702).

An exemption is allowed if the fabrication plant does not utilize any facilities or methods that may alter the physical characteristics or properties of the steel members or components, such as welding, thermal cutting or heating operations. The fabricator would, in any case, need to provide evidence that procedures are used that verify that the proper material specification and grade for the main stress-carrying elements are supplied in accordance with the job specifications and shop drawings.

TABLE 1704.3. See page 17-9.

❖ Table 1704.3 provides a complete list of the types of materials and inspections, the verification required and the referenced standards to be used in evaluating

conformance to and code compliance with main stress-carrying elements of steel construction.

1704.3.1 Welding. Welding inspection shall be in compliance with AWS D1.1. The basis for welding inspector qualification shall be AWS D1.1.

❖ The referenced standard for this section is AWS D1.1 from the American Welding Society (AWS). This standard covers 10 areas of welding: general provisions; design of welded connections; workmanship; technique; qualification; inspection; strengthening and repair of existing structures; design of new buildings and the design of new bridges and new tubular structures. The tables in AWS D1.1 are intended only to provide prequalified joint geometry, such as root opening, angles and clearances, which will permit a qualified welder to deposit sound weld metal.

Weld joints other than those specified by AWS may be qualified, provided that they are tested and qualified in accordance with AWS D1.1.

All prequalified or qualified welds should have a written welding procedure specification prepared by the fabricator following the detail and outline suggested by AWS D1.1. This specification should be available to the engineer, inspector and building official.

AWS D1.1 describes two general categories of inspection: the contractor's quality control (referred to as the "contractor's inspection") and the owner/engineer's quality assurance (referred to as the "verification inspection"). Under the code, the special inspector would typically provide the required verification inspections.

The basis for welding inspector qualification is AWS D1.1. Section 6.1.4 of AWS D 1.1 lists inspector qualification requirements that include AWS QC1 guidelines as well as two other options. Where performing nondestructive testing (NDT), there are additional qualifications given in Section 6.14.6. Section 6.9 of AWS D1.1 requires visual inspection of all welds. AWS D1.1 gives requirements for conducting NDT as well as acceptance criteria; however, the AWS D1.1 document does not generally require NDT (one exception is found in Section 6.11.1).

1704.3.2 Details. The special inspector shall perform an inspection of the steel frame to verify compliance with the details shown on the approved construction documents, such as bracing, stiffening, member locations and proper application of joint details at each connection.

❖ The special inspector is required to perform an inspection of the entire steel frame to verify compliance with the applicable code requirements and the approved engineering drawings.

1704.3.3 High-strength bolts. Installation of high-strength bolts shall be periodically inspected in accordance with AISC specifications.

❖ Installation and periodic inspection of high-strength (ASTM A 325 or ASTM A 490 or equivalent) bolts must be in accordance with AISC 360.

TABLE 1704.3
REQUIRED VERIFICATION AND INSPECTION OF STEEL CONSTRUCTION

VERIFICATION AND INSPECTION	CONTINUOUS	PERIODIC	REFERENCED STANDARD[a]	IBC REFERENCE
1. Material verification of high-strength bolts, nuts and washers:				
a. Identification markings to conform to ASTM standards specified in the approved construction documents.	—	X	Applicable ASTM material specifications; AISC 360, Section A3.3	—
b. Manufacturer's certificate of compliance required.	—	X	—	—
2. Inspection of high-strength bolting:				
a. Bearing-type connections.	—	X	AISC 360, Section M2.5	1704.3.3
b. Slip-critical connections.	X	X		
3. Material verification of structural steel:				
a. Identification markings to conform to ASTM standards specified in the approved construction documents.	—	—	ASTM A 6 or ASTM A 568	1708.4
b. Manufacturers' certified mill test reports.	—	—	ASTM A 6 or ASTM A 568	
4. Material verification of weld filler materials:				
a. Identification markings to conform to AWS specification in the approved construction documents.	—	—	AISC 360, Section A3.5	—
b. Manufacturer's certificate of compliance required.	—	—	—	—
5. Inspection of welding:	—	—		
a. Structural steel:				
1) Complete and partial penetration groove welds.	X	—	AWS D1.1	1704.3.1
2) Multipass fillet welds.	X	—		
3) Single-pass fillet welds > $5/16''$	X	—		
4) Single-pass fillet welds ≤ $5/16''$	—	X		
5) Floor and roof deck welds.	—	X	AWS D1.3	—
b. Reinforcing steel:	—	—		
1) Verification of weldability of reinforcing steel other than ASTM A 706.	—	X	AWS D1.4 ACI 318: 3.5.2	—
2) Reinforcing steel-resisting flexural and axial forces in intermediate and special moment frames, and boundary elements of special reinforced concrete shear walls and shear reinforcement.	X	—		
3) Shear reinforcement.	X	—		
4) Other reinforcing steel.	—	X		
6. Inspection of steel frame joint details for compliance with approved construction documents:		X		1704.3.2
a. Details such as bracing and stiffening.	—	—	—	
b. Member locations.	—	—		
c. Application of joint details at each connection.	—			

For SI: 1 inch = 25.4 mm.
a. Where applicable, see also Section 1707.1, Special inspection for seismic resistance.

1704.3.3.1 General. While the work is in progress, the special inspector shall determine that the requirements for bolts, nuts, washers and paint; bolted parts and installation and tightening in such standards are met. For bolts requiring pretensioning, the special inspector shall observe the preinstallation testing and calibration procedures when such procedures are required by the installation method or by project plans or specifications; determine that all plies of connected materials have been drawn together and properly snugged and monitor the installation of bolts to verify that the selected procedure for installation is properly used to tighten bolts. For joints required to be tightened only to the snug-tight condition, the special inspector need only verify that the connected materials have been drawn together and properly snugged.

❖ Inspection procedures that provide the greatest ensurance that bolts are properly installed and tensioned are provided first by observing the calibration testing of the fasteners using the selected installation procedure, and then by monitoring the work in progress to ensure that the procedure that provided the specified tension is routinely adhered to. When such a program is followed, no further evidence of proper bolt tension is required.

1704.3.3.2 Periodic monitoring. Monitoring of bolt installation for pretensioning is permitted to be performed on a periodic basis when using the turn-of-nut method with matchmarking techniques, the direct tension indicator method or the alternate design fastener (twist-off bolt) method. Joints designated as snug tight need be inspected only on a periodic basis.

❖ Periodic monitoring by a special inspector is permitted for high-strength bolts installed and tightened by the turn-of-nut method with matchmarking techniques, the direct tension indicator method, the alternate design fastener method or joints designated as "snug tight."

1704.3.3.3 Continuous monitoring. Monitoring of bolt installation for pretensioning using the calibrated wrench method or the turn-of-nut method without matchmarking shall be performed on a continuous basis.

❖ Continuous monitoring by a special inspector is required for high-strength bolts installed and tightened by the calibrated wrench method or the turn-of-nut method without matchmarking.

1704.4 Concrete construction. The special inspections and verifications for concrete construction shall be as required by this section and Table 1704.4.

Exception: Special inspections shall not be required for:

1. Isolated spread concrete footings of buildings three stories or less in height that are fully supported on earth or rock.

2. Continuous concrete footings supporting walls of buildings three stories or less in height that are fully supported on earth or rock where:

    2.1. The footings support walls of light-frame construction;

    2.2. The footings are designed in accordance with Table 1805.4.2; or

    2.3. The structural design of the footing is based on a specified compressive strength, f'_c, no greater than 2,500 pounds per square inch (psi) (17.2 MPa), regardless of the compressive strength specified in the construction documents or used in the footing construction.

3. Nonstructural concrete slabs supported directly on the ground, including prestressed slabs on grade, where the effective prestress in the concrete is less than 150 psi (1.03 MPa).

4. Concrete foundation walls constructed in accordance with Table 1805.5(5).

5. Concrete patios, driveways and sidewalks, on grade.

❖ This section establishes criteria for special inspections of elements of buildings and structures of concrete construction. Exceptions to the requirements of this section address concrete components that have little or no load-carrying requirements, such as nonstructural slabs on grade, driveways, patios, etc., or footings and foundations that require no reinforcement and carry relatively low loads.

TABLE 1704.4. See page 17-11.

❖ Required verifications and inspections during concrete construction operations are listed in Table 1704.4. This table provides a complete list of the types of inspections required and the referenced standards for the placing, curing, prestressing and erection of concrete construction.

1704.4.1 Materials. In the absence of sufficient data or documentation providing evidence of conformance to quality standards for materials in Chapter 3 of ACI 318, the building official shall require testing of materials in accordance with the appropriate standards and criteria for the material in Chapter 3 of ACI 318. Weldability of reinforcement, except that which conforms to ASTM A 706, shall be determined in accordance with the requirements of Section 3.5.2 of ACI 318.

❖ Concrete materials, such as cement, aggregates, admixtures and water, must comply with the standards of Chapter 3 of ACI 318, which regulates materials and addresses specific standards. In the absence of sufficient data or documentation, the building official must require testing in accordance with the standards listed in Chapter 3 of ACI 318.

ASTM A 706 is the standard for reinforcing steel that is weldable, meaning that the chemical composition and manufacturing processes are such that the material is well suited for an acceptable quality of weld. Section 3.5.2 of ACI 318 states that any standard other than ASTM A 706 used for reinforcement material would need to be supplemented for weldability requirements. The intent of this provision is that, where welding of reinforcing steel is required, the steel specified and delivered must be checked for weldability.

1704.5 Masonry construction. Masonry construction shall be inspected and evaluated in accordance with the requirements of Sections 1704.5.1 through 1704.5.3, depending on the classification of the building or structure or nature of the occupancy, as defined by this code.

Exception: Special inspections shall not be required for:

1. Empirically designed masonry, glass unit masonry or masonry veneer designed by Section 2109, 2110 or Chapter 14, respectively, or by Chapter 5, 7 or 6 of ACI 530/ASCE 5/TMS 402, respectively, when they

TABLE 1704.4
REQUIRED VERIFICATION AND INSPECTION OF CONCRETE CONSTRUCTION

VERIFICATION AND INSPECTION	CONTINUOUS	PERIODIC	REFERENCED STANDARD[a]	IBC REFERENCE
1. Inspection of reinforcing steel, including prestressing tendons, and placement.	—	X	ACI 318: 3.5, 7.1-7.7	1913.4
2. Inspection of reinforcing steel welding in accordance with Table 1704.3, Item 5b.	—	—	AWS D1.4 ACI 318: 3.5.2	—
3. Inspect bolts to be installed in concrete prior to and during placement of concrete where allowable loads have been increased.	X	—	—	1911.5
4. Verifying use of required design mix.	—	X	ACI 318: Ch. 4, 5.2-5.4	1904.2.2, 1913.2, 1913.3
5. At the time fresh concrete is sampled to fabricate specimens for strength tests, perform slump and air content tests, and determine the temperature of the concrete.	X	—	ASTM C 172 ASTM C 31 ACI 318: 5.6, 5.8	1913.10
6. Inspection of concrete and shotcrete placement for proper application techniques.	X	—	ACI 318: 5.9, 5.10	1913.6, 1913.7, 1913.8
7. Inspection for maintenance of specified curing temperature and techniques.	—	X	ACI 318: 5.11-5.13	1913.9

8. Inspection of prestressed concrete:
  a. Application of prestressing forces.
  b. Grouting of bonded prestressing tendons in the seismic-force-resisting system.
	X
	X
		—	ACI 318: 18.20
			ACI 318: 18.18.4
				—

9. Erection of precast concrete members.	—	X	ACI 318: Ch. 16	—
10. Verification of in-situ concrete strength, prior to stressing of tendons in posttensioned concrete and prior to removal of shores and forms from beams and structural slabs.	—	X	ACI 318: 6.2	—
11. Inspect formwork for shape, location and dimensions of the concrete member being formed.	—	X	ACI 318: 6.1.1	—

For SI: 1 inch = 25.4 mm.

a. Where applicable, see also Section 1707.1, Special inspection for seismic resistance.

are part of structures classified as Occupancy Category I, II or III in accordance with Section 1604.5.

2. Masonry foundation walls constructed in accordance with Table 1805.5(1), 1805.5(2), 1805.5(3) or 1805.5(4).

3. Masonry fireplaces, masonry heaters or masonry chimneys installed or constructed in accordance with Section 2111, 2112 or 2113, respectively.

❖ This section establishes whether special inspection is required, and the necessary level, for masonry construction. This is determined by the design method (empirical versus engineered) and whether or not a structure is classified as Occupancy Category IV in accordance with Section 1604.5. These requirements are summarized in Table 1704.5. Additionally, masonry foundation walls complying with the prescriptive tables in Section 1805.5 are exempt from special inspection.

TABLE 1704.5
SPECIAL INSPECTION OF MASONRY CONSTRUCTION

	TYPE OF FACILITY			
	Nonessential		Essential	
Design method	Empirical	Engineered	Empirical	Engineered
Special inspection	Exempt	Level 1	Level 1	Level 2

1704.5.1 Empirically designed masonry, glass unit masonry and masonry veneer in Occupancy Category IV. The minimum special inspection program for empirically designed masonry, glass unit masonry or masonry veneer designed by Section 2109, 2110 or Chapter 14, respectively, or by Chapter 5, 7 or 6 of ACI 530/ASCE 5/TMS 402, respectively, in structures classified as Occupancy Category IV, in accordance with Section 1604.5, shall comply with Table 1704.5.1.

❖ This section defines the minimum level of special inspection required for empirically designed masonry, glass unit masonry and masonry veneer in Occupancy Category IV buildings (see Table 1704.5.1).

TABLE 1704.5.1. See page 17-13.

❖ The minimum required special inspections and verifications during masonry construction are listed in Table 1704.5.1. This table also lists the required criteria from the code, ACI 530 and ACI 530.1. This table applies to empirically designed masonry in Occupancy Category IV facilities and to engineered masonry in all other occupancy categories.

1704.5.2 Engineered masonry in Occupancy Category I, II or III. The minimum special inspection program for masonry designed by Section 2107 or 2108 or by chapters other than Chapters 5, 6 or 7 of ACI 530/ASCE 5/TMS 402 in structures classified as Occupancy Category I, II or III, in accordance with Section 1604.5, shall comply with Table 1704.5.1.

❖ Engineered masonry in Occupancy Category I, II and III facilities requires special inspection. The minimum special inspections required are listed in Table 1704.5.1.

1704.5.3 Engineered masonry in Occupancy Category IV. The minimum special inspection program for masonry designed by Section 2107 or 2108 or by chapters other than Chapters 5, 6 or 7 of ACI 530/ASCE 5/TMS 402 in structures classified as Occupancy Category IV, in accordance with Section 1604.5, shall comply with Table 1704.5.3.

❖ Engineered masonry in Occupancy Category IV facilities requires special inspection. The minimum special inspections required are listed in Table 1704.5.3.

TABLE 1704.5.3. See page 17-14.

❖ The minimum required special inspections and verifications during masonry construction of engineered design masonry in essential facilities are listed in Table 1704.5.3. This table also lists the required criteria from the code, ACI 530 and ACI 530.1.

1704.6 Wood construction. Special inspections of the fabrication process of prefabricated wood structural elements and assemblies shall be in accordance with Section 1704.2. Special inspections of site-built assemblies shall be in accordance with this section.

❖ The fabrication process of wood structural elements and assemblies (such as wood trusses) that is being performed on the premises of a fabricator's shop must receive special inspection in accordance with Section 1704.2.

1704.6.1 High-load diaphragms. High-load diaphragms designed in accordance with Table 2306.3.2 shall be installed with special inspections as indicated in Section 1704.1. The special inspector shall inspect the wood structural panel sheathing to ascertain whether it is of the grade and thickness shown on the approved building plans. Additionally, the special inspector must verify the nominal size of framing members at adjoining panel edges, the nail or staple diameter and length, the number of fastener lines and that the spacing between fasteners in each line and at edge margins agrees with the approved building plans.

❖ This section requires special inspection of specific portions of diaphragms that are designed in accordance with Table 2306.3.2. By their very nature, these "high-load" diaphragms are likely to carry more significant wind or seismic loads, making it necessary to perform special inspection.

TABLE 1704.5.1
LEVEL 1 SPECIAL INSPECTION

INSPECTION TASK	FREQUENCY OF INSPECTION		REFERENCE FOR CRITERIA		
	Continuous during task listed	Periodically during task listed	IBC section	ACI 530/ASCE 5/TMS 402[a]	ACI 530.1/ASCE 6/TMS 602[a]
1. As masonry construction begins, the following shall be verified to ensure compliance:					
a. Proportions of site-prepared mortar.	—	X	—	—	Art. 2.6A
b. Construction of mortar joints.	—	X	—	—	Art. 3.3B
c. Location of reinforcement, connectors, prestressing tendons and anchorages.	—	X	—	—	Art. 3.4, 3.6A
d. Prestressing technique.	—	X	—	—	Art. 3.6B
e. Grade and size of prestressing tendons and anchorages.	—	X	—	—	Art. 2.4B, 2.4H
2. The inspection program shall verify:					
a. Size and location of structural elements.	—	X	—	—	Art. 3.3G
b. Type, size and location of anchors, including other details of anchorage of masonry to structural members, frames or other construction.	—	X	—	Sec. 1.2.2(e), 2.1.4, 3.1.6	—
c. Specified size, grade and type of reinforcement.	—	X	—	Sec. 1.13	Art. 2.4, 3.4
d. Welding of reinforcing bars.	X	—	—	Sec. 2.1.10.7.2, 3.3.3.4(b)	—
e. Protection of masonry during cold weather (temperature below 40°F) or hot weather (temperature above 90°F).	—	X	Sec. 2104.3, 2104.4	—	Art. 1.8C, 1.8D
f. Application and measurement of prestressing force.	—	X	—	—	Art. 3.6B
3. Prior to grouting, the following shall be verified to ensure compliance:					
a. Grout space is clean.	—	X	—	—	Art. 3.2D
b. Placement of reinforcement and connectors and prestressing tendons and anchorages.	—	X	—	Sec. 1.13	Art. 3.4
c. Proportions of site-prepared grout and prestressing grout for bonded tendons.	—	X	—	—	Art. 2.6B
d. Construction of mortar joints.	—	X	—	—	Art. 3.3B
4. Grout placement shall be verified to ensure compliance with code and construction document provisions.	X	—	—	—	Art 3.5
a. Grouting of prestressing bonded tendons.	X	—	—	—	Art. 3.6C
5. Preparation of any required grout specimens, mortar specimens and/or prisms shall be observed.	X	—	Sec. 2105.2.2, 2105.3	—	Art. 1.4
6. Compliance with required inspection provisions of the construction documents and the approved submittals shall be verified.	—	X	—	—	Art. 1.5

For SI: °C = (°F - 32)/1.8.
a. The specific standards referenced are those listed in Chapter 35.

1704.7 Soils. Special inspections for existing site soil conditions, fill placement and load-bearing requirements shall be as required by this section and Table 1704.7. The approved soils report, required by Section 1802.2, and the documents prepared by the registered design professional in responsible charge shall be used to determine compliance. During fill placement, the special inspector shall determine that proper materials and procedures are used in accordance with the provisions of the approved soils report, as specified in Section 1803.5.

Exception: Special inspection is not required during placement of controlled fill having a total depth of 12 inches (305 mm) or less.

❖ The load-bearing capacity of the supporting soil has a significant impact on the structural integrity of any

TABLE 1704.5.3
LEVEL 2 SPECIAL INSPECTION

INSPECTION TASK	FREQUENCY OF INSPECTION		REFERENCE FOR CRITERIA		
	Continuous during task listed	Periodically during task listed	IBC section	ACI 530/ ASCE 5/ TMS 402[a]	ACI 530.1/ ASCE 6/ TMS 602[a]
1. From the beginning of masonry construction, the following shall be verified to ensure compliance:					
a. Proportions of site-prepared mortar, grout and prestressing grout for bonded tendons.	—	X	—	—	Art. 2.6A
b. Placement of masonry units and construction of mortar joints.	—	X	—	—	Art. 3.3B
c. Placement of reinforcement, connectors and prestressing tendons and anchorages.	—	X	—	Sec. 1.13	Art. 3.4, 3.6A
d. Grout space prior to grouting.	X	—	—	—	Art. 3.2D
e. Placement of grout.	X	—	—	—	Art. 3.5
f. Placement of prestressing grout.	X	—	—	—	Art. 3.6C
2. The inspection program shall verify:					
a. Size and location of structural elements.	—	X	—	—	Art. 3.3G
b. Type, size and location of anchors, including other details of anchorage of masonry to structural members, frames or other construction.	X	—	—	Sec. 1.2.2(e), 2.1.4, 3.1.6	—
c. Specified size, grade and type of reinforcement.	—	X	—	Sec. 1.13	Art. 2.4, 3.4
d. Welding of reinforcing bars.	X	—	—	Sec. 2.1.10.7.2, 3.3.3.4(b)	—
e. Protection of masonry during cold weather (temperature below 40°F) or hot weather (temperature above 90°F).	—	X	Sec. 2104.3, 2104.4	—	Art. 1.8C, 1.8D
f. Application and measurement of prestressing force.	X	—	—	—	Art. 3.6B
3. Preparation of any required grout specimens, mortar specimens and/or prisms shall be observed.	X	—	Sec. 2105.2.2, 2105.3	—	Art. 1.4
4. Compliance with required inspection provisions of the construction documents and the approved submittals shall be verified.	—	X	—	—	Art. 1.5

For SI: °C = (°F - 32)/1.8.

a. The specific standards referenced are those listed in Chapter 35.

building. The increasing use of prepared fill as a load-bearing soil strata has established the need to make specific mention of this construction in the special inspection requirements. The amount of compaction and the methods vary depending on the particular design. Use of proper compaction, lift and density, however, is critical to achieving the desired bearing capacity.

TABLE 1704.7. See below.

❖ Tabular formatting of soil inspections clearly conveys the intended requirements for testing and inspection of soils for controlled fill and determination of soil-bearing capacity. The verifications prior to fill placement include verifying that the site preparation meets specified requirements, including proper excavation depth, removal of all deleterious material and any other special requirements that the soils engineer deems necessary for the design (see Items 1, 2 and 5). Verifications of the fill material and the placement operation are treated in Items 3 and 4. Without observing and documenting that the proper material is used and the specified compaction techniques and lifts are employed, the specified load-bearing capacity may not be achieved. A major factor in the design of the fill is the in-place density. This evaluation is needed so that the compaction methods result in adequate soil-bearing capacity.

1704.8 Pile foundations. Special inspections shall be performed during installation and testing of pile foundations as required by Table 1704.8. The approved soils report, required by Section 1802.2, and the documents prepared by the registered design professional in responsible charge shall be used to determine compliance.

❖ Table 1704.8 lists the special inspections that are required for pile foundations. A soils investigation is required by Section 1802.2.4. Section 1808.2.2 establishes specific criteria that the soils report should include, which would also be the basis for field verifications by a special inspector.

TABLE 1704.8. See page 17-16.

❖ This table provides a concise list of requirements for inspection and verification of pile materials, testing and installation.

1704.9 Pier foundations. Special inspections shall be performed during installation and testing of pier foundations as required by Table 1704.9. The approved soils report, required by Section 1802.2, and the documents prepared by the registered design professional in responsible charge shall be used to determine compliance.

❖ Table 1704.9 lists the special inspections that are required for pier foundations. A soils investigation is required by Section 1802.2.4. Section 1808.2.2 establishes specific criteria that the soils report should include, which would also be the basis for field verifications by a special inspector.

TABLE 1704.9. See page 17-16.

❖ This table provides a concise list of requirements for inspection and verification of pier materials, testing and installation.

1704.10 Sprayed fire-resistant materials. Special inspections for sprayed fire-resistant materials applied to structural elements and decks shall be in accordance with Sections 1704.10.1 through 1704.10.5. Special inspections shall be based on the fire-resistance design as designated in the approved construction documents.

❖ To ensure that a sprayed fire-resistant material, or SFRM as it is commonly referred to, performs as intended by the manufacturer, certain conditions are required to be met during its application. The conditions include: temperature at time of application; substrate conditions; protection provided and thickness and density of material. These conditions must be checked and verified at the time of, or prior to, the application of the SFRM. This section stipulates that special inspections are required in accordance with Sections 1704.10.1 through 1704.10.5.

TABLE 1704.7
REQUIRED VERIFICATION AND INSPECTION OF SOILS

VERIFICATION AND INSPECTION TASK	CONTINUOUS DURING TASK LISTED	PERIODICALLY DURING TASK LISTED
1. Verify materials below footings are adequate to achieve the design bearing capacity.	—	X
2. Verify excavations are extended to proper depth and have reached proper material.	—	X
3. Perform classification and testing of controlled fill materials.	—	X
4. Verify use of proper materials, densities and lift thicknesses during placement and compaction of controlled fill.	X	—
5. Prior to placement of controlled fill, observe subgrade and verify that site has been prepared properly.	—	X

1704.10.1 Structural member surface conditions. The surfaces shall be prepared in accordance with the approved fire-resistance design and the approved manufacturer's written instructions. The prepared surface of structural members to be sprayed shall be inspected before the application of the sprayed fire-resistant material.

❖ The integrity of an SFRM system depends, foremost, on the conditions of the surface of the steel member to which it is to be applied. The system must be fully adhered to the surface for proper performance, in accordance with design values.

1704.10.2 Application. The substrate shall have a minimum ambient temperature before and after application as specified in the approved manufacturer's written instructions. The area for application shall be ventilated during and after application as required by the approved manufacturer's written instructions.

❖ During application of SFRMs, and immediately thereafter during cure of the material, several items must be controlled, including ambient temperature during the application and temperature of the substrate and the SFRMs. Temperature control is important to determine

TABLE 1704.8
REQUIRED VERIFICATION AND INSPECTION OF PILE FOUNDATIONS

VERIFICATION AND INSPECTION TASK	CONTINUOUS DURING TASK LISTED	PERIODICALLY DURING TASK LISTED
1. Verify pile materials, sizes and lengths comply with the requirements.	X	—
2. Determine capacities of test piles and conduct additional load tests, as required.	X	—
3. Observe driving operations and maintain complete and accurate records for each pile.	X	—
4. Verify placement locations and plumbness, confirm type and size of hammer, record number of blows per foot of penetration, determine required penetrations to achieve design capacity, record tip and butt elevations and document any pile damage.	X	—
5. For steel piles, perform additional inspections in accordance with Section 1704.3.	—	—
6. For concrete piles and concrete-filled piles, perform additional inspections in accordance with Section 1704.4.	—	—
7. For specialty piles, perform additional inspections as determined by the registered design professional in responsible charge.	—	—
8. For augered uncased piles and caisson piles, perform inspections in accordance with Section 1704.9.	—	—

TABLE 1704.9
REQUIRED VERIFICATION AND INSPECTION OF PIER FOUNDATIONS

VERIFICATION AND INSPECTION TASK	CONTINUOUS DURING TASK LISTED	PERIODICALLY DURING TASK LISTED
1. Observe drilling operations and maintain complete and accurate records for each pier.	X	—
2. Verify placement locations and plumbness, confirm pier diameters, bell diameters (if applicable), lengths, embedment into bedrock (if applicable) and adequate end bearing strata capacity.	X	—
3. For concrete piers, perform additional inspections in accordance with Section 1704.4.	—	—
4. For masonry piers, perform additional inspections in accordance with Section 1704.5.	—	—

that the necessary chemical reactions needed to make a particular material bond to the steel surfaces and hold together do, in fact, happen. The minimum or maximum temperatures necessary for proper bond and cure depend on the specific type of material. The scope of the special inspection also includes the proper ventilation for curing.

1704.10.3 Thickness. The average thickness of the sprayed fire-resistant materials applied to structural elements shall not be less than the thickness required by the approved fire-resistant design. Individually measured thickness, which exceeds the thickness specified in a design by $1/4$ inch (6.4 mm) or more, shall be recorded as the thickness specified in the design plus $1/4$ inch (6.4 mm). For design thicknesses 1 inch (25 mm) or greater, the minimum allowable individual thickness shall be the design thickness minus $1/4$ inch (6.4 mm). For design thicknesses less than 1 inch (25 mm), the minimum allowable individual thickness shall be the design thickness minus 25 percent. Thickness shall be determined in accordance with ASTM E 605. Samples of the sprayed fire-resistant materials shall be selected in accordance with Sections 1704.10.3.1 and 1704.10.3.2.

❖ For the system to provide the required design fire-resistance rating, it must be applied at the appropriate thickness. The required sampling provided for in Sections 1704.10.3.1 and 1704.10.3.2 is based on ASTM E 605. This standard also provides testing methods that are commonly used by the industry.

1704.10.3.1 Floor, roof and wall assemblies. The thickness of the sprayed fire-resistant material applied to floor, roof and wall assemblies shall be determined in accordance with ASTM E 605 by taking the average of not less than four measurements for each 1,000 square feet (93 m²) of the sprayed area on each floor or part thereof.

❖ Sampling of a SFRM for membrane components (floors, roofs or walls) is based on the square footage of the components. The size of the sample and the amount of area per sample are taken from ASTM E 605.

1704.10.3.2 Structural framing members. The thickness of the sprayed fire-resistant material applied to structural members shall be determined in accordance with ASTM E 605. Thickness testing shall be performed on not less than 25 percent of the structural members on each floor.

❖ Sampling of a SFRM for structural elements is based on the number of structural elements. Again, sample size and number of elements represented by one sample are taken from ASTM E 605.

1704.10.4 Density. The density of the sprayed fire-resistant material shall not be less than the density specified in the approved fire-resistant design. Density of the sprayed

fire-resistant material shall be determined in accordance with ASTM E 605.

❖ The density of a SFRM will have an impact on the fire-resistance rating of the system; therefore, it is important that the density of the material be measured to verify that the product is as designed. The sampling requirements given are the same as those for thickness measurements in Sections 1704.10.3.1 and 1704.10.3.2. The required method of determining density is provided in ASTM E 605.

1704.10.5 Bond strength. The cohesive/adhesive bond strength of the cured sprayed fire-resistant material applied to structural elements shall not be less than 150 pounds per square foot (psf) (7.18 kN/m²). The cohesive/adhesive bond strength shall be determined in accordance with the field test specified in ASTM E 736 by testing in-place samples of the sprayed fire-resistant material selected in accordance with Sections 1704.10.5.1 and 1704.10.5.2.

❖ Bond strength of the material is an essential variable in the design equation for an appropriate fire-resistance rating of a SFRM. A minimum cohesive/adhesive bond strength of 150 pounds per square foot (psf) (7.18 kN/m²) is required in this section based on the American Institute of Architects (AIA) Master Specification and the recommendations of the General Services Administration (GSA) for durability and serviceability of the material.

1704.10.5.1 Floor, roof and wall assemblies. The test samples for determining the cohesive/adhesive bond strength of the sprayed fire-resistant materials shall be selected from each floor, roof and wall assembly at the rate of not less than one sample for every 10,000 square feet (929 m²) or part thereof of the sprayed area in each story.

❖ The sampling rate for bond in this section is less than the samples required for thickness and density, since the thickness and density samples minimize the need for bond strength sampling.

1704.10.5.2 Structural framing members. The test samples for determining the cohesive/adhesive bond strength of the sprayed fire-resistant materials shall be selected from beams, girders, joists, trusses and columns at the rate of not less than one sample for each type of structural framing member for each 10,000 square feet (929 m²) of floor area or part thereof in each story.

❖ The bond strength sampling rate in this section is the same as that indicated in Section 1704.10.5.1 for floor, roof and wall assemblies.

1704.11 Mastic and intumescent fire-resistant coatings. Special inspections for mastic and intumescent fire-resistant coatings applied to structural elements and decks shall be in accordance with AWCI 12-B. Special inspections shall be based on the fire-resistance design as designated in the approved construction documents.

❖ Special inspection of mastic and intumescent fire-resistant coatings is justified because these products are often complex systems that require special expertise from applicators and quality assurance personnel. It is essential to confirm that their installation is in accordance with the manufacturer's instructions and the terms of their listing so that they will perform as expected.

1704.12 Exterior insulation and finish systems (EIFS). Special inspections shall be required for all EIFS applications.

Exceptions:

1. Special inspections shall not be required for EIFS applications installed over a water-resistive barrier with a means of draining moisture to the exterior.

2. Special inspections shall not be required for EIFS applications installed over masonry or concrete walls.

❖ Special inspections are required for all exterior insulation and finish system (EIFS) installations except for the two exceptions in this section. Exception 1 recognizes that EIFS, which are installed over a water-resistive barrier and incorporate flashings at penetrations and terminations and a means of drainage to the exterior, afford a built-in redundancy to water penetration that makes the need for special inspections less critical.

Exception 2 recognizes that concrete and masonry substrates are relatively durable and the exposure to moisture in wall conditions does not necessarily have a detrimental effect on these materials.

1704.13 Special cases. Special inspections shall be required for proposed work that is, in the opinion of the building official, unusual in its nature, such as, but not limited to, the following examples:

1. Construction materials and systems that are alternatives to materials and systems prescribed by this code.

2. Unusual design applications of materials described in this code.

3. Materials and systems required to be installed in accordance with additional manufacturer's instructions that prescribe requirements not contained in this code or in standards referenced by this code.

❖ This section requires special inspections for proposed work that is unique and not specifically addressed in the code or in standards referenced by the code. For example, a designer chooses to utilize a new type of wood-laminated beam system in lieu of standard steel beam construction. Because the laminated beam is a new product, the inspector must rely on load tables, connection details and bearing length charts from the manufacturer, provided that the system is previously approved for installation by the building official.

[F] 1704.14 Special inspection for smoke control. Smoke control systems shall be tested by a special inspector.

❖ This section requires that all smoke control systems be tested by special inspection.

[F] 1704.14.1 Testing scope. The test scope shall be as follows:

1. During erection of ductwork and prior to concealment for the purposes of leakage testing and recording of device location.

2. Prior to occupancy and after sufficient completion for the purposes of pressure difference testing, flow measurements and detection and control verification.

❖ An approved special inspection agency of smoke control systems must have expertise and certification in fire protection engineering, mechanical engineering or air balancing.

[F] 1704.14.2 Qualifications. Special inspection agencies for smoke control shall have expertise in fire protection engineering, mechanical engineering and certification as air balancers.

❖ An approved special inspection agency of smoke control systems must have expertise and certification in fire protection engineering, mechanical engineering or air balancing.

SECTION 1705
STATEMENT OF SPECIAL INSPECTIONS

1705.1 General. Where special inspection or testing is required by Section 1704, 1707 or 1708, the registered design professional in responsible charge shall prepare a statement of special inspections in accordance with Section 1705 for submittal by the permit applicant (see Section 1704.1.1).

❖ Section 1705 outlines the requirements of the basic statement of special inspections as well as the added requirements for seismic and wind resistance, arranging them in logical order. The intent is to document the required inspections and testing in order to foster a systematic approach that helps to achieve the goal of a finished structure that meets or exceeds the minimum performance expectations.

1705.2 Content of statement of special inspections. The statement of special inspections shall identify the following:

1. The materials, systems, components and work required to have special inspection or testing by the building official or by the registered design professional responsible for each portion of the work.

2. The type and extent of each special inspection.

3. The type and extent of each test.

4. Additional requirements for special inspection or testing for seismic or wind resistance as specified in Section 1705.3, 1705.4, 1707 or 1708.

5. For each type of special inspection, identification as to whether it will be continuous special inspection or periodic special inspection.

❖ This section details the areas to be addressed in the statement of special inspection. It requires a list of materials and work subject to special inspection, the types of inspections and inspection frequency.

1705.3 Seismic resistance. The statement of special inspections shall include seismic requirements for the following cases:

1. The seismic-force-resisting systems in structures assigned to Seismic Design Category C, D, E or F in accordance with Section 1613.

2. Designated seismic systems in structures assigned to Seismic Design Category D, E or F.

3. The following additional systems and components in structures assigned to Seismic Design Category C:

   3.1. Heating, ventilating and air-conditioning (HVAC) ductwork containing hazardous materials and anchorage of such ductwork.

   3.2. Piping systems and mechanical units containing flammable, combustible or highly toxic materials.

   3.3. Anchorage of electrical equipment used for emergency or standby power systems.

4. The following additional systems and components in structures assigned to Seismic Design Category D:

   4.1. Systems required for Seismic Design Category C.

   4.2. Exterior wall panels and their anchorage.

   4.3. Suspended ceiling systems and their anchorage.

   4.4. Access floors and their anchorage.

   4.5. Steel storage racks and their anchorage, where the importance factor is equal to 1.5 in accordance with Section 15.5.3 of ASCE 7.

5. The following additional systems and components in structures assigned to Seismic Design Category E or F:

   5.1. Systems required for Seismic Design Categories C and D.

   5.2. Electrical equipment.

Exception: Seismic requirements are permitted to be excluded from the statement of special inspections for structures designed and constructed in accordance with the following:

1. The structure consists of light-frame construction; the design spectral response acceleration at short periods, S_{DS}, as determined in Section 1613.5.4, does not exceed 0.5g; and the height of the structure does not exceed 35 feet (10 668 mm) above grade plane; or

2. The structure is constructed using a reinforced masonry structural system or reinforced concrete

structural system; the design spectral response acceleration at short periods, S_{DS}, as determined in Section 1613.5.4, does not exceed 0.5g; and the height of the structure does not exceed 25 feet (7620 mm) above grade plane; or

3. Detached one- or two-family dwellings not exceeding two stories in height, provided the structure does not have any of the following plan or vertical irregularities in accordance with Section 12.3.2 of ASCE 7:

   3.1. Torsional irregularity.

   3.2. Nonparallel systems.

   3.3. Stiffness irregularity–extreme soft story and soft story.

   3.4. Discontinuity in capacity–weak story.

❖ This section provides inspection requirements for seismic-force-resisting systems, nonstructural components and designated seismic systems. Failures of these components and systems have been well documented in past earthquakes. Figures 1705.3(1) and 1705.3(2) illustrate a ductwork failure. Where the risk of earthquake damage is elevated, the seismic provisions of the code introduce additional quality assurance measures with the intention of verifying that construction is in accordance with the construction documents.

The exceptions to including these additional seismic requirements in the statement of special inspections apply to light-frame construction with a height no greater than 35 feet (10 668 mm) above grade plane that has a relatively low seismic risk (S_{DS} does not exceed 0.50g), or structures constructed of reinforced masonry not more than 25 feet (7620 mm) above grade plane that are located in areas of low seismic risk (S_{DS} does not exceed 0.50g).

Exception 3 applies to detached one- or two-family dwellings not exceeding two stories in height. The exception is also limited to those structures that do not have any of the following irregularities: torsional irregularity; extreme torsional irregularity; nonparallel systems; stiffness irregularity (soft story); stiffness irregularity (extreme soft story) or discontinuity in capacity (weak story). It is important to emphasize that this exception is for the inclusion of additional special inspections, not for the design of the structure in accordance with the requirements of the code.

1705.3.1 Seismic requirements in the statement of special inspections. When Section 1705.3 specifies that seismic requirements be included, the statement of special inspections shall identify the following:

1. The designated seismic systems and seismic-force-resisting systems that are subject to special inspections in accordance with Section 1705.3.

2. The additional special inspections and testing to be provided as required by Sections 1707 and 1708 and other

applicable sections of this code, including the applicable standards referenced by this code.

❖ The important items to be subjected to these added special inspections must be clearly identified along with the inspections and tests for seismic resistance required by Sections 1707 and 1708.

1705.4 Wind resistance. The statement of special inspections shall include wind requirements for structures constructed in the following areas:

1. In wind Exposure Category B, where the 3-second-gust basic wind speed is 120 miles per hour (mph) (52.8 m/s) or greater.

Figure 1705.3(1)
EARTHQUAKE INDUCED FAILURE OF DUCTWORK SUPPORTS

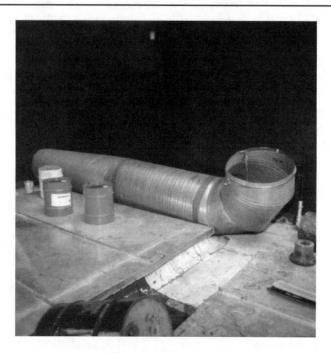

Figure 1705.3(2)
FALLEN DUCTWORK CAUSED BY SUPPORT FAILURE

2. In wind Exposure Category C or D, where the 3-second-gust basic wind speed is 110 mph (49 m/s) or greater.

❖ Where the risk of wind damage is elevated, the code requires added special inspections for key structural components.

1705.4.1 Wind requirements in the statement of special inspections. When Section 1705.4 specifies that wind requirements be included, the statement of special inspections shall identify the main windforce-resisting systems and wind-resisting components subject to special inspections as specified in Section 1705.4.2.

❖ The wind hazard thresholds established in Section 1705.4 identify the conditions that warrant additional inspections to provide for wind resistance. As a minimum, these wind-resistance inspections must cover the items that are listed in Section 1704.5.2.

1705.4.2 Detailed requirements. The statement of special inspections shall include at least the following systems and components:

1. Roof cladding and roof framing connections.

2. Wall connections to roof and floor diaphragms and framing.

3. Roof and floor diaphragm systems, including collectors, drag struts and boundary elements.

4. Vertical windforce-resisting systems, including braced frames, moment frames and shear walls.

5. Windforce-resisting system connections to the foundation.

6. Fabrication and installation of systems or components required to meet the impact-resistance requirements of Section 1609.1.2.

Exception: Fabrication of manufactured systems or components that have a label indicating compliance with the wind-load and impact-resistance requirements of this code.

❖ This section lists the structural elements that are critical to wind performance and are therefore subjected to these additional special inspections. The exception exempts manufactured components and assemblies that are labeled as complying with the required wind load and impact resistance.

SECTION 1706
CONTRACTOR RESPONSIBILITY

1706.1 Contractor responsibility. Each contractor responsible for the construction of a main wind- or seismic-force-resisting system, designated seismic system or a wind- or seismic-resisting component listed in the statement of special inspections shall submit a written statement of responsibility to the building official and the owner prior to the commencement of work on the system or component. The contractor's statement of responsibility shall contain the following:

1. Acknowledgment of awareness of the special requirements contained in the statement of special inspections;

2. Acknowledgment that control will be exercised to obtain conformance with the construction documents approved by the building official;

3. Procedures for exercising control within the contractor's organization, the method and frequency of reporting and the distribution of the reports; and

4. Identification and qualifications of the person(s) exercising such control and their position(s) in the organization.

❖ The statement of contractor responsibility is required wherever the statement of special inspections includes additional wind- or seismic-resistance inspections. This statement by the contractor is separate from the statement of special inspections. It represents the contractor's acknowledgement of these added special inspections that are beyond what is typically required. The exact methods of control are the responsibility of individual contractors, subject to approval by the building official; however, special inspections of the work are required in specific situations to provide the building official reasonable assurance that there is compliance with the approved construction documents.

The qualifications of each contractor and the subcontractors can vary considerably. This statement is an opportunity to identify those areas of special concern that must be addressed during the construction process, including, but not limited to, types of testing, frequency of testing, types of inspections, frequency of inspections and the extent of the structural observations to be performed.

SECTION 1707
SPECIAL INSPECTIONS FOR
SEISMIC RESISTANCE

1707.1 Special inspections for seismic resistance. Special inspections itemized in Sections 1707.2 through 1707.10, unless exempted by the exceptions of Section 1704.1, are required for the following:

1. The seismic-force-resisting systems in structures assigned to Seismic Design Category C, D, E or F, as determined in Section 1613.

2. Designated seismic systems in structures assigned to Seismic Design Category D, E or F.

3. Architectural, mechanical and electrical components in structures assigned to Seismic Design Category C, D, E or F that are required in Sections 1707.7 and 1707.8.

❖ The added special inspection requirements in this section are an important consideration in carrying out the intent of the seismic provisions of the code. A certain amount of inelastic behavior is inherent in a building's response to the design earthquake. Strong ground shaking caused by earthquakes tends to expose any underlying flaws in a building's construction or design. Thus, the code specifies additional special inspections of various structural and nonstructural components for seismic resistance in order to provide further verification that these portions of the finished structure will be in accordance with the construction documents. It is imperative that the professional who designs systems that are critical to the earthquake performance also identifies them and specifies the necessary inspection and testing.

Item 1 states the seismic risk threshold for seismic-force-resisting systems. The inspections of items described in Sections 1707.2 through 1707.4 would apply to these structures. Item 2 covers designated seismic systems that are nonstructural components and systems that are assigned an importance factor greater than 1. The only inspections that apply specifically to designated seismic systems are given in Section 1707.9, but other requirements, such as those in Section 1707.8 for mechanical and electrical components in general, would also be applicable where the component is given an importance factor greater than 1. Item 3 states the seismic design categories for architectural components in Section 1707.7 and for electrical and mechanical components in Section 1707.8. Note that the seismic design categories in Sections 1707.7 and 1707.8 do not always coincide with those in Item 3. Since Item 3 is a general statement, the more specific seismic design categories for the various items in Sections 1707.7 and 1707.8 should be followed.

1707.2 Structural steel. Continuous special inspection is required for structural welding in accordance with AISC 341.

Exceptions:

1. Single-pass fillet welds not exceeding $^5/_{16}$ inch (7.9 mm) in size.

2. Floor and roof deck welding.

❖ Section 1707.1 requires the special inspection of seismic-force-resisting systems of structures classified as Seismic Design Category C, D, E or F. This section specifically requires continuous special inspection of the structural welding in accordance with AISC 341 (with two exceptions). Also note that Section 2205.2.2 requires structural steel seismic-force-resisting systems (BSFRS) to be in accordance with AISC 341 for Seismic Design Category D, E or F structures.

AISC 341, Section 16 states, "The general requirements and responsibilities for performance of a quality assurance plan shall be in accordance with the requirements of the Authority Having Jurisdiction ....," which is, in effect, a reference back to the code as it is adopted and enforced locally. Item 1 requires visual in-spection, as well as NDT in conformance with AWS D1.1, but doesn't specify where. It is likely that this refers to the specific NDT required in Item 2 and it is also a more general caution that NDT does not replace the need to do 100 percent visual inspections, as AWS D1.1 already requires. Item 2 requires NDT per AWS D1.1 for inspection of all complete and partial joint penetration groove welds that are subject to net tensile forces for any BSFRS that is designed under AISC 341. Since no specific method of NDT is prescribed, any type of NDT permitted under AWS D1.1 would appear to suffice (see caution on NDT methods in Section 6.14 of AWS D1.1). Consequently, the statement of special inspection should be specific regarding the type of NDT. Since NDT applies to joints subject to net tensile forces, the extent of NDT that will be necessary is subject to the outcome of the structural analysis of the BSFRS. An exception permits the NDT to be less than 100 percent if both the engineer and the building official approve the lower percentage, although no specific basis is given for such approval.

1707.3 Structural wood. Continuous special inspection is required during field gluing operations of elements of the seismic-force-resisting system. Periodic special inspection is required for nailing, bolting, anchoring and other fastening of components within the seismic-force-resisting system, including wood shear walls, wood diaphragms, drag struts, braces, shear panels and hold-downs.

Exception: Special inspection is not required for wood shear walls, shear panels and diaphragms, including nailing, bolting, anchoring and other fastening to other components of the seismic-force-resisting system, where the fastener spacing of the sheathing is more than 4 inches (102 mm) on center (o.c.).

❖ Continuous special inspection of field gluing operations of structural wood is required, while special inspection of fastenings of components within the seismic-force-resisting system is required on a periodic basis. Since special inspection should only be necessary for complex, highly loaded systems, an exception waives the special inspection of fastening in shear walls and diaphragms that are lightly loaded. Lower design loads means these elements will have less stringent requirements for bolting, anchoring and fastening. The building official who is already inspecting the sheathing will also inspect fastening in these instances. Rather than being based on the component's design shear, the exception is based on the sheathing fastener spacing as a more practical threshold. All things being equal, as the fastener spacing decreases the component's design load is higher and the tighter spacing increases the potential for splitting framing members. The latter concern is a quality issue associated primarily with nailing.

1707.4 Cold-formed steel framing. Periodic special inspections is required during welding operations of elements of the seismic-force-resisting system. Periodic special inspection is required for screw attachment, bolting, anchoring and other fastening of components within the seismic-force-resisting system, including struts, braces, and hold-downs.

❖ Periodic special inspection is required for cold-formed steel framing and its fastening.

1707.5 Pier foundations. Special inspection is required for pier foundations for buildings assigned to Seismic Design Category C, D, E or F in accordance with Section 1613. Periodic special inspection is required during placement of reinforcement and continuous special inspection is required during placement of the concrete.

❖ For structures classified as Seismic Design Category C, D, E or F, the pier foundations will typically be of concrete that is subject to special inspections in accordance with Section 1704.4 as well as the pier verifications required by Section 1704.9. The requirements in this section may be somewhat redundant since they correspond to concrete inspections that are required in Section 1704.4 regardless of seismic design category.

1707.6 Storage racks and access floors. Periodic special inspection is required during the anchorage of access floors and storage racks 8 feet (2438 mm) or greater in height in structures assigned to Seismic Design Category D, E or F.

❖ The anchorage of storage racks and access floors in structures assigned to Seismic Design Category D, E or F requires periodic special inspection.

1707.7 Architectural components. Periodic special inspection is required during the erection and fastening of exterior cladding, interior and exterior nonbearing walls and interior and exterior veneer in structures assigned to Seismic Design Category D, E or F.

Exceptions:

1. Special inspection is not required for architectural components in structures 30 feet (9144 mm) or less in height.

2. Special inspection is not required for cladding and veneer weighing 5 psf (24.5 N/m²) or less.

3. Special inspection is not required for interior nonbearing walls weighing 15 psf (73.5 N/m²) or less.

❖ Periodic special inspection of the installation of the listed architectural components verifies that the intent of the design is carried out in the field. The listed components are often supported at different levels of a structure, meaning the component and connection design must consider relative movements. Where connections are designed to allow for relative displacements, verifying that the installation is in accordance with the construction documents is a prudent measure. The special inspector should verify that the method of anchoring or fastening and the number, spacing and types of fasteners actually used conform with the approved construction documents for the component installed.

1707.8 Mechanical and electrical components. Special inspection for mechanical and electrical equipment shall be as follows:

1. Periodic special inspection is required during the anchorage of electrical equipment for emergency or standby power systems in structures assigned to Seismic Design Category C, D, E or F;

2. Periodic special inspection is required during the installation of anchorage of other electrical equipment in structures assigned to Seismic Design Category E or F;

3. Periodic special inspection is required during installation of piping systems intended to carry flammable, combustible or highly toxic contents and their associated mechanical units in structures assigned to Seismic Design Category C, D, E or F;

4. Periodic special inspection is required during the installation of HVAC ductwork that will contain hazardous materials in structures assigned to Seismic Design Category C, D, E or F; and

5. Periodic special inspection is required during the installation of vibration isolation systems in structures assigned to Seismic Design Category C, D, E or F where the construction documents require a nominal clearance of 0.25 inches (6.4 mm) or less between the equipment support frame and restraint.

❖ It is anticipated that the minimum requirements for mechanical and electrical components will be complied with when the special inspector is satisfied that the method of anchoring or fastening and the number, spacing and types of fasteners actually used conform with the approved construction documents for the components installed. It is noted that such special inspection requirements are for selected electrical, lighting, piping and ductwork components in Seismic Design Category C, D, E or F.

Item 5 requires the inspection of vibration isolation systems where an optional lower clearance is specified by the design. Typically, vibration isolated mechanical equipment requires an increased seismic design force. Note b of Table 13.6-1 of ASCE 7 provides an option to reduce this seismic design force on vibration-isolated components and systems having a clearance of 0.25 inches (6.4 mm) or less between the equipment support frame and the restraint. To confirm that the design intent is carried out in the field, isolated equipment installations utilizing a reduced clearance is subject to special inspection.

1707.9 Designated seismic system verifications. The special inspector shall examine designated seismic systems requiring seismic qualification in accordance with Section 1708.5 and verify that the label, anchorage or mounting conforms to the certificate of compliance.

❖ The special inspector is required to verify that the component label, anchorage or mounting conforms to the certificate of compliance.

1707.10 Seismic isolation system. Periodic special inspection is required during the fabrication and installation of isolator units and energy dissipation devices that are part of the seismic isolation system.

❖ Seismic isolation units and energy dissipation devices must receive periodic special inspection during fabrication and installation. The appropriate threshold for requiring such inspections would be that listed for seismic-force-resisting systems in Item 1 of Section 1707.1.

SECTION 1708
STRUCTURAL TESTING FOR
SEISMIC RESISTANCE

1708.1 Masonry. Testing and verification of masonry materials and assemblies prior to construction shall comply with the requirements of Sections 1708.1.1 through 1708.1.4, depending on the classification of the building or structure or nature of the occupancy, as defined by this code.

❖ This section establishes the level of quality assurance required for masonry construction. This is determined by the design method (empirical versus engineered) and the building's occupancy category. These requirements are summarized in Table 1708.1. The quality assurance levels identified as Level 1 and Level 2 coincide with the level of special inspection required by Section 1704.5.

TABLE 1708.1
QUALITY ASSURANCE OF MASONRY CONSTRUCTION

	TYPE OF FACILITY			
	I, II or III		IV	
Design Method	Empirical	Engineered	Empirical	Engineered
Quality Assurance	Note a	Level 1	Level 1	Level 2

a. Certificate of compliance (see Section 1708.1.1).

1708.1.1 Empirically designed masonry and glass unit masonry in Occupancy Category I, II or III. For masonry designed by Section 2109 or 2110 or by Chapter 5 or 7 of ACI 530/ASCE 5/TMS 402 in structures classified as Occupancy Category I, II or III, in accordance with Section 1604.5, certificates of compliance used in masonry construction shall be verified prior to construction.

❖ The minimum quality assurance required for empirically designed masonry and glass unit masonry in Occupancy Category I, II or III facilities is a certificate of compliance indicating that the masonry materials used comply with the code and contract documents. Certifi-

cates of compliance are a requirement for all masonry construction in order to provide the building official with evidence that the masonry materials and products conform to the code and the construction documents.

1708.1.2 Empirically designed masonry and glass unit masonry in Occupancy Category IV. The minimum testing and verification prior to construction for masonry designed by Section 2109 or 2110 or by Chapter 5 or 7 of ACI 530/ASCE 5/TMS 402 in structures classified as Occupancy Category IV, in accordance with Section 1604.5, shall comply with the requirements of Table 1708.1.2.

❖ The minimum level of testing and verification required for empirically designed masonry and glass unit masonry in Occupancy Category IV facilities is a certificate of compliance indicating that the masonry materials used comply with the code and contract documents plus verification of f'_m prior to construction (see Table 1708.1.2, "Level 1 Quality Assurance").

TABLE 1708.1.2
LEVEL 1 QUALITY ASSURANCE

MINIMUM TESTS AND SUBMITTALS
Certificates of compliance used in masonry construction.
Verification of f'_m and f'_{AAC} prior to construction, except where specifically exempted by this code.

❖ See the commentary to Section 1708.1.4 for discussion.

1708.1.3 Engineered masonry in Occupancy Category I, II or III. The minimum testing and verification prior to construction for masonry designed by Section 2107 or 2108 or by chapters other than Chapter 5, 6 or 7 of ACI 530/ASCE 5/TMS 402 in structures classified as Occupancy Category I, II or III, in accordance with Section 1604.5, shall comply with Table 1708.1.2.

❖ Engineered masonry in Occupancy Category I, II or III facilities requires the same testing and verification as empirically designed in Occupancy Category IV facilities (see Table 1708.1.2, "Level 1 Quality Assurance").

1708.1.4 Engineered masonry in Occupancy Category IV. The minimum testing and verification prior to construction for masonry designed by Section 2107 or 2108 or by chapters other than Chapter 5, 6 or 7 of ACI 530/ASCE 5/TMS 402 in structures classified as Occupancy Category IV, in accordance with Section 1604.5, shall comply with Table 1708.1.4.

❖ Engineered masonry in Occupancy Category IV facilities requires testing and verification prior to construction. The minimum tests and submittals are listed in Table 1708.1.4. The requirements are the same as for engineered masonry in Occupancy Category I, II or III facilities, with the added requirements that f'_m must be verified every 5,000 square feet (465 m²) during construction and the proportions of materials in mortar and grout must be verified as delivered to the site (see Table 1708.1.4, "Level 2 Quality Assurance").

TABLE 1708.1.4
LEVEL 2 QUALITY ASSURANCE

MINIMUM TESTS AND SUBMITTALS
Certificates of compliance used in masonry construction.
Verification of f'_m and f'_{AAC} prior to construction and every 5,000 square feet during construction.
Verification of proportions of materials in mortar and grout as delivered to the site.

For SI: 1 square foot = 0.0929 m^2.

❖ See the commentary to Section 1708.1.4 for discussion.

1708.2 Testing for seismic resistance. The tests specified in Sections 1708.3 through 1708.6 are required for the following:

1. The seismic-force-resisting systems in structures assigned to Seismic Design Category C, D, E or F, as determined in Section 1613.

2. Designated seismic systems in structures assigned to Seismic Design Category D, E or F.

3. Architectural, mechanical and electrical components in structures assigned to Seismic Design Category C, D, E or F that are required in Section 1708.5.

❖ This section specifies when material seismic-resistance tests for seismic-force-resisting systems and designated seismic systems are required. These requirements supplement the test requirements contained in the referenced standards given in other sections of the code. The seismic provisions of the code are written to rely heavily on the concept of material tests such that good quality material is used for seismic-resistant construction.

Some inelastic behavior can be anticipated in a building's response to the design earthquake. Strong ground shaking caused by earthquakes tends to expose any underlying flaws in a building's construction or design. Thus, the code specifies testing of some structural and nonstructural components as additional verification that the seismic resistance provided in the finished structure meets the intent of the design. It is imperative that the professional who designs systems that are critical to the earthquake performance also identifies them and specifies the necessary testing in the statement of special inspections.

Item 1 states the seismic risk threshold for seismic-force-resisting systems. The testing of items in Sections 1708.3 and 1708.4 would apply to these structures. Item 2 covers designated seismic systems that are nonstructural components and systems that are assigned an importance factor greater than 1. The testing applicable to designated seismic systems is given in Section 1708.5. Item 3 is somewhat redundant since there are no testing requirements for architectural, electrical and mechanical components in

Section 1708, other than the aforementioned designated seismic system testing.

1708.3 Reinforcing and prestressing steel. Certified mill test reports shall be provided for each shipment of reinforcing steel used to resist flexural, shear and axial forces in reinforced concrete intermediate frames, special moment frames and boundary elements of special reinforced concrete or reinforced masonry shear walls. Where ASTM A 615 reinforcing steel is used to resist earthquake-induced flexural and axial forces in special moment frames and in wall boundary elements of shear walls in structures assigned to Seismic Design Category D, E or F, as determined in Section 1613, the testing requirements of ACI 318 shall be met. Where ASTM A 615 reinforcing steel is to be welded, chemical tests shall be performed to determine weldability in accordance with Section 3.5.2 of ACI 318.

❖ Certified material test reports are required for rebar. When ASTM A 615 is used in special moment frames and shear walls in structures classified as Seismic Design Category D, E or F, the testing requirements of Section 21.2.5 of ACI 318 must be used (see Section 1908.1.5). Where ASTM A 615 rebar is to be welded, a chemical analysis must be provided to determine weldability of the steel in accordance with AWS D1.4.

1708.4 Structural steel. The testing contained in the quality assurance plan shall be as required by AISC 341 and the additional requirements herein. The acceptance criteria for nondestructive testing shall be as required in AWS D1.1 as specified by the registered design professional.

Base metal thicker than 1.5 inches (38 mm), where subject to through-thickness weld shrinkage strains, shall be ultrasonically tested for discontinuities behind and adjacent to such welds after joint completion. Any material discontinuities shall be accepted or rejected on the basis of ASTM A 435 or ASTM A 898 (Level 1 criteria) and criteria as established by the registered design professional(s) in responsible charge and the construction documents.

❖ Structural steel must be tested as required by AISC 341. Any NDT must use the acceptance criteria in AWS D1.1 and as specified by the registered design professional. Base metal thicker than 1.5 inches (38 mm) is subject to laminations and discontinuities during the steel-making process. Any base metal thicker than 1.5 inches (38 mm) that will be subjected to through-thickness weld shrinkage strains must be ultrasonically tested after welding is completed. The acceptance criteria for ultrasonic testing are required to be on the basis of ASTM A 435 or A 898 (Level 1 criteria) and as established by the registered design professional.

1708.5 Seismic qualification of mechanical and electrical equipment. The registered design professional in responsible charge shall state the applicable seismic qualification requirements for designated seismic systems on the construction documents. Each manufacturer of designated seismic system components shall test or analyze the component and its mounting system or anchorage and submit a certificate of compliance for review and acceptance by the registered design professional in responsible charge of the design of the designated seismic system and for approval by the building official. Qualification shall be by an actual test on a shake table, by three-dimensional shock tests, by an analytical method using dynamic characteristics and forces, by the use of experience data (i.e., historical data demonstrating acceptable seismic performance) or by a more rigorous analysis providing for equivalent safety.

❖ The registered design professional is required to specify the applicable seismic qualification requirements on the construction documents for designated seismic systems. This is necessary for the manufacturer who is required to provide a certificate of compliance. Affected mechanical and electrical components are required to demonstrate seismic performance by testing or analytical methods. The intent is to verify the ability of critical equipment to not only survive an earthquake but to also remain functional.

A certificate of compliance for review and acceptance is also required. The registered design professional responsible must then review and accept the certificate of compliance, and the building official must approve it. ICC Evaluation Service *Acceptance Criteria for Seismic Qualification Testing of Nonstructural Components* (AC 156) provides testing criteria that can be used to comply with this provision.

1708.6 Seismically isolated structures. For required system tests, see Section 17.8 of ASCE 7.

❖ The referenced section of ASCE 7 contains detailed provisions for isolation system testing. This testing provides effective stiffness and effective damping values to be used in the design of a seismically isolated structure. A minimum of two full-size specimens must be tested for each proposed type and size of the isolator unit. These prototypes are not to be used in the construction, unless approved by both the registered design professional and the building official.

SECTION 1709
STRUCTURAL OBSERVATIONS

1709.1 General. Where required by the provisions of Section 1709.2 or 1709.3 the owner shall employ a registered design professional to perform structural observations as defined in Section 1702.

At the conclusion of the work included in the permit, the structural observer shall submit to the building official a written statement that the site visits have been made and identify any reported deficiencies that, to the best of the structural observer's knowledge, have not been resolved.

❖ This section requires that a registered engineer or architect be employed by the owner to provide on-site visits to observe compliance with the structural drawings. Structural observations are required under certain conditions in high-seismic and high-wind locations (see Section 1702 for the definition of "Structural observation"), providing an additional level of inspections beyond those provided by the building inspector or the special inspector. The intent of requiring structural observations by a registered design professional for

these structures is to verify that the seismic-force-resisting systems, the designated seismic systems and the windforce-resisting systems are constructed in general conformance with the construction documents.

1709.2 Structural observations for seismic resistance. Structural observations shall be provided for those structures included in Seismic Design Category D, E or F, as determined in Section 1613, where one or more of the following conditions exist:

1. The structure is classified as Occupancy Category III or IV in accordance with Section 1604.5.

2. The height of the structure is greater than 75 feet (22 860 mm) above the base.

3. The structure is assigned to Seismic Design Category E, is classified as Occupancy Category I or II in accordance with Section 1604.5 and is greater than two stories in height.

4. When so designated by the registered design professional in responsible charge of the design.

5. When such observation is specifically required by the building official.

❖ This section lists the thresholds of seismic risk that necessitate structural observation. Items 1 through 3 are specific conditions that require structural observation, while Items 4 and 5 give discretion to the registered design professional or the building official to require structural observation in other instances (see commentary, Section 1709.1).

1709.3 Structural observations for wind requirements. Structural observations shall be provided for those structures sited where the basic wind speed exceeds 110 mph (49 m/s), determined from Figure 1609, where one or more of the following conditions exist:

1. The structure is classified as Occupancy Category III or IV in accordance with Table 1604.5.

2. The building height is greater than 75 feet (22 860 mm).

3. When so designated by the registered design professional in responsible charge of the design.

4. When such observation is specifically required by the building official.

❖ This section lists the thresholds of high wind exposure that necessitate structural observation. Items 1 and 2 are specific conditions that require structural observation, while Items 3 and 4 give discretion to the registered design professional or the building official to require structural observation in other instances (see commentary, Section 1709.1).

SECTION 1710
DESIGN STRENGTHS OF MATERIALS

1710.1 Conformance to standards. The design strengths and permissible stresses of any structural material that are identified by a manufacturer's designation as to manufacture and grade by mill tests, or the strength and stress grade is otherwise confirmed to the satisfaction of the building official, shall conform to the specifications and methods of design of accepted engineering practice or the approved rules in the absence of applicable standards.

❖ Structural materials must conform to applicable design standards, approved rules and accepted methods of engineering practice. Conformance to these provisions and to the manufacturer's designations provides the building official with the information needed to verify that the materials will perform their intended function satisfactorily.

1710.2 New materials. For materials that are not specifically provided for in this code, the design strengths and permissible stresses shall be established by tests as provided for in Section 1711.

❖ Materials that are not explicitly covered by the code are allowed when subjected to the appropriate testing demonstrating adequate performance (see Section 1701.2).

SECTION 1711
ALTERNATIVE TEST PROCEDURE

1711.1 General. In the absence of approved rules or other approved standards, the building official shall make, or cause to be made, the necessary tests and investigations; or the building official shall accept duly authenticated reports from approved agencies in respect to the quality and manner of use of new materials or assemblies as provided for in Section 104.11. The cost of all tests and other investigations required under the provisions of this code shall be borne by the permit applicant.

❖ Test reports from approved agencies may be used as a basis for approval of materials that are not within the purview of any approved rules (i.e., "new materials" as mentioned in Section 1701.2). This section directly references Section 104.11. It is within the power of the building official to accept reports from an approved agency. In determining the approval, the building official should check that the agency is an independent third-party agency with no financial or fiduciary affiliations with the applicant or material supplier. The capability and competency of the agency must also be examined. It should be noted that this section assigns responsibility for costs of the testing to the applicant.

SECTION 1712
TEST SAFE LOAD

1712.1 Where required. Where proposed construction is not capable of being designed by approved engineering analysis, or where proposed construction design method does not comply with the applicable material design standard, the system of construction or the structural unit and the connections shall be subjected to the tests prescribed in Section 1714. The building official shall accept certified reports of such tests conducted by an approved testing agency, provided that such tests meet the requirements of this code and approved procedures.

❖ Testing to determine safe load is required when a structural component cannot be designed in accordance with approved engineering practices or where the construction design method does not fully comply with the respective material design standard listed in Chapter 35. If either of these situations exist, the structural components are required to be subjected to the prescriptive tests listed in Section 1714, which address loading and deflection criteria.

An example of a structural component that cannot be designed by approved engineering practice is a composite concrete and steel slab in which the shear connector is some type of new configuration called a "widget." The horizontal shear that can be developed is unknown; therefore, a complete analysis cannot be performed. This section also restates the building official's option of accepting data from an approved testing agency, as previously stated in Section 1703.4.

SECTION 1713
IN-SITU LOAD TESTS

1713.1 General. Whenever there is a reasonable doubt as to the stability or load-bearing capacity of a completed building, structure or portion thereof for the expected loads, an engineering assessment shall be required. The engineering assessment shall involve either a structural analysis or an in-situ load test, or both. The structural analysis shall be based on actual material properties and other as-built conditions that affect stability or load-bearing capacity, and shall be conducted in accordance with the applicable design standard. If the structural assessment determines that the load-bearing capacity is less than that required by the code, load tests shall be conducted in accordance with Section 1713.2. If the building, structure or portion thereof is found to have inadequate stability or load-bearing capacity for the expected loads, modifications to ensure structural adequacy or the removal of the inadequate construction shall be required.

❖ The intent of this section is to utilize an engineering analysis to verify the adequacy of the structure, if possible. The load test requirement should only be done if an engineering analysis does not verify structural adequacy. Load tests are last resort options, and the building official should document his or her reasons for any load testing requirement.

An example of the executed structural analysis would be an analysis by a third-party engineering firm acceptable to both the building official and the owner. The structural integrity may be questioned for items such as visible signs of excessive settlement or lateral deflection, such as cracks in concrete foundation walls or excessive vibration when the assembly is loaded.

A load test procedure must simulate the actual load conditions to which the structure is subjected during normal use (see Section 1713.3 for details).

1713.2 Test standards. Structural components and assemblies shall be tested in accordance with the appropriate material standards listed in Chapter 35. In the absence of a standard that contains an applicable load test procedure, the test procedure shall be developed by a registered design professional and approved. The test procedure shall simulate loads and conditions of application that the completed structure or portion thereof will be subjected to in normal use.

❖ When load test procedures for materials are given by the applicable referenced material standard, the test procedure outlined in that specific standard must be adhered to without variation. If a referenced standard lacks a load test procedure or a material or assembly does not have a specific referenced standard, then such a test must be developed by a registered design professional and approved by the building official. The test procedure must be representative of and simulate the actual loading conditions that the completed structure or portion thereof will be subjected to during normal use.

1713.3 In-situ load tests. In-situ load tests shall be conducted in accordance with Section 1713.3.1 or 1713.3.2 and shall be supervised by a registered design professional. The test shall simulate the applicable loading conditions specified in Chapter 16 as necessary to address the concerns regarding structural stability of the building, structure or portion thereof.

❖ The criteria for in-situ load tests are set forth for two categories: procedures specified, which are regulated by Section 1713.3.1, and procedures not specified, which are regulated by Section 1713.3.2. This section further requires that the test be performed under the supervision of a registered design professional and that it simulates the actual loads and conditions of the completed structure or portion thereof.

1713.3.1 Load test procedure specified. Where a standard listed in Chapter 35 contains an applicable load test procedure and acceptance criteria, the test procedure and acceptance criteria in the standard shall apply. In the absence of specific load factors or acceptance criteria, the load factors and acceptance criteria in Section 1713.3.2 shall apply.

❖ The load test must be in accordance with the applicable referenced standard. Section 1713.3.2 must only be utilized in the absence of either a specific standard or specific load factors and acceptance criteria from applicable referenced standards.

1713.3.2 Load test procedure not specified. In the absence of applicable load test procedures contained within a standard referenced by this code or acceptance criteria for a specific material or method of construction, such existing structure shall be subjected to a test procedure developed by a registered design professional that simulates applicable loading and deformation conditions. For components that are not a part of the seismic-load-resisting system, the test load shall be equal to two times the unfactored design loads. The test load shall be left in place for a period of 24 hours. The structure shall be considered to have successfully met the test requirements where the following criteria are satisfied:

1. Under the design load, the deflection shall not exceed the limitations specified in Section 1604.3.

2. Within 24 hours after removal of the test load, the structure shall have recovered not less than 75 percent of the maximum deflection.

3. During and immediately after the test, the structure shall not show evidence of failure.

❖ If the applicable standards do not specify load factor or testing criteria acceptance methods, then the testing criteria listed in this section must be followed. Note that the design load includes design live load and all dead loads that are not yet in place, such as the dead load from tenant walls in a speculative office building.

SECTION 1714
PRECONSTRUCTION LOAD TESTS

1714.1 General. In evaluating the physical properties of materials and methods of construction that are not capable of being designed by approved engineering analysis or do not comply with applicable material design standards listed in Chapter 35, the structural adequacy shall be predetermined based on the load test criteria established in this section.

❖ This section establishes requirements for load testing structural assemblies that are either incapable of being designed or those that, for one reason or another, do not comply with the applicable material design standards. This section does not govern the load testing of existing buildings, which is governed by Section 1713.

The different categories of preconstruction load tests are addressed herein. Specified load test procedures are regulated by Section 1714.2. Load test procedures that are not specified are regulated by Section 1714.3. Wall and partition assemblies are regulated by Section 1714.4. Exterior window and door assemblies are regulated by Section 1714.5, and test specimens are regulated by Section 1714.6.

1714.2 Load test procedures specified. Where specific load test procedures, load factors and acceptance criteria are included in the applicable design standards listed in Chapter 35, such test procedures, load factors and acceptance criteria shall apply. In the absence of specific test procedures, load factors or acceptance criteria, the corresponding provisions in Section 1714.3 shall apply.

❖ This section has priority over Section 1714.3, provided that load factors and acceptance criteria are established in the applicable design standards.

1714.3 Load test procedures not specified. Where load test procedures are not specified in the applicable design standards listed in Chapter 35, the load-bearing and deformation capacity of structural components and assemblies shall be determined on the basis of a test procedure developed by a registered design professional that simulates applicable loading and deformation conditions. For components and assemblies that are not a part of the seismic-load-resisting system, the test shall be as specified in Section 1714.3.1. Load tests shall simulate the applicable loading conditions specified in Chapter 16.

❖ In the absence of load factors and acceptance criteria in the applicable design standards and in accordance with Section 1714.2, this section is to be used by the building official to determine if conformance to the applicable code requirements has been achieved. Additionally, this section provides the design professional and the building official with specific loading and pass/fail criteria.

1714.3.1 Test procedure. The test assembly shall be subjected to an increasing superimposed load equal to not less than two times the superimposed design load. The test load shall be left in place for a period of 24 hours. The tested assembly shall be considered to have successfully met the test requirements if the assembly recovers not less than 75 percent of the maximum deflection within 24 hours after the removal of the test load. The test assembly shall then be reloaded and subjected to an increasing superimposed load until either structural failure occurs or the superimposed load is equal to two and one-half times the load at which the deflection limitations specified in Section 1714.3.2 were reached, or the load is equal to two and one-half times the superimposed design load. In the case of structural components and assemblies for which deflection limitations are not specified in Section 1714.3.2, the test specimen shall be subjected to an increasing superimposed load until structural failure occurs or the load is equal to two and one-half times the desired superimposed design load. The allowable superimposed design load shall be taken as the lesser of:

1. The load at the deflection limitation given in Section 1714.3.2.

2. The failure load divided by 2.5.

3. The maximum load applied divided by 2.5.

❖ Load test criteria relating to superimposed design loads are established herein. These requirements are a compilation of commonly accepted engineering practices to test adequately against structural failure.

In the case of structural components and assemblies for which maximum deflection limitations are not addressed in Section 1714.3.2, the test assemblies must be subjected to increasing superimposed loads until failure occurs or the load is equal to two and one-half times the superimposed design load, whichever occurs first.

1714.3.2 Deflection. The deflection of structural members under the design load shall not exceed the limitations in Section 1604.3.

❖ Acceptance criteria for deflection of structural systems when subjected to the allowable design load are used to demonstrate adequate structural performance and are addressed in Section 1604.3.

1714.4 Wall and partition assemblies. Load-bearing wall and partition assemblies shall sustain the test load both with and without window framing. The test load shall include all design load components. Wall and partition assemblies shall be tested both with and without door and window framing.

❖ Load-bearing wall and partition assemblies must sustain loads with and without window framing. It is not appropriate to assume that a wall will sustain loads better if window framing is involved in the test. Each individual design must be evaluated separately based on the construction of that assembly. All design loads, such as vertical and lateral forces, must be included in the test.

1714.5 Exterior window and door assemblies. The design pressure rating of exterior windows and doors in buildings shall be determined in accordance with Section 1714.5.1 or 1714.5.2.

Exception: Structural wind load design pressures for window units smaller than the size tested in accordance with Section 1714.5.1 or 1714.5.2 shall be permitted to be higher than the design value of the tested unit provided such higher pressures are determined by accepted engineering analysis. All components of the small unit shall be the same as the tested unit. Where such calculated design pressures are used, they shall be validated by an additional test of the window unit having the highest allowable design pressure.

❖ This section allows two methods of load test for exterior window and door assemblies. The first method, provided for in Section 1714.5.1, allows products to be tested and labeled as conforming to AAMA/WMDA/CSA 101/I.S.2/A440. The second method allows products to be tested in accordance with ASTM E 330 and the glazing must comply with Section 2403. The exception allows window units smaller than the size tested to have higher design pressures, provided the higher pressures are determined by accepted engineering analysis, all components of the smaller unit are the same as the tested unit and an additional test of the smaller unit having the highest calculated design pressure is performed in accordance with Section 1714.5.1 or 1714.5.2.

1714.5.1 Exterior windows and doors. Exterior windows and sliding doors shall be tested and labeled as conforming to AAMA/WDMA/CSA101/I.S.2/A440. The label shall state the name of the manufacturer, the approved labeling agency and the product designation as specified in AAMA/WDMA/CSA101/I.S.2/A440. Exterior side-hinged doors shall be tested and labeled as conforming to AAMA/WDMA/CSA101/I.S.2/A440 or comply with Section 1714.5.2. Products tested and labeled as conforming to AAMA/WDMA/CSA 101/I.S.2/A440 shall not be subject to the requirements of Sections 2403.2 and 2403.3.

❖ This section requires exterior windows and doors complying with AAMA/WMDA/CSA 101/I.S.2/A440 to be labeled as such. Products so tested and labeled must not be required to meet the provisions of Sections 2403.2 and 2403.3. By requiring the product to be labeled, the building official does not have to interpret test results and determine load-carrying capacities, or accept the manufacturer's interpretation of tests.

1714.5.2 Exterior windows and door assemblies not provided for in Section 1714.5.1. Exterior window and door assemblies shall be tested in accordance with ASTM E 330. Exterior window and door assemblies containing glass shall comply with Section 2403. The design pressure for testing shall be calculated in accordance with Chapter 16. Each assembly shall be tested for 10 seconds at a load equal to 1.5 times the design pressure.

❖ This section allows an alternative to the provisions of Section 1714.5.1.

This procedure is to be used to verify the integrity of door and window assemblies as a whole, and its results do not supersede, but rather complement, the requirements of Chapter 24. Glass thickness must be determined in accordance with the provisions of Chapter 24. The assemblies must comply with Section 2403, and the design pressure for testing is determined from Chapter 16. Each assembly is required to be tested for 10 seconds at a load equal to 1.5 times the design pressure. When testing a product line that has a variety of sizes, the most critical size can usually be tested and the results used to qualify other similar products within its family. In general, the larger size and the most heavily loaded of each particular design, type, construction or configuration should be tested. The code does not specify how many specimens are to be tested. ASTM E 330 states that if only one sample is tested, it should be selected by the specifying authority.

1714.6 Test specimens. Test specimens and construction shall be representative of the materials, workmanship and details normally used in practice. The properties of the materials used to construct the test assembly shall be determined on the basis of tests on samples taken from the load assembly or on representative samples of the materials used to construct the load test assembly. Required tests shall be conducted or witnessed by an approved agency.

❖ The test specimen must resemble and simulate as much as possible the design being tested using materials and workmanship that could be expected in the actual construction or fabrication. The test itself must be witnessed or conducted by an agency acceptable to and approved by the building official.

SECTION 1715
MATERIAL AND TEST STANDARDS

1715.1 Test standards for joist hangers and connectors.

❖ This section prescribes the test standard and criteria to be used for joist hangers and connectors. This criteria is meant to be used in establishing the load capacity of joist hangers and connectors used in wood construction for which there is no calculated procedure recognized by the code.

The referenced test standard, ASTM D 1761, requires the joist length to be twice the joist depth plus 10 inches (254 mm), but this can result in longer joists that fail before the hanger. In establishing a hanger's allowable load, the objective is to have the hanger fail in the test, rather than the joist. The exception sets a maximum joist length in order to facilitate the appropriate failure mode.

1715.1.1 Test standards for joist hangers. The vertical load-bearing capacity, torsional moment capacity and deflection characteristics of joist hangers shall be determined in accordance with ASTM D 1761 using lumber having a specific gravity of 0.49 or greater, but not greater than 0.55, as determined in accordance with AF&PA NDS for the joist and headers.

Exception: The joist length shall not be required to exceed 24 inches (610 mm).

❖ The specified ASTM standard test method provides a procedure for evaluating the vertical load-carrying capacity, torsional moment capacity and deflection characteristics of joist hangers and similar devices used to connect wood joists to headers of wood or other materials. The lumber used for the test specimen must have a specific gravity equal to or greater than 0.49, but not greater than 0.55.

1715.1.2 Vertical load capacity for joist hangers. The vertical load capacity for the joist hanger shall be determined by testing a minimum of three joist hanger assemblies as specified in ASTM D 1761. If the ultimate vertical load for any one of the tests varies more than 20 percent from the average ultimate vertical load, at least three additional tests shall be conducted. The allowable vertical load of the joist hanger shall be the lowest value determined from the following:

1. The lowest ultimate vertical load for a single hanger from any test divided by three (where three tests are conducted and each ultimate vertical load does not vary more than 20 percent from the average ultimate vertical load).

2. The average ultimate vertical load for a single hanger from all tests divided by three (where six or more tests are conducted).

3. The average from all tests of the vertical loads that produce a vertical movement of the joist with respect to the header of 0.125 inch (3.2 mm).

4. The sum of the allowable design loads for nails or other fasteners utilized to secure the joist hanger to the wood members and allowable bearing loads that contribute to the capacity of the hanger.

5. The allowable design load for the wood members forming the connection.

❖ The method prescribed establishes the allowable load for normal duration, as defined by the American Forest & Paper Association (AF&PA) *National Design Specification (NDS) for Wood Construction*. Additionally, allowable stresses cannot exceed those allowed by the code. For example, published allowable loads cannot contain nail loads higher than those allowed by NDS, nor can tension in steel strapping exceed that allowed by the steel design standards noted in Chapter 22.

For loads of other than normal duration, the stresses or loads may be increased or must be decreased as noted by the appropriate design standard, but in no case can the load exceed that which will produce $^1/_8$-inch (3.2 mm) deflection.

EXAMPLE:

Given: A manufacturer's test results for a particular joist hanger are as follows:

Test 1 Ultimate load = 1,000 pounds with the $^1/_8$-inch deflection occurring at 400 pounds.

Test 2 Ultimate load = 1,100 pounds with the $^1/_8$-inch deflection occurring at 350 pounds.

Test 3 Ultimate load = 900 pounds with the $^1/_8$-inch deflection occurring 375 pounds.

The manufacturer submitted structural calculations indicating that the allowable design load of the nails is 250 pounds (114 kg), and the allowable shear load in the wood joists framing to the hangers is 280 pounds (127 kg). Joist hanger geometry does not allow for any meaningful calculation of stresses in the steel sections of the hanger.

Find: The allowable load for the joist hanger.

Solution:

Average ultimate load = (1,000 + 1,100 + 900) ÷ 3.0 = 1,000 pounds

Since 20 percent of 1,000 is 200, the test scatter is within the allowable range of plus and minus 20 percent of the average ultimate load; therefore, three tests are sufficient to establish allowable load.

Thus, the allowable load is 250 pounds (114 kg) based on the lesser of:

Lowest ultimate load ÷ 3.0 = 300 pounds (137 kg)

$^1/_8$-inch deflection in any test = 350 pounds (159 kg)

Allowable nail load = 250 pounds (114 kg)

Allowable joist shear = 280 pounds (127 kg)

In this case, the allowable vertical load for a normal duration is limited by the calculated allowable design load of the fastener [250 pounds (113 kg)] in accordance with Item 4 of Section 1715.1.2. This value is permitted to be modified by a duration of loading factor; however, the modified value cannot exceed the lowest test value as determined in accordance with Items 1, 2 and 3 of Section 1715.1.2.

1715.1.3 Torsional moment capacity for joist hangers. The torsional moment capacity for the joist hanger shall be determined by testing at least three joist hanger assemblies as specified in ASTM D 1761. The allowable torsional moment of the joist hanger shall be the average torsional moment at which the lateral movement of the top or bottom of the joist with respect to the original position of the joist is 0.125 inch (3.2 mm).

❖ The allowable torsional moment capacity for a joist hanger is determined by testing in accordance with ASTM D 1761 with the limitation that rotational deflection of the top or bottom of the joist with respect to the header must not exceed 0.125 inch (3.2 mm).

1715.1.4 Design value modifications for joist hangers. Allowable design values for joist hangers that are determined by Item 4 or 5 in Section 1715.1.2 shall be permitted to be mod-

ified by the appropriate duration of loading factors as specified in AF&PA NDS but shall not exceed the direct loads as determined by Item 1, 2 or 3 in Section 1715.1.2. Allowable design values determined by Item 1, 2 or 3 in Section 1715.1.2 shall not be modified by duration of loading factors.

❖ The calculated allowable design values, as determined by Item 4 or 5 of Section 1715.1.2 and modified by duration of loading factors, must not exceed the lowest test value as determined by Item 1, 2 or 3 of Section 1715.1.2

1715.2 Concrete and clay roof tiles.

❖ This section prescribes the test standards and criteria to be used to determine the overturning resistance and wind characteristics of concrete and clay roof tiles.

1715.2.1 Overturning resistance. Concrete and clay roof tiles shall be tested to determine their resistance to overturning due to wind in accordance with SBCCI SSTD 11 and Chapter 15.

❖ Section 1715.2.1 requires concrete and clay tiles to be tested to determine their overturning resistance in accordance with SBCCI SSTD 11. SBCCI SSTD 11 prescribes methods for determining the allowable overturning moment for mechanically fastened, adhesive-set and mortar-set tiles. A test procedure is also prescribed for determining the allowable uplift loads on hip/ridge tiles.

1715.2.2 Wind tunnel testing. When roof tiles do not satisfy the limitations in Chapter 16 for rigid tile, a wind tunnel test shall be used to determine the wind characteristics of the concrete or clay tile roof covering in accordance with SBCCI SSTD 11 and Chapter 15.

❖ The wind tunnel test procedures in SBCCI SSTD 11 must be used if the roof tiles do not meet the limitations of Chapter 16 for rigid tile.

Bibliography

The following resource materials are referenced in this chapter or are relevant to the subject matter addressed in this chapter.

AAMA/WMDA/CSA 101/I.S.2/A440-05, *Voluntary Specifications for Aluminum Vinyl (PVC) and Wood Windows and Glass Doors.* Schaumburg, IL: American Architectural Manufacturers Association, 2005.

ACI 318-05, *Building Code Requirements for Structural Concrete.* Farmington Hills, MI: American Concrete Institute, 2005.

ACI 530/ASCE 5/TMS 402-05, *Building Code Requirements for Masonry Structures.* Farmington Hills, MI: American Concrete Institute; New York: American Society of Civil Engineers; Boulder, CO: The Masonry Council, 2005.

ACI 530.1/ASCE 6/TMS 602-05, *Specifications for Masonry Structures.* Farmington Hills, MI: American Concrete Institute; New York: American Society of Civil Engineers; Boulder, CO: The Masonry Council, 2005.

AF&PA NDS-05, *National Design Specification for Wood Construction.* Washington, D.C.: American Forest & Paper Association, 2005.

AISC 341-05, *Seismic Provisions for Structural Steel Buildings.* Chicago: American Institute of Steel Construction, 2005.

AISC 360-05, *Specification for Structural Steel Buildings.* Chicago: American Institute of Steel Construction, 2005.

ASCE 7-05, *Minimum Design Loads for Buildings and Other Structures.* New York: American Society of Civil Engineers, 2005.

ASTM A 6-04a, *Specification for General Requirements for Rolled Structural Bars, Plates, Shapes, and Sheet Piling.* West Conshohocken, PA: ASTM International, 2004.

ASTM A 36/A 36M-04, *Specification for Carbon Structural Steel.* West Conshohocken, PA: ASTM International, 2004.

ASTM A 325-04a, *Specification for Structural Bolts, Steel, Heat Treated, 120/105 ksi Minimum Tensile Strength.* West Conshohocken, PA: ASTM International, 2004.

ASTM A 435-90 (2001), *Specification for Straight Beam Ultrasound Examination of Steel Plates.* West Conshohocken, PA: ASTM International, 2001.

ASTM A 490-04a, *Specification for Structural Bolts, Alloy Steel, Heat Treated, 150 ksi Minimum Tensile Strength.* West Conshohocken, PA: ASTM International, 2004.

ASTM A 568-03, *Specification for Steel Sheet, Carbon and High-Strength, Low-Alloy, Hot-Rolled and Cold-Rolled, General Requirements For.* West Conshohocken, PA: ASTM International, 2003.

ASTM A 615/A 615M-04a, *Specification for Deformed and Plain Billet-Steel Bars for Concrete Reinforcement.* West Conshohocken, PA: ASTM International, 2004.

ASTM A 706-04a, *Specification for Low-Alloy Steel Deformed and Plain Bars for Concrete Reinforcement.* West Conshohocken, PA: ASTM International, 2004.

ASTM A 898-91(2001), *Straight Beam Ultrasound Examination of Rolled Steel Structural Shapes.* West Conshohocken, PA: ASTM International, 2001.

ASTM C 31/C31M-03a, *Standard Practice for Making and Curing Concrete Test Specimens in the Field.* West Conshohocken, PA: ASTM International, 2003.

ASTM C 172-04, *Standard Practice for Sampling Freshly Mixed Concrete.* West Conshohocken, PA: ASTM International, 2004.

ASTM D 1761-88 (2000)e1, *Standard Test Method for Mechanical Fasteners in Wood.* West Conshohocken, PA: ASTM International, 2000.

ASTM E 330-02, *Standard Test Methods for Structural Performance of Exterior Windows, Curtain Walls, and Doors by Uniform Static Air Pressure Difference.* West Conshohocken, PA: ASTM International, 2002.

ASTM E 605-00, *Test Methods for Thickness and Density of Sprayed Fire-Resistive Material Applied to Structural Members.* West Conshohocken, PA: ASTM International, 2000.

AWCI 12-B, *Technical Manual 12-B Standard Practice for the Testing and Inspection of Field Applied Thin-Film Intumescent Fire-Resistive Materials; as Annoted Guide, First Edition.* Falls Church, VA: The Association of the Wall and Ceiling Industries International, 1998

AWS D1.1-04, *Structural Welding Code–Steel.* Miami, FL: American Welding Society, 2004.

AWS D1.3-98, *Structural Welding Code–Sheet Steel.* Miami, FL: American Welding Society, 1998.

AWS D1.4-98, *Structural Welding Code–Reinforced Steel.* Miami, FL: American Welding Society, 1998.

AWS QC1-88, *Standard and Guide for Qualification and Certification of Welding Inspectors.* Miami, FL: American Welding Society, 1988.

FEMA 450, *NEHRP Recommended Provisions for Seismic Regulations for New Buildings and Other Structures.* Washington, D.C.: Federal Emergency Management Agency, 2004.

Henry, John R. "Special Inspection, Structural Observation and Quality Assurance Under the 2000 IBC." *Structural Engineer*, April 2002, pp 24-28.

ICC-ES AC 156, *Acceptance Criteria for Seismic Qualification Testing of Nonstructural Components.* Whittier, CA: International Code Council Evaluation Service, 2000.

IFC-06, *International Fire Code.* Falls Church, VA: International Code Council, 2006.

IMC-06, *International Mechanical Code.* Falls Church, VA: International Code Council, 2006.

IPC-06, *International Plumbing Code.* Falls Church, VA: International Code Council, 2006.

IRC-06, *International Residential Code.* Falls Church, VA: International Code Council, 2006.

Model Program for Special Inspection. Falls Church, VA: International Codes Council, 2002.

Specification for Structural Joints Using A325 or A490 Bolts. Chicago: Research Council on Structural Connections (c/o American Institute of Steel Construction, Inc.), 2004.

SBCCI SSTD 11-97, *Standard for Determining Wind Resistance of Concrete or Clay Roof Tiles*. Birmingham, AL: Southern Building Code Congress International, 1997.

UL 127-99, *Factory-Built Fireplaces*. Northbrook, IL: Underwriters Laboratories Inc., 1999.

Chapter 18:
Soils and Foundations

General Comments

Chapter 18 contains provisions regulating the design and construction of foundations for buildings and other structures and involves geotechnical and structural considerations in the selection and installation of adequate supports for the loads transferred from the structure above.

Section 1801 gives the general scope or purpose of the provisions in Chapter 18.

Section 1802 provides the requirements for foundation and soils investigations that are to be conducted at the site prior to final design and construction.

Section 1803 includes excavation, grading and backfill provisions.

Section 1804 establishes the allowable load-bearing values for soils where site soil tests do not verify that higher soil values are appropriate.

Section 1805 contains the regulations for spread footings and foundations.

Section 1806 provides requirements for retaining walls.

Section 1807 provides specifications for the dampproofing and waterproofing of floor slabs and below-grade walls.

Section 1808 gives the requirements for pier and pile foundations.

Section 1809 provides the specifications for driven pile foundations.

Section 1810 includes the provisions for cast-in-place concrete pile foundations.

Section 1811 contains the regulations for composite piles.

Section 1812 provides the pier foundation specifications that are required by the code.

The proper design and construction of a foundation system is critical to the satisfactory performance of the entire building structure that it supports.

Foundation problems are not uncommon and vary greatly. They may be of a simple or complex nature and may be manageable or without practical remedy. The uncertainties of foundation construction make it extremely difficult to seek and handle every potential for failure within the text of the code. The provisions of the code are meant to set forth and regulate the minimum standards and conventional practices needed for the design and construction of foundation systems so as to provide adequate safety to life and property. Due care must be exercised in the planning and design of foundation systems based on obtaining sufficient soils information, the use of accepted engineering procedures, experience and good technical judgement.

Essentially, there are two parts to the foundation system: the substructure and the soil. The substructure consists of structural components that serve as the medium through which the building loads are transmitted to the supporting earth (soil or rock). The substructure components may consist of shallow foundations, such as basement walls, grade walls, beams, isolated spread footings or combinations of these components. Shallow foundations may also involve the use of mat or raft foundations. As may be required, the substructure can consist of deep foundations involving the use of piles; drilled shafts or piers; caissons or other similar deep foundations. The second part of the foundation system involves the use of soil (including rock) as a structural material to carry the load of the building or any other load transmitted through the substructure.

The substructure and the soil are interdependent elements of the foundation system and must be understood and dealt with as a composite engineering consideration. Indeed, the selection of the kind of substructure to be used, whether it employs any of the different types of shallow foundations commonly used or adopts the use of a deep foundation, is a direct function of the nature of the soil encountered at the project site. Chapter 18 broadly outlines the conventional systems of foundation construction. Although it does not specifically include special or patented systems, it does not preclude their use because most such construction will fall into the general categories of foundation systems prescribed in the provisions and, thus, meet the intent of the code.

In determining the load-bearing capacity and other values of the soil mass, the code provisions address such considerations in terms of "undisturbed" soil. Special provisions are included where prepared fill is to be utilized for foundation support.

Purpose

The provisions of this chapter set forth the minimum requirements for the design, construction and resistance to water intrusion (for areas not subject to scour or water pressure by wind and wave action) of foundation systems for buildings and other structures.

SECTION 1801
GENERAL

1801.1 Scope. The provisions of this chapter shall apply to building and foundation systems in those areas not subject to scour or water pressure by wind and wave action. Buildings and foundations subject to such scour or water pressure loads shall be designed in accordance with Chapter 16.

❖ The provisions contained in this chapter for foundation design and construction do not apply to structures located in flood hazard areas where potential scour or water pressure from wind and wave action exist. Special design considerations are required for foundations subject to scour and water pressure that this chapter does not address. Section 1612 addresses flood loads and the establishment of flood hazard areas. Section 1612.4 requires buildings and structures located in a flood hazard area to be designed in accordance with ASCE 24. This standard provides extensive guidance to be used where scour by wind and wave action is expected to significantly impact the foundation of a building or structure. Once the expected loss of supporting soil has been determined, an effective foundation system can be designed and constructed. Failure to consider scour in areas subject to scour during a design flood event may lead to significant building damage or failure. For additional guidance, refer to FEMA 55, *Coastal Construction Manual.*

1801.2 Design. Allowable bearing pressures, allowable stresses and design formulas provided in this chapter shall be used with the allowable stress design load combinations specified in Section 1605.3. The quality and design of materials used structurally in excavations, footings and foundations shall conform to the requirements specified in Chapters 16, 19, 21, 22 and 23 of this code. Excavations and fills shall also comply with Chapter 33.

❖ Design requirements in Chapter 18 are based on an allowable stress design (ASD) approach. Allowable stresses and service loads should not be used with the load combinations for strength design, and vice versa. This section clarifies the applicable load combinations from Chapter 16 that are to be used for the design of foundations.

1801.2.1 Foundation design for seismic overturning. Where the foundation is proportioned using the load combinations of Section 1605.2, and the computation of the seismic overturning moment is by the equivalent lateral-force method or the modal analysis method, the proportioning shall be in accordance with Section 12.13.4 of ASCE 7.

❖ This provision correlates with ASCE 7 earthquake load requirements that are based on the NEHRP *Recommended Provisions for Seismic Regulations for New Buildings.* When using load and resistance factor load combinations to size foundations the seismic overturning computed by the equivalent lateral force method or the modal analysis method is permitted to be reduced.

SECTION 1802
FOUNDATION AND SOILS INVESTIGATIONS

1802.1 General. Foundation and soils investigations shall be conducted in conformance with Sections 1802.2 through 1802.6. Where required by the building official, the classification and investigation of the soil shall be made by a registered design professional.

❖ Sections 1802.2 through 1802.6 specify the conditions for which foundation and soil investigations are required. Although this section states that the classification and investigation of soils is to be done by a registered design professional only when required by the building official, consultation with geotechnical engineers or individuals with significant experience in soil and foundation analysis is advised.

1802.2 Where required. The owner or applicant shall submit a foundation and soils investigation to the building official where required in Sections 1802.2.1 through 1802.2.7.

Exception: The building official need not require a foundation or soils investigation where satisfactory data from adjacent areas is available that demonstrates an investigation is not necessary for any of the conditions in Sections 1802.2.1 through 1802.2.6.

❖ Soils investigations to determine subsurface conditions should be made prior to the design and construction of new buildings and other structures. Such investigations should also be conducted when additions or alterations to existing facilities are contemplated and are of such scope as would significantly increase or change the distribution of foundation loads.

There are two main objectives for conducting a soils investigation. The first is of a confirmatory nature. Its purpose is to obtain information already known from adjacent structures, such as soil-boring records, field test results, laboratory test data and analyses and any other knowledge useful in the design of the foundation system. The second objective is of an exploratory nature. It is warranted where soils information does not exist or is insufficient or unsatisfactory for use in the design of the foundation system.

Regardless of the objective of the soils investigation, the information generally required includes one or more (or all) of the following items for determining subsurface conditions:

1. The depth, thickness and composition of each soil stratum;

2. For rock, the characteristics of the rock stratum (or strata), including the thickness of the rock to a reasonable depth;

3. The depth of ground water below the site surface; and

4. The engineering properties of the soil and rock strata that are pertinent for the proper design and performance of the foundation system.

For shallow foundations, the soils investigation should yield sufficient information to establish the character and load-bearing capacity of the soil (or rock) at depths that will receive the foundations.

Foundation problems are not uncommon and may vary greatly, ranging from very simple and manageable problems to very complex situations that may be either manageable or without practical remedy. The field of soil mechanics and foundation engineering is diverse and complicated, and since it is not an exact science, its application requires specialized knowledge and judgmental decisions based on experience. Where subsurface conditions are found or suspected to be of a critical nature, the building official is encouraged to seek the professional advice of experienced foundation engineers.

As indicated in the exception, where geotechnical data from adjacent areas are well known, the building official can accept the use of local engineering practices for the design of foundations.

1802.2.1 Questionable soil. Where the classification, strength or compressibility of the soil are in doubt or where a load-bearing value superior to that specified in this code is claimed, the building official shall require that the necessary investigation be made. Such investigation shall comply with the provisions of Sections 1802.4 through 1802.6.

❖ Whenever relevant soil characteristics are in doubt, or where a design is based upon load-bearing values that are greater than those specified in the code, the building official may require investigation and testing of the soil.

One such method of investigation includes construction test pits for field load-bearing tests. Test pits are usually required to be at least 4 square feet (0.37 m²) in area and be excavated down to the elevation of the proposed bearing surface. The typical apparatus for making such tests involves placing the test loads on a platform supported on a post through which the applied loads are transferred to a bearing plate of a specified size and, in turn, to the soil below. A typical setup for field load tests is shown in Figure 1802.2.1.

It is important that the load (weights on platform) is applied such that all of it will be transmitted to the soil as a static load without impact, fluctuation or eccentricity. The load should be applied incrementally, and continuous records of all settlements should be kept. Measurements are usually made by settlement recording devices, such as dial gauges, capable of measuring the settlement of the test bearing plate to an accuracy of at least 0.01 inch (0.25 mm).

The test is continued until either the maximum test load is reached or the ratio of load increment to settlement increment reaches a minimum, steady magnitude sustained for a period of 48 hours. After the load is released, the elastic rebound of the soil is also measured for a period of time.

Load test results are normally presented in a load settlement diagram in which the applied test load measured in tons per square foot is plotted in relation to the settlement readings recorded in fractions of an inch. The bearing capacity of the soil can be computed from the test results.

There are some drawbacks to the use of field load tests for determining soil-bearing capacity. Test results can be misleading if the soil under the footing is not uniform for the full depth of load influence, which is equal to about twice the width of the footing. Also, since a load test is conducted for a short duration, settlements that occur due to the consolidation of the soil over a very long time cannot be predicted. Since this type of field load-bearing testing is relatively expensive, it is not widely used.

This section permits the use of other methods for determining the safe bearing capacity of soils. Standard laboratory tests can usually produce sufficient proof of satisfactory bearing capacity and settlement information. However, there are conditions when standard laboratory tests may not produce reliable results, such as when clay materials contain a pattern of cracks or when stiff clays may have suffered differential movement or expansion.

Figure 1802.2.1
TYPICAL SETUP FOR CONDUCTING STATIC LOAD TESTS

1802.2.2 Expansive soils. In areas likely to have expansive soil, the building official shall require soil tests to determine where such soils do exist.

❖ Expansive soils, often referred to as "swelling soils," contain montmorillonite minerals and have the characteristics of absorbing water and swelling, or shrinking and cracking when drying. Significant volume changes can cause serious damage to buildings and other structures as well as to pavements and sidewalks.

Swelling soils are found throughout the nation, but are more prevalent in regions with dry or moderately arid climates.

1802.2.3 Ground-water table. A subsurface soil investigation shall be performed to determine whether the existing ground-water table is above or within 5 feet (1524 mm) below the elevation of the lowest floor level where such floor is located below the finished ground level adjacent to the foundation.

> **Exception:** A subsurface soil investigation shall not be required where waterproofing is provided in accordance with Section 1807.

❖ There are several reasons for conducting a subsurface investigation to determine the level of ground water at a construction site. If the ground-water table is above subsurface slabs (i.e., basement floors), then walls and floors need to be designed to resist hydrostatic pressures. Foundation walls and basement slabs may need to be dampproofed or waterproofed, depending on the location of the ground-water table. A subsurface investigation will also determine the type of drainage system needed as a permanent installation, whether there will be any major water problems that could affect the excavation operations and construction of the foundation system, and if it is necessary to provide a temporary drainage system of a type and size that will control ground-water seepage.

Ground-water levels can vary significantly over a year's time, as well as from year to year. While it would be ideal to make ground-water table observations that encompass a full annual cycle, the reality is that such an undertaking would not normally facilitate a design/construction program and would be impractical. Ground-water observations must be made in shorter time intervals; however, this situation poses some real problems. For example, measurements of ground-water levels in bore holes taken 24 hours after completion of the soil borings can provide an acceptable indication of the water level in permeable soils such as sand, gravel or sand/gravel mixtures. Fine-grained soils of low permeability, such as clays, require the use of observation tubes (piezometers) and, depending on the specific properties of the soil, a time period of 10 weeks or longer to obtain acceptable readings.

Water levels established by either of the two methods described above are sufficient indication of the water conditions at the time of measurement, but do not necessarily represent the highest possible ground-water levels that can occur. For design purposes, the water levels established by field observations may need to be adjusted with the climatological and hydrological records of the region in order to establish the high and low points.

As indicated in the exception, a subsurface investigation is not required where floors, walls, joints and penetrations are waterproofed as required in Section 1807.3.

1802.2.4 Pile and pier foundations. Pile and pier foundations shall be designed and installed on the basis of a foundation investigation and report as specified in Sections 1802.4 through 1802.6 and Section 1808.2.2.

❖ A foundation investigation is required when pile and pier foundations are used. Such investigations are needed to define as accurately as possible the subsurface conditions of soil and rock materials, establish the soil and rock profiles across the construction site and locate the ground-water table. Sometimes, it may also be necessary to determine specific soil properties, such as shear strength, relative density, compressibility and other such technical data required for analyzing subsurface conditions. Foundation investigations may also be used to render such valuable data as information on existing construction at the site or on neighboring properties (including boring and test records), the type and condition of the existing structures, their age, the type of foundations used and performance over the years. Knowledge of existing deleterious substances in the soils that could affect the durability (as well as the performance) of the piles, data on geologic conditions at the site (including such information as the existence of mines, earth cavities, underground streams or other adverse water conditions) and a history of any seismic activity and hurricanes are also important.

The types of information described above are usually obtained by means of soil and rock borings; laboratory and field tests and engineering analyses. Such information is used for determining design loads; types and lengths of piles; driving criteria and selection of equipment and probable durability of pile materials in relation to subsurface conditions.

1802.2.5 Rock strata. Where subsurface explorations at the project site indicate variations or doubtful characteristics in the structure of the rock upon which foundations are to be constructed, a sufficient number of borings shall be made to a depth of not less than 10 feet (3048 mm) below the level of the foundations to provide assurance of the soundness of the foundation bed and its load-bearing capacity.

❖ Rock may be found at levels near or at the earth's surface, upon which shallow foundations can be supported or will range downward to very low levels, upon which piles and other types of deep foundations can bear.

Most intact rock will have compressive strengths that far exceed the requirements for foundation support. It is most common, however, to find cracks, joints and other defects in rock formations that will increase the compressibility of the material. Depending on the nature and extent of the defects, settlement may become the governing factor in determining allowable load-bearing capacity rather than rock strength.

Where the condition of the rock is in doubt, borings must be made at least 10 feet (305 mm) into the rock stratum below the bottom of the footings to verify the soundness of the material and to determine its load-bearing capacity.

1802.2.6 Seismic Design Category C. Where a structure is determined to be in Seismic Design Category C in accordance with Section 1613, an investigation shall be conducted and shall include an evaluation of the following potential hazards resulting from earthquake motions: slope instability, liquefaction and surface rupture due to faulting or lateral spreading.

❖ The potential for liquefaction, surface rupture or slope instability at a building site is greater in areas of moderate and high seismicity than in areas of low seismicity. Also, the consequences of damage resulting from such hazards are more severe for buildings in higher seismic use groups (such as essential facilities). Thus, this section requires a soils investigation report for sites of buildings assigned to Seismic Design Category C and higher, including an evaluation of specific hazards resulting from earthquake motions. The purpose of this section is to reduce the hazard of large ground movement and the damaging effects on the structure that might result if any of the conditions cited in the code text should occur.

Liquefaction of saturated granular soils has been a major source of building damage during past earthquakes. For example, many structures in Niigata, Japan suffered major damage as a consequence of liquefaction during its 1964 earthquake. Loss of bearing strength, differential settlement and differential horizontal displacement due to lateral spread were the direct causes of damage. Many structures have been similarly damaged by differential ground displacements during U.S. earthquakes, such as the San Fernando Valley Juvenile Hall during the 1971 San Fernando, California earthquake and the Marine Sciences Laboratory at Moss Landing, California during the 1989 Loma Prieta event.

For more information regarding evaluation of slope instability, liquefaction and surface rupture due to faulting or lateral spreading, see Section 7.4 of the 1997 NEHRP provisions commentary.

1802.2.7 Seismic Design Category D, E or F. Where the structure is determined to be in Seismic Design Category D, E or F, in accordance with Section 1613, the soils investigation requirements for Seismic Design Category C, given in Section 1802.2.6, shall be met, in addition to the following. The investigation shall include:

1. A determination of lateral pressures on basement and retaining walls due to earthquake motions.

2. An assessment of potential consequences of any liquefaction and soil strength loss, including estimation of differential settlement, lateral movement or reduction in foundation soil-bearing capacity, and shall address mitigation measures. Such measures shall be given consideration in the design of the structure and can include but are not limited to ground stabilization, selection of appropriate foundation type and depths, selection of appropriate structural systems to accommodate anticipated displacements or any combination of these measures. The potential for liquefaction and soil strength loss shall be evaluated for site peak ground acceleration magnitudes and source characteristics consistent with the design earthquake ground motions. Peak ground acceleration

shall be determined from a site-specific study taking into account soil amplification effects, as specified in Chapter 21 of ASCE 7.

Exception: A site-specific study need not be performed, provided that peak ground acceleration equal to $S_{DS}/2.5$ is used, where S_{DS} is determined in accordance with Section 21.2.1 of ASCE 7.

❖ This section includes additional requirements for the soil investigation report for sites of structures assigned to Seismic Design Category D and higher. The investigation must determine lateral pressures on basement and retaining walls due to earthquake motions. Earthquake motions create increased lateral soil pressure on walls below the ground surface, especially in soft soils in areas of high seismicity. The requirements of this section are necessary to make sure that the dynamic soil pressures are included in the design of basement and retaining walls. Section 7.5.1 of the 1997 NEHRP provisions commentary includes a discussion about how earth-retaining structures have been designed for dynamic loads.

Additionally, a thorough assessment of potential consequences of any liquefaction and soil strength loss needs to be made and considered in the design of the structure. See the commentary to Section 1802.2.6 for a discussion of earthquake damage due to liquefaction. Design to prevent damage due to liquefaction consists of three parts: evaluation of liquefaction hazard; evaluation of potential ground displacement and mitigating the hazard by: designing to resist ground displacement, reducing the potential for liquefaction or choosing an alternative site with less hazard.

The assessment is required to be made for site peak ground acceleration based on a site-specific study unless a peak ground acceleration equal to $S_{DS}/2.5$ is used. The peak ground acceleration needs to be known by the geotechnical engineer in order to evaluate the potential for liquefaction and soil strength loss. The exception allows an estimate of $S_{DS}/2.5$ to be used to represent the peak ground acceleration.

1802.3 Soil classification. Where required, soils shall be classified in accordance with Section 1802.3.1 or 1802.3.2.

❖ This section provides parameters for classification of soils in accordance with the code.

1802.3.1 General. For the purposes of this chapter, the definition and classification of soil materials for use in Table 1804.2 shall be in accordance with ASTM D 2487.

❖ Where required, soils are to be classified in accordance with ASTM D 2487. This standard provides a system for which to classify soils for engineering purposes based on laboratory determination of particle size characteristics, liquid limit and plasticity index. The classification system identifies three major soil divisions—course-grained, fine-grained and highly organic—which are further broken down into 15 basic soil groups. ASTM D 2487 is the ASTM version of the Unified Soil Classification System.

1802.3.2 Expansive soils. Soils meeting all four of the following provisions shall be considered expansive, except that tests to show compliance with Items 1, 2 and 3 shall not be required if the test prescribed in Item 4 is conducted:

1. Plasticity index (PI) of 15 or greater, determined in accordance with ASTM D 4318.

2. More than 10 percent of the soil particles pass a No. 200 sieve (75 μm), determined in accordance with ASTM D 422.

3. More than 10 percent of the soil particles are less than 5 micrometers in size, determined in accordance with ASTM D 422.

4. Expansion index greater than 20, determined in accordance with ASTM D 4829.

❖ Expansive soils, often referred to as "swelling soils," contain montmorillonite minerals that shrink or swell with changes in moisture content. Expansive soils tend to swell when absorbing water or shrink and crack when drying. Significant volume changes can cause serious damage to buildings and other structures as well as to pavements and sidewalks. Swelling soils are found throughout the nation, but are more prevalent in regions with dry or moderately arid climates.

There is a general relationship between the plasticity index (PI) of a soil as determined by the ASTM D 4318 standard test method and the potential for expansion, as shown in Figure 1802.3.2.

This section defines "Expansive soil" as any plastic material with a PI of 15 or greater with more than 10 percent of the soil particles passing a No. 200 sieve and less than 5 micrometers in size. As an alternative, tests in accordance with ASTM D 4829 can be used to determine if a soil is expansive. The expansion index is a measure of the swelling potential of the soil.

The amount and depth of potential swelling that can occur in a clay material are, to some extent, functions of the cyclical moisture content in the soil. In dryer climates where the moisture content in the soil near the ground surface is low because of evaporation, there is a greater potential for extensive swelling than the same soil in wetter climates where the variations of moisture content are not as severe. Volume changes in highly expansive soils range between 7 and 10 percent, but experience has shown that occasionally, under abnormal conditions, they can reach as high as 25 percent.

1802.4 Investigation. Soil classification shall be based on observation and any necessary tests of the materials disclosed by borings, test pits or other subsurface exploration made in appropriate locations. Additional studies shall be made as necessary to evaluate slope stability, soil strength, position and adequacy of load-bearing soils, the effect of moisture variation on soil-bearing capacity, compressibility, liquefaction and expansiveness.

❖ When soils are required to be classified, the classification must be based on observations and tests such as borings or test pits. In addition to the situations specified that require a soils investigation and classification, the evaluation of slope stability, soil strength, position and adequacy of load-bearing soils, moisture effects, compressibility and liquefaction are required to be performed when deemed necessary by the building official or registered design professional.

1802.4.1 Exploratory boring. The scope of the soil investigation including the number and types of borings or soundings, the equipment used to drill and sample, the in-situ testing equipment and the laboratory testing program shall be determined by a registered design professional.

❖ Whenever the allowable load on a soil is in doubt and a field investigation is necessary, exploratory borings are to be made to determine the load-bearing value of the soil. The soil investigation is to be performed by a registered design professional, which in most cases would be a geotechnical engineer. Exploratory borings and their associated tests should be conducted in each area of relatively dissimilar subsoil conditions.

1802.5 Soil boring and sampling. The soil boring and sampling procedure and apparatus shall be in accordance with generally accepted engineering practice. The registered design professional shall have a fully qualified representative on the site during all boring and sampling operations.

❖ A qualified representative is required on site to verify that the information obtained from the investigation will be adequate, valid and acceptable to the building official.

1802.6 Reports. The soil classification and design load-bearing capacity shall be shown on the construction documents. Where required by the building official, a written report of the investigation shall be submitted that includes, but need not be limited to, the following information:

1. A plot showing the location of test borings and/or excavations.

2. A complete record of the soil samples.

3. A record of the soil profile.

SWELLING POTENTIAL PLASTICITY INDEX

Low	0-15
Medium	0-35
High	20-55
Very high	35 and above

Figure 1802.3.2
SWELLING POTENTIAL OF SOILS AND PLASTICITY INDEX

Source: R.B. Peck, W.E. Hanson and T.H. Thornburn. *Foundation Engineering,* 2nd ed. (New York: John Wiley & Sons, 1974).

---

4. Elevation of the water table, if encountered.

5. Recommendations for foundation type and design criteria, including but not limited to: bearing capacity of natural or compacted soil; provisions to mitigate the effects of expansive soils; mitigation of the effects of liquefaction, differential settlement and varying soil strength; and the effects of adjacent loads.

6. Expected total and differential settlement.

7. Pile and pier foundation information in accordance with Section 1808.2.2.

8. Special design and construction provisions for footings or foundations founded on expansive soils, as necessary.

9. Compacted fill material properties and testing in accordance with Section 1803.5.

❖ When soils are required to be classified, their classification and load-bearing capacity must be shown on the construction documents. If a written report is required by the building official, it is required to include at a minimum the nine items listed in this section. These items will establish a retrievable and verifiable record of the soil conditions if problems are encountered in the future. These items also provide the minimum necessary information for compliance with the code and an adequate foundation system.

SECTION 1803
EXCAVATION, GRADING AND FILL

1803.1 Excavations near footings or foundations. Excavations for any purpose shall not remove lateral support from any footing or foundation without first underpinning or protecting the footing or foundation against settlement or lateral translation.

❖ The purpose of this section is to provide soil stability of adjacent foundations when excavations are made. Due to their lack of shear strength, cohesionless soils, such as sand, will slide to the bottom of an excavation until a certain slope of the sides is reached. This slope is known as the angle of repose of natural slope and is independent of the depth of the excavation.

Cohesive (fine-grained) soils, such as clay, behave much differently compared to granular materials. For example, unsupported vertical cuts of 20 feet (6096 mm) or more can be made in stiff plastic clay materials. This is due to a firm bond between the particles of the cohesive soil. But the strength of this bond (cohesiveness) will vary based on the conditions of the soil, such as density, water content, plasticity and sensitivity (loss of shear strength upon disturbance).

In cohesive soils, when a certain critical depth of excavation is reached, the sides of the cut will fail and the soil mass will fall to the bottom. Unlike granular materials, such as sand, the steepest slope at which a cohesive soil will stand decreases as the depth of the excavation increases. Technically, in cohesive soils, the resistance against sliding is a function of the shearing resistance of the material and its corresponding angle

of internal friction (frictional resistance between particles).

The use of the angle of internal friction (and other factors) to calculate slope stability is applicable not only to cohesive soils, but also to granular materials. For example, when the slope angle of an excavation in a bed of sand exceeds the angle of internal friction of the material, the sand will slide down the slope; therefore, the steepest slope that sand can attain is equal to the angle of internal friction. The angle of repose (previously discussed) will be approximately the same value as the angle of internal friction only when the sand is in a dry and loose condition (such as in a stockpile) or is fully immersed in water.

For simplicity, what we have been dealing with in this part of the commentary is slope stability as it relates to homogeneous soils such as sand and clay. In nature, however, soils often occur as mixtures or layers (strata) of different materials, making the determination of slope stability a highly technical and complex subject. Normally for shallow excavations, determination of safe slopes is a matter of applying local experience. In cases of deep cuts, however, slope stability is best determined through tests and analytical methods performed by professionals experienced in foundation engineering.

Some texts dealing with soil mechanics contain tables that indicate the angles of slopes that can be expected for various soil materials commonly found throughout the country. While such tables provide useful information, they should be used only as a guide. They should not be used for design purposes, nor employed when a critical subsurface condition exists or when the type of soil (established by borings) does not closely fit those described in the tables.

1803.2 Placement of backfill. The excavation outside the foundation shall be backfilled with soil that is free of organic material, construction debris, cobbles and boulders or a controlled low-strength material (CLSM). The backfill shall be placed in lifts and compacted, in a manner that does not damage the foundation or the waterproofing or dampproofing material.

Exception: Controlled low-strength material need not be compacted.

❖ Section 1803.2 requires that soils used for backfilling foundation excavations must be free of organic material, construction debris or large rocks. The type of soil used for backfill purposes becomes an important consideration in the design of foundation walls. For example, clean sand, gravel or a mixture of these two granular materials is considered the best kind of backfill to use because each is free draining and generally has frost-free properties. On the other hand, fine-grained soils, such as clays, tend to accumulate moisture and are susceptible to swelling and shrinking, as well as frost action. Such backfill materials, particularly at times when shrinkage cracks occur, can become loaded with rainwater, thus subjecting foundation walls and basement floors to hydrostatic pressures and possible structural damage.

Backfilling and related work should be performed in such a way as to prevent the movement of the earth of adjoining properties or the subsequent caving in of backfilled areas. Backfilling should not be done until retaining walls, foundation walls or other construction against which backfill is to be placed is in a suitable condition to resist lateral pressures. This section requires that backfill be free of organic materials, construction debris, cobbles and boulders.

In addition to carefully selecting the backfill material, the soil should be placed in lifts, usually 9 inches (229 mm) or less, and compacted to prevent significant subsidence due to consolidation under its own weight. While compaction is done by hand-operated tampers or other portable compaction equipment, care should be taken to prevent any possible damage to waterproofing or dampproofing installations and to avoid overcompaction of backfill since it may cause excessive earth pressure against foundation walls. As an alternative to compacted backfill the code also permits controlled low-strength material (CLSM) (see commentary, Section 1803.6).

1803.3 Site grading. The ground immediately adjacent to the foundation shall be sloped away from the building at a slope of not less than one unit vertical in 20 units horizontal (5-percent slope) for a minimum distance of 10 feet (3048 mm) measured perpendicular to the face of the wall. If physical obstructions or lot lines prohibit 10 feet (3048 mm) of horizontal distance, a 5-percent slope shall be provided to an approved alternative method of diverting water away from the foundation. Swales used for this purpose shall be sloped a minimum of 2 percent where located within 10 feet (3048 mm) of the building foundation. Impervious surfaces within 10 feet (3048 mm) of the building foundation shall be sloped a minimum of 2 percent away from the building.

> **Exception:** Where climatic or soil conditions warrant, the slope of the ground away from the building foundation is permitted to be reduced to not less than one unit vertical in 48 units horizontal (2-percent slope).

The procedure used to establish the final ground level adjacent to the foundation shall account for additional settlement of the backfill.

❖ This section requires that the ground immediately adjacent to the foundation be sloped away from the building. The intent is to facilitate water drainage and reduces the potential for water standing under and around the building.

Where a full 10 feet (3048 mm) of slope is not available on a building site, an alternative method of diverting water away from the foundation is permitted by providing a minimum 5-percent slope to an approved diversion structure. The use of swales to convey surface water is recognized, provided the minimum slope is provided where necessary.

The exception permits the slope to be reduced to a rate of 1 unit vertical in 48 units horizontal (2-percent slope) where climatic or soil conditions warrant. This exception would be applicable in arid areas or sites

that are surrounded by free-draining soils, such as sand.

1803.4 Grading and fill in flood hazard areas. In flood hazard areas established in Section 1612.3, grading and/or fill shall not be approved:

1. Unless such fill is placed, compacted and sloped to minimize shifting, slumping and erosion during the rise and fall of flood water and, as applicable, wave action.

2. In floodways, unless it has been demonstrated through hydrologic and hydraulic analyses performed by a registered design professional in accordance with standard engineering practice that the proposed grading or fill, or both, will not result in any increase in flood levels during the occurrence of the design flood.

3. In flood hazard areas subject to high-velocity wave action, unless such fill is conducted and/or placed to avoid diversion of water and waves toward any building or structure.

4. Where design flood elevations are specified but floodways have not been designated, unless it has been demonstrated that the cumulative effect of the proposed flood hazard area encroachment, when combined with all other existing and anticipated flood hazard area encroachment, will not increase the design flood elevation more than 1 foot (305 mm) at any point.

❖ This section puts limitations on fill and grading in flood hazard areas. In many flood hazard areas, it is common to use fill to achieve the appropriate elevation of the building's lowest floor. Item 1 intends to minimize the risk of fills becoming unstable in a flood. Fill materials that become saturated during conditions of flooding may become unstable and fill slopes may be exposed to erosive velocities and waves. When placed in a flood hazard area, fill and grading should be done so as to be stable as floodwaters rise and, in particular, as floodwaters fall and the saturated materials drain.

Item 2 permits grading or fill in a floodway only if it is demonstrated that it will not adversely affect surrounding areas by increasing the design flood elevation. As the definition in Section 1612.2 indicates, a "Floodway" is that portion of a flood hazard area that is reserved for the discharge of the design flood event. The National Flood Insurance Program (NFIP) requires that the impact of development or encroachment on the floodway should be considered.

In flood hazard areas that are subject to high-velocity wave action, fill can divert the flow of floodwaters and increase flood risks on other properties. Item 3 provides coordination with ASCE 24 provisions relating to fill in flood hazard areas subject to high-velocity wave action. These areas are commonly referred to as "coastal high hazard areas" or "V zones," and typically are indicated on flood hazard maps of communities along the open coast. In these areas, changing the shape of the ground through grading or fill can divert erosive flows and increase wave energies that, in turn, increase the forces that affect a building and any adjacent structures.

Although FEMA designates floodways on many rivers and streams shown on Flood Insurance Rate Maps (FIRMs), some riverine flood hazard areas have base flood elevations but do not have delineated floodways. In these areas the effect of floodplain development on flood elevations has not been evaluated. Item 4 provides consistency with the NFIP with respect to development in areas where a base flood elevation has been established, but no floodway has been designated has been. It allows a proposed development to be approved if it is demonstrated that the flood elevations will not be increased by more than 1 foot (305 mm). Development in riverine floodplains can increase flood levels and loads on other properties, especially if it occurs in areas known as floodways that must be reserved to convey flood flows (see the definition of the term "Floodway").

1803.5 Compacted fill material. Where footings will bear on compacted fill material, the compacted fill shall comply with the provisions of an approved report, which shall contain the following:

1. Specifications for the preparation of the site prior to placement of compacted fill material.

2. Specifications for material to be used as compacted fill.

3. Test method to be used to determine the maximum dry density and optimum moisture content of the material to be used as compacted fill.

4. Maximum allowable thickness of each lift of compacted fill material.

5. Field test method for determining the in-place dry density of the compacted fill.

6. Minimum acceptable in-place dry density expressed as a percentage of the maximum dry density determined in accordance with Item 3.

7. Number and frequency of field tests required to determine compliance with Item 6.

Exception: Compacted fill material less than 12 inches (305 mm) in depth need not comply with an approved report, provided it has been compacted to a minimum of 90 percent Modified Proctor in accordance with ASTM D 1557. The compaction shall be verified by a qualified inspector approved by the building official.

❖ Where prepared fill is to be utilized for foundation support, the soils report required by Section 1803.5 is to contain detailed information for approval of the fill. The information on the prepared fill in the soils report is necessary to permit a reasonable prediction as to the load-bearing capacity of the fill material. Additionally, when prepared fill is to be utilized where special inspections are required, the fill operation itself must be performed under the scrutiny of a special inspection. For example, if one (or more) of the exceptions in Section 1704.1 applies, special inspection of the prepared fill operation is not required.

1803.6 Controlled low-strength material (CLSM). Where footings will bear on controlled low-strength material (CLSM), the CLSM shall comply with the provisions of an approved report, which shall contain the following:

1. Specifications for the preparation of the site prior to placement of the CLSM.

2. Specifications for the CLSM.

3. Laboratory or field test method(s) to be used to determine the compressive strength or bearing capacity of the CLSM.

4. Test methods for determining the acceptance of the CLSM in the field.

5. Number and frequency of field tests required to determine compliance with Item 4.

❖ As an alternative to compacted fill in accordance with Section 1803.5, this section permits the use of CLSM for the support of footings. CLSM must be placed in accordance with an approved report that includes requirements for the material strength and field verification.

In Chapter 2, CLSM is defined as "self-compacting cementitious materials." This class of material is commonly referred to by a variety of other names, which include flowable fill, controlled density fill, unshrinkable fill and soil-cement slurry. Guidance on the use of these materials can be found in ACI 229R. Additional documents that may be useful references for sampling and testing these materials include the following ASTM International standards:

- ASTM D 4832, *Standard Test Method for Preparation and Testing of Controlled Low-Strength Material (CLSM) Test Cylinders.*

- ASTM D 5971, *Standard Practice for Sampling Freshly Mixed Controlled Low-Strength Material.*

- ASTM D 6023, *Standard Test Method for Unit Weight, Yield, Cement Content, and Air Content (Gravimetric) of Controlled Low-Strength Material (CLSM).*

- ASTM D 6024, *Standard Test Method for Ball Drop on Controlled Low-Strength Material (CLSM) to Determine Suitability for Load Application.*

- ASTM D 6103, *Standard Test Method for Flow Consistency of Controlled Low-Strength Material (CLSM).*

SECTION 1804
ALLOWABLE LOAD-BEARING VALUES OF SOILS

1804.1 Design. The presumptive load-bearing values provided in Table 1804.2 shall be used with the allowable stress design load combinations specified in Section 1605.3.

❖ This section clarifies the use of the presumptive load-bearing values relating to foundations and footing design. Foundation design is based on the ASD approach. Since Chapter 16 includes provisions, including load combinations, for strength design [load and resistance factor design (LRFD)], it is necessary to clarify that these values are to be used with the ASD load combinations.

1804.2 Presumptive load-bearing values. The maximum allowable foundation pressure, lateral pressure or lateral sliding-resistance values for supporting soils near the surface shall not exceed the values specified in Table 1804.2 unless data to substantiate the use of a higher value are submitted and approved.

Presumptive load-bearing values shall apply to materials with similar physical characteristics and dispositions.

Mud, organic silt, organic clays, peat or unprepared fill shall not be assumed to have a presumptive load-bearing capacity unless data to substantiate the use of such a value are submitted.

Exception: A presumptive load-bearing capacity is permitted to be used where the building official deems the load-bearing capacity of mud, organic silt or unprepared fill is adequate for the support of lightweight and temporary structures.

❖ Where the load-bearing capacity of the soil has not been determined by borings, as specified in Section 1802, the presumptive load-bearing values listed in Table 1804.2 are intended to apply in the design of shallow foundation systems.

While fill material (unconsolidated), mud, muck, peat, organic silt, soft clay and other unprepared fill materials are considered to have no presumptive load-bearing value, soil tests may show that they do have some limited load-bearing capacity and, based on this type of evidence, the building official may approve the construction of lightweight structures upon such soils.

The presumption is that the building official possesses sufficient technical knowledge on the character and behavior of subsurface materials to render a valid judgment on the adequacy of the soil to support satisfactorily the lightweight or temporary structure, or he or she has sought and gained specific advice through consultation with professionals who are competent in the field of foundation engineering. It would be an unwise practice to authorize construction on exceptionally weak soils without the benefit of technical knowledge to make judgmental decisions.

TABLE 1804.2. See below.

❖ The values in Table 1804.2 are intended to be lower-bound allowable pressures to be used where the bearing value is not determined by borings as specified in Section 1802. The classes of soil and rock listed in Table 1804.2 are those materials most commonly found at construction sites around the country. The allowable foundation pressures expressed in terms of pounds per square foot (psf) for each class of subsurface materials listed in Table 1804.2 are based on long experience with the behavior of these materials under loaded conditions.

Should local soil conditions be significantly different than those listed in Table 1804.2, then, with the approval of the building official, local experience can be employed in the design of foundations, particularly where actual load-bearing values are less than the allowable unit pressures given in the table.

Whether the allowable load-bearing values of Table 1804.2 are used or local conditions on soil capacity prevail, many precautions must be taken in the design of foundation systems that are not regulatory functions of the code, but rather are the professional considerations and design applications of those who are engaged in foundation engineering. The building official needs to pay particular attention to proper selection of the allowable load-bearing capacity of the soil because it is the source of many foundation failures. The problem arises when the load-bearing material directly

TABLE 1804.2
ALLOWABLE FOUNDATION AND LATERAL PRESSURE

| CLASS OF MATERIALS | ALLOWABLE FOUNDATION PRESSURE (psf)[d] | LATERAL BEARING (psf/f below natural grade)[d] | LATERAL SLIDING | |
			Coefficient of friction[a]	Resistance (psf)[b]
1. Crystalline bedrock	12,000	1,200	0.70	—
2. Sedimentary and foliated rock	4,000	400	0.35	—
3. Sandy gravel and/or gravel (GW and GP)	3,000	200	0.35	—
4. Sand, silty sand, clayey sand, silty gravel and clayey gravel (SW, SP, SM, SC, GM and GC)	2,000	150	0.25	—
5. Clay, sandy clay, silty clay, clayey silt, silt and sandy silt (CL, ML, MH and CH)	1,500[c]	100	—	130

For SI: 1 pound per square foot = 0.0479 kPa, 1 pound per square foot per foot = 0.157 kPa/m.

a. Coefficient to be multiplied by the dead load.

b. Lateral sliding resistance value to be multiplied by the contact area, as limited by Section 1804.3.

c. Where the building official determines that in-place soils with an allowable bearing capacity of less than 1,500 psf are likely to be present at the site, the allowable bearing capacity shall be determined by a soils investigation.

d. An increase of one-third is permitted when using the alternate load combinations in Section 1605.3.2 that include wind or earthquake loads.

under a foundation overlies a stratum (or strata) of weaker material having a smaller allowable load-bearing capacity. The selection of the load-bearing value to be used in the design of the foundation should take into account the load distribution at the weaker stratum (or strata) so that the pressure on the soil does not exceed its allowable load-bearing capacity. In this respect, it is important to have a soil profile showing the classes of material at the construction site or at properties in the near vicinity of or adjacent to the area of construction. Such information can be a part of the records or other data required by Section 1802.6 or the results of a soil investigation.

With this information, the building official can consult with the registered design professional in charge of the foundation design to ascertain that due care was exercised in adopting proper soil load-bearing values.

1804.3 Lateral sliding resistance. The resistance of structural walls to lateral sliding shall be calculated by combining the values derived from the lateral bearing and the lateral sliding resistance shown in Table 1804.2 unless data to substantiate the use of higher values are submitted for approval.

For clay, sandy clay, silty clay and clayey silt, in no case shall the lateral sliding resistance exceed one-half the dead load.

❖ The lateral sliding resistance of structural walls is determined by combining the lateral bearing and lateral sliding values from Table 1804.2. The lateral sliding value from Table 1804.2 is determined by multiplying the applicable coefficient of friction by the dead load. For clay; sandy clay; silty clay and clayey silt; silt and sandy silt, the lateral sliding value is set at 130 psf (6.23 kPa). However, this section also stipulates that for clay, sandy clay, silty clay and clayey silt, the lateral sliding resistance cannot exceed one-half the dead load.

1804.3.1 Increases in allowable lateral sliding resistance. The resistance values derived from the table are permitted to be increased by the tabular value for each additional foot (305 mm) of depth to a maximum of 15 times the tabular value.

Isolated poles for uses such as flagpoles or signs and poles used to support buildings that are not adversely affected by a 0.5 inch (12.7 mm) motion at the ground surface due to short-term lateral loads are permitted to be designed using lateral-bearing values equal to two times the tabular values.

❖ The lateral sliding resistance is calculated as the sum of the lateral bearing and lateral sliding values from Table 1804.2. The lateral bearing values are determined as the product of the tabular value and the depth below natural grade. This section essentially limits the depth to which the lateral bearing values are added to the lateral sliding values at 15 feet (4572 mm).

For isolated poles used as supports for flagpoles or signs, and poles used to support buildings that are not adversely affected by a 1/2-inch (12.7 mm) motion at the ground surface due to short-term lateral loads, the lateral bearing values in Table 1804.2 are permitted to be multiplied by two.

SECTION 1805
FOOTINGS AND FOUNDATIONS

1805.1 General. Footings and foundations shall be designed and constructed in accordance with Sections 1805.1 through 1805.9. Footings and foundations shall be built on undisturbed soil, compacted fill material or CLSM. Compacted fill material shall be placed in accordance with Section 1803.5. CLSM shall be placed in accordance with Section 1803.6.

The top surface of footings shall be level. The bottom surface of footings is permitted to have a slope not exceeding one unit vertical in 10 units horizontal (10-percent slope). Footings shall be stepped where it is necessary to change the elevation of the top surface of the footing or where the surface of the ground slopes more than one unit vertical in 10 units horizontal (10-percent slope).

❖ Sections 1805.1 through 1805.9 address requirements related to proper design and installation of footings and foundations. It is important that footings and foundations be built on undisturbed soil of known bearing value or properly compacted fill with known bearing capacity. As an alternative to compacted fill the code permits the use of CLSM (see commentary, Section 1803.6). The tops and bottoms of footings are required to be essentially level, with a slope of 1 unit vertical in 10 units horizontal (10-percent slope) permitted for the bottom of footings. Where the slope of the surface of the ground exceeds one unit vertical in 10 units horizontal, footings are required to be stepped. Although not specifically mentioned, crack propagation at the joints should be considered when determining the overlapping and vertical dimensions of the steps.

1805.2 Depth of footings. The minimum depth of footings below the undisturbed ground surface shall be 12 inches (305 mm). Where applicable, the depth of footings shall also conform to Sections 1805.2.1 through 1805.2.3.

❖ Footings are required to extend below the ground surface a minimum of 12 inches (305 mm). This is considered a minimum depth to protect the footing from movement of the soil caused by freezing and thawing in mild climate areas (see Section 1805.2.1 for general frost protection requirements).

1805.2.1 Frost protection. Except where otherwise protected from frost, foundation walls, piers and other permanent supports of buildings and structures shall be protected by one or more of the following methods:

1. Extending below the frost line of the locality;

2. Constructing in accordance with ASCE 32; or

3. Erecting on solid rock.

Exception: Free-standing buildings meeting all of the following conditions shall not be required to be protected:

1. Classified in Occupancy Category I, in accordance with Section1604.5;

2. Area of 600 square feet (56 m²) or less for light-frame construction or 400 square feet (37 m²) or less for other than light-frame construction; and

3. Eave height of 10 feet (3048 mm) or less.

Footings shall not bear on frozen soil unless such frozen condition is of a permanent character.

❖ Shallow foundations must be placed on soil strata with adequate load-bearing capacity and at depths to which freezing cannot penetrate. In winter, frost action can raise the ground level (frost heave), whereas in springtime the same area will soften and settle back. If foundations are built in soil strata that can freeze, then the heave or vertical movement of the ground, which is rarely uniform, can cause serious damage to buildings and other structures. Frost heave can become particularly aggravated in clay soils. Well-drained soils, such as sand and gravel, are not as susceptible to extensive movements.

Unless the exception applies, the foundation is to be protected from frost in accordance with this section. A common method of accomplishing this is by placing the footing bottom below the frost line. The "frost line" is defined as the lowest level below the ground surface to which a temperature of 32°F (0°C) extends. The factors determining the depth of the frost line are air temperature and the length of time it is below freezing [32°F (0°C)], as well as the ability of the soil to conduct heat and its level of thermal conductivity. Frost lines vary significantly throughout the country, ranging from 5 inches (127 mm) in the deep south to 100 inches (2540 mm) in the uppermost northern regions. The frost-free depth for shallow foundations is dependent on the frost line set for the particular locality of construction.

Another form of protection is the use of frost-protected shallow foundations (FPSF) in accordance with ASCE 32. This type of frost protection utilizes slab edge insulation to minimize heat loss at the slab edge. By retaining heat from the building in the ground, it has the effect of raising the frost line around the perimeter of the building.

Foundations are not to be placed on frozen soil because when the ground thaws, uneven settlement of the structure is apt to occur, thereby causing structural damage. This section does, however, permit footings to be constructed on permanently frozen soil. In permafrost areas, special precautions are necessary to prevent heat from the structure from thawing the soil beneath the foundation.

The exception to frost protection applies to low risk structures such as a detached garage. The permitted areas are intended to accommodate garage sizes that are commonly constructed. Note that the minimum footing depth of Section 1805.2 would apply where this exception is invoked.

1805.2.2 Isolated footings. Footings on granular soil shall be so located that the line drawn between the lower edges of adjoining footings shall not have a slope steeper than 30 degrees (0.52 rad) with the horizontal, unless the material supporting the higher footing is braced or retained or otherwise laterally supported in an approved manner or a greater slope has been properly established by engineering analysis.

❖ The bottoms of adjacent footings bearing on granular soil are to be located so that a line drawn between their closest edges would not be steeper than 30 degrees (0.52 rad) from the horizontal. The exceptions are where the soil surrounding the higher footing is laterally braced or retained as approved by the building official or where engineering analysis shows that a greater slope can be tolerated. The purpose of this restriction is to provide conditions of safety against the possible influence of lateral and vertical soil pressures transmitted to the lower footing(s) by the loads of higher adjacent foundations (see Figure 1805.2.2).

For SI: 1 degree = 0.01745 rad.

Figure 1805.2.2
ISOLATED FOOTINGS

1805.2.3 Shifting or moving soils. Where it is known that the shallow subsoils are of a shifting or moving character, footings shall be carried to a sufficient depth to ensure stability.

❖ See the commentary to Sections 1802.2.2 and 1802.3.2 for a discussion regarding expansive soils. Shallow foundations placed on or within a mass of shifting or moving soils must be designed with great rigidity and strength in order to adequately resist soil movement and avoid damage. This section requires footings to be carried to a sufficient depth for adequate stability. Adequate stability implies, among other things, the consideration of uplift and perpendicular forces exerted on the footings due to soil movement.

1805.3 Footings on or adjacent to slopes. The placement of buildings and structures on or adjacent to slopes steeper than one unit vertical in three units horizontal (33.3-percent slope) shall conform to Sections 1805.3.1 through 1805.3.5.

❖ The provisions of this section apply to buildings placed on or adjacent to slopes steeper than 1 unit vertical in 3 units horizontal.

1805.3.1 Building clearance from ascending slopes. In general, buildings below slopes shall be set a sufficient distance from the slope to provide protection from slope drainage, erosion and shallow failures. Except as provided for in Section 1805.3.5 and Figure 1805.3.1, the following criteria will be assumed to provide this protection. Where the existing slope is steeper than one unit vertical in one unit horizontal (100-percent slope), the toe of the slope shall be assumed to be at the intersection of a horizontal plane drawn from the top of the foundation and a plane drawn tangent to the slope at an angle of 45 degrees (0.79 rad) to the horizontal. Where a retaining wall is constructed at the toe of the slope, the height of the slope shall be measured from the top of the wall to the top of the slope.

❖ Figure 1805.3.1 depicts criteria for locating buildings adjacent to slopes. The setback is intended to provide protection to the structure not only from shallow slope failures (sometimes referred to as "sloughing") but also from erosion and slope drainage. At ascending slopes, the setback also provides access around the building and helps create a light and open-air environment. Where an existing slope is steeper than 1:1, the toe of the slope is assumed to be at a distance determined by the intersection of a horizontal line at the top of the foundation and a line drawn at a 45-degree (0.79 rad) angle to the horizontal line, and terminating at the top of the slope. This determination of slope height and measurement of the setback is illustrated in Figure 1805.3.1(1).

FIGURE 1805.3.1. See below.

❖ This figure illustrates required setbacks at slopes. The setback is a function of the height of the slope (see commentary, Sections 1805.3.1 and 1805.3.2).

1805.3.2 Footing setback from descending slope surface. Footings on or adjacent to slope surfaces shall be founded in firm material with an embedment and set back from the slope surface sufficient to provide vertical and lateral support for the footing without detrimental settlement. Except as provided for in Section 1805.3.5 and Figure 1805.3.1, the following setback is deemed adequate to meet the criteria. Where the slope is steeper than 1 unit vertical in 1 unit horizontal (100-percent slope), the required setback shall be measured from an imaginary plane 45 degrees (0.79 rad) to the horizontal, projected upward from the toe of the slope.

❖ The provisions of this section restrict the placement of footings at the top of slopes so that vertical and lateral

For SI: 1 foot = 304.8 mm.

FIGURE 1805.3.1
FOUNDATION CLEARANCES FROM SLOPES

For SI: 1 degree = 0.01745 rad.

Figure 1805.3.1(1)
BUILDINGS ADJACENT TO ASCENDING SLOPE EXCEEDING 1 TO 1

support for the footing is provided. This setback is shown in Figure 1805.3.1. When the slope is greater than one unit vertical in one unit horizontal, the setback is measured from a top of the slope that is established as the intersection of the ground surface and an imaginary plane at 45 degrees (0.79 rad) to the horizontal, projected from the toe of the slope. This requirement is illustrated in Figure 1805.3.2. For most conditions, the required setbacks will provide adequate lateral support for the foundations.

1805.3.3 Pools. The setback between pools regulated by this code and slopes shall be equal to one-half the building footing setback distance required by this section. That portion of the pool wall within a horizontal distance of 7 feet (2134 mm) from the top of the slope shall be capable of supporting the water in the pool without soil support.

❖ This section specifies the required setback distances for pools located near ascending or descending slopes. The minimum setback distance is established as one-half the required building setback from Section 1805.3.2. The pool wall that is within 7 feet (2134 mm) of the top slope is required to be self-supporting without support from the soil and is intended to provide additional safety measures should localized minor sliding and sloughing occur.

1805.3.4 Foundation elevation. On graded sites, the top of any exterior foundation shall extend above the elevation of the street gutter at point of discharge or the inlet of an approved drainage device a minimum of 12 inches (305 mm) plus 2 percent. Alternate elevations are permitted subject to the approval of the building official, provided it can be demonstrated that required drainage to the point of discharge and away from the structure is provided at all locations on the site.

❖ Figure 1805.3.4 depicts the requirements of this section regarding elevation for exterior foundations with respect to the street, gutter or point of inlet of a drain-

For SI: 1 degree = 0.01745 rad.

Figure 1805.3.2
BUILDINGS ADJACENT TO DESCENDING SLOPE EXCEEDING 1 TO 1

For SI: 1 inch = 25.4 mm, 1 foot = 304.8 mm.

Figure 1805.3.4
FOOTING ELEVATION ON GRADED SITES

age device. The elevation of the street or gutter shown is that point at which drainage from the site reaches the street or gutter.

This requirement is intended to protect the building from water encroachment in case of heavy or unprecedented rain and may be modified on the approval of the building official if he or she finds that positive drainage slopes are provided to drain water away from the building and that the drainage pattern is not subject to temporary flooding due to landscaping or other impediments to drainage. This section is related more directly to the provisions for site grading in Section 1803.3 than it is to Section 1805.3 requirements for footings adjacent to slopes.

1805.3.5 Alternate setback and clearance. Alternate setbacks and clearances are permitted, subject to the approval of the building official. The building official is permitted to require an investigation and recommendation of a registered design professional to demonstrate that the intent of this section has been satisfied. Such an investigation shall include consideration of material, height of slope, slope gradient, load intensity and erosion characteristics of slope material.

❖ This section provides the building official the authority to approve alternative setbacks and clearances from slopes, provided he or she is satisfied that the intent of this section has been met. The building official has the authority to require a foundation investigation by a registered design professional to show that the intent of the code has been met. This item also specifies the parameters that must be considered by the registered design professional in the investigation.

1805.4 Footings. Footings shall be designed and constructed in accordance with Sections 1805.4.1 through 1805.4.6.

❖ Sections 1805.4.1 through 1805.4.6 pertain to the design and construction of concrete, masonry-unit, steel grillage and timber footings.

1805.4.1 Design. Footings shall be so designed that the allowable bearing capacity of the soil is not exceeded, and that differential settlement is minimized. The minimum width of footings shall be 12 inches (305 mm).

Footings in areas with expansive soils shall be designed in accordance with the provisions of Section 1805.8.

❖ Regardless of the type of shallow foundation used, the allowable bearing capacity of soil must not be exceeded. There are two premises by which allowable soil-bearing pressures are established. The first premise requires that the safety factor against ultimate shear failure of the soil be adequate. The second premise requires that settlements under allowable bearing pressures not exceed tolerable values. In most cases, settlement governs the value established for allowable soil-bearing capacity. Bearing capacity is usually determined from a soils investigation and engineering analysis.

When the soils profile of a construction site is established by a sufficient number of test borings, and it indicates that a nonuniform soil condition exists where the strata of suitable bearing materials occurs at varying thicknesses or different depths, the foundation design must be adjusted to the subsurface condition to provide for the proper and safe performance of the foundation system.

Under such circumstances, it becomes necessary in the design of shallow foundations to determine the different depths at which isolated or continuous stepped footings need to be placed in order to obtain equal bearing pressures and avoid serious structural damage caused by differential (unequal) settlement of the different parts of the foundation system.

Another design method for obtaining equal bearing pressures and keeping the footings at a common elevation is to size the footings in accordance with the allowable bearing capacity of the soil at each location, thus producing a balanced design of the foundation system and preventing differential settlement.

Section 1805.8 stipulates required methods and design of foundations on expansive soils.

1805.4.1.1 Design loads. Footings shall be designed for the most unfavorable effects due to the combinations of loads specified in Section 1605.2 or 1605.3. The dead load is permitted to include the weight of foundations, footings and overlying fill. Reduced live loads, as specified in Sections 1607.9 and 1607.11, are permitted to be used in the design of footings.

❖ Foundations are permitted to be designed using either ASD or strength design. The appropriate load combinations are investigated to determine the most severe structural effects. Live load reductions permitted in Section 1607 apply equally to the foundation design loads. This section also clarifies what portions of the foundation construction are considered dead loads. See the commentary for Section 1606 for discussion of dead load estimates.

1805.4.1.2 Vibratory loads. Where machinery operations or other vibrations are transmitted through the foundation, consideration shall be given in the footing design to prevent detrimental disturbances of the soil.

❖ Vibrations emanating from machinery operations and transmitted to the soil through the foundation may cause serious settlement to occur, particularly to foundations bearing on granular materials. While granular soils generally have a considerable volume of voids, the foundation pressure is usually carried and distributed by the bearing of grain on grain without detrimental deformations. Granular materials subjected to strong vibrations, however, may result in the particles readjusting and slipping into the void spaces. Essentially, the soil mass is consolidated and reduced in volume, causing vertical settlement.

Vibratory loads that cause a disturbance of the soil will flush water out of the material. In saturated granular soils, such as sand, the flushing action may cause a "quick" condition that allows sudden flow beneath the foundation, sometimes resulting in serious structural damage.

In soils that are rather impermeable, such as clays, vibration may also cause water to flush out, but it will take a very long time. Also, foundation settlement will occur over a prolonged period.

Care should be taken in the design of foundations to eliminate completely or minimize the transmission of vibratory loads to load-bearing soils.

1805.4.2 Concrete footings. The design, materials and construction of concrete footings shall comply with Sections 1805.4.2.1 through 1805.4.2.6 and the provisions of Chapter 19.

Exception: Where a specific design is not provided, concrete footings supporting walls of light-frame construction are permitted to be designed in accordance with Table 1805.4.2.

❖ The design, construction and materials used for concrete footings are to comply with Chapter 19 and ACI 318, with the applicable requirements of Sections 1805.4.2.1 through 1805.4.2.6.

The exception applies to footings supporting walls of light-frame construction when a specific footing design is not provided. The purpose of Table 1805.4.2 is to specify footing sizes that can be used to safely support walls of light-frame construction. The table is based on anticipated loads on foundations due to wall, floor and roof systems.

TABLE 1805.4.2. See below.

❖ Table 1805.4.2 provides prescriptive designs for concrete and masonry-unit footings supporting walls of light-frame construction when a specific design is not provided. The first column of the table is specifically intended to apply to the number of floors supported by the footing, not the number of stories of a building. This table provides for the minimum width and thickness of footings.

1805.4.2.1 Concrete strength. Concrete in footings shall have a specified compressive strength (f'_c) of not less than 2,500 pounds per square inch (psi) (17 237 kPa) at 28 days.

❖ This section of the code requires a minimum concrete compressive strength for footings of 2,500 psi (17 237 kPa) at 28 days. The code's intent is that concrete used as structural material for footing construction will be of sufficient strength and durability to satisfy safety requirements.

1805.4.2.2 Footing seismic ties. Where a structure is assigned to Seismic Design Category D, E or F in accordance with Section 1613, individual spread footings founded on soil defined in Section 1613.5.2 as Site Class E or F shall be interconnected by ties. Ties shall be capable of carrying, in tension or compression, a force equal to the product of the larger footing load times the seismic coefficient, S_{DS}, divided by 10 unless it is demonstrated that equivalent restraint is provided by reinforced concrete beams within slabs on grade or reinforced concrete slabs on grade.

❖ This section requires that spread footings on soft soil profiles be interconnected by ties when they support structures assigned to Seismic Design Category D and higher. The purpose of this section is to preclude excessive movement of one column or wall with respect to another (differential settlement). One of the prerequisites of adequate structural performance during an earthquake is that the foundation of the structure acts as a unit. This is typically accomplished by tying together the pile caps, piers or piles with ties capable of carrying, in tension or compression, a force equal to 10 percent of the larger pile cap or column load multiplied by S_{DS}. S_{DS} is the design spectral response acceleration at short periods as determined in Section 1613.5.4. This can be accomplished through the use of concrete floor slabs or tie beams. The differential movement of the foundation should not exceed that included in the design of the seismic-force-resisting system.

1805.4.2.3 Plain concrete footings. The edge thickness of plain concrete footings supporting walls of other than light-frame construction shall not be less than 8 inches (203 mm) where placed on soil.

Exception: For plain concrete footings supporting Group R-3 occupancies, the edge thickness is permitted to be 6

TABLE 1805.4.2
FOOTINGS SUPPORTING WALLS OF LIGHT-FRAME CONSTRUCTION[a, b, c, d, e]

NUMBER OF FLOORS SUPPORTED BY THE FOOTING[f]	WIDTH OF FOOTING (inches)	THICKNESS OF FOOTING (inches)
1	12	6
2	15	6
3	18	8[g]

For SI: 1 inch = 25.4 mm, 1 foot = 304.8 mm.

a. Depth of footings shall be in accordance with Section 1805.2.

b. The ground under the floor is permitted to be excavated to the elevation of the top of the footing.

c. Interior-stud-bearing walls are permitted to be supported by isolated footings. The footing width and length shall be twice the width shown in this table, and footings shall be spaced not more than 6 feet on center.

d. See Section 1908 for additional requirements for footings of structures assigned to Seismic Design Category C, D, E or F.

e. For thickness of foundation walls, see Section 1805.5.

f. Footings are permitted to support a roof in addition to the stipulated number of floors. Footings supporting roof only shall be as required for supporting one floor.

g. Plain concrete footings for Group R-3 occupancies are permitted to be 6 inches thick.

inches (152 mm), provided that the footing does not extend beyond a distance greater than the thickness of the footing on either side of the supported wall.

❖ An isolated spread footing of square or rectangular shape is the most common type of shallow foundation used in building construction. Basically, the footing is a slab of concrete supporting a column or other concentrated load. The footing thickness or depth is generally a function of shear strength and is established by the more severe of two structural conditions. The thickness of a footing determined by one-way shear action may be compared to the shear in a beam where the vertical shear plane extends across the entire structural section. Thickness established by two-way shear action (punching shear) is based on the consideration that the column (or column pedestal) tends to punch through the footing and failure occurs as a fracture in the form of a truncated pyramid concentrated under the load (see Figure 1805.4.2.3). Foundation designs will normally result in concrete footing thicknesses that are greater than the minimum thicknesses prescribed in the code. The minimum thickness requirements for plain concrete footings are intended to provide adequate construction based on experience.

In compliance with the requirements of ACI 318, the edge thickness for plain concrete footings is not to be less than 8 inches (203 mm). In computing footing stresses (flexure, combined flexure, axial load and shear), however, the overall depth is to be 2 inches (51 mm) less than the actual thickness of the footing for all foundations on soil (see ACI 318 for design of plain concrete members) to allow for the irregularities of excavated earth surfaces and for possible contamination of the concrete by the soil in contact with the construction.

The exception permits the edge thickness of plain concrete footings to be reduced to 6 inches (152 mm) for occupancies in Group R-3 of light-frame construction where the footing does not extend beyond a distance greater than the thickness of the footing on either side of the supported wall. For lightweight construction with the dimensional limitations given, shear stresses in the concrete will not usually govern the design thicknesses of such foundations. Plain concrete is not to be used for footings on piles.

1805.4.2.4 Placement of concrete. Concrete footings shall not be placed through water unless a tremie or other method approved by the building official is used. Where placed under or in the presence of water, the concrete shall be deposited by approved means to ensure minimum segregation of the mix and negligible turbulence of the water.

❖ Placing concrete under water should be avoided wherever possible. The risk of segregation of the concrete mixture is much greater when depositing under water as opposed to air; however, when concrete must be placed under water, it should be done by any one of several accepted methods used for such construction, including tremie. Due care must be exercised in the concreting operations so as to avoid or minimize seg-

Figure 1805.4.2.3
PUNCHING SHEAR
(EXAGGERATED ILLUSTRATION)

regation of the mix and turbulence of the water.

Generally, when concrete is to be placed under water, the mixture should be proportioned to provide a good plastic mix and high workability so that it will flow without segregation. The slump of the concrete should be 5 inches (127 mm) or greater. This desired consistency can be obtained by the use of rounded aggregates, a higher percentage of fines and entrained air. Cement content should be increased by 10 to 15 percent above the quantities required for similar mixtures placed in air to compensate for increases in water-cement ratios (see Chapter 19). In no case should the cement content be less than 600 pounds per cubic yard (355 kg/m³) of concrete.

1805.4.2.5 Protection of concrete. Concrete footings shall be protected from freezing during depositing and for a period of not less than five days thereafter. Water shall not be allowed to flow through the deposited concrete.

❖ Concrete for footings should not be placed during rain, sleet or snow or in freezing weather unless adequate protection, as approved by the building official, is provided. Such protection, when required, is to be provided during the concreting operations and for a period of not less than five days thereafter. Rainwater or water from other sources must not be allowed to flow through freshly deposited concrete so as to increase the mixing water content or to damage the surface finish. For detailed information on materials, methods and procedures used in cold-weather operations, refer to ACI 306R, ACI 306.1 and Chapter 19.

1805.4.2.6 Forming of concrete. Concrete footings are permitted to be cast against the earth where, in the opinion of the building official, soil conditions do not require forming. Where forming is required, it shall be in accordance with Chapter 6 of ACI 318.

❖ Earth cuts are not normally used to form foundation construction, except for footings. Soil used to form the vertical sides of footings is to have sufficient stiffness to maintain the desired shape and dimensions before and during concreting operations. In the event that the soil is deemed to be unstable for such purpose, the building official is to require that formwork be built in accordance with the provisions of ACI 318.

1805.4.3 Masonry-unit footings. The design, materials and construction of masonry-unit footings shall comply with Sections 1805.4.3.1 and 1805.4.3.2, and the provisions of Chapter 21.

> **Exception:** Where a specific design is not provided, masonry-unit footings supporting walls of light-frame construction are permitted to be designed in accordance with Table 1805.4.2.

❖ Sections 1805.4.3.1 and 1805.4.3.2 provide general requirements for the construction of solid masonry footings. This type of foundation was widely used until the middle of the last century when it was replaced by steel or wood grillage and eventually by more economical plain and reinforced concrete foundations. At that time, masonry foundations were built of either stone cut to specific sizes or of rubble stone of random sizes bonded together with mortar. Masonry footings under columns were usually constructed as high piers in the shape of truncated pyramids. While stone footings may still exist in very old structures, they are extinct in modern construction. Today, although scarcely used, masonry footings may be constructed of hard-burned brick in cement mortar to support the walls of lightweight construction, such as for low residential buildings. Brick footings are usually set on a full bed of mortar spread upon the earth or a grout sill. Stepped footing courses are recommended to be built with the brick units laid on edge, which are capable of resisting a greater transverse stress than flat courses.

The exception applies to footings supporting walls of light-frame construction when a specific footing design is not provided. The purpose of Table 1805.4.2 is to specify footing sizes and depths that can be used to safely support walls of light-frame construction. The table is based on anticipated loads on foundations due to wall, floor and roof systems.

1805.4.3.1 Dimensions. Masonry-unit footings shall be laid in Type M or S mortar complying with Section 2103.8 and the depth shall not be less than twice the projection beyond the wall, pier or column. The width shall not be less than 8 inches (203 mm) wider than the wall supported thereon.

❖ Footings made of masonry units are to be laid in Type M or S mortar complying with the requirements of Section 2103.7. Type M is a high-strength mortar suitable for general use and, in particular, where maximum ma-

sonry compressive strength is required. It is also suitable for unreinforced masonry below grade and in construction that is in contact with earth. Type S is a general-use mortar where high lateral strength of masonry is required. It is specifically recommended for use in reinforced masonry.

Projections of footings beyond a wall, pier or column base are not to exceed dimensions by more than one-half the depth of the foundation. For example, a footing supporting a wall and consisting of three courses of brick with a depth of about 8 inches (203 mm) is not to project more than 4 inches (102 mm) ($^{1}/_{2}$ by 8 inches) beyond the base of the wall. This is to keep the shearing stresses in the footing within a safe limitation.

Masonry footings are not to be less than 8 inches (203 mm) wider than the thickness of the supported wall so as to provide adequate distribution of the wall load to the load-bearing soil.

1805.4.3.2 Offsets. The maximum offset of each course in brick foundation walls stepped up from the footings shall be 1.5 inches (38 mm) where laid in single courses, and 3 inches (76 mm) where laid in double courses.

❖ Foundation walls that entail stepping back successive courses of brick (called racking) are not to have horizontal offsets exceeding $1^{1}/_{2}$ inches (38 mm) from the face of the course below. If the steps are made in double course increments, then the offsets must not exceed 3 inches (76 mm).

Sometimes, wide footings are required so as not to exceed the safe load-bearing capacity of the supporting soil. Since footing projections beyond a wall are limited in dimension (see Section 1805.3.1), the wall must inevitably be stepped out to provide the proper base on the footing (see Figure 1805.4.3.2).

1805.4.4 Steel grillage footings. Grillage footings of structural steel shapes shall be separated with approved steel spacers and be entirely encased in concrete with at least 6 inches (152 mm) on the bottom and at least 4 inches (102 mm) at all other points. The spaces between the shapes shall be completely filled with concrete or cement grout.

❖ Steel grillage footings were extensively used during the latter part of the last century, but the development and use of reinforced concrete foundations have made this type of construction all but obsolete. They are, however, still used for underpinning purposes. There are many steel grillage footings in existence in old buildings.

A typical grillage footing consists of two or more tiers of steel beams (usually I sections) with each tier placed at right angles to the one below it. The beams in each tier are usually held together by a system of bolts and pipe spacers. The beams should be clean and unpainted and the whole system completely filled and encased in concrete with at least 6 inches (152 mm) of cover on the bottom and 4 inches (102 mm) at all other points. In lieu of concrete (other than the encasement), the spaces between the steel beams may be filled with cement grout.

For SI: 1 inch = 25.4 mm.

Figure 1805.4.3.2
BRICK FOOTING OFFSET

1805.4.5 Timber footings. Timber footings are permitted for buildings of Type V construction and as otherwise approved by the building official. Such footings shall be treated in accordance with AWPA U1 (Commodity Specification A, Use Category 4B). Treated timbers are not required where placed entirely below permanent water level or where used as capping for wood piles that project above the water level over submerged or marsh lands. The compressive stresses perpendicular to the grain in untreated timber footings supported upon treated piles shall not exceed 70 percent of the allowable stresses for the species and grade of timber as specified in the AF&PA NDS.

❖ The use of timber footings is allowed only in Type V construction, or as otherwise approved by the building official. Such footings are commonly built as grillages of heavy timbers, using large sections, such as railroad ties.

If the footings are to be constructed at depths above the water table, they are to be given a preservative treatment by pressure processes to protect the materials from decay, fungi and harmful insects. Pressure treatment is to be in accordance with the requirements of the U1 standard of the American-Wood Preservers' Association (AWPA). AWPA U1, Specification A, Use Category 4B refers to sawn wood products that are used in contact with the ground. Except for use on cut surfaces, brush or spray applications of preservatives are not acceptable methods of treatment.

Preservative treatment by the pressure process within the limitations specified in AWPA U1 should not significantly affect the strength of the wood. Part of the process, however, involves the conditioning of timbers before treatment. Timbers conditioned by steaming or boiling under vacuum can suffer significant reductions in strength. This condition is recognized in the American Forest and Paper Association National Design Standard (AF&PA NDS) by the use of the untreated factor C_u.

The design values for treated round timber piles given in this specification are adjusted to compensate for strength reductions because of conditioning prior to treatment. It is recommended that the values given for the several species of wood be used in the design of treated timber footings. Design values given in other tables contained in the AF&PA referenced standard are for general structural purposes using untreated lumber. While the section of this specification for pressure-preservative treatment stipulates that the design values apply to wood products that are pressure impregnated by an approved process, it is not apparent that reductions have been made to allow for possible strength loss in the wood because of the treatment process.

Untreated timber may be used when the footings are completely embedded in soil below the water level. Long experience has shown that timber foundations permanently confined in water will stay sound and durable indefinitely. Wood submerged in fresh water cannot decay because the necessary air is excluded. It is not uncommon, however, to have changing ground-water levels because of changes in adjacent drainage systems and a variety of other subterranean conditions. Precautions should be taken so that untreated timber footings are placed at depths sufficiently below the ground-water level such that small drops in the water level will not expose the footings to air. Otherwise, decay will set in, causing settlements and possible structural failures.

The code also prescribes that compressive stresses perpendicular to grain in untreated timber footings supported upon piles are not to exceed 70 percent of the allowable stresses specified for the species and grade of timber given in the AF&PA referenced standard.

1805.4.6 Wood foundations. Wood foundation systems shall be designed and installed in accordance with AF&PA Technical Report No. 7. Lumber and plywood shall be treated in accordance with AWPA U1 (Commodity Specification A, Use

Category 4B and Section 5.2) and shall be identified in accordance with Section 2303.1.8.1.

❖ This section covers the design and installation of a multicomponent wood foundation system complying with the specifications of AF&PA Technical Report No. 7. The construction is essentially a below-grade, load-bearing, wood-frame system that serves as the enclosure for basements and crawl spaces and also as the structural foundation for the support of light-frame structures.

This foundation system consists of several principal components:

1. Walls that consist of plywood panels attached to a wood stud framing system;

2. A composite footing consisting of a continuous wood plate set on a bed of granular materials, such as sand, gravel or crushed stone, which supports the foundation walls and transmits their loads to the bearing soil below;

3. Polyethylene film that serves as a vapor barrier and covers the exterior side of the plywood foundation walls from grade down to the footing plate;

4. Caulking compounds used for sealing joints in plywood walls as well as bonding agents for attaching the polyethylene film to the plywood and for sealing the film joints;

5. Metal fasteners made of silicon, bronze, copper or stainless steel or hot-dipped, zinc-coated steel nails or staples; and

6. Pressure-treated plywood and lumber to protect the foundation material against decay, termites and other insects.

A typical cross section of a wood foundation basement wall is shown in Figure 1805.4.6.

All plywood and lumber used in the foundation system are required to receive a preservative treatment

For SI: 1 inch = 25.4 mm, 1 foot = 304.8 mm.

Figure 1805.4.6
WOOD FOUNDATION

to protect the material against decay, termites and other insects in accordance with AWPA U1, Specification A, Use Category 4B and Section 5.2. Specification A of AWPA U1 pertains to sawn wood products, while Use Category 4B refers to wood that is in contact with the ground. The additional reference to Section 5.2 incorporates requirements under Specification A that are unique to permanent wood foundations. This section also requires the treated wood to be identified in accordance with Section 2303.1.8.1. See the commentary to that section for a discussion of identification requirements.

Ordinary metal fasteners in contact with preservatives used in the treatment of wood products will corrode over time and can cause serious structural problems. For this reason, AF&PA Technical Report No. 7 prescribes that only fasteners made of silicon bronze, copper or Type 304 or 316 stainless steel, as defined by the American Iron and Steel Institute (AISI), are permitted to be used in a wood foundation system. Hot-dipped, zinc-coated steel nails are permitted to be used in a wood foundation system when installed in accordance with the specific conditions contained in AF&PA Technical Report No. 7 for the surface treatment of nails and moisture protection of the foundation.

Certain cold-weather precautions should be taken in the installation of the wood foundation system. The composite footing consisting of a wood plate supported on a bed of stone or sand fill should not be placed on frozen ground. While the bottom of the wood plate footing would normally be placed below the frost line (i.e., basement construction), under certain drainage conditions, AF&PA Technical Report No. 7 permits the wood plate to be set above the frost line level. The building official should ascertain that such an alternative design satisfies the intent of the code. Most important is the use of proper sealants during very cold weather. All manufacturers of sealants and bonding agents impose temperature restrictions on the use of their products. Only sealants and bonding agents specifically produced for cold-weather conditions should be used.

Another cold-weather precaution relates to the use of sealants. All manufacturers of sealants and bonding agents place temperature limitations on their use. In very cold weather, only those sealants that are specially produced for low-temperature applications should be used for the caulking of joints and for bonding purposes. This is an important consideration since sealants are a factor in verifying the water-tight integrity of the foundation installation.

The wood foundation system is an innovative design consisting of many parts and several kinds of materials. It can be expected to perform satisfactorily as a foundation for light construction only if the comprehensive provisions and recommendations of AF&PA Technical Report No. 7 are strictly followed.

1805.5 Foundation walls. Concrete and masonry foundation walls shall be designed in accordance with Chapter 19 or 21,

respectively. Foundation walls that are laterally supported at the top and bottom and within the parameters of Tables 1805.5(1) through 1805.5(5) are permitted to be designed and constructed in accordance with Sections 1805.5.1 through 1805.5.5.

❖ This section covers foundation walls constructed of concrete or of masonry materials that are primarily intended for, but not necessarily limited to, basement construction in residential and light commercial buildings or other light structures.

Foundation walls are usually designed and constructed in accordance with the accepted engineering practices to carry the vertical loads from the structure above, resist wind and any lateral forces transmitted to the foundations and sustain earth pressures exerted against the walls, including any forces that may be imposed by frost action. For reinforced and plain concrete, the physical properties and design criteria are provided in Chapter 19. For masonry construction, the physical properties and design criteria are provided in Chapter 21.

This section also provides the option of using the design tables for plain masonry and concrete foundation walls, and the preengineered designs for reinforced masonry and concrete foundation walls. Foundation walls that are not laterally supported at the top and bottom are required to be designed by the specified standards, since the designs in the tables are only for walls that are laterally supported at the top and bottom.

1805.5.1 Foundation wall thickness. The minimum thickness of concrete and masonry foundation walls shall comply with Sections 1805.5.1.1 through 1805.5.1.3.

❖ Sections 1805.5.1.1 and 1805.5.1.2 prescribe the minimum thickness of foundation walls based on the thickness of the walls supported, and the soil loads and heights for concrete or masonry foundation walls. The walls are to comply with the minimum thicknesses required for either section. Section 1805.5.1.3 prescribes the minimum thickness of foundation walls constructed of rough or random rubble stone.

1805.5.1.1 Thickness at top of foundation wall. The thickness of foundation walls shall not be less than the thickness of the wall supported, except that foundation walls of at least 8-inch (203 mm) nominal width are permitted to support brick-veneered frame walls and 10-inch-wide (254 mm) cavity walls provided the requirements of Section 1805.5.1.2 are met. Corbeling of masonry shall be in accordance with Section 2104.2. Where an 8-inch (203 mm) wall is corbeled, the top corbel shall not extend higher than the bottom of the floor framing and shall be a full course of headers at least 6 inches (152 mm) in length or the top course bed joint shall be tied to the vertical wall projection. The tie shall be W2.8 (4.8 mm) and spaced at a maximum horizontal distance of 36 inches (914 mm); the hollow space behind the corbelled masonry shall be filled with mortar or grout.

❖ The minimum thicknesses in this section are to facilitate the support of the wall above grade. The thickness requirements in this section are empirical and have been used successfully for many years. Where corbeling is utilized to match the width of a masonry cavity wall above, see Section 2104.2 for construction requirements. A header course is provided at the corbel to transfer the load deeper into the supporting wall. An alternative to the header, based on Section 2109.6.3.1, utilizes a filled collar joint and ties to accomplish this load transfer. This option is illustrated in Figure 1805.5.1.1.

1805.5.1.2 Thickness based on soil loads, unbalanced backfill height and wall height. The thickness of foundation walls shall comply with the requirements of Table 1805.5(5) for concrete walls, Table 1805.5(1) for plain masonry walls or Table 1805.5(2), 1805.5(3) or 1805.5(4) for masonry walls with reinforcement. When using the tables, masonry shall be laid in running bond and the mortar shall be Type M or S.

Unbalanced backfill height is the difference in height between the exterior finish ground level and the lower of the top of the concrete footing that supports the foundation wall or the interior finish ground level. Where an interior concrete slab on grade is provided and is in contact with the interior surface of the foundation wall, the unbalanced backfill height is permitted to be measured from the exterior finish ground level to the top of the interior concrete slab.

❖ The thickness of masonry walls is also required to meet the minimum requirements of Tables 1805.5(1), 1805.5(2), 1805.5(3) or 1805.5(4). The thickness of concrete walls is also required to meet the minimum requirements of Tables 1805.5(5). These prescriptive foundation wall tables are based on soil type and wall height as well as the unbalanced backfill height.

As stated, unbalanced backfill height is the difference in height of the exterior and interior finished ground levels. The height of unbalanced backfill identifies the magnitude of the lateral soil load for the different classifications of soils presented in the tables. For an illustration of unbalanced backfill height where an interior slab on grade is in contact with the foundation wall, see Figure 1805.5.1.2.

TABLE 1805.5(1). See page 18-23.

❖ This table provides wall thickness designs for various wall heights and unbalanced backfill depths. The thicknesses in the table are proportional to the lateral soil loads for each classification of soils. The soil class designations are according to ASTM D 2487 (based on the Unified Soil Classification System that is commonly found in textbooks on soil mechanics). The lateral soil loads in the table are consistent with the soil loads in Table 1610.1. The description of the soils associated with each soil symbol (e.g., GW) is given in Table 1610.1.

The thicknesses in the table for plain masonry are based on research by the National Concrete Masonry Association (NCMA) in the report *Research Evaluation of the Flexural Strength of Concrete Masonry,* Project No. 93-172, December 7, 1993.

TABLE 1805.5(2). See page 18-24.

❖ For general comments regarding this table, see the commentary to Table 1805.5(1). The reinforcement size and spacing in the table are based on the design requirements from ACI 530/ASCE 5/TMS 402.

Three different tables are provided for three distinct wall thicknesses. Each table also specifies the re-

W 2.8 TIE x 6 INCHES (MIN.)
SPACED 36 INCHES (MAX.) O.C.

MORTAR
OR GROUT

ANCHOR BOLT
BED JOINT OF TOP COURSE

FOUNDATION WALL

For SI: 1 inch = 25.4 mm.

Figure 1805.5.1.1
TIE AT TOP COURSE BED JOINT

quired effective depth distance. For an explanation of the effective depth, see the commentary to Section 1805.5.2.2.

TABLE 1805.5(3). See page 18-25.

❖ See the commentary to Table 1805.5(2).

TABLE 1805.5(4). See page 18-26.

❖ See the commentary to Table 1805.5(2).

TABLE 1805.5(5). See page 18-27.

❖ For general comments regarding this table, see the commentary to Table 1805.5(1). The thickness for the

walls in the table is based on the design requirements from ACI 318. The tables also specify the required size and spacing of the reinforcement.

In a departure from the masonry foundation wall tables, the soil classes are not included. Note d points out that Section 1610.1 requires that foundation walls greater than 8 feet (2438 mm) in height be designed for at-rest lateral soil loads. Since some soil classes have at-rest design lateral loads of 60 psf per foot (2873 Pa/mm) of depth, that column of the table is applicable for both active pressures and at-rest pressures, depending upon the soil's classification (see Table 1610.1).

**Figure 1805.5.1.2
UNBALANCED BACKFILL HEIGHT**

**TABLE 1805.5(1)
PLAIN MASONRY FOUNDATION WALLS**[a, b, c]

MAXIMUM WALL HEIGHT (feet)	MAXIMUM UNBALANCED BACKFILL HEIGHT[e] (feet)	MINIMUM NOMINAL WALL THICKNESS (inches)		
		Soil classes and lateral soil load[a] (psf per foot below natural grade)		
		GW, GP, SW and SP soils 30	GM, GC, SM, SM-SC and ML soils 45	SC, ML-CL and Inorganic CL soils 60
7	4 (or less)	8	8	8
	5	8	10	10
	6	10	12	10 (solid[c])
	7	12	10 (solid[c])	10 (solid[c])
8	4 (or less)	8	8	8
	5	8	10	12
	6	10	12	12 (solid[c])
	7	12	12 (solid[c])	Note d
	8	10 (solid[c])	12 (solid[c])	Note d
9	4 (or less)	8	8	8
	5	8	10	12
	6	12	12	12 (solid[c])
	7	12 (solid[c])	12 (solid[c])	Note d
	8	12 (solid[c])	Note d	Note d
	9	Note d	Note d	Note d

For SI: 1 inch = 25.4 mm, 1 foot = 304.8 mm, 1 pound per square foot per foot = 0.157kPa/m.

a. For design lateral soil loads, see Section 1610. Soil classes are in accordance with the Unified Soil Classification System and design lateral soil loads are for moist soil conditions without hydrostatic pressure.

b. Provisions for this table are based on construction requirements specified in Section 1805.5.2.2.

c. Solid grouted hollow units or solid masonry units.

d. A design in compliance with Chapter 21 or reinforcement in accordance with Table 1805.5(2) is required.

e. For height of unbalanced backfill, see Section 1805.5.1.2.

TABLE 1805.5(2)
8-INCH MASONRY FOUNDATION WALLS WITH REINFORCEMENT WHERE d ≥ 5 INCHES[a, b, c]

MAXIMUM WALL HEIGHT (feet-inches)	MAXIMUM UNBALANCED BACKFILL HEIGHT[d] (feet-inches)	VERTICAL REINFORCEMENT		
		Soil classes and lateral soil load[a] (psf per foot below natural grade)		
		GW, GP, SW and SP soils 30	GM, GC, SM, SM-SC and ML soils 45	SC, ML-CL and Inorganic CL soils 60
7-4	4-0 (or less)	#4 at 48″ o.c.	#4 at 48″ o.c.	#4 at 48″ o.c.
	5-0	#4 at 48″ o.c.	#4 at 48″ o.c.	#4 at 48″ o.c.
	6-0	#4 at 48″ o.c.	#5 at 48″ o.c.	#5 at 48″ o.c.
	7-4	#5 at 48″ o.c.	#6 at 48″ o.c.	#7 at 48″ o.c.
8-0	4-0 (or less)	#4 at 48″ o.c.	#4 at 48″ o.c.	#4 at 48″ o.c.
	5-0	#4 at 48″ o.c.	#4 at 48″ o.c.	#4 at 48″ o.c.
	6-0	#4 at 48″ o.c.	#5 at 48″ o.c.	#5 at 48″ o.c.
	7-0	#5 at 48″ o.c.	#6 at 48″ o.c.	#7 at 48″ o.c.
	8-0	#5 at 48″ o.c.	#6 at 48″ o.c.	#7 at 48″ o.c.
8-8	4-0 (or less)	#4 at 48″ o.c.	#4 at 48″ o.c.	#4 at 48″ o.c.
	5-0	#4 at 48″ o.c.	#4 at 48″ o.c.	#5 at 48″ o.c.
	6-0	#4 at 48″ o.c.	#5 at 48″ o.c.	#6 at 48″ o.c.
	7-0	#5 at 48″ o.c.	#6 at 48″ o.c.	#7 at 48″ o.c.
	8-8	#6 at 48″ o.c.	#7 at 48″ o.c.	#8 at 48″ o.c.
9-4	4-0 (or less)	#4 at 48″ o.c.	#4 at 48″ o.c.	#4 at 48″ o.c.
	5-0	#4 at 48″ o.c.	#4 at 48″ o.c.	#5 at 48″ o.c.
	6-0	#4 at 48″ o.c.	#5 at 48″ o.c.	#6 at 48″ o.c.
	7-0	#5 at 48″ o.c.	#6 at 48″ o.c.	#7 at 48″ o.c.
	8-0	#6 at 48″ o.c.	#7 at 48″ o.c.	#8 at 48″ o.c.
	9-4	#7 at 48″ o.c.	#8 at 48″ o.c.	#9 at 48″ o.c.
10-0	4-0 (or less)	#4 at 48″ o.c.	#4 at 48″ o.c.	#4 at 48″ o.c.
	5-0	#4 at 48″ o.c.	#4 at 48″ o.c.	#5 at 48″ o.c.
	6-0	#4 at 48″ o.c.	#5 at 48″ o.c.	#6 at 48″ o.c.
	7-0	#5 at 48″ o.c.	#6 at 48″ o.c.	#7 at 48″ o.c.
	8-0	#6 at 48″ o.c.	#7 at 48″ o.c.	#8 at 48″ o.c.
	9-0	#7 at 48″ o.c.	#8 at 48″ o.c.	#9 at 48″ o.c.
	10-0	#7 at 48″ o.c.	#9 at 48″ o.c.	#9 at 48″ o.c.

For SI: 1 inch = 25.4 mm, 1 foot = 304.8 mm, 1 pound per square foot per foot = 0.157 kPa/m.

a. For design lateral soil loads, see Section 1610. Soil classes are in accordance with the Unified Soil Classification System and design lateral soil loads are for moist soil conditions without hydrostatic pressure.

b. Provisions for this table are based on construction requirements specified in Section 1805.5.2.2.

c. For alternative reinforcement, see Section 1805.5.3.

d. For height of unbalanced backfill, see Section 1805.5.1.2.

TABLE 1805.5(3)
10-INCH MASONRY FOUNDATION WALLS WITH REINFORCEMENT WHERE d ≥ 6.75 INCHES[a, b, c]

MAXIMUM WALL HEIGHT (feet-inches)	MAXIMUM UNBALANCED BACKFILL HEIGHT[d] (feet-inches)	VERTICAL REINFORCEMENT		
		Soil classes and lateral soil load[a] (psf per foot below natural grade)		
		GW, GP, SW and SP soils 30	GM, GC, SM, SM-SC and ML soils 45	SC, ML-CL and Inorganic CL soils 60
7-4	4-0 (or less)	#4 at 56″ o.c.	#4 at 56″ o.c.	#4 at 56″ o.c.
	5-0	#4 at 56″ o.c.	#4 at 56″ o.c.	#4 at 56″ o.c.
	6-0	#4 at 56″ o.c.	#4 at 56″ o.c.	#5 at 56″ o.c.
	7-4	#4 at 56″ o.c.	#5 at 56″ o.c.	#6 at 56″ o.c.
8-0	4-0 (or less)	#4 at 56″ o.c.	#4 at 56″ o.c.	#4 at 56″ o.c.
	5-0	#4 at 56″ o.c.	#4 at 56″ o.c.	#4 at 56″ o.c.
	6-0	#4 at 56″ o.c.	#4 at 56″ o.c.	#5 at 56″ o.c.
	7-0	#4 at 56″ o.c.	#5 at 56″ o.c.	#6 at 56″ o.c.
	8-0	#5 at 56″ o.c.	#6 at 56″ o.c.	#7 at 56″ o.c.
8-8	4-0 (or less)	#4 at 56″ o.c.	#4 at 56″ o.c.	#4 at 56″ o.c.
	5-0	#4 at 56″ o.c.	#4 at 56″ o.c.	#4 at 56″ o.c.
	6-0	#4 at 56″ o.c.	#4 at 56″ o.c.	#5 at 56″ o.c.
	7-0	#4 at 56″ o.c.	#5 at 56″ o.c.	#6 at 56″ o.c.
	8-8	#5 at 56″ o.c.	#7 at 56″ o.c.	#8 at 56″ o.c.
9-4	4-0 (or less)	#4 at 56″ o.c.	#4 at 56″ o.c.	#4 at 56″ o.c.
	5-0	#4 at 56″ o.c.	#4 at 56″ o.c.	#4 at 56″ o.c.
	6-0	#4 at 56″ o.c.	#5 at 56″ o.c.	#5 at 56″ o.c.
	7-0	#4 at 56″ o.c.	#5 at 56″ o.c.	#6 at 56″ o.c.
	8-0	#5 at 56″ o.c.	#6 at 56″ o.c.	#7 at 56″ o.c.
	9-4	#6 at 56″ o.c.	#7 at 56″ o.c.	#8 at 56″ o.c.
10-0	4-0 (or less)	#4 at 56″ o.c.	#4 at 56″ o.c.	#4 at 56″ o.c.
	5-0	#4 at 56″ o.c.	#4 at 56″ o.c.	#4 at 56″ o.c.
	6-0	#4 at 56″ o.c.	#5 at 56″ o.c.	#5 at 56″ o.c.
	7-0	#5 at 56″ o.c.	#6 at 56″ o.c.	#7 at 56″ o.c.
	8-0	#5 at 56″ o.c.	#7 at 56″ o.c.	#8 at 56″ o.c.
	9-0	#6 at 56″ o.c.	#7 at 56″ o.c.	#9 at 56″ o.c.
	10-0	#7 at 56″ o.c.	#8 at 56″ o.c.	#9 at 56″ o.c.

For SI: 1 inch = 25.4 mm, 1 foot = 304.8 mm, 1 pound per square foot per foot = 0.157kPa/m.

a. For design lateral soil loads, see Section 1610. Soil classes are in accordance with the Unified Soil Classification System and design lateral soil loads are for moist soil conditions without hydrostatic pressure.
b. Provisions for this table are based on construction requirements specified in Section 1805.5.2.2.
c. For alternative reinforcement, see Section 1805.5.3.
d. For height of unbalanced fill, see Section 1805.5.1.2.

TABLE 1805.5(4)
12-INCH MASONRY FOUNDATION WALLS WITH REINFORCEMENT WHERE d ≥ 8.75 INCHES[a, b, c]

MAXIMUM WALL HEIGHT (feet-inches)	MAXIMUM UNBALANCED BACKFILL HEIGHT[d] (feet-inches)	VERTICAL REINFORCEMENT		
		Soil classes and lateral soil load[a] (psf per foot below natural grade)		
		GW, GP, SW and SP soils 30	GM, GC, SM, SM-SC and ML soils 45	SC, ML-CL and Inorganic CL soils 60
7-4	4-0 (or less)	#4 at 72″ o.c.	#4 at 72″ o.c.	#4 at 72″ o.c.
	5-0	#4 at 72″ o.c.	#4 at 72″ o.c.	#4 at 72″ o.c.
	6-0	#4 at 72″ o.c.	#4 at 72″ o.c.	#5 at 72″ o.c.
	7-4	#4 at 72″ o.c.	#5 at 72″ o.c.	#6 at 72″ o.c.
8-0	4-0 (or less)	#4 at 72″ o.c.	#4 at 72″ o.c.	#4 at 72″ o.c.
	5-0	#4 at 72″ o.c.	#4 at 72″ o.c.	#4 at 72″ o.c.
	6-0	#4 at 72″ o.c.	#4 at 72″ o.c.	#5 at 72″ o.c.
	7-0	#4 at 72″ o.c.	#5 at 72″ o.c.	#6 at 72″ o.c.
	8-0	#5 at 72″ o.c.	#6 at 72″ o.c.	#7 at 72″ o.c.
8-8	4-0 (or less)	#4 at 72″ o.c.	#4 at 72″ o.c.	#4 at 72″ o.c.
	5-0	#4 at 72″ o.c.	#4 at 72″ o.c.	#4 at 72″ o.c.
	6-0	#4 at 72″ o.c.	#4 at 72″ o.c.	#5 at 72″ o.c.
	7-0	#4 at 72″ o.c.	#5 at 72″ o.c.	#6 at 72″ o.c.
	8-8	#5 at 72″ o.c.	#7 at 72″ o.c.	#8 at 72″ o.c.
9-4	4-0 (or less)	#4 at 72″ o.c.	#4 at 72″ o.c.	#4 at 72″ o.c.
	5-0	#4 at 72″ o.c.	#4 at 72″ o.c.	#4 at 72″ o.c.
	6-0	#4 at 72″ o.c.	#5 at 72″ o.c.	#5 at 72″ o.c.
	7-0	#4 at 72″ o.c.	#5 at 72″ o.c.	#6 at 72″ o.c.
	8-0	#5 at 72″ o.c.	#6 at 72″ o.c.	#7 at 72″ o.c.
	9-4	#6 at 72″ o.c.	#7 at 72″ o.c.	#8 at 72″ o.c.
10-0	4-0 (or less)	#4 at 72″ o.c.	#4 at 72″ o.c.	#4 at 72″ o.c.
	5-0	#4 at 72″ o.c.	#4 at 72″ o.c.	#4 at 72″ o.c.
	6-0	#4 at 72″ o.c.	#5 at 72″ o.c.	#5 at 72″ o.c.
	7-0	#4 at 72″ o.c.	#6 at 72″ o.c.	#6 at 72″ o.c.
	8-0	#5 at 72″ o.c.	#6 at 72″ o.c.	#7 at 72″ o.c.
	9-0	#6 at 72″ o.c.	#7 at 72″ o.c.	#8 at 72″ o.c.
	10-0	#7 at 72″ o.c.	#8 at 72″ o.c.	#9 at 72″ o.c.

For SI: 1 inch = 25.4 mm, 1 foot = 304.8 mm, 1 pound per square foot per foot = 0.157kPa/m.

a. For design lateral soil loads, see Section 1610. Soil classes are in accordance with the Unified Soil Classification System and design lateral soil loads are for moist soil conditions without hydrostatic pressure.

b. Provisions for this table are based on construction requirements specified in Section 1805.5.2.2.

c. For alternative reinforcement, see Section 1805.5.3.

d. For height of unbalanced backfill, see Section 1805.5.1.2.

TABLE 1805.5(5)
CONCRETE FOUNDATION WALLS[b, c]

MAXIMUM WALL HEIGHT (feet)	MAXIMUM UNBALANCED BACKFILL HEIGHT[e] (feet)	VERTICAL REINFORCEMENT AND SPACING (inches)								
		Design lateral soil load[a] (psf per foot of depth)								
		30			45			60		
		Minimum wall thickness (inches)								
		7.5	9.5	11.5	7.5	9.5	11.5	7.5	9.5	11.5
5	4	PC	PC	PC	PC	PC	PC	PC	PC	PC
	5	PC	PC	PC	PC	PC	PC	PC	PC	PC
6	4	PC	PC	PC	PC	PC	PC	PC	PC	PC
	5	PC	PC	PC	PC	PC	PC	PC	PC	PC
	6	PC	PC	PC	PC	PC	PC	PC	PC	PC
7	4	PC	PC	PC	PC	PC	PC	PC	PC	PC
	5	PC	PC	PC	PC	PC	PC	PC	PC	PC
	6	PC	PC	PC	PC	PC	PC	#5 at 48″	PC	PC
	7	PC	PC	PC	#5 at 46″	PC	PC	#6 at 48″	PC	PC
8	4	PC	PC	PC	PC	PC	PC	PC	PC	PC
	5	PC	PC	PC	PC	PC	PC	PC	PC	PC
	6	PC	PC	PC	PC	PC	PC	#5 at 43″	PC	PC
	7	PC	PC	PC	#5 at 41″	PC	PC	#6 at 43″	PC	PC
	8	#5 at 47″	PC	PC	#6 at 43″	PC	PC	#6 at 32″	#6 at 44″	PC
9	4	PC	PC	PC	PC	PC	PC	PC	PC	PC
	5	PC	PC	PC	PC	PC	PC	PC	PC	PC
	6	PC	PC	PC	PC	PC	PC	#5 at 39″	PC	PC
	7	PC	PC	PC	#5 at 37″	PC	PC	#6 at 38″	#5 at 37″	PC
	8	#5 at 41″	PC	PC	#6 at 38″	#5 at 37″	PC	#7 at 39″	#6 at 39″	#4 at 48″
	9[d]	#6 at 46″	PC	PC	#7 at 41″	#6 at 41″	PC	#7 at 31″	#7 at 41″	#6 at 39″
10	4	PC	PC	PC	PC	PC	PC	PC	PC	PC
	5	PC	PC	PC	PC	PC	PC	PC	PC	PC
	6	PC	PC	PC	PC	PC	PC	#5 at 37″	PC	PC
	7	PC	PC	PC	#6 at 48″	PC	PC	#6 at 35″	#6 at 48″	PC
	8	#5 at 38″	PC	PC	#7 at 47″	#6 at 47″	PC	#7 at 35″	#7 at 48″	#6 at 45″
	9[d]	#6 at 41″	#4 at 48″	PC	#7 at 37″	#7 at 48″	#4 at 48″	#6 at 22″	#7 at 37″	#7 at 47″
	10[d]	#7 at 45″	#6 at 45″	PC	#7 at 31″	#7 at 40″	#6 at 38″	#6 at 22″	#7 at 30″	#7 at 38″

For SI: 1 inch = 25.4 mm, 1 foot = 304.8 mm, 1 pound per square foot = 0.157 kPa/m.
a. For design lateral soil loads for different classes of soil, see Section 1610.
b. Provisions for this table are based on construction requirements specified in Section 1805.5.2.1.
c. "PC" means plain concrete.
d. Where design lateral soil loads from Table 1610.1 are used, the requirements for 30 and 45 psf per foot of depth are not applicable. See Section 1610.
e. For height of unbalanced backfill, see Section 1805.5.1.2.

1805.5.1.3 Rubble stone. Foundation walls of rough or random rubble stone shall not be less than 16 inches (406 mm) thick. Rubble stone shall not be used for foundations for structures in Seismic Design Category C, D, E or F.

❖ Foundation walls of rough or random rubble stone are limited to a thickness of 16 inches (406 mm). Because of the use of rough stones of irregular size and shape, a larger thickness is required (as compared to concrete or masonry walls) for adequate bonding of the stone and mortar to provide structural stability against soil pressures, as well as protection from water penetration.

1805.5.2 Foundation wall materials. Concrete foundation walls constructed in accordance with Table 1805.5(5) shall comply with Section 1805.5.2.1. Masonry foundation walls constructed in accordance with Table 1805.5(1), 1805.5(2), 1805.5(3) or 1805.5(4) shall comply with Section 1805.5.2.2.

❖ This section specifies minimum requirements for materials and construction of foundation walls constructed in accordance with Tables 1805.5(1) through 1805.5(5). Additional guidance based on a structure's seismic design category is given in Section 1805.5.5.

1805.5.2.1 Concrete foundation walls. Concrete foundation walls shall comply with the following:

1. The size and spacing of vertical reinforcement shown in Table 1805.5(5) is based on the use of reinforcement with a minimum yield strength of 60,000 psi (414 MPa). Vertical reinforcement with a minimum yield strength of 40,000 psi (276 MPa) or 50,000 psi (345 MPa) is permitted, provided the same size bar is used and the spacing shown in the table is reduced by multiplying the spacing by 0.67 or 0.83, respectively.

2. Vertical reinforcement, when required, shall be placed nearest the inside face of the wall a distance, $d,$ from the outside face (soil side) of the wall. The distance, $d,$ is equal to the wall thickness, $t,$ minus 1.25 inches (32 mm) plus one-half the bar diameter, d_b [$d = t - (1.25 + d_b/2)$]. The reinforcement shall be placed within a tolerance of $\pm \frac{3}{8}$ inch (9.5 mm) where d is less than or equal to 8 inches (203 mm) or $\pm \frac{1}{2}$ inch (2.7 mm) where d is greater than 8 inches (203 mm).

3. In lieu of the reinforcement shown in Table 1805.5(5), smaller reinforcing bar sizes with closer spacings that provide an equivalent cross-sectional area of reinforcement per unit length of wall are permitted.

4. Concrete cover for reinforcement measured from the inside face of the wall shall not be less than $\frac{3}{4}$ inch (19.1 mm). Concrete cover for reinforcement measured from the outside face of the wall shall not be less than 1.5 inches (38 mm) for No. 5 bars and smaller and not less than 2 inches (51 mm) for larger bars.

5. Concrete shall have a specified compressive strength, f_c', of not less than 2,500 psi (17.2 MPa) at 28 days.

6. The unfactored axial load per linear foot of wall shall not exceed 1.2 tf_c', where t is the specified wall thickness in inches.

❖ The section contains the construction details and material properties needed to utilize the prescriptive designs of Table 1805.5(5) for concrete foundation walls.

1805.5.2.2 Masonry foundation walls. Masonry foundation walls shall comply with the following:

1. Vertical reinforcement shall have a minimum yield strength of 60,000 psi (414 MPa).

2. The specified location of the reinforcement shall equal or exceed the effective depth distance, d, noted in Tables 1805.5(2), 1805.5(3) and 1805.5(4) and shall be measured from the face of the exterior (soil) side of the wall to the center of the vertical reinforcement. The reinforcement shall be placed within the tolerances specified in ACI 530.1/ASCE 6/TMS 402, Article 3.4 B7 of the specified location.

3. Grout shall comply with Section 2103.12.

4. Concrete masonry units shall comply with ASTM C 90.

5. Clay masonry units shall comply with ASTM C 652 for hollow brick, except compliance with ASTM C 62 or ASTM C 216 is permitted when solid masonry units are installed in accordance with Table 1805.5(1) for plain masonry.

6. Masonry units shall be installed with Type M or S mortar in accordance with Section 2103.8.

7. The unfactored axial load per linear foot of wall shall not exceed 1.2 tf_m' where t is the specified well thickness in inches and f_m' is the specified compressive strength of masonry in pounds per square inch.

❖ This section contains the construction details and material properties needed to utilize the prescriptive designs of Tables 1805.5(1) through 1805.5(4) for masonry foundation walls.

Item 2 provides specific placement requirements regarding the location of reinforcement in connection with Tables 1805.5(2) through 1805.5(4). This dimension is necessary so that the wall is capable of developing the required bending capacity to resist the lateral soil loads.

1805.5.3 Alternative foundation wall reinforcement. In lieu of the reinforcement provisions for masonry foundation walls in Table 1805.5(2), 1805.5(3) or 1805.5(4), alternative reinforcing bar sizes and spacings having an equivalent cross-sectional area of reinforcement per linear foot (mm) of wall are permitted to be used, provided the spacing of reinforcement does not exceed 72 inches (1829 mm) and reinforcing bar sizes do not exceed No. 11.

❖ The purpose of this section is to provide an alternative for the reinforcement requirements in Tables 1805.5(2), 1805.5(3) and 1805.5(4) that would result in the same strength for lateral soil load resistance. For example, if No. 5 reinforcement bars were installed where the table indicates No. 4 at 48 inches (1219 mm) on center (o.c.), the spacing of the No. 5 bars for equivalent reinforcement area per foot of wall would be 72 inches (1829 mm) o.c. The increase in spacing would be allowed, since the cross-sectional area of a No. 5 bar is approximately 50 percent higher than a No. 4 reinforcement bar.

1805.5.4 Hollow masonry walls. At least 4 inches (102 mm) of solid masonry shall be provided at girder supports at the top of hollow masonry unit foundation walls.

❖ Foundation walls made of hollow load-bearing masonry units complying with the requirements of ASTM C 90 are to be reinforced with at least 4 inches (102 mm) of solid masonry at all locations where girders bear on masonry. This provides reinforcement against excessive stresses that may be induced in the wall through the support of heavy concentrated loads.

1805.5.5 Seismic requirements. Tables 1805.5(1) through 1805.5(5) shall be subject to the following limitations in Sections 1805.5.5.1 and 1805.5.5.2 based on the seismic design category assigned to the structure as defined in Section 1613.

❖ This section clarifies the application of the prescriptive foundation wall tables based on a structure's seismic design category.

1805.5.5.1 Seismic requirements for concrete foundation walls. Concrete foundation walls designed using Table 1805.5(5) shall be subject to the following limitations:

1. Seismic Design Categories A and B. No additional seismic requirements, except provide not less than two No. 5 bars around window and door openings. Such bars shall extend at least 24 inches (610 mm) beyond the corners of the openings.

2. Seismic Design Categories C, D, E and F. Tables shall not be used except as allowed for plain concrete members in Section 1908.1.15.

❖ This section applies to concrete foundation walls. For structures classified as either Seismic Design Category A or B, the prescriptive tables apply along with the supplemental reinforcement specified at openings. For all other seismic design categories, use of these prescriptive tables is limited by the referenced code section.

1805.5.5.2 Seismic requirements for masonry foundation walls. Masonry foundation walls designed using Tables 1805.5(1) through 1805.5(4) shall be subject to the following limitations:

1. Seismic Design Categories A and B. No additional seismic requirements.

2. Seismic Design Category C. A design using Tables 1805.5(1) through 1805.5(4) is subject to the seismic requirements of Section 2106.4.

3. Seismic Design Category D. A design using Tables 1805.2(2) through 1805.5(4) is subject to the seismic requirements of Section 2106.5.

4. Seismic Design Categories E and F. A design using Tables 1805.2(2) through 1805.5(4) is subject to the seismic requirements of Section 2106.6.

❖ This section applies to masonry foundation walls. For structures classified as either Seismic Design Category A or B, the prescriptive tables apply. For all other seismic design categories, use of these prescriptive tables is limited by the referenced code sections. For structures classified as Seismic Design Category D or higher, only the tables that require reinforcement can be used.

1805.5.6 Foundation wall drainage. Foundation walls shall be designed to support the weight of the full hydrostatic pressure of undrained backfill unless a drainage system is installed in accordance with Sections 1807.4.2 and 1807.4.3.

❖ This section clarifies that if a foundation drainage system is not provided, the foundation wall has to be designed to support the weight of the full hydrostatic pressure of the undrained backfill. The lateral pressures exerted from undrained backfill will be significantly higher than those exerted from properly drained backfill.

1805.5.7 Pier and curtain wall foundations. Except in Seismic Design Categories D, E and F, pier and curtain wall foundations are permitted to be used to support light-frame construction not more than two stories in height, provided the following requirements are met:

1. All load-bearing walls shall be placed on continuous concrete footings bonded integrally with the exterior wall footings.

2. The minimum actual thickness of a load-bearing masonry wall shall not be less than 4 inches (102 mm) nominal or 3.625 inches (92 mm) actual thickness, and shall be bonded integrally with piers spaced 6 feet (1829 mm) on center (o.c.).

3. Piers shall be constructed in accordance with Chapter 21 and the following:

   3.1. The unsupported height of the masonry piers shall not exceed 10 times their least dimension.

   3.2. Where structural clay tile or hollow concrete masonry units are used for piers supporting beams and girders, the cellular spaces shall be filled solidly with concrete or Type M or S mortar.

   Exception: Unfilled hollow piers are permitted where the unsupported height of the pier is not more than four times its least dimension.

   3.3. Hollow piers shall be capped with 4 inches (102 mm) of solid masonry or concrete or the cavities of the top course shall be filled with concrete or grout.

4. The maximum height of a 4-inch (102 mm) load-bearing masonry foundation wall supporting wood frame walls and floors shall not be more than 4 feet (1219 mm) in height.

5. The unbalanced fill for 4-inch (102 mm) foundation walls shall not exceed 24 inches (610 mm) for solid masonry, nor 12 inches (305 mm) for hollow masonry.

❖ Pier and curtain wall foundation systems are now and have always been a very popular method of construction for many years, particularly throughout the eastern United States. The origin of the design is seen in many precolonial wood-frame buildings. Pier and curtain wall foundations are only permitted to support structures of light-frame construction (wood or light-gage steel framing members) not more than two stories in height and assigned to Seismic Design Category A, B or C. Seismic detailing requirements for higher seismic design categories have not yet been developed. The term "pier" used in this application means "pilaster" rather than a type of footing. The term "curtain wall" refers to minimum 4-inch-thick (102 mm) masonry load-bearing walls. The provisions apply to simple wood-frame buildings where the combined loads are minimal.

This type of foundation system is also addressed in Section R404.1.5.1 of the *International Residential Code®* (IRC®), which includes an accompanying figure illustrating the code requirements. This IRC figure is reproduced here as Figure 1805.5.7.

1805.6 Foundation plate or sill bolting. Wood foundation plates or sills shall be bolted or strapped to the foundation or foundation wall as provided in Chapter 23.

❖ Anchorage of the sill plate in Seismic Design Categories D, E and F is addressed in Section 2305.3.11. Sill plate anchorage for conventional light-frame construction is addressed in Section 2308.6.

1805.7 Designs employing lateral bearing. Designs to resist both axial and lateral loads employing posts or poles as columns embedded in earth or embedded in concrete footings in the earth shall conform to the requirements of Sections 1805.7.1 through 1805.7.3.

❖ The criteria in this section apply to the lateral resistance of posts or poles embedded either directly in earth or in a concrete footing. They were originally developed to reduce the complexity of analyzing the soil-structure interaction, thereby facilitating the design of cantilevered poles used to support outdoor advertising. While more accurate methods are available for determining the resistance of posts or poles to axial and lateral loads, the other methods are much more complex, and the results do not differ significantly from this method.

1805.7.1 Limitations. The design procedures outlined in this section are subject to the following limitations:

1. The frictional resistance for structural walls and slabs on silts and clays shall be limited to one-half of the normal force imposed on the soil by the weight of the footing or slab.

2. Posts embedded in earth shall not be used to provide lateral support for structural or nonstructural materials such as plaster, masonry or concrete unless bracing is provided that develops the limited deflection required.

Wood poles shall be treated in accordance with AWPA U1 for sawn timber posts (Commodity Specification A, Use Category 4B) and for round timber posts (Commodity Specification B, Use Category 4B).

❖ The limitations imposed by this section are intended to address both structural stability and serviceability. The limitation of the frictional resistance for structural walls and slabs on silts and clays is consistent with Section 1804.3 and Table 1804.2, which also limit the sliding resistance to one-half the dead load. The limitations on the types of construction materials that utilize lateral support of poles are based on the brittle nature of the materials. In order to prevent excessive distortions, which would produce cracking of these brittle materials, this section limits the use of poles unless some type of rigid cross bracing is provided to limit the deflections to those that can be tolerated by the materials.

1805.7.2 Design criteria. The depth to resist lateral loads shall be determined by the design criteria established in Sections 1805.7.2.1 through 1805.7.2.3, or by other methods approved by the building official.

❖ The design criteria for the use of poles embedded in the ground or in concrete footings in the ground were developed in the 1940s. The design criteria address conditions where constraint is provided at the ground surface, such as a rigid floor, and where no constraint is provided. The original design criteria established a $^1/_2$-inch (12.7 mm) lateral pole deformation at the surface of the ground. These criteria were also based on field tests conducted in a range of sandy and gravelly soils and of silts and clays.

1805.7.2.1 Nonconstrained. The following formula shall be used in determining the depth of embedment required to resist lateral loads where no constraint is provided at the ground surface, such as rigid floor or rigid ground surface pavement, and where no lateral constraint is provided above the ground surface, such as a structural diaphragm.

$$d = 0.5A\{1 + [1 + (4.36h/A)]^{1/2}\} \qquad \textbf{(Equation 18-1)}$$

where:

A = $2.34P/S_1 b$.

b = Diameter of round post or footing or diagonal dimension of square post or footing, feet (m).

d = Depth of embedment in earth in feet (m) but not over 12 feet (3658 mm) for purpose of computing lateral pressure.

h = Distance in feet (m) from ground surface to point of application of "P."

P = Applied lateral force in pounds (kN).

S_1 = Allowable lateral soil-bearing pressure as set forth in Section 1804.3 based on a depth of one-third the depth of embedment in pounds per square foot (psf) (kPa).

❖ The required embedment depth for posts or poles that are not restrained at or above the ground surface is determined by Equation 18-1. This equation is applicable to posts or poles that have no lateral restraint at the ground surface. Nonconstrained conditions in-

clude flexible pavements such as asphalt. The code further restricts the use of this equation to structures that have no lateral support, such as from a floor diaphragm, provided above the ground level.

Determining the minimum embedment depth of a vertical member resisting lateral loads requires two steps:

1. Assuming an embedment depth, calculate the soil pressure that the embedded member imposes on the soil, and

2. Compare the calculated pressure to an allowable lateral soil bearing pressure.

This two-step process is expressed as a single trial-and-error equation, which requires an iterative pro-

For SI: 1 inch = 25.4 mm, 1 foot = 304.8 mm, 1 degree = 0.01745 rad.

Figure 1805.5.7
FOUNDATION WALL CLAY MASONRY CURTAIN WALL WITH CONCRETE MASONRY PIERS

cess to arrive at a satisfactory depth of embedment.

Note that the allowable lateral soil bearing pressure is in accordance with Section 1804.3. This includes increases permitted in Section 1804.3.1. In particular, the allowable lateral bearing values can be doubled for poles supporting structures that can tolerate 0.5 inches (12.7 mm) of movement at the ground surface – generally considered applicable to isolated poles.

1805.7.2.2 Constrained. The following formula shall be used to determine the depth of embedment required to resist lateral loads where constraint is provided at the ground surface, such as a rigid floor or pavement.

$$d^2 = 4.25(Ph/S_3 b) \qquad \textbf{(Equation 18-2)}$$

or alternatively

$$d^2 = 4.25 (M_g/S_3 b) \qquad \textbf{(Equation 18-3)}$$

where:

M_g = Moment in the post at grade, in foot-pounds (kN-m).

S_3 = Allowable lateral soil-bearing pressure as set forth in Section 1804.3 based on a depth equal to the depth of embedment in pounds per square foot (kPa).

❖ The required embedment depth for posts or poles that are restrained at the ground surface is determined by Equation 18-2 or 18-3.

1805.7.2.3 Vertical load. The resistance to vertical loads shall be determined by the allowable soil-bearing pressure set forth in Table 1804.2.

❖ See the commentary to Section 1804.2 and Table 1804.2. While no specific limit is provided with respect to vertical loading, this methodology was originally developed for cantilevered posts where the primary concern is lateral resistance (see commentary, Section 1805.7). The above determination of embedment length is independent of the vertical load.

1805.7.3 Backfill. The backfill in the annular space around columns not embedded in poured footings shall be by one of the following methods:

1. Backfill shall be of concrete with an ultimate strength of 2,000 psi (13.8 MPa) at 28 days. The hole shall not be less than 4 inches (102 mm) larger than the diameter of the column at its bottom or 4 inches (102 mm) larger than the diagonal dimension of a square or rectangular column.

2. Backfill shall be of clean sand. The sand shall be thoroughly compacted by tamping in layers not more than 8 inches (203 mm) in depth.

3. Backfill shall be of controlled low-strength material (CLSM).

❖ In order for the post or pole to meet the conditions and limitations of research for which the above criteria was established, backfill in the annular space around a column not embedded in a concrete footing must be either 2,000 psi (13.8 MPa) concrete, CLSM (see com-

mentary, Section 1803.6) or clean sand thoroughly compacted by tamping in layers not more than 8 inches (203 mm) in depth.

1805.8 Design for expansive soils. Footings or foundations for buildings and structures founded on expansive soils shall be designed in accordance with Section 1805.8.1 or 1805.8.2.

Footing or foundation design need not comply with Section 1805.8.1 or 1805.8.2 where the soil is removed in accordance with Section 1805.8.3, nor where the building official approves stabilization of the soil in accordance with Section 1805.8.4.

❖ Expansive soils, often referred to as "swelling soils," contain montmorillonite minerals and have the characteristics of absorbing water and swelling or shrinking and cracking when drying. Significant volume changes can cause serious damage to buildings and other structures as well as to pavements and sidewalks. Swelling soils are found throughout the nation, but are more prevalent in regions with dry or moderately arid climates.

There is a general relationship between the PI of a soil as determined by the ASTM D 4318 standard test method and the potential for expansion. This relationship is shown in Figure 1802.3.2.

Section 1802.3.2 defines "Expansive soil" as any plastic material with a PI of 15 or greater with more than 10 percent of the soil particles passing a No. 200 sieve and less than 5 micrometers in size. As an alternative, tests in accordance with ASTM D 4829 can be used to determine if a soil is expansive. The expansion index is a measure of the swelling potential of the soil.

The amount and depth of potential swelling that can occur in a clay material are, to some extent, functions of the cyclical moisture content in the soil. In dryer climates where the moisture content in the soil near the ground surface is low because of evaporation, there is a greater potential for extensive swelling than the same soil in wetter climates where the variations of moisture content are not as severe.

Volume changes in highly expansive soils range between 7 and 10 percent, but experience has shown that occasionally, under abnormal conditions, they can reach as high as 25 percent.

1805.8.1 Foundations. Footings or foundations placed on or within the active zone of expansive soils shall be designed to resist differential volume changes and to prevent structural damage to the supported structure. Deflection and racking of the supported structure shall be limited to that which will not interfere with the usability and serviceability of the structure.

Foundations placed below where volume change occurs or below expansive soil shall comply with the following provisions:

1. Foundations extending into or penetrating expansive soils shall be designed to prevent uplift of the supported structure.

2. Foundations penetrating expansive soils shall be designed to resist forces exerted on the foundation due to

soil volume changes or shall be isolated from the expansive soil.

❖ Shallow foundation systems placed on or within a mass of expansive soil must be designed with great rigidity and strength in order to adequately resist swelling pressures and avoid serious structural damage. Sometimes footings and piers, as well as foundation walls and grade beams, are isolated from the swelling soils by intervening fills or granular materials. While this type of insulation serves to cushion or diminish lateral pressures, it will not prevent structure heaving, except for grade beams constructed on collapsible forms.

The preferable foundation construction in expansive soils is the use of drilled piers with belled footings extending below the zone of swelling activity, or at least to a soil's stratum where the seasonal moisture content of the expansive soil will remain within a tolerable range. This type of construction is made possible in swelling soil's because the material usually consists of stiff clays that do not contain free water, providing excellent conditions for drilling holes in the ground. Piers with belled footings have been widely used for foundations in expansive soils, even in the construction of single-family dwellings.

The concrete shaft must be reinforced for its entire length because the swelling soil is apt to exert high uplift forces and subject the drilled pier construction to tensile stresses. This condition can be prevented or sufficiently mitigated by isolating the concrete from the swelling soil by surrounding the shaft with a vertical layer of granular soil or other suitable materials that possess little or no shearing strength (e.g., vermiculite).

1805.8.2 Slab-on-ground foundations. Moments, shears and deflections for use in designing slab-on-ground, mat or raft foundations on expansive soils shall be determined in accordance with *WRI/CRSI Design of Slab-on-Ground Foundations* or *PTI Standard Requirements for Analysis of Shallow Concrete Foundations on Expansive Soils*. Using the moments, shears and deflections determined above, nonprestressed slabs-on-ground, mat or raft foundations on expansive soils shall be designed in accordance with *WRI/CRSI Design of Slab-on-Ground Foundations* and post-tensioned slab-on-ground, mat or raft foundations on expansive soils shall be designed in accordance with *PTI Standard Requirements for Design of Shallow Post-Tensioned Concrete Foundations on Expansive Soils*. It shall be permitted to analyze and design such slabs by other methods that account for soil-structure interaction, the deformed shape of the soil support, the plate or stiffened plate action of the slab as well as both center lift and edge lift conditions. Such alternative methods shall be rational and the basis for all aspects and parameters of the method shall be available for peer review.

❖ It is not uncommon for concrete slabs on ground to be reinforced (prestressed) with post-tensioned strands so that the magnitude of tensile forces induced in the slab from uplift of the soil is reduced, thus avoiding cracking of the concrete at its top surface. It has also

been a practice to cast the slab on cellular forms made of cardboard or other collapsible materials that will support wet concrete, but will yield when subjected to swelling pressures. Since there is no rebound of the form material after subsidence of the soil, floors must be designed as structural slabs.

This section recognizes the analytical procedures in WRI/CRSI *Design of Slab-on-Ground Foundations* and PTI *Standard Requirements for Analysis of Shallow Concrete Foundations on Expansive soils* for slab-on-ground foundations located on expansive soils. The WRI/CRSI document also provides design requirements for reinforced concrete slabs. The PTI *Standard Requirements for Design of Shallow Post-Tensioned Concrete Foundations on Expansive Soils* addresses design requirements for post-tensioned concrete slabs on expansive soils.

1805.8.3 Removal of expansive soil. Where expansive soil is removed in lieu of designing footings or foundations in accordance with Section 1805.8.1 or 1805.8.2, the soil shall be removed to a depth sufficient to ensure a constant moisture content in the remaining soil. Fill material shall not contain expansive soils and shall comply with Section 1803.5 or 1803.6.

> **Exception:** Expansive soil need not be removed to the depth of constant moisture, provided the confining pressure in the expansive soil created by the fill and supported structure exceeds the swell pressure.

❖ The best way to avoid the problems associated with expansive soils is to remove such soil from the construction site and, where necessary, replace it with suitable compacted fill material. Depending on site conditions, however, this method is not always economically feasible and is not widely used.

1805.8.4 Stabilization. Where the active zone of expansive soils is stabilized in lieu of designing footings or foundations in accordance with Section 1805.8.1 or 1805.8.2, the soil shall be stabilized by chemical, dewatering, presaturation or equivalent techniques.

❖ This section allows expansive soils to be stabilized by either chemical, presaturation or dewatering methods or by other equivalent techniques. These methods, however, have limited use.

Chemical stabilization may be effectively accomplished by the use of lime thoroughly mixed with the soil and compacted at approximately the optimum moisture content. The purpose of using lime is to reduce the plasticity of the soil, which will reduce the swelling potential (see commentary, Section 1805.8). Since this method requires a uniform mix of lime and soil, however, it is generally limited to compacted fills. There is a method of pressure injecting lime slurry into heavily fissured clay materials, but the method is not appropriate for all site conditions.

Presaturation by flooding the site is rarely a totally effective way to stabilize expansive soils, since it takes a very long time for the water to penetrate to any great depth. Dewatering of expansive soils that consist generally of dense clays without free water is not an effective way to control moisture content.

One good method of stabilization is to place sufficient compacted fill material on the site so that the downward pressures of the fill will, to the extent possible, balance the upward swelling pressures of the soil. The effectiveness of this balancing concept depends on the pressures developed by the expansive soil as a function of estimated volume change and the depth of compacted fill material necessary to counteract these pressures. The application of this stabilization method is usually economically feasible when the soil pressures are low, around 500 psf (23.9 kPa); however, pressures occasionally have reached as high as 20 tons per square foot (1915 kPa).

1805.9 Seismic requirements. See Section 1908 for additional requirements for footings and foundations of structures assigned to Seismic Design Category C, D, E or F.

For structures assigned to Seismic Design Category D, E or F, provisions of ACI 318, Sections 21.10.1 to 21.10.3, shall apply when not in conflict with the provisions of Section 1805. Concrete shall have a specified compressive strength of not less than 3,000 psi (20.68 MPa) at 28 days.

Exceptions:

1. Group R or U occupancies of light-framed construction and two stories or less in height are permitted to use concrete with a specified compressive strength of not less than 2,500 psi (17.2 MPa) at 28 days.

2. Detached one- and two-family dwellings of light-frame construction and two stories or less in height are not required to comply with the provisions of ACI 318, Sections 21.10.1 to 21.10.3.

❖ This section first provides an important reference to Section 1908 for additional requirements concerning footings and foundations of buildings assigned to Seismic Design Category C, D, E or F. As specified in Section 1908.1.15, all footings and foundations in these structures require reinforcement except for footings supporting walls in detached one- and two-family dwellings three stories or less in height and constructed with stud-bearing walls. The concern is that plain concrete footings may be susceptible to damage, which can be reduced or avoided if reinforcement is provided. Foundation damage is obviously very expensive to repair and is better avoided.

The second paragraph of this section refers to provisions in ACI 318 concerning the design and construction of foundations of buildings assigned to high seismic performance and design categories. Chapter 21 of ACI 318 is entitled "Special Provisions for Seismic Design," and Section 21.10 contains the requirements pertaining to foundations. This paragraph also requires a minimum concrete compressive strength for footings of 3,000 psi (20.68 MPa) at 28 days. Section 21.2.4.1 of ACI 318 prescribes such a limitation for reinforced concrete structural members that are part of the seismic-force-resisting system of a building. The code's intent is that concrete used as a structural ma-

terial for footing construction will be of sufficient strength and durability. The exception permits the use of 2,500 psi (17.2 MPa) concrete in smaller residential-type structures. Additionally, relatively small, detached dwellings classified as Seismic Design Category D, E or F are exempted from the referenced ACI 318 provisions. These structures would only need to comply with requirements for Seismic Design Category C.

SECTION 1806
RETAINING WALLS

1806.1 General. Retaining walls shall be designed to ensure stability against overturning, sliding, excessive foundation pressure and water uplift. Retaining walls shall be designed for a safety factor of 1.5 against lateral sliding and overturning.

❖ This section provides design considerations for a retaining wall. The lateral pressure of the soil against the retaining wall is greatly influenced by soil moisture. Backfill is usually kept from being saturated for an extended length of time by placing drains near the base of the retaining wall to remove the water in the soil behind it.

SECTION 1807
DAMPPROOFING AND WATERPROOFING

1807.1 Where required. Walls or portions thereof that retain earth and enclose interior spaces and floors below grade shall be waterproofed and dampproofed in accordance with this section, with the exception of those spaces containing groups other than residential and institutional where such omission is not detrimental to the building or occupancy.

Ventilation for crawl spaces shall comply with Section 1203.4.

❖ Section 1807 covers the requirements for waterproofing and dampproofing those parts of substructure construction that need to be provided with moisture protection. Sections 1807.1 through 1807.4.3 identify the locations where moisture barriers are required and specify the materials to be used and the methods of application. The provisions also deal with subsurface water conditions, drainage systems and other protection requirements.

The term "waterproofing" is often used where dampproofing is really meant to take place. Although both terms are intended to apply to the installation and use of moisture barriers, dampproofing does not furnish the same degree of moisture protection.

Dampproofing generally refers to the application of one or more coatings of a compound or other materials that are impervious to water, which are used to prevent the passage of water vapor through walls or other building components, and which restrict the flow of water under slight hydrostatic pressure. Waterproofing, on the other hand, refers to the application of coatings and sealing materials to walls or other building components to prevent moisture from penetrating in either a vapor or liquid form, even under conditions of significant hydrostatic pressure. Hydrostatic pressure is created by the presence of water under pressure. This pressure can occur when the ground-water table rises above the bottom of the foundation wall, or the soil next to the foundation wall becomes saturated with water caused by uncontrolled storm water runoff.

Section 1807.1 is an overall requirement that waterproofing and dampproofing applications are to be made to horizontal and vertical surfaces of below-ground spaces where the occupancy would normally be affected by the intrusion of water or moisture. Moisture or water in a floor below grade can cause damage to structural members such as columns, posts or load-bearing walls, as well as pose a health hazard by promoting the growth of bacteria or fungi and adversely affect any mechanical and electrical appliances that may be located at that level. It can also cause a great deal of damage to goods that may be located or stored in that lower level. These vertical and horizontal surfaces include foundation walls, retaining walls, underfloor spaces and floor slabs. Waterproofing and dampproofing are not required in locations other than residential and institutional occupancies where the omission of moisture barriers would not adversely affect the use of the spaces. An example of a location where waterproofing or dampproofing would not be required is in an open parking structure, as long as the structural components are individually protected against the effects of water. Waterproofing and dampproofing are not permitted to be omitted from residential and institutional occupancies where people may be sleeping or services are provided on the floor below grade. A person walking in a flooded basement may be in a very hazardous situation, particularly if the possibility of an electrical charge in the water exists that is caused by electrical service at that level.

Section 1807.1.1 addresses the type of problem faced when a portion of a story is above grade, while Section 1807.1.2 limits any infiltration of water into crawl spaces so as to protect this area from potential water damage and prevent ponding of water. Both of these sections reference other applicable sections of the code, as well as the exceptions.

1807.1.1 Story above grade plane. Where a basement is considered a story above grade plane and the finished ground level adjacent to the basement wall is below the basement floor elevation for 25 percent or more of the perimeter, the floor and walls shall be dampproofed in accordance with Section 1807.2 and a foundation drain shall be installed in accordance with Section 1807.4.2. The foundation drain shall be installed around the portion of the perimeter where the basement floor is below ground level. The provisions of Sections 1802.2.3, 1807.3 and 1807.4.1 shall not apply in this case.

❖ The provisions of this section, stated in another way, require that where a basement is deemed to be a story above grade plane (see definition, Section 202), the section of the basement floor that occurs below the ex-

terior ground level and the walls that bound that part of the floor are to be dampproofed in accordance with the requirements of Section 1807.2.

The use of dampproofing, rather than waterproofing, is permitted here since hydrostatic pressure will not tend to develop against the walls if the basement is a story above grade plane and the ground level adjacent to the basement wall is below the basement floor elevation for no less than 25 percent of the basement perimeter.

Any water pressure that may occur against the walls below ground or under the basement floor would be relieved by the water drainage system required in this section. The drainage system would be installed at the base of the wall construction in accordance with Section 1807.4.2 for a minimum distance along those portions of the wall perimeter where the basement floor is below ground level.

Because of the relationship of grade to the basement floor and the inclusion of foundation drains, the potential for hydrostatic pressure buildup is not significant; therefore, a ground-water table investigation, waterproofing and the basement floor gravel base course is not required.

The objective of Section 1807.4.1 is to prevent moisture migration in basement spaces. In story-above-grade construction that meets the requirements of this section, the basement floor would be only partly below ground level (sometimes a small part) and the need for moisture protection as required by Section 1807.4.1 would be unnecessary. Dampproofing of the floor slab would be required, however, in accordance with Section 1807.2.1.

1807.1.2 Under-floor space. The finished ground level of an under-floor space such as a crawl space shall not be located below the bottom of the footings. Where there is evidence that the ground-water table rises to within 6 inches (152 mm) of the ground level at the outside building perimeter, or that the surface water does not readily drain from the building site, the ground level of the under-floor space shall be as high as the outside finished ground level, unless an approved drainage system is provided. The provisions of Sections 1802.2.3, 1807.2, 1807.3 and 1807.4 shall not apply in this case.

❖ The requirements of this section are designed to prevent any ponding of water in underfloor spaces, such as crawl spaces. Crawl spaces are particularly susceptible to ponding of water, since they are usually uninhabitable spaces that are observed very infrequently. Water can build up in these spaces and remain for an extended period of time without being noticed by the building occupants. This type of stagnant water under a building, which can harbor disease, mold and disease-carrying insects, such as mosquitoes, can result in a serious health concern. Water buildup in a crawl space can also damage the structural integrity of the building. Wood exposed to water will deteriorate and rot, while concrete and masonry exposed to water will deteriorate with a loss of strength.

Steel exposed to water or high humidity can eventually rust to the extent that effective structural capability is jeopardized. Water buildup in a crawl space can also damage any mechanical or electrical appliances, which may be located in the space, by causing corrosion of electrical parts or metal skins and deterioration to insulation used to protect heating elements.

Where it is known that the water table can rise to within 6 inches (152 mm) of the outside ground level, or where there is evidence that surface water cannot readily drain from the site, then the finished ground surface in underfloor spaces is to be set at an elevation equal to the outside ground level around the perimeter of the building unless an approved drainage system is provided. In order for the drainage system to be approved, it must be demonstrated to be adequate to prevent the infiltration of water into the underfloor space. This is done by determining the maximum possible flow of water near the foundation wall and footing and designing the drainage system to remove that flow of water as it occurs, without permitting the buildup of water at the foundation wall.

To prevent the ponding of water in the underfloor space from a rise in the ground-water table, or from storm water runoff, the finished ground level of an underfloor space is not to be located below the bottom of the foundation footings.

Dampproofing (see Section 1807.2) the foundation walls, waterproofing (see Section 1807.3) and providing subsoil drainage (see Section 1807.4) is not necessary if the ground level of the underfloor space is as high as the ground level at the outside of the building perimeter, as the foundation walls do not enclose an interior space below grade. Compliance with Sections 1807.2, 1807.3 and 1807.4 would still be required where the finished ground surface of the underfloor space is below the outside ground level.

1807.1.2.1 Flood hazard areas. For buildings and structures in flood hazard areas as established in Section 1612.3, the finished ground level of an under-floor space such as a crawl space shall be equal to or higher than the outside finished ground level.

> **Exception:** Under-floor spaces of Group R-3 buildings that meet the requirements of FEMA/ FIA-TB-11.

❖ According to Section 1612.2, the definition of "Basement" for buildings and structures located in designated flood hazard areas is "the portion of a building having its floor subgrade (below grade) on all sides." This definition pertains to enclosed areas below elevated buildings and structures whether or not there is enough clearance for such areas to be occupied. An area without sufficient clearance to be occupied is often referred to as a "crawl space." Local officials may be called on to determine whether a crawl space is a basement under the flood-plain management requirements of the NFIP. Neither the use of the enclosed space nor the clearance height is the controlling factor as to whether an enclosed area below an elevated building is a basement under Section 1612.2. The con-

trolling factor is whether the interior grade is below the exterior grade on all sides, even if the interior grade is established by the footing excavation that was not backfilled.

So that buildings and structures located in designated flood hazard areas receive favorable flood insurance premium rates, enclosed areas below elevated buildings or structures must meet the requirements of Section 1612.4, which references ASCE 24. In the event that a building or structure is not built in accordance with Section 1612.4, especially if it is determined to have a basement that is subgrade on all sides, federal flood insurance premium rates may be significantly higher.

In recognition of common construction practices in some parts of the country, the exception permits crawl spaces to be as much as 2 feet (610 mm) below the lowest adjacent exterior grade in accordance with FEMA Technical Bulletin 11-01, *Crawlspace Construction for Buildings Located in Special Flood Hazard Areas*, provided the additional requirements of that document are met [specifically, the maximum depth below grade is limited to 2 feet (210 mm) and maximum depth of floodwater above grade is limited to 2 feet (210 mm)]. Communities that retain a floodplain management ordinance and that choose to allow this practice must amend their ordinance to include these additional requirements.

1807.1.3 Ground-water control. Where the ground-water table is lowered and maintained at an elevation not less than 6 inches (152 mm) below the bottom of the lowest floor, the floor and walls shall be dampproofed in accordance with Section 1807.2. The design of the system to lower the ground-water table shall be based on accepted principles of engineering that shall consider, but not necessarily be limited to, permeability of the soil, rate at which water enters the drainage system, rated capacity of pumps, head against which pumps are to operate and the rated capacity of the disposal area of the system.

❖ After completion of building construction, it is often necessary to maintain the water table at a level that is at least 6 inches (152 mm) below the bottom of the lowest floor in order to prevent the flow or seepage of water into the basement. Where the site consists of well-draining soil and the highest point of the water table occurs naturally at or lower than the required level stated above, there is no need to provide a site drainage system specifically designated to control the ground-water level. Where the soil characteristics and site topography are such that the water table can rise to a level that will produce a hydrostatic pressure against the basement structure, then a site drainage system may be installed to reduce the water level if there is sufficient land area to accomplish the purpose. When ground-water control in accordance with this section is provided, waterproofing in accordance with Section 1807.3 is not required.

There are many types of site drainage systems that can be employed to control ground-water levels. The most commonly used systems may involve the installation of drainage ditches or trenches filled with pervious materials, sump pits and discharge pumps, well point systems, drainage wells with deep-well pumps, sand-drain installations, etc. This section requires that all such systems be designed and constructed using accepted engineering principles and practices based upon considerations that include the permeability of the soil, amount and rate at which water enters the system, pump capacity, capacity of the disposal area and other such factors that are necessary for the complete design of an operable drainage system.

1807.2 Dampproofing required. Where hydrostatic pressure will not occur as determined by Section 1802.2.3, floors and walls for other than wood foundation systems shall be dampproofed in accordance with this section. Wood foundation systems shall be constructed in accordance with AF&PA Technical Report No. 7.

❖ For a general definition of "Dampproofing," see the commentary to Section 1807. Where a ground-water table investigation made in accordance with the requirements of Section 1802.2.3 (see commentary) has established that the high water table will occur at such a level that the building substructure will not be subjected to hydrostatic pressure, then dampproofing in accordance with this section and a subsoil drain in accordance with Section 1807.4 are sufficient to control moisture in the floor below grade. Since the wall will not be subject to water under pressure, the more restrictive provisions of waterproofing, as outlined in Section 1807.3, are not required. Wood foundation systems specified in Section 1805.4.6 (see commentary) are to be dampproofed as required by AF&PA Technical Report No. 7.

1807.2.1 Floors. Dampproofing materials for floors shall be installed between the floor and the base course required by Section 1807.4.1, except where a separate floor is provided above a concrete slab.

Where installed beneath the slab, dampproofing shall consist of not less than 6-mil (0.006 inch; 0.152 mm) polyethylene with joints lapped not less than 6 inches (152 mm), or other approved methods or materials. Where permitted to be installed on top of the slab, dampproofing shall consist of mopped-on bitumen, not less than 4-mil (0.004 inch; 0.102 mm) polyethylene, or other approved methods or materials. Joints in the membrane shall be lapped and sealed in accordance with the manufacturer's installation instructions.

❖ Floors requiring dampproofing in accordance with Section 1807.2 are to employ materials as specified in Section 1807.2.1. The dampproofing materials must be placed between the floor construction and the supporting gravel or stone base as shown in Figure 1807.4.2. Even if a floor base in accordance with Section 1807.4.1 is not required, dampproofing is still to be placed under the slab unless otherwise specified in Section 1807.2.1.

The installation is intended to provide a moisture barrier against the passage of water vapor or seepage into below-ground spaces.

The dampproofing material most commonly used for underslab installations consists of a polyethylene film no less than 6 mil [0.006 inch; (0.152 mm)] in thickness, which is applied over the gravel or stone base required in Section 1807.4.1. Care must be used in the installation of the material over the rough surface of the base and during the concreting operations so as not to puncture the polyethylene. Joints must be lapped at least 6 inches (152 mm). Other materials used in a similar way are made of neoprene or butyl rubber.

Dampproofing materials can also be applied on top of the base concrete slab if a separate floor is provided above the base slab, since dampproofing is provided to prevent moisture infiltration of the interior space, not the concrete slab.

Materials commonly used for dampproofing floors are listed in Figure 1807.2.2.

1807.2.2 Walls. Dampproofing materials for walls shall be installed on the exterior surface of the wall, and shall extend from the top of the footing to above ground level.

Dampproofing shall consist of a bituminous material, 3 pounds per square yard (16 N/m^2) of acrylic modified cement, 0.125 inch (3.2 mm) coat of surface-bonding mortar complying with ASTM C 887, any of the materials permitted for waterproofing by Section 1807.3.2 or other approved methods or materials.

❖ Walls requiring dampproofing in accordance with Section 1807.2 are first to be prepared as required in Section 1807.2.2.1 and then coated with a bituminous material, cement or mortar as specified in Section 1807.2.2 or other approved materials and methods of application. Approved materials are those that will prevent moisture from penetrating the foundation wall when water is present but not under pressure.

Coatings are applied to cover prepared exterior wall surfaces extending from the top of the wall footings to slightly above ground level so that the entire wall that contacts the ground is protected. Surfaces are usually primed to provide a bond coat and then dampproofed with a protective coat of asphalt or tar pitch. Emulsion-type coatings may be applied directly on "green" concrete or unit masonry walls; however, because

MATERIAL	SPECIFICATION
Asphalt	ASTM D 449
Asphalt primer	ASTM D 41
Coal-tar	ASTM D 450
Concrete and masonry oil primer (for coal-tar applications only)	ASTM D 43
Treated glass fabric	ASTM D 1668

**Figure 1807.2.2
MATERIALS FOR WATERPROOFING AND
DAMPPROOFING INSTALLATIONS**

they are water-soluble materials, their use is not generally recommended. Installation should comply with the manufacturer's instructions.

Any of the materials specified in Section 1807.3.2 for waterproofing are also allowed to be used for dampproofing. Figure 1807.2.2 provides a list of bituminous materials that can be used, including the applicable standards that may be used as the basis of acceptance of such materials. Included in Figure 1807.2.2 is ASTM D 1668 for glass fabric that is treated with asphalt (Type I), coal-tar pitch (Type II) or organic resin (Type III).

Surface-bonding mortar complying with ASTM C 887 may be utilized. This specification covers the materials, properties and packaging of dry, combined materials for use as surface-bonding mortar with concrete masonry units that have not been prefaced, coated or painted. Since this specification does not address design or application, manufacturers' recommendations should be followed. This standard covers proportioning, physical requirements, sampling and testing. The minimum thickness of the coating is $^1/_8$ inch (3.2 mm).

Acrylic-modified cement coatings may be utilized at the rate of 3 pounds per square yard (16 N/m²). These types of materials have been used and performed successfully as dampproofing materials for foundation walls. Surface-bonding mortar and acrylic-modified cement are limited in use to dampproofing. The ability of these two types of products to bridge nonstructural cracks, as required in Section 1807.3.2 for waterproofing materials, is not known; therefore, their use is limited to dampproofing and they are not permitted to be used as waterproofing. Dampproofing may also include other materials and methods of installation acceptable to the building official.

1807.2.2.1 Surface preparation of walls. Prior to application of dampproofing materials on concrete walls, holes and recesses resulting from the removal of form ties shall be sealed with a bituminous material or other approved methods or materials. Unit masonry walls shall be parged on the exterior surface below ground level with not less than 0.375 inch (9.5 mm) of portland cement mortar. The parging shall be coved at the footing.

> **Exception:** Parging of unit masonry walls is not required where a material is approved for direct application to the masonry.

❖ Before applying dampproofing materials, the concrete must be free of any holes or recesses that could affect the proper sealing of the wall surfaces. Air trapped beneath a dampproofing coating or membranes can cause blistering, while rocks and other sharp objects can puncture membranes. Irregular surfaces can also create uneven layering of coatings, which can result in vulnerable areas of dampproofing. Surface irregularities commonly associated with concrete wall construction can be sealed with bituminous materials or filled with portland cement grout or other approved methods.

Unit masonry walls are usually parged (plastered)

with a $^1/_2$-inch-thick (12.7 mm) layer of portland cement and sand mix (1:2$^1/_2$ by volume) or with Type M mortar proportioned in accordance with the requirements of ASTM C 270, and applied in two $^1/_4$-inch-thick (6.4 mm) layers. In no case is parging to result in a final thickness of less than $^3/_8$ inch (9.5 mm). The parging is to be coved at the joint formed by the base of the wall and the top of the wall footing to prevent the accumulation of water at that location.

The moisture protection of unit masonry walls provided by the parging method may not be required where approved dampproofing materials, such as grout coatings, cement-based paints or bituminous coatings, can be applied directly to masonry surfaces.

1807.3 Waterproofing required. Where the ground-water investigation required by Section 1802.2.3 indicates that a hydrostatic pressure condition exists, and the design does not include a ground-water control system as described in Section 1807.1.3, walls and floors shall be waterproofed in accordance with this section.

❖ For a general definition of "Waterproofing," and the distinction between dampproofing and waterproofing, see the commentary to Section 1807.

The significance of waterproofing installations is that they are intended to provide moisture barriers against water seepage that may be forced into below-ground spaces by hydrostatic pressure.

Where a ground-water table investigation made in accordance with the requirements of Section 1802.2.3 (see commentary) has established that the high water table will occur at such a level that the building substructure will be subjected to hydrostatic pressure, and where the water table is not lowered by a water control system, as prescribed in the commentary to Section 1807.1.3, all floors and walls below ground level are to be waterproofed in accordance with Sections 1807.3.1, 1807.3.2 and 1807.3.3.

1807.3.1 Floors. Floors required to be waterproofed shall be of concrete and designed and constructed to withstand the hydrostatic pressures to which the floors will be subjected.

Waterproofing shall be accomplished by placing a membrane of rubberized asphalt, butyl rubber, fully adhered/fully bonded HDPE or polyolefin composite membrane or not less than 6-mil [0.006 inch (0.152 mm)] polyvinyl chloride with joints lapped not less than 6 inches (152 mm) or other approved materials under the slab. Joints in the membrane shall be lapped and sealed in accordance with the manufacturer's installation instructions.

❖ Since floors that are required to be waterproofed are subjected to hydrostatic uplift pressures, such floors must, for all practical purposes, be made of concrete and designed and constructed to resist the maximum hydrostatic pressures possible. It is particularly important that the floor slab be properly designed, since severe cracking or movement of the concrete would allow water seepage into below-ground spaces because the ability of the waterproofing materials to bridge small cracks would be exceeded. Concrete floor construction is to comply with the applicable provisions of Chapter 19.

Below-ground floors subjected to hydrostatic uplift pressures are to be waterproofed with membrane materials placed as underslab or split-slab installations, including such materials as rubberized asphalt, butyl rubber, fully bonded HDPE, fully bonded polyolefin and neoprene or with polyvinyl chloride (PVC) not less than 6 mil [0.006 inch; (0.152 mm)] in thickness, lapped at least 6 inches (152 mm). There are many proprietary membrane products available in the marketplace that are specifically made for waterproofing floors and walls (i.e., polyethylene sheets sandwiched between layers of asphalt), which may be used for that purpose when approved by the building official.

All membrane joints are to be lapped and sealed in accordance with the manufacturer's instructions to form a continuous, impermeable moisture barrier.

1807.3.2 Walls. Walls required to be waterproofed shall be of concrete or masonry and shall be designed and constructed to withstand the hydrostatic pressures and other lateral loads to which the walls will be subjected.

Waterproofing shall be applied from the bottom of the wall to not less than 12 inches (305 mm) above the maximum elevation of the ground-water table. The remainder of the wall shall be dampproofed in accordance with Section 1807.2.2. Waterproofing shall consist of two-ply hot-mopped felts, not less than 6-mil (0.006 inch; 0.152 mm) polyvinyl chloride, 40-mil (0.040 inch; 1.02 mm) polymer-modified asphalt, 6-mil (0.006 inch; 0.152 mm) polyethylene or other approved methods or materials capable of bridging nonstructural cracks. Joints in the membrane shall be lapped and sealed in accordance with the manufacturer's installation instructions.

❖ Walls that are required to be waterproofed in accordance with Section 1807.3 must first be prepared as required in Section 1807.3.2.1.

The walls must be designed to resist the hydrostatic pressure anticipated at the site, as well as any other lateral loads to which the wall will be subjected, such as soil pressures or seismic loads. As with floors required to be waterproofed, it is particularly important that walls required to be waterproofed be properly designed to resist all loads present, since cracking and other damage would allow water seepage into below-ground spaces. Water seepage can lead to deterioration of the foundation as wood rots, concrete and masonry erode and steel rusts. Such deterioration can cause structural failure of the foundation. More importantly, failure of the foundation wall can lead to structural failure of the building, since the foundation supports the building structure. Masonry or concrete construction must comply with the applicable provisions of Chapters 21 and 19, respectively.

Figures 1807.2.2 and 1807.3.2 list materials commonly used for the installation of moisture barriers in wall construction and the related standard that may be used as a basis for acceptance of such materials. Asphalt and coal-tar products are not compatible and should not be used together.

MATERIAL	SPECIFICATION
Asphalt-saturated asbestos felt	ASTM D 250
Asphalt-saturated burlap fabric	ASTM D 1327
Asphalt-saturated cotton fabric	ASTM D 173
Asphalt-saturated organic felt	ASTM D 226
Coal-tar-saturated burlap fabric	ASTM D 1327
Coal-tar-saturated cotton fabric	ASTM D 173
Coal-tar-saturated organic felt	ASTM D 227

**Figure 1807.3.2
MATERIALS FOR WATERPROOFING
INSTALLATIONS**

Waterproofing installations are to extend from the bottom of the wall to a height no less than 12 inches (305 mm) above the maximum elevation of the ground-water table determined in accordance with the requirements of Section 1802.2.3 (see commentary). The remainder of the wall below ground level (if the height is small) may be waterproofed as a continuation of the installation or must be dampproofed in accordance with the requirements of Section 1807.2.2. If the ground-water table investigation is not conducted on the basis of the exception to Section 1802.2.3, then waterproofing should be provided from a point below the footing to above the ground level.

This section requires that waterproofing must consist of two-ply hot-mopped felts. The practice of the waterproofing industry is to select the number of plies of membrane material based on the hydrostatic head (height of water pressure against the wall). As a general practice, if the head of water is between 1 foot (305 mm) and 3 feet (914 mm), two plies of felt or fabric membrane are used; between 4 feet (1219 mm) and 10 feet (3048 mm), three-ply construction is needed and between 11 feet (3353 mm) and 25 feet (7620 mm), four-ply construction is necessary.

Waterproofing installations may also use polyvinyl chloride materials of no less than 6-mil [0.006 inch (0.152 mm)] thick, 40-mil [0.040 inch (1.02 mm)] polymer-modified asphalt or 6-mil [0.006 inch (0.152 mm)] polyethylene. These materials have been widely recognized for their effectiveness in bridging nonstructural cracks. Other approved materials and methods may be used provided that the same performance standards are met. All membrane joints must be lapped and sealed in accordance with the manufacturer's instructions.

1807.3.2.1 Surface preparation of walls. Prior to the application of waterproofing materials on concrete or masonry walls, the walls shall be prepared in accordance with Section 1807.2.2.1.

❖ Before applying waterproofing materials to concrete or masonry walls, the surfaces must be prepared in accordance with the requirements of Section 1807.2.2.1, which requires the sealing of all holes and recesses.

Surfaces to be waterproofed must also be free of any projections that might puncture or tear membrane materials that are applied over the surfaces.

1807.3.3 Joints and penetrations. Joints in walls and floors, joints between the wall and floor and penetrations of the wall and floor shall be made water-tight utilizing approved methods and materials.

❖ This section requires that all joints occurring in floors and walls and at locations where floors and walls meet, as well as all penetrations in floors and walls, be made water tight by approved methods. Sealing joints and penetrations is of primary importance to ensuring the effectiveness of the waterproofing. If the joints or penetrations are not sealed properly, they can develop leaks, which become a passageway for water to enter the building. Since the remainder of the foundation is wrapped in waterproofing, moisture can actually become trapped in the foundation walls or floor slab, and serious damage to these structural components can occur. Such methods may involve the use of construction keys (e.g., between the base of the wall and the top of the footing) or, if there is hydrostatic pressure, floor and wall joints may require the use of manufactured waterstops made of metal, rubber, plastic or mastic materials. Figures 1807.3.3(1) and 1807.3.3(2) illustrate examples of joint treatment and penetration treatment of waterproofing. Floor edges along walls

Figure 1807.3.3(1)
DETAIL—MEMBRANE PLACEMENT THROUGH A KEY JOINT

Figure 1807.3.3(2)
PIPE DETAIL—PROPER PLACEMENT OF WATERPROOFING ELEMENTS

and floor expansion joints may employ the use of any number of preformed expansion joint materials, such as asphalt, polyurethane, sponge rubber, self-expanding cork, cellular fibers bonded with bituminous materials, etc., which all comply with applicable ASTM or AASHTO standards or other federal specifications. A variety of sealants may be used together with the preformed joint materials. Gaskets made of neoprene and other materials are also available for use in concrete and masonry joints. The National Roofing Contractor's *NRCA Roofing and Waterproofing Manual* provides details for the reinforcement of membrane terminations, corners, intersections of slabs and walls, through-wall and slab penetrations and other locations.

Penetrations in walls and floors may be made water tight with grout or manufactured fill materials and sealants made for that purpose.

1807.4 Subsoil drainage system. Where a hydrostatic pressure condition does not exist, dampproofing shall be provided and a base shall be installed under the floor and a drain installed around the foundation perimeter. A subsoil drainage system designed and constructed in accordance with Section 1807.1.3 shall be deemed adequate for lowering the ground-water table.

❖ This section covers subsoil drainage systems in conjunction with dampproofing (see Section 1807.2) to protect below-ground spaces from water seepage. Such systems are not used where basements or other below-ground spaces are subject to hydrostatic pressure because they would not be effective in disposing of the amount of water anticipated if a hydrostatic pressure condition exists. Ground-water tables may be reduced to acceptable levels by methods described in the commentary to Section 1807.1.3.

The details of subsoil drainage systems are covered in the requirements of Sections 1807.4.1 through 1807.4.3.

1807.4.1 Floor base course. Floors of basements, except as provided for in Section 1807.1.1, shall be placed over a floor base course not less than 4 inches (102 mm) in thickness that consists of gravel or crushed stone containing not more than 10 percent of material that passes through a No. 4 (4.75 mm) sieve.

Exception: Where a site is located in well-drained gravel or sand/gravel mixture soils, a floor base course is not required.

❖ This section requires that basement floors, except for story-above-grade construction, must be placed on a gravel or stone base no less than 4 inches (102 mm) thick. Not more than 10 percent of the material is to pass a No. 4 sieve so as to provide a porous installation. Material that passes a No. 4 sieve would be fine silt or clay that would not permit the free movement of water through the floor base.

This requirement serves three purposes. The first is to provide an adjustment to the irregularities of a compacted subgrade so as to produce a level surface upon which to cast a concrete slab. The second is to provide a capillary break so that moisture from the soil

below will not rise to the underside of the floor. Finally, where required, the porous base can act as a drainage system to expel underslab water by means of gravity or the use of a sump pump or other approved methods.

The exception allows for the omission of the floor base when the natural soils beneath the floor slab consist of well-draining granular materials, such as sand, stone or mixtures of these materials. The exception is consistent with the requirements of Section 1807.4.3. Some caution, however, is justified in the use of this exception. If the granular soils contain an excessive percentage of fine materials, the porosity and the ability of the soil to provide a capillary break may be considerably diminished. The exception is only to be applied if the natural base is equivalent to the floor base otherwise required by this section.

1807.4.2 Foundation drain. A drain shall be placed around the perimeter of a foundation that consists of gravel or crushed stone containing not more than 10-percent material that passes through a No. 4 (4.75 mm) sieve. The drain shall extend a minimum of 12 inches (305 mm) beyond the outside edge of the footing. The thickness shall be such that the bottom of the drain is not higher than the bottom of the base under the floor, and that the top of the drain is not less than 6 inches (152 mm) above the top of the footing. The top of the drain shall be covered with an approved filter membrane material. Where a drain tile or perforated pipe is used, the invert of the pipe or tile shall not be higher than the floor elevation. The top of joints or the top of perforations shall be protected with an approved filter membrane material. The pipe or tile shall be placed on not less than 2 inches (51 mm) of gravel or crushed stone complying with Section 1807.4.1, and shall be covered with not less than 6 inches (152 mm) of the same material.

❖ This section describes the materials and features of construction required for the installation of foundation drain systems. Perimeter foundation drains provide a means to remove free ground water and prevent leakage into habitable below-grade spaces. The first part of this provision describes a foundation drain of gravel or crushed stone. The balance of the provision describes the installation of drain tile or perforated pipe where either of those is provided. The use of drain tiles is important in areas having moderate to heavy rainfall as well as in soils that have low percolation rates. The requirements of this section are illustrated in Figure 1807.4.2.

This type of drain system is suitable where the water table occurs at such elevation that there is no hydrostatic pressure exerted against the basement floor and walls and where the amount of seepage from the surrounding soil is so small that the water can be readily discharged by gravity or mechanical means into sewers or ditches. The objective is to combine the protection afforded by the dampproofing of walls and floors (see Section 1807.2) and that given by perimeter drains so as to maintain below-ground spaces in a dry condition.

Gravel or crushed stone drain material is to be

placed in the excavation so that it will extend out from the edge of the wall footing a distance of at least 12 inches (305 mm), with the bottom of the fill being no higher than the bottom of the base under the floor (see Section 1807.4.1) and the top of the filter material being no less than 6 inches (152 mm) above the top of the wall footing so that water will not collect along the top of the footing. Requiring the bottom of the gravel or crushed rock drain to be no higher than the bottom of the floor base is necessary so that if the water table rises into the floor base, it will also be able to rise unobstructed into the foundation drain. The foundation drain will then drain the water away from the building, as required by Section 1807.4.3. The top of the drain must be covered with an approved filter membrane to allow water to pass through without allowing, or at least greatly reducing, the possibility of fine soil materials entering the drainage system.

Where drain tile or perforated drain pipe is utilized, the drain system consists of clay or concrete drain tiles or corrugated metal or nonmetallic pipes installed in a filter bed with at least 2 inches (51 mm) of filter material and covered with at least 6 inches (152 mm) of filter material to maintain good water flow into the drain tile or pipe. The foundation drain is set adjacent to the wall footing and extends around the perimeter of the building. Drain tiles are placed end to end with open joints to permit water to enter the system. Metallic and nonmetallic drains are made with perforations at the invert (bottom) section of the pipe and are installed with connected ends. The drain pipe invert is not to be set higher than the basement floor line such that water conveyed by the drain does not seep into the filter material and create a hydrostatic pressure condition against the foundation wall and footing. The inverts should not be placed below the bottom of the adjacent wall footings so as to avoid carrying away fine soil particles that, in time, could undermine the footing and settlement of the foundation walls.

Tile joints or pipe perforations should be covered with an approved filter membrane material to prevent them from becoming clogged, preventing fine particles that may be contained in the surrounding soil from entering the system and being carried away by water. The filter material around the drain tiles or pipes (not to be confused with filter membrane material) is to consist of selected gravel and crushed stone containing no more than 10 percent of material that passes through a No. 4 sieve. The filter materials should be selected to prevent the movement of particles from the protected soil surrounding the drain installation into the drain.

1807.4.3 Drainage discharge. The floor base and foundation perimeter drain shall discharge by gravity or mechanical means into an approved drainage system that complies with the *International Plumbing Code*.

Exception: Where a site is located in well-drained gravel or sand/gravel mixture soils, a dedicated drainage system is not required.

❖ This section references the *International Plumbing Code®* (IPC®) for the requirements of installing piping

For SI: 1 inch = 25.4 mm.

Figure 1807.4.2
FOUNDATION DRAINAGE SYSTEM

systems for the disposal of water from the floor base (see Section 1807.4.1) and the foundation drains (see Section 1807.4.2). Chapter 11 of the IPC deals with the piping materials, applicable standards and methods of installation of subsurface storm drains to facilitate water discharge either by gravity or mechanical means.

Where the soil at the site consists of well-drained granular materials (i.e., gravel or sand-gravel mixtures) to prevent the occurrence of hydrostatic pressure against the foundation walls and under the floor slab, the use of a dedicated drainage system as prescribed in the IPC is not required, since the site soils would permit natural drainage.

SECTION 1808
PIER AND PILE FOUNDATIONS

1808.1 Definitions. The following words and terms shall, for the purposes of this section, have the meanings shown herein.

❖ Section 1808 addresses the general requirements for the installation of various types of pier and pile foundations, and provides specific provisions related to allowable pier and pile loads, lateral support of piles, stability and other pertinent considerations required for the satisfactory construction of these deep foundation systems.

Definitions are intended to facilitate the understanding of code provisions and to minimize potential confusion. This section contains the definitions of terms associated with piles and piers. Note that these terms are also listed in Chapter 2 with a cross reference to this section. Including definitions that pertain primarily to piles and piers in this section provides convenient access to them without having to refer back to Chapter 2. The use and application of all defined terms, including those defined herein, as well as undefined terms are set forth in Section 201.

FLEXURAL LENGTH. Flexural length is the length of the pile from the first point of zero lateral deflection to the underside of the pile cap or grade beam.

❖ This definition in used in connection with the special detailing requirements for the design of piles.

MICROPILES. Micropiles are 12-inch-diameter (305 mm) or less bored, grouted-in-place piles incorporating steel pipe (casing) and/or steel reinforcement.

❖ See Section 1810.8 and the commentary for additional information on micropiles.

PIER FOUNDATIONS. Pier foundations consist of isolated masonry or cast-in-place concrete structural elements extending into firm materials. Piers are relatively short in comparison to their width, with lengths less than or equal to 12 times the least horizontal dimension of the pier. Piers derive their load-carrying capacity through skin friction, through end bearing, or a combination of both.

❖ See Section 1812 and the commentary for additional information on pier foundations.

Belled piers. Belled piers are cast-in-place concrete piers constructed with a base that is larger than the diameter of the remainder of the pier. The belled base is designed to increase the load-bearing area of the pier in end bearing.

❖ See Section 1812.6 and the commentary for additional information on belled piers.

PILE FOUNDATIONS. Pile foundations consist of concrete, wood or steel structural elements either driven into the ground or cast in place. Piles are relatively slender in comparison to their length, with lengths exceeding 12 times the least horizontal dimension. Piles derive their load-carrying capacity through skin friction, end bearing or a combination of both.

❖ See Sections 1809 through 1811 and the commentary for additional information on pile foundations.

Augered uncased piles. Augered uncased piles are constructed by depositing concrete into an uncased augered hole, either during or after the withdrawal of the auger.

❖ See Section 1810.3 and the commentary for additional information on drilled or augered uncased piles.

Caisson piles. Caisson piles are cast-in-place concrete piles extending into bedrock. The upper portion of a caisson pile consists of a cased pile that extends to the bedrock. The lower portion of the caisson pile consists of an uncased socket drilled into the bedrock.

❖ See Section 1810.7 and the commentary for additional information on caisson piles.

Concrete-filled steel pipe and tube piles. Concrete-filled steel pipe and tube piles are constructed by driving a steel pipe or tube section into the soil and filling the pipe or tube section with concrete. The steel pipe or tube section is left in place during and after the deposition of the concrete.

❖ See Section 1810.6 and the commentary for additional information on concrete-filled steel pipe and tube piles.

Driven uncased piles. Driven uncased piles are constructed by driving a steel shell into the soil to shore an unexcavated hole that is later filled with concrete. The steel casing is lifted out of the hole during the deposition of the concrete.

❖ See Section 1810.4 and the commentary for additional information on driven uncased piles.

Enlarged base piles. Enlarged base piles are cast-in-place concrete piles constructed with a base that is larger than the diameter of the remainder of the pile. The enlarged base is designed to increase the load-bearing area of the pile in end bearing.

❖ See Section 1810.2 and the commentary for additional information on enlarged base piles.

Steel-cased piles. Steel-cased piles are constructed by driving a steel shell into the soil to shore an unexcavated hole. The steel casing is left permanently in place and filled with concrete.

❖ See the commentary to Section 1810.5.

Timber piles. Timber piles are round, tapered timbers with the small (tip) end embedded into the soil.

❖ See Section 1809.1 and the commentary for additional information on timber piles.

1808.2 Piers and piles—general requirements.

❖ This section provides general requirements applicable to the construction and installation of pier and pile foundations.

1808.2.1 Design. Piles are permitted to be designed in accordance with provisions for piers in Section 1808 and Sections 1812.3 through 1812.10 where either of the following conditions exists, subject to the approval of the building official:

1. Group R-3 and U occupancies not exceeding two stories of light-frame construction, or

2. Where the surrounding foundation materials furnish adequate lateral support for the pile.

❖ Subject to the approval of the building official, piles may be designed under the provisions for piers in either of two situations. One is where such piles are given enough lateral support by surrounding foundation materials to prevent or adequately limit flexural (bending) deformations. The second such allowance is for Group R-3 and U occupancies that are built of light-frame construction not exceeding two stories in height.

1808.2.2 General. Pier and pile foundations shall be designed and installed on the basis of a foundation investigation as defined in Section 1802, unless sufficient data upon which to base the design and installation is available.

The investigation and report provisions of Section 1802 shall be expanded to include, but not be limited to, the following:

1. Recommended pier or pile types and installed capacities.

2. Recommended center-to-center spacing of piers or piles.

3. Driving criteria.

4. Installation procedures.

5. Field inspection and reporting procedures (to include procedures for verification of the installed bearing capacity where required).

6. Pier or pile load test requirements.

7. Durability of pier or pile materials.

8. Designation of bearing stratum or strata.

9. Reductions for group action, where necessary.

❖ A foundations investigation is mandatory when pier or pile foundations are used. Such investigations are needed to define as accurately as possible the subsurface conditions of soil and rock materials, establish the soil and rock profiles across the construction site and locate the ground-water table. Sometimes it may also be necessary to determine specific soil properties, such as shear strength, relative density, compressibility and other such technical data required for analyzing subsurface conditions. Foundation investigations may also be used to render such valuable data as information on existing construction at the site or on neighboring properties (including boring and test records), the type and condition of the existing structures, their age, the type of foundations used and performance over the years. Knowledge of existing deleterious substances in the soils that could affect the durability (as well as the performance) of piers or piles, data on geologic conditions at the site (including such information as the existence of mines, earth cavities, underground streams or other adverse water conditions) and a history of any seismic activity, hurricanes, etc., are also important.

The types of information described above are usually obtained by means of soil and rock borings; laboratory and field tests; engineering analyses and research. Such information is used for determining design loads; type, spacing and lengths of piles; suitable bearing strata; driving criteria and selection of equipment; probable durability of pile materials in relation to subsurface conditions, etc.

1808.2.3 Special types of piles. The use of types of piles not specifically mentioned herein is permitted, subject to the approval of the building official, upon the submission of acceptable test data, calculations and other information relating to the structural properties and load capacity of such piles. The allowable stresses shall not in any case exceed the limitations specified herein.

❖ Piles are generally identified according to the materials used (concrete, steel or wood) or the methods of construction or installation. While the most commonly used types of piles are addressed in Sections 1808 through 1811, this section acknowledges that there are many variations of pile types used in the construction of deep foundations, including some special or proprietary types that are beyond the scope of the code.

However, while it is not the intent to preclude the use of such special or proprietary types of piles, it is necessary to ensure their structural integrity by submitting test data, calculations, information on structural properties and load capacity and installation procedures to the building official for approval.

1808.2.4 Pile caps. Pile caps shall be of reinforced concrete, and shall include all elements to which piles are connected, including grade beams and mats. The soil immediately below the pile cap shall not be considered as carrying any vertical load. The tops of piles shall be embedded not less than 3 inches (76 mm) into pile caps and the caps shall extend at least 4 inches (102 mm) beyond the edges of piles. The tops of piles shall be cut back to sound material before capping.

❖ Pile caps include all elements to which the piles are connected and are to be of reinforced concrete and designed in accordance with the requirements of ACI 318. For footings (pile caps) on piles, computations for moments and shears may be based on the assumption that the load reaction from any pile is concentrated at the pile center (see ACI 318 for loads and reactions of footings on piles).

The soil immediately under the pile cap is not considered to provide any support for vertical loads. For a detailed explanation of this requirement, see the commentary to Section 1808.2.9.3.

The heads of all piles are to be embedded no less than 3 inches (76 mm) into pile caps and the edges of the pile caps are to extend at least 4 inches (102 mm) beyond the closest sides of all piles. The degree of fixity between a pile head and the concrete cap depends on the method of connection required to satisfy design considerations.

1808.2.5 Stability. Piers or piles shall be braced to provide lateral stability in all directions. Three or more piles connected by a rigid cap shall be considered braced, provided that the piles are located in radial directions from the centroid of the group not less than 60 degrees (1 rad) apart. A two-pile group in a rigid cap shall be considered to be braced along the axis connecting the two piles. Methods used to brace piers or piles shall be subject to the approval of the building official.

Piles supporting walls shall be driven alternately in lines spaced at least 1 foot (305 mm) apart and located symmetrically under the center of gravity of the wall load carried, unless effective measures are taken to provide for eccentricity and lateral forces, or the wall piles are adequately braced to provide for lateral stability. A single row of piles without lateral bracing is permitted for one- and two-family dwellings and lightweight construction not exceeding two stories or 35 feet (10 668 mm) in height, provided the centers of the piles are located within the width of the foundation wall.

❖ A group of piles designed to support a common load and, as may be required, to resist horizontal forces, must be braced or rigidly tied together to act as a single structural unit that will provide lateral stability in all directions. Piles connected by a rigid, reinforced concrete pile cap are deemed to be braced construction that serves the intent of this provision.

Three or more piles are generally used to support a building column load or other isolated concentrated load. In a three-pile group, lateral stability is assured by requiring that the piles are located such that they will not be less than 60 degrees (1.0 rad) apart as measured from the centroid of the group in a radial direction.

For stability of a pile group supporting a wall structure, the piles are braced by a continuous rigid footing and are alternately staggered in two lines at least 1 foot (305 mm) apart and symmetrically located on each side of the center of gravity of the wall. Other approved pile arrangements may be used to support

walls, provided that the piles are adequately braced and the lateral stability of the foundation construction is ensured.

For lightweight construction, such as one- and two-family dwellings not exceeding two stories or 35 feet (10 668 mm) in height, a single row of piles located within the width of the wall is permitted. In this case, pile group stability is theoretically afforded only in the direction of the line of piles. However, in reality, cross walls, floor slabs and other structural components of building construction may contribute to the lateral stability of the foundation system.

1808.2.6 Structural integrity. Piers or piles shall be installed in such a manner and sequence as to prevent distortion or damage that may adversely affect the structural integrity of piles being installed or already in place.

❖ The placement of piles is often by driving, vibrating, jacking, jetting, direct weight or a combination of these methods. Because of the generally harsh nature of pile installation operations, piles can experience some degree of damage during placement; however, damage can be prevented or minimized by selecting the proper type of pile placement methods and techniques, as well as the right equipment to accomplish the work, all based on adequate knowledge of the soil conditions obtained from a foundation investigation (see Section 1808.2.2).

Piles must be placed in a manner that maintains their structural integrity and installed to such depths as determined by foundation investigation and engineering analysis to safely resist the design loads that are to be imposed upon them. Care must be exercised during pile placement operations to provide for the safety of adjacent piles or other structures-leaving their strength and load capacity unimpaired. Any pile damaged during installation to the extent that its structural integrity is affected must be satisfactorily repaired or rejected.

1808.2.7 Splices. Splices shall be constructed so as to provide and maintain true alignment and position of the component parts of the pier or pile during installation and subsequent thereto and shall be of adequate strength to transmit the vertical and lateral loads and moments occurring at the location of the splice during driving and under service loading. Splices shall develop not less than 50 percent of the least capacity of the pier or pile in bending. In addition, splices occurring in the upper 10 feet (3048 mm) of the embedded portion of the pier or pile shall be capable of resisting at allowable working stresses the moment and shear that would result from an assumed eccentricity of the pier or pile load of 3 inches (76 mm), or the pier or pile shall be braced in accordance with Section 1808.2.5 to other piers or piles that do not have splices in the upper 10 feet (3048 mm) of embedment.

❖ While it is physically and economically better to drive piles in one piece, site conditions sometimes necessitate that piles be driven in spliced sections. For example, when the soil or rock-bearing stratum is located so deep below the ground that the leads on the driving equipment will not receive full-length piles, it becomes necessary to install the piles sectionally or, where possible, to take up the extra length by setting the tip in a preexcavated hole (see commentary, Section 1808.2.13). When piles are installed in areas such as existing buildings with restricted headroom, they are also required to be placed in spliced sections. There are a number of other reasons for field-splicing piles, such as restrictions on shipping lengths, the use of composite piles, etc.

This provision requires that splices be constructed so as to provide and maintain true alignment and position of the pile sections during installation. Furthermore, splices must be of sufficient strength to transmit safely the vertical and lateral loads on the piles, as well as to resist the bending stresses that may occur at splice locations during the driving operations and under long-term service loads. Splices are to develop at least 50 percent of the value of the pile in bending. Consideration should be given to the design of splices at locations where the piles may be subject to tension. Additionally, splices that occur in the upper 10 feet (3048 mm) of pile embedment are to be designed to resist the bending moments and shears at the allowable stress levels of the pile material, based on an assumed pile load eccentricity of 3 inches (76 mm), unless the pile is properly braced. Proper bracing of a spliced pile is deemed to exist if stability of the pile group is present in accordance with the provisions of Section 1808.2.4, provided that the other piles in the group do not have splices in the upper 10 feet (3048 mm) of their embedded length.

There are different methods employed in splicing piles based on the different materials used in pile construction. For example, timber piles are spliced by one of two commonly used methods. The first method uses a pipe sleeve with a length of about four to five times the diameter of the pile. The butting ends of the pile are sawn square for full contact of the two pile sections, and the spliced portions of the timber pile are trimmed smoothly around their periphery so as to fit tightly into the pipe sleeve. The other splicing method involves the use of steel straps and bolts. The butting ends of the pile sections are sawn square for full contact and proper alignment and the four sides are planed flat to receive the splicing straps. This type of splicing can resist some uplift forces.

Splicing of precast concrete piles usually occurs at the head portions of the piles where, after the piles are driven to their required depth, pile heads are cut off or spliced to the desired elevation for proper embedment in the concrete pile caps. Any portion of the pile that is cracked or shattered caused by the driving operations or cutting off of pile heads should be removed and spliced with fresh concrete. To cut off a precast concrete pile section, a deep groove is chiseled around the pile exposing the reinforcing bars, which are then cut off (by torch) to desired heights or extensions. The pile section above the groove is snapped off (by crane) and a new pile section is freshly cast to tie in with the precast pile.

Steel H-piles are spliced in the same manner as steel columns, normally by welding the sections together. Welded splices may be welded-plate or bar splices, butt-welded splices, special welded splice fittings or a combination of these. Spliced materials should be kept on the inner faces of the H-pile sections to avoid forcing a hole in the ground larger than the pile, causing at least a temporary loss in frictional value and lateral support that might result in excessive bending stresses.

Steel pipe piles may be spliced by butt welding, sometimes using straps to guide the sections and provide more strength to the welded joint. Another method is to use inside sleeves having a driving fit, with a flange extending between the pipe sections. By applying bituminous cement or compound on the outside of the ring before driving, a water-tight joint is obtained.

1808.2.8 Allowable pier or pile loads.

❖ This section provides the general requirements for the determination of allowable pile loads based on soil characteristics and capacity, as well as on installation conditions involving driveability; pier and pile stresses; overloads and other similar criteria.

1808.2.8.1 Determination of allowable loads. The allowable axial and lateral loads on piers or piles shall be determined by an approved formula, load tests or method of analysis.

❖ There are two general considerations for determining capacity as required for the design and installation of pier or pile foundations. The first consideration involves the determination of the underlying soil or rock characteristics. The second is the application of approved driving formulas, load tests or accepted methods of analysis to determine the pier or pile capacities required to resist the axial and lateral loads they will be subjected to, as well as to provide the basis for the proper selection of pile-driving equipment.

1808.2.8.2 Driving criteria. The allowable compressive load on any pile where determined by the application of an approved driving formula shall not exceed 40 tons (356 kN). For allowable loads above 40 tons (356 kN), the wave equation method of analysis shall be used to estimate pile driveability of both driving stresses and net displacement per blow at the ultimate load. Allowable loads shall be verified by load tests in accordance with Section 1808.2.8.3. The formula or wave equation load shall be determined for gravity-drop or power-actuated hammers and the hammer energy used shall be the maximum consistent with the size, strength and weight of the driven piles. The use of a follower is permitted only with the approval of the building official. The introduction of fresh hammer cushion or pile cushion material just prior to final penetration is not permitted.

❖ It has been accepted practice for many decades to predict the load capacity of a pile by its resistance to driving as determined by a pile-driving formula. The simple premise upon which a pile-driving formula is founded is that as the resistance of a pile to driving increases, the pile's capacity to support loads also increases. While several pile formulas have been developed over the years, none have been completely dependable.

The Engineering-News formula is the simplest and probably the most widely used in the United States. This calculation method, as well as other formulas in common use today, have generally shown poor correlations with load test results. However, the comparative differences between pile capacities as determined by driving formulas and the results of load tests are much smaller for soils consisting of free-draining, coarse-grained materials, such as sand and gravel, than for soils consisting of silt, clay or fine, dense sand.

The use of pile-driving formulas to determine pile capacities should generally be avoided, except on small jobs where the piles are to be driven in well-drained granular soils, and the cost of load testing cannot be justified.

1808.2.8.3 Load tests. Where design compressive loads per pier or pile are greater than those permitted by Section 1808.2.10 or where the design load for any pier or pile foundation is in doubt, control test piers or piles shall be tested in accordance with ASTM D 1143 or ASTM D 4945. At least one pier or pile shall be test loaded in each area of uniform subsoil conditions. Where required by the building official, additional piers or piles shall be load tested where necessary to establish the safe design capacity. The resulting allowable loads shall not be more than one-half of the ultimate axial load capacity of the test pier or pile as assessed by one of the published methods listed in Section 1808.2.8.3.1 with consideration for the test type, duration and subsoil. The ultimate axial load capacity shall be determined by a registered design professional with consideration given to tolerable total and differential settlements at design load in accordance with Section 1808.2.12. In subsequent installation of the balance of foundation piles, all piles shall be deemed to have a supporting capacity equal to the control pile where such piles are of the same type, size and relative length as the test pile; are installed using the same or comparable methods and equipment as the test pile; are installed in similar subsoil conditions as the test pile; and, for driven piles, where the rate of penetration (e.g., net displacement per blow) of such piles is equal to or less than that of the test pile driven with the same hammer through a comparable driving distance.

❖ The most reliable method for determining pile capacity is by a load test. A load test should be conducted wherever feasible and used where the pile capacity is intended to exceed 40 tons (356 kN) per pile (see Section 1808.2.8.2). Test piles are to be of the same type and size intended for use in the permanent foundation and installed with the same equipment, by the same procedure and in the same soils intended or specified for the work.

Load tests are to be conducted in accordance with the requirements of ASTM D 1143 or D 4945, which covers procedures for testing vertical or batter foundation piles, individually or in groups, to determine the ul-

timate pile load (pile capacity) and whether the pile or pile group is capable of supporting the load(s) without excessive or continuous settlement. Recognition, however, must be given to the fact that load-settlement characteristics and pile capacity determinations are based on data derived at the time and under conditions of the test. The long-term performance of a pile or group of piles supporting actual loads may produce behaviors that are different than those indicated by load test results. Judgement based on experience must be used to predict pile capacity and expected behavior.

The load-bearing capacity of all piles, except those seated on rock, does not reach the ultimate load until after a period of rest. The results of load tests cannot be deemed accurate or reliable unless there is an allowance for a period of adjustment. For piles driven in permeable soils, such as coarse-grained sand and gravel, the waiting period may be as little as two or three days. For test piles driven in silt, clay or fine sand, the waiting period may be 30 days or longer. The waiting period may be determined by testing (i.e., by redriving piles) or from previous experience.

This section requires that at least one pile be tested in each area of uniform subsoil conditions. The statement should not be misconstrued to mean that the tested area is to have only one uniform stratum of subsurface material, but rather that the soil profile, which may consist of several layers (strata) of different materials, must represent a substantially unchanging crosssection in each area to be tested.

The allowable pile load to be used for design purposes is not to be more than one-half of the test pile's ultimate axial load capacity, as determined in Section 1808.2.8.3.1. In establishing the pile capacity, the registered design professional must consider the tolerable settlement that can be structure dependent (see Section 1808.2.12).

The rate of penetration of production piles must be equal to or less than that of the test pile(s). All production piles should be of the same type, size and approximate length as the prototype test pile, as well as installed with comparable equipment and methods. Driven pile capacities are only valid when the same hammer is used, because different hammers of the same model, though comparable, can actually have different efficiencies. Production piles should also be installed in soils similar to those for the test pile.

1808.2.8.3.1 Load test evaluation. It shall be permitted to evaluate pile load tests with any of the following methods:

1. Davisson Offset Limit.

2. Brinch-Hansen 90% Criterion.

3. Butler-Hoy Criterion.

4. Other methods approved by the building official.

❖ This section lists generally accepted methods that can be used to determine the ultimate axial load capacity of test piles. Since no single method applies to all situations that are encountered, this list provides the necessary latitude to select a method of load test evaluation that is appropriate for the type of pile being tested, the test procedure and the subsurface conditions. The Davisson Offset method is perhaps the most widely used method of test load evaluation. It has proven to provide capacities that are conservative, yet reasonable. The Brinch-Hansen 90% Criterion and Butler-Hoy Criterion are considered a little less conservative than Davisson, but are viable methods of evaluating test piles.

1808.2.8.4 Allowable frictional resistance. The assumed frictional resistance developed by any pier or uncased cast-in-place pile shall not exceed one-sixth of the bearing value of the soil material at minimum depth as set forth in Table 1804.2, up to a maximum of 500 psf (24 kPa), unless a greater value is allowed by the building official after a soil investigation, as specified in Section 1802, is submitted or a greater value is substantiated by a load test in accordance with Section 1808.2.8.3. Frictional resistance and bearing resistance shall not be assumed to act simultaneously unless recommended by a soil investigation as specified in Section 1802.

❖ Under certain circumstances, such as when the pile extends through cohesive soils, like clays, to a bearing stratum of compact sand and gravels, both skin friction and end bearing act together to support the pile. However, the nature of load sharing between the two and whether in fact both exist simultaneously can be determined only by a soils investigation. Thus, the code requires that in order to assume that both skin friction and end bearing act simultaneously, the assumption must be justified by means of a soils investigation.

1808.2.8.5 Uplift capacity. Where required by the design, the uplift capacity of a single pier or pile shall be determined by an approved method of analysis based on a minimum factor of safety of three or by load tests conducted in accordance with ASTM D 3689. The maximum allowable uplift load shall not exceed the ultimate load capacity as determined in Section 1808.2.8.3 divided by a factor of safety of two. For pile groups subjected to uplift, the allowable working uplift load for the group shall be the lesser of:

1. The proposed individual pile uplift working load times the number of piles in the group.

2. Two-thirds of the effective weight of the pile group and the soil contained within a block defined by the perimeter of the group and the length of the pile.

❖ Piles and piers subjected to uplift forces act in tension and are actually friction piles. The amount of tension that can be sustained by a pier or pile depends on the strength of the pier or pile material and the frictional or cohesive properties of the soil. Tensile resistance is not necessarily correlated with the bearing capacity of a pier or pile under compressive load. For example, the tensile resistance of a friction pile in clay will usually be about the same value as its bearing capacity because the skin friction developed in cohesive soils is very large. In comparison, a friction pile installed in granular materials (noncohesive), such as sand, will

develop a tensile resistance that is considerably less than its bearing capacity.

Analytical methods can be used to determine the ultimate uplift resistance of a pier or pile, provided that the properties of the soil are well known. When the ultimate uplift resistance is established by analysis, a safety factor of three must be applied to determine the allowable uplift load of the pile.

The response of a pier or vertical or batter pile to an axially applied uplift force is best determined by an extraction test performed in accordance with the requirements of the ASTM D 3689 standard listed in Chapter 35 and in accordance with the provisions of this section.

Piers and piles must be well anchored into the pile cap by adequate connection devices in order to be effective in resisting uplift forces. In turn, the pile cap must also be designed to resist uplift stresses. Sometimes, it is necessary to give special consideration in the design of the pile itself to take the tensile stresses imposed by uplift conditions. For example, a cast-in-place or precast concrete pile must be designed so that the tensile reinforcement will extend the full length of the pile. Special consideration should also be given to the design of pile splices that are to act in tension.

1808.2.8.6 Load-bearing capacity. Piers, individual piles and groups of piles shall develop ultimate load capacities of at least twice the design working loads in the designated load-bearing layers. Analysis shall show that no soil layer underlying the designated load-bearing layers causes the load-bearing capacity safety factor to be less than two.

❖ The bearing capacity of a pier or pile, whether it is a single acting pile or part of a group, is determined as a pier- or pile-soil system. In this respect, pier and pile-bearing capacity is a function of either the strength properties of the pier or pile or the supporting strength of the soil. Obviously, the bearing capacity is controlled by the smaller value obtained in the two considerations.

In most cases, the supporting strength of the soil governs the bearing capacity of a pier or pile foundation. This section requires that the bearing capacity of piers or an individual pile or group of piles must not be more than one-half of the ultimate load capacities of the piers or piles as a function of the bearing capacity of the soil.

Sometimes, soils investigations show that weaker layers of soil underlie the intended bearing strata. To avoid damaging settlements, the weaker soils must have a safety factor of 2 or more as determined by analytical methods. Where the safety factor is less than 2, piers or piles must either be driven to deeper bearing soils to obtain adequate and safe support or the design capacity of the piers or piles must be reduced, thus increasing the total number of piles in the foundation system.

1808.2.8.7 Bent piers or piles. The load-bearing capacity of piers or piles discovered to have a sharp or sweeping bend shall be determined by an approved method of analysis or by load testing a representative pier or pile.

❖ This section requires that piers or piles that are discovered to have sharp or sweeping bends, usually occurring because of obstructions encountered during driving operations, must be analyzed by an approved method or load tested by a representative pier or pile to determine their load-carrying capacity. Where acceptable, such piers or piles may be used at a reduced capacity; otherwise, they should be abandoned and replaced.

1808.2.8.8 Overloads on piers or piles. The maximum compressive load on any pier or pile due to mislocation shall not exceed 110 percent of the allowable design load.

❖ Because of subsurface obstructions or other reasons, it is sometimes necessary to offset piers or piles a small distance from their intended locations or they may be driven out of position. In such cases, the load distribution in a group of piles may be changed from the design requirements and cause some of the piles to be overloaded. This section requires that the maximum compressive load on any pier or pile caused by mislocation should not exceed 110 percent of the allowable design load. Piers or piles exceeding this limitation must be extracted and installed in the proper location or other approved remedies must be applied, such as installing additional piers or piles to balance the group.

1808.2.9 Lateral support.

❖ This section provides specific requirements for lateral support of piers and piles and the allowable lateral load.

1808.2.9.1 General. Any soil other than fluid soil shall be deemed to afford sufficient lateral support to the pier or pile to prevent buckling and to permit the design of the pier or pile in accordance with accepted engineering practice and the applicable provisions of this code.

❖ Long experience with pile installations and performance under loaded conditions has shown that piles surrounded by earth, including even very soft and compressible clays, provide sufficient lateral restraint to prevent buckling.

1808.2.9.2 Unbraced piles. Piles standing unbraced in air, water or in fluid soils shall be designed as columns in accordance with the provisions of this code. Such piles driven into firm ground can be considered fixed and laterally supported at 5 feet (1524 mm) below the ground surface and in soft material at 10 feet (3048 mm) below the ground surface unless otherwise prescribed by the building official after a foundation investigation by an approved agency.

❖ Piles driven in fluid soils, such as saturated silts, or in water, or piles that extend up into open air are not laterally supported and may be subject to high bending stresses and subsequent buckling. Under such conditions, piles must be designed as columns in accordance with the applicable provisions of the code. This section designates a point of fixity [5 feet (1524 mm) below ground surface for firm soils and 10 feet (3048 mm) below ground surface for soft soils] without requiring a complete soils investigation. This provision essentially provides a starting point for analyzing the fixity of the pile.

1808.2.9.3 Allowable lateral load. Where required by the design, the lateral load capacity of a pier, a single pile or a pile group shall be determined by an approved method of analysis or by lateral load tests to at least twice the proposed design working load. The resulting allowable load shall not be more than one-half of that test load that produces a gross lateral movement of 1 inch (25 mm) at the ground surface.

❖ Because of wind loads, unbalanced building loads, earth pressures and the like, it is inevitable that piers, individual piles or groups of vertical piles supporting buildings or other structures will be subjected to lateral forces. The distribution of these lateral forces to piers or piles largely depends on how the loads are carried down through the structural framing system and transferred through the supporting foundation to the piers or piles. The amount of lateral load that can be taken by the pier or pile foundation is a function of the type of pier or pile used; the soil characteristics, particularly in the upper 10 feet (3048 mm) of the piers or piles; the embedment of the pier or pile head (fixity); the magnitude of the axial compressive load on the pier or pile; the nature of the lateral forces and the amount of horizontal pier or pile movement deemed acceptable.

The degree of fixity of the pier or pile head is an important design consideration under very high lateral loading unless some other method, such as the use of batter piles, is employed to resist lateral loads. The fixing of the pier or pile head against rotation reduces the lateral deflection. In general, pile butts are embedded 3 inches (76 mm) to 4 inches (102 mm) into the pile cap (see Section 1808.2.3) with no ties to the cap. These pile heads are neither fixed nor free, but somewhere in the middle. Such construction is satisfactory for most loading conditions.

The magnitude of friction developed between the surfaces of two structural elements in contact with each other is a function of the weight or loads applied. The larger the weight, the greater the frictional resistance developed. In the design of pile foundations, frictional resistance between the soil and the bottom of the pile caps (footings) should not be relied on to provide lateral restraint, since the vertical loads are transmitted through the piles to the supporting soil below and not to the ground immediately under the pile caps. Only the weights of the pile caps can supply some frictional resistance because such footings are constructed by placing fresh concrete on the soil, thus providing a positive contact. The weights of the pile caps in comparison to the magnitude of loads and lateral forces transmitted to the piles is nominal, however, and not too significant from a structural design standpoint. Also, in rare occurrences, soil has been known to settle under pile caps, leaving open spaces and thus eliminating the development of any frictional restraint.

Generally, about $^1/_4$-inch (6.4 mm) horizontal movement of a pier or pile is considered acceptable without tests. Piers or piles with their upper sections embedded in deep strata of very soft or soft clays and silts should not be relied on to resist lateral forces of more than 1,000 pounds (4.45 kN) per pile.

Where vertical piles are subjected to lateral forces exceeding acceptable limitations, the use of batter piles may be required. Lateral forces on many structures are also resisted by the embedded foundation walls and the sides of the pile caps.

The allowable lateral load capacity of a pier, single pile or group of piles is to be determined either by approved analytical methods or load tests. Load tests are to be conducted to produce lateral forces that are twice the proposed design load; however, in no case is the allowable pier or pile load to exceed one-half of the test load, which produces a gross lateral pier or pile movement of 1 inch (25 mm) as measured at the ground surface.

1808.2.10 Use of higher allowable pier or pile stresses. Allowable stresses greater than those specified for piers or for each pile type in Sections 1809 and 1810 are permitted where supporting data justifying such higher stresses is filed with the building official. Such substantiating data shall include:

1. A soils investigation in accordance with Section 1802.

2. Pier or pile load tests in accordance with Section 1808.2.8.3, regardless of the load supported by the pier or pile.

The design and installation of the pier or pile foundation shall be under the direct supervision of a registered design professional knowledgeable in the field of soil mechanics and pier or pile foundations who shall certify to the building official that the piers or piles as installed satisfy the design criteria.

❖ In other parts of this chapter, limitations are specified for the stress values used for design purposes. These allowable stresses are stated as a percentage of some limiting strength property of the pier or pile material. For example, allowable design stresses for piles made of steel are stated as a percentage of the yield strengths of the several grades of steel typically used for pile construction. For concrete, the allowable design stress is prescribed as a percentage of the 28-day specified compression strength. The allowable design stresses permitted for timber piles are based on the natural strengths of the different species of wood used for pile construction. The values have been developed and tabulated by the AF&PA and include reductions in pile strengths because of preservative treatment.

The allowable design stresses stipulated in the code for the different types of piles provide an adequate factor of safety against the dynamic forces of pile driving that may cause damage to the piles and prevent overstresses because of loading and subsoil conditions.

This section allows the use of higher allowable stresses when evidence supporting the values is submitted and approved by the building official. The data submitted to the building official should include analytical evaluations and findings from a foundation investigation as specified in Section 1808.2.1, and the results of load tests performed in accordance with the requirements of Section 1808.2.8.3. The technical data and the recommendation for the use of higher stress values must come from a registered engineer who is knowledgeable in soil mechanics and experienced in the design of pile foundations. This engineer is to supervise the pile design work and witness the installation of the pile foundation so as to certify to the building official that the construction satisfies the design criteria. In any case, the use of greater design stresses is not to result in permitting design loads that are larger than one-half of the test loads (see Section 1808.2.8.3).

1808.2.11 Piles in subsiding areas. Where piles are installed through subsiding fills or other subsiding strata and derive support from underlying firmer materials, consideration shall be given to the downward frictional forces that may be imposed on the piles by the subsiding upper strata.

Where the influence of subsiding fills is considered as imposing loads on the pile, the allowable stresses specified in this chapter are permitted to be increased where satisfactory substantiating data are submitted.

❖ Where compacted fill is placed over compressible soils, the underlying material will consolidate because of the added weight. The depth to suitable bearing material will, over a period of time, be shifted downward by the forces of the subsiding soil. Such forces caused by the weight of the fill are transmitted to the piles by skin friction and, in effect, serve as added loads on the piles. The magnitude of such loads must be determined by accepted analytical methods and be taken into account in the design of pile foundations.

1808.2.12 Settlement analysis. The settlement of piers, individual piles or groups of piles shall be estimated based on approved methods of analysis. The predicted settlement shall cause neither harmful distortion of, nor instability in, the structure, nor cause any stresses to exceed allowable values.

❖ A settlement analysis is performed to design a pier or pile foundation system that will maintain the stability and structural integrity of the supported building or structure. Foundation systems that suffer serious settlements, particularly differential settlements, can cause great structural damage to the supported structure as well as to the foundation itself.

The settlement analysis of an individual pier or pile is a complex procedure. In comparison, the analysis of a group of piles is even more complex because of the overlapping soil stresses caused by closely spaced piles. Analytical procedures vary with the type of piers or piles and especially with the type of soil.

Settlements are of two basic types: immediate settlements are those that occur as soon as the load is applied and usually take place within a period of less than 7 days; consolidation settlements are time dependent and take place over a long period of time. All cohesionless soils, such as granular materials consisting of sand, gravel or a mixture of both, which have a large coefficient of permeability (rapid draining properties), undergo immediate settlements. All fine-grained, saturated, cohesive soils, such as clays, undergo time-dependent consolidation settlements.

Settlement analysis would generally include cases involving piers and end-bearing piles on rock or hard soils as well as friction-type piles in both granular and cohesive soils. Load tests are often performed to provide data that will aid the settlement analysis.

The code places no specific limitation on the amount of settlement; instead, it states that the effects of any settlement, particularly differential settlements, should not be harmful to the structure. Thus, the tolerable settlement is largely dependent on the type of structure being considered. Guidance on tolerable settlements can be found in engineering text books such as *Soil Mechanics in Engineering Practice* (Terzaghi and Peck), *Foundation Analysis and Design* (Bowles) and *Foundation Engineering* (Hanson, Peck and Thornburn). As a rule, these references indicate that a total settlement of 1-inch (25 mm) is acceptable for the majority of structures while other structures can tolerate even greater settlements without distress. On the other hand, a more restrictive settlement criterion can be necessary based on the needs of a specific structure. Current engineering practice is often based on an allowable total settlement of 1-inch (25 mm) with the objective of controlling the differential settlements to $^3/_4$ inch (19.1 mm) or less. The effects of differential settlements need to be considered in the structural design as a self-straining force, T, when designing for the load combinations of Section 1605.

1808.2.13 Preexcavation. The use of jetting, augering or other methods of preexcavation shall be subject to the approval of the building official. Where permitted, preexcavation shall be carried out in the same manner as used for piers or piles subject to load tests and in such a manner that will not impair the carrying capacity of the piers or piles already in place or damage adjacent structures. Pile tips shall be driven below the preexcavated depth until the required resistance or penetration is obtained.

❖ The use of preexcavation methods to facilitate the installation of piers or piles is often necessary for a number of reasons, including:
- To drive piles through upper layers of hard soil;
- To penetrate through subsurface obstructions;
- To eliminate or reduce the possibility of ground heave that could result in lifting adjacent piles already driven;
- To reduce ground pressures that could result in the lateral movement of adjacent piles or structures;
- To reduce the amount of driving necessary to seat the piles in the required bearing stratum;
- To reduce the possibility of damaging vibrations;
- To reduce the amount of noise associated with pile-driving operations; and

- To accommodate the placement of piles that are longer than the leads of the driving equipment.

Methods commonly used for preexcavation are jetting and predrilling. The jetting method has been found to be more effective in granular soils, such as sand and fine gravel, than in cohesive materials, such as clay. However, jetting should not be done in granular soils containing very coarse gravel, cobbles and small boulders because such material cannot be removed by the jet stream and will result in a collection of stones at the bottom of the hole, making it very difficult, if not impossible, to drive a pile through the mass of stone material. The jetting operations must be controlled to avoid excessive losses of soil that could affect the stability of adjacent structures or the required bearing capacity of previously installed piles.

Jetting operations must be carefully controlled to avoid excessive loss of soil, which could affect the load-bearing capacity of piles already installed or the stability of adjacent structures.

Piles should be driven below the depth of the jetted hole until the required resistance or penetration is obtained. Before this preexcavation method is used, consideration should be given to the possibility that jetting, unless strictly controlled, can adversely affect load transfer, particularly as it involves the placement of nontapered piles.

In comparison to the jetting method described, a more controllable form of preexcavation is by predrilling or coring. This method greatly reduces the possibility of detrimental effects on adjacent piles or structures and can be performed as a dry operation or a wet rotary process. Dry drilling can be done by the use of a continuous-flight auger or a short-flight auger attached to the end of a drill stem or kelly bar. Wet drilling requires a hollow-stem, continuous-flight auger or a hollow drill stem employing the use of spade bits. When the wet rotary process of predrilling is used, bentonite slurry or plain water is circulated to keep the hole open. As in the case of jetting, piles should be driven with tips below the predrilled hole. This is necessary to prevent any voids or very loose or soft soils from occurring below the pile tip.

There are other methods used for preexcavation purposes, such as the dry tube method and spudding, but such procedures are seldom used. The methods to be employed for preexcavation are subject to the approval of the building official.

1808.2.14 Installation sequence. Piles shall be installed in such sequence as to avoid compacting the surrounding soil to the extent that other piles cannot be installed properly, and to prevent ground movements that are capable of damaging adjacent structures.

❖ As piles are driven sequentially within a group, progressive compaction of the surrounding soil occurs and there can be a buildup of unequal pressures around the piles, possibly causing some of them to be deflected off-line.

This situation is more apt to develop when piles are

closely spaced. More importantly, soil compaction during driving operations can cause variations in pile lengths within a group to the extent that some of the piles may stop considerably short of reaching specified bearing material. Ground heave is another possible effect of soil compaction (see Section 1808.2.19).

To prevent or significantly reduce the problems associated with soil compaction, the driving sequence becomes an important consideration in the planning of pile installation work. For example, in placing a group of piles, if the outer piles within the group are driven first, the inner piles may be short of their required length because of soil compaction that builds up progressively toward the center of the cluster. Starting the pile-driving sequence at the edge of a group makes the piles progressively more difficult to drive, resulting in a one-sided bearing group. The generally accepted driving sequence is to work from the center of a cluster of piles outward. For pile arrangements consisting of rows of widely spaced piles (i.e., under a continuous concrete mat), driving can be done sequentially from one side to the other.

1808.2.15 Use of vibratory drivers. Vibratory drivers shall only be used to install piles where the pile load capacity is verified by load tests in accordance with Section 1808.2.8.3. The installation of production piles shall be controlled according to power consumption, rate of penetration or other approved means that ensure pile capacities equal or exceed those of the test piles.

❖ The use of vibratory drivers for the installation of piles is not applicable to all types of soil conditions. They are most effective in granular soils with the use of nondisplacement piles, such as steel H-piles and pipe piles driven open ended. Vibratory drivers are also used for extracting piles or temporary casings employed in the construction of cast-in-place concrete piles.

Vibratory drivers, either low or high frequency, cause the pile to penetrate the soil by longitudinal vibrations. While this type of pile driver can produce remarkable results in the installation of nondisplacement piles under favorable soil conditions, the greatest difficulty is the lack of a reliable method of estimating the load-bearing capacity. After the pile has been installed with a vibratory driver, pile capacity can best be determined by using an impact-type hammer to set the pile in its final position.

An acceptable means of controlling pile capacity is by determining the power consumption in relation to the rate of penetration. Nonetheless, the use of a vibratory driver is only permitted where the pile load capacity is established by load tests in accordance with the requirements of Section 1808.2.8.3.

1808.2.16 Pile driveability. Pile cross sections shall be of sufficient size and strength to withstand driving stresses without damage to the pile, and to provide sufficient stiffness to transmit the required driving forces.

❖ Additionally, the pile must be capable of transmitting dynamic driving forces to mobilize the required ulti-

mate pile capacity within the soil without severe elastic energy losses. Pile driveability depends on the pile stiffness, which is a function of pile length, cross-sectional area and modulus of elasticity (MOE). It should be noted that the yield strength does not affect stiffness; thus, caution should be observed in the use of high yield-strength steels for higher loads on smaller cross sections requiring higher dynamic driving energy. A wave equation analysis would reflect pile stiffness or driveability.

The selection of pile types and dimensional requirements for driveability is a function of soil characteristics.

1808.2.17 Protection of pile materials. Where boring records or site conditions indicate possible deleterious action on pier or pile materials because of soil constituents, changing water levels or other factors, the pier or pile materials shall be adequately protected by materials, methods or processes approved by the building official. Protective materials shall be applied to the piles so as not to be rendered ineffective by driving. The effectiveness of such protective measures for the particular purpose shall have been thoroughly established by satisfactory service records or other evidence.

❖ Piles are often exposed to the deteriorating effects of biological, chemical and physical actions caused by a hostile underground environment that may exist at the time of their installation or that may later develop at the site. Under such conditions, pile materials must be properly protected to ensure their expected durability.

The problems associated with pile durability relate directly to the type of pile materials used. For example, concrete piles that are entirely embedded in undisturbed soil are deemed to be permanent installations. The level of the ground-water table is generally not a factor affecting the durability of concrete piles. Ground water that contains deleterious substances and readily flows through disturbed or granular soils, such as sand and gravel (regardless of the level of the water table), can have a deteriorating effect on concrete piles. Concrete piles embedded in fine grained, impervious soils, such as clay, generally are not adversely affected by ground water containing harmful substances. Concrete can also be affected by exposure to soils having a high sulfate content, unless Type II or V portland cement is used in making the concrete mixture.

Concrete piles installed in saltwater, such as for buildings or other structures in waterfront construction, are subject to chemical action from polluted waters, frost action on porous concrete, spalling and rusting of steel reinforcement. Spalling action may become particularly critical under tidal conditions where alternate wetting and drying of the concrete occurs in conjunction with cycles of freezing and thawing. Generally, concrete can be protected from damage by such adverse conditions with the use of special cements, dense concrete mixtures rich in cement content, adequate concrete cover over the reinforcement, air entrainment, suitable concrete admixtures or special surface coatings.

Piles made of steel materials and embedded entirely in undisturbed soil (regardless of its type) are not significantly affected by corrosion due to oxidation, mainly because undisturbed soil is so deficient in oxygen that progressive corrosion is repressed. However, steel may be subject to serious corrosion and structural deterioration where ground water contains deleterious substances from coal piles, alkali soils, active cinder fills, chemical waste from manufacturing operations, etc. Under such conditions, steel piles may be protected by encasement in concrete or by applying protective coatings such as coal-tar or other suitable materials. Steel piles installed in saltwater or exposed to a saltwater environment can corrode severely and should be protected by encasement in concrete or by the application of approved coatings. Piles that extend above ground level and are exposed to air should be painted in the same way as any type of structural steel construction to prevent rusting. Corrosion of load-bearing steel can also occur because of electrolytic action, and in such cases, cathodic protection may be required.

Timber piles totally embedded in the earth below the low point of ground-water level or entirely submerged in fresh water will last indefinitely without preservative treatment. However, timber piles that extend above the ground-water level or are exposed to air or saltwater are subject to decay as well as attacks by insects and marine borers. The piles may also be subjected to damage by the percolation of ground water heavily charged with alkali and acids. Under such conditions, timber piles must be pressure treated with preservatives in accordance with AWPA Standard U1 listed in Chapter 35.

Generally, at any site where piles are to be installed and where the soil is suspect or there is sufficient evidence of an adverse underground environment, a soils investigation should be conducted to determine the need and method to protect the piles against possible deterioration.

1808.2.18 Use of existing piers or piles. Piers or piles left in place where a structure has been demolished shall not be used for the support of new construction unless satisfactory evidence is submitted to the building official, which indicates that the piers or piles are sound and meet the requirements of this code. Such piers or piles shall be load tested or redriven to verify their capacities. The design load applied to such piers or piles shall be the lowest allowable load as determined by tests or redriving data.

❖ After the demolition of an existing building, piers or piles remaining in place cannot be reused to support a new structure unless sufficient and reliable information is provided to the building official showing that the new loads to be imposed on the existing pier or pile system will be adequately supported. This requirement is necessary because of the lack of adequate soil data and technical information on the pier or piling material used and the unavailability of pile-driving records made during the construction of the existing pier or pile foundation. The true condition of the piers or piles is not known, since they may have deteriorated over time, possibly reducing their load capacity. Pier or pile capacities may be determined by load tests, or the piles may be retracted and redriven to verify their load capacities.

1808.2.19 Heaved piles. Piles that have heaved during the driving of adjacent piles shall be redriven as necessary to develop the required capacity and penetration, or the capacity of the pile shall be verified by load tests in accordance with Section 1808.2.8.3.

❖ When piles are driven into cohesive soils, particularly saturated plastic clay materials, they often displace a volume of soil equal to that of the piles themselves. Such soil displacement usually occurs as ground heaves and may cause adjacent piles already driven to lift up and become unseated, causing a loss of pile load capacity. When this happens, heaved piles must be redriven to firm bearing in order to regain their required capacity. If heaved piles are not redriven, their capacities must be verified by means of load tests.

It should be noted that not all types of piles can be redriven. For example, heaved uncased cast-in-place concrete piles, or sectional piles whose splices cannot take tension, should be abandoned and replaced.

In redriving heaved piles, the same or comparable driving equipment should be employed as used in the original installation; however, there are exceptions to this rule. For example, during the installation of concrete-filled pipe piles, only the empty pipes were first driven. In redriving this type of pile, the pipes may now be filled with concrete resulting in much stiffer and heavier sections than were driven initially. In such cases, the driving technique must be adjusted to accommodate a considerably lesser driving resistance.

One method commonly used to prevent or reduce objectionable displacements caused by pile installations in soils subject to heaving is to preexcavate pile holes in accordance with the requirements of Section 1808.2.13.

1808.2.20 Identification. Pier or pile materials shall be identified for conformity to the specified grade with this identity maintained continuously from the point of manufacture to the point of installation or shall be tested by an approved agency to determine conformity to the specified grade. The approved agency shall furnish an affidavit of compliance to the building official.

❖ All pier or pile materials must be identified for conformity to construction specifications, providing information such as strength (species and grade for timber piles) and dimensions and other pertinent information as may be required. Such identification must be provided for all piers and piles, whether they are taken from the manufacturer's stock or made for a particular project. Identification is to be maintained from the place of manufacture to the shipment, on-site handling, storage and installation of the piers or piles. Manufacturers, upon request, usually furnish certificates of compliance with construction specifications.

In the absence of adequate data, piers and piles are to be tested to prove conformity to the specified grade.

In addition to mill certificates (steel piles), identification is made through plant manufacturing or inspection reports (precast concrete piers or piles and timber piles) and delivery tickets (concrete). Timber piles are stamped (branded) with information such as producer, species, treatment and length.

Identification is essential when high yield-strength steel is specified. Frequently, pile cutoff lengths are re-used and pile material may come from a jobber, a contractor's yard or a material supplier. In such cases, mill certificates are not available and the steel should be tested to see if it complies with the specifications.

1808.2.21 Pier or pile location plan. A plan showing the location and designation of piers or piles by an identification system shall be filed with the building official prior to installation of such piers or piles. Detailed records for piers or individual piles shall bear an identification corresponding to that shown on the plan.

❖ This section requires that a pier or pile location plan clearly showing the designation of all piers or piles in the foundation system be submitted to the building official prior to installation of piles. Preferably, such plans should be submitted before delivery of the piers or piles to the construction site and prior to staking the pier or pile locations.

The pier or pile location plan is not only an important tool for the installation of piers or piles, but also serves to communicate information between the owner, engineer, contractor, manufacturer, special inspector, building official and other interested persons. The special inspector (see commentary, Section 1808.2.22) must keep piling logs, records and reports based on this identification system. The use of the pier or pile plan is particularly important at sites where the variations in soil profiles are so extensive that it becomes necessary to manufacture piers or piles of different lengths to reach proper bearing levels. The building official should receive revised copies of the pier or pile location plan whenever field changes are made that add, delete or relocate piles.

1808.2.22 Special inspection. Special inspections in accordance with Sections 1704.8 and 1704.9 shall be provided for piles and piers, respectively.

❖ This section requires special inspections (see definition in Section 1702) in accordance with Sections 1704.8 and 1704.9 to verify proper and safe installations. To be approved by the building official, an inspector should be experienced in pier or pile installation work. An inspector must be able to read and understand foundation plans and specifications, maintain accurate records, have a thorough understanding of the scope of the pile work to be performed, give concise and timely reports to the owner or engineer and submit in writing to the building official whatever field information is required.

The duties of the pier or pile inspector include the inspection and approval of pile-driving equipment, the inspection and acceptance of piles furnished for the work and the verification of all pile location stakes and spacings. Duties also include the observation and recording of required information relating to the installation of the pier or pile foundation, including keeping a log for each pile showing either the number of hammer blows for each foot of penetration and the final penetration in blows per inch or the driving record for only the last few feet of penetration. The driving log should also include information on the duration and cause of any delays, such as splicing time, equipment breakdown, changing cushions, etc. The record should also show the final tip and butt (cutoff) elevations, as well as the depths of preexcavation holes (see Section 1808.2.13). Finally, there should also be a record of any pile damage and repair work; pile extractions and replacements; observations on pile heaving and depth of redriving and information on pile alignment.

1808.2.23 Seismic design of piers or piles.

❖ This section includes additional requirements concerning the seismic design of piers and piles (as defined in Section 1808.1) supporting structures assigned to Seismic Design Category C and higher. For more information regarding piers and piles, consult the following publications, which are listed in the bibliography to this chapter: *Recommendations for Design, Manufacture, and Installation of Concrete Piles* (ACI 543R-74), *Design and Construction of Drilled Piers* (ACI 336.3R-93) (also *ACI Manual of Concrete Practice*, Part 4) and "Recommended Practice for Design, Manufacture and installation of Prestressed Concrete Piling," *PCI Journal*.

1808.2.23.1 Seismic Design Category C. Where a structure is assigned to Seismic Design Category C in accordance with Section 1613, the following shall apply. Individual pile caps, piers or piles shall be interconnected by ties. Ties shall be capable of carrying, in tension and compression, a force equal to the product of the larger pile cap or column load times the seismic coefficient, S_{DS}, divided by 10 unless it can be demonstrated that equivalent restraint is provided by reinforced concrete beams within slabs on grade, reinforced concrete slabs on grade, confinement by competent rock, hard cohesive soils or very dense granular soils.

Exception: Piers supporting foundation walls, isolated interior posts detailed so the pier is not subject to lateral loads, lightly loaded exterior decks and patios of Group R-3 and U occupancies not exceeding two stories of light-frame construction, are not subject to interconnection if it can be shown the soils are of adequate stiffness, subject to the approval of the building official.

❖ The purpose of this section is to preclude excessive movement of one pile or pile group with respect to another. The section is similar to Section 1805.4.2.2. One of the prerequisites of adequate structural performance during an earthquake is that the foundation of the structure must act as a unit. This is typically accomplished by tying together the pile caps, piers or piles with ties capable of carrying, both in tension and compression, a force equal to 10 percent of the larger pile cap or column load multiplied by S_{DS}, which is the design spectral response accelerations at short periods as determined in Section 1613.5.4. This can be accomplished through the use of concrete floor slabs or tie beams. Reliance upon lateral soil pressure is typically not a good idea because the motion is imparted from soil to structure (not inversely as is commonly assumed), and if the soil is soft enough to require piles, little reliance can be placed on soft-soil passive pressure to restrain relative pile displacement under dynamic conditions. If soils are shown to be of adequate stiffness, the building official may allow the use of the exception for the following piers in Group R-3 and U occupancies not exceeding two stories of light-frame construction:

1. Piers supporting foundation walls.

2. Piers supporting isolated interior posts detailed so the pier is not subject to lateral loads.

3. Piers supporting lightly loaded exterior decks and patios.

1808.2.23.1.1 Connection to pile cap. Concrete piles and concrete-filled steel pipe piles shall be connected to the pile cap by embedding the pile reinforcement or field-placed dowels anchored in the concrete pile in the pile cap for a distance equal to the development length. For deformed bars, the development length is the full development length for compression or tension, in the case of uplift, without reduction in length for excess area. Alternative measures for laterally confining concrete and maintaining toughness and ductile-like behavior at the top of the pile will be permitted provided the design is such that any hinging occurs in the confined region.

Ends of hoops, spirals and ties shall be terminated with seismic hooks, as defined in Section 21.1 of ACI 318, turned into the confined concrete core. The minimum transverse steel ratio for confinement shall not be less than one-half of that required for columns.

For resistance to uplift forces, anchorage of steel pipe (round HSS sections), concrete-filled steel pipe or H-piles to the pile cap shall be made by means other than concrete bond to the bare steel section.

Exception: Anchorage of concrete-filled steel pipe piles is permitted to be accomplished using deformed bars developed into the concrete portion of the pile.

Splices of pile segments shall develop the full strength of the pile, but the splice need not develop the nominal strength of the pile in tension, shear and bending when it has been designed to resist axial and shear forces and moments from the load combinations of Section 1605.4.

❖ This section prescribes special detailing requirements between the pile and pile cap, including required development of the longitudinal pile reinforcement in the cap and associated transverse reinforcement. This reinforcement is required to extend into the pile cap to tie the elements together and to assist in load transfer at the top of the pile to the pile cap. The connection must consist of embedment of the pile reinforcement in the pile cap for a distance equal to the development length

as specified in ACI 318. Field-placed dowels anchored in the plastic concrete piles are acceptable. The development length to be provided is that for compression or, where uplift is indicated by analysis, tension without reduction in length for excess area. Where seismic confinement reinforcement at the top of the pile is required, alternative measures for laterally confining concrete and maintaining toughness and ductile-like behavior at the top of the pile are permitted, provided the design would force the hinge to occur in the confined region. Splices of pile segments are required to develop the nominal pile strength, or splices can be designed for the special seismic load combinations of Section 1605.4.

1808.2.23.1.2 Design details. Pier or pile moments, shears and lateral deflections used for design shall be established considering the nonlinear interaction of the shaft and soil, as recommended by a registered design professional. Where the ratio of the depth of embedment of the pile-to-pile diameter or width is less than or equal to six, the pile may be assumed to be rigid.

Pile group effects from soil on lateral pile nominal strength shall be included where pile center-to-center spacing in the direction of lateral force is less than eight pile diameters. Pile group effects on vertical nominal strength shall be included where pile center-to-center spacing is less than three pile diameters. The pile uplift soil nominal strength shall be taken as the pile uplift strength as limited by the frictional force developed between the soil and the pile.

Where a minimum length for reinforcement or the extent of closely spaced confinement reinforcement is specified at the top of the pier or pile, provisions shall be made so that those specified lengths or extents are maintained after pier or pile cut-off.

❖ This section addresses miscellaneous issues unique to the seismic design of piers and piles. If the length is less than or equal to six times the least horizontal dimension of the pier, the pier can be assumed rigid and moments, shears and lateral deflections can be calculated accordingly. Where the length exceeds six times the least horizontal dimension, the nonlinear interaction of the pier or pile and soil effects are to be included in the design. The effect of abrupt changes in soil deposits, such as changes from soft to firm or loose to dense soils, should be included in the design.

This section also prescribes conditions under which group effects on the nominal pile strength, lateral as well as vertical, need to be included in the design.

The last paragraph of this section is intended to account for the condition where a pile encounters refusal at a shallower depth than intended and an unanticipated portion of the pile is cut off. It is imperative that the required reinforcement be provided at the top of the pile even when excess pile length is cut off.

1808.2.23.2 Seismic Design Category D, E or F. Where a structure is assigned to Seismic Design Category D, E or F in accordance with Section 1613, the requirements for Seismic Design Category C given in Section 1808.2.23.1 shall be met, in addition to the following. Provisions of ACI 318, Section

21.10.4, shall apply when not in conflict with the provisions of Sections 1808 through 1812. Concrete shall have a specified compressive strength of not less than 3,000 psi (20.68 MPa) at 28 days.

Exceptions:

1. Group R or U occupancies of light-framed construction and two stories or less in height are permitted to use concrete with a specified compressive strength of not less than 2,500 psi (17.2 MPa) at 28 days.

2. Detached one- and two-family dwellings of light-frame construction and two stories or less in height are not required to comply with the provisions of ACI 318, Section 21.10.4.

3. Section 21.10.4.4(a) of ACI 318 need not apply to concrete piles.

❖ This section includes additional requirements applicable to buildings assigned to Seismic Design Category D, E or F. The section is similar to Section 1805.9, except that Section 21.10.4 of ACI 318 is referenced.

Concrete piers and piles in buildings assigned to Seismic Design Category D, E or F are required to have a specified compressive strength of no less than 3,000 psi (20.68 MPa) at 28 days. This same requirement is also in Section 1805.9, which is applicable to foundations in buildings assigned to Seismic Design Category D, E or F.

An exception is provided for the use of 2,500 psi (17.2 MPa) concrete in smaller residential-type structures. In addition, small detached dwellings are exempt from the referenced ACI 318 provisions and merely need to meet the requirements for Seismic Design Category C. Exception 3 eliminates an ACI 318 requirement that is instead incorporated (with modification) in Section 1808.2.23.2.1.

1808.2.23.2.1 Design details for piers, piles and grade beams. Piers or piles shall be designed and constructed to withstand maximum imposed curvatures from earthquake ground motions and structure response. Curvatures shall include free-field soil strains modified for soil-pile-structure interaction coupled with pier or pile deformations induced by lateral pier or pile resistance to structure seismic forces. Concrete piers or piles on Site Class E or F sites, as determined in Section 1613.5.2, shall be designed and detailed in accordance with Sections 21.4.4.1, 21.4.4.2 and 21.4.4.3 of ACI 318 within seven pile diameters of the pile cap and the interfaces of soft to medium stiff clay or liquefiable strata. For precast prestressed concrete piles, detailing provisions as given in Sections 1809.2.3.2.1 and 1809.2.3.2.2 shall apply. Grade beams shall be designed as beams in accordance with ACI 318, Chapter 21. When grade beams have the capacity to resist the forces from the load combinations in Section 1605.4, they need not conform to ACI 318, Chapter 21.

❖ The first paragraph of this section requires special consideration of flexural loads on piers or piles due to earthquake motions. The following discussion taken from the NEHRP provisions commentary, Section 7.5.4, provides justification for these requirements:

Special consideration is required in the design of concrete piles subject to significant bending during earthquake shaking. Bending can become crucial to pile design where portions of the foundation piles are supported in soils, such as loose granular materials or soft soils that are susceptible to large deformations or strength degradation. Severe pile bending problems can result from various combinations of soil conditions during strong ground shaking. For example:

1. Soil settlement at the pile-cap interface either from consolidation of soft soil prior to the earthquake or from soil compaction during the earthquake can create a free-standing short column adjacent to the pile cap.

2. Large deformations, or a reduction in strength or both, resulting from liquefaction of loose granular materials can cause bending or conditions of free-standing columns.

3. Large deformations in soft soils can cause varying degrees of pile bending. The degree of pile bending will depend upon thickness and strength of the soft soil layer(s) and the properties of the soft/stiff soil interface(s).

The designer needs to consider the variation in soil conditions and driven pile lengths in providing for pile ductility at potential high curvature interfaces. Interaction between the geotechnical and structural engineers is essential.

It is prudent to design piles to remain functional during and following earthquakes in view of the fact that it is difficult to repair foundation damage. The desired foundation performance can be accomplished by proper selection and detailing of the pile foundation system. Such design should accommodate bending from both reaction to the building's inertial loads and those induced by the motions of the soils themselves.

The second paragraph of this section allows grade beams to be designed to have the strength to resist the design forces given by the special seismic load combinations instead of designing the grade beams as beams in accordance with ACI 318, Chapter 21. The special seismic load combinations estimate the maximum forces that can realistically develop in the grade beam in an earthquake situation.

1808.2.23.2.2 Connection to pile cap. For piles required to resist uplift forces or provide rotational restraint, design of anchorage of piles into the pile cap shall be provided considering the combined effect of axial forces due to uplift and bending moments due to fixity to the pile cap. Anchorage shall develop a minimum of 25 percent of the strength of the pile in tension. Anchorage into the pile cap shall be capable of developing the following:

1. In the case of uplift, the lesser of the nominal tensile strength of the longitudinal reinforcement in a concrete pile, or the nominal tensile strength of a steel pile, or the

pile uplift soil nominal strength factored by 1.3 or the axial tension force resulting from the load combinations of Section 1605.4.

2. In the case of rotational restraint, the lesser of the axial and shear forces, and moments resulting from the load combinations of Section 1605.4 or development of the full axial, bending and shear nominal strength of the pile.

❖ The requirements of this section address the need for conservatism in the design of connections between piles and pile caps. They are intended to allow energy dissipating mechanisms, such as rocking, to occur in the soil without failure of the pile.

1808.2.23.2.3 Flexural strength. Where the vertical lateral-force-resisting elements are columns, the grade beam or pile cap flexural strengths shall exceed the column flexural strength.

The connection between batter piles and grade beams or pile caps shall be designed to resist the nominal strength of the pile acting as a short column. Batter piles and their connection shall be capable of resisting forces and moments from the load combinations of Section 1605.4.

❖ This section requires that the grade beam or pile cap flexural strength exceed that of the supported column flexural strength if the column is a part of the lateral-force-resisting system.

Additional requirements are specified in this section for batter piles in order to limit earthquake damage to these systems. By their nature, batter pile systems have limited ductility and have performed poorly under strong ground motions. They are required to be designed using the special seismic load combinations of Section 1605.4, which estimate the maximum force expected to be developed in these elements during a seismic event.

SECTION 1809
DRIVEN PILE FOUNDATIONS

1809.1 Timber piles. Timber piles shall be designed in accordance with the AFPA NDS.

❖ Round timber piles are best suited as friction-type piles. They are not recommended as piles to be driven through dense gravel, boulders or till, or for end-bearing piles on rock. While timber piles do not have the high allowable load capacities of structural steel or concrete piles, they are probably the most commonly used type of pile for deep foundation construction throughout the United States, mainly because of their availability, ease of handling and, sometimes, for economic reasons.

Timber piles are shaped from tree trunks and their lengths are dependent on the heights that the various species of trees used for making piles will grow. Piles are made as tapered sections because of the natural taper of tree trunks.

Round timber piles are usually made from Southern pine in lengths up to about 80 feet (24 384 mm) and

from Pacific Coast Douglas fir in lengths up to about 125 feet (38 100 mm). Other species commonly used for piles are red oak and red pine. Timber piles that are 40 feet (12 192 mm) to 60 feet (18 288 mm) in length are common, but longer lengths cannot be obtained economically in all areas of the country.

Untreated timber piles that are embedded below the permanent ground-water level (fresh water only) may last indefinitely. Under conditions where piles are required to extend above ground-water level but still remain totally embedded in the ground, the untreated wood material may be subject to decay. In cases where untreated timber piles extend above the ground surface into the air, they are exposed to decay and insect attack. The durability of timber piles is best served by applying treatment with an approved preservative.

This section requires timber piles to be designed in accordance with the AF&PA's *National Design Specification for Wood Construction* (NDS). Section 6 of the NDS provides the appropriate design values, different species of timber piles and the applicable adjustment factors. While timber piles have to be designed in accordance with the NDS, they are usually only considered for design loads of 10 (89 kN) to 50 tons (445 kN).

One of the significant problems associated with timber pile installations is the possibility of damage caused by overdriving. Overdriving of wood piles may cause failure by bending, brooming at the tip, crushing or brooming at the butt end or splitting or breaking along the pile section (see Figure 1809.1.3). Another problem is the difficulties encountered when splicing timber piles to achieve greater lengths.

1809.1.1 Materials. Round timber piles shall conform to ASTM D 25. Sawn timber piles shall conform to DOC PS-20.

❖ The ASTM D 25 standard referenced in this section covers the physical characteristics of treated and untreated round timber piles. Essentially, the standard divides timber piles into two classifications: friction and end-bearing piles. The dimensional requirements for each classification are tabulated, giving the minimum circumference requirements for pile heads and the corresponding smaller tip circumference based on pile taper and lengths between 20 (6096 mm) and 120 feet (36 576 mm), measured in increments of 10 feet (3048 mm). The ASTM D 25 standard also includes requirements for the quality of wood, tolerances on pile straightness, twist of grain, knots, holes, scars, wood checks, shakes, splits and other necessary information. However, the requirements stated in the standard do not relate to the several species of wood used for making timber piles, nor to their relative strengths. A timber pile with typical dimensions from ASTM D 25 is shown in Figure 1809.1.1.

1809.1.2 Preservative treatment. Timber piles used to support permanent structures shall be treated in accordance with this section unless it is established that the tops of the untreated timber piles will be below the lowest ground-water level assumed to exist during the life of the structure. Preservative and minimum final retention shall be in accordance with

AWPA U1 (Commodity Specification E, Use Category 4C) for round timber piles and AWPA U1 (Commodity Specification A, Use Category 4B) for sawn timber piles. Preservative-treated timber piles shall be subject to a quality control program administered by an approved agency. Pile cutoffs shall be treated in accordance with AWPA M4.

❖ For a general discussion on the need for treating wood with preservatives, see the commentary to Sections 1805.4.5, 1808.2.17 and 1809.1. This section of the code specifically requires that the preservative treatment of round timber piles and sawn timber piles by pressure processes conform to the applicable requirements of AWPA U1. Round timber pile requirements are covered in Commodity Specification E. Use Category 4C refers to conditions of very severe ground contact. Sawn timber piles are covered under Commodity Specification A and Use Category 4B, which refer to ground contact in severe environments. Note that round timber piles conform to ASTM D 25, while sawn timber piles comply with the grading rules of the species for strength.

This section also requires that, for timber piles subjected to saltwater exposure, the treatment of piles with water-borne preservatives and creosote complies with quality control procedures administered by an approved agency. Guides used in establishing these procedures are those contained in the American Wood Preservers' Bureau (AWPB) publications such as:

- MP1, *Quality Control and Inspection Procedures for Dual Treatment of Marine Piling Pressure Treated with Waterborne Preservatives and Creosote for Use in Marine Waters*;

- MP2, *Quality Control and Inspection Procedures for Marine Piling Pressure Treated with Creosote for Use in Marine Waters*; or

- MP4, *Quality Control and Inspection Procedures for Marine Piling Pressure Treated with Waterborne Preservatives for Use in Marine Waters*.

The field treatment of cuts and injuries to timber piles, including the treatment at pile cutoffs, with preservatives made for applications of creosote and creosote mixtures or with water-borne preservatives is covered by AWPA M4.

1809.1.3 Defective piles. Any substantial sudden increase in rate of penetration of a timber pile shall be investigated for possible damage. If the sudden increase in rate of penetration cannot be correlated to soil strata, the pile shall be removed for inspection or rejected.

❖ A small drop in blow count is not usually a cause for concern and is not uncommon in layered soils. It is only a substantial change in the rate of pile penetration that is not related to soil properties, which indicates a potential problem. If the penetration resistance suddenly decreases, serious damage to the pile could have occurred (including breaking), and the condition must be investigated. The pile can be withdrawn and inspected, and if required, abandoned and replaced.

Another significant problem associated with the use

For SI: 1 inch = 25.4 mm, 1 foot = 304.8 mm.

Figure 1809.1.1
TIMBER PILES

of timber piles is the possibility of structural damage caused by overdriving. Overdriving usually occurs upon reaching rock or hard soil stratas and may result in pile bending, brooming at the tip, crushing or brooming at the head or splitting or breaking along the pile section (see Figure 1809.1.3). To avoid such damage, if the penetration resistance suddenly increases, pile driving should be stopped immediately.

1809.1.4 Allowable stresses. The allowable stresses shall be in accordance with the AF&PA NDS.

❖ This section provides a reference to an appropriate material standard for determining allowable stresses.

1809.2 Precast concrete piles.

❖ This section sets forth minimum material and design requirements for precast nonprestressed (conventionally reinforced) and prestressed concrete piles.

1809.2.1 General. The materials, reinforcement and installation of precast concrete piles shall conform to Sections 1809.2.1.1 through 1809.2.1.4.

❖ Precast concrete piles are manufactured as nonprestressed (conventionally reinforced) or prestressed. Both types can be formed by bedcasting, spinning (centrifugal casting), vertical casting, slip forming or extrusion methods. They are usually made in square,

BENDING   BROOMING   BREAKING

Figure 1809.1.3
TYPICAL EFFECTS OF OVERDRIVING TIMBER PILES

12"-24"

12"-24" DIA.

NOTE: REINFORCING MAY
BE PRESTRESSED.

12"-54" DIA.

TAPER MAY
BE OMITTED →

TYPICAL CROSS SECTIONS

For SI: 1 inch = 25.4 mm.

Figure 1809.2.1
PRECAST NONPRESTRESSED AND PRESTRESSED CONCRETE PILES

octagonal or round shapes (see Figure 1809.2.1). Precast piles are manufactured as solid units or may be made with a hollow core. They can also be made with internal jet pipes or inspection ducts.

Precast piles are generally ordered in predetermined lengths based on the findings and analysis of soil exploratory work conducted at the project site.

This type of pile is a displacement pile that is normally installed with pile-driving equipment.

Precast concrete piles are to be designed in accordance with ACI 318. Nonprestressed concrete piles are usually considered for lengths of 40 feet (12 192 mm) to 50 feet (15 240 mm), while prestressed concrete piles are usually considered for lengths of 60 to 100 feet (18 288 to 30 480 mm). The general loading range for precast concrete piles is 40 to 400 tons (355 to 3558 kN).

Advantages of using precast concrete piles include their high load capacities and corrosion resistance. Disadvantages include their vulnerability to damage associated with handling, high breakage rates, particularly when spliced, and the considerable displacement of soil.

1809.2.1.1 Design and manufacture. Piles shall be designed and manufactured in accordance with accepted engineering practice to resist all stresses induced by handling, driving and service loads.

❖ Precast piles must be properly designed to resist the stresses induced both by handling and driving operations, and later imposed by service loads. Care must be given during handling and installation to minimize or avoid possible damage to the piles, such as cracking, crushing or spalling.

This section requires the minimum lateral dimension to be 8 inches (203 mm) and corners of square piles to be chamfered. Chamfered corners consist of rounding off or smoothing the corners of a square pile. The triangular portions of the corners are typically weak spots and chamfering the corners reduces the probability of cracking or breakage of the pile at these points.

1809.2.1.2 Minimum dimension. The minimum lateral dimension shall be 8 inches (203 mm). Corners of square piles shall be chamfered.

❖ This section provides the minimum dimension for precast concrete piles based on the size required to withstand the pile-driving operation. Chamfered corners are required to reduce concrete spalling during driving.

1809.2.1.3 Reinforcement. Longitudinal steel shall be arranged in a symmetrical pattern and be laterally tied with steel ties or wire spiral spaced not more than 4 inches (102 mm) apart, center to center, for a distance of 2 feet (610 mm) from the ends of the pile; and not more than 6 inches (152 mm) elsewhere except that at the ends of each pile, the first five ties or spirals shall be spaced 1 inch (25 mm) center to center. The gage of ties and spirals shall be as follows:

For piles having a diameter of 16 inches (406 mm) or less, wire shall not be smaller than 0.22 inch (5.6 mm) (No. 5 gage).

For piles having a diameter of more than 16 inches (406 mm) and less than 20 inches (508 mm), wire shall not be smaller than 0.238 inch (6 mm) (No. 4 gage).

For piles having a diameter of 20 inches (508 mm) and larger, wire shall not be smaller than 0.25 inch (6.4 mm) round or 0.259 inch (6.6 mm) (No. 3 gage).

❖ For precast piles, the longitudinal steel must be set in a symmetrical arrangement in the pile. To resist high-impact stresses, lateral steel ties used for confining the longitudinal reinforcement are to be provided at each end of the pile for a distance of 2 feet (610 mm) or more and be closely spaced at 3 inches (76 mm) o.c., except that the first five ties from each end are to be spaced at 1-inch (25 mm) centers. Between these two closely tied ends, the longitudinal steel must be similarly tied at spacings not to exceed 6 inches (152 mm) o.c. so as to provide a reinforcing cage that will keep the pile from buckling or cracking during handling.

1809.2.1.4 Installation. Piles shall be handled and driven so as not to cause injury or overstressing, which affects durability or strength.

❖ This section requires that all precast piles be handled in such a manner and driven so as not to cause injury or overstressing, which affects pile durability and structural strength. Sometimes cracking or spalling occurs during the driving of both conventionally reinforced and prestressed concrete piles. Such damage can be classified into four basic types of failure:

1. The spalling of concrete at the butt end (head) of the pile caused by very high or irregular compressive stress concentrations may be the result of a number of causes, such as: insufficient cushioning at the head of the pile; the butt is not square or perpendicular to the longitudinal axis of the pile; the driving hammer and the pile are not aligned, causing eccentric impact forces; the reinforcing steel is not cut flush at the top of the pile, allowing the impact force of the hammer to be transmitted to the concrete through the projecting steel; insufficient spiral reinforcing at the head of the pile, resulting in spalling or splitting of the concrete; pile corners that spall on impact of the hammer because they were not chamfered (see Section 1809.2.1.2) and concrete fatigue caused by a large number of hammer blows at a very high stress level.

2. Spalling at the tip of the pile is usually caused by extremely high driving resistance, particularly when the pile point is seated on rock.

3. The breaking or transverse cracking of a pile is a complex occurrence. When a hammer strikes the cushion or head of a pile, a compressive stress is produced that travels as a wave along the length of the pile down to the tip. This wave action may be reflected from the point of the pile as either a compressive or tensile stress based on the soil resistance at the point, or the wave

may simply pass into the soil. In cases where there is very little or no soil resistance, the compressive stress wave may be reflected back up the pile as a tensile stress and, depending on its magnitude, may cause damage along the pile. This phenomenon usually occurs in piles that are 50 feet (15 240 mm) or longer. When tip resistance is very low, such as in extremely soft soils, voids or in predrilled (or jetted) holes, the driving energy should be reduced to avoid high tensile stresses.

4. Spiral or transverse cracking of precast concrete piles may be caused by a combination of torsional stress (diagonal tension) and reflected tensile stress. Torsion is usually caused by excessive restraint of the pile in the leads, resulting in rotation of the pile during driving. Another reason is that the helmet or drive cap may fit too tightly on the pile, thus preventing the pile from rotating or adjusting slightly because of the soil action on the embedded section of the pile.

Good driving practices will generally eliminate the problems associated with the installation of precast concrete piles, as described. Good practice procedures are given in ACI 543R, which consists of recommendations for the design, manufacture and installation of concrete piles.

1809.2.2 Precast nonprestressed piles. Precast nonprestressed concrete piles shall conform to Sections 1809.2.2.1 through 1809.2.2.5.

❖ Precast nonprestressed (conventionally reinforced) concrete piles are manufactured from concrete and have reinforcement consisting of a steel reinforcing cage made up of several longitudinal bars or tie steel in the form of individual hoops or spirals.

1809.2.2.1 Materials. Concrete shall have a 28-day specified compressive strength (f'_c) of not less than 3,000 psi (20.68 MPa).

❖ Concrete used for precast nonprestressed concrete piles is to have a minimum specified 28-day compressive stress (f'_c) of 3,000 psi (20.68 Mpa).

While ACI 318 specifically excludes concrete piles from its design provisions, the material, quality and mix requirements contained in this standard generally apply to the manufacture of precast concrete piles.

To provide for handling stresses, concrete and reinforcing steel are often designed to allow for an overstress of up to 50 percent. Nonetheless, this section requires that the longitudinal reinforcement consists of not less than four bars and be at least 0.8 percent of the cross-sectional area of the pile, regardless that the calculated lifting stresses may sometimes indicate lesser requirements. Four bars are a practical minimum for symmetrical placement in square piles, which are the most commonly used type.

1809.2.2.2 Minimum reinforcement. The minimum amount of longitudinal reinforcement shall be 0.8 percent of the concrete section and shall consist of at least four bars.

❖ This section specifies the minimum amount of longitudinal reinforcement where seismic effects are minimal or low. See Sections 1809.2.2.2.1 and 1809.2.2.2.2 for reinforcement requirements for moderate and high seismic regions.

1809.2.2.2.1 Seismic reinforcement in Seismic Design Category C. Where a structure is assigned to Seismic Design Category C in accordance with Section 1613, the following shall apply. Longitudinal reinforcement with a minimum steel ratio of 0.01 shall be provided throughout the length of precast concrete piles. Within three pile diameters of the bottom of the pile cap, the longitudinal reinforcement shall be confined with closed ties or spirals of a minimum $^3/_8$ inch (9.5 mm) diameter. Ties or spirals shall be provided at a maximum spacing of eight times the diameter of the smallest longitudinal bar, not to exceed 6 inches (152 mm). Throughout the remainder of the pile, the closed ties or spirals shall have a maximum spacing of 16 times the smallest longitudinal bar diameter not to exceed 8 inches (203 mm).

❖ This section includes moderately ductile detailing requirements for precast nonprestressed piles in buildings assigned to Seismic Design Category C. The minimum longitudinal and transverse reinforcement requirements are so that piles will be able to accommodate seismically induced ground deformations that are expected to affect Seismic Design Category C buildings. The 1-percent minimum longitudinal reinforcement is a standard requirement for reinforced concrete columns. A 6-bar-diameter or 6-inch (152 mm) spacing of transverse reinforcement is a fairly common requirement to prevent buckling of longitudinal compression reinforcement. The transverse reinforcement spacing requirement of this section for the confinement region of the pile is somewhat less than that, allowing an 8-bar-diameter or 6-inch (152 mm) spacing. Outside of this region the spacing requirements for transverse reinforcing is less stringent.

1809.2.2.2.2 Seismic reinforcement in Seismic Design Category D, E or F. Where a structure is assigned to Seismic Design Category D, E or F in accordance with Section 1613, the requirements for Seismic Design Category C in Section 1809.2.2.2.1 shall apply except as modified by this section. Transverse confinement reinforcement consisting of closed ties or equivalent spirals shall be provided in accordance with Sections 21.4.4.1, 21.4.4.2 and 21.4.4.3 of ACI 318 within three pile diameters of the bottom of the pile cap. For other than Site Class E or F, or liquefiable sites and where spirals are used as the transverse reinforcement, a volumetric ratio of spiral reinforcement of not less than one-half that required by Section 21.4.4.1(a) of ACI 318 shall be permitted.

❖ This section requires that the provisions for Seismic Design Category C in Section 1809.2.2.1 must, as a minimum, be met. The confinement region of the pile requires the more stringent and demanding detailing provided in the referenced sections of ACI 318, which are the requirements for special moment frames. These increased transverse reinforcement requirements are intended to provide a high degree of ductility in the upper portion (confinement region) of the pile. Experience has shown that concrete piles tend to hinge or sustain damage immediately below the pile cap; therefore, tie spacing is reduced in this area to better confine the concrete.

1809.2.2.3 Allowable stresses. The allowable compressive stress in the concrete shall not exceed 33 percent of the 28-day specified compressive strength (f'_c) applied to the gross cross-sectional area of the pile. The allowable compressive stress in the reinforcing steel shall not exceed 40 percent of the yield strength of the steel (f_y) or a maximum of 30,000 psi (207 MPa). The allowable tensile stress in the reinforcing steel shall not exceed 50 percent of the yield strength of the steel (f_y) or a maximum of 24,000 psi (165 MPa).

❖ This section specifies the allowable stresses in precast nonprestressed concrete piles. The allowable stresses are significantly less than the yield strength.

1809.2.2.4 Installation. A precast concrete pile shall not be driven before the concrete has attained a compressive strength of at least 75 percent of the 28-day specified compressive strength (f'_c), but not less than the strength sufficient to withstand handling and driving forces.

❖ This section requires that precast reinforced concrete piles must not be driven until the concrete has acquired at least three-quarters of its 28-day specified compressive strength. For example, if a pile is manufactured at the minimum compressive strength of 4,000 psi (27.6 MPa), it must not be driven until it has obtained a strength of at least 3,000 psi (20.68 MPa) ($^3/_4$ x 4,000) (27.6 MPa). In all cases, however, concrete strength must have developed to the point that it is sufficient to sustain the stresses imposed on the pile by handling and driving operations.

1809.2.2.5 Concrete cover. Reinforcement for piles that are not manufactured under plant conditions shall have a concrete cover of not less than 2 inches (51 mm).

Reinforcement for piles manufactured under plant control conditions shall have a concrete cover of not less than 1.25 inches (32 mm) for No. 5 bars and smaller, and not less than 1.5 inches (38 mm) for No. 6 through No. 11 bars except that longitudinal bars spaced less than 1.5 inches (38 mm) clear distance apart shall be considered bundled bars for which the minimum concrete cover shall be equal to that for the equivalent diameter of the bundled bars.

Reinforcement for piles exposed to seawater shall have a concrete cover of not less than 3 inches (76 mm).

❖ The minimum concrete cover is intended to provide protection to the reinforcement for piles exposed to the weather and against the effects of deleterious substances that may be present in the ground. This section permits the reinforcement of precast piles manufactured under plant control conditions to have reduced concrete cover requirements in comparison

to piles produced under other precasting conditions. However, the term "manufactured under plant control conditions" should not be construed as meaning that piles must be produced in a plant. Precast piles made at the job site will also qualify under this section, provided that the control of form dimensions, placing of reinforcement, concrete quality and curing procedures are all equal to the level of control exercised in a plant.

1809.2.3 Precast prestressed piles. Precast prestressed concrete piles shall conform to the requirements of Sections 1809.2.3.1 through 1809.2.3.5.

❖ Precast prestressed concrete piles are either of the pretensioned or post-tensioned type. Pretensioned type of piles are normally cast in a plant in their full lengths, as predetermined by soils investigations and engineering analyses. The reinforcement in the concrete pile consists of tendons (stressing steel) that are tensioned before the concrete is placed. After casting and when the concrete has attained sufficient strength, the stretched tendons are released from their anchorage, thus placing the pile in continuous compression. In comparison, post-tensioned type of piles are made in a plant or on the job site, and the tendons are released after the concrete has hardened. In addition to the tendons, this type of pile also contains mild reinforcing steel to resist the handling stresses before the stressing steel is tensioned.

One of the primary advantages of prestressed over nonprestressed concrete piles is durability. Since the concrete is under continuous compression, hairline cracks are kept tightly closed and, thus, are more durable. Another advantage is that the tensile stresses that can develop in the concrete under certain driving conditions are less critical. Prestressed concrete piles are best suited for friction piles in sand, gravel and clays.

The purpose for prestressing concrete piles is explained in the commentary to Section 1809.2.3.4.

1809.2.3.1 Materials. Prestressing steel shall conform to ASTM A 416. Concrete shall have a 28-day specified compressive strength (f'_c) of not less than 5,000 psi (34.48 MPa).

❖ Rods, strands or wires conforming to the requirements of ASTM A 416 are used as stressing steel in the manufacture of precast, prestressed concrete piles of both the pretensioned and the post-tensioned type. The most commonly used stressing steel is the seven wire, uncoated, stress-relieved strand.

This section requires that concrete used in the manufacture of prestressed piles must have a 28-day specified compressive strength of no less than 5,000 psi (34.48 MPa). The use of higher strength concrete results in smaller volumetric changes with corresponding reductions in prestress losses and higher effective prestress values.

1809.2.3.2 Design. Precast prestressed piles shall be designed to resist stresses induced by handling and driving as well as by loads. The effective prestress in the pile shall not be less than 400 psi (2.76 MPa) for piles up to 30 feet (9144 mm) in length,

550 psi (3.79 MPa) for piles up to 50 feet (15 240 mm) in length and 700 psi (4.83 MPa) for piles greater than 50 feet (15 240 mm) in length.

Effective prestress shall be based on an assumed loss of 30,000 psi (207 MPa) in the prestressing steel. The tensile stress in the prestressing steel shall not exceed the values specified in ACI 318.

❖ For prestressed concrete piles, the effective prestress (the stress remaining in the pile after all losses have occurred—excluding the effect of superimposed loads and the weight of the pile) is not to be less than the specified values for the various lengths. Experience has shown that an effective prestress less than the minimum value prescribed herein is sometimes inadequate in preventing or controlling cracking of the concrete during the handling and installation operations. This section also requires the effective prestress to be based on an assumed loss of 30,000 psi (207 MPa) in the prestressing steel and the tensile stresses in the steel not to exceed the values set forth in ACI 318.

1809.2.3.2.1 Design in Seismic Design Category C. Where a structure is assigned to Seismic Design Category C in accordance with Section 1613, the following shall apply. The minimum volumetric ratio of spiral reinforcement shall not be less than 0.007 or the amount required by the following formula for the upper 20 feet (6096 mm) of the pile.

$$\rho_s = 0.12 f'_c / f_{yh} \qquad \text{(Equation 18-4)}$$

where:

f'_c = Specified compressive strength of concrete, psi (MPa).

f_{yh} = Yield strength of spiral reinforcement ≤ 85,000 psi (586 MPa).

ρ_s = Spiral reinforcement index (vol. spiral/vol. core).

At least one-half the volumetric ratio required by Equation 18-4 shall be provided below the upper 20 feet (6096 mm) of the pile.

The pile cap connection by means of dowels as indicated in Section 1808.2.23.1 is permitted. Pile cap connection by means of developing pile reinforcing strand is permitted provided that the pile reinforcing strand results in a ductile connection.

❖ This section gives requirements for precast prestressed piles supporting structures assigned to Seismic Design Category C. The minimum spiral reinforcement requirement that results in ductile prestressed concrete piles is based on Precast/Prestressed Concrete Institute's (PCI) *Recommended Practice for Design, Manufacture and Installation of Prestressed Concrete Piling*. This document incorporates work done on piles in the United States and New Zealand.

These piles exhibit larger curvatures in the top 20 feet (6096 mm). This section requires (confinement) spiral reinforcement as determined by Equation 18-4. Based on ACI 318, this formula is deemed to provide

for moderate ductility. In the lower portion of the pile, the required reinforcing is reduced by one-half.

1809.2.3.2.2 Design in Seismic Design Category D, E or F. Where a structure is assigned to Seismic Design Category D, E or F in accordance with Section 1613, the requirements for Seismic Design Category C in Section 1809.2.3.2.1 shall be met, in addition to the following:

1. Requirements in ACI 318, Chapter 21, need not apply, unless specifically referenced.

2. Where the total pile length in the soil is 35 feet (10 668 mm) or less, the lateral transverse reinforcement in the ductile region shall occur through the length of the pile. Where the pile length exceeds 35 feet (10 668 mm), the ductile pile region shall be taken as the greater of 35 feet (10 668 mm) or the distance from the underside of the pile cap to the point of zero curvature plus three times the least pile dimension.

3. In the ductile region, the center-to-center spacing of the spirals or hoop reinforcement shall not exceed one-fifth of the least pile dimension, six times the diameter of the longtitudinal strand, or 8 inches (203 mm), whichever is smaller.

4. Circular spiral reinforcement shall be spliced by lapping one full turn and bending the end of the spiral to a 90-degree hook or by use of a mechanical or welded splice complying with Sec. 12.14.3 of ACI 318.

5. Where the transverse reinforcement consists of circular spirals, the volumetric ratio of spiral transverse reinforcement in the ductile region shall comply with the following:

$$\rho_s = 0.25(f'_c / f_{yh})(A_g / A_{ch} - 1.0)[0.5 + 1.4P/(f'_c A_g)]$$

(Equation 18-5)

but not less than:

$$\rho_s = 0.12(f'_c / f_{yh})[0.5 + 1.4P/(f'_c A_g)]$$

(Equation 18-6)

and need not exceed:

$$\rho_s = 0.021 \qquad \text{(Equation 18-7)}$$

where:

A_g = Pile cross-sectional area, square inches (mm²).

A_{ch} = Core area defined by spiral outside diameter, square inches (mm²).

f'_c = Specified compressive strength of concrete, psi (MPa).

f_{yh} = Yield strength of spiral reinforcement ≤ 85,000 psi (586 MPa).

P = Axial load on pile, pounds (kN), as determined from Equations 16-5 and 16-6.

ρ_s = Volumetric ratio (vol. spiral/ vol. core).

This required amount of spiral reinforcement is permitted to be obtained by providing an inner and outer spiral.

6. When transverse reinforcement consists of rectangular hoops and cross ties, the total cross-sectional area of lateral transverse reinforcement in the ductile region with spacings, and perpendicular to dimension, h_c, shall conform to:

$$A_{sh} = 0.3 s h_c (f'_c/f_{yh})(A_g/A_{ch} - 1.0)[0.5 + 1.4P/(f'_c A_g)]$$

(Equation 18-8)

but not less than:

$$A_{sh} = 0.12 s h_c (f'_c/f_{yh})[0.5 + 1.4P/(f'_c A_g)]$$

(Equation 18-9)

where:

f_{yh} = $\leq 70{,}000$ psi (483 MPa).

h_c = Cross-sectional dimension of pile core measured center to center of hoop reinforcement, inch (mm).

s = Spacing of transverse reinforcement measured along length of pile, inch (mm).

A_{sh} = Cross-sectional area of tranverse reinforcement, square inches (mm²).

f'_c = Specified compressive strength of concrete, psi (MPa).

The hoops and cross ties shall be equivalent to deformed bars not less than No. 3 in size. Rectangular hoop ends shall terminate at a corner with seismic hooks.

Outside of the length of the pile requiring transverse confinement reinforcing, the spiral or hoop reinforcing with a volumetric ratio not less than one-half of that required for transverse confinement reinforcing shall be provided.

❖ The provisions for Seismic Design Category C must be followed in addition to these requirements. This section clarifies that Chapter 21 of ACI 318 does not generally apply to precast prestressed concrete piles unless it is explicitly referenced. This chapter was never intended for such piles.

The provisions found in this section are based on PCI's "Recommended Practice for Design, Manufacture and Installation of Prestressed Concrete Piling." This document incorporates work done on piles in the United States and New Zealand.

The maximum volumetric ratio set forth in Equation 18-7 is based on pile testing that has shown the 0.021 maximum is sufficient for the smaller square precast prestressed concrete piles to perform in a ductile manner.

1809.2.3.3 Allowable stresses. The allowable design compressive stress, f_c, in concrete shall be determined as follows:

$$f_c = 0.33 f'_c - 0.27 f_{pc} \qquad \textbf{(Equation 18-10)}$$

where:

f'_c = The 28-day specified compressive strength of the concrete.

f_{pc} = The effective prestress stress on the gross section.

❖ The allowable design compressive stress (f_c) in concrete is determined using Equation 18-10. The term "$0.33 f'_c$" is the same as the allowable compressive stress used for conventionally reinforced concrete piles (see Section 1809.2.2.3). Since prestressing places the concrete in compression, the concrete compressive stress thus induced and designated by the term "$0.27 f_{pc}$" must be subtracted from the gross allowable stress in the pile ($0.33 f'_c$), with the remainder (f_c) representing the net amount of compressive stress value available for the service loads on the pile. The notation f_{pc} is the effective prestress value on the gross area of the pile section, which is the stress remaining in the pile after all losses have occurred, excluding the effect of the service loads and the weight of the piles.

The initial prestressing force can be measured with some accuracy during and immediately after the tensioning operations. The final prestress varies with time because of losses caused by the elastic shortening of concrete, creep, the shrinkage of concrete, relaxation of the steel stress and frictional losses caused by curvature in the stressing tendons. Such losses in prestress are included in the notation $0.27 f_{pc}$ in the formula (equation) first stated above. Furthermore, the allowable design load (P_a) can be expressed by the following equation:

$$P_a = (0.33 f'_c - 0.27 f_{pc}) A_c = f_c A_c$$

where A_c is the area of the concrete section, including the prestressing steel.

1809.2.3.4 Installation. A prestressed pile shall not be driven before the concrete has attained a compressive strength of at least 75 percent of the 28-day specified compressive strength (f'_c), but not less than the strength sufficient to withstand handling and driving forces.

❖ The requirements for the installation of prestressed concrete piles are the same as the installation provisions for conventionally reinforced precast concrete piles (see commentary, Section 1809.2.2.4).

The reason for prestressing concrete piles is to place the concrete under continuous compression so that any hairline cracks that may develop will be kept tightly closed and to prevent possible injury to the pile from tensile stresses that may occur during installation operations. The handling and driving of prestressed piles do not require the same degree of care as needed to install conventionally reinforced concrete piles. As a general rule, prestressed piles are more durable than precast reinforced piles.

1809.2.3.5 Concrete cover. Prestressing steel and pile reinforcement shall have a concrete cover of not less than $1\frac{1}{4}$ inches (32 mm) for square piles of 12 inches (305 mm) or smaller size and $1\frac{1}{2}$ inches (38 mm) for larger piles, except that for piles exposed to seawater, the minimum protective concrete cover shall not be less than $2\frac{1}{2}$ inches (64 mm).

❖ The minimum concrete cover is intended to provide protection to the reinforcement for piles exposed to the weather and against the effects of deleterious substances that may be present in the ground. The concrete cover requirement for piles exposed to seawater is increased to $2\frac{1}{2}$ inches (64 mm) because of the highly corrosive effects of salt.

1809.3 Structural steel piles. Structural steel piles shall conform to the requirements of Sections 1809.3.1 through 1809.3.4.

❖ The piles in this category are made of load-bearing steel materials that can develop high load capacities, and include H-piles and open-ended steel pipe piles. In addition to stating the applicable material specifications for the manufacturing of steel piles, this section also prescribes the allowable design stresses and dimensional requirements (see Figure 1809.3).

Figure 1809.3
STEEL H-PILE SECTIONS

1809.3.1 Materials. Structural steel piles, steel pipe and fully welded steel piles fabricated from plates shall conform to ASTM A 36, ASTM A 252, ASTM A 283, ASTM A 572, ASTM A 588, ASTM A 690, ASTM A 913 or ASTM A 992.

❖ The materials used in the manufacture of steel piles must comply with requirements of one of the listed ASTM International standards.

While H-piles are ordinarily made with steel materials conforming to ASTM A 36 or ASTM A 572 requirements, the employment of other standards is not excluded, provided it can be shown that such steels meet the applicable chemical and mechanical properties of one of the listed specifications to establish the suitability of the material for pile use.

Welded or seamless pipe piles must conform to ASTM A 252 requirements.

1809.3.2 Allowable stresses. The allowable axial stresses shall not exceed 35 percent of the minimum specified yield strength (F_y).

Exception: Where justified in accordance with Section 1808.2.10, the allowable axial stress is permitted to be increased above $0.35F_y$, but shall not exceed $0.5F_y$.

❖ In the elastic design of steel foundation piles, the compressive stresses are limited to a percentage of the steel yield strength (F_y). For structural steel piles, the maximum allowable design stresses are not to exceed $0.35F_y$ (35 percent of the yield strength). Allowable design stresses may, however, be increased to $0.50F_y$ on the basis that justification for such an increase is provided by both approved analytical methods and load tests conducted in accordance with the requirements of Section 1808.2.10.

H-piles consist of rolled wide-flange steel H-beams. They are manufactured in standard sizes with nominal beam depths in the range of 8 inches (203 mm) to 14 inches (355 mm).

H-piles are usually employed as deep end-bearing piles because of their high load-carrying capacity and, with few exceptions, they are nondisplacement piles that can readily penetrate soil strata to reach rock or other suitable point-supporting subsurface materials. Ideally, steel H-piles are driven to hard or medium-hard rock. Many load tests have shown that, for H-piles driven to refusal on rock, the stresses at failure closely equate to the yield strength of the material. Under such conditions, consideration should be given to designing H-pile foundations at a high load capacity per pile based on design stress levels up to $0.50F_y$ and using pile-tip reinforcement.

High load-carrying capacity can also be obtained by H-piles driven to practical refusal in load-bearing strata of soft rock, hardpan overlying a rock layer or in very dense granular soils. Unlike piles seated on hard or medium-hard rock, however, the ability of these strata to support pile loads up to the yield point of the steel may be questionable. A thorough analytical study combined with an extensive load test program is required before considering using design stresses up to $0.50F_y$.

Experience indicates that corrosion is not a practical problem for steel piles driven in natural soil, due primarily to the absence of oxygen in the soil; however, in fill ground at or above the ground-water table, moderate corrosion may occur and protection may be needed. A commonly used protection method includes coatings applied before driving.

H-piles can also be used as friction piles. Because

of the many variables involved in the design of this type of pile, as compared to piles driven to end-bearing on rock, however, it is not possible to produce conclusive information on maximum loads and stresses. It is to be recognized that the soil, rather than the structural strength of the H-pile, will be the controlling factor in the design. Additionally, the design of friction piles is directly influenced by such considerations as group action, driveability and settlement limitations, as opposed to piles driven to end-bearing rock.

As required by this section, an increase in the allowable design stresses from $0.35F_y$ up to $0.50F_y$, representing a maximum step-up of 43 percent in the design load, needs to be justified by approved analytical methods as well as by the results of load testing, particularly where it involves steel pile installations in subsurface load-bearing materials other than hard or medium-hard rock. When possible, building officials should seek the recommendations of registered engineers who are experienced in soil mechanics and in the design of deep foundations before authorizing increases in allowable compressive stresses.

Steel pipe piles are made of seamless, welded or spiral-welded pipes and are most frequently filled with concrete. Pipe piles conforming to ASTM A 252 are used as both friction piles and end-bearing installations. Piles may be driven open ended or closed ended. Open-ended piles are generally specified where the soil investigation indicates that rock load-bearing strata are found at levels reasonably close to the ground surface, especially where the loads to be supported are great. The earth forced into the interior core of the pipe pile by driving operations is cleaned out by a blowing technique or auguring, depending on the type of soil encountered. The pile is then filled with concrete.

The use of closed-ended piles is indicated when soil investigations show the absence of rock formations or when rock is found at excessive depths. Under such conditions, closed-ended piles are driven as friction piles.

The yield strengths (F_y) of welded or seamless pipe piles specified in ASTM A 252 and upon which the allowable compressive stresses are based are as follows: for Grade 1 steel, the yield strength is 30,000 psi (207 MPa); for Grade 2, 35,000 psi (241 MPa) and for Grade 3, 45,000 psi (310 MPa). Grade 1 steel is seldom used for pipe pile fabrication.

1809.3.3 Dimensions of H-piles. Sections of H-piles shall comply with the following:

1. The flange projections shall not exceed 14 times the minimum thickness of metal in either the flange or the web and the flange widths shall not be less than 80 percent of the depth of the section.

2. The nominal depth in the direction of the web shall not be less than 8 inches (203 mm).

3. Flanges and web shall have a minimum nominal thickness of $^3/_8$ inch (9.5 mm).

❖ Dimensional requirements for manufacturing structural steel H-piles are provided in this section. H-piles are proportioned to withstand the large impact stresses imposed on piles during hard driving. The thicknesses of the flanges and the web of a rolled-steel H-pile section are made equal in order to avoid damage that could occur during hard driving if the piles were proportioned with a mixture of thick and thinner parts. Flange widths are proportioned in relation to the depth of the pile section to provide rigidity in the weak axis.

It would seem unnecessary to repeat the dimensioning requirements for H-piles in the code, since they were originally created by the steel industry and have for decades been the industry standard used in the manufacture of hot-rolled steel shapes. The main purpose is to provide the dimensional basis for the fabrication of similar pile products made principally of welded steel plates and other necessary steel parts.

While it is the general preference and practice to use rolled-steel piles, occasionally it becomes necessary to fabricate special pile sections because of: time problems imposed by mill scheduling and delivery; a special need for heavier H-pile sections than are customarily available; an immediate need for replacement piles or for any other reason. The dimensional requirements contained in this section regulate the fabrication of such special piles and provide the building official with a basis for approval. Fabricated pile materials are to comply with the requirements of Section 1809.3.1.

This section requires that the flange projection not exceed 14 times the minimum thickness of metal in either the flange or the web. The measurement of the flange projection is shown in Figure 1809.3 and can be calculated by the indicated formula:

$$P = \frac{W - t_w}{2} \; or \; \frac{W - t_f}{2}$$

where W is the flange width, t_w is the web thickness, t_f is the flange thickness and P is the flange projection.

For example, the dimensions of an HP 14 x 73 pile section are as follows:

The width of flange, W = 14.585 inches.
The thickness of the flange, t_f = 0.505 inch.
The thickness of the web, t_w = .505 inch.
Therefore, the actual flange projection is:

$$P = \frac{14.585 - 0.505}{2} = 7.04 \; inches \, (179 \; mm)$$

The maximum allowable flange projection is 14 × t_f or 14 × t_w (whichever is smaller).

Allowable flange projection: 14 × 0.505 = 7.07 inches (179.5 mm), which is greater than the actual flange projection of 7.04 inches (178.8 mm); therefore, an HP 14 × 73 pile is acceptable.

This section also specifies that flange widths are not to be less than 80 percent of the depth of the pile section. In actuality, practically all H-piles are manufactured so that their flanges are slightly greater in width (by fractions of an inch) than the depths of the sections. In fact, piles are squared so that their flange widths are about equal to their depth.

Flange and web thickness is not to be less than $^3/_8$ inch (9.5 mm). In H-piles, the flange thicknesses are equal to the web thickness.

1809.3.4 Dimensions of steel pipe piles. Steel pipe piles driven open ended shall have a nominal outside diameter of not less than 8 inches (203 mm). The pipe shall have a minimum cross section of 0.34 square inch (219 mm²) to resist each 1,000 foot-pounds (1356 N-m) of pile hammer energy, or shall have the equivalent strength for steels having a yield strength greater than 35,000 psi (241 Mpa) or the wave equation analysis shall be permitted to be used to assess compression stresses induced by driving to evaluate if the pile section is appropriate for the selected hammer. Where pipe wall thickness less than 0.179 inch (4.6 mm) is driven open ended, a suitable cutting shoe shall be provided.

❖ Because of their uniform cross section from butt to tip, steel pipe piles provide unvarying resistance to bending and lateral forces applied in any direction.

A pipe to be driven open ended must have an outside diameter of no less than 8 inches (204 mm). Smaller diameter pipes could allow soil to bridge the sides and cause plugging at their tips. This section also establishes the minimum thickness of the pipe wall.

The minimum area of the pipe cross section can be determined by one of two methods. The first is a long standing rule of thumb that relates pile area to the hammer energy. The second approach recognizes the more up-to-date wave equation. The latter approach is commonly used to evaluate the induced driving stresses in piles.

SECTION 1810
CAST-IN-PLACE CONCRETE PILE FOUNDATIONS

1810.1 General. The materials, reinforcement and installation of cast-in-place concrete piles shall conform to Sections 1810.1.1 through 1810.1.3.

❖ Section 1810 addresses materials, design and installation of cast-in-place concrete piles. Cast-in-place concrete piles are installed by placing concrete into preformed holes in the ground extending to required load-bearing depths. Since this type of pile is not subjected to driving stresses, only the stresses imposed by the service loads are considered in the design. The predetermination of pile lengths is not as important as it is with manufactured piles since adjustments can readily be made in the field to accommodate subsurface conditions. Reinforcement, if required, is installed prior to or during concrete placement.

Cast-in-place concrete piles basically consist of two types: uncased and cased. The design and construction of the various types of uncased piles are covered in Sections 1810.2, 1810.3 and 1810.4. The cased type of pile is covered in Section 1810.5.

Section 1810.6 addresses concrete-filled steel pipe and tube piles. Section 1810.7 addresses caisson piles.

Concrete materials, the proportioning and quality of concrete mixtures and the reinforcement used for cast-in-place piles are to comply with Sections 1810.1.1 and 1810.1.2, and specifically with the applicable provisions of Chapter 19. General installation requirements applicable to all types of cast-in-place concrete pile construction covered by Section 1810 are to comply with Section 1810.1.3.

1810.1.1 Materials. Concrete shall have a 28-day specified compressive strength (f'_c) of not less than 2,500 psi (17.24 MPa). Where concrete is placed through a funnel hopper at the top of the pile, the concrete mix shall be designed and proportioned so as to produce a cohesive workable mix having a slump of not less than 4 inches (102 mm) and not more than 6 inches (152 mm). Where concrete is to be pumped, the mix design including slump shall be adjusted to produce a pumpable concrete.

❖ Concrete used for cast-in-place piles is to have a minimum 28-day compressive strength of 2,500 psi (17.24 MPa). Higher concrete compressive strengths are usually specified to increase pile load capacity as well as for economic reasons.

This section requires that concrete be designed and proportioned to produce a cohesive workable mix with a consistency (slump) ranging between 4 inches (102 mm) and 6 inches (152 mm). This provision is within the bounds required for the proper placement of conventional structural concrete for cased cast-in-place piles. The slump requirements stated here, however, should not be considered so absolute that, with the use of engineering judgement, the consistency of the concrete cannot be adjusted to satisfy prevailing conditions of construction. For example, concrete placement conducted under difficult conditions—such as for piles containing heavy reinforcement, for very long cased piles, for pile shells driven on steep batters or for other demanding situations—may require the employment of special concrete mixes using reduced quantities of coarse aggregates with corresponding increases in sand and cement content. Under such conditions, slumps of 4 inches (102 mm) with a tolerance of plus 2 inches (51 mm) or minus 1 inch (25 mm) are in order.

The consistency of concrete required for uncased cast-in-place concrete piles may be altogether different than the slumps specified and used for cased construction. For example, concrete placed in drilled holes should have a slump of at least 6 inches (152 mm) so that the concrete flows properly and full bond is attained with the reinforcement.

Furthermore, high slump concrete is used so that

the complete volume of the hole is filled, including natural soil crevices and pockets and surface irregularities caused by the drilling operations. High slumps are also used where the concrete must displace drilling slurry.

Concrete slumps as high as 8 inches (203 mm) are sometimes needed where temporary casings are used to facilitate pile construction and are subsequently extracted as the concrete is placed.

Concrete is usually placed through a funnel hopper at the top of the pile opening, but other approved methods, such as pumping or tremie, may also be used.

1810.1.2 Reinforcement. Except for steel dowels embedded 5 feet (1524 mm) or less in the pile and as provided in Section 1810.3.4, reinforcement where required shall be assembled and tied together and shall be placed in the pile as a unit before the reinforced portion of the pile is filled with concrete except in augered uncased cast-in-place piles. Tied reinforcement in augered uncased cast-in-place piles shall be placed after piles are concreted, while the concrete is still in a semifluid state.

❖ The typical procedure for constructing cast-in-place concrete piles is to tie the head of the pile with the pile cap, inserting dowels in the freshly placed concrete pile to obtain an embedment of about 5 feet (1524 mm). Main reinforcement consisting of a cage of longitudinal deformed reinforcing bars tied together with individual steel hoops or a spiral must be placed and securely held in position in the pile opening (cased or uncased) prior to the placement of concrete. This method, however, is not applicable to auger-injected concrete piles (see commentary, Section 1810.3.4).

Reinforcement required for resisting tensile stresses imposed by uplift forces may consist of a single bar or other structural steel unit or a cluster of bars placed at the center for the full length of the pile.

1810.1.2.1 Reinforcement in Seismic Design Category C. Where a structure is assigned to Seismic Design Category C in accordance with Section 1613, the following shall apply. A minimum longitudinal reinforcement ratio of 0.0025 shall be provided for uncased cast-in-place concrete drilled or augered piles, piers or caissons in the top one-third of the pile length, a minimum length of 10 feet (3048 mm) below the ground or that required by analysis, whichever length is greatest. The minimum reinforcement ratio, but no less than that ratio required by rational analysis, shall be continued throughout the flexural length of the pile. There shall be a minimum of four longitudinal bars with closed ties (or equivalent spirals) of a minimum $^3/_8$ inch (9 mm) diameter provided at 16-longitudinal-bar diameter maximum spacing. Transverse confinement reinforcement with a maximum spacing of 6 inches (152 mm) or 8-longitudinal-bar diameters, whichever is less, shall be provided within a distance equal to three times the least pile dimension of the bottom of the pile cap.

❖ Longitudinal and transverse reinforcement requirements prescribed by this section result in moderate ductility in buildings assigned to Seismic Design Category C to withstand seismically induced ground deformations that they can encounter. Separate transverse

reinforcement is specified for the portions of the pile within and outside of the potential plastic hinge zone.

The purpose of this section is to include pile bending, which is a result of ground horizontal movement during an earthquake, in the structural design. The reinforcement in the pile, required to resist the tension caused by pile bending, increases the ductility of the foundation such that bending or shear failure is precluded.

1810.1.2.2 Reinforcement in Seismic Design Category D, E or F. Where a structure is assigned to Seismic Design Category D, E or F in accordance with Section 1613, the requirements for Seismic Design Category C given above shall be met, in addition to the following. A minimum longitudinal reinforcement ratio of 0.005 shall be provided for uncased cast-in-place drilled or augered concrete piles, piers or caissons in the top one-half of the pile length a minimum length of 10 feet (3048 mm) below ground or throughout the flexural length of the pile, whichever length is greatest. The flexural length shall be taken as the length of the pile to a point where the concrete section cracking moment strength multiplied by 0.4 exceeds the required moment strength at that point. There shall be a minimum of four longitudinal bars with transverse confinement reinforcement provided in the pile in accordance with Sections 21.4.4.1, 21.4.4.2 and 21.4.4.3 of ACI 318 within three times the least pile dimension of the bottom of the pile cap. A transverse spiral reinforcement ratio of not less than one-half of that required in Section 21.4.4.1(a) of ACI 318 for other than Class E, F or liquefiable sites is permitted. Tie spacing throughout the remainder of the concrete section shall neither exceed 12-longitudinal-bar diameters, one-half the least dimension of the section, nor 12 inches (305 mm). Ties shall be a minimum of No. 3 bars for piles with a least dimension up to 20 inches (508 mm), and No. 4 bars for larger piles.

❖ The requirements of Section 1810.1.2.1 apply in addition to the requirements of this section. This section is similar in intent, with increased minimum reinforcement requirements for ductility during earthquake ground motion that may be experienced by buildings assigned to Seismic Design Category D and higher. Separate transverse reinforcement requirements are given for portions of a pile within and beyond the potential plastic hinge zone. The specified confinement reinforcing is similar to ACI 318 requirements for special moment frames, except that a reduction is permitted for competent soils in recognition of the confinement attributed to those soils. Note that flexural length is redefined for purposes of this section.

1810.1.3 Concrete placement. Concrete shall be placed in such a manner as to ensure the exclusion of any foreign matter and to secure a full-sized shaft. Concrete shall not be placed through water except where a tremie or other approved method is used. When depositing concrete from the top of the pile, the concrete shall not be chuted directly into the pile but shall be poured in a rapid and continuous operation through a funnel hopper centered at the top of the pile.

❖ The pile opening must be free of all foreign matter prior to placing concrete. If there is excessive water in the hole [about 4 inches (102 mm) is permitted], the concrete may be placed by tremie or pumping methods. Concrete is usually deposited from the top through a funnel hopper, which should have a discharge opening that is sized smaller than the smallest pile cross section in the foundation system. Vibrators should not be used in the concreting operations of cast-in-place, uncased concrete piles because they may damage the earth walls and cause contamination of the concrete. Improper vibration may also cause excessive bleeding.

Other installation problems associated with cast-in-place concrete piles are discussed in the commentary to Section 1810.3.3.

1810.2 Enlarged base piles. Enlarged base piles shall conform to the requirements of Sections 1810.2.1 through 1810.2.5.

❖ The pile opening must be free of all foreign matter prior to placing concrete. If there is excessive water in the hole [about 4 inches (102 mm) is permitted], the concrete may be placed by tremie or pumping methods. Concrete is usually deposited from the top through a funnel hopper, which should have a discharge opening that is sized smaller than the smallest pile cross section in the foundation system. Vibrators should not be used in the concreting operations of cast-in-place, uncased concrete piles because they may damage the earth walls and cause contamination of the concrete. Improper vibration may also cause excessive bleeding.

Other installation problems associated with cast-in-place concrete piles are discussed in the commentary to Section 1810.3.3.

Enlarged base cast-in-place concrete piles are of two types. The first type is the compacted concrete base pile, which consists of a bulb-shaped footing formed after driving a steel casing to its final depth. This type of pile is also known as a pressure-injected footing. The second type of enlarged base pile consists of a truncated cone or pyramid-shaped precast concrete tip larger than the casing diameter, which is driven into the soil together with the steel casing. For more comprehensive descriptions of the two types of enlarged base piles, see the commentary to Section 1810.2.3.

Enlarged base piles are a displacement type of piles that are installed only in granular soils, since the voids between the particles of soil allow densification (compression) of the material surrounding the pile without creating excessive pressures that could damage adjacent piles by causing lateral movements or heaving.

Enlarged base piles generally function as end-bearing piles, deriving their bearing capacity from the strength of the granular materials upon which they are seated. See Figure 1810.2 for illustrations of typical enlarged base cast-in-place concrete piles.

1810.2.1 Materials. The maximum size for coarse aggregate for concrete shall be $^3/_4$ inch (19.1 mm). Concrete to be compacted shall have a zero slump.

❖ Concrete materials used for enlarged base piles are to comply with the applicable requirements of Chapter

Figure 1810.2
ENLARGED BASE PILES—UNCASED OR CASED SHAFTS

19. Coarse aggregate materials used in the concrete mix are not to exceed $^3/_4$ inch (19.1 mm) in size. Concrete used for the bulb type of pile described in Section 1810.2 is to have a zero slump in order to provide a stiff mix capable of being compacted by a heavy drop weight.

1810.2.2 Allowable stresses. The maximum allowable design compressive stress for concrete not placed in a permanent steel casing shall be 25 percent of the 28-day specified compressive strength (f'_c). Where the concrete is place in a permanent steel casing, the maximum allowable concrete stress shall be 33 percent of the 28-day specified compressive strength (f'_c).

❖ For uncased cast-in-place concrete piles formed by driving temporary steel casings, the allowable design stress in the concrete is not to exceed $0.25 f'_c$ (25 percent of the 28-day specified compressive strength). Pile capacity based on concrete strength alone (without consideration for soil capacity) must be calculated on the basis of the internal cross-sectional area of the pile casing (see commentary, Section 1810.4.1).

For cased piles where a steel casing is left permanently in place, the allowable design stress in the concrete is not to exceed $0.33 f'_c$ (33 percent of the 28-day specified compressive strength).

1810.2.3 Installation. Enlarged bases formed either by compacting concrete or driving a precast base shall be formed in or driven into granular soils. Piles shall be constructed in the same manner as successful prototype test piles driven for the project. Pile shafts extending through peat or other organic soil shall be encased in a permanent steel casing. Where a cased shaft is used, the shaft shall be adequately reinforced to resist column action or the annular space around the pile shaft shall be filled sufficiently to reestablish lateral support by the soil. Where pile heave occurs, the pile shall be replaced unless it is demonstrated that the pile is undamaged and capable of carrying twice its design load.

❖ The pressure-injected footing type of pile is normally installed by first driving a steel casing into the ground. This is done by means of a drop weight (hammer) operating inside the casing and impacting against a concrete or gravel plug confined in the tip of the steel casing. Next, after the required depth has been reached, the plug is driven out of the casing and an enlarged base is formed in the granular bearing soil by progressively adding and driving out small batches of zero-slump concrete by the operation of the drop weight. Caged reinforcement, when required, is inserted in the casing and joined to the footing by placing and compacting (by drop weight) a small amount of concrete. In the final step of the installation procedure, the uncased pile shaft is formed by placing zero-slump concrete in small lifts and compacting with the drop weight as the casing is withdrawn. Where caged reinforcement is used, the drop weight operates inside the cage. An illustration of the pressure-injected footing method of installation is shown in Figure 1810.2.3.

The shaft of this type of pile may also be constructed by placing high-slump concrete into the core as the drive casing is withdrawn.

A permanent cased shaft pile may be formed by inserting a steel shell in the drive casing after forming the footing, as described above, then filling the steel shell with conventional concrete and removing the drive casing.

The second type of enlarged base pile generally described in the introductory remarks of Section 1810.4 is installed with an enlarged precast concrete base at the end of a mandrel-driven steel casing. A major problem associated with this type of pile installation is that, in driving the permanent casing into the ground, the precast base creates a hole that is larger than the diameter of the pile shaft, leaving an open space around the pile and thus losing the lateral support usually provided by the surrounding soil. The same problem also applies to the pressure-injected footing type of pile where a permanent shell is used to form the pile shaft (as described in the previous paragraph), creating an open space around the shaft after the drive casing is withdrawn. In such cases, the annular space around the pile shaft must be filled to provide the necessary lateral support. The customary practice is to fill the annular space by pumping grout or washing in granular material.

(A) DRIVING PILING TUBE
(B) COMPACTING CONCRETE IN SHAFT
(C) PLACING CONCRETE IN PILING TUBE
(D) COMPLETED PILE

SOURCE: "PRACTICAL GUIDELINES FOR THE SELECTION, DESIGN AND INSTALLATION OF PILES" BY THE ASCE COMMITTEE ON DEEP FOUNDATION.

Figure 1810.2.3
PRESSURE-INJECTED FOOTING

The spacings of enlarged base piles are normally greater than most other types of conventional piles in order to avoid base interferences and the close over-lapping of soil-bearing areas that would serve to reduce pile capacity.

1810.2.4 Load-bearing capacity. Pile load-bearing capacity shall be verified by load tests in accordance with Section 1808.2.8.3.

❖ Due to irregularities in pile shapes, specifically as they relate to pressure-injected footing types of piles, the bearing capacity must be ascertained by means of load tests performed in accordance with the require-ments of Section 1808.2.8.3.

1810.2.5 Concrete cover. The minimum concrete cover shall be $2^1/_2$ inches (64 mm) for uncased shafts and 1 inch (25 mm) for cased shafts.

❖ In lined shafts, the clearances required for the installa-tion of reinforcement in the pile casings can be closely controlled with spacers; therefore, this section permits the minimum concrete cover dimension to be as small as 1 inch (25 mm). Due to surface irregularities in the walls of unlined piles, however, greater dimensional tolerance is needed to install the reinforcement; there-fore, the minimum cover requirement is set at $2^1/_2$ inches (63 mm).

1810.3 Drilled or augered uncased piles. Drilled or augered uncased piles shall conform to Sections 1810.3.1 through 1810.3.5.

❖ Holes for the drilled type of piles are formed by ma-chine drilling with augers or bucket-type drills (with or without protective casings) and then filling them with concrete by conventional methods, which include fun-nel hoppers, tremies or pumping. In comparison, holes for the augered type of piles are formed by a hol-low-stem auger, and subsequently, concrete is pres-sure injected into the holes through the hollow stem as the auger is withdrawn.

There are many problems associated with drilled hole piles that are discussed in the commentary to Section 1810.3.3; however, there are several advan-tages to using this type of pile, which include economy, complete nondisplacement, minimal driving vibration, high skin friction, good contact on rock for end bearing and no splicing is required.

1810.3.1 Allowable stresses. The allowable design stress in the concrete of drilled or augered uncased piles shall not exceed 33 percent of the 28-day specified compressive strength (f'_c). The allowable compressive stress of reinforcement shall not exceed 40 percent of the yield strength of the steel or 25,500 psi (175.8 MPa).

❖ The allowable design stresses for drilled or augered uncased piles are not to exceed $0.33 f'_c$ (33 percent of the 28-day specified concrete compressive strength). Installation methods of drilled cast-in-place concrete piles and augured cast-in-place piles are discussed in the previous section.

1810.3.2 Dimensions. The pile length shall not exceed 30 times the average diameter. The minimum diameter shall be 12 inches (305 mm).

Exception: The length of the pile is permitted to exceed 30 times the diameter, provided that the design and installation of the pile foundation are under the direct supervision of a registered design professional knowledgeable in the field of soil mechanics and pile foundations. The registered design professional shall certify to the building official that the piles were installed in compliance with the approved construction documents.

❖ The dimensional relationship between the diameter and length of a pile has been established for many de-cades and is based on the premise that a pile under axial load acts as a column and is subject to buckling. However, it has also been established through techno-logical advancements, research and experience that piles completely contained in soils, even in soft materi-als, do not behave as free-standing columns and the danger of pile buckling is extremely remote. Not with-standing this kind of evidence, the code limitations placed on dimensional requirements based on diame-ter-to-length ratios have become accepted practices.

1810.3.3 Installation. Where pile shafts are formed through unstable soils and concrete is placed in an open-drilled hole, a steel liner shall be inserted in the hole prior to placing the con-crete. Where the steel liner is withdrawn during concreting, the level of concrete shall be maintained above the bottom of the liner at a sufficient height to offset any hydrostatic or lateral soil pressure.

Where concrete is placed by pumping through a hol-low-stem auger, the auger shall be permitted to rotate in a clockwise direction during withdrawal. The auger shall be withdrawn in continuous increments. Concreting pumping pressures shall be measured and maintained high enough at all times to offset hydrostatic and lateral earth pressures. Concrete volumes shall be measured to ensure that the volume of con-crete placed in each pile is equal to or greater than the theoreti-cal volume of the hole created by the auger. Where the installation process of any pile is interrupted or a loss of con-creting pressure occurs, the pile shall be redrilled to 5 feet (1524 mm) below the elevation of the tip of the auger when the installation was interrupted or concrete pressure was lost and reformed. Augered cast-in-place piles shall not be installed within six pile diameters center to center of a pile filled with concrete less than 12 hours old, unless approved by the build-ing official. If the concrete level in any completed pile drops due to installation of an adjacent pile, the pile shall be replaced.

❖ There are many problems associated with drilled hole piles. Most of these problems are related to soil condi-tions, including soil or rock debris accumulating at the base of the pile or occurring in the pile shaft; reduc-tions in the shaft cross section caused by the necking of soil walls because of soft materials or earth pres-sures; discontinuities in the pile shaft; hollows on the surface of the shaft and other problems linked to drill-ing operations. Such problems are usually addressed by the use of proper installation techniques.

Whenever unstable soils are encountered in drilling operations, such as loose granular soils, organic soils, very soft silts or clays and water-bearing subsurface materials, a temporary steel liner is to be placed in the hole to prevent the collapse of the earth walls or sloughing off of the soil during concrete placement.

In placing concrete in temporarily lined holes, the top of the concrete should be kept well above the bottom edge of the steel liner as it is withdrawn in order to offset any hydrostatic or lateral soil pressures. Stiff (low slump) concrete should not be used, so as to avoid the potential problem of concrete arching in the liner tube and causing discontinuities or voids to occur in the pile shaft as the liner is withdrawn. For a discussion on concrete consistency (slump values), see the commentary to Section 1810.1.1.

While many of the problems associated with auger-placed concrete piles are similar to those described above for drilled hole piles, other problems may also be introduced because of the particular installation technique. The pile installation procedure using a hollow stem continuous-flight auger, both for drilling and concrete placement purposes, does not permit the use of temporary steel liners. Damage to this type of pile, because of improper installation procedures, can result in the incomplete filling of the pile hole, concrete discontinuities along the pile shaft, reductions in the cross-sectional area of the shaft (necking) and soil inclusions. Also, the loss of side support of the drilled hole or vertical displacement caused by ground pressures or soil movement may cause serious damage to the pile shaft.

In drilling the pile hole, the hollow-stem auger should be rotated and advanced continuously until the required tip elevation is reached. At that point in the drilling operation, rotation of the auger should be stopped to avoid removing excess soils that could result in damaging adjacent piles and to keep the auger flights full of soil as a means of retaining the hole walls. Concrete is then pumped through the hollow stem, filling the hole from the bottom up as the auger is withdrawn.

Concrete should be pumped under continuous pressure, and the rate of withdrawal of the auger should be carefully controlled for a continuous and full-sized shaft. The auger should be withdrawn in a smooth, continuous operation to prevent the surrounding soil from squeezing into the hole (necking) and possibly contaminating the concrete.

Discontinuities in the concrete shaft can develop if the auger is improperly withdrawn.

Concrete pumping pressures at the auger outlet should be greater than any hydrostatic or lateral pressures occurring in the hole. Excessively high pumping pressures, however, should be avoided in placing concrete surrounded by soft soils because it could cause upward or lateral movement of adjacent piles.

The volume of concrete placed in the hole is to be measured and should normally exceed the theoretical volume of the augered hole by approximately 10 percent. If the amount of the concrete pumped into the hole is considerably more than the quantity indicated above, however, then the cause should be investigated. It could mean, for instance, that concrete (grout) is being pumped into soft soil strata, solution cavities, underground pipelines or other underground structures.

If, for any reason, the pile-concreting operation is interrupted, a pressure drop occurs in pumping the concrete, the auger is improperly handled and is raised too fast or for any other cause that could damage the pile or cause a reduction in the required size of the pile section, then the hollow-stem auger must be redrilled to the required tip elevation and the pile reformed from the bottom up.

Unless approved by the building official, no new pile holes are to be drilled or injected with concrete if they are located within a center-to-center distance of six pile diameters from other adjacent piles containing fresh concrete that have not been allowed to set for a period of 12 hours or more. This is to reduce the possibility of damage to adjacent piles.

If the concrete surface of a completed pile is observed to drop below its cast elevation as a result of the drilling of adjacent piles, then the completed pile should be rejected and replaced. Such an occurrence could indicate some deformation along the pile shaft that could affect its structural integrity.

1810.3.4 Reinforcement. For piles installed with a hollow-stem auger where full-length longitudinal steel reinforcement is placed without lateral ties, the reinforcement shall be placed through the hollow stem of the auger prior to filling the pile with concrete. All pile reinforcement shall have a concrete cover of not less than 2.5 inches (64 mm).

Exception: Where physical constraints do not allow the placement of the longitudinal reinforcement prior to filling the pile with concrete or where partial-length longitudinal reinforcement is placed without lateral ties, the reinforcement is allowed to be placed after the piles are completely concreted but while concrete is still in a semifluid state.

❖ Unlike drilled cast-in-place concrete piles, caged reinforcement for auger-placed piles cannot be installed prior to filling the pile hole with concrete because the hollow-stem auger must be positioned in the hole at all times during drilling and concreting operations. To facilitate this, the cage must be pushed through the concrete in fluid form after the auger has been withdrawn. The problem associated with the placement of caged reinforcement in this way, particularly where it involves long cages, is that it cannot be determined if the assembly has been positioned appropriately within the filled hole such that the reinforcement will have a minimum $2\frac{1}{2}$-inch (63 mm) concrete cover at all places (see Section 1810.1.2 for a description of caged reinforcement).

As specified in this section, where pile reinforcement consists of individual longitudinal reinforcing bars without lateral ties it must be placed through the hollow stem of the auger prior to filling the hole with concrete.

The exception recognizes that it is not possible to always place full-length longitudinal steel reinforcement in the hollow stem of the auger prior to placing the concrete. In low headroom piles, the connector or splice in the reinforcing is often too large to fit in the hollow stem. Many specified single-center bars are too large to fit in the hollow stem and allow placement of concrete containing standard aggregates. Additionally, many large-center bars have a 90- (1.57 rad) or 180-degree (3.14 rad) hook, which prevents the bars from going into the hollow stem. This exception permits the placement of the longitudinal steel into the pile after placing concrete is complete.

1810.3.5 Reinforcement in Seismic Design Category C, D, E or F. Where a structure is assigned to Seismic Design Category C, D, E or F in accordance with Section 1613, the corresponding requirements of Sections 1810.1.2.1 and 1810.1.2.2 shall be met.

❖ This section provides a convenient cross reference to Sections 1810.1.2.1 and 1810.1.2.2 for seismic reinforcement detailing requirements for drilled or augered uncased piles. Augered uncased piles are defined in Section 1808.1.

1810.4 Driven uncased piles. Driven uncased piles shall conform to Sections 1810.4.1 through 1810.4.4.

❖ A driven uncased pile is a displacement type of cast-in-place concrete pile formed by driving a temporary steel pile casing into the ground, filling the opening with concrete and then extracting the casing. Pile casings are driven either with a closed end using a detachable tip made of steel or concrete that remains permanently in the ground upon withdrawal of the casing or by using a mandrel that seals off the tip of the casing, keeping soil from entering the core during the driving operation.

A reinforcing cage can be installed in the casing before placing concrete. Sometimes the top part of the pile is formed with a short section of corrugated steel shell that is left permanently in place to protect the pile from contamination by loose soil or entry of surface water when the driven casing is withdrawn and the level of concrete subsides to fill in the spaces previously occupied by the casing. The space created at the top of the pile caused by the drop in the concrete level is subsequently filled with concrete up to the pile cutoff elevation.

Sections 1810.4.1 through 1810.4.4 provide detailed requirements for the design and construction of driven uncased piles.

1810.4.1 Allowable stresses. The allowable design stress in the concrete shall not exceed 25 percent of the 28-day specified compressive strength (f'_c) applied to a cross-sectional area not greater than the inside area of the drive casing or mandrel.

❖ The allowable stress in the concrete is not to exceed $0.25f'_c$ (25 percent of the 28-day specified concrete compressive strength). Pile capacity based on concrete strength is to be calculated on the basis of a cross-sectional area no greater than the inside area of the drive casing or the area of the mandrel. The reason for this requirement is that the exact or average area of the pile hole cannot be established with any certainty because of the dimensional irregularities of the pile hole after removal of the casing.

1810.4.2 Dimensions. The pile length shall not exceed 30 times the average diameter. The minimum diameter shall be 12 inches (305 mm).

Exception: The length of the pile is permitted to exceed 30 times the diameter, provided that the design and installation of the pile foundation is under the direct supervision of a registered design professional knowledgeable in the field of soil mechanics and pile foundations. The registered design professional shall certify to the building official that the piles were installed in compliance with the approved design.

❖ See the commentary to Section 1810.3.2 for a discussion of pile length-to-diameter limits.

1810.4.3 Installation. Piles shall not be driven within six pile diameters center to center in granular soils or within one-half the pile length in cohesive soils of a pile filled with concrete less than 48 hours old unless approved by the building official. If the concrete surface in any completed pile rises or drops, the pile shall be replaced. Piles shall not be installed in soils that could cause pile heave.

❖ Pile casings driven in granular soils are to be spaced at least six pile diameters center to center. Pile casings driven in cohesive soils are not to be spaced less than one-half of the pile length when the concrete (in adjacent piles) is less than 48 hours old. This latter provision should not be construed to mean that piles in clay materials cannot be centered at spacings closer than 15 times their length (one-half by 30 times the diameter; see Section 1810.4.2). Indeed, they can be installed at the same spacings as piles in granular materials. Piles driven in cohesive materials, however, can cause significant soil displacement that can induce high lateral pressures on adjacent piles. Unless the concrete in these adjacent piles has had sufficient time to set (48 hours or more) and acquire some of its ultimate strength, considerable pile damage can be caused by the earth pressures. Because of the voids between the particles in granular soils, the effects of soil displacement are not as serious as for cohesive soils, such as clay.

The 48-hour concrete time-set requirement as stated in this section should not be confused with the 12-hour requirement referred to in Section 1810.3.3 for drilled or augered piles. While both types are cast-in-place concrete piles, unlike the driven uncased piles specified in Section 1810.4, drilled or augered piles are not of the displacement type, which essentially compacts the soil and may cause great lateral pressures and earth movement. The soil in the drilled method of installation is removed rather than displaced, and the subsurface influence on adjacent piles is far less than on the driven pile, particularly in cohesive soils.

Driven uncased piles are exposed to various types of potential damage. Besides the possibility of concrete contamination by the surrounding soil and surface water at the top of the pile, as described in Section 1810.4, the pile cross section may be subjected to squeezing or necking from lateral soil pressures and intrusions of displaced soils or other obstructions. Also, the concrete may be damaged by loss of support caused by removal of adjacent pile casings from the soil surrounding the pile. Driven uncased piles are not recommended under conditions where significant ground heave could occur or where highly unstable soils exist.

Subsurface investigations and load testing may be required to establish pile capacity, since there can be no correlation between the driving resistance of the pile casing and pile capacity. This is because the casing is eventually removed.

1810.4.4 Concrete cover. Pile reinforcement shall have a concrete cover of not less than 2.5 inches (64 mm), measured from the inside face of the drive casing or mandrel.

❖ For driven uncased piles, pile reinforcement, where required, is to be caged and installed so that the concrete cover over the reinforcement is no less than $2\frac{1}{2}$ inches (64 mm), measured from the inside face of the pile casing or inward from the periphery of the mandrel.

1810.5 Steel-cased piles. Steel-cased piles shall comply with the requirements of Sections 1810.5.1 through 1810.5.4.

❖ Steel-cased piles are the most widely used type of cast-in-place concrete piles. Essentially, they consist of mandrel-driven, light-gage steel shells or thin-walled pipes that are left permanently in place, reinforced as required by the design and filled with concrete. The shell is either a constant section for the full length of the pile or a step-tapered shape. A typical illustration is shown in Figure 1810.5.

1810.5.1 Materials. Pile shells or casings shall be of steel and shall be sufficiently strong to resist collapse and sufficiently water tight to exclude any foreign materials during the placing of concrete. Steel shells shall have a sealed tip with a diameter of not less than 8 inches (203 mm).

❖ Steel pipe shells or casings driven with a mandrel can be torn or otherwise damaged because of underground obstructions, such as rock crevices. If this happens, the water tightness and structural integrity of the pile may be affected. Sometimes the buildup of ground pressures during driving may cause pile shells to squeeze or even collapse after the mandrel has been withdrawn. This section requires that steel-cased piles

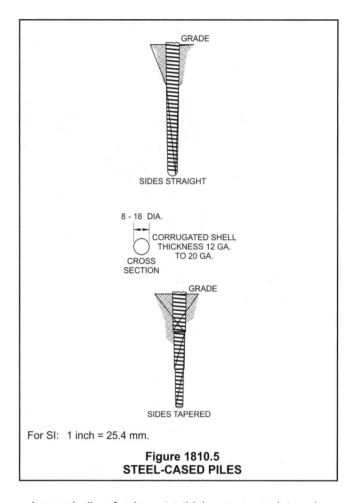

For SI: 1 inch = 25.4 mm.

Figure 1810.5
STEEL-CASED PILES

have shells of adequate thickness to resist serious damage or collapse and to maintain sufficient water tightness so as to prevent any foreign materials from entering the pile opening during the placing of concrete.

Pile tips are to have a closed end and be no less than 8 inches (203 mm) in diameter.

1810.5.2 Allowable stresses. The allowable design compressive stress in the concrete shall not exceed 33 percent of the 28-day specified compressive strength (f'_c). The allowable concrete compressive stress shall be 0.40 (f'_c) for that portion of the pile meeting the conditions specified in Sections 1810.5.2.1 through 1810.5.2.4.

❖ For cased cast-in-place concrete piles formed by driving permanent steel casings, the allowable design compressive stress in the concrete is not to exceed 0.33 f'_c (33 percent of the 28-day specified compressive strength). Except for the minimum pile tip diameter of 8 inches (203 mm), the pile section designed for a maximum allowable stress of 0.33 f'_c may be of any diameter, limited only by the standard sizes available from the manufacturers of steel pile casings.

The allowable concrete compressive stress may be increased to 0.40 f'_c, provided that the piles comply with the limiting provisions contained in Sections 1810.5.2.1 through 1810.5.2.4. The basis for this increase in allowable concrete stress is the added strength given to the concrete by the confining action of the steel shell. The general formula for increased allowable stress caused by confinement is:

$$f_c = 0.33\, f'_c \left(\frac{1 + 7.5tf_y}{Df'_c} \right)$$

where:

F_c = Allowable concrete stress.

f'_c = Specified 28-day concrete strength.

T = Thickness of steel shell.

F_y = Yield strength of steel.

D = Diameter of steel shell.

This formula is from the Portland Cement Association's (PCA) *Report on Allowable Stresses in Concrete Piles*.

When values for the various terms as set forth in Sections 1810.5.2.1 through 1810.5.2.4 are inserted in the given formula, the resulting allowable stress is $0.40f'_c$. Higher allowable stresses would result if, for example, the shell thickness was increased or the shell diameter was decreased. This increased allowable stress caused by confinement applies only to nonaxial load-bearing steel where the stress in the steel is taken in hoop tension instead of axial compression.

1810.5.2.1 Shell thickness. The thickness of the steel shell shall not be less than manufacturer's standard gage No. 14 gage (0.068 inch) (1.75 mm) minimum.

❖ Steel pile shells are to be No. 14 gage (U.S. standard) or thicker, but are not to be considered in the design of the pile to carry a portion of the pile load. The equivalent thickness for No.14-gage material is approximately 0.068 inches (1.7 mm).

1810.5.2.2 Shell type. The shell shall be seamless or provided with seams of strength equal to the basic material and be of a configuration that will provide confinement to the cast-in-place concrete.

❖ The shell for this type of pile must be seamless or have spirally welded seams and be of the strength and configuration required to provide structural confinement of the concrete fill. "Confinement" is the technical qualification that permits the use of increased allowable compressive stresses. Simply stated, the pile shell restrains the concrete in directions perpendicular to the applied stresses (see commentary, Section 1810.5.2).

1810.5.2.3 Strength. The ratio of steel yield strength (f_y) to 28-day specified compressive strength (f'_c) shall not be less than six.

❖ This section requires that the ratio of the yield strength (f_y) of the steel used in pile casings to the design compressive strength of concrete (f'_c) is not to be less than six. The yield strength of the steel used for pile casings of the type specified in this section is normally 30,000

psi (207 MPa) or greater. For example, in selecting a casing with a yield strength (f_y) of 30,000 psi (207 MPa) and a concrete compressive strength of 3,000 psi (21 MPa), the resulting ratio (f_y / f'_c) would be 10, which is greater than the minimum required ratio of six; therefore, the strengths of the pile materials are acceptable. For comparison, use the same steel casing and a specified concrete compressive strength (f'_c) of 5,000 psi (34 MPa). The resulting ratio would be exactly six. It can readily be seen that for concrete strengths greater than 5,000 psi (34 MPa), the type of steel used for the casing material would need yield strengths greater than 30,000 psi (207 MPa). For example, in order to meet the minimum ratio of six as required by this section, if the 28-day concrete compressive strength (f'_c) was specified at 6,000 psi (41 MPa), the material to be used for the casing would require a steel yield strength of at least 36,000 psi (248 MPa) conforming to ASTM A 36.

1810.5.2.4 Diameter. The nominal pile diameter shall not be greater than 16 inches (406 mm).

❖ The nominal pile diameter is not to exceed 16 inches (406 mm) in order to qualify for the "confined concrete" provision (see Section 1810.5.2) that allows for an increase in the allowable design compressive stress.

1810.5.3 Installation. Steel shells shall be mandrel driven their full length in contact with the surrounding soil.

The steel shells shall be driven in such order and with such spacing as to ensure against distortion of or injury to piles already in place. A pile shall not be driven within four and one-half average pile diameters of a pile filled with concrete less than 24 hours old unless approved by the building official. Concrete shall not be placed in steel shells within heave range of driving.

❖ For a general description of steel-cased cast-in-place concrete piles, refer to the commentary to Section 1810.5.

One advantage to this type of pile is that after the driving and removal of the mandrel before the concrete is placed, the steel casing can be inspected internally along its full length; therefore, any damage that has resulted from the pile-driving operation can be readily discovered and corrected.

Concrete placement is to comply with the applicable provisions of Section 1810.1.3.

Steel casings are not to be driven closer than four and one-half pile diameters of any adjacent piles filled with concrete until the concrete in the shells has cured for at least 24 hours and achieved an early strength. The basic reason for requiring four and one-half pile diameters between driving and concreting is a general reluctance to expose fresh concrete to vibrations as it sets. Several independent tests have shown, however, that there is no detrimental effect caused by vibrations on setting concrete. In some cases, there was a strength gain. The code requirement of four and one-half pile diameters is reasonable. It is often impractical to place concrete in pile shells right next to piles being driven. Pile casings can be driven earlier than the minimum 24-hour period when approved by the building official.

Driven light-gage steel pile shells that are left open can also be damaged by earth pressures when driving other piles in the close vicinity. Under such circumstances, pile shells are sometimes protected by inserting dummy mandrels.

Pile shells that are in place, adjacent to and within heave range of a pile being driven are normally left open if the soil condition is such that it can cause pile heave. This condition is particularly critical in cohesive soils, such as clay. Heaved piles must be redriven before filling with concrete. If a concreted pile has heaved, however, it can be safely redriven if proper techniques and a suitably designed pile cushion are used.

1810.5.4 Reinforcement. Reinforcement shall not be placed within 1 inch (25 mm) of the steel shell. Reinforcing shall be required for unsupported pile lengths or where the pile is designed to resist uplift or unbalanced lateral loads.

❖ Reinforcing steel is required in steel-cased, cast-in-place concrete piles for those lengths of pile that are laterally unsupported (such as piles projecting above the ground surface) and piles that must be designed for uplift forces (tension loads) or unbalanced lateral loads. Such reinforcement is to be designed to resist the bending and tensile stresses induced on the pile unit.

Steel-reinforcing cages must be installed at least 1 inch (25 mm) clear of the inside surface of the pile casing to allow sufficient space for the proper bonding of the steel reinforcing bars and the fill concrete.

1810.5.4.1 Seismic reinforcement. Where a structure is assigned to Seismic Design Category C, D, E or F in accordance with Section 1613, the reinforcement requirements for drilled or augered uncased piles in Section 1810.3.5 shall be met.

Exception: A spiral-welded metal casing of a thickness no less than the manufacturer's standard gage No. 14 gage [0.068 inch (1.7 mm)] is permitted to provide concrete confinement in lieu of the closed ties or equivalent spirals required in an uncased concrete pile. Where used as such, the metal casing shall be protected against possible deleterious action due to soil constituents, changing water levels or other factors indicated by boring records of site conditions.

❖ The purpose of this section is to include pile bending, which is a result of ground horizontal movement during an earthquake, in the structural design. The reinforcement in the pile, required to resist tension caused by pile bending, increases the ductility of the foundation such that bending or shear failure is precluded. The shear strength and confining ability of spiral-welded metal casing eliminates the need for special pile ties, which is the basis for the exception in this section.

1810.6 Concrete-filled steel pipe and tube piles. Concrete-filled steel pipe and tube piles shall conform to the requirements of Sections 1810.6.1 through 1810.6.5.

❖ This section covers concrete-filled steel piles consisting of either seamless or welded pipe or closed-ended

8"-18" DIA.

CROSS SECTION

CORRUGATED STEEL THICKNESS 12 GA. TO 20 GA.

SIDES STRAIGHT OR TAPERED

For SI: 1 inch = 25.4 mm.

Figure 1810.6
CONCRETE-FILLED STEEL PIPE PILE

tubular steel piles with either straight or tapered sections. These piles may be installed as friction or end-bearing units. If driven open ended, the pipe is cleaned out (earth core removed) before being filled with concrete. The steel pipe may be driven with an internal mandrel (see Figure 1810.6).

1810.6.1 Materials. Steel pipe and tube sections used for piles shall conform to ASTM A 252 or ASTM A 283. Concrete shall conform to Section 1810.1.1. The maximum coarse aggregate size shall be $^3/_4$ inch (19.1 mm).

❖ Seamless or welded-steel pipe piles must conform to the requirements of ASTM A 252. Steel tube piles are to be made of materials conforming to the requirements of ASTM A 283, or such other approved specifications suitable for use in pile fabrication.

The concrete used for filling the piles is to consist of $^3/_4$-inch (19.1 mm) maximum-sized coarse aggregates and conform to the requirements of Section 1810.1.1. Concrete should not contain any admixtures (e.g., calcium chloride) that would be corrosive to steel.

1810.6.2 Allowable stresses. The allowable design compressive stress in the concrete shall not exceed 33 percent of the 28-day specified compressive strength (f'_c). The allowable design compressive stress in the steel shall not exceed 35 percent of the minimum specified yield strength of the steel (F_y), provided F_y shall not be assumed greater than 36,000 psi (248 MPa) for computational purposes.

Exception: Where justified in accordance with Section 1808.2.10, the allowable stresses are permitted to be increased to 0.50 F_y.

❖ The allowable design compressive stress in concrete must not exceed 0.33 f'_c. The allowable design compressive stress in steel must not exceed 0.35F_y, except that the allowable design compressive stress may be increased up to 50 percent of the minimum specified yield strength of steel where substantiated by Section 1808.2.10.

The allowable stress provisions for both concrete and steel materials used for this type of pile are adequately covered by this section and are self-explanatory. For concrete-filled pipe piles, the full design load is shared by both steel and concrete; however, in driving an empty pipe pile, the full dynamic stresses resulting from the required ultimate capacity (twice design load) must be taken only by steel. Increasing allowable stresses to 0.50F_y could lead to pile deformations during driving.

In comparison, for mandrel-driven piles, the full design load is taken by concrete fill because the pile walls are so thin [$^1/_{10}$-inch (2.5 mm) minimum or 12 gage] that the strength of the steel shells is totally disregarded, with one exception. In accordance with the recommendations of a report on the design, manufacture and installation of concrete piles (ACI 543R), plain or fluted casings of the minimum thickness stated in Section 1810.6.3 may be designed as a load-carrying pile component, provided the cross-sectional area of the casing is at least 3 percent of the gross pile area.

Steel pipe piles are made of welded or seamless pipe in three grades of steel having yield strengths of 30,000, 35,000 and 45,000 psi (207, 242 and 311 MPa) in accordance with ASTM A 252. They are used as either end-bearing or friction-type piles and may be driven with open or closed ends. Open-ended piles are generally used where rock-bearing materials are found at levels reasonably close to the surface, and especially where very large loads are to be supported. In driving an open-ended pipe pile, the earth is forced into the core of the pipe and must be cleaned out either by a blowing technique or augering, depending on the type of soils encountered. As may be specified, the pipe is then filled with concrete.

When soils investigations show the absence of rock formations, or where rock occurs at excessive depths and it is not practical or economically feasible to reach such depths, closed-end pipe piles are driven as friction piles.

1810.6.3 Minimum dimensions. Piles shall have a nominal outside diameter of not less than 8 inches (203 mm) and a minimum wall thickness in accordance with Section 1809.3.4. For mandrel-driven pipe piles, the minimum wall thickness shall be $^1/_{10}$ inch (2.5 mm).

❖ This section requires that seamless or welded-steel pipe piles and tube piles have a nominal outside diameter of no less than 8 inches (203 mm). The 8-inch (203 mm) minimum diameter is necessary in order to inspect the inside of the pile to observe any damage that may have occurred during the driving process. Additionally, the minimum diameter is necessary to provide stiffness for driving purposes. The nominal wall thicknesses of steel pipe piles are to be in accordance with Section 1809.3.4.

1810.6.4 Reinforcement. Reinforcement steel shall conform to Section 1810.1.2. Reinforcement shall not be placed within 1 inch (25 mm) of the steel casing.

❖ Steel reinforcement placed in concrete-filled cores of steel pipe or tube piles to resist the bending stresses induced by lateral forces acting upon the piles or by other structural loads is to comply with the requirements of Section 1810.1.2.

Internal pile reinforcement, where required, consists of a cage made of several longitudinal deformed reinforcing bars tied together with lateral steel in the form of individual hoops or a spiral. Metal reinforcement is to comply with the applicable ASTM specifications referenced in ACI 318.

Preassembled cages should be provided with spacer bars to facilitate accurate placement and allow for a minimum clearance of 1 inch (25 mm) between the reinforcement and the pile casing. Longitudinal bars may be discontinued along the pile when no longer needed because of load transfer into the soil; no more than two bars should be stopped off at any one point along the pile. This provision is especially critical in thin-walled piles where the casings do not carry any of the axial or bending loads.

1810.6.4.1 Seismic reinforcement. Where a structure is assigned to Seismic Design Category C, D, E or F in accordance with Section 1613, the following shall apply. Minimum reinforcement no less than 0.01 times the cross-sectional area of the pile concrete shall be provided in the top of the pile with a length equal to two times the required cap embedment anchorage into the pile cap, but not less than the tension development length of the reinforcement. The wall thickness of the steel pipe shall not be less than $^3/_{16}$ inch (5 mm).

❖ This section prescribes seismic reinforcement requirements for concrete-filled steel pipe and tube piles in structures assigned to Seismic Design Category C, D, E or F. These types of piles are defined in Section 1808.1. The minimum wall thickness is based on testing and practice that has resulted in successful installations in moderate to high seismic regions.

1810.6.5 Placing concrete. The placement of concrete shall conform to Section 1810.1.3, but is permitted to be chuted directly into smooth-sided pipes and tubes without a centering funnel hopper.

❖ The placement of concrete in the core of a pipe pile or thin-walled tube pile must be performed in accordance with the requirements of Section 1810.1.3 to verify that the required design strength of the concrete is fully developed throughout the cross section and length of the pile. Section 1810.1.3 specifically requires a funnel hopper centered at the top of the pile. The primary purpose of the funnel hopper is to prevent the concrete from encountering the soil at the perimeter of the hole for a drilled pile . Smooth sided pipe or tube piles act as the hopper in themselves. The limitation to smooth-sided pipes is to rule out pipe sections having ridges or corrugations that could cause segregation.

Overdriving steel pipe, thin-walled tube piles or encountering damaging subsurface obstructions can result in bent piles, torn pile walls and distorted or ruptured pile tips. Such damage can allow the entry of soils or water into the pile cores. Pile walls may also collapse because of the buildup of excessive soil pressures during driving operations, especially in the case of thin-walled piles after removal of the mandrel.

Piles should be inspected internally to determine their condition after driving. If serious damage is discovered that would affect the structural performance of the pile, appropriate measures should be taken to correct the condition before the placement of concrete.

1810.7 Caisson piles. Caisson piles shall conform to the requirements of Sections 1810.7.1 through 1810.7.6.

❖ The caisson pile, commonly known as a "drilled-in caisson," is installed as a special type of high load-capacity pile. Its construction features are described in Section 1810.7.1. The design and allowable stress provisions, as well as the material, steel core and installation requirements, are contained in Sections 1810.7.2 through 1810.7.6.

Some advantages of using caisson piles include the control of construction during installation; nondisplacement; the ability to clean out the shaft and drive

the pile further; its high load capacities and ease in splicing. The main disadvantage of caisson piles is their significant cost.

An illustration of a typical caisson pile is shown in Figure 1810.7.

For SI: 1 inch = 25.4 mm.

Figure 1810.7
CAISSON PILES

1810.7.1 Construction. Caisson piles shall consist of a shaft section of concrete-filled pipe extending to bedrock with an uncased socket drilled into the bedrock and filled with concrete. The caisson pile shall have a full-length structural steel core or a stub core installed in the rock socket and extending into the pipe portion a distance equal to the socket depth.

❖ The caisson pile is a high-load capacity, end-bearing type of pile. Essentially, it is constructed as a cased, cast-in-place concrete pile formed by driving a thick-walled, open-ended steel pipe down to suitable rock material, cleaning out the soil materials within the pipe, drilling a socket into the rock, inserting a structural steel core into the pipe and then filling the pipe and drilled socket with concrete. The core material used in the pipe casing is of hot rolled structural shapes such as wide flange or I-beam sections. Steel rails are also used. Core material is installed to extend from the bottom of the rock socket up to the head of the pile or part of the way up the casing, depending on design requirements.

1810.7.2 Materials. Pipe and steel cores shall conform to the material requirements in Section 1809.3. Pipes shall have a minimum wall thickness of $^3/_8$ inch (9.5 mm) and shall be fitted with a suitable steel-driving shoe welded to the bottom of the pipe. Concrete shall have a 28-day specified compressive strength (f'_c) of not less than 4,000 psi (27.58 MPa). The concrete mix shall be designed and proportioned so as to produce a cohesive workable mix with a slump of 4 inches to 6 inches (102 mm to 152 mm).

❖ The pipe used in the construction of caisson piles must have a minimum thickness of $^3/_8$ inch (9.5 mm) and comply with the requirements of the ASTM A 252 standard specification for welded and seamless pipe piles. While this specification tabulates pipe pile sections up to 20 inches (508 mm) in outside diameter, larger sections may be ordered from the manufacturers of such products. Caisson piles as large as 36 inches (914 mm) in diameter have been used.

Caisson pipes must be fitted with a hardened steel drive shoe to facilitate hard driving to the rock and to sufficiently seat the tip of the pipe into the rock and form a seal that will prevent water and foreign matter from entering the pile hole during the concrete filling operations.

The materials used for the concrete fill must be proportioned to provide a workable mix that will have a minimum 28-day specified compressive strength (f'_c) of 4,000 psi (27.6 MPa) and a consistency (slump) of 4 inches (102 mm) to 6 inches (152 mm). Concrete is to comply with the applicable provisions of Chapter 19.

1810.7.3 Design. The depth of the rock socket shall be sufficient to develop the full load-bearing capacity of the caisson pile with a minimum safety factor of two, but the depth shall not be less than the outside diameter of the pipe. The design of the rock socket is permitted to be predicated on the sum of the allowable load-bearing pressure on the bottom of the socket plus bond along the sides of the socket. The minimum outside diameter of the caisson pile shall be 18 inches (457 mm), and the diameter of the rock socket shall be approximately equal to the inside diameter of the pile.

❖ The principal structural feature of the caisson type of pile is the rock socket, which is filled with concrete and designed to take the full load of the pile by end bearing and the frictional resistance offered by the rough walls of the socket.

The relationship between the depth of the socket and the bearing capacity of the rock cannot be determined with any great accuracy because of the natural joints, bedding planes and fissures generally found in rock formations. Experience has shown, however, that the frictional resistance in hard rock is often sufficient to carry the load. In the case of soft rock, the shear strength of the socket walls should be determined by testing rock samples.

1810.7.4 Structural core. The gross cross-sectional area of the structural steel core shall not exceed 25 percent of the gross area of the caisson. The minimum clearance between the structural core and the pipe shall be 2 inches (51 mm). Where cores are to be spliced, the ends shall be milled or ground to provide full contact and shall be full-depth welded.

❖ Core material made of structural steel or rails must comply with ASTM A 36. Other steels used as permanent load-bearing structural cores in concrete piles are referenced in ACI 543R, which deals with the design, manufacture and installation of concrete piles. The purpose of the steel core is to bond with the concrete fill to act as a composite section and provide additional strength and load-bearing capacity.

1810.7.5 Allowable stresses. The allowable design compressive stresses shall not exceed the following: concrete, $0.33 f'_c$; steel pipe, $0.35 F_y$ and structural steel core, $0.50 F_y$.

❖ Load-carrying capacity is determined as a composite of the strengths of the three components of the caisson piles: the driven steel pipe, the concrete and the structural core. As in other types of cased cast-in-place concrete piles, the concrete stress is not to exceed 0.33 f'_c (33 percent of the 28-day specified concrete compressive strength). Steel pipe stresses are not to exceed $0.35 F_y$ (35 percent of the steel yield strength) and the structural steel core material stresses are not to exceed $0.50 F_y$.

1810.7.6 Installation. The rock socket and pile shall be thoroughly cleaned of foreign materials before filling with concrete. Steel cores shall be bedded in cement grout at the base of the rock socket. Concrete shall not be placed through water except where a tremie or other approved method is used.

❖ A brief description of the installation sequence of caisson piles is given in the commentary to Section 1810.7.1. One of the problems in the construction of caisson piles is that some water will seep into the bottom of the pile opening through the rock joints or fissures in the socket. Sometimes water may seep into the opening at the end of the pipe casing because of incomplete seating of the pipe into the rock material. If the water in the hole cannot be controlled by ejection or other methods rendering the placement of concrete under reasonably dry conditions, the concrete may be placed by a tremie or other methods when approved by the building official.

1810.8 Micropiles. Micropiles shall conform to the requirements of Sections 1810.8.1 through 1810.8.5.

❖ The name "micropile" is a reference to its relative size compared to conventional piles. They are also referred to as "minipiles" or "pin piles." These provisions provide minimum requirements for the design and installation of micropiles. This type of pile has been a popular alternative to more conventional pile types where headroom or equipment access is otherwise limited.

1810.8.1 Construction. Micropiles shall consist of a grouted section reinforced with steel pipe or steel reinforcing. Micropiles shall develop their load-carrying capacity through a bond zone in soil, bedrock or a combination of soil and bedrock. The full length of the micropile shall contain either a steel pipe or steel reinforcement.

❖ Section 1808.1 defines micropiles as bored or grouted in-place piles no more than 12 inches (305 mm) in diameter and incorporating steel pipe casing or steel reinforcement. This section elaborates on that definition to provide a description of micropile construction. Either the steel pipe casing or the steel reinforcement is required to extend the full length of the pile. Note the further requirement in Section 1810.8.4.1 for a permanent steel casing in structures that are classified as Seismic Design Category C. Steel pipe casings are typically manufactured in segments with threaded ends.

1810.8.2 Materials. Grout shall have a 28-day specified compressive strength (f'_c) of not less than 4,000 psi (27.58 MPa). The grout mix shall be designed and proportioned so as to produce a pumpable mixture. Reinforcement steel shall be deformed bars in accordance with ASTM A 615 Grade 60 or 75 or ASTM A 722 Grade 150.

Pipe/casing shall have a minimum wall thickness of $^3/_{16}$ inch (4.8 mm) and as required to meet Section 1808.2.7. Pipe/casing shall meet the tensile requirements of ASTM A 252 Grade 3, except the minimum yield strength shall be as used in the design submittal [typically 50,000 psi to 80,000 psi (345 MPa to 552 MPa)] and minimum elongation shall be 15 percent.

❖ This section provides the minimum specifications for the component materials of the micropile, grout, reinforcing steel and steel pipe casing. It also establishes a minimum wall thickness for the steel pipe that is virtually identical to the limits for other steel pipe piles (see Section 1809.3.4).

1810.8.3 Allowable stresses. The allowable design compressive stress on grout shall not exceed $0.33 f'_c$. The allowable design compressive stress on steel pipe and steel reinforcement shall not exceed the lesser of $0.4 F_y$, or 32,000 psi (220 MPa). The allowable design tensile stress for steel reinforcement shall not exceed $0.60 F_y$. The allowable design tensile stress for the cement grout shall be zero.

❖ The code limits the allowable stress on grout to $0.33 f'_c$. The allowable stress on steel is limited to $0.4F_y$, but not greater than 32,000 psi (220 Mpa), which is based on 0.4×80 ksi, recognizing that high-strength steel reinforcing bars and pipes are frequently used in micropiles at these design stresses. Net tension must be resisted entirely by reinforcing steel.

1810.8.4 Reinforcement. For piles or portions of piles grouted inside a temporary or permanent casing or inside a hole drilled into bedrock or a hole drilled with grout, the steel pipe or steel reinforcement shall be designed to carry at least 40 percent of the design compression load. Piles or portions of piles grouted in an open hole in soil without temporary or permanent casing and without suitable means of verifying the hole diameter during grouting shall be designed to carry the entire compression load in the reinforcing steel. Where a steel pipe is used for reinforcement, the portion of the cement grout enclosed within the pipe is permitted to be included at the allowable stress of the grout.

❖ Micropiles with steel pile casing in place at the time grout is placed are designed with 40 percent of the compression load carried by the steel casing or steel reinforcing. Micropiles with grout that is placed without a casing in place must be designed so that 100 percent of the compression load is carried by the reinforcing steel.

1810.8.4.1 Seismic reinforcement. Where a structure is assigned to Seismic Design Category C, a permanent steel casing shall be provided from the top of the pile down 120 percent times the flexural length. The flexural length shall be determined in accordance with Section 1808.1. Where a structure is assigned to Seismic Design Category D, E or F, the pile shall be considered as an alternative system. In accordance with Section 104.11, the alternative pile system design, supporting documentation and test data shall be submitted to the building official for review and approval.

❖ In structures that are classified as Seismic Design Category A or B, there are no additional micropile requirements. In structures that are classified as Seismic Design Category C, permanent steel pipe casing must be provided for the length of the micropile noted. In structures that are classified as Seismic Design Category D, or higher, these micropile provisions are not applicable; approval of micropiles can only be as an alternative method of design and construction.

1810.8.5 Installation. The pile shall be permitted to be formed in a hole advanced by rotary or percussive drilling methods, with or without casing. The pile shall be grouted with a fluid cement grout. The grout shall be pumped through a tremie pipe extending to the bottom of the pile until grout of suitable quality returns at the top of the pile. The following requirements apply to specific installation methods:

1. For piles grouted inside a temporary casing, the reinforcing steel shall be inserted prior to withdrawal of the casing. The casing shall be withdrawn in a controlled manner with the grout level maintained at the top of the pile to ensure that the grout completely fills the drill hole. During withdrawal of the casing, the grout level inside the casing shall be monitored to check that the flow of grout inside the casing is not obstructed.

2. For a pile or portion of a pile grouted in an open drill hole in soil without temporary casing, the minimum design diameter of the drill hole shall be verified by a suitable device during grouting.

3. For piles designed for end bearing, a suitable means shall be employed to verify that the bearing surface is properly cleaned prior to grouting.

4. Subsequent piles shall not be drilled near piles that have been grouted until the grout has had sufficient time to harden.

5. Piles shall be grouted as soon as possible after drilling is completed.

6. For piles designed with casing full length, the casing must be pulled back to the top of the bond zone and reinserted or some other suitable means shall be employed to verify grout coverage outside the casing.

❖ This section provides the basic micropile installation requirements. Micropile boreholes are typically advanced by either of the listed methods, rotary drilling or rotary percussive drilling. Installation requirements differ based on whether a steel casing is permanent or temporary or not provided (Item 6, 1 or 2, respectively).

SECTION 1811
COMPOSITE PILES

1811.1 General. Composite piles shall conform to the requirements of Sections 1811.2 through 1811.5.

❖ Composite piles are made of two or more sections of different materials or of pile types that are spliced together to form a single pile unit. These piles are usually used when significant pile lengths are needed. However, due to economic considerations and problems with splicing, composite piles are seldom used today.

An illustration of typical pile combinations is shown in Figure 1811.1.

Sections 1811.2 through 1811.5 set forth the minimum design, construction and installation requirements for composite piles.

TYPICAL COMBINATION

Figure 1811.1
COMPOSITE PILES

1811.2 Design. Composite piles consisting of two or more approved pile types shall be designed to meet the conditions of installation.

❖ Like any other type of pile, the design of composite piles should consider: the installation requirements involving the handling and storage of piling materials; the selection of pile-driving equipment, with special regard for the physical properties of the weaker section; and the application of procedures that would prevent possible installation problems and prevent any adverse effect on the structural integrity of the pile during driving and its long-term performance. The driveability of a pile consisting of two or more different materials or pile types should be given special consideration to produce a pile design that will have sufficient stiffness and rigidity at the splice points to transmit the driving energy adequately, as well as withstand the driving forces.

1811.3 Limitation of load. The maximum allowable load shall be limited by the capacity of the weakest section incorporated in the pile.

❖ The strength of a composite pile consisting of two or more sections of different materials or pile types is governed by the weaker section of the spliced components. This section requires that the maximum load-carrying capacity of a composite pile be established by the physical properties of the weakest section. This limitation, while necessary from an engineering standpoint, provides for the inefficient use of the stronger pile sections. It is an inherent disadvantage in the use of composite piles.

1811.4 Splices. Splices between concrete and steel or wood sections shall be designed to prevent separation both before and after the concrete portion has set, and to ensure the alignment and transmission of the total pile load. Splices shall be designed to resist uplift caused by upheaval during driving of adjacent piles, and shall develop the full compressive strength and not less than 50 percent of the tension and bending strength of the weaker section.

❖ The strength of a composite pile is governed not only by the weaker section, but also by the strength and details of the splice that joins and holds the pile sections together.

There are several problems associated with the splicing of sections comprising a composite pile. For example, if the length of a composite pile is of such dimension that it will fit in the leaders of the driving equipment, then the full length of the spliced pile can be driven as a continuous operation. However, if the pile is too long to fit in the leaders as a single unit, the pile must be installed in sections, and the driving would have to be interrupted in order to make the splice in the leaders.

It is most important that splicing devices be made in such a way that their installation will be simple and quick. When pile driving must be stopped to make a splice in the leaders, the time expended should not be of a duration that would allow the soil to set up (soil freeze) so that continued driving would produce excessive stresses in the pile.

Sometimes when the predetermined length of a composite pile is too long to fit in the leaders, but not overly long, the difference in length between the pile and the leaders may be made up by inserting the lower end of the spliced unit in a preexcavated hole, thus allowing for continuous driving of the pile (see Section 1808.2.13).

Another problem related to splicing occurs in trying to accurately align the pile sections. When the lower section of a composite pile is driven nearly full length into the ground, it is very difficult to determine its true direction, particularly if the pile has drifted off line because of soil pressures, subsurface obstructions, improper driving or for any other reason. When a pile section is connected to a section already driven in the ground, the upper section should be installed in the same direction as the longitudinal axis of the lower section, even though the direction may not be vertical. It is better to maintain a misdirection rather than to try to correct a situation and create a bend at the pile joint.

The selection of the splicing method to be used for assembling a composite pile should be based on the driving and service load stresses that must be resisted by the pile. Splices should be designed to prevent separation of the pile sections during construction and thereafter. Special consideration must be given to the design of splices to resist uplift forces, whether the piles are subject to heaving or specifically designed as tension piles. Designing the splice strong enough to develop about one and one-half to two times the design uplift force is proper engineering practice. Pile splices must also be designed to resist compressive, bending and shear stresses imposed by construction and service loads.

1811.5 Seismic reinforcement. Where a structure is assigned to Seismic Design Category C, D, E or F in accordance with Section 1613, the following shall apply. Where concrete and steel are used as part of the pile assembly, the concrete reinforcement shall comply with that given in Sections 1810.1.2.1 and 1810.1.2.2 or the steel section shall comply with Section 1810.6.4.1.

❖ This section includes seismic requirements for composite piles, which are piles constructed as composites of two or more of the pile types defined in Section 1808.1. As such, these piles must comply with the applicable seismic requirements for the different types of piles used as indicated by the specific sections referenced.

SECTION 1812
PIER FOUNDATIONS

1812.1 General. Isolated and multiple piers used as foundations shall conform to the requirements of Sections 1812.2 through 1812.10, as well as the applicable provisions of Section 1808.2.

❖ Section 1812 addresses general requirements for the construction of isolated pier foundations (drilled or excavated shafts with or without belled bottoms) using plain or reinforced concrete or piers encased by steel shafts.

Piers can be used for both shallow and deep foundation construction, but are principally employed as deep foundations. There are basically two methods for constructing piers. The first method involves hand excavation of pits or holes that must be lined to prevent caving of the earth sides. The openings are subsequently filled with concrete. The "Chicago Caisson" method is an example of this type of construction. A circular hole is excavated at least $3^1/_2$ feet (1067 mm) in diameter to allow for working space; a larger dimension may be necessary as required by the design. On the way down, the excavation is lined with vertical boards (lagging) that are held in place by horizontal steel rings. The bottom of the shaft may be belled out to increase the bearing area. The hole is then filled with concrete, leaving the lagging and steel rings in place as part of the permanent foundation. In the past, it was not uncommon to carry foundations of this type down 50 or 60 feet (15 240 or 18 288 mm) below ground. Hand-excavated sheeted pits or cased holes are seldom used today, except for shallow foundations or for underpinning purposes.

The second and most common method used for constructing piers consists of drilling a hole in the earth (by machine), belling out the bottom where necessary to obtain the required bearing area and placing concrete in the hole. Pier holes may be cased or left without such protection based on the stability of the soil. Piers may contain reinforcement or be constructed without reinforcement, depending on design requirements. Drilled piers are also known by other names, such as "drilled shafts" or "drilled caissons."

Piers, unlike piles, are not normally used in groups, but are usually constructed as isolated foundations. Piers are generally designed as end-bearing units to carry loads greater than the capacity of single piles.

Sections 1812.2 through 1812.10 provide the minimum design, construction and installation requirements for piers. This section also references Section 1808.2, which addresses general requirements for piers and piles, including the required foundation investigation.

1812.2 Lateral dimensions and height. The minimum dimension of isolated piers used as foundations shall be 2 feet (610 mm), and the height shall not exceed 12 times the least horizontal dimension.

❖ Pier foundations are not to have widths (or diameters) less than 2 feet (610 mm) and should be proportioned so that the height (length) of the pier is no more than 12 times the least lateral dimension. This is an empirical requirement to prevent effects similar to those experienced in slender columns where, because of the flexibility of slender compression members, excessive bending stresses that cause structural failure, including buckling, may be induced.

1812.3 Materials. Concrete shall have a 28-day specified compressive strength (f'_c) of not less than 2,500 psi (17.24 MPa). Where concrete is placed through a funnel hopper at the top of the pier, the concrete mix shall be designed and proportioned so as to produce a cohesive workable mix having a slump of not less than 4 inches (102 mm) and not more than 6 inches (152 mm). Where concrete is to be pumped, the mix design including slump shall be adjusted to produce a pumpable concrete.

❖ Concrete used for piers is to have a minimum 28-day compressive strength of 2,500 psi (17.24 MPa). Higher concrete compressive strengths are usually specified to increase pier load capacity as well as for economic reasons.

This section requires that concrete be designed and proportioned to produce a cohesive workable mix with a consistency (slump) of no less than 4 inches (102 mm) and 6 inches (152 mm). The slump requirements stated here, however, should not be considered so absolute that, with the use of engineering judgement, the consistency of the concrete cannot be adjusted to satisfy prevailing conditions of construction. For example, concrete placement conducted under difficult conditions—such as for piles containing heavy reinforcement, very long cased piles, pile shells driven on steep batters or other demanding situations—may require the employment of special concrete mixes using reduced quantities of coarse aggregates with corresponding increases in sand and cement content. Under such conditions, slumps of 4 inches (102 mm) with a tolerance of plus 2 inches (51 mm) or minus 1 inch (25 mm) are in order.

High slump concrete is used so that the complete volume of the hole is filled, including natural soil crevices and pockets and surface irregularities caused by the drilling operations. High slumps are also used where the concrete must displace drilling slurry.

Concrete is usually placed through a funnel hopper at the top of the pile opening, but other approved methods such as pumping or tremie may also be used.

1812.4 Reinforcement. Except for steel dowels embedded 5 feet (1524 mm) or less in the pier, reinforcement where required shall be assembled and tied together and shall be placed in the pier hole as a unit before the reinforced portion of the pier is filled with concrete.

Exception: Reinforcement is permitted to be wet set and the $2^1/_2$- inch (64 mm) concrete cover requirement be reduced to 2 inches (51 mm) for Group R-3 and U occupancies not exceeding two stories of light-frame construction, provided the construction method can be demonstrated to the satisfaction of the building official.

Reinforcement shall conform to the requirements of Sections 1810.1.2.1 and 1810.1.2.2.

Exceptions:

1. Isolated piers supporting posts of Group R-3 and U occupancies not exceeding two stories of light-frame construction are permitted to be reinforced as required by rational analysis but not less than a minimum of one No. 4 bar, without ties or spirals, when detailed so the pier is not subject to lateral loads and the soil is determined to be of adequate stiffness.

2. Isolated piers supporting posts and bracing from decks and patios appurtenant to Group R-3 and U occupancies not exceeding two stories of light-frame construction are permitted to be reinforced as required by rational analysis but not less than one No. 4 bar, without ties or spirals, when the lateral load, E, to the top of the pier does not exceed 200 pounds (890 N) and the soil is determined to be of adequate stiffness.

3. Piers supporting the concrete foundation wall of Group R-3 and U occupancies not exceeding two stories of light-frame construction are permitted to be reinforced as required by rational analysis but not less than two No. 4 bars, without ties or spirals, when it can be shown the concrete pier will not rupture when designed for the maximum seismic load, E_m, and the soil is determined to be of adequate stiffness.

4. Closed ties or spirals where required by Section 1810.1.2.2 are permitted to be limited to the top 3 feet (914 mm) of the piers 10 feet (3048 mm) or less in depth supporting Group R-3 and U occupancies of Seismic Design Category D, not exceeding two stories of light-frame construction.

❖ This section prescribes the necessary seismic reinforcement for piers. Piers are defined in Section 1808.1 and are relatively short in comparison to their width. Piers are required to comply with the same seismic reinforcement requirements as cast-in-place concrete piles in Sections 1810.1.2.1 and 1810.1.2.2. However, there are quite a number of exceptions applicable to Group R-3 and U occupancies. Often piers are used in residential applications because of expansive soils and high ground-water tables. The requirement for transverse reinforcement is excessive for the interior piers of wood-frame residential structures that only support vertical loads imposed by wood posts and for lightly loaded piers. Exception 1 exempts such interior piers from the transverse reinforcement requirements of this section. Exceptions 2 through 4 allow lesser reinforcement than what is required by this section for residential structures, given certain conditions.

1812.5 Concrete placement. Concrete shall be placed in such a manner as to ensure the exclusion of any foreign matter and to secure a full-sized shaft. Concrete shall not be placed through water except where a tremie or other approved method is used. When depositing concrete from the top of the pier, the concrete shall not be chuted directly into the pier but shall be poured in a rapid and continuous operation through a funnel hopper centered at the top of the pier.

❖ The drilled hole must be free of all foreign matter prior to placing concrete. If there is excessive water in the hole [about 4 inches (102 mm) is permitted], the concrete may be placed by tremie or pumping methods. Concrete is usually deposited from the top through a funnel hopper that should have a discharge opening that is sized smaller than the smallest pier cross section in the foundation system. Vibrators should not be used in the concreting operations because they may damage the earth walls and cause contamination of the concrete. Improper vibration may also cause excessive bleeding.

1812.6 Belled bottoms. Where pier foundations are belled at the bottom, the edge thickness of the bell shall not be less than that required for the edge of footings. Where the sides of the bell slope at an angle less than 60 degrees (1 rad) from the horizontal, the effects of vertical shear shall be considered.

❖ Piers constructed as deep foundations are sometimes belled out at their bottoms to increase bearing areas. Belled bottoms can be made by either hand excavation or a mechanical method involving underreaming with a belling bucket, as is commonly used in the construction of drilled piers.

The soil should be sufficiently cohesive so the roof of the bell will not collapse during excavation, cleanout operations and the placement of fresh concrete. Where the character of the soil is such that attempts at belling out pier bottoms might not produce acceptable results, it would be preferable to carry the piers down as straight shafts into soil strata with better load-bearing values, allowing the loads to be carried by the smaller pier bottoms or to such depths that would permit the soils to carry the loads by side friction. The

building official should ascertain the suitability of the soil for bell construction through soil investigations and reports and by the recommendations of the architect or engineer.

Section 1812.6 requires bells to have vertical edges at their bottoms that are equal to the thickness requirements for concrete footings. The purpose of this specification is to prevent shear breaks in angled edges caused by soil pressures. Bell slopes (sides) are not to have sides that are less than 60 degrees (1 rad) from the horizontal, unless the effects of vertical shear are considered in the design (see Figure 1812.6).

1812.7 Masonry. Where the unsupported height of foundation piers exceeds six times the least dimension, the allowable working stress on piers of unit masonry shall be reduced in accordance with ACI 530/ASCE 5/TMS 402.

❖ ACI 530 contains provisions for reducing the allowable working stress on masonry columns based on the height to radius of gyration of the column.

1812.8 Concrete. Where adequate lateral support is not provided, and the unsupported height to least lateral dimension does not exceed three, piers of plain concrete shall be designed and constructed as pilasters in accordance with ACI 318. Where the unsupported height to least lateral dimension exceeds three, piers shall be constructed of reinforced concrete, and shall conform to the requirements for columns in ACI 318.

Exception: Where adequate lateral support is furnished by the surrounding materials as defined in Section 1808.2.9, piers are permitted to be constructed of plain or reinforced concrete. The requirements of ACI 318 for bearing on concrete shall apply.

❖ The slenderness ratio (unsupported height to least lateral dimension) is limited to three for piers constructed of plain concrete. When the slenderness ratio exceeds three, the pier is required to be constructed of reinforced concrete in accordance with ACI 318. Where

Figure 1812.6
BELLED BOTTOMS OF PIER FOUNDATIONS

the surrounding soil provides lateral support for the pier, this section permits the pier to be of plain or reinforced concrete. The bearing on concrete requirements of ACI 318 will govern.

1812.9 Steel shell. Where concrete piers are entirely encased with a circular steel shell, and the area of the shell steel is considered reinforcing steel, the steel shall be protected under the conditions specified in Section 1808.2.17. Horizontal joints in the shell shall be spliced to comply with Section 1808.2.7.

❖ The use of permanent steel shells as reinforcement for concrete and as may be required to facilitate excavation conditions is generally associated with drilled pier construction. A drilled concrete pier consisting of a steel shell is analogous to a pipe pile. The engineering significance is the same for both types of construction, except that pipe piles are driven in groups of two or more for the support of a building column while drilled piers serving the same purpose and of larger diameters are designed as single, isolated foundations. Shells may be installed during or after excavation operations, depending on the stability of the soils.

Steel materials must be protected from corrosive soils and shell joints must be properly spliced, all in accordance with the applicable provisions of Section 1808.2.7.

1812.10 Dewatering. Where piers are carried to depths below water level, the piers shall be constructed by a method that will provide accurate preparation and inspection of the bottom, and the depositing or construction of sound concrete or other masonry in the dry.

❖ Where pier foundations are carried to depths below water level, the construction operations should be conducted under dry conditions. Operating in dry conditions becomes particularly important where it involves inspections of drilled excavations to ascertain the accuracy of shaft alignment and dimensions; when necessary, to verify the load-bearing capacity of the supporting soil or rock stratum; to verify the removal of loose material from pier bottoms; to inspect bell-outs for possible failures and to ascertain visually that the opening will remain stable during concreting operations.

Where water inflow is expected or encountered in deep excavations and dewatering becomes necessary, it can be accomplished by any of a number of commonly used methods, ranging from the simple practice of pumping out excavations to the more complex systems employed by lowering water tables below foundation bottoms.

The choice of the dewatering method depends mainly on the site conditions, including the soil characteristics and soil profile, the level of the water table and the type and depth of the foundation construction. Serious water problems occur in granular soils rather than in cohesive clay materials.

Pier foundations are best used in soils that will permit construction in dry conditions for the reasons stated above. When the findings of soil investigations indicate a high probability of encountering serious water problems and it is apparent that the control of water inflow will be most difficult and, perhaps, economically impractical, it is preferable to select another type of foundation construction (i.e., driven piles or other suitable foundations).

One of the most common methods used for stopping the inflow of water from pockets or strata of cohesionless soils (granular materials) encountered during the drilling of deep pier holes is to introduce a slurry consisting of soil, bentonite and water. The fluid is mixed by rotating, lifting and lowering the drilling auger. When the walls of the pier opening at the locations of water inflow have been sealed by the fluid mixture, a pipe casing is inserted to protect the hole and the slurry is then bailed out, allowing the drilling to proceed under dry conditions.

Bibliography

The following resource materials are referenced in this chapter or are relevant to the subject matter addressed in this chapter.

ACI 229R-99, *Controlled Low-Strength Materials.* Farmington Hills, MI: American Concrete Institute, 1999.

ACI 306R-88, *Cold Weather Concreting.* Farmington Hills, MI: American Concrete Institute, 1988.

ACI 306.1-90, *Standard Specification for Cold Weather Concrete.* Farmington Hills, MI: American Concrete Institute, 1990.

ACI 318-05, *Building Code Requirements for Structural Concrete.* Farmington Hills, MI: American Concrete Institute, 2005.

ACI 530/ASCE 5/TMS 402-05, *Building Code Requirements for Masonry Structures.* Farmington Hills, MI: American Concrete Institute; Reston, VA: American Society of Civil Engineers; Boulder, CO: The Masonry Society, 2005.

ACI 336.3R-93. *Design and Construction of Drilled Piers.* Famington, MI: American Concrete Institute, 1993.

ACI 530.1/ASCE 6/TMS 602-05, *Specification for Masonry Structures.* Farmington Hills, MI: American Concrete Institute; Reston, VA: American Society of Civil Engineers; Boulder, CO: The Masonry Society, 2005.

ACI 543R-74 (1980), *Recommendations for Design, Manufacture and Installation of Concrete Piles.* Farmington Hills, MI: American Concrete Institute, 1980.

ACI 2006 Manual of Concrete Practice. Farmington Hills, MI: American Concrete Institute, 2006.

AF&PA NDS-05, *National Design Specification for Wood Construction—with 2005 Supplement.* Washington, D.C.: American Forest and Paper Association, 2005.

AF&PA Technical Report 7-87, *Basic Requirements for Permanent Wood Foundation Systems*. Washington, D.C.: American Forest and Paper Association, 1987.

ASCE 20-96, *Standard Guidelines for the Design and Installation of Pile Foundations*. Reston, VA: American Society of Civil Engineers, 1997.

ASCE 24-05, *Flood Resistant Design and Construction Standard*. Reston, VA: American Society of Civil Engineers, 2005.

ASCE 32-01, *Design of Frost Protected Shallow Foundations*. Reston, VA: American Society of Civil Engineers, 2001.

ASTM A 36M-04, *Specification for Carbon Structural Steel*. West Conshohocken, PA: ASTM International, 2004.

ASTM A 252-98(2002), *Specification for Welded and Seamless Steel Pipe Piles*. West Conshohocken, PA: ASTM International, 2002.

ASTM A 283/A 283M-03, *Specification for Low and Intermediate Tensile Strength Carbon Steel Plates*. West Conshohocken, PA: ASTM International, 2003.

ASTM A 416/416M-02, *Specification for Steel Strand, Uncoated Seven-Wire for Prestressed Concrete*. West Conshohocken, PA: ASTM International, 2002.

ASTM A 572/A 572M-04, *Specification for High-Strength Low-Alloy Columbian-Vanadium Structural Steel*. West Conshohocken, PA: ASTM International, 2004.

ASTM A 588/A 588M-04, *Specification for High-Strength Low-Alloy Structural Steel with 50 ksi (345 Mpa) Minimum Yield Point to 4 inches Thick*. West Conshohocken, PA: ASTM International, 2004.

ASTM A 913/A 913M-04, *Specification for High-Strength Low-Alloy Steel Shapes of Structural Quality, Produced by Quenching and Self-Tempering Process (QST)*. West Conshohocken, PA: ASTM International, 2004.

ASTM C 90-03, *Specification for Load-bearing Concrete Masonry Units*. West Conshohocken, PA: ASTM International, 2003.

ASTM C 270-04, *Specification for Mortar for Unit Masonry*. West Conshohocken, PA: ASTM International, 2004.

ASTM C 887-79a(2001), *Specification for Packaged, Dry, Combined Materials for Surface Bonding Mortar*. West Conshohocken, PA: ASTM International, 2001.

ASTM D 25-99E01, *Specification for Round Timber Piles*. West Conshohocken, PA: ASTM International, 1999.

ASTM D 422-63(2002), *Standard Test Method for Particle Size Analysis of Soils*. West Conshohocken, PA: ASTM International, 2002.

ASTM D 1143-81 (1994)e1, *Standard Test Method for Piles Under Static Axial Compressive Load*. West Conshohocken, PA: ASTM International, 1994.

ASTM D 1557-e01, *Standard Test Method for Laboratory Compaction Characteristics of Soil Using Modified Effort*. West Conshohocken, PA: ASTM International, 2001.

ASTM D 1668-97a, *Standard Specification for Glass Fabrics (Woven and Treated) for Roofing and Waterproofing*. West Conshohocken, PA: ASTM International, 1997.

ASTM D 2487-00, *Standard Practice for Classification of Soils for Engineering Purposes*. West Conshohocken, PA: ASTM International, 2000.

ASTM D 3689-90 (1995), *Specification of Individual Piles Under Static Axial Tensile Load*. West Conshohocken, PA: ASTM International, 1995.

ASTM D 4318-00, *Standard Test Method for Liquid Limit, Plastic Limit, and Plasticity Index of Soils*. West Conshohocken, PA: ASTM International, 2000.

ASTM D 4829-03, *Test Method for Expansion Index of Soils*. West Conshohocken, PA: ASTM International, 2003.

ASTM D 4832-02, *Standard Test Method for Preparation and Testing of Controlled Low Strength Material (CLSM) Test Cylinders*. West Conshohocken, PA: ASTM International, 2002.

ASTM D 4945-00, *Test Method for High-Strain Dynamic Testing of Piles*. West Conshohocken, PA: ASTM International, 2000.

ASTM D 5971-01, *Standard Practice for Sampling Freshly Mixed Controlled Low-Strength Material*. West Conshohocken, PA: ASTM International, 2001.

ASTM D 6023-02, *Standard Test Method for Unit Weight, Yield, Cement Content, and Air Content (Gravimetric) of Controlled Low-Strength Material (CLSM)*. West Conshohocken, PA: ASTM International, 2002.

ASTM D 6024-02, *Standard Test Method for Ball Drop on Controlled Low-Strength Material (CLSM) to Determine Suitability for Load Application*. West Conshohocken, PA: ASTM International, 2002.

ASTM D 6103-97, *Standard Test Method for Flow Consistency of Controlled Low Strength Material (CLSM)*. West Conshohocken, PA: ASTM International, 1997.

AWPA M4-01, *Standard for the Care of Preservative-Treated Wood Products*. Grandbury, TX: American Wood-Preservers' Association, 2001.

AWPA U1-04, *Use Category System: User Specification for Treated Wood, except* Section 6, Commodity Specification H. Grandbury, TX: American Wood-Preservers' Association, 2004

AWPB MP1, *Quality Control and Inspection Procedures for Dual Treatment of Marine Piling Pressure Treated with Waterborne Preservatives and Creosote for Use in Marine Waters.* American Wood Preservers Bureau.

AWPB MP2, *Quality Control and Inspection Procedures for Dual Treatment of Marine Piling Pressure Treated with Waterborne Preservatives and Creosote for Use in Marine Waters.* American Wood Preservers Bureau.

AWPB MP4, *Quality Control and Inspection Procedures for Dual Treatment of Marine Piling Pressure Treated with Waterborne Preservatives and Creosote for Use in Marine Waters.* American Wood Preservers Bureau.

Bowles, J. E. *Foundation Analysis and Design, 5th edition* New York: McGraw Hill, 1996.

DOC PS-20-99, *American Softwood Lumber Standard.* Gaithersburg, MD: U.S. Department of Commerce; National Institute of Standards and Technology, 1999.

FEMA 55, *Coastal Construction Manual.* Washington, D.C.: Federal Emergency Management Agency, 1999.

FEMA TB11-01, *Crawlspace Construction for Buildings Located in Special Flood Hazard Areas.* Washington, D.C.: Federal Emergency Management Agency, 2001.

Hanson, W.E., R.B. Peck, and T.H. Thornburn. *Foundation Engineering, 2nd edition.* New York: John Wiley & Sons, 1974.

IPC-06, *International Plumbing Code.* Falls Church, VA: International Code Council, 2006.

IRC-06, *International Residential Code.* Falls Church, VA: International Code Council, 2006.

NCMA Report, *Research Evaluation of the Flexural Strength of Concrete Masonry, Project No. 93-172, Dec. 7, 1993.* Herndon VA: National Concrete Masonry Association, 1993.

NRCA Roofing and Waterproofing Manual, 3rd edition. Chicago: National Roofing Contractors Association, 1989.

PTI, *Design of Post-Tensioned Slabs-on-Ground, 3rd edition.* Phoenix: Post-Tensioning Institute, 2004.

PTI, *Standard Requirements for Analysis of Shallow Concrete Foundations on Expansive Soils, 1st ed.* Phoenix: Post-Tensioning Institute, 2004.

PTI, *Standard Requirements for Design of Shallow Post-Tensioned Concrete Foundations on Expansive Soils, 1st ed.* Phoenix: Post-Tensioning Institute, 2004.

Recommended Design Specifications for Driven Bearing Piles 3rd ed. Boulder, CO: Pile Driving Contractors Association, 2001.

"Recommended Practice for Design, Manufacture and installation of Prestressed Concrete Piling." *PCI Journal,* V. 38, No. 2, March-April 1993, pp. 14-41.

Terzaghi, Karl and R. B. Peck. *Soil Mechanics in Engineering Practice, 2nd edition.* New York: John Wiley & Sons, 1967.

WRI/CRSI-96, *Design of Slab-on-Ground Foundations.* Leesburg, VA: Wire Reinforcing Institute, 1996.

Chapter 19:
Concrete

General Comments

The words "cement" and "concrete" are used in several sections in this chapter. At the outset of this commentary, the distinction between cement and concrete should be made. Often, one may hear of references to cement floors, cement sidewalks and driveways, or of cement walls in describing walls made of cast-in-place or precast concrete or of concrete masonry, and the like. Cement and concrete are not interchangeable words. Cement is only one of four basic ingredients (cement, water, sand and stone) that make up the concrete material used in construction. In the end, there are only concrete structures and concrete building components. Structures of cement alone simply do not exist.

While concrete is the most versatile of the structural materials, it is also one of the most variable and complex materials used in building construction. Concrete is used in a number of different ways. It is used for building substructures, spread footings, mats, piles and foundation walls. Above-ground concrete is used for building walls and enclosures, for floors and roof construction, and for structural framework. Concrete also has a number of ancillary uses in building facilities: the construction of sidewalks, patios and driveways; for sewers and basins; for equipment bases; for fire protection purposes; as an insulating material; and so on. Concrete can be cast into any desired shape that can be structurally formed in the field or in a precasting plant with a variety of architectural finishes and surface configurations to satisfy aesthetic requirements. Mixtures can be varied to achieve strengths [2,500 to 20,000 pounds per square inch (psi) (17 238 to 137 908 kPa)] that meet the structural needs of the design. Concrete mixes can also be made to achieve good workability of the fresh material and to provide durability of the hardened product exposed to adverse environmental conditions, as well as to the conditions of wear imposed by the use of the facilities.

Because of the variable properties of the material and the numerous design and construction options available in the uses of concrete, due care and control throughout the entire construction process is necessary. This process includes: selection of materials and design of the mixture; mixing and transport of the concrete; proper construction and use of formwork; setting of the reinforcement and, in prestressed construction, the tensioning of the tendons and finally, the placement, finishing, curing and testing of the concrete material.

Purpose

The provisions of this chapter are intended to set the minimum accepted practices that apply to the design and construction of buildings or structural components using concrete.

SECTION 1901
GENERAL

1901.1 Scope. The provisions of this chapter shall govern the materials, quality control, design and construction of concrete used in structures.

❖ Section 1901 provides information on the application and organization of the concrete provisions of Chapter 19. For additional background information on the provisions of this chapter, particularly as they relate to the inspection of concrete construction, consult PCA-EB115, *Concrete Inspection Handbook,* listed in the bibliography. Section 1901.1 provides a definitive statement on the important aspects of concrete construction covered by Chapter 19. If proper attention is given to these aspects, it will improve the chances of a quality concrete structure that is code compliant.

1901.2 Plain and reinforced concrete. Structural concrete shall be designed and constructed in accordance with the requirements of this chapter and ACI 318 as amended in Section 1908 of this code. Except for the provisions of Sections 1904 and 1910, the design and construction of slabs on grade shall not be governed by this chapter unless they transmit vertical loads or lateral forces from other parts of the structure to the soil.

❖ The structural concrete provisions of the code comply strictly with the provisions of ACI 318/318R, except for the modifications contained in Section 1908. Slabs on grade are another important exception. Slabs on grade typically only convey the gravity dead load of the slab and the gravity live load superimposed on the slab directly to the soil; thus, slabs are not considered to be the "structural concrete" that must comply with the design provisions of ACI 318. Slabs on grade, however, must comply with the durability provisions of Section 1904 and the minimum thickness and vapor barrier provisions of Section 1911.

The last sentence of Section 1901.2 clarifies that

slabs on grade used to transmit "vertical loads," "lateral forces" or both from other parts of the structure to the soil must comply with all the applicable portions of Chapter 19 and ACI 318. In many cases, slabs on grade are used to transmit lateral forces due to wind or seismic into the ground. A common example is a typical preengineered metal building in which the shear at the bottom of the columns is resisted by placing "hairpins" around the column anchor bolts and extending the hairpins into the slab on grade. Although this may appear to be a new requirement, codes have for many years required that structures be designed so that they do not slide or overturn.

1901.3 Source and applicability. The format and subject matter of Sections 1902 through 1907 of this chapter are patterned after, and in general conformity with, the provisions for structural concrete in ACI 318.

❖ Section 1901.3 references ACI 318; thus, all of the requirements in ACI 318 become an enforceable part of the code. Sections 1902 through 1907 of the code parallel Chapters 2 through 7 of ACI 318. Architects, engineers, building officials and others who are responsible for the design and construction of concrete must use ACI 318 as supplemented by this chapter to fully comply with the code.

ACI 318 includes requirements for all structural concrete. The term "structural concrete" refers to all concrete used for structural purposes, including both plain and reinforced. By definition, "Plain concrete" is either unreinforced or contains less reinforcement than would be required for "reinforced concrete" by ACI 318. The definition of "Reinforced concrete" includes conventionally reinforced and prestressed concrete. ACI 318 is equally applicable to cast-in-place concrete and precast concrete. Additionally, it has provisions for composite concrete and steel compression members utilizing structural steel shapes and steel pipe or tubing.

The general purpose of ACI 318 is to provide a minimum acceptable standard for use in governing concrete material properties and the design and construction of concrete structural elements. While engineering knowledge, experience and sound judgement must prevail, nonetheless, structural elements of concrete must comply with ACI 318 unless preconstruction load tests are conducted in accordance with Section 1714 of the code.

ACI 318 generally does not govern special concrete structures, such as arches, tanks, reservoirs, bins and silos, blast-resistant structures and chimneys; however, portions of it may be applicable (e.g., durability requirements).

ACI 318 does not govern the design and construction of composite concrete slabs on steel deck, as this type of construction is regulated by ASCE 3 (see Section 2209.2). However, ASCE 3 references the applicable provisions of ACI 318 (i.e., Chapters 2 through 7 of ACI 318, which are also the basis for Sections 1902

through 1907 of this chapter). The design and construction of concrete slabs cast on noncomposite, stay-in-place steel deck are regulated by this chapter and ACI 318.

ACI 318 does not govern portions of concrete piles, caissons and drilled piers embedded in soil capable of providing lateral support, except in buildings assigned to Seismic Design Category D, E or F (see Section 1808.2.23). In general, portions of concrete pier and pile foundations embedded in soil capable of providing lateral support to the foundation element must comply with Sections 1808 through 1812. Portions of concrete piles, caissons and drilled piers in soil not capable of providing lateral support, projecting above the ground surface or in water must comply with this chapter and ACI 318 (see Section 1808.2.9.2).

Generally, foundation walls and shallow footings must comply with Chapter 19 and ACI 318; however, the code user should be aware of requirements located in Section 1805 that may provide an alternative to those in this chapter.

ACI 318 cites numerous material and test standards of ASTM International (ASTM) that are applicable to concrete construction. These ASTM standards are declared to be part of the ACI document and, therefore, also become enforceable under the code.

Chapter 1 of ACI 318 covers general requirements that may be in conflict with the administration and enforcement sections of the code, particularly those provisions that relate to drawings and inspections. In cases of conflict, the provisions in the code will govern in accordance with Section 102.4.

1901.4 Construction documents. The construction documents for structural concrete construction shall include:

1. The specified compressive strength of concrete at the stated ages or stages of construction for which each concrete element is designed.

2. The specified strength or grade of reinforcement.

3. The size and location of structural elements, reinforcement, and anchors.

4. Provision for dimensional changes resulting from creep, shrinkage and temperature.

5. The magnitude and location of prestressing forces.

6. Anchorage length of reinforcement and location and length of lap splices.

7. Type and location of mechanical and welded splices of reinforcement.

8. Details and location of contraction or isolation joints specified for plain concrete.

9. Minimum concrete compressive strength at time of posttensioning.

10. Stressing sequence for posttensioning tendons.

11. For structures assigned to Seismic Design Category D, E or F, a statement if slab on grade is designed as a

structural diaphragm (see Section 21.10.3.4 of ACI 318).

❖ This section provides a list of important information that must be included in construction documents. These items that are specifically related to concrete construction, are in addition to the structural design information required by Section 1603, and additional requirements of Section 106. The list is not necessarily all-inclusive and may be expanded at the discretion of the building official.

1901.5 Special inspection. The special inspection of concrete elements of buildings and structures and concreting operations shall be as required by Chapter 17.

❖ This section provides a cross reference to the code's special inspection provisions for concrete found in Section 1704, which specifies that most structural concrete, with the exception of some foundations, needs to have a special inspection. ACI 318 does not contain provisions for the special inspection of concrete, except that Section 1.3.5 requires continuous inspection of the placement of reinforcement and concrete in special moment frames that resist seismic forces in regions of high seismic risk, which correspond to Seismic Design Category D, E or F of the code.

SECTION 1902
DEFINITIONS

1902.1 General. The words and terms defined in ACI 318 shall, for the purposes of this chapter and as used elsewhere in this code for concrete construction, have the meanings shown in ACI 318.

❖ Section 1902 contains definitions of terms that are associated with the subject matter of this chapter. It is important to emphasize that these terms are not exclusively related to this chapter but are applicable everywhere the term is used in the code. The code has transcribed only those definitions for concrete construction that are used in Chapter 19. Additional definitions and related commentary for Chapter 19 are contained in ACI 318, Chapter 2.

Definitions of terms can help in understanding and applying the code requirements. The purpose for including these definitions within this chapter is to provide more convenient access to them without having to refer back to Chapter 2 of the code. Also, for convenience, these terms are listed in Chapter 2 with a cross reference to this section. The use and application of all defined terms, including those defined herein, are set forth in Section 201.

SECTION 1903
SPECIFICATIONS FOR TESTS AND MATERIALS

1903.1 General. Materials used to produce concrete, concrete itself and testing thereof shall comply with the applicable stan-

dards listed in ACI 318. *Where required, special inspections and tests shall be in accordance with Chapter 17.*

❖ Section 1903 regulates the materials commonly used for both reinforced and plain concrete construction and requires the selection and acceptance of such materials to conform to the provisions of ACI 318. Chapter 3 of ACI 318 establishes the basis for the selection, specification and acceptance of materials used in concrete construction, including cement, aggregate, water, steel reinforcement and admixtures.

Cement: The words "cement" and "concrete" are used in several sections of Chapter 19. It is important to understand the distinction between cement and concrete. Often, one may hear of references to cement floors, or cement sidewalks and driveways, or of cement walls in describing walls made of cast-in-place or precast concrete or of concrete masonry. Cement and concrete are not interchangeable words. Cement is only one of three basic ingredients (cement, water and aggregate) that make up the concrete material used in construction. In the end use, there are only concrete structures and building components. Structures of cement alone simply do not exist. The code permits the use of cements that meet ASTM C 150, ASTM C 595, ASTM C 845 and ASTM C 1157. General information on each of these cements is provided below.

Portland Cement—ASTM C 150:
ASTM C 150 specifies the chemical composition, physical properties and test methods for producing eight types of portland cement as follows:

1. Type I portland cement is for general use where the special properties of other types of cement are not required. While Type I is the most widely used type of portland cement produced, it should not be used in concrete structures that will be subject to sulfate attack, such as from soil or ground water, or in large concrete placements where the temperature rises from the heat of hydration produced by the chemical reaction of cement and water and can adversely affect the construction process.

2. Type IA portland cement is an air-entraining cement for general use, as is Type I portland cement, but it is used specifically where the benefits of air entrainment are desired. The purpose of air entrainment is discussed in other parts of this commentary.

3. Type II portland cement is for general use when moderate sulfate resistance or moderate heat of hydration is desired.

4. Type IIA portland cement is for general use, as is Type II portland cement, but is used where the benefits of air entrainment are also desired.

5. Type III portland cement is a high early strength cement especially used to accelerate the hard-

ening of wet concrete to facilitate construction scheduling.

6. Type IIIA portland cement has the same properties and is used for the same purposes as Type III portland cement, but is also used where the benefits of air entrainment are desired.

7. Type IV portland cement is used especially where a low heat of hydration is required. It is used principally in mass concrete construction such as dams, bridge abutments, heavy foundations, etc. It is seldom used for building construction.

8. Type V portland cement is for special use where high sulfate resistance is desired, and is principally used in areas where the soils or ground waters have high concentrations of sulfate.

Blended Hydraulic Cement—ASTM C 595:

ASTM C 595 specifies five types of blended cements as follows:

1. Portland blast-furnace slag cement, Type IS, is for use in general concrete construction. Portland cement (ASTM C 150) may be modified by blending fine-granulated blast-furnace slag to provide moderate sulfate resistance; Type IS (MS); air entrainment, Type IS (A); moderate heat of hydration, Type IS (MH) or any combination by adding the indicated suffixes.

2. Portland-pozzolan cement, Type IP, is for general use and will perform similar to Type I portland cement, except that it can have slightly lower 28-day concrete compressive strength and can be less resistant to deicer scaling. The cement can be modified to provide moderate sulfate resistance, Type IP (MS); air entrainment, Type IP (A); moderate heat of hydration, Type IP (MH) or any combination by adding the indicated suffixes.

3. Slag cement, Type S, is used in combination with portland cement in making concrete or in combination with hydrated lime in making masonry mortar. It can be modified to provide air entrainment, Type S (A). Section 3.2.1 of ACI 318 prohibits the use of slag cement in structural concrete.

4. Pozzolan-modified portland cement, Type I (PM), is used in general concrete construction and can be modified to provide moderate sulfate resistance, Type I (PM)(MS); air entrainment, Type I (PM)(A); or moderate heat of hydration, Type I (PM)(MH), respectively.

5. Slag-modified portland cement, Type I (SM), is used for general concrete construction and may be modified to provide moderate sulfate resistance, air entrainment or moderate heat of hydration by adding the suffixes (MS), (A) or (MH), respectively.

Expansive Hydraulic Cements—ASTM C 845:

Expansive hydraulic cements expand during the early hardening period after the concrete has set. The cements are essentially composed of hydraulic calcium aluminates and calcium sulfates. These cements are used on projects where shrinkage-compensating concrete is specified. When restrained, the expansion of shrinkage-compensating concrete induces compressive stresses that can offset tensile stresses due to drying shrinkage in the concrete member. Typically, this type of concrete is used for floor slabs and pavements to minimize or eliminate drying-shrinkage cracking. The cement is classified as Type E-1 and will include the letter-designating suffix (K), (M) or (S), depending on the chemical composition of the cement.

Performance Blended Hydraulic Cement—ASTM C 1157:

ASTM C 1157 differs from ASTM C 150 and ASTM C 595 in that it does not establish the chemical composition of the different types of cements; however, individual constituents used to manufacture ASTM C 1157 cements must comply with the requirements specified in the standard. The standard also provides for several optional requirements, including one for cement with low reactivity to alkali-reactive aggregates.

Aggregates: Section 3.3 of ACI 318 requires that concrete aggregates conform to ASTM C 33 or ASTM C 330. All fine and coarse aggregates used in producing normal-weight concrete [135 to 160 pounds per square foot (psf) (6.5 to 7.6 kNm^2)] must comply with the provisions of ASTM C 33. All aggregates used in making lightweight structural concrete [85 to 115 psf (4 to 5.5 kNm^2)] must comply with the requirements of ASTM C 330. Both ASTM C 33 and ASTM C 330 specify physical property requirements and provide for sampling and testing of aggregate materials.

Fine aggregates used in normal-weight concrete consist of natural sand, manufactured sand or a combination of both materials. Coarse aggregates consist of either gravel, crushed gravel, crushed stone, air-cooled blast-furnace slag, crushed hydraulic-cement concrete or combinations of any of these materials. Aggregates used for making lightweight structural concrete are expanded shale, clay, slate and slag. Lightweight aggregates, such as pumice, pearlite and vermiculite, are generally used in making insulating concrete, but their use is prohibited in structural concrete.

The performance characteristics of concrete are greatly affected by the selection and proportioning of aggregates in the concrete mixture. Aggregates not only affect the workability of fresh concrete, but in the hardened state they are a factor in providing adequate resistance to abrasion; freezing and thawing, drying shrinkage and fire exposure. They also affect the strength of concrete.

Harmful substances may be present in aggregates

in such quantities that the performance of the concrete could be seriously affected. ASTM C 33 and ASTM C 330 set content limitations for such deleterious materials.

Sometimes the practicalities of construction necessitate deviations from normal practices. Aggregates that conform to ASTM material standards are not always economically or readily available, and local, nonconforming aggregates may be successfully used in concrete. In such cases, ACI 318 permits evidence of a long history of satisfactory performance or special tests conducted to show adequate strength and durability of the concrete to serve as the basis for acceptance of nonconforming aggregate materials.

Water: The water used in concrete mixtures is required to be clear and clean. It must not contain injurious amounts of oils, acids, alkalis, salts, organic materials or other substances that are harmful to concrete or concrete reinforcement.

Potable water obtained from municipal water supply systems is normally suitable for use as mixing water in concrete. However, this does not mean that nonpotable water cannot be used in making concrete. ACI 318 permits the use of nonpotable water under specific conditions of acceptance. Water that appears discolored or has an unusual or repulsive taste or odor must undergo an analysis using applicable ASTM test methods. Evidence of its acceptability can be provided by checking service records of concrete made with such water, or from other information that demonstrates the water will not be detrimental to the concrete. In the absence of such evidence, water suspected of being of questionable quality can be subject to the acceptance criteria for concrete strength and time set contained in the ASTM C 94.

Environmental concerns have caused many ready-mixed producers to reuse washout water (i.e., water used to reclaim aggregate or clean out returning trucks) as mix water. This is permitted by ASTM C 94, provided the water meets the criteria outlined in the standard. In addition, the standard permits the concrete purchaser to require optional testing to determine chloride ion, sulfate, alkali and total solids in the water. This is particularly important when chloride ion content of the concrete must be limited for corrosion protection, and when aggregate susceptible to alkali reactivity will be used, or both (see Section 1904.4 or ACI 318, Section 4.4.1).

Steel reinforcement: Steel reinforcement conforming to the requirements of ACI 318 must be of the deformed type (with lugs or protrusions), except that plain steel reinforcement can be used for spirals and prestressing tendons. Under certain conditions, reinforcement can also consist of structural steel, steel pipe or steel tubing.

ACI 318, Section 3.5, requires that reinforcement for concrete must comply with one of the following ASTM material standards:

1. Deformed reinforcement;

2. Deformed and plain billet-steel bars, ASTM A 615;3.

3. Deformed and plain low-alloy steel bars, ASTM A 706. (This special reinforcement is used primarily in structures assigned to Seismic Design Category D, E or F and is not readily available in all parts of the country.); or4.

4. Rail-steel and axle-steel deformed bars, ASTM A996.

Bar mats for concrete reinforcement must conform to ASTM A 184 and be fabricated from reinforcing bars that conform to either ASTM A 615, ASTM A 616, ASTM A 617 or ASTM A 706.

Deformed wire must conform to ASTM A 496 but the wire must not be smaller than Size D-4; and welded deformed wire fabric for concrete reinforcement must conform to ASTM A 497. Welded intersections must not be spaced farther apart than 16 inches (406 mm) in the direction of the principal reinforcement, except when used as stirrups.

Welded plain wire fabric for concrete reinforcement must conform to ASTM A 185. The code considers welded plain fabric to be deformed reinforcement.

Reinforcing bars may be zinc coated (galvanized) or epoxy coated to provide corrosion resistance.

Galvanized bars must conform to ASTM A 767. Epoxy-coated bars must conform to ASTM A 775.

Steel wire and welded wire fabric may be epoxy coated in accordance with ASTM A 884 to provide corrosion resistance.

ASTM A 996 is a replacement for ASTM A 616 and ASTM A 617. Unlike its predecessors, ASTM A 996 does not provide for plain bars. Although manufacturers will be producing bars in accordance with the new standard, there are no additional changes that will affect those involved in the design, construction and inspection of reinforced concrete structures.

Plain reinforcement:
Plain bars for spiral reinforcement must conform to either ASTM A 615 or ASTM A 706. Smooth wire for spiral reinforcement must conform to ASTM A 82.

Prestressing steel:
Steel used for prestressed concrete must conform to one of the following ASTM standards:

1. Wire, including low-relaxation wire, must be uncoated, stress-relieved steel wire, ASTM A 421;

2. Strand, including low-relaxation wire, must be uncoated, seven-wire, stress-relieved steel, ASTM A 416;

3. High-strength steel bars, ASTM A 722; or

4. Wire, strands and bars that are not specifically listed in ASTM A 421, ASTM A 416 or ASTM A 722 can be used, provided they conform to the minimum requirements of these specifications and do not have properties that make the materi-

als less satisfactory than intended by these standard specifications.

Structural steel:

Structural steel used with reinforcing bars in composite compression members must conform to one of the following ASTM standards:

1. Carbon steel shapes of structural quality, ASTM A 36;

2. High-strength, low-alloy structural steel shapes, ASTM A 242;

3. High-strength, low-alloy columbium-vanadium steels of structural quality, ASTM A 572; or

4. High-strength, low-alloy structural steel with 50,000 psi (344 750 kPa) minimum yield point to 4 inches (102 mm) thick, ASTM A 588.

Steel pipe or tubing:

Steel pipe or tubing for composite compression members consisting of a steel-encased concrete core must comply with one of the following ASTM standards:

1. Grade B seamless and welded black and hot-dipped galvanized steel pipe, ASTM A 53;

2. Cold-formed welded and seamless carbon steel structural tubing of round, square, rectangular or special shape, ASTM A 500; or

3. Hot-formed welded and seamless carbon steel structural tubing of round, square, rectangular or special shape, ASTM A 501.

See Section 1916 for concrete-filled pipe columns.

The most important considerations before welding reinforcing bars are the weldability of the steel and the compatibility of the welding process. The American Welding Society (AWS) specifies the proper welding practices in the D1.4 standard.

The ASTM standards for concrete reinforcement permitted in the code each contain the material properties of the specific types of metal. The steel chemistry set forth in any of these standards must be restricted to a given range to provide compatibility with the welding requirements of AWS D1.4. Therefore, to assure the weldability of the steel and conformance with AWS requirements, this section stipulates that the ASTM reinforcing bar specifications must be supplemented with a report addressing the material properties of the particular type of steel to be used in the welding process. However, the steel produced using ASTM A 706 is exempt from the requirement for a supplemental report because it was purposely developed for welding and already has a restricted chemistry.

Where there is a need to weld existing reinforcing bars in making alterations to existing buildings, and mill reports of the steel are not available, AWS D1.4 requires that a chemical analysis be made to determine the composition of the steel. As a practical alternative, an assumption can be made that the carbon equivalent is above 0.75 percent for the reinforcement, and the steel can be welded after a preheat temperature of 500°F (260°C) is applied. While this procedure solves the welding problem, consideration must be given to the possible effects of heat damage to the concrete and the stress levels in the reinforcing steel.

Admixtures: By definition, an admixture is a material other than water, aggregates or hydraulic cement, used as an ingredient in concrete (or mortar) and added to the concrete immediately before or during mixing. There are four reasons for using admixtures:

1. To achieve certain desired properties in concrete more readily and effectively than can be obtained by other means;

2. To insure the quality of concrete during the mixing, transporting, placing and curing operations under adverse weather conditions;

3. To overcome certain adverse conditions during concreting operations; and

4. To reduce the cost of concrete construction.

While each of the several different types of admixtures used in concrete is intended to have a beneficial effect on certain properties of concrete, at the same time, the admixture can have an adverse effect on other properties. Therefore, to make sure the overall quality of concrete intended by the design engineer and the code is met, the use of admixtures must have prior approval of the engineer.

The effectiveness of an admixture in concrete is dependent on such factors as the type and the amount of cement; the shape, gradation and proportions of the aggregate used in the concrete mixture; the water content and the mixing time, slump and temperatures of the surrounding air and concrete. To insure proper proportioning of the concrete in accordance with the requirements of Section 1905.2, trial mixtures should be made with the admixture and the job materials at temperatures and humidities anticipated at the job site. Trial mixtures will establish the compatibility of the admixture with the other admixtures that may be used in the concrete, as well as determine the effects of the admixture on the properties of fresh and hardened concrete. The amount of admixture used in concrete should not exceed the amount recommended by the manufacturer or the optimum amount determined by laboratory tests.

Because concentrations of chloride ions can cause severe corrosive effects on metallic materials, admixtures containing chloride as part of the formulation, including the use of calcium chloride as an antifreeze admixture during cold weather, are not allowed in prestressed concrete, concrete with aluminum embedment or concrete that is in contact with permanent installations of galvanized formwork.

Admixtures used in concrete must comply with the applicable ASTM material specifications referenced in

ACI 318. There are several classes of admixtures covered by the ASTM specifications. Each class of admixture, as well as each type of admixture within a class, serve different purposes and affect the properties of concrete in different ways.

Air-entraining Admixtures—ASTM C 260:

Air-entraining admixtures incorporate microscopic air bubbles in the concrete mixture. Air entrainment serves three main purposes:

1. It improves the resistance of moist concrete to the effects of freezing and thawing cycles;

2. It improves the resistance of concrete to the actions of deicing chemicals; and

3. It improves the workability of fresh concrete.

Entrained air can be introduced in concrete by means of air-entraining cements specified in Section 1903.2, or as a separate ingredient (i.e., an admixture) in the concrete mixture (ASTM C 260) introduced before or during the mixture operations. If desired, both methods can be used together, but difficulties may arise in controlling the amount of air entrainment.

Water-reducing, Retarding and Accelerating Admixtures—ASTM C 494 and ASTM C 1017:

The ASTM C 494 material specification for chemical admixtures gives the physical requirements for each of seven types of admixture—Types A through G listed below—and prescribes the applicable test methods and other information necessary for the acceptance of admixture materials:

Type A—Water-reducing admixtures
Type B—Retarding admixtures
Type C—Accelerating admixtures
Type D—Water-reducing and retarding admixtures
Type E—Water-reducing and accelerating admixtures
Type F—Water-reducing and high-range admixtures
Type G—Water-reducing, high-range and retarding admixtures.

Type A water-reducing (and set-controlling) admixtures can be used in several ways in proportioning concrete mixtures to achieve certain desired results:

1. The cement content can be maintained while reducing the amount of water, thus lowering the w/cm (w/cm) ratio and increasing the strength of concrete, as well as its durability.

2. The w/cm ratio can be maintained by reducing both the water and cement contents of the concrete resulting in a cost savings (less cement used); and

3. The slump of the concrete can be increased without increasing the w/cm ratio, thus maintaining the same strength level while improving the workability of the concrete.

The purpose of Type B chemical admixtures is to retard the rate of the setting of concrete. Such admixtures are normally used in hot weather when an accelerated rate of hardening of concrete occurs, making placing and finishing operations very difficult. On the negative side, the use of retarding admixtures will cause reductions in concrete strength at very early ages. More important, the effects of retarding admixtures on other properties of concrete, particularly shrinkage, are unpredictable.

Type C accelerating admixtures are used mainly in cold-weather operations to shorten the setting time of concrete and accelerate the strength development at early ages. The problem is that calcium chloride is the prevalent ingredient used in producing these admixtures. Calcium chloride can cause an increase in drying shrinkage, reinforcement corrosion, discoloration of concrete (darkens concrete), loss of strength at later ages and possibly scaling. This type of admixture is prohibited in prestressed concrete construction.

Type D chemical admixtures perform as a combination of the properties of Types A and B. They are used to reduce the quantity of mixing water required to produce concrete of a given consistency (slump) and to retard the setting time of the concrete.

Type E chemical admixtures perform as a combination of the properties of Type A and C admixtures. They serve to reduce the quantity of mixing water required to produce concrete of a given consistency (slump) and accelerate the setting time and early strength development of the concrete.

Type F chemical admixtures are water-reducing admixtures similar to Type A, except that they can reduce the quantity of mixing water by 12 percent or more to produce concrete of a given consistency. Up to 30-percent water reduction can be achieved with these admixtures.

Type G chemical admixtures are high-range, water-reducing and retarding admixtures similar in function to Types F and B, except they can reduce the quantity of mixing water by 12 percent or more to produce concrete of a given consistency and retard the setting time of the concrete.

ASTM C 1017 regulates admixtures that are similar to a high-range water reducer, but are used in a different manner. These admixtures are typically referred to as "superplasticizers." Instead of being used to reduce the amount of mix water and maintain the same consistency (i.e., as a water reducer), the superplasticizer is used with the same amount of mix water that would be necessary to produce relatively low-slump concrete [e.g., $3^1/_2$ inches (89 mm)]. The superplasticizer causes an increase in the slump, which makes the concrete flowable. Two types of admixtures are provided for under ASTM C 1017: Type I, plasticizing; and Type II, plasticizing and retarding. The requirements of ASTM C 1017 are similar to those of ASTM C 494 (Types F and G); however, in order to comply with ASTM C 1017, the admixture must increase the slump at least $3^1/_2$ inches (89 mm) compared to the control specimen.

Pozzolans—ASTM C 618:

Admixtures made of raw or natural pozzolans and fly ash are covered by ASTM C 618 material specifications. While these mineral admixtures will benefit as well as adversely affect the properties of concrete, they are widely used as partial replacements for portland cement and often result in more economical concrete construction.

By definition, a pozzolan is a siliceous or aluminosiliceous material that in itself possesses little or no cementitious value, but will, in finely divided form and in the presence of water, chemically react with the calcium hydroxide released by the hydration of portland cement to form compounds possessing cementitious properties. Pozzolanic materials used for making admixtures come from a number of natural sources, such as diatomaceous earth, opaline cherts, clays, shales, volcanic tuffs and pumicites. The ASTM C 618 standard designates these materials as Class N pozzolans.

Fly ash is the most widely used of all the mineral admixtures in concrete. Essentially, it is a finely divided residue (a powder) that results from the combustion of pulverized coal. Large quantities of fly ash are produced in coal-burning electric power-generating plants. Fly ash is primarily silicate glass containing silica, aluminum, iron and calcium. It also contains small amounts of magnesium, sulfur, sodium, potassium and carbon. ASTM C 618 designates these materials as Class F and C fly ashes, depending on the percentages of calcium and carbon contents. Each class of fly ash affects the properties of concrete in different ways. It is a complex subject, and for more information on fly ash and other mineral admixtures, refer to the Portland Cement Association publications cited in the bibliography.

Aside from economic reasons, pozzolan and fly ash admixtures are used in concrete for any one or more of the following purposes: increase sulfate resistance, reduce expansions in concrete due to alkali-silica reactions, reduce the permeability of concrete, decrease heat generation or improve the workability and finishing qualities of the concrete.

The properties of pozzolans and fly ash may vary widely. Fly ash, in particular, can have varying amounts of carbon, silica, sulfur, alkalis and other ingredients that can adversely affect the strength, air content and durability of concrete. Because it is a residue and not a controlled product, the properties of fly ash can vary widely, not only between productions coming from different electric power-generating plants, but also between batches coming from the same plant.

Slag—ASTM C 989:

Ground-granulated blast-furnace slag complying with the requirements of ASTM C 989 provides characteristics similar to those exhibited by pozzolan ad-

mixtures. Ground-granulated blast-furnace slag and fly ash are often used in combination as admixtures in portland cement (ASTM C 150) concrete. The practice has been growing in recent years due to energy conservation considerations, but mainly because of savings in the cost of concrete when fly ash or slag or both fly ash and slag are used to partially replace cement.

Silica Fume—ASTM C 1240:

Silica fume is a byproduct of the manufacture of elemental silicon or ferro-silicon alloys in electric arc furnaces. When used in conjunction with a high-range water-reducing admixture, it is possible to produce concrete with compressive strengths of 20,000 psi (138 MPa) or higher. It is also used to achieve a very dense cement paste matrix to reduce the permeability of concrete, thus providing better corrosion protection to reinforcing steel in concrete subjected to deicing chemicals (e.g., parking garages). Tests have shown that silica fume greatly improves the sulfate resistance of concrete. For this reason, the code requires that concrete to be subjected to very severe sulfate exposure be made with Type V ASTM C 150 cement plus a pollozan, such as silica fume, that has shown through tests or service records to improve sulfate resistance (see Table 4.3.1 of ACI 318).

Storage of materials: Section 3.7 of ACI 318 regulates the storage of concrete ingredients and prohibits use of such materials if they have suffered contamination or deterioration.

Cement stored in contact with damp air or moisture sets more slowly and has less strength than cement that is maintained in a dry condition. Bulk cement can be stored in weather-tight containers, such as steel bins or concrete silos.

Bags of cement are typically stored in warehouses or sheds where the relative humidity is kept as low as possible. Bags of cement must not be stored on damp floors or on earth (job-site conditions; dirt), but should rest on pallets and be stacked close together to reduce air circulation. Bags should be covered with a waterproof covering to prevent deterioration when exposed to weather.

Aggregates are normally stockpiled on the ground in areas that have been stripped of all vegetation and debris and have been leveled to facilitate handling operations. When two or more types or sizes of aggregates are involved, the stockpiles must be so arranged as to avoid crowding or overlapping with other materials or other sizes of like aggregates. In removing material from a stockpile, a layer of aggregate of sufficient thickness must be left on the ground to keep the handling equipment, such as a front-end loader or clamshell, from picking up earth along with the aggregate. Stockpiles typically are built up in centric uniform layers to prevent coarse aggregates from rolling down the sides and segregating the material.

Ordinarily, cement does not remain in storage long, but it can be stored for long periods without deterioration. At the time of use, however, cement is required to be free flowing and relatively free of lumps. If the lumps do not readily break up or there is other reason to doubt the quality of the cement, the material can be tested for strength or loss on ignition in accordance with the ASTM standard under which it was manufactured before it is used in concrete.

If there is reason to doubt that any aggregate taken from a stockpile is not clean or appears to be contaminated, any of the applicable standard ASTM tests that would detect impurities should be made before the material is used for making concrete.

1903.2 Glass fiber reinforced concrete. *Glass fiber reinforced concrete (GFRC) and the materials used in such concrete shall be in accordance with the PCI MNL 128 standard.*

❖ This section references a document that addresses the quality of materials used in glass fiber reinforced concrete panels. Glass fiber reinforced concrete is utilized in the prescriptive fire-resistance ratings for wall assemblies [see Table 720.1(2), Item 15].

SECTION 1904
DURABILITY REQUIREMENTS

1904.1 Water-cementitious materials ratio. Where maximum w/cm ratios are specified in ACI 318, they shall be calculated in accordance with ACI 318, Section 4.1.

❖ The provisions of Section 1904 specify the proportioning of concrete that is necessary to obtain the required strengths and to provide the properties of concrete that will resist the effects of adverse environmental conditions. This includes concrete that will be subjected to freeze-thaw cycles; in contact with deicer chemicals or exposed to the deteriorating effects of sulfates; corrosive actions of chloride ions on steel reinforcement or other harmful exposures.

Section 4.1 of ACI 318 clarifies that in addition to including traditional cements in the weight of cement used to calculate the w/cm ratio, it is permissible to include the weights of fly ash or other pozzolan, slag and silica fume. The user is also alerted to the fact that Section 4.2.3 of ACI 318 limits the amount of fly ash or other pozzolan, slag and silica fume in concrete to be exposed to deicer chemicals.

1904.2 Freezing and thawing exposures. Concrete that will be exposed to freezing and thawing, deicing chemicals or other exposure conditions as defined below shall comply with Sections 1904.2.1 through 1904.2.3.

❖ This section requires that concrete be proportioned so that the required strength, and other properties of hardened concrete needed to resist the effects of freezing and thawing, will be achieved.

1904.2.1 Air entrainment. Concrete exposed to freezing and thawing or deicing chemicals shall be air entrained in accordance with ACI 318, Section 4.2.1:

❖ Freezing and thawing cycles can be one of the most destructive weathering factors for concrete when it is wet. For nonair-entrained concrete, deterioration is caused by the freezing of water in the cement matrix, in the aggregate or both. Studies have documented that concrete provided with proper air entrainment is highly resistant to this deterioration. Table 4.2.1 of ACI 318 specifies the required air content based on the aggregate size and exposure condition. An exposure condition would be considered severe if the concrete is located in a cold climate and could be continuously in contact with moisture prior to freezing or exposed to deicing chemicals. A parking structure is an example of a severe exposure because vehicles may track in deicing chemicals from the streets. The exposure condition would be considered moderate if concrete is in a cold climate and in contact with moisture prior to freezing only occasionally, and no deicing chemicals are used. Exterior walls, beams, girders and slabs not in direct contact with the ground are examples of moderate exposure.

Normal-weight and lightweight concrete subjected to freezing and thawing cycles or deicer chemicals must be air entrained in compliance with the air content requirements set forth in Table 4.2.1 of ACI 318.

Concrete with compressive strengths in excess of 5,000 psi (34.47 MPa) has characteristics that improve resistance to freezing and thawing (e.g., lower w/cm ratio, lower porosity), and therefore are permitted to have 1 percent less air content. For example, the required total air content for a 5,500 psi (37.9 MPa) mix that will be made with $^3/_8$-inch (9.5 mm) maximum size aggregate and subjected to a "severe exposure" is $7^1/_2$ - 1 = $6^1/_2$ percent.

Air is introduced in concrete by the use of air-entraining cements, air-entraining admixtures or a combination of both.

The required air content in Table 4.2.1 of ACI 318, expressed as a percentage of the volume of concrete, is based on the size of coarse aggregates used in the mix. For smaller aggregate sizes, the percentage is increased to account for the increase in mortar volume that results when aggregate size decreases.

The code requires that all concrete exposed to freezing and thawing conditions or to deicer chemicals must be air entrained. While air entrainment is mandatory, to do otherwise would result in exposed concrete that is afforded less protection than is necessary to provide durability in cold-weather climates. For comprehensive information on both air content and selecting proportions for normal-weight concrete, refer to ACI 211.1.

1904.2.2 Concrete properties. Concrete that will be subject to the following exposures shall conform to the corresponding maximum w/cm ratios and minimum specified concrete compressive strength requirements of ACI 318, Section 4.2.2:

1. Concrete intended to have low permeability where exposed to water;

2. Concrete exposed to freezing and thawing in a moist condition or deicer chemicals; or

3. Concrete with reinforcement where the concrete is exposed to chlorides from deicing chemicals, salt, salt water, brackish water, seawater or spray from these sources.

> **Exception:** *For occupancies and appurtenances thereto in Group R occupancies that are in buildings less than four stories in height, normal-weight aggregate concrete shall comply with the requirements of Table 1904.2.2 based on the weathering classification (freezing and thawing) determined from Figure 1904.2.2.*

In addition, concrete exposed to deicing chemicals shall conform to the limitations of Section 1904.2.3.

❖ Concrete subject to the exposures listed in Table 4.2.2 of ACI 318 must comply with the limitations on water-cementious materials (w/cm) ratio, or specified compressive strength, or both. The intent of the upper limit on the w/cm ratio and lower limit on specified compressive strength is to achieve dense, impermeable or water-tight concrete. Note that the minimum specified compressive strength applies to both normal-weight and lightweight aggregate concrete, while the maximum w/cm ratio applies only to normal-weight concrete. The variability of moisture absorption in lightweight aggregate makes calculation of the w/cm ratio uncertain.

In the exception, in climatic areas where concrete construction is exposed to freeze-thaw cycles or deicer chemicals, concrete must be of the quality necessary to resist the harmful effects of such weather conditions. In all buildings that are subject to special exposure conditions, except those of Group R occupancies that are less than four stories in height, the w/cm ratio and specified compressive strength of the concrete is required to be designed to conform to the requirements of Table 4.2.2 of ACI 318. The specified compressive strengths of concrete in low-rise Group R occupancies (less than four stories) need only comply with Table 1904.2.2, which mandates a minimum specified compressive strength as a function of the concrete element and exposure.

Table 4.2.2 of ACI 318 provides w/cm ratios for normal-weight aggregate concrete in conjunction with various exposure conditions. The values for w/cm ratios are the maximum permissible without regard for the relative strength of concrete. A control on the w/cm ratio is provided in addition to the minimum specified compressive strength values (f'_c) for normal-weight concrete, because the w/cm ratio is a more reliable indication of durability and watertightness of concrete. The table only gives a minimum specified compressive strength value for proportioning concrete mixtures using lightweight aggregates because the amount of water absorbed by lightweight aggregates is difficult to determine, making the calculation of w/cm ratios unreliable.

TABLE 1904.2.2. See below.

❖ This table gives the minimum specified compressive strength values (f'_c) required for concrete used in various components for construction of Group R occupancy buildings that are less than four stories, and which are subject to either freezing and thawing while wet or exposure to deicer chemicals, such as walls, slabs and foundations. The main purpose of this table is to assure adequate performance of the concrete by requiring a minimum compressive strength (f'_c) based on the classifications of weathering determined from Figure 1904.2.2.

Note b requires air-entrained concrete for garage floor slabs located in moderate and severe weathering

TABLE 1904.2.2
MINIMUM SPECIFIED COMPRESSIVE STRENGTH (f'_c)

TYPE OR LOCATION OF CONCRETE CONSTRUCTION	MINIMUM SPECIFIED COMPRESSIVE STRENGTH (f'_c at 28 days, psi)		
	Negligible exposure	Moderate exposure	Severe exposure
Basement walls[c] and foundations not exposed to the weather	2,500	2,500	2,500[a]
Basement slabs and interior slabs on grade, except garage floor slabs	2,500	2,500	2,500[a]
Basement walls[c], foundation walls, exterior walls and other vertical concrete surfaces exposed to the weather	2,500	3,000[b]	3,000[b]
Driveways, curbs, walks, patios, porches, carport slabs, steps and other flatwork exposed to the weather, and garage floor slabs	2,500	3,000[b, d]	3,500[b, d]

For SI: 1 pound per square inch = 0.00689 MPa.

a. Concrete in these locations that can be subjected to freezing and thawing during construction shall be of air-entrained concrete in accordance with Section 1904.2.1.

b. Concrete shall be air entrained in accordance with Section 1904.2.1.

c. Structural plain concrete basement walls are exempt from the requirements for exposure conditions of Section 1904.2.2 (see Section 1909.6.1).

d. For garage floor slabs where a steel trowel finish is used, the total air content required by Section 1904.2.1 is permitted to be reduced to not less than 3 percent, provided the minimum specified compressive strength of the concrete is increased to 4,000 psi.

regions. The purpose of this air-entrainment is to protect the slab from deterioration by deicer chemicals that may be carried into the garage on vehicle tires and the underside of the vehicle. Most building owners prefer a garage slab with a smooth trowel finish to facilitate cleaning, but experience has shown that it is difficult to get a smooth, durable finish by using a steel trowel on concrete with total air content in the range of 5 to 7 percent. Note d provides an option to air-entrained concrete that has been successfully applied to garage floors in severe and moderate weathering regions. Field experience has shown that durable concrete may be obtained with a lesser amount of air-entrainment if the specified compressive strength of the concrete is increased to 4,000 psi (27.6 MPa). The higher concrete strength is accompanied by a denser cement-paste matrix, which results in the concrete being less permeable.

FIGURE 1904.2.2. See below.

❖ Figure 1904.2.2 shows the geographic regions within the U.S. mainland, which are classified as having negligible, moderate and severe weather exposures. The sole purpose of the map is to locate those weathering areas in the country where the requirements for minimum specified compressive strength of concrete of Table 1904.2.2 apply. Note that the boundary lines defining the weather regions are only approximate. Therefore, building officials in communities located near a boundary line must establish applicable exposure classifications (severe, moderate or negligible) based on their knowledge of weather conditions in the area and the relative performance of concrete.

1904.2.3 Deicing chemicals. For concrete exposed to deicing chemicals, the maximum weight of fly ash, other pozzolans, silica fume or slag that is included in the concrete shall not exceed the percentages of the total weight of cementitious materials permitted by ACI 318, Section 4.2.3.

❖ The purpose of this section is to limit the quantity of certain materials in a concrete mixture to assure adequate durability of the concrete that will be exposed to deicing chemicals. Although research has shown that the use of fly ash, slag and silica fume produces a finer pore structure in concrete resulting in a lower permeability and improved durability, their effectiveness is limited by the relative amounts of these materials and portland or blended cement. The limitations of this section do not apply if the concrete will not be exposed to deicing chemicals.

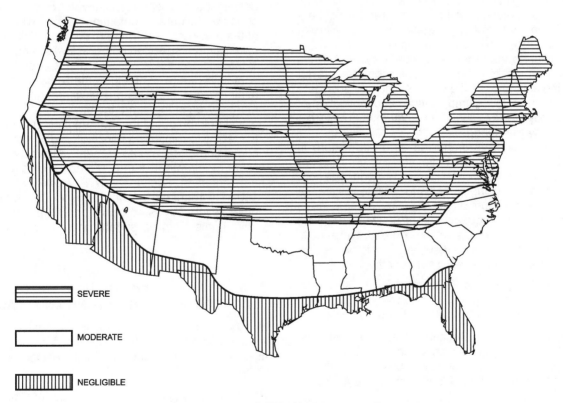

SEVERE

MODERATE

NEGLIGIBLE

FIGURE 1904.2.2
WEATHERING PROBABILITY MAP FOR CONCRETE[a, b, c]

a. Lines defining areas are approximate only. Local areas can be more or less severe than indicated by the region classification.

b. A "severe" classification is where weather conditions encourage or require the use of deicing chemicals or where there is potential for a continuous presence of moisture during frequent cycles of freezing and thawing. A "moderate" classification is where weather conditions occasionally expose concrete in the presence of moisture to freezing and thawing, but where deicing chemicals are not generally used. A "negligible" classification is where weather conditions rarely expose concrete in the presence of moisture to freezing and thawing.

c. Alaska and Hawaii are classified as severe and negligible, respectively.

1904.3 Sulfate exposures. Concrete that will be exposed to sulfate-containing solutions or soils shall comply with the maximum w/cm ratios, minimum specified compressive strength and be made with the appropriate type of cement in accordance with the provisions of ACI 318, Section 4.3.

❖ Section 4.3 of ACI 318 regulates the proportioning and types of cement permitted in concrete that will be exposed to sulfate solutions.

Many naturally occurring soils contain sufficient amounts of sulfates to cause deterioration of concrete that does not possess certain physical and chemical qualities. Referred to as alkaline soils, they contain concentrations of sodium and magnesium sulfates (salts). Water or solutions coming in contact with concrete can also contain sulfates (e.g., seawater and sewage). Concrete made with sulfate-resisting cement and having limitations on its maximum w/cm ratio minimum specified compressive strength (f'_c) or both provides resistance to sulfate attack.

Table 4.3.1 of ACI 318 provides the requirements for concrete exposed to sulfate-containing soils and solutions based on four exposure levels: negligible, moderate, severe and very severe. The table specifies the types of sulfate-resisting cements to be used in concrete mixtures, the maximum w/cm ratio for normal-weight aggregate concrete and the minimum specified compressive strengths for normal-weight aggregate and lightweight aggregate concrete—all relative to the four exposure conditions stated above.

In addition to the proper selection of cement, w/cm ratio or strength of concrete, other requirements for producing sulfate-resisting concrete include adequate air entrainment, low slump, sufficient consolidation of the concrete mixture, adequate cover of the reinforcement and proper curing.

Calcium chloride is prohibited as an admixture in concrete that is exposed to severe and very severe sulfate-containing soil or solutions. The principal consideration in producing cements for sulfate resistance is to limit the tricalcium aluminate (C_3A) content in accordance with the requirements of ASTM C 150. The use of calcium chloride as an admixture, because of its calcium content, would adversely affect the performance of cement in resisting sulfate attack.

1904.4 Corrosion protection of reinforcement. Reinforcement in concrete shall be protected from corrosion and exposure to chlorides in accordance with ACI 318, Section 4.4.

❖ Section 4.4 of ACI 318 prescribes measures that must be taken to protect reinforcement from corrosion where exposed to high concentrations of chloride ion (Cl-) present in calcium chloride and other materials having chloride ion.

The maximum concentration of water-soluble chloride ion permitted in hardened concrete from all sources is given in Table 4.4.1 of ACI 318. Chloride in concrete could be in a water-soluble state or be chemically combined with other ingredients. The soluble chlorides induce corrosion while chlorides in combined form have little or no corrosive effect. Sampling and testing for water-soluble chloride ions in concrete must comply with the requirements of ASTM C 1218.

Additional information on the effects of chlorides on the corrosion of reinforcing steel is given in the ACI 201.2R and ACI 222R.

For corrosion protection, Table 4.4.1 of ACI 318 limits the maximum soluble chloride ion concentration in hardened concrete that is contributed by the ingredients (i.e., mix water, aggregates, cementitious materials and admixtures). Tests for water-soluble chloride ion concentrations can be performed on hardened concrete in accordance with ASTM C 1218. An initial evaluation of the chloride ion content is possible by testing the individual concrete ingredients and comparing the summation, based on the concrete proportions, to values in Table 4.4.1 of ACI 318. If the total chloride ion content exceeds the table values, it is advisable to test samples of the hardened concrete. The total value for the ingredients can be higher than that permitted by Table 4.4.1 of ACI 318, because some of the chloride ions present in the individual ingredients are insoluble or will have reacted with the cement during hydration, and, therefore will not be available in hardened concrete.

To provide the properties of concrete that will resist the harmful effects of exposure to chlorides from deicing chemicals, salts, seawater or brackish water, concrete mixtures containing normal-weight aggregates must conform to the maximum w/cm ratios and the minimum specified compressive strengths (f'_c) indicated in Table 4.2.2 of ACI 318. Concrete mixtures containing lightweight aggregates must conform to the minimum specified compressive strength (f'_c) requirements given in the same table.

It should be noted that the requirements of this section apply to reinforced concrete and to plain concrete containing steel reinforcement. Minimum concrete cover requirements for the protection of reinforcement are stated in Section 1907.7. Additionally, Section 1907.7.5 requires that in corrosive environments or other severe exposure conditions, minimum cover requirements must be suitably increased (see Section 1907.7.5).

SECTION 1905
CONCRETE QUALITY, MIXING AND PLACING

1905.1 General. The required strength and durability of concrete shall be determined by compliance with the proportioning, testing, mixing and placing provisions of Sections 1905.1.1 through 1905.13.

❖ Section 1905 specifies the methods for proportioning concrete mixtures to obtain the proper workability of fresh concrete and to secure properties of strength, durability, watertightness and abrasion resistance in hardened concrete as necessary to meet performance requirements. The section also regulates the mixing; handling and conveying; depositing and curing of concrete to insure the intended performance of the finished product.

Section 1905.1 covers the methods used in proportioning mixtures to provide adequate workability of fresh concrete and to render those qualities of strength and durability in hardened concrete that are necessary to meet performance requirements. This section includes methods for testing concrete and criteria for its evaluation and acceptance based on strength requirements, as well as requirements for mixing, handling, transporting, placing and curing of concrete for field quality.

1905.1.1 Strength. Concrete shall be proportioned to provide an average compressive strength as prescribed in Section 1905.3 and shall satisfy the durability criteria of Section 1904. Concrete shall be produced to minimize the frequency of strengths below f'_c as prescribed in Section 1905.6.3. *For concrete designed and constructed in accordance with this chapter, f'_c shall not be less than 2,500 psi (17.22 MPa).* No maximum specified compressive strength shall apply unless restricted by a specific provision of this code or ACI 318.

❖ Concrete mix designs are proportioned for strength based on probabilistic concepts so that adequate strength will be developed in the concrete. This section requires that the average strength of concrete produced must exceed the larger of (1) the value of f'_c specified by the registered design professional for the structural design requirements and (2) the minimum strength required for any special exposure condition prescribed in Section 1904. Concrete proportioned to the code requirements may produce strength tests that fall below the specified compressive strength, f'_c. This section introduces this concept by noting that it is the code's intent to "minimize frequency of strength below f'_c." The acceptability of concrete based on the measurement of strength is provided for in Section 1905.6. If a concrete strength test falls below f'_c, the acceptability of this lower-strength concrete is provided for in Section 5.6.3.3 of ACI 318.

This section also prescribes that all concrete designed and constructed in accordance with Chapter 19 must have a minimum specified compressive strength, f'_c, of 2,500 psi (17.22 MPa). This limit has been in model building codes for many years. Where compliance with Chapter 21 of ACI 318 is necessary, a higher minimum concrete strength is specified for members resisting earthquake induced forces (see Section 21.2.4 of ACI 318).

f'_c must be verified by making and testing cylinders in accordance with Section 1905.6.3. Typical test specimens for evaluating compressive strength of concrete are cylinders 6 inches (152 mm) in diameter by 12 inches (305 mm) high (see commentary, Section 1905.6.3).

Most laboratory tests to determine compressive strength levels are based on 28-day cured specimens. If the specified compressive strength of concrete used in the structural design is other than a 28-day age, the specific age must be stated in the construction documents and the concrete must be tested accordingly.

For example, certain conditions may require a concrete structure to receive its service load at an age earlier than 28 days. Under such a circumstance, the compressive strength of the concrete must be specified and obtained in a time frame that will accommodate the loading requirement. Another example is when early concrete strength is provided to facilitate a faster schedule for removal of forms and shores and other similar situations, obtaining earlier strength in concrete is desirable. On some projects, the architect or engineer may require that the specified compressive strength of concrete, f'_c, be based on a period of time longer than 28 days. A typical example is concrete in columns in the lower stories of a high-rise building. Since these columns will not be subjected to their full design load for several months, a longer period of time can be allowed since the concrete continues to gain strength after the first 28 days. In these cases, 56-day and 84-day strengths are commonly specified.

Sections 9.5.2.3 (modulus of rupture), 11.2 (concrete shear strength) and 12.2.4 (development of reinforcement) of ACI 318 require that design criteria be modified for lightweight aggregate concrete based on a relationship between splitting tensile strength, f_{ct}, and specified compressive strength, f'_c. Testing for splitting tensile strength must be in compliance with ASTM C 330.

Section 5.1.5 of ACI 318 prohibits the use of the splitting tensile strength test (ASTM C 330) as a substitute for the more rigorous testing and acceptance criteria required in Section 1905.6.

1905.2 Selection of concrete proportions. Concrete proportions shall be determined in accordance with the provisions of ACI 318, Section 5.2.

❖ Section 5.2 of ACI 318 provides the basic criteria that must be met when proportioning concrete mixes. The objective in designing concrete mixtures is to determine the most economical combination of available ingredients to produce concrete that will satisfy the performance requirements under particular conditions of use.

Materials must be properly proportioned to produce concrete mixtures that meet minimum workability (and consistency), special exposure and strength requirements. Workability is the measure of the ease of placing, consolidating and finishing freshly mixed concrete. Consistency is the measure of its ability to flow. The code requires that concrete must be readily workable into forms and around reinforcement and to have a plastic consistency that will prevent segregation and excessive bleeding. Segregation occurs when the aggregates in freshly mixed concrete are not held in suspension, which produces a mixture that is not homogeneous. Bleeding is the migration of water to the top surface of freshly placed concrete. Excessive bleeding increases the w/cm ratio near the top surface of the concrete, creating a potential for poor durability of the top surface.

Resistance to special exposure conditions provides

minimum safeguards to protect against deterioration of concrete and reinforcement due to freeze-thaw exposure and exposure to chemicals found in concrete and the surrounding environment (see commentary, Section 1904).

The requirement that proportioning provide "conformance with the strength test requirements of Section 1905.6" means that after the mixture proportions have been selected and the job has started, concrete delivered to the job must be evaluated and found acceptable in accordance with the provisions of Section 1905.6. This clarifies that the ultimate goal of concrete proportioning is to provide quality concrete throughout the term of construction. Testing and evaluation of concrete during the construction process is necessary to insure that the selected mixture proportions continue to perform should materials and job-related conditions vary during the construction process.

When different materials like aggregate or cement are to be used in different portions of the concrete work, a separate evaluation of the concrete mixture for conformance to the code and job specifications must be performed.

The preferred methods for selecting concrete mixture proportions are those using field experience or laboratory trial mixtures in accordance with Section 1905.3. Estimation of the w/cm ratio in accordance with Section 1905.4 is permitted when no prior field experience or trial mixture data is available, and only when approved by the engineer or architect as Section 5.4 of ACI 318 indicates.

1905.3 Proportioning on the basis of field experience and/or trial mixtures. Concrete proportioning determined on the basis of field experience and/or trial mixtures shall be done in accordance with ACI 318, Section 5.3.

❖ In proportioning concrete on the basis of field experience, a concrete production facility (e.g., a ready-mixed concrete supplier) that has a suitable record of at least 30 consecutive strength tests, or two groups of 30 consecutive tests of similar materials and conditions that will compare closely with the work to be performed, may use the standard deviation methods of ACI 318, Section 5.3.1, to calculate the required average compressive strength. The concrete produced must meet a compressive strength that does not deviate more than 1,000 psi (6.9 MPa) from the required specified compressive strength of concrete, (f'_c).

Experience has shown that there is enough variation in concrete mixtures, even if they are made of the same materials and proportions, to render variations in test results. This happens even when test specimens are made from the same batch of concrete. Therefore, in using field experience as the basis for proportioning new concrete, it is necessary to account for such differences by measuring how a group of test results varies from the average. This is done by applying ACI 318 calculation methods for determining standard deviations.

The standard deviation of strength tests of a concrete mixture with at least 30 consecutive tests can be determined as follows:

$$s = \left[\frac{\Sigma (x_i - x)^2}{n-1} \right]^{1/2}$$

where:

s = Standard deviation, psi.

x_i = Individual strength tests as defined in Section 5.6.2.4 of ACI 318.

x = Average number of consecutive strength test results.

n = Number of consecutive strength tests.

If two records are used to obtain at least 30 tests, the standard deviation used should be the statistical average of the values calculated from each test record in accordance with the following formula:

$$s = \left[\frac{(n_1 - 1)(s_1)^2 + (n_2 - 1)(s_2)^2}{(n_1 + n_2 - 2)} \right]^{1/2}$$

where:

s = Statistical average standard deviation where two test records are used to estimate the standard deviation.

s_1, s_2 = Standard deviations calculated from two test records, 1 and 2, respectively.

n_1, n_2 = Number of tests in each test record, respectively.

Example:
Calculations for determining the standard deviation are as follows:

In this example, a record of 30 consecutive strength tests is available. Each test represents the average of two cylinders tested at 28-day strengths. The specified compressive strength (f'_c) required for the design is 4,000 psi (27.6 MPa) and therefore, from experience, the average compressive strength (f_{cr}) would be in the vicinity of 4,500 psi (31.0 MPa) more or less. The record of 30 tests to be used in the analysis would have to be at about that level of strength.

Step 1: Calculate x, which is the average strength of the entire group of 30 tests. Add all the individual test strengths (x_i), and divide the total by the number of tests (n). For simplicity, each of the individual tests is not shown here, and only the totals are used in this example.

Assume the strength value for the entire group of 30 tests adds up to 132,510 psi (913.5 MPa).

Therefore, $x = \dfrac{132,510}{30} = 4,417$

Step 2: Determine the value of the term $(x_i - x)^2$. This value needs to be calculated for each of the 30 tests: x was calculated in Step 1 above. x_i represents the value of each individual test.

Therefore, as an example:

Test 1 = $(4,760 - 4,417)^2 = 343^2 = 117,649$ psi
Test 2 = $(4,300 - 4,417)^2 = -117^2 = 13,689$ psi
Test 3 = $(4,840 - 4,417)^2 = 423^2 = 178,929$ psi

and so on until all 30 expressions are determined.

Step 3: Add up all the numbers obtained in Step 2 above. The total figure represents the expression, $\Sigma (x_i - x)^2$. For this example, assume the total adds up to 5,241,720 psi.

Step 4: Determine the standard deviation.

$$s = \left[\frac{5,241,720}{30-1} \right]^{1/2} = 425 \text{ psi}$$

When available records consist of less than the required number of 30 consecutive tests [as specified in ACI 318, Section 5.3.1.1(c)], but are between 15 and 29 tests, the standard deviation must be increased by the applicable modification factor listed in ACI 318, Table 5.3.1.2. This will result in obtaining a more conservative required average strength as a protection against the possibility that the use of a smaller number of sample tests may cause an underestimation of the standard deviation.

When the record of strength tests is insufficient (less than 15 tests), and the procedures given in ACI 318, Section 5.3.1.1 cannot be applied, the required strength must exceed the specified design strength in accordance with ACI 318, Table 5.3.2.2, and documented in accordance with ACI 318, Section 5.3.3.

1905.4 Proportioning without field experience or trial mixtures. Concrete proportioning determined without field experience or trial mixtures shall be done in accordance with ACI 318, Section 5.4.

❖ The cost of determining trial mixture data may not be justifiable on projects involving small quantities of concrete. Also, on occasion, the concrete supply to a project may be interrupted and there is not sufficient time to perform tests and evaluation on a new concrete supply. This provision permits concrete mixtures with no prior experience or trial mixture data, as required in Section 1905.3, to be based on other experience or information, subject to the approval of the engineer or architect. The concrete produced must have an average compressive strength at least 1,200 psi (8.3 MPa) greater than the specified compressive strength, f'_c, and meet minimum durability prescribed in Section 1904. In addition, the provisions should not be used where the specified compressive strength, f'_c, of the concrete exceeds 4,000 psi (27.6 MPa).

1905.5 Average strength reduction. As data become available during construction, it is permissible to reduce the amount by which the average compressive strength (f'_c) is required to

exceed the specified value of f'_c in accordance with ACI 318, Section 5.5.

❖ During a construction project, if the strength test results of the concrete are running appreciably higher than the required average strength established for the work in accordance with Section 1905.3, the engineer may order a reduction in the required strength. Such adjustment must be based on average test results, as required by ACI 318, Section 5.5, and the procedures for proportioning concrete on the basis of field experience as specified in ACI 318, Sections 5.3.1 and 5.3.2. The reduced value is still subject to the special exposure requirements of Sections 1904.2 through 1904.4. Normally, strength reductions are warranted on construction projects that use large quantities of concrete.

1905.6 Evaluation and acceptance of concrete. The criteria for evaluation and acceptance of concrete shall be as specified in Sections 1905.6.2 through 1905.6.5.

❖ The specified frequency of testing and application of minimum acceptance criteria for concrete used in the construction process is covered in this section.

1905.6.1 Qualified technicians. Concrete shall be tested in accordance with the requirements in Sections 1905.6.2 through 1905.6.5. Qualified field testing technicians shall perform tests on fresh concrete at the job site, prepare specimens required for curing under field conditions, prepare specimens required for testing in the laboratory and record the temperature of the fresh concrete when preparing specimens for strength tests. Qualified laboratory technicians shall perform all required laboratory tests.

❖ This section contains a new provision in ACI 318 that requires technicians performing code-related tests on fresh concrete in the field, including fabricating strength test specimens (concrete cylinders) and code-related tests on concrete in the laboratory, to be qualified to perform such tests. Technicians desiring to perform field tests can demonstrate their qualifications to the satisfaction of the building official by becoming certified by ACI as a "Concrete Field Test Technician–Grade I," or by meeting the requirements of ASTM C 1077, or a comparable program. Similarly, laboratory technicians can demonstrate their competence by becoming certified by ACI as a "Concrete Laboratory Testing Technician," "Concrete Strength Testing Technician" or by meeting the requirements of ASTM C 1077 or a comparable program.

1905.6.2 Frequency of testing. The frequency of conducting strength tests of concrete and the minimum number of tests shall be as specified in ACI 318, Section 5.6.2.

Exception: When the total volume of a given class of concrete is less than 50 cubic yards (38 m³), strength tests are not required when evidence of satisfactory strength is submitted to and approved by the building official.

❖ Frequency of testing is a significant factor in the effectiveness of quality control of concrete. To ensure that test results will be representative of the concrete in the structure, samples of fresh concrete to determine its compressive strength must strictly follow the provisions of this section. On a given project, if the quantity of a specific class (i.e., mix design) of concrete totals less than 50 cubic yards (38 m³), the exception permits strength tests to be waived with the approval of the building official if evidence is provided to show that proper concrete strength can be obtained. Such evidence may include certification from the concrete supplier that the material meets strength requirements based on field experience of previous projects or by trial batches in accordance with ACI 318 (see commentary, Section 1905.3).

Section 5.6.2.1 of ACI 318 prescribes three minimum sampling criteria for strength tests of each class of concrete (i.e., each mix design) based on the amount of concrete placed on a given day. For test samples to provide a true measure of concrete strength, they must be taken on a random basis to provide a fair representation of the quality of concrete used in the construction. To be representative, the choice of both times and batches of concrete to be sampled must be made on the basis of chance alone. Samples should not be selected on the basis of the appearance of the concrete, convenient times or for any other reason that would cause the test results to lose their statistical validity. Procedures for random selection of batches are described in ASTM D 3665. Procedures for sampling freshly mixed concrete are provided in ASTM C 172.

The representative nature of strength test data becomes less reliable when there are fewer than five tests. The criterion in Section 5.6.2.2 of ACI 318 is intended to provide the largest possible sample size for projects that use small volumes of concrete.

The statistical approach used in ACI 318 to proportion concrete mixtures for strength is based on strength test results that are the average of two cylinders. Likewise, the statistical approach is based on 28-day cylinder breaks, or other test age as specified in the construction documents. Using criteria other than that prescribed in this section will result in strength data that is not representative of concrete in the structure.

Example:
The quantity of concrete required to cast a 4-inch-thick (102 mm) slab for a floor, 3,000 square feet (279 m²) in area, would be about 37 cubic yards (28 m³). This represents the total volume of concrete involved in the project. About four truckloads of ready-mixed concrete using 10 cubic yard (7.6 m³) capacity mixers (or agitators) are needed. How many strength tests are required?

According to the provisions of Section 5.6.2.1 of ACI 318, only one strength test is required because less than 150 cubic yards (115 m³) of concrete and less than 5,000 square feet (465 m²) of floor are to be cast in one day. Section 5.6.2.2 of ACI 318 modifies this requirement by specifying that when less than five strength tests for the total volume of a particular class of concrete are involved on a given project, at least

one must be made for each batch of concrete. In this example, each truckload of concrete is a "batch" and, therefore, at least four strength tests are required. Section 5.6.2.4 of ACI 318 prescribes that a strength test be the average strengths of two cylinders made from the same sample (batch) of concrete. This means that a minimum of eight cylinders must be tested for the 37 cubic yards (28 m³) of concrete used in this project.

However, the exception in Section 1905.6.2 allows the building official to waive the strength tests when the quantity of concrete to be cast is less than 50 cubic yards (38 m³), but only if, in his or her judgment, there is sufficient evidence that the concrete will meet strength requirements.

1905.6.3 Strength test specimens. Specimens prepared for acceptance testing of concrete in accordance with Section 1905.6.2 and strength test acceptance criteria shall comply with the provisions of ACI 318, Section 5.6.3.

❖ Section 5.6.3 of ACI 318 provides criteria for determining the acceptability of concrete strength. As soon as test results become known during the course of the work, they can be immediately evaluated and the concrete accepted if it meets specified strength requirements. However, should concrete fail to meet the strength requirements in the acceptance criteria, changes should be made to the concrete to increase the average strength of subsequent test results.

ASTM C 172 provides a procedure for obtaining a representative sample of fresh concrete as delivered to the project site. The sample of concrete is then used to make strength test specimens (cylinders) that are tested to determine compliance with the quality control requirements of the code. The timing and manner of taking samples of freshly mixed concrete are critical to the production of quality concrete. Unless the sample is truly representative, the test results may be misleading.

After the sample is taken, specimens for strength testing must be prepared in accordance with ASTM C 31, which provides a standardized procedure for making test cylinders under field conditions, as well as for proper curing, protection and transporting to the laboratory for testing. In the laboratory the cylinders are tested in accordance with ASTM C 39 to determine the compressive strength of concrete. ASTM C 31 requires that whenever cylinders are fabricated for strength tests, the slump, air content and temperature of the sample of concrete must be determined. It should be pointed out that the requirement for determining the air content applies to all concrete, including nonair-entrained concrete.

Concrete strength is deemed satisfactory as long as the average of any three consecutive strength tests (average of two cylinders) equals or exceeds the specified compressive strength (f'_c), and no individual strength test (average of two cylinders) falls below the specified compressive strength of concrete, f'_c, by the amount specified.

When concrete fails to satisfy the strength requirements in Section 5.6.3.3 of ACI 318, changes to the concrete mix must be made to increase the average of subsequent strength test results. The changes that must be made to improve the concrete mixture will depend on particular circumstances and job specifications, but could include changes to the content of cement, water and air. Modifications that can be made to the concrete mixture to increase strength include changes in cement content, slump and air entrainment. Changes that only involve modifying mix proportions do not require resubmittal or approval of the concrete mixture. However, if the modifications involve changes to the ingredients of concrete, such as the use of a different type of cement, materials obtained from different sources or the addition of admixtures, evidence of an acceptable improvement in the concrete strength must be presented.

1905.6.4 Field-cured specimens. Where required by the building official to determine adequacy of curing and protection of concrete in the structure, specimens shall be prepared, cured, tested and test results evaluated for acceptance in accordance with ACI 318, Section 5.6.4.

❖ Section 5.6.4 of ACI 318 provides requirements for making, protecting and curing concrete cylinders in the field to determine if protection and curing of concrete in the structure is adequate in unusual weather conditions. They are also required to determine when forms may be removed (see Section 6.2.2.1 of ACI 318).

Normally, the requirement for field-cured cylinders is found in the construction specifications. Although field-cured cylinders are not required by the code to verify adequacy of curing and protection, this section authorizes the building official to require them when necessary.

The building official should consider several things when deciding whether or not field-cured cylinders are necessary. One reason for testing field-cured cylinders is to determine if the concrete in the structure is being properly cured. Typically, when concrete is kept in a moist condition at a temperature above 50°F (10°C) seven days (three days for high early strength concrete), it will undergo proper curing as required. Another reason is to confirm the concrete strength prior to removal of forms and shores. Section 6.2.2.1(c) of ACI 318 requires confirmation of concrete strength by the testing of field-cured cylinders or other approved methods. Deciding whether or not to require field-cured cylinders will usually be based on one or more of the following:

1. Prior experience with the contractor;

2. Prior test results of field-cured cylinders;

3. Temperature conditions during construction;

4. Cold-weather protection method(s) for the concrete;

5. Protection method(s) to minimize moisture loss from the concrete; or

6. Method(s) for supplying additional moisture to the concrete.

ASTM C 31 provides procedures not only for making and curing concrete test specimens that are to be field cured, but also for laboratory curing specimens as prescribed in Section 1905.6.3. Field-cured cylinders are stripped from their molds after a minimum initial curing and are stored in or on the structure as near as possible to the representative concrete. All surfaces of the cylinder are protected from the elements in as near as possible the same way as the formed work, thus providing the cylinders with the same temperature and moisture environments as the structural work.

Section 5.6.4.3 of ACI 318 stipulates that test cylinders to be field cured must be molded from the same sample of concrete as laboratory-cured test cylinders, thus providing companion cylinders for comparison testing as required in the following section.

The purpose of field-cured specimens is to simulate the curing of concrete under field conditions and to compare the test results of the field cylinders with those of laboratory-cured specimens made and cured under ideal moisture conditions. The comparison is made between the actual strengths of the field-cured cylinders and the laboratory-cured specimens, and not between the test results of the field specimens and the specified compressive strength of concrete, f'_c. Experience has shown that field-cured specimens generally do not test less than 85 percent of the standard laboratory moist cured cylinders. However, in cases when the field specimens fail to meet the 85-percent criterion specified in Section 5.6.4.4 of ACI 318, the test results are still acceptable if the strengths exceed the specified compressive strength by more than 500 psi (3.45 MPa).

1905.6.5 Low-strength test results. Where any strength test (see ACI 318, Section 5.6.2.4) falls below the specified value of f'_c, the provisions of ACI 318, Section 5.6.5, shall apply.

❖ Section 5.6.5 of ACI 318 prescribes the procedures that must be applied for structural safety when test results fail to meet the acceptance criteria. The building official must apply judgment as to the true significance of low-strength test results and whether they indicate need for concern. For example, strength tests failing to meet the acceptance criteria of the following section can be expected to occur in approximately one in every 100 tests, even though concrete strength and uniformity may be satisfactory. Allowance should be made for such statistically expected variations in deciding whether the strength level being produced is adequate.

When test results for laboratory or field-cured cylinders indicate a strength value lower than permitted in this section, an assessment of the significance of the difference must be made. It is imperative that the investigation follows a logical sequence of possible cause and effect. First, test reports should be reviewed and results analyzed before any action is taken. Of special interest in the early investigation should be the procedures used in making, handling and testing the cylinders since almost all deficiencies in failure to follow prescribed procedures will lower cylinder strength. Examples include: extra days in the field; curing over 80°F (26.7°C); frozen cylinders; impact during transporting to lab; delay in moist curing at the lab; improper caps and insufficient care in breaking cylinders. Second, engineering computations should be made to determine if the design of the concrete in question, including the effects of reinforcement, provides sufficient safety factors to support the intended service loads at the lower strength indicated by the test results. Nondestructive tests, such as the Schimdt rebound hammer (ASTM C 805), the Windor probe (ASTM C 803), or an ultrasonic pulse velocity test may be used to determine whether or not a portion of the structure actually contains low-strength concrete. In extreme cases, compressive strength tests of cores, as provided for in the following section, may be necessary. For in-place concrete investigations, it is essential to know where in the structure the questionable concrete was deposited. This information should be part of the data recorded at the time the test cylinders were molded. This information is both important and necessary when determining the location to core drill, should the need arise.

If a preliminary investigation determines that core tests should be required, core drilling in the area in question should be performed according to procedures outlined in ASTM C 42. The testing of cores requires great care in the operation itself and in the interpretation of the test results; thus, it is critical that ASTM C 42 is followed in detail.

Section 6.5.3 of ACI 318 provides conditioning requirements for core samples that are intended to produce strength tests that are indicative of the in-situ concrete strength for the area in question.

In evaluating core test results, the fact that core strengths may not equal the specified compressive strength of concrete, f'_c, should not be a cause for concern, provided they are within the limits specified. It is realistic to expect the compressive strength of cores to be less than f'_c because the size of the core differs from that of a standard concrete test cylinder, and procedures for obtaining and curing the cores are different. Therefore, the concrete can be considered acceptable from the standpoint of strength if test results from the cores are at least 85 percent of f'_c for the average of three cores and 75 percent of f'_c for a single core. Cores must be obtained and tested in accordance with ASTM C 42. Cores for compressive strength testing must have a length-to-diameter ratio between 1.0 and 2.10. For ratios less than 1.94, the compressive strength is multiplied by a correction factor (less than one) given in ASTM C 42 to correlate the strength of the core to that of a standard concrete test cylinder that has a length-to-diameter ratio of 2. Because of the less than delicate nature of the coring and sawing process, this section also permits erratic core and cube strengths to be discarded from the sample under consideration.

As a last resort, load tests may be required to check the adequacy of structural members that are seriously in doubt. Generally, such tests are suited only for flexural members, such as floors or beams, but they may sometimes apply to other members. Load testing is a highly specialized procedure that must be performed and interpreted by an engineer fully qualified in the proper techniques. Another "appropriate action" may be approving the structural element for a reduced live load.

1905.7 Preparation of equipment and place of deposit. Prior to concrete being placed, the space to receive the concrete and the equipment used to deposit it shall comply with ACI 318, Section 5.7.

❖ Section 5.7 of ACI 318 specifies the preparatory work on formwork, reinforcement and concrete equipment that is necessary prior to the placement of concrete. The section addresses the need for using clean equipment and for cleaning forms and reinforcement before placing concrete. All debris—including sawdust, nails, wood pieces and other litter—that usually collects inside the forms must be removed. The surface condition of reinforcement must comply with Section 7.4 of ACI 318. All standing water, snow or ice must also be removed from the forms.

1905.8 Mixing. Mixing of concrete shall be performed in accordance with ACI 318, Section 5.8.

❖ Section 5.8 of ACI 318 provides basic requirements for concrete mixing.

Concrete is to be mixed thoroughly until it is uniform in appearance and all ingredients are evenly distributed. If concrete has been adequately mixed, samples taken from different portions of a batch will have essentially the same unit weight, air content, slump and coarse aggregate content.

Ready-mixed concrete is commonly mixed in a central plant and transported to the construction site in agitating or nonagitating trucks, mixed entirely in transit or after reaching the job or mixed partially in a central plant and completed in transit or at the job site. ASTM C 94 and ASTM C 685 cover methods of mixing and transporting concrete.

Concrete is sometimes mixed at the construction site in a stationary mixer. Job-mixed concrete must conform to the requirements of ACI 318, Section 5.8.3, which provides general provisions for materials handling, batching and the mixing of concrete, and includes requirements for job records.

1905.9 Conveying. The method and equipment for conveying concrete to the place of deposit shall comply with ACI 318, Section 5.9.

❖ The handling and transporting of concrete needs to be carefully controlled in each step of the operation to maintain uniformity of the materials. Section 5.9 of ACI 318 requires the equipment for handling and transporting concrete to be capable of supplying concrete to the place of deposit continuously and reliably under all conditions and methods of placement, including by crane buckets; wheelbarrows or power buggies; chutes; belt conveyors or pumps or by any other acceptable means.

To maintain uniformity and an even distribution of ingredients in the concrete mixture, it is important to handle and transport the material in such a way as to avoid segregation of the coarse aggregates from the mortar or the water from the other ingredients comprising the mixture.

Section 5.9.2 of ACI 318 requires that equipment used for handling and transporting concrete be capable of supplying concrete to the place of deposit continuously and reliably under all conditions. Equipment that may be used to convey concrete includes pumps, belt conveyors, pneumatic systems, wheelbarrows, power buggies, crane buckets and tremies.

1905.10 Depositing. The depositing of concrete shall comply with the provisions of ACI 318, Section 5.10.

❖ Section 5.10 of ACI 318 addresses the placement of fresh concrete in its final position, as well as concerns, such as placement timing, concrete that has partially hardened, continuous concreting, retempering, depositing in walls, construction joints and consolidation.

Segregation is the separation of coarse aggregate from mortar. Concrete should be deposited at or near its final position in the structure. There is a tendency for aggregates to segregate when concrete is flowed laterally into place or deposited in locations that necessitate rehandling.

At all times during the placement operations, concrete is required to remain in a plastic state to allow it to flow readily into the spaces between reinforcement. Placing should be carried on at such a rate that the concrete, which is being integrated with fresh concrete, is still plastic. Concrete is deemed to be in a "plastic" or semifluid state when it is capable of being easily molded by hand. A very wet or soupy concrete mixture is not within the definition of "Plastic." A plastic mix keeps all grains of sand and pieces of coarse aggregate encased and held in place. The parts of such mixtures are not apt to segregate while being transported, placed or worked in the forms.

Concrete that has become partially hardened (i.e., advanced beyond an initial set) prior to placement will have significantly less strength than concrete that has been placed while in a plastic state and allowed to gain strength under appropriate curing conditions. Concrete that has been contaminated by foreign materials, such as dirt or sawdust, will have lower strength, be less durable and have other less desirable properties depending on the amount of contaminants.

Concrete may be remixed after first achieving its initial set without losing appreciable strength; however, it should be done infrequently and with caution. Because of the potential for lower strength, it must be approved by the registered design professional.

Concrete should be deposited continuously or in layers of such thickness that it will not be deposited on concrete that has hardened sufficiently to cause flow lines, seams or planes of weakness within the section. If a section cannot be placed continuously, preplanned construction joints (in accordance with Section 6.4 of ACI 318) should be provided as called for on the drawings or as may otherwise be permitted.

Section 5.10.6 of ACI 318 requires that the top surface of concrete being placed in a vertical member, such as walls, be kept as level as practical. The top surface of a concrete lift provides a potential location for seams or planes of weakness. The structural consequence of these potential planes of weakness is less if they are approximately level or horizontal.

A construction joint occurs where freshly mixed concrete is placed in contact with existing hardened concrete. Special details and procedures are necessary to enable construction joints to provide structural integrity and other desirable characteristics, as outlined in Section 6.4 of ACI 318.

Consolidation is the process of compacting fresh concrete to mold it within forms and around embedded items and reinforcement to eliminate stone pockets and entrapped air. Consolidation is accomplished by hand rodding or mechanical methods, such as vibration. Consolidation around reinforcement is particularly important to develop bond strength between the concrete and the reinforcement.

1905.11 Curing. The length of time, temperature and moisture conditions for curing of concrete shall be in accordance with ACI 318, Section 5.11.

❖ Section 5.11 of ACI 318 governs the curing of concrete. Concrete's strength increases over time. Curing requires the maintenance of an appropriate moisture content and temperature in fresh concrete for a specific period of time. The goal of the curing process is to minimize the loss of water by evaporation while the concrete is gaining strength. Proper curing is needed in order to achieve the required field strength of the concrete and to provide a durable finished product.

Immediately after placement and finishing, concrete must be protected and maintained with minimal moisture loss at a relatively constant temperature for the period necessary for the hydration of cement and the hardening of concrete so that the desired properties of strength and durability may develop. Concrete (other than high early strength) that is maintained in a moist condition and at a temperature above 50°F (10°C) for seven days can be expected to obtain its specified compressive strength. While the amount of mixing water in the concrete mixture is normally sufficient for curing purposes, excessive water loss by evaporation should be prevented by properly applying curing materials and methods. Evaporation may reduce the amount of water below what is necessary for the development of the desired properties of concrete.

Concrete made with high early strength cement develops strength much earlier than concrete made with normal (Type I) or moderate sulfate-resistance (Type II) cements. Earlier development of strength reduces the length of time (from seven to three days) that it is necessary to provide moist, temperate curing conditions.

The principal reason for using accelerated methods for curing concrete is to speed up strength development and reduce the time of curing. Accelerated curing is commonly used in the manufacturing of precast structural concrete elements. ACI 318, Section 5.11.3, prescribes acceptable methods, testing and acceptance criteria for accelerated curing. Curing by these special methods can produce compressive strengths that are lower than concrete moist cured by conventional methods. As a consequence, concrete mixtures should be proportioned to meet the specified compressive strength requirements under such curing conditions and be provided with supplementary strength testing.

1905.12 Cold weather requirements. Concrete to be placed during freezing or near-freezing weather shall comply with the requirements of ACI 318, Section 5.12.

❖ Concrete placed in cold weather can adversely affect the strength and durability of the hardened material unless proper protection is provided to prevent the fresh concrete from freezing. This section describes the preparatory work necessary before concrete is placed in cold weather. Cold weather is defined as a period when for more than three consecutive days, the mean daily temperature drops below 40°F (4.4°C). When temperatures are above 50°F (10°C) during more than half of any 24-hour period, the concrete should not be regarded as coming under the cold weather requirements of this section. For more information, refer to ACI 306R, which provides requirements for protection, temperatures during and after concrete placement, curing methods and other detailed information needed for cold weather concreting.

1905.13 Hot weather requirements. Concrete to be placed during hot weather shall comply with the requirements of ACI 318, Section 5.13.

❖ Hot weather cannot be simply defined in terms of time/temperature limits as is used to define cold weather (see commentary, Section 1905.12) because of the influence of other environmental factors. Hot weather, as it affects concrete, can best be described as any combination of high temperature, low relative humidity and wind velocity that tends to adversely affect the properties of fresh or hardened concrete.

Hot weather can create difficulties in fresh concrete such as:

1. Increased water demand;

2. Accelerated slump loss;

3. Increased rate of setting and the ensuing problems of placing and finishing concrete;

4. Increased tendency for plastic cracking and

5. Control of air content of air-entrained concrete.

Although adding water to concrete during hot weather would seem to be the appropriate solution, it can have adverse effects on the properties of hardened concrete, such as:

1. Decreased strength;

2. Decreased durability and watertightness;

3. Nonuniform surface appearance and

4. Increased tendency for drying shrinkage.

Detailed information on the materials, methods and procedures necessary to alleviate the difficulties described above is contained in ACI 305R.

SECTION 1906
FORMWORK, EMBEDDED PIPES AND CONSTRUCTION JOINTS

1906.1 Formwork. The design, fabrication and erection of forms shall comply with ACI 318, Section 6.1.

❖ These provisions give the minimum requirements for the design, construction and materials for formwork to provide installations that are adequate and safe to receive the concrete work as shown on the plans and specified in the construction documents.

Section 6.1.1 of ACI 318 provides minimum performance requirements for formwork with respect to shape, alignment and dimension of concrete members. Formwork is very seldom constructed to exactly the dimensions shown on the construction documents, nor is it perfectly straight, level and plumb. Fortunately, it seldom needs to be perfect. Tolerances for alignment and dimensions are normally found in the construction documents. In the absence of such information, allowable tolerances can be found in ACI 117 and ACI 347. The recommended practices covered in this document include preparation of construction documents; design criteria for vertical and horizontal forces and lateral pressures; design considerations (including capacities of formwork accessories); preparation of formwork design drawings, construction and use of forms (including safety considerations); materials for formwork and formwork for special structures and special methods of construction.

While the information contained in ACI 347 is pertinent to the work of engineers, architects and building officials, it is important to understand that the recommended practices are based on the premise that the layout, design and construction of formwork are the responsibility of the contractor.

Forms must have adequate strength and be tightly fitted to prevent gaps from developing at the joints and seams, which would allow the loss of cement paste (mortar). Loss of mortar can cause honeycombing and rock pockets that diminish concrete appearance and may adversely affect strength and durability.

Formwork should be straight, free of warping and adequately braced and tied together to resist concrete pressure without bulging or failing.

Shoring and formwork supports must be designed so they do not impose undue loading or are not anchored in a manner that will cause damage to previously placed concrete.

Although forms have a temporary function, they are engineered structures that must be designed to withstand the loads and vibrations encountered during concrete placement. ACI 318, Section 6.1.5, requires that formwork be designed to consider rate and method of concrete placement, construction loads (i.e., vertical, horizontal and impact) and special forming requirements (e.g., shells, folded plates, domes and architectural elements).

When prestressing forces are induced in post-tensioned concrete members, the members tend to shorten and, in the case of beams and slabs, also tend to "lift." The shortening of the elements may damage forming if it is not designed and constructed to accommodate this movement. Because prestressing forces generally induce compression in portions of beams and slabs that are resisting flexural tension stresses when loaded (i.e., the bottom portion of the mid-span region of a member), the member tends to "lift" when the prestressing forces are applied to the concrete. This tendency to lift may temporarily remove all gravity loads from the shores that have been supporting the formwork and element prior to application of the prestressing. If the shores are not mechanically attached to the formwork and braced laterally, they may partially or completely fall out, which will leave the post-tensioned member with no support for construction loads that may be applied later.

1906.2 Removal of forms, shores and reshores. The removal of forms and shores, including from slabs and beams (except where cast on the ground), and the installation of reshores shall comply with ACI 318, Section 6.2.

❖ Section 6.2 of ACI 318 provides general requirements for the removal of forms and the installation of reshores that are necessary to provide for worker safety and prevent damage to the concrete during the construction process.

Forms must not be removed until the concrete has developed sufficient tensile strength to resist cracking, spalling or, worse yet, structural failure. Usually, the side (vertical) forms of beams, girders, columns and walls can be removed after 12 hours. Before the removal of formwork supporting the bottom (horizontal)

portions of structural elements, such as slabs, beams and girders, additional evaluation is required to determine the appropriate time for removal to assure that the integrity of the structure is maintained.

Section 6.2.2 of ACI 318 provides general requirements for the removal of forms and shores and installation of reshores for structural slabs and beams (i.e., those not supported on the ground) to provide a safe structure during construction and prevent damage to the concrete during the construction process.

Section 6.2.2.1 of ACI 318 requires the contractor to develop a procedure and schedule for the removal of forms and placement of reshores supporting slabs and beams (other than those supported by the ground) before construction is started. The time for removal of the slab and beam forms must be determined by structural analysis while giving consideration to the effect of construction loads and the actual strength of the concrete. Before forms are stripped, concrete must acquire sufficient strength to avoid collapse or excessive deflection or distortions that would cause permanent structural damage. The strength of concrete must be determined prior to form removal by testing field-cured cylinders in accordance with ASTM C 39 or by other procedures approved by the building official such as:

1. Compressive strength tests, ASTM C 873 [These tests are limited to slabs where the depth of concrete is 5 to 12 inches (127 to 305 mm)];

2. Penetration resistance of hardened concrete, ASTM C 803;

3. Pullout strength tests, ASTM C 900; and

4. Maturity factor determinations in accordance with ASTM C 1074.

The strength analysis and concrete data used to develop the procedures and time schedule must be furnished to the building official, if requested.

For safety purposes, special loading conditions, such as unsymmetrical placement of concrete, impact loads, uplift and concentrated loads must be considered in the design of formwork, including the loads transmitted from the floors above in multistory construction.

Construction loads include combinations of dead loads, such as the weights of formwork and fresh concrete, and live loads, such as workmen, equipment, stored materials, runways and other loads imposed by the concreting operations. These loads are often greater than the design live load and must be supported by shores unless structural analysis and test data show that shores may be omitted.

Before the dead load and any superimposed construction loads can be safely supported by an unshored, post-tensioned, prestressed member, the concrete must attain the compressive strength established by the engineer, and prestressing must be applied to the concrete by tensioning the tendons to the value specified in the construction documents. If shores are removed from such a member prior to the concrete being prestressed or from one in which inadequate prestress has been applied to the concrete, the member may be damaged or experience structural failure.

1906.3 Conduits and pipes embedded in concrete. Conduits, pipes and sleeves of any material not harmful to concrete and within the limitations of ACI 318, Section 6.3, are permitted to be embedded in concrete with approval of the registered design professional.

❖ Conduits, pipes and sleeves may be embedded in concrete, provided that the properties of the material or the size and location of embedments will not damage the concrete. In the design of structural concrete, embedments must not be considered as structurally replacing any of the displaced concrete, except in the case of concrete under compression in accordance with ACI 318, Section 6.3.6.

The use of bare aluminum in structural concrete is prohibited because the material reacts with concrete, and in the presence of chloride ions, may produce an electrolytic reaction with steel reinforcement causing corrosion of the steel that may result in cracking or spalling of the concrete or both. Further, in conjunction with the use of aluminum materials in concrete, particularly aluminum conduits, stray electric currents may accelerate the electrolytic reaction. Therefore, any aluminum embedments in concrete must be suitably coated or covered to prevent direct contact with the concrete.

Care must be taken in the design and construction of slabs, walls or beams to ensure that any conduits, pipes or sleeves passing through these structural elements will not be located in places that would create planes of weakness or distress and compromise the strength of the concrete construction. The locations of conduits, pipes and sleeves passing through structural elements should be shown and properly located on the project plans. In the absence of such information, the building official should seek the approval of the design engineer before allowing installation.

Conduits and pipes are permitted to be embedded within a column, provided that such installations do not displace more than 4 percent of the column cross-sectional area on which the strength is calculated or that is required for fire protection purposes. This allowance is based on the premise that a column is a compression member, and that a nominal displacement of concrete with steel will not seriously affect structural integrity.

ACI 318, Section 6.3.5, gives empirical rules for acceptance of conduits and pipes embedded in slabs, walls and beams (other than those merely passing through) that apply unless the structural engineer specifically approves drawings showing the size and location of the conduit and pipe.

1906.4 Construction joints. Construction joints, including their location, shall comply with the provisions of ACI 318, Section 6.4.

❖ There are three basic types of joints used in concrete construction: construction joints, contraction or control joints and isolation joints. Section 6.4 of ACI 318 deals only with construction joints, which are mainly used in structural floor and roof systems. Construction joints are stopping places in the process of concrete placement. A true construction joint should bond new concrete to existing concrete and permit no movement. This condition of construction is essential in structural concrete. See the commentary to Section 1909.3 for information on contraction and isolation joints.

The surface of previously placed concrete must be clean and free of laitance before concrete placement is continued. Laitance is the light gray or nearly white substance that may form during the finishing operation. The laitance layer has little strength, and if not removed will reduce the bond between hardened and fresh concrete.

If the hardened concrete joint is not lightly wetted prior to placement of fresh concrete, its dry surface may absorb water from the fresh concrete that is needed for proper strength development, thus creating a weakened plane. Overwetting, especially when it creates standing water, will increase the w/cm ratio of the concrete at the joint, which can also reduce localized strength and bonding potential.

The location of construction joints is an important structural consideration that must be specified by the designer either in the construction documents or on a case-by-case basis. Where shear due to gravity is not significant, which typically occurs in the middle of the span of flexural members such as slabs and beams, a vertical joint (without shear key, diagonal dowels or other shear transfer method) may be acceptable. ACI 318, Section 11.7.9, requires that the construction joint be clean and free of laitance to provide proper bond between the hardened joint and the fresh concrete. Joint surfaces (where the designer has assumed a coefficient of friction for shear transfer) must be roughened to an amplitude of approximately $^1/_4$ inch (6.4 mm).

As a general rule, construction joints should be located where shear, due to gravity loads, is at a minimum. In slabs, beams and girders, this location is within the middle third of the span, preferably nearer the center where the shear stress approaches zero. Where a beam intersects a girder, the joint in the girder must be offset a distance equal to twice the width of the beam.

Where horizontal members, such as beams, girders and slabs are placed monolithically with their supporting columns or walls, cracking may develop at the interface of the horizontal members and the vertical supporting member due to settlement and bleeding of the plastic concrete in the supporting member. To lessen the potential for such cracking, concrete in walls and columns supporting slabs, beams or girders must be allowed to harden before additional concrete is placed.

When beams, girders, brackets, column capitals, haunches and drop panels are not placed monolithically with the floor or roof slab, there is potential for development of a weakened plane that will reduce the flexural capacity of the composite member. By definition, monolithic concrete contains no joints other than construction joints.

SECTION 1907
DETAILS OF REINFORCEMENT

1907.1 Hooks. Standard hooks on reinforcing bars used in concrete construction shall comply with ACI 318, Section 7.1.

❖ The ends of most reinforcing bars, except at lap splices, must be bent to form a hook in order to provide better anchorage to the concrete. ACI 318 requires that hooks be bent 90 degrees (1.6 rad) ("L" shape), 135 degrees (2.4 rad) or 180 degrees (3.1 rad) depending on their function and loading condition.

1907.2 Minimum bend diameters. Minimum reinforcement bend diameters utilized in concrete construction shall comply with ACI 318, Section 7.2.

❖ If reinforcement is bent too sharply, the outside surface of the bend may develop cracks that will reduce anchorage strength and increase the potential for corrosion. To prevent this from occurring, the code restricts the minimum inside diameter of reinforcing bar bends based on the bar size; requiring larger minimum-diameter bends for larger bar sizes.

1907.3 Bending. The bending of reinforcement shall comply with ACI 318, Section 7.3.

❖ Section 7.3 of ACI 318 identifies the scope of the code requirements regarding bending of reinforcement. Proper bending is needed so that the reinforcement will not have a reduction in tensile strength.

Normally, bends in reinforcement complying with the minimum diameters specified in Section 7.2 of ACI 318 can be accomplished with the reinforcement at ambient temperatures. In some unusual construction conditions, normal-bend diameters may not be practical and bends with smaller diameters than specified in Section 7.2 of ACI 318 may be necessary. In order to accomplish such bends, the reinforcement may need to be heated prior to bending. This section permits this practice, provided it is approved by the registered design professional.

Reinforcement that is partially embedded in the concrete must be bent in the field on occasion. Under such circumstances, the registered design professional must decide whether the bars are to be bent cold or heated. Heating is usually ordered for large-sized bars that cannot be easily bent cold. Reinforcing bars can be bent successfully if they are first preheated to between 1,100 and 1,200°F (593 and 649°C) and then gently bent in as gradual an arc as possible. The decision to heat partially embedded reinforcement is normally based on the proximity of the bar bend to the adjacent concrete. Heating, where necessary, must be controlled at all times to prevent damage to the bars that would adversely affect structural strength. In the case of partially embedded reinforcement, heating

must also be done in a manner to avoid spalling or damaging the concrete.

1907.4 Surface conditions of reinforcement. The surface conditions of reinforcement shall comply with the provisions of ACI 318, Section 7.4.

❖ Section 7.4 of ACI 318 identifies the code requirements regarding surface conditions of reinforcement. Surface conditions are important for proper bond strength to the concrete.

At the time of concrete placement, all metal reinforcement must be free from mud, oil or other deleterious materials that may adversely affect or reduce the bond between the reinforcement and the concrete. Bonding of the two materials is necessary in order for the concrete to take full advantage of the tensile strength provided by the reinforcement.

Research has shown that the amount of rust and mill scale that normally remains on reinforcement after handling actually improves the bond between concrete and the reinforcement. Rust becomes a concern when it has progressed to a state where there is a measurable loss in height of deformations and weight of a specimen subjected to a hand-wire-brushed test.

Excessive rust, pitting or both must not be permitted to develop on the surface of prestressing steel because these highly stressed elements are more susceptible to corrosion. Oil and dirt must also be removed to maintain a proper bond between the steel and the concrete.

1907.5 Placing reinforcement. The placement of reinforcement, including tolerances on depth and cover, shall comply with the provisions of ACI 318, Section 7.5. Reinforcement shall be accurately placed and adequately supported before concrete is placed.

❖ Section 7.5 of ACI 318 identifies the scope of the code requirements regarding reinforcement placement.

Reinforcement, tendons and post-tensioning ducts must be accurately positioned, supported and secured to prevent displacement by construction loads or the placing of concrete beyond the tolerances permitted by Section 7.5.2 of ACI 318. Section 7.5.1 of ACI 318 recognizes the practical necessity of placing some items in precast elements after the concrete is in place and relies on the registered design professional to determine proper installation.

Reinforcement, including bars and welded wire fabric, tendons and post-tensioning ducts must be placed and held in position as shown on the project drawings or as otherwise specified by the engineer. The strength of a concrete member can be adversely affected by the improper positioning of reinforcement that exceeds the tolerances permitted by Sections 7.5.2.1 and 7.5.2.2 of ACI 318.

Section 7.5.2.1 of ACI 318 provides tolerances that apply simultaneously to concrete cover and member effective depth (d). Cover is measured from the outside of the reinforcement to the face of the concrete. "Effective depth" or (d) is defined as the distance from the extreme compression fiber to the centroid of tension reinforcement. The effective depth (d) is a very important dimension since deviation in this dimension, especially for members of lesser depth, can have an adverse effect on the strength of the member. The permitted tolerances take this into account with a smaller permitted variation [i.e., $^3/_8$ inch (9.5 mm)] for smaller members; for example, members where d is 8 inches (203 mm) or less. Note that a plus (+) tolerance increases the dimension and a minus (-) tolerance decreases the dimension. Where only a minus (-) tolerance is indicated for cover, there is no limit in the other direction; however, normally when cover is increased beyond that specified, effective depth (d) is reduced. The last sentence of the text allows a reduced cover dimension for formed soffits, because they provide additional exposure protection. Measurement of effective depth (d) and cover is illustrated in Figure 1907.5.

The locations of bends and ends of reinforcement as required by the construction documents are important to the structural integrity of concrete members. In addition to the code-prescribed rebar placing tolerances, job specifications may contain more stringent reinforcement tolerances that must be followed during construction.

Section 7.5.3 of ACI 318 permits the draping of welded wire fabric in short-span slabs [10 feet (3048 mm) or less] where the fabric consists of smaller sizes of wire that are flexible enough to curve readily over the supports. "A point near the top of the slab" and "a point near the bottom of the slab" should be interpreted as distances from the top and bottom of the slab that provide the cover required by Section 1907.7. Provisions for anchorage requirements at supports are found in Sections 12.11 and 12.12 of ACI 318.

Unless otherwise approved by the engineer, Section 7.5.4 of ACI 318 prohibits tack welding of crossing bars. Metallurgical notches are often created as a re-sult of tack welding at the welded point in which the material is quickly heated at a high temperature and then allowed to cool rapidly. Welding crossbars can seriously weaken the bar at the point of welding.

1907.6 Spacing limits for reinforcement. The clear distance between reinforcing bars, bundled bars, tendons and ducts shall comply with ACI 318, Section 7.6.

❖ The code requires that reinforcement be separated by a minimum distance to allow concrete to flow readily around it and to prevent too close a spacing between bars on a line that can cause shear and shrinkage cracking. The exception would be bars that are bundled in accordance with ACI 318 Section 7.6.6 (see Section 1907.7.4).

1907.7 Concrete protection for reinforcement. The minimum concrete cover for reinforcement shall comply with Sections 1907.7.1 through 1907.7.7.

❖ A minimum thickness of concrete cover must be provided to protect reinforcement against weather and corrosive environments and may be required to protect against exposure to fire. In some extremely corrosive atmospheres, such as in the presence of chemical fumes, or in the vicinity of saltwater or where there are severe moisture and temperature changes, the amount of specified cover may need to be increased. Galvanized or epoxy-coated reinforcement is also used to provide corrosion protection.

1907.7.1 Cast-in-place concrete (nonprestressed). Minimum concrete cover shall be provided for reinforcement in nonprestressed, cast-in-place concrete construction in accordance with ACI 318, Section 7.7.1.

❖ Section 7.7.1 of ACI 318 specifies the minimum concrete cover generally required for cast-in-place concrete. Other requirements, such as fire protection or

Figure 1907.5
TYPICAL SLAB & BEAM CONSTRUCTION
(NOT TO SCALE)

exposure to corrosive environments could necessitate cover in excess of that required by this section.

Concrete cover is measured from the concrete surface to the outermost surface of the steel to which the cover requirement applies. The values contained in Table 7.7.1 of ACI 318 provide minimum concrete cover for reinforcement in cast-in-place concrete exposed to most environments. Section 4.4.2 of ACI 318 requires that if concrete with reinforcement will be exposed to chlorides from deicing chemicals, salts, saltwater, brackish water, seawater or spray from these sources, it must comply with Table 4.2.2 and the minimum concrete cover requirements of this section. The intent is that the provisions of Section 1907.7.5 apply (see commentary, Section 1907.7.5). For concrete elements required to achieve a fire-resistance rating, see the commentary to Section 1907.7.7.

1907.7.2 Cast-in-place concrete (prestressed). The minimum concrete cover for prestressed and nonprestressed reinforcement, ducts and end fittings in cast-in-place prestressed concrete shall comply with ACI 318, Section 7.7.2.

❖ This section references ACI 318 for minimum cover requirements for reinforcement in prestressed concrete. The minimum concrete cover values equally apply to prestressed and nonprestressed reinforcement, ducts and end fittings. See the commentaries to Section 1907.7.5 for prestressed concrete exposed to corrosive environments and Section 1907.7.7 for prestressed concrete elements required to be fire-resistance rated.

1907.7.3 Precast concrete (manufactured under plant control conditions). The minimum concrete cover for prestressed and nonprestressed reinforcement, ducts and end fittings in precast concrete manufactured under plant control conditions shall comply with ACI 318, Section 7.7.3.

❖ This section references ACI 318 for minimum cover requirements for reinforcement for precast concrete manufactured under plant control conditions. The phrase "manufactured under plant control conditions" does not necessarily mean that precast members must be produced in a plant. Precast structural elements can also be made at the job site and may qualify under this section of the code, provided the control of form dimensions, placing of reinforcement, quality control of concrete and curing procedures are equal to practices normally expected in a plant. These tighter controls permit lesser cover thickness requirements for precast concrete versus cast-in-place construction. See the commentaries to Section 1907.7.5 for precast concrete exposed to corrosive environments and Section 1907.7.7 for precast concrete elements required to be fire-resistance rated.

1907.7.4 Bundled bars. The minimum concrete cover for bundled bars shall comply with ACI 318, Section 7.7.4.

❖ In construction that requires heavy concentrations of reinforcement, bundles of standard bar sizes can save space and reduce congestion, thus allowing for proper placement and consolidation of concrete. Bundling of

bars is permitted, but only if such bundles are in conformance with ACI 318, Section 7.6.6. The minimum cover for bundled bars must be equal to the equivalent diameter of the bundle, but needs to be greater than 2 inches (51 mm), except for concrete cast against and permanently exposed to earth, where the minimum cover is 3 inches (76 mm). Equivalent diameter is determined from the total cross-sectional area of the bundled bars.

1907.7.5 Corrosive environments. In corrosive environments or other severe exposure conditions, prestressed and nonprestressed reinforcement shall be provided with additional protection in accordance with ACI 318, Section 7.7.5.

❖ Where severe corrosive environments are encountered, the amount of concrete protection must be increased in accordance with Section 7.7.5 of ACI 318. Section 1904.2.2 cites several examples of concrete that will be exposed to external sources of chloride in service. The amount of increase for the cover over reinforcing steel is generally based on the judgement of the registered design professional. The commentary to this section of ACI 318 recommends (but does not require) a minimum cover for reinforcement of 2 inches (51 mm) for walls and slabs and 2$\frac{1}{2}$ inches (64 mm) for other structural members. These two dimensions can be reduced to 1$\frac{1}{2}$ and and 2 inches (38.1 and 64 mm), respectively, for precast concrete manufactured under plant control conditions. The provisions also require that the denseness and nonporosity of the protecting concrete also be considered. Section 4.4.2 and Table 4.2.2 of ACI 318 mandate a lower w/cm ratio, a minimum specified compressive strength, or both for concrete exposed to these environments, thus providing a more impermeable (i.e., denser) cement paste matrix. Even higher densities can be achieved with the use of some admixtures, such as silica fume. As an alternative to providing increased cover and denser cement paste, other protection may be provided, such as the use of galvanized or epoxy-coated reinforcement, or cathodically protecting the reinforcement.

1907.7.6 Future extensions. Exposed reinforcement, inserts and plates intended for bonding with future extensions shall be protected from corrosion.

❖ The exposed surfaces of reinforcement, inserts and plates that will be used for extensions of a structure must be protected to prevent corrosion that will cause them to fail the criteria of Section 1907.4 and otherwise reduce their ability to bond with future concrete.

1907.7.7 Fire protection. When this code requires a thickness of cover for fire protection greater than the minimum concrete cover specified in Section 1907.7, such greater thickness shall be used.

❖ The cover requirements in Section 1907.7.1 for reinforcement in cast-in-place concrete construction will generally equal or exceed the requirements for fire resistance ratings. There are, however, a few exceptions for higher fire-resistance ratings, particularly those for members having 3- and 4-hour fire endurance ratings. Three of these exceptions under the prescriptive fire-resistance-rated assemblies should be noted:

1. The reinforcement in unrestrained, reinforced concrete floor and roof slabs requiring a fire-resistance rating of 3 hours must have a 1 1/4-inch (32 mm) minimum concrete cover [see Table 721.2.3(1)].

2. The main reinforcement in unrestrained beams less than 8 inches (203 mm) in width and requiring a fire-resistance rating of 3 hours may require a cover of more than 1 1/2 inches (38.1 mm) [see Table 721.2.3(2)].

3. The main reinforcement in columns having a required fire-resistance rating of 2 hours or greater may require a cover of 2 inches (51 mm). This should be achieved with No. 4 ties (see Section 721.2.4.2).

Monolithic cast-in-place concrete is usually considered restrained construction in accordance with Appendix X3 of ASTM E 119 and, therefore, the exceptions listed above may not totally apply or may apply on a limited basis under certain conditions of construction. An unrestrained condition is one in which the load-carrying element is free to expand and rotate at the supports. As an example, this may apply to precast concrete elements. For a more comprehensive guide to determining these conditions, refer to Appendix X3 of ASTM E 119. For additional information on fire resistance of concrete, see Sections 721.1 and 721.2.

1907.8 Special reinforcement details for columns. Offset bent longitudinal bars in columns and load transfer in structural steel cores of composite compression members shall comply with the provisions of ACI 318, Section 7.8.

❖ The provisions of ACI 318, Section 7.8, only apply to special reinforcement details in columns where they change in dimension and columns of composite construction that contain structural steel cores. Offset bends in column longitudinal reinforcement are permitted only where column faces are offset less than 3 inches (76 mm). Otherwise, the longitudinal bars must be discontinued and spliced as illustrated in Figure 1907.8.

1907.9 Connections. Connections between concrete framing members shall comply with the provisions of ACI 318, Section 7.9.

❖ The provisions of ACI 318, Section 7.9, require that at connections of principal framing elements (such as beam-to-column connections), ties, spirals, stirrups or other forms of confinement must be provided for splices of continuing reinforcement and anchorage of reinforcement that terminates in the connection. This is necessary so the joint will maintain its flexural capacity without deteriorating under repeated loadings.

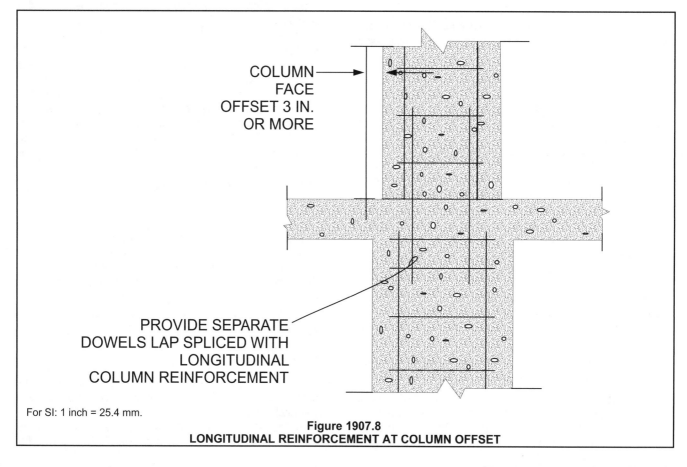

COLUMN FACE OFFSET 3 IN. OR MORE

PROVIDE SEPARATE DOWELS LAP SPLICED WITH LONGITUDINAL COLUMN REINFORCEMENT

For SI: 1 inch = 25.4 mm.

**Figure 1907.8
LONGITUDINAL REINFORCEMENT AT COLUMN OFFSET**

1907.10 Lateral reinforcement for compression members. Lateral reinforcement for concrete compression members shall comply with the provisions of ACI 318, Section 7.10.

❖ Section 7.10 of ACI 318 provides minimum design and detailing requirements that apply to lateral reinforcement used in compression members such as columns. Lateral reinforcement may consist of ties or spirals or both that encircle and confine the longitudinal reinforcement. In some cases, the lateral reinforcement may be called upon to provide resistance to shear or torsion or both, in which case the ties are referred to as "stirrups."

1907.11 Lateral reinforcement for flexural members. Lateral reinforcement for compression reinforcement in concrete flexural members shall comply with the provisions of ACI 318, Section 7.11.

❖ Section 7.11 of ACI 318 provides minimum design and detailing requirements that apply to lateral reinforcement used to confine compression reinforcement in flexural members, such as beams and girders. In some cases, lateral reinforcement may be called upon to provide resistance to shear or torsion or both, in which case the ties are referred to as "stirrups."

1907.12 Shrinkage and temperature reinforcement. Reinforcement for shrinkage and temperature stresses in concrete members shall comply with the provisions of ACI 318, Section 7.12.

❖ Section 7.12 of ACI 318 requires that a minimum amount of reinforcement be placed in one-way structural slabs to minimize cracking caused by concrete shrinkage and large temperature fluctuations and to tie the structure together. One-way structural slabs have flexural reinforcement extending in one direction only. Shrinkage and temperature reinforcement is placed at a right angle to the principal reinforcement. This requirement does not apply to soil-supported slabs.

1907.13 Requirements for structural integrity. The detailing of reinforcement and connections between concrete members shall comply with the provisions of ACI 318, Section 7.13, to improve structural integrity.

❖ The intent of ACI 318, Section 7.13, is to improve structural redundancy and ductility The provisions require minimum reinforcement detailing so that concrete members of a structure are effectively tied together to improve overall structural integrity should it suffer severe local damage from abnormal loads.

SECTION 1908
MODIFICATIONS TO ACI 318

1908.1 General. The text of ACI 318 shall be modified as indicated in Sections 1908.1.1 through 1908.1.16.

❖ Section 1908 contains modifications to ACI 318; the standard adopted by reference in Section 1901.2.

Many of the modifications are related to the seismic design of concrete structures. Text that differs from or is not found in ACI 318 is printed in italics to identify the changes.

1908.1.1 ACI 318, Section 10.5. Modify ACI 318, Section 10.5, by adding new Section 10.5.5 to read as follows:

10.5.5 – In structures assigned to Seismic Design Category B, beams in ordinary moment frames forming part of the seismic-force-resisting system shall have at least two main flexural reinforcing bars continuously top and bottom throughout the beam and continuous through or developed within exterior columns or boundary elements.

❖ The ACI 318 modification in this section adds a requirement for continuity reinforcement in order to increase the ductility of ordinary reinforced concrete moment frames. Ordinary moment frames of reinforced concrete are defined in Section 21.1 of ACI 318 as complying only with the provisions of Chapters 1 through 18 and are limited to structures having a relatively low seismic risk. There are no applicable seismic design provisions from Chapter 21 of ACI 318 since there is no need to provide for increased energy dissipation and ductility as is required for intermediate or special reinforced concrete moment frames. In the 1994 edition of the *NEHRP Recommended Provisions for Seismic Regulations for New Buildings,* the response modification coefficient, *R*, for ordinary reinforced concrete moment frames was increased to its current value of three. This increased *R*-value was agreed to after consideration of analytical and experimental findings, provided the continuity reinforcement herein described is also required for these frames.

1908.1.2 ACI 318, Section 11.11. Modify ACI 318, Section 11.11, by changing its title to read as shown below and by adding new Section 11.11.3 to read as follows:

11.11– Special provisions for columns.

11.11.3 – In structures assigned to Seismic Design Category B, columns of ordinary moment frames having a clear height-to-maximum-plan-dimension ratio of five or less shall be designed for shear in accordance with 21.12.3.

❖ This modification to ACI 318 adds a requirement that is intended to reduce the risk of a shear failure in the columns of ordinary moment frames. Investigation of earthquake damage has shown that failures of short, stiff columns are a concern. This is not surprising since the shear force generated in a column is inversely proportional to the clear height of the column. To minimize the chance of a shear failure that would inhibit the inelastic deformability anticipated in these systems, short stiff columns of ordinary moment frames must be designed for shear in accordance with the requirements for intermediate moment frames.

1908.1.3 ACI 318, Section 21.1. Modify existing definitions and add the following definitions to ACI 318, Section 21.1.

DESIGN DISPLACEMENT. Total lateral displacement expected for the design-basis earthquake, *as specified by Section 12.8.6 of ASCE 7.*

❖ This IBC modification to the definition given in ACI 318 provides a more specific code reference. Design displacement is defined to be the same as δ_x (or interstory drift based on δ_x) as given in Section 12.8.6 of ASCE 7.

DETAILED PLAIN CONCRETE STRUCTURAL WALL. *A wall complying with the requirements of Chapter 22, including 22.6.7.*

❖ The classification of shear wall systems is indicative of the amount of reinforcement and the type of detailing to be provided. Shear wall system classifications are also used to identify the appropriate earthquake design coefficients and system limitations in ASCE 7. This definition describes a type of shear wall system permitted in the earthquake load requirements of ASCE 7, but not defined under ACI 318. The definition provides a specific reference to the requirements of Section 1908.1.14, which supplements ACI 318 provisions for plain concrete by providing additional requirements for "detailed plain concrete structural walls." Note that ACI 318 refers to a "shear wall" as a "structural wall."

ORDINARY PRECAST STRUCTURAL WALL. *A precast wall complying with the requirements of Chapters 1 through 18.*

❖ This definition adds a shear wall system that is not explicitly defined in ACI 318. The classification of shear wall systems reflects the amount of reinforcement and the type of detailing to be provided. Shear wall system classifications are also used to identify the appropriate earthquake design coefficients and limitations in ASCE 7. The ACI 318 definition of "Ordinary reinforced concrete structural wall" includes both cast-in-place and precast walls. Because ordinary reinforced concrete structural walls constructed of precast elements are not expected to perform as well as a cast-in-place wall when subjected to the same earthquake ground motion, a precast ordinary reinforced concrete structural wall does not warrant the same response modification coefficient, *R*, as a comparable cast-in-place shear wall system wall. In order to assign a distinct *R*-value to precast shear wall systems, they must be defined separately. Note that ACI 318 refers to a "shear wall" as a "structural wall."

ORDINARY REINFORCED CONCRETE STRUCTURAL WALL. A *cast-in-place* wall complying with the requirements of Chapters 1 through 18.

❖ The classification of shear wall systems is indicative of the amount of reinforcement and the type of detailing to be provided. Shear wall system classifications are also used to identify the appropriate earthquake design coefficients and limitations in ASCE 7. The definition of "Ordinary reinforced concrete structural wall" in ACI 318 includes both cast-in-place and precast walls. This modification clarifies that under the code the term "ordinary reinforced concrete structural wall" applies only to cast-in-place shear wall systems. Note that ACI 318 refers to a "shear wall" as a "structural wall."

ORDINARY STRUCTURAL PLAIN CONCRETE WALL. A wall complying with the requirements of Chapter 22, *excluding 22.6.7.*

❖ The classification of shear wall systems reflects the amount of reinforcement and the type of detailing to be provided. Shear wall system classifications are also used to identify the appropriate earthquake design coefficients and limitations in ASCE 7. This modification clarifies that the additional code requirements intended for "detailed plain concrete shear walls" do not apply to "ordinary plain concrete shear walls." Note that ACI 318 refers to a "shear wall" as a "structural wall."

WALL PIER. *A wall segment with a horizontal length-to-thickness ratio of at least 2.5, but not exceeding 6, whose clear height is at least two times its horizontal length.*

❖ See Figure 1908.1.3.

* CLEAR HEIGHT OF WALL PIER IS AT LEAST 2 TIMES THE HORIZONTAL LENGTH ℓ

Figure 1908.1.3
PLAN OF WALL WITH OPENINGS

1908.1.4 ACI 318, Section 21.2.1. Modify ACI 318 Sections 21.2.1.2, 21.2.1.3 and 21.2.1.4, to read as follows:

21.2.1.2 – For structures assigned to Seismic Design Category A or B, provisions of Chapters 1 through 18 and 22 shall apply except as modified by the provisions of this chapter. Where the seismic design loads are computed using provisions for intermediate or special concrete systems, the requirements of Chapter 21 for intermediate or special systems, as applicable, shall be satisfied.

21.2.1.3 – For structures assigned to Seismic Design Category C, intermediate or special moment frames, intermediate precast structural walls or ordinary or special reinforced concrete structural walls shall be used to resist seismic forces induced by earthquake motions. Where the design seismic loads are computed using provisions for special concrete systems, the requirements of Chapter 21 for special systems, as applicable, shall be satisfied.

21.2.1.4 – For structures assigned to Seismic Design Category D, E or F, special moment frames, special reinforced concrete structural walls, diaphragms and trusses and foundations complying with 21.2 through 21.10 or intermediate precast structural walls complying with 21.13 shall be used to resist forces induced by earthquake motions. Members not proportioned to resist earthquake forces shall comply with 21.11.

❖ This modification provides a clarification that is necessary for correct application of the earthquake design provisions.

1908.1.5 ACI 318, Section 21.2.5. Modify ACI 318, Section 21.2.5, by renumbering as Section 21.2.5.1 and adding new Section 21.2.5.2 to read as follows:

21.2.5 – Reinforcement in members resisting earthquake-induced forces.

21.2.5.1 – Except as permitted in 21.2.5.2, reinforcement resisting earthquake-induced flexural and axial forces in frame members and in structural wall boundary elements shall comply with ASTM A 706. ASTM A 615, Grades 40 and 60 reinforcement, shall be permitted in these members if (a) the actual yield strength based on mill tests does not exceed the specified yield, f_y, strength by more than 18,000 psi (124 MPa) [retests shall not exceed this value by more than an additional 3,000 psi (21 MPa)], and (b) the ratio of the actual tensile strength to the actual yield strength is not less than 1.25.

For computing shear strength, the value of f_{yt} for transverse reinforcement, including spiral reinforcement, shall not exceed 60,000 psi (414 MPa).

21.2.5.2 – Prestressing steel shall be permitted in flexural members of frames, provided the average prestress, f_{pc}, calculated for an area equal to the member's shortest cross-sectional dimension multiplied by the perpendicular dimension shall be the lesser of 700 psi (4.83 MPa) or $f'_c/6$

at locations of nonlinear action where prestressing steel is used in members of frames.

❖ This ACI 318 modification is based on the 2003 NEHRP provisions. It renumbers ACI 318, Section 21.2.5, as 21.2.5.1, and adds what amounts to an exception in Section 21.2.5.2.

21.2.5.1: ACI 318 requires reinforcement resisting earthquake-induced flexural and axial forces in frame members and in shear wall boundary elements to comply with ASTM A 706. ASTM A 615, Grade 40 or 60 reinforcement is permitted in these members subject to two additional conditions. Under the code the use of prestressing steel to resist earthquake-induced flexural and axial forces in flexural members of frames is permitted subject to the conditions in Sections 21.2.5.2.

The modification to the ACI 318 limitation on the yield stress of transverse reinforcement clarifies that the limitation is only for the purpose of computing shear strength. Sections 9.4 and 10.9.3 of ACI 318 have increased the maximum specified yield strength of spiral reinforcement to allow up to 100 ksi (6 890 MPa). This had been limited to 60 ksi (414 MPa) previously. The ACI 318 limitation is based on concerns that high-strength spiral reinforcement might be less ductile than conventional mild reinforcement and that spiral failure has in fact been observed in earthquakes. It has been pointed out that spiral failures have typically been the result of insufficient spiral reinforcement, rather than a lack of ductility of the spiral reinforcement.

21.2.5.2: This ACI 318 modification allows prestressing steel in the flexural members of frames within the stated limitations. The significance of this provision is that it permits the use of prestressing steel in flexural members of frames that are designed using the design coefficients for a reinforced concrete frame, either intermediate or special. The limitations originated from research carried out by Ishizuka and Hawkins at the University of Washington, Seattle in the mid-1980s. F_{pc} is the average compressive stress in concrete due to effective prestress force only (after allowance for all prestress losses). Based on the noted research, the 1994 NEHRPs provisions introduced the use of prestressing steel in conjunction with deformed reinforcing bars in frames resisting earthquake forces, provided f_{pc} is kept below the greater of 350 psi (2.4 MPa and $f'_c/12$. When the 1997 *Uniform Building Code*® (UBC) added this provision, the limit on f_{pc} was worded more restrictively as the lesser of 350 psi (2.4 MPa) or $f'_c/12$. In the 1997 NEHRP provisions, the limit was increased to the lesser of 700 psi (4.8 MPa) or $f'_c/6$ as the IBC currently permits, based on newer research by Stanton, Stone and Cheok.

1908.1.6 ACI 318, Section 21.2. Modify ACI 318, Section 21.2, by adding new Section 21.2.9 to read as follows:

21.2.9 – Anchorages for unbonded post-tensioning tendons resisting earthquake induced forces in structures assigned to Seismic Design Category C, D, E or F shall withstand, without failure, 50 cycles of loading ranging between 40 and 85 percent of the specified tensile strength of the prestressing steel.

❖ This ACI 318 modification is based on the 2003 NEHRP provisions. The requirement to demonstrate the satisfactory seismic performance of tendon anchorages was first introduced in the 1994 NEHRP provisions and is now part of the code.

1908.1.7 ACI 318, Section 21.3. Modify ACI 318, Section 21.3, by adding new Section 21.3.2.5 to read as follows:

21.3.2.5 – Unless the special moment frame is qualified for use through structural testing as required by 21.6.3, for flexural members prestressing steel shall not provide more than one-quarter of the strength for either positive or negative moment at the critical section in a plastic hinge location and shall be anchored at or beyond the exterior face of a joint.

❖ This ACI 318 modification is based on the 2003 NEHRP provisions. The added restrictions on special moment frames in this section are based directly on research by Ishizuka and Hawkins discussed under Section 1908.1.5. If the limitations are exceeded, qualifying a moment frame system by structural testing that complies with the criteria of ACI T1.1 is another option allowed under Section 21.6.3 of ACI 318. This referenced acceptance criteria allows the use of reinforced concrete moment frames not satisfying Chapter 21 requirements if testing demonstrates performance that is consistent with the intent of the code. It is intended for structures having a high seismic risk and could be applied to frames utilizing precast elements, precast prestressed elements or post-tensioned reinforcement.

1908.1.8 ACI 318, Section 21.7. Modify ACI 318, Section 21.7, by adding new Section 21.7.10 to read as follows:

21.7.10 – Wall piers and wall segments.

21.7.10.1 – Wall piers not designed as a part of a special moment frame shall have transverse reinforcement designed to satisfy the requirements in 21.7.10.2.

Exceptions:

1. *Wall piers that satisfy 21.11.*

2. *Wall piers along a wall line within a story where other shear wall segments provide lateral support to the wall piers and such segments have a total stiffness of at least six times the sum of the stiffness of all the wall piers.*

21.7.10.2 – Transverse reinforcement with seismic hooks at both ends shall be designed to resist the shear forces determined from 21.4.5.1. Spacing of transverse reinforcement shall not exceed 6 inches (152 mm). Transverse reinforce-

ment shall be extended beyond the pier clear height for at least 12 inches (305 mm).

21.7.10.3 – Wall segments with a horizontal length-to-thickness ratio less than 2.5 shall be designed as columns.

❖ This ACI 318 modification is based on the 2003 NEHRP provisions. It adds Section 21.7.10 to the ACI 318 provisions for special reinforced concrete structural walls. As stated in Section 1908.1.4, this type of system is intended for structures assigned to Seismic Design Category D, E or F (see modified Section 21.2.1.4 of ACI 318).

21.7.10: This added section is titled "Wall Piers and Wall Segments." Wall piers are defined in Section 1908.1.3 and illustrated in Figure 1908.3. Wall segments that meet the definition of wall pier must be designed required by this section.

21.7.10.1: Wall pier detailing requirements are stated in Section 21.7.10.2, but three alternatives are provided. The first alternative is that the wall pier may be designed as part of a special moment frame. Two other alternatives are provided by the exceptions. The first exception applies to wall piers that are not part of the lateral-force-resisting system and that satisfy the requirement for deformation compatibility. Wall piers laterally supported by much stiffer shear walls along the same line within a story.

21.7.10.2: Wall pier requirements were first introduced in the 1988 UBC out of concern that thin column-like elements between openings in shear walls were being designed without proper transverse reinforcement. In order to prevent premature shear failure, a wall pier must be designed for the shear force computed in accordance with ACI 318 requirements for special moment frame members subjected to axial load and bending. Transverse reinforcement is required to be spaced at no more than 6 inches (152 mm), and is required to be extended beyond the clear pier height by at least 12 inches (305 mm).

21.7.10.3: This section clarifies that a segment of wall with a horizontal length-to-thickness ratio of less than $2^1/_2$ is considered a column and must be designed accordingly.

1908.1.9 ACI 318, Section 21.8. Modify Section 21.8.1 to read as follows:

21.8.1 – Special structural walls constructed using precast concrete shall satisfy all the requirements of 21.7 for cast-in-place special structural walls in addition to Sections 21.13.2 *through* 21.13.4.

❖ The modification made to ACI 318 in this section coordinates with the sections that are referenced—and are themselves modified in Section 1908.1.13.

1908.1.10 ACI 318, Section 21.10.1.1. Modify ACI 318, Section 21.10.1.1, to read as follows:

21.10.1.1 – Foundations resisting earthquake-induced forces or transferring earthquake-induced forces between a structure and the ground shall comply with the requirements of Section 21.10 and other applicable provisions of ACI 318 *unless modified by Chapter 18 of the International Building Code.*

❖ This ACI 318 modification points out that there are foundation requirements in the code that, in accordance with Section 102.4, would take precedence over those of ACI 318.

1908.1.11 ACI 318, Section 21.11. Modify ACI 318, Section 21.11.2.2 to read as follows:

21.11.2.2 – Members with factored gravity axial forces exceeding ($A_g f'_c$ /10) shall satisfy 21.4.3, 21.4.4.1(c), 21.4.4.3 and 21.4.5. The maximum longitudinal spacing of ties shall be s_o for the full column height. Spacing, s_o, shall not exceed the smaller of six diameters of the smallest longitudinal bar enclosed and 6 inches (152 mm). *Lap splices of longitudinal reinforcement in such members need not satisfy 21.4.3.2 in structures where the seismic-force-resisting system does not include special moment frames.*

❖ This ACI 318 modification is based on the 2003 NEHRP provisions. It modifies Section 21.11 requirements for frame members not proportioned to resist forces induced by earthquake motions in structures assigned to Seismic Design Category D, E or F.

ACI 318 requires members with factored gravity axial forces exceeding $A_g f'_c$ /10 to satisfy specific requirements for members of special moment frames, including ACI 318, Section 21.4.3. The code modification excludes the requirement of ACI 318, Section 21.4.3.2, for lap splices to be in the center half of the column where the seismic-force-resisting system does not include special reinforced concrete moment frames. The amendment regarding the location of lap splices is considered reasonable because gravity columns are required to sustain only their gravity load-carrying capacity under the design earthquake displacements, whereas columns forming part of the seismic-force-resisting system are required to sustain full gravity, as well as lateral loads under the same displacements. While a requirement to keep lap splices away from regions of potential plastic hinging near the ends of the column may be justified in the latter case, it is not necessary for a gravity column.

1908.1.12 ACI 318, Section 21.12.5. Modify ACI 318, Section 21.12.5, by adding new Section 21.12.5.6 to read as follows:

21.12.5.6 – *Columns supporting reactions from discontinuous stiff members, such as walls, shall be designed for the special load combinations in Section 1605.4 of the International Building Code and shall be provided with transverse reinforcement at the spacing, s_o, as defined in 21.12.5.2 over their full height beneath the level at which the discontinuity occurs. This transverse reinforcement shall be extended above and below the column as required in 21.4.4.5.*

❖ This section adds a provision to the ACI 318 requirements under columns of intermediate moment frames. Because of a desire on the part of building owners for an open ground story (for stores, ballrooms in hotels and so forth), shear walls are sometimes discontinued

above that story. Columns supporting discontinued shear walls are particularly vulnerable to damage, as has been observed in past earthquakes. Thus, there are two safeguards concerning these elements built into the code—one related to strength and the other related to inelastic deformability or detailing. The strength of columns supporting discontinued stiff members must be augmented by designing them using the special load combinations of Section 1605.4, with E_m defined in Section 1602. Additionally, these columns must have transverse reinforcement not only within the regions of potential plastic hinging at the ends, but also throughout their height. The transverse reinforcement must continue above and below the column in accordance with the requirements for special moment frame members.

1908.1.13 ACI 318, Section 21.13. Modify ACI 318, Section 21.13, by renumbering Section 21.13.3 to become 21.13.4 and adding new Sections 21.13.3, 21.13.5 and 21.13.6 to read as follows:

21.13.3 – Except for Type 2 mechanical splices, connection elements that are designed to yield shall be capable of maintaining 80 percent of their design strength at the deformation induced by the design displacement.

21.13.4 – Elements of the connection that are not designed to yield shall develop at least 1.5 S_y.

21.13.5 – Wall piers not designed as part of a moment frame shall have transverse reinforcement designed to resist the shear forces determined from 21.12.3. Spacing of transverse reinforcement shall not exceed 8 inches (203 mm). Transverse reinforcement shall be extended beyond the pier clear height for at least 12 inches (305 mm).

Exceptions:

1. *Wall piers that satisfy 21.11.*

2. *Wall piers along a wall line within a story where other shear wall segments provide lateral support to the wall piers and such segments have a total stiffness of at least six times the sum of the stiffnesses of all the wall piers.*

21.13.6 – Wall segments with a horizontal length-to-thickness ratio less than 2.5 shall be designed as columns.

❖ This modification to ACI 318 affects connections between precast concrete wall panels in intermediate precast structural walls. ACI 318 Section 21.13.2 restricts yielding to steel elements or reinforcement in connections between wall panels, or between wall panels and the foundation of intermediate precast structural walls. This restriction is also made applicable by Section 21.8.1 to special precast structural walls as well – see Section 1908.1.9.

21.13.3: This ACI 318 modification is consistent with a clarification of ductile elements that is made in the 2003 NEHRP provisions. Connection elements that are designed to yield must be capable of maintaining 80 percent of their design strength up to the stage where a structure is subjected to the design displacement as defined in ACI 318 (see Section 1908.1.3).

21.13.5: Provisions for wall piers are added for intermediate structural walls that parallel those for special structural walls in Section 1908.1.8.

1908.1.14 ACI 318, Section 22.6. Modify ACI 318, Section 22.6, by adding new Section 22.6.7 to read:

22.6.7 – Detailed plain concrete structural walls.

22.6.7.1 – Detailed plain concrete structural walls are walls conforming to the requirements of ordinary structural plain concrete walls and 22.6.7.2.

22.6.7.2 - Reinforcement shall be provided as follows:

(a) Vertical reinforcement of at least 0.20 square inch (129 mm²) in cross-sectional area shall be provided continuously from support to support at each corner, at each side of each opening and at the ends of walls. The continuous vertical bar required beside an opening is permitted to substitute for one of the two No. 5 bars required by 22.6.6.5.

(b) Horizontal reinforcement at least 0.20 square inch (129 mm²) in cross-sectional area shall be provided:

1. Continuously at structurally connected roof and floor levels and at the top of walls;

2. At the bottom of load-bearing walls or in the top of foundations where doweled to the wall; and

3. At a maximum spacing of 120 inches (3048 mm).

Reinforcement at the top and bottom of openings, where used in determining the maximum spacing specified in Item 3 above, shall be continuous in the wall.

❖ This ACI 318 modification provides requirements for detailed plain concrete structural walls that are based on the 2003 NEHRP provisions (see definition in Section 1908.1.3). In addition to conforming to the requirements for ordinary plain concrete shear walls, detailed plain concrete shear walls must have some reinforcement to provide limited ductility and to keep the concrete from disintegrating should cracking occur before or during a seismic event. Detailed plain concrete shear walls may only be used in structures assigned to Seismic Design Category A or B. The area of vertical reinforcement of 0.20 square inch (129 mm²) corresponds to the area of a No. 4 reinforcing bar. The provision requiring vertical reinforcement at the "ends of walls" is intended to also require a bar on each side of a contraction (i.e., control) joint.

Item 1 requires that a bar be located at the top of a wall. Where the provisions of Item 1 require a closer spacing of horizontal reinforcement than Item 3 [i.e., 120 inches or 10 feet (3048 mm)], then the closer spacing applies.

1908.1.15 ACI 318, Section 22.10. Delete ACI 318, Section 22.10, and replace with the following:

22.10 – Plain concrete in structures assigned to Seismic Design Category C, D, E or F.

22.10.1 – Structures assigned to Seismic Design Category C, D, E or F shall not have elements of structural plain concrete, except as follows:

(a) Structural plain concrete basement, foundation or other walls below the base are permitted in detached one- and two-family dwellings three stories or less in height constructed with stud-bearing walls. In dwellings assigned to Seismic Design Category D or E, the height of the wall shall not exceed 8 feet (2438 mm), the thickness shall not be less than 7¹/₂ inches (190 mm), and the wall shall retain no more than 4 feet (1219 mm) of unbalanced fill. Walls shall have reinforcement in accordance with 22.6.6.5.

(b) Isolated footings of plain concrete supporting pedestals or columns are permitted, provided the projection of the footing beyond the face of the supported member does not exceed the footing thickness.

> *Exception: In detached one- and two-family dwellings three stories or less in height, the projection of the footing beyond the face of the supported member is permitted to exceed the footing thickness.*

(c) Plain concrete footings supporting walls are permitted, provided the footings have at least two continuous longitudinal reinforcing bars. Bars shall not be smaller than No. 4 and shall have a total area of not less than 0.002 times the gross cross-sectional area of the footing. For footings that exceed 8 inches (203 mm) in thickness, a minimum of one bar shall be provided at the top and bottom of the footing. Continuity of reinforcement shall be provided at corners and intersections.

> *Exceptions:*
>
> *1. In detached one- and two-family dwellings three stories or less in height and constructed with stud-bearing walls, plain concrete footings without longitudinal reinforcement supporting walls are permitted.*
>
> *2. For foundation systems consisting of a plain concrete footing and a plain concrete stemwall, a minimum of one bar shall be provided at the top of the stemwall and at the bottom of the footing.*
>
> *3. Where a slab on ground is cast monolithically with the footing, one No. 5 bar is permitted to be located at either the top of the slab or bottom of the footing.*

❖ Structural plain concrete walls are prohibited in structures assigned to Seismic Design Category C due to their inability to respond inelastically, which is considered to be essential if they are to perform satisfactory during the design earthquake. Item (a) permits this in one- and two-family dwellings where the stated conditions are met.

Plain concrete footings supporting columns or pedestals are permitted if the projection of the footing beyond the face of the column or pedestal does not exceed the footing thickness. The limitation on the projection was judged to be prudent because of the empirical manner in which the vertical component of earthquake ground motion is considered. With the limitation on the projection, a flexural or shear failure in the footing is virtually impossible.

Wall footings of structural plain concrete are permitted in all structures assigned to Seismic Design Category C. Some longitudinal reinforcement is specified in order to keep the footing together should it crack before or during an earthquake. The reinforcement will also allow the footing to act as a "reinforced" concrete member should the earthquake-induced ground motion cause the supported structure to be subjected to differential vertical movement. For footings 8 inches (203 mm) or less in thickness, the bars will typically be located with 3 inches (76 mm) of concrete cover to the bottom as required by Section 1907.7.1. For footings thicker than 8 inches (203 mm), one bar will be located with respect to the bottom as noted previously, and the other with $1^1/_2$ or 2 inches (38 or 51 mm) of cover from the top of the footing as required under Section 1907.7.1. Typically, two No. 4 [total of 0.40 square inches (258 mm^2)] or two No. 5 [total of 0.62 square inches (400 mm^2)] bars will suffice, with the former being adequate for footings up to 200 square inches (0.129 m^2) in cross-sectional area [0.40 in.^{2}/0.002 = 200 in.^{2} (258 mm^2/.002 × 10^6 = 0.129 m^2)], and the latter being suitable for footings up to 310 square inches (0.200 m^2) in cross-sectional area [0.62 in.^{2}/0.002 = 310 in.^{2} (400 mm^2/.002 × 10^6 = 0.200 m^2)]. Where one bar will be placed in the bottom of the footing and the other will be placed in the top of the stem wall, the concrete surface of the top of the footing should be intentionally roughened so that a good bond develops between the two concrete placements.

Exception 1 permits structural plain concrete footings without longitudinal reinforcement in certain one- and two-family dwellings. Where a slab-on-ground is cast monolithically with the footing, typically referred to as a "turned-down slab," Exception 3 allows one No. 5 bar (versus two No. 4 bars) to be located in either the top or bottom of the footing. In this case, the top of the footing is the top of the slab.

1908.1.16 ACI 318, Section D.3.3. Modify ACI 318, Sections D.3.3.2 through D.3.3.5, to read as follows:

D.3.3.2 – *In structures assigned to Seismic Design Category C, D, E or F,* post-installed anchors for use under D.2.3 shall have passed the Simulated Seismic Tests of ACI 355.2.

D.3.3.3 – *In structures assigned to Seismic Design Category C, D, E or F,* the design strength of anchors shall be taken as $0.75\varphi N_n$ and $0.75\varphi V_n$, where φ is given in D.4.4 or D.4.5, and N_n and V_n are determined in accordance with D.4.1.

D.3.3.4 – *In structures assigned to Seismic Design Category C, D, E or F,* anchors shall be designed to be governed by tensile or shear strength of a ductile steel element, unless D.3.3.5 is satisfied.

D.3.3.5 – Instead of D.3.3.4, the attachment that the anchor is connecting to the structure shall be designed so that the attachment will undergo ductile yielding at a load level corresponding to anchor forces no greater than the design strength of anchors specified in D.3.3.3, *or the minimum design strength of the anchors shall be at least 2.5 times the factored forces transmitted by the attachment.*

❖ This modification to Appendix D of ACI 318 is based on the 2003 NEHRP provisions. References to seismic risk have been changed to the corresponding seismic design category for consistency with the earthquake load requirements of Chapter 16.

SECTION 1909
STRUCTURAL PLAIN CONCRETE

1909.1 Scope. The design and construction of structural plain concrete, both cast-in-place and precast, shall comply with the minimum requirements of Section 1909 and ACI 318, Chapter 22, as modified in Section 1908.

❖ Section 1909 covers the minimum requirements for structural members regulated by the code and constructed of plain concrete. By definition, plain concrete contains no reinforcement, or has less reinforcement than necessary to meet the requirements for reinforced concrete in ACI 318. In addition to the provisions of Section 1909, there are other provisions in Chapter 22 of ACI 318 that must be met for plain concrete used for structural purposes.

1909.1.1 Special structures. For special structures, such as arches, underground utility structures, gravity walls and shielding walls, the provisions of this section shall govern where applicable.

❖ Special structures regulated by the code, such as an underground utility vault with plain concrete walls, plain concrete retaining walls designed to resist overturning by gravity, plain concrete walls constructed for shielding purposes and plain concrete walls used in conjunction with outdoor handball courts are also required to meet the applicable requirements of this section.

1909.2 Limitations. The use of structural plain concrete shall be limited to:

1. Members that are continuously supported by soil, such as walls and footings, or by other structural members capable of providing continuous vertical support.

2. Members for which arch action provides compression under all conditions of loading.

3. Walls and pedestals.

The use of structural plain concrete columns and structural plain concrete footings on piles is not permitted. See Section 1908.1.15 for additional limitations on the use of structural plain concrete.

❖ Since the structural integrity of plain concrete members depends solely on the properties of the concrete, their use is limited to members that are generally under compression. In addition, such members must be able to tolerate random cracking. For these reasons, such members are limited to uses where (1) they are continuously supported by soil, such as footings, or by other structural members capable of providing continuous vertical support; (2) arch action provides compression under all conditions of loading or (3) as walls and pedestals. Examples of structural members permitted to be of plain concrete include footings, foundation and basement walls, above-grade walls, arches over openings in a concrete wall in lieu of a reinforced lintel and pedestals. Pedestals are frequently used in crawl spaces or under decks to support floor framing, and when used as such they are usually referred to as piers. A "pedestal" must be proportioned so the height does not exceed three times the least lateral dimension. If the height-to-least-lateral dimension exceeds three, the member is defined as a "column" and must be designed as a reinforced member according to ACI 318.

Although pile caps (i.e., footings supported on piles) are generally cast on the soil, they are designed to be supported by the piles, much like a structural floor or roof slab is supported by beams that are in turn supported by columns. For this reason, footings supported by piles are not permitted to be of plain concrete.

It is generally desirable that structural plain concrete members be under compression at all times; however, this is not always possible, especially in the case of walls of plain concrete. When a wall is subjected to lateral loads from wind or soil, or both, a portion of the wall cross section may be under net tension. The tension is a result of lateral loads causing flexural tension exceeding the compressive stress from the axial load on the wall, if any.

A common type of construction in areas of the country where expansive soils are present is drilled pier foundations supporting reinforced concrete grade beams, which in turn support concrete basement walls. Under this scenario, the use of a structural plain concrete basement wall is permitted, provided the wall is proportioned to resist the axial and lateral loads to which it will be subjected.

Structural plain concrete members are generally restricted to buildings classified as Seismic Design Category A or B. In some cases, Section 1908.1.15 allows plain concrete members in higher seismic design categories if some reinforcement is provided to give the members some ductility and hold the concrete together should cracking occur during a seismic event. The presence of this nominal amount of reinforcement does not change the classification of a member from plain to reinforced concrete.

1909.3 Joints. Contraction or isolation joints shall be provided to divide structural plain concrete members into flexurally discontinuous elements in accordance with ACI 318, Section 22.3.

❖ Generally, considerably more water is added to a concrete mix than is necessary to react with the cement. The additional water that is added for workability must eventually evaporate and will continue to do so until the moisture content of the hardened concrete reaches equilibrium with the surrounding environment. As the concrete dries, it tries to shrink. If the member, such as a slab or basement wall, is restrained so it cannot shrink, tensile stresses develop in the concrete. If the tensile strength of the concrete is exceeded, the slab or wall will crack. Temperature changes and creep have the same effect as moisture changes.

Section 22.3 provides two alternatives for addressing shrinkage of concrete. In the first case, the concrete element can be isolated from other elements that may restrain it by using isolation joint material, typically referred to as "expansion joint material." This is not always feasible; therefore, a second alternative allows the element to be subdivided into flexurally discontinuous smaller elements with contraction joints, often called "control joints." A contraction joint is a weakened cross section of the concrete member where tensile stresses in the restrained concrete will be concentrated as it tries to shrink. Because the concrete is intentionally weakened at the contraction joint, the crack will occur at that location. The contraction joint can be formed while the concrete is still plastic, such as with a groover, or in the case of a slab, the hardened concrete can be sawed before it has dried sufficiently enough to crack. In the case of a wall, the joint is typically created by installing strips on the inside surfaces of the wall forms. To be effective, the joint should be at least one-quarter the slab or wall thickness. This means for walls, each strip located on opposing form surfaces should be at least one-eighth the wall thickness. To be effective, joints need to be located where the tensile stress will be greatest; typically at offsets in slabs and at openings in walls.

1909.4 Design. Structural plain concrete walls, footings and pedestals shall be designed for adequate strength in accordance with ACI 318, Sections 22.4 through 22.8.

> **Exception:** For Group R-3 occupancies and buildings of other occupancies less than two stories in height of light-frame construction, the required edge thickness of ACI 318 is permitted to be reduced to 6 inches (152 mm), provided that the footing does not extend more than 4 inches (102 mm) on either side of the supported wall.

❖ Sections 22.4 through 22.8 of ACI 318 contain design procedures for elements permitted to be of structural plain concrete, including walls, footings and pedestals. The procedures do not allow credit to be given for any reinforcement that may be present in the member.

This means that if the required strength of a member exceeds its design strength when calculated without benefit of any reinforcement present, it must be designed as reinforced concrete.

ACI 318 does not permit columns of plain concrete; however, plain concrete pedestals are permitted, provided the ratio of unsupported height-to-average-least-lateral dimension does not exceed three. For example, if the unsupported height of a 10-inch (254 mm) square pier is 28 inches (711 mm), the ratio is 2.8 [28 inches ÷ 10 inches = 2.8 (711 mm ÷ 254 mm = 2.8)]; therefore, the pier qualifies as a "pedestal."

The exception applies to Section 22.7.4 of ACI 318 which requires the thickness of a structural plain concrete footing to be a minimum of 8 inches (203 mm). This exception permits the thickness of the footing to be reduced to 6 inches (152 mm). The reduced footing thickness applies to any Group R-3 occupancy and to buildings of any other occupancy that are one story in height and of light-frame construction. An additional limitation is that the footing cannot extend more than 4 inches (102 mm) beyond the face of the supported wall. The 4-inch (102 mm) limitation is to reduce the likelihood that the footing will experience a flexural failure at the junction of the footing and the wall supported above.

1909.5 Precast members. The design, fabrication, transportation and erection of precast, structural plain concrete elements shall be in accordance with ACI 318, Section 22.9.

❖ ACI 318 requires that the design of precast members must include all loading conditions that the member will be subjected to from fabrication to in-place use in the completed structure. The load considerations must include form removal, storage, transportation and erection at the job site. The end use of precast members is subject to the same limitations as previously cited in the commentary to Section 1909.2.

1909.6 Walls. In addition to the requirements of this section, structural plain concrete walls shall comply with the applicable requirements of ACI 318, Chapter 22.

❖ Section 22.6 of ACI 318 has two design procedures for structural plain concrete walls. For walls where the resultant of all factored loads (axial and lateral) falls in the middle one-third of the wall, an empirical procedure that considers only the axial load is permitted. Where the resultant of all factored loads falls outside the middle one-third of the wall, a rational procedure specified in ACI 318 that considers both axial and lateral loads must be used. Regardless of which one of the two design approaches is used, the provisions of Section 1909.6.1 through 1909.6.3 apply. These provisions only apply to walls that are laterally supported at the top and bottom with elements, such as slabs on ground and structural floors/roofs. Walls designed in accordance with these provisions are intended for use in low-rise buildings assigned to Seismic Design Category A or B Chapter 32 of PCA-EB070 contains numerous aids to facilitate the design of structural plain concrete walls and footings.

1909.6.1 Basement walls. The thickness of exterior basement walls and foundation walls shall be not less than $7^1/_2$ inches (191 mm). Structural plain concrete exterior basement walls shall be exempt from the requirements for special exposure conditions of Section 1904.2.2.

❖ This section requires exterior basement and foundation walls to be a minimum of $7^1/_2$ inches (191 mm) in thickness. Though the code states exterior walls, the intent of the section is for all basement and foundation walls retaining earth to meet the $7^1/_2$ inch (191 mm) minimum dimension. Such a situation may occur in a building with a partial basement. The provisions also exempt basement walls from the special exposure conditions described in Section 1904.2.2.

1909.6.2 Other walls. Except as provided for in Section 1909.6.1, the thickness of bearing walls shall be not less than $1/_{24}$ the unsupported height or length, whichever is shorter, but not less than $5^1/_2$ inches (140 mm).

❖ This section applies to interior and exterior bearing walls above grade, and to interior bearing walls in a basement. It requires the thickness to be a minimum of $5^1/_2$ inches (140 mm), but in no case less than $1/_{24}$ of the unsupported height or length of the wall, whichever is less. For example, a concrete wall with a height of 8 feet (2438 mm) is required to be a minimum of $5^1/_2$ inches (140 mm), since $1/_{24}$ of the height of the wall is less than $5^1/_2$ inches [8 feet × 12 inches/feet × $1/_{24}$ = 4 inches (2438 mm × $1/_{24}$ = 102 mm)]. For a 12-foot-high (3658 mm) wall, the minimum thickness required is 6 inches [12 feet × 12 inches/feet × $1/_{24}$ = 6 inches (3658 mm × $1/_{24}$ = 152 mm)].

1909.6.3 Openings in walls. Not less than two No. 5 bars shall be provided around window and door openings. Such bars shall extend at least 24 inches (610 mm) beyond the corners of openings.

❖ All openings for windows and doors in structural plain concrete walls must be bound by two No. 5 (16) bars as follows. For openings not extending to the bottom of the wall, the bars must be placed on all four sides of the opening. For openings extending to the bottom of the wall, the bars are only necessary along the sides and top. To provide continuity at the corners of the openings, the bars must extend 24 inches (610 mm) past the corners. These extensions should occur at all four corners of an opening not extending to the bottom of the wall, and at the two top corners for other openings.

SECTION 1910
MINIMUM SLAB PROVISIONS

1910.1 General. The thickness of concrete floor slabs supported directly on the ground shall not be less than $3^1/_2$ inches (89 mm). A 6-mil (0.006 inch; 0.15 mm) polyethylene vapor retarder with joints lapped not less than 6 inches (152 mm) shall be placed between the base course or subgrade and the concrete floor slab, or other approved equivalent methods or materials shall be used to retard vapor transmission through the floor slab.

Exception: A vapor retarder is not required:

1. For detached structures accessory to occupancies in Group R-3, such as garages, utility buildings or other unheated facilities.

2. For unheated storage rooms having an area of less than 70 square feet (6.5 m²) and carports attached to occupancies in Group R-3.

3. For buildings of other occupancies where migration of moisture through the slab from below will not be detrimental to the intended occupancy of the building.

4. For driveways, walks, patios and other flatwork which will not be enclosed at a later date.

5. Where approved based on local site conditions.

❖ Section 1910 requires that slabs on grade be a minimum thickness so superimposed loads are transmitted to the subgrade without causing structural distress of the slab. Additionally, slabs must be protected on the underside with material that will retard the transmission of water vapor through the floor slab into the enclosed space so moisture levels are not excessive.

The coarse aggregate most commonly used in concrete slabs supported directly on the ground is allowed up to 5 percent (by weight) of the stone larger than 1 inch (25 mm) in size, with all coarse aggregates in this grading classification passing a $1^1/_2$-inch (38 mm) sieve. Section 3.3.2 of ACI 318 requires that the nominal maximum size of coarse aggregate be no larger than one-third the depth of the slab, thus making it necessary that slabs be a minimum thickness of $3^1/_2$ inches (89 mm) to accommodate coarse aggregate stones between 1 and $1^1/_2$ inches (25 and 38 mm) in size. Also, as a matter of experience, a $3^1/_2$-inch (89 mm) minimum slab thickness is needed to support typical concentrated loads without damage to the concrete, particularly at locations where small voids or pockets in the subgrade may develop because of improper compaction during construction or from other causes.

An approved 6-mil vapor retarder (see Section 202 for definition) placed between the concrete slab and the supporting ground is required to prevent the migration of water vapor through the slab into the space above. The five exceptions acknowledge conditions where the migration of limited moisture through the slab will not adversely affect the occupancy of the structure. Exception 3 is intended to give relief to occupancies in which additional moisture will not be detrimental, such as indoor swimming pools, industrial processes generating large amounts of moisture and certain storage facilities. Exception 5 is intended to give relief based on environmental or geological conditions, such as locations either with arid climates where the relative humidity is typically low or with well-drained soils and low water tables. Vapor retarders, while preventing moisture from migrating through

the slab from below, can adversely affect the construction and performance of concrete slabs on grade. Finishing operations may be delayed since it will normally take longer for the bleed water to evaporate from the surface. Additionally, the tendency for concrete to develop both plastic and drying shrinkage cracking may be exacerbated. Another problem that occurs is the tendency for the slab to curl or warp upward at the edges. Curling occurs because the bottom of the slab remains moist longer than the top portion due to the vapor retarder; therefore, the top shortens more than the bottom, causing curling. If none of the exceptions apply and the vapor retarder must be provided, one solution to the above problems is to place a 2- or 3-inch (50 or 76 mm) layer of sand on top of the vapor retarder prior to concrete placement. The sand will absorb some of the excess water from the concrete and allow earlier finishing. This will also partially or wholly mitigate some of the other adverse effects. If a layer of sand is used, it must be compacted by wetting the day before the concrete is to be placed. However, if the sand is to function effectively as a "blotter," it must be free of drainable water at the time of concrete placement.

SECTION 1911
ANCHORAGE TO CONCRETE—
ALLOWABLE STRESS DESIGN

1911.1 Scope. The provisions of this section shall govern the allowable stress design of headed bolts and headed stud anchors cast in normal-weight concrete for purposes of transmitting structural loads from one connected element to the other. These provisions do not apply to anchors installed in hardened concrete or where load combinations include earthquake loads or effects. The bearing area of headed anchors shall be not less than one and one-half times the shank area. Where strength design is used, or where load combinations include earthquake loads or effects, the design strength of anchors shall be determined in accordance with Section 1912. Bolts shall conform to ASTM A 307 or an approved equivalent.

❖ Section 1911 contains requirements for design of anchorage to concrete by the ASD method, also called "working stress design." Provisions for strength design of anchorage to concrete are given in Section 1912. This section must not be used for load combinations that include seismic forces or their effects. In those cases, the provisions of Section 1912 are to be used.

The code provides two methods of design for anchors cast in concrete. Section 1911 provides for limited use of the ASD method and Section 1912 provides an expanded approach for the strength design method. The ASD provisions of this section are adapted from the 1997 edition of the UBC. They are based on limited test data of headed bolts cast in normal-weight concrete and subjected to static loading. Therefore, its application may not be used for hooked (J- or L-) bolts, lightweight concrete, post-installed anchors (i.e., anchors installed in hardened concrete) or when load combinations include earthquake loads.

1911.2 Allowable service load. The allowable service load for headed anchors in shear or tension shall be as indicated in Table 1911.2. Where anchors are subject to combined shear and tension, the following relationship shall be satisfied:

$$(P_s / P_t)^{5/3} + (V_s / V_t)^{5/3} \leq 1 \qquad \textbf{(Equation 19-1)}$$

where:

P_s = Applied tension service load, pounds (N).

P_t = Allowable tension service load from Table 1911.2, pounds (N).

V_s = Applied shear service load, pounds (N).

V_t = Allowable shear service load from Table 1911.2, pounds (N).

❖ Table 1911.2 gives allowable service loads for tension or shear, but not at the same time. In real-world applications, both forces are frequently acting at the same time. This section provides a formula for determining the maximum values that may be assigned to bolts that are subjected to combinations of shear and tension loading. The term "allowable service load" signifies that unfactored shear and tension loads are to be used. They may include live, dead, wind and other loading conditions, except for earthquake loads.

TABLE 1911.2. See below.

❖ Table 1911.2 provides shear and tension loads that may be assigned to headed bolts or stud anchors that are cast in normal-weight concrete. Note that the table assigns values for either shear or tension. Both values

cannot be assigned to the same bolt at the same time. The table values are based on a particular combination of bolt size [from $^1/_4$ to $1^1/_4$ inches (6.4 to 32 mm)] and concrete strength [i.e., 2,500, 3,000 and 4,000 psi (17, 21 and 28 MPa)]. When using the table values, minimum bolt embedments, edge distances and spacing between bolts must be observed. Edge distances are measured from the face of the bolt to the sides and ends of concrete members. Decreases up to 50 percent in edge distance or spacing or both are provided for in Section 1911.3.

1911.3 Required edge distance and spacing. The allowable service loads in tension and shear specified in Table 1911.2 are for the edge distance and spacing specified. The edge distance and spacing are permitted to be reduced to 50 percent of the values specified with an equal reduction in allowable service load. Where edge distance and spacing are reduced less than 50 percent, the allowable service load shall be determined by linear interpolation.

❖ This section expands the applicability of Table 1911.2 when bolt locations cannot meet the required minimum edge distances or spacing or both specified by the table. Tabulated minimum edge distances and spacings may be reduced up to 50 percent for a corresponding 50-percent reduction of service loads. For reductions in edge distance or spacing or both less than 50 percent, corresponding service loads may be determined by interpolation. For example, if $^5/_8$-inch (15.9 mm) bolts are embedded $4^1/_2$ inches (114 mm) in 2,500 psi (17 MPa) concrete, the table assigns shear and tension values of 2,750 and 1,500 pounds (12 100 and 6600 N), respectively, if at least $3^3/_4$-inch (95 mm) edge distance and $7^1/_2$-inch (191 mm) spacing be-

TABLE 1911.2
ALLOWABLE SERVICE LOAD ON EMBEDDED BOLTS (pounds)

BOLT DIAMETER (inches)	MINIMUM EMBEDMENT (inches)	EDGE DISTANCE (inches)	SPACING (inches)	MINIMUM CONCRETE STRENGTH (psi)					
				$f'_c = 2,500$		$f'_c = 3,000$		$f'_c = 4,000$	
				Tension	Shear	Tension	Shear	Tension	Shear
$^1/_4$	$2^1/_2$	$1^1/_2$	3	200	500	200	500	200	500
$^3/_8$	3	$2^1/_4$	$4^1/_2$	500	1,100	500	1,100	500	1,100
$^1/_2$	4	3	6	950	1,250	950	1,250	950	1,250
	4	5	5	1,450	1,600	1,500	1,650	1,550	1,750
$^5/_8$	$4^1/_2$	$3^3/_4$	$7^1/_2$	1,500	2,750	1,500	2,750	1,500	2,750
	$4^1/_2$	$6^1/_4$	$7^1/_2$	2,125	2,950	2,200	3,000	2,400	3,050
$^3/_4$	5	$4^1/_2$	9	2,250	3,250	2,250	3,560	2,250	3,560
	5	$7^1/_2$	9	2,825	4,275	2,950	4,300	3,200	4,400
$^7/_8$	6	$5^1/_4$	$10^1/_2$	2,550	3,700	2,550	4,050	2,550	4,050
1	7	6	12	3,050	4,125	3,250	4,500	3,650	5,300
$1^1/_8$	8	$6^3/_4$	$13^1/_2$	3,400	4,750	3,400	4,750	3,400	4,750
$1^1/_4$	9	$7^1/_2$	15	4,000	5,800	4,000	5,800	4,000	5,800

For SI: 1 inch = 25.4 mm, 1 pound per square inch = 0.00689 MPa, 1 pound = 4.45 N.

tween bolts are provided. If only a 25-percent reduction in edge distance or spacing or both is desired, the corresponding shear and tension values (determined by linear interpolation) are 2,063 and 1,125 pounds (9077 and 4950 N), respectively.

1911.4 Increase in allowable load. Increase of the values in Table 1911.2 by one-third is permitted where the provisions of Section 1605.3.2 permit an increase in allowable stress for wind loading.

❖ This section follows a time-honored principle of permitting a one-third increase in allowable stress for wind. The rationale for this increase is that design wind loads occur infrequently and their maximum effects last only a few seconds. Therefore, an overstress of one-third for a very short time at infrequent intervals is considered acceptable.

1911.5 Increase for special inspection. Where special inspection is provided for the installation of anchors, a 100-percent increase in the allowable tension values of Table 1911.2 is permitted. No increase in shear value is permitted.

❖ The tension and shear values in Table 1911.2 are based on load test data that has been reduced by a safety factor. The code acknowledges that special inspection improves construction practices sufficiently to allow a 100-percent increase in the tension values provided in the table, thus allowing a lower safety factor when special inspection is provided.

The shear values of the table are not permitted to be increased when special inspection is provided, because the table values reflect a safety factor that is not overly conservative. The provisions of this section were taken from the footnotes of Table 19D (Allowable Service Load on Embedded Bolts) of the 1997 UBC. The footnotes first appeared in the 1976 UBC and were accompanied by a 50-percent increase in the table's shear values, but no increase in the table's tensile values. The footnotes also enabled the tensile values to be increased 50 percent, but only when special inspection is provided.

SECTION 1912
ANCHORAGE TO CONCRETE—
STRENGTH DESIGN

1912.1 Scope. The provisions of this section shall govern the strength design of anchors installed in concrete for purposes of transmitting structural loads from one connected element to the other. Headed bolts, headed studs and hooked (J- or L-) bolts cast in concrete and expansion anchors and undercut anchors installed in hardened concrete shall be designed in accordance with Appendix D of ACI 318 as modified by Section 1908.1.16, provided they are within the scope of Appendix D.

Exception: Where the basic concrete breakout strength in tension of a single anchor, N_b, is determined in accordance with Equation (D-7), the concrete breakout strength requirements of Section D.4.2.2 shall be considered satisfied by the design procedures of Sections D.5.2 and D.6.2

for anchors exceeding 2 inches (51 mm) in diameter or 25 inches (635 mm) tensile embedment depth.

The strength design of anchors that are not within the scope of Appendix D of ACI 318, and as amended above, shall be in accordance with an approved procedure.

❖ Section 1912 is restricted in scope to anchors that transmit structural loads from attachments into concrete members and vice versa under typical in-service conditions. Other standards, such as those applicable to construction and material handling, for instance, can require more stringent safety levels and must be followed where they apply.

The first edition of the code included provisions for the strength design of anchorage to concrete. At that time, the provisions that were being developed for inclusion in ACI 318 were not yet available for reference by the code. With the publication of the 2002 edition of ACI 318, requirements for the strength design of concrete anchorage were included in Appendix D. The code provisions were then replaced with a reference to Appendix D of ACI 318, since it was the most up-to-date version of the same requirements. Based on the content of the newer requirements in Appendix D, the scope of this section also includes post-installed anchors, such as expansion anchors and undercut anchors, provided they are evaluated and categorized in accordance with ACI 355.2. Design of anchors that are not within the scope of this section must be in accordance with an approved procedure. Examples of anchors that fall into the latter category are specialty inserts, through-bolts, adhesive anchors, grouted anchors and powder-actuated fasteners.

Section D.2.4 of the referenced ACI 318 provisions excludes load applications producing high-cycle fatigue or extremely short duration impact (such as blast or shock wave). This statement does not exclude seismic loads, and, in fact, Section 1911 for design of anchorage to concrete by the ASD method requires the use of strength design for load combinations that include seismic forces or their effects.

Section D.4.2.2 of ACI 318 allows determination of the concrete breakout strength in both tension and shear to be determined according to the referenced sections that are based on the concrete capacity design (CCD) method. Consistent with the range of test data that was available, this section restricts the use of the option to compute concrete breakout strength of the anchorage based on an anchor's diameter or embedment length. Instead, it requires testing of anchors exceeding 2 inches (51 mm) in diameter or with a depth of embedment greater than 25 inches (635 mm) in order to determine concrete breakout strength. The exception is an important modification to this anchor size and embedment limit that is based on subsequent analysis of test data. It indicates that if Equation D-7 is used to establish the concrete breakout strength of a single anchor, then these limits on size and embedment depth do not apply.

SECTION 1913
SHOTCRETE

1913.1 General. Shotcrete is mortar or concrete that is pneumatically projected at high velocity onto a surface. Except as specified in this section, shotcrete shall conform to the requirements of this chapter for plain or reinforced concrete.

❖ Section 1913 regulates the materials and test procedures for shotcrete construction. Shotcrete is the commonly accepted and generic name for pneumatically projected mortar or concrete. Terms, such as gunite, sprayed concrete, spraycrete, pneumatically applied mortar and concrete are often used to refer to shotcrete.

ACI 506R, the guide for shotcreting, defines "Shotcrete" as mortar or concrete conveyed through a hose and pneumatically projected at high velocity onto a surface. The force of the jet impacting on the shotcrete surface compacts the material, and when applied properly, the shotcrete surface will support itself without sagging or sloughing, even for vertical and overhead applications. Shotcrete must conform to the requirements for structural concrete specified in this chapter and ACI 318 unless modified by Section 1913.

1913.2 Proportions and materials. Shotcrete proportions shall be selected that allow suitable placement procedures using the delivery equipment selected and shall result in finished in-place hardened shotcrete meeting the strength requirements of this code.

❖ Although the compressive strength of shotcrete must be verified prior to beginning work, its resulting strength and acceptability relies more on the suitability of placement procedures and the delivery equipment. Prior to construction, strength is established by the submittal of a mix design that is supported by test results. Mix designs from previous jobs can be used. Product data sheets for any admixtures, and graduations for the sand and aggregate, need to be included in the submittal. Verification of in-place hardened shotcrete strength is determined by testing in accordance with Section 1913.10.

1913.3 Aggregate. Coarse aggregate, if used, shall not exceed $^3/_4$ inch (19.1 mm).

❖ For construction sections that are several inches thick, and where adequate gunning equipment is available, the use of shotcrete mixtures containing coarse aggregates may be advantageous. Coarse aggregate must not exceed $^3/_4$ inch (19.1 mm) in size and the gradation limitations for a combination of fine (sand) and coarse aggregates should be in accordance with the requirements of ACI 506.2. Normal-weight coarse aggregate for concrete must comply with the requirements of ASTM C 33. Lightweight coarse aggregate for concrete must comply with the requirements of ASTM C 330 (see commentary, Section 1903).

1913.4 Reinforcement. Reinforcement used in shotcrete construction shall comply with the provisions of Sections 1913.4.1 through 1913.4.4.

❖ The best results will be obtained when the reinforcing steel is designed and placed to cause the least interference with the high-velocity placement of shotcrete. When proper techniques are used, shotcrete can be successfully placed through two or three layers of reinforcing steel. The size, spacing and splicing of reinforcement is specified in Sections 1913.4.1, 1913.4.2 and 1913.4.3, respectively.

1913.4.1 Size. The maximum size of reinforcement shall be No. 5 bars unless it is demonstrated by preconstruction tests that adequate encasement of larger bars will be achieved.

❖ Depending on the thickness and nature of the work, reinforcement will consist of either deformed bars or welded wire fabric. Bar sizes are limited to a No. 5 (16) bar. Larger bar sizes can be used if it can be shown that adequate bonding with and encasement in the concrete can be accomplished. Preconstruction tests may be required by the building official to provide verification that proper bonding can be achieved (see Section 1913.5).

1913.4.2 Clearance. When No. 5 or smaller bars are used, there shall be a minimum clearance between parallel reinforcement bars of $2^1/_2$ inches (64 mm). When bars larger than No. 5 are permitted, there shall be a minimum clearance between parallel bars equal to six diameters of the bars used. When two curtains of steel are provided, the curtain nearer the nozzle shall have a minimum spacing equal to 12 bar diameters and the remaining curtain shall have a minimum spacing of six bar diameters.

Exception: Subject to the approval of the building official, required clearances shall be reduced where it is demonstrated by preconstruction tests that adequate encasement of the bars used in the design will be achieved.

❖ Experience has shown that a $2^1/_2$-inch (64 mm) space between No. 5 (16) bars, and a six-bar diameter space between larger bars, is sufficient clearance for placement of shotcrete where there is one layer of reinforcement. Experience has also shown that shotcrete application can be effective when the spacing between bars in the closer of two layers (nearer to nozzle) of steel is 12 bar diameters, and in the remaining layer the bars have a spacing of six bar diameters. The "exception" clarifies that lesser clearance dimensions are permitted if they can be verified by preconstruction tests.

For welded wire fabric, ACI 506R recommends a minimum spacing of 2 inches (51 mm) between wires in both directions. The minimum clearance between the reinforcement and the form or other back-up surface may vary depending on whether a mortar mix (fine aggregate only) or concrete (fine and coarse aggregate) is used for shotcrete and whether conventional reinforcing steel or welded wire fabric is used. This spacing should be specified or shown on the construction documents. ACI 506R recommends that no less than a 2-inch (51 mm) clearance should be used for No. 5 (16) reinforcing bars. The minimum cover for reinforcement must comply with the requirements of Section 1907.7.

1913.4.3 Splices. Lap splices of reinforcing bars shall utilize the noncontact lap splice method with a minimum clearance of 2 inches (51 mm) between bars. The use of contact lap splices necessary for support of the reinforcing is permitted when approved by the building official, based on satisfactory preconstruction tests that show that adequate encasement of the bars will be achieved, and provided that the splice is oriented so that a plane through the center of the spliced bars is perpendicular to the surface of the shotcrete.

❖ Bars and welded wire fabric must be lapped in such a manner so as not to create weak sections in the shotcrete. Lapped reinforcing bars should not be tied together, but rather be separated by at least 2 inches (51 mm) wherever possible. The building official should only permit contact lap splices when it can be shown by preconstruction tests that the reinforcement will be completely encased. Welded wire fabric must be lapped 1 $^1/_2$ squares in all directions. Continuous chair support should not be placed parallel and directly underneath bars because chairs may prevent full shotcrete encasement or cause sand pockets. Individual chairs should also to be offset to prevent a line of weakness in the shotcrete.

1913.4.4 Spirally tied columns. Shotcrete shall not be applied to spirally tied columns.

❖ Typically, plain reinforcement (i.e., no surface deformations) is used for spiral reinforcement in columns. Shotcrete is not permitted to be applied to spirally tied columns because of the lack of these deformations. Additionally, ACI 318 requires that the clear spacing between spirals must not exceed 3 inches (76 mm). These conditions make it difficult to achieve adequate bond between the concrete and the reinforcement.

1913.5 Preconstruction tests. When required by the building official, a test panel shall be shot, cured, cored or sawn, examined and tested prior to commencement of the project. The sample panel shall be representative of the project and simulate job conditions as closely as possible. The panel thickness and reinforcing shall reproduce the thickest and most congested area specified in the structural design. It shall be shot at the same angle, using the same nozzleman and with the same concrete mix design that will be used on the project. The equipment used in preconstruction testing shall be the same equipment used in the work requiring such testing, unless substitute equipment is approved by the building official.

❖ When required by the building official, preconstruction test specimens must be made for each type of construction application (flat, vertical or overhead) using the materials, mix proportions, equipment and the noz-

zle operator proposed for the project. Test panels should be at least 30 inches (762 mm) square, and must have the same reinforcement and thickness [but not less than 3 inches (76 mm)] as specified and shown on the construction documents for permanent construction. Drilled cores should be taken from the panels and tested in accordance with ASTM C 42 procedures for compliance with strength requirements. Preconstruction testing may not be necessary if it can be shown to the satisfaction of the building official that mix, materials, equipment and construction personnel have produced satisfactory results on similar work.

1913.6 Rebound. Any rebound or accumulated loose aggregate shall be removed from the surfaces to be covered prior to placing the initial or any succeeding layers of shotcrete. Rebound shall not be used as aggregate.

❖ Rebound is loose aggregate and cement paste that bounces off application surfaces after colliding with formwork, reinforcement or the shotcrete surface itself. Factors that affect the amount of rebound include the position of the work (flat, vertical or overhead applications), air pressure, cement content, water content, maximum size and grading of aggregates, amount of reinforcement and thickness of the layer. Rebound must not be reused as aggregate or in any way worked back into the construction. If rebound does not fall clear of the work, it must be removed.

1913.7 Joints. Except where permitted herein, unfinished work shall not be allowed to stand for more than 30 minutes unless edges are sloped to a thin edge. For structural elements that will be under compression and for construction joints shown on the approved construction documents, square joints are permitted. Before placing additional material adjacent to previously applied work, sloping and square edges shall be cleaned and wetted.

❖ Construction joints are generally tapered to a thin edge over a width of approximately 12 inches (305 mm). Square construction joints (nontapered) should be avoided wherever possible in shotcrete construction because they form a trap for rebound. Where the joint will be subjected to compressive stresses, however, square joints are frequently used and, in such cases, care must be taken to avoid or remove trapped rebound at the joint. All joints, tapered or square, must be thoroughly cleaned of loose material and wetted prior to the application of additional shotcrete.

1913.8 Damage. In-place shotcrete that exhibits sags, sloughs, segregation, honeycombing, sand pockets or other obvious defects shall be removed and replaced. Shotcrete above sags and sloughs shall be removed and replaced while still plastic.

❖ After placement, any shotcrete that lacks uniformity or exhibits segregation, honeycombing or delamination, or that contains dry patches, slugs, voids or sand pockets (porous areas low in cement content), or that sags or sloughs should be removed and replaced. As a practical matter, small holes created by cores taken for testing purposes cannot be filled solidly by shooting mortar at high velocity. The patching mixture to be applied by hand should be made of the same materials and approximately of the same proportions as used for the shotcrete, except that coarse aggregate, if used, should be omitted. Chapter 9 of ACI 301 provides additional recommendations for repairing shotcrete, including filling of holes created where cores were obtained for testing.

1913.9 Curing. During the curing periods specified herein, shotcrete shall be maintained above 40°F (4°C) and in moist condition.

❖ Like cast-in-place concrete, shotcrete must be cured in a moist condition to achieve its potential strength and durability. Good curing practice for shotcrete construction generally requires that surfaces be kept continuously moist for a period of at least 7 days. During curing, the air that is in contact with shotcrete surfaces must be maintained at temperatures above freezing, with the temperature of the shotcrete maintained above 40°F (4°C). For recommendations on winter protection, see the commentary to Section 1905.12 or refer to ACI 306R.

1913.9.1 Initial curing. Shotcrete shall be kept continuously moist for 24 hours after shotcreting is complete or shall be sealed with an approved curing compound.

❖ Immediately after finishing operations are complete, shotcrete must be kept continuously moist for a period of 24 hours by one of the following methods:

1. Ponding or continuous sprinkling;

2. Use of absorptive mat or fabric, sand or other protective covering kept continuously moist;

3. Continuous steam not exceeding 150°F (66°C) or vapor mist bath; or

4. Membrane-forming curing compounds and application conforming to ASTM C 309.

1913.9.2 Final curing. Final curing shall continue for seven days after shotcreting, or for three days if high- early-strength cement is used, or until the specified strength is obtained. Final curing shall consist of the initial curing process or the shotcrete shall be covered with an approved moisture-retaining cover.

❖ Immediately following the initial curing specified in Section 1913.9.1, additional curing must be provided using one of the following materials or methods:

1. Continue the method used in the initial curing process (see commentary, Section 1914.9.1);

2. Use sheet materials conforming to ASTM C 171; or

3. Use other moisture-retaining covers acceptable to the registered design professional and approved by the building official.

Curing must be continued for at least the first seven days after shotcreting, or for the first three days if

high-early-strength cement is used (Type III or IIIA portland cement), or until the specified strength is obtained. During the curing period, shotcrete must be maintained at a temperature above 40°F (4°C) and in a moist condition.

1913.9.3 Natural curing. Natural curing shall not be used in lieu of that specified in this section unless the relative humidity remains at or above 85 percent, and is authorized by the registered design professional and approved by the building official.

❖ Natural curing is exposure to the atmosphere without any cover or water spray. If the atmospheric conditions surrounding the shotcrete construction are satisfactory, such as when the relative humidity is at or above 85 percent, the registered design professional, with the approval of the building official, may authorize natural curing. Where natural curing is used, the shotcrete must be maintained above 40°F (4°C) for the period specified in Section 1913.9.2.

1913.10 Strength tests. Strength tests for shotcrete shall be made by an approved agency on specimens that are representative of the work and which have been water soaked for at least 24 hours prior to testing. When the maximum-size aggregate is larger than $^3/_8$ inch (9.5 mm), specimens shall consist of not less than three 3-inch-diameter (76 mm) cores or 3-inch (76 mm) cubes. When the maximum-size aggregate is $^3/_8$ inch (9.5 mm) or smaller, specimens shall consist of not less than 2-inch-diameter (51 mm) cores or 2-inch (51 mm) cubes.

❖ This section provides basic strength test criteria for shotcrete. Experience has shown that aggregate size has a direct bearing on the accuracy of compressive tests. Therefore, core and cube specimens containing aggregates that are larger than $^3/_8$ inch (9.5 mm) must be a minimum of 3 inches (76 mm) in diameter (or cross section). Test specimens are required to be soaked in water for at least 24 hours prior to testing to maintain a common environmental condition for all specimens and to insure uniformity between test results. Cores should be obtained and tested in accordance with ASTM C 42.

1913.10.1 Sampling. Specimens shall be taken from the in-place work or from test panels, and shall be taken at least once each shift, but not less than one for each 50 cubic yards (38.2 m³) of shotcrete.

❖ Frequency of testing is a significant factor in the effectiveness of quality control of concrete. To ensure that test results will be representative of the concrete in the structure, specimens must be taken at the minimum frequency required by this section.

1913.10.2 Panel criteria. When the maximum-size aggregate is larger than $^3/_8$ inch (9.5 mm), the test panels shall have minimum dimensions of 18 inches by 18 inches (457 mm by 457 mm). When the maximum size aggregate is $^3/_8$ inch (9.5 mm) or smaller, the test panels shall have minimum dimensions of 12 inches by 12 inches (305 mm by 305 mm). Panels shall be shot in the same position as the work, during the course of the work and by the nozzlemen doing the work. The conditions under which the panels are cured shall be the same as the work.

❖ When test panels are used, they are required to meet the criteria in this section. Experience has shown that aggregate size has a direct bearing on the ability to produce test panels that accurately portray the condition of in-place shotcrete. Evidence suggests that test panels containing aggregates larger than $^3/_8$ inch (9.5 mm) must be a minimum of 18 inches (457 mm) square, and that test panels with aggregate $^3/_8$ inch (9.5 mm) or smaller must be a minimum 12 inches (305 mm) square. The ability of the nozzle operator is a significant factor in achieving quality shotcrete. Therefore, it is critical that the nozzle operator doing the work also prepares the test panels. Since the position of the work (flat, vertical or overhead) and the method of curing have an impact on the quality of work, they must also be accounted for when making test panels.

1913.10.3 Acceptance criteria. The average compressive strength of three cores from the in-place work or a single test panel shall equal or exceed $0.85 f'_c$ with no single core less than $0.75 f'_c$. The average compressive strength of three cubes taken from the in-place work or a single test panel shall equal or exceed f'_c with no individual cube less than $0.88 f'_c$. To check accuracy, locations represented by erratic core or cube strengths shall be retested.

❖ In evaluating core test results, the fact that core strengths may not equal the specified compressive strength of concrete, f'_c, should not be a cause for concern, provided they are within the limits specified. It is realistic to expect the compressive strength of cores to be less than f'_c because the size of the core differs from that of a standard concrete test cylinder, and procedures for obtaining and curing the cores are different. Therefore, the shotcrete can be considered acceptable from the standpoint of strength if test results from the cores are at least 85 percent of f'_c for the average of three cores and 75 percent of f'_c for a single core. Cores should be obtained and tested in accordance with ASTM C 42. Cores for compressive strength testing must have a length-to-diameter ratio between 1.0 and 2.10. For ratios less than 1.94, the compressive strength is multiplied by a correction factor (less than 1) given in ASTM C 42 to correlate the strength of the core to that of a standard concrete test cylinder that has a length-to-diameter ratio of 2. Experience has shown that test results from cubes can be held to a higher standard, thus requiring the average of three cubes to equal or exceed f'_c with no individual cube less than $0.88f'_c$. Since the height-to-least-lateral dimension of a cube is 1, ACI 506R indicates that the compressive strength of a cube can be correlated to that of a standard test cylinder by multiplying the test results by 0.85. Either correlated or uncorrelated results for cubes can be used to determine if the specified acceptance criteria are met. Because of the less than delicate nature of the coring and sawing process, this section also permits erratic core and cube

strengths to be discarded from the sample under consideration.

SECTION 1914
REINFORCED GYPSUM CONCRETE

1914.1 General. Reinforced gypsum concrete shall comply with the requirements of ASTM C 317 and ASTM C 956.

❖ The provisions of Section 1914 apply to poured-in-place reinforced gypsum concrete that is used to construct structural roof decks (when placed over permanent formboards). Gypsum concrete is a manufactured product consisting of calcined gypsum to which the producer may add aggregates complying with ASTM C 35, wood chips or wood shavings. Only water is added at the job site.

Similar proprietary products manufactured under various trade names are intended to be used in nonstructural applications such as floor toppings, where increased sound resistance is desired. These products can also be used as floor levers, and this application is prevalent in rehabilitation of existing structures. This section is not intended to govern the use of these materials where used in nonstructural applications.

Gypsum concrete (the product) used in reinforced gypsum concrete construction must conform to the requirements of ASTM C 317. This standard provides for two classes (i.e., A and B) of gypsum concrete based on minimum compressive strength. Class A must achieve a minimum compressive strength of 500 psi (3.5 MPa) and have a density not exceeding 60 lb/ft³ (960 kg/m³). Class B must achieve a compressive strength of 1,000 psi (6.9 MPa) with no restriction on density. Before the development of newer insulating materials suitable for installation on roof decks, reinforced gypsum concrete roof decks were used where energy efficiency was a concern. To increase the energy efficiency of the material, an aggregate complying with ASTM C 35, such as perlite or vermiculite, is used to reduce the density of the gypsum concrete, thus increasing its *R*-value.

The design and installation of reinforced gypsum concrete roof decks must comply with the requirements of ASTM C 956. Gypsum concrete roof decks must be reinforced with zinc-coated (galvanized) welded or woven wire mesh or fabric and must be designed to support anticipated loads without exceeding the allowable stresses specified in Appendix Section X2 of ASTM C 956. ASTM C 956 indicates that "methods of design [of reinforced gypsum concrete] shall follow established principles of mechanics and principles of design for reinforced concrete in accordance with ACI 318." Requirements of Section 7.12 of ACI 318 for shrinkage and temperature reinforcement in reinforced concrete do not apply to reinforced gypsum concrete.

1914.2 Minimum thickness. The minimum thickness of reinforced gypsum concrete shall be 2 inches (51 mm) except the minimum required thickness shall be reduced to 1¹/₂ inches (38 mm), provided the following conditions are satisfied:

1. The overall thickness, including the formboard, is not less than 2 inches (51 mm).

2. The clear span of the gypsum concrete between supports does not exceed 33 inches (838 mm).

3. Diaphragm action is not required.

4. The design live load does not exceed 40 pounds per square foot (psf) (1915 Pa).

❖ This section provides an exception to the ASTM C 956 minimum 2-inch (51 mm) thickness requirement for re-inforced gypsum concrete roof decks. The design live load mentioned in Item 4 should be interpreted to mean the larger of the design roof live load and the de-sign snow load.

SECTION 1915
CONCRETE-FILLED PIPE COLUMNS

1915.1 General. Concrete-filled pipe columns shall be manu-factured from standard, extra-strong or double-extra-strong steel pipe or tubing that is filled with concrete so placed and manipulated as to secure maximum density and to ensure com-plete filling of the pipe without voids.

❖ Section 1915 regulates the design and construction of concrete-filled steel pipe or tube columns that rely on the concrete and steel acting compositely to resist loads.

Although the code requires concrete-filled pipe col-umns to be standard, extra-strong or double ex-tra-strong steel pipe, this language is out of date since it only refers to pipe manufactured in accordance with ASTM A 53. Both ACI 318 and AISC 360 require that steel tube or pipe used in composite concrete-filled pipe columns comply with ASTM A 53 Grade B, ASTM A 500 or ASTM A 501. In addition, both ACI 318 and AISC 360 provide limitations on minimum steel wall thickness and either of these referenced standards should be consulted for up-to-date design information.

During placement, the concrete must be thoroughly consolidated with a mechanical vibrator to develop a cohesive mass that fills all voids in the steel pipe or tube. This assures a complete bonding of the concrete to the steel shell, and any reinforcing steel and other embedded items that may be present in the pipe or tube.

1915.2 Design. The safe supporting capacity of concrete-filled pipe columns shall be computed in accordance with the approved rules or as determined by a test.

❖ The load-carrying capacity of concrete-filled pipe col-umns should be calculated in accordance with the pro-visions contained in ACI 318 or AISC 360 for compos-ite compression members. In lieu of using analytical procedures, the load-carrying capacity may be deter-mined by preconstruction load tests in accordance with Section 1714.

1915.3 Connections. Caps, base plates and connections shall be of approved types and shall be positively attached to the shell and anchored to the concrete core. Welding of brackets without mechanical anchorage shall be prohibited. Where the pipe is slotted to accommodate webs of brackets or other con-nections, the integrity of the shell shall be restored by welding to ensure hooping action of the composite section.

❖ In general terms, this section prescribes conditions for making structural connections to concrete-filled pipe columns. For the column to act compositely, all loads must be transferred to the concrete core, not just to the steel shell, as would be the case with a bracket welded to the exterior surface of the shell. Since increased structural capacity is achieved with concrete fill, the in-tegrity of the shell must be preserved so it can function to laterally confine the concrete core. If a connection requires welding to the steel shell, it must be done prior to filling the core with concrete unless it can be shown that the concrete will not be damaged from the heat of welding.

1915.4 Reinforcement. To increase the safe load-supporting capacity of concrete-filled pipe columns, the steel reinforce-ment shall be in the form of rods, structural shapes or pipe embedded in the concrete core with sufficient clearance to ensure the composite action of the section, but not nearer than 1 inch (25 mm) to the exterior steel shell. Structural shapes used as reinforcement shall be milled to ensure bearing on cap and base plates.

❖ This section does not require steel reinforcement in the concrete core; however, if it is added it must com-ply with the requirements of this section and the appli-cable provisions of ACI 318 or AISC 360. Since the concrete fill and the steel must serve as a composite section to support the loads on the column, construc-tion must be such that reinforcement placed in the core of the pipe has sufficient clearance to allow the concrete to flow around it for proper encasement and bonding.

1915.5 Fire-resistance-rating protection. Pipe columns shall be of such size or so protected as to develop the required fire-resistance ratings specified in Table 601. Where an outer steel shell is used to enclose the fire-resistant covering, the shell shall not be included in the calculations for strength of the col-umn section. The minimum diameter of pipe columns shall be 4 inches (102 mm) except that in structures of Type V construc-tion not exceeding three stories or 40 feet (12 192 mm) in height, pipe columns used in the basement and as secondary steel members shall have a minimum diameter of 3 inches (76 mm).

❖ Composite pipe columns required to be fire-resistant rated may be designed to achieve the required rating with the steel left unprotected on the exterior. This is accomplished by allowing the concrete to act as a heat sink, and Kodur and MacKinnon have developed a de-tailed methodology for providing the required fire resis-tance in this manner.

Some prefabricated steel pipe columns consist of an inner pipe filled with concrete and an outer steel shell encasing conventional concrete or a proprietary insulating concrete that enables the column assembly to achieve a fire-resistance rating. Where such a column is required to have a fire-resistance rating, the structural calculations for the load-carrying capacity of the column must not include the outer steel shell unless it is protected with materials that provide a fire-resistance rating as required in Table 601.

Since the concrete in these columns is completely encased, if it is heated sufficiently by a fire, free moisture in the concrete will be turned into steam. If provisions are not made to relieve the resulting pressure, the pipe or tube may burst causing premature failure of the column. To prevent this, Kodur and MacKinnon recommend that a 1-inch-diameter (25 mm) vent hole be provided near the top and bottom of the column. If the column is continuous through more than one story of the structure, pairs of holes as described previously should be provided in each story. Listed columns may have other venting arrangements. Care must be taken to ensure that the holes are not inadvertently or unintentionally plugged during the construction process.

1915.6 Approvals. Details of column connections and splices shall be shop fabricated by approved methods and shall be approved only after tests in accordance with the approved rules. Shop-fabricated concrete-filled pipe columns shall be inspected by the building official or by an approved representative of the manufacturer at the plant.

❖ Concrete-filled pipe columns, including their connection details and splices, which are shop fabricated as preengineered sections, are to be inspected at the plant by the building official or an approved manufacturer's representative. Approvals of such manufactured products are to be based on accepted fabrication practices and tests.

Bibliography

The following resource materials are referenced in this chapter or are relevant to the subject matter addressed in this chapter.

ACI 116R-00, *Cement and Concrete Terminology*. Farmington Hills, MI: American Concrete Institute, 1990.

ACI 117-90/117R-90, *Standard Specifications for Tolerances for Concrete Construction and Materials and Commentary*. Farmington Hills, MI: American Concrete Institute, 1990.

ACI 201.2R-92 (Reapproved 1997), *Guide to Durable Concrete*. Farmington Hills, MI: American Concrete Institute, 1992.

ACI 211.1-98, *Standard Practice for Selecting Proportions for Normal, Heavyweight, and Mass Concrete*. Farmington Hills, MI: American Concrete Institute, 1998.

ACI 211.2-98, *Standard Practice for Selecting Proportions for Structural Lightweight Concrete*. Farmington Hills, MI: American Concrete Institute, 1998.

ACI 212.3R-04, *Chemical Admixtures for Concrete*. Farmington Hills, MI: American Concrete Institute, 2004.

ACI 222.1-96, *Provisional Standard Test Method for Water-Soluble Chloride Available for Corrosion of Embedded Steel in Mortar and Concrete Using the Soxhlet Extractor*. Farmington Hills, MI: American Concrete Institute, 1996.

ACI 301-05, *Specifications for Structural Concrete*. Farmington Hills, MI: American Concrete Institute, 2005.

ACI 304R-00, *Guide for Measuring, Mixing, Transporting and Placing Concrete*. Farmington Hills, MI: American Concrete Institute, 2000.

ACI 305R-99, *Report on Hot Weather Concreting*. Farmington Hills, MI: American Concrete Institute, 1999.

ACI 306R-88 (Reapproved 1997), *Cold Weather Concreting*. Farmington Hills, MI: American Concrete Institute, 1988.

ACI 308R-01, *Standard Practice for Curing Concrete*. Farmington Hills, MI: American Concrete Institute, 2001.

ACI 309R-96, *Guide for Consolidation of Concrete*. Farmington Hills, MI: American Concrete Institute, 1996.

ACI 318-05/318R-05, *Building Code Requirements for Structural Concrete and Commentary*. Farmington Hills, MI: American Concrete Institute, 2005.

ACI 347-04, *Guide to Formwork for Concrete*. Farmington Hills, MI: American Concrete Institute, 2004.

ACI 349-01/349R-01, *Code Requirements for Nuclear Safety Related Concrete Structures and Commentary*. Farmington Hills, MI: American Concrete Institute, 2001.

ACI 355.2-04, *Evaluating the Performance of Post-Installed Mechanical Anchors in Concrete*. Farmington Hills, MI: American Concrete Institute, 2004.

ACI 355.2R-04, *Commentary on Evaluating the Performance of Post-Installed Mechanical Anchors in Concrete*. Farmington Hills, MI: American Concrete Institute, 2004.

ACI 423.6-01/423.6R-01, *Specification for Unbonded Single-Strand Tendons and Commentary*. Farmington Hills, MI: American Concrete Institute, 2001.

ACI 506R-05, *Guide to Shotcrete*. Farmington Hills, MI: American Concrete Institute, 2005.

ACI 506.2-95, *Specification for Shotcrete*. Farmington Hills, MI: American Concrete Institute, 1995.

ACI T1.1-01, *Acceptance Criteria for Moment Frames Based on Structural Testing*. Farmington Hills, MI: American Concrete Institute, 2001.

ACI T1.1R-01, *Commentary on Acceptance Criteria for Moment Frames Based on Structural Testing*. Farmington Hills, MI: American Concrete Institute, 2001.

AISC 360-05, *Specification for Structural Steel Buildings*. Chicago: American Institute of Steel Construction, 2005.

ANSI A58.1-1982, *Minimum Design Loads for Buildings and Other Structures* (Also, ANSI A58.1-1972). New York: American National Standards Institute, 1972, 1982.

ASCE 3-91, *Standard for the Structural Design of Composite Slabs*. New York: American Society of Civil Engineers, 1992.

ASCE 7-05, *Minimum Design Loads for Buildings and Other Structures* (also ASCE 7-88, ASCE 7-93, ASCE 7-95). New York: American Society of Civil Engineers, 1990, 1993, 1995, ASCE: Reston, VA, 2005.

ASME B1.1-89, *Unified Inch Screw Threads (UN and UNR Thread Form)*. Fairfield, NJ: American Society of Mechanical Engineers, 1989.

ASME B18.2.1-96, *Square and Hex Bolts and Screws, Inch Series*. Fairfield, NJ: American Society of Mechanical Engineers, 1996.

ASME B18.2.6-96, *Fasteners for Use in Structural Applications*. Fairfield, NJ: American Society of Mechanical Engineers, 1996.

ASTM A 36/A 36M-04, *Standard Specification for Carbon Structural Steel*. West Conshohocken, PA: ASTM International, 2004.

ASTM A 53-97, *Standard Specification for Pipe, Steel, Black and Hot-Dipped, Zinc-Coated Welded and Seamless*. West Conshohocken, PA: ASTM International, 1997.

ASTM A 82-02, *Standard Specification for Steel Wire, Plain, for Concrete Reinforcement*. West Conshohocken, PA: ASTM International, 2002.

ASTM A 184-96, *Standard Specification for Fabricated Deformed Steel Bar Mats for Concrete Reinforcement*. West Conshohocken, PA: ASTM International, 1996.

ASTM A 185-02, *Standard Specification for Steel Welded Wire Fabric, Plain, for Concrete Reinforcement*. West Conshohocken, PA: ASTM International, 2002.

ASTM A 242-93a/A 242M-93a, *Standard Specification for High-Strength Low-Alloy Structural Steel*. West Conshohocken, PA: ASTM International, 1993.

ASTM A 416-02, *Standard Specification for Steel Strand, Uncoated Seven-Wire for Prestressed Concrete*. West Conshohocken, PA: ASTM International, 2002.

ASTM A 421/A 421M-02, *Standard Specification for Uncoated Stress-Relieved Steel Wire for Prestressed Concrete*. West Conshohocken, PA: ASTM International, 2002.

ASTM A 496-02, *Standard Specification for Steel Wire, Deformed, for Concrete Reinforcement*. West Conshohocken, PA: ASTM International, 2002.

ASTM A 497-01, *Standard Specification for Steel Welded Wire Fabric, Deformed, for Concrete Reinforcement*. West Conshohocken, PA: ASTM International, 2001.

ASTM A 500-96, *Standard Specification for Cold-Formed Welded and Seamless Carbon Steel Structural Tubing in Rounds and Shapes*. West Conshohocken, PA: ASTM International, 1996.

ASTM A 501-96, *Standard Specification for Hot-Formed Welded and Seamless Carbon Steel Structural Tubing*. West Conshohocken, PA: ASTM International, 1996.

ASTM A 572/A 572M-04, *Standard Specification for High-Strength Low-Alloy Columbium-Vanadium Steels of Structural Quality*. West Conshohocken, PA: ASTM International, 2004.

ASTM A 588/A 588M-04, *Standard Specification for High-Strength Low-Alloy Structural Steel with 50 ksi [345 MPa] Minimum Yield Point to 4 inches [100 mm] Thick*. West Conshohocken, PA: ASTM International, 2004.

ASTM A 615-04a, *Standard Specification for Deformed and Plain Billet-Steel Bars for Concrete Reinforcement*. West Conshohocken, PA: ASTM International, 2004.

ASTM A 706/A 706M-04a, *Standard Specification for Low-Alloy Steel Deformed and Plain Bars for Concrete Reinforcement*. West Conshohocken, PA: ASTM International, 2004.

ASTM A 722/A 722M-98, *Standard Specification for Uncoated High-Strength Steel Bar for Prestressing Concrete*. West Conshohocken, PA: ASTM International, 1998.

ASTM A 767/A 767M-00b, *Standard Specification for Zinc-Coated (Galvanized) Steel Bars for Concrete Reinforcement*. West Conshohocken, PA: ASTM International, 2000.

ASTM A 775-04, *Standard Specification for Epoxy-Coated Reinforcing Steel Bars*. West Conshohocken, PA: ASTM International, 2004.

ASTM A 884-99, *Standard Specification for Epoxy-Coated Steel Wire and Welded Wire Fabric for Reinforcement*. West Conshohocken, PA: ASTM International, 1999.

ASTM A 996/A 996M-04, *Specification for Rail-Steel and Axle-Steel Deformed Bars for Concrete Reinforcement*. West Conshohocken, PA: ASTM International, 2004.

ASTM C 31/C 31M-03a, *Standard Practice for Making and Curing Concrete Test Specimens in the Field*. West Conshohocken, PA: ASTM International, 2003.

ASTM C 33-03, *Standard Specification for Concrete Aggregates*. West Conshohocken, PA: ASTM International, 2003.

ASTM C 35-95 (2001), *Standard Specification for Inorganic Aggregates for Use in Gypsum Plaster*. West Conshohocken, PA: ASTM International, 1995.

ASTM C 39-99ae1, *Standard Test Method for Compressive Strength of Cylindrical Concrete Specimens*. West Conshohocken, PA: ASTM International, 1999.

ASTM C 42/C 42M-04, *Standard Test Method for Obtaining and Testing Drilled Cores and Sawed Beams of Concrete*. West Conshohocken, PA: ASTM International, 2004.

ASTM C 94/C 94M-04, *Standard Specification for Ready-Mixed Concrete*. West Conshohocken, PA: ASTM International, 2004.

ASTM C 142-78(1990), *Standard Test Method for Clay Lumps and Friable Particles in Aggregates*. West Conshohocken, PA: ASTM International, 1990.

ASTM C 150-04, *Standard Specification for Portland Cement*. West Conshohocken, PA: ASTM International, 2004.

ASTM C 171-92, *Standard Specification for Sheet Materials for Curing Concrete*. West Conshohocken, PA: ASTM International, 1992.

ASTM C 172-04, *Standard Practice For Sampling Freshly Mixed Concrete*. West Conshohocken, PA: ASTM International, 2004.

ASTM C 231-97, *Standard Test Method for Air Content of Freshly Mixed Concrete by the Pressure Method*. West Conshohocken, PA: ASTM International, 1997.

ASTM C 260-95, *Standard Specification for Air-Entraining Admixtures for Concrete*. West Conshohocken, PA: ASTM International, 1995.

ASTM C 309-93, *Standard Specification for Liquid Membrane-Forming Compounds for Curing Concrete*. West Conshohocken, PA: ASTM International, 1993.

ASTM C 317-00, *Standard Specification for Gypsum Concrete*. West Conshohocken, PA: ASTM International, 2000.

ASTM C 330-04, *Standard Specification for Lightweight Aggregates for Structural Concrete*. West Conshohocken, PA: ASTM International, 2004.

ASTM C 494-92, *Standard Specification for Chemical Admixtures for Concrete*. West Conshohocken, PA: ASTM International, 1992.

ASTM C 595-01, *Standard Specification for Blended Hydraulic Cements*. West Conshohocken, PA: ASTM International, 2001.

ASTM C 618-99, *Standard Specification for Fly Ash and Raw or Calcined Natural Pozzolan for Use as a Mineral Admixture in Portland Cement Concrete*. West Conshohocken, PA: ASTM International, 1999.

ASTM C 685/C 685M-98a, *Standard Specification for Concrete Made by Volumetric Batching and Continuous Mixing*. West Conshohocken, PA: ASTM International, 1998.

ASTM C 803-90, *Standard Test Method for Penetration Resistance of Hardened Concrete*. West Conshohocken, PA: ASTM International, 1990.

ASTM C 805-94, *Standard Test Method for Rebound Number of Hardened Concrete*. West Conshohocken, PA: ASTM International, 1994.

ASTM C 845-04, *Standard Specification for Expansive Hydraulic Cement*. West Conshohocken, PA: ASTM International, 2004.

ASTM C 873-94, *Standard Test Method for Compressive Strength of Concrete Cylinders Cast in Place in Cylindrical Molds*. West Conshohocken, PA: ASTM International, 1994.

ASTM C 900-87(1993), *Standard Test Method of Pullout Strength of Hardened Concrete*. West Conshohocken, PA: ASTM International, 1993.

ASTM C 956-04, *Standard Specification for Installation of Cast-in-Place Reinforced Gypsum Concrete*. West Conshohocken, PA: ASTM International, 2004.

ASTM C 989-04, *Standard Specification for Ground Granulated Blast-Furnace Slag for Use in Concrete and Mortars*. West Conshohocken, PA: ASTM International, 2004.

ASTM C 1017-92, *Standard Specification for Chemical Admixtures for Use in Producing Flowing Concrete*. West Conshohocken, PA: ASTM International, 1992.

ASTM C 1074-93, *Standard Practice of Estimating Concrete Strength by the Maturity Method*. West Conshohocken, PA: ASTM International, 1993.

ASTM C 1077-96, *Standard Practice for Laboratories Testing Concrete and Concrete Aggregates for Use in Construction and Criteria for Laboratory Evaluation*. West Conshohocken, PA: ASTM International, 1996.

ASTM C 1157M-03, *Standard Performance Specification for Blended Hydraulic Cement*. West Conshohocken, PA: ASTM International, 2003.

ASTM C 1218/C 1218M-99, *Standard Test Method for Water-Soluble Chloride in Mortar and Concrete*. West Conshohocken, PA: ASTM International, 1999.

ASTM C 1240-04, *Standard Specification for Silica Fume for Use as a Mineral Admixture in Hydraulic-Ce-*

ment Concrete, Mortar, and Grout. West Conshohocken, PA: ASTM International, 2004.

ASTM D 3665-93, Standard Practice for Random Sampling of Construction Materials. West Conshohocken, PA: ASTM International, 1993.

ASTM E 119-00, Standard Test Methods for Fire Tests of Building Construction Materials. West Conshohocken, PA: ASTM International, 2000.

ATC 3-06, Tentative Provisions for the Development of Seismic Regulations for Buildings. Applied Technology Council. NBS Special Publication 510, NSF Publication 78-8. Washington, DC: U.S. Government Printing Office, 1978.

AWS D1.1-04, Structural Welding Code — Steel. Miami: American Welding Society, 2004.

AWS D1.4-98, Structural Welding Code — Reinforcing Steel. Miami: American Welding Society, 1998.

BOCA National Building Code/1999. Country Club Hills, IL: Building Officials and Code Administrators International, Inc., 1999.

Cook, R. A. and R. E. Klingner. "Behavior of Ductile Multiple-Anchor Steel-to-Concrete Connections with Surface-Mounted Baseplates." Anchors in Concrete: Design and Behavior. American Concrete Institute Special Publication SP-130, February 1992, pp. 61–122.

Cook, R. A. and R. E. Klingner. "Ductile Multiple-Anchor Steel-to-Concrete Connections." Journal of Structural Engineering. ASCE, Vol. 118, No. 6, June 1992, pp. 1645-1665.

Design of Fasteners in Concrete. Comite Euro-International du Beton (CEB). London: Thomas Telford Services Ltd., Jan. 1997.

DeVries, Richard A. Anchorage of Headed Reinforcement in Concrete. Ph.D. Dissertation. The University of Texas at Austin, December 1996.

Eligehausen, R., W. Fuchs, and B. Mayer. "Load Bearing Behavior of Anchor Fastenings in Tension." Betonwerk + Fertigteiltechnik, December 1987, pp. 826–832, and January 1988, pp. 29–35.

Eligehausen, R. and W. Fuchs. "Load Bearing Behavior of Anchor Fastenings under Shear, Combined Tension and Shear or Flexural Loadings." Betonwerk + Fertigteiltechnik. pp. 48-56, February 1988.

Eligehausen, R. and T. Balogh. "Behavior of Fasteners Loaded in Tension in Cracked Reinforced Concrete." ACI Structural Journal, Vol. 92, No. 3, May-June 1995, pp. 365-379.

Farrow, C. B. and R. E. Klingner. "Tensile Capacity of Anchors with Partial or Overlapping Failure Surfaces: Evaluation of Existing Formulas on an LRFD Basis." ACI Structural Journal, Vol. 92, No. 6, November-December 1995, pp. 698-710.

Fastenings to Concrete and Masonry Structures, State of the Art Report. Comite Euro-International du Beton, (CEB), Bulletin No. 216. London: Thomas Telford Services Ltd., 1994.

FEMA 450-1, NEHRP Recommended Provisions for Seismic Regulations for New Buildings and Other Structures, Part 1. Washington, DC: Prepared by the Building Seismic Safety Council for the Federal Emergency Management Agency, 2004.

FEMA 450-2, NEHRP Recommended Provisions for Seismic Regulations for New Buildings and Other Structures, Part 2 Commentary. Washington, DC: Prepared by the Building Seismic Safety Council for the Federal Emergency Management Agency, 2004.

Fuchs, W., R. Eligehausen, and J. Breen. "Concrete Capacity Design (CCD) Approach for Fastening to Concrete." ACI Structural Journal, Vol. 92, No. 1, January-February 1995, pp. 73-93. Discussion—ACI Structural Journal, Vol. 92, No. 6, November-December 1995, pp. 787-802.

Furche, J. and R. Eligehausen. "Lateral Blow-out Failure of Headed Studs Near a Free Edge." Anchors in Concrete: Design and Behavior. American Concrete Institute Special Publication SP-130, February 1992, pp. 235-252.

Ghosh, S.K. "Design of Reinforced Concrete Buildings Under the 1997 UBC." Building Standards. International Conference of Building Officials, May-June 1998, pp. 20-24.

Ghosh, S.K. "Needed Adjustments in 1997 UBC." Proceedings, 1998 Convention, Structural Engineers Association of California, October 7-10, 1998, Reno-Sparks, NV, pp. T9.1-T9.15.

Ghosh, S.K., S.D. Nakaki, and K. Krishnan. "Precast Structures in Regions of High Seismicity: 1997 UBC Design Provisions." PCI Journal, Vol. 42, No. 6, November-December 1997, pp. 76-93.

Ishizuka, T. and N.M. Hawkins. "Effect of Bond Deterioration on the Seismic Response of Reinforced and Partially Prestressed Concrete and Ductile Moment Resistant Frames." Report SM87-2. Seattle: Department of Civil Engineering, University of Washington, 1987.

Klingner, R., J. Mendonca, and J. Malik. "Effect of Reinforcing Details on the Shear Resistance of Anchor Bolts under Reversed Cyclic Loading." Journal of the American Concrete Institute, Vol. 79, No. 1, 1982, pp. 3-12.

Kodur, V. and D. MacKinnon. "Design of Concrete-Filled Hollow Structural Steel Columns for Fire Endurance." AISC Engineering Journal.

Kuhn, D. and F. Shaikh. "Slip-Pullout Strength of Hooked Anchors." Research Report. University of Wisconsin—Milwaukee, 1996.

Lotze, D. and R.E. Klingner. "Behavior of Multiple-Anchor Attachments to Concrete from the Perspective of Plastic Theory." Report PMFSEL 96-4. Ferguson Structural Engineering Laboratory, the University of Texas at Austin, March 1997.

Lutz, L. "Discussion to Concrete Capacity Design (CCD) Approach for Fastening to Concrete." *ACI Structural Journal*, November/December 1995, pp. 791-792 and author's closure, pp. 798-799.

Manual of Standard Practice, 26th edition. Schaumburg, IL: Concrete Reinforcing Steel Institute, 1997.

Matlock, A. H., J. Yamazaki, and B. T. Kattula. "Comparative Study of Prestressed Concrete Beams, with and without Bond." *ACI Journal, Proceedings*, Vol. 68, No. 2, February 1971, pp. 116-125.

Muspratt, M. A. "Behavior of a Prestressed Concrete Waffle Slab with Unbonded Tendons." *ACI Journal, Proceedings*, Vol. 66, No. 12, December 1969, pp. 1001-1004.

Odello, R. J. and B. M. Mehta. "Behavior of a Continuous Prestressed Concrete Slab with Drop Panels." Report. Division of Structural Engineering and Structural Mechanics, University of California, Berkeley, 1967.

PCA-EB001, *Design and Control of Concrete Mixtures*, 13th edition. Skokie, IL: Portland Cement Association, 1988.

PCA-EB070, *Notes on ACI 318-99 Building Code Requirements for Structural Concrete — with Design Applications*. Skokie, IL: Portland Cement Association, 1999.

PCA-EB080, *Strength Design of Anchorage to Concrete*. Skokie, IL: Portland Cement Association, 2002.

PCA-EB115, *Concrete Inspection Handbook*. Skokie, IL: Portland Cement Association, 2000.

PCA-IS521, *Strength Design Load Combinations for Concrete Elements*. Skokie, IL: Portland Cement Association, 1998.

PCA-SP040, *Fly Ash in Cement and Concrete*. Skokie, IL: Portland Cement Association, 1987.

PCI Design Handbook, 4th edition. Chicago: Precast/Prestressed Concrete Institute, 1992.

Placing Reinforcing Bars, 7th edition. Schaumburg, IL: Concrete Reinforcing Steel Institute, 1997.

Primavera, E.J ., J. P. Pinelli, and E. H. Kalajian. "Tensile Behavior of Cast-in-Place and Undercut Anchors in High-Strength Concrete." *ACI Structural Journal*, Vol. 94, No. 5, September-October 1997, pp. 583-594.

Shaikh, A. F. and W. Yi. "In-Place Strength of Welded Studs." *PCI Journal*, Vol. 30 (2), March-April 1985.

Standard Building Code, 1999 ed. Birmingham, AL: Southern Building Code Congress, International, 1999.

Stanton, J., W. C. Stone, and G. S. Cheok. "A Hybrid Reinforced Precast Frame for Seismic Regions."*PCI Journal*, Vol. 42, No. 2, March-April 1997, pp. 20-32.

Uniform Building Code, 1997 ed. Whittier, CA: International Conference of Building Officials, 1997.

Wong, T. L. "Stud Groups Loaded in Shear." M.S. Thesis, Oklahoma State University, 1988.

Zhang, Y. "Dynamic Behavior of Multiple Anchor Connections in Cracked Concrete." Ph.D. Dissertation, the University of Texas at Austin, August 1997.

Chapter 20:
Aluminum

General Comments

Chapter 20 contains standards for the use of aluminum in building construction. Only provisions for structural applications of these materials are included.

This chapter does not seek to establish standards for aluminum specialty products, such as storefront framing, architectural hardware, etc. The use of aluminum in heating, ventilating and air-conditioning (HVAC) and plumbing systems is addressed in the *International Mechanical Code®* (IMC®) and the *International Plumbing Code®* (IPC®), respectively. This chapter applies to the structural requirements.

Purpose

Aluminum has certain physical properties, structural characteristics and nonstructural characteristics that set it apart as a building material. By utilizing the standards set forth in this chapter, a proper application of this material can be obtained.

SECTION 2001
GENERAL

2001.1 Scope. This chapter shall govern the quality, design, fabrication and erection of aluminum.

❖ The scope for Chapter 20 is provided in this section.

SECTION 2002
MATERIALS

2002.1 General. Aluminum used for structural purposes in buildings and structures shall comply with AA ASM 35 and Parts 1-A and 1-B of the Aluminum Design Manual. The nominal loads shall be the minimum design loads required by Chapter 16.

❖ The referenced standards to be applied in the utilization of aluminum in buildings and other structures are established. Parts 1-A and 1-B of the Aluminum Association's (AA) *Aluminum Design Manual* apply to the design of aluminum building-type structural load-carrying members and elements according to allowable stress design (ASD) and load and resistance factor design (LRFD) criteria, respectively. ASM 35 sets criteria for the use and installation of flashings, sheet roofing and similar applications.

Bibliography

The following resource materials are referenced in this chapter or are relevant to the subject matter addressed in this chapter.

AA ADM1-00, *Aluminum Design Manual*; "Part 1-A: Specifications for Aluminum Structures, Allowable Stress Design" and "Part 1-B: Specifications for Aluminum Structures, Load and Resistance Factor Design of Buildings and Similar Type Structures." Washington, D.C.: The Aluminum Association, 2000.

AA ASM 35-80, *Specification for Aluminum Sheet Metal Work in Building Construction*. Washington, D.C.: The Aluminum Association, 1980.

IMC-2003, *International Mechanical Code*. Falls Church, VA: International Code Council, 2003.

IPC-2003, *International Plumbing Code*. Falls Church, VA: International Code Council, 2003.

Chapter 21:
Masonry

General Comments

Masonry construction has been used for at least 10,000 years in a variety of structures—homes, private and public buildings and historical monuments. The masonry of ancient times involved two major materials: brick manufactured from sun-dried mud or burned clay and shale; and natural stone.

The first masonry structures were unreinforced and intended to support mainly gravity loads. The weight of these structures stabilized them against lateral loads from wind and earthquakes.

Masonry construction has progressed through several stages of development. Fired clay brick became the principal building material in the United States during the middle 1800s. Concrete masonry was introduced to construction during the early 1900s and, along with clay masonry, expanded in use to all types of structures.

Historically, "rules of thumb" (now termed "empirical design") were the only available methods of masonry design. Only in recent times have masonry structures been engineered using structural calculations. In the last 45 years, the introduction of engineered reinforced masonry has resulted in structures that are stronger and more stable against lateral loads, such as wind and seismic.

Masonry consists of a variety of materials. Raw materials are made into masonry units of different sizes and shapes, each having specific physical and mechanical properties. Both the raw materials and the method of manufacture affect masonry unit properties.

The word "masonry" is a general term that applies to construction using hand-placed units of clay, concrete, structural clay tile, glass block, natural stones and the like. One or more types of masonry units are bonded together with mortar, metal ties, reinforcement and accessories to form walls and other structural elements.

Proper masonry construction depends on correct design, materials, handling, installation and workmanship.

With a fundamental understanding of the functions and properties of the materials that comprise masonry construction and with proper design and construction, quality masonry structures are not difficult to obtain.

During the pioneer era of U.S. history, the fireplace was the central focus of residential cooking and heating. Today, the fireplace is essentially a decorative feature of residential construction. For energy conservation, existing fireplaces are sometimes converted and new fireplaces are designed to provide supplemental heat.

Of the many types of fireplaces, the most common are single face. Multifaced fireplaces, such as a corner fireplace with two adjacent open sides, fireplaces with two opposite faces open (common exposure to two rooms) or fireplaces with three or all faces open also occur, but are less common.

While the provisions of this chapter are for single-faced fireplaces, almost all types of masonry fireplaces include the same basic construction features: the base assembly, which consists of a foundation and hearth support; the firebox assembly, which consists of a fireplace opening, a hearth, a firebox or combustion chamber; and the throat and the smoke chamber, which supports the chimney liner.

Masonry fireplaces are made primarily of clay brick or natural stones, but also of concrete masonry or cast-in-place concrete. Chimneys for medium- and high-heat appliances require special attention for fire safety.

Purpose

Chapter 21 provides comprehensive and practical requirements for masonry construction, based on the latest state of technical knowledge. The provisions of Chapter 21 require minimum accepted practices and the use of standards for the design and construction of masonry structures and elements of structures. The provisions address: material specifications and test methods; types of wall construction; criteria for engineered design (by working stress and strength design methods); criteria for empirical design; required details of construction and other aspects of masonry, including execution of construction. The provisions are intended to result in safe and durable masonry. The provisions of Chapter 21 are also intended to prescribe minimum accepted practices for the design and construction of glass unit masonry, masonry fireplaces, masonry heaters and masonry chimneys.

SECTION 2101
GENERAL

2101.1 Scope. This chapter shall govern the materials, design, construction and quality of masonry.

❖ Section 2101 prescribes general requirements for masonry designed in accordance with Chapter 21 of the code. It identifies masonry design methods and the conditions required for the use of each method. The methods are intended as a practical means for safety under a variety of potential service conditions.

Minimum requirements for construction documents and fireplace drawings are also included in Section 2101.

Chapter 21 contains the minimum code requirements for acceptance of masonry design and construction by the building official. Compliance with these requirements is intended to result in masonry construction with the minimum required structural adequacy and durability. Requirements more stringent than these are appropriate where mandated by sound engineering and judgment. Less restrictive requirements, however, are not permitted.

2101.2 Design methods. Masonry shall comply with the provisions of one of the following design methods in this chapter as well as the requirements of Sections 2101 through 2104. Masonry designed by the allowable stress design provisions of Section 2101.2.1, the strength design provisions of Section 2101.2.2 or the prestressed masonry provisions of Section 2101.2.3 shall comply with Section 2105.

❖ This section requires masonry to comply with one of six design methods and the requirements contained in Sections 2101 through 2104 for construction documents, materials and construction.

Other provisions of the code also apply to masonry. For example, fire-resistant construction using masonry is required to comply with Chapter 7. Design loads and related requirements, including seismic forces and detailing, are required to comply with Chapter 16. Masonry foundations are required to comply with the provisions of Chapter 18. Special inspections of masonry construction are required in Chapter 17. Masonry veneer is addressed in Chapter 14.

2101.2.1 Allowable stress design. Masonry designed by the allowable stress design method shall comply with the provisions of Sections 2106 and 2107.

❖ This section requires that masonry designed by the allowable stress design method meets both the allowable stress design requirements in Section 2107 and the seismic design requirements in Section 2106. Section 2107 requires allowable stress design to comply with Chapters 1 and 2 of ACI 530/ASCE 5/TMS 402 with several exceptions. Additional information on these procedures is given in the commentaries to Section 2107 and ACI 530/ASCE 5/TMS 402.

ACI 530/ASCE 5/TMS 402 and ACI 530.1/ASCE 6/TMS 602 are referenced throughout Chapter 21. A description of these standards is warranted here. Both are joint publications of the Masonry Standards Joint Committee (MSJC), which includes the American Concrete Institute (ACI), the Structural Engineering Institute of the American Society of Civil Engineers (ASCE) and The Masonry Society (TMS). These standards are typically referred to as the MSJC Code and MSJC Specification, respectively, to reflect their joint authorship and sponsorship of the committee that oversees their development. The standards are developed through an ANSI-regulated consensus process and reflect the current state of technical knowledge on masonry design and construction.

The MSJC Code (ACI 530/ASCE 5/TMS 402) contains minimum requirements for masonry elements of structures. Topics include: construction documents; quality assurance; materials; analysis and design; strength and serviceability; flexural and axial stresses; shear; reinforcement; walls; columns; pilasters; beams and lintels and empirical design.

The allowable stress design method in MSJC Code assumes linearly elastic material behavior and properties and uses nominal loads (see Chapter 16). The standard also includes provisions for strength design, prestressed masonry design, empirical design, veneer, glass unit masonry and design of autoclaved aerated concrete masonry.

The MSJC Specification (ACI 530.1/ASCE 6/TMS 602) sets minimum acceptable levels of construction. It includes minimum requirements for composition; preparation and placement of materials; quality assurance for materials and masonry; execution of masonry construction; inspection and verification of quality. MSJC Specification contains both mandatory requirements as well as optional requirements that may be project specific. The mandatory requirements are enforceable code requirements; the optional requirements may be invoked by the design professional. The specification is meant to be modified for use with the particular project under design.

2101.2.2 Strength design. Masonry designed by the strength design method shall comply with the provisions of Sections 2106 and 2108, except that autoclaved aerated concrete (AAC) masonry shall comply with the provisions of Section 2106 and Chapter 1 and Appendix A of ACI 530/ASCE 5/TMS 402. AAC masonry shall not be used in the seismic-force-resisting system of structures classified as Seismic Design Category B, C, D, E or F.

❖ Masonry is required to meet the strength design provisions referenced in Section 2108 and the seismic design requirements in Section 2106. Additional information on the strength design requirements is given in the commentaries to Section 2108 and to Chapter 3 of the MSJC Code. The 2005 edition of the MSJC Code, Appendix A introduced strength design provisions for autoclaved aerated concrete (AAC) masonry. This section also prohibits the use of AAC masonry for the seismic-force-resisting system in structures classified as Seismic Design Category B, C, D, E or F. This limitation is not included in the MSJC Code, nor does it apply to structures permitted to be designed under the *International Residential Code®* (IRC®).

2101.2.3 Prestressed masonry. Prestressed masonry shall be designed in accordance with Chapters 1 and 4 of ACI 530/ASCE 5/TMS 402 and Section 2106. Special inspection during construction shall be provided as set forth in Section 1704.5.

❖ Prestressed masonry must comply with Chapters 1 and 4 of the MSJC Code as well as seismic detailing requirements in Section 2106. Additional information on prestressed masonry design is given in the MSJC Code commentary for Chapter 4.

2101.2.4 Empirical design. Masonry designed by the empirical design method shall comply with the provisions of Sections 2106 and 2109 or Chapter 5 of ACI 530/ASCE 5/TMS 402.

❖ This section permits the empirical design of masonry either by the provisions of Section 2109 or Chapter 5 of the MSJC Code. The provisions of Section 2109 are identical in intent to those of the MSJC Code, with some additional provisions added to Section 2109 that address surface-bonded masonry construction and adobe masonry that are not included in the MSJC Code document. Additional information on these provisions is given in the commentaries to Section 2109 and the MSJC Code.

2101.2.5 Glass unit masonry. Glass unit masonry shall comply with the provisions of Section 2110 or Chapter 7 of ACI 530/ASCE 5/ TMS 402.

❖ Glass masonry must comply with either the provisions of Section 2110 or Chapter 7 of the MSJC Code. The provisions in Section 2110 are based on the requirements in Chapter 7 of the MSJC Code. Additional information on these provisions is given in the commentaries to Section 2109 and the MSJC Code.

2101.2.6 Masonry veneer. Masonry veneer shall comply with the provisions of Chapter 14 or Chapter 6 of ACI 530/ASCE 5/TMS 402.

❖ This section requires masonry veneer to comply with the provisions of Chapter 14; specifically, Sections 1405.5 for anchored masonry veneer and 1405.9 for adhered masonry veneer. These sections reference the provisions in Chapter 6 of the MSJC Code. Additional information on these provisions is given in the commentaries to Chapter 14 and the MSJC Code.

2101.3 Construction documents. The construction documents shall show all of the items required by this code including the following:

1. Specified size, grade, type and location of reinforcement, anchors and wall ties.

2. Reinforcing bars to be welded and welding procedure.

3. Size and location of structural elements.

4. Provisions for dimensional changes resulting from elastic deformation, creep, shrinkage, temperature and moisture.

❖ Construction requirements must be clearly identified in the contract documents so that the structure is properly constructed using appropriate materials and methods. This section requires that, as a minimum, critical items required by the code and by the particular design be shown in the construction documents. The list is a minimum and should not be considered all-inclusive by the design professional. Both the design professional and the building official are permitted to require additional items as needed for a particular structure.

2101.3.1 Fireplace drawings. The construction documents shall describe in sufficient detail the location, size and construction of masonry fireplaces. The thickness and characteristics of materials and the clearances from walls, partitions and ceilings shall be clearly indicated.

❖ This section requires the submission of construction documents for all fireplaces so that compliance with appropriate code sections can be properly determined during plan review. The type of information and its format for plan review are established in this section. Construction documents are required showing relationships of components, as well as details related to the specific characteristics of the materials and techniques to be used when erecting the fireplace and chimney system. Such details are to include the type of brick or stone; refractory brick; concrete masonry; mortar requirements; wall thicknesses; clearances; dimensions of openings and dimensions of the firebox and the hearth extension.

SECTION 2102
DEFINITIONS AND NOTATIONS

2102.1 General. The following words and terms shall, for the purposes of this chapter and as used elsewhere in this code, have the meanings shown herein.

❖ Certain terms are used in the code to describe something very specific to the related subject matter while other terms may have multiple meanings that can vary based on the subject matter and context. To avoid misunderstandings, it is preferable to reach a consensus on the meaning of these terms. Definitions are intended to facilitate the understanding of code provisions and to minimize potential confusion.

 This section contains definitions of terms associated with masonry construction. Note that these terms are also listed in Chapter 2 with a cross reference to this section. Definitions that pertain primarily to masonry construction are included within this chapter to provide convenient access to them without having to refer back to Chapter 2. The use and application of all defined terms, including those defined herein, as well as undefined terms are set forth in Section 201.

AAC MASONRY. Masonry made of autoclaved aerated concrete (AAC) units, manufactured without internal reinforcement and bonded together using thin- or thick-bed mortar.

❖ This definition establishes that requirements in this chapter apply to masonry units manufactured from AAC. AAC masonry units are a low-density cementitious product first introduced into the 2005 edition of the MSJC Code and MSJC Specification, although its use has been prevalent in other countries for several decades. Besides being a relatively light-weight product, AAC is also considered to provide good thermal and acoustic insulation. AAC masonry units are bonded together using a thin-bed polymer mortar specifically manufactured for use with AAC masonry.

ADOBE CONSTRUCTION. Construction in which the exterior load-bearing and nonload-bearing walls and partitions are of unfired clay masonry units, and floors, roofs and interior framing are wholly or partly of wood or other approved materials.

❖ Adobe masonry was popular in the southwest United States due to the availability of soil for units, the frequent exposure to intense sunlight to dry the units, the thermal mass provided by the completed adobe structure and the low cost of this form of construction. This form of construction has relatively low strength, a lack of formalized design procedures and labor-intensive manufacture of units and construction of the building; thus, it has not been used as much in recent years.

The two types of adobe masonry, stabilized and unstabilized, are briefly described below. Prescriptive design requirements for adobe masonry are contained in Section 2109.8.

Adobe, stabilized. Unfired clay masonry units to which admixtures, such as emulsified asphalt, are added during the manufacturing process to limit the units' water absorption so as to increase their durability.

❖ Admixtures are added to the soil to produce more durable units.

Adobe, unstabilized. Unfired clay masonry units that do not meet the definition of "Adobe, stabilized."

❖ Unstabilized adobe does not contain stabilizers in the soil and is, therefore, not as durable as stabilized adobe.

ANCHOR. Metal rod, wire or strap that secures masonry to its structural support.

❖ Anchors are fasteners connecting two components. Figure 2102.1(1) shows examples of anchor bolts that can be used in masonry and Figure 2102.1(3) shows some uses of anchor bolts to connect wood floors and masonry walls.

Anchors are often used to connect masonry walls to

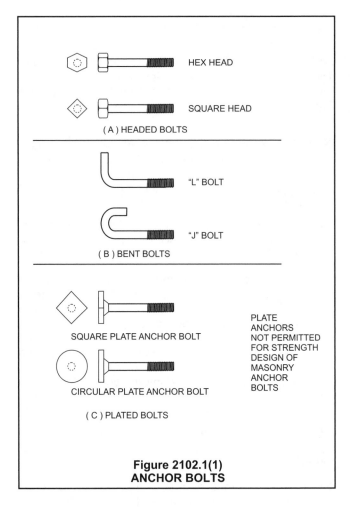

**Figure 2102.1(1)
ANCHOR BOLTS**

adjoining walls, floors, roofs or the foundation below. Requirements for strength and durability of metal anchors are given in Section 2103.13.

ARCHITECTURAL TERRA COTTA. Plain or ornamental hard-burned modified clay units, larger in size than brick, with glazed or unglazed ceramic finish.

❖ Architectural terra cotta refers to fired clay units with architectural shape and fired glazed coating. While rarely used today in new construction, repairs to terra cotta in existing building construction are not uncommon. These clay masonry units are usually produced for custom-made, anchored, ornamental veneers.

AREA.

❖ Different areas are used in different calculations throughout this chapter. It is important to use the correct area, as each different area may give dramatically different results that may not be appropriate.

Bedded. The area of the surface of a masonry unit that is in contact with mortar in the plane of the joint.

❖ The bedded area is simply the area of the unit's surface on which mortar is placed and through which stresses are transferred to the adjacent work.

Gross cross-sectional. The area delineated by the out-to-out specified dimensions of masonry in the plane under consideration.

❖ The gross cross-sectional area of the masonry is the specified masonry width (thickness) multiplied by the specified length, as illustrated in Figure 2102.1(2). While subtraction of core areas of the masonry unit is not required, subtraction of the space between wythes is required in noncomposite walls. Empirical compressive stress design is based on the gross cross-sectional area of the masonry.

Net cross-sectional. The area of masonry units, grout and mortar crossed by the plane under consideration based on out-to-out specified dimensions.

❖ The net cross-sectional area encompasses the area of units, grout and mortar contained within the plane under consideration. For ungrouted masonry, this area is sometimes equal to the bedded area, or more often to the minimum specified area of the face shells. For grouted masonry, this also includes that area of cores, cells or spaces filled with grout.

Figure 2102.1(2)
MASONRY/FLOOR ANCHOR BOLT CONNECTOR

GROSS AREA = SPECIFIED LENGTH × SPECIFIED THICKNESS

Figure 2102.1(3)
GROSS CROSS-SECTIONAL AREA FOR SINGLE WYTHE WALL UNDER LOAD

AUTOCLAVED AERATED CONCRETE (AAC). Low-density cementitious product of calcium silicate hydrates, whose material specifications are defined in ASTM C 1386.

❖ AAC is a relatively lightweight concrete product that is manufactured in panels that can be cut into smaller blocks to use as masonry units. The referenced material standard establishes the minimum compressive strength based on the class of the AAC. The definition of AAC masonry given above establishes that requirements in this chapter apply to masonry units that consist of AAC.

BED JOINT. The horizontal layer of mortar on which a masonry unit is laid.

❖ This is a horizontal mortar joint [see Figure 2102.1(4)] that separates a course of masonry units from the ones above and supports the weight of the masonry. Unlike the head or collar joint, it is easily closed when the masonry unit is placed. For a masonry unit in the typical (stretcher) orientation, the bed joint faces are the top and bottom, while the bed surface of the masonry unit is the underside. A special type of bed joint is the base-course joint or starting joint placed over foundations. See Section 2104.1.2 for requirements for thicknesses, placement and permitted tolerances for bed joints.

BOND BEAM. A horizontal grouted element within masonry in which reinforcement is embedded.

❖ Bond beams permit horizontal reinforcement to be placed in masonry. For hollow masonry unit walls, special units can either be manufactured or the webs and end shells can be reduced by saw cutting to allow horizontal reinforcing bars to be placed in the wall.

BOND REINFORCING. The adhesion between steel reinforcement and mortar or grout.

❖ This term describes the adhesion between reinforcing steel and grout or mortar that transfers stresses between those elements.

BRICK.

❖ Brick is composed of masonry units that are generally prismatic (rectangular) in shape.

Calcium silicate (sand lime brick). A masonry unit made of sand and lime.

❖ This solid brick unit is made principally from high-silica sand and lime.

Clay or shale. A masonry unit made of clay or shale, usually formed into a rectangular prism while in the plastic state and burned or fired in a kiln.

❖ These masonry units are manufactured from surface clay, shale or fire clay. Different manufacturing processes and physical properties are associated with each material. Surface clays are found in sedimentary layers near the surface. Shales are clays subjected to geologic pressure, resulting in a solid state similar to slate. Fire clays are mined from deeper layers, resulting in more uniform properties. These units are formed into the desired shape by extrusion, molding or pressing. They are then fired in a kiln to increase their strength and durability.

Concrete. A masonry unit having the approximate shape of a rectangular prism and composed of inert aggregate particles embedded in a hardened cementitious matrix.

❖ Concrete brick units are made from a zero-slump mix of portland cement (and possibly other cementitious materials), aggregates, water and admixtures. These units are solid or have a shallow depression called a "frog." Slump brick, for example, is a decorative concrete brick with bulged sides resulting from the consistency of the mix and the manufacturing process.

BUTTRESS. A projecting part of a masonry wall built integrally therewith to provide lateral stability.

❖ These elements serve one or more purposes and are sometimes called "pilasters." As secondary structural members, they are used to limit the unbraced horizontal length of walls (and their corresponding length-to-thickness ratios) by providing horizontally spaced points of lateral support. As lateral-load-resisting beam elements, or as vertical-load-resisting beam columns built integrally with the wall, they are primary structural members.

Figure 2102.1(4)
MASONRY CONSTRUCTION TERMS

CAST STONE. A building stone manufactured from portland cement concrete precast and used as a trim, veneer or facing on or in buildings or structures.

❖ Cast stone is a simulated stone precast from portland cement concrete. This material is typically used for veneer, but can also be used in other applications.

CELL. A void space having a gross cross-sectional area greater than $1\frac{1}{2}$ square inches (967 mm²).

❖ This term defines a large intentional void within a masonry unit. Grout and reinforcing steel are often placed in cells to form reinforced masonry.

CHIMNEY. A primarily vertical enclosure containing one or more passageways for conveying flue gases to the outside atmosphere.

❖ The *International Fuel Gas Code®* (IFGC®) and the *International Mechanical Code®* (IMC®) address the installation of chimneys and venting systems that are required to convey products of combustion from fuel-burning appliances to the atmosphere. A chimney is a primarily vertical enclosure containing one or more flues. This chapter regulates the construction of masonry chimneys in Section 2113. Chimneys differ from metal vents in the materials from which they are constructed and the type of appliance they are designed to serve. Chimneys can vent much hotter flue gases than metal vents.

CHIMNEY TYPES.

❖ Provisions for several types of chimneys are contained in Chapter 21, as described below. The construction of the chimney is affected by the type of appliance and the temperature of the flue gases that are generated.

High-heat appliance type. An approved chimney for removing the products of combustion from fuel-burning, high-heat appliances producing combustion gases in excess of 2,000°F (1093°C) measured at the appliance flue outlet (see Section 2113.11.3).

❖ High-heat chimneys are used in industrial applications, such as incinerators, kilns and blast furnaces. Section 2113.11.3 contains requirements for the construction and installation of chimneys for high-heat appliances.

Low-heat appliance type. An approved chimney for removing the products of combustion from fuel-burning, low-heat appliances producing combustion gases not in excess of 1,000°F (538°C) under normal operating conditions, but capable of producing combustion gases of 1,400°F (760°C) during intermittent forces firing for periods up to 1 hour. Temperatures shall be measured at the appliance flue outlet.

❖ Most domestic fuel-burning appliances are low-heat appliances. Low-heat appliances include solid-fuel-burning appliances, such as room heaters and wood stoves. Section 2113 contains requirements for the construction and installation of chimneys for low-heat appliances.

Masonry type. A field-constructed chimney of solid masonry units or stones.

❖ Masonry chimneys can have one or more flues and are field constructed of masonry units, stone, concrete and fired-clay materials. Masonry chimneys can stand alone or be part of a masonry fireplace. Section 2113 contains requirements for the construction and installation of masonry chimneys.

Most masonry chimneys require a chimney liner that is resistant to heat and the corrosive action of the products of combustion. Chimney liners are generally made of fired-clay tile, refractory brick, poured-in-place refractory materials or stainless steel.

Medium-heat appliance type. An approved chimney for removing the products of combustion from fuel-burning, medium-heat appliances producing combustion gases not exceeding 2,000°F (1093°C) measured at the appliance flue outlet (see Section 2113.11.2).

❖ Some examples of medium-heat appliances are annealing furnaces, galvanizing furnaces, pulp dryers and charcoal furnaces. Section 2113.11.2 contains requirements for the construction and installation of chimneys for medium-heat appliances.

CLEANOUT. An opening to the bottom of a grout space of sufficient size and spacing to allow the removal of debris.

❖ These openings allow debris to be removed from a space to be grouted. The code references the MSJC Specifications for minimum construction requirements for masonry, including the minimum size and maximum spacing of cleanouts for grouted masonry.

COLLAR JOINT. Vertical longitudinal joint between wythes of masonry or between masonry and backup construction that is permitted to be filled with mortar or grout.

❖ A collar joint is a filled space between masonry wythes [see Figure 2102.1(4)]. Care is necessary for proper construction of collar joints, especially where solid filling is required.

COLUMN, MASONRY. An isolated vertical member whose horizontal dimension measured at right angles to its thickness does not exceed three times its thickness and whose height is at least four times its thickness.

❖ Masonry columns typically resist moment and axial compression and sometimes axial tension from uplift. Masonry elements falling within the dimensional limits for columns must be designed and detailed accordingly, with minimum column ties and minimum vertical reinforcement.

The requirements for masonry columns vary. Some members meeting the geometric requirements for columns do not have significant structural demands placed on them. In light of this, Section 2107.4 exempts some column-type elements from these column detailing requirements.

COMPOSITE ACTION. Transfer of stress between components of a member designed so that in resisting loads, the combined components act together as a single member.

❖ When the prescriptive detailing requirements of Section 2.1.5.2.1 of the MSJC Code are used to bond multiple wythes of masonry together, all wythes are assumed to act in unison in resisting applied loads.

COMPOSITE MASONRY. Multiwythe masonry members acting with composite action.

❖ When masonry wythes are connected or bonded together so that stresses can be transferred adequately between them, the masonry is considered as composite masonry. Section 2.1.5.2.1 of the MSJC Code provides minimum bonding requirements for masonry to be considered as composite.

COMPRESSIVE STRENGTH OF MASONRY. Maximum compressive force resisted per unit of net cross-sectional area of masonry, determined by the testing of masonry prisms or a function of individual masonry units, mortar and grout.

❖ The specified compressive strength of masonry, f'_m, is used for the engineered design of masonry (allowable stress design of Section 2107 and strength design of Section 2108). The average compressive strength of masonry, determined by the prism test method or the unit strength method (Section 2105.2), must equal or exceed the specified compressive strength.

CONNECTOR. A mechanical device for securing two or more pieces, parts or members together, including anchors, wall ties and fasteners.

❖ A few types of steel connectors are illustrated in Figures 2102.1(2) and 2102.1(4). Masonry connectors attach intersecting components and also act as bonding elements. The specified size, grade, type and location of connectors are required on construction documents, in accordance with Section 2101.3.

COVER. Distance between surface of reinforcing bar and edge of member.

❖ An adequate thickness of masonry materials is needed between reinforcing steel and the surface of the masonry for two important reasons. The first reason is for proper transfer of stresses between the reinforcing steel and the masonry. The second reason is so that the reinforcement is protected from corrosion or degradation. Accordingly, larger cover is required in the MSJC standards for masonry with more severe exposure. The cover includes all masonry materials, including grout, mortar and units, between the outside surface of the reinforcement and the nearest outside surface of the masonry assembly.

DIAPHRAGM. A roof or floor system designed to transmit lateral forces to shear walls or other lateral-load-resisting elements.

❖ A diaphragm is a planar horizontal structural element (for example, a floor or roof) designed to transmit horizontal forces to vertical resisting elements (for example, shear walls or frames). Diaphragms are essential elements in the lateral-load-resisting system of a structure. Diaphragms are considered as rigid or flexible in their own planes. Flexible diaphragms deflect more than rigid diaphragms under imposed loads.

DIMENSIONS.

❖ Different dimensions are used to designate sizes of masonry units and masonry elements. The terms below denote the common meanings of various types of dimensions used in the chapter.

SPECIFIED DIMENSIONS

NOMINAL DIMENSIONS
(i.e., ³/₈″ MORTAR JOINT INCLUDED)

For SI: 1 inch = 25.4 mm.

Figure 2102.1(5)
SPECIFIED AND NOMINAL DIMENSIONS FOR NOMINAL 8 × 8 × 16 CONCRETE MASONRY UNITS

Actual. The measured dimension of a masonry unit or element.

❖ The actual dimensions of a masonry unit are its measured dimensions. Actual dimensions should equal the specified dimensions within the construction or manufacturing tolerances.

Nominal. The specified dimension plus an allowance for the joints with which the units are to be laid. Thickness is given first, followed by height and then length.

❖ The nominal dimensions of a masonry unit are the specified dimensions, plus the specified thickness of one mortar joint. Nominal dimensions are used for architectural layout of masonry structures. Figure 2102.1(5) shows nominal dimensions for a specific concrete masonry unit.

Specified. The dimensions specified for the manufacture or construction of masonry, masonry units, joints or any other component of a structure.

❖ The specified dimensions are prescribed in the construction documents. Figure 2102.1(5) shows specified and nominal dimensions for a concrete masonry unit.

EFFECTIVE HEIGHT. For braced members, the effective height is the clear height between lateral supports and is used for calculating the slenderness ratio. The effective height for unbraced members is calculated in accordance with engineering mechanics.

❖ Effective height is a theoretical distance used to predict the buckling load (compressive capacity as governed by stability) of a wall or column. For a braced condition (no sidesway), effective height is conservatively required to be assumed as the clear height between points of lateral support. For an unbraced condition (sidesway permitted), the effective height is greater than the clear height and must be calculated.

FIREPLACE. A hearth and fire chamber or similar prepared place in which a fire may be made and which is built in conjunction with a chimney.

❖ Requirements for masonry fireplaces are contained in Section 2111. Note that a chimney must be provided, as that is the means of conveying the products of combustion to the exterior of the building.

FIREPLACE THROAT. The opening between the top of the firebox and the smoke chamber.

❖ This definition is necessary for proper understanding of the code criteria for location and minimum cross-sectional area. This criterion is based on many years of successful performance and is needed to provide proper construction requirements (see Section 2111.7).

FOUNDATION PIER. An isolated vertical foundation member whose horizontal dimension measured at right angles to its thickness does not exceed three times its thickness and whose height is equal to or less than four times its thickness.

❖ Masonry foundation piers are similar to masonry columns, except that they are shorter and do not need to comply with the prescriptive reinforcement detailing requirements for masonry columns.

GROUTED MASONRY.

❖ Masonry with grout, either in the cells of hollow units or in the collar joint, is considered grouted masonry. Grouted masonry has a greater surface area to resist loads and a better transfer of stresses to reinforcing steel.

Grouted hollow-unit masonry. That form of grouted masonry construction in which certain designated cells of hollow units are continuously filled with grout.

❖ Hollow masonry units are often reinforced and grouted to provide stronger elements. Table 7 of the MSJC Specification provides requirements on fine or coarse grout based on the dimensions of the cell to be grouted.

Grouted multiwythe masonry. That form of grouted masonry construction in which the space between the wythes is solidly or periodically filled with grout.

❖ The space between wythes of multiwythe masonry can be grouted to provide stronger elements. Table 7 of the MSJC Specification, provides requirements on fine or coarse grout, based on the dimensions of the space to be grouted.

HEAD JOINT. Vertical mortar joint placed between masonry units within the wythe at the time the masonry units are laid.

❖ Vertically oriented joints between masonry units are head joints [see Figure 2102.1(4)].

HEADER (Bonder). A masonry unit that connects two or more adjacent wythes of masonry.

❖ Masonry bond between adjacent masonry wythes, and masonry bond anchorage between intersecting masonry walls, are occasionally accomplished with connecting units called "headers" or "bonders." The units may be visible on the outside of either wythe, or may not be visible on one or more wythes. If not visible, they are referred to as "blind headers." Headers must have a minimum embedment in each wythe.

HEIGHT, WALLS. The vertical distance from the foundation wall or other immediate support of such wall to the top of the wall.

❖ This term means the actual height, measured from the bottom to the top of the wall, for free-standing cantilever walls, or as the vertical distance between points of lateral support for walls spanning between floor or roof levels.

MASONRY. A built-up construction or combination of building units or materials of clay, shale, concrete, glass, gypsum, stone or other approved units bonded together with or without mortar or grout or other accepted methods of joining.

❖ The materials (other than gypsum) and elements constructed as stated in this definition are considered masonry construction and are regulated by Chapter 21. This term identifies the building elements of plain (unreinforced) masonry, reinforced masonry, grouted masonry, glass unit masonry and masonry veneer.

Ashlar masonry. Masonry composed of various-sized rectangular units having sawed, dressed or squared bed surfaces, properly bonded and laid in mortar.

❖ Units for ashlar masonry construction are rectangular in shape but variable in size.

Coursed ashlar. Ashlar masonry laid in courses of stone of equal height for each course, although different courses shall be permitted to be of varying height.

❖ In coursed ashlar masonry, all units in one course are the same height, although different courses may have different heights.

Glass unit masonry. Masonry composed of glass units bonded by mortar.

❖ Glass unit masonry consists of nonclad-bearing assemblies designed in accordance with Sections 2101.2.5 and 2110.

Plain masonry. Masonry in which the tensile resistance of the masonry is taken into consideration and the effects of stresses in reinforcement are neglected.

❖ Plain masonry has historically been referred to as "unreinforced masonry." Since such masonry may actually contain some reinforcement, however, the term "unreinforced" has fallen out of favor. When reinforcement is contained in plain masonry, its contribution to the strength of the system is required to be ignored. The bond between the masonry units and mortar is critical in the performance of plain masonry.

Random ashlar. Ashlar masonry laid in courses of stone set without continuous joints and laid up without drawn patterns. When composed of material cut into modular heights, discontinuous but aligned horizontal joints are discernible.

❖ Random ashlar has discontinuous bed joints because units have different heights.

Reinforced masonry. Masonry construction in which reinforcement acting in conjunction with the masonry is used to resist forces.

❖ Reinforced masonry contains reinforcement (currently limited to steel reinforcement) and is designed considering the tensile strength of that reinforcement. Not all masonry containing reinforcement is considered reinforced masonry. Some plain masonry contains reinforcement (usually to reduce the size of any cracks that may form), but the contribution of that reinforcement is required to be neglected.

Solid masonry. Masonry consisting of solid masonry units laid contiguously with the joints between the units filled with mortar.

❖ This term describes single- or multiwythe walls composed of solid masonry units, including the thickness of the collar joint if it is filled with mortar or grout.

Unreinforced (plain) masonry. Masonry in which the tensile resistance of masonry is taken into consideration and the resistance of the reinforcing steel, if present, is neglected.

❖ See the discussion under the definition of "Plain masonry."

MASONRY UNIT. Brick, tile, stone, glass block or concrete block conforming to the requirements specified in Section 2103.

❖ Masonry units are natural stone units or manufactured units of fired clay, shale, cementitious materials or glass.

Clay. A building unit larger in size than a brick, composed of burned clay, shale, fired clay or mixtures thereof.

❖ Clay masonry units are manufactured from fired clay or shale (also see the definition of "Brick, clay or shale").

Concrete. A building unit or block larger in size than 12 inches by 4 inches by 4 inches (305 mm by 102 mm by 102 mm) made of cement and suitable aggregates.

❖ Concrete masonry units are manufactured from a zero-slump mixture of portland cement (and possibly other cementitious materials), aggregates, water and sometimes admixtures.

Hollow. A masonry unit whose net cross-sectional area in any plane parallel to the load-bearing surface is less than 75 percent of its gross cross-sectional area measured in the same plane.

❖ Hollow masonry units are those having a specified net cross-sectional area less than 75 percent of their corresponding gross cross-sectional area. Where the specified net cross-sectional area is equal to or greater than 75 percent of the gross cross-sectional area, the unit is considered to be solid.

Solid. A masonry unit whose net cross-sectional area in every plane parallel to the load-bearing surface is 75 percent or more of its gross cross-sectional area measured in the same plane.

❖ Solid masonry units have a specified net cross-sectional area 75 percent or greater of their corresponding gross cross-sectional area. Where the specified net cross-sectional area is less than 75 percent of the gross cross-sectional area, the unit is considered to be hollow.

MEAN DAILY TEMPERATURE. The average daily temperature of temperature extremes predicted by a local weather bureau for the next 24 hours.

❖ This is the predicted average daily temperature to confirm when cold-weather and hot-weather construction techniques are required to be followed.

MORTAR. A plastic mixture of approved cementitious materials, fine aggregates and water used to bond masonry or other structural units.

❖ Mortar is the material that bonds units and accessories together and compensates for dimensional variations of the units. Both the plastic and hardened properties of mortar are important for strong, durable, water-tight construction. Material requirements and referenced standards for several permitted mortar types are given in Section 2103.8.

MORTAR, SURFACE-BONDING. A mixture to bond concrete masonry units that contains hydraulic cement, glass fiber reinforcement with or without inorganic fillers or organic modifiers and water.

❖ This mortar is a packaged, dry, combined material permitted for use in the surface bonding of concrete masonry units that have not been prefaced, coated or painted. Masonry units are stacked without mortar joints and surface-bonding mortar is then applied to both sides of the wall surface, creating a structural element.

PLASTIC HINGE. The zone in a structural member in which the yield moment is anticipated to be exceeded under loading combinations that include earthquakes.

❖ The portion of a member where the yield moment is expected to be exceeded under seismic loads is considered a plastic hinge zone. Location and detailing of plastic hinge zones are a critical part of strength design for seismic loads.

PRESTRESSED MASONRY. Masonry in which internal stresses have been introduced to counteract potential tensile stresses in masonry resulting from applied loads.

❖ This definition provides an understanding of the term "prestressed masonry," which is used to define particular types of shear wall systems recognized under the code.

PRISM. An assemblage of masonry units and mortar with or without grout used as a test specimen for determining properties of the masonry.

❖ Compliance with the specified compressive strength of masonry, f'_m, can be verified by prism tests. The prism configuration and construction methods for such tests are prescribed in ASTM C 1314.

RUBBLE MASONRY. Masonry composed of roughly shaped stones.

❖ Rubble consists of pieces of stone that are irregular in shape and size. Rubble is often laid to form walls, foundations and paving.

   Coursed rubble. Masonry composed of roughly shaped stones fitting approximately on level beds and well bonded.

❖ Coursed rubble consists of roughly shaped stones that are laid with continuous bed joints.

   Random rubble. Masonry composed of roughly shaped stones laid without regularity of coursing but well bonded and fitted together to form well-divided joints.

❖ Random rubble has approximately level beds, but discontinuous bed joints because of the varying heights of the individual units.

TYPICAL RUNNING BOND BRICK UNITS

TYPICAL RUNNING BOND CONCRETE MASONRY UNITS

UNIT LENGTH — ¼ UNIT OVERLAP

UNIT LENGTH — ¼ UNIT OVERLAP

MASONRY THAT IS NOT OVERLAPPED A MINIMUM OF ONE-FOURTH OF THE UNIT LENGTH IS CONSIDERED TO BE LAID IN STACK BOND.

Figure 2102.1(6)
RUNNING BOND MASONRY

Rough or ordinary rubble. Masonry composed of unsquared field stones laid without regularity of coursing but well bonded.

❖ This type of rubble is laid without regular coursing.

RUNNING BOND. The placement of masonry units such that head joints in successive courses are horizontally offset at least one-quarter the unit length.

❖ Figure 2102.1(6) illustrates the required overlap for running bonds. The minimum overlap is necessary to provide strength between units when masonry spans horizontally. Bed joint reinforcement or reinforced bond beams are required in Section 1.11 of the MSJC Code for masonry in other than running bond. Masonry not laid in running bond is often referred to as "stack bond" construction.

SHEAR WALL.

❖ Shear walls are vertical diaphragms that resist the effects of in-plane lateral loads. They are part of the lateral-force-resisting system (also see the definition of "Seismic-force-resisting system" in Section 1613). The various types of shear wall systems are defined in order to correctly characterize their expected performance level in resisting earthquake forces. These system descriptions correspond to the basic seismic-force-resisting system categories that are used to comply with the earthquake load provisions. Shear walls must be adequately connected to floor and roof diaphragms and to the foundation, so that loads can be transferred effectively between these elements.

Detailed plain masonry shear wall. A masonry shear wall designed to resist lateral forces neglecting stresses in reinforcement, and designed in accordance with Section 2106.1.1.

❖ Such shear walls are designed as plain (unreinforced) masonry, but contain a minimum amount of reinforcement in the horizontal and vertical directions. Because of this reinforcement, these walls have more favorable seismic design parameters, including higher response modification factors, R, than ordinary plain masonry shear walls.

Intermediate prestressed masonry shear wall. A prestressed masonry shear wall designed to resist lateral forces considering stresses in reinforcement, and designed in accordance with Section 2106.1.1.2.

❖ This definition distinguishes intermediate prestressed masonry shear walls from the other types of prestressed masonry shear walls that are recognized by the code in order to classify the lateral-force-resisting system in determining earthquake loads.

Intermediate reinforced masonry shear wall. A masonry shear wall designed to resist lateral forces considering stresses in reinforcement, and designed in accordance with Section 2106.1.1.

❖ These shear walls are designed as reinforced masonry and also must contain a minimum amount of prescriptive reinforcement. Because they contain reinforcement, their seismic performance will be better than that of plain masonry shear walls in seismic events and they are accordingly permitted in areas of moderate as well as low seismic risk. These walls have more favorable seismic design parameters, including higher response modification factors, R, than plain masonry shear walls and ordinary reinforced masonry shear walls.

Ordinary plain masonry shear wall. A masonry shear wall designed to resist lateral forces neglecting stresses in reinforcement, and designed in accordance with Section 2106.1.1.

❖ Such shear walls meet only minimum requirements, without minimum amounts of horizontal and vertical reinforcement. Thus, they may be used only in areas of low seismic risk. Plain masonry walls are designed as unreinforced masonry (by the noted sections), although they may in fact contain reinforcement.

Ordinary plain prestressed masonry shear wall. A prestressed masonry shear wall designed to resist lateral forces considering stresses in reinforcement, and designed in accordance with Section 2106.1.1.1.

❖ This definition distinguishes ordinary plain prestressed masonry shear walls from the other types of prestressed masonry shear walls that are recognized by the code in order to classify the lateral-force-resisting system in determining earthquake loads.

Ordinary reinforced masonry shear wall. A masonry shear wall designed to resist lateral forces considering stresses in reinforcement, and designed in accordance with Section 2106.1.1.

❖ These shear walls are designed as reinforced masonry. Because they contain reinforcement, their seismic performance is expected to be better than that of plain masonry shear walls and they are accordingly permitted in areas of moderate as well as low seismic risk. These walls have more favorable seismic design parameters, including higher response modification factors, R, than plain masonry shear walls. When used in areas of moderate seismic risk (Seismic Design Category C), however, minimum reinforcement is required as noted in Section 1.14.2.2.3 of the MSJC Code (see Section 2106.4).

Special prestressed masonry shear wall. A prestressed masonry shear wall designed to resist lateral forces considering stresses in reinforcement and designed in accordance with Section 2106.1.1.3 except that only grouted, laterally restrained tendons are used.

❖ This definition distinguishes special prestressed masonry shear walls from the other types of prestressed masonry shear walls that are recognized by the code in order to classify the lateral-force-resisting system in determining earthquake loads.

Special reinforced masonry shear wall. A masonry shear wall designed to resist lateral forces considering stresses in reinforcement, and designed in accordance with Section 2106.1.1.

❖ These shear walls are designed as reinforced masonry and must also meet prescriptive reinforcement limits and material limitations. Because of these requirements, they are permitted to be used in all seismic risk areas. These walls have the most favorable seismic design parameters, including the highest response modification factors, R, of any of the masonry shear wall types.

SHELL. The outer portion of a hollow masonry unit as placed in masonry.

❖ The shells of a masonry unit are defined by how the unit is used in construction. They are the portions of a hollow masonry unit exposed on the faces of elements and may include face shells and end webs.

SPECIFIED. Required by construction documents.

❖ The construction documents contain material and construction requirements essential to the proper performance of the structure. These requirements are considered minimums and material or construction that does not comply is not permitted by the code.

SPECIFIED COMPRESSIVE STRENGTH OF MASONRY, f'_m. Minimum compressive strength, expressed as force per unit of net cross-sectional area, required of the masonry used in construction by the construction documents, and upon which the project design is based. Whenever the quantity f'_m is under the radical sign, the square root of numerical value only is intended and the result has units of pounds per square inch (psi) (MPa).

❖ Engineered design of structural masonry is based on the specified compressive strength of the masonry, f'_m. This strength is required to be shown on the contract documents. Strength of the constructed masonry, determined by the unit strength method or prism strength method, is required to equal or exceed the specified compressive strength of masonry.

STACK BOND. The placement of masonry units in a bond pattern is such that head joints in successive courses are vertically aligned. For the purpose of this code, requirements for stack bond shall apply to masonry laid in other than running bond.

❖ Figure 2102.1(6) illustrates the required overlap for running bonds. If this required overlap is not provided, the wall is considered to be laid in stack bond. Reinforcement, or reinforced bond beams, are required by Section 1.11 of the MSJC Code for stack bond masonry (also see Section 2109.6.5.2).

STONE MASONRY. Masonry composed of field, quarried or cast stone units bonded by mortar.

❖ Stone masonry is comprised of natural marble, limestone, granite, sandstone and slate for building purposes. Ashlar stone is further distinguished as coursed or random. Rubble stone masonry is further distinguished as coursed, random or rough.

Ashlar stone masonry. Stone masonry composed of rectangular units having sawed, dressed or squared bed surfaces and bonded by mortar.

❖ Units for ashlar masonry construction are rectangular in shape but variable in size.

Rubble stone masonry. Stone masonry composed of irregular-shaped units bonded by mortar.

❖ Unlike ashlar stone masonry, which has rectangular units, rubble stone masonry units are irregular in shape. Rubble stone masonry is further distinguished as coursed, random or rough.

STRENGTH.

❖ The term "strength" is used in both general and specific senses. In the general sense, the strength of a member is its capacity to resist internal forces and moments. In the specific sense, strength is further categorized by type. The tensile strength of a member, for instance, refers to how much tensile force the member can support.

In the context of strength design, the force resulting from factored design actions is referred to as the "required strength." An approximation to the "minimum expected" strength of the member is referred to as the "nominal strength" (see the commentary to "Nominal strength"). This nominal strength is then multiplied by a strength reduction factor to account for material, design and construction variabilities to determine the design strength. The design strength must equal or exceed the required strength.

In the context of allowable stress design, the force resulting from unfactored design loads is referred to as the "applied (actual) force." The anticipated strength would then be reduced by appropriate safety factors to either an allowable stress or an allowable strength (for example, for anchor bolts). This allowable stress or strength is required to equal or exceed the applied (actual) stress or force.

"Design strength," "Nominal strength" and "Required strength" are defined in more detail immediately below. Allowable stresses and strengths are defined in the MSJC Code.

Strength can also refer to the load at which a test specimen fails (for example, prism strength).

Design strength. Nominal strength multiplied by a strength reduction factor.

❖ The strength design procedures of Section 2108 use the term "design strength" to indicate a realistic capacity of a member considering material, design and construction variabilities. The design strength is obtained by multiplying the nominal strength by a strength reduction factor. The design strength must equal or exceed the required strength (see separate definition in Section 2102.1).

Nominal strength. Strength of a member or cross section calculated in accordance with these provisions before application of any strength-reduction factors.

❖ The strength design procedures of Section 2108 use the term "nominal strength" to refer to the capacity of a masonry member, which is determined based on the assumptions contained in Section 2108. It is sometimes referred to as the "expected" strength of the member; however, this is a misnomer. The nominal strength would equal the expected strength if: the masonry member were constructed of materials complying exactly with the minimum material requirements; design equations were perfect and construction tolerances were zero. Since material strengths commonly exceed minimum requirements, expected strength is often much higher than nominal strength. For instance, Grade 60 reinforcement often has a yield strength of around 66,000 pounds per square inch (psi) (455 MPa), even though the minimum specified yield strength requirement is 60,000 psi (414 MPa). Therefore, the nominal strength is not the expected strength of the member, although it can be grossly classified as the minimum expected strength. Expected strength design is not included in Chapter 21 and is currently beyond the scope of the code. To determine the design strength, the nominal strength is multiplied by a strength reduction factor to account for material, design and construction variability. The design strength is required to equal or exceed the required strength.

Required strength. Strength of a member or cross section required to resist factored loads.

❖ In the strength design of masonry (see Section 2108), the required strength is that which corresponds to the factored design loads on the structure. The design strength (nominal strength times the appropriate strength reduction factor) must equal or exceed the required strength.

THIN-BED MORTAR. Mortar for use in construction of AAC unit masonry with joints 0.06 inch (1.5 mm) or less.

❖ Thin-bed mortar used in the construction of AAC masonry is not conventional mortar, but is instead a mortar manufactured specifically for use in AAC masonry construction.

TIE, LATERAL. Loop of reinforcing bar or wire enclosing longitudinal reinforcement.

❖ Lateral ties enclose longitudinal reinforcement. They are typically used in columns to support the compression reinforcement and masonry core so that these can support extreme loads even with some degradation of the masonry. Reinforcement can be assumed to be effective in carrying compressive forces only when supported by lateral ties.
Lateral ties also resist shear loads. Ties must form a closed rectangle or loop to completely surround the longitudinal reinforcement. Stirrups, in contrast, can be open.

TIE, WALL. A connector that connects wythes of masonry walls together.

❖ Ties are used to connect adjacent wythes and are subject to requirements for strength, durability and installation. Ties are adjustable or nonadjustable. A typical nonadjustable "Z" wire tie is shown in Figure 2109.6.3(1).

TILE. A ceramic surface unit, usually relatively thin in relation to facial area, made from clay or a mixture of clay or other ceramic materials, called the body of the tile, having either a "glazed" or "unglazed" face and fired above red heat in the course of manufacture to a temperature sufficiently high enough to produce specific physical properties and characteristics.

❖ Ceramic tile units are manufactured from nonmetallic materials and fired at high temperatures to obtain specific properties. Tile is considered a thin, nonstructural finish that must be supported by a strong, stiff, dimensionally stable backing.

TILE, STRUCTURAL CLAY. A hollow masonry unit composed of burned clay, shale, fire clay or mixture thereof, and having parallel cells.

❖ These clay masonry units are produced as end tiles and side tiles and differ from clay brick by having required cells with thinner webs between them.

WALL. A vertical element with a horizontal length-to-thickness ratio greater than three, used to enclose space.

❖ Masonry walls typically enclose space. They are generally required to be designed and installed for weather resistance, durability and adequate structural strength. The given dimensional requirements differentiate walls from columns.

Cavity wall. A wall built of masonry units or of concrete, or a combination of these materials, arranged to provide an airspace within the wall, and in which the inner and outer parts of the wall are tied together with metal ties.

❖ Cavity walls are made up of solid or hollow masonry units separated by a continuous airspace or cavity. This continuous airspace adds insulating value and acts as a barrier to moisture when detailed with flashing and weep holes. In many cavity walls, thermal insulation is placed between the wythes to further enhance thermal efficiency.

Composite wall. A wall built of a combination of two or more masonry units bonded together, one forming the backup and the other forming the facing elements.

❖ A composite wall is a multiwythe wall with wythes that act together to resist loads. The distinction of having wythes with different mechanical properties is important in engineering design. Walls constructed of different materials must be evaluated for lateral- and vertical-load-bearing performance and for differential movement between the wythes.

Dry-stacked, surface-bonded walls. A wall built of concrete masonry units where the units are stacked dry, without mortar on the bed or head joints, and where both sides of the wall are coated with a surface-bonding mortar.

❖ Although this type of wall is dry stacked, a leveling course set in a full bed of mortar is often used. Dry-stacked walls are also required to be placed in a running bond pattern (see the commentary to the definition of "Mortar, surface-bonding").

Masonry-bonded hollow wall. A wall built of masonry units so arranged as to provide an airspace within the wall, and in which the facing and backing of the wall are bonded together with masonry units.

❖ Hollow walls are similar to cavity walls in that they are made up of solid or hollow units separated by an airspace. Unlike a cavity wall, however, the wythes are bonded together by masonry units, which cause the wythes to act together under load.

Parapet wall. The part of any wall entirely above the roof line.

❖ These portions of masonry walls project above the roof. A parapet wall is exposed to weather on both sides and is laterally unsupported at the top. Parapets often have copings.

WEB. An interior solid portion of a hollow masonry unit as placed in masonry.

❖ The webs of hollow units are provided to support and strengthen the face shells. Web heights are permitted to be reduced so that horizontal reinforcement can be placed in the element.

WYTHE. Each continuous, vertical section of a wall, one masonry unit in thickness.

❖ Sometimes referred to as a "leaf" or "tier," each wythe is one thickness of a masonry unit.

NOTATIONS.

A_n = Net cross-sectional area of masonry, square inches (mm^2).

b = Effective width of rectangular member or width of flange for T and I sections, inches (mm).

d_b = Diameter of reinforcement, inches (mm).

F_s = Allowable tensile or compressive stress in reinforcement, psi (MPa).

f_r = Modulus of rupture, psi (MPa).

f_y = Specified yield stress of the reinforcement or the anchor bolt, psi (MPa).

f'_{AAC} = Specified compressive strength of AAC masonry, the minimum compressive strength for a class of AAC masonry as specified in ASTM C 1386, psi (MPa).

f'_m = Specified compressive strength of masonry at age of 28 days, psi (MPa).

f'_{mi} = Specified compressive strength of masonry at the time of prestress transfer, psi (MPa).

K = The lesser of the masonry cover, clear spacing between adjacent reinforcement, or five times d_b, inches (mm).

L_s = Distance between supports, inches (mm).

L_w = Length of wall, inches (mm).

l_d = Required development length or lap length of reinforcement, inches (mm).

l_{de} = Embedment length of reinforcement, inches (mm).

P_w = Weight of wall tributary to section under consideration, pounds (N).

t = Specified wall thickness dimension or the least lateral dimension of a column, inches (mm).

V_n = Nominal shear strength, pounds (N).

V_u = Required shear strength due to factored loads, pounds (N).

W = Wind load, or related internal moments in forces.

γ = Reinforcement size factor.

ρ_n = Ratio of distributed shear reinforcement on plane perpendicular to plane of A_{mv}.

ρ_{max} = Maximum reinforcement ratio.

ϕ = Strength reduction factor.

❖ Explanations of the notations used in Chapter 21 that are listed above clarify the differences among these terms and the appropriate units, if any, that apply to them.

SECTION 2103
MASONRY CONSTRUCTION MATERIALS

2103.1 Concrete masonry units. Concrete masonry units shall conform to the following standards: ASTM C 55 for concrete brick; ASTM C 73 for calcium silicate face brick; ASTM C 90 for load-bearing concrete masonry units or ASTM C 744 for prefaced concrete and calcium silicate masonry units.

❖ Proper selection of materials is essential to produce masonry with adequate strength and durability. This section sets forth prescriptive and performance-based requirements (referenced standards) for masonry materials. Test procedures and criteria for establishing and verifying quality are included. Concrete masonry refers to solid and hollow concrete units, including concrete brick, concrete block, split-face block, slump block and other special units. This section requires conformance to ASTM standards for each specific type of concrete masonry unit. The standards include requirements for materials; manufacture; physical properties; strength; absorption; minimum dimensions and permissible variations; inspection; testing and rejection.

Concrete masonry units are selected based on the desired use and appearance. Units are typically specified by density class and compressive strength.

Hollow load-bearing concrete masonry units are manufactured in accordance with the requirements of ASTM C 90 using portland cement, water and mineral aggregates. Other suitable materials, such as approved admixtures, are permitted in accordance with ASTM C 90.

Hollow load-bearing concrete masonry units have three weight classifications: normal-weight units of 125 pounds per cubic foot (pcf) (2000 kg/m³) or more; medium-weight units of between 105 pcf (1680 kg/m³) and 125 pcf (2000 kg/m³) and lightweight units of less than 105 pcf (1680 kg/m³). Lightweight aggregates generally are expanded blast-furnace slag or similar suitable materials. Normal-weight units are made from sand, gravel, crushed stone or air-cooled, blast-furnace-slag aggregates.

Concrete brick is required to comply with ASTM C 55. Concrete building brick and other solid concrete veneer and facing units are typically smaller than concrete masonry units conforming to ASTM C 90. They are made from portland cement, water and mineral aggregates, with or without inclusion of other approved materials.

Concrete brick is manufactured as lightweight, medium-weight and normal-weight units as described above for hollow concrete masonry units. Concrete brick is manufactured in two grades: Grade N and Grade S. Grade N units are used as architectural veneer and facing units in exterior walls, where high strength and resistance to moisture penetration and freeze-thaw cycling are required. Grade S units are used when moderate strength and resistance to freeze-thaw action and moisture penetration are required.

Calcium silicate face brick is a solid masonry unit complying with ASTM C 73. These units are manufactured principally from silica sand, hydrated lime and water.

Calcium silicate face brick is manufactured in two grades: Grade SW and Grade MW. Grade SW is required where exposure to moisture in the presence of freezing temperatures is anticipated. Grade MW is permitted where the anticipated exposure has freezing temperatures, but without water saturation.

Grade SW units have a minimum permitted compressive strength on the gross area of 4,500 psi (31.0 MPa) (average of three units), but not less than 3,500 psi (24.1MPa) for any individual unit. Grade MW units have a minimum permitted compressive strength on the gross area of 2,500 psi (17.2 MPa) (average of three units), but not less than 2,000 psi (13.8 MPa) for any individual unit.

ASTM C 744 is referenced for the manufacture of prefaced concrete and calcium silicate masonry units, commonly referred to as "glazed" concrete masonry units. The specified exposed surfaces of these units are covered during their manufacture with resin, resin and inert filler or cement and inert filler to produce a smooth resinous tile-like facing.

Facing requirements of that standard address resistance to chemicals; failure of adhesion of the facing material; abrasion surface-burning characteristics, color and color change; soiling and cleansability.

The standard also covers dimensional tolerances, including face dimensions and distortions.

2103.2 Clay or shale masonry units. Clay or shale masonry units shall conform to the following standards: ASTM C 34 for structural clay load-bearing wall tile; ASTM C 56 for structural clay nonload-bearing wall tile; ASTM C 62 for building brick (solid masonry units made from clay or shale); ASTM C 1088 for solid units of thin veneer brick; ASTM C 126 for ceramic-glazed structural clay facing tile, facing brick and solid masonry units; ASTM C 212 for structural clay facing tile; ASTM C 216 for facing brick (solid masonry units made from clay or shale); ASTM C 652 for hollow brick (hollow masonry units made from clay or shale); and ASTM C 1405 for glazed brick (single-fired solid brick units).

> **Exception:** Structural clay tile for nonstructural use in fireproofing of structural members and in wall furring shall not be required to meet the compressive strength specifications. The fire-resistance rating shall be determined in accordance with ASTM E 119 and shall comply with the requirements of Table 602.

❖ Section 2103.2 requires conformance with ASTM standards for masonry units manufactured from clay or shale. The various standards also include requirements for materials; manufacture; physical properties; minimum dimensions and permissible variations; inspections and testing.

Clay or shale masonry units are manufactured from clay and shale, which are compounds of silica or alumina. Shale is simply a hardened clay. The raw materials are formed into the desired shape by extrusion and cutting, molding or pressing while in the plastic state. The units are then fired in a kiln. The raw materials and the manufacturing process influence the physical properties of the manufactured unit.

Clay or shale masonry units are selected for their intended use from a variety of shapes, sizes and strengths. Units are specified based on grades and type.

Solid face brick units are required to conform to ASTM C 216. This standard covers clay brick intended to be used in masonry and supplying structural or facing components, or both, to the structure. These units are available in a variety of sizes, textures, colors and shapes.

ASTM C 216 contains requirements for two grades of durability: Grade SW and Grade MW. Grade SW brick is intended for use where high and uniform resistance to damage caused by freeze-thaw cycling is desired and where the brick may be subjected to such cycling while saturated with water. Grade MW brick is intended for use where moderate resistance to cyclic freeze-thaw damage is permissible or where the brick may be damp but not saturated with water when such cycling occurs. ASTM C 216 further classifies face brick into three types of appearance: Types FBS, FBX and FBA. Type FBS (face brick standard) units are permitted for general use in masonry. Type FBX (face brick select) units are also for general use, but have a higher degree of precision and lower permissible variation in size than Type FBS units. Type FBA (face brick architectural) units are also for general use, but are intentionally manufactured to produce characteristic architectural effects resulting from nonuniformity in size and texture of the units.

Solid units of building brick are required to conform to ASTM C 62. This specification covers brick intended for use in both structural and nonstructural masonry where external appearance is not critical. This brick was formerly called "common brick," and the standard does not contain appearance requirements.

ASTM C 62 contains requirements for three grades of durability: Grade SW, Grade MW and Grade NW. Grade SW brick is intended for use where high and uniform resistance to freeze-thaw damage is desired and where the brick may be subjected to such cycling while saturated with water. Grade MW brick is intended for use where moderate resistance to cyclic freeze-thaw damage is permissible or where the brick may be damp but not saturated with water when such cycling occurs. Grade NW is used where little resistance to cyclic freeze-thaw damage is required and is acceptable for applications protected from water absorption and freezing.

Hollow brick is required to conform to ASTM C 652, which regulates hollow building and hollow facing brick. Hollow brick differs from structural clay tile in that it has more stringent physical property requirements, such as thicker shell and web dimensions and higher minimum compressive strengths.

Hollow brick has two grades of durability: Grade SW and Grade MW. Grade SW is used where a high and uniform degree of resistance to freeze-thaw degradation and disintegration by weathering is desired and when the brick may be saturated while frozen. Grade MW is intended for use where a moderate and somewhat nonuniform degree of resistance to freeze-thaw degradation is permissible.

ASTM C 652 also classifies hollow brick into four types of appearance: Types HBS, HBX, HBA and HBB. Type HBS (hollow brick standard) units are permitted for general use. Type HBX (hollow brick select) units are also for general use, but have a higher degree of dimensional precision than Type HBS units. Type HBA (hollow brick architectural) units are also for general use, but are intentionally manufactured to produce characteristic architectural effects resulting from nonuniformity in size and texture of the units. Type HBB (hollow building brick) units are for general use in masonry where a particular color, texture, finish, uniformity or limits on cracks, warpage or other imperfections detracting from their appearance are not a consideration.

ASTM C 1088 regulates thin brick used in adhered veneer having a maximum actual thickness of 1³/₄ inches (44 mm). Thin veneer brick units are specified in three types of appearance and two grades of durability.

Exterior-grade units are for exposure to weather; interior-grade units are not. Type TBS (standard) units

are permitted for general use. Type TBX (select) units are also for general use, but have a higher degree of precision and lower permissible variation in size than Type TBS units. Type TBA (architectural) units are also for general use, but are intentionally manufactured to produce characteristic effects in variations of size, color and texture.

ASTM C 34 is referenced for the manufacture of structural clay load-bearing wall tile. This type of tile has two grades: Grade LBX and LB. Grade LBX is suitable for general use in masonry construction and can be used in masonry exposed to weathering, provided that the units meet durability requirements for Grade SW of ASTM C 216 for solid units of face brick. This tile grade is also suitable for the direct application of stucco.

Grade LB is suitable for general use in masonry not exposed to freeze-thaw action, or for exposed masonry where protected with a minimum 3-inch (76 mm) facing of stone, brick or other masonry materials.

Fire-resistant tile intended for use in load-bearing masonry is required to conform to ASTM C 34. Tile intended for use in fireproofing structural members is required to be of such sizes and shapes as to cover completely the exposed surfaces of the members.

ASTM C 212, the referenced standard regulating structural clay facing tile, covers two types of structural clay load-bearing facing tile: Types FTX and FTS.

Type FTX clay facing tile is a smooth-face tile suitable for use in exposed exterior and interior masonry walls and partitions where low absorption, easy cleaning and resistance to staining are required. The physical characteristics of this tile require a high degree of mechanical perfection, narrow color range and minimum variation in face dimensions.

Type FTS clay facing tile is a smooth- or rough-textured face tile suitable for general use in exposed exterior and interior masonry walls and partitions where moderate absorption, variation in face dimensions, minor defects in surface finish and moderate color variations are permissible.

There are two classes of tile for the types stated above: standard (for general use) and special duty (having superior resistance to impact and moisture transmission and supporting greater lateral and vertical loads).

ASTM C 56 is the standard regulating the manufacture of nonload-bearing tile, which is made from the same materials as other types of tile previously mentioned and is used for partitions, fireproofing and furring.

ASTM C 126 prescribes requirements for ceramic-glazed structural clay load-bearing facing tile and brick and other solid masonry units made from clay, shale, fire clay or combinations thereof. The standard specifies two grades and two types of ceramic-glazed masonry: Grade S is used with comparatively narrow mortar joints, while Grade SS is used where variations in face dimensions are very small; Type I units are used where only one finished face is to be exposed, while Type II units are used where both finished faces are to be exposed.

The finish of glazed units is important. Requirements and test methods are prescribed for imperviousness, opacity, resistance to fading and crazing, hardness, abrasion resistance and flame spread smoke density. Compressive strength requirements, dimensional tolerances and permissible distortions are also prescribed by the standard.

The exception to this section allows structural clay tile that does not meet the compressive strength specifications to be used as nonstructural fireproofing for structural members. The fire-resistance rating must be determined in accordance with ASTM E 119.

2103.3 AAC masonry. AAC masonry units shall conform to ASTM C 1386 for the strength class specified.

❖ AAC masonry units are a low-density cementitious product first introduced into the 2005 edition of the MSJC Code and MSJC Specification, although its use has been prevalent in other countries for several decades. AAC masonry units must comply with ASTM C 1386 for the strength class specified.

2103.4 Stone masonry units. Stone masonry units shall conform to the following standards: ASTM C 503 for marble building stone (exterior); ASTM C 568 for limestone building stone; ASTM C 615 for granite building stone; ASTM C 616 for sandstone building stone; or ASTM C 629 for slate building stone.

❖ Many types of natural stone, including marble, granite, slate, limestone and sandstone, are used in building construction. This section covers natural stones that are sawed, cut, split or otherwise shaped for masonry purposes. Natural stones for building purposes are specified in various grades, textures and finishes and are required to have physical properties appropriate to their intended use.

Where applicable to specific kinds of natural stone, the finished units are required to be sound and free from spalls, cracks, open seams, pits and other defects that would impair structural strength and durability and, when applicable, fire resistance.

This section requires conformance to the following ASTM standards for natural stone masonry: ASTM C 503 for marble building stone (exterior); ASTM C 568 for limestone building stone; ASTM C 615 for granite building stone; ASTM C 616 for sandstone building stone or ASTM C 629 for slate building stone. The standards contain requirements for absorption, density (except for slate), compressive strength (except for slate), modulus of rupture, abrasion resistance (except for granite) and acid resistance (slate only).

2103.5 Ceramic tile. Ceramic tile shall be as defined in, and shall conform to the requirements of, ANSI A137.1.

❖ Ceramic tile is made from clay, possibly mixed with other ceramic materials. Metallic oxides may be included for glaze coloring. Ceramic tile products are available in a broad range of sizes, appearances, characteristics and functions.

ANSI A137.1 is the recognized industry standard for the manufacture, testing and labeling of ceramic tile. According to this standard, tile should be shipped in sealed cartons with the grade of contents indicated by grade seals with a distinctive coloring: blue for standard grade (units as perfect and free from defects as is possible in the manufacturing process) and yellow for "seconds" (units having slight defects, but free from structural defects and cracks).

ANSI A137.1 groups ceramic tile into four major types: glazed wall tile, mosaic tile, quarry tile and paving tile.

2103.6 Glass unit masonry. Hollow glass units shall be partially evacuated and have a minimum average glass face thickness of $^3/_{16}$ inch (4.8 mm). Solid glass-block units shall be provided when required. The surfaces of units intended to be in contact with mortar shall be treated with a polyvinyl butyral coating or latex-based paint. Reclaimed units shall not be used.

❖ Consensus national standards have not been written to establish minimum material properties or test methods for glass blocks. Reliance must be placed on the manufacturer's specifications and glass block is required to meet the minimum specified dimensions of those specifications. Glass block has generally performed well in service.

2103.7 Second-hand units. Second-hand masonry units shall not be reused unless they conform to the requirements of new units. The units shall be of whole, sound materials and free from cracks and other defects that will interfere with proper laying or use. Old mortar shall be cleaned from the unit before reuse.

❖ This section allows for the use of salvaged brick and other second-hand masonry units, provided that their quality and condition meet the requirements for new masonry units. Second-hand units must be whole, of sound material, clean and free from defects that would interfere with proper laying or use.

Most second-hand masonry units come from the demolition of old buildings. Masonry units manufactured in the past do not generally compare with the quality of masonry made by modern manufacturing methods under controlled conditions. Therefore, designers should expect salvaged masonry units to have lower strength and durability than new units.

Generally, the difference between walls laid up with new masonry units and second-hand units of the same type is the adhesion of the mortar to the masonry surfaces. When new masonry units are laid in fresh mortar, water and fine cementitious particles are absorbed into the masonry, thereby improving bond strength. In contrast, pores in the bed faces of second-hand masonry units, regardless of cleaning, are filled with particles of cement, lime and deleterious substances that impede adequate absorption, thereby adversely affecting bond between the mortar and the masonry.

2103.8 Mortar. Mortar for use in masonry construction shall conform to ASTM C 270 and shall conform to the proportion specifications of Table 2103.8(1) or the property specifications of Table 2103.8(2). Type S or N mortar shall be used for glass unit masonry. The amount of water used in mortar for glass unit masonry shall be adjusted to account for the lack of absorption. Retempering of mortar for glass unit masonry shall not be permitted after initial set. Unused mortar shall be discarded within $2^1/_2$ hours after initial mixing, except that unused mortar for glass unit masonry shall be discarded within $1^1/_2$ hours after initial mixing.

❖ Masonry mortar bonds masonry units to form an integral structure. Mortar provides a tight and weather-resistant seal between units; bonds with steel joint reinforcement, ties and other metal accessories and compensates for dimensional variations in masonry units. It can also serve aesthetic purposes through contrasts of color, texture and shadow lines created by different types of tooled joints.

This section of the code establishes ASTM C 270 as the standard regulating mortar to be used in masonry construction. The standard covers mortars for use in the construction of nonreinforced and reinforced unit masonry structures. It specifies types of mortar and two alternative specifications—the proportion specification and the property specification.

Type M mortar has the highest compressive and tensile bond strength. Its use should be limited only to those conditions where it is needed, such as high compressive loads or extreme environments.

Type S mortar is a general-purpose mortar with high compressive and tensile bond strengths. It is often used in reinforced masonry and in unreinforced masonry that requires high strength to resist out-of-plane lateral loads.

Type N mortar is a general-purpose mortar with medium compressive and tensile bond strengths. It is used where high vertical and lateral loads are not expected.

Type O is a low-strength mortar suitable for use in nonload-bearing construction and where the masonry will not be subject to severe weathering or freeze-thaw cycling.

The materials used in mortar are required to conform to the following standard specifications referenced in ASTM C 5, ASTM C 91, ASTM C 144, ASTM C 150, ASTM C 207, ASTM C 270, ASTM C 595 and ASTM C 1329.

In addition to type, mortar for masonry is further classified according to its primary cementitious materials. In portland cement-lime mortar, the cementitious materials are portland cement and hydrated mason's lime. In masonry-cement mortar and mortar-cement mortar, the principal cementitious material is portland cement, which is contained in the masonry cement and the mortar cement.

Regardless of a masonry mortar's principal cementitious constituents, each ingredient—cement, lime (when used), sand and water—contributes to the mortar's overall performance. Portland cement, mortar cement, masonry cement or blended hydraulic cement make the hardened mortar strong and durable. Lime contributes to the workability and water retention of fresh mortar and to the flexibility, compressive strength and tensile bond strength of hardened mortar. Sand serves as a filler material, increases strength and reduces shrinkage. Water determines the consistency of fresh mortar and also hydrates the cementitious materials, resulting in hardening of the mortar.

Tables 2103.8(1) and 2103.8(2) are based on Tables 1 and 2 of ASTM C 270. These tables provide requirements for the production of mortar by the proportion method (by volume) and by the property method (by compressive strength and other properties), respectively. Whether the proportion or property specification governs depends on the contract documents. When neither proportion nor property specification is prescribed, the proportion specification governs, unless data are presented to and accepted by the specifier to show that mortar meets the requirements of the property specification.

ASTM C 270 also covers materials and testing requirements for water retention, compressive strength and air content.

The types of mortar recommended for use in various types of masonry construction are shown in the appendix to ASTM C 270. The performance of masonry is influenced by mortar workability, water retentivity, bond strength, compressive strength and long-term deformability (creep). Each mortar type has corresponding use limitations. For example, allowable compressive stresses permitted in Section 2109.3.2 for empirical design are greater for Types M and S mortar than for Type N mortar.

Glass-block units are required to be laid in Type S or N mortar. Admixtures containing set accelerators or antifreeze compounds are not permitted in mortars for glass-block masonry.

TABLE 2103.8(1). See page 21-20.

❖ This table prescribes proportions for each mortar type when mortar is specified by proportion. The table covers portland cement-lime mortars, masonry-cement mortars and mortar-cement mortars. Materials are measured by volume. The required volume of mason's sand to be used is based on the combined volume of all cementitious materials (including lime, if applicable).

TABLE 2103.8(2). See page 21-20.

❖ This table prescribes minimum required physical properties of plastic and hardened mortar for each type when mortar is specified by property. The table covers portland cement-lime mortars, masonry-cement mortars and mortar-cement mortars. The specified properties are compressive strength, water retention and air content. Values are for mortars prepared in a laboratory in accordance with ASTM C 270. These values do not directly relate to specimens of field mortar tested in accordance with ASTM C 780.

TABLE 2103.8(1)
MORTAR PROPORTIONS

MORTAR	TYPE	PROPORTIONS BY VOLUME (cementitious materials)								AGGREGATE MEASURED IN A DAMP, LOOSE CONDITION
		Portland cement[a] or blended cement[b]	Masonry cement[c]			Mortar cement[d]			HYDRATED LIME[e] OR LIME PUTTY	
			M	S	N	M	S	N		
Cement-lime	M	1	—	—	—	—	—	—	$^1/_4$	Not less than $2^1/_4$ and not more than 3 times the sum of the separate volumes of cementitious materials
	S	1	—	—	—	—	—	—	over $^1/_4$ to $^1/_2$	
	N	1	—	—	—	—	—	—	over $^1/_2$ to $1^1/_4$	
	O	1	—	—	—	—	—	—	over $1^1/_4$ to $2^1/_2$	
Mortar cement	M	1	—	—	—	—	—	1	—	
	M	—	—	—	—	1	—	—	—	
	S	$^1/_2$	—	—	—	—	—	1	—	
	S	—	—	—	—	—	1	—	—	
	N	—	—	—	—	—	—	1	—	
	O	—	—	—	—	—	—	1	—	
Masonry cement	M	1	—	—	1	—	—	—	—	
	M	—	1	—	—	—	—	—	—	
	S	$^1/_2$	—	—	1	—	—	—	—	
	S	—	—	1	—	—	—	—	—	
	N	—	—	—	1	—	—	—	—	
	O	—	—	—	1	—	—	—	—	

a. Portland cement conforming to the requirements of ASTM C 150.
b. Blended cement conforming to the requirements of ASTM C 595.
c. Masonry cement conforming to the requirements of ASTM C 91.
d. Mortar cement conforming to the requirements of ASTM C 1329.
e. Hydrated lime conforming to the requirements of ASTM C 207.

TABLE 2103.8(2)
MORTAR PROPERTIES[a]

MORTAR	TYPE	AVERAGE COMPRESSIVE[b] STRENGTH AT 28 DAYS minimum (psi)	WATER RETENTION minimum (%)	AIR CONTENT maximum (%)
Cement-lime	M	2,500	75	12
	S	1,800	75	12
	N	750	75	14[c]
	O	350	75	14[c]
Mortar cement	M	2,500	75	12
	S	1,800	75	12
	N	750	75	14[c]
	O	350	75	14[c]
Masonry cement	M	2,500	75	18
	S	1,800	75	18
	N	750	75	20[d]
	O	350	75	20[d]

For SI: 1 inch = 25.4 mm, 1 pound per square inch = 6.895 kPa.

a. This aggregate ratio (measured in damp, loose condition) shall not be less than $2^1/_4$ and not more than 3 times the sum of the separate volumes of cementitious materials.
b. Average of three 2-inch cubes of laboratory-prepared mortar, in accordance with ASTM C 270.
c. When structural reinforcement is incorporated in cement-lime or mortar cement mortars, the maximum air content shall not exceed 12 percent.
d. When structural reinforcement is incorporated in masonry cement mortar, the maximum air content shall not exceed 18 percent.

2103.9 Surface-bonding mortar. Surface-bonding mortar shall comply with ASTM C 887. Surface bonding of concrete masonry units shall comply with ASTM C 946.

❖ Specifications for materials to be used in premixed surface-bonding mortar and for the properties of the mortar are contained in ASTM C 887. Requirements for masonry units and other materials used in constructing dry-stacked, surface-bonded masonry for walls and for the construction itself are contained in ASTM C 946.

In addition to its primary function of bonding masonry units, surface-bonding mortar can provide resistance to penetration by wind-driven rain.

2103.10 Mortars for ceramic wall and floor tile. Portland cement mortars for installing ceramic wall and floor tile shall comply with ANSI A108.1A and ANSI A108.1B and be of the compositions indicated in Table 2103.10.

❖ This section pertains to cement mortars and organic adhesives used for setting ceramic wall and floor tiles. Each mortar type has certain qualities that make it suitable for installing tile over different kinds of backing materials or under a certain set of conditions.

Cement mortars are "thick-bed" mortars applied in thicknesses of $^3/_4$ to $1^1/_4$ inches (19.1 to 32 mm) on floors and $^3/_4$ to 1 inch (19.1 to 25 mm) on walls to achieve the specified slopes and flatness in the finished tile work. Portland cement mortars are suitable for setting ceramic tile in most installations. They can be applied over properly prepared backings of clay or concrete masonry; concrete; wood frame; rough wood floors and plywood floors; foam insulation board; gypsum wallboard and portland cement or gypsum plaster. Cement mortars can be reinforced with metal lath or wire mesh; in such cases, however, additional mortar thickness may be required. Cement mortars have good structural strength and are not affected by prolonged contact with water.

Complete material and installation specifications are contained in ANSI A108.1A and A108.1B, which include required specifications for: installation of wire lath and scratch coats; mortar mixes, bond coat mixes and mortar application; installation methods on floors, walls and countertops; grouting of tile and general requirements for tile installations.

TABLE 2103.10. See next column.

❖ Portland cement-lime mortars are a mixture of portland cement (ASTM C 150), hydrated lime (ASTM C 207) and damp mason's sand (ASTM C 144). Proportions of these ingredients are given in Table 2103.10, dealing with the portland cement-lime mortars most commonly used for setting ceramic tile on wall, floor and ceiling construction and measured in parts by volume. Complete material and installation specifications are contained in ANSI A108.1.

TABLE 2103.10
CERAMIC TILE MORTAR COMPOSITIONS

LOCATION	MORTAR	COMPOSITION
Walls	Scratchcoat	1 cement; $^1/_5$ hydrated lime; 4 dry or 5 damp sand
Walls	Setting bed and leveling coat	1 cement; $^1/_2$ hydrated lime; 5 damp sand to 1 cement 1 hydrated lime, 7 damp sand
Floors	Setting bed	1 cement; $^1/_{10}$ hydrated lime; 5 dry or 6 damp sand; or 1 cement; 5 dry or 6 damp sand
Ceilings	Scratchcoat and sand bed	1 cement; $^1/_2$ hydrated lime; $2^1/_2$ dry sand or 3 damp sand

2103.10.1 Dry-set portland cement mortars. Premixed prepared portland cement mortars, which require only the addition of water and are used in the installation of ceramic tile, shall comply with ANSI A118.1. The shear bond strength for tile set in such mortar shall be as required in accordance with ANSI A118.1. Tile set in dry-set portland cement mortar shall be installed in accordance with ANSI A108.5.

❖ Dry-set portland cement mortar for ceramic tile is required to comply with ANSI A118.1, which describes test methods and minimum requirements for dry-set mortar, including sampling, free water content, setting characteristics, shrinkage, shear strength and staining.

Dry-set mortar is a mixture of portland cement, sand and perhaps water-retention admixtures. Dry-set mortars are suitable for use over properly prepared backings of clay or concrete masonry; concrete; cut-cell expanded polystyrene or rigid closed-cell urethane insulation board; gypsum wallboard; lean portland cement mortar and hardened wall and floor setting beds.

Installation must comply with ANSI A108.5. The "thin-bed" mortars are applied in a single layer as thin as $^3/_{32}$ inch (2.4 mm), but usually in thicknesses of $^1/_8$ to $^1/_4$ inch (3.2 to 6.4 mm). The use of dry-set mortar for leveling work is limited to a maximum thickness of $^1/_4$ inch (6.4 mm).

This material has excellent water and impact resistance. It is also water-cleanable, nonflammable, good for exterior use and requires no presoaking of the tile. Dry-set mortar is intended for use with glazed wall tile, ceramic mosaics, pavers and quarry tile.

Shear bond strength (of tile set in mortar) is required to be tested in accordance with standards applicable to the mortar used. Tile set with dry-set mortar must be installed in accordance with ANSI A108.5.

2103.10.2 Latex-modified portland cement mortar. Latex-modified portland cement thin-set mortars in which latex is added to dry-set mortar as a replacement for all or part of the gauging water that are used for the installation of ceramic tile shall comply with ANSI A118.4. Tile set in latex-modified portland cement shall be installed in accordance with ANSI A108.5.

❖ Latex-modified portland cement mortar is a mixture of portland cement, sand and special latex admixtures. It is applied as a thin-bed material, like dry-set portland cement mortar (see commentary, Section 2103.10.1). Since the latex used in these mortars varies among manufacturers, instructions for mixing and use must be followed carefully. Applicable material and installation standards are ANSI A108.5 and A118.4.

2103.10.3 Epoxy mortar. Ceramic tile set and grouted with chemical-resistant epoxy shall comply with ANSI A118.3. Tile set and grouted with epoxy shall be installed in accordance with ANSI A108.6.

❖ Epoxy mortar is a two-part system consisting of epoxy resin and hardener. It is applied as a layer of $^1/_{16}$ to $^1/_8$ inch (1.6 to 3.2 mm) in thickness and is suitable for use on properly prepared floors of concrete, wood, plywood, steel plate or ceramic tile. It is particularly suitable where chemical resistance or high bond strength is required. Deflection control is critical for the successful use of this material, and floor systems should not deflect more than $^1/_{360}$ of their span. Epoxy mortar is recommended for setting ceramic mosaics, quarry tile and paver tile. Applicable material and installation standards are ANSI A108.6 and A118.3.

2103.10.4 Furan mortar and grout. Chemical-resistant furan mortar and grout that are used to install ceramic tile shall comply with ANSI A118.5. Tile set and grouted with furan shall be installed in accordance with ANSI A108.8.

❖ Furan mortar is a two-part system consisting of a furan resin and a hardener. It is suitable where chemical resistance is critical. It is used primarily on floors in laboratories and industrial plants. Acceptable subfloors include concrete, steel plate and ceramic tile. Furan grout is intended for quarry tile and pavers, mainly in industrial areas requiring maximum chemical resistance.

ANSI A118.5 is the standard regulating both furan mortar and grout. Installation must comply with ANSI A108.8.

2103.10.5 Modified epoxy-emulsion mortar and grout. Modified epoxy-emulsion mortar and grout that are used to install ceramic tile shall comply with ANSI A118.8. Tile set and grouted with modified epoxy-emulsion mortar and grout shall be installed in accordance with ANSI A108.9.

❖ Modified epoxy-emulsion mortar and grout used to install ceramic tile are required to comply with ANSI A118.8 and to be installed in accordance with ANSI A108.9.

ANSI A118.8 describes the test methods and the minimum requirements for modified epoxy-emulsion mortar and grout. The chemical and solvent resistance of these mortars and grouts tends to exceed those of organic adhesives and equal those of latex-modified portland cement mortars. They are not, however, de-signed to meet the requirements of ANSI A108.6 or A118.3.

These types of mortars and grouts are three-part systems that include emulsified epoxy resins and hardeners, preblended portland cement and silica sand. They are used as a bond-coat setting mortar or grout, and can be cleaned from wall and floor surfaces using a wet sponge prior to initial set.

ANSI A118.8 regulates water absorption, flexural strength, thermal expansion, linear shrinkage, tensile strength and compressive strength.

2103.10.6 Organic adhesives. Water-resistant organic adhesives used for the installation of ceramic tile shall comply with ANSI A136.1. The shear bond strength after water immersion shall not be less than 40 psi (275 kPa) for Type I adhesive and not less than 20 psi (138 kPa) for Type II adhesive when tested in accordance with ANSI A136.1. Tile set in organic adhesives shall be installed in accordance with ANSI A108.4.

❖ Organic adhesives are prepared materials that cure or set by evaporation and that are ready to use without adding liquid or powder. Adhesives are suitable for installing tiles on prepared wall and floor surfaces, including: brick and concrete masonry; concrete; gypsum wallboard; portland cement or gypsum plaster and wood-flooring systems.

Organic adhesives (mastics) are applied as a thin layer approximately $^1/_{16}$ inch (1.6 mm) thick. An underlayment is used to level and true surfaces. Organic adhesive does not permit the soaking of tiles and is not suitable for exterior use. Bond strength varies greatly among the numerous brands of organic adhesives available for use in construction. Adhesives must meet minimum bond-strength requirements of this section of ANSI A136.1 for Type I and II adhesives. The installation of tile with organic adhesives is required to conform to ANSI A108.4.

2103.10.7 Portland cement grouts. Portland cement grouts used for the installation of ceramic tile shall comply with ANSI A118.6. Portland cement grouts for tile work shall be installed in accordance with ANSI A108.10.

❖ Portland cement grouts are the most commonly used grouts for tile walls. The mixture of portland cement and other ingredients is water resistant and uniform in color. The water in the grout is essential to promote a good bond and to develop full grout strength. This type of grout is required to comply with ANSI A118.6 and to be installed in accordance with ANSI A108.10.

2103.11 Mortar for AAC masonry. Thin-bed mortar for AAC masonry shall comply with Section 2103.11.1. Mortar for leveling courses of AAC masonry shall comply with Section 2103.11.2.

❖ Except for the initial leveling courses of AAC masonry, which uses Type M or S masonry mortar, a special thin-bed mortar is used to bind AAC masonry units.

2103.11.1 Thin-bed mortar for AAC masonry. Thin-bed mortar for AAC masonry shall be specifically manufactured for use with AAC masonry. Testing to verify mortar properties shall be conducted by the thin-bed mortar manufacturer and confirmed by an independent testing agency:

1. The compressive strength of thin-bed mortar, as determined by ASTM C 109, shall meet or exceed the strength of the AAC masonry units.

2. The shear strength of thin-bed mortar shall meet or exceed the shear strength of the AAC masonry units for wall assemblages tested in accordance with ASTM E 519.

3. The flexural tensile strength of thin-bed mortar shall not be less than the modulus of rupture of the masonry units. Flexural strength shall be determined by testing in accordance with ASTM E 72 (transverse load test), ASTM E 518 Method A (flexural bond strength test) or ASTM C 1072 (flexural bond strength test).

   3.1. For conducting flexural strength tests in accordance with ASTM E 518, at least five test specimens shall be constructed as stack-bonded prisms at least 32 inches (810 mm) high. The type of mortar specified by the AAC unit manufacturer shall be used.

   3.2. For flexural strength tests in accordance with ASTM C 1072, test specimens shall be constructed as stack-bonded prisms comprised with at least three bed joints. A total of at least five joints shall be tested using the type of mortar specified by the AAC unit manufacturer.

4. The splitting tensile strength of AAC masonry assemblages composed of two AAC masonry units bonded with one thin-bed mortar joint shall be determined in accordance with ASTM C 1006 and shall equal or exceed $2.4\sqrt{f'_{AAC}}$.

❖ Thin-bed mortar is required to meet the strength requirements of this section, which are consistent with those in the MSJC Specification.

2103.11.2 Mortar for leveling courses of AAC masonry. Mortar used for the leveling courses of AAC masonry shall conform to Section 2103.8 and shall be Type M or S.

❖ Traditional masonry mortar, either Type M or S, is used for the leveling course of AAC masonry to account for any variances in the foundation, and thus provide a level surface for other courses of AAC masonry. Thin-bed mortar is used for subsequent courses of AAC masonry.

2103.12 Grout. Grout shall conform to Table 2103.12 or to ASTM C 476. When grout conforms to ASTM C 476, the grout shall be specified by proportion requirements or property requirements.

❖ Grout intended for use in the construction of engineered and empirically designed masonry structures is required to comply with ASTM C 476, which regulates materials, measurement, mixing and storage of materials.

Two types of grout are used in masonry: fine and coarse. Fine grout is made of cement, sand and water, with optional small quantities of lime. Coarse grout includes the same ingredients, plus pea gravel, which is limited to $3/8$ inch (9.5 mm) in size.

Whether fine or coarse grout is used depends on the size of the grout space to be filled and the height of the grout pour to be used in construction. Table 1.16.1 of the MSJC Code contains the limiting cavity and cell sizes for a given grout pour height for both fine and coarse grouts.

The materials used in masonry grout are required to be listed in ASTM C 476 and must conform to the following standard specifications: ASTM C 5; ASTM C 150; ASTM C 207; ASTM C 404 and ASTM C 595.

Grout may be specified either by proportion or property. When the proportion method is specified, Table 2103.12 provides required proportions of grout materials by volume. Information on property requirements is given in ASTM C 476.

TABLE 2103.12
GROUT PROPORTIONS BY VOLUME FOR MASONRY CONSTRUCTION

TYPE	PARTS BY VOLUME OF PORTLAND CEMENT OR BLENDED CEMENT	PARTS BY VOLUME OF HYDRATED LIME OR LIME PUTTY	AGGREGATE, MEASURED IN A DAMP, LOOSE CONDITION	
			Fine	Coarse
Fine grout	1	$0-1/10$	$2^1/_4$-3 times the sum of the volumes of the cementitious materials	—
Coarse grout	1	$0-1/10$	$2^1/_4$-3 times the sum of the volumes of the cementitious materials	1-2 times the sum of the volumes of the cementitious materials

❖ This table, which is based on ASTM C 476, prescribes volume proportions of grout for masonry. Information on fine and coarse grout is given in the commentary to the MSJC Specification. Additionally, the MSJC Code provides specific requirements for the maximum pour height for both fine and coarse grout.

2103.13 Metal reinforcement and accessories. Metal reinforcement and accessories shall conform to Sections 2103.13.1 through 2103.13.8.

❖ This section contains standards and material requirements for anchors, joint reinforcement, wire accessories, ties, wire fabric and for required corrosion protection of these items.

2103.13.1 Deformed reinforcing bars. Deformed reinforcing bars shall conform to one of the following standards: ASTM A 615 for deformed and plain billet-steel bars for concrete reinforcement; ASTM A 706 for low-alloy steel deformed bars for concrete reinforcement; ASTM A 767 for zinc-coated reinforcing steel bars; ASTM A 775 for epoxy-coated reinforcing steel bars; and ASTM A 996 for rail and axle steel-deformed bars for concrete reinforcement.

❖ ASTM A 615 regulates the manufacture of plain and deformed reinforcement, Grades 40 and 60, for new or recycled steel.

ASTM A 706 regulates the manufacture of plain and deformed reinforcement, Grade 60 only, from mold-cast or strand-cast steel. This steel is intended for welding, in accordance with the procedures of AWS D1.4. It is more ductile than other steel and is especially suitable for structures in zones of high seismicity (see commentary, Section 2108.3).

Where masonry is expected to be exposed to especially corrosive conditions, galvanized reinforcement meeting ASTM A 767 or epoxy-coated reinforcement meeting ASTM A 775 can be used.

ASTM A 996 regulates rail steel and axle steel deformed bars.

Deformed reinforcement is recommended for nearly every application in reinforced masonry and is required in most instances.

2103.13.2 Joint reinforcement. Joint reinforcement shall comply with ASTM A 951. The maximum spacing of crosswires in ladder-type joint reinforcement and point of connection of cross wires to longitudinal wires of truss-type reinforcement shall be 16 inches (400 mm).

❖ ASTM A 951 contains material requirements, mechanical properties and tolerances for joint reinforcement. Longitudinal wires are required to be deformed in accordance with this standard to provide a mechanical bond with the surrounding mortar. Cross wires are plain.

ASTM A 82 regulates the manufacture of cold-drawn, plain steel wire for joint reinforcement, wire anchors and ties. Joint reinforcement can be epoxy coated to provide added corrosion protection (see commentary, Section 2103.13.7).

2103.13.3 Deformed reinforcing wire. Deformed reinforcing wire shall conform to ASTM A 496.

❖ ASTM A 496 regulates the manufacture of cold-worked, deformed wire having yield stresses of 70 ksi (483 MPa) (welded wire fabric) and 75 ksi (517 MPa). This wire is used as reinforcement and in the manufacture of welded, deformed wire fabric. It is not commonly used in masonry.

2103.13.4 Wire fabric. Wire fabric shall conform to ASTM A 185 for plain steel-welded wire fabric for concrete reinforcement or ASTM A 497 for welded deformed steel wire fabric for concrete reinforcement.

❖ ASTM A 185 regulates the manufacture of plain welded wire fabric, using ASTM A 82 steel. ASTM A 497 regulates the manufacture of welded deformed wire fabric using steels conforming to ASTM A 82 and A 496, alone or in combination. Wire fabric is seldom used in masonry.

2103.13.5 Anchors, ties and accessories. Anchors, ties and accessories shall conform to the following standards: ASTM A 36 for structural steel; ASTM A 82 for plain steel wire for concrete reinforcement; ASTM A 185 for plain steel-welded wire fabric for concrete reinforcement; ASTM A 240 for chromium and chromium-nickle stainless steel plate, sheet and strip; ASTM A 307 Grade A for anchor bolts; ASTM A 480 for flat rolled stainless and heat-resisting steel plate, sheet and strip; and ASTM A 1008 for cold-rolled carbon steel sheet.

❖ Anchors, ties and accessories are manufactured in a variety of ways with a number of materials. Applicable ASTM specifications prescribe minimum requirements for these materials.

ASTM A 36 regulates the manufacture of hot-rolled steel with a minimum specified yield stress of 36 ksi (248 MPa), for general structural purposes. This material is suitable for welding.

ASTM A 82 regulates the manufacture of cold-drawn, plain-steel wire for joint reinforcement, wire anchors and ties.

ASTM A 185 regulates the manufacture of plain-steel welded-wire fabric, using ASTM A 82 steel or ASTM A 167 Type 304, corrosion-resistant steel.

ASTM A 366 regulates the manufacture of sheet steel used for metal anchors and ties. The specification covers cold-rolled carbon sheet steel of commercial quality, in coils or cut lengths. This material is intended for exposed or unexposed parts made by bending, moderate drawing, forming or welding. This standard also specifies mechanical properties and tests relative to bending, hardness and moderate deformability and sets requirements for chemical composition. See Figure 2109.6.3(1) for typical masonry accessories.

2103.13.6 Prestressing tendons. Prestressing tendons shall conform to one of the following standards:

1. Wire . . . . ASTM A 421

2. Low-relaxation wire . . . . ASTM A 421

3. Strand . . . . ASTM A 416

4. Low-relaxation strand . . . . ASTM A 416

5. Bar . . . . ASTM A 722

Exceptions:

1. Wire, strands and bars not specifically listed in ASTM A 421, ASTM A 416 or ASTM A 722 are permitted, provided they conform to the minimum requirements in ASTM A 421, ASTM A 416 or ASTM A 722 and are approved by the architect/engineer.

2. Bars and wires of less than 150 kips per square inch (ksi) (1034 MPa) tensile strength and conforming to ASTM A 82, ASTM A 510, ASTM A 615, ASTM A 996 or ASTM A 706 are permitted to be used as prestressed tendons, provided that:

    2.1. The stress relaxation properties have been assessed by tests according to ASTM E 328 for the maximum permissible stress in the tendon.

2.2. Other nonstress-related requirements of ACI 530/ASCE 5/TMS 402, Chapter 4, addressing prestressing tendons are met.

❖ This section specifies the materials for components of the prestressing system. They are similar to those used in prestressed concrete.

2103.13.7 Corrosion protection. Corrosion protection for prestressing tendons shall comply with the requirements of ACI 530.1/ASCE 6/TMS 602, Article 2.4G. Corrosion protection for prestressing anchorages, couplers and end blocks shall comply with the requirements of ACI 530.1/ASCE 6/TMS 602, Article 2.4H. Corrosion protection for carbon steel accessories used in exterior wall construction or interior walls exposed to a mean relative humidity exceeding 75 percent shall comply with either Section 2103.13.7.2 or 2103.13.7.3. Corrosion protection for carbon steel accessories used in interior walls exposed to a mean relative humidity equal to or less than 75 percent shall comply with either Section 2103.13.7.1, 2103.13.7.2 or 2103.13.7.3.

❖ Joint reinforcement, anchors, wall ties and accessories are required to be galvanized to protect the steel from deterioration and corrosion, unless the steel is inherently corrosion resistant, such as ASTM A 167 Type 304 stainless steel. The application and thickness of the galvanizing depends on the intended location or severity of exposure. Carbon steel accessories located in exterior masonry wall construction are required to be either hot-dipped galvanized in accordance with Section 2103.13.7.2 or epoxy coated in accordance with Section 2103.13.7.3. Carbon steel accessories for use in interior-wall construction are permitted to comply with less-restrictive mill galvanized requirements as noted in Section 2103.13.7.1.

2103.13.7.1 Mill galvanized. Mill galvanized coatings shall be applied as follows:

1. For joint reinforcement, wall ties, anchors and inserts, a minimum coating of 0.1 ounce per square foot ($31g/m^2$) complying with the requirements of ASTM A 641 shall be applied.

2. For sheet metal ties and sheet metal anchors, a minimum coating complying with Coating Designation G-60 according to the requirements of ASTM A 653 shall be applied.

3. For anchor bolts, steel plates or bars not exposed to the earth, weather or a mean relative humidity exceeding 75 percent, a coating is not required.

❖ Carbon steel accessories for use in interior wall construction are permitted to comply with less restrictive material specifications (ASTM A 641), as noted.

2103.13.7.2 Hot-dipped galvanized. Hot-dipped galvanized coatings shall be applied after fabrication as follows:

1. For joint reinforcement, wall ties, anchors and inserts, a minimum coating of 1.5 ounces per square foot (458 g/m²) complying with the requirements of ASTM A 153, Class B shall be applied.

2. For sheet metal ties and anchors, the requirements of ASTM A 153, Class B shall be met.

3. For steel plates and bars, the requirements of either ASTM A 123 or ASTM A 153, Class B shall be met.

❖ Carbon steel accessories for use in interior-wall construction are permitted to comply with less-restrictive material specifications (ASTM A 641), as noted.

2103.13.7.3 Epoxy coatings. Carbon steel accessories shall be epoxy coated as follows:

1. For joint reinforcement, the requirements of ASTM A 884, Class A, Type 1 having a minimum thickness of 7 mils (175 μm) shall be met.

2. For wire ties and anchors, the requirements of ASTM A 899, Class C having a minimum thickness of 20 mils (508 μm) shall be met.

3. For sheet metal ties and anchors, a minimum thickness of 20 mils (508 μm) per surface shall be provided or a minimum thickness in accordance with the manufacturer's specification shall be provided.

❖ As an alternative to hot-dipped galvanizing, carbon steel accessories in exterior walls can be epoxy coated. This section specifies the appropriate coating type and minimum coating thickness applicable to each type of accessory.

2103.13.8 Tests. Where unidentified reinforcement is approved for use, not less than three tension and three bending tests shall be made on representative specimens of the reinforcement from each shipment and grade of reinforcing steel proposed for use in the work.

❖ Reinforcing bars are to be rolled with raised symbols or letters impressed on the metal identifying the manufacturing mill. When required by the building official, the grade of material is to be identified by a satisfactory mill test. When the manufacturing mill is not identified but the reinforcement is approved for use under ordinary material procedures, at least three tension and three bending tests are required on representative specimens of the reinforcement from each shipment and grade of reinforcing steel proposed for use. The tension and bending tests are required to be performed by an approved testing agency.

SECTION 2104
CONSTRUCTION

2104.1 Masonry construction. Masonry construction shall comply with the requirements of Sections 2104.1.1 through 2104.5 and with ACI 530.1/ASCE 6/TMS 602.

❖ This section establishes the requirements, based on accepted practice and referenced standards, regulating materials and construction methods used in engineered and empirically designed masonry construction. Engineered masonry construction is further regulated by Sections 2101.2.1, 2101.2.2 and 2101.2.3.

2104.1.1 Tolerances. Masonry, except masonry veneer, shall be constructed within the tolerances specified in ACI 530.1/ASCE 6/TMS 602.

❖ The MSJC Specification, ACI 530.1/ASCE 6/IMS 602 prescribes tolerances for placement of reinforcement and masonry units. Specific tolerances listed include cross sections of elements; thicknesses of mortar joints; widths of grout spaces; variation from level; variation from plumb; trueness to a line; alignment of columns and walls; location of elements and placement of reinforcement and accessories. These required tolerances are intended to inhibit corrosion; provide placement locations reflecting those assumed in design; provide clearance for mortar and grout and maintain compatible lateral deflections of parallel wythes. Tolerances are relatively strict since masonry is a structural material and is also exposed to weather. Aesthetics are not a factor in these requirements.

2104.1.2 Placing mortar and units. Placement of mortar and clay and concrete units shall comply with Sections 2104.1.2.1, 2104.1.2.2, 2104.1.2.3 and 2104.1.2.5. Placement of mortar and glass unit masonry shall comply with Sections 2104.1.2.4 and 2104.1.2.5. Placement of thin-bed mortar and AAC masonry shall comply with Section 2104.1.2.6.

❖ Workmanship is of primary importance in providing strong and durable masonry construction. This is especially true for the placement of mortar and masonry units. Water penetration can often be traced to bond breaks at the mortar-unit interface. Additionally, masonry strength relies on the mortar bedded areas required in this section and the MSJC Specification.

2104.1.2.1 Bed and head joints. Unless otherwise required or indicated on the construction documents, head and bed joints shall be $^3/_8$ inch (9.5 mm) thick, except that the thickness of the bed joint of the starting course placed over foundations shall not be less than $^1/_4$ inch (6.4 mm) and not more than $^3/_4$ inch (19.1 mm).

❖ The required thickness for bed joints has a permitted tolerance of plus or minus $^1/_8$ inch (3 mm). This thickness is intended to provide bonding of reinforcement and ties and to allow for the dimensional tolerances of the masonry units. The greater permitted thickness of the starting mortar course is intended to accommodate variations in the elevation of the top surface of the foundation.

2104.1.2.1.1 Open-end units. Open-end units with beveled ends shall be fully grouted. Head joints of open-end units with beveled ends need not be mortared. The beveled ends shall form a grout key that permits grouts within $^5/_8$ inch (15.9 mm) of the face of the unit. The units shall be tightly butted to prevent leakage of the grout.

❖ Open-end units can be placed around vertical reinforcing steel, rather than having to be lifted up and over it, as closed-end units must be. Such units are manufactured with one or no end shells.

The special type of open-end unit described in this section is manufactured with beveled ends of the faceshells, so that when the unit is grouted, grout fills the head joint seam (see Figure 2104.1.2.1.1). Because of this, head joints of such construction need not be filled with mortar. Open-end units without this bevel must be laid with mortared head joints.

Figure 2104.1.2.1.1
MORTARLESS HEAD JOINT UNIT

2104.1.2.2 Hollow units. Hollow units shall be placed such that face shells of bed joints are fully mortared. Webs shall be fully mortared in all courses of piers, columns, pilasters, in the starting course on foundations where adjacent cells or cavities are to be grouted, and where otherwise required. Head joints shall be mortared a minimum distance from each face equal to the face shell thickness of the unit.

❖ Figure 2104.1.2.2 shows the typical placement of mortar for hollow-unit masonry walls. Except for the initial bed joint at the starter course, the cross webs of hollow units in walls are not usually mortared except at the cross webs of cells that are to be grouted in partially grouted walls. Cross webs of hollow units in piers, columns and pilasters, however, are required to be mortared.

2104.1.2.3 Solid units. Unless otherwise required or indicated on the construction documents, solid units shall be placed in fully mortared bed and head joints. The ends of the units shall be completely buttered. Head joints shall not be filled by slushing with mortar. Head joints shall be constructed by shoving mortar tight against the adjoining unit. Bed joints shall not be furrowed deep enough to produce voids.

❖ Solid units result in fully mortared bed joints since the 25-percent core area is effectively crossed with mortar. Height-to-thickness ratios and calculated stresses are based on the fully mortared bed joint.

Filling head joints by slushing from above is not permitted, since this could result in voids and lead to water penetration. The head-joint mortar may be placed on the end of the masonry unit prior to placing the unit in the wall. The furrow typically formed during bed joint placement should not be too deep since mortar voids would result (see Figure 2104.1.2.3).

2104.1.2.4 Glass unit masonry. Glass units shall be placed so head and bed joints are filled solidly. Mortar shall not be furrowed.

Unless otherwise required, head and bed joints of glass unit masonry shall be $1/4$ inch (6.4 mm) thick, except that vertical joint thickness of radial panels shall not be less than $1/8$ inch (3.2 mm). The bed joint thickness tolerance shall be minus $1/16$ inch (1.6 mm) and plus $1/8$ inch (3.2 mm). The head joint thickness tolerance shall be plus or minus $1/8$ inch (3.2 mm).

❖ Glass unit masonry is required to be placed with full head and bed joints. Glass units are manufactured to be modular with a joint thickness of $1/4$ inch (6.4 mm), smaller than the typical $3/8$ inch (9.6 mm) for concrete masonry and the $3/8$ inch (9.6 mm) or $1/2$ inch (12.8 mm) joint thickness for clay brick masonry. Designers should recognize this difference and use correct nominal dimensions when laying out glass unit masonry.

Tolerances for glass unit masonry are much tighter than for other types of masonry because glass unit masonry is manufactured to very tight tolerances and is more dimensionally stable than other types of masonry.

Head joints on radial panels of glass unit masonry are permitted to be as thin as $1/8$ inch (3.2 mm), because placing rectangular units in a curved wall requires wider joints on one face of the wall and narrower joints on the other face.

2104.1.2.5 Placement in mortar. Units shall be placed while the mortar is soft and plastic. Any unit disturbed to the extent that the initial bond is broken after initial positioning shall be removed and relaid in fresh mortar.

❖ Mortar should not be spread too far ahead of units, as it will stiffen and lose plasticity, especially in hot weather. Mortar that has stiffened should not be used. ASTM C 270 requires that mortar be used within 2½

Figure 2104.1.2.2
MORTAR PLACEMENT ON HOLLOW UNITS IN WALLS

hours of initial mixing.

A test for mortar workmanship includes removal of a unit. Bonded mortar will stick to the removed unit as well as the remaining masonry; stiff, unbonded mortar will not. Broken joints should be remortared.

2104.1.2.6 Thin-bed mortar and AAC masonry units. AAC masonry construction shall begin with a leveling course of masonry meeting the requirements of Section 2104.1.2. Subsequent courses of AAC masonry units shall be laid with thin-bed mortar using a special notched trowel manufactured for use with thin-bed mortar to spread the mortar so that it completely fills the bed joints. Unless otherwise specified, the head joints shall be similarly filled. Joints in AAC masonry shall be approximately $^1/_{16}$ inch (1.5 mm) and shall be formed by striking on the ends and tops of AAC masonry units with a rubber mallet. Minor adjustments in unit position shall be made while the mortar is still soft and plastic by tapping it into the proper position. Minor sanding of the exposed faces of AAC masonry shall be permitted to provide a smooth and plumb surface.

❖ Placement of AAC masonry on thin-bed mortar differs from placement of brick, concrete masonry or stone units, in that a traditional masonry trowel, mortar and laying procedures do not effectively work for AAC masonry and thin-bed mortar. This section, therefore, describes how AAC units must be laid to adequately develop bond between the units and the thin-bed mortar.

Traditional masonry mortar, either Type M or S, is used for the leveling course of AAC masonry to account for any variances in the foundation, and thus provide a level surface for other courses of AAC masonry. Thin-bed mortar is used for subsequent courses of masonry.

2104.1.2.7 Grouted masonry. Between grout pours, a horizontal construction joint shall be formed by stopping all wythes at the same elevation and with the grout stopping a minimum of $1^1/_2$ inches (38 mm) below a mortar joint, except at the top of the wall. Where bond beams occur, the grout pour shall be stopped a minimum of $^1/_2$ inch (12.7 mm) below the top of the masonry.

❖ When there is a possibility of a cold joint forming between grout lifts, a grout key between subsequent grout lifts that does not coincide with the horizontal mortar joint is preferred.

2104.1.3 Installation of wall ties. The ends of wall ties shall be embedded in mortar joints. Wall tie ends shall engage outer face shells of hollow units by at least $^1/_2$ inch (12.7 mm). Wire wall ties shall be embedded at least $1^1/_2$ inches (38 mm) into the mortar bed of solid masonry units or solid-grouted hollow units. Wall ties shall not be bent after being embedded in grout or mortar.

❖ Installation of wall ties requires adequate embedment in mortar, and adequate strength and durability of ties.

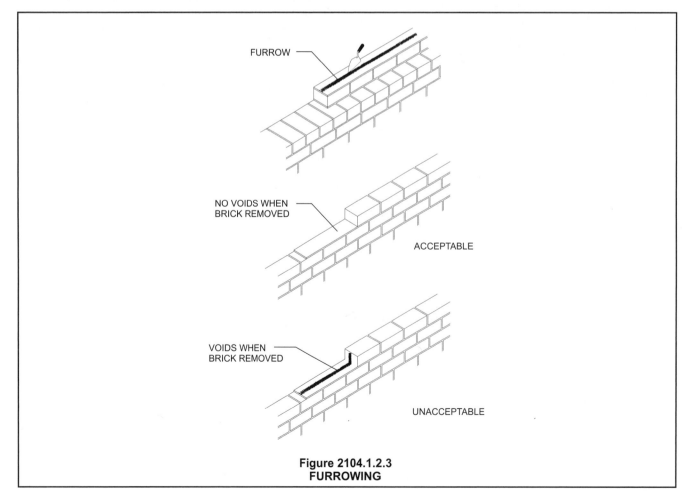

FURROW

NO VOIDS WHEN BRICK REMOVED

ACCEPTABLE

VOIDS WHEN BRICK REMOVED

UNACCEPTABLE

Figure 2104.1.2.3
FURROWING

Wall ties are not permitted to be bent after being embedded in grout or mortar, since movement of the ties will reduce the effectiveness of the embedment and subsequent bonding.

2104.1.4 Chases and recesses. Chases and recesses shall be constructed as masonry units are laid. Masonry directly above chases or recesses wider than 12 inches (305 mm) shall be supported on lintels.

❖ Chases and recesses are designed so that they do not reduce the required strength of the walls or affect the required fire-resistance rating of the wall. Any required chases must be formed during construction of the masonry wall and not by cutting out a section of the finished wall, which could create a weak point and jeopardize the wall strength.

This empirical limitation requires lintels for chases and recesses wider than 12 inches (305 mm). Although arching action reduces lintel moments, the potential for weaknesses created by chases and recesses makes them subject to engineered design. The supporting lintel is permitted to be of masonry, concrete or steel and must be engineered as required in Section 2104.1.5.

2104.1.5 Lintels. The design for lintels shall be in accordance with the masonry design provisions of either Section 2107 or 2108. Minimum length of end support shall be 4 inches (102 mm).

❖ Masonry lintels are required to be engineered as load-bearing beam elements by either the allowable stress design method of Section 2107 or the strength design method of Section 2108. This requirement is consistent with the empirical provisions, which do not include horizontally spanning elements such as beams. Bearing on masonry is required to be a minimum of 4 inches (102 mm) to adequately distribute stresses.

2104.1.6 Support on wood. Masonry shall not be supported on wood girders or other forms of wood construction except as permitted in Section 2304.12.

❖ Masonry is only permitted to be supported on wood construction, in accordance with Section 2304.12, because of both the danger of collapse in case of fire and the serviceability considerations noted below.

Masonry is brittle and relatively weak in tension. It can crack when subjected to deformations. Wood is flexible and usually exhibits elastic deformation and creep (additional deflection from long-term loads). Because of this, masonry supported on wood tends to crack. Therefore, masonry that is supported by wood members must consider long-term effects in accordance with Section 2304.12.

Glass-block panels are typically limited in size and are often used in windows as decoration. For this limited use, support on wood is permitted. The deflection limitation in Section 2110.4.2 is consistent with the requirements for masonry. In calculating the probable deflection of the supporting wood member, consideration of creep is recommended.

2104.1.7 Masonry protection. The top of unfinished masonry work shall be covered to protect the masonry from the weather.

❖ Water can accumulate within masonry walls and it evaporates relatively slowly. Moisture migrates to the surface, dissolves salts in the masonry and deposits them on the surface as a white powder called "efflorescence." To prevent this, the tops of walls exposed to weather are required to be covered.

2104.1.8 Weep holes. Weep holes provided in the outside wythe of masonry walls shall be at a maximum spacing of 33 inches (838 mm) on center (o.c.). Weep holes shall not be less than $3/_{16}$ inch (4.8 mm) in diameter.

❖ Masonry is not water tight; water can penetrate into and through it. Weep holes are to be provided in all masonry wall construction to help water leave the wall. Without weep holes, moisture would remain within the masonry construction and evaporation would be its only means of exit.

Many methods for providing weep holes are available, including open head joints; $3/_8$-inch-diameter (9.6 mm) holes in joints and wicking cord laid in the joint and behind the brick.

The 33-inch (838 mm) maximum spacing of weep holes, while apparently odd since it is not modular for typical masonry, was selected because weep holes are often spaced at 32 inches (813 mm) on center, plus or minus typical construction tolerances.

2104.2 Corbeled masonry. Except for corbels designed per Section 2107 or 2108, the following shall apply:

1. Corbels shall be constructed of solid masonry units.

2. The maximum corbeled projection beyond the face of the wall shall not exceed:

    2.1. One-half of the wall thickness for multiwythe walls bonded by mortar or grout and wall ties or masonry headers or

    2.2. One-half the wythe thickness for single wythe walls, masonry bonded hollow walls, multiwythe walls with open collar joints and veneer walls.

3. The maximum projection of one unit shall not exceed:

    3.1. One-half the nominal unit height of the unit or

    3.2. One-third the nominal thickness of the unit or wythe.

4. The back surface of the corbelled section shall remain within 1 inch (25 mm) of plane.

❖ This section covers the limitations and the construction features of corbeled masonry.

A corbel is formed by projecting courses of masonry, with the first or lowest course projecting out from the face of the wall and each successive course projecting out from the supporting course below. While corbels may be constructed using various load-bearing masonry units, they are most commonly applied to clay masonry and are used for aesthetic purposes and to support offset or thickened walls above them.

The permitted extent of corbeling is illustrated in Figure 2104.2. The total horizontal projection of a corbel from the face of the wall is limited to no more than one-half the wall thickness of a solid wall or more than one-half the thickness of the wythe in a cavity wall. Furthermore, the projection of a single course is not to exceed one-half of the unit height, nor one-third of the unit bed depth, whichever is less.

These limitations establish a maximum angle of corbelled brick masonry, measured from the plane of the wall, of about 26.5 degrees (0.46 rad). A smaller angle can be achieved by decreasing the brick projections to less than the maximum allowed by this section.

Corbels produce and often transfer eccentric loads, which must be considered in the design of masonry walls.

Corbels constructed in walls of hollow units or in hollow masonry walls are required to be of solid masonry. This requirement can be satisfied by using solid units, grouting the cores of hollow units or grouting open spaces in hollow walls.

2104.2.1 Molded cornices. Unless structural support and anchorage are provided to resist the overturning moment, the center of gravity of projecting masonry or molded cornices shall lie within the middle one-third of the supporting wall. Terra cotta and metal cornices shall be provided with a structural frame of approved noncombustible material anchored in an approved manner.

❖ Figure 2104.2.1 illustrates this requirement. Unless cornices or other projections are specifically designed

For SI: 1 degree = 0.01745 rad.

Figure 2104.2
CORBELED MASONRY

Figure 2104.2.1
MOLDED CORNICE

to be supported and anchored to the wall or other adequate structural supports, such as columns, beams, spandrels or floor slabs, the center of gravity of the projecting element is required to fall within the middle third (kern) of the supporting wall.

Cornices are usually ornamental and may or may not support other loads. They can be made of cast-in-place or precast concrete; cast or natural stone; terra cotta; brick or even metal.

Metal and terra-cotta cornices are commonly furnished with a structural frame of noncombustible material that can be anchored to the masonry wall or to other building supports as approved by the building official.

2104.3 Cold weather construction. The cold weather construction provisions of ACI 530.1/ASCE 6/TMS 602, Article 1.8 C, or the following procedures shall be implemented when either the ambient temperature falls below 40°F (4°C) or the temperature of masonry units is below 40°F (4°C).

❖ Provisions in this Section are based on Article 1.8 C of the MSJC Specification. Readers are referred to the commentary on the MSJC Specification for additional information on these provisions.

When masonry construction is conducted in temperatures below 40°F (4°C), both the masonry components and the structure are required to be protected in accordance with this section. One of the main goals of these provisions is to prevent fresh mortar from freezing.

During cold-weather construction, particularly in temperatures below freezing, the curing and subsequent performance of masonry are influenced by the temperature and properties of the mortar and the masonry units, as well as the severity of the exposure (temperature and wind).

2104.3.1 Preparation.

1. Temperatures of masonry units shall not be less than 20°F (-7°C) when laid in the masonry. Masonry units containing frozen moisture, visible ice or snow on their surface shall not be laid.

2. Visible ice and snow shall be removed from the top surface of existing foundations and masonry to receive new construction. These surfaces shall be heated to above freezing, using methods that do not result in damage.

❖ Under cold-weather conditions masonry construction should not proceed if these requirements are not met. Masonry units to be used during cold-weather construction should be dry. Masonry must not be laid on snow or ice-covered surfaces, because bond cannot be developed between the units and the mortar bed under those conditions. Additionally, the units can move when the snow or ice thaws.

2104.3.2 Construction. The following requirements shall apply to work in progress and shall be based on ambient temperature.

❖ This section sets minimum requirements to be met as masonry work proceeds based on the ambient temperature. Mortars (or grout) mixed at low temperatures using cold (not frozen) materials have plastic properties that are significantly different from those mixed at temperatures above 40°F (4°C). If the water content of the mortar is more than 6 to 8 percent when it freezes, expansion and possible damage could occur. If the water content of the mortar is less than 6 to 8 percent, the expected mortar expansion is less severe.

Antifreeze admixtures should not be permitted as a means of lowering the freezing point of mortar. Simply keeping the mortar from freezing does not result in proper hydration. Some commercially available mortar admixtures that claim to have antifreeze qualities may be improperly described or labeled because they do not significantly lower the freezing point of the mortar, but rather serve as accelerators of cement hydration. Calcium chloride is the most commonly used accelerator and is the main ingredient in many commercially available mortar admixtures. Because chlorides enhance the corrosion of embedded steel, however, admixtures containing chlorides should not be used in masonry containing metal ties, joint reinforcement or other metal accessories in contact with mortar.

Mortar mixed at a low temperature requires longer curing times and gains strength more slowly than mortar mixed at higher temperatures. To counteract the slow strength development of mortar at low temperatures, mixing water is required to be heated. In some instances, the sand used in the mortar is also heated. To accelerate early strength development, Type III (high-early-strength portland cement) can be substituted for Type I cement.

Cold-weather conditions do not require any changes to the mortar or grout mix proportions of cement, lime and sand (and coarse aggregate for grout).

2104.3.2.1 Construction requirements for temperatures between 40°F (4°C) and 32°F (0°C). The following construction requirements shall be met when the ambient temperature is between 40°F (4°C) and 32°F (0°C):

1. Glass unit masonry shall not be laid.

2. Water and aggregates used in mortar and grout shall not be heated above 140°F (60°C).

3. Mortar sand or mixing water shall be heated to produce mortar temperatures between 40°F (4°C) and 120°F (49°C) at the time of mixing. When water and aggregates for grout are below 32°F(0°C), they shall be heated.

❖ See the commentary to Section 2104.3.2.

2104.3.2.2 Construction requirements for temperatures between 32°F (0°C) and 25°F (-4°C). The requirements of Section 2104.3.2.1 and the following construction requirements shall be met when the ambient temperature is between 32°F (0°C) and 25°F (-4°C):

1. The mortar temperature shall be maintained above freezing until used in masonry.

2. Aggregates and mixing water for grout shall be heated to produce grout temperature between 70°F (21°C) and

120°F (49°C) at the time of mixing. Grout temperature shall be maintained above 70°F (21°C) at the time of grout placement.

3. Heat AAC masonry units to a minimum temperature of 40°F (4°C) before installing thin-bed mortar.

❖ See the commentary to Section 2104.3.2.

2104.3.2.3 Construction requirements for temperatures between 25°F (-4°C) and 20°F (-7°C). The requirements of Sections 2104.3.2.1 and 2104.3.2.2 and the following construction requirements shall be met when the ambient temperature is between 25°F (-4°C) and 20°F (-7°C):

1. Masonry surfaces under construction shall be heated to 40°F (4°C).

2. Wind breaks or enclosures shall be provided when the wind velocity exceeds 15 miles per hour (mph) (24 km/h).

3. Prior to grouting, masonry shall be heated to a minimum of 40°F (4°C).

❖ Low temperatures do not significantly affect the performance characteristics of masonry units. The absorption (suction) characteristics of cold, wet and frozen masonry units are decreased, however, which may affect mortar bond. Preheating masonry units prevents the sudden cooling of warm mortar in contact with cold units. A heated unit will absorb more water from the mortar because of the absorptive characteristics of a cooling body.

In selecting masonry units for cold-weather construction, consideration should be given to the absorption rates (see ASTM C 67). Units with initial rates of absorption of 30 grams per 30 square inches (19 355 mm²) per minute are preferable for cold-weather construction because they greatly reduce the possibility of disruptive expansion of the mortar due to freezing. Units with an initial rate of absorption of 5 to 6 grams per 30 square inches (19 355 mm²) per minute or less may not absorb sufficient water to prevent disruptive expansion if the mortar freezes (also see commentary, Section 2104.3.2).

2104.3.2.4 Construction requirements for temperatures below 20°F (-7°C). The requirements of Sections 2104.3.2.1, 2104.3.2.2 and 2104.3.2.3 and the following construction requirement shall be met when the ambient temperature is below 20°F (-7°C): Enclosures and auxiliary heat shall be provided to maintain air temperature within the enclosure to above 32°F (0°C).

❖ In addition to all of the above requirements, ambient temperatures below 20°F (-7°C) require the masonry construction to be enclosed and the use of auxiliary heat (also see commentary, Section 2104.3.2).

2104.3.3 Protection. The requirements of this section and Sections 2104.3.3.1 through 2104.3.3.5 apply after the masonry is placed and shall be based on anticipated minimum daily tem-

perature for grouted masonry and anticipated mean daily temperature for ungrouted masonry.

❖ This section sets minimum requirements to be met after masonry work is completed based on the anticipated temperature. Lower expected temperatures require more extensive means of protection.

2104.3.3.1 Glass unit masonry. The temperature of glass unit masonry shall be maintained above 40°F (4°C) for 48 hours after construction.

❖ See the commentary to Sections 2104.3 and 2104.3.3.

2104.3.3.2 AAC masonry. The temperature of AAC masonry shall be maintained above 32°F (0°C) for the first 4 hours after thin-bed mortar application.

❖ As with the other provisions of Section 2104.3, this section is based on the MSJC Specification. As noted in the commentary on the MSJC Specification construction experience indicates that AAC thin-bed mortar reaches strength faster than masonry mortar, although it is more sensitive to severe cold weather.

2104.3.3.3 Protection requirements for temperatures between 40°F (4°C) and 25°F (-4°C). When the temperature is between 40°F (4°C) and 25°F (-4°C), newly constructed masonry shall be covered with a weather-resistive membrane for 24 hours after being completed.

❖ See the commentary to Sections 2104.3 and 2104.3.3.

2104.3.3.4 Protection requirements for temperatures between 25°F (-4°C) and 20°F (-7°C). When the temperature is between 25°F (-4°C) and 20°F (-7°C), newly constructed masonry shall be completely covered with weather-resistive insulating blankets, or equal protection, for 24 hours after being completed. The time period shall be extended to 48 hours for grouted masonry, unless the only cement in the grout is Type III portland cement.

❖ See the commentary to Sections 2104.3 and 2104.3.3.

2104.3.3.5 Protection requirements for temperatures below 20°F (-7°C). When the temperature is below 20°F (-7°C), newly constructed masonry shall be maintained at a temperature above 32°F (0°C) for at least 24 hours after being completed by using heated enclosures, electric heating blankets, infrared lamps or other acceptable methods. The time period shall be extended to 48 hours for grouted masonry, unless the only cement in the grout is Type III portland cement.

❖ When temperatures are less than 20°F (-7°C), temporary heating is required to protect the masonry (also see commentary, Sections 2104.3 and 2104.3.3).

2104.4 Hot weather construction. The hot weather construction provisions of ACI 530.1/ASCE 6/TMS 602, Article 1.8 D, or the following procedures shall be implemented when the temperature or the temperature and wind-velocity limits of this section are exceeded.

❖ Provisions in this Section are based on Article 1.8 C of the MSJC Specification for Masonry Structures. Readers are referred to the commentary on the MSJC Specification for additional information on these provisions.

Temperature, solar radiation, humidity and wind influence the rate of absorption of masonry units and the rate of mortar set. During hot-weather conditions, special precautions must be taken for adequate strength gain of the masonry.

2104.4.1 Preparation. The following requirements shall be met prior to conducting masonry work.

❖ This section imposes minimum required precautions before masonry work proceeds under hot-weather conditions. The purpose is to maintain mortar temperatures less than that which causes a flash set

2104.4.1.1 Temperature. When the ambient temperature exceeds 100°F (38°C), or exceeds 90°F (32°C) with a wind velocity greater than 8 mph (3.5 m/s):

1. Necessary conditions and equipment shall be provided to produce mortar having a temperature below 120°F (49°C).

2. Sand piles shall be maintained in a damp, loose condition.

❖ When hot-weather conditions exceed those specified, procedures are required to keep mortar and sand in an acceptable condition.

2104.4.1.2 Special conditions. When the ambient temperature exceeds 115°F (46°C), or 105°F (40°C) with a wind velocity greater than 8 mph (3.5 m/s), the requirements of Section 2104.4.1.1 shall be implemented, and materials and mixing equipment shall be shaded from direct sunlight.

❖ When hot-weather conditions are extreme, materials and mixing equipment are required to be shaded, in addition to the requirements in Section 2104.4.1.1. Unshaded equipment becomes so hot that mortar can experience flash set.

2104.4.2 Construction. The following requirements shall be met while masonry work is in progress.

❖ This section imposes minimum precautions to be taken as masonry work proceeds under hot-weather conditions. Their purpose is to keep mortar temperatures lower than that which causes flash set. Ice is prohibited because of the potential for weak mortar and grout.

2104.4.2.1 Temperature. When the ambient temperature exceeds 100°F (38°C), or exceeds 90°F (32°C) with a wind velocity greater than 8 mph (3.5 m/s):

1. The temperature of mortar and grout shall be maintained below 120°F (49°C).

2. Mixers, mortar transport containers and mortar boards shall be flushed with cool water before they come into contact with mortar ingredients or mortar.

3. Mortar consistency shall be maintained by retempering with cool water.

4. Mortar shall be used within 2 hours of initial mixing.

5. Thin-bed mortar shall be spread no more than 4 feet (1219 mm) ahead of AAC masonry units.

6. AAC masonry units shall be placed within one minute after spreading thin-bed mortar.

❖ When hot-weather conditions exceed the limitations of this section, procedures are required to keep mortar and grout cool. These procedures include flushing mortar board, containers and other surfaces in contact with mortar and grout with water and retempering.

2104.4.2.2 Special conditions. When the ambient temperature exceeds 115°F (46°C), or exceeds 105°F (40°C) with a wind velocity greater than 8 mph (3.5 m/s), the requirements of Section 2104.4.2.1 shall be implemented and cool mixing water shall be used for mortar and grout. The use of ice shall be permitted in the mixing water prior to use. Ice shall not be permitted in the mixing water when added to the other mortar or grout materials.

❖ In extreme hot-weather conditions, ice can be added to the mixing water for mortar and grout to maintain their temperatures at acceptable levels. Ice cannot, however, be present in the mixing water when it is added to the dry materials in the mortar or grout.

2104.4.3 Protection. When the mean daily temperature exceeds 100°F (38°C) or exceeds 90°F (32°C) with a wind velocity greater than 8 mph (3.5 m/s), newly constructed masonry shall be fog sprayed until damp at least three times a day until the masonry is three days old.

❖ When hot-weather conditions are extended (as defined by the mean daily temperature), masonry is required to be fogsprayed to help hydrate it. Excessive wetting of the wall, however, can cause adverse effects such as efflorescence and moisture expansion in concrete masonry.

2104.5 Wetting of brick. Brick (clay or shale) at the time of laying shall require wetting if the unit's initial rate of water absorption exceeds 30 grams per 30 square inches (19 355 mm²) per minute or 0.035 ounce per square inch (1 g/645 mm²) per minute, as determined by ASTM C 67.

❖ Clay brick laid on fresh mortar generally absorbs water from the mortar along with fine particles of cementitious materials, which help to bond the mortar with the brick. If the brick absorbs too much water, the mortar may have insufficient water for hydration. It is, therefore, necessary to determine the initial rate of absorption (IRA) of the brick.

Mortar bonds best with brick that has an IRA an between 5 and 30 grams per 30 square inches (19 355 mm²) of surface per minute. Where the water absorption rate exceeds 30 grams per 30 square inches (19 355 mm²) per minute (as determined by ASTM C 67), brick is required to be wetted before placement, but should be surface dry when laid.

SECTION 2105
QUALITY ASSURANCE

2105.1 General. A quality assurance program shall be used to ensure that the constructed masonry is in compliance with the construction documents.

The quality assurance program shall comply with the inspection and testing requirements of Chapter 17.

❖ This section requires that a quality assurance program be used so that the masonry is constructed according to the contract documents. In addition, inspection and testing complying with Chapter 17 also requires verification of the masonry compressive strength.

The quality assurance program shall comply with the inspection and testing requirements of Chapter 17.

2105.2 Acceptance relative to strength requirements.

❖ The quality assurance provisions in this section emphasize verification of masonry compressive strengths. This is accomplished by comparing conservatively estimated strengths (based on unit strength and mortar type) or tested prism strengths to the specified compressive strength of the masonry, f'_m, and, when required, by mortar, grout or both to see that they are in compliance.

These quality assurance methods are for general consistency of the constructed masonry. Masonry is relatively strong in compression and thus rarely fails in that manner, but rather in flexural tension. Compression tests are required, however, because they are simple ways of assessing quality.

As implied in the above paragraph, two methods are prescribed in Section 2105.2 to judge acceptance relative to the compressive strength of the masonry assemblage—the unit strength method and the prism test method. These are described in Section 2105.2 and in this commentary. When the strength of constructed masonry is questioned, testing of prisms that have been saw cut from the masonry is permitted in accordance with Section 2105.3.

2105.2.1 Compliance with f'_m and f'_{AAC}. Compressive strength of masonry shall be considered satisfactory if the compressive strength of each masonry wythe and grouted collar joint equals or exceeds the value of f'_m for clay and concrete masonry and f'_{AAC} for AAC masonry. For partially grouted clay and concrete masonry, the compressive strength of both the grouted and ungrouted masonry shall equal or exceed the applicable f'_m. At the time of prestress, the compressive strength of the masonry shall equal or exceed f'_{mi}, which shall be less than or equal to f'_m.

❖ Design of structural masonry is based on the specified compressive strength of the masonry, f'_m, and the specified strength of AAC masonry, f'_{AAC}. This strength is required to be shown on the contract documents, since structural design is based on it. Strength of the constructed masonry determined by the unit or the prism strength method is required to equal or exceed the specified compressive strength of the masonry.

In a multiwythe wall designed as a composite wall, the compressive strength of masonry for each wythe or grouted collar joint must equal or exceed the specified compressive strength.

2105.2.2 Determination of compressive strength. The compressive strength for each wythe shall be determined by the unit strength method or by the prism test method as specified herein.

❖ There are two means of determining the compressive strength of masonry: the unit strength method and the prism test method. The first eliminates the expense of prism tests, but is more conservative.

2105.2.2.1 Unit strength method.

❖ This section describes a prescriptive procedure to estimate the expected compressive strength of the masonry based on the compressive strength of masonry units and the mortar type. This so-called unit strength method was generated using prism test data. Mortar joint thickness is limited because it influences the compressive strength of masonry.

2105.2.2.1.1 Clay masonry. The compressive strength of masonry shall be determined based on the strength of the units and the type of mortar specified using Table 2105.2.2.1.1, provided:

1. Units conform to ASTM C 62, ASTM C 216 or ASTM C 652 and are sampled and tested in accordance with ASTM C 67.

2. Thickness of bed joints does not exceed $^5/_8$ inch (15.9 mm).

3. For grouted masonry, the grout meets one of the following requirements:

   3.1. Grout conforms to ASTM C 476.

   3.2. Minimum grout compressive strength equals or exceeds f'_m but not less than 2,000 psi (13.79 MPa). The compressive strength of grout shall be determined in accordance with ASTM C 1019.

❖ This section prescribes the conditions under which the unit strength method can be used for clay masonry. These conditions include requirements for clay masonry units, maximum mortar joint thickness and grout.

TABLE 2105.2.2.1.1
COMPRESSIVE STRENGTH OF CLAY MASONRY

NET AREA COMPRESSIVE STRENGTH OF CLAY MASONRY UNITS (psi)		NET AREA COMPRESSIVE STRENGTH OF MASONRY (psi)
Type M or S mortar	Type N mortar	
1,700	2,100	1,000
3,350	4,150	1,500
4,950	6,200	2,000
6,600	8,250	2,500
8,250	10,300	3,000
9,900	—	3,500
13,200	—	4,000

For SI: 1 pound per square inch = 0.00689 MPa.

❖ Table 2105.2.2.1.1 lists the compressive strength of masonry in terms of the strength of the clay masonry unit and the mortar type. This table is based on the research results cited in the commentary to the MSJC Specification. A similar table has been used successfully in both the MSJC Specification and the *Uniform Building Code®* (UBC™) since 1988.

The designer can use this table to estimate a specified compressive strength of masonry to use in design, based on the expected strength of the clay masonry units and the specified mortar type. The contractor can use the table to find what clay unit masonry strength and mortar type are needed to comply with the specified strength of the masonry, f'_m, given in the contract documents. The column entitled "Net Area Compressive Strength of Masonry" must equal or exceed the specified strength of the masonry, f'_m.

2105.2.2.1.2 Concrete masonry. The compressive strength of masonry shall be determined based on the strength of the unit and type of mortar specified using Table 2105.2.2.1.2, provided:

1. Units conform to ASTM C 55 or ASTM C 90 and are sampled and tested in accordance with ASTM C 140.

2. Thickness of bed joints does not exceed $5/8$ inch (15.9 mm).

3. For grouted masonry, the grout meets one of the following requirements:

    3.1. Grout conforms to ASTM C 476.

    3.2. Minimum grout compressive strength equals or exceeds f'_m but not less than 2,000 psi (13.79 MPa). The compressive strength of grout shall be determined in accordance with ASTM C 1019.

❖ This section prescribes the conditions under which the unit strength method can be used for concrete masonry. These conditions include requirements for the concrete masonry units, maximum mortar joint thickness and grout. Concrete masonry units must be tested in accordance with ASTM C 140.

TABLE 2105.2.2.1.2
COMPRESSIVE STRENGTH OF CONCRETE MASONRY

NET AREA COMPRESSIVE STRENGTH OF CONCRETE MASONRY UNITS (psi)		NET AREA COMPRESSIVE STRENGTH OF MASONRY (psi)[a]
Type M or S mortar	Type N mortar	
1,250	1,300	1,000
1,900	2,150	1,500
2,800	3,050	2,000
3,750	4,050	2,500
4,800	5,250	3,000

For SI: 1 inch = 25.4 mm, 1 pound per square inch = 0.00689 MPa.
a. For units less than 4 inches in height, 85 percent of the values listed.

❖ Table 2105.2.2.1.2 lists the compressive strength of masonry in terms of the strengths of the concrete masonry units and the mortar type. This table is based on the research cited in the commentary to the MSJC Specification. A similar table has been used successfully in both the MSJC Specification and the UBC since 1988.

The designer can use this table to estimate a specified compressive strength of masonry for use in de-

sign, based on the expected strength of the concrete masonry units and the specified mortar type. The contractor can use the table to find what concrete masonry strength and mortar type are needed to comply with the specified strength of the masonry, f'_m, given in the contract documents. The column entitled "Net Area Compressive Strength of Masonry" must equal or exceed the specified strength of the masonry, f'_m.

2105.2.2.1.3 AAC masonry. The compressive strength of AAC masonry shall be based on the strength of the AAC masonry unit only and the following shall be met:

1. Units conform to ASTM C 1386.

2. Thickness of bed joints does not exceed $1/8$ inch (3.2 mm).

3. For grouted masonry, the grout meets one of the following requirements:

    3.1. Grout conforms to ASTM C 476.

    3.2. Minimum grout compressive strength equals or exceeds f'_{AAC} but not less than 2,000 psi (13.79 MPa). The compressive strength of grout shall be determined in accordance with ASTM C 1019.

❖ Due to the relative difference between the physical properties of AAC masonry units and the thin-bed mortar used to construct AAC masonry, the mortar does not have a significant contribution to the compressive strength of AAC masonry assemblies. Therefore, the compressive strength of AAC masonry is considered equal to the compressive strength of the AAC masonry units alone.

2105.2.2.2 Prism test method.

❖ The prism test method is used when required in the project specifications or when the restrictions of Section 2105.2.2.1 do not apply. Prisms are required to be constructed in accordance with ASTM C 1314 using the same materials and workmanship as in the structure.

ASTM C 1314 replaced ASTM E 447 for field-constructed prism specimens, which was referenced in editions of the specification prior to 1999. The use of ASTM C 1314 is intended to address many of the concerns over the difficulty and imprecision of ASTM E 447 for large prisms.

2105.2.2.2.1 General. The compressive strength of clay and concrete masonry shall be determined by the prism test method:

1. Where specified in the construction documents.

2. Where masonry does not meet the requirements for application of the unit strength method in Section 2105.2.2.1.

❖ Prism tests are required whenever specified and whenever the masonry does not meet the restrictions for the unit strength method.

2105.2.2.2.2 Number of prisms per test. A prism test shall consist of three prisms constructed and tested in accordance with ASTM C 1314.

❖ Whenever prism testing is specified or used, three prism specimens must be constructed and tested in accordance with ASTM C 1314.

2105.3 Testing prisms from constructed masonry. When approved by the building official, acceptance of masonry that does not meet the requirements of Section 2105.2.2.1 or 2105.2.2.2 shall be permitted to be based on tests of prisms cut from the masonry construction in accordance with Sections 2105.3.1, 2105.3.2 and 2105.3.3.

❖ While uncommon, there are times when the strength of masonry determined by the unit strength method or prism test method may be questioned or may be lower than the specified strength. Because low strengths could result from inappropriate testing procedures or unintentional damage to the test specimens, prisms may be saw cut from the completed masonry wall and tested. This section prescribes procedures for such tests.

Such testing is difficult, requires at least 28 days and requires replacement of the affected wall area. Therefore, every effort should be taken so that strengths determined by the unit strength method or the prism test method are adequate.

2105.3.1 Prism sampling and removal. A set of three masonry prisms that are at least 28 days old shall be saw cut from the masonry for each 5,000 square feet (465 m²) of the wall area that is in question but not less than one set of three masonry prisms for the project. The length, width and height dimensions of the prisms shall comply with the requirements of ASTM C 1314. Transporting, preparation and testing of prisms shall be in accordance with ASTM C 1314.

❖ Removal of prisms from a constructed wall requires care so that the prism is not damaged and that damage to the wall is minimal. Prisms must be representative of the wall, yet not contain reinforcing steel, which would bias the results. As with a prism test of newly constructed masonry, a prism test from existing masonry requires three prism specimens.

2105.3.2 Compressive strength calculations. The compressive strength of prisms shall be the value calculated in accordance ASTM C 1314, except that the net cross-sectional area of the prism shall be based on the net mortar bedded area.

❖ Compressive strength calculations from saw-cut specimens must be based on the net mortar bedded area, which must be determined before the prism is tested. The testing agency must determine this area accurately.

2105.3.3 Compliance. Compliance with the requirement for the specified compressive strength of masonry, f'_m, shall be considered satisfied provided the modified compressive strength equals or exceeds the specified f'_m. Additional testing of specimens cut from locations in question shall be permitted.

❖ Strengths determined from saw-cut prisms must equal or exceed the specified strength of masonry, f'_m.

SECTION 2106
SEISMIC DESIGN

2106.1 Seismic design requirements for masonry. Masonry structures and components shall comply with the requirements in Section 1.14.2.2 and Section 1.14.3, 1.14.4, 1.14.5, 1.14.6 or 1.14.7 of ACI 530/ASCE 5/TMS 402 depending on the structure's seismic design category as determined in Section 1613. All masonry walls, unless isolated on three edges from in-plane motion of the basic structural systems, shall be considered to be part of the seismic-force-resisting system. In addition, the following requirements shall be met.

❖ Section 2106 contains minimum requirements for masonry structures based upon their seismic design category. This section requires the use of the MSJC Code seismic design criteria. Requirements established for various seismic risk categories are cumulative from lower to higher categories. These prescriptive and design-oriented provisions have been established to improve the performance of masonry during seismic events by providing additional structural strength, ductility and stability against the dynamic effects of earthquakes. As seismic demand increases, the provisions require more positive connection between structural elements, increased ductility and greater material reliability.

Section 1613 contains criteria for determining a structure's seismic design category and references ASCE 7 criteria for calculating seismic loading. The requirements for seismic resistance (for example, design forces and detailing) remain applicable and are discussed in the Chapter 16 commentary. Compliance with Chapter 21 is not a substitute for compliance with the seismic provisions required by Section 1613. More information on seismic design is contained in the commentaries to ASCE 7 and the NEHRP Recommended Provisions for Seismic Regulations for New Buildings.

To comply with the provisions in Section 2106, the seismic design category must be determined for the building or structure under consideration. Refer to the commentary to Section 1613 for information on determining the seismic design category. Based on a structure's seismic design category, the designer needs to meet the minimum requirements of Section 2106.2, 2106.3, 2106.4, 2106.5 or 2106.6 as well as the referenced MSJC Code sections.

The reader is also referred to the commentary to the MSJC Code for additional background on seismic design requirements for masonry.

2106.1.1 Basic seismic-force-resisting system. Buildings relying on masonry shear walls as part of the basic seismic-force-resisting system shall comply with Section 1.14.2.2 of ACI 530/ASCE 5/TMS 402 or with Section 2106.1.1.1, 2106.1.1.2 or 2106.1.1.3.

❖ A basic seismic-force-resisting system must be defined for all buildings. Most masonry buildings use shear walls to serve as the basic seismic-force-resisting system, although other systems are sometimes used (such as concrete or steel frames with masonry infill). Such shear walls must be designed by the engineering methods in Section 2107 or 2108, unless the structure is assigned to Seismic Design Category A, in which case the empirical provisions of Section 2109 may be used.

There are three types of masonry shear wall systems that are recognized by the code, but are not specifically listed in the MSJC Code. They are ordinary plain prestressed shear walls (Section 2106.1.1.1), intermediate prestressed masonry shear walls (Section 2106.1.1.2) and special prestressed masonry shear walls (Section 2106.1.1.3). The shear wall systems recognized by the MSJC Code are discussed below.

Ordinary plain masonry shear walls (see Section 1.14.2.2.1 of the MSJC Code) meet minimum requirements only and thus may be used only in areas of low seismic risk. Plain masonry walls are designed as unreinforced masonry (by the noted section), although they may in fact contain reinforcement.

Ordinary reinforced masonry shear walls (see Section 1.14.2.2.3 of the MSJC Code) are required to meet minimum requirements for reinforced masonry as noted in the referenced section. Because they contain reinforcement, their performance is expected to be better than that of plain masonry shear walls and they are accordingly permitted in areas of both low and moderate seismic risk. Additionally, these walls have more favorable seismic design parameters, including higher response modification factors, R, than plain masonry shear walls. When assigned to moderate seismic risk areas (Seismic Design Category C), however, minimum reinforcement is required as noted in Section 2106.4.

Detailed plain masonry shear walls (see Section 1.14.2.2.2 of the MSJC Code) are designed as unreinforced masonry in accordance with the section noted, but contain minimum reinforcement in the horizontal and vertical directions. Because of this reinforcement, these walls have more favorable seismic design parameters, including higher response modification factors, R, than ordinary plain masonry shear walls.

Intermediate reinforced masonry shear walls (see Section 1.14.2.2.4 of the MSJC Code) are designed as reinforced masonry as noted in the referenced section and are also required to contain a minimum amount of prescriptive reinforcement. Because they contain reinforcement, their seismic performance is better than that of plain masonry shear walls and they are accordingly permitted in both areas of low and moderate seismic risk. Additionally, these walls have more favorable seismic design parameters, including higher response

modification factors, R, than plain masonry shear walls and ordinary reinforced masonry shear walls.

Special reinforced masonry shear walls (see Section 1.14.2.2.5 of the MSJC Code) are designed as reinforced masonry as noted in the referenced section and are also required to meet restrictive reinforcement and material requirements. Because of these reinforcement and material requirements, they are permitted to be used in all seismic risk areas. Additionally, these walls have the most favorable seismic design parameters, including the highest response modification factors, R, of any of the masonry shear wall types.

2106.1.1.1 Ordinary plain prestressed masonry shear walls. Ordinary plain prestressed masonry shear walls shall comply with the requirements of Chapter 4 of ACI 530/ASCE 5/TMS 402.

❖ This type of shear wall is recognized as a basic seismic-force-resisting system under the code and it must comply with the limitations for these systems in Section 12.2 of ASCE 7. The only other stipulation is that it comply with the prestressed masonry requirements of the MSJC Code.

2106.1.1.2 Intermediate prestressed masonry shear walls. Intermediate prestressed masonry shear walls shall comply with the requirements of Section 1.14.2.2.4 of ACI 530/ASCE 5/TMS 402 and shall be designed by Chapter 4, Section 4.4.3, of ACI 530/ASCE 5/TMS 402 for flexural strength and by Section 3.3.4.1.2 of ACI 530/ASCE 5/TMS 402 for shear strength. Sections 1.14.2.2.5, 3.3.3.5 and 3.3.4.3.2(c) of ACI 530/ASCE 5/TMS 402 shall be applicable for reinforcement. Flexural elements subjected to load reversals shall be symmetrically reinforced. The nominal moment strength at any section along a member shall not be less than one-fourth the maximum moment strength. The cross-sectional area of bonded tendons shall be considered to contribute to the minimum reinforcement in Section 1.14.2.2.4 of ACI 530/ASCE 5/TMS 402. Tendons shall be located in cells that are grouted the full height of the wall.

❖ This type of shear wall is recognized as a basic seismic-force-resisting system under the code and it must comply with the limitations for these systems in Section 12.2 of ASCE 7. Besides requiring that it comply with the prestressed masonry provisions of the MSJC Code, additional requirements must be met that are consistent with the results of research and testing carried out in New Zealand. These additional requirements intend that the walls develop their flexural capacity prior to shear failure. Tendons are limited to cells that are fully grouted, since testing has not substantiated that laterally unrestrained tendons are satisfactory for moderate seismic hazards.

2106.1.1.3 Special prestressed masonry shear walls. Special prestressed masonry shear walls shall comply with the requirements of Section 1.14.2.2.5 of ACI 530/ASCE 5/TMS 402 and shall be designed by Chapter 4, Section 4.4.3, of ACI 530/ASCE 5/TMS 402 for flexural strength and by Section 3.3.4.1.2 of ACI 530/ASCE 5/TMS 402 for shear strength. Sections 1.14.2.2.5(a), 3.3.3.5 and 3.3.4.3.2(c) of ACI

530/ASCE 5/TMS 402 shall be applicable for reinforcement. Flexural elements subjected to load reversals shall be symmetrically reinforced. The nominal moment strength at any section along a member shall not be less than one-fourth the maximum moment strength. The cross-sectional area of bonded tendons shall be considered to contribute to the minimum reinforcement in Section 1.14.2.2.5 of ACI 530/ASCE 5/TMS 402.

❖ This type of shear wall is recognized as a basic seismic-force-resisting system under the code and it must comply with the limitations for these systems in Section 12.2 of ASCE 7. Besides requiring that it comply with the prestressed masonry provisions of the MSJC Code, additional limitations apply that are based on the testing used to substantiate these systems. These additional requirements intend that the walls develop their flexural capacity prior to shear failure.

2106.1.1.3.1 Prestressing tendons. Prestressing tendons shall consist of bars conforming to ASTM A 722.

❖ Test specimens for prestressed masonry shear wall systems used only high-strength bar tendons, indicating that these bars are an appropriate restriction for these systems that are intended for exposure to high seismic hazards.

2106.1.1.3.2 Grouting. All cells of the masonry wall shall be grouted.

❖ Testing of prestressed masonry shear wall systems indicates that exposure to high seismic hazard is only permissible if all cells of the hollow concrete masonry unit wall are grouted solid. This is because partially grouted walls have demonstrated somewhat brittle behavior.

2106.2 Anchorage of masonry walls. Masonry walls shall be anchored to the roof and floors that provide lateral support for the wall in accordance with Section 1604.8.2.

❖ This provision applies to all masonry structures, regardless of seismic design category. Because masonry walls typically depend on lateral support from floors and roofs, they are required to be anchored directly to those elements. Section 1604.8.2 prohibits reliance on friction (from dead load) alone to hold walls, floors and roofs together and prescribes minimum forces for design of anchorage.

2106.3 Seismic Design Category B. Structures assigned to Seismic Design Category B shall conform to the requirements of Section 1.14.4 of ACI 530/ASCE 5/TMS 402 and to the additional requirements of this section.

❖ Requirements in Seismic Design Category B are slightly more restrictive than in Seismic Design Category A. Since the requirements are cumulative with each successive seismic design category, masonry assigned to Seismic Design Category B must meet the requirements for Seismic Design Category A (Section 1.14.3 of the MSJC Code) as well as Section 1.14.4 of the MSJC Code and the requirements in Section 2106.3. Therefore, besides those requirements in Section 2106.2, masonry shear walls must also be one of the types discussed in the commentary to Section

2106.1.1 and be rationally designed in this seismic design category and above.

2106.3.1 Masonry walls not part of the lateral-force-resisting system. Masonry partition walls, masonry screen walls and other masonry elements that are not designed to resist vertical or lateral loads, other than those induced by their own mass, shall be isolated from the structure so that the vertical and lateral forces are not imparted to these elements. Isolation joints and connectors between these elements and the structure shall be designed to accommodate the design story drift.

❖ So that seismic loads are not inadvertently transferred into elements such as masonry partition and screen walls, they are required to be isolated from the seismic-force-resisting system. However, these elements need out-of-plane support. Appropriate connectors are available and should be used. This is, in effect, a modification to the seismic requirements of the MSJC Code, since in that standard this is a requirement for SDC C structures and higher.

2106.4 Additional requirements for structures in Seismic Design Category C. Structures assigned to Seismic Design Category C shall conform to the requirements of Section 2106.3, Section 1.14.5 of ACI 530/ASCE 5/TMS 402 and the additional requirements of this section.

❖ In addition to the requirements of seismic design categories, minimum levels of reinforcement and detailing are required to enhance ductility and load-transfer capability in masonry structures assigned to Seismic Design Category C. The minimum provisions for improved performance of masonry construction in Seismic Design Category C must be met, regardless of the method of design.

2106.4.1 Design of discontinuous members that are part of the lateral-force-resisting system. Columns and pilasters that are part of the lateral-force-resisting system and that support reactions from discontinuous stiff members such as walls shall be provided with transverse reinforcement spaced at no more than one-fourth of the least nominal dimension of the column or pilaster. The minimum transverse reinforcement ratio shall be 0.0015. Beams supporting reactions from discontinuous walls or frames shall be provided with transverse reinforcement spaced at no more than one-half of the nominal depth of the beam. The minimum transverse reinforcement ratio shall be 0.0015.

❖ The requirements in this section are intended to reduce the chance of local failure or collapse in elements supporting discontinuous portions of the lateral-force-resisting system by increasing the strength and toughness of those elements. Elements used to redistribute or transfer the effects of seismic overturning are susceptible to local inelastic response that can significantly impair the ability of the lateral-force-resisting system to achieve the required overall ductility. The provisions in this section are intended to increase the ability of elements to resist inelastic deformations. This allows inelastic deformations to be distributed throughout the buildings, as is implied by the large response modification factors used in design.

2106.5 Additional requirements for structures in Seismic Design Category D. Structures assigned to Seismic Design Category D shall conform to the requirements of Section 2106.4, Section 1.14.6 of ACI 530/ASCE 5/TMS 402 and the additional requirements of this section.

❖ Requirements in this section parallel most of the requirements in the MSJC Code.

2106.5.1 Loads for shear walls designed by the allowable stress design method. When calculating in-plane shear or diagonal tension stresses by the allowable stress design method, shear walls that resist seismic forces shall be designed to resist 1.5 times the seismic forces required by Chapter 16. The 1.5 multiplier need not be applied to the overturning moment.

❖ This provision is based on a similar provision from the UBC. It requires that the in-plane shear stresses due to seismic loading be increased by 50 percent for design purposes and is intended to provide adequate strength and ductility in shear walls of structures in seismically active areas.

2106.5.2 Shear wall shear strength. For a shear wall whose nominal shear strength exceeds the shear corresponding to development of its nominal flexural strength, two shear regions exist.

For all cross sections within a region defined by the base of the shear wall and a plane at a distance L_w above the base of the shear wall, the nominal shear strength shall be determined by Equation 21-1.

$$V_n = A_n \rho_n f_y \qquad \text{(Equation 21-1)}$$

The required shear strength for this region shall be calculated at a distance $L_w/2$ above the base of the shear wall, but not to exceed one-half story height.

For the other region, the nominal shear strength of the shear wall shall be determined from Section 2108.

❖ The intent of this provision is to provide a ductile flexural limit state. The plastic hinge region is considered to extend vertically from the base of the wall to a distance equal to the plan length of the wall. In this region, the shear strength of the wall is based on the transverse reinforcement only. Above the plastic hinge, the shear strength of the wall is based on both the masonry and the transverse reinforcement.

2106.6 Additional requirements for structures in Seismic Design Category E or F. Structures assigned to Seismic Design Category E or F shall conform to the requirements of Section 2106.5 and Section 1.14.7 of ACI 530/ASCE 5/TMS 402.

❖ Additional restrictions are imposed on buildings assigned to the highest seismic risk categories.

SECTION 2107
ALLOWABLE STRESS DESIGN

2107.1 General. The design of masonry structures using allowable stress design shall comply with Section 2106 and the requirements of Chapters 1 and 2 of ACI 530/ASCE 5/TMS 402 except as modified by Sections 2107.2 through 2107.8.

❖ Section 2107 adopts the allowable stress design method of Chapters 1 and 2 of the MSJC Code with modifications that the International Building Code Structural Committee feels are needed. This method of engineered masonry design has been used successfully for years and remains the dominant method of designing masonry structures.

Refer to the commentary to Chapters 1 and 2 of the MSJC Code for additional information on the allowable stress design method for masonry. This section also requires conformance to the seismic design provisions of Section 2106.

2107.2 ACI 530/ASCE 5/TMS 402, Section 2.1.2, load combinations. Delete Section 2.1.2.1.

❖ The sections of the MSJC Code specifically not adopted exclude redundant or conflicting provisions such as the MSJC load combinations, which are only intended to apply where no such load combinations are included in the code.

2107.3 ACI 530/ASCE 5/TMS 402, Section 2.1.3, design strength. Delete Sections 2.1.3.4 through 2.1.3.4.3.

❖ The pseudo-strength design provision of the MSJC Code, Section 2.1.3.4, is also excluded. This method allows masonry to be designed using strength-based seismic loads under the allowable stress provisions. Since Chapter 16 includes service-level load combinations, this pseudo-strength design is not needed and is therefore not permitted.

2107.4 ACI 530/ASCE 5/TMS 402, Section 2.1.6, columns. Add the following text to Section 2.1.6:

2.1.6.6 Light-frame construction. Masonry columns used only to support light-frame roofs of carports, porches, sheds or similar structures with a maximum area of 450 square feet (41.8 m²) assigned to Seismic Design Category A, B or C are permitted to be designed and constructed as follows:

1. Concrete masonry materials shall be in accordance with Section 2103.1 of the *International Building Code*. Clay or shale masonry units shall be in accordance with Section 2103.2 of the *International Building Code*.

2. The nominal cross-sectional dimension of columns shall not be less than 8 inches (203 mm).

3. Columns shall be reinforced with not less than one No. 4 bar centered in each cell of the column.

4. Columns shall be grouted solid.

5. Columns shall not exceed 12 feet (3658 mm) in height.

6. Roofs shall be anchored to the columns. Such anchorage shall be capable of resisting the design loads specified in Chapter 16 of the *International Building Code*.

7. Where such columns are required to resist uplift loads, the columns shall be anchored to their footings

with two No. 4 bars extending a minimum of 24 inches (610 mm) into the columns and bent horizontally a minimum of 15 inches (381 mm) in opposite directions into the footings. One of these bars is permitted to be the reinforcing bar specified in Item 3 above. The total weight of a column and its footing shall not be less than 1.5 times the design uplift load.

❖ This section exempts lightly loaded columns (such as those that support carport roofs, that experience primarily axial tension and flexure in high-wind events) from the prescriptive requirements of Section 2.1.6 of the MSJC Code. All other columns need to comply with Section 2.1.6 of the MSJC Code. This provision is similar to one in the *Standard Building Code®* (SBC®) and is intended to relax an aspect of the MSJC Code, which defines columns by geometry rather than function. According to that document, masonry members of a certain geometry, even though they act primarily in flexure, are classified as columns and thus must meet minimum reinforcement requirements for columns.

2107.5 ACI 530/ASCE 5/TMS 402, Section 2.1.10.7.1.1, lap splices. Modify Section 2.1.10.7.1.1 as follows:

2.1.10.7.1.1 The minimum length of lap splices for reinforcing bars in tension or compression, l_d, shall be

$$l_d = 0.002 d_b f_s \qquad \textbf{(Equation 21-2)}$$

For SI: $l_d = 0.29 d_b f_s$

but not less than 12 inches (305 mm). In no case shall the length of the lapped splice be less than 40 bar diameters.

where:

d_b = Diameter of reinforcement, inches (mm).

f_s = Computed stress in reinforcement due to design loads, psi (MPa).

In regions of moment where the design tensile stresses in the reinforcement are greater than 80 percent of the allowable steel tension stress, F_s, the lap length of splices shall be increased not less than 50 percent of the minimum required length. Other equivalent means of stress transfer to accomplish the same 50 percent increase shall be permitted.

Where epoxy coated bars are used, lap length shall be increased by 50 percent.

❖ This provision is based on allowable stress detailing requirements that were used in the UBC. Lap splice and development length requirements for allowable stress and strength design of masonry differ, due in part to the slightly more conservative nature of the allowable stress design (ASD) requirements. Note that Section 2107.6 prohibits the lap splicing of reinforcement larger than No. 9 (32 mm) bars.

2107.6 ACI 530/ASCE 5/TMS 402, Section 2.1.10.7, splices of reinforcement. Modify Section 2.1.10.7 as follows:

2.1.10.7 Splices of reinforcement. Lap splices, welded splices or mechanical splices are permitted in accordance with the provisions of this section. All welding shall conform to AWS D1.4. Reinforcement larger than No. 9 (M

#29) shall be spliced using mechanical connections in accordance with Section 2.1.10.7.3.

❖ Research has shown that effectively lap splicing large reinforcing bars in masonry is difficult and impractical because of the excessive lap lengths required. This section adds a provision not in the MSJC Code requiring mechanical splices for all bars larger than No. 9 (32 mm) in diameter.

2107.7 ACI 530/ASCE 5/TMS 402, Section 2.3.6, maximum bar size. Add the following to Chapter 2:

2.3.6 Maximum bar size. The bar diameter shall not exceed one-eighth of the nominal wall thickness and shall not exceed one-quarter of the least dimension of the cell, course or collar joint in which it is placed.

❖ Allowing large bars in the masonry walls can result in an overreinforced section that increases the risk of a brittle failure. This section specifies limits on the diameter of reinforcement that are not in the MSJC Code in order to prevent potential problems with overreinforcement and congestion of reinforcement. These requirements are based on practical guidelines and empirical judgment rather than research.

2107.8 ACI 530/ASCE 5/TMS 402, Section 2.3.7, maximum reinforcement percentage. Add the following text to Chapter 2:

2.3.7 Maximum reinforcement percentage. Special reinforced masonry shear walls having a shear span ratio, M/V_d, equal to or greater than 1.0 and having an axial load, P, greater than $0.05 f'_m A_n$ that are subjected to in-plane forces shall have a maximum reinforcement ratio, ρ_{max}, not greater than that computed as follows:

$$\rho_{max} = \frac{n f'_m}{2 f_y \left(n + \dfrac{f_y}{f'_m} \right)} \qquad \textbf{(Equation 21-3)}$$

The maximum reinforcement ratio does not apply in the out-of-plane direction.

❖ The ASD provisions of the MSJC Code have no limit on the maximum reinforcement ratio. This section places such a limit on the maximum reinforcement ratio in special reinforced masonry shear walls. It is necessary to account for the relatively high potential for inelastic response in such systems.

This requirement is based on the recommendations of a blue-ribbon panel appointed by the Masonry Alliance for Codes and Standards (MACS). It was the panel's conclusion that these requirements should only apply to the design of shear walls for in-plane forces that have significant axial load ($P > 0.05 f'm A_n$) and that are controlled by flexure (i.e., $M/V_d \geq 1.0$).

SECTION 2108
STRENGTH DESIGN OF MASONRY

2108.1 General. The design of masonry structures using strength design shall comply with Section 2106 and the requirements of Chapters 1 and 3 of ACI 530/ASCE 5/TMS 402, except as modified by Sections 2108.2 through 2108.4.

Exception: AAC masonry shall comply with the requirements of Chapter 1 and Appendix A of ACI 530/ASCE 5/TMS 402.

❖ The first edition of the code contained extensive strength design requirements, since a consensus standard for the strength design of masonry did not yet exist. Starting with the 2003 edition of the code, this section has required that strength design be in accordance with Chapters 1 and 3 of the MSJC Code with some modifications to specific sections. This section also invokes the minimum seismic requirements of Section 2106 for all masonry designed by the strength method. Additionally, masonry designed by this method (as with any engineered masonry construction) must be inspected during construction in accordance with the special inspection provisions of Section 1704.5.

2108.2 ACI 530/ASCE 5/TMS 402, Section 3.3.3.3 development. Add the following text to Section 3.3.3.3:

The required development length of reinforcement shall be determined by Equation (3-15), but shall not be less than 12 inches (305 mm) and need not be greater than 72 d_b.

❖ This section modifies the MSJC Code by placing a limit on the calculated development length, which in turn limits the maximum required splice length. The limit is based upon historical limits used in the UBC. It is intended to address concerns that the MSJC Code, at times, requires excessive lap splice lengths.

2108.3 ACI 530/ASCE 5/TMS 402, Section 3.3.3.4, splices. Modify items (b) and (c) of Section 3.3.3.4 as follows:

3.3.3.4 (b). A welded splice shall have the bars butted and welded to develop at least 125 percent of the yield strength, f_y, of the bar in tension or compression, as required. Welded splices shall be of ASTM A 706 steel reinforcement. Welded splices shall not be permitted in plastic hinge zones of intermediate or special reinforced walls or special moment frames of masonry.

3.3.3.4 (c). Mechanical splices shall be classified as Type 1 or 2 according to Section 21.2.6.1 of ACI 318. Type 1 mechanical splices shall not be used within a plastic hinge zone or within a beam-column joint of intermediate or special reinforced masonry shear walls or special moment frames. Type 2 mechanical splices are permitted in any location within a member.

❖ This section modifies the strength design splice requirements for consistency with the 2000 edition of the NEHRP provisions. Splices for reinforcement can be achieved by lapping the reinforcement, welding the reinforcement or mechanical splicing.

Two modifications are made to welded splice requirements [Item (b)]. Splices in reinforcing steel used in the lateral-force-resisting system subjected to high seismic strains must be able to develop the strength of the steel in order to achieve the required performance. To be successfully welded, the chemistry of the steel must be controlled to limit carbon content as well as other elements, such as sulfur and phosphorus. Since the chemistry of reinforcing steel conforming to ASTM A 615, for example, is not controlled and is likely to be unknown, a weld that develops the strength of the steel is not guaranteed. ASTM A 706 steel, on the other hand, has controlled chemistry and can be reliably welded; therefore, if splices are to be accomplished by welding, the use of ASTM A 706 reinforcing steel is mandatory.

Welded splices are required to be able to develop at least 125 percent of the yield strength of the spliced rebars; however, reinforcing steel conforming to ASTM A 706 or for that matter ASTM A 615 or A 996 can have an actual yield strength greater than 125 percent of the minimum required yield strength. This means a code-conforming welded splice may conceivably fail before the spliced bars have yielded, thereby limiting the inelastic deformability of that structural member. The use of welded splices is, therefore, prohibited at locations of potential plastic hinging of members in structural systems that are expected to undergo significant inelastic response in resisting forces due to earthquakes.

The modification to mechanical splices in Item (c) requires that the splice be classified in accordance with ACI 318 as Class 1 or Class 2. This is due to the fact that reinforcing steel is predominantly produced from remelted steel scrap, making it difficult to control the strength. The resulting products tend to have a strength considerably higher than the specified yield strength. This is similar to the situation that has occurred in structural steel where the actual yield strength can be much greater than the specified yield strength. Since there is no upper limit on the yield strength (except for ASTM A 706) and only a minimum required yield strength, most reinforcing steel will have a higher yield point than that specified.

Testing by the California Department of Transportation (CALTRANS) indicates the overstrength can be as much as 60 percent over the specified strength. Cyclic tests of splices meeting only the 125-percent criterion (i.e. Class 1) show that, in many cases, they cannot survive several excursions in the post yield range as imposed by cyclic testing. Splices in reinforcing steel used in the lateral-force-resisting system in plastic hinge zones and beam-column joints are subjected to high seismic strains, and they must be able to develop the strength of the steel in order to achieve the required performance. Hence, the requirement for a Type 2 splice that develops the specified tensile strength of the bar.

2108.4 ACI 530/ASCE 5/TMS 402, Section 3.3.3.5, maximum areas of flexural tensile reinforcement. Add the following text to Section 3.3.3.5:

3.3.3.5.5 For special prestressed masonry shear walls, strain in all prestressing steel shall be computed to be compatible with a strain in the extreme tension reinforcement equal to five times the strain associated with the reinforcement yield stress, f_y. The calculation of the maximum reinforcement shall consider forces in the prestressing steel that correspond to these calculated strains.

❖ This MSJC Code section limits the percentages of flexural reinforcement in order to provide ductile behavior. Overreinforced flexural members can fail in a brittle mode by the crushing of the masonry. Such failures are sudden and catastrophic and, therefore, must be avoided.

Section 2106.1.1.3, covering special prestressed masonry shear walls, requires compliance with this code section, which includes a modification of the MSJC Code section specifically for those shear wall types. This modification clarifies how the reinforcing limitation is to be met for this structural system.

SECTION 2109
EMPIRICAL DESIGN OF MASONRY

2109.1 General. Empirically designed masonry shall conform to this chapter or Chapter 5 of ACI 530/ASCE 5/TMS 402.

❖ This section permits empirical design of masonry by either the provisions of Section 2109 or Chapter 5 of the MSJC Code. This is because nearly all of the requirements in Section 2109 are based on the requirements in Chapter 5 of the MSJC Code with minor modifications. Additional information on these provisions can be found in the commentary to the MSJC Code.

Empirical provisions are design rules developed by experience rather than engineering analysis. The empirical rules in these provisions are based on records dating back as far as 1889 in *A Treatise on Masonry Construction* by Ira Baker. The most recent publication providing the basis for these empirical provisions is ANSI A41.1.

This empirical design method is based on several premises: gravity loads are reasonably centered on bearing walls; effects of reinforcement are neglected; buildings have limited height, seismic risk and wind loading. The requirements of and limitations regarding the use of empirical design reflect these assumptions.

2109.1.1 Limitations. The use of empirical design of masonry shall be limited as follows:

1. Empirical design shall not be used for buildings assigned to Seismic Design Category D, E or F as specified in Section 1613, nor for the design of the seismic-force-resisting system for buildings assigned to Seismic Design Category B or C.

2. Empirical design shall not be used for masonry elements that are part of the lateral-force-resisting system where the basic wind speed exceeds 110 mph (79 m/s).

3. Empirical design shall not be used for interior masonry elements that are not part of the lateral- force-resisting system in buildings other than enclosed buildings as defined in Chapter 6 of ASCE 7 in:

   3.1. Buildings over 180 feet (55 100 mm) in height.

   3.2. Buildings over 60 feet (18 400 mm) in height where the basic wind speed exceeds 90 mph (40 m/s).

   3.3. Buildings over 35 feet (10 700 mm) in height where the basic wind speed exceeds 100 mph (45 m/s).

   3.4. Where the basic wind speed exceeds 110 mph (79 m/s).

4. Empirical design shall not be used for exterior masonry elements that are not part of the lateral- force-resisting system and that are more than 35 feet (10 700 mm) above ground:

   4.1. Buildings over 180 feet (55 100 mm) in height.

   4.2. Buildings over 60 feet (18 400 mm) in height where the basic wind speed exceeds 90 mph (40 m/s).

   4.3. Buildings over 35 feet (10 700 mm) in height where the basic wind speed exceeds 100 mph (45 m/s).

5. Empirical design shall not be used for exterior masonry elements that are less than or equal to 35 feet (10 700 mm) above ground where the basic wind speed exceeds 110 mph (79 m/s).

6. Empirical design shall only be used when the resultant of gravity loads is within the center third of the wall thickness and within the central area bounded by lines at one-third of each cross-sectional dimension of foundation piers.

7. Empirical design shall not be used for AAC masonry.

In buildings that exceed one or more of the above limitations, masonry shall be designed in accordance with the engineered design provisions of Section 2107 or 2108 or the foundation wall provisions of Section 1805.5.

❖ Empirical design is permitted for structures having limited seismic risk, wind loading and height. These limitations are justified, since buildings that were representative of the historically based empirical provisions are uncommon today. For example, buildings of the past were smaller, had more interior masonry walls and typically had different floor construction than modern buildings.

Where any one of the stated limitations is exceeded, the masonry structure is not permitted to be empirically designed. Engineered design in accordance with Section 2107 or 2108 is required in such instances. Foundation walls complying with Section 1805.5 are also acceptable.

The empirical provisions of Section 2109 or Chapter 5 of the MSJC Code are adequate for the level of risk associated with the seismic-force-resisting systems of buildings located in Seismic Design Category A. Where masonry is used for purposes other than the seismic-force-resisting system, the empirical design method may be used for buildings assigned to Seismic Design Category A, B or C. Engineered design is required for buildings in higher seismic design categories.

Item 2 applies specifically to the lateral-load-resisting system. For example, empirical design of the lateral-load-resisting system is permitted only where the basic wind speed does not exceed 110 mph (145 km/hr). Otherwise, engineered design is required. These are similar to the wind load restrictions in Chapter 5 of the MSJC Code.

Items 3 through 5 place limits on various building elements based on the element and building height as well as the basic wind speed. Empirical design is permitted for elements not effectively participating in the lateral-load-resisting system. Engineered design is required where the lateral-load-resisting system of a building has substantial height.

The use of empirical design for AAC masonry is not permitted. Additional discussion of the empirical design method can be found in the MSJC Code commentary.

2109.2 Lateral stability.

❖ The lateral stability requirements of this section are required for buildings using empirical design for lateral load resistance. The limitations of this section do not apply when one of the engineered masonry design methods is employed. This section contains requirements for empirical design of lateral-load-resisting systems composed of diaphragms and shear walls.

Requirements include minimum lengths of masonry shear walls in both principal plan directions of the building and maximum span-to-width (depth) ratios of floor and roof diaphragms. Requirements for roofs and dry-stacked, surface-bonded walls are also part of this section

Lateral load resistance is a basic requirement for structural design. The distribution of loads within the lateral-load-resisting system is a function of the relative rigidities of diaphragms and shear walls. The lateral-load-resisting system transfers lateral wind and seismic forces to the foundation in the form of base shear and overturning moment and maintains stability of the structure under gravity loads.

2109.2.1 Shear walls. Where the structure depends upon masonry walls for lateral stability, shear walls shall be provided parallel to the direction of the lateral forces resisted.

❖ The shear wall requirements in this section apply where empirically designed masonry is used for lateral load resistance. Shear walls are required in both principal plan directions of the structure, parallel to the lateral loads required in Chapter 16. Load-bearing walls serve as shear walls.

2109.2.1.1 Cumulative length of shear walls. In each direction in which shear walls are required for lateral stability, shear walls shall be positioned in two separate planes. The minimum cumulative length of shear walls provided shall be 0.4 times the long dimension of the building. Cumulative length of shear walls shall not include openings or any element with a length that is less than one-half its height.

❖ Figure 2109.2.1.1 diagrams the shear wall lengths in each direction of the building, parallel to the lateral loads. The term "cumulative" refers to the sum of the lengths of shear wall segments in a single direction. The required length of shear wall in each direction is 0.4 times the long dimension of the building.

Walls above and below openings such as windows and doors are not considered to be shear wall segments. Neither are portions of a wall that have a length less than one-half the wall height.

2109.2.1.2 Maximum diaphragm ratio. Masonry shear walls shall be spaced so that the length-to-width ratio of each diaphragm transferring lateral forces to the shear walls does not exceed the values given in Table 2109.2.1.2.

❖ For empirical design, this section limits length- (or span) to-width ratios of floor and roof diaphragms between shear walls, based on the inherent rigidity of the diaphragm construction. This effectively distributes shear walls throughout the structure at adequate intervals for the diaphragm span.

The assumed deflected shape of a diaphragm is illustrated in Figure 2109.2.1.2(1). Flexible diaphragms deflect more than rigid ones, possibly resulting in detachment of components. A rigid diaphragm deflects little and transfers loads to shear walls in proportion to their relative rigidities (an important engineered masonry consideration). Symmetrical distribution of shear walls in the building plan is desirable, reducing torsional rotation of the building under lateral loads.

TABLE 2109.2.1.2
DIAPHRAGM LENGTH-TO-WIDTH RATIOS

FLOOR OR ROOF DIAPHRAGM CONSTRUCTION	MAXIMUM LENGTH-TO-WIDTH RATIO OF DIAPHRAGM PANEL
Cast-in-place concrete	5:1
Precast concrete	4:1
Metal deck with concrete fill	3:1
Metal deck with no fill	2:1
Wood	2:1

❖ Table 2109.2.1.2 sets the maximum span-to-width ratio of diaphragms constructed of the listed materials. See Figure 2109.2.1.2(2) for an illustration of the terms "diaphragm span" (length) and "width" (depth). Concrete diaphragms are permitted to have high span- (length-) to-width ratios because of their rigidity, whereas wood diaphragms are required to have low span- (length-) to-width ratios because of their flexibility.

MINIMUM CUMULATIVE SHEAR WALL LENGTH = 0.4 x LONG DIMENSION
MINIMUM L = 0.4(60.7') = 24.3'
X - DIRECTION: L = 2(5.7 + 6.0 + 4.0 + 4.0 + 4.0 + 4.0 + 6.0 + 5.7) = 78.8 ft. > 24.3 ft. OK
Y - DIRECTION: L = 2(24.0 + 10.0 + 10.0 + 12.7 + 6.0) = 125.4 ft. > 24.3 ft. OK

For SI: 1 inch = 25.4 mm. 1 foot = 304.8 mm.

Figure 2109.2.1.1
CUMULATIVE LENGTH OF SHEAR WALL REQUIREMENTS FOR EMPIRICALLY DESIGNED MASONRY

VERTICAL RESISTING ELEMENT

LATERAL FORCES

WIDTH (DEPTH)

DIAPHRAGM

DEFLECTION

DEFLECTED SHAPE

SPAN

Figure 2109.2.1.2(1)
DIAPHRAGM ACTION (PLAN VIEW)

$\dfrac{\text{SPAN}}{\text{WIDTH}} \leq$ TABLE 2109.2.1.3 VALUE

MASONRY SHEAR WALL

Figure 2109.2.1.2(2)
EMPIRICAL MAXIMUM DIAPHRAGM RATIO

2109.2.2 Roofs. The roof construction shall be designed so as not to impart out-of-plane lateral thrust to the walls under roof gravity load.

❖ Roofs are not permitted to rely on masonry walls to resist thrust perpendicular to the wall. The low tensile capacity of masonry and the lack of engineered design for these loads in the empirical provisions result in this requirement. Connections that apply thrust perpendicular to the wall are not permitted.

As another example, a wood-frame cathedral ceiling does not, by design, have ceiling joists to resist outward thrust. If the ridge beam is not designed for vertical support and limited deflection, opposing sloped rafters usually transfer thrust to the outside walls.

Consideration of horizontal thrust is also necessary for truss connections, especially scissor trusses. In a truss, the bottom chord is usually in tension. Chord elongation can impart significant lateral thrust perpendicular to masonry bearing walls. In general, the lower the roof slope, the greater the lateral thrust.

2109.2.3 Surface-bonded walls. Dry-stacked, surface-bonded concrete masonry walls shall comply with the requirements of this code for masonry wall construction, except where otherwise noted in this section.

❖ Dry-stacked, surface-bonded masonry walls consist of courses of concrete masonry units without mortar joints assembled to form unreinforced walls. Both sides of the walls are coated with a $^1/_{16}$- to $^1/_8$-inch-thick layer (1.6 to 3.2 mm) of cementitious mortar reinforced with glass fibers capable of increasing the tensile strength of the masonry and unifying the construction.

2109.2.3.1 Strength. Dry-stacked, surface-bonded concrete masonry walls shall be of adequate strength and proportions to support all superimposed loads without exceeding the allowable stresses listed in Table 2109.2.3.1. Allowable stresses not specified in Table 2109.2.3.1 shall comply with the requirements of ACI 530/ASCE 5/TMS 402.

❖ In surface-bonded masonry construction, strength is lower than conventional masonry construction because of the lack of solid contact between the units.

Where stresses are not specified in Table 2109.2.3.1, engineered design in accordance with Section 2107 or 2108 is required.

TABLE 2109.2.3.1
ALLOWABLE STRESS GROSS CROSS-SECTIONAL
AREA FOR DRY-STACKED, SURFACE-BONDED
CONCRETE MASONRY WALLS

DESCRIPTION	MAXIMUM ALLOWABLE STRESS (psi)
Compression standard block	45
Flexural tension Horizontal span Vertical span	30 18
Shear	10

For SI: 1 pound per square inch = 0.006895 MPa.

❖ The values for allowable stresses based on the gross cross-sectional area of masonry units are given in this table. The gross cross-sectional area is the actual area of a section perpendicular to the direction of the load, without subtraction of the core areas of hollow masonry units.

The flexural strength of surface-bonded walls is about the same as conventional masonry with mortar joints. In the vertical direction, where walls are supported top and bottom, Table 2109.2.3.1 allows a maximum flexural tensile stress of 18 psi (0.12 MPa) based on the gross area. In the horizontal direction, where walls span laterally between supports, a maximum flexural stress of 30 psi (0.21 MPa) is permitted based on the gross area when units are dry stacked in running bond.

The shear strength of surface-bonded walls is less than that of conventional, mortar-jointed walls. This table allows a shear strength of 10 psi (0.069 MPa) based on the gross area.

2109.2.3.2 Construction. Construction of dry- stacked, surface-bonded masonry walls, including stacking and leveling of units, mixing and application of mortar and curing and protection shall comply with ASTM C 946.

❖ The construction of dry-stacked, surface-bonded walls must conform to the requirements of ASTM C 946.

It is not practical to construct the horizontal surface of a wall footing or foundation level enough to receive the base (first) course of masonry without additional leveling. Therefore, it is customary to lay the base course of masonry on a mortar bed so that the remainder of the dry-stacked units will be erected level. As the units are erected, their ends should be butted together as tightly as possible.

If the bearing surfaces of the concrete units are not ground smooth and flat, shims may be required between the units to erect the wall plumb and level. Such shims should be of metal, mortar or plastic.

Because dry-stacked walls have no mortar joints, it is not possible to use horizontal steel joint reinforcement to reduce the size of cracks associated with temperature and moisture movements. Reinforced bond beams and sufficient control joints in surface-bonded masonry can be used to reduce the widths of such cracks.

The joints of dry-stacked units are tight, with no space for connectors to be embedded in the wall. The face shells or cross webs of such concrete masonry units, therefore, should be notched or depressed to accommodate ties and anchors that must be embedded in grout.

Packaged dry-bonding mortar should be mixed with water at the job site in accordance with ASTM C 946 or the manufacturer's recommendations, including curing and protection procedures after application of the material. Surface-bonding mortars are usually applied by hand troweling to thicknesses between $^1/_{16}$ and $^1/_8$ inch (1.6 to 3.2 mm). While they may also be sprayed

on, this is usually followed by hand or mechanical troweling to obtain the desired finish.

2109.3 Compressive stress requirements.

❖ This section applies to empirically designed masonry and, as with the other empirical design requirements of Section 2109, not to masonry designed by the engineered approaches of Sections 2107 and 2108.

Vertical dead and live loads, as required in Chapter 16, encompass a wide range of possibilities, as does the conceivable configuration of supported floor spans. Consequently, specifying empirical minimum sizes that would account for all vertical loading and span conditions would be impractical. Section 2109.3 contains an empirical compressive stress design procedure for single- and multiwythe masonry addressing actual required loading and resulting in minimum areas of masonry to resist vertical loads. The design is based on an average compressive stress on the gross cross-sectional area, using specified instead of nominal dimensions.

The result is a way of sizing and proportioning masonry without complete engineering analysis and design. The required areas are intended to be conservative with respect to engineered design and, along with the other empirical design provisions, to adequately address buildings and elements permitted to be empirically designed by the provisions of Section 2109.

2109.3.1 Calculations. Compressive stresses in masonry due to vertical dead plus live loads, excluding wind or seismic loads, shall be determined in accordance with Section 2109.3.2.1. Dead and live loads shall be in accordance with Chapter 16, with live load reductions as permitted in Section 1607.9.

❖ This section identifies the vertical design loads that must be used in calculating average compressive stresses. The required dead plus live loads are intended to include all loads except wind and seismic loads. Design roof loads given in Chapter 16 are required, including applicable design snow loads.

Effects of wind and seismic forces on empirically designed masonry structures are addressed by the lateral stability requirements of Section 2109.2 and by the seismic, wind and building height limitations of Section 2109.1.1.

2109.3.2 Allowable compressive stresses. The compressive stresses in masonry shall not exceed the values given in Table 2109.3.2. Stress shall be calculated based on specified rather than nominal dimensions.

❖ The maximum permitted compressive stresses on the gross cross-sectional area are given in Table 2109.3.2. Calculation of gross-area compressive stresses is discussed in the commentary to Section 2109.3.2.1. Section 2109.3.2 requires the use of specified dimensions (see also the commentary for the definition of "Dimensions" in Section 2102.1).

TABLE 2109.3.2. See below.

❖ The compressive strength of the unit, as well as the mortar type used, limit the allowable compressive stress or the gross area.

2109.3.2.1 Calculated compressive stresses. Calculated compressive stresses for single wythe walls and for multiwythe composite masonry walls shall be determined by dividing the design load by the gross cross-sectional area of the member. The area of openings, chases or recesses in walls shall not be included in the gross cross-sectional area of the wall.

❖ Using the vertical design load required by Section 2109.3.1, the compressive stress on the gross

cross-sectional area of the masonry must be calculated.

Gross cross-sectional area is illustrated in Figure 2102.1(3) for two single-wythe wall examples. Specified (not nominal) dimensions are used and mortared head joints are included. Cores are not required to be subtracted. Other openings in the wall, including chases and recesses, are required to be subtracted.

The calculated compressive stress is the total required design load divided by the gross cross-sectional area, as illustrated in Figure 2109.3.2.1(1) for a single-wythe wall. The dead weight of the masonry units is part of the total required design load. The calculated compressive stresses are not to exceed the allowable values listed in Table 2109.3.2.

The allowable compressive stresses for masonry directly under concentrated loads (bearing stresses) are

TABLE 2109.3.2
ALLOWABLE COMPRESSIVE STRESSES FOR EMPIRICAL DESIGN OF MASONRY

CONSTRUCTION; COMPRESSIVE STRENGTH OF UNIT GROSS AREA (psi)	ALLOWABLE COMPRESSIVE STRESSES[a] GROSS CROSS-SECTIONAL AREA (psi)	
	Type M or S mortar	Type N mortar
Solid masonry of brick and other solid units of clay or shale; sand-lime or concrete brick:		
8,000 or greater	350	300
4,500	225	200
2,500	160	140
1,500	115	100
Grouted masonry, of clay or shale; sand-lime or concrete:		
4,500 or greater	225	200
2,500	160	140
1,500	115	100
Solid masonry of solid concrete masonry units:		
3,000 or greater	225	200
2,000	160	140
1,200	115	100
Masonry of hollow load-bearing units:		
2,000 or greater	140	120
1,500	115	100
1,000	75	70
700	60	55
Hollow walls (noncomposite masonry bonded)[b] Solid units:		
2,500 or greater	160	140
1,500	115	100
Hollow units	75	70
Stone ashlar masonry:		
Granite	720	640
Limestone or marble	450	400
Sandstone or cast stone	360	320
Rubble stone masonry Coursed, rough or random	120	100

For SI: 1 pound per square inch = 0.006895 MPa.

a. Linear interpolation for determining allowable stresses for masonry units having compressive strengths which are intermediate between those given in the table is permitted.

b. Where floor and roof loads are carried upon one wythe, the gross cross-sectional area is that of the wythe under load; if both wythes are loaded, the gross cross-sectional area is that of the wall minus the area of the cavity between the wythes. Walls bonded with metal ties shall be considered as noncomposite walls unless collar joints are filled with mortar or grout.

recommended in the commentary to the MSJC Code: 125 percent of the Table 2109.3.2 value if the load acts on the full wall thickness; or 150 percent of the Table 2109.3.2 value if the load acts on concentrically placed bearing plates greater than one-half, but less than the full supporting area. Concentrated loads on load-bearing walls transmitted from beams, girders or other structural elements normally bear on units of solid ma-

sonry or on hollow masonry units with grout-filled cores at least 4 inches (102 mm) high.

Bearing plates are often used to distribute concentrated loads and to prevent damage to the bearing areas of the supporting masonry. Masonry bond beams can also be used for this purpose.

Concentrated loads must be distributed over an area with a length that cannot exceed the cen-

$$\frac{\text{DESIGN LOAD}}{\text{GROSS CROSS-SECTIONAL AREA}} \leq \text{TABLE 2109.3.2 ALLOWABLE COMPRESSIVE STRESS}$$

GROSS CROSS-SECTIONAL AREA = SPECIFIED LENGTH x SPECIFIED THICKNESS

Figure 2109.3.2.1(1)
ALLOWABLE COMPRESSIVE STRESS FOR AN EMPIRICALLY DESIGNED MASONRY WALL

For SI: 1 degree = .01745 rad.

Figure 2109.3.2.1(2)
ASSUMED DISTRIBUTION OF COMPRESSIVE STRESS UNDER CONCENTRATED LOAD

ter-to-center distance between loads, nor one-half of the wall height [see Figure 2109.3.2.1(2)]. Loads are not to be distributed across continuous vertical joints, such as expansion or control joints in masonry walls, or across head joints in stack bond construction. For large concentrated loads, masonry pilasters may be required.

2109.3.2.2 Multiwythe walls. The allowable stress shall be as given in Table 2109.3.2 for the weakest combination of the units used in each wythe.

❖ Multiwythe masonry walls are often constructed with units having different mechanical properties. An example of a multiwythe wall is a composite 8-inch (204 mm) nominal thickness wall using Type S mortar, consisting of one wythe of 4-inch-wide (102 mm) common brick and one wythe of 4-inch-wide (102 mm) lightweight concrete block, bonded together with required joint reinforcement and having completely filled collar joints.

In this example, if the compressive strength of the concrete masonry unit is less than that of the common brick, the allowable compressive stress for both wythes is to be based on the strength of the concrete masonry unit.

For loading on only one wythe of a composite (grouted collar joint), multiwythe masonry wall, the wythes and the collar joint are to be considered part of the gross cross-sectional area. For loading on only one wythe of a noncomposite (ungrouted collar joint) multiwythe masonry wall, the gross cross-sectional area is limited to the area of the wythe under load, and the allowable compressive stresses in Table 2109.3.2 are based on the units and mortar comprising the wythe under load.

The use of the term "composite" refers to multicomponent masonry members that act as a unit. Note b of Table 2109.3.2 considers a multiwythe masonry wall to act compositely (monolithically) if it is bonded with metal ties (or joint reinforcement) and has completely filled collar joints.

Openings in the wall, including chases and recesses, are required to be subtracted from the gross wall area, as stated in Section 2109.3.2.1.

2109.4 Lateral support.

❖ This section contains empirical design provisions for the spacing of lateral support locations of masonry walls. Requirements include maximum ratios of wall length or wall height to wall thickness. The requirements of this section do not apply when an engineered masonry design method is utilized.

Compression elements such as columns and walls may have axial capacities limited by buckling, based on slenderness effects (which depends on their stiffness and unsupported length). Limits are also included for nonbearing walls to account for resistance to out-of-plane loads.

The registered design professional is responsible for indicating in the construction documents the method of lateral support of masonry walls.

2109.4.1 Intervals. Masonry walls shall be laterally supported in either the horizontal or vertical direction at intervals not exceeding those given in Table 2109.4.1.

❖ The requirements of this section prescribe maximum length or height-to-thickness ratios for locations of lateral support (see Figure 2109.4.1). Providing lateral support at the resulting maximum heights increases the buckling capacity to an acceptable level. Spacing of lateral support must be designed either vertically or horizontally, but not in both directions.

TABLE 2109.4.1
WALL LATERAL SUPPORT REQUIREMENTS

CONSTRUCTION	MAXIMUM WALL LENGTH TO THICKNESS OR WALL HEIGHT TO THICKNESS
Bearing walls	
Solid units or fully grouted	20
All others	18
Nonbearing walls	
Exterior	18
Interior	36

❖ This table shows maximum length- or height-to-thickness ratios between locations of lateral support. Lateral support is required to be provided either vertically or horizontally, but not in both directions.

Figure 2109.4.1 illustrates use of the table for an empirically designed masonry building. As shown, the height is measured between points of lateral support.

2109.4.2 Thickness. Except for cavity walls and cantilever walls, the thickness of a wall shall be its nominal thickness measured perpendicular to the face of the wall. For cavity walls, the thickness shall be determined as the sum of the nominal thicknesses of the individual wythes. For cantilever walls, except for parapets, the ratio of height-to-nominal thickness shall not exceed 6 for solid masonry or 4 for hollow masonry. For parapets, see Section 2109.5.4.

❖ The nominal thickness or the sum of the nominal thicknesses is used in determining the wall thickness for use in Table 2109.4.1. Figure 2102.1(5) illustrates nominal versus specified thickness of a particular concrete masonry unit. Nominal dimensions are discussed in the commentary to Section 2102.1 (see the definition of "Dimensions").

The use of the nominal thickness is permitted for calculating thickness ratios for empirical design. In contrast, the engineered design method uses the rational Euler buckling equation and the accuracy of specified dimensions is appropriately required. Table 2109.4.1 is intended, however, to produce conservative results with respect to the engineered method in most cases.

For cavity walls, the adjacent wythes do not act together because of the absence of a shear connection. Thus, the wall thickness consists of only the sum of the wythes.

This section does not include retaining walls, which must comply with the requirements of Chapter 18. For definitions of "Solid" and "Hollow" masonry, refer to Section 2102.1.

Free-standing cantilever walls have a lower allowable height-to-thickness ratio because of the lack of lateral support at the free (unsupported) end. Parapets are subject to the more restrictive requirements of Section 2109.5.4 because of their location and exterior exposure.

2109.4.3 Support elements. Lateral support shall be provided by cross walls, pilasters, buttresses or structural frame members when the limiting distance is taken horizontally, or by floors, roofs acting as diaphragms or structural frame members when the limiting distance is taken vertically.

❖ Lateral support points complying with the requirements of this section are achieved by required anchorage complying with Section 2109.7.

The code requires only one direction of span for lateral support, either vertical or horizontal. The lateral support of masonry walls can be achieved by floors or roofs when the limiting distance is measured vertically, or by columns, buttresses (pilasters) or cross walls when the limiting distance is measured horizontally.

A masonry pier is an isolated column designed to support vertical loads. A pilaster is a masonry column integrally bonded to a wall.

A buttress is a pilaster with an outside edge (face running in the same direction as the wall) that slants toward the wall and has a horizontal cross section that increases from top to bottom. Buttresses are used mainly for out-of-plane lateral support of high walls. Both pilasters and buttresses may project out from one

Figure 2109.4.1
LATERAL SUPPORT REQUIREMENTS FOR EMPIRICALLY DESIGNED MASONRY WALLS

or both faces of the wall.

For empirical design, solid-bearing walls are permitted to span 20 times their thickness between supports, while hollow walls or walls of hollow masonry units can span 18 times their thickness. For example, a 16-inch (406 mm) hollow block wall or wall constructed with hollow units can have an unsupported height or length of 24 feet (7315 mm) [18 times 1.33 feet (405 mm)].

2109.5 Thickness of masonry. Minimum thickness requirements shall be based on nominal dimensions of masonry.

❖ Section 2109.5 provides requirements for minimum nominal thicknesses of empirically designed walls of masonry, including rubble stone (see the definition of "Dimensions"). Changes in thickness and parapet walls are also included in this section.

2109.5.1 Thickness of walls. The thickness of masonry walls shall conform to the requirements of Section 2109.5.

❖ Section 2109.5.1 contains provisions for minimum nominal thicknesses of empirically designed masonry walls. The requirements of this section do not apply when engineered masonry design is used.

2109.5.2 Minimum thickness.

❖ The minimum thicknesses required in this section are nominal thicknesses (in accordance with the definition in Section 2102.1) and apply to bearing walls. Like the measurement of thickness in Section 2109.4, the space between cavity walls and multiwythe, noncomposite walls is to be excluded when determining wall thickness.

2109.5.2.1 Bearing walls. The minimum thickness of masonry bearing walls more than one story high shall be 8 inches (203 mm). Bearing walls of one-story buildings shall not be less than 6 inches (152 mm) thick.

❖ These minimum thicknesses for load-bearing walls are based upon successful historical use. Single-story buildings experience limited lateral loads and, therefore, limited base shears and overturning moments. This is the justification for reducing the minimum nominal wall thickness in one-story structures.

2109.5.2.2 Rubble stone walls. The minimum thickness of rough, random or coursed rubble stone walls shall be 16 inches (406 mm).

❖ Rubble stone walls are composed of stone masonry having irregularly shaped units bonded by mortar. The greater thickness required for these walls is justified by this irregularity. Accordingly, Table 2109.3.2 permits relatively low compressive stresses for this material. Nominal thickness in this case is the average thickness. This section is not applicable to ashlar masonry (rectangular units).

2109.5.2.3 Shear walls. The minimum thickness of masonry shear walls shall be 8 inches (203 mm).

❖ An empirical minimum nominal shear wall thickness of 8 inches (203 mm) is required to transfer lateral loads to the foundation. This minimum is based on experience. Shear walls that are also load-bearing walls are required to comply with the compressive stress requirements of Section 2109.3.2 and the lateral support requirements of Section 2109.4. Those requirements may govern.

2109.5.2.4 Foundation walls. The minimum thickness of foundation walls shall be 8 inches (203 mm) and as required by Section 2109.5.3.1.

❖ See the commentary for Section 2109.5.3.

2109.5.2.5 Foundation piers. The minimum thickness of foundation piers shall be 8 inches (203 mm).

❖ See the commentary for Section 2109.5.3.

2109.5.2.6 Parapet walls. The minimum thickness of parapet walls shall be 8 inches (203 mm) and as required by Section 2109.5.4.1.

❖ Because of their exposure conditions and the hazard associated with unreinforced parapets, a minimum thickness of 8 inches (203 mm) and a maximum height-to-thickness ratio of 3:1 is required. These requirements are minimums and thicker parapets or smaller height-to-thickness ratios may be necessary.

2109.5.2.7 Change in thickness. Where walls of masonry of hollow units or masonry bonded hollow walls are decreased in thickness, a course or courses of solid masonry shall be interposed between the wall below and the thinner wall above, or special units or construction shall be used to transmit the loads from face shells or wythes above to those below.

❖ Where hollow walls are decreased in thickness, one or more courses of solid masonry are to be placed between the thicker wall below and the thinner wall above. Alternatively, special construction can be introduced to transmit the load from the wall above to the supporting wall below. For walls constructed with concrete masonry units, a bond beam as thick as the lower wall may be placed between the two wall sections.

2109.5.3 Foundation walls. Foundation walls shall comply with the requirements of Section 2109.5.3.1 or 2109.5.3.2.

❖ Masonry foundation walls must comply with Section 1805.5 if the requirements of this section cannot be met.

2109.5.3.1 Minimum thickness. Minimum thickness for foundation walls shall comply with the requirements of Table 2109.5.3.1. The provisions of Table 2109.5.3.1 are only applicable where the following conditions are met:

1. The foundation wall does not exceed 8 feet (2438 mm) in height between lateral supports;

2. The terrain surrounding foundation walls is graded to drain surface water away from foundation walls;

3. Backfill is drained to remove ground water away from foundation walls;

4. Lateral support is provided at the top of foundation walls prior to backfilling;

5. The length of foundation walls between perpendicular masonry walls or pilasters is a maximum of three times the basement wall height;

6. The backfill is granular and soil conditions in the area are nonexpansive; and

7. Masonry is laid in running bond using Type M or S mortar.

❖ This section provides empirical criteria for foundation walls that are based on similar requirements in the MSJC Code. It is necessary to satisfy the seven listed conditions in order to use this approach.

TABLE 2109.5.3.1
FOUNDATION WALL CONSTRUCTION

WALL CONSTRUCTION	NOMINAL WALL THICKNESS (inches)	MAXIMUM DEPTH OF UNBALANCED BACKFILL (feet)
Fully grouted masonry	8	7
	10	8
	12	8
Hollow unit masonry	8	5
	10	6
	12	7
Solid unit masonry	8	5
	10	7
	12	7

For SI: 1 inch = 25.4 mm, 1 foot = 304.8 mm.

❖ The minimum foundation wall thickness can be established from this table based on the depth of fill that is supported, as well as the proposed wall construction.

2109.5.3.2 Design requirements. Where the requirements of Section 2109.5.3.1 are not met, foundation walls shall be designed in accordance with Section 1805.5.

❖ If the conditions of Section 2109.5.3.1 cannot be met, the more general requirements for foundation walls in Chapter 18 must be complied with.

2109.5.4 Parapet walls.

❖ Parapet walls project above the roof line, exposing them to weather on both faces that may require special consideration.

2109.5.4.1 Minimum thickness. The minimum thickness of unreinforced masonry parapets shall meet Section 2109.5.2.6 and their height shall not exceed three times their thickness.

❖ Because of their exposure conditions and the hazard associated with unreinforced parapets, a minimum thickness of 8 inches (203 mm) and a maximum height-to-thickness ratio of 3:1 is required. These requirements are minimums and thicker parapets or smaller height-to-thickness ratios may be necessary.

2109.5.4.2 Additional provisions. Additional provisions for parapet walls are contained in Sections 1503.2 and 1503.3.

❖ See the commentary to Sections 1503.2 and 1503.3 regarding the application of these provisions to parapets.

2109.6 Bond.

❖ Section 2109.6 contains provisions for bonding adjacent wythes of empirically designed multiwythe masonry walls. The requirements of this section do not apply to engineered masonry design methods of Section 2107 or 2108.

2109.6.1 General. The facing and backing of multiwythe masonry walls shall be bonded in accordance with Section 2109.6.2, 2109.6.3 or 2109.6.4.

❖ This section establishes the requirements and methods of bonding together the facing and backing of adjacent wythes of multiwythe masonry walls. The use of masonry units, nonadjustable metal ties, adjustable metal ties and metal joint reinforcement as bonding elements is provided for in this section.

Bonding increases structural integrity, including load transfer between adjacent wythes and across head and collar joints. The requirements include both transverse (through-wall) and longitudinal (in-wall) bonding. Longitudinal bonding is provided for in Section 2109.5, with requirements for running bond and stack bond.

Adjacent wythes of multiwythe masonry walls are usually brick-to-brick, brick-to-block or block-to-block. Collar joints are usually $^3/_8$ to 4 inches (9.5 to 102 mm) thick.

Adjacent wythes of multiwythe masonry walls are considered composite (monolithic) when connected by metal ties or joint reinforcement and by a collar joint solidly filled with mortar or grout. Accordingly, Note b of Table 2109.3.2 requires completely filled collar joints if adjacent wythes bonded with metal ties are to be considered composite (monolithic) for empirical compressive stress design.

2109.6.2 Bonding with masonry headers.

❖ Masonry headers (bonders) are permitted to be used to connect adjacent wythes, in accordance with this section. Differential thermal movement, especially at exterior walls, can crack masonry bonders. Metal ties or joint reinforcement are more ductile and are recommended at these locations. Walls having the specified masonry bonders are considered composite (monolithic). Header requirements are based on successful past performance.

2109.6.2.1 Solid units. Where the facing and backing (adjacent wythes) of solid masonry construction are bonded by means of masonry headers, no less than 4 percent of the wall surface of each face shall be composed of headers extending not less than 3 inches (76 mm) into the backing. The distance between adjacent full-length headers shall not exceed 24 inches (610 mm) either vertically or horizontally. In walls in which a single header does not extend through the wall, headers from the opposite sides shall overlap at least 3 inches (76 mm), or headers from opposite sides shall be covered with another header course overlapping the header below at least 3 inches (76 mm).

❖ Solid masonry units are defined in Section 2102.1. The 4-percent requirement applies to each surface of the wall and 24-inch (610 mm) spacing is required in both horizontal and vertical directions. Spacing is measured between the nearest surfaces of masonry units. The 3-inch (76 mm) overlap is a projected measurement, since headers are spaced a maximum of 24 inches (610 mm) vertically.

Bonding requirements for solid units are less restrictive than for hollow units because of the greater cross-sectional area crossing the collar joint.

2109.6.2.2 Hollow units. Where two or more hollow units are used to make up the thickness of a wall, the stretcher courses shall be bonded at vertical intervals not exceeding 34 inches (864 mm) by lapping at least 3 inches (76 mm) over the unit below, or by lapping at vertical intervals not exceeding 17 inches (432 mm) with units that are at least 50 percent greater in thickness than the units below.

❖ Hollow masonry units are defined in Section 2102.1. An entire course of headers is required at the stated vertical spacing. Spacing is measured between the nearest surfaces of masonry units. The halved spacing of 17 inches (432 mm) is permitted for units having twice the thickness, because of the greater cross-sectional area crossing the collar joint. Bonding require-

ments for hollow units are more restrictive than for solid units because of the smaller cross-sectional area crossing the collar joint.

2109.6.2.3 Masonry bonded hollow walls. In masonry bonded hollow walls, the facing and backing shall be bonded so that not less than 4 percent of the wall surface of each face is composed of masonry bonded units extending not less than 3 inches (76 mm) into the backing. The distance between adjacent bonders shall not exceed 24 inches (610 mm) either vertically or horizontally.

❖ This section prescribes procedures for bonding hollow walls using masonry units.

2109.6.3 Bonding with wall ties or joint reinforcement.

❖ Figure 2109.6.3(1) illustrates typical masonry accessories. Shown are ladder-type and truss-type joint reinforcement and a "Z" wire wall tie. The spacing requirements given in this section are illustrated in Figure 2109.6.3(2). Strength and durability requirements for metal ties and joint reinforcement are given in Section 2103.13. Wall ties are required to be placed in accordance with Section 2104.1.3.

Metal wall ties or joint reinforcement provide a more ductile connection between adjacent wythes than do masonry headers. The size, spacing and number of ties required in this section are based on experience.

Adjacent wythes bonded with metal ties or joint reinforcement are considered composite (monolithic) only if the collar joint between them is completely filled with mortar or grout.

2109.6.3.1 Bonding with wall ties. Except as required by Section 2109.6.3.1.1, where the facing and backing (adjacent

Figure 2109.6.3(1)
TYPICAL MASONRY ACCESSORIES

wythes) of masonry walls are bonded with wire size W2.8 (MW18) wall ties or metal wire of equivalent stiffness embedded in the horizontal mortar joints, there shall be at least one metal tie for each $4^1/_2$ square feet (0.42 m²) of wall area. The maximum vertical distance between ties shall not exceed 24 inches (610 mm), and the maximum horizontal distance shall not exceed 36 inches (914 mm). Rods or ties bent to rectangular shape shall be used with hollow masonry units laid with the cells vertical. In other walls, the ends of ties shall be bent to 90-degree (1.57 rad) angles to provide hooks no less than 2 inches (51 mm) long. Wall ties shall be without drips. Additional bonding ties shall be provided at all openings, spaced not more than 36 inches (914 mm) apart around the perimeter and within 12 inches (305 mm) of the opening.

❖ Requirements for spacing of nonadjustable wall ties are shown in Figure 2109.6.3(2). The minimum wire size is W2.8. Where wythes of hollow units are bonded and cores are required to be vertical, rectangular-shaped ties are required. "Z" ties, illustrated in Figure 2109.6.3(1), are used in other applications. Drips on wall ties can adversely affect the tie strength and are, therefore, prohibited.

2109.6.3.1.1 Bonding with adjustable wall ties. Where the facing and backing (adjacent wythes) of masonry are bonded with adjustable wall ties, there shall be at least one tie for each 1.77 square feet (0.164 m²) of wall area. Neither the vertical nor horizontal spacing of the adjustable wall ties shall exceed 16 inches (406 mm). The maximum vertical offset of bed joints from one wythe to the other shall be $1^1/_4$ inches (32 mm). The maximum clearance between connecting parts of the ties shall be $^1/_{16}$ inch (1.6 mm). When pintle legs are used, ties shall have at least two wire size W2.8 (MW18) legs.

❖ Spacing requirements for bonding with adjustable ties are shown in Figure 2109.6.3(2) and are more restrictive because of the lower stiffness of those ties. A single tie of the pintle- and eye-type is required to have two legs. The maximum clearance between the eye and pintle is $^1/_{16}$ inch (1.6 mm). Figures 2109.6.3.1.1(1), (2) and (3) show requirements for adjustableanchors.

2109.6.3.2 Bonding with prefabricated joint reinforcement. Where the facing and backing (adjacent wythes) of masonry are bonded with prefabricated joint reinforcement,

For SI: 1 inch = 25.4 mm, 1 square foot = 0.0929 m².

Figure 2109.6.3(2)
EMPIRICAL REQUIREMENTS FOR BONDING MULTIWYTHE WALLS WITH TIES OR JOINT REINFORCEMENT

Figure 2109.6.3.1.1(1)
ADJUSTABLE JOINT REINFORCEMENT ASSEMBLIES

A) LADDER TYPES   B) TRUSS TYPES

SECTION

⅝ INCH MIN.
OR T/2 MAX.

A) TRUSS TYPE

⅝ INCH MIN.
OR T/2 MAX.

B) LADDER TYPE

For SI: 1 inch = 25.4 mm.

Figure 2109.6.3.1.1(2)
ADJUSTABLE ASSEMBLY DETAILS

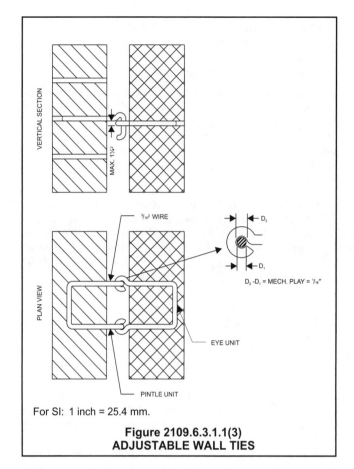

VERTICAL SECTION

MAX. 1¼"

PLAN VIEW

5/16" WIRE

D_2

D_1

$D_2 - D_1$ = MECH. PLAY = 1/16"

EYE UNIT

PINTLE UNIT

For SI: 1 inch = 25.4 mm.

Figure 2109.6.3.1.1(3)
ADJUSTABLE WALL TIES

there shall be at least one cross wire serving as a tie for each $2^2/_3$ square feet (0.25 m²) of wall area. The vertical spacing of the joint reinforcing shall not exceed 24 inches (610 mm). Cross wires on prefabricated joint reinforcement shall not be less than W1.7 (MW11) and shall be without drips. The longitudinal wires shall be embedded in the mortar.

❖ Truss and ladder-type joint reinforcement is illustrated in Figures 2102.1(4), 2109.6.3(1), 2109.6.3.1.1(1) and 2109.6.3.1.1(2). Ladder-type joint reinforcement better accommodates differential movement between wythes. Cross wires are required to have a minimum wire size of W1.7 for the transfer of stresses between adjacent wythes across the collar joint. Cross wires are permitted to be plain, while longitudinal wires are required to be deformed for increased bond strength.

2109.6.4 Bonding with natural or cast stone.

❖ Because of the irregularity of mortar joints permitted with coursed, random or rough stone masonry, as well as the need for peripheral mortar area in through-wall bonding units, metal ties do not adequately bond and are not required in this type of construction.

2109.6.4.1 Ashlar masonry. In ashlar masonry, bonder units, uniformly distributed, shall be provided to the extent of not less than 10 percent of the wall area. Such bonder units shall extend not less than 4 inches (102 mm) into the backing wall.

❖ Ashlar is often classified as either coursed or random. Bonding requirements apply equally to both types. The

10-percent requirement applies to one wall surface only. The "uniform distribution" requirement is intended to allow for slight deviations necessary for the nonmodular units typically encountered.

2109.6.4.2 Rubble stone masonry. Rubble stone masonry 24 inches (610 mm) or less in thickness shall have bonder units with a maximum spacing of 36 inches (914 mm) vertically and 36 inches (914 mm) horizontally, and if the masonry is of greater thickness than 24 inches (610 mm), shall have one bonder unit for each 6 square feet (0.56 m²) of wall surface on both sides.

❖ Rubble stone masonry is required to be through bonded using stones placed at a maximum spacing of 3 feet (914 mm) vertically and horizontally. Where the wall is more than 24 inches (610 mm) thick, at least one bond unit is required for each 6 square feet (0.56 m²) of wall surface on both sides.

2109.6.5 Masonry bonding pattern.

❖ Masonry may be constructed with a variety of bond patterns to provide a variety of decorative appearances. Because some of these bond patterns are stronger than others, two broad categories have been established to describe them. Section 2109.6.5.1 applies to masonry laid in running bond. Section 2109.6.5.2 applies to masonry laid in "other than running bond," which in that section is referred to as "stack bond," although that term can also be used to describe a very specific bond pattern in which the head joints are aligned continuously from course to course, rather than being offset. See the definitions of "Running bond" and "Stack bond" in Section 2102.1 and related commentary.

2109.6.5.1 Masonry laid in running bond. Each wythe of masonry shall be laid in running bond, head joints in successive courses shall be offset by not less than one-fourth the unit length or the masonry walls shall be reinforced longitudinally as required in Section 2109.6.5.2.

❖ The requirement for running bond is illustrated in Figure 2102.1(6). The minimum overlap is intended to provide structural continuity across head joints.

2109.6.5.2 Masonry laid in stack bond. Where unit masonry is laid with less head joint offset than in Section 2109.6.5.1, the minimum area of horizontal reinforcement placed in mortar bed joints or in bond beams spaced not more than 48 inches (1219 mm) apart, shall be 0.0003 times the vertical cross-sectional area of the wall.

❖ Where the overlap required for running bond, as illustrated in Figure 2102.1(6), is not provided, additional structural continuity must be provided across head joints in accordance with this section. Each mortar joint or bond beam is required to have the minimum stated reinforcement to provide distributed reinforcement in the wall.

2109.7 Anchorage.

❖ This section contains provisions for anchorage of empirically designed masonry elements at locations of lateral support, including intersecting walls, floors, roofs and adjoining structural framing. The requirements of this section do not apply to the engineered masonry design methods of Sections 2107 and 2108.

2109.7.1 General. Masonry elements shall be anchored in accordance with Sections 2109.7.2 through 2109.7.4.

❖ This section establishes the requirements and methods of anchoring masonry walls to intersecting walls, floors and roofs. This section covers anchoring requirements for masonry bonders and for metal ties and anchors.

Anchorage in accordance with this section is required for empirically designed masonry. Anchorage provisions address empirical lateral load resistance, but not the vertical load resistance of the connection, floor or roof system. The vertical load resistance of the connection, floor or roof system, including permitted spans, bearing requirements and bracing, must comply with applicable code requirements. The empirical provisions in this chapter do not address elements with horizontal spans, such as wood, floor decks, roof systems or concrete or steel beams.

2109.7.2 Intersecting walls. Masonry walls depending upon one another for lateral support shall be anchored or bonded at locations where they meet or intersect by one of the methods indicated in Sections 2109.7.2.1 through 2109.7.2.5.

❖ Where required by the lateral bracing requirements in Section 2109.4, intersecting walls and partitions must be anchored together by any of the methods described in the sections that follow (see Figure 2109.7.2). These requirements, however, do not prohibit other methods for connecting intersecting walls.

For example, long walls, by design, often require control joints or expansion joints at cross walls. The construction details for these walls differ considerably from walls that are required to be anchored or bonded monolithically.

Intersecting walls, such as tees and corners, may be tied together using interlocking masonry units or heavy metal ties placed in the bed joints.

2109.7.2.1 Bonding pattern. Fifty percent of the units at the intersection shall be laid in an overlapping masonry bonding pattern, with alternate units having a bearing of not less than 3 inches (76 mm) on the unit below.

❖ Masonry bonding requires that at least 50 percent of the masonry units cross at the juncture of the walls as bonding units. The units must be laid with a 3-inch (76 mm) minimum bearing. Figure 2109.7.2 shows typical brick walls bonded at a tee intersection and a corner. While these coursings are common, others are possible.

2109.7.2.2 Steel connectors. Walls shall be anchored by steel connectors having a minimum section of $^1/_4$ inch (6.4 mm) by $1^1/_2$ inches (38 mm), with ends bent up at least 2 inches (51 mm) or with cross pins to form anchorage. Such anchors shall be at least 24 inches (610 mm) long and the maximum spacing shall be 48 inches (1219 mm).

❖ Rigid steel connectors must be used at vertical intervals of 4 feet (1219 mm) or less. For this type of anchorage, the most commonly used fastener is a steel strip with cross-sectional dimensions not less than $^1/_4$ by $1^1/_2$ inches (6.4 by 38 mm) in lengths of 2 feet (610 mm) or more and with 2-inch (51 mm) hooked ends [see Figure 2109.6.3(1)]. Figure 2109.7.2 shows a tee and corner connection made with steel connectors.

2109.7.2.3 Joint reinforcement. Walls shall be anchored by joint reinforcement spaced at a maximum distance of 8 inches (203 mm). Longitudinal wires of such reinforcement shall be at least wire size W1.7 (MW 11) and shall extend at least 30 inches (762 mm) in each direction at the intersection.

❖ Figure 2109.7.2 shows a tee and corner connection made with joint reinforcement. Prefabricated joint reinforcement made for that purpose is commonly used. Joint reinforcement is spaced not more than 8 inches (204 mm) vertically.

The ladder-type joint reinforcement shown is made with $^3/_{16}$-inch-diameter (4.8 mm) deformed side bars cross connected with a minimum wire size of W1.7.

2109.7.2.4 Interior nonload-bearing walls. Interior nonload-bearing walls shall be anchored at their intersection, at vertical intervals of not more than 16 inches (406 mm) with joint reinforcement or $^1/_4$-inch (6.4 mm) mesh galvanized hardware cloth.

❖ Where interior nonload-bearing masonry walls or partitions intersect, they may be connected by bonders described in Section 2109.7.2.1 by joint reinforcement (if constructed separately) or $^1/_4$-inch (6.4 mm) galvanized steel mesh spaced vertically at a maximum interval of 16 inches (406 mm).

2109.7.2.5 Ties, joint reinforcement or anchors. Other metal ties, joint reinforcement or anchors, if used, shall be spaced to provide equivalent area of anchorage to that required by this section.

❖ Truss-type, prefabricated metal ties and steel strip anchors bent at the ends can be hooked into vertical mortar joints or embedded in grout-filled cores of concrete block units. Other methods of reinforcement can be used if the cross-sectional area and distribution are equivalent.

2109.7.3 Floor and roof anchorage. Floor and roof diaphragms providing lateral support to masonry shall comply with the live loads in Section 1607.3 and shall be connected to the masonry in accordance with Sections 2109.7.3.1 through 2109.7.3.3. Roof loading shall be determined in accordance with Chapter 16 and, when net uplift occurs, uplift shall be

For SI: 1 inch = 25.4 mm.

Figure 2109.7.2
ANCHORAGE OF INTERSECTING WALLS

resisted entirely by an anchorage system designed in accordance with the provisions of Sections 2.1 and 2.3, Sections 3.1 and 3.3 or Chapter 4 of ACI 530/ASCE 5/TMS 402.

❖ Where the lateral support locations required in Section 2109.4 are provided by intersecting floors or roofs, the minimum anchorage requirements of this section apply. The specified anchorage is intended to transfer shear between a floor or roof diaphragm and the wall, as well as provide locations of lateral support.

This section does not address vertical load resistance of floors or roofs. The connections required by the loads transferred from floors and roofs must be in accordance with applicable requirements for those elements, including required design loads in Chapter 16. The minimum requirements stated here do not address floor and roof loads to be resisted by the connection.

2109.7.3.1 Wood floor joists. Wood floor joists bearing on masonry walls shall be anchored to the wall at intervals not to exceed 72 inches (1829 mm) by metal strap anchors. Joists parallel to the wall shall be anchored with metal straps spaced not more than 72 inches (1829 mm) o.c. extending over or under and secured to at least three joists. Blocking shall be provided between joists at each strap anchor.

❖ Anchorage with metal straps is specified by this section. Blocking is required for stability against rotation of joists parallel to the wall. Strength and material requirements for metal straps are specified in Section 2103.13.

2109.7.3.2 Steel floor joists. Steel floor joists bearing on masonry walls shall be anchored to the wall with $^3/_8$-inch (9.5 mm) round bars, or their equivalent, spaced not more than 72 inches (1829 mm) o.c. Where joists are parallel to the wall, anchors shall be located at joist bridging.

❖ The strength and material requirements for $^3/_8$-inch (9.5 mm) round bars are specified in Section 2103.13. The requirements for steel joist construction, including materials, design, load-bearing capacity and cross bridging, are in Chapter 22.

2109.7.3.3 Roof diaphragms. Roof diaphragms shall be anchored to masonry walls with $^1/_2$-inch-diameter (12.7 mm) bolts, 72 inches (1829 mm) o.c. or their equivalent. Bolts shall extend and be embedded at least 15 inches (381 mm) into the masonry, or be hooked or welded to not less than 0.20 square inch (129 mm²) of bond beam reinforcement placed not less than 6 inches (152 mm) from the top of the wall.

❖ Roofs must be anchored to bond beams and to the tops of other walls. The material requirements for the bolts are given in Section 2103.13.5.

2109.7.4 Walls adjoining structural framing. Where walls are dependent upon the structural frame for lateral support, they shall be anchored to the structural members with metal anchors or otherwise keyed to the structural members. Metal anchors shall consist of $^1/_2$-inch (12.7 mm) bolts spaced at 48

inches (1219 mm) o.c. embedded 4 inches (102 mm) into the masonry, or their equivalent area.

❖ Selection of connectors for anchoring masonry to steel beams and columns is based on the loads that must be resisted. Lateral restraint perpendicular to the wall is required as a location of lateral support. Diaphragm transfer of shear forces is also required parallel to the wall, if such forces occur. Alternative methods of reinforcement can be used if the cross-sectional area and spacing are equivalent.

2109.8 Adobe construction. Adobe construction shall comply with this section and shall be subject to the requirements of this code for Type V construction.

❖ Adobe masonry was popular in the southwest United States due to the availability of soil for units, the limited rainfall and low humidity to dry the units, the thermal mass provided by the completed adobe structure and the low cost of this form of construction. Requirements for adobe construction are based on previous requirements in the SBC and the UBC. They are a combination of empirical provisions and rudimentary engineering. Since there are no ASTM standards for adobe materials, test methods have been included in the code. Design is based on gross cross-sectional dimensions.

Requirements for unstabilized adobe are contained in Section 2109.8.1. Requirements for stabilized adobe are contained in Section 2109.8.2. Requirements in Sections 2109.8.3 and 2109.8.4 apply to both unstabilized and stabilized adobe. This is one of the few sources for such design information.

2109.8.1 Unstabilized adobe.

❖ Unstabilized adobe does not contain stabilizers and is generally not as durable or dimensionally stable as stabilized adobe.

2109.8.1.1 Compressive strength. Adobe units shall have an average compressive strength of 300 psi (2068 kPa) when tested in accordance with ASTM C 67. Five samples shall be tested and no individual unit is permitted to have a compressive strength of less than 250 psi (1724 kPa).

❖ Average compressive strength, based on five specimens tested in accordance with ASTM C 67, must be at least 300 psi (2068 kPa).

2109.8.1.2 Modulus of rupture. Adobe units shall have an average modulus of rupture of 50 psi (345 kPa) when tested in accordance with the following procedure. Five samples shall be tested and no individual unit shall have a modulus of rupture of less than 35 psi (241 kPa).

❖ Average modulus of rupture, based on five specimens tested in accordance with Sections 2109.8.1.2.1 through 2109.8.1.2.4, must be at least 50 psi (345 kPa).

2109.8.1.2.1 Support conditions. A cured unit shall be simply supported by 2-inch-diameter (51 mm) cylindrical supports located 2 inches (51 mm) in from each end and extending the full width of the unit.

❖ These required support conditions are typical for modulus of rupture tests.

2109.8.1.2.2 Loading conditions. A 2-inch-diameter (51 mm) cylinder shall be placed at midspan parallel to the supports.

❖ Loading through a hydraulic cylinder at midspan is common for these tests.

2109.8.1.2.3 Testing procedure. A vertical load shall be applied to the cylinder at the rate of 500 pounds per minute (37 N/s) until failure occurs.

❖ The required application of vertical load is easily controlled in testing laboratories.

2109.8.1.2.4 Modulus of rupture determination. The modulus of rupture shall be determined by the equation:

$$f_r = 3WL_s/2bt^2 \qquad \text{(Equation 21-4)}$$

where, for the purposes of this section only:

b = Width of the test specimen measured parallel to the loading cylinder, inches (mm).

f_r = Modulus of rupture, psi (MPa).

L_s = Distance between supports, inches (mm).

t = Thickness of the test specimen measured parallel to the direction of load, inches (mm).

W = The applied load at failure, pounds (N).

❖ Equation 21-4 is based on simple engineering mechanics and is valid for all rectangular specimens tested in this fashion.

2109.8.1.3 Moisture content requirements. Adobe units shall have a moisture content not exceeding 4 percent by weight.

❖ This section limits the moisture content of unstabilized adobe units to acceptable levels.

2109.8.1.4 Shrinkage cracks. Adobe units shall not contain more than three shrinkage cracks and any single shrinkage crack shall not exceed 3 inches (76 mm) in length or $^1/_8$ inch (3.2 mm) in width.

❖ As adobe units dry, they shrink and can crack. This section places limits on those potential cracks to keep the masonry structurally sound and reasonably water resistant.

2109.8.2 Stabilized adobe.

❖ This type of adobe is manufactured with stabilizers to increase its durability and decrease its water absorption.

2109.8.2.1 Material requirements. Stabilized adobe shall comply with the material requirements of unstabilized adobe in addition to Sections 2109.8.2.1.1 and 2109.8.2.1.2.

❖ Stabilized adobe must comply with the few material requirements for unstabilized adobe in Section

2109.8.1. Stabilized units must also comply with soil compatibility and absorption requirements in Sections 2109.8.2.1.1 and 2109.8.2.1.2.

2109.8.2.1.1 Soil requirements. Soil used for stabilized adobe units shall be chemically compatible with the stabilizing material.

❖ The soil and stabilizing materials must be chemically compatible, so that the stabilized units will be durable.

2109.8.2.1.2 Absorption requirements. A 4-inch (102 mm) cube, cut from a stabilized adobe unit dried to a constant weight in a ventilated oven at 212°F to 239°F (100°C to 115°C), shall not absorb more than $2^1/_2$ percent moisture by weight when placed upon a constantly water-saturated, porous surface for seven days. A minimum of five specimens shall be tested and each specimen shall be cut from a separate unit.

❖ This section prescribes a test method to verify that stabilized adobe units meet absorption limits.

2109.8.3 Allowable stress. The allowable compressive stress based on gross cross-sectional area of adobe shall not exceed 30 psi (207 kPa).

❖ This section prescribes the allowable compressive stress of adobe based on its gross cross-sectional area.

2109.8.3.1 Bolts. Bolt values shall not exceed those set forth in Table 2109.8.3.1.

❖ This section requires the capacity of bolts to be based on Table 2109.8.3.1. Specific types of bolts are not identified, but headed, bent-bar and plate anchors should all be acceptable.

TABLE 2109.8.3.1
ALLOWABLE SHEAR ON BOLTS IN ADOBE MASONRY

DIAMETER OF BOLTS (inches)	MINIMUM EMBEDMENT (inches)	SHEAR (pounds)
$^1/_2$	—	—
$^5/_8$	12	200
$^3/_4$	15	300
$^7/_8$	18	400
1	21	500
$1^1/_8$	24	600

For SI: 1 inch = 25.4 mm, 1 pound = 4.448 N.

❖ The allowable shear values in this table are based on the capacity of the adobe masonry. The capacity of the anchor-bolt steel is much higher, so the lower strength of the adobe controls.

2109.8.4 Construction.

❖ This section contains general construction requirements for height restrictions, mortar restrictions, mortar joint construction and water-resistance requirements for parapet walls and also specific requirements for wall thickness; foundations; isolated piers and columns; tie beams; exterior finish and lintels.

2109.8.4.1 General.

❖ This section contains general construction requirements for height restrictions, mortar restrictions, mortar joint construction and water-resistance requirements for parapet walls.

2109.8.4.1.1 Height restrictions. Adobe construction shall be limited to buildings not exceeding one story, except that two-story construction is allowed when designed by a registered design professional.

❖ Because of the low strength of adobe masonry, it is limited to use in single-story buildings, unless a registered design professional is hired, in which case two-story buildings are permitted.

2109.8.4.1.2 Mortar restrictions. Mortar for stabilized adobe units shall comply with Chapter 21 or adobe soil. Adobe soil used as mortar shall comply with material requirements for stabilized adobe. Mortar for unstabilized adobe shall be portland cement mortar.

❖ A variety of mortars are acceptable for stabilized adobe as noted; however, portland cement-lime mortars are required for unstabilized adobe. Selection of a relatively weak mortar that is compatible with the units is appropriate.

2109.8.4.1.3 Mortar joints. Adobe units shall be laid with full head and bed joints and in full running bond.

❖ Full mortar joints, the same as for other solid units, are required for adobe construction. Units are required to be laid in running bond.

2109.8.4.1.4 Parapet walls. Parapet walls constructed of adobe units shall be waterproofed.

❖ Waterproofing parapets reduces moisture infiltration into the adobe.

2109.8.4.2 Wall thickness. The minimum thickness of exterior walls in one-story buildings shall be 10 inches (254 mm). The walls shall be laterally supported at intervals not exceeding 24 feet (7315 mm). The minimum thickness of interior load-bearing walls shall be 8 inches (203 mm). In no case shall the unsupported height of any wall constructed of adobe units exceed 10 times the thickness of such wall.

❖ Because of the low strength of adobe masonry walls, thicker walls with more closely spaced supports are required.

2109.8.4.3 Foundations.

❖ This section prescribes foundation requirements for adobe masonry.

2109.8.4.3.1 Foundation support. Walls and partitions constructed of adobe units shall be supported by foundations or footings that extend not less than 6 inches (152 mm) above adjacent ground surfaces and are constructed of solid masonry (excluding adobe) or concrete. Footings and foundations shall comply with Chapter 18.

❖ So that adobe masonry is properly supported, solid masonry or concrete foundations are required.

2109.8.4.3.2 Lower course requirements. Stabilized adobe units shall be used in adobe walls for the first 4 inches (102 mm) above the finished first-floor elevation.

❖ Because of their greater durability, stabilized adobe units are required at the base of adobe walls. Conventional masonry units can also be used to satisfy this requirement.

2109.8.4.4 Isolated piers or columns. Adobe units shall not be used for isolated piers or columns in a load-bearing capacity. Walls less than 24 inches (610 mm) in length shall be considered isolated piers or columns.

❖ Adobe units are not strong enough to carry significant loads and are, therefore, not permitted to be used as isolated piers or columns.

2109.8.4.5 Tie beams. Exterior walls and interior load-bearing walls constructed of adobe units shall have a continuous tie beam at the level of the floor or roof bearing and meeting the following requirements.

❖ To distribute loads more evenly into the adobe, tie beams are required at the floor or roof levels. Tie beams can be constructed of concrete or wood as described in Sections 2109.8.4.5.1 and 2108.8.4.5.2, respectively.

2109.8.4.5.1 Concrete tie beams. Concrete tie beams shall be a minimum depth of 6 inches (152 mm) and a minimum width of 10 inches (254 mm). Concrete tie beams shall be continuously reinforced with a minimum of two No. 4 reinforcing bars. The ultimate compressive strength of concrete shall be at least 2,500 psi (17.2 MPa) at 28 days.

❖ This section provides requirements for concrete tie beams to be cast above adobe masonry walls to distribute loads from floors and roofs.

2109.8.4.5.2 Wood tie beams. Wood tie beams shall be solid or built up of lumber having a minimum nominal thickness of 1 inch (25 mm), and shall have a minimum depth of 6 inches (152 mm) and a minimum width of 10 inches (254 mm). Joints in wood tie beams shall be spliced a minimum of 6 inches (152 mm). No splices shall be allowed within 12 inches (305 mm) of an opening. Wood used in tie beams shall be approved naturally decay-resistant or pressure-treated wood.

❖ This section provides requirements for wood tie beams to be constructed above adobe masonry walls to distribute loads from floors and roofs.

2109.8.4.6 Exterior finish. Exterior walls constructed of unstabilized adobe units shall have their exterior surface covered with a minimum of two coats of portland cement plaster having a minimum thickness of $^3/_4$ inch (19.1 mm) and conforming to ASTM C 926. Lathing shall comply with ASTM C 1063. Fasteners shall be spaced at 16 inches (406 mm) o.c.

maximum. Exposed wood surfaces shall be treated with an approved wood preservative or other protective coating prior to lath application.

❖ Unstabilized adobe must be coated with plaster to increase its durability.

2109.8.4.7 Lintels. Lintels shall be considered structural members and shall be designed in accordance with the applicable provisions of Chapter 16.

❖ Lintels over door and window openings are required to be structurally designed to carry imposed loads and to distribute those loads into the supporting adobe.

SECTION 2110
GLASS UNIT MASONRY

2110.1 Scope. This section covers the empirical requirements for nonload-bearing glass unit masonry elements in exterior or interior walls.

❖ Section 2110 contains provisions for glass unit masonry walls, which are nearly identical to the glass unit masonry provisions in Chapter 7 of the MSJC Code. Because those provisions are essentially the same, Section 2101.2.5 permits glass unit masonry to comply with the provisions of Chapter 7 of the MSJC Code or of Section 2110.

Glass unit masonry panels are permitted to be used in interior or exterior walls, provided that they are nonload bearing and comply with the requirements of Section 2110, which are partly empirical and partly based on tests.

2110.1.1 Limitations. Solid or hollow approved glass block shall not be used in fire walls, party walls, fire barriers or fire partitions, or for load-bearing construction. Such blocks shall be erected with mortar and reinforcement in metal channel-type frames, structural frames, masonry or concrete recesses, embedded panel anchors as provided for both exterior and interior walls or other approved joint materials. Wood strip framing shall not be used in walls required to have a fire-resistance rating by other provisions of this code.

Exceptions:

1. Glass-block assemblies having a fire protection rating of not less than $^3/_4$ hour shall be permitted as opening protectives in accordance with Section 715 in fire barriers and fire partitions that have a required fire-resistance rating of 1 hour or less and do not enclose exit stairways or exit passageways.

2. Glass-block assemblies as permitted in Section 404.5, Exception 2.

❖ Structural glass blocks are not permitted in fire walls, party walls, fire barrier walls or fire partitions, with two exceptions. Exception 1 permits glass blocks that have been tested and classified for a $^3/_4$-hour fire protection rating in openings to be used in fire barrier walls or fire partitions with a required fire-resistance rating of 1 hour or less. Since 1-hour fire barrier walls can be utilized to enclose interior exit stairways and

exit ramps (see Section 1020.1), as well as exit passageways (see Section 1021.3), the exception does not apply to those locations. This is consistent with Sections 1020.1.1 and 1021.4, which limit openings in these exit components to those that are necessary for egress purposes. Exception 2 permits glass blocks to be installed in accordance with the requirements in Section 404.5, Exception 2 for the enclosure of atriums. Because Section 404.5 requires a 1-hour fire barrier wall to enclose an atrium, this exception is redundant since Exception 1 already permits this.

2110.2 Units. Hollow or solid glass-block units shall be standard or thin units.

❖ This section contains minimum requirements for glass masonry units, since a corresponding ASTM standard does not exist. Units are permitted to be either hollow or solid and are required to meet requirements for either standard or thin units.

Glass units are usually factory coated at their edges. Uncoated glass-block units can be field coated by following the manufacturer's instructions.

2110.2.1 Standard units. The specified thickness of standard units shall be at least $3^7/_8$ inches (98 mm).

❖ This section requires a specified thickness for standard glass masonry units of $3^7/_8$ inches (98 mm).

2110.2.2 Thin units. The specified thickness of thin units shall be $3^1/_8$ inches (79 mm) for hollow units or 3 inches (76 mm) for solid units.

❖ Thicknesses for thin glass unit masonry are given in this section.

2110.3 Panel size.

❖ This section provides limits on the size of exterior standard-unit and thin-unit panels, interior panels, solid glass-block panels and curved glass-block panels.

The glass-block panels must be restrained laterally and be capable of resisting horizontal forces. Panels exceeding these size limits require intermediate structural supports so that loads can be adequately resisted.

2110.3.1 Exterior standard-unit panels. The maximum area of each individual exterior standard-unit panel shall be 144 square feet (13.4 m²) when the design wind pressure is 20 psf (958 N/m²). The maximum panel dimension between structural supports shall be 25 feet (7620 mm) in width or 20 feet (6096 mm) in height. The panel areas are permitted to be adjusted in accordance with Figure 2110.3.1 for other wind pressures.

❖ Single panels of glass block are limited to a maximum length of 25 feet (7620 mm) and a maximum height of 20 feet (6096 mm) between structural supports. When subjected to a wind pressure of 20 psf (958 N/m²), any single panel cannot exceed 144 square feet (13 m²) in area. This maximum panel permissible area can be adjusted, however, for other wind pressures by the use of Figure 2110.3.1.

The wind load resistance curve represents capacities for a variety of panel conditions. The 144-square-foot (13 m²) area limit is based on a safety factor of 2.7 when the design wind pressure is 20 psf (958 N/m²).

2110.3.2 Exterior thin-unit panels. The maximum area of each individual exterior thin-unit panel shall be 85 square feet (7.9 m²). The maximum dimension between structural supports shall be 15 feet (4572 mm) in width or 10 feet (3048 mm) in height. Thin units shall not be used in applications where the design wind pressure exceeds 20 psf (958 N/m²).

❖ The limitations on the use of exterior panels with thin-unit glass block are more restrictive than those in Section 2110.3.1 for exterior panels with standard units, since thin units are not as strong as standard ones.

2110.3.3 Interior panels. The maximum area of each individual standard-unit panel shall be 250 square feet (23.2 m²). The maximum area of each thin-unit panel shall be 150 square feet (13.9 m²). The maximum dimension between structural supports shall be 25 feet (7620 mm) in width or 20 feet (6096 mm) in height.

❖ Interior panels can be larger than exterior ones since wind pressures are expected to be lower.

2110.3.4 Solid units. The maximum area of solid glass-block wall panels in both exterior and interior walls shall not be more than 100 square feet (9.3 m²).

❖ Panels constructed of solid glass block must also meet the requirements of this section.

2110.3.5 Curved panels. The width of curved panels shall conform to the requirements of Sections 2110.3.1, 2110.3.2 and 2110.3.3, except additional structural supports shall be provided at locations where a curved section joins a straight section, and at inflection points in multicurved walls.

❖ Curved panels of glass block must meet the appropriate requirements of Sections 2110.3.1 through 2110.3.3, plus the requirements for additional supports at critical locations as defined in this section.

2110.4 Support.

❖ This section requires that glass-unit masonry panels be laterally supported by panel anchors, channel-type restraints or a combination of both. Channel-type restraints can be made of concrete, masonry, metal, wood or other materials, provided that adequate lateral support is achieved.

2110.4.1 General requirements. Glass unit masonry panels shall be isolated so that in-plane loads are not imparted to the panel.

❖ Isolation joints are needed at the top and sides of glass-unit masonry panels so that in-plane forces are not transferred to the panels.

2110.4.2 Vertical. Maximum total deflection of structural members supporting glass unit masonry shall not exceed $1/_{600}$.

❖ The sizes of structural members supporting glass-block panels must be determined by structural analysis in order to avoid excessive deflections that could damage the glass-block construction. Deflections of supporting members are limited to a maximum of $l/600$, where l is the span of the supporting member.

2110.4.2.1 Support on wood construction. Glass unit masonry having an installed weight of 40 psf (195 kg/m²) or

For SI: 1 square foot = 0.0929 m², 1 pound per square foot = 47.9 N/m².

FIGURE 2110.3.1
GLASS MASONRY DESIGN WIND LOAD RESISTANCE

less and a maximum height of 12 feet (3658 mm) shall be permitted to be supported on wood construction.

❖ Glass unit masonry is commonly used in windows and other similar applications whereby it is supported on wood construction. The limits stipulated are intended to address safety concerns associated with fire safety of glass unit masonry supported on combustible construction.

2110.4.2.2 Expansion joint. A vertical expansion joint in glass unit masonry shall be provided to isolate the glass unit masonry supported by wood construction from that supported by other types of construction.

❖ Due to the differences in materials and associated expansion/contraction, deflection glass unit masonry that is supported on wood construction is required to be isolated from adjacent sections of glass unit masonry that is supported on more rigid construction.

2110.4.3 Lateral. Glass unit masonry panels more than one unit wide or one unit high shall be laterally supported along their tops and sides. Lateral support shall be provided by panel anchors along the top and sides spaced not more than 16 inches (406 mm) o.c. or by channel-type restraints. Glass unit masonry panels shall be recessed at least 1 inch (25 mm) within channels and chases. Channel-type restraints shall be oversized to accommodate expansion material in the opening and packing and sealant between the framing restraints and the glass unit masonry perimeter units. Lateral supports for glass unit masonry panels shall be designed to resist applied loads, or a minimum of 200 pounds per lineal feet (plf) (2919 N/m) of panel, whichever is greater.

Exceptions:

1. Lateral support at the top of glass unit masonry panels that are no more than one unit wide shall not be required.

2. Lateral support at the sides of glass unit masonry panels that are no more than one unit high shall not be required.

❖ Glass-block panels in exterior masonry walls or in openings of structural framing systems (curtain walls) are required to be restrained laterally to resist both external and internal pressures caused by wind and horizontal forces from earthquakes. Lateral support can be provided by channel-type or panel anchors.

Adhering to the dimensional limitations imposed on glass-block panels and complying with the requirements for construction will generally produce glass-block elements that adequately resist normal wind conditions. In regions with very high winds or high seismic risk, however, glass-block panels and their anchorage to supporting structural elements should be checked for adequacy.

The exceptions in this section recognize that loads and expected movement of small glass-unit panels are small enough to permit installation without isolation joints.

2110.4.3.1 Single unit panels. Single unit glass unit masonry panels shall conform to the requirements of Section 2110.4.3, except lateral support shall not be provided by panel anchors.

❖ Single-unit panels generally have the same requirements as multiple-unit panels described in Section 2110.4.3.

2110.5 Expansion joints. Glass unit masonry panels shall be provided with expansion joints along the top and sides at structural supports. Expansion joints shall have sufficient thickness to accommodate displacements of the supporting structure, but shall not be less than $^3/_8$ inch (9.5 mm) in thickness. Expansion joints shall be entirely free of mortar or other debris and shall be filled with resilient material. The sills of glass-block panels shall be coated with approved water-based asphaltic emulsion, or other elastic waterproofing material, prior to laying the first mortar course.

❖ Sills supporting glass-block panels on a mortar bed are required to be first made water resistant with a heavy coat of asphalt emulsion or other approved water-resistant material. When the emulsion on the sill has dried, the full mortar bed can be placed, followed by the lowest course of glass block.

Before any glass-block units are placed, however, the head and jamb areas over the full height and width of the glass-block panel should be provided with expansion strips [$^3/_8$ inches (9.5 mm) thick] made for this purpose.

Glass-block units should be placed in successive courses on mortar beds containing panel reinforcement as required by the project documents, either directly or through reference to the manufacturer's instructions. Panel reinforcement should not be placed across expansion joints. Joints should be tooled smooth and concave before the mortar sets and joints around the perimeter of the panel should be raked out sufficiently to receive filler and caulking materials.

After the mortar sets, expansion joints are typically closed with a sealant.

2110.6 Mortar. Mortar for glass unit masonry shall comply with Section 2103.8.

❖ See the commentary to Section 2103.7. Glass-unit masonry panels are to be laid in Type N or S mortar.

2110.7 Reinforcement. Glass unit masonry panels shall have horizontal joint reinforcement spaced not more than 16 inches (406 mm) on center, located in the mortar bed joint, and extending the entire length of the panel but not across expansion joints. Longitudinal wires shall be lapped a minimum of 6 inches (152 mm) at splices. Joint reinforcement shall be placed in the bed joint immediately below and above openings in the panel. The reinforcement shall have not less than two parallel longitudinal wires of size W1.7 (MW11), and have welded cross wires of size W1.7 (MW11).

❖ Panel reinforcement made especially for glass block is typically of the ladder type, formed of two W1.7 galvanized wires spaced 2 inches (51 mm) apart with W1.7 galvanized cross wires welded at 16-inch (406 mm) intervals. Where placed continuously, the sections are required to lap at least 6 inches (152 mm). Such reinforcement must not extend across expansion joints because it would compromise their effectiveness.

SECTION 2111
MASONRY FIREPLACES

2111.1 Definition. A masonry fireplace is a fireplace constructed of concrete or masonry. Masonry fireplaces shall be constructed in accordance with this section.

❖ The provisions of this section apply to the design and construction of concrete and masonry fireplaces, which for simplicity are referred to as "masonry fireplaces." Figure 2111.1 shows a cross section of a typical masonry fireplace. In accordance with the definition in Section 2102.1, a fireplace consists of a hearth and fire chamber and it is built in conjunction with a chimney.

2111.2 Footings and foundations. Footings for masonry fireplaces and their chimneys shall be constructed of concrete or solid masonry at least 12 inches (305 mm) thick and shall extend at least 6 inches (153 mm) beyond the face of the fireplace or foundation wall on all sides. Footings shall be founded on natural undisturbed earth or engineered fill below frost

depth. In areas not subjected to freezing, footings shall be at least 12 inches (305 mm) below finished grade.

❖ Masonry fireplaces and chimneys must be supported on adequate footings due to their weight and the forces imposed on them by wind, earthquakes and other effects. This section prescribes minimum footing requirements that are typically adequate for standard fireplaces and chimneys. Extremely large, tall or heavy fireplaces and chimneys, however, may need more substantial foundations (see Figure 2111.1).

2111.2.1 Ash dump cleanout. Cleanout openings, located within foundation walls below fireboxes, when provided, shall be equipped with ferrous metal or masonry doors and frames constructed to remain tightly closed, except when in use. Cleanouts shall be accessible and located so that ash removal will not create a hazard to combustible materials.

❖ Noncombustible, tightly sealed cleanout doors are required to reduce the danger of fire spread through the cleanout openings. Cleanout openings are required to be easily accessible to allow ash to be readily removed.

2111.3 Seismic reinforcing. Masonry or concrete fireplaces shall be constructed, anchored, supported and reinforced as required in this chapter. In Seismic Design Category D, masonry and concrete fireplaces shall be reinforced and

*EXCEPT AS PERMITTED FOR RUMFORD FIREPLACES [SEE FIGURE 2111.11(2)]

For SI: 1 inch = 25.4 mm.

Figure 2111.1
SECTION THROUGH FIREPLACE

anchored as detailed in Sections 2111.3.1, 2111.3.2, 2111.4 and 2111.4.1 for chimneys serving fireplaces. In Seismic Design Category A, B or C, reinforcement and seismic anchorage is not required. In Seismic Design Category E or F, masonry and concrete chimneys shall be reinforced in accordance with the requirements of Sections 2101 through 2108.

❖ Unreinforced fireplaces and chimneys subjected to strong ground motion have sustained damage in past earthquakes. The requirements in this section provide minimum reinforcement in an effort to increase structural integrity in structures classified as Seismic Design Category D. More substantial reinforcement may be indicated in structures having a higher seismic risk or for atypical fireplaces and chimneys.

2111.3.1 Vertical reinforcing. For fireplaces with chimneys up to 40 inches (1016 mm) wide, four No. 4 continuous vertical bars, anchored in the foundation, shall be placed in the concrete between wythes of solid masonry or within the cells of hollow unit masonry and grouted in accordance with Section 2103.12. For fireplaces with chimneys greater than 40 inches (1016 mm) wide, two additional No. 4 vertical bars shall be provided for each additional 40 inches (1016 mm) in width or fraction thereof.

❖ These requirements are traditional minimum prescriptive provisions to help maintain the integrity of fireplaces and chimneys during earthquakes. To resist strong earthquakes, however, more substantial reinforcement may be required.

2111.3.2 Horizontal reinforcing. Vertical reinforcement shall be placed enclosed within $^1/_4$-inch (6.4 mm) ties or other reinforcing of equivalent net cross-sectional area, spaced not to exceed 18 inches (457 mm) on center in concrete; or placed in the bed joints of unit masonry at a minimum of every 18 inches (457 mm) of vertical height. Two such ties shall be provided at each bend in the vertical bars.

❖ These requirements are traditional minimum prescriptive provisions to help maintain the integrity of fireplaces and chimneys during earthquakes. The vertical reinforcement required by Section 2111.3.1 must be enclosed within the horizontal reinforcement required by this section. To resist strong earthquakes, however, more substantial reinforcement may be required.

2111.4 Seismic anchorage. Masonry and concrete chimneys in Seismic Design Category D shall be anchored at each floor, ceiling or roof line more than 6 feet (1829 mm) above grade, except where constructed completely within the exterior walls. Anchorage shall conform to the following requirements.

❖ Fireplace chimneys can fail by overturning during earthquakes. This section duplicates chimney anchorage requirements that are in Section 2113.4. Seismic anchorage to floors and roof diaphragms is required for exterior chimneys in order to prevent detachment from the structure. While it is not required for a chimney that is enclosed within the exterior walls, anchorage could be utilized to minimize the risk of damage to floor or roof framing due to pounding.

2111.4.1 Anchorage. Two $^3/_{16}$-inch by 1-inch (4.8 mm by 25.4 mm) straps shall be embedded a minimum of 12 inches (305 mm) into the chimney. Straps shall be hooked around the outer bars and extend 6 inches (152 mm) beyond the bend. Each strap shall be fastened to a minimum of four floor joists with two $^1/_2$-inch (12.7 mm) bolts.

❖ The prescriptive requirements in this section are traditional for typical chimney construction. More substantial anchorage may be required in structures having an elevated seismic risk or for chimneys where the distance between floor and roof diaphragms is large.

2111.5 Firebox walls. Masonry fireboxes shall be constructed of solid masonry units, hollow masonry units grouted solid, stone or concrete. When a lining of firebrick at least 2 inches (51 mm) in thickness or other approved lining is provided, the minimum thickness of back and sidewalls shall each be 8 inches (203 mm) of solid masonry, including the lining. The width of joints between firebricks shall not be greater than $^1/_4$ inch (6.4 mm). When no lining is provided, the total minimum thickness of back and sidewalls shall be 10 inches (254 mm) of solid masonry. Firebrick shall conform to ASTM C 27 or ASTM C 1261 and shall be laid with medium-duty refractory mortar conforming to ASTM C 199.

❖ This section specifies the minimum thicknesses of refractory brick or solid masonry necessary to contain the generated heat.
  Solid masonry walls forming the firebox are required to have a minimum total thickness of 8 inches (204 mm), including the refractory lining.
  The refractory lining is to consist of a low-duty, fire-clay refractory brick with a minimum thickness of 2 inches (52 mm), laid with medium-duty refractory mortar. Mortar joints are generally $^1/_{16}$ to $^3/_{16}$ inch (1.6 to 4.8 mm) thick, but not thicker than $^1/_4$ inch (6.4 mm), to reduce thermal movements and prevent joint deterioration.
  Where a firebrick lining is not used in firebox construction, the wall thickness is not to be less than 10 inches (254 mm) of solid masonry. Firebrick is required to conform to ASTM C 27 or ASTM C 1261. Firebrick must be laid with medium-duty refractory mortar conforming to ASTM C 199.

2111.5.1 Steel fireplace units. Steel fireplace units are permitted to be installed with solid masonry to form a masonry fireplace provided they are installed according to either the requirements of their listing or the requirements of this section. Steel fireplace units incorporating a steel firebox lining shall be constructed with steel not less than $^1/_4$ inch (6.4 mm) in thickness, and an air-circulating chamber which is ducted to the interior of the building. The firebox lining shall be encased with solid masonry to provide a total thickness at the back and sides of not less than 8 inches (203 mm), of which not less than 4 inches (102 mm) shall be of solid masonry or concrete. Circulating air ducts employed with steel fireplace units shall be constructed of metal or masonry.

❖ This section provides minimum requirements for steel linings used in masonry fireplaces to improve heat flow.

2111.6 Firebox dimensions. The firebox of a concrete or masonry fireplace shall have a minimum depth of 20 inches (508 mm). The throat shall not be less than 8 inches (203 mm) above the fireplace opening. The throat opening shall not be less than 4 inches (102 mm) in depth. The cross-sectional area of the passageway above the firebox, including the throat, damper and smoke chamber, shall not be less than the cross-sectional area of the flue.

> **Exception:** Rumford fireplaces shall be permitted provided that the depth of the fireplace is at least 12 inches (305 mm) and at least one-third of the width of the fireplace opening, and the throat is at least 12 inches (305 mm) above the lintel, and at least $^1/_{20}$ the cross-sectional area of the fireplace opening.

❖ The proper functioning of the fireplace depends on the size of the face opening and the chimney dimensions, which in turn are related to the room size [see Figure 2111.1]. This section specifies a minimum depth of 20 inches (508 mm) for the combustion chamber because that depth influences the draft requirement. The dimensions of the firebox (depth, opening size and shape) are usually based on two considerations: aesthetics and the need to prevent the room from over-heating. Suggested dimensions for single-opening fireboxes are given in technical publications of the Brick Institute of America (BIA) and the National Concrete Masonry Association (NCMA).

This section also provides additional criteria for the throat's location and minimum cross-sectional area. Those criteria are based on many years of construc-tion of successfully functioning fireplaces.

The exception permits the use of Rumford fireplaces and the dimensions associated with this design style. Rumford fireplaces are tall, shallow fireplaces that can radiate a large amount of heat into a room.

The code reference to the depth of the fireplace is interpreted as the depth of the firebox [see Figure 2111.6]. The throat is required to be made at least 12 inches (305 mm) above the lintel and at least one-twentieth of the cross-sectional area of the fire-place opening. Smoke chambers and flues for Rumford fireplaces should be sized and built like other masonry fireplaces. While those who build Rumford fireplaces do not completely agree about how they work, many books and guides address their construction.

2111.7 Lintel and throat. Masonry over a fireplace opening shall be supported by a lintel of noncombustible material. The minimum required bearing length on each end of the fireplace opening shall be 4 inches (102 mm). The fireplace throat or damper shall be located a minimum of 8 inches (203 mm) above the top of the fireplace opening.

❖ Permanent support for the masonry above the fire-place opening is provided by noncombustible lintels of steel, masonry or concrete. Combustible lintels (for example, those made from wood) are not appropriate due to the risk of fire damage and the probable col-lapse of the masonry above the opening.

The minimum bearing requirement of 4 inches (102 mm) is empirical, based on typical masonry fireplace

For SI: 1 inch = 25.4 mm.

Figure 2111.6
RUMFORD FIREPLACE

openings. Lintels that support more than the typical weight of masonry above the fireplace opening may require a larger bearing area.

The throat of a fireplace is the slot-like opening above the firebox, through which flames, smoke and hot combustion gases pass into the smoke chamber. The throat is as wide as the combustion chamber and is required to be located at least 8 inches (204 mm) above the lintel for conventional fireplaces. The back wall of the combustion chamber extends up to the throat, which is provided with a metal damper.

2111.7.1 Damper. Masonry fireplaces shall be equipped with a ferrous metal damper located at least 8 inches (203 mm) above the top of the fireplace opening. Dampers shall be installed in the fireplace or at the top of the flue venting the fireplace, and shall be operable from the room containing the fireplace. Damper controls shall be permitted to be located in the fireplace.

❖ A damper is used to close the chimney flue when the fireplace is not in use. This section provides guidance on its location and construction.

2111.8 Smoke chamber walls. Smoke chamber walls shall be constructed of solid masonry units, hollow masonry units grouted solid, stone or concrete. Corbeling of masonry units shall not leave unit cores exposed to the inside of the smoke chamber. The inside surface of corbeled masonry shall be parged smooth. Where no lining is provided, the total minimum thickness of front, back and sidewalls shall be 8 inches (203 mm) of solid masonry. When a lining of firebrick at least 2 inches (51 mm) thick, or a lining of vitrified clay at least $^5/_8$ inch (15.9 mm) thick, is provided, the total minimum thickness of front, back and sidewalls shall be 6 inches (152 mm) of solid masonry, including the lining. Firebrick shall conform to ASTM C 27 or ASTM C 1261 and shall be laid with refractory mortar conforming to ASTM C 199.

❖ The minimum wall thickness specified for the throat and smoke chamber is required to provide support for the flue construction, as well as adequate thermal insulation for the adjacent combustible construction.

In conventional fireplace construction, the smoke chamber is a tapering section whose vertical dimension is measured from the damper or throat to the bottom of the chimney flue. The actual height is a function of the fireplace opening.

2111.8.1 Smoke chamber dimensions. The inside height of the smoke chamber from the fireplace throat to the beginning of the flue shall not be greater than the inside width of the fireplace opening. The inside surface of the smoke chamber shall not be inclined more than 45 degrees (0.76 rad) from vertical when prefabricated smoke chamber linings are used or when the smoke chamber walls are rolled or sloped rather than corbeled. When the inside surface of the smoke chamber is formed by corbeled masonry, the walls shall not be corbeled more than 30 degrees (0.52 rad) from vertical.

❖ This section specifies the smoke chamber configuration that is needed for the proper function of the masonry chimney (see Figure 2111.1).

2111.9 Hearth and hearth extension. Masonry fireplace hearths and hearth extensions shall be constructed of concrete or masonry, supported by noncombustible materials, and reinforced to carry their own weight and all imposed loads. No combustible material shall remain against the underside of hearths or hearth extensions after construction.

❖ The fireplace hearth consists of two parts. The hearth, commonly called the "inner hearth," is the floor of the combustion chamber and is obviously constructed of noncombustible material. The outer hearth, commonly known as the "hearth extension," projects beyond the face of the fireplace into the room and also consists of noncombustible materials, such as brick or concrete masonry, concrete, floor tile or stone (see Figure 2111.1). The hearth extension may continue out from the inner hearth at the same level or be stepped down to a lower level. Minimum dimensions for the hearth extension are prescribed in Section 2111.10 and are discussed in the commentary to that section.

Combustible forms and centers could ignite from exposure to heat from the adjacent fireplace and from burning embers that escape the firebox; these and other similar concealed, combustible components must be removed.

2111.9.1 Hearth thickness. The minimum thickness of fireplace hearths shall be 4 inches (102 mm).

❖ The required minimum thickness of 4 inches (102 mm) is an empirical requirement that has historically been successful.

2111.9.2 Hearth extension thickness. The minimum thickness of hearth extensions shall be 2 inches (51 mm).

Exception: When the bottom of the firebox opening is raised at least 8 inches (203 mm) above the top of the hearth extension, a hearth extension of not less than $^3/_8$-inch-thick (9.5 mm) brick, concrete, stone, tile or other approved noncombustible material is permitted.

❖ These requirements are empirical and have historically been successful.

2111.10 Hearth extension dimensions. Hearth extensions shall extend at least 16 inches (406 mm) in front of, and at least 8 inches (203 mm) beyond, each side of the fireplace opening. Where the fireplace opening is 6 square feet (0.557 m²) or larger, the hearth extension shall extend at least 20 inches (508 mm) in front of, and at least 12 inches (305 mm) beyond, each side of the fireplace opening.

❖ The hearth extension is required to extend the full width of the fireplace opening and at least 8 inches (203 mm) beyond each side of the opening. It is also required to extend at least 16 inches (406 mm) out from the face of the fireplace. For fireplace openings larger than 6 square feet (0.557 m²) in area, the hearth extension is required to extend at least 20 inches (508 mm) beyond the face of the fireplace and at least 12 inches (305 mm) beyond each side of the fireplace opening (see Figure 2111.10).

The hearth extension is intended to serve as a fire-protective separation between the firebox and adjacent combustible flooring or furnishings and to prevent accidental spills of hot embers or logs from the fire.

2111.11 Fireplace clearance. Any portion of a masonry fireplace located in the interior of a building or within the exterior wall of a building shall have a clearance to combustibles of not less than 2 inches (51 mm) from the front faces and sides of masonry fireplaces and not less than 4 inches (102 mm) from the back faces of masonry fireplaces. The airspace shall not be filled, except to provide fireblocking in accordance with Section 2111.12.

Exceptions:

1. Masonry fireplaces listed and labeled for use in contact with combustibles in accordance with UL 127 and installed in accordance with the manufacturer's installation instructions are permitted to have combustible material in contact with their exterior surfaces.

2. When masonry fireplaces are constructed as part of masonry or concrete walls, combustible materials shall not be in contact with the masonry or concrete walls less than 12 inches (306 mm) from the inside surface of the nearest firebox lining.

3. Exposed combustible trim and the edges of sheathing materials, such as wood siding, flooring and drywall, are permitted to abut the masonry fireplace sidewalls and hearth extension, in accordance with Figure 2111.11, provided such combustible trim or sheathing is a minimum of 12 inches (306 mm) from the inside surface of the nearest firebox lining.

4. Exposed combustible mantels or trim is permitted to be placed directly on the masonry fireplace front sur-

rounding the fireplace opening, provided such combustible materials shall not be placed within 6 inches (153 mm) of a fireplace opening. Combustible material directly above and within 12 inches (305 mm) of the fireplace opening shall not project more than $^1/_8$ inch (3.2 mm) for each 1-inch (25 mm) distance from such opening. Combustible materials located along the sides of the fireplace opening that project more than $1^1/_2$ inches (38 mm) from the face of the fireplace shall have an additional clearance equal to the projection.

❖ Combustible materials, such as framing studs and joists, must not be installed closer than 2 inches (51 mm) from the front or side surfaces of fireplace walls because of the fire hazard to materials in this location. Heat transmitted through fireplace walls can ignite combustible structural materials in contact with the walls. For this reason, a minimum required clearance has been established from the fireplace to combustibles.

The exceptions cover specific instances where the clearance requirements vary. For a discussion and illustration of Exception 3, see Figure 2111.11 and commentary.

Exception 4 regulates the location of mantels and trim adjacent to fireplace openings. A less restrictive projection limit is applicable to combustible materials along the side of the fireplace opening, because radiant heat and convection heat are a greater problem for combustible material above the fireplace opening than for combustible material at the side of the fireplace opening. The reduced side clearance criterion has been permitted under the *Standard Building Code* for many years.

MASONRY

20" MIN. FOR FIREPLACE OPENINGS 6 SQ.FT. AND LARGER

16" MIN. FOR LESS THAN 6 SQ.FT. FIREPLACE OPENING

FIREPLACE OPENING

HEARTH

12" MIN. FOR FIREPLACE OPENINGS 6 SQ.FT. AND LARGER

8" MIN. FOR LESS THAN 6 SQ.FT. FIREPLACE OPENING

PLAN VIEW

For SI: 1 inch = 25.4 mm,
1 square foot = 0 .0929 m².

Figure 2111.10
HEARTH EXTENSION

FIGURE 2111.11. See below.

❖ This figure depicts the provisions of Exception 4 to Section 2111.11 for combustible trim and sheathing adjacent to masonry fireplaces and hearth extensions.

2111.12 Fireplace fireblocking. All spaces between fireplaces and floors and ceilings through which fireplaces pass shall be fireblocked with noncombustible material securely fastened in place. The fireblocking of spaces between wood joists, beams or headers shall be to a depth of 1 inch (25 mm) and shall only be placed on strips of metal or metal lath laid across the spaces between combustible material and the chimney.

❖ Fireblocking is required to prevent the travel of flames, smoke or hot gases to other areas of the building through gaps between the chimney and the floor or ceiling assemblies. The 1-inch (25 mm) depth requirement is intended to be both a minimum and a maximum.

2111.13 Exterior air. Factory-built or masonry fireplaces covered in this section shall be equipped with an exterior air supply to ensure proper fuel combustion unless the room is mechanically ventilated and controlled so that the indoor pressure is neutral or positive.

❖ Adequate airflow is needed to provide exterior oxygen for the fire and to maintain draft through the chimney so that toxic combustion gases can be exhausted. Adequate airflow is especially important in modern, tightly sealed buildings. Note that this requirement applies to factory-built fireplaces as well as masonry fireplaces.

2111.13.1 Factory-built fireplaces. Exterior combustion air ducts for factory-built fireplaces shall be listed components of the fireplace, and installed according to the fireplace manufacturer's instructions.

❖ Factory-built fireplaces are required to include exterior combustion air ducts. To function properly, these units need to be installed according to the manufacturer's recommendations.

2111.13.2 Masonry fireplaces. Listed combustion air ducts for masonry fireplaces shall be installed according to the terms of their listing and manufacturer's instructions.

❖ For proper functioning and airflow, air ducts for masonry fireplaces must be installed according to the manufacturer's recommendations (see also commentary, Section 2111.13).

2111.13.3 Exterior air intake. The exterior air intake shall be capable of providing all combustion air from the exterior of the dwelling. The exterior air intake shall not be located within the garage, attic, basement or crawl space of the dwelling nor shall the air intake be located at an elevation higher than the firebox. The exterior air intake shall be covered with a corrosion-resistant screen of $1/4$-inch (6.4 mm) mesh.

❖ Air intakes are required to provide air from outside of the building. The air intakes are not permitted to be placed in garages, basements, crawl spaces or other areas where gases from the exhaust of automobiles, furnaces and other sources could be brought into the building.
  Air intakes should be covered with screens to prevent pests from entering the building when the fireplace is not functioning.

2111.13.4 Clearance. Unlisted combustion air ducts shall be installed with a minimum 1-inch (25 mm) clearance to combustibles for all parts of the duct within 5 feet (1524 mm) of the duct outlet.

❖ The 1-inch (25 mm) clearance is required to reduce the risk of fire through the air intakes.

2111.13.5 Passageway. The combustion air passageway shall be a minimum of 6 square inches (3870 mm²) and not more than 55 square inches (0.035 m²), except that combustion air systems for listed fireplaces or for fireplaces tested for emissions shall be constructed according to the fireplace manufacturer's instructions.

❖ These minimum requirements are intended to provide adequate airflow through the fireplace.

For SI: 1 inch = 25.4 mm

FIGURE 2111.11
ILLUSTRATION OF EXCEPTION TO FIREPLACE CLEARANCE PROVISION

2111.13.6 Outlet. The exterior air outlet is permitted to be located in the back or sides of the firebox chamber or within 24 inches (610 mm) of the firebox opening on or near the floor. The outlet shall be closable and designed to prevent burning material from dropping into concealed combustible spaces.

❖ The requirements on the location of the outlet in the firebox chamber are needed to ensure that adequate airflow to the fire is provided, while avoiding the direct flow of air into the adjacent room. Such openings must not become clogged with ash, which would restrict airflow. Even more important, burning material must not drop into concealed combustible spaces due to the risk of fire damage to the building. Outlets must therefore be both closable and adequately designed to prevent this.

SECTION 2112
MASONRY HEATERS

2112.1 Definition. A masonry heater is a heating appliance constructed of concrete or solid masonry, hereinafter referred to as "masonry," which is designed to absorb and store heat from a solid fuel fire built in the firebox by routing the exhaust gases through internal heat exchange channels in which the flow path downstream of the firebox may include flow in a horizontal or downward direction before entering the chimney and which delivers heat by radiation from the masonry surface of the heater.

❖ Masonry heaters are appliances designed to absorb and store heat from a relatively small fire and to radiate that heat into the building interior. They are thermally more efficient than traditional fireplaces because of their design. Interior passageways through the heater allow hot exhaust gases from the fire to transfer heat into the masonry, which then radiates into the building.

2112.2 Installation. Masonry heaters shall be installed in accordance with this section and comply with one of the following:

1. Masonry heaters shall comply with the requirements of ASTM E 1602; or

2. Masonry heaters shall be listed and labeled in accordance with UL 1482 and installed in accordance with the manufacturer's installation instructions.

❖ The listed reference standards provide guidelines for installing masonry heaters.

2112.3 Footings and foundation. The firebox floor of a masonry heater shall be a minimum thickness of 4 inches (102 mm) of noncombustible material and be supported on a noncombustible footing and foundation in accordance with Section 2113.2.

❖ This section refers the user to the foundation requirements for masonry chimneys.

2112.4 Seismic reinforcing. In Seismic Design Category D, E and F, masonry heaters shall be anchored to the masonry foundation in accordance with Section 2113.3. Seismic reinforcing

shall not be required within the body of a masonry heater with a height that is equal to or less than 3.5 times its body width and where the masonry chimney serving the heater is not supported by the body of the heater. Where the masonry chimney shares a common wall with the facing of the masonry heater, the chimney portion of the structure shall be reinforced in accordance with Section 2113.

❖ Because of the large bulk and squat geometry of these heaters, seismic reinforcement is not typically required. Flexural tensile stresses, which typically cause damage to unreinforced masonry, rarely occur. Where chimneys extend above these heaters, however, seismic reinforcement is required by Section 2113. See the commentary to that section for additional information on this requirement.

2112.5 Masonry heater clearance. Combustible materials shall not be placed within 36 inches (765 mm) of the outside surface of a masonry heater in accordance with NFPA 211, Section 8-7 (clearances for solid fuel-burning appliances), and the required space between the heater and combustible material shall be fully vented to permit the free flow of air around all heater surfaces.

Exceptions:

1. When the masonry heater wall thickness is at least 8 inches (203 mm) thick of solid masonry and the wall thickness of the heat exchange channels is at least 5 inches (127 mm) thick of solid masonry, combustible materials shall not be placed within 4 inches (102 mm) of the outside surface of a masonry heater. A clearance of at least 8 inches (203 mm) shall be provided between the gas-tight capping slab of the heater and a combustible ceiling.

2. Masonry heaters listed and labeled in accordance with UL 1482 and installed in accordance with the manufacturer's instructions.

❖ Heat conducted through masonry heater walls can ignite combustible structural materials in contact with these walls. For this reason, a minimum required clearance to combustibles from masonry heaters has been established. Because masonry heaters typically generate more heat for a longer period of time than traditional fireplaces, greater clearances to combustible materials are needed to reduce the risk of fire.

SECTION 2113
MASONRY CHIMNEYS

2113.1 Definition. A masonry chimney is a chimney constructed of concrete or masonry, hereinafter referred to as "masonry." Masonry chimneys shall be constructed, anchored, supported and reinforced as required in this chapter.

❖ A masonry chimney is a field-constructed assembly that can consist of masonry units, grout, reinforced concrete, rubble stone, fire-clay liners and mortars.

The IFGC and the IMC address the installation of chimneys and venting systems that are required to convey products of combustion from fuel-burning appliances to the outdoors so that the operation of the appliance does not adversely affect building occupants. A masonry chimney is permitted to serve residential (low-heat), medium- and high-heat appliances. A masonry chimney is required for masonry fireplaces constructed in accordance with Section 2111. This section provides the construction requirements for all masonry chimneys.

2113.2 Footings and foundations. Footings for masonry chimneys shall be constructed of concrete or solid masonry at least 12 inches (305 mm) thick and shall extend at least 6 inches (152 mm) beyond the face of the foundation or support wall on all sides. Footings shall be founded on natural undisturbed earth or engineered fill below frost depth. In areas not subjected to freezing, footings shall be at least 12 inches (305 mm) below finished grade.

❖ Masonry chimneys must be supported on adequate foundations due to their weight and the forces imposed on them by wind, earthquakes and other effects. This section prescribes minimum foundation requirements that are typically adequate for standard chimneys.

Extremely large, tall or heavy chimneys, however, may need more substantial foundations. Also, a chimney foundation probably will support a larger load than the adjacent building foundations. For this reason, chimney footings and foundations are often separated from the building foundation. A chimney foundation monolithic with the building foundation is permitted, provided the soil pressure and anticipated settlement are approximately uniform.

2113.3 Seismic reinforcing. Masonry or concrete chimneys shall be constructed, anchored, supported and reinforced as required in this chapter. In Seismic Design Category D, masonry and concrete chimneys shall be reinforced and anchored as detailed in Sections 2113.3.1, 2113.3.2 and 2113.4. In Seismic Design Category A, B or C, reinforcement and seismic anchorage is not required. In Seismic Design Category E or F, masonry and concrete chimneys shall be reinforced in accordance with the requirements of Sections 2101 through 2108.

❖ Unreinforced fireplaces and chimneys subjected to strong ground motion have been severely damaged in past earthquakes. The requirements in this section provide minimum reinforcement in an effort to keep these structures together during such events. More substantial reinforcement, however, may be required in areas of high seismicity or for atypical chimneys.

2113.3.1 Vertical reinforcing. For chimneys up to 40 inches (1016 mm) wide, four No. 4 continuous vertical bars anchored in the foundation shall be placed in the concrete between wythes of solid masonry or within the cells of hollow unit masonry and grouted in accordance with Section 2103.12. Grout shall be prevented from bonding with the flue liner so

that the flue liner is free to move with thermal expansion. For chimneys greater than 40 inches (1016 mm) wide, two additional No. 4 vertical bars shall be provided for each additional 40 inches (1016 mm) in width or fraction thereof.

❖ These requirements are traditional minimum prescriptive provisions to help maintain the structural integrity of fireplaces and chimneys during earthquakes. More reinforcement may be required in areas of high seismicity or for atypical chimneys.

2113.3.2 Horizontal reinforcing. Vertical reinforcement shall be placed enclosed within $^1/_4$-inch (6.4 mm) ties, or other reinforcing of equivalent net cross-sectional area, spaced not to exceed 18 inches (457 mm) o.c. in concrete, or placed in the bed joints of unit masonry, at a minimum of every 18 inches (457 mm) of vertical height. Two such ties shall be provided at each bend in the vertical bars.

❖ These requirements are traditional minimum prescriptive provisions intended to help maintain the integrity of fireplaces and chimneys during earthquakes. The vertical reinforcement required by Section 2113.3.1 must be enclosed within the horizontal reinforcement required by this section. More reinforcement may be required in areas of high seismicity or for atypical chimneys.

2113.4 Seismic anchorage. Masonry and concrete chimneys and foundations in Seismic Design Category D shall be anchored at each floor, ceiling or roof line more than 6 feet (1829 mm) above grade, except where constructed completely within the exterior walls. Anchorage shall conform to the following requirements.

❖ Chimneys must be connected to floor and roof diaphragms to prevent overturning during earthquakes. Chimneys must be anchored at the ceiling line of roof or ceiling assemblies and at floor levels below the roof. Such anchorage is of lesser importance where the floor assembly is 6 feet (1829 mm) or less above grade.

2113.4.1 Anchorage. Two $^3/_{16}$-inch by 1-inch (4.8 mm by 25 mm) straps shall be embedded a minimum of 12 inches (305 mm) into the chimney. Straps shall be hooked around the outer bars and extend 6 inches (152 mm) beyond the bend. Each strap shall be fastened to a minimum of four floor joists with two $^1/_2$-inch (12.7 mm) bolts.

❖ The prescriptive requirements in this section are traditional for typical fireplaces and chimneys. More substantial anchorage may be required in areas of high seismicity, for large fireplaces or where the distance between floor and roof diaphragms is large.

2113.5 Corbeling. Masonry chimneys shall not be corbeled more than half of the chimney's wall thickness from a wall or foundation, nor shall a chimney be corbeled from a wall or foundation that is less than 12 inches (305 mm) in thickness unless it projects equally on each side of the wall, except that on the second story of a two-story dwelling, corbeling of chimneys on the exterior of the enclosing walls is permitted to equal the wall thickness. The projection of a single course shall not exceed one-half the unit height or one-third of the unit bed depth, whichever is less.

❖ A corbel is formed by projecting courses of masonry with the first or lowest course projecting out from the face of the wall and each successive course projecting out from the supporting course below. The angle of a corbel is limited so that the structural capacities of the masonry are not exceeded.

Figure 2113.5 illustrates the limitations of corbeling in an eccentrically loaded masonry chimney. The total corbeled projection cannot exceed 6 inches (152 mm). Where corbeling projects from the wall on one side only, the minimum allowable thickness of the wall is 12 inches (305 mm). The maximum allowable horizontal projection of an individual course of brick cannot exceed one-half the height and one-third the thickness of the masonry unit.

A chimney that projects equally on each side of the wall loads the wall concentrically. The requirement for a minimum wall thickness of 12 inches (305 mm) does not apply to this chimney configuration.

This section contains an exception to the limitations on corbel projection. In single-family homes, traditional chimneys originated in the second story and commonly served heating appliances in bedrooms, studies and sewing rooms. The exception allows corbeling to project from the exterior masonry wall a distance equivalent to the masonry wall thickness. This exception, however, does not affect the limitation on minimum wall thickness stated in the second sentence of this section. For example, a chimney in the second story of a two-story building cannot be corbeled at all if the wall is less than 12 inches (305

mm) thick and the corbeling is to project from one side only. If the corbeling is to project equally from both sides of the wall, the minimum wall thickness does not apply.

2113.6 Changes in dimension. The chimney wall or chimney flue lining shall not change in size or shape within 6 inches (152 mm) above or below where the chimney passes through floor components, ceiling components or roof components.

❖ At changes in shape or direction, masonry chimneys can have thinner walls, making them susceptible to leakage of water from the outside and to hot spots and leakage of combustion gases from the inside. This is a fire hazard. The intent of this provision is to prohibit changes in shape and size near combustible construction. This section prohibits any change in dimension or direction of a masonry chimney from 6 inches (152 mm) below the combustible construction to 6 inches (152 mm) above it (see Figure 2113.6).

2113.7 Offsets. Where a masonry chimney is constructed with a fireclay flue liner surrounded by one wythe of masonry, the maximum offset shall be such that the centerline of the flue above the offset does not extend beyond the center of the chimney wall below the offset. Where the chimney offset is supported by masonry below the offset in an approved manner, the maximum offset limitations shall not apply. Each individual corbeled masonry course of the offset shall not exceed the projection limitations specified in Section 2113.5.

❖ An offset requires two changes in direction and causes two vertical sections of a chimney to be misaligned (offset) from each other. The intent of this section is to provide an upper portion of an offset that is structurally stable with respect to the lower portion.

For SI: 1 inch = 25.4 mm.

Figure 2113.5
CORBELING LIMITATIONS

The offset limitation does not apply when the inclined portion of the chimney and the portion above the offset are supported in an approved manner by underlying masonry construction.

2113.8 Additional load. Chimneys shall not support loads other than their own weight unless they are designed and constructed to support the additional load. Masonry chimneys are permitted to be constructed as part of the masonry walls or concrete walls of the building.

❖ Because the requirements for chimneys in Section 2113 are based on past performance, chimneys should not be used to support other loads unless they are specifically designed to do so.

2113.9 Termination. Chimneys shall extend at least 2 feet (610 mm) higher than any portion of the building within 10 feet (3048 mm), but shall not be less than 3 feet (914 mm) above the highest point where the chimney passes through the roof.

❖ Chimneys must be terminated well above adjacent portions of the building so that flue gases are exhausted away from combustible materials and for proper airflow through the chimney. The termination requirements are illustrated in Figure 2113.9. These requirements are applicable to chimneys serving masonry fireplaces or residential-type appliances. More restrictive termination requirements are specified for medium-heat and high-heat appliances.

2113.9.1 Spark arrestors. Where a spark arrestor is installed on a masonry chimney, the spark arrestor shall meet all of the following requirements:

1. The net free area of the arrestor shall not be less than four times the net free area of the outlet of the chimney flue it serves.

2. The arrestor screen shall have heat and corrosion resistance equivalent to 19-gage galvanized steel or 24-gage stainless steel.

3. Openings shall not permit the passage of spheres having a diameter greater than $^{1}/_{2}$ inch (13 mm) nor block the passage of spheres having a diameter less than $^{3}/_{8}$ inch (11 mm).

4. The spark arrestor shall be accessible for cleaning and the screen or chimney cap shall be removable to allow for cleaning of the chimney flue.

❖ This section provides specifications for spark arrestors, if they are provided. Their use is not mandated by the code, but owners or builders often install them.

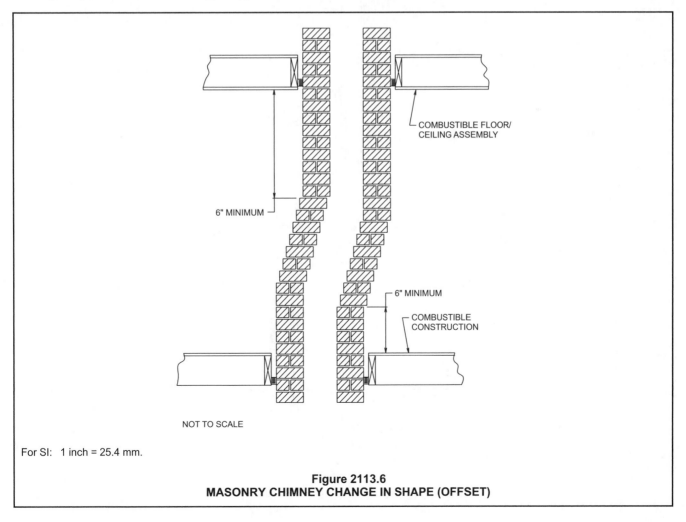

6" MINIMUM

COMBUSTIBLE FLOOR/ CEILING ASSEMBLY

6" MINIMUM

COMBUSTIBLE CONSTRUCTION

NOT TO SCALE

For SI: 1 inch = 25.4 mm.

Figure 2113.6
MASONRY CHIMNEY CHANGE IN SHAPE (OFFSET)

2113.10 Wall thickness. Masonry chimney walls shall be constructed of concrete, solid masonry units or hollow masonry units grouted solid with not less than 4 inches (102 mm) nominal thickness.

❖ The minimum chimney wall thickness is necessary to achieve a thermal mass that will predictably control heat transmission through the walls of the chimney. This thickness is applicable to chimneys serving masonry fireplaces or residential-type appliances. More restrictive thicknesses are specified for medium-heat and high-heat appliances.

2113.10.1 Masonry veneer chimneys. Where masonry is used as veneer for a framed chimney, through flashing and weep holes shall be provided as required by Chapter 14.

❖ While the definition of "Masonry chimney" would not necessarily include a framed chimney with masonry veneer, this cross reference to Chapter 14 is provided as a precaution.

2113.11 Flue lining (material). Masonry chimneys shall be lined. The lining material shall be appropriate for the type of appliance connected, according to the terms of the appliance listing and the manufacturer's instructions.

❖ The liner forms the flue passageway and is the actual conduit for all products of combustion. The temperature and composition of flue gases will vary based on the type of appliance being vented. The flue lining is required to withstand exposure to high temperatures and corrosive chemicals. It protects the masonry construction of the chimney walls and allows the chimney to be gas tight. A mismatch between an appliance and the type of chimney can result in a hazardous operating condition.

2113.11.1 Residential-type appliances (general). Flue lining systems shall comply with one of the following:

1. Clay flue lining complying with the requirements of ASTM C 315, or equivalent.

2. Listed chimney lining systems complying with UL 1777.

3. Factory-built chimneys or chimney units listed for installation within masonry chimneys.

CROSS SECTION — WHEN 10'-0" OR LESS FROM RIDGE

CROSS SECTION — WHEN MORE THAN 10'-0" FROM RIDGE

For SI: 1 foot = 304.8 mm.

Figure 2113.9
MINIMUM TERMINATION OF CHIMNEYS

4. Other approved materials that will resist corrosion, erosion, softening or cracking from flue gases and condensate at temperatures up to 1,800°F (982°C).

❖ This section requires that the lining for residential-type appliances comply with ASTM C 315 or other approved equivalent and be capable of resisting degradation from flue gas. Chimney liner systems that are tested and labeled by an approved agency in accordance with UL 1777 are also permitted.

2113.11.1.1 Flue linings for specific appliances. Flue linings other than those covered in Section 2113.11.1 intended for use with specific appliances shall comply with Sections 2113.11.1.2 through 2113.11.1.4 and Sections 2113.11.2 and 2113.11.3.

❖ This section identifies flue lining materials to be used for specific appliances.

2113.11.1.2 Gas appliances. Flue lining systems for gas appliances shall be in accordance with the *International Fuel Gas Code*.

❖ The IFGC must be used to determine appropriate flue-lining systems for gas appliances.

2113.11.1.3 Pellet fuel-burning appliances. Flue lining and vent systems for use in masonry chimneys with pellet fuel-burning appliances shall be limited to flue lining systems complying with Section 2113.11.1 and pellet vents listed for installation within masonry chimneys (see Section 2113.11.1.5 for marking).

❖ Flue-lining and vent systems in masonry chimneys of pellet fuel-burning appliances can either conform to Section 2113.11.1 or be an approved listed system.

2113.11.1.4 Oil-fired appliances approved for use with L-vent. Flue lining and vent systems for use in masonry chimneys with oil-fired appliances approved for use with Type L vent shall be limited to flue lining systems complying with Section 2113.11.1 and listed chimney liners complying with UL 641 (see Section 2113.11.1.5 for marking).

❖ Flue lining and vent systems in masonry chimneys of oil-fired appliances with Type L vents can either conform to Section 2113.11.1 or be an approved listed system (UL 641).

2113.11.1.5 Notice of usage. When a flue is relined with a material not complying with Section 2113.11.1, the chimney shall be plainly and permanently identified by a label attached to a wall, ceiling or other conspicuous location adjacent to where the connector enters the chimney. The label shall include the following message or equivalent language: "This chimney is for use only with (type or category of appliance) that burns (type of fuel). Do not connect other types of appliances."

❖ Clearly displayed information on the appropriate use and fuel for appliances is required to protect against the use of improper types of fuel that could cause fire.

2113.11.2 Concrete and masonry chimneys for medium-heat appliances.

❖ This section establishes requirements for masonry chimneys serving medium-heat appliances, including the chimney materials, lining, termination height and proper clearances to combustibles.

2113.11.2.1 General. Concrete and masonry chimneys for medium-heat appliances shall comply with Sections 2113.1 through 2113.5.

❖ Chimneys serving a medium-heat appliance must comply with the general chimney requirements in Sections 2113.1 through 2113.5. These include minimum requirements for footings and foundations; seismic reinforcement; anchorage and corbeling.

2113.11.2.2 Construction. Chimneys for medium-heat appliances shall be constructed of solid masonry units or of concrete with walls a minimum of 8 inches (203 mm) thick, or with stone masonry a minimum of 12 inches (305 mm) thick.

❖ Masonry chimneys for medium-heat appliances must be constructed of concrete in accordance with Chapter 19, or of solid masonry units in accordance with this chapter. Chimneys constructed of solid masonry units or reinforced concrete are required to have a minimum wall thickness of 8 inches (203 mm). The minimum thickness requirement is necessary to achieve a thermal mass that will predictably control heat transmission through the walls of the chimney. Chimneys constructed of rubble stone masonry (irregular or roughly shaped stones) are required to have a wall thickness of no less than 12 inches (305 mm). The rate of heat transmission through stone walls is not predictable because of the varying thickness of the individual stones and the nonuniform mortar joints between the stones. For these reasons, stone wall chimneys are required to be thicker to provide a reasonable margin of safety.

2113.11.2.3 Lining. Concrete and masonry chimneys shall be lined with an approved medium-duty refractory brick a minimum of 4$^1/_2$ inches (114 mm) thick laid on the 4$^1/_2$-inch bed (114 mm) in an approved medium-duty refractory mortar. The lining shall start 2 feet (610 mm) or more below the lowest chimney connector entrance. Chimneys terminating 25 feet (7620 mm) or less above a chimney connector entrance shall be lined to the top.

❖ A medium-heat appliance produces flue-gas temperatures up to 2,000°F (1093°C). The chimney lining reduces heat transmission to the walls of the chimney and contains flue gases within a continuous duct until they are away from the building. This section requires at least a medium-duty refractory brick. A 4$^1/_2$-inch (114 mm) medium-duty refractory brick lining tested and classified in accordance with ASTM C 27 meets this criterion. Each course of the refractory brick liner is required to be installed on a full bed of refractory mortar to form a structurally stable, continuous, gas-tight liner.

Figure 2113.11.2.3 depicts a masonry chimney serving a medium-heat appliance. The liner extends from 2 feet (610 mm) below the lowest inlet to the top of the chimney, which is located less than 25 feet (7620 mm) above the highest inlet.

A chimney liner is required in all portions of the chimney that are exposed to flue gases. The liner is required to extend below the lowest inlet to provide protection to the masonry from flue gases, which can deteriorate the masonry and the mortar joints.

2113.11.2.4 Multiple passageway. Concrete and masonry chimneys containing more than one passageway shall have the liners separated by a minimum 4-inch-thick (102 mm) concrete or solid masonry wall.

❖ When a chimney requires multiple passageways, a 4-inch (102 mm) partition of solid masonry is required between them to act as a barrier and to enhance structural integrity.

2113.11.2.5 Termination height. Concrete and masonry chimneys for medium-heat appliances shall extend a minimum of 10 feet (3048 mm) higher than any portion of any building within 25 feet (7620 mm).

❖ Chimneys for medium-heat appliances are required to extend at least 10 feet (3048 mm) above the highest portion of the building within 25 feet (7620 mm) horizontally. This is intended to provide an acceptable height to carry away the flue gases safely and to provide adequate clearance to the roof and surrounding structures to allow any burning embers to extinguish before landing.

2113.11.2.6 Clearance. A minimum clearance of 4 inches (102 mm) shall be provided between the exterior surfaces of a concrete or masonry chimney for medium-heat appliances and combustible material.

❖ A 4-inch (102 mm) minimum airspace clearance to combustibles is required for a medium-heat masonry chimney. This large clearance is needed because of the higher temperatures produced by medium-heat appliances.

2113.11.3 Concrete and masonry chimneys for high-heat appliances.

❖ This section establishes the requirements for concrete and masonry chimneys serving high-heat appliances, including chimney materials, lining, termination height and proper clearances to combustibles.

WHEN DISTANCE IS LESS THAN 25'-0", LINER IS REQUIRED TO EXTEND TO TOP OF CHIMNEY

10'-0" MINIMUM ABOVE ANY PORTION OF BUILDING WITHIN 25'-0"

APPLIANCE FLUE CONNECTION

2'-0" MIN. BELOW LOWEST CHIMNEY INLET

CLEANOUT

For SI: 1 foot = 304.8 mm.

Figure 2113.11.2.3
MASONRY CHIMNEY FOR MEDIUM-HEAT APPLIANCE

2113.11.3.1 General. Concrete and masonry chimneys for high-heat appliances shall comply with Sections 2113.1 through 2113.5.

❖ Chimneys serving high-heat appliances are required to comply with the general chimney requirements in Sections 2113.1 through 2113.5, including minimum provisions for footings and foundations; seismic reinforcement; anchorage and corbeling.

2113.11.3.2 Construction. Chimneys for high-heat appliances shall be constructed with double walls of solid masonry units or of concrete, each wall to be a minimum of 8 inches (203 mm) thick with a minimum airspace of 2 inches (51 mm) between the walls.

❖ The flue gases produced by a high-heat appliance can have temperatures above 2,000°F (1,093°C). To prevent fire, such chimneys must be enclosed by two wythes of masonry with an airspace between them. Thus, the chimney liner is required to be protected by two solid masonry unit walls or reinforced concrete walls, each a minimum of 8 inches (204 mm) thick with an airspace between them of at least 2 inches (51 mm). The airspace is intended to provide thermal insulation and to allow for thermal expansion of the chimney components.

2113.11.3.3 Lining. The inside of the interior wall shall be lined with an approved high-duty refractory brick, a minimum of $4^1/_2$ inches (114 mm) thick laid on the $4^1/_2$-inch bed (114 mm) in an approved high-duty refractory mortar. The lining shall start at the base of the chimney and extend continuously to the top.

❖ High-heat appliances can produce flue gases with temperatures exceeding 2,000°F (1,093°C). The chimney lining reduces heat transmission to the walls of the chimney and contains flue gases within a continuous passageway until they are outside the building. This section permits only a high-duty refractory brick having a minimum thickness of $4^1/_2$ inches (114 mm) to be used in a masonry chimney serving a high-heat appliance. An approved high-duty refractory lining usually consists of brick tested and classified in accordance with ASTM C 27. To provide a structurally stable and continuous gas-tight liner, each course of the refractory brick liner must be laid on a full bed of refractory mortar approved for high-heat appliances.

2113.11.3.4 Termination height. Concrete and masonry chimneys for high-heat appliances shall extend a minimum of 20 feet (6096 mm) higher than any portion of any building within 50 feet (15 240 mm).

❖ Chimneys for high-heat appliances must terminate at least 20 feet (6096 mm) above the highest portion of the building within 50 feet (15 240 mm) horizontally. This allows flue gases to be safely carried away and provides enough clearance to allow any burning embers to extinguish before landing on combustibles.

2113.11.3.5 Clearance. Concrete and masonry chimneys for high-heat appliances shall have approved clearance from buildings and structures to prevent overheating combustible materials, permit inspection and maintenance operations on the chimney and prevent danger of burns to persons.

❖ The extremely high-temperature flue gas produced by a high-heat appliance is a major fire safety concern, and requiring sufficient clearance from buildings protects nearby combustible material, permits inspection and maintenance of the chimney and minimizes danger to persons.

2113.12 Clay flue lining (installation). Clay flue liners shall be installed in accordance with ASTM C 1283 and extend from a point not less than 8 inches (203 mm) below the lowest inlet or, in the case of fireplaces, from the top of the smoke chamber to a point above the enclosing walls. The lining shall be carried up vertically, with a maximum slope no greater than 30 degrees (0.52 rad) from the vertical.

Clay flue liners shall be laid in medium-duty refractory mortar conforming to ASTM C 199 with tight mortar joints left smooth on the inside and installed to maintain an air space or insulation not to exceed the thickness of the flue liner separating the flue liners from the interior face of the chimney masonry walls. Flue lining shall be supported on all sides. Only enough mortar shall be placed to make the joint and hold the liners in position.

❖ Section 2113.11.1 lists clay flue liners as an acceptable flue-lining system for masonry chimneys serving residential-type appliances as well as masonry fireplaces. This section provides the installation requirements for clay flue liners. The liner forms the flue passageway and is the actual conduit of all products of combustion. It must withstand exposure to high temperatures and corrosive chemicals from the flue gases. The chimney lining protects the masonry construction of the chimney walls and allows the chimney to be constructed gas tight.

Installation must comply with ASTM C 1283. The liner is required to extend below the lowest inlet to provide protection to the masonry from flue gases, which can deteriorate the masonry and mortar joints.

Refractory mortar must comply with ASTM C 199.

2113.13 Additional requirements.

❖ This section contains requirements for listed materials to be used as flue linings and for spaces surrounding chimney lining systems.

2113.13.1 Listed materials. Listed materials used as flue linings shall be installed in accordance with the terms of their listings and the manufacturer's instructions.

❖ Listed materials for flue linings must be installed in accordance with the manufacturer's instructions. Such instructions are not listed in the code since they vary with each system.

2113.13.2 Space around lining. The space surrounding a chimney lining system or vent installed within a masonry chimney shall not be used to vent any other appliance.

> **Exception:** This shall not prevent the installation of a separate flue lining in accordance with the manufacturer's instructions.

❖ Where a listed flue-lining system, such as a flexible metallic pipe, is installed in a chimney flue, an annular space can be created. This annular space will have an irregular shape that is not conducive to the flow of flue gas. In addition, the annular space surrounding a chimney lining system provides a thermal buffer between the flue and the surrounding masonry. Using it for another purpose is prohibited.

2113.14 Multiple flues. When two or more flues are located in the same chimney, masonry wythes shall be built between adjacent flue linings. The masonry wythes shall be at least 4 inches (102 mm) thick and bonded into the walls of the chimney.

> **Exception:** When venting only one appliance, two flues are permitted to adjoin each other in the same chimney with only the flue lining separation between them. The joints of the adjacent flue linings shall be staggered at least 4 inches (102 mm).

❖ A single masonry chimney can contain any number of flue-gas passageways. When more than two flues are contained in the same chimney, the flue liners are required to be subdivided by masonry wythes into groups of not more than two (see Figure 2113.14). The purpose of the masonry wythes is to unify the chimney structurally and to isolate pairs of flues serving dissimilar appliances.

2113.15 Flue area (appliance). Chimney flues shall not be smaller in area than the area of the connector from the appliance. Chimney flues connected to more than one appliance shall not be less than the area of the largest connector plus 50 percent of the areas of additional chimney connectors.

> **Exceptions:**
>
> 1. Chimney flues serving oil-fired appliances sized in accordance with NFPA 31.
>
> 2. Chimney flues serving gas-fired appliances sized in accordance with the *International Fuel Gas Code*.

❖ Flues are sized to match the opening at the top of the appliance so that appliances function properly. Smaller flues are not permitted since they may constrict the passage of gases out from the appliance.

2113.16 Flue area (masonry fireplace). Flue sizing for chimneys serving fireplaces shall be in accordance with Section 2113.16.1 or 2113.16.2.

❖ This section provides requirements for the net cross-sectional area of the flue and throat between the

MIN. 4″ SOLID MASONRY

PLAN

STAGGER JOINTS OF ADJOINING FLUE LINERS A MIN. OF 4″

SECTION

For SI: 1 inch = 25.4 mm.

Figure 2113.14
LOW-HEAT CHIMNEY WITH MORE THAN TWO FLUES

firebox and the smoke chamber. Airflow through the fireplace is affected by the dimensions of the firebox opening, the shape and cross-sectional area of the flue and the height of the chimney (see Figure 2113.16). For proper fireplace operation, the required cross-sectional area is about one-tenth that of the fireplace opening. This ratio may vary somewhat with the height of the chimney and the configuration (round or rectangular) of the chimney flue.

This section states that either of the two options may be used to determine the flue area. Section 2113.16.1 prescribes the minimum flue area based on the fireplace opening alone, while Section 2113.16.2 prescribes it based on the height of the fireplace, the fire-

place opening area and flue type.

This figure prescribes minimum flue sizes as a function of chimney height and the fireplace opening area. For example, for a 20-foot-high (6096 mm) chimney and a fireplace opening area of 2,000 square inches (1.3 m^2), the minimum cross-sectional area is 164 square inches (1.05 m^2) for a round flue or 190 square inches (0.12 m^2) for a square or rectangular flue. Using Table 2113.16(1), that minimum required cross-sectional area can be met by a circular flue with a diameter of 15 inches (381 mm). Using Table 2113.16(2), the minimum cross-sectional area would require a rectangular flue with a nominal outside dimension of 16 by 20 inches (405 by 508 mm).

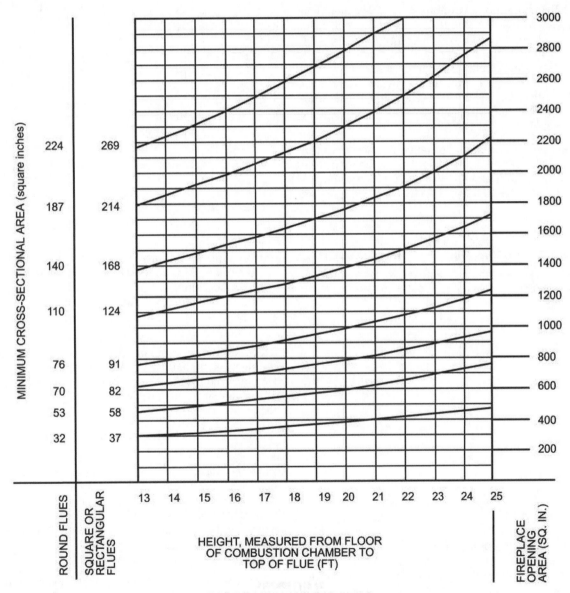

For SI: 1 inch = 25.4 mm, 1 square inch = 645 mm^2.

FIGURE 2113.16
FLUE SIZES FOR MASONRY CHIMNEYS

2113.16.1 Minimum area. Round chimney flues shall have a minimum net cross-sectional area of at least $^1/_{12}$ of the fireplace opening. Square chimney flues shall have a minimum net cross-sectional area of at least $^1/_{10}$ of the fireplace opening. Rectangular chimney flues with an aspect ratio less than 2 to 1 shall have a minimum net cross-sectional area of at least $^1/_{10}$ of the fireplace opening. Rectangular chimney flues with an aspect ratio of 2 to 1 or more shall have a minimum net cross-sectional area of at least $^1/_8$ of the fireplace opening.

❖ This section prescribes the minimum flue area based on the fireplace opening and the shape of the flue. For a given size fireplace opening, circular chimney flues are permitted to have a smaller area than rectangular flues. Wide, narrow flues require a larger area for adequate airflow.

2113.16.2 Determination of minimum area. The minimum net cross-sectional area of the flue shall be determined in accordance with Figure 2113.16. A flue size providing at least the equivalent net cross-sectional area shall be used. Cross-sectional areas of clay flue linings are as provided in Tables 2113.16(1) and 2113.16(2) or as provided by the manufacturer or as measured in the field. The height of the chimney shall be measured from the firebox floor to the top of the chimney flue.

❖ This section provides an alternative method to determine flue size based on the height of the chimney, the fireplace opening area and the flue type [see commentary, Tables 2113.16(1) and 2113.16(2) and Figure 2113.16].

TABLE 2113.16(1)
NET CROSS-SECTIONAL AREA OF ROUND FLUE SIZES[a]

FLUE SIZE, INSIDE DIAMETER (inches)	CROSS-SECTIONAL AREA (square inches)
6	28
7	38
8	50
10	78
10 $^3/_4$	90
12	113
15	176
18	254

For SI: 1 inch = 25.4 mm, 1 square inch = 645.16 mm².
a. Flue sizes are based on ASTM C 315.

❖ This table gives the areas of circular flues of standard sizes so that users of the code can easily determine a flue size complying with Figure 2113.16. The flue areas shown in this table are determined by the equation $A = \pi \times (d/2)^2$, where A is the cross-sectional area, d is the inside diameter of the flue and π is approximated as 3.14. Alternatively, a minimum diameter can be calculated as $d = [(A/\pi)^{1/2}]/4$.

TABLE 2113.16(2)
NET CROSS-SECTIONAL AREA OF SQUARE AND RECTANGULAR FLUE SIZES

FLUE SIZE, OUTSIDE NOMINAL DIMENSIONS (inches)	CROSS-SECTIONAL AREA (square inches)
4.5 × 8.5	23
4.5 × 13	34
8 × 8	42
8.5 × 8.5	49
8 × 12	67
8.5 × 13	76
12 × 12	102
8.5 × 18	101
13 × 13	127
12 × 16	131
13 × 18	173
16 × 16	181
16 × 20	222
18 × 18	233
20 × 20	298
20 × 24	335
24 × 24	431

For SI: 1 inch = 25.4 mm, 1 square inch = 645.16 mm².

❖ Clay flue liners are designated according to their nominal outside dimensions. This table gives the net cross-sectional area of standard size square and rectangular flues so that code users can readily determine a flue size that provides the minimum required area.

2113.17 Inlet. Inlets to masonry chimneys shall enter from the side. Inlets shall have a thimble of fireclay, rigid refractory material or metal that will prevent the connector from pulling out of the inlet or from extending beyond the wall of the liner.

❖ The inlet must be noncombustible and strong enough not to be pulled out.

2113.18 Masonry chimney cleanout openings. Cleanout openings shall be provided within 6 inches (152 mm) of the base of each flue within every masonry chimney. The upper edge of the cleanout shall be located at least 6 inches (152 mm) below the lowest chimney inlet opening. The height of the opening shall be at least 6 inches (152 mm). The cleanout shall be provided with a noncombustible cover.

Exception: Chimney flues serving masonry fireplaces, where cleaning is possible through the fireplace opening.

❖ This section requires a cleanout to be installed in a chimney to facilitate cleaning and inspection. A fireplace inherently provides access to its chimney through the firebox, throat and smoke chamber. The cleanout cover and opening frame are to be of an approved noncombustible material, such as cast iron or precast concrete and must be arranged to remain tightly closed. The requirement for placing the cleanout at least 6 inches (152 mm) below the lowest connection to the chimney is intended to minimize the possibility of combustion products exiting the chimney through the cleanout.

2113.19 Chimney clearances. Any portion of a masonry chimney located in the interior of the building or within the exterior wall of the building shall have a minimum airspace clearance to combustibles of 2 inches (51 mm). Chimneys located entirely outside the exterior walls of the building, including chimneys that pass through the soffit or cornice, shall have a minimum airspace clearance of 1 inch (25 mm). The airspace shall not be filled, except to provide fireblocking in accordance with Section 2113.20.

Exceptions:

1. Masonry chimneys equipped with a chimney lining system listed and labeled for use in chimneys in contact with combustibles in accordance with UL 1777, and installed in accordance with the manufacturer's instructions, are permitted to have combustible material in contact with their exterior surfaces.

2. Where masonry chimneys are constructed as part of masonry or concrete walls, combustible materials shall not be in contact with the masonry or concrete wall less than 12 inches (305 mm) from the inside surface of the nearest flue lining.

3. Exposed combustible trim and the edges of sheathing materials, such as wood siding, are permitted to abut the masonry chimney sidewalls, in accordance with Figure 2113.19, provided such combustible trim or

sheathing is a minimum of 12 inches (305 mm) from the inside surface of the nearest flue lining. Combustible material and trim shall not overlap the corners of the chimney by more than 1 inch (25 mm).

❖ Clearance between the external surfaces of the masonry chimney and all combustible materials must be maintained. The intent of this section is to require a 2-inch (51 mm) minimum airspace clearance between combustibles and surfaces of all chimneys located in the interior of a building or within an exterior wall assembly. A 1-inch (25 mm) airspace clearance is allowed only where the chimney is located entirely outside of the building.

If any portion of a chimney is located in an exterior wall, that chimney must be considered as an interior chimney and must have a 2-inch (51 mm) minimum airspace clearance. The 1-inch (25 mm) clearance is allowed because the exterior surface of an outdoor chimney can dissipate heat. An outdoor chimney is exposed to outdoor ambient temperatures, allowing it to operate with cooler surface temperatures. Like all airspace clearances, the 1-inch (25 mm) airspace clearance is not permitted to be filled with any material except the noncombustible material necessary for fireblocking. The required clearance for exterior chimneys applies to all combustible materials, including sheathing, siding, insulation, framing and trim.

The exception recognizes a variety of chimney liners that are tested and labeled in accordance with UL 1777.

UL 1777 covers metallic and nonmetallic chimney liners intended for field installation into new and existing masonry chimneys used for natural draft venting of gas, oil and solid-fuel-burning appliances having maximum continuous flue-gas temperatures not exceeding

For SI: 1 inch = 25.4 mm.

FIGURE 2113.19
ILLUSTRATION OF EXCEPTION THREE CHIMNEY CLEARANCE PROVISION

1,000°F (538°C). Some lining systems are labeled for reduced clearance applications and could allow for the construction or rehabilitation of chimneys in contact with combustibles, without compromising the safety normally provided by the code-prescribed airspace clearances.

Chimney liner systems are typically metal or poured-in-place concrete and incorporate insulation to retard the transfer of heat to the surrounding masonry walls of the chimney.

This figure clarifies Exception 3 to the clearance requirements for masonry chimneys. The edge abutment of combustible sheathing or trim where there is an adequate thickness of masonry is a long-standing practice that is considered safe, provided the minimum clearance to the flue lining is maintained.

2113.20 Chimney fireblocking. All spaces between chimneys and floors and ceilings through which chimneys pass shall be fireblocked with noncombustible material securely fastened in place. The fireblocking of spaces between wood joists, beams or headers shall be to a depth of 1 inch (25 mm) and shall only be placed on strips of metal or metal lath laid across the spaces between combustible material and the chimney.

❖ Fireblocking is required to prevent the travel of flames, smoke and hot gases to other areas of the building through the gaps between the chimney and the floor or ceiling assemblies. The 1-inch (25 mm) depth requirement is intended to be both the minimum and the maximum [see Figures 2113.20(1) and (2)].

Bibliography

The following resource materials are referenced in this chapter or are relevant to the subject matter addressed in this chapter.

ACI 530-05/ASCE 5-05/TMS 402-05, *Building Code Requirements for Masonry Structures.* Farmington Hills, MI: American Concrete Institute; Reston, VA: Structural Engineering Institute of the American Society of Civil Engineers; Boulder, CO: The Masonry Society, 2005.

ACI 530.1-05/ASCE 6-05/TMS 602-05, *Specifications for Masonry Structures.* Farmington Hills, MI: American Concrete Institute; Reston, VA: Structural Engineering Institute of the American Society of Civil Engineers; Boulder, CO: The Masonry Society, 2005.

ANSI A41.1-53, *American Standard Building Code Requirements for Masonry.* New York: American National Standards Institute, 1953.

ANSI A108.1A-99, *Glazed Wall Tile, Ceramic Mosaic Tile, Quarry Tile and Paver Tile Installed with Portland Cement Mortar.* New York: American National Standards Institute, 1999.

ANSI A108.1B-99, *Glazed Wall Tile, Ceramic Mosaic Tile, Quarry Tile and Paver Tile Installed with Portland Cement Mortar.* New York: American National Standards Institute, 1999.

NONCOMBUSTIBLE FIREBLOCK

Figure 2113.20(1)
FIREBLOCKING

ANSI A108.4-99, *Ceramic Tile Installed with Organic Adhesives or Water Cleanable Tile Setting Epoxy Adhesive*. New York: American National Standards Institute, 1999.

ANSI A108.5-99, *Ceramic Tile Installed with Dry-Set Portland Cement Mortar or Latex-Portland Cement Mortar*. New York: American National Standards Institute, 1999.

ANSI A108.6-99, *Ceramic Tile Installed with Chemical Resistant, Water Cleanable Tile Setting and Grouting Epoxy*. New York: American National Standards Institute, 1999.

ANSI A108.8-99, *Ceramic Tile Installed with Chemical Resistant Furan Mortar and Grout*. New York: American National Standards Institute, 1999.

ANSI A108.9-99, *Ceramic Tile Installed with Modified Epoxy Emulsion Mortar/Grout*. New York: American National Standards Institute, 1999.

ANSI A108.10-99, *Installation of Grout in Tilework*. New York: American National Standards Institute, 1999.

ANSI A118.1-99, *Dry-Set Portland Cement Mortar*. New York: American National Standards Institute, 1999.

ANSI A118.2-99, *Conductive Dry-Set Portland Cement Mortar*. New York: American National Standards Institute, 1999.

ANSI A118.3-99, *Chemical Resistant Water Cleanable Tile Setting and Grouting Epoxy and Water Cleanable Tile Setting Epoxy Adhesive*. New York: American National Standards Institute, 1999.

ANSI A118.4-99, *Latex-Portland Cement Mortar*. New York: American National Standards Institute, 1999.

ANSI A118.5-99, *Chemical Resistant Furan*. New York: American National Standards Institute, 1999.

ANSI A118.6-99, *Ceramic Tile Grouts*. New York: American National Standards Institute, 1999.

ANSI A118.8-99, *Modified Epoxy Emulsion Mortar/Grout*. New York: American National Standards Institute, 1999.

ANSI A136.1-99, *Organic Adhesives for Installation of Ceramic Tile*. New York: American National Standards Institute, 1999.

ANSI A137.1-88, *Ceramic Tile*. New York: American National Standards Institute, 1988.

ASCE 7-05, *Minimum Design Loads for Buildings and Other Structures*. Reston, VA: American Society of Civil Engineers, 2005.

ASTM A 36/A 36 M-04, *Specification for Structural Steel*. West Conshohocken, PA: ASTM International, 2004.

ASTM A 82-02, *Specification for Steel Wire, Plain, for Concrete Reinforcement*. West Conshohocken, PA: ASTM International, 2002.

ASTM A 153-03, *Specification for Zinc Coating (Hot Dip) on Iron and Steel Hardware*. West Conshohocken, PA: ASTM International, 2003.

ASTM A 167-99, *Specification for Stainless and Heat-Resisting Chromium-Nickel Steel Plate, Sheet and Strip*. West Conshohocken, PA: ASTM International, 1999.

ASTM A 185-02, *Specification for Steel Welded Wire Fabric, Plain, for Concrete Reinforcement*. West Conshohocken, PA: ASTM International, 2002.

For SI: 1 inch = 25.4 mm.

Figure 2113.20(2)
FIREBLOCKING (SECTION)

ASTM A 496-02, *Specification for Steel Wire, Deformed, for Concrete Reinforcement*. West Conshohocken, PA: ASTM International, 2002.

ASTM A 497-01, *Specification for Steel Welded Reinforcement, Deformed, for Concrete*. West Conshohocken, PA: ASTM International, 2001.

ASTM A 615-04a, *Specification for Deformed and Plain Billet-Steel Bars for Concrete Reinforcement*. West Conshohocken, PA: ASTM International, 2004.

ASTM A 641-03, *Specification for Zinc-Coated (Galvanized) Carbon Steel Wire*. West Conshohocken, PA: ASTM International, 2003.

ASTM A 706/A706M-04a, *Specification for Low-Alloy Steel Deformed Bars for Concrete Reinforcement*. West Conshohocken, PA: ASTM International, 2004.

ASTM A 767/A 767M-00b, *Specification for Zinc-Coated (Galvanized) Bars for Concrete Reinforcement*. West Conshohocken, PA: ASTM International, 2000.

ASTM A 775/A 775M-04, *Specification for Epoxy-Coated Reinforcing Steel Bars*. West Conshohocken, PA: ASTM International, 2004.

ASTM A 951-02, *Specification for Masonry Joint Reinforcement*. West Conshohocken, PA: ASTM International, 2002.

ASTM C 5-03, *Specification for Quicklime for Structural Purposes*. West Conshohocken, PA: ASTM International, 2003.

ASTM C 27-98 (2002), *Standard Classification of Fireclay and High-Alumina Refractory Brick*. West Conshohocken, PA: ASTM International, 2002.

ASTM C 34-03, *Specification for Structural Clay Load-Bearing Wall Tile*. West Conshohocken, PA: ASTM International, 2003.

ASTM C 55-03, *Specification for Concrete Building Brick*. West Conshohocken, PA: ASTM International, 2003.

ASTM C 56-96 (2001), *Specification for Structural Clay Non-Load-Bearing Tile*. West Conshohocken, PA: ASTM International, 2001.

ASTM C 62-04, *Specification for Building Brick (Solid Masonry Units Made from Clay or Shale)*. West Conshohocken, PA: ASTM International, 2004.

ASTM C 67-03ae01, *Test Methods of Sampling and Testing Brick and Structural Clay Tile*. West Conshohocken, PA: ASTM International, 2003.

ASTM C 73-99a, *Specification for Calcium Silicate Face Brick (Sand Lime Brick)*. West Conshohocken, PA: ASTM International, 1999.

ASTM C 90-03, *Specification for Load-Bearing Concrete Masonry Units*. West Conshohocken, PA: ASTM International, 2003.

ASTM C 91-03a, *Standard Specification for Masonry Cement*. West Conshohocken, PA: ASTM International, 2003.

ASTM C 97-83, *Standard Test Methods for Absorption and Bulk Specific Gravity of Natural Stone*. West Conshohocken, PA: ASTM International, 1983.

ASTM C 99-87, *Standard Test Method for Modulus of Rupture of Natural Building Stone*. West Conshohocken, PA: ASTM International, 1987.

ASTM C 120-85, *Standard Test Method for Flexural Testing of Slate (Modulus of Rupture, Modulus of Elasticity)*. West Conshohocken, PA: ASTM International, 1985.

ASTM C 121-85, *Standard Test Method for Water Absorption of Slate*. West Conshohocken, PA: ASTM International, 1985.

ASTM C 126-99, *Specification for Ceramic Glazed Structural Clay Facing Tile, Facing Brick and Solid Masonry Units*. West Conshohocken, PA: ASTM International, 1999.

ASTM C 140-02c, *Methods of Sampling and Testing Concrete Masonry Units*. West Conshohocken, PA: ASTM International, 2002.

ASTM C 144-89, *Standard Specification for Aggregate for Masonry Mortar*. West Conshohocken, PA: ASTM International, 1989.

ASTM C 150-04, *Specification for Portland Cement*. West Conshohocken, PA: ASTM International, 2004.

ASTM C 170-87, *Standard Test Method for Compressive Strength of Natural Building Stone*. West Conshohocken, PA: ASTM International, 1987.

ASTM C 199-84 (2000), *Standard Test Method for Pier Test for Refractory Mortars*. West Conshohocken, PA: ASTM International, 2000.

ASTM C 207-04, *Standard Specification for Hydrated Lime for Masonry Purposes*. West Conshohocken, PA: ASTM International, 2004.

ASTM C 212-00, *Specification for Structural Clay Facing Tile*. West Conshohocken, PA: ASTM International, 2000.

ASTM C 216-04a, *Specification for Facing Brick (Solid Masonry Units Made from Clay or Shale)*. West Conshohocken, PA: ASTM International, 2004.

ASTM C 217-01A, *Standard Test Method for Weather Resistance of Natural Slate*. West Conshohocken, PA: ASTM International, 2001.

ASTM C 270-04, *Specification for Mortar for Unit Masonry*. West Conshohocken, PA: ASTM International, 2004.

ASTM C 305-82, *Standard Method for Mechanical Mixing of Hydraulic Cement Pastes and Mortars for Plastic Consistency*. West Conshohocken, PA: ASTM International, 1982.

ASTM C 315-02, *Specification for Clay Flue Linings*. West Conshohocken, PA: ASTM International, 2002.

ASTM C 331-04, *Standard Specification for Lightweight Aggregates for Concrete Masonry Units*. West Conshohocken, PA: ASTM International, 2004.

ASTM C 404-87, *Standard Specification for Aggregates for Masonry Grout*. West Conshohocken, PA: ASTM International, 1987.

ASTM C 476-02, *Specification for Grout for Masonry*. West Conshohocken, PA: ASTM International, 2002.

ASTM C 503-03, *Specification For Marble Dimension Stone (Exterior)*. West Conshohocken, PA: ASTM International, 2003.

ASTM C 568-03, *Specification for Limestone Dimension Stone*. West Conshohocken, PA: ASTM International, 2003.

ASTM C 595-01, *Specification for Blended Hydraulic Cements*. West Conshohocken, PA: ASTM International, 2001.

ASTM C 615-03, *Specification for Granite Dimension Stone*. West Conshohocken, PA: ASTM International, 2003.

ASTM C 616-03, *Specification for Quartz-Based Dimension Stone*. West Conshohocken, PA: ASTM International, 2003.

ASTM C 629-03, *Specification for Slate Dimension Stone*. West Conshohocken, PA: ASTM International, 2003.

ASTM C 652-04a, *Specification for Hollow Brick (Hollow Masonry Units Made from Clay or Shale)*. West Conshohocken, PA: ASTM International, 2004.

ASTM C 744-99, *Specification for Prefaced Concrete and Calcium Silicate Masonry Units*. West Conshohocken, PA: ASTM International, 1999.

ASTM C 887-79a(2001), *Specification for Packaged, Dry, Combined Materials for Surface Bonding Mortar*. West Conshohocken, PA: ASTM International, 2001.

ASTM C 946-91 (2001), *Practice for Construction of Dry-Stacked, Surface-Bonded Walls*. West Conshohocken, PA: ASTM International, 2001.

ASTM C 1019-03, *Test Method for Sampling and Testing Grout*. West Conshohocken, PA: ASTM International, 2003.

ASTM C 1088-04, *Specification for Thin Veneer Brick Units Made from Clay or Shale*. West Conshohocken, PA: ASTM International, 2004.

ASTM C 1261-04, *Firebox Brick for Residential Fireplaces*. West Conshohocken, PA: ASTM International, 2004.

ASTM C 1283-03e01, *Standard Practice for Installing Clay Flue Liners*. West Conshohocken, PA: ASTM International, 2003.

ASTM C 1314-03b, *Standard Method for Constructing and Testing Masonry Prisms Used to Determine Compliance with Specified Compressive Strength of Masonry*. West Conshohocken, PA: ASTM International, 2003.

ASTM C 1327-97, *Specification for Mortar Cement*. West Conshohocken, PA: ASTM International, 1997.

ASTM E 72-02, *Standard Test Methods of Conducting Strength Tests of Panels for Building Construction*. West Conshohocken, PA: ASTM International, 2002.

ASTM E 84-04, *Test Method for Surface Burning Characteristics of Building Materials*. West Conshohocken, PA: ASTM International, 2004.

ASTM E 119-00, *Test Methods for Fire Tests of Building Construction and Materials*. West Conshohocken, PA: ASTM International, 2000.

ASTM E 447-92b, *Tests Methods for Compressive Strength of Masonry Prisms*. West Conshohocken, PA: ASTM International, 1992.

ASTM E 518-03, *Standard Test Methods for Flexural Bond Strength of Masonry*. West Conshohocken, PA: ASTM International, 2003.

ASTM E 519-02, *Standard Test Method for Diagonal Tension (Shear) in Masonry Assemblages*. West Conshohocken, PA: ASTM International, 2002.

ASTM E 1602-03, *Standard Guide for Construction of Solid Fuel Burning Masonry Heaters*. West Conshohocken, PA: ASTM International, 2003.

AWS D 1.4-98, *Structural Welding Code—Reinforced Steel*. Miami, FL: American Welding Society, 1998.

Baker, Ira. *A Treatise on Masonry Construction*. Chicago: University of Illinois, 1889.

Commentary on Building Code Requirements for Masonry Structures (ACI 530-05/ASCE 5-05/TMS 402-05). Farmington Hills, MI: American Concrete Institute; Reston, VA: Structural Engineering Institute of the American Society of Civil Engineers; Boulder, CO: The Masonry Society, 2005.

Commentary on Specifications for Masonry Structures (ACI 530.1-05/ASCE 6-05/TMS 602-05). Farmington Hills, MI: American Concrete Institute; Reston, VA: Structural Engineering Institute of the American Society of Civil Engineers; Boulder, CO: The Masonry Society, 2005.

Compressive Strength Testing of Masonry Mortar, A TMS Monograph. Boulder, CO: The Masonry Society, 1996.

FEMA 450-1, *NEHRP Recommended Provisions for Seismic Regulations for New Buildings and Other Structures*. Washington, D.C.: Federal Emergency Management Agency, 2004.

FEMA 450-2, *NEHRP Recommended Provisions for Seismic Regulations for New Buildings and Other*

Structures Part 2: Commentary. Washington, D.C.: Federal Emergency Management Agency, 2004.

IFC-06, *International Fire Code.* Falls Church, VA: International Code Council, 2006.

IFGC-06, *International Fuel Gas Code.* Falls Church, VA: International Code Council, 2006.

IMC-06, *International Mechanical Code.* Falls Church, VA: International Code Council, 2006.

IRC-06, *International Residential Code.* Falls Church, VA: International Code Council, 2006.

UBC-97, *Uniform Building Code.* Whittier, CA: International Conference of Building Officials, 1997.

UL 641-95. *Type L Low-Temperature Venting Systems.* Northbrook, IL: Underwriters Laboratories Inc., 1995.

UL 1777-98, *Chimney Liners.* Northbrook, IL: Underwriters Laboratories Inc., 1998.

Chapter 22:
Steel

General Comments

Chapter 22 contains provisions governing the materials, design, construction and quality of steel structural members.

Section 2201 contains the general requirements.

Section 2202 includes the definitions of terms related to structural steel construction.

Section 2203 includes provisions for the identification and protection of steel for structures.

Section 2204 addresses requirements for welded and bolted connections.

Section 2205 references the design standards for structural steel construction.

Section 2206 addresses steel joists.

Section 2207 specifies the design standard for steel cable structures.

Section 2208 references the specification for the design and installation of steel storage racks.

Section 2209 specifies the appropriate referenced standard for cold-formed steel.

Section 2210 references standards for light-framed construction utilizing cold-formed steel.

Section 2211 provides detailed provisions for light-frame cold-formed steel shear walls.

Steel is a noncombustible material commonly associated with Type 1 and 2 construction; however, it is permitted to be used and commonly found in all types of construction. There are two main families of steel members: the first is hot-rolled structural shapes and members that are made up of a combination of rolled shapes and plates, and the second is composed of sections that are cold formed from steel sheets, strips, plates or flat bars in roll-forming machines, or by bending in brake-press operations.

The code requires that materials used in the design of structural steel members conform to designated national standards. Chapter 22 is involved largely with the design and use of steel materials using the specifications and standards of the American Institute of Steel Construction (AISC), the American Iron and Steel Institute (AISI), the Steel Joist Institute (SJI) and the American Society of Civil Engineers (ASCE).

Purpose

The purpose of Chapter 22 is to provide the requirements necessary for the design and construction of structural steel, cold-formed steel, steel joists, steel cable structures, steel storage racks and composite construction. This chapter specifies the appropriate design and construction standards for these types of structures. It also provides a road map of the applicable technical requirements for steel structures. Detailed provisions are included for the wind and seismic requirements for light- frame cold-formed steel walls.

SECTION 2201
GENERAL

2201.1 Scope. The provisions of this chapter govern the quality, design, fabrication and erection of steel used structurally in buildings or structures.

❖ Chapter 22 is formatted in a very similar fashion to the legacy model codes, since the code is essentially a road map for the use of design standards appropriate for various types of steel construction.

While this section indicates that Chapter 22 governs "quality, design, fabrication and erection," this is a general statement and the code user should take note of the scope of reference used to adopt specific standards since it may be different. Each section is intended to reflect the scope of the adopted standard and it is this specific scope of reference that establishes how each standard must be applied. The following text will discuss the background of each section.

SECTION 2202
DEFINITIONS

2202.1 Definitions. The following words and terms shall, for the purposes of this chapter and as used elsewhere in this code, have the meaning shown herein.

❖ Certain terms are used in the code to describe something very specific to the related subject matter while other terms may have multiple meanings that can vary based on the subject matter and context. To avoid misunderstandings, it is preferable to reach a consensus on the meaning of these terms. Definitions are intended to facilitate the understanding of code provisions and to minimize potential confusion.

This section contains definitions of terms associated with steel construction. Note that these terms are also listed in Chapter 2 with a cross reference to this section. Definitions that pertain primarily to steel construction are included within this chapter to provide convenient access to them without having to refer back to Chapter 2. The use and application of all defined terms, including those defined herein, as well as undefined terms are set forth in Section 201.

STEEL CONSTRUCTION, COLD-FORMED. That type of construction made up entirely or in part of steel structural members cold formed to shape from sheet or strip steel such as roof deck, floor and wall panels, studs, floor joists, roof joists and other structural elements.

❖ This definition is necessary to distinguish structural members of cold-formed steel from other types of steel elements. A distinguishing characteristic is that the forming operation takes place at room temperature without the addition of heat. Specific references to AISI standards and specifications applicable to cold-formed steel construction are provided in Sections 2209 and 2210.

STEEL JOIST. Any steel structural member of a building or structure made of hot-rolled or cold-formed solid or open-web sections, or riveted or welded bars, strip or sheet steel members, or slotted and expanded, or otherwise deformed rolled sections.

❖ This definition is necessary to distinguish steel joists from other types of steel elements. References to SJI specifications are provided in Section 2206.

STEEL MEMBER, STRUCTURAL. Any steel structural member of a building or structure consisting of a rolled steel structural shape other than cold-formed steel, or steel joist members.

❖ This definition is necessary to distinguish structural steel members from other types of steel elements. References to AISC standards and specifications for structural steel are provided in Section 2205.

SECTION 2203
IDENTIFICATION AND PROTECTION
OF STEEL FOR STRUCTURAL PURPOSES

2203.1 Identification. Steel furnished for structural load-carrying purposes shall be properly identified for conformity to the ordered grade in accordance with the specified ASTM standard or other specification and the provisions of this chapter. Steel that is not readily identifiable as to grade from marking and test records shall be tested to determine conformity to such standards.

❖ A wide variety of steel types and grades are available and important properties, such as tensile strength and weldability, vary accordingly. For that reason, it is imperative that the properties of the steel that is furnished be verifiable. This section requires that all steel must be identified so that both the engineer and inspector can verify conformance with the approved plans as well as the code. All steel that is not marked for these identification purposes shall be tested to verify conformance.

2203.2 Protection. Painting of structural steel shall comply with the requirements contained in AISC 360. Individual structural members and assembled panels of cold-formed steel construction, except where fabricated of approved corrosion-resistant steel or of steel having a corrosion-resistant or other approved coating, shall be protected against corrosion with an approved coat of paint, enamel or other approved protection.

❖ This section cautions the user that protection of structural steel members is a consideration. The need to provide protection, however, is dependent upon the potential for corrosion in a given installation. Typically, unprotected carbon steel can perform adequately in a dry, interior exposure while weathering steel can tolerate atmospheric exposure without the need for additional protection. Therefore, the AISC provisions require consideration of corrosion where it could actually impair the strength or serviceability of structural steel. A design that can tolerate the anticipated corrosion is permitted as an alternative to providing protection against corrosion. Many types of coating systems are available, yet painting remains the most common type of corrosion protection. Where painting of structural steel is called for, it must be in accordance with the AISC specification. Also, cold-formed steel structural members need to be of corrosion-resistant steel, or have a corrosion-resistant metallic coating, paint or other approved coating.

SECTION 2204
CONNECTIONS

2204.1 Welding. The details of design, workmanship and technique for welding, inspection of welding and qualification of welding operators shall conform to the requirements of the specifications listed in Sections 2205, 2206, 2207, 2209 and 2210. Special inspection of welding shall be provided where required by Section 1704.

❖ This section does not directly adopt specifications for the design and details of welding; rather, it requires that welding be accomplished in accordance with the requirements of the appropriate design specification. This is done because in the case of structural steel and cold-formed steel, the referenced standards adopt different American Welding Society (AWS) specifications, while in the case of steel joists, welding requirements are contained within the SJI specifications.

The processes and techniques for welding heavy structural steel are different from those necessary for welding thinner cold-formed steel. Therefore, the AISC specification for structural steel construction requires welding to be in accordance with AWS D1.1, *Structural Welding Code—Steel*, which has a thickness limitation of not less than 0.125 inches (3.2 mm). The AISI standards on cold-formed steel refer to AWS D1.3, *Structural Welding Code—Sheet Steel*, which is limited to thicknesses of not more than 0.188 inches (4.8 mm). The overlapping thickness limitations provide the flexibility to allow the sole use of AWS D1.1 or D1.3 on projects where all welded steel meets the corresponding thickness limitations.

Reference is made to the special inspection requirements in Section 1704. Note the exception to continuous inspection of welding in Section 1704.3 where the applicable conditions are met. Although they are not directly referenced here, Sections 1707.2 and 1707.4 address special inspections for welding of structural steel and cold-formed steel that are triggered by the requirements for seismic resistance.

2204.2 Bolting. The design, installation and inspection of bolts shall be in accordance with the requirements of the specifications listed in Sections 2205, 2206, 2209 and 2210. Special inspection of the installation of high-strength bolts shall be provided where required by Section 1704.

❖ This section is very similar to the previous section on welding in that it does not adopt bolt installation standards separately. The requirements for the design, installation and inspection for bolting are found in standards referenced in the sections that are noted.

2204.2.1 Anchor rods. Anchor rods shall be set accurately to the pattern and dimensions called for on the plans. The protrusion of the threaded ends through the connected material shall be sufficient to fully engage the threads of the nuts, but shall not be greater than the length of the threads on the bolts.

❖ This section provides the requirements for the setting and projection of anchor rods. The end of the anchor rod threads must not project above the face of the connected part as is indicated in Figure 2204.2.1.

Figure 2204.2.1
PROJECTION OF ANCHOR ROD THREADS

SECTION 2205
STRUCTURAL STEEL

2205.1 General. The design, fabrication and erection of structural steel for buildings and structures shall be in accordance with AISC 360. Where required, the seismic design of steel structures shall be in accordance with the additional provisions of Section 2205.2.

❖ This section requires that structural steel (see the definition in Section 2202) be designed in accordance with

the AISC *Specification for Structural Steel Buildings* (AISC 360). It is the successor to the separate AISC specifications previously published and combines design provisions for both allowable stress design (ASD) and load and resistance factor design (LRFD) in what is referred to as a unified format. AISC 360 also incorporates requirements that were previously published in a separate standard for the design of round, square and rectangular steel tubular structural sections.

2205.2 Seismic requirements for steel structures. The design of structural steel structures to resist seismic forces shall be in accordance with the provisions of Section 2205.2.1 or 2205.2.2 for the appropriate seismic design category.

❖ The code adopts state-of-the-art seismic requirements based on AISC 341, *Seismic Provisions for Structural Steel Buildings*. It incorporates significant research results from the SAC Research Program, which began as a result of the damage to steel moment frames that was caused by the Northridge earthquake. AISC 341 is divided into two parts. Part I provides the requirements for designing seismic-force-resisting systems of structural steel. It incorporates provisions for ASD and LRFD. Part II is applicable to composite structural steel and reinforced concrete seismic-force-resisting systems.

2205.2.1 Seismic Design Category A, B or C. Structural steel structures assigned to Seismic Design Category A, B or C shall be of any construction permitted in Section 2205. An *R* factor as set forth in Section 12.2.1 of ASCE 7 for the appropriate steel system is permitted where the structure is designed and detailed in accordance with the provisions of AISC 341, Part I. Systems not detailed in accordance with the above shall use the *R* factor in Section 12.2.1 of ASCE 7 designated for "structural steel systems not specifically detailed for seismic resistance."

❖ The earthquake provisions in Section 1613 require all structures to be assigned to a Seismic Design Category, which is a function of the building occupancy as well as the design earthquake ground motions. Seismic Design Category is an indicator of the seismic risk and it is used to establish, amongst other things, the complexity of the seismic analysis and the types of structural systems that may be utilized to resist the effects of earthquakes. Structures classified as Seismic Design Category A, B or C are considered to have a low to moderate seismic risk and this section of the code clarifies the options that are available in designing a seismic-force-resisting system utilizing structural steel. These options are as follows:

Option 1: The engineer is permitted to use a "structural steel system not specifically detailed for seismic resistance," as stated in ASCE 7, which is assigned a response modification coefficient, *R*, of 3 in Table 12.2-1 of ASCE 7. The system is then designed for the resulting earthquake load effects using AISC 360 without the detailing required by AISC 341. Another way of looking at this is that the ductility inherent in a structural steel system that is designed and constructed under AISC 360 warrants use of an *R*-3.

Option 2: If the engineer wishes to use a coefficient

greater than R-3 in order to lower the design earthquake forces, the AISC 341 seismic detailing provisions are mandatory. For example, a special steel concentrically braced frame (SCBF) could be utilized. ASCE 7 Table 12.2-1 specifies an R-6 for this system. All other factors being equal, the design base shear for this system is reduced to half that required under Option 1 above. However, the design and construction requirements for SCBF members and connections in Section 13 of AICS 341 must be satisfied.

It is also worth noting that this section actually states the minimum requirements for structural steel systems in buildings that are assigned to Seismic Design Category B or C only. For structures classified as Seismic Design Category A (i.e., low earthquake risk), compliance with this section is more correctly an alternative to the minimum earthquake requirements that are specified in Section 11.7 of ASCE 7. That document merely requires design for a minimum lateral earthquake load that is independent of the type of seismic-force-resisting system that is utilized (i.e., no R-value is used).

2205.2.2 Seismic Design Category D, E or F. Structural steel structures assigned to Seismic Design Category D, E or F shall be designed and detailed in accordance with AISC 341, Part I.

❖ In buildings classified as Seismic Design Category D, E and F, the provisions of AISC 341, Part I are mandatory for seismic-force-resisting systems of structural steel. Where the seismic risk is considered high, the earthquake load requirements that are referenced in Section 1613 limit the permitted seismic-force-resisting systems to those considered to be capable of inelastic response. The design earthquake forces are a function of the response modification coefficient, R, for the seismic-force-resisting system that is utilized. The design of the steel members and connections in accordance with AISC 341 requirements for the selected seismic-force-resisting system provides the appropriate level of energy dissipation by inelastic response.

2205.3 Seismic requirements for composite construction. The design, construction and quality of composite steel and concrete components that resist seismic forces shall conform to the requirements of the AISC 360 and ACI 318. An R factor as set forth in Section 12.2.1 of ASCE 7 for the appropriate composite steel and concrete system is permitted where the structure is designed and detailed in accordance with the provisions of AISC 341, Part II. In Seismic Design Category B or above, the design of such systems shall conform to the requirements of AISC 341, Part II.

❖ Composite construction consists of steel and concrete members that act in unison to resist the effects of applied loads. Composite action used in gravity load applications has been well established. The use of composite lateral-force-resisting systems has been growing in low-risk seismic areas, but has been hampered in high-risk seismic areas due to the lack of design requirements. The code contains the first concise technical provisions for combining structural steel and concrete in a composite seismic-force-resisting system, while requiring further substantiation where the seismic risk is high.

Composite structures classified as Seismic Design Category A are permitted to be designed in accordance with ACI 318 and AISC 360. In structures classified as Seismic Design Category B or above a composite seismic-force-resisting system is required to be one of the types listed in Table 12.2-1 of ASCE 7. The corresponding response reduction coefficient, R, is used to determine the seismic base shear and the system is designed in accordance with Part II of AISC 341.

2205.3.1 Seismic Design Categories D, E and F. Composite structures are permitted in Seismic Design Categories D, E and F, subject to the limitations in Section 12.2.1 of ASCE 7, where substantiating evidence is provided to demonstrate that the proposed system will perform as intended by AISC 341, Part II. The substantiating evidence shall be subject to building official approval. Where composite elements or connections are required to sustain inelastic deformations, the substantiating evidence shall be based on cyclic testing.

❖ Since these are new requirements where the seismic risk is high, such as in buildings that are classified as Seismic Design Category D, E or F, composite seismic-force-resisting sytems are permitted, provided test data or other evidence that the connections have the necessary rotational capacity to resist the expected deformations is submitted to the building official for approval. Where the composite elements are expected to undergo inelastic demands, the evidence must be based on cyclic tests in accordance with the test protocol and acceptance criteria established by AISC 341. In addition to this data, the use of Part II of AISC 341 is required.

SECTION 2206
STEEL JOISTS

2206.1 General. The design, manufacture and use of open web steel joists and joist girders shall be in accordance with one of the following Steel Joist Institute (SJI) specifications:

1. SJI K-1.1

2. SJI LH/DLH-1.1

3. SJI JG-1.1

Where required, the seismic design of buildings shall be in accordance with the additional provisions of Section 2205.2 or 2210.5.

❖ Steel joists are required to be designed and manufactured in accordance with the specifications published by the SJI. Seismic loading must always be evaluated and the code user is referred to the sections governing design of steel seismic-force-resisting systems. Open web steel joists are primarily used as gravity-load-carrying members, and where a steel joist transfers earthquake loads or carries other unique loads, this information needs to be provided to the joist manufacturer.

2206.2 Design. The registered design professional shall indicate on the construction documents the steel joist and/or steel joist girder designations from the specifications listed in Section 2206.1 and shall indicate the requirements for joist and joist girder design, layout, end supports, anchorage, non-SJI standard bridging, bridging termination connections and bearing connection design to resist uplift and lateral loads. These documents shall indicate special requirements as follows:

1. Special loads including:

    1.1. Concentrated loads;

    1.2. Nonuniform loads;

    1.3. Net uplift loads;

    1.4. Axial loads;

    1.5. End moments; and

    1.6. Connection forces.

2. Special considerations including:

    2.1. Profiles for nonstandard joist and joist girder configurations (standard joist and joist girder configurations are as indicated in the SJI catalog);

    2.2. Oversized or other nonstandard web openings; and

    2.3. Extended ends.

3. Deflection criteria for live and total loads for non-SJI standard joists.

❖ Sections 2206.2 through 2206.5 are meant to delineate the responsibilities of the steel joist manufacturer and the registered design professional who performs the overall structural design. Like precast concrete components and wood I-joists, the design of steel joists is typically relegated to the component manufacturer by contractual arrangements on any given project. Since the steel joist manufacturer designs the steel joists, this section lists the necessary steel joist design criteria that the registered design professional must furnish in the construction documents.

2206.3 Calculations. The steel joist and joist girder manufacturer shall design the steel joists and/or steel joist girders in accordance with the current SJI specifications and load tables to support the load requirements of Section 2206.2. The registered design professional may require submission of the steel joist and joist girder calculations as prepared by a registered design professional responsible for the product design. If requested by the registered design professional, the steel joist manufacturer shall submit design calculations with a cover letter bearing the seal and signature of the joist manufacturer's registered design professional. In addition to standard calculations under this seal and signature, submittal of the following shall be included:

1. Non-SJI standard bridging details (e.g. for cantilevered conditions, net uplift, etc.).

2. Connection details for:

    2.1. Non-SJI standard connections (e.g. flush-framed or framed connections);

    2.2. Field splices; and

    2.3. Joist headers.

❖ This section requires the steel joist manufacturer to design the steel joists in accordance with the design criteria provided by the registered design professional and also the applicable SJI specification(s). Where a more complete accounting of the steel joist design is desired, the registered design professional is permitted to request sealed design calculations from the steel joist manufacturer. The list of "nonstandard" items that are required for submittal is intended to minimize the chance that key details could be overlooked.

2206.4 Steel joist drawings. Steel joist placement plans shall be provided to show the steel joist products as specified on the construction documents and are to be utilized for field installation in accordance with specific project requirements as stated in Section 2206.2. Steel placement plans shall include, at a minimum, the following:

1. Listing of all applicable loads as stated in Section 2206.2 and used in the design of the steel joists and joist girders as specified in the construction documents.

2. Profiles for nonstandard joist and joist girder configurations (standard joist and joist girder configurations are as indicated in the SJI catalog).

3. Connection requirements for:

    3.1. Joist supports;

    3.2. Joist girder supports;

    3.3. Field splices; and

    3.4. Bridging attachments.

4. Deflection criteria for live and total loads for non-SJI standard joists.

5. Size, location and connections for all bridging.

6. Joist headers.

Steel joist placement plans do not require the seal and signature of the joist manufacturer's registered design professional.

❖ The steel joist placement plans are provided by the steel joist manufacturer and illustrate how the steel joists are to be installed in more detail than the project's construction documents that are prepared by the registered design professional. The list of information that must be included is intended to minimize the possibility of overlooking critical details. The steel joist placement plan is considered to be a type of "shop drawing" that does not require the seal of a registered design professional.

2206.5 Certification. At completion of fabrication, the steel joist manufacturer shall submit a certificate of compliance in accordance with Section 1704.2.2 stating that work was performed in accordance with approved construction documents and with SJI standard specifications.

❖ The certificate of compliance provides a written declaration from the steel joist manufacturer that attests to its compliance with the SJI specifications as well as the construction documents.

SECTION 2207
STEEL CABLE STRUCTURES

2207.1 General. The design, fabrication and erection including related connections, and protective coatings of steel cables for buildings shall be in accordance with ASCE 19.

❖ The code adopts ASCE 19 for the design of steel cable structures. This standard addresses the use of carbon steel or stainless steel cables formed from coated or uncoated helically twisted wire strand, wire rope or a parallel wire strand.

2207.2 Seismic requirements for steel cable. The design strength of steel cables shall be determined by the provisions of ASCE 19 except as modified by these provisions.

1. A load factor of 1.1 shall be applied to the prestress force included in T_3 and T_4 as defined in Section 3.12.

2. In Section 3.2.1, Item (c) shall be replaced with "1.5 T_3" and Item (d) shall be replaced with "1.5 T_4."

❖ This section provides the user with appropriate modifications for structural cables that are intended for use in seismic applications. These modifications were first developed by the Building Seismic Safety Council for incorporation into the first edition of the *NEHRP Recommended Provisions for Seismic Regulations for New Buildings and Other Structures.*

SECTION 2208
STEEL STORAGE RACKS

2208.1 Storage racks. The design, testing and utilization of industrial steel storage racks shall be in accordance with the *RMI Specification for the Design, Testing and Utilization of Industrial Steel Storage Racks.* Racks in the scope of this specification include industrial pallet racks, movable shelf racks and stacker racks and does not apply to other types of racks, such as drive-in and drive-through racks, cantilever racks, portable racks or rack buildings. Where required, the seismic design of storage racks shall be in accordance with the provisions of Section 15.5.3 of ASCE 7.

❖ This section requires that the design of steel storage racks be in accordance with the Rack Manufactures Institute (RMI) standard. The types of racks that are included in the scope of that standard are noted. Other types of racks, such as drive-in or drive-through racks or cantilever racks, portable racks or rack buildings, that are beyond the scope of the RMI standard, should be designed in accordance with applicable referenced specifications such as AISC 360. Where the racks are located in structures classified as Seismic Design Category C or higher, they also must be designed in accordance with the earthquake requirements of Chapter 16. According to the referenced ASCE 7 provision, racks supported at or below the base of the structure must comply with the requirements for nonbuilding structures, while racks supported above the base of the structure must comply with the requirements for architectural components.

SECTION 2209
COLD-FORMED STEEL

2209.1 General. The design of cold-formed carbon and low-alloy steel structural members shall be in accordance with AISI-NAS. The design of cold-formed stainless-steel structural members shall be in accordance with ASCE 8. Cold-formed steel light-framed construction shall comply with Section 2210.

❖ This section adopts AISI-NAS, which has been developed by the AISI in cooperation with the Canadian Standards Association (CSA) and Camara Nacional de la Industria del Hierro y del Acero (CANACERO). It supercedes the standards previously published individually by AISI and CSA and is intended for use throughout North America. The latter is accomplished by including three appendix chapters with the country-specific provisions applicable in either the United States, Canada or Mexico. Like its predecessor, it incorporates LRFD and ASD into a single document.

  Cold-formed stainless steel is required to comply with ASCE 8. This standard applies to structural members that are cold formed from annealed and cold-rolled stainless steels. ASCE 8 includes both the LRFD and ASD methods.

2209.2 Composite slabs on steel decks. Composite slabs of concrete and steel deck shall be designed and constructed in accordance with ASCE 3.

❖ This section adopts ASCE 3, which covers the design and testing of both normal and lightweight concrete placed over a cold-formed steel deck. The steel deck performs the dual function of acting as a form for the concrete during construction and as positive reinforcement for the slab during service.

SECTION 2210
COLD-FORMED STEEL
LIGHT-FRAMED CONSTRUCTION

2210.1 General. The design, installation and construction of cold-formed carbon or low-alloy steel, structural and nonstructural steel framing shall be in accordance with AISI-General and AISI-NAS.

❖ Section 2210 applies to light-framed construction utilizing cold-formed steel. Light-framed construction is characterized by vertical (i.e., walls) and horizontal (i.e., floors and roofs) structural elements formed by a system of repetitive members (see Chapter 2 definition). This section references AISI-NAS in addition to AISI-General. As its name implies, AISI-General covers general requirements for the design and installation of cold-formed steel framing members in walls, floors and roof assemblies and it is intended to be a supplement to AISI-NAS. For specific design information, AISI-General refers the reader to AISI-NAS.

2210.2 Headers. The design and installation of cold-formed steel box headers, back-to-back headers and single and double L-headers used in single-span conditions for load-carrying purposes shall be in accordance with AISI-Header, subject to the limitations therein.

❖ The code references AISI *Standard for Cold-Formed Steel Framing-Header Design*, which has design and installation requirements for the commonly used header cross sections depicted in Figure 2210.2. Box and back-to-back headers consist of two cold-formed steel C sections. Single and double L-headers consist of cold-formed steel angle sections connected to the top track.

Figure 2210.2
HEADER TYPES

2210.3 Trusses. The design, quality assurance, installation and testing of cold-formed steel trusses shall be in accordance with AISI-Truss, subject to the limitations therein.

❖ The code references AISI *Standard for Cold-Formed Steel Framing–Truss Design,* or AISI-Truss. This document covers truss design, quality criteria, installation and bracing as well as test methods. Since it is not uncommon for the design of cold-formed steel trusses to be performed by someone other than the registered design professional responsible for the building structure, AISI-Truss spells out the responsibilities of the truss designer and the building designer.

2210.4 Wall stud design. The design and installation of cold-formed steel studs for structural and nonstructural walls shall be in accordance with AISI-WSD.

❖ AISI-WSD provides design and installation requirements for cold-formed steel studs.

2210.5 Lateral design. The design of light-framed cold-formed steel walls and diaphragms to resist wind and seismic loads shall be in accordance with AISI-Lateral.

❖ Where shear walls, diagonal strap bracing or diaphragms of light-framed cold-formed steel provide resistance to the effects of earthquakes and wind loads, AISI-Lateral provides the design requirements for the lateral-force-resisting system. Shear wall design provisions include both segmented shear walls (Type I) and perforated shear walls (Type II).

2210.6 Prescriptive framing. Detached one- and two-family dwellings and townhouses, up to two stories in height, shall be permitted to be constructed in accordance with AISI-PM, subject to the limitations therein.

❖ AISI-PM contains prescriptive requirements for light-framed cold-formed steel construction that can be applied in one- and two-family dwellings as well as townhouses no more that two stories. The provisions are similar to those found in the *International Residential Code®* (IRC®).

Bibliography

The following resource material is referenced in this chapter or is relevant to the subject matter addressed in this chapter.

AISC 341-05, *Seismic Provisions for Structural Steel Buildings*. Chicago: American Institute of Steel Construction, 2005.

AISC 360-05, *Specification for Structural Steel Buildings*. Chicago: American Institute of Steel Construction, 2005.

AISI General-04, *Standard for Cold-Formed Steel Framing—General Provisions*. Washington, D.C.: American Iron and Steel Institute, 2004.

AISI Lateral-04, *Standard for Cold-Formed Steel Framing—Lateral Design*. Washington, D.C.: American Iron and Steel Institute, 2004.

AISI Header-04, *Standard for Cold-Formed Steel Framing—Header Design*. Washington, D.C.: American Iron and Steel Institute, 2004.

AISI NAS-01, *North American Specification for the Design of Cold-Formed Steel Structural Members, including 2004 Supplement*. Washington, D.C.: American Iron and Steel Institute, 2001.

AISI PM-01, *Standard for Cold-Formed Steel Framing-Prescriptive Method for One- and Two-Family Dwellings, including 2004 Supplement*. Washington, D.C.: American Iron and Steel Institute, 2001.

AISI Truss-04, *Standard for Cold-Formed Steel Framing-Truss Design*. Washington, DC: American Iron and Steel Institute, 2004.

AISI WSD-04, *Standard for Cold-Formed Steel Framing-Wall Stud Design*. Washington, D.C.: American Iron and Steel Institute, 2004.

ASCE 3-91, *Standard for the Construction and Inspection of Composite Slabs*. Reston, VA: American Society of Civil Engineers, 1991.

ASCE 7-05, *Minimum Design Loads for Buildings and Other Structures*. New York: American Society of Civil Engineers, 2005.

ASCE 8-02, *Standard Specification for the Design of Cold-Formed Stainless Steel Structural Members*. Reston, VA: American Society of Civil Engineers, 2002.

ASCE 19-96, *Structural Applications of Steel Cables for Buildings*. Reston, VA: American Society of Civil Engineers, 1996.

AWS D1.1-04, *Structural Welding Code—Steel*. Miami, FL: American Welding Society, 2004.

AWS D1.3-98, *Structural Welding Code—Sheet Steel*. Miami, FL: American Welding Society, 1998.

FEMA 450, *NEHRP Recommended Provisions for Seismic Regulations for New Buildings and Other Structures*. Washington, D.C.: Federal Emergency Management Agency, 2004.

IRC-2006, *International Residential Code*. Falls Church, VA: International Code Council, 2006

Naeim, Farzad. *The Seismic Design Handbook*. Norwell, MA: Kluwer Academic Publishers, 2001.

RMI-02, *Specification for the Design, Testing and Utilization of Industrial Steel Storage Racks*. Charlotte, NC: Rack Manufacturers Institute, 2002.

SJI JG-1.1, *Standard Specification for Joist Girders*. Myrtle Beach, SC: Steel Joist Institute, 2005.

SJI K-1.1, *Standard Specification for Open Web Steel Joists, K Series*. Myrtle Beach, SC: Steel Joist Institute, 2005.

SJI LH/DLH-1.1, *Standard Specification for Longspan Steel Joists LH Series and Deep Longspan Steel Joists, DLH Series*. Myrtle Beach, SC: Steel Joist Institute, 2005.

Williams, Alan. *Seismic and Wind Forces Structural Design Examples*. Falls Church, VA: International Code Council, 2005.

Williams, Alan. *Structural Steel Design LRFD*. Falls Church, VA: International Code Council, 2005.

Chapter 23:
Wood

General Comments

This chapter contains information required to design and construct buildings or structures that include wood or wood-based structural elements, and is organized around the application of three design methodologies: allowable stress design (ASD), load and resistance factor design (LRFD) and conventional construction. Included are references to design and manufacturing standards for various wood and wood-based products; general construction requirements; design criteria for lateral-force-resisting systems and specific requirements for the application of the three design methods (ASD, LRFD and conventional construction). Chapter 23 includes elements of all three previous regional model codes—the *BOCA® National Building Code* (BNBC), the *Standard Building Code* (SBC) and the *Uniform Building Code™* (UBC™). It most closely follows the format of the UBC.

Acceptable standards for the manufacture of wood or wood-based products include provisions for sizes, grades (labels), quality control and certification programs, or similar methods of identification. Specific requirements and tables have been developed using referenced design methods to provide a minimum level of safety. This chapter also contains requirements both for the use of products in conjunction with wood and wood-based structural elements, and for prevention of decay.

In general, only Type III, IV or V buildings may be constructed of wood. Accordingly, Chapter 23 is referenced when the combination of the occupancy (determined in Chapter 3) and the height and area of the building or structure (determined in Chapter 5) indicate that the construction (specified in Chapter 6) can be Type III, IV or V. Another basis for referencing Chapter 23 is when wood elements are used in Type I or II structures as permitted in Section 603. This chapter gives information on the application of fire-retardant-treated wood, interior wood elements and trim in these structures. All structural criteria for application of referenced standards and procedures included in Chapter 23 are based on the loading requirements of Chapter 16 or on historical performance.

Chapter 23 is not a textbook on construction. It is assumed that the reader has both the training and experience needed to understand the principles and practices of wood design and construction. Without such understanding, some sections may be misunderstood and misapplied. This commentary should help to promote better understanding of the structure and application of the methods specified in Chapter 23.

Section 2301 identifies three methods of design, and compliance with one or more is required.

Section 2302 contains the most commonly used terms to describe wood and wood-based products. Some engineering terms are also included.

Section 2303 provides reference to manufacturing standards, necessary specification criteria and use and application provisions.

Section 2304 contains provisions for the proper design and construction of all wood structures and the use of all wood products. This section also includes the fastening schedule.

Section 2305 contains general design requirements for lateral-force-resisting systems. Whether the structure is engineered using ASD or LRFD, the provisions of this section apply.

Section 2306 contains provisions for the design of structures using ASD. Historically, all of the industry publications have been developed for ASD. More recently, LRFD has been introduced, making it necessary to distinguish clearly which provisions are appropriate for ASD or LRFD. The provisions of this section are not appropriate for LRFD.

Section 2307 references the consensus standard, which governs the design of structures using the LRFD methodology.

Although fairly limited in application, Section 2308 contains the provisions for conventional construction that may be used to construct certain wood-frame structures. Limitations on the use of this section are provided in Section 2308.2 for a quick determination.

Purpose

This chapter provides minimum guidance for the design of buildings and structures that use wood and wood-based products in their framing and fabrication. Alternative methods and materials can be used where justified by engineering analysis and testing. In all cases, the provisions of Section 2304 apply to all elements of wood-frame construction.

SECTION 2301
GENERAL

2301.1 Scope. The provisions of this chapter shall govern the materials, design, construction and quality of wood members and their fasteners.

❖ Section 2301 includes specifications for use of and standards for production of wood and wood-based products such as boards, dimensional lumber and engineered wood products, such as I-joists, glued-laminated timber, structural panels, trusses, particleboard, fiberboard and hardboard. Also included are criteria and specifications for the use of other materials such as connectors used in conjunction with wood or wood-based products. Other chapters of the code also affect the use of wood materials in buildings and should be referenced prior to making final decisions on the use of any product.

The scope of this chapter is established in Section 2301.1, and broadly encompasses wood products and the limitations placed on them and their various applications within the code.

2301.2 General design requirements. The design of structural elements or systems, constructed partially or wholly of wood or wood-based products, shall be in accordance with one of the following methods:

1. Allowable stress design in accordance with Sections 2304, 2305 and 2306.

2. Load and resistance factor design in accordance with Sections 2304, 2305 and 2307.

3. Conventional light-frame construction in accordance with Sections 2304 and 2308.

> **Exception:** Buildings designed in accordance with the provisions of the AF&PA WFCM shall be deemed to meet the requirements of the provisions of Section 2308.

❖ This chapter includes three methods of designing with wood or wood-based products. This section limits designs to one of these three methods unless an alternative method has been proven to be acceptable as permitted in Section 104.11. It is not uncommon for only one element of a structure to require engineered design. This section recognizes "partial" design.

2301.3 Nominal sizes. For the purposes of this chapter, where dimensions of lumber are specified, they shall be deemed to be nominal dimensions unless specifically designated as actual dimensions (see Section 2304.2).

❖ The use of nominal sizes for lumber is part of the traditional nomenclature for grading and identification of manufactured pieces. "Nominal" simply refers to the short-hand term such as "2 × 4" when the actual piece of lumber has a real dimension of 1 $^1/_2$ inches by 3 $^1/_2$ inches (38 mm by 89 mm)—not 2 inches by 4 inches (51 mm by 102 mm). Section 2304.2, however, prescribes that in determining the required size for design

purposes, computations must be based on the actual size, rather than the nominal size, of the lumber.

SECTION 2302
DEFINITIONS

2302.1 Definitions. The following words and terms shall, for the purposes of this chapter, have the meanings shown herein.

❖ Certain terms are used in the code to describe something very specific to the related subject matter while other terms may have multiple meanings that can vary based on the subject matter and context. To avoid misunderstandings, it is preferable to reach a consensus on the meaning of these terms. Definitions are intended to facilitate the understanding of code provisions and to minimize potential confusion.

This section contains definitions of terms associated with wood construction. Note that these terms are also listed in Chapter 2 with a cross reference to this section. Definitions that pertain primarily to wood construction are included within this chapter to provide convenient access to them without having to refer back to Chapter 2. The use and application of all defined terms, including those defined herein, as well as undefined terms are set forth in Section 201.

ACCREDITATION BODY. An approved, third-party organization that is independent of the grading and inspection agencies, and the lumber mills, and that initially accredits and subsequently monitors, on a continuing basis, the competency and performance of a grading or inspection agency related to carrying out specific tasks.

❖ The process of determining the grade of lumber not only includes the actual method of applying the grade stamp to the product, but also the certification of the grading agency and its methods of quality control, its rules and the work of its agents. For example, an independent third-party quality control program meeting the accreditation requirements of the American Lumber Standards Committee, Inc. (ALSC) or an equivalent process is required for all softwood lumber that is to be graded for use in construction in the United States.

BRACED WALL LINE. A series of braced wall panels in a single story that meets the requirements of Section 2308.3 or 2308.12.4.

❖ A braced wall line consists of conventionally framed shear walls that provide lateral force resistance for buildings constructed in accordance with Section 2308.

BRACED WALL PANEL. A section of wall braced in accordance with Section 2308.9.3 or 2308.12.4.

❖ A braced wall panel is a segment of a braced wall line. Braced wall panels are discussed in Section 2308.9.3.

COLLECTOR. A horizontal diaphragm element parallel and in line with the applied force that collects and transfers diaphragm shear forces to the vertical elements of the lateral-force-resisting system and/or distributes forces within the diaphragm.

❖ The collector is a portion of the diaphragm that transfers the load from the point of application into the adjoining elements, usually the braced wall lines.

CONVENTIONAL LIGHT-FRAME WOOD CONSTRUCTION. A type of construction whose primary structural elements are formed by a system of repetitive wood-framing members. See Section 2308 for conventional light-frame wood construction provisions.

❖ Conventional light-frame wood construction is one of the three methods for designing wood buildings or structures that is recognized in this chapter. Section 2308 provides limits for use of prescriptive provisions and specifications for construction when this method is chosen.

CRIPPLE WALL. A framed stud wall extending from the top of the foundation to the underside of floor framing for the lowest occupied floor level.

❖ Cripple walls are built on the top of footings or foundation walls. They can typically be found along the top of stepped foundation walls where the grade adjoining the structure changes height. Cripple walls must be properly braced to resist lateral forces. They are often treated the same as a first-story wall. Provisions for the bracing of cripple walls are in Section 2308.9.4.

DIAPHRAGM, UNBLOCKED. A diaphragm that has edge nailing at supporting members only. Blocking between supporting structural members at panel edges is not included. Diaphragm panels are field nailed to supporting members.

❖ Unblocked diaphragms can resist low and moderate shear forces. Diaphragm sheathing may be applied with the long dimension of the sheathing either perpendicular or parallel to the main framing members. When the edge of the sheathing is not supported by the main framing member, the diaphragm is considered to be unblocked. Various tables in this chapter provide different values for structural sheathing when it is blocked and when it is not blocked.

DRAG STRUT. See "Collector."

❖ A drag strut is a type of collector for lateral loads when the diaphragm is not continuous. This can be found when there are openings in a floor, or the floor is not all on one level.

FIBERBOARD. A fibrous, homogeneous panel made from lignocellulosic fibers (usually wood or cane) and having a density of less than 31 pounds per cubic foot (pcf) (497 kg/m^3) but more than 10 pcf (160 kg/m^3).

❖ Fiberboard is used primarily as an insulating board and for decorative purposes, but may also be used as wall or roof sheathing. The cellulosic components of fiberboard are broken down to individual fibers and molded to create the bond between the fibers. Other ingredients may be added during processing to provide or improve certain properties such as strength and water resistance, in addition to surface finishes for

decorative products. Fiberboard is used in all locations where panels are desirable, including wall sheathing, insulation of walls and roofs, roof decking, doors and interior finish.

GLUED BUILT-UP MEMBER. A structural element, the section of which is composed of built-up lumber, wood structural panels or wood structural panels in combination with lumber, all parts bonded together with structural adhesives.

❖ Typically, these built-up beam and column sections consist of one or more webs with glued-lumber flanges and stiffeners. They are custom-designed structural elements that require the application of ASD or LRFD standards.

GRADE (LUMBER). The classification of lumber in regard to strength and utility in accordance with American Softwood Lumber Standard DOC PS 20 and the grading rules of an approved lumber rules-writing agency.

❖ The grade identifies how capable a particular piece of lumber is to resist applied loads. The mark is applied to solid sawn pieces of wood and includes the species, grade and whether it was finished (surfaced) green or dry. The species and grade designation on the grade mark, when used in conjunction with the design value in the AF&PA National Design Specifications® Supplement (AF&PA NDS®), provides all the information necessary to determine what load the piece of lumber is capable of holding. The ALSC provides facsimile sheets for agencies accredited to grade lumber. Although rare, fake grade stamps can be found in the marketplace.

HARDBOARD. A fibrous-felted, homogeneous panel made from lignocellulosic fibers consolidated under heat and pressure in a hot press to a density not less than 31 pcf (497 kg/m^3).

❖ Hardboard is used for various interior applications, as well as siding applications. Other ingredients may be added during processing to provide or improve properties such as strength, water resistance and general utility.

NAILING, BOUNDARY. A special nailing pattern required by design at the boundaries of diaphragms.

❖ When designing a diaphragm or a shear wall, the sheathing for those elements must be attached to provide the design resistance to the applied load. Shear capacity tables for wood structural panels specify the nail size and spacing to be used along the edge and in the field of each panel. Boundary nailing is the required nailing pattern for panels located along the edges of diaphragms where the stresses are typically high.

NAILING, EDGE. A special nailing pattern required by design at the edges of each panel within the assembly of a diaphragm or shear wall.

❖ When designing a diaphragm or shear panel, the sheathing for those elements must be attached in a fashion that will provide design resistance to the applied load. Shear capacity tables for wood structural panels used as sheathing specify nail size and spacing. For example, Table 2306.3.1 requires 6 penny (6d) nails to penetrate $1^1/_4$ inches (32 mm) when a $^5/_{16}$-inch (7.9 mm) panel is used on 2-inch-wide (51 mm) framing. To use a shear capacity of 250 pounds per linear feet (plf) (3648 N/m), the nail spacing at boundary conditions where the diaphragm is blocked is recommended to be 4 inches (102 mm) on center (o.c.).

NAILING, FIELD. Nailing required between the sheathing panels and framing members at locations other than boundary nailing and edge nailing.

❖ When designing a diaphragm or shear panel, the sheathing for those elements must be attached in a fashion that will provide design resistance to the applied load. Shear capacity tables for wood structural panels used as sheathing specify nail size and spacing. For example, Note b in Table 2306.3.1 indicates that the maximum spacing for nails along intermediate framing members must be 12 inches (305 mm) o.c.

NATURALLY DURABLE WOOD. The heartwood of the following species with the exception that an occasional piece with corner sapwood is permitted if 90 percent or more of the width of each side on which it occurs is heartwood.

❖ Because of their natural ability to resist deterioration, the harder portions of some species of wood are considered to be naturally durable. The code specifies that "occasional" sapwood is permitted if heartwood constitutes 90 percent of each side.

Decay resistant. Redwood, cedar, black locust and black walnut.

❖ Redwood, cedar, black locust and black walnut lumber are known to resist deterioration due to the action of microbes that enter the wood fibers. The code defines these species of lumber as being decay resistant.

Termite resistant. Redwood and Eastern red cedar.

❖ Redwood and eastern red cedar are considered to be resistant to infestation by termites and are thus listed as naturally durable. The Formosan Termite, however, is capable of destroying all naturally durable species of wood.

NOMINAL SIZE (LUMBER). The commercial size designation of width and depth, in standard sawn lumber and glued-laminated lumber grades; somewhat larger than the standard net size of dressed lumber, in accordance with DOC PS 20 for sawn lumber and with the AF&PA NDS for glued-laminated lumber.

❖ Unless specifically required by the code, all dimensions listed in the code for wood member sizes are nominal, not actual. DOC PS 20-99 specifies the required minimum dimension of lumber for each stated nominal size. The process used to smooth and finish the surface of the lumber and to dry the wood removes a certain thickness of wood on each side of the piece;

therefore, a 2 × 4 is approximately $1^1/_2$ inches by $3^1/_2$ inches (38 mm by 89 mm). The actual dimension of the lumber, otherwise known as dressed size, is limited to ensure consistency in lumber sizes, which is essential to the satisfactory construction of a wood-frame building.

PARTICLEBOARD. A generic term for a panel primarily composed of cellulosic materials (usually wood), generally in the form of discrete pieces or particles, as distinguished from fibers. The cellulosic material is combined with synthetic resin or other suitable bonding system by a process in which the interparticle bond is created by the bonding system under heat and pressure.

❖ Particleboard is one of the family of wood-based products that can be used as wood panels. The definition describes the characteristics unique to particleboard as compared to other types of wood-based composite panels. Particleboard intended for use as wood structural panels is subject to the requirements of Section 2303.1.7. Particleboard intended for use in shear panels is regulated by the requirements of Section 2306.4.3.

PREFABRICATED WOOD I-JOIST. Structural member manufactured using sawn or structural composite lumber flanges and wood structural panel webs bonded together with exterior exposure adhesives, which forms an "I" cross-sectional shape.

❖ Wood I-joists are structural members typically used in floor and roof construction manufactured out of sawn or structural composite lumber flanges and structural panel webs, bonded together with exterior adhesives forming an "I" cross section. This definition explains which engineered wood products are considered prefabricated wood I-joists in order to clarify the requirements in Section 2303.1.2.

PRESERVATIVE-TREATED WOOD. Wood (including plywood) pressure treated with preservatives in accordance with Section 2303.1.8.

❖ Wood that is exposed to high levels of moisture or heat is susceptible to decay both from fungus and other organisms and insect attack. The damage caused by decay or insects can jeopardize the performance of wood members so as to reduce the performance below that required by the code. Section 2304.11 identifies the locations where the use of preservative-treated wood is required. It is important to note that preservative treatment requires a pressure-treatment process. Painting, coating or other surface treatment does not produce preservative-treated wood that will perform as required. The treatment process uses pressure to achieve the depth of penetration of preservative into the wood that is needed to verify that the wood will be resistant to decay and insects over time. Surface treatments may be washed away by rain or ground water, or may chip or peel. The American Wood Preservers'

Association (AWPA) is the consensus standards writing organization for treated wood.

SHEAR WALL. A wall designed to resist lateral forces parallel to the plane of a wall.

❖ Shear walls are the transfer element for the lateral forces that are applied to a floor or roof diaphragm. They function structurally to provide a path for these loads to the foundation. Typically in wood framing these walls are framed with studs and have a structural wood panel attached to the outside using specific nailing patterns to provide the necessary stiffness. They also may have a tie-down attached at both ends of the wall to prevent overturning of the wall.

Shear wall, perforated. A wood structural panel sheathed wall with openings, that has not been specifically designed and detailed for force transfer around openings.

❖ This term is used to describe shear walls designed by using the provisions of Section 2305.3.8.2 as an alternative to segmented shear wall design methodology (see commentary, Section 2305.3.8.2).

Shear wall segment, perforated. A section of shear wall with full-height sheathing that meets the height-to-width ratio limits of Section 2305.3.4.

❖ This term refers to a fully sheathed portion of a perforated shear wall that also meets the limit on height-to-width ratio. Only these portions of the perforated shear wall are considered in determining the percentage of full-height sheathing as described in Section 2305.3.8.2.2, Item 1.

STRUCTURAL COMPOSITE LUMBER. Structural member manufactured using wood elements bonded together with exterior adhesives. Examples of structural composite lumber are:

❖ This definition identifies which engineered wood products are considered structural composite lumber. This facilitates the application of the requirements of Section 2303.1.9.

Laminated veneer lumber (LVL). A composite of wood veneer sheet elements with wood fibers primarily oriented along the length of the member.

❖ See the commentary to "Structural composite lumber" above.

Parallel strand lumber (PSL). A composite of wood strand elements with wood fibers primarily oriented along the length of the member.

❖ See the commentary to "Structural composite lumber" above.

STRUCTURAL GLUED-LAMINATED TIMBER. An engineered, stress-rated product of a timber laminating plant, comprised of assemblies of specially selected and prepared wood laminations in which the grain of all laminations is approximately parallel longitudinally and the laminations are bonded with adhesives.

❖ Structural wood members can be created by laminating smaller pieces of lumber together in a specific way

using glues to create high-performing structural members. Typically, these 2-inch-wide (51 mm) members are arranged in a specific order so that their strength characteristics are used most efficiently. Although these members are made of smaller pieces of lumber, they act together to create sizes that are included in the requirements for heavy timber framing members. Section 2303.1.3 specifies the method of manufacturing these members, and the means of identifying their performance capabilities.

SUBDIAPHRAGM. A portion of a larger wood diaphragm designed to anchor and transfer local forces to primary diaphragm struts and the main diaphragm.

❖ Subdiaphragms can be used by the designer to distribute large diaphragm forces into walls. They are especially common in buildings that incorporate concrete and masonry walls, and are subject to seismic forces.

TIE-DOWN (HOLD-DOWN). A device used to resist uplift of the chords of shear walls.

❖ A shear wall collects lateral forces, which are typically applied at the top of and parallel to the wall. This load will rack the wall and attempt to turn it over. If the strength of the sheathing and framing is correct for the load and appropriately attached, the sheathing will resist the tendency to rack the wall. To prevent the wall from overturning, it must be adequately anchored to the structure below. Anchor bolts are not typically designed to resist these loads. Hold-downs are often necessary at the ends of the shear wall designed to resist high lateral loads due to wind or seismic events. Section 2305.3.7 specifies the conditions where hold-downs are required.

TREATED WOOD. Wood impregnated under pressure with compounds that reduce its susceptibility to flame spread or to deterioration caused by fungi, insects or marine borers.

❖ Wood is treated to reduce its ability to propagate flame or resist damage caused by fungus or insects. Several AWPA standards are prescribed in Section 2303.1.8. These include water-borne as well as several oil-borne standards for treatment, testing and quality control of various wood and wood-based products. The ALSC also accredits third-party inspection agencies, in a manner similar to their grading rules certification, for the quality control of preservative-treated wood products. Facsimiles of accredited agencies' quality marks for treatment are available from ALSC.

WOOD SHEAR PANEL. A wood floor, roof or wall component sheathed to act as a shear wall or diaphragm.

❖ Wood shear panels are horizontal (diaphragms) or vertical (wood shear walls) panels, which transmit forces in the plane of the sheathing caused by lateral loads transmitted from one component of the structure to another. Wood shear panels usually consist of wood framing (studs or joists) either singly or doubly sheathed. The code gives design provisions for wood shear panels that are sheathed in 4-foot by 8-foot (1219 mm by 2438 mm) wood structural panels.

WOOD STRUCTURAL PANEL. A panel manufactured from veneers, wood strands or wafers or a combination of veneer and wood strands or wafers bonded together with waterproof synthetic resins or other suitable bonding systems. Examples of wood structural panels are:

❖ The code defines wood structural panels in terms of the materials commonly used in their manufacture. Wood structural panels intended for structural use must comply with DOC PS1 or PS 2 in accordance with Section 2303.1.4. Both of these documents specify the required structural performance characteristics of wood structural panels, and detail the requirements for the third-party label each panel is to bear.

Composite panels. A wood structural panel that is comprised of wood veneer and reconstituted wood-based material and bonded together with waterproof adhesive;

❖ Composite panels are structural panels that are made up of a combination of wood veneers and wood-based materials. The wood veneer usually forms the two outer layers and may also be used in the core of the panel. These layers of veneer and wood flakes or strands, called "furnish," may be cross laminated.

Oriented strand board (OSB). A mat-formed wood structural panel comprised of thin rectangular wood strands arranged in cross-aligned layers with surface layers normally arranged in the long panel direction and bonded with waterproof adhesive; or

❖ Oriented strand board (OSB) is one of three types of wood structural panels. OSB panels are fabricated out of multiple layers made up of wood flakes or strands called "furnish." Like plywood, these layers of furnish are oriented at 90 degrees (1.57 rad) to each other. This gives OSB panels properties very similar to those of plywood.

Plywood. A wood structural panel comprised of plies of wood veneer arranged in cross-aligned layers. The plies are bonded with waterproof adhesive that cures on application of heat and pressure.

❖ Plywood is one of three types of wood structural panels. It is manufactured by gluing together three or more cross-laminated layers of wood. The code does not differentiate between OSB and plywood, since manufacturing is done according to either the DOC PS 1 standard for industrial plywood or the DOC PS 2 standard for OSB and performance-rated plywood.

SECTION 2303
MINIMUM STANDARDS AND QUALITY

2303.1 General. Structural sawn lumber; end-jointed lumber; prefabricated wood I-joists; structural glued-laminated timber; wood structural panels, fiberboard sheathing (when used structurally); hardboard siding (when used structurally); particleboard; preservative-treated wood; structural log members; structural composite lumber; round timber poles and piles; fire-retardant-treated wood; hardwood plywood; wood trusses; joist hangers; nails; and staples shall conform to the applicable provisions of this section.

❖ When designed and built in accordance with the standards listed in Section 2303, a building or structure is deemed to comply with the code. These standards contain most of the information needed to adequately design a structure. It is necessary for the designer to have a working knowledge of engineering principles and experience with construction to properly interpret the recommendations and meet the provisions of other applicable sections of the code.

Section 2303.1 lists the various materials that have production and quality control standards. The use of these standards is fundamental for manufacturers in producing products and maintaining quality control procedures. Designers, owners and building officials must understand these standards and the methods prescribed in them to be able to identify products that have been produced in accordance with their criteria. Without the knowledge that the products meet the standard, there is little assurance that a safe and efficient building or structure will be constructed.

2303.1.1 Sawn lumber. Sawn lumber used for load-supporting purposes, including end-jointed or edge-glued lumber, machine stress-rated or machine-evaluated lumber, shall be identified by the grade mark of a lumber grading or inspection agency that has been approved by an accreditation body that complies with DOC PS 20 or equivalent. Grading practices and identification shall comply with rules published by an agency approved in accordance with the procedures of DOC PS 20 or equivalent procedures. In lieu of a grade mark on the material, a certificate of inspection as to species and grade issued by a lumber grading or inspection agency meeting the requirements of this section is permitted to be accepted for precut, remanufactured or rough-sawn lumber and for sizes larger than 3 inches (76 mm) nominal thickness.

Approved end-jointed lumber is permitted to be used interchangeably with solid-sawn members of the same species and grade.

❖ All lumber used to support loads in a building or structure is required to be properly identified. Every species and grade of lumber has a unique inherent strength value. These values are further modified in sawn timber by the presence of growth characteristics that vary from piece to piece, such as knots, slope of grain, checks, etc. Without adequate identification, it would be impossible to verify the material used in the field. The required grade mark must identify the species or species grouping, grade and moisture content at the time of surfacing, the grading agency and the mill name or grader's number. Figure 2303.1.1 illustrates typical grade mark labels.

Certification is an acceptable alternative to a grade mark from both United States and Canadian grading agencies. Grading agencies are certified by the ALSC.

Design values are published by lumber grade rules-writing agencies for both individual and grouped species. A grouped species is lumber that is cut and marketed in lots containing two or more species, such as Spruce-Pine-Fir. These species grow together in large areas. It is more economical to market the lumber as a species group than attempt segregation. The assigned strength values include those applicable to the weaker species in the group.

The code also allows certain types of structural lumber to have a certificate of inspection instead of a grade mark. A certificate of inspection is acceptable for precut, remanufactured or rough-sawn lumber and for sizes larger than 3 inches (76 mm) nominal in thickness. It is industry practice to place only one label (grade mark) on a piece of lumber, which may be removed on precut and remanufactured lumber. Each piece of lumber is graded after it has been cut to a standard size. The grade of the piece is determined based on its size, number and location of strength-reducing characteristics. Therefore, one log may produce lumber of two or more different grades.

It is also industry practice not to label lumber having a nominal thickness larger than 3 inches (76 mm), or rough-sawn material where the label may be illegible. A certificate of inspection from an approved agency is acceptable instead of the label for these types of lumber. The certificate should be filed with the permanent records of the building or structure.

If defects exceeding those permitted for the grade allegedly installed are visible, then a grader would be able to determine that the wood is definitely not of a suitable grade. To determine if the wood in question is definitely of a suitable grade, the grader must inspect all four faces of the piece. This cannot happen once the lumber is installed in the building, as other components of the building will be covering up some of the faces of the pieces.

End-joined or edge-glued lumber is acceptable when identified by an appropriate grade mark. The

NDS permits the use of such lumber in Section 4.1.6 of the standard.

2303.1.2 Prefabricated wood I-joists. Structural capacities and design provisions for prefabricated wood I-joists shall be established and monitored in accordance with ASTM D 5055.

❖ This section specifies that the shear, moment and stiffness capacities of prefabricated wood I-joists be established and monitored by ASTM D 5055. This standard also specifies that application details such as bearing length and web openings are to be considered in determining structural capacity. Wood I-joists are structural members typically used in floor and roof construction manufactured out of sawn or structural composite lumber flanges and structural panel webs, bonded together with exterior adhesives forming an "I" cross section (see definition of "Prefabricated wood I-joist" in Section 2302.1). The standard requires I-joist manufacturers to employ an independent inspection agency to monitor the procedures for quality assurance. Finally, the standard specifies that proper installation instructions accompany the product to the job site. The instructions are required to include weather protection, handling requirements and, where required, web reinforcement, connection details, lateral support, bearing details, web hole-cutting limitations and any special situation.

2303.1.3 Structural glued-laminated timber. Glued-laminated timbers shall be manufactured and identified as required in AITC A190.1 and ASTM D 3737.

❖ Glued-laminated timbers are required by this section to be manufactured following ANSI/AITC 90.1 and ASTM D 3737. Knowing the standard these products must meet makes it easier to determine that the product found in the field will meet the design requirements.

2303.1.4 Wood structural panels. Wood structural panels, when used structurally (including those used for siding, roof and wall sheathing, subflooring, diaphragms and built-up members), shall conform to the requirements for their type in DOC PS 1 or PS 2. Each panel or member shall be identified for grade and glue type by the trademarks of an approved testing and grading agency. Wood structural panel components shall be designed and fabricated in accordance with the applicable standards listed in Section 2306.1 and identified by the trademarks of an approved testing and inspection agency indicating conformance with the applicable standard. In addition, wood structural panels when permanently exposed in outdoor applications shall be of exterior type, except that wood structural panel roof sheathing exposed to the outdoors on the underside is permitted to be interior type bonded with exterior glue, Exposure 1.

❖ "Wood structural panels" is a collective term referring to plywood, OSB and other composite panels of wood-based materials (see definition of "Wood structural panel" in Section 2302). Identification of grade for plywood includes N, A, B, C Plugged, C and D (from no knots or patches to large knots and knotholes).

Plywood is manufactured with an odd number of lay-

Figure 2303.1.1
TYPICAL LABELS
(Grade Marks)
American Lumber Standards Committee, U.S.D.C.

ers, which have their grain direction placed perpendicular to each other and are bonded together with a strong adhesive. Each individual layer may consist of an assembly of several different plies laminated together with their grains running in the same direction. This alternation of grain direction in successive layers gives a plywood panel dimensional stability across its width. Stamped on top of typically all panels is a span rating that indicates the maximum roof and floor joist spacings that can be accommodated. Additionally, an exposure rating is assigned as Exterior (designed for applications subject to permanent exposure to the weather or moisture), Exposure 1 (designed for applications where long construction delays may be expected prior to providing protection against moisture or weather extremes) or Exposure 2 (intended solely for protected construction applications where only moderate delays in providing protection from moisture are expected).

APA Rated Sturd-I-Floor panels are commonly used for thicker sheathing applications where added stiffness is desired. This type of panel has a single span rating, indicating that it is specifically engineered for floors, as the name would also suggest. Edges may be blocked or may be ordered as tongue-and-groove, which provides added stability in an unblocked diaphragm. If a diaphragm can handle the applied loading without the addition of blocks around all panel edges, such a specification will lower cost and leave fewer avenues for framing mistakes.

2303.1.5 Fiberboard. Fiberboard for its various uses shall conform to ASTM C 208. Fiberboard sheathing, when used structurally, shall be identified by an approved agency as conforming to ASTM C 208.

❖ All fiberboard must meet the requirements of ASTM C 208, as well as being verminproof, resistant to rot-producing fungi and water repellent. These standards give physical requirements for construction grades of fiberboard, including sheathing grade and roof-insulating grade. The sheathing grade of fiberboard is further broken down into regular and intermediate densities.

2303.1.5.1 Jointing. To ensure tight-fitting assemblies, edges shall be manufactured with square, shiplapped, beveled, tongue-and-groove or U-shaped joints.

❖ Tight-fitting joints in the fiberboard are required for all applications, including insulation, siding and wall sheathing.

2303.1.5.2 Roof insulation. Where used as roof insulation in all types of construction, fiberboard shall be protected with an approved roof covering.

❖ Fiberboard is not intended for prolonged exposure to sunlight, wind, rain or snow. Where fiberboard is used as roof insulation, it must be protected with an approved roof covering to prevent water saturation and subsequent delamination and to avoid decay and destruction of the glue bond by moisture.

2303.1.5.3 Wall insulation. Where installed and fireblocked to comply with Chapter 7, fiberboards are permitted as wall insulation in all types of construction. In fire walls and fire barriers, unless treated to comply with Section 803.1 for Class A materials, the boards shall be cemented directly to the concrete, masonry or other noncombustible base and shall be protected with an approved noncombustible veneer anchored to the base without intervening airspaces.

❖ Fiberboard is permitted without any fire-resistance treatment in the walls of all types of construction (see Section 603.1). When used in fire walls and fire barrier walls, fiberboard must be either treated to comply with Class A flame spread or adhered directly to a noncombustible base and protected by a tight-fitting, noncombustible veneer that is fastened through the fiberboard to the base. This will prevent the fiberboard from contributing to the spread of fire.

2303.1.5.3.1 Protection. Fiberboard wall insulation applied on the exterior of foundation walls shall be protected below ground level with a bituminous coating.

❖ Fiberboard insulation applied to the exterior side of foundation walls is required to be protected from the weather to improve its service life and maintain its performance characteristics. Of particular concern is foundation insulation that is in close proximity to grade and runs the risk of being damaged by a lawn mower, rocks or soil that is kicked up against it, water from a garden hose or rainwater splash back, etc. Protection is required for all fiberboard insulation on the exterior face of foundation walls.

2303.1.6 Hardboard. Hardboard siding used structurally shall be identified by an approved agency conforming to AHA A135.6. Hardboard underlayment shall meet the strength requirements of $^7/_{32}$-inch (5.6 mm) or $^1/_4$-inch (6.4 mm) service class hardboard planed or sanded on one side to a uniform thickness of not less than 0.200 inch (5.1 mm). Prefinished hardboard paneling shall meet the requirements of AHA A135.5. Other basic hardboard products shall meet the requirements of AHA A135.4. Hardboard products shall be installed in accordance with manufacturer's recommendations.

❖ Hardboard siding that is to be used structurally must be manufactured in accordance with AHA A135.6 and marked to indicate conformance with the standard, whether primed or unprimed, and to identify the producer and the type, either lap or panel. Hardboard products are produced primarily from inter-felted ligno-cellulosic fibers. There are five classes based on strength values. Underlayments are limited to $^7/_{32}$-inch (5.6 mm) or $^1/_4$-inch (6.4 mm) service class.

Prefinished hardboard is required to be manufactured to the AHA A135.5 standard, and must be marked to indicate the standard and to identify the producer, flame spread index, finish class, type of gloss and type of substrate, or must be accompanied by written certification of the same information.

2303.1.7 Particleboard. Particleboard shall conform to ANSI A208.1. Particleboard shall be identified by the grade mark or certificate of inspection issued by an approved agency. Particleboard shall not be utilized for applications other than indicated in this section unless the particleboard complies with the provisions of Section 2306.4.3.

❖ Sponsored by the Composite Panel Association (CPA), ANSI A208.1 is the basic specification for the manufacture of particleboard, which establishes a system of grade marks for the boards' grade, density and strength.

Particleboard used in construction is medium density and is first designated by an "M." The second digit or letter in the designation is related to grade. The designations range from 1 to 3, with higher designations being the strongest. Grade M-S refers to medium density, "special grade" particleboard. This grade was added to ANSI A208.1 after the M-1, M-2 and M-3 grades had been established. Grade M-S falls between M-1 and M-2 in physical properties.

An optional third part of the grade designation indicates that the particleboard has a special characteristic. The grades of particleboard specified in Table 2306.4.3, M-S and M-2 "Exterior Glue," are manufactured with exterior glue to increase their durability characteristics.

While ANSI A208.1 has provisions for Grade M-3 particleboard, panels that meet the requirements of this higher grade are more commonly evaluated and used as wood structural panels, in accordance with DOC PS-2. Therefore, Grade M-3 material is not addressed in Table 2306.4.3.

2303.1.7.1 Floor underlayment. Particleboard floor underlayment shall conform to Type PBU of ANSI A208.1. Type PBU underlayment shall not be less than $^1/_4$-inch (6.4 mm) thick and shall be installed in accordance with the instructions of the Composite Panel Association.

❖ Although similar to medium density, Grade 1 particleboard—particleboard intended for use as floor underlayment—is designated "PBU" and has stricter limits on levels of formaldehyde emission permitted than those placed on Grade "M" particleboard. Particleboard intended for use as floor underlayment is not commonly manufactured with exterior glue, which could emit higher levels of formaldehyde than that permitted by ANSI A208.1 for Grade "PBU" floor underlayment.

Particleboard underlayment is often applied over a structural subfloor to provide a smooth surface for resilient-finish or textile floor coverings. The minimum $^1/_4$-inch (6.4 mm) thickness is applicable over panel-type subflooring. Particleboard underlayment installed over board or deck subflooring that has multiple joints should have a thickness of $^3/_8$ inch (9.5 mm). Joints in the underlayment should not be over joints in the subflooring.

All particleboard underlayment with thicknesses of $^1/_4$ through $^3/_4$ inch (6.4 through 19.1 mm) should be attached with a minimum of 6d annular threaded nails spaced 6 inches (152 mm) o.c. on the edges and 10 inches (254 mm) o.c. for intermediate supports.

2303.1.8 Preservative-treated wood. Lumber, timber, plywood, piles and poles supporting permanent structures required by Section 2304.11 to be preservative treated shall conform to the requirements of the applicable AWPA Standard U1 and M4 for the species, product, preservative and end use. Preservatives shall be listed in Section 4 of AWPA U1. Lumber and plywood used in wood foundation systems shall conform to Chapter 18.

❖ Wood is able to absorb chemicals because of its cellular characteristics. Preservative treatment procedures that will repel termites and destroy decay-causing fungus utilize this capability. The process includes placing wood in a large cylinder and applying a vacuum to remove as much air as possible from the wood. The chemical solution is then introduced and pressure is applied to force the solution into the wood. The pressure is maintained until the desired absorption is obtained. A final vacuum is often applied to remove as much excess preservative as possible.

There are several variations of this basic process, all of which must provide the required retention of preservatives. AWPA U1, Section 4 prescribes the requirements for the preservatives that are used, while AWPA U1 specifies the minimum results of the treatment process based on a commodity specification and use category. These requirements include the depth of penetration of the chemical into the wood, and the amount of chemical retained by the wood once the process is completed. Additional requirements for wood used in foundation systems are given in Section 1805.4.6.

2303.1.8.1 Identification. Wood required by Section 2304.11 to be preservative treated shall bear the quality mark of an inspection agency that maintains continuing supervision, testing and inspection over the quality of the preservative-treated wood. Inspection agencies for preservative-treated wood shall be listed by an accreditation body that complies with the requirements of the American Lumber Standards Treated Wood Program, or equivalent. The quality mark shall be on a stamp or label affixed to the preservative-treated wood, and shall include the following information:

1. Identification of treating manufacturer.

2. Type of preservative used.

3. Minimum preservative retention (pcf).

4. End use for which the product is treated.

5. AWPA standard to which the product was treated.

6. Identity of the accredited inspection agency.

❖ Quality marks are necessary to determine that preservative-treated wood conforms to applicable standards. The identifying mark must be by an approved inspection agency that has continuous follow-up services. Additionally, the inspection agency must be listed and certified as being competent by an approved organization. The ALSC provides certification of treating agencies. Facsimiles of its quality marks are available by contacting the ALSC. The required quality mark is not a substitute for a grade mark. When wood or wood-based materials are being used structurally, both the quality mark and grade mark must be displayed on the piece.

2303.1.8.2 Moisture content. Where preservative-treated wood is used in enclosed locations where drying in service cannot readily occur, such wood shall be at a moisture content of 19 percent or less before being covered with insulation, interior wall finish, floor covering or other materials.

❖ Water-borne preservatives are subject to leaching unless properly dried and protected. The requirement for reducing the moisture content to 19 percent or less is intended to aid in preventing such leaching. Also, all structural members are presumed to have a moisture content of 19 percent or less.

   Preservative treatment does not require any adjustment of design values. Some species have a high surface tension, which means that it is difficult to penetrate the surface of the lumber. These species are often incised to break the surface tension and allow the chemicals to penetrate the member. Incising requires a reduction in the design values for the member.

2303.1.9 Structural composite lumber. Structural capacities for structural composite lumber shall be established and monitored in accordance with ASTM D 5456.

❖ The purpose of this section is to specify the appropriate standard for structural composite lumber. Included within the standard are criteria for laminated veneer lumber and parallel strand lumber. The ASTM standard includes requirements for testing, criteria for determining allowable stresses, requirements for independent inspection and quality assurance procedures (also see definition in Section 2302).

2303.1.10 Structural log members. Stress grading of structural log members of nonrectangular shape, as typically used in log buildings, shall be in accordance with ASTM D 3957. Such structural log members shall be identified by the grade mark of an approved lumber grading or inspection agency. In lieu of a grade mark on the material, a certificate of inspection as to species and grade issued by a lumber grading or inspection agency meeting the requirements of this section shall be permitted.

❖ This section address grading requirements for logs used as structural members by reference the ASTM D 3957 methods for establishing structural capacities of logs. It also specifies the requirement for a grading stamp or an alternative certification on structural logs.

2303.1.11 Round timber poles and piles. Round timber poles and piles shall comply with ASTM D 3200 and ASTM D 25, respectively

❖ This section provides references to the appropriate material standards for round timber poles and piles.

2303.2 Fire-retardant-treated wood. Fire-retardant-treated wood is any wood product which, when impregnated with chemicals by a pressure process or other means during manufacture, shall have, when tested in accordance with ASTM E 84, a listed flame spread index of 25 or less and show no evidence of significant progressive combustion when the test is continued for an additional 20-minute period. In addition, the flame front shall not progress more than 10.5 feet (3200 mm) beyond the centerline of the burners at any time during the test.

❖ Fire-retardant-treated wood is plywood and lumber that has been pressure impregnated with chemicals to improve its flame spread characteristics beyond that of untreated wood. The effectiveness of the pressure-impregnated fire-retardant treatment is determined by subjecting the material to tests conducted in accordance with ASTM E 84, with the modification that the test is extended to 30 minutes rather than 15 minutes. Using this procedure, a flame spread index is established during the standard 10-minute test period. The test is continued for an additional 20 minutes. During this added time period, there must not be any significant flame spread. At no time must the flame spread more than 10 $^{1}/_{2}$ feet (3200 mm) past the centerline of the burners.

   The result of impregnating wood with fire-retardant chemicals is a chemical reaction at certain temperature ranges. This reaction reduces the release of certain intermediate products that contribute to the flaming of wood, and also results in the formation of a greater percentage of charcoal and water. Some chemicals are also effective in reducing the oxidation rate for charcoal residue. Fire-retardant chemicals also reduce the heat release rate of fire-retardant-treated wood when burning over a wide range of temperatures. This section gives provisions for the treatment and use of fire-retardant-treated wood.

2303.2.1 Labeling. Fire-retardant-treated lumber and wood structural panels shall be labeled. The label shall contain the following items:

1. The identification mark of an approved agency in accordance with Section 1703.5.

2. Identification of the treating manufacturer.

3. The name of the fire-retardant treatment.

4. The species of wood treated.

5. Flame spread and smoke-developed index.

6. Method of drying after treatment.

7. Conformance with appropriate standards in accordance with Sections 2303.2.2 through 2303.2.5.

8. For fire-retardant-treated wood exposed to weather, damp or wet locations, include the words "No increase in the listed classification when subjected to the Standard Rain Test" (ASTM D 2898).

❖ For continued quality, each piece of fire-retardant-treated wood must be identified by an approved agency having a reinspection service. The identification must show the performance rating of the material, including the 30-minute ASTM E 84 test results determined in Section 2303.2, and design adjustment values determined in Section 2303.2.2. The third-party agency that provides the fire-retardant-treated wood label is also required to state on the label that the fire-retardant-treated wood complies with the requirements of Section 2303.2, and that design adjustment values have been determined for the fire-retardant-treated wood in compliance with the provisions of Section 2303.2.2.

The fire-retardant-treated wood label must be distinct from the grading label to avoid confusion between the two. The grading label provides information about the properties of wood before it is fire-retardant treated. The fire-retardant-treated wood label provides properties of the wood after fire-retardant-treated wood treatment. It is imperative that the fire-retardant-treated wood label be presented in such a manner that it complements the grading label, and does not create confusion over which label takes precedence.

2303.2.2 Strength adjustments. Design values for untreated lumber and wood structural panels, as specified in Section 2303.1, shall be adjusted for fire-retardant-treated wood. Adjustments to design values shall be based on an approved method of investigation that takes into consideration the effects of the anticipated temperature and humidity to which the fire-retardant-treated wood will be subjected, the type of treatment and redrying procedures.

❖ Experience has shown that certain factors can affect the physical properties of fire-retardant-treated wood. Among these factors are the pressure treatment and redrying processes used, and the extremes of temperature and humidity that the fire-retardant-treated wood will be subjected to once installed. The design values for all fire-retardant-treated wood must be adjusted for the effects of the treatment and environmental conditions, such as high temperature and humidity in attic installations. This section requires the determination of these design adjustment values, based on an investigation procedure that includes subjecting the fire-retardant-treated wood to similar temperatures and humidities, and has been approved by the building official. The fire-retardant-treated wood tested must be identical to that which is produced. Items to be considered by the building official reviewing the test procedure include species and grade of the untreated wood, and conditioning of wood, such as drying before the fire-retardant treatment process. A fire-retardant wood treater may choose to have its treatment process evaluated by model code evaluation services.

The fire-retardant-treated wood is required by Section 2303.2.1 to be labeled with the design adjustment values. These can take the form of factors that are multiplied by the original design values of the untreated wood to determine its allowable stresses, or new allowable stresses that have already been factored down in consideration of the fire-retardant treatment.

2303.2.2.1 Wood structural panels. The effect of treatment and the method of redrying after treatment, and exposure to high temperatures and high humidities on the flexure properties of fire-retardant-treated softwood plywood shall be determined in accordance with ASTM D 5516. The test data developed by ASTM D 5516 shall be used to develop adjustment factors, maximum loads and spans, or both, for untreated plywood design values in accordance with ASTM D 6305. Each manufacturer shall publish the allowable maximum loads and spans for service as floor and roof sheathing for its treatment.

❖ This section references the test standard developed to evaluate the flexural properties of fire-retardant-treated plywood that is exposed to high temperatures. Note that while the section title refers to wood structural panels, the referenced standard is limited to softwood plywood. Therefore, judgement is required in determining the effects of elevated temperature and humidity on other types of wood structural panels.

2303.2.2.2 Lumber. For each species of wood that is treated, the effects of the treatment, the method of redrying after treatment and exposure to high temperatures and high humidities on the allowable design properties of fire-retardant-treated lumber shall be determined in accordance with ASTM D 5664. The test data developed by ASTM D 5664 shall be used to develop modification factors for use at or near room temperature and at elevated temperatures and humidity in accordance with ASTM D 6841. Each manufacturer shall publish the modification factors for service at temperatures of not less than 80°F (27°C) and for roof framing. The roof framing modification factors shall take into consideration the climatological location.

❖ This section references the test standard developed to determine the necessary adjustments to design values for lumber that has been fire-retardant treated and includes the effects of elevated temperatures.

2303.2.3 Exposure to weather, damp or wet locations. Where fire-retardant-treated wood is exposed to weather, or damp or wet locations, it shall be identified as "Exterior" to indicate there is no increase in the listed flame spread index as defined in Section 2303.2 when subjected to ASTM D 2898.

❖ Some fire-retardant treatments are soluble when exposed to the weather or used under high-humidity conditions. Note that the humidity threshold established for interior applications in Section 2303.2.4 is 92 percent. Therefore, fire-retardant-treated wood used in an interior location that will exceed this threshold should comply with this section. When a fire-retardant-treated wood product is to be exposed to any of the conditions noted in this section, it must be further tested in accordance with ASTM D 2898. Testing requires the material to meet the performance criteria listed in Section 2303.2. The material is then subjected to the ASTM weathering test and retested after drying. There must not be any significant differences in the performance recorded before and after the weathering test.

2303.2.4 Interior applications. Interior fire-retardant-treated wood shall have moisture content of not over 28 percent when tested in accordance with ASTM D 3201 procedures at 92-percent relative humidity. Interior fire-retardant-treated wood shall be tested in accordance with Section 2303.2.2.1 or 2303.2.2.2. Interior fire-retardant-treated wood designated as Type A shall be tested in accordance with the provisions of this section.

❖ The environment in which the fire-retardant-treated wood is used can affect its performance. In order to make sure that the performance will be adequate in a humid interior condition, testing in accordance with ASTM D 3201 as well as that specified in Section 2303.2.2.1 or 2303.2.2.2 is required for all interior wood. Requiring all interior wood to be tested for the effects of high temperature and humidity reduces the chance of premature failure due to improper use of fire-retardant-treated wood.

2303.2.5 Moisture content. Fire-retardant-treated wood shall be dried to a moisture content of 19 percent or less for lumber and 15 percent or less for wood structural panels before use. For wood kiln dried after treatment (KDAT), the kiln temperatures shall not exceed those used in kiln drying the lumber and plywood submitted for the tests described in Section 2303.2.2.1 for plywood and 2303.2.2.2 for lumber.

❖ These moisture content thresholds are necessary to prevent leaching of the fire retardant from the wood. Section 2303.2.2 requires that the strength adjustments consider the redrying procedure, and this section clarifies that the drying temperatures for wood kiln dried after treatment must be consistent with those on which the adjustment factors were based.

2303.2.6 Type I and II construction applications. See Section 603.1 for limitations on the use of fire-retardant-treated wood in buildings of Type I or II construction.

❖ Although Type I and II construction typically would only allow noncombustible materials, Section 603.1 allows various parts of a building to be combustible. Specifically, there are 22 items listed that are exceptions to the strict limits on Types I and II construction. There are three conditions listed in that section that will allow fire-retardant-treated wood to be used: in nonbearing partitions where their fire resistance is 2 hours or less; in nonbearing exteriors that do not require a fire-resistance rating and in roof construction of buildings not over two stories high.

2303.3 Hardwood and plywood. Hardwood and decorative plywood shall be manufactured and identified as required in HPVA HP-1.

❖ Hardwood plywood, like any other product that undergoes a change in the process of being configured for use in construction, has standard criteria for the manufacturing process and the quality of the materials that are to be used. ANSI/HPVA HP-1 is the standard for hardwood plywood, and is one of the family of standards for this type of product.

2303.4 Trusses.

❖ Trusses are engineered components. They can also be incorporated into a building of otherwise conven-

tional light-frame construction in accordance with Section 2308.4. All types of trusses are included within these provisions, including those with parallel, pitched or curved chords; bowstring, sawtooth or camelback profiles; heavy-bolted timber or light-gage metal plate connected elements. Other trusses that are somewhat common in the industry include those with wood chords and metal tubular webs connected with bolts or pins.

2303.4.1 Design. Wood trusses shall be designed in accordance with the provisions of this code and accepted engineering practice. Members are permitted to be joined by nails, glue, bolts, timber connectors, metal connector plates or other approved framing devices.

❖ The complete design of trusses can be a complex process. The designer must keep the entire system in mind, as trusses do not typically exist as single members, but rather work together to form a complete structural system. Truss designs should address connections such as the truss-to-wall plate since they are not covered in the fastening schedule provided by Table 2304.9.1. A variety of references are available to assist in the design of wood trusses, including the *Timber Construction Manual* and *Structural Design in Wood*.

Heavy timber trusses may be typically provided in new construction for appearance purposes, whereas trusses that will be concealed within the ceiling/attic space are typically built with smaller dimension lumber joined together with light-gage metal plates (see Section 2303.4.2).

2303.4.1.1 Truss designer. The individual or organization responsible for the design of trusses.

❖ Truss design is a specialty and the designer should be thoroughly familiar with member forces, connection capabilities and behavior as well as what is necessary to cause the entire truss system to function in a structurally-sound fashion. This section defines a term unique to this section, truss designer. Engineers and contractors often depend on the fabricator to design the truss. Sometimes special connections for trusses to the supporting structure may be required, such as those necessary to resist high uplift forces, and the truss designer must coordinate these with the building designer. The truss designer is familiar with the manner in which those connectors can be applied to the truss (i.e., to engage the whole section, attach only to the chord or attach only to the web).

2303.4.1.2 Truss design drawings. The written, graphic and pictorial depiction of each individual truss shall be provided to the building official and approved prior to installation. Truss design drawings shall also be provided with the shipment of trusses delivered to the job site. Truss design drawings shall include, at a minimum, the information specified below:

1. Slope or depth, span and spacing;

2. Location of joints;

3. Required bearing widths;

4. Design loads as applicable;

5. Top chord live load (including snow loads);

6. Top chord dead load;

7. Bottom chord live load;

8. Bottom chord dead load;

9. Concentrated loads and their points of application as applicable;

10. Controlling wind and earthquake loads as applicable;

11. Adjustments to lumber and metal connector plate design value for conditions of use;

12. Each reaction force and direction;

13. Metal connector plate type, size, thickness or gage, and the dimensioned location of each metal connector plate except where symmetrically located relative to the joint interface;

14. Lumber size, species and grade for each member;

15. Connection requirements for:

    15.1. Truss to truss;

    15.2. Truss ply to ply; and

    15.3. Field splices.

16. Calculated deflection ratio and maximum vertical and horizontal deflection for live and total load as applicable;

17. Maximum axial tension and compression forces in the truss members; and

18. Required permanent individual truss member bracing and method per Section 2303.4.1.5, unless a specific truss member permanent bracing plan for the roof or floor structural system is provided by a registered design professional.

Where required by one of the following, each individual truss design drawing shall bear the seal and signature of the truss designer:

1. Registered design professional; or

2. Building official; or

3. Statutes of the jurisdiction in which the project is to be constructed.

> **Exceptions:**
>
> 1. When a cover sheet/truss index sheet combined into a single cover sheet is attached to the set of truss design drawings for the project, the single sheet/truss index sheet is the only document that needs to be signed and sealed within the truss submittal package.
>
> 2. When a cover sheet and a truss index sheet are separately provided and attached to the set of

truss design drawings for the project, both the cover sheet and the truss index sheet are the only documents that need to be signed and sealed within the truss submittal package.

❖ The intent of this section is to minimize the confusion that exists when the term "shop drawings" and "construction documents" are used as they relate to trusses. The code requires this information to document what the truss is supposed to do and how the trusses are intended to function together as a complete structural system. Engineers and contractors frequently depend on the fabricator to design the truss. Sometimes special connections for trusses to the building below may be required, such as those necessary to resist high uplift forces, and the truss designer must coordinate these with the building designer. The truss engineer knows and understands how connectors can be applied to resist imposed forces and that knowledge is critical to providing an adequate design to transfer the forces to the building below. The building engineer can then use that information to properly complete the connection and provide a continuous load path to the foundation.

The majority of the items listed are readily available from any designer of trusses. The code specifies the information needed based on the difficulty in determining the appropriateness of the trusses in the field. Since documentation is included with the submitted plans, the inspector should be able to readily verify the field installation meets the intent of the truss engineer.

The requirements for signing and sealing each truss design drawing as well as allowing for a cover or index sheet to be used in lieu of signing and sealing each individual truss design drawing is the approach that is often used in current construction practice.

2303.4.1.3 Truss placement diagram. The truss manufacturer shall provide a truss placement diagram that identifies the proposed location for each individually designated truss and references the corresponding truss design drawing. The truss placement diagram shall be provided as part of the truss submittal package, and with the shipment of trusses delivered to the job site. Truss placement diagrams shall not be required to bear the seal or signature of the truss designer.

> **Exception:** When the truss placement diagram is prepared under the direct supervision of a registered design professional, it is required to be signed and sealed.

❖ This section constitutes a definition of the term "truss placement diagram." It is intended to minimize the confusion that exists in the construction industry between a variety of terms that are often used interchangeably, such as "installation documents," "construction documents," "shop drawings," etc. The term "truss placement diagram" is preferred by the truss industry and is very specific.

2303.4.1.4 Truss submittal package. The truss submittal package shall consist of each individual truss design drawing, the truss placement diagram for the project, the truss member permanent bracing specification and, as applicable, the cover sheet/truss index sheet.

❖ This section lists the minimum information that must be provided as a submittal by the truss manufacturer.

2303.4.1.5 Truss member permanent bracing. Where permanent bracing of truss members is required on the truss design drawings, it shall be accomplished by one of the following methods:

1. The trusses shall be designed so that the buckling of any individual truss member can be resisted internally by the structure (e.g. buckling member T-bracing, L-bracing, etc.) of the individual truss. The truss individual member buckling reinforcement shall be installed as shown on the truss design drawing or on supplemental truss member buckling reinforcement diagrams provided by the truss designer.

2. Permanent bracing shall be installed using standard industry bracing details that conform with generally accepted engineering practice. Individual truss member continuous lateral bracing location(s) shall be shown on the truss design drawing.

❖ There are two different types of lateral bracing for trusses and their systems: temporary and permanent. Temporary lateral bracing is typically required for safe installation before the sheathing is applied and nailed to the top and bottom chords. All forms of lateral bracing are critical and are well-described within common truss publications. Permanent lateral bracing will remain in place for the life of the structure and is intended to stabilize entire trusses so that they can function as the truss designer intends. Item 18 of Section 2303.4.1.2 requires that permanent individual truss member bracing be provided on the truss design drawing. This section emphasizes the need for permanent truss bracing to be installed and lists the options for permanent truss member bracing.

The truss designer provides engineering for the entire truss system, which must include complete the design of individual trusses, girder trusses, truss-to-truss connections and temporary and permanent lateral bracing. In some cases, permanent lateral bracing applied to the webs or the bottom chord may require a connection to the building, such as at a balloon-framed wall, which would then require the involvement of the building designer. In these special cases, the truss and building engineers must work together to provide an acceptable detail that will fulfill the truss bracing requirements in such a way that it is compatible with both the truss system and the building system.

Another type of permanent bracing commonly recognized is the plywood sheathing applied to the top chords or the gypsum wallboard that may be applied to the bottom chords (provided the stiffness of the gypsum wallboard is adequate enough to meet the needs of the truss designer). Since the truss designer is thoroughly familiar with the lateral bracing forces and capabilities of the system that he or she has designed,

that engineer is typically the one who must specify whether the sheathing can be used for permanent lateral bracing. This is not a difficult step, since that truss designer is given a set of drawings on which to base his or her design, which would show any sheathing that is intended by the building designer for other purposes. If there are any questions, the truss designer should contact the building designer to discuss the lateral bracing system that the truss designer is planning to design into his or her system.

Publication BCSI 1-03, *Guide to Good Practice for Handling, Installing and Bracing of Metal Plate Connected Wood Trusses*, published by the Wood Truss Council of America (WTCA) and the Truss Plate Institute (TPI), describes different methods that may be used to provide permanent lateral bracing for web members, including continuous lateral bracing, T-reinforcement, L-reinforcement, scab reinforcement, stacked web reinforcement and proprietary metal reinforcement products.

2303.4.1.6 Anchorage. All transfer of loads and anchorage of each truss to the supporting structure is the responsibility of the registered design professional.

❖ This section clarifies that the transfer of all design loads through the building and the connections of the trusses to the supporting structure to resist those loads remain the responsibility of the building's registered design professional.

2303.4.1.7 Alterations to trusses. Truss members and components shall not be cut, notched, drilled, spliced or otherwise altered in any way without written concurrence and approval of a registered design professional. Alterations resulting in the addition of loads to any member (e.g., HVAC equipment, water heater) shall not be permitted without verification that the truss is capable of supporting such additional loading.

❖ In any structural member, the existence of holes or notches can affect its ability to carry load. The effect of alterations on the capacity of truss members must be assessed by a registered design professional. This requirement applies not only to the truss chords and webs, but also to permanent lateral bracing components. Additionally, trusses are designed to carry a specific load and any loads that are added must be assessed by a registered design professional. The roof live load must also be considered in any review of existing trusses, as that live load also covers reroofing operations that may take place in the future.

2303.4.2 Metal-plate-connected trusses. In addition to Sections 2303.4.1 through 2303.4.1.7, the design, manufacture and quality assurance of metal-plate-connected wood trusses shall be in accordance with TPI 1. Manufactured trusses shall comply with Section 1704.6 as applicable.

❖ Metal-plate-connected wood trusses have been a popular product for construction of wood-frame buildings for many years. There are significant benefits to building with these engineered systems of wood and metal connectors; however, there are also criteria associated with the manufacturing, transportation and placement of trusses that are critical to their productive use in a structure. TPI 1 is the industry-developed standard for managing this process.

A manufacturer of trusses is also required by this section to have an agency review its operations to determine compliance with this standard. This type of evaluation by a qualified person is necessary for the quality of the product and its ability to perform as designed.

2303.5 Test standard for joist hangers and connectors. For the required test standards for joist hangers and connectors, see Section 1715.1.

❖ Tests for joist hangers and connections are included in the criteria for special inspections for materials and test standards in Section 1715. These criteria reference ASTM D 1761 and establish the criteria for use of lumber with a specific gravity of 0.49 or larger, but not more than 0.55 in accordance with the AF&PA NDS. Additionally, Section 1715.2 adds limits on the vertical load capacity for joist hangers by requiring that there be consistency of the tested capacity within 20 percent of the average ultimate load. If this cannot be achieved, additional testing is required.

2303.6 Nails and staples. Nails and staples shall conform to requirements of ASTM F 1667. Nails used for framing and sheathing connections shall have minimum average bending yield strengths as follows: 80 kips per square inch (ksi) (551 MPa) for shank diameters larger than 0.177 inch (4.50 mm) but not larger than 0.254 inch (6.45 mm), 90 ksi (620 MPa) for shank diameters larger than 0.142 inch (3.61 mm) but not larger than 0.177 inch (4.50 mm) and 100 ksi (689 MPa) for shank diameters of at least 0.099 inch (2.51 mm) but not larger than 0.142 inch (3.61 mm).

❖ The code specifies the material standard for nails and staples that provides appropriate fastener requirements. ASTM F 1667 specifies dimensions, etc. for nails and staples. Minimum nail strengths are specified based upon the stiffness of the nail shank. Minimum fastening is provided in the nailing schedule in Table 2304.9.1.

Just as the capacity of nailed connections is dependent upon the stiffness of the nail shank, the capacity of stapled connections is dependent upon the stiffness of the staple legs. Research was performed at Virginia Tech on stapled connections paralleling the research that was performed there on nailed connections. The research on nailed connections was incorporated into the AF&PA NDS. The research on stapled connections has been incorporated into the ICC Evaluation Services (ICC ES) Acceptance Criteria 201. The determination of staple bending moments is in accordance ASTM F 1575 with the modifications of that standard that are made in Sections 3.5.1 through 3.5.4 of ICC ES Acceptance Criteria 201.

2303.7 Shrinkage. Consideration shall be given in design to the possible effect of cross-grain dimensional changes considered vertically which may occur in lumber fabricated in a green condition.

❖ Lumber described as in a "green condition" has a moisture content higher than that used to define a dry condition under the applicable grading rules, which is reflected on the grade mark (see Figure 2303.1.1). Because it will shrink more than dry lumber, use of lumber that is fabricated in a green condition requires consideration of the effects of cross-grain shrinkage. The extent of cross-grain shrinkage is a function of the lumber's moisture at the time of construction and the amount of drying that occurs after construction.

SECTION 2304
GENERAL CONSTRUCTION REQUIREMENTS

2304.1 General. The provisions of this section apply to design methods specified in Section 2301.2.

❖ This section is intended to clarify some of the more important details of wood-frame construction and, as a result, provide consistency in building inspections. The methods of design are noted in Section 2301.2 and in the industry documents specified therein. Regardless of the design method chosen, there are certain construction requirements that need to be enforced. Section 2304 is related more to construction issues than it is to design issues.

Wood-framed construction has not changed much over the years, which is beneficial for a variety of reasons. First of all, many framers prefer doing things the way they've always done them in the past, most of which are detailed in carpentry manuals or architectural detailing books. To the extent such practices do not violate the need to support or transfer loads, it is preferable to allow them. Another reason that it is important to reach some type of agreement with these framers is to help maintain a consistent product that a building inspector would be familiar with from building to building. This section helps to achieve this goal.

2304.2 Size of structural members. Computations to determine the required sizes of members shall be based on the net dimensions (actual sizes) and not nominal sizes.

❖ ASD and LRFD methods require the use of the actual lumber dimensions that occur after the wood has dried. Typically, dimension lumber will shrink $^1/_2$ inch (12.7 mm) in thickness and $^1/_2$ inch to $^3/_4$ inch (12.7 mm to 19 mm) in face width as noted in the American Softwood Lumber Standard DOC PS 20. This provision is an attempt to make sure that if design principles are applied to a specific element, the designer uses design values consistent with the actual load-carrying capacity of the element, as reflected in the referenced design standards.

2304.3 Wall framing. The framing of exterior and interior walls shall be in accordance with the provisions specified in Section 2308 unless a specific design is furnished.

❖ The conventional construction criteria in Section 2308 provide great detail and specifications for the construction of walls. Under the majority of circumstances, the methods and procedures for framing a wall in accordance with Section 2308.9 will provide the necessary resistance to all conventional-level loads for a typical wall. Although special loading conditions need to be considered in the design of nonconventional walls, as directed by Section 2308.4, the completed product will likely be similar in shape and form as an otherwise conventionally framed wall.

2304.3.1 Bottom plates. Studs shall have full bearing on a 2-inch-thick (actual $1^1/_2$-inch, 38 mm) or larger plate or sill having a width at least equal to the width of the studs.

❖ It is necessary to provide a continuous wood plate at the bottom of the wall with a minimum nominal thickness of 2 inches (51 mm), so that the loads supported by a stud in a wood-framed wall are adequately transferred to the supporting structural members below. The plate is required to be at least the same width as the studs so that the full capacity of each stud is realized. This requirement is commonly known in the industry and will typically be met unless there is some type of problem in the field or such a condition occurs in existing construction. If the bottom plate of the wall is found to be narrower than the width of the studs for some reason, the issue that needs to be checked structurally is the perpendicular-to-grain bearing stress in that plate. The AF&PA NDS published by the American Forest & Paper Association, provides guidelines for calculating and checking this stress.

2304.3.2 Framing over openings. Headers, double joists, trusses or other approved assemblies that are of adequate size to transfer loads to the vertical members shall be provided over window and door openings in load-bearing walls and partitions.

❖ Walls depend upon the studs to be continuous in order to carry the imposed loads from the top plate to the bottom plate and eventually to the foundation. When there is an opening in the wall, this continuity is interrupted. It is necessary to transfer the loads from above the opening to adjacent structural members, which could be a single stud, multiple studs or a solid post. The loads above the opening are typically carried by a solid wood beam, a combination of cripple studs and plates or some other combination of members that provides the needed capacity to span adequately between the sides of the opening.

An important structural consideration is the connection between the member spanning the opening and the support. Whether that connection is a simple direct support, a hanger or some other type, it is necessary to provide a mechanism that will transfer not only the vertical loads but also forces that occur normal to the wall plane where such a consideration is warranted.

2304.3.3 Shrinkage. Wood walls and bearing partitions shall not support more than two floors and a roof unless an analysis satisfactory to the building official shows that shrinkage of the wood framing will not have adverse effects on the structure or any plumbing, electrical or mechanical systems, or other equipment installed therein due to excessive shrinkage or differential movements caused by shrinkage. The analysis shall also show that the roof drainage system and the foregoing systems or equipment will not be adversely affected or, as an alternate, such systems shall be designed to accommodate the differential shrinkage or movements.

❖ The adverse effects of shrinkage are greatest in multistory buildings having nonstructural components and supported equipment that may not be able to tolerate differential movements caused by shrinkage. This section states conditions under which shrinkage of the wood-framing must be considered in the construction of the wall system. An analysis of this phenomenon is required in order to determine the effect of shrinkage on the building system that is supported by the wall below. A typical wood-framed building will be constructed of mostly dry lumber, which has a moisture content of 19 percent or less. When supply of dry lumber is limited, the contractor may need to use green lumber which has a moisture content greater than 19 percent. The overall effect of lumber shrinkage on a multistory wood-framed building is described in *Design of Wood Structures–ASD*.

Another example of the adverse effects of shrinkage is the impact it can have on shear wall performance. Shrinkage has been recognized as a factor in loose hold-down anchor bolt nuts in earthquake-damaged buildings. In this situation the hold-down cannot resist tension until the slack is taken up, resulting in excessive movement of the shear wall under lateral loading. Since in most cases the shear wall sheathing conceals the hold-downs, this problem often goes undetected.

2304.4 Floor and roof framing. The framing of wood-joisted floors and wood framed roofs shall be in accordance with the provisions specified in Section 2308 unless a specific design is furnished.

❖ Similar to the criteria for wall construction, floors may be constructed using the methods prescribed in Section 2308, subject to the limitations described therein. One such limitation is a restriction on the magnitude of the applied dead, live and snow loads that may be applied to the members described in the rafter span tables. Span Tables 2308.8(1) and (2); 2308.9.5, 2308.9.6, 2308.10.2(1) and (2); and 2308.10.3(1), (2), (3), (4), (5) and (6) each describe a different loading condition for joists, rafters, headers and girders at the ceiling, floor and roof. The maximum floor load covered by Table 2308.8 is 40 pounds per square foot (psf) (1.9 kN/m²) live load. Table 2308.10.2 covers only dead loads of 5 psf (0.24 kN/m²) for ceiling rafters, and Table 2308.10.3 only provides for up to a 50 psf (2.4 kN/m²) ground snow load. The range of required loading is much larger, as demonstrated in Table 1607.1. The only structures for which a 40 psf (1.9 kN/m²) live load on the floor is allowed are catwalks, fire escapes on single-family dwellings, private rooms and wards in a hospital, cell blocks, school classrooms and all areas of residential occupancies other than public areas. All other uses require a higher design floor load, in which case these tables do not apply and an engineered solution per Section 2308.4 is required.

2304.5 Framing around flues and chimneys. Combustible framing shall be a minimum of 2 inches (51 mm), but shall not be less than the distance specified in Sections 2111 and 2113 and the *International Mechanical Code*, from flues, chimneys and fireplaces, and 6 inches (152 mm) away from flue openings.

❖ This provision is duplicated in the *International Mechanical Code*® (IMC®), but must be included in both places in order to alert the framer, installer and inspector to the requirement. When exposed to heat for a prolonged period of time the characteristics of wood may change. In particular, the temperature at which wood will ignite can be lowered. Repeated exposure to temperature extremes has a cumulative effect on the mechanical properties of wood, whereby sustained heating at temperatures above 212°F (100°C) may cause negative nonreversible effects on these properties. In order to help mitigate this phenomenon, a separation is provided so that the wood will not be directly exposed or noncombustible fire-stopping material is placed between the wood and the heat source. The requirement for some type of separation has been required in legacy model codes and is detailed in commonly used construction manuals.

2304.6 Wall sheathing. Except as provided for in Section 1405 for weatherboarding or where stucco construction that complies with Section 2510 is installed, enclosed buildings shall be sheathed with one of the materials of the nominal thickness specified in Table 2304.6 or any other approved material of equivalent strength or durability.

❖ Wall sheathing material can be any of those listed in Table 2304.6. There are two main reasons for providing sheathing to at least one face of wood-framed walls: (1) to stabilize the studs against racking when the wall carries vertical loads and (2) to provide a nominal ability to transfer lateral forces from the roof or floor deck to the sill plate and anchor bolts. Studs nailed to the top and bottom plate are stable to a degree, but will collapse with significant force placed on them unless there is some type of stabilizing membrane or brace. Sheathing, whether it is a heavy structural membrane found on exterior bearing walls or a light one that would be used for nonbearing interior partitions, provides necessary stability for the walls. Some sheathing may serve as insulation for the building, in addition to providing lateral bracing. Other sheathing, may provide a finish for the wall. Sheathing is not the only means of bracing a wall.

This section establishes that only the sheathing types listed in Table 2304.6, which includes board and various panel products (structural panels, boards, gypsum and reinforced cement mortar) are to be used

for walls. There is an allowance that other materials may be used if it is shown they are capable of providing equivalent strength and durability. Let-in braces are common for temporary bracing of the wall against racking forces, but they provide only minimal resistance lateral forces. Many let-in braces failed in tension or at the bottom plate connection during the 1971 San Fernando earthquake. Accordingly, their use is limited to structures with low seismic risk for conventional braced wall panels per Section 2308.9.3.

2304.6.1 Wood structural panel sheathing. Where wood structural panel sheathing is used as the exposed finish on the exterior of outside walls, it shall have an exterior exposure durability classification. Where wood structural panel sheathing is used on the exterior of outside walls but not as the exposed finish, it shall be of a type manufactured with exterior glue (Exposure 1 or Exterior). Where wood structural panel sheathing is used elsewhere, it shall be of a type manufactured with intermediate or exterior glue.

❖ Wood structural panels are defined by this chapter as being plywood, OSB and composite panels. Each of these types of panel products has identical structural characteristics when manufactured according to the PS 2 standard. This section requires equivalent exposure durability performance. It requires the classification to be Exposure 1 or Exterior when the material is likely to be exposed to the weather. According to the American Plywood Association, "Exterior" type panels are suitable for applications permanently exposed to weather, "Exposure 1" type panels may be used for extended temporary exposure to weather (such as during construction when long delays are expected prior to providing permanent protection), "Exposure 2" type panels apply to protected applications not exposed to high humidity and may be briefly subjected to the weather while "Interior" type panels are to be used only where permanent protection against the weather is provided.

TABLE 2304.6. See page 23-18.

❖ Table 2304.6 reinforces Section 2304.6, which requires all sheathing to meet a minimum set of requirements, including minimum material thickness and maximum wall stud spacing depending on the type of sheathing material chosen for a particular project. Because of loading conditions or the design of the assembly to which the wall sheathing is attached, the sheathing may have to be thicker or the wall studs may need to be placed closer together.

2304.6.2 Interior paneling. Softwood wood structural panels used for interior paneling shall conform with the provisions of Chapter 8 and shall be installed in accordance with Table 2304.9.1. Panels shall comply with DOC PS 1 or PS 2. Prefinished hardboard paneling shall meet the requirements of AHA A135.5. Hardwood plywood shall conform to HPVA HP-1.

❖ Interior paneling, similar to exterior paneling, must meet minimum standards of acceptance. DOC PS 1 and PS 2 establish the general standards for all softwood veneer panel products. AHA A135.5 and HPVA HP-1 address the criteria for hardboard paneling and hardwood veneered plywood, respectively.

2304.7 Floor and roof sheathing.

❖ Floor and roof sheathing perform several functions. They are the primary base for both the walking surface for the floor and the main support of the roofing materials. Additionally, the sheathing transfers horizontally applied forces to the vertical lateral-force-resisting elements. Floor sheathing is the structural element in a wood floor that transfers live loads to the floor joists.

Roof sheathing is typically covered by the roofing material. The sheathing typically provides the only means by which loads from wind and snow are transferred to the structural framing members and the fastening that completes this load path is critical. Panels are typically "span-rated," which provides the designer guidance as to the maximum spacing that the framing members can be placed at to adequately support imposed loads.

2304.7.1 Structural floor sheathing. Structural floor sheathing shall be designed in accordance with the general provisions of this code and the special provisions in this section.

Floor sheathing conforming to the provisions of Table 2304.7(1), 2304.7(2), 2304.7(3) or 2304.7(4) shall be deemed to meet the requirements of this section.

❖ The tables referenced are used for both floor and roof applications (see commentary, Section 2304.7.2). In a two-layer floor sheathing system, the structural panels are placed at the bottom ("subfloor") and fastened directly to the framing members, and the top layer ("underlayment") is typically a different grade of material, occasionally specified to resist indentations depending on what floor finish material is to be installed.

2304.7.2 Structural roof sheathing. Structural roof sheathing shall be designed in accordance with the general provisions of this code and the special provisions in this section.

Roof sheathing conforming to the provisions of Table 2304.7(1), 2304.7(2), 2304.7(3) or 2304.7(5) shall be deemed to

meet the requirements of this section. Wood structural panel roof sheathing shall be bonded by exterior glue.

❖ There are four tables referenced in this section that establish either allowable load or allowable span of structural sheathing typically produced by mills in North America. Using the tables is straightforward: simply locate with the span between structural framing members, and depending on the direction of the framing, determine the minimum thickness required by the code. It may be necessary to install thicker material if the loads require additional support.

Lumber and structural sheathing are required to be marked with grade stamps indicating the capabilities of the lumber or panels, including a span rating and structural quality identification, based on the testing and conditions of the material.

TABLE 2304.7(1). See page 23-19.

❖ Table 2304.7(1) applies to "surfaced dry" or "surfaced unseasoned" lumber. These terms refer to the condition of the lumber at the time it was surfaced, or finished. This is only critical when determining the actual dimension of the lumber. A piece that is surfaced dry is narrower than a piece that is surfaced unseasoned (often referred to as surfaced green). The grade stamp will indicate whether it is surfaced dry or surfaced green.

TABLE 2304.7(2). See page 23-19.

❖ The different agencies that publish grading rules are listed and are an important part of the criteria used in determining the adequacy of structural lumber. All of these agencies are regulated by the U.S. Department of Commerce through the ALSC in order to maintain some type of standard by which stress-grading principles are used to create grading rules. These requirements are outlined in ALSC standard DOC PS 20. The rules-writing agencies included in this table are the National Lumber Grades Authority (NLGA), West Coast Lumber Inspection Bureau (WCLIB), Western Wood Products Association (WWPA), Northern Soft-

TABLE 2304.6
MINIMUM THICKNESS OF WALL SHEATHING

SHEATHING TYPE	MINIMUM THICKNESS	MAXIMUM WALL STUD SPACING
Wood boards	$^5/_8$ inch	24 inches on center
Fiberboard	$^1/_2$ inch	16 inches on center
Wood structural panel	In accordance with Tables 2308.9.3(2) and 2308.9.3(3)	—
M-S "Exterior Glue" and M-2 "Exterior Glue" Particleboard	In accordance with Tables 2306.4.3 and 2308.9.3(5)	—
Gypsum sheathing	$^1/_2$ inch	16 inches on center
Gypsum wallboard	$^1/_2$ inch	24 inches on center
Reinforced cement mortar	1 inch	24 inches on center

For SI: 1 inch = 25.4 mm.

wood Lumber Bureau (NSLB), Northeastern Lumber Manufacturers Association (NELMA), Southern Pine Inspection Bureau (SPIB) and Redwood Inspection Service (RIS).

TABLE 2304.7(3). See page 23-20.

❖ Structural wood panels are also graded, but the type of grade stamp used identifies the thickness and the span capabilities of the piece. The panel span rating (first column of the table), also known as a "panel index," will appear on the grade stamp. The first number on the panel span rating indicates the allowable spacing of support members (rafters or trusses) in a roof application. The second number indicates the allowable spacing of floor joists or trusses in a floor application. For instance, if the span rating is 24/16, the panel may be used on roof joists spaced 24 inches (610 mm) o.c. and on floor joists spaced 16 inches (406 mm) o.c. Span ratings were developed for easy identification and to assist in verifying the correct use of panels in the field. According to *Design of Wood Structures – ASD*, typical wood structural panel applications are not controlled so much by the uniform load requirements, but more so on the basis of panel deflection under concentrated loads and judged by the panel's performance under passing traffic.

For the panel span rating to be applicable, other conditions of the table for loading and use must be met. For instance, for sheathing with a panel index of 24/16 to be used on roof rafters spaced 16 inches (406 mm) o.c., the snow load for the roof cannot exceed 40 psf (1.92 MPa) (listed in the sixth column of the table). The allowable spacing of support members is also dependent on the type of edge support provided for the panels (see Note f). The table is applicable only when the strength axis of the panel is perpendicular to the supports, in other words, when the long edge of the panel is perpendicular to the joists or rafters.

TABLE 2304.7(4). See page 23-21.

❖ Often a combination of a wood subfloor and a finish floor is used and there is additional capacity for that composite floor action. This table provides the additional capacity associated with that installation. It is necessary to go to DOC PS 1 to determine what species groups are represented by the numbers in the table. One of the reasons that a combination subfloor-underlayment floor paneling system may be chosen is to provide a more rigid fastening surface for the finish materials, which can improve long-term performance. In fact, some floor finishing manufacturers or organization require a two-layer floor system for certain assemblies.

TABLE 2304.7(5). See page 23-21.

❖ The table provides criteria for panels that may be used on roofs with significant live loads, the maximum being listed at 90 psf (4.31 MPa). That being the case, it is clear that the live loads in excess of 20 psf (0.958 MPa) (which is generally the maximum roof live loading determined in Section 1607.11.2) is referring primarily to the roof snow load rather than a live load, strictly speaking. For further guidance on required snow loads, see the commentary to Section 1608. The allowable spacing of rafters or trusses and the allowable load are determined by the grade and thickness of the panel.

2304.8 Lumber decking.

TABLE 2304.7(1)
ALLOWABLE SPANS FOR LUMBER FLOOR AND ROOF SHEATHING[a,b]

SPAN (inches)	MINIMUM NET THICKNESS (inches) OF LUMBER PLACED			
	Perpendicular to supports		Diagonally to supports	
	Surfaced dry[c]	Surfaced unseasoned	Surfaced dry[c]	Surfaced unseasoned
Floors				
24	$3/4$	$25/32$	$3/4$	$25/32$
16	$5/8$	$11/16$	$5/8$	$11/16$
Roofs				
24	$5/8$	$11/16$	$3/4$	$25/32$

For SI: 1 inch = 25.4 mm.

a. Installation details shall conform to Sections 2304.6.1 and 2304.6.2 for floor and roof sheathing, respectively.

b. Floor or roof sheathing conforming with this table shall be deemed to meet the design criteria of Section 2304.6.

c. Maximum 19-percent moisture content.

TABLE 2304.7(2)
SHEATHING LUMBER, MINIMUM GRADE REQUIREMENTS: BOARD GRADE

SOLID FLOOR OR ROOF SHEATHING	SPACED ROOF SHEATHING	GRADING RULES
Utility	Standard	NLGA, WCLIB, WWPA
4 common or utility	3 common or standard	NLGA, WCLIB, WWPA, NSLB or NELMA
No. 3	No. 2	SPIB
Merchantable	Construction common	RIS

2304.8.1 General. Lumber decking shall be designed and installed in accordance with the general provisions of this code and the provisions of this section. Each piece shall be square-end trimmed. When random lengths are furnished, each piece shall be square-end trimmed across the face so that at least 90 percent of the pieces will be within 0.5 degrees (0.00873 rad) of square. The ends of the pieces shall be permitted to be beveled up to 2 degrees (0.0349 rad) from vertical with the exposed face of the piece slightly longer than the back of the piece. Tongue-and-groove decking shall be installed with the tongues up on sloped or pitched roofs with pattern faces down.

❖ Along with design requirements in Section 2306.1.4, this section provides installation requirements for all lumber decking, including tongue-and-groove as well as mechanically laminated decking.

2304.8.2 Layup patterns. Lumber decking is permitted to be laid up following one of five standard patterns as defined in Sections 2304.8.2.1 through 2304.8.2.5. Other patterns are permitted to be used if justified by engineering analysis.

❖ This section describes typically used layup patterns for lumber decking. It also allows other patterns to be used where the engineer provides substantiation. These layup patterns are determined, in part, by the location of end joints and the pattern is the basis for determining the allowable deck load in Section 2306.1.4.

TABLE 2304.7(3)
ALLOWABLE SPANS AND LOADS FOR WOOD STRUCTURAL PANEL SHEATHING AND SINGLE-FLOOR GRADES CONTINUOUS OVER TWO OR MORE SPANS WITH STRENGTH AXIS PERPENDICULAR TO SUPPORTS[a,b]

SHEATHING GRADES		ROOF[c]				FLOOR[d]
Panel span rating roof/floor span	Panel thickness (inches)	Maximum span (inches)		Load[e] (psf)		Maximum span (inches)
		With edge support[f]	Without edge support	Total load	Live load	
12/0	$^5/_{16}$	12	12	40	30	0
16/0	$^5/_{16}$, $^3/_8$	16	16	40	30	0
20/0	$^5/_{16}$, $^3/_8$	20	20	40	30	0
24/0	$^3/_8$, $^7/_{16}$, $^1/_2$	24	20[g]	40	30	0
24/16	$^7/_{16}$, $^1/_2$	24	24	50	40	16
32/16	$^{15}/_{32}$, $^1/_2$, $^5/_8$	32	28	40	30	16[h]
40/20	$^{19}/_{32}$, $^5/_8$, $^3/_4$, $^7/_8$	40	32	40	30	20[h,i]
48/24	$^{23}/_{32}$, $^3/_4$, $^7/_8$	48	36	45	35	24
54/32	$^7/_8$, 1	54	40	45	35	32
60/32	$^7/_8$, 1 $^1/_8$	60	48	45	35	32
SINGLE FLOOR GRADES		ROOF[c]				FLOOR[d]
Panel span rating	Panel thickness (inches)	Maximum span (inches)		Load[e] (psf)		Maximum span (inches)
		With edge support[f]	Without edge support	Total load	Live load	
16 o.c.	$^1/_2$, $^{19}/_{32}$, $^5/_8$	24	24	50	40	16[h]
20 o.c.	$^{19}/_{32}$, $^5/_8$, $^3/_4$	32	32	40	30	20[h,i]
24 o.c.	$^{23}/_{32}$, $^3/_4$	48	36	35	25	24
32 o.c.	$^7/_8$, 1	48	40	50	40	32
48 o.c.	1$^3/_{32}$, 1$^1/_8$	60	48	50	40	48

For SI: 1 inch = 25.4 mm, 1 pound per square foot = 0.0479 kN/m².

a. Applies to panels 24 inches or wider.

b. Floor and roof sheathing conforming with this table shall be deemed to meet the design criteria of Section 2304.7.

c. Uniform load deflection limitations $^1/_{180}$ of span under live load plus dead load, $^1/_{240}$ under live load only.

d. Panel edges shall have approved tongue-and-groove joints or shall be supported with blocking unless $^1/_4$-inch minimum thickness underlayment or 1 $^1/_2$ inches of approved cellular or lightweight concrete is placed over the subfloor, or finish floor is $^3/_4$-inch wood strip. Allowable uniform load based on deflection of $^1/_{360}$ of span is 100 pounds per square foot except the span rating of 48 inches on center is based on a total load of 65 pounds per square foot.

e. Allowable load at maximum span.

f. Tongue-and-groove edges, panel edge clips (one midway between each support, except two equally spaced between supports 48 inches on center), lumber blocking or other. Only lumber blocking shall satisfy blocked diaphragm requirements.

g. For $^1/_2$-inch panel, maximum span shall be 24 inches.

h. Span is permitted to be 24 inches on center where $^3/_4$-inch wood strip flooring is installed at right angles to joist.

i. Span is permitted to be 24 inches on center for floors where 1$^1/_2$ inches of cellular or lightweight concrete is applied over the panels.

TABLE 2304.7(4)
ALLOWABLE SPAN FOR WOOD STRUCTURAL PANEL COMBINATION SUBFLOOR-UNDERLAYMENT (SINGLE FLOOR)[a,b]
(Panels Continuous Over Two or More Spans and Strength Axis Perpendicular to Supports)

IDENTIFICATION	MAXIMUM SPACING OF JOISTS (inches)				
	16	20	24	32	48
Species group[c]	Thickness (inches)				
1	$^1/_2$	$^5/_8$	$^3/_4$	—	—
2, 3	$^5/_8$	$^3/_4$	$^7/_8$	—	—
4	$^3/_4$	$^7/_8$	1	—	—
Single floor span rating[d]	16 o.c.	20 o.c.	24 o.c.	32 o.c.	48 o.c.

For SI: 1 inch = 25.4 mm, 1 pound per square foot = 0.0479 kN/m².

a. Spans limited to value shown because of possible effects of concentrated loads. Allowable uniform loads based on deflection of $^1/_{360}$ of span is 100 pounds per square foot except allowable total uniform load for $1^1/_8$-inch wood structural panels over joists spaced 48 inches on center is 65 pounds per square foot. Panel edges shall have approved tongue-and-groove joints or shall be supported with blocking, unless $^1/_4$-inch minimum thickness underlayment or $1^1/_2$ inches of approved cellular or lightweight concrete is placed over the subfloor, or finish floor is $^3/_4$-inch wood strip.

b. Floor panels conforming with this table shall be deemed to meet the design criteria of Section 2304.7.

c. Applicable to all grades of sanded exterior-type plywood. See DOC PS 1 for plywood species groups.

d. Applicable to Underlayment grade, C-C (Plugged) plywood, and Single Floor grade wood structural panels.

TABLE 2304.7(5)
**ALLOWABLE LOAD (PSF) FOR WOOD STRUCTURAL PANEL ROOF SHEATHING CONTINUOUS OVER
TWO OR MORE SPANS AND STRENGTH AXIS PARALLEL TO SUPPORTS**
(Plywood Structural Panels Are Five-Ply, Five-Layer Unless Otherwise Noted)[a,b]

PANEL GRADE	THICKNESS (inch)	MAXIMUM SPAN (inches)	LOAD AT MAXIMUM SPAN (psf)	
			Live	Total
Structural I sheathing	$^7/_{16}$	24	20	30
	$^{15}/_{32}$	24	35[c]	45[c]
	$^1/_2$	24	40[c]	50[c]
	$^{19}/_{32}, ^5/_8$	24	70	80
	$^{23}/_{32}, ^3/_4$	24	90	100
Sheathing, other grades covered in DOC PS 1 or DOC PS 2	$^7/_{16}$	16	40	50
	$^{15}/_{32}$	24	20	25
	$^1/_2$	24	25	30
	$^{19}/_{32}$	24	40[c]	50[c]
	$^5/_8$	24	45[c]	55[c]
	$^{23}/_{32}, ^3/_4$	24	60[c]	65[c]

For SI: 1 inch = 25.4 mm, 1 pound per square foot = 0.0479 kN/m².

a. Roof sheathing conforming with this table shall be deemed to meet the design criteria of Section 2304.7.

b. Uniform load deflection limitations $^1/_{180}$ of span under live load plus dead load, $^1/_{240}$ under live load only. Edges shall be blocked with lumber or other approved type of edge supports.

c. For composite and four-ply plywood structural panel, load shall be reduced by 15 pounds per square foot.

2304.8.2.1 Simple span pattern. All pieces shall be supported by two supports.

❖ See the commentary to Section 2304.8.2 and Figure 2304.8.2.1.

2304.8.2.2 Two-span continuous pattern. All pieces shall be supported by three supports, and all end joints shall occur in line on every other support. Supporting members shall be designed to accommodate the load redistribution caused by this pattern.

❖ See the commentary to Section 2304.8.2 and Figure 2304.8.2.2.

2304.8.2.3 Combination simple and two-span continuous pattern. Courses in end spans shall be alternating simple span and two span continuous. End joints are staggered in adjacent courses and occur only over supports.

❖ See the commentary to Section 2304.8.2 and Figure 2304.8.2.3.

2304.8.2.4 Cantilevered pieces intermixed pattern. The decking shall cover a minimum of three spans. Pieces in the starter course and every third course shall be simple span. Pieces in other courses shall be cantilevered over the supports with end joints at alternate quarter or third points of the spans, and each piece shall bear on at least one support.

❖ See commentary to Section 2304.8.2 and Figure 2304.8.2.4.

2304.8.2.5 Controlled random pattern. The decking shall cover a minimum of three spans. End joints within 6 inches (152 mm) of being in line in either direction shall be separated by at least two intervening courses. In the end bays each piece shall bear on at least one support. Where an end joint occurs in an end bay, the next piece in the same course shall continue over the first inner support for at least 24 inches (610 mm). The details of the controlled random pattern shall be as described for each decking material in Section 2304.8.3.3, 2304.8.4.3 or 2304.8.5.3.

ℓ = SPAN OF DECKING

Figure 2304.8.2.1
SIMPLE SPAN PATTERN

END JOINTS OCCUR IN LINE
ON EVERY OTHER SUPPORT

ℓ = SPAN OF DECKING

Figure 2304.8.2.2
TWO-SPAN CONTINUOUS PATTERN

For cantilevered spans with the controlled random pattern, special considerations shall be made when the overhang exceeds 18 inches (457 mm), 24 inches (610 mm) or 36 inches (914 mm) for 2-inch (51 mm), 3-inch (76 mm) or 4-inch (102 mm) nominal thickness decking, respectively. The maximum cantilevered length for the controlled random pattern shall be 30 percent of the length of the first adjacent interior span. For cantilever overhangs within these limits, a structural fascia shall be fastened to each decking piece to maintain a continuous, straight roof line. There shall be no end joints in the cantilevered portion or within one-half of the first adjacent interior span.

❖ General requirements for controlled random layup are stated in this section. They apply along with the additional specific layup requirements subsequently given under mechanically laminated and tongue-and-groove decking. The second paragraph states requirements where there is a cantilevered span such as at a roof overhang (also see commentary, Section 2304.8.2).

2304.8.3 Mechanically laminated decking.

2304.8.3.1 General. Mechanically laminated decking consists of square-edged dimension lumber laminations set on edge and nailed to the adjacent pieces and to the supports.

❖ One method of floor construction is the use of individual wood members set on edge, connected with nails and mechanically laminated to produce a monolithic surface.

This section gives the fastening requirements for floors that are constructed of individual wood members set on edge and mechanically laminated to produce a continuous surface. For the floor to act as a unit, fastening is critical, and therefore it is prescribed in this section. If the floor is constructed in accordance with these fastening requirements, then for purposes of design it can be considered a solid wood floor acting as one element.

ℓ = SPAN OF DECKING

Figure 2304.8.2.3
COMBINATION SIMPLE AND TWO-SPAN CONTINUOUS PATTERN

ℓ = SPAN OF DECKING
SS = SIMPLE SPAN

Figure 2304.8.2.4
CANTILEVERED PIECES INTERMIXED PATTERN

2304.8.3.2 Nailing. The length of nails connecting laminations shall not be less than two and one-half times the net thickness of each lamination. Where deck supports are 48 inches (1219 mm) on center (o.c.) or less, side nails shall be spaced not more than 30 inches (762 mm) o.c. alternately near top and bottom edges, and staggered one-third of the spacing in adjacent laminations. Where supports are spaced more than 48 inches (1219 mm) o.c., side nails shall be spaced not more than 18 inches (457 mm) o.c. alternately near top and bottom edges and staggered one-third of the spacing in adjacent laminations. Two side nails shall be used at each end of butt-jointed pieces.

Laminations shall be toenailed to supports with 20d or larger common nails. Where the supports are 48 inches (1219 mm) o.c. or less, alternate laminations shall be toenailed to alternate supports; where supports are spaced more than 48 inches (1219 mm) o.c., alternate laminations shall be toenailed to every support.

❖ See the commentary to Section 2304.8.3.1.

2304.8.3.3 Controlled random pattern. There shall be a minimum distance of 24 inches (610 mm) between end joints in adjacent courses. The pieces in the first and second courses shall bear on at least two supports with end joints in these two courses occurring on alternate supports. A maximum of seven intervening courses shall be permitted before this pattern is repeated.

❖ These requirements are specific to the layup of mechanically laminated decking and are in addition to the general limitations stated in Section 2304.8.2.5. They are similar to the requirements for 2-inch (51 mm) tongue-and-groove decking that are illustrated in Figure 2304.8.4.3.

2304.8.4 Two-inch sawn tongue-and-groove decking.

2304.8.4.1 General. Two-inch (51 mm) decking shall have a maximum moisture content of 15 percent. Decking shall be machined with a single tongue-and-groove pattern. Each deck piece shall be nailed to each support as required.

❖ Another method of floor construction is the use of sawn lumber tongue-and-groove decking. The projecting tongue on one side fits into the matching groove of the adjacent board, causing these two elements to deflect as a unit when subjected to loads. As Figure 2304.8.4.1 illustrates, 2-inch thick (51 mm) tongue-and-groove boards have one tongue and one groove per board.

For SI: 1 inch = 25.4 mm.

Figure 2304.8.4.1
TWO-INCH TONGUE-AND-GROOVE DECKING

For SI: 1 inch = 25.4 mm.

Figure 2304.8.4.3
CONTROLLED RANDOM PATTERN TWO-INCH TONGUE-AND-GROOVE DECKING

2304.8.4.2 Nailing. Each piece of decking shall be toenailed at each support with one 16d common nail through the tongue and face-nailed with one 16d common nail.

❖ For the floor to act as a unit, fastening is critical, and therefore it is prescribed in this section.

2304.8.4.3 Controlled random pattern. There shall be a minimum distance of 24 inches (610 mm) between end joints in adjacent courses. The pieces in the first and second courses shall bear on at least two supports with end joints in these two courses occurring on alternate supports. A maximum of seven intervening courses shall be permitted before this pattern is repeated.

❖ These requirements are specific to the layup of tongue-and-groove decking and are in addition to the general limitations stated in Section 2304.8.2.5 (see Figure 2304.8.4.3).

2304.8.5 Three- and 4-inch sawn tongue-and-groove decking.

2304.8.5.1 General. Three-inch (76 mm) and 4-inch (102 mm) decking shall have a maximum moisture content of 19 percent. Decking shall be machined with a double tongue-and-groove pattern. Deck pieces shall be interconnected and fastened to the supports as required.

❖ As Figure 2304.8.5.1 illustrates, 3- and 4-inch thick (76 and 102 mm) tongue-and-groove boards have a pair of tongues and a pair of grooves per board.

2304.8.5.2 Nailing. Each piece shall be toenailed at each support with one 40d common nail and face-nailed with one 60d common nail. Courses shall be spiked to each other with 8-inch (203 mm) spikes at intervals not to exceed 30 inches (762 mm) through predrilled edge holes penetrating to a depth of approximately 4 inches (102 mm) and with one spike at a distance not exceeding 10 inches (254 mm) from the end of each piece.

❖ For the floor to act as a unit, fastening is critical, and therefore it is prescribed in this section.

2304.8.5.3 Controlled random pattern. There shall be a minimum distance of 48 inches (1219 mm) between end joints in adjacent courses. Pieces not bearing over a support are permitted to occur in interior bays, provided the adjacent pieces in the same course continue over the support for at least 24 inches (610 mm). This condition shall not occur more than once in every six courses in each interior bay.

❖ These requirements are specific to the layup of tongue-and-groove decking and are in addition to the general limitations stated in Section 2304.8.2.5.

2304.9 Connections and fasteners.

2304.9.1 Fastener requirements. Connections for wood members shall be designed in accordance with the appropriate methodology in Section 2301.2. The number and size of fasteners connecting wood members shall not be less than that set forth in Table 2304.9.1.

❖ Section 2301.2 gives alternative design methodologies for wood construction: (1) the ASD method; (2) the LRFD method; and (3) conventional light-frame methodologies that are embodied in Section 2308 of the code. If conventional provisions are used and there is no design being conducted by means of the ASD or LRFD methodologies, then the prescriptive fastening of Table 2304.9.1 is all that is required. However, when the ASD or LRFD methods are used, fastening must comply with the associated design standard as well as the prescriptive fastenings specified in Table 2304.9.1.

TABLE 2304.9.1. See page 23-26.

❖ This table provides the minimum fastening required for all lightweight wood-frame construction, regardless of the design method used. Familiarity with the terms and configurations of conventional frame construction is necessary to apply this table. All members identified within this table are commonly described in any available wood design or construction manual. See Figure 2304.9.1 for an illustration of certain nail types.

PAIR OF TONGUES FIT INTO PAIR OF GROOVES IN ADJACENT BOARD

For SI: 1 inch = 25.4 mm.

Figure 2304.8.5.1
THREE- AND FOUR-INCH TONGUE-AND-GROOVE DECKING

TABLE 2304.9.1
FASTENING SCHEDULE

CONNECTION	FASTENING[a,m]	LOCATION
1. Joist to sill or girder	3 - 8d common ($2^1/_2$″ × 0.131″) 3 - 3″ × 0.131″ nails 3 - 3″ 14 gage staples	toenail
2. Bridging to joist	2 - 8d common ($2^1/_2$″ × 0.131″) 2 - 3″ × 0.131″ nails 2 - 3″ 14 gage staples	toenail each end
3. 1″ × 6″ subfloor or less to each joist	2 - 8d common ($2^1/_2$″ × 0.131″)	face nail
4. Wider than 1″ × 6″ subfloor to each joist	3 - 8d common ($2^1/_2$″ × 0.131″)	face nail
5. 2″ subfloor to joist or girder	2 - 16d common ($3^1/_2$″ × 0.162″)	blind and face nail
6. Sole plate to joist or blocking	16d ($3^1/_2$″ × 0.135″) at 16″ o.c. 3″ × 0.131″ nails at 8″ o.c. 3″ 14 gage staples at 12″ o.c.	typical face nail
Sole plate to joist or blocking at braced wall panel	3″ - 16d ($3^1/_2$″ × 0.135″) at 16″ 4 - 3″ × 0.131″ nails at 16″ 4 - 3″ 14 gage staples per 16″	braced wall panels
7. Top plate to stud	2 - 16d common ($3^1/_2$″ × 0.162″) 3 - 3″ × 0.131″ nails 3 - 3″ 14 gage staples	end nail
8. Stud to sole plate	4 - 8d common ($2^1/_2$″ × 0.131″) 4 - 3″ × 0.131″ nails 3 - 3″ 14 gage staples	toenail
	2 - 16d common ($3^1/_2$″ × 0.162″) 3 - 3″ × 0.131″ nails 3 - 3″ 14 gage staples	end nail
9. Double studs	16d ($3^1/_2$″ × 0.135″) at 24″ o.c. 3″ × 0.131″ nail at 8″ o.c. 3″ 14 gage staple at 8″ o.c.	face nail
10. Double top plates	16d ($3^1/_2$″ × 0.135″) at 16″ o.c. 3″ × 0.131″ nail at 12″ o.c. 3″ 14 gage staple at 12″ o.c.	typical face nail
Double top plates	8-16d common ($3^1/_2$″ × 0.162″) 12-3″ × 0.131″ nails 12-3″ 14 gage staples	lap splice
11. Blocking between joists or rafters to top plate	3 - 8d common ($2^1/_2$″ × 0.131″) 3 - 3″ × 0.131″ nails 3 - 3″ 14 gage staples	toenail
12. Rim joist to top plate	8d ($2^1/_2$″ × 0.131″) at 6″ o.c. 3″ × 0.131″ nail at 6″ o.c. 3″ 14 gage staple at 6″ o.c.	toenail
13. Top plates, laps and intersections	2 - 16d common ($3^1/_2$″ × 0.162″) 3 - 3″ × 0.131″ nails 3 -3″ 14 gage staples	face nail
14. Continuous header, two pieces	16d common ($3^1/_2$″ × 0.162″)	16″ o.c. along edge
15. Ceiling joists to plate	3 - 8d common ($2^1/_2$″ × 0.131″) 5 - 3″ × 0.131″ nails 5 - 3″ 14 gage staples	toenail
16. Continuous header to stud	4 - 8d common ($2^1/_2$″ × 0.131″)	toenail

(continued)

TABLE 2304.9.1—continued
FASTENING SCHEDULE

CONNECTION	FASTENING[a,m]	LOCATION
17. Ceiling joists, laps over partitions (see Section 2308.10.4.1, Table 2308.10.4.1)	3 - 16d common ($3^1/_2$″ × 0.162″) minimum, Table 2308.10.4.1 4 - 3″ × 0.131″ nails 4 - 3″ 14 gage staples	face nail
18. Ceiling joists to parallel rafters (see Section 2308.10.4.1, Table 2308.10.4.1)	3 - 16d common ($3^1/_2$″ × 0.162″) minimum, Table 2308.10.4.1 4 - 3″ × 0.131″ nails 4 - 3″ 14 gage staples	face nail
19. Rafter to plate (see Section 2308.10.1, Table 2308.10.1)	3 - 8d common ($2^1/_2$″ × 0.131″) 3 - 3″ × 0.131″ nails 3 - 3″ 14 gage staples	toenail
20. 1″ diagonal brace to each stud and plate	2 - 8d common ($2^1/_2$″ × 0.131″) 2 - 3″ × 0.131″ nails 3 - 3″ 14 gage staples	face nail
21. 1″ × 8″ sheathing to each bearing	3 - 8d common ($2^1/_2$″ × 0.131″)	face nail
22. Wider than 1″ × 8″ sheathing to each bearing	3 - 8d common ($2^1/_2$″ × 0.131″)	face nail
23. Built-up corner studs	16d common ($3^1/_2$″ × 0.162″) 3″ × 0.131″ nails 3″ 14 gage staples	24″ o.c. 16″ o.c. 16″ o.c.
24. Built-up girder and beams	20d common (4″ × 0.192″) 32″ o.c. 3″ × 0.131″ nail at 24″ o.c. 3″ 14 gage staple at 24″ o.c. 2 - 20d common (4″ × 0.192″) 3 - 3″ × 0.131″ nails 3 - 3″ 14 gage staples	face nail at top and bottom staggered on opposite sides face nail at ends and at each splice
25. 2″ planks	16d common ($3^1/_2$″ × 0.162″)	at each bearing
26. Collar tie to rafter	3 - 10d common (3″ × 0.148″) 4 - 3″ x 0.131″ nails 4 - 3″ 14 gage staples	face nail
27. Jack rafter to hip	3 - 10d common (3″ x 0.148″) 4 - 3″ × 0.131″ nails 4 - 3″ 14 gage staples 2 - 16d common ($3^1/_2$″ × 0.162″) 3 -3″ × 0.131″ nails 3 - 3″ 14 gage staples	toenail face nail
28. Roof rafter to 2-by ridge beam	2 - 16d common ($3^1/_2$″ × 0.162″) 3 - 3″ × 0.131″ nails 3 - 3″ 14 gage staples 2-16d common ($3^1/_2$″ × 0.162″) 3 - 3″ × 0.131″ nails 3 - 3″ 14 gage staples	toenail face nail
29. Joist to band joist	3 - 16d common ($3^1/_2$″ × 0.162″) 4 - 3″ × 0.131″ nails 4 - 3″ 14 gage staples	face nail

(continued)

TABLE 2304.9.1—continued
FASTENING SCHEDULE

CONNECTION	FASTENING[a,m]		LOCATION
30. Ledger strip	3 - 16d common ($3^1/_2'' \times 0.162''$) 4 - 3" x 0.131" nails 4 - 3" 14 gage staples		face nail
31. Wood structural panels and particleboard[b] Subfloor, roof and wall sheathing (to framing)	$^1/_2''$ and less	6d[c,1] $2^3/_8'' \times 0.113''$ nail[n] $1^3/_4''$ 16 gage[o]	
	$^{19}/_{32}''$ to $^3/_4''$	8d[d] or 6d[e] $2^3/_8'' \times 0.113''$ nail[p] 2" 16 gage[p]	
	$^7/_8''$ to 1"	8d[c]	
Single Floor (combination subfloor-underlayment to framing)	$1^1/_8''$ to $1^1/_4''$	10d[d] or 8d[d]	
	$^3/_4''$ and less	6d[e]	
	$^7/_8''$ to 1"	8d[e]	
	$1^1/_8''$ to $1^1/_4''$	10d[d] or 8d[e]	
32. Panel siding (to framing)	$^1/_2''$ or less	6d[f]	
	$^5/_8''$	8d[f]	
33. Fiberboard sheathing[g]	$^1/_2''$	No. 11 gage roofing nail[h] 6d common nail ($2'' \times 0.113''$) No. 16 gage staple[i]	
	$^{25}/_{32}''$	No. 11 gage roofing nail[h] 8d common nail ($2^1/_2'' \times 0.131''$) No. 16 gage staple[i]	
34. Interior paneling	$^1/_4''$	4d[j]	
	$^3/_8''$	6d[k]	

For SI: 1 inch = 25.4 mm.

a. Common or box nails are permitted to be used except where otherwise stated.

b. Nails spaced at 6 inches on center at edges, 12 inches at intermediate supports except 6 inches at supports where spans are 48 inches or more. For nailing of wood structural panel and particleboard diaphragms and shear walls, refer to Section 2305. Nails for wall sheathing are permitted to be common, box or casing.

c. Common or deformed shank (6d - $2'' \times 0.113''$; 8d - $2^1/_2'' \times 0.131''$; 10d - $3'' \times 0.148''$).

d. Common (6d - $2'' \times 0.113''$; 8d - $2^1/_2'' \times 0.131''$; 10d - $3'' \times 0.148''$).

e. Deformed shank (6d - $2'' \times 0.113''$; 8d - $2^1/_2'' \times 0.131''$; 10d - $3'' \times 0.148''$).

f. Corrosion-resistant siding (6d - $1^7/_8'' \times 0.106''$; 8d - $2^3/_8'' \times 0.128''$) or casing (6d - $2'' \times 0.099''$; 8d - $2^1/_2'' \times 0.113''$) nail.

g. Fasteners spaced 3 inches on center at exterior edges and 6 inches on center at intermediate supports, when used as structural sheathing. Spacing shall be 6 inches on center on the edges and 12 inches on center at intermediate supports for nonstructural applications.

h. Corrosion-resistant roofing nails with $^7/_{16}$-inch-diameter head and $1^1/_2$-inch length for $^1/_2$-inch sheathing and $1^3/_4$-inch length for $^{25}/_{32}$-inch sheathing.

i. Corrosion-resistant staples with nominal $^7/_{16}$-inch crown and $1^1/_8$-inch length for $^1/_2$-inch sheathing and $1^1/_2$-inch length for $^{25}/_{32}$-inch sheathing. Panel supports at 16 inches (20 inches if strength axis in the long direction of the panel, unless otherwise marked).

j. Casing ($1^1/_2'' \times 0.080''$) or finish ($1^1/_2'' \times 0.072''$) nails spaced 6 inches on panel edges, 12 inches at intermediate supports.

k. Panel supports at 24 inches. Casing or finish nails spaced 6 inches on panel edges, 12 inches at intermediate supports.

l. For roof sheathing applications, 8d nails ($2^1/_2'' \times 0.113''$) are the minimum required for wood structural panels.

m. Staples shall have a minimum crown width of $^7/_{16}$ inch.

n. For roof sheathing applications, fasteners spaced 4 inches on center at edges, 8 inches at intermediate supports.

o. Fasteners spaced 4 inches on center at edges, 8 inches at intermediate supports for subfloor and wall sheathing and 3 inches on center at edges, 6 inches at intermediate supports for roof sheathing.

p. Fasteners spaced 4 inches on center at edges, 8 inches at intermediate supports.

Figure 2304.9.1
STANDARD NAILS AND STAPLES
Independent Nail Packing Co., Bridgewater, MA

2304.9.2 Sheathing fasteners. Sheathing nails or other approved sheathing connectors shall be driven so that their head or crown is flush with the surface of the sheathing.

❖ This requirement is a matter of workmanship (see Figure 2304.9.2 for an illustration of a nail driven to fasten sheathing properly). Protruding nails do not provide the intended connecting capacity and could be hazardous. Likewise, nails overdriven into structural sheathing may not perform as expected. Framing installation is often less than perfect and fasteners are overdriven to a point where the top layer of sheathing is crushed beneath the nail head. An occasional overdriven nail may not be significant. As the percentage of overdriven fasteners increases, the issue raised is at what point does this adversely affect the shear capacity of a diaphragm or shear wall element. The American Plywood Association has recognized that this is a common occurrence and has made a guideline available at no cost on its website (www.apawood.org). Another condition to be aware of in sheathing nailing is where the depth of the supporting member is less than the length of a commonly used fastener, such as in a case where sheathing will be applied over the top of flat decking. Shorter nails are available for these situations, but the holding and shear capacities are typically reduced.

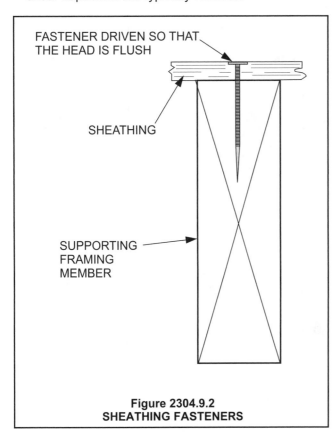

FASTENER DRIVEN SO THAT THE HEAD IS FLUSH

SHEATHING

SUPPORTING FRAMING MEMBER

Figure 2304.9.2
SHEATHING FASTENERS

2304.9.3 Joist hangers and framing anchors. Connections depending on joist hangers or framing anchors, ties and other mechanical fastenings not otherwise covered are permitted where approved. The vertical load-bearing capacity, torsional moment capacity and deflection characteristics of joist hangers shall be determined in accordance with Section 1715.1.

❖ This section recognizes for the use of joist and framing anchors as an alternative to conventional fastening. These items are engineered components and must be used in accordance with the manufacturer's instructions and design specifications. The capacity for each hanger must be specified by the manufacturer and is determined in accordance with the test methods outlined in Section 1715.1. All fasteners required to make the anchor or hanger support load are specified by the manufacturer, who also describes any limitations that the designer or contractor need to consider when using the product.

2304.9.4 Other fasteners. Clips, staples, glues and other approved methods of fastening are permitted where approved.

❖ In addition to joist hangers and typical fastening per Table 2304.9.1, there are other fastening methods that can be approved by the building official, provided there is sufficient evidence to prove their adequacy. This section is included in the code to address the fact that new fasteners and connection methods are continually being developed, which necessitates that the building official give consideration to viable alternatives.

2304.9.5 Fasteners in preservative-treated and fire-retardant-treated wood. Fasteners for preservative- treated and fire-retardant-treated wood shall be of hot dipped zinc-coated galvanized steel, stainless steel, silicon bronze or copper. The coating weights for zinc-coated fasteners shall be in accordance with ASTM A 153.

> **Exception:** Fasteners other than nails, timber rivets, wood screws and lag screws shall be permitted to be of mechanically deposited zinc coated steel with coating weights in accordance with ASTM B 695, Class 55 minimum.

Fastenings for wood foundations shall be as required in AF&PA Technical Report No. 7.

❖ Corrosion-resistant fasteners are required for treated wood, since treated wood is associated with conditions where wood and fasteners could be subjected to moisture. This applies to fire-retardant-treated wood as well as pressure-preservative-treated wood. The base requirements limit galvanized steel to the hot-dipped process. Because mechanical galvanizing is a preferred method for certain fastener types, an exception recognizes this process where it is appropriate.

If treated wood is used for purposes of termite resistance and in locations where moisture accumulation is not likely, there is no practical necessity for the use of corrosion-resistant fasteners. AF&PA Technical Report No. 7 requires specific types of corrosion-resistant nails for locations below grade and above grade in wood foundations, based on the increased exposure to moisture in below-grade applications. This report is available for free download from the American Wood Council's website, www.awc.org.

2304.9.6 Load path. Where wall framing members are not continuous from foundation sill to roof, the members shall be secured to ensure a continuous load path. Where required, sheet metal clamps, ties or clips shall be formed of galvanized steel or other approved corrosion-resistant material not less than 0.040 inch (1.01 mm) nominal thickness.

❖ In conventional construction, the fastening required by Table 2304.9.1 provides for a continuous connection between framing elements as required by this section. If the loads on the building exceed the parameters established in Section 2308.2, the building, in total or in part, must be designed in accordance with the ASD or LRFD methodologies and the connections must be designed in accordance with those standards as well to provide continuity of load support throughout the structure. The designed connections may incorporate engineered tie or strap components with capacities specified by the manufacturer. This section specifies the minimum thickness of sheet metal for fabricated straps or ties to 0.04 inch (1.01 mm) nominal, which is common among manufacturers. Designers and contractors must keep in mind that a building is simply a huge assembly of multiple independent parts. When a building is subject to external forces, each independent part has a tendency to move into a different position and this tendency must be controlled by completing the path to deliver the imposed forces into the ground. This load path may be comprised of hangers from a beam to a column, metal straps from an upper-level stud to a lower-level stud or from hold-down anchors into the concrete. There are countless applications and possibilities, but the overall concept is that every individual piece must be fastened to some adjacent piece.

2304.9.7 Framing requirements. Wood columns and posts shall be framed to provide full end bearing. Alternatively, column-and-post end connections shall be designed to resist the full compressive loads, neglecting end-bearing capacity. Column-and-post end connections shall be fastened to resist lateral and net induced uplift forces.

❖ This section requires that the framing workmanship be such that the ends of wood columns and posts, including studs, bear completely atop the supporting member, whether it be a wood sill plate or a metal base connector. If it does not bear fully on the entire cross section, then design calculations are necessary in order to show that the bearing provided will be capable of supporting the loads. It also requires a positive connection (not merely a friction connection) to resist incidental lateral forces and anticipated uplift forces. For instance, a wood column supporting beams in the basement of a structure could not simply bear on the concrete pad beneath it without a means of positive anchorage to the concrete, to prevent it from accidently being displaced on impact by a person or an object. If the combination of loads on the structure could result in uplift forces, then a connection to resist uplift

would also be required. This section is a further extension of the philosophy outlined in Section 2304.9.6.

2304.10 Heavy timber construction.

❖ Heavy timber construction originated in New England to serve the needs of the growing textile industry. As the industry modernized, the need for larger open space, unobstructed by columns, gradually reduced the demand for this type of construction.

Today, due primarily to its aesthetic value, heavy timber construction is used in many other occupancies. It is commonly used for assembly and mercantile buildings, such as schools and churches, auditoriums, gymnasiums and supermarkets.

The provisions of this section must be used with the minimum dimension requirements in Section 602.4, where heavy timber construction is also classified as "Type IV Construction" per Table 601. A structural analysis must be performed to verify that the minimum dimensions are adequate. These final dimensions will always be based on a shrunken section since timbers are ordinarily supplied in a green condition, unless they are special products manufactured to specific tolerances.

2304.10.1 Columns. Columns shall be continuous or superimposed throughout all stories by means of reinforced concrete or metal caps with brackets, or shall be connected by properly designed steel or iron caps, with pintles and base plates, or by timber splice plates affixed to the columns by metal connectors housed within the contact faces, or by other approved methods.

❖ The AF&PA publication, *Heavy Timber Construction Details—Wood Construction Data 5*, provides details for the proper connection of columns and beams. Traditionally, heavy timber structures were designed and constructed using a prescriptive approach, much like today's conventional construction for light-frame buildings. The number and size of bolts, lag screws or connectors should be determined through analysis of the loads to be supported.

Pintles provide a method to fasten the butt ends of beams or columns. A pintle acts like a short column, connecting the ends of the structural elements.

2304.10.1.1 Column connections. Girders and beams shall be closely fitted around columns and adjoining ends shall be cross tied to each other, or intertied by caps or ties, to transfer horizontal loads across joints. Wood bolsters shall not be placed on tops of columns unless the columns support roof loads only.

❖ Where columns and girders or beams intersect, care must be taken for load transfer. Proper detailing at the intersection of columns and girders will limit the effects of shrinkage. Elements that are not continuous must be properly fastened together to transfer design forces. Today's engineered wood products are available in multiple depths and widths, and in lengths up to 60 feet (18 288 mm), so continuous span structural elements are readily available. Bolster blocks are used to reduce the compression perpendicular to grain stresses where roof beams bear on columns. The shrinkage of the bolster block must be considered in the roof design. Roof beams must have a positive connection to the column to resist uplift.

2304.10.2 Floor framing. Approved wall plate boxes or hangers shall be provided where wood beams, girders or trusses rest on masonry or concrete walls. Where intermediate beams are used to support a floor, they shall rest on top of girders, or shall be supported by ledgers or blocks securely fastened to the sides of the girders, or they shall be supported by an approved metal hanger into which the ends of the beams shall be closely fitted.

❖ Wood members must not bear directly on masonry or concrete, otherwise they will absorb moisture from these elements and lose strength in bearing due to decay. Typically, a steel bearing plate or custom steel hanger is used to support heavy timber elements. Bearing plates provide an even distribution of load on the wall. Wall plate boxes should be designed and detailed to provide a positive connection to the masonry. Where direct bearing on top of the girders is not possible or desirable, a metal hanger designed to transfer all induced loads is permitted. An additional advantage to using a hanger is less shrinkage. Beams consisting of green lumber should be installed so their top edge is slightly above the top edge of the girder.

2304.10.3 Roof framing. Every roof girder and at least every alternate roof beam shall be anchored to its supporting member; and every monitor and every sawtooth construction shall be anchored to the main roof construction. Such anchors shall consist of steel or iron bolts of sufficient strength to resist vertical uplift of the roof.

❖ Anchorage of roof beams and girders should be in accordance with the engineered design. This section provides minimum acceptable anchorage practices that are especially important in unusual or discontinuous types of roof framing, such as the "sawtooth" arrangement. Sawtooth roof construction is used to provide large amounts of daylight into the floor area of the building. The roof line looks much like the cutting edge of a saw, with regularly spaced peaks and valleys. The vertical or near vertical line connecting the peaks and valleys is filled with glazing. This unique framing pattern must be anchored to the main roof supports.

2304.10.4 Floor decks. Floor decks and covering shall not extend closer than $^1/_2$ inch (12.7 mm) to walls. Such $^1/_2$-inch (12.7 mm) spaces shall be covered by a molding fastened to the wall either above or below the floor and arranged such that the molding will not obstruct the expansion or contraction movements of the floor. Corbeling of masonry walls under floors is permitted in place of such molding.

❖ Flooring will expand and contract as the relative humidity in the building changes. To accommodate expansion, a $^1/_2$-inch (12.7 mm) gap is required at the wall. The gap will also prevent moisture in the walls from migrating into the floor boards. The fire integrity of the floor assembly must be maintained, so either molding or corbeling of the wall below is required to prevent the passage of smoke and hot gases.

2304.10.5 Roof decks. Where supported by a wall, roof decks shall be anchored to walls to resist uplift forces determined in

accordance with Chapter 16. Such anchors shall consist of steel or iron bolts of sufficient strength to resist vertical uplift of the roof.

❖ The roof deck must be anchored frequently enough to resist uplift forces as prescribed in Chapter 16. This type of anchorage is not covered by the conventional construction provisions of the code; therefore, engineering is required per the approved methods, ASD or LRFD.

2304.11 Protection against decay and termites.

2304.11.1 General. Where required by this section, protection from decay and termites shall be provided by the use of naturally durable or preservative-treated wood.

❖ Conditions favorable for decay and fungus attack are discussed in Sections 2304.11.2 and 2304.11.4, along with preservative treatments. This section sets forth specific locations of concern and minimum construction practices to prevent such an attack. Termite protection is also defined. According to *The Complete Book of Home Inspection*, termite infestation actually should not be an alarming discovery since termites do their damage very slowly. It will take a colony of 60,000 termites an entire year to destroy only 4 feet (1219 mm) of boards that measure 2 inches by 4 inches (51 by 102 mm). Chemically treated lumber provided as described in this section will generally control infestation possibilities.

2304.11.2 Wood used above ground. Wood used above ground in the locations specified in Sections 2304.11.2.1 through 2304.11.2.7, 2304.11.3 and 2304.11.5 shall be naturally durable wood or preservative-treated wood using water-borne preservatives, in accordance with AWPA U1 (Commodity Specifications A or F) for above-ground use.

❖ The AWPA is the principal standards writing body for the wood preserving industry in the United States. AWPA U1 covers aspects of treated wood from the treating process to appropriate treatment retention levels for specific applications. Wood products used in locations identified in this section are particularly susceptible to deterioration. Deterioration can result from continuous or intermittent contact with moisture or termites.

Decay is caused by fungi, which are low forms of plant life that attack the cell walls of wood in order to feed on the contents of the cells. For fungi to attack wood in service, the following conditions must be present: (1) temperature in the range of 35° to 100°F (1° to 38°C); (2) adequate supply of oxygen; and (3) wood moisture content in excess of 20 percent. A properly-constructed and maintained building will typically remain at a moisture content of less than 15 percent throughout its life, but some special conditions might cause fluctuations leading to problems (see Section 2304.11.2.2).

2304.11.2.1 Joists, girders and subfloor. Where wood joists or the bottom of a wood structural floor without joists are closer than 18 inches (457 mm), or wood girders are closer than 12 inches (305 mm) to the exposed ground in crawl spaces or unexcavated areas located within the perimeter of the building foundation, the floor assembly (including posts, girders, joists and subfloor) shall be of naturally durable or preservative-treated wood.

❖ The clearances specified in this section are recommended for all buildings and are considered necessary to: (1) maintain wood elements in permanent structures at safe moisture contents for decay protection; (2) provide a termite barrier; and (3) facilitate periodic inspection (see Figure 2304.11.2.1). When it is not possible or practical to comply with the clearance specified, the use of naturally durable or pressure-preservative-treated wood is recommended. However, when termites are known to exist, space must be provided for periodic inspection. Often termite infestations are the result of poor construction or improper clearance between wood framing and the surrounding soil. Unfortunately, termites are very determined insects and will often build small, moisture-controlled tubes or tunnels from the ground surface to the source of food, bypassing inedible barriers, which is one of the reasons it is important to provide adequate space below floor framing for periodic visual inspections of wood framing.

2304.11.2.2 Wood supported by exterior foundation walls. Wood framing members, including wood sheathing, that rest on exterior foundation walls and are less than 8 inches (203 mm) from exposed earth shall be of naturally durable or preservative-treated wood.

❖ Traditionally, light-gage metal termite shields were used to obstruct the construction of tunnels from the soil, up the face of the foundation and into the frame. The required clearance, as shown in Figure 2304.11.2.1, provides an exposed foundation surface area that allows for a visual inspection for termite tubes.

2304.11.2.3 Exterior walls below grade. Wood framing members and furring strips attached directly to the interior of exterior masonry or concrete walls below grade shall be of approved naturally durable or preservative-treated wood.

❖ Wood attached directly to masonry or concrete will draw moisture from these materials, and may reach a moisture content greater than 20 percent, depending on the relative humidity of the basement.

2304.11.2.4 Sleepers and sills. Sleepers and sills on a concrete or masonry slab that is in direct contact with earth shall be of naturally durable or preservative-treated wood.

❖ Wood in contact with either the earth or material that retains moisture must be naturally durable or preservative treated. The moisture content of wood will change based on the relative humidity of the air and the moisture of materials with which it is in contact. Interior slabs that are protected from the earth by an impervious moisture barrier should not be considered in direct contact with the earth for purposes of this requirement. Sometimes this moisture barrier becomes damaged in the course of construction, but will remain intact for the most part. Moisture-sensitive flooring materials typically require special measures or care to be taken during placement of the concrete slab and preparation of the subbase, but interior wall sill plates will typically maintain an acceptable moisture content even if special drying measures are not taken for the slab.

2304.11.2.5 Girder ends. The ends of wood girders entering exterior masonry or concrete walls shall be provided with a $^1/_2$-inch (12.7 mm) air space on top, sides and end, unless naturally durable or preservative-treated wood is used.

❖ Clearance around the ends of girders prevents moisture from migrating from the concrete into the wood. The minimum clearance that is shown in Figure 2304.11.2.5 allows for air circulation and periodic in-

For SI: 1 inch = 25.4 mm.

**Figure 2304.11.2.1
WOOD-FRAMING CLEARANCES**

spection for termites. This requirement has been in the model codes for many years, proving it to be an effective step in controlling wood deterioration.

2304.11.2.6 Wood siding. Clearance between wood siding and earth on the exterior of a building shall not be less than 6 inches (152 mm) except where siding, sheathing and wall framing are of naturally durable or preservative-treated wood.

❖ Exterior wood siding should extend at least 1 inch (25 mm) below the top of the foundation to provide a drip line protecting the sill from rainwater. The siding is typically held away from the foundation with a starter strip attached to the sheathing that ends at the bottom of the plate. The 1-inch (25 mm) overhang causes the rainwater to drip off the bottom edge of the siding, rather than being absorbed by the underlying sheathing. A minimum clearance illustrated in Figure 2304.11.2.6 must be provided for wood siding that is neither naturally durable nor preservative treated.

2304.11.2.7 Posts or columns. Posts or columns supporting permanent structures and supported by a concrete or masonry slab or footing that is in direct contact with the earth shall be of naturally durable or preservative-treated wood.

Exceptions:

1. Posts or columns that are either exposed to the weather or located in basements or cellars, supported

For SI: 1 inch = 25.4 mm.

Figure 2304.11.2.5
GIRDER END CLEARANCE

For SI: 1 inch = 25.4 mm.

Figure 2304.11.2.6
SIDING CLEARANCES

by concrete piers or metal pedestals projected at least 1 inch (25 mm) above the slab or deck and 6 inches (152 mm) above exposed earth, and are separated therefrom by an impervious moisture barrier.

2. Posts or columns in enclosed crawl spaces or unexcavated areas located within the periphery of the building, supported by a concrete pier or metal pedestal at a height greater than 8 inches (203 mm) from exposed ground, and are separated therefrom by an impervious moisture barrier.

❖ When the post or column is directly in concrete that is in direct contact with earth, it must be naturally durable or preservative treated. The exceptions illustrated in Figure 2304.11.2.7 permit the use of common framing lumber when the post or column is isolated from the earth or concrete. Generally, structural elements in basements or cellars are easier to inspect, which explains the lower clearances provided by Exception 1. Posts or columns in crawl spaces are required to have a greater clearance to safeguard against termites between inspections. In both exceptions, an impervious moisture barrier restricts the absorption of water from the earth or concrete pier.

2304.11.3 Laminated timbers. The portions of glued-laminated timbers that form the structural supports of a building or other structure and are exposed to weather and not fully protected from moisture by a roof, eave or similar covering shall be pressure treated with preservative or be manufactured from naturally durable or preservative-treated wood.

❖ It is common practice to design large glued-laminated arches that "spring" or are connected to foundations near ground level. Exterior walls are built inside the span of these arches, leaving the initial few feet of the wood arches exposed to the weather. Experience has shown that covering the tops of these arches with metal or other water seals is not sufficient to prevent decay; therefore, arches and other exposed wood members not protected by roofs or similar covers must be laminated of naturally durable or pressure-treated wood. Creosote solutions are typically used for the most severe outdoor exposures, but they are quite dark, oily and pungent and are, therefore, typically restricted to uses where direct human contact is avoided. Oil-borne and water-borne treatments are a bit friendlier, but they have different effects depending on the timing of the application, either before or after laminating the timbers together.

2304.11.4 Wood in contact with the ground or fresh water. Wood used in contact with the ground (exposed earth) in the locations specified in Sections 2304.11.4.1 and 2304.11.4.2 shall be naturally durable (species for both decay and termite resistance) or preservative treated using water-borne preservatives in accordance with AWPA U1 (Commodity Specifications A or F) for soil or fresh water use.

Exception: Untreated wood is permitted where such wood is continuously and entirely below the ground-water level or submerged in fresh water.

❖ Wood that has been completely submerged below the ground-water table or in freshwater has been known to remain in good condition for many years. This method of preservation is the only exception to requiring preservative treatment of all wood in contact with or embedded in the ground.

2304.11.4.1 Posts or columns. Posts and columns supporting permanent structures that are embedded in concrete that is in direct contact with the earth, embedded in concrete that is exposed to the weather or in direct contact with the earth shall be of preservative-treated wood.

❖ This requirement is similar to Section 2304.11.2.7, but emphasizes that posts and columns encased or em-

For SI: 1 inch = 25.4 mm.

Figure 2304.11.2.7
POST AND COLUMN PROTECTION

bedded in concrete must be of preservative-treated wood. The condition is generally more severe than wood placed directly on concrete or embedded in the earth. The embedded portion of the column will most likely be at a moisture content in excess of 20 percent. At some point very near where the concrete ends, the conditions will be ideal for decay—having both a high moisture content and the presence of air. Unless treated wood is used, decay will occur very near where the encasing begins.

2304.11.4.2 Wood structural members. Wood structural members that support moisture-permeable floors or roofs that are exposed to the weather, such as concrete or masonry slabs, shall be of naturally durable or preservative-treated wood unless separated from such floors or roofs by an impervious moisture barrier.

❖ The framing condition described is uncommon, but the requirement for naturally durable or preservative-treated wood is appropriate. Concrete or masonry can contain sufficient moisture to raise the moisture content in supporting wood structural elements above 20 percen; therefore, an impervious moisture barrier or naturally durable or preservative-treated wood is required.

2304.11.5 Supporting member for permanent appurtenances. Naturally durable or preservative-treated wood shall be utilized for those portions of wood members that form the structural supports of buildings, balconies, porches or similar permanent building appurtenances where such members are exposed to the weather without adequate protection from a roof, eave, overhang or other covering to prevent moisture or water accumulation on the surface or at joints between members.

Exception: When a building is located in a geographical region where experience has demonstrated that climatic conditions preclude the need to use durable materials where the structure is exposed to the weather.

❖ Decks and porches exposed to rain and snow are subject to deterioration. In desert areas where rainfall is slight, there may not be enough moisture to cause any deterioration; therefore, the exception is provided. It is important to confirm climatic conditions with the weather bureau or the building official to determine the proper course of action in protecting these appurtenant structures.

2304.11.6 Termite protection. In geographical areas where hazard of termite damage is known to be very heavy, wood floor framing shall be of naturally durable species (termite resistant) or preservative treated in accordance with AWPA U1 for the species, product preservative and end use or provided with approved methods of termite protection.

❖ This provision applies where termite damage has been determined to be very heavy. The termite infestation probability map used in the *International Residential Code*® (IRC®) is included as Figure 2304.11.6. This map provides guidance, but the local jurisdiction

should be consulted as they will be more familiar with the threat posed by termites. Most termites found in the United States are classified as subterranean termites, which indicates their need to live in the ground. For many years, metal termite shields were considered the best method of preventing infestation. When properly installed, such shields are still a good defense. Soil poisoning is considered to be a more efficient method of controlling termites; however, there are obvious negative environmental implications and some regions may forbid this method altogether.

Additionally, all possible locations of infestation must be poisoned; therefore, the work should be carefully monitored and performed by professionals who are competent and understand applicable standards, laws and procedures. Under no circumstances should soil poisoning be attempted by those unfamiliar with toxic products and their application.

2304.11.7 Wood used in retaining walls and cribs. Wood installed in retaining or crib walls shall be preservative treated in accordance with AWPA U1 (Commodity Specifications A or F) for soil and fresh water use.

❖ Where the structure is regulated by the code, wood used to retain soil is to be preservative treated. Presumably, where the failure of the retaining wall or crib may threaten life safety or result in property damage, the durability provided by preservative-treated wood is necessary.

2304.11.8 Attic ventilation. For attic ventilation, see Section 1203.2.

❖ Adequate ventilation allows moisture-laden air to readily escape from the attic space in order to avoid exposure of the framing to excessive moisture (see Section 1203.2 for further discussion).

2304.11.9 Under-floor ventilation (crawl space). For under-floor ventilation (crawl space), see Section 1203.3.

❖ Enclosed spaces under floors of buildings are referred to as crawl spaces when not designed and constructed as basements. The earth is usually exposed or covered with a vapor retarder. Regardless, there is always the potential for a large amount of moisture vapor from the ground being present. This moisture must be removed to prevent wetting and drying cycles that can cause decay and fungus attack in wood members. Experience has shown that crawl spaces constructed in conformance with this section and ventilated to the exterior as required by Section 1203.3 do not develop decay problems.

Where a high ground-water table exists or moisture is abnormally high, the crawl space must be adequately drained in accordance with Section 1807.1.2. It should be recognized that the moisture problem may be exaggerated in colder months; therefore, operable vents should be used with caution. Once insulation in such spaces becomes wet, it seldom dries out, thus negating its performance.

Figure 2304.11.6
TERMITE INFESTATION PROBABILITY MAP

VERY HEAVY

MODERATE TO HEAVY

SLIGHT TO MODERATE

NONE TO SLIGHT

2304.12 Long-term loading. Wood members supporting concrete, masonry or similar materials shall be checked for the effects of long-term loading using the provisions of the AF&PA NDS. The total deflection, including the effects of long-term loading, shall be limited in accordance with Section 1604.3.1 for these supported materials.

> **Exception:** Horizontal wood members supporting masonry or concrete nonstructural floor or roof surfacing not more than 4 inches (102 mm) thick need not be checked for long-term loading.

❖ It is common for wood structural elements to support masonry and concrete construction. When properly designed, taking into consideration long-term deflection, wood can adequately support these products. The deflection limitations on wood supporting masonry and concrete are recommended by that industry. Masonry and concrete are brittle materials, which do not tolerate movement. Wood is hygroscopic, meaning it changes dimensions as a result of absorbing or releasing (drying) water. Dimensional changes in wood due to improper detailing can result in damage to the supported masonry and concrete.

The exception provides some relief. Flooring and roof coverings meeting specific thickness criteria may be supported by wood. The wood must be designed to carry the weight of the supported material, but it is not necessary to consider the effect of long-term loading.

Tabulated modulus of elasticity design values, *E*, in the AF&PA NDS are intended for the calculation of immediate deformation under load. Under sustained loading, wood members exhibit additional time dependent formation (creep), which develops at a slow but steady rate over time. Where dead loads or sustained live loads represent a relatively high percentage of total design load, creep is a design consideration and is handled by these provisions in the AF&PA NDS. In such situations the total deflection under long-term loading is estimated by increasing the initial deflection associated with the long-term load component by 1.5 for seasoned or 2.0 for unseasoned or wet lumber or glued-laminated timber.

SECTION 2305
GENERAL DESIGN REQUIREMENTS FOR LATERAL-FORCE-RESISTING SYSTEMS

2305.1 General. Structures using wood shear walls and diaphragms to resist wind, seismic and other lateral loads shall be designed and constructed in accordance with the provisions of this section. Alternatively, compliance with the AF&PA SDPWS shall be permitted subject to the limitations therein and the limitations of this code.

❖ General design requirements for lateral-force-resisting systems are described in this section and are applicable to engineered structures. Elements discussed in this section are designed using either ASD or LRFD methodologies.

When elements within structures constructed according to Section 2308 need to be engineered, provisions of Section 2305 can be applied without engineering the entire structure. The extent of design to be provided must be determined by the design professional and building official. However, the minimum acceptable extent is often taken to be force transfer into the element, design of the element and force transfer out of the element. When more than one braced wall line or diaphragm in any area of conventional structures requires engineering analysis, the assumptions used to develop Section 2308 may have changed, and a complete analysis may be appropriate for the entire lateral-force-resisting system. For example, the absence of a ceiling diaphragm may create a nonconventional configuration.

2305.1.1 Shear resistance based on principles of mechanics. Shear resistance of diaphragms and shear walls are permitted to be calculated by principles of mechanics using values of fastener strength and sheathing shear resistance.

❖ In lieu of using the tabular values for shear capacity, such as those in Table 2306.3.1, this section permits calculation of shear wall and diaphragm resistance by principles of engineering mechanics. Methodologies are described in the technical literature that allow for an analytical determination of shear capacity. Design values for wood members and fastener strength in wood members can be determined in accordance with the AF&PA NDS.

2305.1.2 Framing. Boundary elements shall be provided to transmit tension and compression forces. Perimeter members at openings shall be provided and shall be detailed to distribute the shearing stresses. Diaphragm and shear wall sheathing shall not be used to splice boundary elements. Diaphragm chords and collectors shall be placed in, or tangent to, the plane of the diaphragm framing unless it can be demonstrated that the moments, shears and deformations, considering eccentricities resulting from other configurations can be tolerated without exceeding the adjusted resistance and drift limits.

❖ The transfer of forces into and out of diaphragms and shear walls is necessary in the analysis of load path continuity. Boundary elements must be sized and connected to the diaphragm to ensure force transfer. This section provides basic framing requirements for boundary elements in shear walls and diaphragms. Good construction practice for boundary elements utilizes framing members in the plane or tangent to the plane of the diaphragm. Where splices occur, transfer of forces is usually through addition of framing members or metal connectors—not the diaphragm or shear wall sheathing.

2305.1.2.1 Framing members. Framing members shall be at least 2 inch (51 mm) nominal width. In general, adjoining panel edges shall bear and be attached to the framing members and butt along their centerlines. Nails shall be placed not less than $^3/_8$ inch (9.5 mm) from the panel edge, not more than 12 inches (305 mm) apart along intermediate supports, and 6 inches (152 mm) along panel edge bearings, and shall be firmly driven into the framing members.

❖ The prescriptive requirements in this section for the use of dimension lumber and minimum panel edge distance for nailing are intended to provide an adequate connection between sheathing and framing in shear wall and diaphragm construction. Edge spacing is intended to prevent splitting of the wood structural panel. The fastener spacing is consistent with the minimum spacing provided in Table 2304.9.1.

2305.1.3 Openings in shear panels. Openings in shear panels that materially affect their strength shall be fully detailed on the plans, and shall have their edges adequately reinforced to transfer all shearing stresses.

❖ Openings occur in shear walls and diaphragms for stairs, windows, doors, shafts and other purposes. The transfer of shear forces around openings must be effectively addressed in the design to provide adequate capacity exists around the perimeter of the opening to transfer shear forces.

2305.1.4 Shear panel connections. Positive connections and anchorages capable of resisting the design forces shall be provided between the shear panel and the attached components. In Seismic Design Category D, E or F, the capacity of toenail connections shall not be used when calculating lateral load resistance to transfer lateral earthquake forces in excess of 150 pounds per foot (2189 N/m) from diaphragms to shear walls, drag struts (collectors) or other elements, or from shear walls to other elements.

❖ This section states the requirement for connections and anchorage capable of resisting design forces with a limitation on the use of toenail connections (see left side of Figure 2305.1.4) for shear transfer in Seismic Design Category D, E or F. Toenails are driven through the end grain of a side member and are used where it is not possible to face nail into the holding member. For example, Item 12 of Table 2304.9.1 provides requirements for a rim joist that is toenailed to a top plate. The toenailed connection does not provide a lateral load resistance equivalent to that of a face nailed connection. Where the seismic risk is elevated, this type of connection cannot be utilized to provide the requisite lateral load path for seismic load effects. Where the transfer of lateral forces from seismic loads is in excess of 150 pounds per foot (2188 N/m), based on calculated lateral load resistance, other connection methods such as framing anchors or face nailing to perpendicular blocking (see right side of Figure 2305.1.4) should be utilized. Although it is not explicit, this limitation should be based on the effects of nominal earthquake loads, as opposed to factored earthquake loading.

2305.1.5 Wood members resisting horizontal seismic forces contributed by masonry and concrete walls. Wood shear walls, diaphragms, horizontal trusses and other members shall not be used to resist horizontal seismic forces contributed by masonry or concrete walls in structures over one story in height.

Exceptions:

1. Wood floor and roof members are permitted to be used in horizontal trusses and diaphragms to resist horizontal seismic forces contributed by masonry or concrete walls, provided such forces do not result in torsional force distribution through the truss or diaphragm.

RIM JOIST

RIM JOIST

TOP PLATE

WALL STUDS

TOE NAILED

FACE NAILED ALTERNATIVE

Figure 2305.1.4
TOENAILED VERSUS FACE-NAILED CONNECTION

2. Wood structural panel sheathed shear walls are permitted to be used to provide resistance to seismic forces contributed by masonry or concrete walls in two-story structures of masonry or concrete walls, provided the following requirements are met:

2.1. Story-to-story wall heights shall not exceed 12 feet (3658 mm).

2.2. Diaphragms shall not be designed to transmit lateral forces by rotation and shall not cantilever past the outermost supporting shear wall.

2.3. Combined deflections of diaphragms and shear walls shall not permit story drift of supported masonry or concrete walls to exceed the limit of Section 12.12.1 in ASCE 7.

2.4. Wood structural panel sheathing in diaphragms shall have unsupported edges blocked. Wood structural panel sheathing for both stories of shear walls shall have unsupported edges blocked and, for the lower story, shall have a minimum thickness of $^{15}/_{32}$ inch (11.9 mm).

2.5. There shall be no out-of-plane horizontal offsets between the first and second stories of wood structural panel shear walls.

❖ The use of wood diaphragms with masonry or concrete walls is a common construction practice. The compatibility of wood and other less ductile construction materials is regulated in this section. Many of the provisions of this section are subjective and based on field observations following major seismic events. This provision clarifies that limitations on wood construction resisting horizontal seismic forces apply when wood systems are resisting seismic forces contributed by masonry or concrete walls and not other types of masonry or concrete construction, such as masonry veneer and concrete floors. For cases where horizontal seismic forces are caused by masonry veneer or concrete floors, elements of the wood lateral-force-resisting system can be designed using standard engineering methods.

2305.1.6 Wood members resisting seismic forces from nonstructural concrete or masonry. Wood members shall be permitted to resist horizontal seismic forces from nonstructural concrete, masonry veneer or concrete floors.

❖ This provision clarifies that limitations on wood construction resisting horizontal seismic forces apply when wood systems are resisting seismic forces contributed by masonry or concrete walls and not other types of masonry or concrete construction such as masonry veneer and concrete floors. Section 2305 provisions apply only to engineered wood buildings designed in accordance with Section 2306 or 2307.

2305.2 Design of wood diaphragms.

2305.2.1 General. Wood diaphragms are permitted to be used to resist horizontal forces provided the deflection in the plane of the diaphragm, as determined by calculations, tests or anal-

ogies drawn therefrom, does not exceed the permissible deflection of attached distributing or resisting elements. Connections shall extend into the diaphragm a sufficient distance to develop the force transferred into the diaphragm.

❖ General requirements for wood diaphragms include consideration of diaphragm strength and deflection. The transfer of forces into and out of the diaphragm must be included in the design. General requirements for diaphragms can be found in APA Research Report 138, *Plywood Diaphragms*.

2305.2.2 Deflection. Permissible deflection shall be that deflection up to which the diaphragm and any attached distributing or resisting element will maintain its structural integrity under design load conditions, such that the resisting element will continue to support design loads without danger to occupants of the structure. Calculations for diaphragm deflection shall account for the usual bending and shear components as well as any other factors, such as nail deformation, which will contribute to deflection.

The deflection (Δ) of a blocked wood structural panel diaphragm uniformly nailed throughout is permitted to be calculated by using the following equation. If not uniformly nailed, the constant 0.188 (For SI: $^1/_{1627}$) in the third term must be modified accordingly.

$$\Delta = \frac{5vL^3}{8EAb} + \frac{vL}{4Gt} + 0.188Le_n + \frac{\Sigma(\Delta_c X)}{2b} \qquad \textbf{(Equation 23-1)}$$

For SI: $\Delta = \dfrac{0.052vL^3}{EAb} + \dfrac{vL}{4Gt} + \dfrac{Le_n}{1627} + \dfrac{\Sigma(\Delta_c X)}{2b}$

where:

A = Area of chord cross section, in square inches (mm²).

b = Diaphragm width, in feet (mm).

E = Elastic modulus of chords, in pounds per square inch (N/mm²).

e_n = Nail or staple deformation, in inches (mm) [see Table 2305.2.2(1)].

Gt = Panel rigidity through the thickness, in pounds per inch (N/mm) of panel width or depth [see Table 2305.2.2(2)].

L = Diaphragm length, in feet (mm).

v = Maximum shear due to design loads in the direction under consideration, in pounds per linear foot (plf) (N/mm).

Δ = The calculated deflection, in inches (mm).

$\Sigma(\Delta_c X)$ = Sum of individual chord-splice slip values on both sides of the diaphragm, each multiplied by its distance to the nearest support.

❖ Equation 23-1 provides the estimate of the deflection of a blocked, uniformly nailed wood structural panel diaphragm. Total deflection represents the sum of four components contributing to the overall deflection: bending deflection, shear deflection, nail slip and chord splice slip. Guidance on the selection of various equation inputs such as nail deformation, effective panel thickness and sum of individual chord-splice values is provided in Tables 2305.2.2(1) and 2305.2.2(2). These tables provide the required parameters needed in the four-part equation. For ease of calculation, the "modulus of rigidity" term (G) is combined with the "effective thickness" term (t) to form a "panel rigidity capacity" term (Gt). Capacity terms such as Gt are used by the panel industry in lieu of section properties and allowable stresses as it makes it easier to tabulate and compare different panel types:

As noted above, the deflection computation is applicable only to blocked diaphragms. For unblocked diaphragms *SEAOC Blue Book*, 1999, Section 805.3.2 provides the following guidance:

> Limited testing of diaphragms (APA, 1952, 1954, 1955, 1967) suggests that the deflection of an unblocked diaphragm at its tabulated allowable shear capacity will be about 2.5 times the calculated deflection of a blocked diaphragm of similar construction and dimensions, at the same shear capacity. If diaphragm framing is spaced more than 24 inches (610 mm) o.c., testing indicates a further increase in deflection of about 20 percent for unblocked diaphragms (e.g., to 3 times the deflection on a comparable blocked diaphragm). This relationship can be used to develop an estimate of the deflection of unblocked diaphragms.

Note that the unblocked diaphragm testing conducted by the APA mentioned above was conducted on diaphragms with aspect ratios in the 2:1 to 3:1 range. As the aspect ratio decreases, it is reasonable to expect the adjustment factor to similarly decrease.

As noted above, the equation for diaphragm deflection is intended for uniform nailing. The 0.188 constant in the "nail-slip deflection-contribution" term was derived based on panel-edge nailing being the same for the entire length of the diaphragm. When nail spacing becomes less dense near the center of the diaphragm span, *ATC 7, Guidelines for the Design of Horizontal Wood Diaphragms*, recommends that the 0.188 constant should be increased in proportion to the average load on each nail with nonuniform nailing, compared to the average load that would be present if a uniform nail spacing had been maintained. The adjusted constant can be expressed as follows:

$0.188 \times (v_n'/v_n)$

where

v_n' = The average non-uniform load per nail.

v_n = The average uniform load per nail.

Graphically, this can be shown and calculated in terms of areas. In Figure 2305.2.2, Area 1 is proportional to v_n and the sum of Areas 2 and 3 is propor-

tional to v_n'.

The adjusted value for the constant would become:

$$0.188\left(\frac{Area_2 + Area_3}{Area_1}\right) =$$

$$0.188\left[\frac{0.5(100+75)20 + 0.5(125)50}{0.5(100)70}\right] =$$

$$0.188 \times 139 = 0.262$$

TABLE 2305.2.2(1). See below.

❖ Tables 2305.2.2(1) and 2305.2.2(2) contain values for the terms (e_n, G, and t) for use in Equations 23-1 and 23-2. The source of these tabulated values is the 1997 UBC, Volume 3, which is the source for Equations 23-1 and 23-2 as well.

G and t are combined into a single table that gives the panel rigidity-through-the-thickness capacity, Gt, which allows the use of similar values for wood structural panels other than plywood. These values are recognized by the industry for all products meeting the criteria of either DOC PS 1 or PS 2, both of which are referenced standards in the code.

TABLE 2305.2.2(2). See page 23-42.

❖ See the commentary for Table 2305.2.2(1).

TABLE 2305.2.2(1)
e_n VALUES (inches) FOR USE IN CALCULATING DIAPHRAGM DEFLECTION DUE TO FASTENER SLIP (Structural I)[a,d]

LOAD PER FASTENER[c] (pounds)	FASTENER DESIGNATIONS[b]			
	6d	8d	10d	14-Ga staple x 2 inches long
60	0.01	0.00	0.00	0.011
80	0.02	0.01	0.01	0.018
100	0.03	0.01	0.01	0.028
120	0.04	0.02	0.01	0.04
140	0.06	0.03	0.02	0.053
160	0.10	0.04	0.02	0.068
180	—	0.05	0.03	—
200	—	0.07	0.47	—
220	—	0.09	0.06	—
240	—	—	0.07	—

For SI: 1 inch = 25.4 mm, 1 foot = 304.8 mm, 1 pound = 4.448 N.

a. Increase e_n values 20 percent for plywood grades other than Structural I.

b. Nail values apply to common wire nails or staples identified.

c. Load per fastener = maximum shear per foot divided by the number of fasteners per foot at interior panel edges.

d. Decrease e_n values 50 percent for seasoned lumber (moisture content < 19 percent).

TABLE 2305.2.2(2)
VALUES OF *Gt* FOR USE IN CALCULATING DEFLECTION OF WOOD STRUCTURAL PANEL SHEAR WALLS AND DIAPHRAGMS

PANEL TYPE	SPAN RATING	VALUES OF *Gt* (lb/in. panel depth or width)							
		OTHER				STRUCTURAL I			
		3-ply Plywood	4-ply Plywood	5-ply Plywood[a]	OSB	3-ply Plywood	4-ply Plywood	5-ply Plywood[a]	OSB
Sheathing	24/0	25,000	32,500	37,500	77,500	32,500	42,500	41,500	77,500
	24/16	27,000	35,000	40,500	83,500	35,000	45,500	44,500	83,500
	32/16	27,000	35,000	40,500	83,500	35,000	45,500	44,500	83,500
	40/20	28,500	37,000	43,000	88,500	37,000	48,000	47,500	88,500
	48/24	31,000	40,500	46,500	96,000	40,500	52,500	51,000	96,000
Single Floor	16 o.c.	27,000	35,000	40,500	83,500	35,000	45,500	44,500	83,500
	20 o.c.	28,000	36,500	42,000	87,000	36,500	47,500	46,000	87,000
	24 o.c.	30,000	39,000	45,000	93,000	39,000	50,500	49,500	93,000
	32 o.c.	36,000	47,000	54,000	110,000	47,000	61,000	59,500	110,000
	48 o.c.	50,500	65,500	76,000	155,000	65,500	85,000	83,500	155,000

	Thickness (in.)	OTHER			STRUCTURAL I		
		A-A, A-C	Marine	All Other Grades	A-A, A-C	Marine	All Other Grades
Sanded Plywood	$1/4$	24,000	31,000	24,000	31,000	31,000	31,000
	$11/32$	25,500	33,000	25,500	33,000	33,000	33,000
	$3/8$	26,000	34,000	26,000	34,000	34,000	34,000
	$15/32$	38,000	49,500	38,000	49,500	49,500	49,500
	$1/2$	38,500	50,000	38,500	50,000	50,000	50,000
	$19/32$	49,000	63,500	49,000	63,500	63,500	63,500
	$5/8$	49,500	64,500	49,500	64,500	64,500	64,500
	$23/32$	50,500	65,500	50,500	65,500	65,500	65,500
	$3/4$	51,000	66,500	51,000	66,500	66,500	66,500
	$7/8$	52,500	68,500	52,500	68,500	68,500	68,500
	1	73,500	95,500	73,500	95,500	95,500	95,500
	$1 1/8$	75,000	97,500	75,000	97,500	97,500	97,500

For SI: 1 inch = 25.4 mm, 1 pound/inch = 0.1751 N/mm.

a. Applies to plywood with five or more layers; for five-ply/three-layer plywood, use values for four ply.

2305.2.3 Diaphragm aspect ratios. Size and shape of diaphragms shall be limited as set forth in Table 2305.2.3.

❖ This section simply references the table for maximum diaphragm ratios. Maximum diaphragm ratios are based on their performance during high wind and seismic events.

TABLE 2305.2.3
MAXIMUM DIAPHRAGM DIMENSION RATIOS
HORIZONTAL AND SLOPED DIAPHRAGM

TYPE	MAXIMUM LENGTH - WIDTH RATIO
Wood structural panel, nailed all edges	4:1
Wood structural panel, blocking omitted at intermediate joints	3:1
Diagonal sheathing, single	3:1
Diagonal sheathing, double	4:1

❖ This table provides maximum aspect ratios for floor and roof diaphragms using wood structural panel or diagonal board sheathing. Additional requirements for open-front structures and cantilevered diaphragms are described in Section 2305.2.5 and Figures 2305.2.5(1) and (2).

2305.2.4 Construction. Wood diaphragms shall be constructed of wood structural panels manufactured with exterior glue and not less than 4 feet by 8 feet (1219 mm by 2438 mm), except at boundaries and changes in framing where minimum sheet dimension shall be 24 inches (610 mm) unless all edges of the undersized sheets are supported by and fastened to framing members or blocking. Wood structural panel thickness for horizontal diaphragms shall not be less than the valves set forth in Tables 2304.7(3), 2304.7(4) and 2304.7(5) for corresponding joist spacing and loads.

❖ Basic prescriptive requirements for diaphragm construction specify the use of 4-foot by 8-foot (1219 mm by 2438 mm) sheets except at boundaries and changes in framing. No sheet less than 24 inches (610 mm) may be used unless all edges of the undersized sheet are supported and fastened to framing or blocking. Panel sections less than 24 inches (610 mm) have been permitted by the UBC for 25 years with no reported problems. This practice has been verified via

Figure 2305.2.2
ADJUSTMENTS FOR NONUNIFORM NAILING

For SI: 1 foot = 304.8 mm

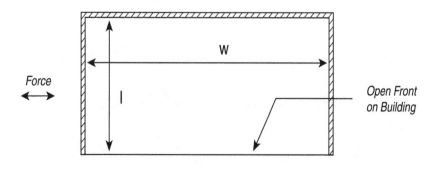

FIGURE 2305.2.5(1)
DIAPHRAGM LENGTH AND WIDTH FOR PLAN VIEW OF OPEN-FRONT BUILDING

full-size testing at the American Plywood Association laboratory.

2305.2.4.1 Seismic Design Category F. Structures assigned to Seismic Design Category F shall conform to the additional requirements of this section.

Wood structural panel sheathing used for diaphragms and shear walls that are part of the seismic-force-resisting system shall be applied directly to the framing members.

> **Exception:** Wood structural panel sheathing in a diaphragm is permitted to be fastened over solid lumber planking or laminated decking, provided the panel joints and lumber planking or laminated decking joints do not coincide.

❖ As an additional precaution in structures assigned to Seismic Design Category F, requirements are provided for the attachment of wood structural panel sheathing to framing members, solid lumber planking or laminated decking.

2305.2.5 Rigid diaphragms. Design of structures with rigid diaphragms shall conform to the structure configuration requirements of Section 12.3.2 of ASCE 7 and the horizontal shear distribution requirements of Section 12.8.4 of ASCE 7.

Open-front structures with rigid wood diaphragms resulting in torsional force distribution are permitted, provided the length, l, of the diaphragm normal to the open side does not exceed 25 feet (7620 mm), the diaphragm sheathing conforms to Section 2305.2.4 and the l/w ratio [as shown in Figure 2305.2.5(1)] is less than 1 for one-story structures or 0.67 for structures over one story in height.

> **Exception:** Where calculations show that diaphragm deflections can be tolerated, the length, l, normal to the open end is permitted to be increased to a l/w ratio not greater than 1.5 where sheathed in compliance with Section 2305.2.4 or to 1 where sheathed in compliance with Section 2306.3.4 or 2306.3.5.

Rigid wood diaphragms are permitted to cantilever past the outermost supporting shearwall (or other vertical resisting element) a length, l, of not more than 25 feet (7620 mm) or two-thirds of the diaphragm width, w, whichever is smaller. Figure 2305.2.5(2) illustrates the dimensions of l and w for a cantilevered diaphragm.

Structures with rigid wood diaphragms having a torsional irregularity in accordance with Table 12.3-1, Item 1, of ASCE 7 shall meet the following requirements: the l/w ratio shall not exceed 1 for one-story structures or 0.67 for structures over one story in height, where l is the dimension parallel to the load direction for which the irregularity exists.

> **Exception:** Where calculations demonstrate that the diaphragm deflections can be tolerated, the width is permitted to be increased and the l/w ratio is permitted to be increased to 1.5 where sheathed in compliance with Section 2305.2.4 or 1 where sheathed in compliance with Section 2306.3.4 or 2306.3.5.

❖ For diaphragms defined as rigid, rotational or torsional behavior is expected and results in redistribution of shear to the vertical-force-resisting elements. Additional prescriptive limits are placed on rigid diaphragms based on the type of irregularity and diaphragm construction. Further discussion of these limits is available in Chapter 12 of the *NEHRP Recommended Provisions Commentary for Seismic Regulations for New Buildings*.

2305.3 Design of wood shear walls.

FIGURES 2305.2.5(1) and 2305.2.5(2). See pages 23-43 and 23-44.

❖ The source of Figures 2305.2.5(1) and 2305.2.5(2) is Chapter 12 of the NEHRP provisions. The figures provide the meaning of the *l* and *w* terms that are specified in Section 2305.2.5.

2305.3.1 General. Wood shear walls are permitted to resist horizontal forces in vertical distributing or resisting elements, provided the deflection in the plane of the shear wall, as determined by calculations, tests or analogies drawn therefrom, does not exceed the more restrictive of the permissible deflection of attached distributing or resisting elements or the drift limits of Section 12.12.1 of ASCE 7. Shear wall sheathing other than wood structural panels shall not be permitted in Seismic Design Category E or F (see Section 1613).

❖ General requirements for wood shear walls include consideration of shear wall strength and deflection. General requirements for wood shear walls can be

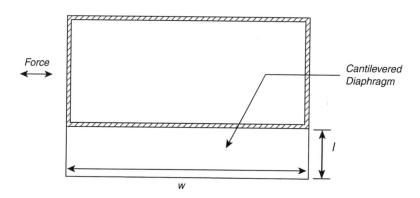

FIGURE 2305.2.5(2)
DIAPHRAGM LENGTH AND WIDTH FOR PLAN VIEW OF CANTILEVERED DIAPHRAGM

found in APA Research Report 154, *Wood Structural Panel Shear Wall*.

2305.3.2 Deflection. Permissible deflection shall be that deflection up to which the shear wall and any attached distributing or resisting element will maintain its structural integrity under design load conditions, i.e., continue to support design loads without danger to occupants of the structure.

The deflection (Δ) of a blocked wood structural panel shear wall uniformly fastened throughout is permitted to be calculated by the use of the following equation:

$$\Delta = \frac{8vh^3}{EAb} + \frac{vh}{Gt} + 0.75he_n + d_a\frac{h}{b} \qquad \textbf{(Equation 23-2)}$$

For SI: $\Delta = \dfrac{vh^3}{3EAb} + \dfrac{vh}{Gt} + \dfrac{he_n}{407.6} + d_a\dfrac{h}{b}$

where:

A = Area of boundary element cross section in square inches (mm²) (vertical member at shear wall boundary).

b = Wall width, in feet (mm).

d_a = Vertical elongation of overturning anchorage (including fastener slip, device elongation, anchor rod elongation, etc.) at the design shear load (v).

E = Elastic modulus of boundary element (vertical member at shear wall boundary), in pounds per square inch (N/mm²).

e_n = Nail or staple deformation, in inches (mm) [see Table 2305.2.2(1)].

Gt = Panel rigidity through the thickness, in pounds per inch (N/mm) of panel width or depth [see Table 2305.2.2(2)].

h = Wall height, in feet (mm).

v = Maximum shear due to design loads at the top of the wall, in pounds per linear foot (N/mm).

Δ = The calculated deflection, in inches (mm).

❖ The equation in this section provides an estimate of the deflection of a blocked wood structural panel shear wall. Total deflection represents the sum of four components contributing to overall deflection: bending deflection, shear deflection, fastener slip and anchorage deformation. Guidance is provided in Tables 2305.2.2(1) and 2305.2.2(2), which provide the required parameters needed for the four-part equation. For ease of calculation the "modulus of rigidity" term (G) was combined with the "effective thickness" term (t) to form a "panel rigidity capacity" term (Gt). Capacity terms such as Gt are used by the panel industry in lieu of section properties and allowable stresses as it makes it easier to tabulate and compare different panel types.

2305.3.3 Construction. Wood shear walls shall be constructed of wood structural panels manufactured with exterior glue and not less than 4 feet by 8 feet (1219 mm by 2438 mm), except at boundaries and at changes in framing. All edges of all panels shall be supported by and fastened to framing members or blocking. Wood structural panel thickness for shear walls shall not be less than set forth in Table 2304.6.1 for corresponding framing spacing and loads, except that $^1/_4$ inch (6.4 mm) is permitted to be used where perpendicular loads permit.

❖ Basic prescriptive requirements for shear wall construction specify the use of 4-foot by 8-foot (1219 mm by 2438 mm) sheets except at boundaries and changes in framing.

2305.3.4 Shear wall aspect ratios. Size and shape of shear walls, perforated shear wall segments within perforated shear walls and wall piers within shear walls that are designed for force transfer around openings shall be limited as set forth in Table 2305.3.4. The height, h, and the width, w, shall be determined in accordance with Sections 2305.3.5 through 2305.3.5.2 and 2305.3.6 through 2305.3.6.2, respectively.

❖ See the commentary to Table 2305.3.4.

TABLE 2305.3.4
MAXIMUM SHEAR WALL DIMENSION RATIOS

TYPE	MAXIMUM HEIGHT-WIDTH RATIO
Wood structural panels or particleboard, nailed edges	For other than seismic: $3^1/_2$:1 For seismic: 2:1[a]
Diagonal sheathing, single	2:1
Fiberboard	$1^1/_2$:1
Gypsum board, gypsum lath, cement plaster	$1^1/_2$:1[b]

a. For design to resist seismic forces, shear wall height-width ratios greater than 2:1, but not exceeding $3^1/_2$:1, are permitted provided the allowable shear values in Table 2306.4.1 are multiplied by $2w/h$.

b. Ratio shown is for unblocked construction. Height-to-width ratio is permitted to be 2:1 where the wall is installed as blocked construction in accordance with Section 2306.4.5.1.2.

❖ Table 2305.3.4 provides maximum shear wall height-to-width ratios, also known as "aspect ratios," for wood-frame shear walls sheathed with wood structural panels, particleboard, fiberboard, diagonal sheathing, gypsum board, gypsum lath or cement plaster.

For shear walls with wood structural panels or particleboard sheathing that resist seismic loads, the maximum aspect ratio is 2:1 unless the tabulated allowable shear values are adjusted in accordance with Note a. This approach for addressing larger aspect ratios is based on the *Recommended Provisions for Seismic Regulations for New Buildings and Other Structures* (NEHRP 2000). Based on review of the test data for larger aspect ratio shear walls, this adjustment addresses the concern of the reduced stiffness associated with these larger aspect ratios. Also note that the resistance to loading other than seismic permits the use of a larger aspect ratio.

2305.3.5 Shear wall height definition. The height of a shear wall, h, shall be defined as:

1. The maximum clear height from the top of the foundation to the bottom of the diaphragm framing above; or

2. The maximum clear height from the top of the diaphragm to the bottom of the diaphragm framing above [see Figure 2305.3.5(a)].

❖ The shear wall height definition is needed for the evaluation of the maximum aspect ratios in Table 2305.3.4. This height definition applies to shear walls that have not been designed for force transfer around openings in accordance with Section 2305.8.1.

FIGURE 2305.3.5. See below.

❖ This figure illustrates the determination of the height and width of a shear wall.

2305.3.5.1 Perforated shear wall segment height definition. The height of a perforated shear wall segment, h, shall be defined as specified in Section 2305.3.5 for shear walls.

❖ See Figure 2305.3.5(a) and the commentary to Section 2305.3.5.

2305.3.5.2 Force transfer shear wall pier height definition. The height, h, of a wall pier in a shear wall with openings designed for force transfer around openings shall be defined as the clear height of the pier at the side of an opening [see Figure 2305.3.5(b)].

❖ See Figure 2305.3.5(b) and the commentary to Section 2305.3.8.1.

2305.3.6 Shear wall width definition. The width of a shear wall, w, shall be defined as the sheathed dimension of the shear wall in the direction of application of force [see Figure 2305.3.5(a)].

❖ Shear wall width, as defined in this section, is provided for the purpose of calculating maximum height-to-width ratios for shear walls. When used to calculate shear wall aspect ratio, it is intended that a standard 4-foot by 8-foot (1219 mm by 2438 mm) panel used in standard 8-foot (2438 mm) wall construction should qualify as meeting an aspect ratio of 2:1.

2305.3.6.1 Perforated shear wall segment width definition. The width of a perforated shear wall segment, w, shall be defined as the width of full-height sheathing adjacent to openings in the perforated shear wall [see Figure 2305.3.5(a)].

❖ Shear wall segments refer to those portions of a shear wall on either side of an opening. Only portions of the wall with full-height sheathing can be considered as contributing to the width of a segment and in satisfying the aspect ratio limits of Section 2305.3.4.

2305.3.6.2 Force transfer shear wall pier width definition. The width, w, of a wall pier in a shear wall with openings designed for force transfer around openings shall be defined as the sheathed width of the pier at the side of an opening [see Figure 2305.3.5(b)].

❖ See Figure 2305.3.5(b) and the commentary to Section 2305.3.8.1.

(a) HEIGHT-TO-WIDTH RATIO FOR SHEAR WALLS AND PERFORATED SHEAR WALLS

(b) HEIGHT-TO-WIDTH RATIO WITH DESIGN FOR FORCE TRANSFER AROUND OPENINGS

FIGURE 2305.3.5
GENERAL DEFINITION OF SHEAR WALL HEIGHT, WIDTH AND HEIGHT-TO-WIDTH RATIO

2305.3.7 Overturning restraint. Where the dead load stabilizing moment in accordance with Chapter 16 allowable stress design load combinations is not sufficient to prevent uplift due to overturning moments on the wall, an anchoring device shall be provided. Anchoring devices shall maintain a continuous load path to the foundation.

❖ This section provides a general requirement to maintain a continuous load path to the foundation for uplift due to wall overturning forces. The location or frequency of this restraint depends on the design method used. The segmented shear wall design approach requires that each segment be evaluated individually for overturning. The perforated shear wall method requires that anchorage be provided at each end of the shear wall. Typically, these tie-downs, or hold-downs, consist of metal straps or other hardware attached to the shear wall end posts.

In recent years, manufacturers of hold-down anchors have worked on producing devices that will transfer the uplift forces to the foundation in a concentric manner, primarily in response to field observations of reduced anchor capacities due to the inherent eccentricity of loading from the center of a stud assembly or post. Nelson and Hamburger reported observations of brittle failure in wood posts through combined tension and flexure failure at the top bolt in a bolt-type hold-down device. These failures occurred at approximately two times the manufacturer's allowable hold-down loads, whereas concentric applications achieved factors of four or five times published loads. These conditions need to be considered in the design of walls for overturning resistance.

2305.3.8 Shear walls with openings. The provisions of this section shall apply to the design of shear walls with openings. Where framing and connections around the openings are designed for force transfer around the openings, the provisions of Section 2305.3.8.1 shall apply. Where framing and connections around the openings are not designed for force transfer around the openings, the provisions of Section 2305.3.8.2 shall apply.

❖ This section suggests that there are only two options for designing shear walls with openings. If one considers the commonly used segmented shear wall design approach, there are, in fact, three methods that could be used.

When framing and connections around the openings are specifically designed and detailed for force transfer around the opening, the provisions of Section 2305.3.8.1 apply. This method is an alternative to segmented shear wall design wherein the shear wall segments are as shown in Figure 2305.3.5(a). In Section 2305.3.8.1, design and detailing around openings are based on rigid body behavior and utilize the full shear resistance developed in the wood structural panel shear wall. The benefit of this additional design and detailing is that the height-to-width ratio of wall piers can be determined as indicated in Figure 2305.3.5(b), potentially allowing a narrower width pier to comply

with the shear wall aspect ratio limits of Section 2305.3.4.

As an option to either the segmented shear wall design method or design for force transfer around openings, the perforated shear wall provisions of Section 2305.3.8.2 can be utilized. In the perforated shear wall method, the wall is not assumed to act as a rigid body and shear capacity is reduced, accounting for the reduced strength and stiffness of unrestrained shear wall segments within the perforated shear wall.

2305.3.8.1 Force transfer around openings. Where shear walls with openings are designed for force transfer around the openings, the limitations of Table 2305.3.4 shall apply to the overall shear wall, including openings, and to each wall pier at the side of an opening. Design for force transfer shall be based on a rational analysis. Detailing of boundary elements around the opening shall be provided in accordance with the provisions of this section[see Figure 2305.3.5(b)].

❖ Shear walls designed using provisions for force transfer around openings utilize aspect ratio limits defined in Table 2305.3.4; however, definitions of wall pier height and width are used to calculate the aspect ratio. Wall pier width and height are shown in Figure 2305.3.5(b). In this approach for design of shear walls with openings, design force transfer around the openings involves developing a system of piers and coupling beams within the shear wall. Load paths for the shear and flexure developed in the piers and coupling beams generally require blocking and strapping extending from each corner of the opening to some distance beyond. This approach is often utilized in highly loaded walls when limited amounts of sheathing are present around openings and often results in shear wall detailing involving many metal tension straps around the openings.

2305.3.8.2 Perforated shear walls. The provisions of Section 2305.3.8.2 shall be permitted to be used for the design of perforated shear walls. For the determination of the height and width of perforated shear wall segments, see Sections 2305.3.5.1 and 2305.3.6.1, respectively.

❖ Shear walls meeting the provisions of this section are commonly referred to as perforated shear walls. Aspect ratio limits of Section 2305.3.4 apply to shear wall segments within the perforated shear wall as described by the definition of "Perforated shear wall segment" in Section 2302.

This method of shear wall design utilizes empirically determined reductions (see Table 2305.3.8.2) of wood structural panel shear wall capacities that are based on the maximum opening height as well as the percentage of a particular shear wall that qualifies as a perforated shear wall segment(s). This method recognizes the strength and stiffness of sheathed areas above and below openings without the need to specifically design and detail for force transfer as required under Section 2305.3.8.1. It is not expected that sheathed wall areas above and below openings behave as coupling beams acting end to end, but rather that they provide local restraint at their ends. As a consequence, significantly reduced capacities are attributed to interior perforated shear wall segments with limited overturning restraint. Further background on the development of provisions for perforated shear walls and example problems are provided in the NEHRP provisions commentary.

TABLE 2305.3.8.2. See below.

❖ See the commentary to Section 2305.3.8.2.

2305.3.8.2.1 Limitations. The following limitations shall apply to the use of Section 2305.3.8.2:

1. A perforated shear wall segment shall be located at each end of a perforated shear wall. Openings shall be permitted to occur beyond the ends of the perforated shear wall, provided the width of such openings is not be included in the width of the perforated shear wall.

2. The allowable shear set forth in Table 2306.4.1 shall not exceed 490 plf (7150 N/m).

3. Where out-of-plane offsets occur, portions of the wall on each side of the offset shall be considered as separate perforated shear walls.

4. Collectors for shear transfer shall be provided through the full length of the perforated shear wall.

5. A perforated shear wall shall have uniform top of wall and bottom of wall elevations. Perforated shear walls not having uniform elevations shall be designed by other methods.

6. Perforated shear wall height, h, shall not exceed 20 feet (6096 mm).

❖ This section limits the configuration of perforated shear walls. Item 2 limits the allowable shear values of the shear wall assemblies in order to rule out high-demand shear wall assemblies that are not considered suitable for this method.

2305.3.8.2.2 Perforated shear wall resistance. The resistance of a perforated shear wall shall be calculated in accordance with the following:

1. The percentage of full-height sheathing shall be calculated as the sum of the widths of perforated shear wall segments divided by the total width of the perforated shear wall, including openings.

2. The maximum opening height shall be taken as the maximum opening clear height. Where areas above and below an opening remain unsheathed, the height of the opening shall be defined as the height of the wall.

3. The unadjusted shear resistance shall be the allowable shear set forth in Table 2306.4.1 for height-to-width ratios of perforated shear wall segments that do not exceed 2:1 for seismic forces and $3^1/_2$:1 for other than seismic forces. For seismic forces, where the height-to-width ratio of any perforated shear wall segment used in the calculation of the sum of the widths of perforated shear wall segments, $\sum L_i$, is greater than 2:1 but does not exceed $3^1/_2$:1, the unadjusted shear resistance shall be multiplied by 2 w/h.

TABLE 2305.3.8.2
SHEAR RESISTANCE ADJUSTMENT FACTOR, C_o

WALL HEIGHT, H	MAXIMUM OPENING HEIGHT[a]				
	H/3	H/2	2H/3	5H/6	H
8' wall	2'-8"	4'-0"	5'-4"	6'-8"	8'-0"
10' wall	3'-4"	5'-0"	6'-8"	8'-4"	10'-0"
Percentage of full-height sheathing[b]	Shear resistance adjustment factor				
10%	1.00	0.69	0.53	0.43	0.36
20%	1.00	0.71	0.56	0.45	0.38
30%	1.00	0.74	0.59	0.49	0.42
40%	1.00	0.77	0.63	0.53	0.45
50%	1.00	0.80	0.67	0.57	0.50
60%	1.00	0.83	0.71	0.63	0.56
70%	1.00	0.87	0.77	0.69	0.63
80%	1.00	0.91	0.83	0.77	0.71
90%	1.00	0.95	0.91	0.87	0.83
100%	1.00	1.00	1.00	1.00	1.00

For SI: 1 inch = 25.4 mm, 1 foot = 304.8 mm.
a. See Section 2305.3.8.2.2, Item 2.
b. See Section 2305.3.8.2.2, Item 1.

4. The adjusted shear resistance shall be calculated by multiplying the unadjusted shear resistance by the shear resistance adjustment factors of Table 2305.3.8.2. For intermediate percentages of full-height sheathing, the values in Table 2305.3.8.2 are permitted to be interpolated.

5. The perforated shear wall resistance shall be equal to the adjusted shear resistance times the sum of the widths of the perforated shear wall segments.

❖ This section describes the determination of perforated shear wall resistance. First, however, it is necessary to determine the unadjusted shear resistance as described in Item 3 of Section 2305.3.8.2.2. For the wall under consideration, determine the percentage of full-height sheathing (see Item 1) and the maximum opening height ratio (see Item 2). Based on these two parameters, determine the shear resistance adjustment factor from Table 2305.3.8.2 and then calculate the adjusted shear resistance according to Item 3. The product of the adjusted shear resistance and the total widths of perforated shear wall segments in the wall under consideration is the shear resistance of that shear wall.

2305.3.8.2.3 Anchorage and load path. Design of perforated shear wall anchorage and load path shall conform to the requirements of Sections 2305.3.8.2.4 through 2305.3.8.2.8, or shall be calculated using principles of mechanics. Except as modified by these sections, wall framing, sheathing, sheathing attachment and fastener schedules shall conform to the requirements of Section 2305.2.4 and Table 2306.4.1.

❖ This section prescribes minimum design forces for uplift anchorage and compression chords in perforated shear walls. These requirements conservatively address the nonuniform shear distribution occurring in perforated shear walls. Alternative designs based on principles of mechanics are permitted.

2305.3.8.2.4 Uplift anchorage at perforated shear wall ends. Anchorage for uplift forces due to overturning shall be provided at each end of the perforated shear wall. The uplift anchorage shall conform to the requirements of Section 2305.3.7, except that for each story the minimum tension chord uplift force, T, shall be calculated in accordance with the following:

$$T = \frac{Vh}{C_o \Sigma L_i}$$ (Equation 23-3)

where:

T = Tension chord uplift force, pounds (N).

V = Shear force in perforated shear wall, pounds (N).

h = Perforated shear wall height, feet (mm).

C_o = Shear resistance adjustment factor from Table 2305.3.8.2.

ΣL_i = Sum of widths of perforated shear wall segments, feet (mm).

❖ Anchorage against uplift is required only at the ends of the perforated shear wall as opposed to providing full

restraint of each segment as is the case when designing with a segmented shear wall approach. The uplift force due to overturning should be computed using Equation 23-3.

2305.3.8.2.5 Anchorage for in-plane shear. The unit shear force, v, transmitted into the top of a perforated shear wall, out of the base of the perforated shear wall at full height sheathing and into collectors connecting shear wall segments shall be calculated in accordance with the following:

$$v = \frac{V}{C_o \Sigma L_i}$$ (Equation 23-4)

where:

v = Unit shear force, pounds per lineal feet (N/m).

V = Shear force in perforated shear wall, pounds (N).

C_o = Shear resistance adjustment factor from Table 2305.3.8.2.

ΣL_i = Sum of widths of perforated shear wall segments, feet (mm).

❖ This section prescribes the unit shear force, v, to be applied to the top and bottom of each perforated shear wall segment. Designing for a unit shear, as well as the uniform uplift force of Section 2305.3.8.2.6, that are based on the unadjusted shear capacity provides a sheathing-to-bottom plate attachment such that the capacity of the shear wall will not be governed by this connection.

2305.3.8.2.6 Uplift anchorage between perforated shear wall ends. In addition to the requirements of Section 2305.3.8.2.4, perforated shear wall bottom plates at full-height sheathing shall be anchored for a uniform uplift force, t, equal to the unit shear force, v, determined in Section 2305.3.8.2.5.

❖ The unit uplift force, t, is taken equal to the unit shear force, v (see commentary, Section 2305.3.8.2.5).

2305.3.8.2.7 Compression chords. Each end of each perforated shear wall segment shall be designed for a compression chord force, C, equal to the tension chord uplift force, T, calculated in Section 2305.3.8.2.4.

❖ The vertical members on either side of all perforated shear wall segments must be designed for the required compression chord force.

2305.3.8.2.8 Load path. Load path. A load path to the foundation shall be provided for each uplift force, T and t, for each shear force, V and v, and for each compression chord force, C. Elements resisting shear wall forces contributed by multiple stories shall be designed for the sum of forces contributed by each story.

❖ A load path for the forces prescribed in this section is necessary. Consideration must be given to shear wall forces from a story, or stories, above in multistory buildings.

2305.3.8.2.9 Deflection of shear walls with openings. The controlling deflection of a blocked shear wall with openings uniformly fastened throughout shall be taken as the maximum individual deflection of the shear wall segments calculated in accordance with Section 2305.3.2, divided by the appropriate shear resistance adjustment factors of Table 2305.3.8.2.

❖ The provision for the calculation of deflection of a perforated shear wall is based on the observation from testing that the reduction in strength is comparable to the reduction in stiffness.

2305.3.9 Summing shear capacities. The shear values for shear panels of different capacities applied to the same side of the wall are not cumulative except as allowed in Table 2306.4.1.

The shear values for material of the same type and capacity applied to both faces of the same wall are cumulative. Where the material capacities are not equal, the allowable shear shall be either two times the smaller shear capacity or the capacity of the stronger side, whichever is greater.

Summing shear capacities of dissimilar materials applied to opposite faces or to the same wall line is not allowed.

Exception: For wind design, the allowable shear capacity of shear wall segments sheathed with a combination of wood structural panels and gypsum wallboard on opposite faces, fiberboard structural sheathing and gypsum wallboard on opposite faces or hardboard panel siding and gypsum wallboard on opposite faces shall equal the sum of the sheathing capacities of each face separately.

❖ This section provides general requirements for summing shear capacities of sheathing materials applied to either one side of a wall or both sides of the same wall. The capacity of a wall that is sheathed on opposite sides with the same material type but different capacities is taken as the greater of either the capacity of the stronger side or twice the shear value of the lesser capacity.

Generally, summing shear capacities of different sheathing materials is not permitted. However, the exception reflects the results of testing conducted to confirm that the summing of shear resistance of different sheathing materials is justifiable for certain combinations of materials. Test results for wood structural panels in combination with gypsum wallboard were provided in APA Report 157.

2305.3.10 Adhesives. Adhesive attachment of shear wall sheathing is not permitted as a substitute for mechanical fasteners, and shall not be used in shear wall strength calculations alone, or in combination with mechanical fasteners in Seismic Design Category D, E or F.

❖ The restriction against adhesive attachment of shear wall sheathing in high seismic areas addresses a concern that the increased stiffness associated with use of adhesives will attract more load to the shear wall than anticipated and could cause brittle failure of anchoring devices.

2305.3.11 Sill plate size and anchorage in Seismic Design Category D, E or F. Anchor bolts for shear walls shall include steel plate washers, a minimum of 0.229 inch by 3 inches by 3 inches (5.82 mm by 76 mm by 76 mm) in size, between the sill plate and nut. The hole in the plate washer is permitted to be diagonally slotted with a width of up to $^3/_{16}$ inch (4.76 mm) larger than the bolt diameter and a slot length not to exceed $1^3/_4$ inches (44 mm), provided a standard cut washer is placed between the plate washer and the nut. Sill plates resisting a design load greater than 490 plf (7154 N/m) using load and resistance factor design or 350 plf (5110 N/m) using allowable stress design shall not be less than a 3-inch (76 mm) nominal member. Where a single 3-inch (76 mm) nominal sill plate is used, 2- 20d box end nails shall be substituted for 2-16d common end nails found in line 8 of Table 2304.9.1.

Exception: In shear walls where the design load is greater than 490 plf (7151 N/m) but less than 840 plf (12 264 N/m) using load and resistance factor design or greater than 350 plf (5110 N/m) but less than 600 plf (8760 N/m) using allowable stress design, the sill plate is permitted to be a 2-inch (51 mm) nominal member if the sill plate is anchored by two times the number of bolts required by design and 0.229-inch by 3-inch by 3-inch (5.82 mm by 76 mm by 76 mm) plate washers are used.

❖ This provision addresses a concern over earthquake damage to highly loaded sill plates. The code requires use of the plate washers between the sill plate and nut when attaching the sill plate to the foundation for engineered shear walls. Even where tie-down devices are provided, tests have demonstrated that the uplift movement at ends of shear walls with tie-down devices is large enough to permit splitting of the sills. It is expected that this plate washer will reduce cross-grain bending forces in the sill plate. In addition, a 3-inch (76 mm) sill plate is required when the effects of nominal earthquake loads exceed 350 plf (5110 N/m) as an added measure of resistance to cross-grain bending forces. Where the single 3-inch (76 mm) sill plate is used, longer end nails should be substituted to provide sufficient penetration into the stud.

This provision permits diagonal slotted holes in the plate washers to address construction tolerance issues. The Curee Woodframe Project found that in order to reduce sill splitting, it is critical to get the edge of the plate washer very close to the edge of the sill at which shear wall sheathing is fastened. Tests with larger distances between the edge of the sill and the edge of the washer experienced splitting at significantly lower force levels. The 3-inch by 3-inch (76 mm by 76 mm) plate washer size is needed, as a minimum, to minimize this distance.

The exception enables the use of 2-inch (51 mm) sill plates for design loads up to 600 plf (8760 N/m) due to the effects of nominal earthquake loads, provided that twice the number of anchor bolts required by design are used to anchor the sill plate. While this requirement may not be practical for new construction, it is expected that this option would be used to upgrade existing structures with 2-inch (51 mm) sill plates.

SECTION 2306
ALLOWABLE STRESS DESIGN

2306.1 Allowable stress design. The structural analysis and construction of wood elements in structures using allowable stress design shall be in accordance with the following applicable standards:

❖ This section contains a number of specifications intended to provide guidance to nonprofessional as well as professional users of the code. These specifications, based on accepted engineering practices and experiences, are considered to be the minimum acceptable methods for constructing wood elements in structures. When designed and built in accordance with the standards listed in this section, a building or structure is deemed to comply with the code. These standards contain most of the information needed to adequately design a structure in accordance with the ASD method. It is necessary for the designer to have a working knowledge of engineering principles and experience to properly interpret these recommendations and meet the provisions of other applicable sections of the code.

The most common and applicable practices are summarized in the standards listed in Section 2306.1. These standards contain references to additional technical publications that address special problems and design criteria for wood structures. The users of these standards should have training and expertise in engineering and construction in order to properly interpret the standards' requirements. These standards employ ASD—the traditional format for presenting safety checking equations in structural design standards. The basic design equation for ASD requires that the specified product allowable design value meet or exceed the actual (applied) stress or other effect imposed by the specified loads. In ASD, the allowable design values are set very low and the nominal load magnitudes are set at "once-in-a-lifetime" levels. This combination produces designs that maintain high safety levels yet remain economically feasible.

While preventing structural collapse remains the primary purpose of structural design, the designer must also consider how the design will perform from the perspective of serviceability, durability and fire safety. The ASD approach is to provide adequate resistance to certain limit states for serviceability. Serviceability limit states are those that restrict the normal use and occupancy of the structure, such as excessive deflection and vibration. The code defines these limits as a ratio of the member span; for example, $L/360$ computed under live load or $L/240$ under total load for floors.

American Forest & Paper Association.

NDS   National Design Specification for Wood Construction

❖ The AF&PA NDS is promulgated and distributed by the American Forest & Paper Association (AF&PA), listed in Chapter 35. The NDS was first adopted in 1944 and has been updated periodically to reflect new knowledge under the auspices of AF&PA and its predeces-

sor organizations, the National Lumber Manufacturers Association (NLMA) and the National Forest Products Association (NFPA). A supplement to the NDS collates all of the recommended working stress values published by the approved grade rules-writing agencies. The NDS is included in AF&PA's *Allowable Stress Design Manual for Wood Construction* (ASD Manual). The ASD Manual provides guidance for selection of most wood-based structural products used in the construction of wood buildings.

AF&PA also publishes *Wood Construction Data (WCD) No. 5, Heavy Timber Construction Details*. It is written in code language for easy reference. WCD No. 5 contains detailed specifications for sizes of members required and other provisions, which collectively provide a definition of heavy timber construction. Recommendations are also included for the fabrication and erection of this unique construction. It should be used in conjunction with the NDS and the American Institute of Timber Construction (AITC) specifications when the members are glued laminated.

The *Supplement to the National Design Specification* (NDS) is a collation of the allowable working stress values for sawn lumber as published by seven grade rules-writing agencies: National Lumber Grades Authority (Canada), Northern Softwood Lumber Bureau, Northeastern Lumber Manufacturers Association, Redwood Inspection Bureau, Southern Pine Inspection Bureau, West Coast Lumber Inspection Bureau and Western Wood Products Association.

These allowable working stress values, often called design values, have been approved by the Board of Review of the ALSC with advice from the U.S. Forest Products Laboratory of the U.S. Department of Agriculture. They have been certified for conformance with the U.S. Department of Commerce Voluntary Standard PS 20. Regional grading agencies formulate and publish grading rules, and ALSC approves these rules as conforming with DOC PS 20.

DOC PS 20 requires that design values for visually graded lumber be developed in accordance with appropriate ASTM standards or other technically sound criteria. The design values reflected in the NDS Supplement are derived from ASTM D 1990, which is utilized to derive strength values from full-sized samples of lumber.

Design values for visually graded timbers, decking and some species and grades of dimension lumber are based on the provisions of ASTM D 245. The methods in ASTM D 245 involve adjusting the strength properties of small clear specimens of wood for the effects of knots, slope of grain, splits, checks, size, duration of load, moisture content and other influencing factors to obtain design values applicable to normal conditions of service.

The design values given in the supplement are for normal loading conditions and under dry service conditions in a covered building. If other conditions of loadings or wet end-use conditions exist, these values must be adjusted in accordance with the recommendations of the NDS.

American Institute of Timber Construction.

AITC 104   Typical Construction Details

AITC 110   Standard Appearance Grades for Structural Glued Laminated Timber

AITC 113   Standard for Dimensions of Structural Glued Laminated Timber

AITC 117   Standard Specifications for Structural Glued Laminated Timber of Softwood Species

AITC 119   Structural Standard Specifications for Glued Laminated Timber of Hardwood Species

AITC A190.1   Structural Glued Laminated Timber

AITC 200   Inspection Manual

❖ The American Institute of Timber Construction (AITC) standards listed are primarily intended for glued-laminated structural members, but also include recommendations for sawn timber members. These standards supplement the provisions of the AF&PA NDS and WCD No. 5 with no conflicting provisions. All the standards should be used collectively.

AITC A190.1 is intended to provide nationally recognized requirements for the production, inspection, testing and certification of structural glued-laminated timber. Section 6 of the standard provides detailed requirements for in-plant quality control and the qualification of third-party inspection and testing agencies. These specifications include provisions for required testing during and after fabrication. Section 7 outlines the requirements for marking, which must be distinctive and must identify the edition of AITC A190.1 used, the inspection and testing agency and the laminating plant.

The marking on other than custom members meeting specific job specifications must also include information on: species or species group of lumber, design values used (if not standard), intended use of member (simple beam, tension or compression member, etc.), type of adhesive (dry-wet-use), appearance grade (industrial-architectural-premium), treatment (if any) and proofloading (if applicable).

American Society of Agricultural Engineers.

ASAE EP 484.2   Diaphragm Design of Metal-Clad, Post-Frame Rectangular Buildings

ASAE EP 486.1   Shallow Post Foundation Design

ASAE 559   Design Requirements and Bending Properties for Mechanically Laminated Columns

❖ American Society of Agricultural Engineers (ASAE) develops standards that are needed for the construction of agricultural and utility buildings.

APA—The Engineered Wood Association.

Panel Design Specification

Plywood Design Specification Supplement 1 - Design & Fabrication of Plywood Curved Panel

Plywood Design Specification Supplement 2 - Design & Fabrication of Glued Plywood-Lumber Beams

Plywood Design Specification Supplement 3 - Design & Fabrication of Plywood Stressed-Skin Panels

Plywood Design Specification Supplement 4 - Design & Fabrication of Plywood Sandwich Panels

Plywood Design Specification Supplement 5 - Design & Fabrication of All-Plywood Beams

EWS T300   Glulam Connection Details

EWS S560   Field Notching and Drilling of Glued Laminated Timber Beams

EWS S475   Glued Laminated Beam Design Tables

EWS X450   Glulam in Residential Construction

EWS X440   Product and Application Guide: Glulam

EWS R540   Builders Tips: Proper Storage and Handling of Glulam Beams

❖ American Plywood Association (APA) recommendations and design standards are considered to provide the best criteria for the design and erection of structures incorporating plywood and other structural-use panel products.

Truss Plate Institute, Inc.

TPI 1   National Design Standard for Metal Plate Connected Wood Truss Construction

❖ TPI 1 is the standard for the design, fabrication, quality control and erection of light-frame metal-plate-connected wood trusses. The standard was prepared through a joint effort by the American National Standards Institute (ANSI) and TPI. This standard should be used in conjunction with other standards for wood construction.

2306.1.1 Joists and rafters. The design of rafter spans is permitted to be in accordance with the *AF&PA Span Tables for Joists and Rafters.*

❖ Spans for joists and rafters may be designed in accordance with the AF&PA *Span Tables for Joists and Rafters* or in accordance with approved engineering standards. When sizes of floor joists are other than standard sizes conforming to DOC PS 20, they must conform to the NDS. When applicable, other span tables may be used, such as those from the IRC, Southern Forest Products Association, Western Wood Products Association and Canadian Wood Council. The spans given in these tables are the same as those published in AF&PA *Span Tables for Joists and Rafters.* The AF&PA span tables were developed to provide uniformity in use of wood joists and rafters for light wood-frame structures where the live loads do not exceed those for which the tables were designed. The tables for floor joists were designed with the live load, dead load and deflection criteria shown in Chapter 16.

2306.1.2 Plank and beam flooring. The design of plank and beam flooring is permitted to be in accordance with the *AF&PA Wood Construction Data No. 4.*

❖ The plank-and-beam method for framing floors and roofs has been used in buildings for many years. The adaptation of this system to residential construction has raised technical issues concerning details of application. AF&PA's *Wood Construction Data No. 4, Plank and Beam Framing for Residential Buildings*, contains information pertaining to design principles, advantages and limitations, construction details and structural requirements for the plank-and-beam method of framing.

2306.1.3 Treated wood stress adjustments. The allowable unit stresses for preservative-treated wood need no adjustment for treatment, but are subject to other adjustments.

The allowable unit stresses for fire-retardant-treated wood, including fastener values, shall be developed from an approved method of investigation that considers the effects of anticipated temperature and humidity to which the fire-retardant-treated wood will be subjected, the type of treatment and the redrying process. Other adjustments are applicable except that the impact load duration shall not apply.

❖ The allowable stresses for fire-retardant-treated wood products must be determined by testing that takes into account anticipated temperatures and humidity levels that will be encountered during service. The allowable stresses for untreated products are not applicable to fire-retardant-treated products, because the treatment chemicals can reduce the strength of materials. This is particularly true for products such as roof sheathing, which is subject to conditions of high temperature and humidity.

Before approving the use of fire-retardant-treated wood products, the building official should require the manufacturer of each subject product to submit its recommended allowable stresses and provide documentation on how those values were determined. This documentation should be required for each chemical that is to be used, because performance will vary between formulations. Some manufacturers maintain evaluation reports that can be used by the building official in reviewing these products.

Additionally, the building official should enforce any restriction on the use and location of any such treated product recommended by the manufacturer or justified based on the nature of the testing performed. For example, when these products are proposed for use in attics, the building official should make sure that they have been tested under conditions that will ensure suitable performance under conditions of high temperature and humidity typically found in attics.

The building official should also be alerted to the fact that the ventilation requirement in the code takes on added importance when such products are used in attics. Ventilation requirements should be strictly enforced, and the building official should consult the manufacturer to determine if any additional ventilation

above the code requirements might be necessary. Ventilation is needed to control temperature and humidity levels to which the products will be subjected. A lack of adequate ventilation can change the service conditions to which the products are subjected and, as a result, reduce the strength of the products below the values published by the manufacturer.

2306.1.4 Lumber decking. The capacity of lumber decking arranged according to the patterns described in Section 2304.8.2 shall be the lesser of the capacities determined for flexure and deflection according to the formulas in Table 2306.1.4.

❖ Along with installation requirements in Section 2304.8, this section incorporates provisions for design of lumber decking.

TABLE 2306.1.4
ALLOWABLE LOADS FOR LUMBER DECKING

PATTERN	ALLOWABLE AREA LOAD[a,b]	
	Flexure	Deflection
Simple span	$\sigma_b = \dfrac{8F_b'}{l^2}\dfrac{d^2}{6}$	$\sigma_\Delta = \dfrac{384\Delta E'}{5l^4}\dfrac{d^3}{12}$
Two-span continuous	$\sigma_b = \dfrac{8F_b'}{l^2}\dfrac{d^2}{6}$	$\sigma_\Delta = \dfrac{185\Delta E'}{l^4}\dfrac{d^3}{12}$
Combination simple- and two-span continuous	$\sigma_b = \dfrac{8F_b'}{l^2}\dfrac{d^2}{6}$	$\sigma_\Delta = \dfrac{131\Delta E'}{l^4}\dfrac{d^3}{12}$
Cantilevered pieces intermixed	$\sigma_b = \dfrac{20F_b'}{3l^2}\dfrac{d^2}{6}$	$\sigma_\Delta = \dfrac{105\Delta E'}{l^4}\dfrac{d^3}{12}$
Controlled random layup		
Mechanically laminated decking	$\sigma_b = \dfrac{20F_b'}{3l^2}\dfrac{d^2}{6}$	$\sigma_\Delta = \dfrac{100\Delta E'}{l^4}\dfrac{d^3}{12}$
2-inch decking	$\sigma_b = \dfrac{20F_b'}{3l^2}\dfrac{d^2}{6}$	$\sigma_\Delta = \dfrac{100\Delta E'}{l^4}\dfrac{d^3}{12}$
3-inch and 4-inch decking	$\sigma_b = \dfrac{20F_b'}{3l^2}\dfrac{d^2}{6}$	$\sigma_\Delta = \dfrac{116\Delta E'}{l^4}\dfrac{d^3}{12}$

For SI: 1 inch = 25.4 mm.

a. σ_b = Allowable total uniform load limited by bending.

  σ_Δ = Allowable total uniform load limited by deflection.

b. d = Actual decking thickness.

  l = Span of decking.

  F_b' = Allowable bending stress adjusted by applicable factors.

  E' = Modulus of elasticity adjusted by applicable factors.

❖ Based on the decking layup patterns in Section 2304.8.2, the applicable equations for computing allowable uniform load are selected from this table. Note that the allowable loading will be the most restrictive value based on bending stresses and deflection. Also note that units are not specified in the equation and units merely need to be consistent in carrying out the load computation.

2306.2 Wind provisions for walls.

2306.2.1 Wall stud bending stress increase. The AF&PA NDS fiber stress in bending (F_b) design values for sawn lumber wood studs resisting out of plane wind loads shall be increased by the factors in Table 2306.2.1, in lieu of the 1.15 repetitive member factor. These increases take into consideration the load sharing and composite actions provided by the wood structural panels as defined in Section 2302.1. The increases shall apply where the studs are designed for bending and are spaced no more than 16 inches (406 mm) o.c., covered on the inside with a minimum of $^1/_2$-inch (12.7 mm) gypsum board fastened in accordance with Table 2306.4.5 and sheathed on the exterior with a minimum of $^3/_8$-inch (9.5 mm) wood structural panel sheathing. All panel joints shall occur over studs or blocking and shall be attached using a minimum of 8d common nails spaced a maximum of 6 inches o.c. (152 mm) at panel edges and 12 inches o.c. (305 mm) at intermediate framing members.

❖ See the commentary to Table 2306.2.1.

TABLE 2306.2.1
WALL STUD BENDING STRESS INCREASE FACTORS

STUD SIZE	SYSTEM FACTOR
2 × 4	1.5
2 × 6	1.35
2 × 8	1.25
2 × 10	1.2
2 × 12	1.15

❖ Design values for structural lumber are published in the AF&PA NDS Supplement. The tabulated design values are for normal load duration under the moisture conditions specified. Bending design values, F_b, for dimension lumber may be multiplied by the repetitive member factor, C_r = 1.15, when such members are used as joists, truss chords, rafters, studs, planks, decking or for similar members that are in contact or spaced not more than 24 inches (610 mm) o.c., are not less than three in number and are joined by floor, roof or other load-distributing elements adequate to support the design load.

For wall systems composed of 2-inch by 4-inch (51 mm by 102 mm) to 2-inch by 10-inch (51 mm by 254 mm) studs faced on one side a minimum of $^3/_8$-inch (9.5 mm) wood structural panel and with a minimum of $^1/_2$-inch (12.7 mm) gypsum board on the other side, the repetitive member factor may be increased to the values in Table 2306.2.1 as a result of load sharing and composite actions of the wall-framing system in a high-wind condition.

2306.3 Wood diaphragms.

❖ A diaphragm is a relatively thin structural element, (e.g., floor or roof assembly), usually rectangular in plan and capable of resisting shear parallel to its edges. The three major elements of a diaphragm are the web (wood structural panels and joists/rafters), the chords (edge members) and the fasteners (such as nails). The web is usually composed of several panels. The chords are also almost always one piece. Therefore, fasteners and connection chord splices must be considered as major factors in the performance of a structural diaphragm.

Most floors, roofs and walls can function as a diaphragm. Full diaphragm action can be obtained by detailing the sheathing fasteners and framing connections. Diaphragm construction is used extensively to resist horizontal forces of wind and earthquakes. Diaphragms may be either vertical (often referred to as "shear walls") or horizontal. In special designs, they may be oriented in any direction. A diaphragm acts in the same manner as an I-beam. The plywood panel skin acts as the web and the edge members act as the flanges, which are called chords. However, due to the relatively greater depth of a diaphragm, the reactions to stress are somewhat different. The diaphragm design assumes that stresses are uniformly distributed across the panel and the web does not contribute to the resistance of tension and compression stresses as it does in a shallow beam.

A series of diaphragms may be tied together to develop a very rigid structure. Their ability to resist large shear stresses has led to their use in innovative construction methods such as folded plate roof elements, geodesic domes and space frames.

Diaphragms must be securely connected to supporting members and to foundations that are capable of resisting all design forces. Anchorage for floors and roof generally presents no special problems. It is only required that the stresses be transferred to the chords, assuming they are adequately supported. Shear walls, however, require special attention to their anchorage to the foundation. The foundation and the connections thereto must be designed to resist the uplift, compression and shear forces applied.

Openings in diaphragms, such as skylights, doors and windows, disrupt the uniform distribution of shear stresses across the diaphragm. Provisions must be made for continuity and compensation for this disruption. For instance, reinforcement may be necessary at windows.

2306.3.1 Wood structural panel diaphragms. Wood structural panel diaphragms are permitted to resist horizontal forces using the allowable shear capacities set forth in Table 2306.3.1 or 2306.3.2. The allowable shear capacities are permitted to be calculated by principles of mechanics without limitations by using values for fastener strength in the AF&PA NDS, structural design properties for wood structural panels based on DOC PS-1 and DOC PS-2 or wood structural panel design properties given in the *APA Panel Design Specification* (PDS).

❖ Wood structural panel diaphragms may be designed using the information in Table 2306.3.1 or 2306.3.2. Note that the panel thickness and joist spacings must also comply with the requirements for sheathing of floors and roofs. Wood structural panel diaphragms may also be designed using accepted engineering principles. AF&PA NDS provides information on shear resistance of mechanical fasteners and chord design. Working stresses for wood structural panels and other materials are contained in other references in the code.

The design of horizontal diaphragms to resist shear stresses depends on the direction of continuous panel joints relative to the direction of load and not on the direction of the long dimension of the panel or framing. Six cases of panel orientation are illustrated in Table 2306.3.1. The nailing schedule and the resultant shear resistance will be affected by the panel layout (relative to load) selected. A diaphragm usually must resist loads in both transverse and longitudinal directions. Each direction must be considered separately with a different case of panel orientation; thus, the allowable load will be applicable to each direction.

Blocked diaphragms have all panel edges supported by and fastened to framing members. Blocking is used at the edges of panels that are not over joists to provide for connecting units and to assist in the transfer of shear forces to the adjacent panel. Unblocked diaphragms may be used, provided the applicable nailing schedule in Table 2304.9.1 is used and the shear stresses do not exceed those allowed for unblocked diaphragms in Table 2306.3.1. Unblocked diaphragms are often controlled by the buckling of the panel (skin). With the same nail spacing, allowable design loads for blocked diaphragms are $1 \frac{1}{2}$ to 2 times the design loads for unblocked diaphragms. In addition, the maximum loads for which blocked diaphragms may be designed are many times greater. The forces that must be resisted usually serve as the basis for choosing blocked or unblocked construction.

2306.3.2 Shear capacities modifications. The allowable shear capacities in Tables 2306.3.1 and 2306.3.2 for horizontal wood structural panel diaphragms shall be increased 40 percent for wind design.

❖ The 40-percent increase for wind is based on reducing the load factor from 2.8, which was originally used to develop shear capacities for diaphragms, down to 2.0 for wind resistance.

TABLE 2306.3.1. See page 23-56.

❖ See the commentary to Section 2306.3.1.
   The table's fifth column specifies the minimum nominal width of framing members at adjoining panel edges and boundaries. The intent is to provide thicker framing where the fastener size, spacing or number of rows at adjoining panel edges, or boundaries, is dictated. The purpose of using larger framing at these locations is to allow the connection details specified to transfer the shear forces from one panel to the next without splitting the framing members. No matter what the edge-nailing requirement is for a specific panel,

the minimum nailing requirement at intermediate framing (framing that does not support a panel edge) is always a single row of nails at 6 or 12 inches (152 or 305 mm) o.c. as this connection is not part of the shear transfer system. As such, from a shear-transfer perspective there is no reason to use the "at adjoining panel edges" framing requirements at intermediate framing locations. The normal (usually 2x) framing specified for the vertical load or other engineering considerations is sufficient at the intermediate framing locations.

Note h clarifies the calculation method required to convert table values for applications other than seismic or wind loading. These adjustments reflect the fact that the values in the tables have a built-in load duration factor (LDF) of 1.6. As such, to convert to a normal load duration (LDF = 1.0) the tabular value must be multiplied by 0.63 (1.0/1.6 = 0.63), and to convert to a permanent load duration (LDF = 0.90) a factor of 0.56 must be used (0.9/1.6 = 0.56).

Note f requires staples to be installed with the crowns of the staple parallel to the long dimension of the framing members. This makes the installation requirement for staples consistent with past practice for staples used in wood-framed diaphragms (see National Evaluation Service, Inc. National Evaluation Report NER-272).

As Figure 2306.3.1(1) illustrates, splitting occurs parallel to the grain and staggering the nails alleviates this problem. Notes c and d address the concern of splitting of framing members where tighter nail spacing is specified. Figure 2306.3.1(2) illustrates staggered nailing that meets the intent of these footnotes.

TABLE 2306.3.2. See page 23-60.

❖ This table contains wood structural panel diaphragm assemblies that have higher capacities compared to those of Table 2306.3.1. These can be useful for larger flat-roofed structures subjected to high wind or seismic loads. The capacities reflected in this table are based upon calculations using the European yield method and were also confirmed by testing.
   Also see the commentary to Table 2306.3.1.

2306.3.3 Diagonally sheathed lumber diaphragms. Diagonally sheathed lumber diaphragms shall be nailed in accordance with Table 2306.3.3.

❖ See the commentary to Table 2306.3.3.

TABLE 2306.3.3. See page 23-61.

❖ This table presents a prescriptive schedule for nailing of nominal 1-inch (25.4 mm) and 2-inch (51 mm) lumber sheathing placed diagonally to the structural framing. The nail schedule is intended to be used in accordance with the requirements of Section 2306.3.4 for single diagonally sheathed lumber diaphragms and Section 2306.3.5 for double diagonally sheathed lumber diaphragms.

TABLE 2306.3.1
ALLOWABLE SHEAR (POUNDS PER FOOT) FOR WOOD STRUCTURAL PANEL DIAPHRAGMS WITH FRAMING OF DOUGLAS FIR-LARCH, OR SOUTHERN PINE[a] FOR WIND OR SEISMIC LOADING[h]

PANEL GRADE	COMMON NAIL SIZE OR STAPLE[f] LENGTH AND GAGE	MINIMUM FASTENER PENETRATION IN FRAMING (inches)	MINIMUM NOMINAL PANEL THICKNESS (inch)	MINIMUM NOMINAL WIDTH OF FRAMING MEMBERS AT ADJOINING PANEL EDGES AND BOUNDARIES[g] (inches)	BLOCKED DIAPHRAGMS				UNBLOCKED DIAPHRAGMS	
					Fastener spacing (inches) at diaphragm boundaries (all cases) at continuous panel edges parallel to load (Cases 3, 4), and at all panel edges (Cases 5, 6)[b] — and Fastener spacing (inches) at other panel edges (Cases 1, 2, 3 and 4)[b]				Fasteners spaced 6" max. at supported edges[b]	
					6 / 6	4 / 6	2½[c] / 4	2[c] / 3	Case 1 (No unblocked edges or continuous joints parallel to load)	All other configurations (Cases 2, 3, 4, 5 and 6)
Structural I Grades	6d[e] (2" × 0.113")	1¼	5/16	2	185	250	375	420	165	125
				3	210	280	420	475	185	140
	1½ 16 Gage	1	5/16	2	155	205	310	350	135	105
				3	175	230	345	390	155	115
	8d (2½" × 0.131)	1⅜	3/8	2	270	360	530	600	240	180
				3	300	400	600	675	265	200
	1½ 16 Gage	1	3/8	2	175	235	350	400	155	115
				3	200	265	395	450	175	130
	10d[d] (3" × 0.148")	1½	15/32	2	320	425	640	730	285	215
				3	360	480	720	820	320	240
	1½ 16 Gage	1	15/32	2	175	235	350	400	155	120
				3	200	265	395	450	175	130
Sheathing, single floor and other grades covered in DOC PS 1 and PS 2	6d[e] (2" × 0.113)	1¼	5/16	2	170	225	335	380	150	110
				3	190	250	380	430	170	125
	1½ 16 Gage	1	5/16	2	140	185	275	315	125	90
				3	155	205	310	350	140	105
	6d[e] (2" × 0.113)	1¼	3/8	2	185	250	375	420	165	125
				3	210	280	420	475	185	140
	8d (2½" × 0.131)	1⅜	3/8	2	240	320	480	545	215	160
				3	270	360	540	610	240	180

(continued)

TABLE 2306.3.1—continued
ALLOWABLE SHEAR (POUNDS PER FOOT) FOR WOOD STRUCTURAL PANEL DIAPHRAGMS WITH FRAMING OF DOUGLAS FIR-LARCH, OR SOUTHERN PINE[a] FOR WIND OR SEISMIC LOADING[h]

PANEL GRADE	COMMON NAIL SIZE OR STAPLE[f] LENGTH AND GAGE	MINIMUM FASTENER PENETRATION IN FRAMING (inches)	MINIMUM NOMINAL PANEL THICKNESS (inch)	MINIMUM NOMINAL WIDTH OF FRAMING MEMBERS AT ADJOINING PANEL EDGES AND BOUNDARIES[g] (inches)	BLOCKED DIAPHRAGMS				UNBLOCKED DIAPHRAGMS	
					Fastener spacing (inches) at diaphragm boundaries (all cases), at continuous panel edges parallel to load (Cases 3, 4), and at all panel edges (Cases 5, 6)[b]				Fasteners spaced 6" max. at supported edges[b]	
					6	4	2½[c]	2[c]	Case 1 (No unblocked edges or continuous joints parallel to load)	All other configurations (Cases 2, 3, 4, 5 and 6)[b]
					Fastener spacing (inches) at other panel edges (Cases 1, 2, 3 and 4)[b]					
					6	6	4	3		
Sheathing, single floor and other grades covered in DOC PS 1 and PS 2 (continued)	1½ 16 Gage	1	3/8	2	160	210	315	360	140	105
				3	180	235	355	400	160	120
	8d (2½"× 0.131")	1 3/8		2	255	340	505	575	230	170
				3	285	380	570	645	255	190
	1½ 16 Gage	1	7/16	2	165	225	335	380	150	110
				3	190	250	375	425	165	125
	8d (2½"× 0.131")	1 3/8		2	270	360	530	600	240	180
				3	300	400	600	675	265	200
	10d (3"× 0.148")	1½	15/32	2	290	385	575	655	255	190
				3	325	430	650	735	290	215
	1½ 16 Gage	1	19/32	2	160	210	315	360	140	105
				3	180	235	355	405	160	120
	10d (3"× 0.148")	1½		2	320	425	640	730	285	215
				3	360	480	720	820	320	240
	1¾ 16 Gage	1		2	175	235	350	400	155	115
				3	200	265	395	450	175	130

(continued)

TABLE 2306.3.1—continued
ALLOWABLE SHEAR (POUNDS PER FOOT) FOR WOOD STRUCTURAL
PANEL DIAPHRAGMS WITH FRAMING OF DOUGLAS FIR-LARCH,
OR SOUTHERN PINE[a] FOR WIND OR SEISMIC LOADING[h]

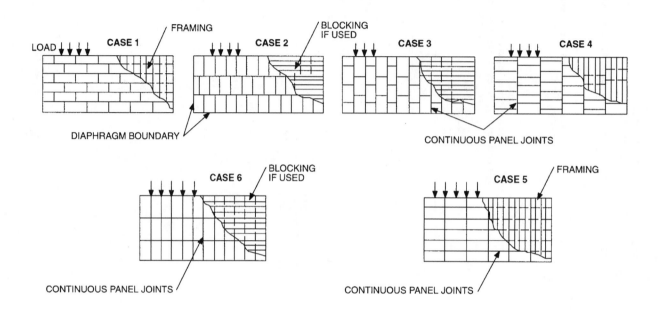

For SI: 1 inch = 25.4 mm, 1 pound per foot = 14.5939 N/m.

a. For framing of other species: (1) Find specific gravity for species of lumber in AF&PA NDS. (2) For staples find shear value from table above for Structural I panels (regardless of actual grade) and multiply value by 0.82 for species with specific gravity of 0.42 or greater, or 0.65 for all other species. (3) For nails find shear value from table above for nail size for actual grade and multiply value by the following adjustment factor: Specific Gravity Adjustment Factor = [1-(0.5 - SG)], where SG = Specific Gravity of the framing lumber. This adjustment factor shall not be greater than 1.

b. Space fasteners maximum 12 inches o.c. along intermediate framing members (6 inches o.c. where supports are spaced 48 inches o.c.).

c. Framing at adjoining panel edges shall be 3 inches nominal or wider, and nails shall be staggered where nails are spaced 2 inches o.c. or $2^{1}/_{2}$ inches o.c.

d. Framing at adjoining panel edges shall be 3 inches nominal or wider, and nails shall be staggered where both of the following conditions are met: (1) 10d nails having penetration into framing of more than $1^{1}/_{2}$ inches and (2) nails are spaced 3 inches o.c. or less.

e. 8d is recommended minimum for roofs due to negative pressures of high winds.

f. Staples shall have a minimum crown width of $^{7}/_{16}$ inch and shall be installed with their crowns parallel to the long dimension of the framing members.

g. The minimum nominal width of framing members not located at boundaries or adjoining panel edges shall be 2 inches.

h. For shear loads of normal or permanent load duration as defined by the AF&PA NDS, the values in the table above shall be multiplied by 0.63 or 0.56, respectively.

Figure 2306.3.1(1)
HOW WOOD SPLITS AND HOW STAGGERING MINIMIZES SPLITTING

For Si: 1 inch = 25.4 mm.

Figure 2306.3.1(2)
STAGGERED NAILING

TABLE 2306.3.2
ALLOWABLE SHEAR (POUNDS PER FOOT) FOR WOOD STRUCTURAL PANEL BLOCKED DIAPHRAGMS
UTILIZING MULTIPLE ROWS OF FASTENERS (HIGH LOAD DIAPHRAGMS) WITH FRAMING OF
DOUGLAS FIR-LARCH OR SOUTHERN PINE[a] FOR WIND OR SEISMIC LOADING[b, g, h]

						BLOCKED DIAPHRAGMS					
						Cases 1 and 2[d]					
						Fastener Spacing Per Line at Boundaries (inches)					
						4		2½		2	
PANEL GRADE[c]	COMMON NAIL SIZE OR STAPLE[f] GAGE	MINIMUM FASTENER PENETRATION IN FRAMING (inches)	MINIMUM NOMINAL PANEL THICKNESS (inch)	MINIMUM NOMINAL WIDTH OF FRAMING MEMBER AT ADJOINING PANEL EDGES AND BOUNDARIES[e]	LINES OF FASTENERS	Fastener Spacing Per Line at Other Panel Edges (inches)					
						6	4	4	3	3	2
Structural I grades	10d common nails	1½	15/32	3	2	605	815	875	1,150	—	—
				4	2	700	915	1,005	1,290	—	—
				4	3	875	1,220	1,285	1,395	—	—
			19/32	3	2	670	880	965	1,255	—	—
				4	2	780	990	1,110	1,440	—	—
				4	3	965	1,320	1,405	1,790	—	—
			23/32	3	2	730	955	1,050	1,365	—	—
				4	2	855	1,070	1,210	1,565	—	—
				4	3	1,050	1,430	1,525	1,800	—	—
	14 gage staples	2	15/32	3	2	600	600	860	960	1,060	1,200
				4	3	860	900	1,160	1,295	1,295	1,400
			19/32	3	2	600	600	875	960	1,075	1,200
				4	3	875	900	1,175	1,440	1,475	1,795
Sheathing single floor and other grades covered in DOC PS 1 and PS 2	10d common nails	1½	15/32	3	2	525	725	765	1,010	—	—
				4	2	605	815	875	1,105	—	—
				4	3	765	1,085	1,130	1,195	—	—
			19/32	3	2	650	860	935	1,225	—	—
				4	2	755	965	1,080	1,370	—	—
				4	3	935	1,290	1,365	1,485	—	—
			23/32	3	2	710	935	1,020	1,335	—	—
				4	2	825	1,050	1,175	1,445	—	—
				4	3	1,020	1,400	1,480	1,565	—	—
	14 gage staples	2	15/32	3	2	540	540	735	865	915	1,080
				4	3	735	810	1,005	1,105	1,105	1,195
			19/32	3	2	600	600	865	960	1,065	1,200
				4	3	865	900	1,130	1,430	1,370	1,485
			23/32	4	3	865	900	1,130	1,490	1,430	1,545

For SI: 1 inch = 25.4 mm, 1 pound per foot = 14.5939 N/m.

a. For framing of other species: (1) Find specific gravity for species of framing lumber in AF&PA NDS. (2) For staples, find shear value from table above for Structural I panels (regardless of actual grade) and multiply value by 0.82 for species with specific gravity of 0.42 or greater, or 0.65 for all other species. (3) For nails, find shear value from table above for nail size of actual grade and multiply value by the following adjustment factor: Specific Gravity Adjustment Factor = [1- (0.5 - SG)], where SG = Specific gravity of the framing lumber. This adjustment factor shall not be greater than 1.

b. Fastening along intermediate framing members: Space fasteners a maximum of 12 inches on center, except 6 inches on center for spans greater than 32 inches.

c. Panels conforming to PS 1 or PS 2.

d. This table gives shear values for Cases 1 and 2 as shown in Table 2306.3.1. The values shown are applicable to Cases 3, 4, 5 and 6 as shown in Table 2306.3.1, providing fasteners at all continuous panel edges are spaced in accordance with the boundary fastener spacing.

e. The minimum nominal depth of framing members shall be 3 inches nominal. The minimum nominal width of framing members not located at boundaries or adjoining panel edges shall be 2 inches.

f. Staples shall have a minimum crown width of 7/16 inch, and shall be installed with their crowns parallel to the long dimension of the framing members.

g. High load diaphragms shall be subject to special inspection in accordance with Section 1704.6.1.

h. For shear loads of normal or permanent load duration as defined by the AF&PA NDS, the values in the table above shall be multiplied by 0.63 or 0.56, respectively.

TABLE 2306.3.3
DIAGONALLY SHEATHED LUMBER DIAPHRAGM NAILING SCHEDULE

SHEATHING NOMINAL DIMENSION	NAILING TO INTERMEDIATE AND END-BEARING STUDS		NAILING AT THE SHEAR PANEL BOUNDARIES	
	Type, size and number of nails per board			
	Common nails	Box nails	Common nails	Box nails
1 × 6	2 - 8d	3 - 8d	3 - 8d	5 - 8d
1 × 8	3 - 8d	4 - 8d	4 - 8d	6 - 8d
2 × 6	2 - 16d	3 - 16d	3 - 16d	5 - 16d
2 × 8	3 - 16d	4 - 16d	4 - 16d	6 - 16d

2306.3.4 Single diagonally sheathed lumber diaphragms. Single diagonally sheathed lumber diaphragms shall be constructed of minimum 1-inch (25 mm) thick nominal sheathing boards laid at an angle of approximately 45 degrees (0.78 rad) to the supports. The shear capacity for single diagonally sheathed lumber diaphragms of southern pine or Douglas fir-larch shall not exceed 300 plf (4378 N/m) of width. The shear capacities shall be adjusted by reduction factors of 0.82 for framing members of species with a specific gravity equal to or greater than 0.42 but less than 0.49 and 0.65 for species with a specific gravity of less than 0.42, as contained in the AF&PA NDS.

❖ The reduction factors are based on common wire nail shear design values for three classifications of lumber as follows:

1. Specific gravity, G, ≥ 0.49 (for example, Southern pine and Douglas fir-larch);

2. 0.42 ≤ G < 0.49 (for example, spruce-pine-fir and hem-fir); and

3. 0.42 > G (for example, aspen, balsam fir and cedars).

The specific gravity of lumber species can be found in Table 12A of AF&PA NDS.

2306.3.4.1 End joints. End joints in adjacent boards shall be separated by at least one stud or joist space and there shall be at least two boards between joints on the same support.

❖ End joints of individual boards are required to be staggered in location as described in this section to maintain the shear capacity specified in Section 2306.3.4.2.

2306.3.4.2 Single diagonally sheathed lumber diaphragms. Single diagonally sheathed lumber diaphragms made up of 2-inch (51 mm) nominal diagonal lumber sheathing fastened with 16d nails shall be designed with the same shear capacities as shear panels using 1-inch (25 mm) boards fastened with 8d nails, provided there are not splices in adjacent boards on the same support and the supports are not less than 4 inch (102 mm) nominal depth or 3 inch (76 mm) nominal thickness.

❖ The shear capacity referenced in this section is conservative, since tests of wood shear panels with 2-inch (51 mm) nominal lumber sheathing are not readily available.

2306.3.5 Double diagonally sheathed lumber diaphragms. Double diagonally sheathed lumber diaphragms shall be constructed of two layers of diagonal sheathing boards at 90 degrees (1.57 rad) to each other on the same face of the supporting members. Each chord shall be considered as a beam with uniform load per foot equal to 50 percent of the unit shear due to diaphragm action. The load shall be assumed as acting normal to the chord in the plan of the diaphragm in either direction. The span of the chord or portion thereof shall be the distance between framing members of the diaphragm, such as the joists, studs and blocking that serve to transfer the assumed load to the sheathing. The shear capacity of double diagonally sheathed diaphragms of Southern pine or Douglas fir-larch shall not exceed 600 plf (8756 kN/m) of width. The shear capacity shall be adjusted by reduction factors of 0.82 for framing members of species with a specific gravity equal to or greater than 0.42 but less than 0.49 and 0.65 for species with a specific gravity of less than 0.42, as contained in the AF&PA NDS. Nailing of diagonally sheathed lumber diaphragms shall be in accordance with Table 2306.3.3.

❖ This section provides the shear capacity for various configurations of double diagonal board-sheathed shear panels. This type of construction is seldom used because of the labor costs of individual board installation. Double diagonal construction has a higher shear capacity than the single diagonal construction described in Section 2306.3.4. For discussion on the reduction factor, see the commentary to Section 2306.3.4.

2306.3.6 Gypsum board diaphragm ceilings. Gypsum board diaphragm ceilings shall be in accordance with Section 2508.5.

❖ See the commentary to Section 2508.5.

2306.4 Shear walls. Panel sheathing joints in shear walls shall occur over studs or blocking. Adjacent panel sheathing joints shall occur over and be nailed to common framing members (see Section 2305.3.1 for limitations on shear wall bracing materials).

❖ A shear wall is a vertical diaphragm designed to resist lateral forces parallel to the plane of the wall. Therefore, much of the commentary in Section 2306.3 on diaphragms is also applicable to this section.

Detailing is critical in the design of shear walls so

that forces are properly distributed to top plates, ledgers, bond beams or any other continuous element at the perimeter to the diaphragm. Therefore, it is important that splices be designed to transmit tension or compression forces occurring at the location of a splice. Panel sheathing joints must occur over studs or blocking and common framing members.

2306.4.1 Wood structural panel shear walls. The allowable shear capacities for wood structural panel shear walls shall be in accordance with Table 2306.4.1. These capacities are permitted to be increased 40 percent for wind design. Shear walls are permitted to be calculated by principles of mechanics without limitations by using values for nail strength given in the AF&PA NDS and wood structural panel design properties given in the *APA Panel Design Specification*.

❖ See the commentary to Table 2306.4.1.

TABLE 2306.4.1. See page 23-63.

❖ The allowable shear capacities for wood structural panel shear walls in Table 2306.4.1 were developed on engineering principles and monotonic testing.

The 40-percent increase in shear capacities for wind loads is based on engineering principles. Full-scale monotonic testing was conducted to validate the model. The increase takes into effect the short time duration of loading associated with a wind event.

Shear walls may also be designed without limitations by a rational method using the values for nail strength in the AF&PA NDS or by using wood structural panel design properties in APA *Panel Design Specification*. For shear walls utilizing wood framing, energy dissipation is almost entirely due to nail bending.

Note i permits the use of a pair of nominal 2-inch members to be used in lieu of a single 3-inch nominal member, provided the two 2-inch members are fastened together to transfer the design shear value. This provision is based on testing by APA, which showed that two 2-inch members fastened to transfer the shear performed as well as shear walls with a single 3-inch nominal member. See the commentary to Table 2306.3.1 for additional information on the footnotes.

2306.4.2 Lumber sheathed shear walls. Single and double diagonally sheathed lumber diaphragms are permitted using the construction and allowable load provisions of Sections 2306.3.4 and 2306.3.5.

❖ Although wood structural panel sheathed shear walls are most prevalent in use today, diagonally sheathed lumber shear walls are still used for new construction in some regions and are encountered in the alteration of existing structures. The shear resistance values are

applicable to buildings exposed to seismic or wind forces.

2306.4.3 Particleboard shear walls. The design shear capacity of particleboard shear walls shall be in accordance with Table 2306.4.3. Shear panels shall be constructed with particleboard sheets not less than 4 feet by 8 feet (1219 mm by 2438 mm), except at boundaries and changes in framing. Particleboard panels shall be designed to resist shear only, and chords, collector members and boundary elements shall be connected at all corners. Panel edges shall be backed with 2-inch (51 mm) nominal or wider framing. Sheets are permitted to be installed either horizontally or vertically. For $^3/_8$-inch (9.5 mm) particleboard sheets installed with the long dimension parallel to the studs spaced 24 inches (610 mm) o.c, nails shall be spaced at 6 inches (152 mm) o.c. along intermediate framing members. For all other conditions, nails of the same size shall be spaced at 12 inches (305 mm) o.c. along intermediate framing members. Particleboard panels less than 12 inches (305 mm) wide shall be blocked. Particleboard shall not be used to resist seismic forces in structures in Seismic Design Category D, E or F.

❖ See the commentary to Table 2306.4.3.

TABLE 2306.4.3. See page 23-64.

❖ Limitations of use in shear walls are greater for particleboard than for wood structural panels and lumber. The design shear capacity of particleboard shear walls is limited to the values listed in Table 2306.4.3, and particleboard shear walls may only be used to resist shear forces. Further, this section specifies that special care must be taken so that all panel edges are properly attached to and fully blocked with 2-inch (51 mm) nominal framing for transfer of shear forces to boundary elements.

2306.4.4 Fiberboard shear walls. The design shear capacity of fiberboard shear walls shall be in accordance with Table 2306.4.4. The fiberboard sheathing shall be applied vertically or horizontally to wood studs not less than 2 inch (51 mm) in nominal thickness spaced 16 inches (406 mm) o.c. Blocking not less than 2 inch (51 mm) nominal in thickness shall be provided at horizontal joints. Fiberboard shall not be used to resist seismic forces in structures in Seismic Design Category D, E or F.

❖ For structural fiberboard to be used in the construction of shear walls, the long dimension of the panel must be parallel to the studs. A 2:1 aspect ratio is established. The shear capacity of fiberboard shear walls is limited to the values listed in Table 2306.4.4. Further, this section specifies that special care is to be taken so that all panel edges are properly attached to and fully blocked with 2-inch (51 mm) nominal framing for transfer of shear forces to boundary elements.

TABLE 2306.4.4. See page 23-64.

❖ See the commentary to Section 2306.4.4.

TABLE 2306.4.1
ALLOWABLE SHEAR (POUNDS PER FOOT) FOR WOOD STRUCTURAL PANEL SHEAR WALLS WITH FRAMING OF DOUGLAS FIR-LARCH OR SOUTHERN PINE[a] FOR WIND OR SEISMIC LOADING[b,h,i,j,l]

PANEL GRADE	MINIMUM NOMINAL PANEL THICKNESS (inch)	MINIMUM FASTENER PENETRATION IN FRAMING (inches)	PANELS APPLIED DIRECT TO FRAMING — NAIL (common or galvanized box) or staple size[x]	Direct 6	Direct 4	Direct 3	Direct 2[e]	PANELS APPLIED OVER 1/2" OR 5/8" GYPSUM SHEATHING — NAIL (common or galvanized box) or staple size[x]	Gyp 6	Gyp 4	Gyp 3	Gyp 2[e]
Structural I Sheathing	5/16	1 1/4	6d (2 × 0.113" common, 2" × 0.099" galvanized box)	200	300	390	510	8d (2 1/2" × 0.131" common, 2 1/2" × 0.113" galvanized box)	200	300	390	510
		1	1 1/2 16 Gage	165	245	325	415	2 16 Gage	125	185	245	315
	3/8	1 3/8	8d (2 1/2" × 0.131" common, 2 1/2" × 0.113" galvanized box)	230[d]	360[d]	460[d]	610[d]	10d (3" × 0.148" common, 3" × 0.128" galvanized box)	280	430	550[f]	730
		1	1 1/2 16 Gage	155	235	315	400	2 16 Gage	155	235	310	400
	7/16	1 3/8	8d (2 1/2" × 0.131" common, 2 1/2" × 0.113" galvanized box)	255[d]	395[d]	505[d]	670[d]	10d (3" × 0.148" common, 3" × 0.128" galvanized box)	280	430	550[f]	730
		1	1 1/2 16 Gage	170	260	345	440	2 16 Gage	155	235	310	400
	15/32	1 3/8	8d (2 1/2" × 0.131" common, 2 1/2" × 0.113" galvanized box)	280	430	550	730	10d (3" × 0.148" common, 3" × 0.1218" galvanized box)	280	430	550[f]	730
		1	1 1/2 16 Gage	185	280	375	475	2 16 Gage	155	235	300	400
	15/32	1 1/2	10d (3" × 0.148" common, 3" × 0.128" galvanized box)	340	510	665[f]	870	—	—	—	—	—
Sheathing, plywood siding[g] except Group 5 Species	5/16 or 1/4[c]	1 1/4	6d (2" × 0.113" common, 2" × 0.099" galvanized box)	180	270	350	450	8d (2 1/2" × 0.131" common, 2 1/2" × 0.113" galvanized box)	180	270	350	450
		1	1 1/2 16 Gage	145	220	295	375	2 16 Gage	110	165	220	285
	3/8	1 1/4	6d (2" × 0.113" common, 2" × 0.099" galvanized box)	200	300	390	510	8d (2 1/2" × 0.131" common, 2 1/2" × 0.113" galvanized box)	200	300	390	510
		1 3/8	8d (2 1/2" × 0.131" common, 2 1/2" × 0.113" galvanized box)	220[d]	320[d]	410[d]	530[d]	10d (3" × 0.148" common, 3" × 0.128" galvanized box)	260	380	490[f]	640
		1	1 1/2 16 Gage	140	210	280	360	2 16 Gage	140	210	280	360
	7/16	1 3/8	8d (2 1/2" × 0.131" common, 2 1/2" × 0.113" galvanized box)	240[d]	350[d]	450[d]	585[d]	10d (3" × 0.148" common, 3" × 0.128" galvanized box)	260	380	490[f]	640
		1	1 1/2 16 Gage	155	230	310	395	2 16 Gage	140	210	280	360
	15/32	1 3/8	8d (2 1/2" × 0.131" common, 2 1/2" × 0.113" galvanized box)	260	380	490	640	10d (3" × 0.148" common, 3" × 0.128" galvanized box)	260	380	490[f]	640
		1 1/2	10d (3" × 0.148" common, 3" × 0.128" galvanized box)	310	460	600[f]	770	—	—	—	—	—
		1	1 1/2 16 Gage	170	255	335	430	2 16 Gage	140	210	280	360
	19/32	1 1/2	10d (3" × 0.148" common, 3" × 0.128" galvanized box)	340	510	665[f]	870	—	—	—	—	—
		1	1 3/4 16 Gage	185	280	375	475	—	—	—	—	—
			Nail Size (galvanized casing)					**Nail Size (galvanized casing)**				
	5/16[c]	1 1/4	6d (2" × 0.099")	140	210	275	360	8d (2 1/2" × 0.113")	140	210	275	360
	3/8	1 3/8	8d (2 1/2" × 0.113")	160	240	310	410	10d (3" × 0.128")	160	240	310[f]	410

(continued)

Notes to Table 2306.4.1

For SI: 1 inch = 25.4 mm, 1 pound per foot = 14.5939 N/m.

a. For framing of other species: (1) Find specific gravity for species of lumber in AF&PA NDS. (2) For staples find shear value from table above for Structural I panels (regardless of actual grade) and multiply value by 0.82 for species with specific gravity of 0.42 or greater, or 0.65 for all other species. (3) For nails find shear value from table above for nail size for actual grade and multiply value by the following adjustment factor: Specific Gravity Adjustment Factor = [1-(0.5 - SG)], where SG = Specific Gravity of the framing lumber. This adjustment factor shall not be greater than 1.

b. Panel edges backed with 2-inch nominal or wider framing. Install panels either horizontally or vertically. Space fasteners maximum 6 inches on center along intermediate framing members for $^3/_8$-inch and $^7/_{16}$-inch panels installed on studs spaced 24 inches on center. For other conditions and panel thickness, space fasteners maximum 12 inches on center on intermediate supports.

c. $^3/_8$-inch panel thickness or siding with a span rating of 16 inches on center is the minimum recommended where applied direct to framing as exterior siding.

d. Allowable shear values are permitted to be increased to values shown for $^{15}/_{32}$-inch sheathing with same nailing provided (a) studs are spaced a maximum of 16 inches on center, or (b) panels are applied with long dimension across studs.

e. Framing at adjoining panel edges shall be 3 inches nominal or wider, and nails shall be staggered where nails are spaced 2 inches on center.

f. Framing at adjoining panel edges shall be 3 inches nominal or wider, and nails shall be staggered where both of the following conditions are met: (1) 10d (3″ × 0.148″) nails having penetration into framing of more than 1 $^1/_2$ inches and (2) nails are spaced 3 inches on center.

g. Values apply to all-veneer plywood. Thickness at point of fastening on panel edges governs shear values.

h. Where panels applied on both faces of a wall and nail spacing is less than 6 inches o.c. on either side, panel joints shall be offset to fall on different framing members, or framing shall be 3-inch nominal or thicker at adjoining panel edges and nails on each side shall be staggered.

i. In Seismic Design Category D, E or F, where shear design values exceed 350 pounds per linear foot, all framing members receiving edge nailing from abutting panels shall not be less than a single 3-inch nominal member, or two 2-inch nominal members fastened together in accordance with Section 2306.1 to transfer the design shear value between framing members. Wood structural panel joint and sill plate nailing shall be staggered in all cases. See Section 2305.3.11 for sill plate size and anchorage requirements.

j. Galvanized nails shall be hot dipped or tumbled.

k. Staples shall have a minimum crown width of $^7/_{16}$ inch and shall be installed with their crowns parallel to the long dimension of the framing members.

l. For shear loads of normal or permanent load duration as defined by the AF&PA NDS, the values in the table above shall be multiplied by 0.63 or 0.56, respectively.

TABLE 2306.4.3
ALLOWABLE SHEAR FOR PARTICLEBOARD SHEAR WALL SHEATHING[b]

PANEL GRADE	MINIMUM NOMINAL PANEL THICKNESS (inch)	MINIMUM NAIL PENETRATION IN FRAMING (inches)	Nail size (common or galvanized box)	PANELS APPLIED DIRECT TO FRAMING			
				Allowable shear (pounds per foot) nail spacing at panel edges (inches)[a]			
				6	4	3	2
M-S "Exterior Glue" and M-2 "Exterior Glue"	$^3/_8$	1 $^1/_2$	6d	120	180	230	300
	$^3/_8$	1 $^1/_2$	8d	130	190	240	315
	$^1/_2$			140	210	270	350
	$^1/_2$	1 $^5/_8$	10d	185	275	360	460
	$^5/_8$			200	305	395	520

For SI: 1 inch = 25.4 mm, 1 pound per foot = 14.5939 N/m.

a. Values are not permitted in Seismic Design Category D, E or F.

b. Galvanized nails shall be hot-dipped or tumbled.

TABLE 2306.4.4
ALLOWABLE SHEAR VALUES (plf) FOR WIND OR SEISMIC LOADING ON
SHEAR WALLS OF FIBERBOARD SHEATHING BOARD CONSTRUCTION FOR TYPE V CONSTRUCTION ONLY[a,b,c,d,e,f,g,h]

THICKNESS AND GRADE	FASTENER SIZE	SHEAR VALUE (pounds per linear foot) 3-INCH NAIL SPACING AROUND PERIMETER AND 6-INCH AT INTERMEDIATE POINTS
$^1/_2$″ Structural	No. 11 gage galvanized roofing nail 1$^1/_2$″ long, $^7/_{16}$″ head	125[g]
$^{25}/_{32}$″ Structural	No. 11 gage galvanized roofing nail 1$^3/_4$″ long, $^7/_{16}$″ head	175[g]

For SI: 1 inch = 25.4 mm, 1 pound per foot = 14.5939 N/m.

a. Fiberboard sheathing diaphragms shall not be used to brace concrete or masonry walls.

b. Panel edges shall be backed with 2 inch or wider framing of Douglas fir-larch or Southern pine.

c. Fiberboard sheathing on one side only.

d. Fiberboard panels are installed with their long dimension parallel or perpendicular to studs.

e. Fasteners shall be spaced 6 inches on center along intermediate framing members.

f. For framing of other species: (1) Find specific gravity for species of lumber in AF&PA NDS and (2) Multiply the shear value from the above table by 0.82 for species with specific gravity of 0.42 or greater, or 0.65 for all other species.

g. The same values can be applied when staples are used as described in Table 2304.9.1.

h. Values are not permitted in Seismic Design Category D, E or F.

2306.4.5 Shear walls sheathed with other materials. Shear capacities for walls sheathed with lath, plaster or gypsum board shall be in accordance with Table 2306.4.5. Shear walls sheathed with lath, plaster or gypsum board shall be constructed in accordance with Chapter 25 and Section 2306.4.5.1. Walls resisting seismic loads shall be subject to the limitations in Section 12.2.1 of ASCE 7.

❖ See the commentary to Table 2306.4.5.

TABLE 2306.4.5. See below.

❖ This table recognizes the shear capacities of gypsum board sheathing, expanded metal or woven wire lath and portland cement plaster in shear walls. Because of the brittleness of these materials, their use to resist seismic forces is limited for structures classified as Seismic Design Category D and prohibited in structures classified as Seismic Design Category E or F. In

TABLE 2306.4.5
ALLOWABLE SHEAR FOR WIND OR SEISMIC FORCES FOR SHEAR WALLS OF LATH
AND PLASTER OR GYPSUM BOARD WOOD FRAMED WALL ASSEMBLIES

TYPE OF MATERIAL	THICKNESS OF MATERIAL	WALL CONSTRUCTION	FASTENER SPACING[b] MAXIMUM (inches)	SHEAR VALUE[a,e] (plf)	MINIMUM FASTENER SIZE[c,d,j,k]
1. Expanded metal or woven wire lath and portland cement plaster	$7/_8''$	Unblocked	6	180	No. 11 gage $1^1/_2''$ long, $7/_{16}''$ head 16 Ga. Galv. Staple, $7/_8''$ legs
2. Gypsum lath, plain or perforated	$3/_8''$ lath and $1/_2''$ plaster	Unblocked	5	100	No. 13 gage, $1^1/_8''$ long, $19/_{64}''$ head, plasterboard nail 16 Ga. Galv. Staple, $1^1/_8''$ long 0.120'' Nail, min. $3/_8''$ head, $1^1/_4''$ long
3. Gypsum sheathing	$1/_2'' \times 2' \times 8'$	Unblocked	4	75	No. 11 gage, $1^3/_4''$ long, $7/_{16}''$ head, diamond-point, galvanized
	$1/_2'' \times 4'$	Blocked[f]	4	175	
		Unblocked	7	100	16 Ga. Galv. Staple, $1^3/_4''$ long
	$5/_8'' \times 4'$	Blocked	4'' edge/ 7'' field	200	6d galvanized 0.120'' Nail, min. $3/_8''$ head, $1^3/_4''$ long
4. Gypsum board, gypsum veneer base or water-resistant gypsum backing board	$1/_2''$	Unblocked[f]	7	75	5d cooler ($1^5/_8'' \times .086''$) or wallboard 0.120'' nail, min. $3/_8''$ head, $1^1/_2''$ long 16 Gage Staple, $1^1/_2''$ long
		Unblocked[f]	4	110	
		Unblocked	7	100	
		Unblocked	4	125	
		Blocked[g]	7	125	
		Blocked[g]	4	150	
		Unblocked	8/12[h]	60	No. 6 $1^1/_4''$ screws[i]
		Blocked[g]	4/16[h]	160	
		Blocked[g]	4/12[h]	155	
		Blocked[f, g]	8/12[h]	70	
		Blocked[g]	6/12[h]	90	
	$5/_8''$	Unblocked[f]	7	115	6d cooler ($1^7/_8'' \times 0.092''$) or wallboard 0.120'' Nail, min. $3/_8''$ head, $1^3/_4''$ long 16 Gage Staple, $1^1/_2''$ legs, $1^5/_8''$ long
			4	145	
		Blocked[g]	7	145	
			4	175	
		Blocked[g] Two-ply	Base ply: 9 Face ply: 7	250	Base ply-6d cooler ($1^7/_8'' \times 0.092''$) or wallboard $1^3/_4'' \times 0.120''$ Nail, min. $3/_8''$ head $1^5/_8''$ 16 Ga. Galv. Staple $1^5/_8''$ 16 Gage Galv. Staple Face ply-8d cooler ($2^3/_8'' \times 0.113''$) or wallboard 0.120'' Nail, min. $3/_8''$ head, $2^3/_8''$ long 15 Ga. Galv. Staple, $2^1/_4''$ long
		Unblocked	8/12[h]	70	No. 6-$1^1/_4''$ screws[i]
		Blocked[g]	8/12[h]	90	

For SI: 1 inch = 25.4 mm, 1 foot = 304.8 mm, 1 pound per foot = 14.5939 N/m.

a. These shear walls shall not be used to resist loads imposed by masonry or concrete construction (see Section 2305.1.5). Values shown are for short-term loading due to wind or seismic loading. Walls resisting seismic loads shall be subject to the limitations in Section 12.2.1 of ASCE 7. Values shown shall be reduced 25 percent for normal loading.

b. Applies to fastening at studs, top and bottom plates and blocking.

c. Alternate fasteners are permitted to be used if their dimensions are not less than the specified dimensions. Drywall screws are permitted to substitute for the 5d ($1^5/_8'' \times 0.086''$), and 6d ($1^7/_8'' \times 0.092''$)(cooler) nails listed above, and No. 6 $1^1/_4$ inch Type S or W screws for 6d ($1^7/_8'' \times 0.092$) (cooler) nails.

d. For properties of cooler nails, see ASTM C 514.

e. Except as noted, shear values are based on a maximum framing spacing of 16 inches on center.

f. Maximum framing spacing of 24 inches on center.

g. All edges are blocked, and edge fastening is provided at all supports and all panel edges.

h. First number denotes fastener spacing at the edges; second number denotes fastener spacing at intermediate framing members.

i. Screws are Type W or S.

j. Staples shall have a minimum crown width of $7/_{16}$ inch, measured outside the legs, and shall be installed with their crowns parallel to the long dimension of the framing members.

k. Staples for the attachment of gypsum lath and woven-wire lath shall have a minimum crown width of $3/_4$ inch, measured outside the legs.

wind design, the shear capacities of these materials are often added to the shear capacity of the wood structural panel or wood sheathing to determine the total shear capacity of the shear wall to resist wind forces.

Note j requires staples to be installed with the crowns of the staple parallel to the long dimension of the framing members. This makes the installation requirement for staples consistent with what has been required for staples used in wood framed diaphragms and shear walls (see National Evaluation Service, Inc. National Evaluation Report NER-272).

2306.4.5.1 Application of gypsum board or lath and plaster to wood framing.

❖ This section presents general prescriptive details on the application of gypsum board and lath and plaster to wood framing. It is deemed that compliance with these will develop a structurally sound and serviceable wall surface.

2306.4.5.1.1 Joint staggering. End joints of adjacent courses of gypsum board shall not occur over the same stud.

❖ Staggering the joints is common practice, and serves to add to the overall shear capacity of the wall.

2306.4.5.1.2 Blocking. Where required in Table 2306.4.5, wood blocking having the same cross-sectional dimensions as the studs shall be provided at joints that are perpendicular to the studs.

❖ Full-sized blocking is required at unsupported edges. This increases the overall shear capacity by allowing direct transfer of shear forces from the gypsum board to the framing elements.

2306.4.5.1.3 Fastening. Studs, top and bottom plates and blocking shall be fastened in accordance with Table 2304.9.1.

❖ Table 2304.9.1 is the fastener schedule for all wood connections in conventional framing. This section simply emphasizes that the proper connections are essential to the wall's performance in resisting shear.

2306.4.5.1.4 Fasteners. The size and spacing of fasteners shall be set forth in Table 2306.4.5. Fasteners shall be spaced not less than $^3/_8$ inch (9.5 mm) from edges and ends of gypsum boards or sides of studs, blocking and top and bottom plates.

❖ The minimum edge spacing of fasteners is necessary so that they develop their full load-resisting capacity without splitting the material at the edge.

2306.4.5.1.5 Gypsum lath. Gypsum lath shall be applied perpendicular to the studs. Maximum allowable shear values shall be as set forth in Table 2306.4.5.

❖ The values in Table 2306.4.5 assume a perpendicular application of the gypsum lath.

2306.4.5.1.6 Gypsum sheathing. Four-foot-wide (1219 mm) pieces of gypsum sheathing shall be applied parallel or perpendicular to studs. Two-foot-wide (610 mm) pieces of gypsum

sheathing shall be applied perpendicular to the studs. Maximum allowable shear values shall be as set forth in Table 2306.4.5.

❖ For gypsum sheathing, the values in Table 2306.4.5 assume application in either parallel or perpendicular orientation, except for sheets 2 feet (610 mm) wide in the first row of the gypsum sheathing portion of the table.

2306.4.5.1.7 Other gypsum boards. Gypsum board shall be applied parallel or perpendicular to studs. Maximum allowable shear values shall be as set forth in Table 2306.4.5.

❖ For typical gypsum board, the values in Table 2306.4.5 assume application in either parallel or perpendicular orientation.

SECTION 2307
LOAD AND RESISTANCE FACTOR DESIGN

2307.1 Load and resistance factor design. The structural analysis and construction of wood elements and structures using load and resistance factor design shall be in accordance with AF&PA NDS.

❖ The 2005 Edition of the AF&PA NDS is a dual format specification that contains provisions for both ASD and LRFD.

Theoretical reliability-based analysis has been used for many years in the electronics and aerospace industries with great success. The extension of theoretical reliability concepts to building applications has proven to be somewhat more difficult, but still achievable. LRFD has evolved to become the preferred format for converting structural design standards to a so-called limit states approach. The LRFD standard is intended for use in conjunction with competent engineering design, accurate fabrication and adequate supervision of construction.

The basic design equation for LRFD requires that the specified product strength or resistance meet or exceed the stress or other effect imposed by the specified loads. In ASD, the permissible stress levels are set very low and the load magnitudes are set at "once-in-a-lifetime" levels. This combination produces designs that maintain high safety levels yet remain economically feasible. In LRFD, the basic design equation follows a similar format, in which factored resistance must be greater than or equal to the factored load effects.

2307.1.1 Wood structural panel shear walls. In Seismic Design Category D, E or F, where shear design values exceed 490 pounds per foot (7154 N/m), all framing members receiving edge nailing from abutting panels shall not be less than a single 3-inch (76 mm) nominal member or two 2-inch (51 mm) nominal members fastened together in accordance with AF&PA NDS to transfer the design shear value between framing members. Wood structural panel joint and sill plate nailing shall be staggered in all cases. See Section 2305.3.11 for sill plate size and anchorage requirements.

❖ A 3-inch (76 mm) nominal framing member is required above the stated threshold where abutting panels are both fastened to that member. The use of a pair of nominal 2-inch (51 mm) members is permitted in lieu of a single 3-inch (76 mm) nominal member-provided the two 2-inch (51 mm) members are fastened together to transfer the design shear value. This provision is based on testing at APA, which showed that two 2-inch (51 mm) members fastened to transfer the shear performed as well as shear walls with a single 3-inch (76 mm) nominal member.

SECTION 2308
CONVENTIONAL LIGHT-FRAME CONSTRUCTION

2308.1 General. The requirements of this section are intended for conventional light-frame construction. Other methods are permitted to be used, provided a satisfactory design is submitted showing compliance with other provisions of this code. Interior nonload-bearing partitions, ceilings and curtain walls of conventional light-frame construction are not subject to the limitations of this section. Alternatively, compliance with AF&PA WFCM shall be permitted subject to the limitations therein and the limitations of this code. Detached one- and two-family dwellings and multiple single-family dwellings (townhouses) not more than three stories above grade plane in height with a separate means of egress and their accessory structures shall comply with the *International Residential Code.*

❖ The provisions of Section 2308 recognize that light-frame wood construction for a limited range of buildings will perform to the extent required by the code based on many years of research into issues such as engineering reliability, redundancy of systems and good historic performance. The stated intent of the code is to provide minimum requirements necessary to safeguard the health, safety and general welfare of the public (see the preface and Section 101.3), not to ensure a complete lack of structural damage after a major wind or seismic event. A structure should not collapse, but may require extensive repairs following a major disaster. The details and specifications for conventional construction are deemed to meet these standards. The development of conventional means of construction is not new. Prescriptive methods have been followed ever since wood has been used as a building material, but in 1990 the Conventional Construction Task Force of the Structural Engineers Association of California was formed for the purpose of reviewing existing code provisions and construction practices in order to develop better and more uniformly enforceable provisions. Its work was integrated into the 1994 UBC and much of that work remains in the codes to this day.

These specifications are generally only applicable to light wood-frame building construction having closely spaced framing, using studs up to 2 inches by

6 inches (51 mm by 152 mm) in size and joists and rafters up to 2 inches by 12 inches (51 mm by 305 mm) in size. With a few exceptions, a "conventional" structure is framed in the "platform" style. Repetitive, closely spaced framing is not specifically defined by the code; however, it is usually assumed to include framing that does not exceed a spacing of 24 inches (610 mm) o.c. This is the greatest spacing commonly found in these types of structures, the greatest spacing covered by AF&PA *Span Tables for Joists and Rafters* and the maximum spacing for which referenced standards allow increases in bending stresses for repetitive member use.

For structures within the scope of conventional construction, however, other methods of construction are permitted, provided that they are designed to comply with either ASD or LRFD methodology, as discussed in the commentary to Section 2301.2.

Some interior nonbearing applications are permitted to use the provisions of Section 2308, even though the remainder of the building does not fall within the limitations of Section 2308.2. These buildings would be of light wood-frame construction as well as other types of construction in which wood framing is permitted under other portions of the code.

2308.1.1 Portions exceeding limitations of conventional construction. When portions of a building of otherwise conventional construction exceed the limits of Section 2308.2, these portions and the supporting load path shall be designed in accordance with accepted engineering practice and the provisions of this code. For the purposes of this section, the term "portions" shall mean parts of buildings containing volume and area such as a room or a series of rooms.

❖ In recognition of the complexity of many wood-framed structures, this section of the code allows for a combined use of prescriptive detailing and engineered systems where the necessary limitations of conventional construction are exceeded. It is the intention that engineered portions within an otherwise conventionally-framed building will be completely integrated in such a way that a complete load path is provided for the engineered system and that the forces carried will be properly delivered there, even across an area of the building that is not engineered. For example, an engineered shear wall line requires collectors that might extend over the top of conventional beams, studs and columns, which is perfectly acceptable, provided the lateral forces that this engineered line of shear wall is to resist are properly transferred there through an engineered load path and ultimately carried to the ground. The applicable code requirements in terms of loads, drag elements, diaphragm limitations, etc. form the basis for design of engineered portions.

2308.2 Limitations. Buildings are permitted to be constructed in accordance with the provisions of conventional light-frame construction, subject to the following limitations, and to further limitations of Sections 2308.11 and 2308.12.

1. Buildings shall be limited to a maximum of three stories above grade plane. For the purposes of this section, for

buildings in Seismic Design Category D or E as determined in Section 1613, cripple stud walls shall be considered to be a story.

> **Exception:** Solid blocked cripple walls not exceeding 14 inches (356 mm) in height need not be considered a story.

2. Bearing wall floor-to-floor heights shall not exceed a stud height of 10 feet (3048 mm) plus a height of floor framing not to exceed 16 inches (406 mm).

3. Loads as determined in Chapter 16 shall not exceed the following:

   3.1. Average dead loads shall not exceed 15 psf (718 N/m²) for combined roof and ceiling, exterior walls, floors and partitions.

   Exceptions:

   1. Subject to the limitations of Sections 2308.11.2 and 2308.12.2, stone or masonry veneer up to the lesser of 5 inches (127 mm) thick or 50 psf (2395 N/m²) and installed in accordance with Chapter 14 is permitted to a height of 30 feet (9144 mm) above a noncombustible foundation, with an additional 8 feet (2438 mm) permitted for gable ends.

   2. Concrete or masonry fireplaces, heaters and chimneys shall be permitted in accordance with the provisions of this code.

   3.2. Live loads shall not exceed 40 psf (1916 N/m²) for floors.

   3.3. Ground snow loads shall not exceed 50 psf (2395 N/m²).

4. Wind speeds shall not exceed 100 miles per hour (mph) (44 m/s) (3-second gust).

   > **Exception:** Wind speeds shall not exceed 110 mph (48.4 m/s) (3-second gust) for buildings in Exposure Category B.

5. Roof trusses and rafters shall not span more than 40 feet (12 192 mm) between points of vertical support.

6. The use of the provisions for conventional light-frame construction in this section shall not be permitted for Occupancy Category IV buildings assigned to Seismic Design Category B, C, D, E or F, as determined in Section 1613.

7. Conventional light-frame construction is limited in irregular structures in Seismic Design Category D or E, as specified in Section 2308.12.6.

❖ It is acknowledged that conventional construction provisions concerning framing members and sheathing that carry gravity loads are adequate due to the fact that the tables for selecting members have been developed from engineering calculations. For resistance to lateral loads (wind and seismic), however, experience has shown that additional requirements—or limits on the application of conventional construction provisions—are needed.

1. The requirements of conventional construction are based on anticipated loads, both gravity and lateral. Buildings greater than three stories load the lower stories to a level higher than addressed by these provisions.

2. Studs in bearing walls have a tendency to bow, or move laterally, along the plane of the wall. Although this lateral movement can be restrained or otherwise accounted for by design, the prescriptive provisions of Section 2308 do not attempt to address the problem for studs within bearing walls that are taller than 10 feet (3048 mm). This height is assumed to be the maximum stable height using the typical grades and species (type) of wood in general use in conventional construction. Refer to Table 2308.9.1.

3. As the requirements of Section 2308 are based on anticipated loads, this item limits both dead and live loads in the structure. The prescriptive provisions of conventional construction are based on relatively light dead loads and do not address support of masonry or concrete other than veneer with a total weight of not greater than 50 psf (2395 N/m²). Additionally, the span tables for joists are based on live loads of residential uses, and other elements that support floor loads assume residential loading.

4. While some elements of conventional construction are acceptable for use in buildings in high-wind areas (as will be seen in the documents referenced in Section 2308.2.1), the general framing, sheathing and connection requirements are recognized to be insufficient. Buildings in Exposure Category B, as defined in Section 1609.4, are somewhat sheltered by other buildings as well as by trees and other obstructions. For that reason, the exception recognizes that conventional construction will provide acceptable lateral resistance in buildings in those areas.

5. In buildings with roof framing spans in excess of 40 feet (12 192 mm), the horizontal thrust of that framing on the top plate on which it rests is greater than can be resisted by the ceiling joist and rafter connections specified in Section 2308.10.4.1. Note that the limitation is on the span of the truss or rafter and not on the width of the building. The building width could exceed 40 feet (12 192 mm) as long as the actual span of the roof framing is no more than 40 feet (12 192 mm).

6. Occupancy Category IV structures that are not classified as Seismic Design Category A require an engineered design. The requirements for seismic design category assignment would

mean this exclusion applies to essential facilities that are classified as Seismic Design Category C, D or F in accordance with Section 1613.5.6. These are considered essential facilities and the design seismic forces in such structures are greater than can be resisted by conventional construction. Because of their role in emergency preparedness, essential facility building plans and calculations are often reviewed more extensively by state and local agencies before a building permit is issued and other required design requirements may apply.

7. Highly irregular portions of buildings (see commentary, Section 2308.12.6) are not permitted to use conventional construction in the higher seismic design categories. One of the reasons for this is that the behavior patterns of irregular building types are more difficult to predict. Conventional methods are allowed where the irregularities are limited that building's performance would be somewhat predictable. Some minor irregularities are allowed and are defined within the entire section.

2308.2.1 Basic wind speed greater than 100 mph (3-second gust). Where the basic wind speed exceeds 100 mph (3-second gust), the provisions of either AF&PA WFCM, or the SBCCI SSTD 10 are permitted to be used.

❖ Where the basic wind speed exceeds the limitation in Section 2308.2, Item 4, the provisions of the section cannot be used in constructing the building. Engineering design, however, is not always required. The engineered yet prescriptive provisions of AF&PA's *Wood Frame Construction Manual for One- and Two-Family Dwellings* (WFCM) and SBCCI SSTD-10 may be used on buildings that are within the scope of those documents.

2308.2.2 Buildings in Seismic Design Category B, C, D or E. Buildings of conventional light-frame construction in Seismic Design Category B or C, as determined in Section 1613, shall comply with the additional requirements in Section 2308.11.

Buildings of conventional light-frame construction in Seismic Design Category D or E, as determined in Section 1613, shall comply with the additional requirements in Section 2308.12.

❖ In recognition of additional lateral forces anticipated in Seismic Design Category B or C and those in D or E, additional limits are placed on the application of conventional construction (see commentary, Sections 2308.11 and 2308.12). One of the reasons that additional care is given to structures having an elevated seismic risk is that the actual loading effect is cyclic in nature rather than a statically applied force. Connections must resist a different character of force that needs to be recognized in the provisions, which would typically resulting in more stringent requirements.

2308.3 Braced wall lines. Buildings shall be provided with exterior and interior braced wall lines as described in Section

2308.9.3 and installed in accordance with Sections 2308.3.1 through 2308.3.4.

❖ Exterior walls of buildings that do not require a seismic analysis in accordance with Section 1614.1 and are located in regions where hurricane winds are not anticipated (see Section 2308.2) are required to be braced in accordance with this section. In addition, if the spacing between the braced exterior walls exceeds 35 feet (10 668 mm) (see Section 2308.3.1), interior braced wall lines are also required.

The requirements for wall bracing vary, depending upon the level of seismic activity for the region in which the building under consideration is located and whether the story in question is supporting other stories. The bracing requirements increase for higher seismic design categories and for walls that support the weight of additional stories above. Various types of structural sheathing materials are permitted, although not every material is permitted in every instance. For example, the let-in brace is not typically permitted in high seismic risk buildings, but may still be used to prevent temporary racking of the wall during construction.

Nailing requirements are critical to the performance of wall bracing. In order for the bracing material to be effective in resisting racking of the wall, it must be installed in the manner prescribed.

2308.3.1 Spacing. Spacing of braced wall lines shall not exceed 35 feet (10 668 mm) o.c. in both the longitudinal and transverse directions in each story.

❖ This section establishes the maximum spacing of braced wall lines and thereby limits the lateral force acting on individual vertical elements within that line. Refer to Figure 2308.9.3 for further understanding of this principle. The philosophy of conventional construction regarding the resistance of lateral forces is to: (1) limit the magnitude of force that individual elements must resist; (2) streamline the effects of force on the entire structure by restricting the type and extent of irregularity; and (3) strengthen resisting elements beyond what might otherwise be installed without requirements. All of these concepts are addressed within Section 2308.

2308.3.2 Braced wall panel connections. Forces shall be transferred from the roofs and floors to braced wall panels and from the braced wall panels in upper stories to the braced wall panels in the story below by the following:

1. Braced wall panel top and bottom plates shall be fastened to joists, rafters or full-depth blocking. Braced wall panels shall be extended and fastened to roof framing at intervals not to exceed 50 feet (15 240 mm) between parallel braced wall lines.

   Exception: Where roof trusses are used, lateral forces shall be transferred from the roof diaphragm to the braced wall by blocking of the ends of the trusses or by other approved methods.

2. Bottom plate fastening to joist or blocking below shall be with not less than 3-16d nails at 16 inches (406 mm) o.c.

3. Blocking shall be nailed to the top plate below with not less than 3-8d toenails per block.

4. Joists parallel to the top plates shall be nailed to the top plate with not less than 8d toenails at 6 inches (152 mm) o.c.

In addition, top plate laps shall be nailed with not less than 8-16d face nails on each side of each break in the top plate.

❖ The means of connection described in this section serves to effectively tie the structure together, so that forces can be transferred from walls and roofs to the braced wall lines. This helps to provide some continuity of load path for braced wall panels which are the conventional lateral-force-resisting elements.

In regard to Item 1, if the maximum dimension of a roof exceeds 50 feet (15 240 mm) without being fastened to a braced wall line, the force during an earthquake due to the mass of the roof would exceed the resistance capability of the braced wall panels. Since the spacing of the walls cannot exceed 35 feet (10 668 mm) in accordance with Section 2308.3.1, there may occur some interior braced wall lines that do not need to be extended and fastened to the roof or floor above. This requirement meets the specifications for braced wall line foundations noted in Section 2308.3.4.

2308.3.3 Sill anchorage. Where foundations are required by Section 2308.3.4, braced wall line sills shall be anchored to concrete or masonry foundations. Such anchorage shall conform to the requirements of Section 2308.6 except that such anchors shall be spaced at not more than 4 feet (1219 mm) o.c. for structures over two stories in height. The anchors shall be distributed along the length of the braced wall line. Other anchorage devices having equivalent capacity are permitted.

❖ Many interior braced wall lines are typically secured a slab with pneumatic fasteners ("shot-pins"), but this section is intended to clarify that something more substantial is needed where foundations are required below these wall lines. Anchor bolts are described in Section 2308.6, but this section places a more restrictive spacing requirement in buildings greater than two stories in height due to the dynamic effects of an earthquake. Although interior braced wall lines in a building with plan dimensions less than or equal to 50 feet (15 240 mm) do not require a foundation or anchor bolts, full-scale testing has shown that shot pins may not perform as had been expected, and the designer or contractor may want to consider providing anchors drilled and secured into a concrete floor slab at interior braced wall panels.

2308.3.3.1 Anchorage to all-wood foundations. Where all-wood foundations are used, the force transfer from the braced wall lines shall be determined based on calculation and shall have a capacity greater than or equal to the connections required by Section 2308.3.3.

❖ There are no prescriptive provisions currently in the code for anchorage of braced wall lines to wood foundations; therefore, transfer of the forces from the wall line to the foundation that is equivalent to the required

sill anchorage must be provided. Connections must be designed to provide capacity equivalent to that of the bolts described in Section 2308.6: ½ inch (12.7 mm) diameter bolts embedded into concrete, spaced at 6 feet (1829 mm) o.c. maximum (two stories or less) or 4 feet (1219 mm) o.c. maximum (over two stories in height).

2308.3.4 Braced wall line support. Braced wall lines shall be supported by continuous foundations.

Exception: For structures with a maximum plan dimension not over 50 feet (15 240 mm), continuous foundations are required at exterior walls only.

❖ An exception allows horizontal dimensions of up to 50 feet (15 240 mm) between continuous foundations. This is intended to permit residential cripple walls to be braced at exterior walls only, within the limitations described, but applies to other conditions as well. At one time in the development of the conventional construction provisions, this requirement was only necessary in the highest seismic regions.

Connections between horizontal and vertical-resisting elements and transferring of forces to the foundations are prescribed in this section. Some suggested details are available in Chapter 12 of the NEHRP provisions commentary.

2308.4 Design of elements. Combining of engineered elements or systems and conventionally specified elements or systems is permitted subject to the following limits:

❖ The conventional construction provisions are intended to not only control the amount of irregularity within a building, but also to make things easier for plans examiners, builders, inspectors and designers. Buildings can be very complex; simple, "box-like" buildings are no longer the norm. This section makes it possible to design a "hybrid" type of building that is a mixture of conventional and engineered elements, provided appropriate attention is given to each.

2308.4.1 Elements exceeding limitations of conventional construction. When a building of otherwise conventional construction contains structural elements exceeding the limits of Section 2308.2, these elements and the supporting load path shall be designed in accordance with accepted engineering practice and the provisions of this code.

❖ The intent of this section is to permit individual framing members that are used in applications that exceed the limitations of Section 2308 (load, span, etc.) to be engineered and placed within an otherwise conventional system and not to require the entire structural system to be engineered. Methods that must be used for all elements or systems that require an engineered design are to comply with the requirements of the code. The code also requires that the load path supporting these elements must be engineered. It should be noted that it is not the intent of this section to allow situations that result in load transfer beyond the underlying assumptions of the tables for other conventional members.

2308.4.2 Structural elements or systems not described herein. When a building of otherwise conventional construction contains structural elements or systems not described in Section 2308, these elements or systems shall be designed in accordance with accepted engineering practice and the provisions of this code. The extent of such design need only demonstrate compliance of the nonconventional elements with other applicable provisions of this code and shall be compatible with the performance of the conventionally framed system.

❖ This section permits individual framing members that are not specifically mentioned in the code (new or proprietary products) to be engineered and placed within otherwise conventional construction and not to require the entire structural system to be engineered. The intent of this section is to recognize that there is a large volume of preengineered products available on the market that might fit into a conventionally framed building, but not necessarily follow code sections for their individual designs. Most elements that have an approval report have published design values based on laboratory testing and should be allowed as alternative materials or methods of construction, provided it is demonstrated that they will adequately fit into the building system without disrupting other requirements of Section 2308. It is not the intent of this provision to require engineered load path reevaluation for new or proprietary products used in conventional applications, even though the products may have a greater potential capacity than the conventional element that it replaces.

2308.5 Connections and fasteners. Connections and fasteners used in conventional construction shall comply with the requirements of Section 2304.9.

❖ As discussed in the commentary to Section 2304.9, the section on fasteners and connectors applies to all types of wood construction addressed in Chapter 23, and the number of fasteners stipulated in Table 2304.9.1 is the minimum to be used. This section emphasizes that requirement.

2308.6 Foundation plates or sills. Foundations and footings shall be as specified in Chapter 18. Foundation plates or sills resting on concrete or masonry foundations shall comply with Section 2304.3.1. Foundation plates or sills shall be bolted or anchored to the foundation with not less than $^1/_2$-inch-diameter (12.7 mm) steel bolts or approved anchors. Bolts shall be embedded at least 7 inches (178 mm) into concrete or masonry, and spaced not more than 6 feet (1829 mm) apart. There shall be a minimum of two bolts or anchor straps per piece with one bolt or anchor strap located not more than 12 inches (305 mm) or less than 4 inches (102 mm) from each end of each piece. A properly sized nut and washer shall be tightened on each bolt to the plate.

❖ This section prescribes the size and spacing of foundation bolts or anchors in structures that comply with conventional construction, but the minimum anchorage requirements indicated would also apply to engineered wood structures unless a design is provided (see Section 2304.3). Note that regulations pertaining to the foundation or footing are found in Chapter 18. Anchor types other than bolts, such as anchor straps, are also permitted when approved. Anchor straps come in different shapes and sizes and the allowable shear values that can be used with them is specified by the respective manufacturer in its literature.

2308.7 Girders. Girders for single-story construction or girders supporting loads from a single floor shall not be less than 4 inches by 6 inches (102 mm by 152 mm) for spans 6 feet (1829 mm) or less, provided that girders are spaced not more than 8 feet (2438 mm) o.c. Spans for built-up 2-inch (51 mm) girders shall be in accordance with Table 2308.9.5 or 2308.9.6. Other girders shall be designed to support the loads specified in this code. Girder end joints shall occur over supports.

Where a girder is spliced over a support, an adequate tie shall be provided. The ends of beams or girders supported on masonry or concrete shall not have less than 3 inches (76 mm) of bearing.

❖ Girders constructed of a single piece of wood must meet the size requirement of this section. Those that are built up—constructed of two or more pieces of 2-inch (51 mm) nominal wide lumber—must comply with the specified tables. All of the span lengths listed in Tables 2308.9.5 and 2308.9.6 are short enough to allow individual pieces of lumber to be continuous between supports (i.e. no splices). Splicing of individual pieces together would require an engineering analysis to design the connectors for shear flow. Other girders are permitted if properly designed.

The minimum bearing length for girders supported by concrete or masonry walls is 3 inches (76 mm). This is based on the anticipated loads for the construction typical in this chapter, the allowable compressive stresses perpendicular to the grain for beam sizes and grades typical of the requirements in this chapter and the consideration of shear failure of the masonry.

2308.8 Floor joists. Spans for floor joists shall be in accordance with Table 2308.8(1) or 2308.8(2). For other grades and or species, refer to the *AF&PA Span Tables for Joists and Rafters*.

❖ See the commentary to Tables 2308.8(1) and (2).

TABLES 2308.8(1) and (2). See pages 23-73 and 23-75.

❖ The spans shown in Tables 2308.8(1) and (2) were derived from the AF&PA *Span Tables for Joists and Rafters*. For simplicity, spans of only the most common species and grades of wood are shown. When other species or grades of wood are used, the AF&PA tables should be consulted to determine the proper span. Additionally, span tables published by the Southern Forest Products Association, the Western Wood Products Association and the Canadian Wood Council can be consulted. The spans published in those tables were derived in the same manner as those published in the AF&PA tables.

2308.8.1 Bearing. Except where supported on a 1-inch by 4-inch (25.4 mm by 102 mm) ribbon strip and nailed to the adjoining stud, the ends of each joist shall not have less than $1^1/_2$

inches (38 mm) of bearing on wood or metal, or less than 3 inches (76 mm) on masonry.

❖ This section is needed so that floor joists have adequate bearing at each end. A minimum bearing distance is stipulated for bearing on wood or steel. The 3 inches (76 mm) on masonry, which should also apply to bearing on concrete, is to minimize the potential for shear failure of the masonry or concrete. As an exception to the minimum 1¹/₂-inch (38 mm) bearing surface on wood, joists are permitted to bear on a ribbon strip nailed to the narrow face of the studs, as long as the ends of the joists are also nailed to the adjoining studs.

2308.8.2 Framing details. Joists shall be supported laterally at the ends and at each support by solid blocking except where the ends of the joists are nailed to a header, band or rim joist or to an adjoining stud or by other means. Solid blocking shall not be less than 2 inches (51mm) in thickness and the full depth of the joist. Notches on the ends of joists shall not exceed one-fourth the joist depth. Holes bored in joists shall not be within 2 inches (51 mm) of the top or bottom of the joist, and the diameter of any such hole shall not exceed one-third the depth of the joist. Notches in the top or bottom of joists shall not exceed one-sixth the depth and shall not be located in the middle third of the span.

Joist framing from opposite sides of a beam, girder or partition shall be lapped at least 3 inches (76 mm) or the opposing joists shall be tied together in an approved manner.

Joists framing into the side of a wood girder shall be supported by framing anchors or on ledger strips not less than 2 inches by 2 inches (51 mm by 51 mm).

❖ The general requirement of this section is that joists be laterally supported by solid blocking at each end and at any intermediate support. The blocking is intended to prevent the joists from rolling over at the supports. The blocking may be omitted where the ends of the joists are restrained either by nailing or by the use of some other means, such as a joist hanger, to accomplish the same action. Most blocking is 2 inches (51 mm) nominal in thickness.

Wood is comprised of relatively small elongated cells oriented parallel to each other and bound together physically and chemically by lignin. The cells are continuous from end to end of a piece of lumber. When a beam or joist is notched or cut, the cell strands are interrupted. In notched beams, the effective depth is reduced equal to the depth of the notch (see Figure 2308.8.2).

Some designs and installation practices require that limited notching and cutting occur. Notching should be avoided when possible. Holes bored in beams and joists create the same problems as notches. When necessary, the holes should be located in areas with the least stress concentration, generally along the neutral axis of the joist and, ideally, at a minimum distance away from the supports.

Joists must be adequately tied together in order to provide continuity and must be adequately supported when not bearing on the top of girders.

2308.8.2.1 Engineered wood products. Cuts, notches and holes bored in trusses, structural composite lumber, structural glue-laminated members or I-joists are not permitted except where permitted by the manufacturer's recommendations or where the effects of such alterations are specifically considered in the design of the member by a registered design professional.

❖ The notching and boring allowances that are permitted for solid sawn joist framing cannot be applied to engineered wood products. Because these types of products are commonly incorporated into conventional construction, this section points out that they must be treated differently. However, it does permit notching and boring that is considered in the design of the member. Each manufacturer typically publishes guidelines to establish safe notching limitations.

2308.8.3 Framing around openings. Trimmer and header joists shall be doubled, or of lumber of equivalent cross section, where the span of the header exceeds 4 feet (1219 mm). The ends of header joists more than 6 feet (1829 mm) long shall be supported by framing anchors or joist hangers unless bearing on a beam, partition or wall. Tail joists over 12 feet (3658 mm) long shall be supported at the header by framing anchors or on ledger strips not less than 2 inches by 2 inches (51 mm by 51 mm).

❖ Figure 2308.8.3 illustrates the criteria for framing around openings.

For SI: 1 inch = 25.4 mm.

Figure 2308.8.2
CUTTING, NOTCHING AND BORING JOISTS

TABLE 2308.8(1)
FLOOR JOIST SPANS FOR COMMON LUMBER SPECIES
(Residential Sleeping Areas, Live Load = 30 psf, L/Δ = 360)

JOIST SPACING (inches)	SPECIES AND GRADE		DEAD LOAD = 10 psf				DEAD LOAD = 20 psf			
			2x6	2x8	2x10	2x12	2x6	2x8	2x10	2x12
			Maximum floor joist spans							
			(ft. - in.)	(ft. - in.)	(ft. - in.)	(ft. - in.)	(ft. - in.)	(ft. - in.)	(ft. - in.)	(ft. - in.)
12	Douglas Fir-Larch	SS	12-6	16-6	21-0	25-7	12-6	16-6	21-0	25-7
	Douglas Fir-Larch	#1	12-0	15-10	20-3	24-8	12-0	15-7	19-0	22-0
	Douglas Fir-Larch	#2	11-10	15-7	19-10	23-0	11-6	14-7	17-9	20-7
	Douglas Fir-Larch	#3	9-8	12-4	15-0	17-5	8-8	11-0	13-5	15-7
	Hem-Fir	SS	11-10	15-7	19-10	24-2	11-10	15-7	19-10	24-2
	Hem-Fir	#1	11-7	15-3	19-5	23-7	11-7	15-2	18-6	21-6
	Hem-Fir	#2	11-0	14-6	18-6	22-6	11-0	14-4	17-6	20-4
	Hem-Fir	#3	9-8	12-4	15-0	17-5	8-8	11-0	13-5	15-7
	Southern Pine	SS	12-3	16-2	20-8	25-1	12-3	16-2	20-8	25-1
	Southern Pine	#1	12-0	15-10	20-3	24-8	12-0	15-10	20-3	24-8
	Southern Pine	#2	11-10	15-7	19-10	24-2	11-10	15-7	18-7	21-9
	Southern Pine	#3	10-5	13-3	15-8	18-8	9-4	11-11	14-0	16-8
	Spruce-Pine-Fir	SS	11-7	15-3	19-5	23-7	11-7	15-3	19-5	23-7
	Spruce-Pine-Fir	#1	11-3	14-11	19-0	23-0	11-3	14-7	17-9	20-7
	Spruce-Pine-Fir	#2	11-3	14-11	19-0	23-0	11-3	14-7	17-9	20-7
	Spruce-Pine-Fir	#3	9-8	12-4	15-0	17-5	8-8	11-0	13-5	15-7
16	Douglas Fir-Larch	SS	11-4	15-0	19-1	23-3	11-4	15-0	19-1	23-0
	Douglas Fir-Larch	#1	10-11	14-5	18-5	21-4	10-8	13-6	16-5	19-1
	Douglas Fir-Larch	#2	10-9	14-1	17-2	19-11	9-11	12-7	15-5	17-10
	Douglas Fir-Larch	#3	8-5	10-8	13-0	15-1	7-6	9-6	11-8	13-6
	Hem-Fir	SS	10-9	14-2	18-0	21-11	10-9	14-2	18-0	21-11
	Hem-Fir	#1	10-6	13-10	17-8	20-9	10-4	13-1	16-0	18-7
	Hem-Fir	#2	10-0	13-2	16-10	19-8	9-10	12-5	15-2	17-7
	Hem-Fir	#3	8-5	10-8	13-0	15-1	7-6	9-6	11-8	13-6
	Southern Pine	SS	11-2	14-8	18-9	22-10	11-2	14-8	18-9	22-10
	Southern Pine	#1	10-11	14-5	18-5	22-5	10-11	14-5	17-11	21-4
	Southern Pine	#2	10-9	14-2	18-0	21-1	10-5	13-6	16-1	18-10
	Southern Pine	#3	9-0	11-6	13-7	16-2	8-1	10-3	12-2	14-6
	Spruce-Pine-Fir	SS	10-6	13-10	17-8	21-6	10-6	13-10	17-8	21-4
	Spruce-Pine-Fir	#1	10-3	13-6	17-2	19-11	9-11	12-7	15-5	17-10
	Spruce-Pine-Fir	#2	10-3	13-6	17-2	19-11	9-11	12-7	15-5	17-10
	Spruce-Pine-Fir	#3	8-5	10-8	13-0	15-1	7-6	9-6	11-8	13-6

(continued)

TABLE 2308.8(1)—continued
FLOOR JOIST SPANS FOR COMMON LUMBER SPECIES
(Residential Sleeping Areas, Live Load = 30 psf, L/Δ = 360)

JOIST SPACING (inches)	SPECIES AND GRADE		DEAD LOAD = 10 psf				DEAD LOAD = 20 psf			
			2x6	2x8	2x10	2x12	2x6	2x8	2x10	2x12
			(ft. - in.)	(ft. - in.)	(ft. - in.)	(ft. - in.)	(ft. - in.)	(ft. - in.)	(ft. - in.)	(ft. - in.)
			Maximum floor joist spans							
19.2	Douglas Fir-Larch	SS	10-8	14-1	18-0	21-10	10-8	14-1	18-0	21-0
	Douglas Fir-Larch	#1	10-4	13-7	16-9	19-6	9-8	12-4	15-0	17-5
	Douglas Fir-Larch	#2	10-1	12-10	15-8	18-3	9-1	11-6	14-1	16-3
	Douglas Fir-Larch	#3	7-8	9-9	11-10	13-9	6-10	8-8	10-7	12-4
	Hem-Fir	SS	10-1	13-4	17-0	20-8	10-1	13-4	17-0	20-7
	Hem-Fir	#1	9-10	13-0	16-4	19-0	9-6	12-0	14-8	17-0
	Hem-Fir	#2	9-5	12-5	15-6	17-1	8-11	11-4	13-10	16-1
	Hem-Fir	#3	7-8	9-9	11-10	13-9	6-10	8-8	10-7	12-4
	Southern Pine	SS	10-6	13-10	17-8	21-6	10-6	13-10	17-8	21-6
	Southern Pine	#1	10-4	13-7	17-4	21-1	10-4	13-7	16-4	19-6
	Southern Pine	#2	10-1	13-4	16-5	19-3	9-6	12-4	14-8	17-2
	Southern Pine	#3	8-3	10-6	12-5	14-9	7-4	9-5	11-1	13-2
	Spruce-Pine-Fir	SS	9-10	13-0	16-7	20-2	9-10	13-0	16-7	19-6
	Spruce-Pine-Fir	#1	9-8	12-9	15-8	18-3	9-1	11-6	14-1	16-3
	Spruce-Pine-Fir	#2	9-8	12-9	15-8	18-3	9-1	11-6	14-1	16-3
	Spruce-Pine-Fir	#3	7-8	9-9	11-10	13-9	6-10	8-8	10-7	12-4
24	Douglas Fir-Larch	SS	9-11	13-1	16-8	20-3	9-11	13-1	16-2	18-9
	Douglas Fir-Larch	#1	9-7	12-4	15-0	17-5	8-8	11-0	13-5	15-7
	Douglas Fir-Larch	#2	9-1	11-6	14-1	16-3	8-1	10-3	12-7	14-7
	Douglas Fir-Larch	#3	6-10	8-8	10-7	12-4	6-2	7-9	9-6	11-0
	Hem-Fir	SS	9-4	12-4	15-9	19-2	9-4	12-4	15-9	18-5
	Hem-Fir	#1	9-2	12-0	14-8	17-0	8-6	10-9	13-1	15-2
	Hem-Fir	#2	8-9	11-4	13-10	16-1	8-0	10-2	12-5	14-4
	Hem-Fir	#3	6-10	8-8	10-7	12-4	6-2	7-9	9-6	11-0
	Southern Pine	SS	9-9	12-10	16-5	19-11	9-9	12-10	16-5	19-11
	Southern Pine	#1	9-7	12-7	16-1	19-6	9-7	12-4	14-7	17-5
	Southern Pine	#2	9-4	12-4	14-8	17-2	8-6	11-0	13-1	15-5
	Southern Pine	#3	7-4	9-5	11-1	13-2	6-7	8-5	9-11	11-10
	Spruce-Pine-Fir	SS	9-2	12-1	15-5	18-9	9-2	12-1	15-0	17-5
	Spruce-Pine-Fir	#1	8-11	11-6	14-1	16-3	8-1	10-3	12-7	14-7
	Spruce-Pine-Fir	#2	8-11	11-6	14-1	16-3	8-1	10-3	12-7	14-7
	Spruce-Pine-Fir	#3	6-10	8-8	10-7	12-4	6-2	7-9	9-6	11-0

Check sources for availability of lumber in lengths greater than 20 feet.
For SI: 1 inch = 25.4 mm, 1 foot = 304.8 mm, 1 pound per square foot = 47.8 N/m².

TABLE 2308.8(2)
FLOOR JOIST SPANS FOR COMMON LUMBER SPECIES
(Residential Living Areas, Live Load = 40 psf, L/Δ = 360)

JOIST SPACING (inches)	SPECIES AND GRADE		DEAD LOAD = 10 psf				DEAD LOAD = 20 psf			
			2x6	2x8	2x10	2x12	2x6	2x8	2x10	2x12
			(ft. - in.)	(ft. - in.)	(ft. - in.)	(ft. - in.)	(ft. - in.)	(ft. - in.)	(ft. - in.)	(ft. - in.)
			Maximum floor joist spans							
12	Douglas Fir-Larch	SS	11-4	15-0	19-1	23-3	11-4	15-0	19-1	23-3
	Douglas Fir-Larch	#1	10-11	14-5	18-5	22-0	10-11	14-2	17-4	20-1
	Douglas Fir-Larch	#2	10-9	14-2	17-9	20-7	10-6	13-3	16-3	18-10
	Douglas Fir-Larch	#3	8-8	11-0	13-5	15-7	7-11	10-0	12-3	14-3
	Hem-Fir	SS	10-9	14-2	18-0	21-11	10-9	14-2	18-0	21-11
	Hem-Fir	#1	10-6	13-10	17-8	21-6	10-6	13-10	16-11	19-7
	Hem-Fir	#2	10-0	13-2	16-10	20-4	10-0	13-1	16-0	18-6
	Hem-Fir	#3	8-8	11-0	13-5	15-7	7-11	10-0	12-3	14-3
	Southern Pine	SS	11-2	14-8	18-9	22-10	11-2	14-8	18-9	22-10
	Southern Pine	#1	10-11	14-5	18-5	22-5	10-11	14-5	18-5	22-5
	Southern Pine	#2	10-9	14-2	18-0	21-9	10-9	14-2	16-11	19-10
	Southern Pine	#3	9-4	11-11	14-0	16-8	8-6	10-10	12-10	15-3
	Spruce-Pine-Fir	SS	10-6	13-10	17-8	21-6	10-6	13-10	17-8	21-6
	Spruce-Pine-Fir	#1	10-3	13-6	17-3	20-7	10-3	13-3	16-3	18-10
	Spruce-Pine-Fir	#2	10-3	13-6	17-3	20-7	10-3	13-3	16-3	18-10
	Spruce-Pine-Fir	#3	8-8	11-0	13-5	15-7	7-11	10-0	12-3	14-3
16	Douglas Fir-Larch	SS	10-4	13-7	17-4	21-1	10-4	13-7	17-4	21-0
	Douglas Fir-Larch	#1	9-11	13-1	16-5	19-1	9-8	12-4	15-0	17-5
	Douglas Fir-Larch	#2	9-9	12-7	15-5	17-10	9-1	11-6	14-1	16-3
	Douglas Fir-Larch	#3	7-6	9-6	11-8	13-6	6-10	8-8	10-7	12-4
	Hem-Fir	SS	9-9	12-10	16-5	19-11	9-9	12-10	16-5	19-11
	Hem-Fir	#1	9-6	12-7	16-0	18-7	9-6	12-0	14-8	17-0
	Hem-Fir	#2	9-1	12-0	15-2	17-7	8-11	11-4	13-10	16-1
	Hem-Fir	#3	7-6	9-6	11-8	13-6	6-10	8-8	10-7	12-4
	Southern Pine	SS	10-2	13-4	17-0	20-9	10-2	13-4	17-0	20-9
	Southern Pine	#1	9-11	13-1	16-9	20-4	9-11	13-1	16-4	19-6
	Southern Pine	#2	9-9	12-10	16-1	18-10	9-6	12-4	14-8	17-2
	Southern Pine	#3	8-1	10-3	12-2	14-6	7-4	9-5	11-1	13-2
	Spruce-Pine-Fir	SS	9-6	12-7	16-0	19-6	9-6	12-7	16-0	19-6
	Spruce-Pine-Fir	#1	9-4	12-3	15-5	17-10	9-1	11-6	14-1	16-3
	Spruce-Pine-Fir	#2	9-4	12-3	15-5	17-10	9-1	11-6	14-1	16-3
	Spruce-Pine-Fir	#3	7-6	9-6	11-8	13-6	6-10	8-8	10-7	12-4

(continued)

TABLE 2308.8(2)—continued
FLOOR JOIST SPANS FOR COMMON LUMBER SPECIES
(Residential Living Areas, Live Load = 40 psf, L/Δ = 360)

JOIST SPACING (inches)	SPECIES AND GRADE		DEAD LOAD = 10 psf				DEAD LOAD = 20 psf			
			2x6	2x8	2x10	2x12	2x6	2x8	2x10	2x12
			Maximum floor joist spans							
			(ft. - in.)	(ft. - in.)	(ft. - in.)	(ft. - in.)	(ft. - in.)	(ft. - in.)	(ft. - in.)	(ft. - in.)
19.2	Douglas Fir-Larch	SS	9-8	12-10	16-4	19-10	9-8	12-10	16-4	19-2
	Douglas Fir-Larch	#1	9-4	12-4	15-0	17-5	8-10	11-3	13-8	15-11
	Douglas Fir-Larch	#2	9-1	11-6	14-1	16-3	8-3	10-6	12-10	14-10
	Douglas Fir-Larch	#3	6-10	8-8	10-7	12-4	6-3	7-11	9-8	11-3
	Hem-Fir	SS	9-2	12-1	15-5	18-9	9-2	12-1	15-5	18-9
	Hem-Fir	#1	9-0	11-10	14-8	17-0	8-8	10-11	13-4	15-6
	Hem-Fir	#2	8-7	11-3	13-10	16-1	8-2	10-4	12-8	14-8
	Hem-Fir	#3	6-10	8-8	10-7	12-4	6-3	7-11	9-8	11-3
	Southern Pine	SS	9-6	12-7	16-0	19-6	9-6	12-7	16-0	19-6
	Southern Pine	#1	9-4	12-4	15-9	19-2	9-4	12-4	14-11	17-9
	Southern Pine	#2	9-2	12-1	14-8	17-2	8-8	11-3	13-5	15-8
	Southern Pine	#3	7-4	9-5	11-1	13-2	6-9	8-7	10-1	12-1
	Spruce-Pine-Fir	SS	9-0	11-10	15-1	18-4	9-0	11-10	15-1	17-9
	Spruce-Pine-Fir	#1	8-9	11-6	14-1	16-3	8-3	10-6	12-10	14-10
	Spruce-Pine-Fir	#2	8-9	11-6	14-1	16-3	8-3	10-6	12-10	14-10
	Spruce-Pine-Fir	#3	6-10	8-8	10-7	12-4	6-3	7-11	9-8	11-3
24	Douglas Fir-Larch	SS	9-0	11-11	15-2	18-5	9-0	11-11	14-9	17-1
	Douglas Fir-Larch	#1	8-8	11-0	13-5	15-7	7-11	10-0	12-3	14-3
	Douglas Fir-Larch	#2	8-1	10-3	12-7	14-7	7-5	9-5	11-6	13-4
	Douglas Fir-Larch	#3	6-2	7-9	9-6	11-0	5-7	7-1	8-8	10-1
	Hem-Fir	SS	8-6	11-3	14-4	17-5	8-6	11-3	14-4	16-10[a]
	Hem-Fir	#1	8-4	10-9	13-1	15-2	7-9	9-9	11-11	13-10
	Hem-Fir	#2	7-11	10-2	12-5	14-4	7-4	9-3	11-4	13-1
	Hem-Fir	#3	6-2	7-9	9-6	11-0	5-7	7-1	8-8	10-1
	Southern Pine	SS	8-10	11-8	14-11	18-1	8-10	11-8	14-11	18-1
	Southern Pine	#1	8-8	11-5	14-7	17-5	8-8	11-3	13-4	15-11
	Southern Pine	#2	8-6	11-0	13-1	15-5	7-9	10-0	12-0	14-0
	Southern Pine	#3	6-7	8-5	9-11	11-10	6-0	7-8	9-1	10-9
	Spruce-Pine-Fir	SS	8-4	11-0	14-0	17-0	8-4	11-0	13-8	15-11
	Spruce-Pine-Fir	#1	8-1	10-3	12-7	14-7	7-5	9-5	11-6	13-4
	Spruce-Pine-Fir	#2	8-1	10-3	12-7	14-7	7-5	9-5	11-6	13-4
	Spruce-Pine-Fir	#3	6-2	7-9	9-6	11-0	5-7	7-1	8-8	10-1

Check sources for availability of lumber in lengths greater than 20 feet.

For SI: 1 inch = 25.4 mm, 1 foot = 304.8 mm, 1 pound per square foot = 47.8 N/m².

a. End bearing length shall be increased to 2 inches.

FIRST HEADER

SECOND HEADER
REQUIRED FOR
HEADER SPANS
MORE THAN 4 FT

HEADER SPAN

FIRST TRIMMER JOIST

TAIL JOIST

SECOND TRIMMER JOIST REQUIRED FOR
HEADER SPANS MORE THAN 4 FT

TAIL JOIST MUST BE SUPPORTED ON METAL
HANGERS OR MIN. 2x2 WOOD LEDGER WHEN TAIL
JOIST EXCEEDS 12 FT LONG

HEADER

For SI: 1 foot = 304.8 mm.

Figure 2308.8.3
FRAMING AROUND OPENING

2308.8.4 Supporting bearing partitions. Bearing partitions parallel to joists shall be supported on beams, girders, doubled joists, walls or other bearing partitions. Bearing partitions perpendicular to joists shall not be offset from supporting girders, walls or partitions more than the joist depth unless such joists are of sufficient size to carry the additional load.

❖ The floor joist spans provided in Tables 2308.8(1) and (2) assume uniform loading conditions and will support only limited concentrated loads. Unless the floor joists are designed using an engineering analysis which considers these loads, bearing partitions must be supported by other bearing partitions, beams or girders.

A major problem that frequently occurs is the orientation of heavy loads, such as special types of bathtubs, parallel with the floor joists. Additional joists should be installed to support these loads (see Figure 2308.8.4), which brings the design of this portion into an engineering analysis.

Bearing partitions oriented perpendicular to joists cannot be offset from supporting girders, walls or partitions more than the depth of the joists, unless the joists are of sufficient size to carry the additional load (see Figure 2308.8.4).

Although not specifically required, it is typically advantageous to double up floor joists supporting nonload-bearing partitions oriented parallel to the joists or to provide solid blocking between the joists to transfer the wall load to the supporting joists (see Figure 2308.8.4) unless the floor sheathing is of sufficient strength to carry the load. APA literature addresses

common framing issues such as these that can be integrated into a conventionally framed floor without an engineering analysis.

2308.8.5 Lateral support. Floor, attic and roof framing with a nominal depth-to-thickness ratio greater than or equal to 5:1 shall have one edge held in line for the entire span. Where the nominal depth-to-thickness ratio of the framing member exceeds 6:1, there shall be one line of bridging for each 8 feet (2438 mm) of span, unless both edges of the member are held in line. The bridging shall consist of not less than 1-inch by 3-inch (25 mm by 76 mm) lumber, double nailed at each end, of equivalent metal bracing of equal rigidity, full-depth solid blocking or other approved means. A line of bridging shall also be required at supports where equivalent lateral support is not otherwise provided.

❖ When the depth-to-thickness ratio of joists and rafters exceeds 5:1, as would be the case in members larger than 2 inches by 10 inches (51 mm by 254 mm), the lateral support required by Section 2308.8.2 is not sufficient to prevent lateral buckling between supports. Additional resistance is required. Sheathing, subflooring, decking and similar materials attached to each joist or rafter are considered to provide edge restraint.

These requirements are cumulative. The support required by Section 2308.8.2 applies to all joists. Additionally, members greater than 2 inches by 10 inches (51 mm by 254 mm) must have one edge held in line, and members greater than 2 inches by 12 inches (51 mm by 305 mm) must have one edge held in line as well as a line of bridging at each 8 feet (2438 mm) of span (which may be omitted if both edges are held in line).

2308.8.6 Structural floor sheathing. Structural floor sheathing shall comply with the provisions of Section 2304.7.1.

❖ Subflooring is considered to be a structural element, whereas underlayment is related to serviceability. Both must comply with the appropriate tables for the maximum span referenced in Section 2304.7.1. Refer to the commentary for Sections 2304.6 and 2304.7.

2308.8.7 Under-floor ventilation. For under-floor ventilation, see Section 1203.3.

❖ Enclosed spaces under floors of buildings are referred to as crawl spaces when not designed and constructed as basements. The earth is usually exposed or covered with a vapor retarder. Regardless, there is always the potential for a large amount of moisture vapor from the ground. This moisture must be removed to prevent wetting and drying cycles that can cause decay and fungus attack in wood members. Experience has shown that crawl spaces constructed in conformance with Section 2308 and ventilated to the exterior as required by Section 1203.3 do not develop decay problems. Access to crawl spaces is required as specified in Section 1209.1.

It should be recognized that moisture problems may be exaggerated in colder months; therefore, operable vents should be used with caution.

2308.9 Wall framing.

2308.9.1 Size, height and spacing. The size, height and spacing of studs shall be in accordance with Table 2308.9.1 except that utility-grade studs shall not be spaced more than 16 inches (406 mm) o.c., or support more than a roof and ceiling, or exceed 8 feet (2438 mm) in height for exterior walls and load-bearing walls or 10 feet (3048 mm) for interior nonload-bearing walls.

❖ The specifications in these sections are empirical interpretations of accepted engineering design practices and traditional designs for light wood-frame buildings. When bearing wall stud heights exceed 10 feet (3048 mm) or the structure is outside the scope of conventional construction requirements, studs must be designed in accordance with accepted engineering practice using the forces prescribed in the code.

TABLE 2308.9.1. See page 79.

❖ Wood studs in conventional construction are limited to the maximum height and spacing as indicated in this table. The height and spacing limits are based on the dead and live loads associated with conventional construction that are specified in Section 2308.2 along with a nominal lateral load to account for wind or earthquakes acting perpendicular to the wall.

2308.9.2 Framing details. Studs shall be placed with their wide dimension perpendicular to the wall. Not less than three studs shall be installed at each corner of an exterior wall.

Exception: At corners, two studs are permitted, provided wood spacers or backup cleats of $^3/_8$-inch-thick (9.5 mm) wood structural panel, $^3/_8$-inch (9.5 mm) Type M "Exterior

Figure 2308.8.4
JOISTS SUPPORTING PARTITIONS AND HEAVY LOADS

TABLE 2308.9.1
SIZE, HEIGHT AND SPACING OF WOOD STUDS

STUD SIZE (inches)	BEARING WALLS				NONBEARING WALLS	
	Laterally unsupported stud height[a] (feet)	Supporting roof and ceiling only	Supporting one floor, roof and ceiling	Supporting two floors, roof and ceiling	Laterally unsupported stud height[a] (feet)	Spacing (inches)
		Spacing (inches)				
2 × 3[b]	—	—	—	—	10	16
2 × 4	10	24	16	—	14	24
3 × 4	10	24	24	16	14	24
2 × 5	10	24	24	—	16	24
2 × 6	10	24	24	16	20	24

For SI: 1 inch = 25.4 mm, 1 foot = 304.8 mm.

a. Listed heights are distances between points of lateral support placed perpendicular to the plane of the wall. Increases in unsupported height are permitted where justified by an analysis.

b. Shall not be used in exterior walls.

Glue" particleboard, 1-inch-thick (25 mm) lumber or other approved devices that will serve as an adequate backing for the attachment of facing materials are used. Where fire-resistance ratings or shear values are involved, wood spacers, backup cleats or other devices shall not be used unless specifically approved for such use.

❖ Wood-frame exterior walls support all vertically imposed loads and also must act as braced walls to resist lateral loads, such as from wind and earthquakes. Adequately secured and sufficient corner posts are required to assist in this function, in addition to providing a degree of continuity in the exterior walls. Traditionally, three studs were installed. Experience has shown that two studs will adequately perform this function, provided other means are employed to support any interior finish that may be installed. Note that these other means may not be acceptable if they affect the fire resistance or shear resisting capacity of the walls. Most books on architectural standards contain common corner framing details. One issue with corner framing that becomes sensitive for engineered construction is when the designer chooses to attach one hold-down in such a way that it will provide uplift resistance for both of the perpendicular walls at that corner. A continuous load path is necessary to deliver uplift forces to the foundation and there is an eccentricity for one of the walls that should be considered.

2308.9.2.1 Top plates. Bearing and exterior wall studs shall be capped with double top plates installed to provide overlapping at corners and at intersections with other partitions. End joints in double top plates shall be offset at least 48 inches (1219 mm), and shall be nailed with not less than eight 16d face nails on each side of the joint. Plates shall be a nominal 2 inches (51 mm) in depth and have a width at least equal to the width of the studs.

Exception: A single top plate is permitted, provided the plate is adequately tied at joints, corners and intersecting walls by at least the equivalent of 3-inch by 6-inch (76 mm by 152 mm) by 0.036-inch-thick (0.914 mm) galvanized steel that is nailed to each wall or segment of wall by six 8d

nails or equivalent, provided the rafters, joists or trusses are centered over the studs with a tolerance of no more than 1 inch (25 mm).

❖ Double top plates serve three major functions:

1. They overlap at corners and interior wall intersections, thus tying the building together;

2. They serve as beams to support joists and rafters that are not located directly over the studs (see Section 2308.9.2.2); and

3. They serve as chords for floor and roof diaphragms (see Section 2305.2).

The tolerance of 1 inch (25 mm) with which the rafters or joists are to be centered over the studs is required to prevent overstress or excessive deflection in the single top plate. The steel strap provides a minimum level of continuity.

2308.9.2.2 Top plates for studs spaced at 24 inches (610 mm). Where bearing studs are spaced at 24-inch (610 mm) intervals and top plates are less than two 2-inch by 6-inch (51 mm by 152 mm) or two 3-inch by 4-inch (76 mm by 102 mm) members and where the floor joists, floor trusses or roof trusses that they support are spaced at more than 16-inch (406 mm) intervals, such joists or trusses shall bear within 5 inches (127 mm) of the studs beneath or a third plate shall be installed.

❖ Ideally, joists, rafters or trusses should be located directly over bearing studs; this is typically not planned for unless the drawings clearly state that it is required. The empirical provisions of this section allow joists or trusses to bear within 5 inches (127 mm) of the stud below when the joists or trusses are spaced more than 16 inches (406 mm) o.c., and the studs are spaced 24 inches (610 mm) o.c. This provision assumes there are double top (bearing) plates that have all joints located over the studs. Additional support is required if the offset is more than 5 inches (127 mm), such as the addition of a third plate.

2308.9.2.3 Nonbearing walls and partitions. In nonbearing walls and partitions, studs shall be spaced not more than 28 inches (711 mm) o.c. and are permitted to be set with the long dimension parallel to the wall. Interior nonbearing partitions shall be capped with no less than a single top plate installed to provide overlapping at corners and at intersections with other walls and partitions. The plate shall be continuously tied at joints by solid blocking at least 16 inches (406 mm) in length and equal in size to the plate or by $1/_2$-inch by $1^1/_2$-inch (12.7 mm by 38 mm) metal ties with spliced sections fastened with two 16d nails on each side of the joint.

❖ Nonbearing partitions are not intended to add to the rigidity or structural performance of the building. They are, however, expected to stay in place and have the ability to support applied finished membranes. The code limits stud spacing to 28 inches (711 mm) o.c., but allows the stud to be oriented with the wide dimension parallel to the wall length. Partitions must be able to support the lateral loads to which they are subjected, but not less than 5 psf (0.240 kN/m²) lateral load, in accordance with Section 1607.13. This section is not intended to apply to interior nonbearing walls that are serving as braced wall lines.

Lateral support of partitions can also be achieved by the use of light-gage metal clips produced by wood connector manufacturers. Allowable loads per clip are published in their respective literature. These clips typically call out for a minimum and maximum gap between the top of the wall and the bottom of the framing.

2308.9.2.4 Plates or sills. Studs shall have full bearing on a plate or sill not less than 2 inches (51 mm) in thickness having a width not less than that of the wall studs.

❖ The single bottom plate serves to anchor the wall to the floor. Studs are attached to the sill by end nailing or toe nailing specified in Table 2304.9.1.

2308.9.3 Bracing. Braced wall lines shall consist of braced wall panels that meet the requirements for location, type and amount of bracing as shown in Figure 2308.9.3, specified in Table 2308.9.3(1) and are in line or offset from each other by not more than 4 feet (1219 mm). Braced wall panels shall start not more than $12^1/_2$-feet (3810 mm) from each end of a braced wall line. Braced wall panels shall be clearly indicated on the plans. Construction of braced wall panels shall be by one of the following methods:

1. Nominal 1-inch by 4-inch (25 mm by 102 mm) continuous diagonal braces let into top and bottom plates and intervening studs, placed at an angle not more than 60 degrees (1.0 rad) or less than 45 degrees (0.79 rad) from the horizontal and attached to the framing in conformance with Table 2304.9.1.

2. Wood boards of $5/_8$ inch (15.9 mm) net minimum thickness applied diagonally on studs spaced not over 24 inches (610 mm) o.c.

3. Wood structural panel sheathing with a thickness not less than $5/_{16}$ inch (7.9 mm) for a 16-inch (406 mm) stud spacing and not less than $3/_8$ inch (9.5 mm) for a 24-inch (610

mm) stud spacing in accordance with Tables 2308.9.3(2) and 2308.9.3(3).

4. Fiberboard sheathing panels not less than $1/_2$ inch (12.7 mm) thick applied vertically or horizontally on studs spaced not over 16 inches (406 mm) o.c. where installed with fasteners in accordance with Section 2306.4.4 and Table 2306.4.4.

5. Gypsum board [sheathing $1/_2$-inch-thick (12.7 mm) by 4-feet-wide (1219 mm) wallboard or veneer base] on studs spaced not over 24 inches (610 mm) o.c. and nailed at 7 inches (178 mm) o.c. with nails as required by Table 2306.4.5.

6. Particleboard wall sheathing panels where installed in accordance with Table 2308.9.3(4).

7. Portland cement plaster on studs spaced 16 inches (406 mm) o.c. installed in accordance with Section 2510.

8. Hardboard panel siding where installed in accordance with Section 2303.1.6 and Table 2308.9.3(5).

For cripple wall bracing, see Section 2308.9.4.1. For Methods 2, 3, 4, 6, 7 and 8, each panel must be at least 48 inches (1219 mm) in length, covering three stud spaces where studs are spaced 16 inches (406 mm) apart and covering two stud spaces where studs are spaced 24 inches (610 mm) apart.

For Method 5, each panel must be at least 96 inches (2438 mm) in length where applied to one face of a panel and 48 inches (1219 mm) where applied to both faces.

All vertical joints of panel sheathing shall occur over studs and adjacent panel joints shall be nailed to common framing members. Horizontal joints shall occur over blocking or other framing equal in size to the studding except where waived by the installation requirements for the specific sheathing materials.

Sole plates shall be nailed to the floor framing and top plates shall be connected to the framing above in accordance with Section 2308.3.2. Where joists are perpendicular to braced wall lines above, blocking shall be provided under and in line with the braced wall panels.

❖ This section continues the provisions that were started in Section 2308.3 for bracing of walls to resist lateral forces. Braced wall panels are portions of walls, as required by Table 2308.9.3(1), composed of the bracing materials discussed below. Portions of braced wall lines are permitted to be offset from each other, and braced wall panels are not required to extend to the end of the braced wall line (see Figure 2308.9.3). The philosophy of this lateral force resisting system is to divide the building into smaller "boxes," similar to the use of stiffening baffles within a storage tank. The roof system delivers the lateral forces to the wall system below and historical analysis has shown that the more regularly defined these "boxes" are, the better the building will perform even though there may not be a continuous load path connecting every single line together.

FIGURE 2308.9.3. See page 23-81.

❖ See the commentary to Section 2308.9.3.

SEISMIC DESIGN CATEGORY	MAXIMUM WALL SPACING (feet)	REQUIRED BRACING LENGTH, b
A, B and C	35'-0"	Table 2308.9.3(1) and Section 2308.9.3
D and E	25'-0"	Table 2308.12.4

For SI: 1 foot = 304.8 mm.

FIGURE 2308.9.3
BASIC COMPONENTS OF THE LATERAL BRACING SYSTEM

TABLE 2308.9.3(1). See below.

❖ See the commentary to Section 2308.9.3.

TABLES 2308.9.3(2) through 2308.9.3(5). See below and page 23-83.

❖ Tables 2308.9.3(2) through 2308.9.3(5) are for the materials that are prescriptively permitted to be used to form the braced wall panels in braced wall lines. The location and minimum acceptable length of braced wall panels are specified in Table 2308.9.3(1). Detailed requirements for various materials—grades, thickness, nailing schedule, stud spacing and similar requirements—are also contained in the tables referenced for each material. Also see the commentary to Section 2303 concerning the minimum standards and quality of these materials.

TABLE 2308.9.3(1)
BRACED WALL PANELS[a]

SEISMIC DESIGN CATEGORY	CONDITION	CONSTRUCTION METHODS[b,c]								BRACED PANEL LOCATION AND LENGTH[d]
		1	2	3	4	5	6	7	8	
A and B	One story, top of two or three story	X	X	X	X	X	X	X	X	Each end and not more than 25 feet on center
	First story of two story or second story of three story	X	X	X	X	X	X	X	X	
	First story of three story	—	X	X	X	X[e]	X	X	X	
C	One story, top of two or three story	—	X	X	X	X	X	X	X	Each end and not more than 25 feet on center
	First story of two story or second story of three story	—	X	X	X	X[e]	X	X	X	Each end and not more than 25 feet on center but total length shall not be less than 25% of building length[f]
	First story of three story	—	X	X	X	X[e]	X	X	X	Each end and not more than 25 feet on center but total length shall not be less than 40% of building length[f]

For SI: 1 inch = 25.4 mm, 1 foot = 304.8 mm.

a. This table specifies minimum requirements for braced panels that form interior or exterior braced wall lines.

b. See Section 2308.9.3 for full description.

c. See Section 2308.9.3.1 for alternative braced panel requirement.

d. Building length is the dimension parallel to the braced wall length.

e. Gypsum wallboard applied to framing supports that are spaced at 16 inches on center.

f. The required lengths shall be doubled for gypsum board applied to only one face of a braced wall panel.

TABLE 2308.9.3(2)
EXPOSED PLYWOOD PANEL SIDING

MINIMUM THICKNESS[a] (inch)	MINIMUM NUMBER OF PLIES	STUD SPACING (inches) Plywood siding applied directly to studs or over sheathing
$^3/_8$	3	16[b]
$^1/_2$	4	24

For SI: 1 inch = 25.4 mm.

a. Thickness of grooved panels is measured at bottom of grooves.

b. Spans are permitted to be 24 inches if plywood siding applied with face grain perpendicular to studs or over one of the following: (1) 1-inch board sheathing, (2) $^7/_{16}$ -inch wood structural panel sheathing or (3) $^3/_8$ -inch wood structural panel sheathing with strength axis (which is the long direction of the panel unless otherwise marked) of sheathing perpendicular to studs.

TABLE 2308.9.3(3)
WOOD STRUCTURAL PANEL WALL SHEATHING[b]
(Not Exposed to the Weather, Strength Axis Parallel or Perpendicular to Studs Except as Indicated Below)

MINIMUM THICKNESS (inch)	PANEL SPAN RATING	STUD SPACING (inches)		
			Nailable sheathing	
		Siding nailed to studs	Sheathing parallel to studs	Sheathing perpendicular to studs
$^5/_{16}$	12/0, 16/0, 20/0 Wall–16″ o.c.	16	—	16
$^3/_8$, $^{15}/_{32}$, $^1/_2$	16/0, 20/0, 24/0, 32/16 Wall–24″ o.c.	24	16	24
$^7/_{16}$, $^{15}/_{32}$, $^1/_2$	24/0, 24/16, 32/16 Wall–24″ o.c.	24	24[a]	24

For SI: 1 inch = 25.4 mm.

a. Plywood shall consist of four or more plies.

b. Blocking of horizontal joints shall not be required except as specified in Sections 2306.4 and 2308.12.4.

TABLE 2308.9.3(4)
ALLOWABLE SPANS FOR PARTICLEBOARD WALL SHEATHING
(Not Exposed to the Weather, Long Dimension of the Panel Parallel or Perpendicular to Studs)

GRADE	THICKNESS (inch)	STUD SPACING (inches)	
		Siding nailed to studs	Sheathing under coverings specified in Section 2308.9.3 parallel or perpendicular to studs
M-S "Exterior Glue" and M-2 "Exterior Glue"	$^3/_8$	16	—
	$^1/_2$	16	16

For SI: 1 inch = 25.4 mm.

TABLE 2308.9.3(5)
HARDBOARD SIDING

SIDING	MINIMUM NOMINAL THICKNESS (inch)	2 × 4 FRAMING MAXIMUM SPACING	NAIL SIZE[a,b,d]	NAIL SPACING	
				General	Bracing panels[c]
1. Lap siding					
Direct to studs	$^3/_8$	16″ o.c.	8d	16″ o.c.	Not applicable
Over sheathing	$^3/_8$	16″ o.c.	10d	16″ o.c.	Not applicable
2. Square edge panel siding					
Direct to studs	$^3/_8$	24″ o.c.	6d	6″ o.c. edges; 12″ o.c. at intermediate supports	4″ o.c. edges; 8″ o.c. at intermediate supports
Over sheathing	$^3/_8$	24″ o.c.	8d	6″ o.c. edges; 12″ o.c. at intermediate supports	4″ o.c. edges; 8″ o.c. at intermediate supports
3. Shiplap edge panel siding					
Direct to studs	$^3/_8$	16″ o.c.	6d	6″ o.c. edges; 12″ o.c. at intermediate supports	4″ o.c. edges; 8″ o.c. at intermediate supports
Over sheathing	$^3/_8$	16″ o.c.	8d	6″ o.c. edges; 12″ o.c. At intermediate supports	4″ o.c. edges; 8″ o.c. at intermediate supports

For SI: 1 inch = 25.4 mm.

a. Nails shall be corrosion resistant.

b. Minimum acceptable nail dimensions:

	Panel Siding (inch)	Lap Siding (inch)
Shank diameter	0.092	0.099
Head diameter	0.225	0.240

c. Where used to comply with Section 2308.9.3.

d. Nail length must accommodate the sheathing and penetrate framing $1^1/_2$ inches.

2308.9.3.1 Alternative bracing. Any bracing required by Section 2308.9.3 is permitted to be replaced by the following:

1. In one-story buildings, each panel shall have a length of not less than 2 feet 8 inches (813 mm) and a height of not more than 10 feet (3048 mm). Each panel shall be sheathed on one face with $^3/_8$-inch-minimum-thickness (9.5 mm) wood structural panel sheathing nailed with 8d common or galvanized box nails in accordance with Table 2304.9.1 and blocked at wood structural panel edges. Two anchor bolts installed in accordance with Section 2308.6 shall be provided in each panel. Anchor bolts shall be placed at each panel outside quarter points. Each panel end stud shall have a tie-down device fastened to the foundation, capable of providing an approved uplift capacity of not less than 1,800 pounds (8006 N). The tie-down device shall be installed in accordance with the manufacturer's recommendations. The panels shall be supported directly on a foundation or on floor framing supported directly on a foundation that is continuous across the entire length of the braced wall line. This foundation shall be reinforced with not less than one No. 4 bar top and bottom.

   Where the continuous foundation is required to have a depth greater than 12 inches (305 mm), a minimum 12-inch by 12-inch (305 mm by 305 mm) continuous footing or turned down slab edge is permitted at door openings in the braced wall line. This continuous footing or turned down slab edge shall be reinforced with not less than one No. 4 bar top and bottom. This reinforcement shall be lapped 15 inches (381 mm) with the reinforcement required in the continuous foundation located directly under the braced wall line.

2. In the first story of two-story buildings, each wall panel shall be braced in accordance with Section 2308.9.3.1, Item 1, except that the wood structural panel sheathing shall be provided on both faces, three anchor bolts shall be placed at one-quarter points, and tie-down device uplift capacity shall not be less than 3,000 pounds (13 344 N).

❖ This provision allows a minimum 32-inch (813 mm) wall bracing length in lieu of the 48-inch (1219 mm) length required elsewhere. This bracing detail allows more flexibility in constructing braced wall panels adjacent to garage doors and other similar openings. This detail was developed by APA and is supported in APA Research Report No. 156. Item 1 describes the requirements for a one-story building (see Figure 2308.9.3.1).

An alternative panel is allowed on the ground level of a two-story structure as well, but the requirements are more stringent—almost double the resistance required for the single-story panel. One important issue is the specification of a continuous foundation for the length of the entire braced wall line that contains these panels, which would even include those interior lines at a spacing less than 50 feet (15 240 mm) from an adjacent braced wall line as described in Section 2308.3.4.

For SI: 1 inch = 25.4 mm,
         1 foot = 304.8 mm.

Figure 2308.9.3.1
ALTERNATIVE BRACED WALL PANELS—ONE-STORY

2308.9.3.2 Alternate bracing wall panel adjacent to a door or window opening. Any bracing required by Section 2308.9.3 is permitted to be replaced by the following when used adjacent to a door or window opening with a full-length header:

1. In one-story buildings, each panel shall have a length of not less than 16 inches (406 mm) and a height of not more than 10 feet (3048 mm). Each panel shall be sheathed on one face with a single layer of $^3/_8$ inch (9.5 mm) minimum thickness wood structural panel sheathing nailed with 8d common or galvanized box nails in accordance with Figure 2308.9.3.2. The wood structural panel sheathing shall extend up over the solid sawn or glued-laminated header and shall be nailed in accordance with Figure 2308.9.3.2. A built-up header consisting of at least two 2 x 12s and fastened in accordance with Item 24 of Table 2304.9.1 shall be permitted to be used. A spacer, if used, shall be placed on the side of the built-up beam opposite the wood structural panel sheathing. The header shall extend between the inside faces of the first full-length outer studs of each panel. The clear span of the header between the inner studs of each panel shall be not less than 6 feet (1829 mm) and not more than 18 feet (5486 mm) in length. A strap with an uplift capacity of not less than 1,000 pounds (4,400 N) shall fasten the header to the inner studs opposite the sheathing. One anchor bolt not less than $^5/_8$ inch (15.9 mm) diameter and installed in accordance with Section 2308.6 shall be provided in the center of each sill plate. The studs at each end of the panel shall have a tie-down device fastened to the foundation with an uplift capacity of not less than 4,200 pounds (18 480 N).

   Where a panel is located on one side of the opening, the header shall extend between the inside face of the first full-length stud of the panel and the bearing studs at the other end of the opening. A strap with an uplift capacity of not less than 1,000 pounds (4400 N) shall fasten the header to the bearing studs. The bearing studs shall also have a tie-down device fastened to the foundation with an uplift capacity of not less than 1,000 pounds (4400 N).

   The tie-down devices shall be an embedded strap type, installed in accordance with the manufacturer's recommendations. The panels shall be supported directly on a foundation that is continuous across the entire length of the braced wall line. This foundation shall be reinforced with not less than one No. 4 bar top and bottom.

   Where the continuous foundation is required to have a depth greater than 12 inches (305 mm), a minimum 12-inch by 12-inch (305 mm by 305 mm) continuous footing or turned down slab edge is permitted at door openings in the braced wall line. This continuous footing or turned down slab edge shall be reinforced with not less than one No. 4 bar top and bottom. This reinforcement shall be lapped not less than 15 inches (381 mm) with the reinforcement required in the continuous foundation located directly under the braced wall line.

2. In the first story of two-story buildings, each wall panel shall be braced in accordance with Item 1 above, except that each panel shall have a length of not less than 24 inches (610 mm).

❖ This provision allows an additional alternative bracing method for use adjacent to a window or door opening. It includes specifications for creating a type of wood-frame system with built-up "columns" at each end that are fastened to header with a nominal amount of rigidity. In this alternative, the header runs the length of the sheathed bracing panel—up to the first full-length stud—and is attached to the full-height sheathing with a grid pattern of nails where the sheathing and header overlap. This provides a partially moment-resistant connection. At the base of the sheathed section embedded, anchors are nailed to the framing and they provide a more conventional moment resisting connection. These framing anchors are sized to provide uplift in addition to shear capacity, reducing the plate anchoring requirements. Additional strap anchors, as shown in Figure 2308.9.3.2, are required. The result of these provisions is a reduced width of the full height leg of the assembly from 32 to 16 inches (813 to 406 mm) for a one-story building, and to 24 inches (610 mm) in the first floor of a two-story building compared to the bracing option permitted in Section 2308.9.3.1. The maximum allowable garage (or other room) size for this application would have to comply with the restrictions for braced wall line spacing and framing member spans noted elsewhere in Section 2308.

FIGURE 2308.9.3.2. See page 23-86.

❖ Narrow wall details such as these have been used successfully in the Pacific Northwest for almost two decades. This alternative was first developed in 1988 and was based on monotonic testing. The popularity of this detail grew until it was used in a number of different jurisdictions in the Pacific Northwest. During a series of cyclic tests conducted by the APA, the single-sided/single-story 32-inch (813 mm) wide alternative bracing in Section 2308.9.3.1 was tested along with the 16-inch (406 mm) wide alternative. Similarly, the double-sided/two-story 32-inch (813 mm) wide assembly was compared with the 24-inch (610 mm) wide alternative. In both cases, testing demonstrated that the Section 2308.9.3.2 alternative bracing performs significantly better in both strength and stiffness than the 32-inch (813 mm) wide bracing in Section 2308.9.3.1.

2308.9.4 Cripple walls. Foundation cripple walls shall be framed of studs not less in size than the studding above with a minimum length of 14 inches (356 mm), or shall be framed of solid blocking. Where exceeding 4 feet (1219 mm) in height, such walls shall be framed of studs having the size required for an additional story.

❖ Cripple walls (sometimes referred to as foundation stud walls or knee walls) are stud walls usually less than 8 feet (2438 mm) in height that rest on the foundation plate and support the first immediate floor above. The minimum stud length of 14 inches (356 mm) is based on the length necessary to properly fasten the studs to the foundation wall plate and the double plate above. Where the studs are less than 14 inches (356 mm) in length, they should be installed with wall plates and with the solid blocking tightly fit between each stud. This blocking performs two purposes. It provides a level, uniform bearing surface for the support of the floor above and transfers lateral forces from the floor to the foundation.

2308.9.4.1 Bracing. For the purposes of this section, cripple walls having a stud height exceeding 14 inches (356 mm) shall be considered a story and shall be braced in accordance with Table 2308.9.3(1) for Seismic Design Category A, B or C. See Section 2308.12.4 for Seismic Design Category D or E.

❖ When the cripple wall stud height is more than 14 inches (356 mm), the wall is to be considered as a first-story wall and the bracing requirements of Table 2308.9.3(1) will apply for buildings in lower seismic design categories. In Seismic Design Category D or E, the increased bracing requirements of Table 2308.12.4 will apply.

2308.9.4.2 Nailing of bracing. Spacing of edge nailing for required wall bracing shall not exceed 6 inches (152 mm) o.c. along the foundation plate and the top plate of the cripple wall. Nail size, nail spacing for field nailing and more restrictive boundary nailing requirements shall be as required elsewhere in the code for the specific bracing material used.

❖ Reasons that cripple wall bracing and nailing are so important are to avoid the creation of a soft story and to provide a continuous load path for lateral forces on the structure above. Nailing specified in this section provides a level of shear resistance at the bottom of the structure that is at least the equivalent to that above.

2308.9.5 Openings in exterior walls.

TABLE 2308.9.5. See page 23-88.

❖ The header support conditions utilized in this table are shown in Figure 2308.9.5 (see commentary, Section 2308.9.5.1).

For SI: 1 foot = 304.8 mm; 1 inch = 25.4 mm; 1 pound = 4.448 N.

FIGURE 2308.9.3.2
ALTERNATE BRACED WALL PANEL ADJACENT TO A DOOR OR WINDOW OPENING

2308.9.5.1 Headers. Headers shall be provided over each opening in exterior-bearing walls. The spans in Table 2308.9.5 are permitted to be used for one- and two-family dwellings. Headers for other buildings shall be designed in accordance with Section 2301.2, Item 1 or 2. Headers shall be of two pieces of nominal 2-inch (51 mm) framing lumber set on edge as permitted by Table 2308.9.5 and nailed together in accordance with Table 2304.9.1 or of solid lumber of equivalent size.

❖ Headers are required to carry loads from the wall, floors and roof above. Header spans for exterior bearing walls of one- and two-family dwellings may be sized and supported in accordance with Table 2308.9.5. Headers for other structures must be designed.

Table 2308.9.5, as well as Table 2308.9.6, provides an empirical procedure for determining header sizes in exterior walls. The tables were developed using the following design load criteria:

Roof LL	=	20 psf	(0.96 kN/m^2)
Roof DL	=	10 psf	(0.48 kN/m^2)
Ceiling LL	=	5 psf	(0.24 kN/m^2)
Ceiling DL	=	5 psf	(0.24 kN/m^2)
Floor LL	=	40 psf	(1.92 kN/m^2)
Floor DL	=	10 psf	(0.48 kN/m^2)
Wall DL	=	11 psf	(0.53 kN/m^2)

The spans were developed by using the given loads and performing moment and deflection calculations to determine the maximum span length for the specific size headers shown in the tables. The span length calculations also included a load combination reduction factor. Current ASD design loads do not address load combination reduction factors. The load combination reduction factors were developed in accordance with LRFD load combinations, the smaller of $1.6L + 0.5S$ or $1.6S + 0.5L$, where S is the snow load and L is the live load. The appropriate factor is then divided by 1.6 to adjust from LRFD to ASD.

The header span in the tables is determined based on the ground snow load in accordance with Figure 1608.2, the building width, the supporting conditions and the size of the header to be used. The header sizes must comprise the required lumber sizes shown in the tables, or may be of solid lumber of equivalent size.

2308.9.5.2 Header support. Wall studs shall support the ends of the header in accordance with Table 2308.9.5. Each end of a lintel or header shall have a length of bearing of not less than $1^1/_2$ inches (38 mm) for the full width of the lintel.

❖ Trimmer or jack studs that support the header at each end, should be continuous from the header to the bottom plate (or sill plate). Cutting the trimmers to support a sill is not allowed. Headers should be adequately nailed together and to the wall studs. Minimum required nailing is specified in Table 2304.9.1 [see Figures 2308.9.5.2(1) and (2)].

2308.9.6 Openings in interior bearing partitions. Headers shall be provided over each opening in interior bearing partitions as required in Section 2308.9.5. The spans in Table 2308.9.6 are permitted to be used. Wall studs shall support the ends of the header in accordance with Table 2308.9.5 or 2308.9.6, as appropriate.

❖ It may not be immediately clear that interior bearing walls will require headers to carry loads over openings in the same fashion that headers within exterior bearing walls must; therefore, this section has been included in the code. At interior bearing walls, the tributary width of loading carried from above extends to both sides of the wall instead of just one side, so care is needed to determine whether an interior header meets conventional restrictions or if it requires engineering per Section 2308.4.

TABLE 2308.9.6. See page 23-91.

❖ The header support conditions that are used in this table are illustrated in Figure 2308.9.6 (see commentary, Section 2308.9.5.1).

2308.9.7 Openings in interior nonbearing partitions. Openings in nonbearing partitions are permitted to be framed with single studs and headers. Each end of a lintel or header shall have a length of bearing of not less than $1^1/_2$ inches (38 mm) for the full width of the lintel.

❖ Because interior nonbearing partitions do not carry gravity loads, the prescriptive header requirements over openings are greatly reduced.

Figure 2308.9.5
HEADER SUPPORT CONDITIONS FOR EXTERIOR LOAD BEARING WALLS

TABLE 2308.9.5
HEADER AND GIRDER SPANS[a] FOR EXTERIOR BEARING WALLS
(Maximum Spans for Douglas Fir-Larch, Hem-Fir, Southern Pine and Spruce-Pine-Fir[b] and Required Number of Jack Studs)

GROUND SNOW LOAD (psf)[e]

HEADERS SUPPORTING	SIZE	30 — 20 ft Span	30 — 20 ft NJ[d]	30 — 28 ft Span	30 — 28 ft NJ[d]	30 — 36 ft Span	30 — 36 ft NJ[d]	50 — 20 ft Span	50 — 20 ft NJ[d]	50 — 28 ft Span	50 — 28 ft NJ[d]	50 — 36 ft Span	50 — 36 ft NJ[d]
Roof & Ceiling	2-2×4	3-6	1	3-2	1	2-10	1	3-2	1	2-9	1	2-6	1
	2-2×6	5-5	1	4-8	1	4-2	1	4-8	1	4-1	1	3-8	2
	2-2×8	6-10	1	5-11	2	5-4	2	5-11	2	5-2	2	4-7	2
	2-2×10	8-5	2	7-3	2	6-6	2	7-3	2	6-3	2	5-7	2
	2-2×12	9-9	2	8-5	2	7-6	2	8-5	2	7-3	2	6-6	2
	3-2×8	8-4	1	7-5	1	6-8	1	7-5	1	6-5	2	5-9	2
	3-2×10	10-6	1	9-1	2	8-2	2	9-1	2	7-10	2	7-0	2
	3-2×12	12-2	2	10-7	2	9-5	2	10-7	2	9-2	2	8-2	2
	4-2×8	9-2	1	8-4	1	7-8	1	8-4	1	7-5	1	6-8	1
	4-2×10	11-8	1	10-6	1	9-5	2	10-6	1	9-1	2	8-2	2
	4-2×12	14-1	1	12-2	2	10-11	2	12-2	2	10-7	2	9-5	2
Roof, Ceiling & 1 Center-Bearing Floor	2-2×4	3-1	1	2-9	1	2-5	1	2-9	1	2-5	1	2-2	1
	2-2×6	4-6	1	4-0	1	3-7	2	4-1	1	3-7	2	3-3	2
	2-2×8	5-9	2	5-0	2	4-6	2	5-2	2	4-6	2	4-1	2
	2-2×10	7-0	2	6-2	2	5-6	2	6-4	2	5-6	2	5-0	2
	2-2×12	8-1	2	7-1	2	6-5	2	7-4	2	6-5	2	5-9	3
	3-2×8	7-2	1	6-3	2	5-8	2	6-5	2	5-8	2	5-1	2
	3-2×10	8-9	2	7-8	2	6-11	2	7-11	2	6-11	2	6-3	2
	3-2×12	10-2	2	8-11	2	8-0	2	9-2	2	8-0	2	7-3	2
	4-2×8	8-1	1	7-3	1	6-7	1	7-5	1	6-6	1	5-11	1
	4-2×10	10-1	1	8-10	2	8-0	2	9-1	2	8-0	2	7-2	2
	4-2×12	11-9	2	10-3	2	9-3	2	10-7	2	9-3	2	8-4	2
Roof, Ceiling & 1 Clear Span Floor	2-2×4	2-8	1	2-4	1	2-1	1	2-7	1	2-3	1	2-0	1
	2-2×6	3-11	1	3-5	2	3-0	2	3-10	2	3-4	2	3-0	2
	2-2×8	5-0	2	4-4	2	3-10	2	4-10	2	4-2	2	3-9	2
	2-2×10	6-1	2	5-3	2	4-8	2	5-11	2	5-1	2	4-7	3
	2-2×12	7-1	2	6-1	3	5-5	3	6-10	3	5-11	3	5-4	3
	3-2×8	6-3	2	5-5	2	4-10	2	6-1	2	5-3	2	4-8	2
	3-2×10	7-7	2	6-7	2	5-11	2	7-5	2	6-5	2	5-9	2
	3-2×12	8-10	2	7-8	2	6-10	2	8-7	2	7-5	2	6-8	2
	4-2×8	7-2	1	6-3	2	5-7	2	7-0	1	6-1	2	5-5	2
	4-2×10	8-9	2	7-7	2	6-10	2	8-7	2	7-5	2	6-7	2
	4-2×12	10-2	2	8-10	2	7-11	2	9-11	2	8-7	2	7-8	2

(continued)

TABLE 2308.9.5—continued
HEADER AND GIRDER SPANS[a] FOR EXTERIOR BEARING WALLS
(Maximum Spans for Douglas Fir-Larch, Hem-Fir, Southern Pine and Spruce-Pine-Fir[b] and Required Number of Jack Studs)

HEADERS SUPPORTING	SIZE	GROUND SNOW LOAD (psf)[e]											
		30						50					
		Building width[c] (feet)											
		20		28		36		20		28		36	
		Span	NJ[d]	Span	NJ[d]	Span	NJ[d]	Span	NJ[d]	Span	NJ[d]	Span	NJ[d]
Roof, Ceiling & 2 Center-Bearing Floors	2-2 × 4	2-7	1	2-3	1	2-0	1	2-6	1	2-2	1	1-11	1
	2-2 × 6	3-9	2	3-3	2	2-11	2	3-8	2	3-2	2	2-10	2
	2-2 × 8	4-9	2	4-2	2	3-9	2	4-7	2	4-0	2	3-8	2
	2-2 × 10	5-9	2	5-1	2	4-7	3	5-8	2	4-11	2	4-5	3
	2-2 × 12	6-8	2	5-10	3	5-3	3	6-6	2	5-9	3	5-2	3
	3-2 × 8	5-11	2	5-2	2	4-8	2	5-9	2	5-1	2	4-7	2
	3-2 × 10	7-3	2	6-4	2	5-8	2	7-1	2	6-2	2	5-7	2
	3-2 × 12	8-5	2	7-4	2	6-7	2	8-2	2	7-2	2	6-5	3
	4-2 × 8	6-10	1	6-0	2	5-5	2	6-8	1	5-10	2	5-3	2
	4-2 × 10	8-4	2	7-4	2	6-7	2	8-2	2	7-2	2	6-5	2
	4-2 × 12	9-8	2	8-6	2	7-8	2	9-5	2	8-3	2	7-5	2
Roof, Ceiling & 2 Clear Span Floors	2-2 × 4	2-1	1	1-8	1	1-6	2	2-0	1	1-8	1	1-5	2
	2-2 × 6	3-1	2	2-8	2	2-4	2	3-0	2	2-7	2	2-3	2
	2-2 × 8	3-10	2	3-4	2	3-0	3	3-10	2	3-4	2	2-11	3
	2-2 × 10	4-9	2	4-1	3	3-8	3	4-8	2	4-0	3	3-7	3
	2-2 × 12	5-6	3	4-9	3	4-3	3	5-5	3	4-8	3	4-2	3
	3-2 × 8	4-10	2	4-2	2	3-9	2	4-9	2	4-1	2	3-8	2
	3-2 × 10	5-11	2	5-1	2	4-7	3	5-10	2	5-0	2	4-6	3
	3-2 × 12	6-10	2	5-11	3	5-4	3	6-9	2	5-10	3	5-3	3
	4-2 × 8	5-7	2	4-10	2	4-4	2	5-6	2	4-9	2	4-3	2
	4-2 × 10	6-10	2	5-11	2	5-3	2	6-9	2	5-10	2	5-2	2
	4-2 × 12	7-11	2	6-10	2	6-2	2	7-9	2	6-9	2	6-0	3

For SI: 1 inch = 25.4 mm, 1 foot = 304.8 mm, 1 pound per square foot = 47.8 N/m².

a. Spans are given in feet and inches (ft-in).
b. Tabulated values are for No. 2 grade lumber.
c. Building width is measured perpendicular to the ridge. For widths between those shown, spans are permitted to be interpolated.
d. NJ - Number of jack studs required to support each end. Where the number of required jack studs equals one, the header is permitted to be supported by an approved framing anchor attached to the full-height wall stud and to the header.
e. Use 30 pounds per square foot ground snow load for cases in which ground snow load is less than 30 pounds per square foot and the roof live load is equal to or less than 20 pounds per square foot.

Figure 2308.9.5.2(1)
STUDS AND HEADERS AROUND WALL OPENINGS

Figure 2308.9.5.2(2)
STUDS AND HEADERS AROUND WALL OPENINGS—BAY WINDOWS
(Source: AF&PA *Wood Frame Construction Manual for One- and Two-Family Dwellings, SBC High Wind Edition*, 1995.)

Figure 2308.9.6
HEADER SUPPORT CONDITIONS FOR INTERIOR LOAD-BEARING WALLS
(Source: AF&PA WOOD-FRAME CONSTRUCTION MANUAL)

TABLE 2308.9.6
HEADER AND GIRDER SPANS[a] FOR INTERIOR BEARING WALLS
(Maximum Spans for Douglas Fir-Larch, Hem-Fir, Southern Pine and Spruce-Pine-Fir[b] and Required Number of Jack Studs)

HEADERS AND GIRDERS SUPPORTING	SIZE	BUILDING WIDTH[c] (feet)					
		20		28		36	
		Span	NJ[d]	Span	NJ[d]	Span	NJ[d]
One Floor Only	2-2 × 4	3-1	1	2-8	1	2-5	1
	2-2 × 6	4-6	1	3-11	1	3-6	1
	2-2 × 8	5-9	1	5-0	2	4-5	2
	2-2 ×10	7-0	2	6-1	2	5-5	2
	2-2 ×12	8-1	2	7-0	2	6-3	2
	3-2 × 8	7-2	1	6-3	1	5-7	2
	3-2 ×10	8-9	1	7-7	2	6-9	2
	3-2 ×12	10-2	2	8-10	2	7-10	2
	4-2 × 8	9-0	1	7-8	1	6-9	1
	4-2 ×10	10-1	1	8-9	1	7-10	2
	4-2 ×12	11-9	1	10-2	2	9-1	2
Two Floors	2-2 × 4	2-2	1	1-10	1	1-7	1
	2-2 × 6	3-2	2	2-9	2	2-5	2
	2-2 × 8	4-1	2	3-6	2	3-2	2
	2-2 ×10	4-11	2	4-3	2	3-10	3
	2-2 ×12	5-9	2	5-0	3	4-5	3
	3-2 × 8	5-1	2	4-5	2	3-11	2
	3-2 ×10	6-2	2	5-4	2	4-10	2
	3-2 ×12	7-2	2	6-3	2	5-7	3
	4-2 × 8	6-1	1	5-3	2	4-8	2
	4-2 ×10	7-2	2	6-2	2	5-6	2
	4-2 ×12	8-4	2	7-2	2	6-5	2

For SI: 1 inch = 25.4 mm, 1 foot = 304.8 mm.

a. Spans are given in feet and inches (ft-in).

b. Tabulated values are for No. 2 grade lumber.

c. Building width is measured perpendicular to the ridge. For widths between those shown, spans are permitted to be interpolated.

d. NJ - Number of jack studs required to support each end. Where the number of required jack studs equals one, the headers are permitted to be supported by an approved framing anchor attached to the full-height wall stud and to the header.

2308.9.8 Pipes in walls. Stud partitions containing plumbing, heating or other pipes shall be so framed and the joists underneath so spaced as to give proper clearance for the piping. Where a partition containing such piping runs parallel to the floor joists, the joists underneath such partitions shall be doubled and spaced to permit the passage of such pipes and shall be bridged. Where plumbing, heating or other pipes are placed in or partly in a partition, necessitating the cutting of the soles or plates, a metal tie not less than 0.058 inch (1.47 mm) (16 galvanized gage) and 1¹/₂ inches (38 mm) wide shall be fastened to each plate across and to each side of the opening with not less than six 16d nails.

❖ This section requires that installation of piping in partitions and through floors does not compromise the strength or serviceability of the framing. This section recognizes that the plumbing trades are not always coordinated with the work of the carpenters and some form of regulation is necessary to guide the installation of plumbing so that it does not affect the behavior of a conventionally framed system. If there are piping installations proposed that are different from what is described in this section, an engineering analysis and design will be required.

2308.9.9 Bridging. Unless covered by interior or exterior wall coverings or sheathing meeting the minimum requirements of this code, stud partitions or walls with studs having a height-to-least-thickness ratio exceeding 50 shall have bridging not less than 2 inches (51 mm) in thickness and of the same width as the studs fitted snugly and nailed thereto to provide adequate lateral support. Bridging shall be placed in every stud cavity and at a frequency such that no stud so braced shall have a height-to-least-thickness ratio exceeding 50 with the height of the stud measured between horizontal framing and bridging or between bridging, whichever is greater.

❖ In the rare cases where a stud wall will not be covered with a sheathing material defined within the Code,

solid blocking is required at certain intervals to help assure that the studs act together as a system to maintain stability and prevent racking. The Code requires fire blocking in other sections, which may also be used to meet the requirement stated here.

2308.9.10 Cutting and notching. In exterior walls and bearing partitions, any wood stud is permitted to be cut or notched to a depth not exceeding 25 percent of its width. Cutting or notching of studs to a depth not greater than 40 percent of the width of the stud is permitted in nonbearing partitions supporting no loads other than the weight of the partition.

❖ Studs should not be cut, notched or bored when possible to maintain the cross-sectional bearing area. When the damage exceeds that specified in Section 2308.9.10, the studs should be doubled or otherwise reinforced to provide the required strength (see Figure 2308.9.10). Such reinforcing will require an engineered design.

2308.9.11 Bored holes. A hole not greater in diameter than 40 percent of the stud width is permitted to be bored in any wood stud. Bored holes not greater than 60 percent of the width of the stud are permitted in nonbearing partitions or in any wall where each bored stud is doubled, provided not more than two such successive doubled studs are so bored.

In no case shall the edge of the bored hole be nearer than ⁵/₈ inch (15.9 mm) to the edge of the stud.

Bored holes shall not be located at the same section of stud as a cut or notch.

❖ Due to the redundancy of this type of construction, limited notching and hole boring are allowed (see commentary, Section 2308.9.10).

2308.10 Roof and ceiling framing. The framing details required in this section apply to roofs having a minimum slope of three units vertical in 12 units horizontal (25-percent slope) or greater. Where the roof slope is less than three units vertical

For SI: 1 inch = 25.4 mm.

Figure 2308.9.10
CUTTING, NOTCHING AND BORED HOLES IN STUDS

in 12 units horizontal (25-percent slope), members supporting rafters and ceiling joists such as ridge board, hips and valleys shall be designed as beams.

❖ The code intends that the framing details for roofs in Section 2308.10 apply only to roofs having a minimum slope of three units vertical in 12 units horizontal. For roofs flatter than 3:12, the side thrust becomes so large as to exceed the capabilities of conventional construction. In that situation, members that support rafters must be designed as beams and must be supported by exterior or interior bearing walls or posts.

2308.10.1 Wind uplift. Roof assemblies shall have rafter and truss ties to the wall below. Resultant uplift loads shall be transferred to the foundation using a continuous load path. The rafter or truss to wall connection shall comply with Tables 2304.9.1 and 2308.10.1.

❖ The subject of resisting wind uplift on roofs in locations other than high-wind areas was not addressed in previous building codes. However, in developing the provisions of Section 2308, it was believed that there was justification for addressing this situation since significant uplift may also occur in other areas as well.

This section requires compliance with both Table 2304.9.1, which is the fastening schedule, and Table 2308.10.1, which specifies the minimum uplift resistance to be provided between the roof framing and the wall below. Although the toe-nail connections required by Table 2304.9.1 could conceivably provide enough uplift resistance to satisfy Table 2308.10.1 for short roof spans in the lower wind speed areas, in almost all instances, additional uplift resistance is required in the form of approved connectors. When uplift connectors are required by the table, one is required on every rafter or truss to the stud below, assuming the roof framing is spaced 24 inches (610 mm) o.c. in accordance with Note b. The intent, again, is to complete the load path to the foundation.

TABLE 2308.10.1. See below.

❖ See the commentary to Section 2308.10.1.

2308.10.2 Ceiling joist spans. Allowable spans for ceiling joists shall be in accordance with Table 2308.10.2(1) or 2308.10.2(2). For other grades and species, refer to the *AF&PA Span Tables for Joists and Rafters.*

❖ Allowable spans for ceiling joists are to be in accordance with Table 2308.10.2(1) or 2308.10.2(2). For other grades and species, refer to the AF&PA *Span Tables for Joists and Rafters* (see commentary, Section 2308.10.3).

TABLES 2308.10.2(1) and (2). See pages 23-94 and 23-96.

❖ For Tables 2308.10.2(1) and (2), see the commentary to Sections 2308.10.3 and 2304.4.

TABLE 2308.10.1
REQUIRED RATING OF APPROVED UPLIFT CONNECTORS (pounds)[a,b,c,e,f,g,h]

BASIC WIND SPEED (3-second gust)	ROOF SPAN (feet)							OVERHANGS (pounds/feet)[d]
	12	20	24	28	32	36	40	
85	-72	-120	-145	-169	-193	-217	-241	-38.55
90	-91	-151	-181	-212	-242	-272	-302	-43.22
100	-131	-281	-262	-305	-349	-393	-436	-53.36
110	-175	-292	-351	-409	-467	-526	-584	-64.56

For SI: 1 inch = 25.4 mm, 1 foot = 304.8 mm, 1 mile per hour = 1.61 km/hr, 1 pound = 0.454 Kg, 1 pound/foot = 14.5939 N/m.

a. The uplift connection requirements are based on a 30-foot mean roof height located in Exposure B. For Exposure C or D and for other mean roof heights, multiply the above loads by the adjustment coefficients below.

EXPOSURE	Mean Roof Height (feet)									
	15	20	25	30	35	40	45	50	55	60
B	1.00	1.00	1.00	1.00	1.05	1.09	1.12	1.16	1.19	1.22
C	1.21	1.29	1.35	1.40	1.45	1.49	1.53	1.56	1.59	1.62
D	1.47	1.55	1.61	1.66	1.70	1.74	1.78	1.81	1.84	1.87

b. The uplift connection requirements are based on the framing being spaced 24 inches on center. Multiply by 0.67 for framing spaced 16 inches on center and multiply by 0.5 for framing spaced 12 inches on center.

c. The uplift connection requirements include an allowance for 10 pounds of dead load.

d. The uplift connection requirements do not account for the effects of overhangs. The magnitude of the above loads shall be increased by adding the overhang loads found in the table. The overhang loads are also based on framing spaced 24 inches on center. The overhang loads given shall be multiplied by the overhang projection and added to the roof uplift value in the table.

e. The uplift connection requirements are based upon wind loading on end zones as defined in Figure 6-2 of ASCE 7. Connection loads for connections located a distance of 20 percent of the least horizontal dimension of the building from the corner of the building are permitted to be reduced by multiplying the table connection value by 0.7 and multiplying the overhang load by 0.8.

f. For wall-to-wall and wall-to-foundation connections, the capacity of the uplift connector is permitted to be reduced by 100 pounds for each full wall above. (For example, if a 500-pound rated connector is used on the roof framing, a 400-pound rated connector is permitted at the next floor level down).

g. Interpolation is permitted for intermediate values of basic wind speeds and roof spans.

h. The rated capacity of approved tie-down devices is permitted to include up to a 60-percent increase for wind effects where allowed by material specifications.

TABLE 2308.10.2(1)
CEILING JOIST SPANS FOR COMMON LUMBER SPECIES
(Uninhabitable Attics Without Storage, Live Load = 10 pounds psf, L/Δ = 240)

CEILING JOIST SPACING (inches)	SPECIES AND GRADE		DEAD LOAD = 5 pounds per square foot			
			Maximum ceiling joist spans			
			2 × 4	2 × 6	2 × 8	2 × 10
			(ft. - in.)	(ft. - in.)	(ft. - in.)	(ft. - in.)
12	Douglas Fir-Larch	SS	13-2	20-8	Note a	Note a
	Douglas Fir-Larch	#1	12-8	19-11	Note a	Note a
	Douglas Fir-Larch	#2	12-5	19-6	25-8	Note a
	Douglas Fir-Larch	#3	10-10	15-10	20-1	24-6
	Hem-Fir	SS	12-5	19-6	25-8	Note a
	Hem-Fir	#1	12-2	19-1	25-2	Note a
	Hem-Fir	#2	11-7	18-2	24-0	Note a
	Hem-Fir	#3	10-10	15-10	20-1	24-6
	Southern Pine	SS	12-11	20-3	Note a	Note a
	Southern Pine	#1	12-8	19-11	Note a	Note a
	Southern Pine	#2	12-5	19-6	25-8	Note a
	Southern Pine	#3	11-6	17-0	21-8	25-7
	Spruce-Pine-Fir	SS	12-2	19-1	25-2	Note a
	Spruce-Pine-Fir	#1	11-10	18-8	24-7	Note a
	Spruce-Pine-Fir	#2	11-10	18-8	24-7	Note a
	Spruce-Pine-Fir	#3	10-10	15-10	20-1	24-6
16	Douglas Fir-Larch	SS	11-11	18-9	24-8	Note a
	Douglas Fir-Larch	#1	11-6	18-1	23-10	Note a
	Douglas Fir-Larch	#2	11-3	17-8	23-0	Note a
	Douglas Fir-Larch	#3	9-5	13-9	17-5	21-3
	Hem-Fir	SS	11-3	17-8	23-4	Note a
	Hem-Fir	#1	11-0	17-4	22-10	Note a
	Hem-Fir	#2	10-6	16-6	21-9	Note a
	Hem-Fir	#3	9-5	13-9	17-5	21-3
	Southern Pine	SS	11-9	18-5	24-3	Note a
	Southern Pine	#1	11-6	18-1	23-1	Note a
	Southern Pine	#2	11-3	17-8	23-4	Note a
	Southern Pine	#3	10-0	14-9	18-9	22-2
	Spruce-Pine-Fir	SS	11-0	17-4	22-10	Note a
	Spruce-Pine-Fir	#1	10-9	16-11	22-4	Note a
	Spruce-Pine-Fir	#2	10-9	16-11	22-4	Note a
	Spruce-Pine-Fir	#3	9-5	13-9	17-5	21-3

(continued)

TABLE 2308.10.2(1)—continued
CEILING JOIST SPANS FOR COMMON LUMBER SPECIES
(Uninhabitable Attics Without Storage, Live Load = 10 pounds psf, L/Δ = 240)

CEILING JOIST SPACING (inches)	SPECIES AND GRADE		DEAD LOAD = 5 pounds per square foot			
			Maximum ceiling joist spans			
			2 × 4	2 × 6	2 × 8	2 × 10
			(ft. - in.)	(ft. - in.)	(ft. - in.)	(ft. - in.)
19.2	Douglas Fir-Larch	SS	11-3	17-8	23-3	Note a
	Douglas Fir-Larch	#1	10-10	17-0	22-5	Note a
	Douglas Fir-Larch	#2	10-7	16-7	21-0	25-8
	Douglas Fir-Larch	#3	8-7	12-6	15-10	19-5
	Hem-Fir	SS	10-7	16-8	21-11	Note a
	Hem-Fir	#1	10-4	16-4	21-6	Note a
	Hem-Fir	#2	9-11	15-7	20-6	25-3
	Hem-Fir	#3	8-7	12-6	15-10	19-5
	Southern Pine	SS	11-0	17-4	22-10	Note a
	Southern Pine	#1	10-10	17-0	22-5	Note a
	Southern Pine	#2	10-7	16-8	21-11	Note a
	Southern Pine	#3	9-1	13-6	17-2	20-3
	Spruce-Pine-Fir	SS	10-4	16-4	21-6	Note a
	Spruce-Pine-Fir	#1	10-2	15-11	21-0	25-8
	Spruce-Pine-Fir	#2	10-2	15-11	21-0	25-8
	Spruce-Pine-Fir	#3	8-7	12-6	15-10	19-5
24	Douglas Fir-Larch	SS	10-5	16-4	21-7	Note a
	Douglas Fir-Larch	#1	10-0	15-9	20-1	24-6
	Douglas Fir-Larch	#2	9-10	14-10	18-9	22-11
	Douglas Fir-Larch	#3	7-8	11-2	14-2	17-4
	Hem-Fir	SS	9-10	15-6	20-5	Note a
	Hem-Fir	#1	9-8	15-2	19-7	23-11
	Hem-Fir	#2	9-2	14-5	18-6	22-7
	Hem-Fir	#3	7-8	11-2	14-2	17-4
	Southern Pine	SS	10-3	16-1	21-2	Note a
	Southern Pine	#1	10-0	15-9	20-10	Note a
	Southern Pine	#2	9-10	15-6	20-1	23-11
	Southern Pine	#3	8-2	12-0	15-4	18-1
	Spruce-Pine-Fir	SS	9-8	15-2	19-11	25-5
	Spruce-Pine-Fir	#1	9-5	14-9	18-9	22-11
	Spruce-Pine-Fir	#2	9-5	14-9	18-9	22-11
	Spruce-Pine-Fir	#3	7-8	11-2	14-2	17-4

For SI: 1 inch = 25.4 mm, 1 foot = 304.8 mm, 1 pound per square foot = 47.8 N/m².

a. Span exceeds 26 feet in length. Check sources for availability of lumber in lengths greater than 20 feet.

TABLE 2308.10.2(2)
CEILING JOIST SPANS FOR COMMON LUMBER SPECIES
(Uninhabitable Attics With Limited Storage, Live Load = 20 pounds per square foot, $L/\Delta = 240$)

CEILING JOIST SPACING (inches)	SPECIES AND GRADE		DEAD LOAD = 10 pounds per square foot			
			Maximum ceiling joist spans			
			2 × 4	2 × 6	2 × 8	2 × 10
			(ft. - in.)	(ft. - in.)	(ft. - in.)	(ft. - in.)
12	Douglas Fir-Larch	SS	10-5	16-4	21-7	Note a
	Douglas Fir-Larch	#1	10-0	15-9	20-1	24-6
	Douglas Fir-Larch	#2	9-10	14-10	18-9	22-11
	Douglas Fir-Larch	#3	7-8	11-2	14-2	17-4
	Hem-Fir	SS	9-10	15-6	20-5	Note a
	Hem-Fir	#1	9-8	15-2	19-7	23-11
	Hem-Fir	#2	9-2	14-5	18-6	22-7
	Hem-Fir	#3	7-8	11-2	14-2	17-4
	Southern Pine	SS	10-3	16-1	21-2	Note a
	Southern Pine	#1	10-0	15-9	20-10	Note a
	Southern Pine	#2	9-10	15-6	20-1	23-11
	Southern Pine	#3	8-2	12-0	15-4	18-1
	Spruce-Pine-Fir	SS	9-8	15-2	19-11	25-5
	Spruce-Pine-Fir	#1	9-5	14-9	18-9	22-11
	Spruce-Pine-Fir	#2	9-5	14-9	18-9	22-11
	Spruce-Pine-Fir	#3	7-8	11-2	14-2	17-4
16	Douglas Fir-Larch	SS	9-6	14-11	19-7	25-0
	Douglas Fir-Larch	#1	9-1	13-9	17-5	21-3
	Douglas Fir-Larch	#2	8-9	12-10	16-3	19-10
	Douglas Fir-Larch	#3	6-8	9-8	12-4	15-0
	Hem-Fir	SS	8-11	14-1	18-6	23-8
	Hem-Fir	#1	8-9	13-5	16-10	20-8
	Hem-Fir	#2	8-4	12-8	16-0	19-7
	Hem-Fir	#3	6-8	9-8	12-4	15-0
	Southern Pine	SS	9-4	14-7	19-3	24-7
	Southern Pine	#1	9-1	14-4	18-11	23-1
	Southern Pine	#2	8-11	13-6	17-5	20-9
	Southern Pine	#3	7-1	10-5	13-3	15-8
	Spruce-Pine-Fir	SS	8-9	13-9	18-1	23-1
	Spruce-Pine-Fir	#1	8-7	12-10	16-3	19-10
	Spruce-Pine-Fir	#2	8-7	12-10	16-3	19-10
	Spruce-Pine-Fir	#3	6-8	9-8	12-4	15-0

(continued)

TABLE 2308.10.2(2)—continued
CEILING JOIST SPANS FOR COMMON LUMBER SPECIES
(Uninhabitable Attics With Limited Storage, Live Load = 20 pounds per square foot, $L/\Delta = 240$)

CEILING JOIST SPACING (inches)	SPECIES AND GRADE		DEAD LOAD = 10 pounds per square foot			
			Maximum ceiling joist spans			
			2 × 4	2 × 6	2 × 8	2 × 10
			(ft. - in.)	(ft. - in.)	(ft. - in.)	(ft. - in.)
19.2	Douglas Fir-Larch	SS	8-11	14-0	18-5	23-4
	Douglas Fir-Larch	#1	8-7	12-6	15-10	19-5
	Douglas Fir-Larch	#2	8-0	11-9	14-10	18-2
	Douglas Fir-Larch	#3	6-1	8-10	11-3	13-8
	Hem-Fir	SS	8-5	13-3	17-5	22-3
	Hem-Fir	#1	8-3	12-3	15-6	18-11
	Hem-Fir	#2	7-10	11-7	14-8	17-10
	Hem-Fir	#3	6-1	8-10	11-3	13-8
	Southern Pine	SS	8-9	13-9	18-1	23-1
	Southern Pine	#1	8-7	13-6	17-9	21-1
	Southern Pine	#2	8-5	12-3	15-10	18-11
	Southern Pine	#3	6-5	9-6	12-1	14-4
	Spruce-Pine-Fir	SS	8-3	12-11	17-1	21-8
	Spruce-Pine-Fir	#1	8-0	11-9	14-10	18-2
	Spruce-Pine-Fir	#2	8-0	11-9	14-10	18-2
	Spruce-Pine-Fir	#3	6-1	8-10	11-3	13-8
24	Douglas Fir-Larch	SS	8-3	13-0	17-1	20-11
	Douglas Fir-Larch	#1	7-8	11-2	14-2	17-4
	Douglas Fir-Larch	#2	7-2	10-6	13-3	16-3
	Douglas Fir-Larch	#3	5-5	7-11	10-0	12-3
	Hem-Fir	SS	7-10	12-3	16-2	20-6
	Hem-Fir	#1	7-6	10-11	13-10	16-11
	Hem-Fir	#2	7-1	10-4	13-1	16-0
	Hem-Fir	#3	5-5	7-11	10-0	12-3
	Southern Pine	SS	8-1	12-9	16-10	21-6
	Southern Pine	#1	8-0	12-6	15-10	18-10
	Southern Pine	#2	7-8	11-0	14-2	16-11
	Southern Pine	#3	5-9	8-6	10-10	12-10
	Spruce-Pine-Fir	SS	7-8	12-0	15-10	19-5
	Spruce-Pine-Fir	#1	7-2	10-6	13-3	16-3
	Spruce-Pine-Fir	#2	7-2	10-6	13-3	16-3
	Spruce-Pine-Fir	#3	5-5	7-11	10-0	12-3

For SI: 1 inch = 25.4 mm, 1 foot = 304.8 mm, 1 pound per square foot = 47.8 N/m².

a. Span exceeds 26 feet in length. Check sources for availability of lumber in lengths greater than 20 feet.

2308.10.3 Rafter spans. Allowable spans for rafters shall be in accordance with Table 2308.10.3(1), 2308.10.3(2), 2308.10.3(3), 2308.10.3(4), 2308.10.3(5) or 2308.10.3(6). For other grades and species, refer to the *AF&PA Span Tables for Joists and Rafters*.

❖ The spans shown in the referenced tables were derived from the AF&PA *Span Tables for Joists and Rafters*. For simplicity, spans of only the most common species, or types, and grades of wood are shown. When other species or grades of wood are used, the AF&PA tables should be consulted to determine the proper span. Additionally, span tables published by the Southern Forest Products Association, the Western Wood Products Association and the Canadian Wood Council can be consulted. The spans in those tables were derived in the same manner as those published in the AF&PA tables.

Spans are based on species of wood, grade, roof live load or ground snow load, dead load and deflection criteria. With respect to the indicated ground snow loads, the spans shown reflect consideration of unbalanced snow loads, in accordance with established roof snow load criteria. Other considerations such as drifting snow loads are not accounted for in these tables. Where conditions such as a low roof adjacent to a higher roof, projections, or roof-mounted equipment are encountered the need to consider drifting must be assessed in accordance with the snow load criteria in ASCE 7.

TABLES 2308.10.3(1) through (6). See page 23-100 through 23-110.

❖ For Tables 2308.10.3(1) through (6), see the commentary to Sections 2308.10.3 and 2304.4.

2308.10.4 Ceiling joist and rafter framing. Rafters shall be framed directly opposite each other at the ridge. There shall be a ridge board at least 1-inch (25 mm) nominal thickness at ridges and not less in depth than the cut end of the rafter. At valleys and hips, there shall be a single valley or hip rafter not less than 2-inch (51 mm) nominal thickness and not less in depth than the cut end of the rafter.

❖ Traditional practice is to provide both a ridge board between opposite rafters as a nailing base and a full bearing for the rafter. Rafters must be placed directly opposite each other, and the ridge board must have a depth equal to or greater than the cut end of the rafter. The rafter ends must be flush against the ridge board to avoid excessive horizontal shear in the rafters (see Figure 2308.10.4). The ridge board only works in conjunction with thrust-resisting elements to maintain rafter support, which is handled by the ceiling joists and connections. See further discussion in the commentary to Section 2308.10.4.1. Additionally, a prescriptively framed roof system will have a series of kickers or interior bearing walls that serve to reduce the span of the rafters to the tabulated values. A lot of wood goes into the framing of a conventional system, but experienced contractors have adequately managed it for many years.

2308.10.4.1 Ceiling joist and rafter connections. Ceiling joists and rafters shall be nailed to each other and the assembly shall be nailed to the top wall plate in accordance with Tables 2304.9.1 and 2308.10.1. Ceiling joists shall be continuous or securely joined where they meet over interior partitions and fastened to adjacent rafters in accordance with Tables 2308.10.4.1 and 2304.9.1 to provide a continuous rafter tie across the building where such joists are parallel to the rafters.

RIDGE BOARD FULL DEPTH OF RAFTER CUT

RAFTER END MUST BE FLUSH AGAINST RIDGE BOARD TO AVOID HORIZONTAL SHEAR IN RAFTER

Figure 2308.10.4
RAFTER END CUT AT RIDGE

Ceiling joists shall have a bearing surface of not less than $1^1/_2$ inches (38 mm) on the top plate at each end.

Where ceiling joists are not parallel to rafters, an equivalent rafter tie shall be installed in a manner to provide a continuous tie across the building, at a spacing of not more than 4 feet (1219 mm) o.c. The connections shall be in accordance with Tables 2308.10.4.1 and 2304.9.1, or connections of equivalent capacities shall be provided. Where ceiling joists or rafter ties are not provided at the top of the rafter support walls, the ridge formed by these rafters shall also be supported by a girder conforming to Section 2308.4.

Rafter ties shall be spaced not more than 4 feet (1219 mm) o.c. Rafter tie connections shall be based on the equivalent rafter spacing in Table 2308.10.4.1. Where rafter ties are spaced at 32 inches (813 mm) o.c., the number of 16d common nails shall be two times the number specified for rafters spaced 16 inches (406 mm) o.c., with a minimum of 4-16d common nails where no snow loads are indicated. Where rafter ties are spaced at 48 inches (1219 mm) o.c., the number of 16d common nails shall be two times the number specified for rafters spaced 24 inches (610 mm) o.c., with a minimum of 6-16d common nails where no snow loads are indicated. Rafter/ceiling joist connections and rafter/tie connections shall be of sufficient size and number to prevent splitting from nailing.

❖ Upon review of a simple free body diagram, it should be apparent that sloped rafters exert a thrusting force on the walls of buildings in a system that is supported by a ridge board. Therefore, unless the ridge is otherwise supported, as in post and beam framing, the building must provide horizontal ties to prevent spreading of walls (see Figure 2308.10.4.1). When ceiling joists are parallel to the rafters and adequately connected to the top plate and rafters, the continuous tie requirements of this section are met. If ceiling joists are placed perpendicular to the span of the rafters, the continuity cannot be established as described, so the code simply requires an "equivalent rafter tie" to be installed. Some publications include details or descriptions of acceptable systems that make use of blocking, straps and adequate sheathing to resist these thrusting forces from the rafters. It could be argued that this condition requires an engineered design in order to establish what is an "equivalent" rafter tie.

Ceiling joists are sometimes raised above the wall support of the rafters to provide additional ceiling height. Only when such design is accompanied by engineering analysis or tests should this be allowed.

Structural problems are often found in houses having cathedral ceilings. Since no joists are present to tie the walls together across the building, properly designed beams and posts must be used to support the rafters to satisfy code requirements.

Wind blowing across the roof develops a negative pressure on the rafters. Rafter ties are intended to prevent separation of the rafters at the ridge under such conditions in addition to the possible use as continuous ties explained in this section. Rafter ties must never be considered as a means of reducing the span of rafters.

TABLE 2308.10.4.1. See page 23-112.

❖ See the commentary to Section 2308.10.4.1.

1x6 COLLAR BEAM EVERY THIRD RAFTER PAIR

CEILING JOIST (RAFTER TIE)

COLLAR BEAMS AND CEILING JOISTS PREVENT LATERAL SPREADING OF THE ROOF SYSTEM DUE TO GRAVITY AND WIND LOADS.

Figure 2308.10.4.1
RAFTER TIES

TABLE 2308.10.3(1)
RAFTER SPANS FOR COMMON LUMBER SPECIES
(Roof Live Load = 20 pounds per square foot, Ceiling Not Attached to Rafters, L/Δ = 180)

RAFTER SPACING (inches)	SPECIES AND GRADE		DEAD LOAD = 10 pounds per square foot					DEAD LOAD = 20 pounds per square foot				
			2 × 4	2 × 6	2 × 8	2 × 10	2 × 12	2 × 4	2 × 6	2 × 8	2 × 10	2 × 12
			Maximum rafter spans									
			(ft.-in.)	(ft.-in.)	(ft.-in.)	(ft.-in.)	(ft.-in.)	(ft.-in.)	(ft.-in.)	(ft.-in.)	(ft.-in.)	(ft.-in.)
12	Douglas Fir-Larch	SS	11-6	18-0	23-9	Note a	Note a	11-6	18-0	23-5	Note a	Note a
	Douglas Fir-Larch	#1	11-1	17-4	22-5	Note a	Note a	10-6	15-4	19-5	23-9	Note a
	Douglas Fir-Larch	#2	10-10	16-7	21-0	25-8	Note a	9-10	14-4	18-2	22-3	25-9
	Douglas Fir-Larch	#3	8-7	12-6	15-10	19-5	22-6	7-5	10-10	13-9	16-9	19-6
	Hem-Fir	SS	10-10	17-0	22-5	Note a	Note a	10-10	17-0	22-5	Note a	Note a
	Hem-Fir	#1	10-7	16-8	21-10	Note a	Note a	10-3	14-11	18-11	23-2	Note a
	Hem-Fir	#2	10-1	15-11	20-8	25-3	Note a	9-8	14-2	17-11	21-11	25-5
	Hem-Fir	#3	8-7	12-6	15-10	19-5	22-6	7-5	10-10	13-9	16-9	19-6
	Southern Pine	SS	11-3	17-8	23-4	Note a	Note a	11-3	17-8	23-4	Note a	Note a
	Southern Pine	#1	11-1	17-4	22-11	Note a	Note a	11-1	17-3	21-9	25-10	Note a
	Southern Pine	#2	10-10	17-0	22-5	Note a	Note a	10-6	15-1	19-5	23-2	Note a
	Southern Pine	#3	9-1	13-6	17-2	20-3	24-1	7-11	11-8	14-10	17-6	20-11
	Spruce-Pine-Fir	SS	10-7	16-8	21-11	Note a	Note a	10-7	16-8	21-9	Note a	Note a
	Spruce-Pine-Fir	#1	10-4	16-3	21-0	25-8	Note a	9-10	14-4	18-2	22-3	25-9
	Spruce-Pine-Fir	#2	10-4	16-3	21-0	25-8	Note a	9-10	14-4	18-2	22-3	25-9
	Spruce-Pine-Fir	#3	8-7	12-6	15-10	19-5	22-6	7-5	10-10	13-9	16-9	19-6
16	Douglas Fir-Larch	SS	10-5	16-4	21-7	Note a	Note a	10-5	16-0	20-3	24-9	Note a
	Douglas Fir-Larch	#1	10-0	15-4	19-5	23-9	Note a	9-1	13-3	16-10	20-7	23-10
	Douglas Fir-Larch	#2	9-10	14-4	18-2	22-3	25-9	8-6	12-5	15-9	19-3	22-4
	Douglas Fir-Larch	#3	7-5	10-10	13-9	16-9	19-6	6-5	9-5	11-11	14-6	16-10
	Hem-Fir	SS	9-10	15-6	20-5	Note a	Note a	9-10	15-6	19-11	24-4	Note a
	Hem-Fir	#1	9-8	14-11	18-11	23-2	Note a	8-10	12-11	16-5	20-0	23-3
	Hem-Fir	#2	9-2	14-2	17-11	21-11	25-5	8-5	12-3	15-6	18-11	22-0
	Hem-Fir	#3	7-5	10-10	13-9	16-9	19-6	6-5	9-5	11-11	14-6	16-10
	Southern Pine	SS	10-3	16-1	21-2	Note a	Note a	10-3	16-1	21-2	Note a	Note a
	Southern Pine	#1	10-0	15-9	20-10	25-10	Note a	10-0	15-0	18-10	22-4	Note a
	Southern Pine	#2	9-10	15-1	19-5	23-2	Note a	9-1	13-0	16-10	20-1	23-7
	Southern Pine	#3	7-11	11-8	14-10	17-6	20-11	6-10	10-1	12-10	15-2	18-1
	Spruce-Pine-Fir	SS	9-8	15-2	19-11	25-5	Note a	9-8	14-10	18-10	23-0	Note a
	Spruce-Pine-Fir	#1	9-5	14-4	18-2	22-3	25-9	8-6	12-5	15-9	19-3	22-4
	Spruce-Pine-Fir	#2	9-5	14-4	18-2	22-3	25-9	8-6	12-5	15-9	19-3	22-4
	Spruce-Pine-Fir	#3	7-5	10-10	13-9	16-9	19-6	6-5	9-5	11-11	14-6	16-10

(continued)

TABLE 2308.10.3(1)—continued
RAFTER SPANS FOR COMMON LUMBER SPECIES
(Roof Live Load = 20 pounds per square foot, Ceiling Not Attached to Rafters, L/Δ = 180)

RAFTER SPACING (inches)	SPECIES AND GRADE		DEAD LOAD = 10 pounds per square foot					DEAD LOAD = 20 pounds per square foot				
			2 × 4	2 × 6	2 × 8	2 × 10	2 × 12	2 × 4	2 × 6	2 × 8	2 × 10	2 × 12
			(ft.-in.)	(ft.-in.)	(ft.-in.)	(ft.-in.)	(ft.-in.)	(ft.-in.)	(ft.-in.)	(ft.-in.)	(ft.-in.)	(ft.-in.)
					Maximum rafter spans							
19.2	Douglas Fir-Larch	SS	9-10	15-5	20-4	25-11	Note a	9-10	14-7	18-6	22-7	Note a
	Douglas Fir-Larch	#1	9-5	14-0	17-9	21-8	25-2	8-4	12-2	15-4	18-9	21-9
	Douglas Fir-Larch	#2	8-11	13-1	16-7	20-3	23-6	7-9	11-4	14-4	17-7	20-4
	Douglas Fir-Larch	#3	6-9	9-11	12-7	15-4	17-9	5-10	8-7	10-10	13-3	15-5
	Hem-Fir	SS	9-3	14-7	19-2	24-6	Note a	9-3	14-4	18-2	22-3	25-9
	Hem-Fir	#1	9-1	13-8	17-4	21-1	24-6	8-1	11-10	15-0	18-4	21-3
	Hem-Fir	#2	8-8	12-11	16-4	20-0	23-2	7-8	11-2	14-2	17-4	20-1
	Hem-Fir	#3	6-9	9-11	12-7	15-4	17-9	5-10	8-7	10-10	13-3	15-5
	Southern Pine	SS	9-8	15-2	19-11	25-5	Note a	9-8	15-2	19-11	25-5	Note a
	Southern Pine	#1	9-5	14-10	19-7	23-7	Note a	9-3	13-8	17-2	20-5	24-4
	Southern Pine	#2	9-3	13-9	17-9	21-2	24-10	8-4	11-11	15-4	18-4	21-6
	Southern Pine	#3	7-3	10-8	13-7	16-0	19-1	6-3	9-3	11-9	13-10	16-6
	Spruce-Pine-Fir	SS	9-1	14-3	18-9	23-11	Note a	9-1	13-7	17-2	21-0	24-4
	Spruce-Pine-Fir	#1	8-10	13-1	16-7	20-3	23-6	7-9	11-4	14-4	17-7	20-4
	Spruce-Pine-Fir	#2	8-10	13-1	16-7	20-3	23-6	7-9	11-4	14-4	17-7	20-4
	Spruce-Pine-Fir	#3	6-9	9-11	12-7	15-4	17-9	5-10	8-7	10-10	13-3	15-5
24	Douglas Fir-Larch	SS	9-1	14-4	18-10	23-4	Note a	8-11	13-1	16-7	20-3	23-5
	Douglas Fir-Larch	#1	8-7	12-6	15-10	19-5	22-6	7-5	10-10	13-9	16-9	19-6
	Douglas Fir-Larch	#2	8-0	11-9	14-10	18-2	21-0	6-11	10-2	12-10	15-8	18-3
	Douglas Fir-Larch	#3	6-1	8-10	11-3	13-8	15-11	5-3	7-8	9-9	11-10	13-9
	Hem-Fir	SS	8-7	13-6	17-10	22-9	Note a	8-7	12-10	16-3	19-10	23-0
	Hem-Fir	#1	8-4	12-3	15-6	18-11	21-11	7-3	10-7	13-5	16-4	19-0
	Hem-Fir	#2	7-11	11-7	14-8	17-10	20-9	6-10	10-0	12-8	15-6	17-11
	Hem-Fir	#3	6-1	8-10	11-3	13-8	15-11	5-3	7-8	9-9	11-10	13-9
	Southern Pine	SS	8-11	14-1	18-6	23-8	Note a	8-11	14-1	18-6	22-11	Note a
	Southern Pine	#1	8-9	13-9	17-9	21-1	25-2	8-3	12-3	15-4	18-3	21-9
	Southern Pine	#2	8-7	12-3	15-10	18-11	22-2	7-5	10-8	13-9	16-5	19-3
	Southern Pine	#3	6-5	9-6	12-1	14-4	17-1	5-7	8-3	10-6	12-5	14-9
	Spruce-Pine-Fir	SS	8-5	13-3	17-5	21-8	25-2	8-4	12-2	15-4	18-9	21-9
	Spruce-Pine-Fir	#1	8-0	11-9	14-10	18-2	21-0	6-11	10-2	12-10	15-8	18-3
	Spruce-Pine-Fir	#2	8-0	11-9	14-10	18-2	21-0	6-11	10-2	12-10	15-8	18-3
	Spruce-Pine-Fir	#3	6-1	8-10	11-3	13-8	15-11	5-3	7-8	9-9	11-10	13-9

For SI: 1 inch = 25.4 mm, 1 foot = 304.8 mm, 1 pound per square foot = 47.9 N/m².

a. Span exceeds 26 feet in length. Check sources for availability of lumber in lengths greater than 20 feet.

TABLE 2308.10.3(2)
RAFTER SPANS FOR COMMON LUMBER SPECIES
(Roof Live Load = 20 pounds per square foot, Ceiling Not Attached to Rafters, $L/\Delta = 240$)

RAFTER SPACING (inches)	SPECIES AND GRADE		DEAD LOAD = 10 pounds per square foot					DEAD LOAD = 20 pounds per square foot				
			2 × 4	2 × 6	2 × 8	2 × 10	2 × 12	2 × 4	2 × 6	2 × 8	2 × 10	2 × 12
			Maximum rafter spans									
			(ft. - in.)	(ft. - in.)	(ft. - in.)	(ft. - in.)	(ft. - in.)	(ft. - in.)	(ft. - in.)	(ft. - in.)	(ft. - in.)	(ft. - in.)
12	Douglas Fir-Larch	SS	10-5	16-4	21-7	Note a	Note a	10-5	16-4	21-7	Note a	Note a
	Douglas Fir-Larch	#1	10-0	15-9	20-10	Note a	Note a	10-0	15-4	19-5	23-9	Note a
	Douglas Fir-Larch	#2	9-10	15-6	20-5	25-8	Note a	9-10	14-4	18-2	22-3	25-9
	Douglas Fir-Larch	#3	8-7	12-6	15-10	19-5	22-6	7-5	10-10	13-9	16-9	19-6
	Hem-Fir	SS	9-10	15-6	20-5	Note a	Note a	9-10	15-6	20-5	Note a	Note a
	Hem-Fir	#1	9-8	15-2	19-11	25-5	Note a	9-8	14-11	18-11	23-2	Note a
	Hem-Fir	#2	9-2	14-5	19-0	24-3	Note a	9-2	14-2	17-11	21-11	25-5
	Hem-Fir	#3	8-7	12-6	15-10	19-5	22-6	7-5	10-10	13-9	16-9	19-6
	Southern Pine	SS	10-3	16-1	21-2	Note a	Note a	10-3	16-1	21-2	Note a	Note a
	Southern Pine	#1	10-0	15-9	20-10	Note a	Note a	10-0	15-9	20-10	25-10	Note a
	Southern Pine	#2	9-10	15-6	20-5	Note a	Note a	9-10	15-1	19-5	23-2	Note a
	Southern Pine	#3	9-1	13-6	17-2	20-3	24-1	7-11	11-8	14-10	17-6	20-11
	Spruce-Pine-Fir	SS	9-8	15-2	19-11	25-5	Note a	9-8	15-2	19-11	25-5	Note a
	Spruce-Pine-Fir	#1	9-5	14-9	19-6	24-10	Note a	9-5	14-4	18-2	22-3	25-9
	Spruce-Pine-Fir	#2	9-5	14-9	19-6	24-10	Note a	9-5	14-4	18-2	22-3	25-9
	Spruce-Pine-Fir	#3	8-7	12-6	15-10	19-5	22-6	7-5	10-10	13-9	16-9	19-6
16	Douglas Fir-Larch	SS	9-6	14-11	19-7	25-0	Note a	9-6	14-11	19-7	24-9	Note a
	Douglas Fir-Larch	#1	9-1	14-4	18-11	23-9	Note a	9-1	13-3	16-10	20-7	23-10
	Douglas Fir-Larch	#2	8-11	14-1	18-2	22-3	25-9	8-6	12-5	15-9	19-3	22-4
	Douglas Fir-Larch	#3	7-5	10-10	13-9	16-9	19-6	6-5	9-5	11-11	14-6	16-10
	Hem-Fir	SS	8-11	14-1	18-6	23-8	Note a	8-11	14-1	18-6	23-8	Note a
	Hem-Fir	#1	8-9	13-9	18-1	23-1	Note a	8-9	12-11	16-5	20-0	23-3
	Hem-Fir	#2	8-4	13-1	17-3	21-11	25-5	8-4	12-3	15-6	18-11	22-0
	Hem-Fir	#3	7-5	10-10	13-9	16-9	19-6	6-5	9-5	11-11	14-6	16-10
	Southern Pine	SS	9-4	14-7	19-3	24-7	Note a	9-4	14-7	19-3	24-7	Note a
	Southern Pine	#1	9-1	14-4	18-11	24-1	Note a	9-1	14-4	18-10	22-4	Note a
	Southern Pine	#2	8-11	14-1	18-6	23-2	Note a	8-11	13-0	16-10	20-1	23-7
	Southern Pine	#3	7-11	11-8	14-10	17-6	20-11	6-10	10-1	12-10	15-2	18-1
	Spruce-Pine-Fir	SS	8-9	13-9	18-1	23-1	Note a	8-9	13-9	18-1	23-0	Note a
	Spruce-Pine-Fir	#1	8-7	13-5	17-9	22-3	25-9	8-6	12-5	15-9	19-3	22-4
	Spruce-Pine-Fir	#2	8-7	13-5	17-9	22-3	25-9	8-6	12-5	15-9	19-3	22-4
	Spruce-Pine-Fir	#3	7-5	10-10	13-9	16-9	19-6	6-5	9-5	11-11	14-6	16-10

(continued)

TABLE 2308.10.3(2)—continued
RAFTER SPANS FOR COMMON LUMBER SPECIES
(Roof Live Load = 20 pounds per square foot, Ceiling Not Attached to Rafters, $L/\Delta = 240$)

RAFTER SPACING (inches)	SPECIES AND GRADE		DEAD LOAD = 10 pounds per square foot					DEAD LOAD = 20 pounds per square foot				
			2 × 4	2 × 6	2 × 8	2 × 10	2 × 12	2 × 4	2 × 6	2 × 8	2 × 10	2 × 12
			\ Maximum rafter spans									
			(ft. - in.)	(ft. - in.)	(ft. - in.)	(ft. - in.)	(ft. - in.)	(ft. - in.)	(ft. - in.)	(ft. - in.)	(ft. - in.)	(ft. - in.)
19.2	Douglas Fir-Larch	SS	8-11	14-0	18-5	23-7	Note a	8-11	14-0	18-5	22-7	Note a
	Douglas Fir-Larch	#1	8-7	13-6	17-9	21-8	25-2	8-4	12-2	15-4	18-9	21-9
	Douglas Fir-Larch	#2	8-5	13-1	16-7	20-3	23-6	7-9	11-4	14-4	17-7	20-4
	Douglas Fir-Larch	#3	6-9	9-11	12-7	15-4	17-9	5-10	8-7	10-10	13-3	15-5
	Hem-Fir	SS	8-5	13-3	17-5	22-3	Note a	8-5	13-3	17-5	22-3	25-9
	Hem-Fir	#1	8-3	12-11	17-1	21-1	24-6	8-1	11-10	15-0	18-4	21-3
	Hem-Fir	#2	7-10	12-4	16-3	20-0	23-2	7-8	11-2	14-2	17-4	20-1
	Hem-Fir	#3	6-9	9-11	12-7	15-4	17-9	5-10	8-7	10-10	13-3	15-5
	Southern Pine	SS	8-9	13-9	18-1	23-1	Note a	8-9	13-9	18-1	23-1	Note a
	Southern Pine	#1	8-7	13-6	17-9	22-8	Note a	8-7	13-6	17-2	20-5	24-4
	Southern Pine	#2	8-5	13-3	17-5	21-2	24-10	8-4	11-11	15-4	18-4	21-6
	Southern Pine	#3	7-3	10-8	13-7	16-0	19-1	6-3	9-3	11-9	13-10	16-6
	Spruce-Pine-Fir	SS	8-3	12-11	17-1	21-9	Note a	8-3	12-11	17-1	21-0	24-4
	Spruce-Pine-Fir	#1	8-1	12-8	16-7	20-3	23-6	7-9	11-4	14-4	17-7	20-4
	Spruce-Pine-Fir	#2	8-1	12-8	16-7	20-3	23-6	7-9	11-4	14-4	17-7	20-4
	Spruce-Pine-Fir	#3	6-9	9-11	12-7	15-4	17-9	5-10	8-7	10-10	13-3	15-5
24	Douglas Fir-Larch	SS	8-3	13-0	17-2	21-10	Note a	8-3	13-0	16-7	20-3	23-5
	Douglas Fir-Larch	#1	8-0	12-6	15-10	19-5	22-6	7-5	10-10	13-9	16-9	19-6
	Douglas Fir-Larch	#2	7-10	11-9	14-10	18-2	21-0	6-11	10-2	12-10	15-8	18-3
	Douglas Fir-Larch	#3	6-1	8-10	11-3	13-8	15-11	5-3	7-8	9-9	11-10	13-9
	Hem-Fir	SS	7-10	12-3	16-2	20-8	25-1	7-10	12-3	16-2	19-10	23-0
	Hem-Fir	#1	7-8	12-0	15-6	18-11	21-11	7-3	10-7	13-5	16-4	19-0
	Hem-Fir	#2	7-3	11-5	14-8	17-10	20-9	6-10	10-0	12-8	15-6	17-11
	Hem-Fir	#3	6-1	8-10	11-3	13-8	15-11	5-3	7-8	9-9	11-10	13-9
	Southern Pine	SS	8-1	12-9	16-10	21-6	Note a	8-1	12-9	16-10	21-6	Note a
	Southern Pine	#1	8-0	12-6	16-6	21-1	25-2	8-0	12-3	15-4	18-3	21-9
	Southern Pine	#2	7-10	12-3	15-10	18-11	22-2	7-5	10-8	13-9	16-5	19-3
	Southern Pine	#3	6-5	9-6	12-1	14-4	17-1	5-7	8-3	10-6	12-5	14-9
	Spruce-Pine-Fir	SS	7-8	12-0	15-10	20-2	24-7	7-8	12-0	15-4	18-9	21-9
	Spruce-Pine-Fir	#1	7-6	11-9	14-10	18-2	21-0	6-11	10-2	12-10	15-8	18-3
	Spruce-Pine-Fir	#2	7-6	11-9	14-10	18-2	21-0	6-11	10-2	12-10	15-8	18-3
	Spruce-Pine-Fir	#3	6-1	8-10	11-3	13-8	15-11	5-3	7-8	9-9	11-10	13-9

For SI: 1 inch = 25.4 mm, 1 foot = 304.8 mm, 1 pound per square foot = 47.9 N/m².

a. Span exceeds 26 feet in length. Check sources for availability of lumber in lengths greater than 20 feet.

TABLE 2308.10.3(3)
RAFTER SPANS FOR COMMON LUMBER SPECIES
(Ground Snow Load = 30 pounds per square foot, Ceiling Not Attached to Rafters, L/Δ = 180)

RAFTER SPACING (inches)	SPECIES AND GRADE		DEAD LOAD = 10 pounds per square foot					DEAD LOAD = 20 pounds per square foot				
			Maximum rafter spans									
			2 × 4	2 × 6	2 × 8	2 × 10	2 × 12	2 × 4	2 × 6	2 × 8	2 × 10	2 × 12
			(ft. - in.)	(ft. - in.)	(ft. - in.)	(ft. - in.)	(ft. - in.)	(ft. - in.)	(ft. - in.)	(ft. - in.)	(ft. - in.)	(ft. - in.)
12	Douglas Fir-Larch	SS	10-0	15-9	20-9	Note a	Note a	10-0	15-9	20-1	24-6	Note a
	Douglas Fir-Larch	#1	9-8	14-9	18-8	22-9	Note a	9-0	13-2	16-8	20-4	23-7
	Douglas Fir-Larch	#2	9-5	13-9	17-5	21-4	24-8	8-5	12-4	15-7	19-1	22-1
	Douglas Fir-Larch	#3	7-1	10-5	13-2	16-1	18-8	6-4	9-4	11-9	14-5	16-8
	Hem-Fir	SS	9-6	14-10	19-7	25-0	Note a	9-6	14-10	19-7	24-1	Note a
	Hem-Fir	#1	9-3	14-4	18-2	22-2	25-9	8-9	12-10	16-3	19-10	23-0
	Hem-Fir	#2	8-10	13-7	17-2	21-0	24-4	8-4	12-2	15-4	18-9	21-9
	Hem-Fir	#3	7-1	10-5	13-2	16-1	18-8	6-4	9-4	11-9	14-5	16-8
	Southern Pine	SS	9-10	15-6	20-5	Note a	Note a	9-10	15-6	20-5	Note a	Note a
	Southern Pine	#1	9-8	15-2	20-0	24-9	Note a	9-8	14-10	18-8	22-2	Note a
	Southern Pine	#2	9-6	14-5	18-8	22-3	Note a	9-0	12-11	16-8	19-11	23-4
	Southern Pine	#3	7-7	11-2	14-3	16-10	20-0	6-9	10-0	12-9	15-1	17-11
	Spruce-Pine-Fir	SS	9-3	14-7	19-2	24-6	Note a	9-3	14-7	18-8	22-9	Note a
	Spruce-Pine-Fir	#1	9-1	13-9	17-5	21-4	24-8	8-5	12-4	15-7	19-1	22-1
	Spruce-Pine-Fir	#2	9-1	13-9	17-5	21-4	24-8	8-5	12-4	15-7	19-1	22-1
	Spruce-Pine-Fir	#3	7-1	10-5	13-2	16-1	18-8	6-4	9-4	11-9	14-5	16-8
16	Douglas Fir-Larch	SS	9-1	14-4	18-10	23-9	Note a	9-1	13-9	17-5	21-3	24-8
	Douglas Fir-Larch	#1	8-9	12-9	16-2	19-9	22-10	7-10	11-5	14-5	17-8	20-5
	Douglas Fir-Larch	#2	8-2	11-11	15-1	18-5	21-5	7-3	10-8	13-6	16-6	19-2
	Douglas Fir-Larch	#3	6-2	9-0	11-5	13-11	16-2	5-6	8-1	10-3	12-6	14-6
	Hem-Fir	SS	8-7	13-6	17-10	22-9	Note a	8-7	13-6	17-1	20-10	24-2
	Hem-Fir	#1	8-5	12-5	15-9	19-3	22-3	7-7	11-1	14-1	17-2	19-11
	Hem-Fir	#2	8-0	11-9	14-11	18-2	21-1	7-2	10-6	13-4	16-3	18-10
	Hem-Fir	#3	6-2	9-0	11-5	13-11	16-2	5-6	8-1	10-3	12-6	14-6
	Southern Pine	SS	8-11	14-1	18-6	23-8	Note a	8-11	14-1	18-6	23-8	Note a
	Southern Pine	#1	8-9	13-9	18-1	21-5	25-7	8-8	12-10	16-2	19-2	22-10
	Southern Pine	#2	8-7	12-6	16-2	19-3	22-7	7-10	11-2	14-5	17-3	20-2
	Southern Pine	#3	6-7	9-8	12-4	14-7	17-4	5-10	8-8	11-0	13-0	15-6
	Spruce-Pine-Fir	SS	8-5	13-3	17-5	22-1	25-7	8-5	12-9	16-2	19-9	22-10
	Spruce-Pine-Fir	#1	8-2	11-11	15-1	18-5	21-5	7-3	10-8	13-6	16-6	19-2
	Spruce-Pine-Fir	#2	8-2	11-11	15-1	18-5	21-5	7-3	10-8	13-6	16-6	19-2
	Spruce-Pine-Fir	#3	6-2	9-0	11-5	13-11	16-2	5-6	8-1	10-3	12-6	14-6

(continued)

TABLE 2308.10.3(3)—continued
RAFTER SPANS FOR COMMON LUMBER SPECIES
(Ground Snow Load = 30 pounds per square foot, Ceiling Not Attached to Rafters, $L/\Delta = 180$)

RAFTER SPACING (inches)	SPECIES AND GRADE		DEAD LOAD = 10 pounds per square foot					DEAD LOAD = 20 pounds per square foot				
			2×4	2×6	2×8	2×10	2×12	2×4	2×6	2×8	2×10	2×12
			Maximum rafter spans									
			(ft.-in.)	(ft.-in.)	(ft.-in.)	(ft.-in.)	(ft.-in.)	(ft.-in.)	(ft.-in.)	(ft.-in.)	(ft.-in.)	(ft.-in.)
19.2	Douglas Fir-Larch	SS	8-7	13-6	17-9	21-8	25-2	8-7	12-6	15-10	19-5	22-6
	Douglas Fir-Larch	#1	7-11	11-8	14-9	18-0	20-11	7-1	10-5	13-2	16-1	18-8
	Douglas Fir-Larch	#2	7-5	10-11	13-9	16-10	19-6	6-8	9-9	12-4	15-1	17-6
	Douglas Fir-Larch	#3	5-7	8-3	10-5	12-9	14-9	5-0	7-4	9-4	11-5	13-2
	Hem-Fir	SS	8-1	12-9	16-9	21-4	24-8	8-1	12-4	15-7	19-1	22-1
	Hem-Fir	#1	7-9	11-4	14-4	17-7	20-4	6-11	10-2	12-10	15-8	18-2
	Hem-Fir	#2	7-4	10-9	13-7	16-7	19-3	6-7	9-7	12-2	14-10	17-3
	Hem-Fir	#3	5-7	8-3	10-5	12-9	14-9	5-0	7-4	9-4	11-5	13-2
	Southern Pine	SS	8-5	13-3	17-5	22-3	Note a	8-5	13-3	17-5	22-0	25-9
	Southern Pine	#1	8-3	13-0	16-6	19-7	23-4	7-11	11-9	14-9	17-6	20-11
	Southern Pine	#2	7-11	11-5	14-9	17-7	20-7	7-1	10-2	13-2	15-9	18-5
	Southern Pine	#3	6-0	8-10	11-3	13-4	15-10	5-4	7-11	10-1	11-11	14-2
	Spruce-Pine-Fir	SS	7-11	12-5	16-5	20-2	23-4	7-11	11-8	14-9	18-0	20-11
	Spruce-Pine-Fir	#1	7-5	10-11	13-9	16-10	19-6	6-8	9-9	12-4	15-1	17-6
	Spruce-Pine-Fir	#2	7-5	10-11	13-9	16-10	19-6	6-8	9-9	12-4	15-1	17-6
	Spruce-Pine-Fir	#3	5-7	8-3	10-5	12-9	14-9	5-0	7-4	9-4	11-5	13-2
24	Douglas Fir-Larch	SS	7-11	12-6	15-10	19-5	22-6	7-8	11-3	14-2	17-4	20-1
	Douglas Fir-Larch	#1	7-1	10-5	13-2	16-1	18-8	6-4	9-4	11-9	14-5	16-8
	Douglas Fir-Larch	#2	6-8	9-9	12-4	15-1	17-6	5-11	8-8	11-0	13-6	15-7
	Douglas Fir-Larch	#3	5-0	7-4	9-4	11-5	13-2	4-6	6-7	8-4	10-2	11-10
	Hem-Fir	SS	7-6	11-10	15-7	19-1	22-1	7-6	11-0	13-11	17-0	19-9
	Hem-Fir	#1	6-11	10-2	12-10	15-8	18-0	6-2	9-1	11-6	14-0	16-3
	Hem-Fir	#2	6-7	9-7	12-2	14-10	17-3	5-10	8-7	10-10	13-3	15-5
	Hem-Fir	#3	5-0	7-4	9-4	11-5	13-2	4-6	6-7	8-4	10-2	11-10
	Southern Pine	SS	7-10	12-3	16-2	20-8	25-1	7-10	12-3	16-2	19-8	23-0
	Southern Pine	#1	7-8	11-9	14-9	17-6	20-11	7-1	10-6	13-2	15-8	18-8
	Southern Pine	#2	7-1	10-2	13-2	15-9	18-5	6-4	9-2	11-9	14-1	16-6
	Southern Pine	#3	5-4	7-11	10-1	11-11	14-2	4-9	7-1	9-0	10-8	12-8
	Spruce-Pine-Fir	SS	7-4	11-7	14-9	18-0	20-11	7-1	10-5	13-2	16-1	18-8
	Spruce-Pine-Fir	#1	6-8	9-9	12-4	15-1	17-6	5-11	8-8	11-0	13-6	15-7
	Spruce-Pine-Fir	#2	6-8	9-9	12-4	15-1	17-6	5-11	8-8	11-0	13-6	15-7
	Spruce-Pine-Fir	#3	5-0	7-4	9-4	11-5	13-2	4-6	6-7	8-4	10-2	11-10

For SI: 1 inch = 25.4 mm, 1 foot = 304.8 mm, 1 pound per square foot = 47.9 N/m².

a. Span exceeds 26 feet in length. Check sources for availability of lumber in lengths greater than 20 feet.

TABLE 2308.10.3(4)
RAFTER SPANS FOR COMMON LUMBER SPECIES
(Ground Snow Load = 50 pounds per square foot, Ceiling Not Attached to Rafters, $L/\Delta = 180$)

RAFTER SPACING (inches)	SPECIES AND GRADE		DEAD LOAD = 10 pounds per square foot					DEAD LOAD = 20 pounds per square foot				
			2 × 4	2 × 6	2 × 8	2 × 10	2 × 12	2 × 4	2 × 6	2 × 8	2 × 10	2 × 12
			Maximum rafter spans									
			(ft. - in.)	(ft. - in.)	(ft. - in.)	(ft. - in.)	(ft. - in.)	(ft. - in.)	(ft. - in.)	(ft. - in.)	(ft. - in.)	(ft. - in.)
12	Douglas Fir-Larch	SS	8-5	13-3	17-6	22-4	26-0	8-5	13-3	17-0	20-9	24-10
	Douglas Fir-Larch	#1	8-2	12-0	15-3	18-7	21-7	7-7	11-2	14-1	17-3	20-0
	Douglas Fir-Larch	#2	7-8	11-3	14-3	17-5	20-2	7-1	10-5	13-2	16-1	18-8
	Douglas Fir-Larch	#3	5-10	8-6	10-9	13-2	15-3	5-5	7-10	10-0	12-2	14-1
	Hem-Fir	SS	8-0	12-6	16-6	21-1	25-6	8-0	12-6	16-6	20-4	23-7
	Hem-Fir	#1	7-10	11-9	14-10	18-1	21-0	7-5	10-10	13-9	16-9	19-5
	Hem-Fir	#2	7-5	11-1	14-0	17-2	19-11	7-0	10-3	13-0	15-10	18-5
	Hem-Fir	#3	5-10	8-6	10-9	13-2	15-3	5-5	7-10	10-0	12-2	14-1
	Southern Pine	SS	8-4	13-0	17-2	21-11	Note a	8-4	13-0	17-2	21-11	Note a
	Southern Pine	#1	8-2	12-10	16-10	20-3	24-1	8-2	12-6	15-9	18-9	22-4
	Southern Pine	#2	8-0	11-9	15-3	18-2	21-3	7-7	10-11	14-1	16-10	19-9
	Southern Pine	#3	6-2	9-2	11-8	13-9	16-4	5-9	8-5	10-9	12-9	15-2
	Spruce-Pine-Fir	SS	7-10	12-3	16-2	20-8	24-1	7-10	12-3	15-9	19-3	22-4
	Spruce-Pine-Fir	#1	7-8	11-3	14-3	17-5	20-2	7-1	10-5	13-2	16-1	18-8
	Spruce-Pine-Fir	#2	7-8	11-3	14-3	17-5	20-2	7-1	10-5	13-2	16-1	18-8
	Spruce-Pine-Fir	#3	5-10	8-6	10-9	13-2	15-3	5-5	7-10	10-0	12-2	14-1
16	Douglas Fir-Larch	SS	7-8	12-1	15-10	19-5	22-6	7-8	11-7	14-8	17-11	20-10
	Douglas Fir-Larch	#1	7-1	10-5	13-2	16-1	18-8	6-7	9-8	12-2	14-11	17-3
	Douglas Fir-Larch	#2	6-8	9-9	12-4	15-1	17-6	6-2	9-0	11-5	13-11	16-2
	Douglas Fir-Larch	#3	5-0	7-4	9-4	11-5	13-2	4-8	6-10	8-8	10-6	12-3
	Hem-Fir	SS	7-3	11-5	15-0	19-1	22-1	7-3	11-5	14-5	17-8	20-5
	Hem-Fir	#1	6-11	10-2	12-10	15-8	18-2	6-5	9-5	11-11	14-6	16-10
	Hem-Fir	#2	6-7	9-7	12-2	14-10	17-3	6-1	8-11	11-3	13-9	15-11
	Hem-Fir	#3	5-0	7-4	9-4	11-5	13-2	4-8	6-10	8-8	10-6	12-3
	Southern Pine	SS	7-6	11-10	15-7	19-11	24-3	7-6	11-10	15-7	19-11	23-10
	Southern Pine	#1	7-5	11-7	14-9	17-6	20-11	7-4	10-10	13-8	16-2	19-4
	Southern Pine	#2	7-1	10-2	13-2	15-9	18-5	6-7	9-5	12-2	14-7	17-1
	Southern Pine	#3	5-4	7-11	10-1	11-11	14-2	4-11	7-4	9-4	11-0	13-1
	Spruce-Pine-Fir	SS	7-1	11-2	14-8	18-0	20-11	7-1	10-9	13-8	16-8	19-4
	Spruce-Pine-Fir	#1	6-8	9-9	12-4	15-1	17-6	6-2	9-0	11-5	13-11	16-2
	Spruce-Pine-Fir	#2	6-8	9-9	12-4	15-1	17-6	6-2	9-0	11-5	13-11	16-2
	Spruce-Pine-Fir	#3	5-0	7-4	9-4	11-5	13-2	4-8	6-10	8-8	10-6	12-3

(continued)

TABLE 2308.10.3(4)—continued
RAFTER SPANS FOR COMMON LUMBER SPECIES
(Ground Snow Load = 50 pounds per square foot, Ceiling Not Attached to Rafters, L/Δ = 180)

RAFTER SPACING (inches)	SPECIES AND GRADE		DEAD LOAD = 10 pounds per square foot					DEAD LOAD = 20 pounds per square foot				
			Maximum rafter spans									
			2 × 4	2 × 6	2 × 8	2 × 10	2 × 12	2 × 4	2 × 6	2 × 8	2 × 10	2 × 12
			(ft. - in.)	(ft. - in.)	(ft. - in.)	(ft. - in.)	(ft. - in.)	(ft. - in.)	(ft. - in.)	(ft. - in.)	(ft. - in.)	(ft. - in.)
19.2	Douglas Fir-Larch	SS	7-3	11-4	14-6	17-8	20-6	7-3	10-7	13-5	16-5	19-0
	Douglas Fir-Larch	#1	6-6	9-6	12-0	14-8	17-1	6-0	8-10	11-2	13-7	15-9
	Douglas Fir-Larch	#2	6-1	8-11	11-3	13-9	15-11	5-7	8-3	10-5	12-9	14-9
	Douglas Fir-Larch	#3	4-7	6-9	8-6	10-5	12-1	4-3	6-3	7-11	9-7	11-2
	Hem-Fir	SS	6-10	10-9	14-2	17-5	20-2	6-10	10-5	13-2	16-1	18-8
	Hem-Fir	#1	6-4	9-3	11-9	14-4	16-7	5-10	8-7	10-10	13-3	15-5
	Hem-Fir	#2	6-0	8-9	11-1	13-7	15-9	5-7	8-1	10-3	12-7	14-7
	Hem-Fir	#3	4-7	6-9	8-6	10-5	12-1	4-3	6-3	7-11	9-7	11-2
	Southern Pine	SS	7-1	11-2	14-8	18-9	22-10	7-1	11-2	14-8	18-7	21-9
	Southern Pine	#1	7-0	10-8	13-5	16-0	19-1	6-8	9-11	12-5	14-10	17-8
	Southern Pine	#2	6-6	9-4	12-0	14-4	16-10	6-0	8-8	11-2	13-4	15-7
	Southern Pine	#3	4-11	7-3	9-2	10-10	12-11	4-6	6-8	8-6	10-1	12-0
	Spruce-Pine-Fir	SS	6-8	10-6	13-5	16-5	19-1	6-8	9-10	12-5	15-3	17-8
	Spruce-Pine-Fir	#1	6-1	8-11	11-3	13-9	15-11	5-7	8-3	10-5	12-9	14-9
	Spruce-Pine-Fir	#2	6-1	8-11	11-3	13-9	15-11	5-7	8-3	10-5	12-9	14-9
	Spruce-Pine-Fir	#3	4-7	6-9	8-6	10-5	12-1	4-3	6-3	7-11	9-7	11-2
24	Douglas Fir-Larch	SS	6-8	10-3	13-0	15-10	18-4	6-6	9-6	12-0	14-8	17-0
	Douglas Fir-Larch	#1	5-10	8-6	10-9	13-2	15-3	5-5	7-10	10-0	12-2	14-1
	Douglas Fir-Larch	#2	5-5	7-11	10-1	12-4	14-3	5-0	7-4	9-4	11-5	13-2
	Douglas Fir-Larch	#3	4-1	6-0	7-7	9-4	10-9	3-10	5-7	7-1	8-7	10-0
	Hem-Fir	SS	6-4	9-11	12-9	15-7	18-0	6-4	9-4	11-9	14-5	16-8
	Hem-Fir	#1	5-8	8-3	10-6	12-10	14-10	5-3	7-8	9-9	11-10	13-9
	Hem-Fir	#2	5-4	7-10	9-11	12-1	14-1	4-11	7-3	9-2	11-3	13-0
	Hem-Fir	#3	4-1	6-0	7-7	9-4	10-9	3-10	5-7	7-1	8-7	10-0
	Southern Pine	SS	6-7	10-4	13-8	17-5	21-0	6-7	10-4	13-8	16-7	19-5
	Southern Pine	#1	6-5	9-7	12-0	14-4	17-1	6-0	8-10	11-2	13-3	15-9
	Southern Pine	#2	5-10	8-4	10-9	12-10	15-1	5-5	7-9	10-0	11-11	13-11
	Southern Pine	#3	4-4	6-5	8-3	9-9	11-7	4-1	6-0	7-7	9-0	10-8
	Spruce-Pine-Fir	SS	6-2	9-6	12-0	14-8	17-1	6-0	8-10	11-2	13-7	15-9
	Spruce-Pine-Fir	#1	5-5	7-11	10-1	12-4	14-3	5-0	7-4	9-4	11-5	13-2
	Spruce-Pine-Fir	#2	5-5	7-11	10-1	12-4	14-3	5-0	7-4	9-4	11-5	13-2
	Spruce-Pine-Fir	#3	4-1	6-0	7-7	9-4	10-9	3-10	5-7	7-1	8-7	10-0

For SI: 1 inch = 25.4 mm, 1 foot = 304.8 mm, 1 pound per square foot = 47.9 N/m².

a. Span exceeds 26 feet in length. Check sources for availability of lumber in lengths greater than 20 feet.

TABLE 2308.10.3(5)
RAFTER SPANS FOR COMMON LUMBER SPECIES
(Ground Snow Load = 30 pounds per square foot, Ceiling Attached to Rafters, $L/\Delta = 240$)

RAFTER SPACING (inches)	SPECIES AND GRADE		DEAD LOAD = 10 pounds per square foot					DEAD LOAD = 20 pounds per square foot				
			2 × 4	2 × 6	2 × 8	2 × 10	2 × 12	2 × 4	2 × 6	2 × 8	2 × 10	2 × 12
			\(Maximum rafter spans\)									
			(ft. - in.)	(ft. - in.)	(ft. - in.)	(ft. - in.)	(ft. - in.)	(ft. - in.)	(ft. - in.)	(ft. - in.)	(ft. - in.)	(ft. - in.)
12	Douglas Fir-Larch	SS	9-1	14-4	18-10	24-1	Note a	9-1	14-4	18-10	24-1	Note a
	Douglas Fir-Larch	#1	8-9	13-9	18-2	22-9	Note a	8-9	13-2	16-8	20-4	23-7
	Douglas Fir-Larch	#2	8-7	13-6	17-5	21-4	24-8	8-5	12-4	15-7	19-1	22-1
	Douglas Fir-Larch	#3	7-1	10-5	13-2	16-1	18-8	6-4	9-4	11-9	14-5	16-8
	Hem-Fir	SS	8-7	13-6	17-10	22-9	Note a	8-7	13-6	17-10	22-9	Note a
	Hem-Fir	#1	8-5	13-3	17-5	22-2	Note a	8-5	12-10	16-3	19-10	23-0
	Hem-Fir	#2	8-0	12-7	16-7	21-0	24-4	8-0	12-2	15-4	18-9	21-9
	Hem-Fir	#3	7-1	10-5	13-2	16-1	18-8	6-4	9-4	11-9	14-5	16-8
	Southern Pine	SS	8-11	14-1	18-6	23-8	Note a	8-11	14-1	18-6	23-8	Note a
	Southern Pine	#1	8-9	13-9	18-2	23-2	Note a	8-9	13-9	18-2	22-2	Note a
	Southern Pine	#2	8-7	13-6	17-10	22-3	24-8	8-7	12-11	16-8	19-11	23-4
	Southern Pine	#3	7-7	11-2	14-3	16-10	20-0	6-9	10-0	12-9	15-1	17-11
	Spruce-Pine-Fir	SS	8-5	13-3	17-5	22-3	Note a	8-5	13-3	17-5	22-3	Note a
	Spruce-Pine-Fir	#1	8-3	12-11	17-0	21-4	24-8	8-3	12-4	15-7	19-1	22-1
	Spruce-Pine-Fir	#2	8-3	12-11	17-0	21-4	24-8	8-3	12-4	15-7	19-1	22-1
	Spruce-Pine-Fir	#3	7-1	10-5	13-2	16-1	18-8	6-4	9-4	11-9	14-5	16-8
16	Douglas Fir-Larch	SS	8-3	13-0	17-2	21-10	Note a	8-3	13-0	17-2	21-3	24-8
	Douglas Fir-Larch	#1	8-0	12-6	16-2	19-9	22-10	7-10	11-5	14-5	17-8	20-5
	Douglas Fir-Larch	#2	7-10	11-11	15-1	18-5	21-5	7-3	10-8	13-6	16-6	19-2
	Douglas Fir-Larch	#3	6-2	9-0	11-5	13-11	16-2	5-6	8-1	10-3	12-6	14-6
	Hem-Fir	SS	7-10	12-3	16-2	20-8	25-1	7-10	12-3	16-2	20-8	24-2
	Hem-Fir	#1	7-8	12-0	15-9	19-3	22-3	7-7	11-1	14-1	17-2	19-11
	Hem-Fir	#2	7-3	11-5	14-11	18-2	21-1	7-2	10-6	13-4	16-3	18-10
	Hem-Fir	#3	6-2	9-0	11-5	13-11	16-2	5-6	8-1	10-3	12-6	14-6
	Southern Pine	SS	8-1	12-9	16-10	21-6	Note a	8-1	12-9	16-10	21-6	Note a
	Southern Pine	#1	8-0	12-6	16-6	21-1	25-7	8-0	12-6	16-2	19-2	22-10
	Southern Pine	#2	7-10	12-3	16-2	19-3	22-7	7-10	11-2	14-5	17-3	20-2
	Southern Pine	#3	6-7	9-8	12-4	14-7	17-4	5-10	8-8	11-0	13-0	15-6
	Spruce-Pine-Fir	SS	7-8	12-0	15-10	20-2	24-7	7-8	12-0	15-10	19-9	22-10
	Spruce-Pine-Fir	#1	7-6	11-9	15-1	18-5	21-5	7-3	10-8	13-6	16-6	19-2
	Spruce-Pine-Fi	#2	7-6	11-9	15-1	18-5	21-5	7-3	10-8	13-6	16-6	19-2
	Spruce-Pine-Fi	#3	6-2	9-0	11-5	13-11	16-2	5-6	8-1	10-3	12-6	14-6

(continued)

TABLE 2308.10.3(5)—continued
RAFTER SPANS FOR COMMON LUMBER SPECIES
(Ground Snow Load = 30 pounds per square foot, Ceiling Attached to Rafters, L/Δ = 240)

RAFTER SPACING (inches)	SPECIES AND GRADE		DEAD LOAD = 10 pounds per square foot					DEAD LOAD = 20 pounds per square foot				
			Maximum rafter spans									
			2 × 4	2 × 6	2 × 8	2 × 10	2 × 12	2 × 4	2 × 6	2 × 8	2 × 10	2 × 12
			(ft. - in.)	(ft. - in.)	(ft. - in.)	(ft. - in.)	(ft. - in.)	(ft. - in.)	(ft. - in.)	(ft. - in.)	(ft. - in.)	(ft. - in.)
19.2	Douglas Fir-Larch	SS	7-9	12-3	16-1	20-7	25-0	7-9	12-3	15-10	19-5	22-6
	Douglas Fir-Larch	#1	7-6	11-8	14-9	18-0	20-11	7-1	10-5	13-2	16-1	18-8
	Douglas Fir-Larch	#2	7-4	10-11	13-9	16-10	19-6	6-8	9-9	12-4	15-1	17-6
	Douglas Fir-Larch	#3	5-7	8-3	10-5	12-9	14-9	5-0	7-4	9-4	11-5	13-2
	Hem-Fir	SS	7-4	11-7	15-3	19-5	23-7	7-4	11-7	15-3	19-1	22-1
	Hem-Fir	#1	7-2	11-4	14-4	17-7	20-4	6-11	10-2	12-10	15-8	18-2
	Hem-Fir	#2	6-10	10-9	13-7	16-7	19-3	6-7	9-7	12-2	14-10	17-3
	Hem-Fir	#3	5-7	8-3	10-5	12-9	14-9	5-0	7-4	9-4	11-5	13-2
	Southern Pine	SS	7-8	12-0	15-10	20-2	24-7	7-8	12-0	15-10	20-2	24-7
	Southern Pine	#1	7-6	11-9	15-6	19-7	23-4	7-6	11-9	14-9	17-6	20-11
	Southern Pine	#2	7-4	11-5	14-9	17-7	20-7	7-1	10-2	13-2	15-9	18-5
	Southern Pine	#3	6-0	8-10	11-3	13-4	15-10	5-4	7-11	10-1	11-11	14-2
	Spruce-Pine-Fir	SS	7-2	11-4	14-11	19-0	23-1	7-2	11-4	14-9	18-0	20-11
	Spruce-Pine-Fir	#1	7-0	10-11	13-9	16-10	19-6	6-8	9-9	12-4	15-1	17-6
	Spruce-Pine-Fir	#2	7-0	10-11	13-9	16-10	19-6	6-8	9-9	12-4	15-1	17-6
	Spruce-Pine-Fir	#3	5-7	8-3	10-5	12-9	14-9	5-0	7-4	9-4	11-5	13-2
24	Douglas Fir-Larch	SS	7-3	11-4	15-0	19-1	22-6	7-3	11-3	14-2	17-4	20-1
	Douglas Fir-Larch	#1	7-0	10-5	13-2	16-1	18-8	6-4	9-4	11-9	14-5	16-8
	Douglas Fir-Larch	#2	6-8	9-9	12-4	15-1	17-6	5-11	8-8	11-0	13-6	15-7
	Douglas Fir-Larch	#3	5-0	7-4	9-4	11-5	13-2	4-6	6-7	8-4	10-2	11-10
	Hem-Fir	SS	6-10	10-9	14-2	18-0	21-11	6-10	10-9	13-11	17-0	19-9
	Hem-Fir	#1	6-8	10-2	12-10	15-8	18-2	6-2	9-1	11-6	14-0	16-3
	Hem-Fir	#2	6-4	9-7	12-2	14-10	17-3	5-10	8-7	10-10	13-3	15-5
	Hem-Fir	#3	5-0	7-4	9-4	11-5	13-2	4-6	6-7	8-4	10-2	11-10
	Southern Pine	SS	7-1	11-2	14-8	18-9	22-10	7-1	11-2	14-8	18-9	22-10
	Southern Pine	#1	7-0	10-11	14-5	17-6	20-11	7-0	10-6	13-2	15-8	18-8
	Southern Pine	#2	6-10	10-2	13-2	15-9	18-5	6-4	9-2	11-9	14-1	16-6
	Southern Pine	#3	5-4	7-11	10-1	11-11	14-2	4-9	7-1	9-0	10-8	12-8
	Spruce-Pine-Fir	SS	6-8	10-6	13-10	17-8	20-11	6-8	10-5	13-2	16-1	18-8
	Spruce-Pine-Fir	#1	6-6	9-9	12-4	15-1	17-6	5-11	8-8	11-0	13-6	15-7
	Spruce-Pine-Fir	#2	6-6	9-9	12-4	15-1	17-6	5-11	8-8	11-0	13-6	15-7
	Spruce-Pine-Fir	#3	5-0	7-4	9-4	11-5	13-2	4-6	6-7	8-4	10-2	11-10

For SI: 1 inch = 25.4 mm, 1 foot = 304.8 mm, 1 pound per square foot = 47.9 N/m^2.

a. Span exceeds 26 feet in length. Check sources for availability of lumber in lengths greater than 20 feet.

TABLE 2308.10.3(6)
RAFTER SPANS FOR COMMON LUMBER SPECIES
(Ground Snow Load = 50 pounds per square foot, Ceiling Attached to Rafters, L/Δ = 240)

RAFTER SPACING (inches)	SPECIES AND GRADE		DEAD LOAD = 10 pounds per square foot					DEAD LOAD = 20 pounds per square foot				
			2 × 4	2 × 6	2 × 8	2 × 10	2 × 12	2 × 4	2 × 6	2 × 8	2 × 10	2 × 12
			Maximum rafter spans									
			(ft. - in.)	(ft. - in.)	(ft. - in.)	(ft. - in.)	(ft. - in.)	(ft. - in.)	(ft. - in.)	(ft. - in.)	(ft. - in.)	(ft. - in.)
12	Douglas Fir-Larch	SS	7-8	12-1	15-11	20-3	24-8	7-8	12-1	15-11	20-3	24-0
	Douglas Fir-Larch	#1	7-5	11-7	15-3	18-7	21-7	7-5	11-2	14-1	17-3	20-0
	Douglas Fir-Larch	#2	7-3	11-3	14-3	17-5	20-2	7-1	10-5	13-2	16-1	18-8
	Douglas Fir-Larch	#3	5-10	8-6	10-9	13-2	15-3	5-5	7-10	10-0	12-2	14-1
	Hem-Fir	SS	7-3	11-5	15-0	19-2	23-4	7-3	11-5	15-0	19-2	23-4
	Hem-Fir	#1	7-1	11-2	14-8	18-1	21-0	7-1	10-10	13-9	16-9	19-5
	Hem-Fir	#2	6-9	10-8	14-0	17-2	19-11	6-9	10-3	13-0	15-10	18-5
	Hem-Fir	#3	5-10	8-6	10-9	13-2	15-3	5-5	7-10	10-0	12-2	14-1
	Southern Pine	SS	7-6	11-0	15-7	19-11	24-3	7-6	11-10	15-7	19-11	24-3
	Southern Pine	#1	7-5	11-7	15-4	19-7	23-9	7-5	11-7	15-4	18-9	22-4
	Southern Pine	#2	7-3	11-5	15-0	18-2	21-3	7-3	10-11	14-1	16-10	19-9
	Southern Pine	#3	6-2	9-2	11-8	13-9	16-4	5-9	8-5	10-9	12-9	15-2
	Spruce-Pine-Fir	SS	7-1	11-2	14-8	18-9	22-10	7-1	11-2	14-8	18-9	22-4
	Spruce-Pine-Fir	#1	6-11	10-11	14-3	17-5	20-2	6-11	10-5	13-2	16-1	18-8
	Spruce-Pine-Fir	#2	6-11	10-11	14-3	17-5	20-2	6-11	10-5	13-2	16-1	18-8
	Spruce-Pine-Fir	#3	5-10	8-6	10-9	13-2	15-3	5-5	7-10	10-0	12-2	14-1
16	Douglas Fir-Larch	SS	7-0	11-0	14-5	18-5	22-5	7-0	11-0	14-5	17-11	20-10
	Douglas Fir-Larch	#1	6-9	10-5	13-2	16-1	18-8	6-7	9-8	12-2	14-11	17-3
	Douglas Fir-Larch	#2	6-7	9-9	12-4	15-1	17-6	6-2	9-0	11-5	13-11	16-2
	Douglas Fir-Larch	#3	5-0	7-4	9-4	11-5	13-2	4-8	6-10	8-8	10-6	12-3
	Hem-Fir	SS	6-7	10-4	13-8	17-5	21-2	6-7	10-4	13-8	17-5	20-5
	Hem-Fir	#1	6-5	10-2	12-10	15-8	18-2	6-5	9-5	11-11	14-6	16-10
	Hem-Fir	#2	6-2	9-7	12-2	14-10	17-3	6-1	8-11	11-3	13-9	15-11
	Hem-Fir	#3	5-0	7-4	9-4	11-5	13-2	4-8	6-10	8-8	10-6	12-3
	Southern Pine	SS	6-10	10-9	14-2	18-1	22-0	6-10	10-9	14-2	18-1	22-0
	Southern Pine	#1	6-9	10-7	13-11	17-6	20-11	6-9	10-7	13-8	16-2	19-4
	Southern Pine	#2	6-7	10-2	13-2	15-9	18-5	6-7	9-5	12-2	14-7	17-1
	Southern Pine	#3	5-4	7-11	10-1	11-11	14-2	4-11	7-4	9-4	11-0	13-1
	Spruce-Pine-Fir	SS	6-5	10-2	13-4	17-0	20-9	6-5	10-2	13-4	16-8	19-4
	Spruce-Pine-Fir	#1	6-4	9-9	12-4	15-1	17-6	6-2	9-0	11-5	13-11	16-2
	Spruce-Pine-Fir	#2	6-4	9-9	12-4	15-1	17-6	6-2	9-0	11-5	13-11	16-2
	Spruce-Pine-Fir	#3	5-0	7-4	9-4	11-5	13-2	4-8	6-10	8-8	10-6	12-3

(continued)

TABLE 2308.10.3(6)—continued
RAFTER SPANS FOR COMMON LUMBER SPECIES
(Ground Snow Load = 50 pounds per square foot, Ceiling Attached to Rafters, $L/\Delta = 240$)

RAFTER SPACING (inches)	SPECIES AND GRADE		DEAD LOAD = 10 pounds per square foot					DEAD LOAD = 20 pounds per square foot				
			Maximum rafter spans									
			2 × 4	2 × 6	2 × 8	2 × 10	2 × 12	2 × 4	2 × 6	2 × 8	2 × 10	2 × 12
			(ft. - in.)	(ft. - in.)	(ft. - in.)	(ft. - in.)	(ft. - in.)	(ft. - in.)	(ft. - in.)	(ft. - in.)	(ft. - in.)	(ft. - in.)
19.2	Douglas Fir-Larch	SS	6-7	10-4	13-7	17-4	20-6	6-7	10-4	13-5	16-5	19-0
	Douglas Fir-Larch	#1	6-4	9-6	12-0	14-8	17-1	6-0	8-10	11-2	13-7	15-9
	Douglas Fir-Larch	#2	6-1	8-11	11-3	13-9	15-11	5-7	8-3	10-5	12-9	14-9
	Douglas Fir-Larch	#3	4-7	6-9	8-6	10-5	12-1	4-3	6-3	7-11	9-7	11-2
	Hem-Fir	SS	6-2	9-9	12-10	16-5	19-11	6-2	9-9	12-10	16-1	18-8
	Hem-Fir	#1	6-1	9-3	11-9	14-4	16-7	5-10	8-7	10-10	13-3	15-5
	Hem-Fir	#2	5-9	8-9	11-1	13-7	15-9	5-7	8-1	10-3	12-7	14-7
	Hem-Fir	#3	4-7	6-9	8-6	10-5	12-1	4-3	6-3	7-11	9-7	11-2
	Southern Pine	SS	6-5	10-2	13-4	17-0	20-9	6-5	10-2	13-4	17-0	20-9
	Southern Pine	#1	6-4	9-11	13-1	16-0	19-1	6-4	9-11	12-5	14-10	17-8
	Southern Pine	#2	6-2	9-4	12-0	14-4	16-10	6-0	8-8	11-2	13-4	15-7
	Southern Pine	#3	4-11	7-3	9-2	10-10	12-11	4-6	6-8	8-6	10-1	12-0
	Spruce-Pine-Fir	SS	6-1	9-6	12-7	16-0	19-1	6-1	9-6	12-5	15-3	17-8
	Spruce-Pine-Fir	#1	5-11	8-11	11-3	13-9	15-11	5-7	8-3	10-5	12-9	14-9
	Spruce-Pine-Fir	#2	5-11	8-11	11-3	13-9	15-11	5-7	8-3	10-5	12-9	14-9
	Spruce-Pine-Fir	#3	4-7	6-9	8-6	10-5	12-1	4-3	6-3	7-11	9-7	11-2
24	Douglas Fir-Larch	SS	6-1	9-7	12-7	15-10	18-4	6-1	9-6	12-0	14-8	17-0
	Douglas Fir-Larch	#1	5-10	8-6	10-9	13-2	15-3	5-5	7-10	10-0	12-2	14-1
	Douglas Fir-Larch	#2	5-5	7-11	10-1	12-4	14-3	5-0	7-4	9-4	11-5	13-2
	Douglas Fir-Larch	#3	4-1	6-0	7-7	9-4	10-9	3-10	5-7	7-1	8-7	10-0
	Hem-Fir	SS	5-9	9-1	11-11	15-2	18-0	5-9	9-1	11-9	14-5	16-8
	Hem-Fir	#1	5-8	8-3	10-6	12-10	14-10	5-3	7-8	9-9	11-10	13-9
	Hem-Fir	#2	5-4	7-10	9-11	12-1	14-1	4-11	7-3	9-2	11-3	13-0
	Hem-Fir	#3	4-1	6-0	7-7	9-4	10-9	3-10	5-7	7-1	8-7	10-0
	Southern Pine	SS	6-0	9-5	12-5	15-10	19-3	6-0	9-5	12-5	15-10	19-3
	Southern Pine	#1	5-10	9-3	12-0	14-4	17-1	5-10	8-10	11-2	13-3	15-9
	Southern Pine	#2	5-9	8-4	10-9	12-10	15-1	5-5	7-9	10-0	11-11	13-11
	Southern Pine	#3	4-4	6-5	8-3	9-9	11-7	4-1	6-0	7-7	9-0	10-8
	Spruce-Pine-Fir	SS	5-8	8-10	11-8	14-8	17-1	5-8	8-10	11-2	13-7	15-9
	Spruce-Pine-Fir	#1	5-5	7-11	10-1	12-4	14-3	5-0	7-4	9-4	11-5	13-2
	Spruce-Pine-Fir	#2	5-5	7-11	10-1	12-4	14-3	5-0	7-4	9-4	11-5	13-2
	Spruce-Pine-Fir	#3	4-1	6-0	7-7	9-4	10-9	3-10	5-7	7-1	8-7	10-0

For SI: 1 inch = 25.4 mm, 1 foot = 304.8 mm, 1 pound per square foot = 47.9 N/m².

TABLE 2308.10.4.1
RAFTER TIE CONNECTIONS[g]

RAFTER SLOPE	TIE SPACING (inches)	NO SNOW LOAD				GROUND SNOW LOAD (pound per square foot)							
						30 pounds per square foot				50 pounds per square foot			
						Roof span (feet)							
		12	20	28	36	12	20	28	36	12	20	28	36
		Required number of 16d common ($3^1/_2''$ x 0.162'') nails[a,b] per connection[c,d,e,f]											
3:12	12	4	6	8	10	4	6	8	11	5	8	12	15
	16	5	7	10	13	5	8	11	14	6	11	15	20
	24	7	11	15	19	7	11	16	21	9	16	23	30
	32	10	14	19	25	10	16	22	28	12	27	30	40
	48	14	21	29	37	14	32	36	42	18	32	46	60
4:12	12	3	4	5	6	3	5	6	8	4	6	9	11
	16	3	5	7	8	4	6	8	11	5	8	12	15
	24	4	7	10	12	5	9	12	16	7	12	17	22
	32	6	9	13	16	8	12	16	22	10	16	24	30
	48	8	14	19	24	10	18	24	32	14	24	34	44
5:12	12	3	3	4	5	3	4	5	7	3	5	7	9
	16	3	4	5	7	3	5	7	9	4	7	9	12
	24	4	6	8	10	4	7	10	13	6	10	14	18
	32	5	8	10	13	6	10	14	18	8	14	18	24
	48	7	11	15	20	8	14	20	26	12	20	28	36
7:12	12	3	3	3	4	3	3	4	5	3	4	5	7
	16	3	3	4	5	3	4	5	6	3	5	7	9
	24	3	4	6	7	3	5	7	9	4	7	10	13
	32	4	6	8	10	4	8	10	12	6	10	14	18
	48	5	8	11	14	6	10	14	18	9	14	20	26
9:12	12	3	3	3	3	3	3	3	4	3	3	4	5
	16	3	3	3	4	3	3	4	5	3	4	5	7
	24	3	3	5	6	3	4	6	7	3	6	8	10
	32	3	4	6	8	4	6	8	10	5	8	10	14
	48	4	6	9	11	5	8	12	14	7	12	16	20
12:12	12	3	3	3	3	3	3	3	3	3	3	3	4
	16	3	3	3	3	3	3	3	4	3	3	4	5
	24	3	3	3	4	3	3	4	6	3	4	6	8
	32	3	3	4	5	3	5	6	8	4	6	8	10
	48	3	4	6	7	4	7	8	12	6	8	12	16

For SI: 1 inch = 25.4 mm, 1 foot = 304.8 mm, 1 pound per square foot = 47.8 N/m².

a. 40d box (5'' × 0.162'') or 16d sinker ($3^1/_4'' × 0.148''$) nails are permitted to be substituted for 16d common ($3^1/_2'' × 0.16''$) nails.

b. Nailing requirements are permitted to be reduced 25 percent if nails are clinched.

c. Rafter tie heel joint connections are not required where the ridge is supported by a load-bearing wall, header or ridge beam.

d. When intermediate support of the rafter is provided by vertical struts or purlins to a load-bearing wall, the tabulated heel joint connection requirements are permitted to be reduced proportionally to the reduction in span.

e. Equivalent nailing patterns are required for ceiling joist to ceiling joist lap splices.

f. Connected members shall be of sufficient size to prevent splitting due to nailing.

g. For snow loads less than 30 pounds per square foot, the required number of nails is permitted to be reduced by multiplying by the ratio of actual snow load plus 10 divided by 40, but not less than the number required for no snow load.

2308.10.4.2 Notches and holes. Notching at the ends of rafters or ceiling joists shall not exceed one-fourth the depth. Notches in the top or bottom of the rafter or ceiling joist shall not exceed one-sixth the depth and shall not be located in the middle one-third of the span, except that a notch not exceeding one-third of the depth is permitted in the top of the rafter or ceiling joist not further from the face of the support than the depth of the member.

Holes bored in rafters or ceiling joists shall not be within 2 inches (51 mm) of the top and bottom and their diameter shall not exceed one-third the depth of the member.

❖ Wood is comprised of relatively small elongated cells oriented parallel to each other and bound together physically and chemically by lignin. When a rafter is notched or cut, the cell strands are interrupted and the loss in strength needs to be accounted for. This section of the code defines where notches and holes may occur without having to perform an engineering analysis on the wood member.

Some designs and installation practices require that limited notching and cutting occur. Although it is permissible to engineer notches into wood framing, notching should be avoided when possible. Holes bored in rafters create the same problems as notches and have a more significant effect on the strength of the member if they are located near the top or bottom edge. When necessary, the holes should be located in areas with the least stress concentration, generally along the neutral axis of the rafter. Beams subject to high horizontal shear stress (short span/heavy load) should not be cut. Wood members having a thickness of over 4 inches (102 mm) should not be notched, except at the ends of the members.

2308.10.4.3 Framing around openings. Trimmer and header rafters shall be doubled, or of lumber of equivalent cross section, where the span of the header exceeds 4 feet (1219 mm). The ends of header rafters more than 6 feet (1829 mm) long shall be supported by framing anchors or rafter hangers unless bearing on a beam, partition or wall.

❖ See the commentary to Figure 2308.8.3 for an illustration of the criteria for framing around openings.

2308.10.5 Purlins. Purlins to support roof loads are permitted to be installed to reduce the span of rafters within allowable limits and shall be supported by struts to bearing walls. The maximum span of 2-inch by 4-inch (51 mm by 102 mm) purlins shall be 4 feet (1219 mm). The maximum span of the 2-inch by 6-inch (51 mm by 152 mm) purlin shall be 6 feet (1829 mm), but in no case shall the purlin be smaller than the supported rafter. Struts shall not be smaller than 2-inch by 4-inch (51 mm by 102 mm) members. The unbraced length of struts shall not exceed 8 feet (2438 mm) and the minimum slope of the struts shall not be less than 45 degrees (0.79 rad) from the horizontal.

❖ This section permits the use of purlins and struts to reduce the span of rafters. Where the roof slope is less than three units vertical in 12 units horizontal, the code requires that members supporting rafters such as

ridge boards, for example, be designed as beams. However, struts can be installed from a partition up to the ridge board to reduce the span (see Figure 2308.10.5).

2308.10.6 Blocking. Roof rafters and ceiling joists shall be supported laterally to prevent rotation and lateral displacement in accordance with the provisions of Section 2308.8.5.

❖ Solid blocking must be provided at the ends of rafters to prevent twisting or rolling over. There are other common ways of providing resistance to this type of action, but solid blocking is always preferred since it also provides a nominal amount of lateral force transfer from the roof sheathing to the wall or beam below.

2308.10.7 Engineered wood products. Prefabricated wood I-joists, structural glued-laminated timber and structural composite lumber shall not be notched or drilled except where permitted by the manufacturer's recommendations or where the effects of such alterations are specifically considered in the design of the member by a registered design professional.

❖ Engineered wood products should not be cut, notched or bored unless those alterations are considered in the design of the member. This section makes it clear that any alterations to an engineered wood product must be considered in the design of the member. The consideration is to be made either by the manufacturer and reflected in use recommendations (which is common in I-joists, permitting some limited alterations to webs) or made by a registered design professional.

2308.10.8 Roof sheathing. Roof sheathing shall be in accordance with Tables 2304.7(3) and 2304.7(5) for wood structural panels, and Tables 2304.7(1) and 2304.7(2) for lumber and shall comply with Section 2304.7.2.

❖ See the commentary to Section 2304.7.2.

2308.10.8.1 Joints. Joints in lumber sheathing shall occur over supports unless approved end-matched lumber is used, in which case each piece shall bear on at least two supports.

❖ When board sheathing is used, the ends of the boards must be located over the joists or rafters. Such support is not required when end-matched tongue-and-groove lumber is used or when the ends are otherwise prevented from moving relative to each other.

2308.10.9 Roof planking. Planking shall be designed in accordance with the general provisions of this code.

In lieu of such design, 2-inch (51 mm) tongue-and-groove planking is permitted in accordance with Table 2308.10.9. Joints in such planking are permitted to be randomly spaced, provided the system is applied to not less than three continuous spans, planks are center matched and end matched or splined, each plank bears on at least one support, and joints are separated by at least 24 inches (610 mm) in adjacent pieces.

❖ Lumber planking on roofs must conform to the requirements of Section 2304.7.2 and Table 2304.7(1) for typical lumber roof sheathing. This section provides an alternative using 2-inch (51 mm) tongue-and-groove planking.

TABLE 2308.10.9. See page 23-115.

❖ Table 2308.10.9 gives allowable span (spacing of roof supports) based on the live load and deflection limits (from the code) and the design values for the planking used.

2308.10.10 Wood trusses. Wood trusses shall be designed in accordance with Section 2303.4.

❖ Trusses are to be designed in accordance with Section 2303.4 (see the commentary to that section). They are engineered elements that can be incorporated into a building of otherwise conventional light-frame construction in accordance with Section 2308.4.

2308.10.11 Attic ventilation. For attic ventilation, see Section 1203.2.

❖ Large amounts of water vapor migrate by movement of air-carrying water or by diffusion through the building envelope materials because of a vapor pressure difference. The sources of water vapor include cooking, laundering, bathing and breathing and perspiration of persons. These can account for an average daily production of 25 pounds (11.3 kg) of water vapor for a family of four in a typical dwelling. This average can be much higher where appliances such as humidi-

fiers, washers and dryers are used.

As the vapor moves into the attic, it may reach its dew point, thus condensing on wood roof components. This wetting and drying action will cause rotting and decay. To avoid this, the attic must be ventilated to prevent the accumulation of water on building components. The installation of a vapor retarder acts to prevent the passage of moisture to the attic. An effective vapor retarder allows for a decrease in ventilation requirements. Vapor retarders are ineffective when openings in the barrier are allowed where moisture can be carried by air into the attic. Care should be exercised so that attic vent openings remain unobstructed. Attic ventilation is also required for the removal of excess heat. Access to attic spaces is required by Section 1208.2.

Truss manufacturers may provide special blocking that contains openings along the centerline and a galvanized screen to be shipped with their product that will effectively meet the attic ventilation requirements at the eaves. Structural calculations have shown that the strength of this blocking with openings is not reduced significantly but allows it to effectively transfer forces from the roof diaphragm to the wall below.

For SI: 1 inch = 25.4 mm, 1 foot = 304.8 mm.

**Figure 2308.10.5
BRACED RAFTER CONSTRUCTION**

TABLE 2308.10.9
ALLOWABLE SPANS FOR 2-INCH TONGUE-AND-GROOVE DECKING

SPAN[a] (feet)	LIVE LOAD (pound per square foot)	DEFLECTION LIMIT	BENDING STRESS (f) (pound per square inch)	MODULUS OF ELASTICITY (E) (pound per square inch)
		Roofs		
4	20	1/240 1/360	160	170,000 256,000
	30	1/240 1/360	210	256,000 384,000
	40	1/240 1/360	270	340,000 512,000
4.5	20	1/240 1/360	200	242,000 305,000
	30	1/240 1/360	270	363,000 405,000
	40	1/240 1/360	350	484,000 725,000
5.0	20	1/240 1/360	250	332,000 500,000
	30	1/240 1/360	330	495,000 742,000
	40	1/240 1/360	420	660,000 1,000,000
5.5	20	1/240 1/360	300	442,000 660,000
	30	1/240 1/360	400	662,000 998,000
	40	1/240 1/360	500	884,000 1,330,000
6.0	20	1/240 1/360	360	575,000 862,000
	30	1/240 1/360	480	862,000 1,295,000
	40	1/240 1/360	600	1,150,000 1,730,000
6.5	20	1/240 1/360	420	595,000 892,000
	30	1/240 1/360	560	892,000 1,340,000
	40	1/240 1/360	700	1,190,000 1,730,000
7.0	20	1/240 1/360	490	910,000 1,360,000
	30	1/240 1/360	650	1,370,000 2,000,000
	40	1/240 1/360	810	1,820,000 2,725,000

(continued)

TABLE 2308.10.9–continued
ALLOWABLE SPANS FOR 2-INCH TONGUE-AND-GROOVE DECKING

SPAN[a] (feet)	LIVE LOAD (pound per square foot)	DEFLECTION LIMIT	BENDING STRESS (f) (pound per square inch)	MODULUS OF ELASTICITY (E) (pound per square inch)
Roofs				
7.5	20	1/240 1/360	560	1,125,000 1,685,000
7.5	30	1/240 1/360	750	1,685,000 2,530,000
7.5	40	1/240 1/360	930	2,250,000 3,380,000
8.0	20	1/240 1/360	640	1,360,000 2,040,000
8.0	30	1/240 1/360	850	2,040,000 3,060,000
Floors				
4	40	1/360	840	1,000,000
4.5	40	1/360	950	1,300,000
5.0	40	1/360	1,060	1,600,000

For SI: 1 inch = 25.4 mm, 1 foot = 304.8 mm, 1 pound per square foot = 0.0479kN/m^2, 1 pound per square inch = 0.00689 N/mm^2.

a. Spans are based on simple beam action with 10 pounds per square foot dead load and provisions for a 300-pound concentrated load on a 12-inch width of decking. Random layup is permitted in accordance with the provisions of Section 2308.10.9. Lumber thickness is $1^1/_2$ inches nominal.

2308.11 Additional requirements for conventional construction in Seismic Design Category B or C. Structures of conventional light-frame construction in Seismic Design Category B or C, as determined in Section 1613, shall comply with Sections 2308.11.1 through 2308.11.3, in addition to the provisions of Sections 2308.1 through 2308.10.

❖ This section places additional construction limits on structures where the seismic risk is moderate. The limits on conventional construction are based on FEMA 450.

2308.11.1 Number of stories. Structures of conventional light-frame construction shall not exceed two stories in height in Seismic Design Category C.

❖ This section limits conventional construction to two stories in height in Seismic Design Category C because of amplified dynamic effects in taller structures.

When conventional construction provisions are used for one- and two-family dwellings in Seismic Design Category C, the allowable number of stories is increased to three.

2308.11.2 Concrete or masonry. Concrete or masonry walls or masonry veneer shall not extend above the basement.

Exceptions:

1. Masonry veneer is permitted to be used in the first two stories above grade plane or the first three stories above grade plane where the lowest story has concrete or masonry walls in Seismic Design Category B, provided that structural use panel wall bracing is used and the length of bracing provided is one- and one-half times the required length as determined in Table 2308.9.3(1).

2. Masonry veneer is permitted to be used in the first story above grade plane or the first two stories above grade plane where the lowest story has concrete or masonry walls in Seismic Design Category B or C.

3. Masonry veneer is permitted to be used in the first two stories above grade plane in Seismic Design Categories B and C, provided the following criteria are met:

   3.1. Type of brace per Section 2308.9.3 shall be Method 3 and the allowable shear capacity in accordance with Table 2306.4.1 shall be a minimum of 350 plf (5108 N/m).

   3.2. The bracing of the top story shall be located at each end and at least every 25 feet (7620 mm) o.c. but not less than 40 percent of the braced wall line. The bracing of the first story shall be located at each end and at least every 25 feet (7620 mm) o.c. but not less than 35 percent of the braced wall line.

   3.3. Hold-down connectors shall be provided at the ends of braced walls for the second floor to first floor wall assembly with an allowable design of 2,000 pounds (8896 N). Hold-down connectors shall be provided at the ends of each wall segment of the braced walls for the first floor to foundation with an allowable design of 3,900 pounds (17 347 N). In all cases, the hold-down connector force shall be transferred to the foundation.

   3.4. Cripple walls shall not be permitted.

❖ Because of their weight, concrete walls, masonry walls and masonry veneer can impose lateral loads from seismic events that are beyond the loading considered acceptable for conventional light-frame construction. For this reason, the code requires an engineered design for concrete or masonry walls above the basement. The use of masonry veneer above grade is permitted in conventionally framed buildings in accordance with the exceptions. Exception 1 allows masonry veneer either two or three stories above the basement of structures classified as Seismic Design Category B under certain conditions that include increasing the amount of wall bracing. It would not be necessary to meet these conditions for masonry veneer on the first story only, since the second exception permits this. Exception 2 allows one story of masonry veneer above grade (i.e., the basement level) and two stories above grade plane if the lower story walls are constructed of concrete or masonry, which are engineered for seismic effects. Exception 3 allows up to two stories of masonry veneer above grade plane, provided the added requirements for wall bracing and hold-downs are complied with in order to increase the resistance of the lateral-force-resisting system.

2308.11.3 Framing and connection details. Framing and connection details shall conform to Sections 2308.11.3.1 through 2308.11.3.3.

❖ Sections 2308.11.3.1 through 2308.11.3.3 contain specific connection requirements for Seismic Design Category B or C.

2308.11.3.1 Anchorage. Braced wall lines shall be anchored in accordance with Section 2308.6 at foundations.

❖ Braced wall lines must be anchored to the foundation so that they will not overturn because of a moment created by lateral loads. Anchorage for overturning will nearly always be more critical for the individual shear panel than for the entire wall, though the effectiveness of the entire system cannot be ignored. Conventional construction assumes an entire system that is adequately tied together by all of the requirements noted throughout the code to provide a complete load path for imposed forces.

2308.11.3.2 Stepped footings. Where the height of a required braced wall panel extending from foundation to floor above varies more than 4 feet (1219 mm), the following construction shall be used:

1. Where the bottom of the footing is stepped and the lowest floor framing rests directly on a sill bolted to the footings, the sill shall be anchored as required in Section 2308.3.3.

2. Where the lowest floor framing rests directly on a sill bolted to a footing not less than 8 feet (2438 mm) in length along a line of bracing, the line shall be considered to be braced. The double plate of the cripple stud wall beyond the segment of footing extending to the lowest framed floor shall be spliced to the sill plate with metal ties, one on each side of the sill and plate. The metal ties shall not be less than 0.058 inch [1.47 mm (16 galvanized gage)] by 1.5 inches (38 mm) wide by 48 inches (1219 mm) with eight 16d common nails on each side of the

splice location (see Figure 2308.11.3.2). The metal tie shall have a minimum yield of 33,000 pounds per square inch (psi) (227 MPa).

3. Where cripple walls occur between the top of the footing and the lowest floor framing, the bracing requirements for a story shall apply.

❖ The primary objectives of footing design are to provide a level surface for construction of the foundation wall; to provide adequate transfer and distribution of the building loads to the underlying soil; to provide adequate strength to prevent differential settlement of the building and to provide adequate anchorage or mass to resist potential uplift and overturning forces from lateral loads placed on the structure. The most common footing type in residential construction is a continuous concrete spread footing. These concrete footings are recognized in prescriptive footing size tables for most typical conditions. In contrast, special conditions such as stepped footings, which might be required in steeply sloped sites, require special consideration and possibly an engineered design.

FIGURE 2308.11.3.2. See page 23-118.

❖ This figure illustrates the code requirements for stepped footing details (see Section 2308.11.3.2 for further discussion). Care should be taken to install the metal strap shown so that the nailing does not interfere with anchor bolts at the sill plate. Additionally, nails used to fasten the strap to pressure-treated plates must be galvanized or otherwise protected as described in Section 2304.9.5.

2308.11.3.3 Openings in horizontal diaphragms. Openings in horizontal diaphragms with a dimension perpendicular to the joist that is greater than 4 feet (1.2 m) shall be constructed in accordance with the following:

1. Blocking shall be provided beyond headers.

2. Metal ties not less than 0.058 inch [1.47 mm (16 galvanized gage)] by 1.5 inches (38 mm) wide with eight 16d common nails on each side of the header-joist intersection shall be provided (see Figure 2308.11.3.3). The metal ties shall have a minimum yield of 33,000 psi (227 MPa).

❖ Horizontal diaphragms are floor and roof assemblies that are usually clad with structural wood sheathing panels, such as plywood or OSB. Though more complicated and difficult to visualize, lateral forces that are applied to a building from wind or seismic events follow a load path that distributes and transfers shear and overturning forces from the lateral loads. When openings are built into the diaphragm, they disrupt the continuity of load across the diaphragm and they must be reinforced to compensate. Another concern is the stiffness of the diaphragm. These provisions are a prescriptive solution for openings not greater than 4 feet (1219 mm) in dimension and provide a general means for a load path in these specific cases in lieu of an engineered design.

2x SILL PLATE

2'-0" MIN.

SPLICE

WHERE FOOTING SECTION "A" IS MORE THAN 8'-0",
PROVIDE METAL TIE 16GA x 1 1/2" x 4'-0" MIN., EACH SIDE
W/ 8-16d COMMON NAILS
EACH SIDE OF SPLICE

2-2x PLATE

CONCRETE
STEPPED FOOTING

2x CRIPPLE
STUD WALL

FOOTING SECTION "A"

NOTE: WHERE FOOTING SECTION "A"
IS LESS THAN 8'-0" LONG IN A
25'-0" TOTAL LENGTH WALL, PROVIDE
BRACING AT CRIPPLE STUD WALL

For SI: 1 inch = 25.4 mm, 1 foot = 304.8 mm.

FIGURE 2308.11.3.2
STEPPED FOOTING CONNECTION DETAILS

FIGURE 2308.3.3. See page 23-119.

❖ See the commentary to Section 2308.11.3.3.

2308.12 Additional requirements for conventional construction in Seismic Design Category D or E. Structures of conventional light-frame construction in Seismic Design Category D or E, as determined in Section 1613, shall conform to Sections 2308.12.1 through 2308.12.9, in addition to the requirements for Seismic Design Category B or C in Section 2308.11.

❖ Conventional construction is a type of construction that has been used successfully for many years and relies on standard practice as governed by prescriptive building code requirements. For conditions where the lateral loading on the building is greater, such as in higher seismic areas of the country with Seismic Design Categories D and E, additional prescriptive methods or limitations are considered necessary. The basis for the requirements for Seismic Design Category D or E is FEMA 450.

2308.12.1 Number of stories. Structures of conventional light-frame construction shall not exceed one story in height in Seismic Design Category D or E.

❖ Similar to buildings classified as Seismic Design Category C, prescriptive structures having a higher seismic risk need to be limited in height due to historic observations of the effects of severe earthquakes. Structural failures were observed in the aftermath of the 1994 Northridge, California earthquake of multistory residential buildings that were sheathed with the weakest of materials allowed under the prescriptive methods. As a result of these observations, newer regulations

for conventional construction have included more restrictive use of materials and a more restrictive placement of braced wall lines and panels.

2308.12.2 Concrete or masonry. Concrete or masonry walls or masonry veneer shall not extend above the basement.

Exception: Masonry veneer is permitted to be used in the first story above grade plane in Seismic Design Category D, provided the following criteria are met:

1. Type of brace in accordance with Section 2308.9.3 shall be Method 3 and the allowable shear capacity in accordance with Table 2306.4.1 shall be a minimum of 350 plf (5108 N/m).

2. The bracing of the first story shall be located at each end and at least every 25 feet (7620 mm) o.c. but not less than 45 percent of the braced wall line.

3. Hold-down connectors shall be provided at the ends of braced walls for the first floor to foundation with an allowable design of 2,100 pounds (9341 N).

4. Cripple walls shall not be permitted.

❖ Because of their weight, concrete walls, masonry walls and masonry veneer can impose lateral loads from seismic events in excess of the loading considered acceptable for conventional light-frame construction. For this reason, the code prohibits the use of concrete or masonry walls above the basement. The use of masonry veneer one story above grade plane is permitted for structures classified as Seismic Design Category D in accordance with the exception, provided the requirements for wall bracing and hold-downs are met.

2308.12.3 Braced wall line spacing. Spacing between interior and exterior braced wall lines shall not exceed 25 feet (7620 mm).

❖ This provision mandates the maximum spacing between lines of bracing elements and limiting offsets with a line. It limits the lateral forces supported by the elements by limiting the tributary area dn the building area. A series of braced wall panels comprise a braced wall line. A two-dimensional grid of interior and exterior braced wall lines breaks up the building into a series of boxes as illustrated in Figure 2308.9.3. The size of the box is either 25 or 35 feet (7620 or 10 668 mm), depending on seismic design category and wind speed. The 25-foot (7620 mm) dimension is related to the maximum framing spans, which are included in other sections of this chapter. For additional information on the development and philosophy of the conventional construction method, see the commentary for Section 2308.1.

2308.12.4 Braced wall line sheathing. Braced wall lines shall be braced by one of the types of sheathing prescribed by Table 2308.12.4 as shown in Figure 2308.9.3. The sum of lengths of braced wall panels at each braced wall line shall conform to Table 2308.12.4. Braced wall panels shall be distributed along the length of the braced wall line and start at not more than 8 feet (2438 mm) from each end of the braced wall line. Panel sheathing joints shall occur over studs or blocking. Sheathing shall be fastened to studs, top and bottom plates and at panel edges occurring over blocking. Wall framing to which sheathing used for bracing is applied shall be nominal 2 inch wide [actual $1^1/_2$ inch (38 mm)] or larger members.

Cripple walls having a stud height exceeding 14 inches (356 mm) shall be considered a story for the purpose of this section and shall be braced as required for braced wall lines in accordance with Table 2308.12.4. Where interior braced wall lines occur without a continuous foundation below, the length of parallel exterior cripple wall bracing shall be one and one-half times the lengths required by Table 2308.12.4. Where the cripple wall sheathing type used is Type S-W and this additional length of bracing cannot be provided, the capacity of Type S-W sheathing shall be increased by reducing the spacing of fasteners along the perimeter of each piece of sheathing to 4 inches (102 mm) o.c.

❖ Braced wall lines are required to contain braced wall panels sheathed and placed in accordance with one of the methods specified in Table 2308.12.4, as illustrated in Figure 2308.9.3. All vertical joints of the panel are required to occur over studs. Horizontal joints are required to occur over blocking equal in size to the studding unless waived by the installation requirements for the specific sheathing materials. A collector or drag strut, which is usually a system of members in light-frame construction, collects and transfers loads by tension or compression to the shear-resisting segments of a wall line. In a typical home, special design of chord members may involve some modest detailing of splices at the diaphragm boundary. Compressive forces are rarely a concern within the conventional construction limits where at least a double top plate is used as a collector, particularly where the collector is braced against lateral buckling by attachment to other construction.

TABLE 2308.12.4. See page 23-120.

❖ See the commentary to Section 2308.12.4.

METAL TIE 16GA. x 1 1/2" x 4'-0" MIN., (4 TOTAL)
W/ 16-16d COMMON NAILS AS SHOWN

-OR-

METAL TIE 16GA. x 1 1/2" x (OPENING WIDTH + 4'-0") MIN.,
(2 TOTAL) W/ 24-16d COMMON NAILS

For SI: 1 inch = 25.4 mm, 1 foot = 304.8 mm.

FIGURE 2308.11.3.3
OPENINGS IN HORIZONTAL DIAPHRAGMS

TABLE 2308.12.4
WALL BRACING IN SEISMIC DESIGN CATEGORIES D AND E
(Minimum Length of Wall Bracing per each 25 Linear Feet of Braced Wall Line[a])

CONDITION	SHEATHING TYPE[b]	$S_{DS} < 0.50$	$0.50 \leq S_{DS} < 0.75$	$0.75 \leq S_{DS} \leq 1.00$	$S_{DS} > 1.00$
One story	G-P[c]	10 feet 8 inches	14 feet 8 inches	18 feet 8 inches	25 feet 0 inches
	S-W	5 feet 4 inches	8 feet 0 inches	9 feet 4 inches	12 feet 0 inches

For SI: 1 inch = 25.4 mm, 1 foot = 304.8 mm.

a. Minimum length of panel bracing of one face of the wall for S-W sheathing or both faces of the wall for G-P sheathing; h/w ratio shall not exceed 2:1. For S-W panel bracing of the same material on two faces of the wall, the minimum length is permitted to be one-half the tabulated value but the h/w ratio shall not exceed 2:1 and design for uplift is required.

b. G-P = gypsum board, fiberboard, particleboard, lath and plaster or gypsum sheathing boards; S-W = wood structural panels and diagonal wood sheathing.

c. Nailing as specified below shall occur at all panel edges at studs, at top and bottom plates and, where occurring, at blocking:

For $^1/_2$-inch gypsum board, 5d (0.113 inch diameter) cooler nails at 7 inches on center;

For $^5/_8$-inch gypsum board, No. 11 gage (0.120 inch diameter) at 7 inches on center;

For gypsum sheathing board, $1^3/_4$ inches long by $^7/_{16}$-inch head, diamond point galvanized nails at 4 inches on center;

For gypsum lath, No. 13 gage (0.092 inch) by $1^1/_8$ inches long, $^{19}/_{64}$-inch head, plasterboard at 5 inches on center;

For Portland cement plaster, No. 11 gage (0.120 inch) by $1^1/_2$ inches long, $^7/_{16}$- inch head at 6 inches on center;

For fiberboard and particleboard, No. 11 gage (0.120 inch) by $1^1/_2$ inches long, $^7/_{16}$-inch head, galvanized nails at 3 inches on center.

2308.12.5 Attachment of sheathing. Fastening of braced wall panel sheathing shall not be less than that prescribed in Table 2308.12.4 or 2304.9.1. Wall sheathing shall not be attached to framing members by adhesives.

❖ The nailing required by Table 2308.12.4 is specified in Note c. The connections in Table 2304.9.1 are the minimum allowed for conventional construction. Adhesives are not ordinarily used to fasten panels to wall studs and plates and therefore have not been thoroughly tested for such installations. Floor sheathing panels are often glued down to the framing members prior to final nailing in order to help prevent squeaks, but that need does not exist with wall panels.

2308.12.6 Irregular structures. Conventional light-frame construction shall not be used in irregular portions of structures in Seismic Design Category D or E. Such irregular portions of structures shall be designed to resist the forces specified in Chapter 16 to the extent such irregular features affect the performance of the conventional framing system. A portion of a structure shall be considered to be irregular where one or more of the conditions described in Items 1 through 6 below are present.

1. Where exterior braced wall panels are not in one plane vertically from the foundation to the uppermost story in which they are required, the structure shall be considered to be irregular [see Figure 2308.12.6(1)].

   Exception: Floors with cantilevers or setbacks not exceeding four times the nominal depth of the floor joists [see Figure 2308.12.6(2)] are permitted to support braced wall panels provided:

   1. Floor joists are 2 inches by 10 inches (51 mm by 254 mm) or larger and spaced not more than 16 inches (406 mm) o.c.

   2. The ratio of the back span to the cantilever is at least 2:1.

   3. Floor joists at ends of braced wall panels are doubled.

   4. A continuous rim joist is connected to the ends of cantilevered joists. The rim joist is permitted to be spliced using a metal tie not less than 0.058 inch (1.47 mm) (16 galvanized gage) and $1^1/_2$ inches (38 mm) wide fastened with six 16d common nails on each side. The metal tie shall have a minimum yield of 33,000 psi (227 MPa).

   5. Joists at setbacks or the end of cantilevered joists shall not carry gravity loads from more than a single story having uniform wall and roof loads, nor carry the reactions from headers having a span of 8 feet (2438 mm) or more.

2. Where a section of floor or roof is not laterally supported by braced wall lines on all edges, the structure shall be considered to be irregular [see Figure 2308.12.6(3)].

   Exception: Portions of roofs or floors that do not support braced wall panels above are permitted to extend up to 6 feet (1829 mm) beyond a braced wall line [see Figure 2308.12.6(4)].

3. Where the end of a required braced wall panel extends more than 1 foot (305 mm) over an opening in the wall below, the structure shall be considered to be irregular. This requirement is applicable to braced wall panels offset in plane and to braced wall panels offset out of plane as permitted by the exception to Item 1 above in this section [see Figure 2308.12.6(5)].

   Exception: Braced wall panels are permitted to extend over an opening not more than 8 feet (2438 mm) in width where the header is a 4-inch by 12-inch (102 mm by 305 mm) or larger member.

4. Where portions of a floor level are vertically offset such that the framing members on either side of the offset cannot be lapped or tied together in an approved manner, the structure shall be considered to be irregular [see Figure 2308.12.6(6)].

   Exception: Framing supported directly by foundations need not be lapped or tied directly together.

5. Where braced wall lines are not perpendicular to each other, the structure shall be considered to be irregular [see Figure 2308.12.6(7)].

6. Where openings in floor and roof diaphragms having a maximum dimension greater than 50 percent of the distance between lines of bracing or an area greater than 25 percent of the area between orthogonal pairs of braced wall lines are present, the structure shall be considered to be irregular [see Figure 2308.12.6(8)].

❖ Irregular structures are those that require engineered design in Seismic Design Category D or E because of their unusual shape or discontinuities in the lateral-force-resisting system. Irregular conditions identified within the code are known to create torsion and unbalanced load distribution to the shear walls that necessitates an engineered design. The design professional is left to judge the extent of the portion required to be designed as a result of an irregularity. This often involves design of the nonconforming element, force transfer into the element and a load path from the element to the foundation. A nonconforming portion will sometimes have enough of an impact on the behavior of a structure to warrant that the entire lateral-force-resisting system receive an engineered design. Buildings may have slightly irregular features, as identified in this section and in Figures 2308.12.6(1) through 2308.12.6(8), and a designer who wishes to avoid an engineering design must lay out their building in such a way as to meet these restrictions.

1. This limit applies when braced wall panels are offset out of plane from floor to floor. In-plane offsets are discussed in another item. Ideally, braced wall panels would always stack above each other from floor to floor.

   Because cantilevers and setbacks are often incorporated into structures, the exception to this provision defines the restriction on cantilever and set-back length in order to maintain regular building behavior.

2. This limitation applies to open-front structures or portions of structures. The conventional construction bracing concept is based on using braced wall lines to divide a structure into a series of "boxes" of limited dimension, with the seismic force to each box being limited by the size. The intent is that each box be supported by braced wall lines on all four sides, limiting the amount of torsion that can occur. The exception, which permits portions of roofs or floors to extend past the braced wall line, is intended to permit construction such as porch roofs and bay windows. Walls with no lateral resistance are allowed in areas where braced walls are prohibited.

3. This limitation applies when braced wall panels are offset in plane. Ends of braced wall panels supported on window or door headers can be calculated to transfer large vertical reactions to headers that may not be of adequate size to resist these reactions. This section restricts the

portion of an upper braced wall panel that may extend over an opening below where a header of limited size may exist. This is allowed on the basis that the vertical reaction is within an approximate 45-degree (0.78 rad) line of the header support and, therefore, will not result in critical shear or flexure. Where a 4-inch by 12-inch nominal (102 mm by 305 mm) header is framed over the opening below a braced wall panel, the panel may extend completely over this header provided the opening is not greater than 8 feet (2438 mm) in length. All other header conditions require an engineered design.

4. This limitation results from observation of damage that is somewhat unique to split-level wood-frame construction. If floors on either side of an offset move in opposite directions due to earthquake or wind loading, the short bearing wall in the middle becomes unstable and vertical support for the upper joists can be lost, resulting in a collapse. If the vertical offset is limited to a dimension equal to or less than the joist depth, then a simple strap tie directly connecting joists on different levels can be provided, making the irregularity manageable.

5. This limitation applies to nonperpendicular braced wall lines. When braced wall lines are not perpendicular to each other, further evaluation is needed to determine force distributions and required bracing.

6. This limitation attempts to place a practical limit on openings in floors and roofs. Because stair openings are essential to residential construction and have long been used without any report of life safety hazards after a large earthquake, these have been found to be acceptable in conventional construction.

FIGURES 2308.12.6(1) through (8). See pages 23-122 through 23-124.

❖ For Figures 2308.12.6(1) through (8), see the commentary to Section 2308.12.6.

2308.12.7 Anchorage of exterior means of egress components. Exterior egress balconies, exterior exit stairways and similar means of egress components shall be positively anchored to the primary structure at not over 8 feet (2438 mm) o.c. or shall be designed for lateral forces. Such attachment shall not be accomplished by use of toenails or nails subject to withdrawal.

❖ Because of the importance of moving people out of a building after a catastrophic event, the exterior means of egress components identified in this section must be positively anchored (i.e. no toe-nails or nails in withdrawal) to the structure in buildings classified as Seismic Design Category D or E to provide resistance to lateral forces.

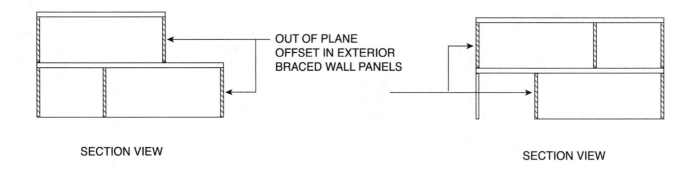

SECTION VIEW   SECTION VIEW

FIGURE 2308.12.6(1)
BRACED WALL PANELS OUT OF PLANE

SECTION THRU CANTILEVER   SECTION THRU SET BACK

For SI: 1 foot = 304.8 mm.

FIGURE 2308.12.6(2)
BRACED WALL PANELS SUPPORTED BY CANTILEVER OR SET BACK

PLAN VIEW

FIGURE 2308.12.6(3)
FLOOR OR ROOF NOT SUPPORTED ON ALL EDGES

PLAN VIEW

ROOF OR FLOOR SHALL BE PERMITTED
TO EXTEND UP TO 6' BEYOND
THE BRACED WALL LINE

NO BRACED WALL PANEL ABOVE
PERMITTED AT THIS LOCATION

For SI: 1 foot = 304.8 mm.

FIGURE 2308.12.6(4)
ROOF OR FLOOR EXTENSION BEYOND BRACED WALL LINE

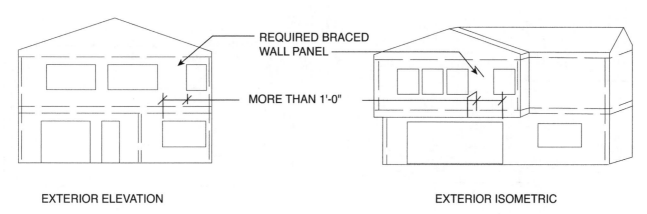

REQUIRED BRACED
WALL PANEL

MORE THAN 1'-0"

EXTERIOR ELEVATION

EXTERIOR ISOMETRIC

For SI: 1 foot = 304.8 mm.

FIGURE 2308.12.6(5)
BRACED WALL PANEL EXTENSION OVER OPENING

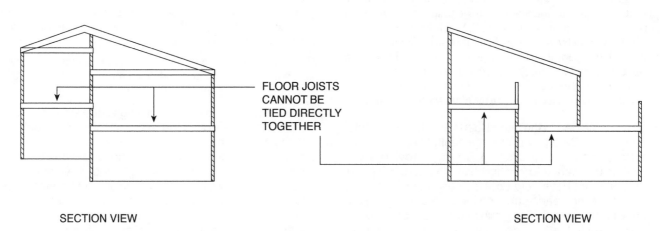

FLOOR JOISTS
CANNOT BE
TIED DIRECTLY
TOGETHER

SECTION VIEW

SECTION VIEW

FIGURE 2308.12.6(6)
PORTIONS OF FLOOR LEVEL OFFSET VERTICALLY

PLAN VIEW

FIGURE 2308.12.6(7)
BRACED WALL LINES NOT PERPENDICULAR

FIGURE 2308.12.6(8)
OPENING LIMITATIONS FOR FLOOR AND ROOF DIAPHRAGMS

2308.12.8 Steel plate washers. Steel plate washers shall be placed between the foundation sill plate and the nut. Such washers shall be a minimum of 0.229 inch by 3 inches by 3 inches (5.82 mm by 76 mm by 76 mm) in size. The hole in the plate washer is permitted to be diagonally slotted with a width of up to $^3/_{16}$ inch (4.76 mm) larger than the bolt diameter and a slot length not to exceed $1^3/_4$ inches (44 mm), provided a standard cut washer is placed between the plate washer and the nut.

❖ Walls subjected to the lateral loads anticipated in Seismic Design Category D or E, particularly tall walls, will have a tendency to overturn. To help provide some level of resistance to this action in a braced wall panel that may not have hold-downs at each end, the code increases the bearing area on the sill plate at each anchor bolt by stipulating a minimum-size plate washer that spreads the load over a wider area. Although an engineering analysis may indicate that this does not provide enough overturning resistance, and the fact that there are sill plate splices and other discontinuities that would affect the system, laboratory testing has shown that these methods add some measure of resistance to a prescriptive system.

2308.12.9 Anchorage in Seismic Design Category E. Steel bolts with a minimum nominal diameter of $^5/_8$ inch (15.9 mm) shall be used in Seismic Design Category E.

❖ Lateral seismic loads on buildings cause the bottom of the walls to slide, and the tendency of sill plate bolts is to rip through the wood. Dowel-type connectors (bolts, nails, screws and pins) rely on metal-to-wood bearing for the transfer of lateral loads. One of the factors that determines dowel bearing strength is the diameter of the bolt. By increasing the minimum diameter of the bolts from ½ inch (12.7 mm) as specified in Section 2308.6, the bearing surface area of the bolt is increased to provide a connection to resist the increased seismic forces of Seismic Design Category E.

Bibliography

The following resource materials are referenced in this chapter or are relevant to the subject matter addressed in this chapter.

1999 Handbook for Ceramic Tile Installation. Anderson, SC: TCA1999.

AF&PA, *Allowable Stress Design Manual for Wood Construction (ASD Manual).* Washington, D.C.: American Forest & Paper Association.

AF&PA, *Heavy Timber Construction Details—Wood Construction Data 5.* Washington, D.C.: American Forest & Paper Association.

AF&PA NDS—05, *National Design Specification® for Wood Construction with 2005 Supplement.* Washington, D.C.: American Forest & Paper Association, 2005.

AF&PA No.4—89, *Plank and Beam Framing for Residential Buildings.* Washington, D.C.: American Forest & Paper Association, 1989.

AF&PA, *Span Tables for Joists and Rafters.* Washington, D.C.: American Forest & Paper Association, 1993.

AF&PA Technical Report 7—87, *Basic Requirements for Permanent Wood Foundation System.* Washington, D.C.: American Forest & Paper Association, 1987.

AF&PA, *Wood Frame Construction Manual for One- and Two-Family Dwellings (WFCM).* Washington, D.C.: American Forest and Paper Association, 2001.

AHA A135.4—04, *Basic Hardboard.* Palatine, IL: American Hardwood Association, 2004.

AHA A135.5—04, *Prefinished Hardboard Paneling.* Palatine, IL: American Hardwood Association, 2004.

AHA A135.6—98, *Hardboard Siding.* Palatine, IL: American Hardwood Association, 1998.

AITC 104—2003, *Typical Construction Details.* Englewood, CO: American Institute of Timber Construction, 2003.

AITC 110—01, *Standard Appearance Grades for Structural Glued Laminated Timber.* Englewood, CO: American Institute of Timber Construction, 2001.

AITC 112—93, *Standard for Tongue-and-Groove Heavy Timber Roof Decking.* Englewood, CO: American Institute of Timber Construction, 1993.

AITC 113—01, *Standard for Dimensions of Structural Glued Laminated Timber.* Englewood, CO: American Institute of Timber Construction, 2001.

AITC 117—04, *Structural Glued-Laminated Timber.* Englewood, CO: American Institute of Timber Construction, 2004.

AITC 119—96, *Standard Specifications for Hardwood Glued-Laminated Timber.* Englewood, CO: American Institute of Timber Construction, 1996.

AITC 200—2004, *Manufacturing Quality Control Systems Manual for Structural Glued Laminated Timber.* Englewood, CO: American Institute of Timber Construction, 2004.

AITC A 190.1—02, *Structural Glued-Laminated Timber.* Englewood, CO: American Institute of Timber Construction, 2002.

American Wood Council. "Details for Conventional Wood Frame Construction." AF&PA, pp. 10 & 42. 2001.

Andreason, Kenneth R. and John D. Rose. *APA Report T94-5: Northridge, California Earthquake.* American Plywood Association, pp. 2 & 5. March 1994.

ANSI A 208.1—99, *Particleboard.* New York: American National Standards Institute, 1999.

APA PDS—04, *Panel Design Specification.* Tacoma, WA: APA—Engineered Wood Association, 2004.

APA PDS—90, Supplement 1—*Design and Fabrication of Plywood Curved Panels.* Tacoma, WA: APA—Engineered Wood Association, 1995.

APA PDS—90, Supplement 3—*Design and Fabrication of Plywood Stressed-Skin Panels.* Tacoma, WA: APA—Engineered Wood Association, 1996.

APA PDS—90, Supplement 4—*Design and Fabrication of Plywood Sandwich Panels.* Tacoma, WA: APA—Engineered Wood Association, 1993.

APA PDS—92, Supplement 2—*Design and Fabrication of Plywood-Lumber Beams.* Tacoma, WA: APA—Engineered Wood Association, 1998.

APA PDS—95, Supplement 5—*Design and Fabrication of All-Plywood Beams.* Tacoma, WA: APA—Engineered Wood Association, 1995.

APA Report 157—95, *Wood Structural Panel Shear Walls with Gypsum Wallboard and Window/Door Openings.* Tacoma, WA: APA—The Engineered Wood Association, 1995.

APA Research Report 138, *Plywood Diaphragms.* Tacoma, WA: APA—Engineered Wood Association.

APA Research Report 154, *Wood Structural Panel Shear Wall.* Tacoma, WA: APA—Engineered Wood Association.

American Institute of Timber Construction. *Timber Construction Manual, 3rd Edition.* New York: John Wiley and Sons, Inc. 1985

ASAE EP 484.2 (1998), *Diaphragm Design of Metal-Clad, Post-Frame Rectangular Buildings.* St. Joseph, MI: American Society of Agricultural Engineers, 1998.

ASAE EP 486.1 (2000), *Shallow-Post Foundation Design.* St. Joseph, MI: American Society of Agricultural Engineers, 2000.

ASAE EP 559 (1997), *Design Requirements and Bending Properties for Mechanically Laminated Columns.*

St. Joseph, MI: American Society of Agricultural Engineers, 1997.

ASCE 7—05, *Minimum Design Loads for Buildings and Other Structures.* Reston, VA: American Society of Civil Engineers, 2005.

ASTM D 2898-94(1999), *Test Methods for Accelerated Weathering of Fire-Retardant-Treated Wood for Fire Testing.* West Conshohocken, PA: ASTM International, 1999.

ASTM D 3200-74(2000), *Standard Specification and Test Method for Establishing Recommended Design Stresses for Round Timber Construction Poles.* West Conshohocken, PA: ASTM International, 2000.

ASTM D 3201-94(2003), *Test Method for Hygroscopic Properties of Fire-Retardant-Treated Wood and Wood-base Products.* West Conshohocken, PA: ASTM International, 2003.

ASTM D 3737-03, *Practice for Establishing Allowable Properties for Structural Glued Laminated Timber (Glulam).* West Conshohocken, PA: ASTM International, 2003.

ASTM D 3957-74(2000), *Standard Practices for Establishing Stress Grades for Structural Members Used in Log Buildings.* West Conshohocken, PA: ASTM International, 2000.

ASTM D 5055-04, *Specification for Establishing and Monitoring Structural Capacities of Prefabricated Wood I-Joists.* West Conshohocken, PA: ASTM International, 2004.

ASTM D 6841-03, Standard Practice for Calculating Design Value Treatment Adjustment Factors for Fire-Retardant-Treated Wood. West Conshohocken, PA: ASTM International, 2003.

ATC 7, *Guidelines for the Design of Horizontal Wood Diaphragms.* Redwood City, CA: Applied Technology Council, 1981.

AWPA U1—04, *Use Category System: User Specification for Treated Wood.* Grandbury, TX: American Wood-Preservers' Association, 2004.

Becker, Norman. *The Complete Book of Home Inspection, 2nd Edition.* McGraw-Hill, New York: 1993.

Bossi, Robert J. and Kelly E. Cobeen. "Conventional Construction for Today's Buildings." *ICBO Building Standards*, Sept.-Oct. 1996, p. 14.

Breyer, Donald E., Kenneth J. Fridley, David G. Pollock, Jr. and Kelly E. Cobeen. *Design of Wood Structures – ASD (Fifth Edition).* New York: McGraw-Hill, 2003.

Cobeen, Kelly, James Russell and J. Daniel Dolan. *CUREE W30a & 30b: Recommendations for Earthquake Resistance in the Design and Construction of Woodframe Buildings.* Consortium of Universities for Research in Earthquake Engineering (CUREE), 2004.

DOC PS 1—95, *Construction and Industrial Plywood.* Gaithersburg, MD: U.S. Department of Commerce, 1995.

DOC PS 2—92, *Performance Standard for Wood-Based Structural-Use Panels.* Gaithersburg, MD: U.S. Department of Commerce, 1992.

DOC PS 20—99, *American Softwood Lumber Standard.* Gaithersburg, MD: U.S. Department of Commerce, 1999.

EWS R540—96, *Builders Tips: Proper Storage and Handling of Glulam Beams.* Tacoma, WA: APA—Engineered Wood Association, 1996.

EWS S475—99, *Glued Laminated Beam Design Tables.* Tacoma, WA: APA—Engineered Wood Association, 1999.

EWS S560—99, *Field Notching and Drilling of Glued Laminated Timber Beams.* Tacoma, WA: APA—Engineered Wood Association, 1999.

EWS T300—99, *Glulam Connection Details.* Tacoma, WA: APA—Engineered Wood Association, 1999.

EWS X440—00, *Product and Application Guide: Gluman.* Tacoma, WA: APA—Engineered Wood Association, 2000.

EWS X450—97, *Glulam in Residential Construction.* Tacoma, WA: APA—Engineered Wood Association, 1997.

FEMA 450-1, *NEHRP Recommended Provisions for Seismic Regulations for New Buildings and Other Structures.* Washington, D.C.: Federal Emergency Management Agency, 2004.

FEMA 450-2, *NEHRP Recommended Provisions for Seismic Regulations for New Buildings and Other Structures Part 2: Commentary.* Washington, D.C.: Federal Emergency Management Agency, 2004.

Henry, John R. "A Better Way to Think: LRFD for Engineered Wood Construction." *Structural Engineer*, February 2003, pp 20-25.

HPVA HP-1—00, *The American National Standard for Hardwood and Decorative Plywood.* Reston, VA: Hardwood Plywood Veneer Association, 2000.

Hoke, John Ray, Jr., editor. *Ramsay/Sleeper Architectural Graphic Standards, 8th Edition.* New York: John Wiley & Sons, 1988.

ICC-ES Acceptance Criteria for Staples. Whittier, CA: ICC Evaluation Service, Inc., 2005. AC 201.

Publication BCSI 1-03, *Guide to Good Practice for Handling, Installing and Bracing of Metal Plate Connected Wood Trusses.* Madison, WI: Wood Truss Council of America (WTCA) & Truss Plate Institute (TPI), 2003

Recommended Lateral Force Requirements and Commentary (SEAOC Blue Book). Sacramento, CA: Structural Engineers Association of California, 1999.

SBCCI SSTD 10—99, *Standard for Hurricane Resistant Construction.* Birmingham, AL: Southern Building Code Congress International, 1999.

Stalnaker, Judith J. and Ernest Harris. *Structural Design in Wood.* New York: Van Nostrand Reinhold, 1989.

Technical Note B429F: *Non-Load Bearing Partitions on APA Structural Panel Floors.* Washington, D.C.: American Plywood Association, September 1999.

TPI 1—02, *National Design Standard for Metal Plate Connected Wood Trusses Construction.* New York: American National Standards Institute, 2002.

Williams, T. Jeff. *Basic Carpentry Techniques* San Ramon, CA: Orth Books (Chevron Chemical Company), 1981.

Wood Handbook, Wood as an Engineering Material. Washington, DC: USDA Forest Products Laboratory, U.S. Government Printing Office, 1999.

The Wood Truss Handbook. Miami: Gang-Nail Systems, Inc., 1981.

Chapter 24:
Glass and Glazing

General Comments

Chapter 24 includes design provisions and quality requirements for glazing consisting of glass and, to a limited extent, light-transmitting plastics.

Section 2401 contains the general contents and scope of this chapter.

Section 2402 provides meanings for words and phrases necessary for a complete understanding of the requirements of this chapter.

Section 2403 establishes the identification requirements for safety and nonsafety glazing, glass support and framing systems, glass thicknesses and requirements for glass at interior locations. This section also addresses special provisions for louvered and jalousie windows.

Section 2404 provides the design requirements and limitations for vertical glass exposed to wind loads, and sloped glass exposed to wind, snow and dead loads. Design factors for the more common types of glass are included in the design tables.

Section 2405 defines the allowable glazing types and related requirements for sloped glazing and skylights.

Section 2406 provides a detailed listing of the applications where safety glazing is required. Test standards category classifications for safety glazing are also included in this section.

Section 2407 considers glass used as either structural balustrades or in-fill panels in handrail and guardrail systems.

Section 2408 clarifies the special requirements for glass areas exposed to extraordinary impact from people. These are applicable to walls and doors in certain athletic facilities.

Section 2409 covers glass in elevator enclosures.

As is true with other chapters in the code that deal with building materials and components, this chapter evolved from the considerations of safety, public welfare and the integrity of a building's construction. There is an ever-growing demand from building owners and architects to incorporate larger quantities of glass in buildings. This in turn increases the concerns for such things as safety and energy conservation. The glass industry is being called on to meet these demands. Model building codes are changing almost daily to stay abreast of the ever-changing technology being employed in the glass industry.

Recognizing the risks should sloped glass fail, the model code groups and the glass and glazing industry have developed changes that only allow safety glazing materials over occupiable areas. Complying materials include laminated, fully tempered and wired glasses and acrylic and polycarbonate plastics. Except in residential occupancies, fully tempered glass is required to have a screen below the glass.

In addition to the concern for overhead glazing, a need developed to codify and address glazing materials in locations that are exposed to human impact, including but not necessarily limited to: sliding patio doors; side-swinging storm and primary use doors; tub and shower enclosures; and doors and certain fixed and operable glass panels. Building officials and the glass industry joined together to develop and enact safety glazing requirements to fill this need.

As a result, safety glazing laws have been adopted in many states. Based on the recommendation of the Consumer Product Safety Commission (CPSC), a federal law was enacted in 1977 that mandated safety glazing in all areas that might reasonably be exposed to impact from people. The requirements of this federal law were similar to the laws adopted by these various states.

This chapter also addresses the design and selection of glass to withstand wind loads, snow loads, dead loads and other imposed loads.

To assist in evaluating the requirements for glass in buildings, properties of the various types are listed below:

Annealed glass—has been subjected to the annealing process, which consists of controlled cooling in a cooling or annealing oven. Types of annealed glass include float, plate and sheet.

Float glass—is produced by floating a continuous ribbon of molten glass on molten tin. As the glass flows downstream, the temperature of the glass and tin is reduced. When the glass ribbon solidifies, it is lifted from the tin onto rollers and slowly cooled in a controlled cooling oven or annealing lehr. There is no further treatment of the glass. Most clear, gray, bronze and green-tinted glass for architectural use is presently produced by the float process.

Plate glass—is produced by grinding and polishing rough glass blanks. The glass blanks are formed by drawing semimolten glass through textured rollers, which are then fed through an annealing lehr prior to grinding and polishing. It should be noted that there is limited domestic production of plate glass material.

Sheet glass—is manufactured by continuously drawing molten glass from a bath, horizontally or vertically over rollers, through an annealing lehr. Because of the manufacturing process, sheet glass has an inherent waviness characteristic, and is, therefore, not typically used in applications where good optical quality is desired, such as architectural glazing, reflective glazing or show windows. The most common use of sheet glass is in residential applications.

Tempered or fully tempered glass—is annealed glass that has been heated and quenched in a controlled operation to provide a high level of surface compression. For most types of loading conditions, it has approximately four times the strength of regular annealed glass. Tempered glass breaks into small, relatively harmless fragments, and is used as safety glazing material. After tempering, this glass cannot be cut, drilled or otherwise altered without breaking.

Heat-strengthened glass—is produced in much the same way as tempered glass, except the level of surface compression is less. Heat-strengthened glass has approximately twice the strength of regular annealed glass and breaks into relatively large pieces. It is not a safety glazing material.

Laminated glass—is a sandwich of two or more plies of glass bonded together with resilient plastic interlayers, usually polyvinyl butyral. The glass plies may be regular, tempered or heat-strengthened glass. The plastic interlayer thickness is generally 0.030 inch (0.76 mm), but may range up to 0.090 inch (2.3 mm) for some applications. When this glass breaks, the fragments are held together by the plastic interlayer.

Wired glass—is a monolithic glass with wire mesh embedded in approximately the center of the glass thickness. Coupled with a suitable framing system, some wire glass is fire resistant. The surface of wire glass may be smooth or textured. Wire glass cannot be tempered or heat strengthened.

Insulating glass—is a factory-assembled glazing unit consisting of two or more panes of glass separated by airspaces. The edges of the finished glazing unit, including the airspaces, are hermetically sealed. The entrapped air is removed, creating a vacuum, which is necessary to prevent condensation within the sealed airspaces.

The airspaces between the glass can be filled with specialized gas to add to the energy efficiency of the glazing unit. The panes of glass may be any type, except for patterned glasses with deep patterns.

Patterned glass—has a texture or pattern embossed on one or both surfaces during the manufacturing process. Some glass with shallow patterns can be tempered and heat strengthened.

Spandrel glass—is manufactured from either float glass or plate glass that has an opaque coating of fired-on ceramic glaze on the interior surface (or room side). This glass is typically heat strengthened or tempered to resist the thermal stresses (movement) caused by the high absorption of solar radiation by the opaque coating.

Purpose

The primary purpose of this chapter is to provide information to establish the adequacy of glazing from the standpoint of life safety and performance. The sections on glass in handrails and guards, glazing in athletic facilities and safety glazing dictate a performance standard that will minimize risks to a level consistent with other provisions of the code.

SECTION 2401
GENERAL

2401.1 Scope. The provisions of this chapter shall govern the materials, design, construction and quality of glass, light-transmitting ceramic and light-transmitting plastic panels for exterior and interior use in both vertical and sloped applications in buildings and structures.

❖ This chapter establishes regulations for glass and glazing used in buildings and structures that, when installed, are subjected to wind loads, snow loads and dead loads. Engineering and design requirements in the form of tables, formulas and design loads are contained in this chapter. Additional structural requirements are found in Chapter 16.

A second concern of this chapter is glass and glazing used in areas where it is likely to be broken by human contact. Requirements for identification of safety glass, human impact loads, identification of hazardous locations and the types of glass that are allowed in railings and walls are included in this chapter.

This chapter details certain prescriptive and performance requirements for glazing used in buildings, including thickness limitations, design of glass supports and allowable sizes based on wind, snow and dead loads. The limitations on glass used in sloped glazing and areas exposed to human impact are also included. The information contained in this chapter on light-transmitting plastics is limited. For additional information on plastic materials, referenced standards and other code requirements for plastics, refer to Chapter 26.

2401.2 Glazing replacement. The installation of replacement glass shall be as required for new installations.

❖ Any glazing installed in existing construction is required to meet all the requirements and standards of the code. This includes installing new glass in an existing window, door or other type of opening, even where the glass being replaced did not comply with the standards of the code.

Relocating any type of manufactured unit containing glass, such as a door or window, from one location in an existing building or structure to another location in the same building or structure, or even a different building or structure, is to be considered new glazing, even though the unit containing the glass is existing.

SECTION 2402
DEFINITIONS

2402.1 Definitions. The following words and terms shall, for the purposes of this chapter and as used elsewhere in this code, have the meanings shown herein.

❖ No attempt has been made in the code or this chapter to define commonly used words or terms that are used in accordance with their established and accepted dictionary meanings, except where a word or term has been loosely used and it is necessary to define its meaning to avoid misunderstanding. For compatibility, when one of the *International Codes®* is referenced to establish the meaning of a word, term, phrase or abbreviation, the referenced code must bear the same publication date as the *International Building Code®* (IBC®) to which it is applied.

Definitions of terms can help in the understanding and application of code requirements. The purpose for including these definitions in this chapter is to provide more convenient access without having to refer back to Chapter 2. For convenience, these terms are also listed in Chapter 2 with a cross reference to this section. The use and application of all defined terms, including those defined herein, are set forth in Chapter 2.

DALLE GLASS. A decorative composite glazing material made of individual pieces of glass that are embedded in a cast matrix of concrete or epoxy.

❖ Dalle glass is a form of "stained" or "decorative" glass, and is more commonly known as "faceted" glass. Dalle glass is an outdated term and is rarely used. It can be prefabricated in panels that are set into framed openings, or it can be constructed in place, similar to the way a mason would construct a single wythe masonry wall. The actual glazing can be heavy pieces of flat or textured glass, irregular-shaped chunks of glass, glass shapes or translucent minerals cut to form geometric patterns or artistic scenes. Knowing that dalle glass or faceted glass is a type of decorative glass is necessary because decorative glass is not required to be safety glazed in hazardous locations as required in Section 2406.3 (see Section 2406.3.1, Exception 2, for specific exceptions).

DECORATIVE GLASS. A carved, leaded or Dalle glass or glazing material whose purpose is decorative or artistic, not

functional; whose coloring, texture or other design qualities or components cannot be removed without destroying the glazing material and whose surface, or assembly into which it is incorporated, is divided into segments.

❖ This definition is needed to understand and properly apply Section 2406.3.1, Exception 2. Decorative glass is not to be included in the requirements for specific hazardous locations, and thus is not required to be safety glazing.

The code uses the terms "glass" and "glazing" interchangeably as if there is no difference between the two. This can, and quite often does, cause confusion when talking about "safety glazing" or "safety glass."

Glass is a hard, brittle, inorganic material that is manufactured from a mixture of silica, flux and a stabilizer to form a transparent, translucent or opaque material. While in its molten (semiliquid) form, it can be blown, drawn, rolled, pressed or cast into a variety of shapes. Most of the glass referenced in this code is in sheet form. Glass, safety or not, as well as other materials can be used as glazing.

Glazing is to furnish or fit with glass, such as to install glass in an opening in a door. Glass installed in an exterior wall would be properly identified as a glazed opening, or more commonly, a window. An enclosure around a shower can be constructed of many materials, but it is considered to be a glazed enclosure if the enclosing material is glass. This code does permit the limited use of some plastics as glazing materials.

SECTION 2403
GENERAL REQUIREMENTS FOR GLASS

2403.1 Identification. Each pane shall bear the manufacturer's mark designating the type and thickness of the glass or glazing material. The identification shall not be omitted unless approved and an affidavit is furnished by the glazing contractor certifying that each light is glazed in accordance with approved construction documents that comply with the provisions of this chapter. Safety glazing shall be identified in accordance with Section 2406.2.

Each pane of tempered glass, except tempered spandrel glass, shall be permanently identified by the manufacturer. The identification mark shall be acid etched, sand blasted, ceramic fired, laser etched, embossed or of a type that, once applied, cannot be removed without being destroyed.

Tempered spandrel glass shall be provided with a removable paper marking by the manufacturer.

❖ Basic marking for identifying type and thickness of glass, as well as requirements and precautionary statements regarding the glass support system, are prescribed in this section. Detailed performance requirements are addressed in other sections.

Glass is a product that is manufactured and fabricated as assemblies at locations other than the building site. Identification of glass and glazing with a mark bearing the type and thickness is needed by the building inspector to perform a visual inspection (see Figure 2403.1). The inspection is needed to verify that the glazing is installed in the proper location in the building in compliance with this chapter for hazardous locations and structural loads. The code requires that each light, which is an individual pane of glass, be marked. Only tempered glass is required by the code to be permanently marked. Laminated, annealed, float and spandrel, unlike tempered glass, can be cut and fabricated without fracturing. Laminated glass is stockpiled in standard-size sheets by manufacturers. It consists of two or more sheets of glass held together by an intervening layer of plastic material. The stock sheets are marked by the manufacturer in one corner, which is cut off in the fabrication process. A permanent mark is therefore impractical, and the building official relies on an affidavit from the installer of the glazing.

The glazing contractor is responsible for the final installation of glass and glazing in the building, and must submit, with the approval of the building official, an affidavit as an alternative to marking. The affidavit must certify that each light is glazed under the approved plans and specifications. The affidavit relieves the building inspector of the responsibility for inspecting the glass and glazing for type and thickness. The affidavit does not relieve the plans examiner from the responsibility of reviewing the plans and specifications for compliance with the code. An affidavit may not be submitted for tempered glass that must be permanently identified in compliance with the requirements of this section.

Only tempered glass is required by the code to be permanently identified; the identification is to be visible after installation. Tempered glass, both heat strengthened and fully tempered, is produced by reheating and rapidly cooling annealed glass. Heat-strengthened and fully tempered glass have increased mechanical strength and resistance to thermal stresses. Fully tempered glass is twice as strong as heat-strengthened glass. Fully tempered glass will pulverize into innumerable small, cube-like fragments when broken at any point. Heat-strengthened glass breaks similarly to annealed glass. Heat-strengthened and fully tempered glass cannot be altered after fabrication. Tempered glass cannot be cut, drilled, ground or polished without fracture. The manufacturer of tempered glass must furnish the exact size and shape.

Permanent identification of spandrel glass is not required by the code, since a permanent label may alter the performance of the glass and cause breakage. Spandrel glass is an opaque glass that is heat strengthened through the process of fire fusing an opaque ceramic color to the interior surface of the glass. Spandrel glass is used as exterior windows and curtain walls to conceal internal construction and to provide solar control and energy conservation. An exposed permanent mark on tinted and reflective spandrel glass could change the coefficient of expansion of the glass. The label reacts differently to the heating and cooling effects of the atmosphere than the coated glass surface, which can cause increased incidence of glass breakage. The code, therefore, permits a removable paper mark for identification of spandrel glass.

For SI: 1 inch = 25.4 mm, 1 square foot = 0.0929 m^2.

a. "I" after "16 CFR 1201" indicates sizes 9 square feet and less are covered; "II" indicates all sizes are covered.
b. "U" indicates all sizes are covered; "L" indicates only sizes listed by the Safety Glazing Certification Council (SGCC) are covered.
c. Numbers or letters for date coding and plant identification may also be included.

Figure 2403.1
EXAMPLE IDENTIFYING MARK

2403.2 Glass supports. Where one or more sides of any pane of glass are not firmly supported, or are subjected to unusual load conditions, detailed construction documents, detailed shop drawings and analysis or test data assuring safe performance for the specific installation shall be prepared by a registered design professional.

❖ Most glass in buildings is firmly supported on all four edges. With many contemporary designs, glass is installed with support at the top and bottom edges only and, in some cases, with support on three edges. The free edges usually abut adjacent glass panels. When one or more edges of the glass are unsupported, the resistance to wind load is reduced. In this case, or in other cases where glass is subjected to unusual loading conditions, the code requires special engineering analysis and detailed construction documents to demonstrate compliance.

2403.3 Framing. To be considered firmly supported, the framing members for each individual pane of glass shall be designed so the deflection of the edge of the glass perpendicular to the glass pane shall not exceed $^1/_{175}$ of the glass edge length or $^3/_4$ inch (19.1 mm), whichever is less, when subjected to the larger of the positive or negative load where loads are combined as specified in Section 1605.

❖ Framing for glazing systems includes, but is not necessarily limited to: aluminum storefront framing; hollow metal framing; job-fabricated custom wood framing; and factory-manufactured window units constructed of wood, metal, vinyl or a combination of these materials, most of which are industry standards acknowledged and approved by the code. Regardless of the framing system, it is to comply with the structural requirements of Chapter 16.

It is important to note that "all edges" of each piece of glass must be supported within the parameters described in this section and elsewhere in the code. If not, detailed engineering analysis must be done as indicated in Section 2403.2.

2403.4 Interior glazed areas. Where interior glazing is installed adjacent to a walking surface, the differential deflection of two adjacent unsupported edges shall not be greater than the thickness of the panels when a force of 50 pounds per linear foot (plf) (730 N/m) is applied horizontally to one panel at any point up to 42 inches (1067 mm) above the walking surface.

❖ This provision addresses the installation of glass panels in interior locations adjacent to walking surfaces, such as storefronts within covered mall buildings. Unlike exterior, interior installations do not require the placement of mullions between adjacent panels to provide weather resistance. Without intermediate mullions, individual glass panels are free to move independently of each other. When only one panel is subjected to horizontal pressure, the deflection of the panel may cause gaps between adjacent panels not subjected to the same horizontal pressure. These gaps pose a potential pinching hazard to anyone who may contact the glass panels and cause the differential deflection, especially children. This hazard is greater when the glass panels are not mounted in the same plane and the meeting edges form an angle.

The design criteria provided in this section limit the use of glass panels to those that will not deflect more than the thickness of the glass panel itself when the specified force is applied at a height up to 42 inches (1067 mm) above the walking surface. This limitation protects against the occurrence of gaps between adjacent panels that would otherwise create a pinching hazard. The 42-inch (1067 mm) dimension represents the height range in which a person's body will likely contact the glass. The load must be assumed to occur at the point anywhere within the 42-inch (1067 mm) range that will cause the greatest differential deflection between the glass panels.

2403.5 Louvered windows or jalousies. Float, wired and patterned glass in louvered windows and jalousies shall be no thinner than nominal $^3/_{16}$ inch (4.8 mm) and no longer than 48 inches (1219 mm). Exposed glass edges shall be smooth.

Wired glass with wire exposed on longitudinal edges shall not be used in louvered windows or jalousies.

Where other glass types are used, the design shall be submitted to the building official for approval.

❖ Louvered and jalousie windows consist of a series of overlapping horizontal glass louvers that pivot simultaneously in the window frame. During opening, the bottom edge of each louver swings toward the exterior and the top edge swings toward the interior. The glass panels are supported on two vertical sides with the horizontal edges of the glass exposed.

For safety reasons, the exposed edges of the louvered panels are required to be smooth; thus, wire glass panels, if used, must have no exposed projecting wire. To provide structural strength, the panels are limited to a minimum thickness of $^3/_{16}$ inch (4.8 mm) and a maximum length of 48 inches (1219 mm).

Louvered and jalousie windows are exempt from safety glazing requirements in all applications, including those where a flat pane of glass is otherwise required to be safety glass. This exemption is based on records that show the injuries associated with this use of glass are primarily from persons impacting the glass edge with no cutting or piercing injuries resulting from glass breakage. Safety glass would not have an effect on this type of injury. There are also practical production reasons associated with fabricating safety glazing for the relatively long, thin slats.

SECTION 2404
WIND, SNOW, SEISMIC
AND DEAD LOADS ON GLASS

2404.1 Vertical glass. Glass sloped 15 degrees (0.26 rad) or less from vertical in windows, curtain and window walls, doors and other exterior applications shall be designed to resist the wind loads in Section 1609 for components and cladding. Glass in glazed curtain walls, glazed storefronts and glazed partitions shall meet the seismic requirements of ASCE 7, Section 13.5.9. The load resistance of glass under uniform load shall be determined in accordance with ASTM E 1300.

The design of vertical glazing shall be based on the following equation:

$$F_{gw} \leq F_{ga} \qquad \text{(Equation 24-1)}$$

where:

F_{gw} = Wind load on the glass computed in accordance with Section 1609.

F_{ga} = Short duration load on the glass as determined in accordance with ASTM E 1300.

❖ The procedures for determining compliance of the more common types of glass used in vertical applications are explained in this section. For a given glass type, thickness, plate length and plate width, allowable loads are determined in accordance with ASTM E 1300 using nonfactored loads and established factors that are based on glass type. ASTM E 1300 also considers the influence of width-to-height ratios, effects of weathering and exposure and statistical influence of glass size. The standard contains more than 40 nonfactored charts covering one-, two-, three- and four-sided support conditions. Application of ASTM E 1300 results in an allowable uniform wind load for a specified glass type, thickness and size. The allowable wind load must be at least the required wind loads determined from Section 1609 for components and cladding. If the glass allowable wind load is less than the required wind load from Section 1609, thicker glass is needed.

In buildings having an elevated seismic risk, glazed partitions, glazed storefronts and glazed curtain walls must be capable of withstanding the relative seismic displacements between the points of attachment to the supporting structure(s). The objective of the referenced ASCE 7 provision is to limit the hazard of falling glass during and after an earthquake.

2404.2 Sloped glass. Glass sloped more than 15 degrees (0.26 rad) from vertical in skylights, sunrooms, sloped roofs and other exterior applications shall be designed to resist the most critical of the following combinations of loads.

$$F_g = W_o - D \qquad \text{(Equation 24-2)}$$

$$F_g = W_i + D + 0.5\,S \qquad \text{(Equation 24-3)}$$

$$F_g = 0.5\,W_i + D + S \qquad \text{(Equation 24-4)}$$

where:

D = Glass dead load psf (kN/m²).

For glass sloped 30 degrees (0.52 rad) or less from horizontal,

= 13 t_g (For SI: 0.0245 t_g).

For glass sloped more than 30 degrees (0.52 rad) from horizontal,

= 13 $t_g \cos \theta$ (For SI: 0.0245 $t_g \cos \theta$).

F_g = Total load, psf (kN/m²) on glass.

S = Snow load, psf (kN/m²) as determined in Section 1608.

t_g = Total glass thickness, inches (mm) of glass panes and plies.

W_i = Inward wind force, psf (kN/m²) as calculated in Section 1609.

W_o = Outward wind force, psf (kN/m²) as calculated in Section 1609.

θ = Angle of slope from horizontal.

Exception: Unit skylights shall be designed in accordance with Section 2405.5.

The design of sloped glazing shall be based on the following equation:

$$F_g \leq F_{ga} \qquad \text{(Equation 24-5)}$$

where:

F_g = Total load on the glass determined from the load combinations above.

F_{ga} = Short duration load resistance of the glass as determined according to ASTM E 1300 for Equations 24-2 and 24-3; or the long duration load resistance of the glass as determined according to ASTM E 1300 for Equation 24-4.

❖ This section provides the procedure for determining the maximum allowable load that may be supported by a sloped glazing system. The maximum allowable load must be greater than the required design loads. The required design loads are first determined from the three load combinations listed as Equations 24-2, 24-3 and 24-4. The applicable wind loads from the components and cladding requirements of Section 1609 and the snow loads from Section 1608 are used to determine the required design loads from the three load combinations.

After the required design loads are determined, the maximum allowable load resistance of the specified glass is determined in accordance with ASTM E 1300. See the commentary to Section 2404.1 for a discussion of ASTM E 1300.

If the determined maximum allowable load resistance of the glass is not greater than that required from the load combinations, an option would be to choose a greater glass thickness and determine the maximum allowable load resistance of the glass in accordance with ASTM E 1300.

For load combinations given by Equations 24-2 and 24-3 wind is the primary loading, and the short-term load resistance from ASTM E 1300 applies. In Equation 24-4, however, snow is a primary load and wind becomes a "companion load" (factor of 0.5 is applied to wind) requiring use of the long duration load resistance in satisfying Equation 24-5. In the latter case, the nominal snow loads should consider drifts, partial snow load, etc., in accordance with the requirements under Section 1608.

The exception refers to design provisions that are specific to unit skylights (i.e., single panel, factory-assembled units).

2404.3 Wired, patterned and sandblasted glass.

2404.3.1 Vertical wired glass. Wired glass sloped 15 degrees (0.26 rad) or less from vertical in windows, curtain and window walls, doors and other exterior applications shall be designed to resist the wind loads in Section 1609 for components and cladding according to the following equation:

$$F_{gw} < 0.5 F_{ge} \qquad \text{(Equation 24-6)}$$

where:

F_{gw} = Is the wind load on the glass computed per Section 1609.

F_{ge} = Nonfactored load from ASTM E 1300 using a thickness designation for monolithic glass that is not greater than the thickness of wired glass.

❖ Wired glass is not within the scope of ASTM E 1300 and the requirements for vertical wired glass are therefore specified in this code. As with other vertical glazing, the wind loading on the glass is required to be determined in accordance with Section 1609 of the code. The allowable load on vertical wired glass is required to be that required for monolithic glass, as contained in ASTM E 1300, of a thickness less than or equal to that specified for the wired glass. ASTM E 1300 provides the method for determining values of short-duration and long-duration load resistance for specific glazing types by taking a nonfactored load from the load chart and adjusting that value for the glass type as well as the load duration. For wired glass, patterned glass, or sandblasted glass the nonfactored load from ASTM E 1300 is utilized. The factors in the corresponding code equations adjust the nonfactored load to establish a corresponding resistance value for these glass types. Note that a factor of 0.5 is used with the nonfactored load determined in accordance with ASTM E 1300. This equates to a factor of 2 on the wind loads determined in Section 1609.

2404.3.2 Sloped wired glass. Wired glass sloped more than 15 degrees (0.26 rad) from vertical in skylights, sunspaces, sloped roofs and other exterior applications shall be designed to resist the most critical of the combinations of loads from Section 2404.2.

For Equations 24-2 and 24-3:

$$F_g < 0.5 F_{ge} \qquad \text{(Equation 24-7)}$$

For Equation 24-4:

$$F_g < 0.3 F_{ge} \qquad \text{(Equation 24-8)}$$

where:

F_g = Total load on the glass.

F_{ge} = Nonfactored load from ASTM E 1300.

❖ Wired glass is not within the scope of ASTM E 1300 and the requirements for sloped wired glass are therefore specified in this code. As with other sloped glazing, the maximum allowable load must be greater than the required loads. The required design loads are first determined from the three load combinations listed as Equations 24-2, 24-3 and 24-4. The applicable wind loads from the components and cladding requirements of Section 1609 and the snow loads from Section 1608 are used to determine the required design loads from the three load combinations.

After the required design loads are determined, the maximum allowable load resistance of the specified glass is determined in accordance with ASTM E 1300. ASTM E 1300 provides the method for determining values of short-duration and long-duration load resistance for specific glazing types by taking a nonfactored load from the load chart and adjusting that value for the glass type as well as the load duration. For wired glass, patterned glass, or sandblasted glass the nonfactored load from ASTM E 1300 is utilized. The factors in the corresponding code equations adjust the nonfactored load to establish a corresponding resistance value for these glass types. Note that factors of either 0.5 or 0.3 are used with the nonfactored load determined in accordance with ASTM E 1300. This equates to a factor of 2 or 3.33 on the loads determined in the corresponding load combination. The resistance for sloped wired glass, patterned glass or sandblasted glass is to be determined using the nonfactored load from ASTM E 1300. The code fails to specify which type of glazing is appropriate to establish this resistance, but Section 2404.3.1 uses the values for monolithic glass; therefore, it is likely that the intent is to use the nonfactored load for monolithic glass in this section as well.

2404.3.3 Vertical patterned glass. Patterned glass sloped 15 degrees (0.26 rad) or less from vertical in windows, curtain and window walls, doors and other exterior applications shall be designed to resist the wind loads in Section 1609 for components and cladding according to the following equation:

$$F_{gw} < 1.0 F_{ge} \qquad \text{(Equation 24-9)}$$

where:

F_{gw} = Wind load on the glass computed per Section 1609.

F_{ge} = Nonfactored load from ASTM E 1300. The value for patterned glass shall be based on the thinnest part of the glass. Interpolation between nonfactored load charts in ASTM E 1300 shall be permitted.

❖ Patterned glass is not within the scope of ASTM E 1300 and the requirements for vertical patterned glass are therefore specified in this code. As with other vertical glazing, the wind loading on the glass is required to be determined in accordance with Section 1609 of the code. The allowable load on vertical patterned glass is to be as contained in ASTM E 1300 for a thickness less than or equal to the thinnest portion specified for the patterned glass. ASTM E 1300 provides the method for determining values of short-duration and long-duration load resistance for specific glazing types by taking a nonfactored load from the load chart and adjusting that value for the glass type as well as the load duration. For wired glass, patterned glass or sandblasted glass the nonfactored load from ASTM E 1300 is utilized. The factors in the corresponding code equations adjust the nonfactored load to establish a corresponding resistance value for these glass types. The resistance for sloped wired glass, patterned glass or sandblasted glass is to be determined using the nonfactored load from ASTM E 1300. The code fails to specify which type of glazing is appropriate to establish this resistance, but Section 2404.3.1 uses the values for monolithic glass; therefore, it is likely that the intent is to use the nonfactored load for monolithic glass in this section as well.

2404.3.4 Sloped patterned glass. Patterned glass sloped more than 15 degrees (0.26 rad) from vertical in skylights, sunspaces, sloped roofs and other exterior applications shall be designed to resist the most critical of the combinations of loads from Section 2404.2.

For Equations 24-2 and 24-3:

$$F_g < 1.0\,F_{ge} \qquad \textbf{(Equation 24-10)}$$

For Equation 24-4:

$$F_g < 0.6 F_{ge} \qquad \textbf{(Equation 24-11)}$$

where

F_g = Total load on the glass.

F_{ge} = Nonfactored load from ASTM E 1300. The value for patterned glass shall be based on the thinnest part of the glass. Interpolation between the nonfactored load charts in ASTM E 1300 shall be permitted.

❖ Patterned glass is not within the scope of ASTM E 1300 and the requirements for sloped patterned glass are therefore specified in this code. As with other sloped glazing, the maximum allowable load must be greater than the required loads. The required design loads are first determined from the three load combinations listed as Equations 24-2, 24-3 and 24-4. The applicable wind loads from the components and cladding requirements of Section 1609 and the snow loads from Section 1608 are used to determine the required design loads from the three load combinations.

The allowable load on vertical patterned glass is to be as contained in ASTM E 1300 for a thickness less than or equal to the thinnest portion specified for the patterned glass. ASTM E 1300 provides the method for determining values of short-duration and long-duration load resistance for specific glazing types by taking a nonfactored load from the load chart and adjusting that value for the glass type as well as the load duration. For wired glass, patterned glass or sand-

blasted glass the nonfactored load from ASTM E 1300 is utilized. The factors in the corresponding code equations adjust the nonfactored load to establish a corresponding resistance value for these glass types. Note that factors of either 1.0 or 0.6 are used with the nonfactored load determined in accordance with ASTM E 1300. This equates to a factor of 1 or 1.66 on the loads determined in the corresponding load combinations. The resistance for sloped wired glass, patterned glass or sandblasted glass is to be determined using the nonfactored load from ASTM E 1300. The code fails to specify which type of glazing is appropriate to establish this resistance, but Section 2404.3.1 uses the values for monolithic glass; therefore, it is likely that the intent is to use the nonfactored load for monolithic glass in this section as well.

2404.3.5 Vertical sandblasted glass. Sandblasted glass sloped 15 degrees (0.26 rad) or less from vertical in windows, curtain and window walls, doors, and other exterior applications shall be designed to resist the wind loads in Section 1609 for components and cladding according to the following equation:

$$F_g < 0.5\,F_{ge} \qquad \textbf{(Equation 24-12)}$$

where:

F_g = Total load on the glass.

F_{ge} = Nonfactored load from ASTM E 1300. The value for sandblasted glass is for moderate levels of sandblasting.

❖ Sandblasted glass is not within the scope of ASTM E 1300 and the requirements for vertical sandblasted glass are therefore specified in this code. As with other vertical glazing, the wind loading on the glass is required to be determined in accordance with Section 1609 of the code. The allowable load on vertical patterned glass is to be as contained in ASTM E 1300. ASTM E 1300 provides the method for determining values of short duration and long-duration load resistance for specific glazing types by taking a nonfactored load from the load chart and adjusting that value for the glass type as well as the load duration. For wired glass, patterned glass or sandblasted glass the nonfactored load from ASTM E 1300 is utilized. The factors in the corresponding code equations adjust the nonfactored load to establish a corresponding resistance value for these glass types. Note that a factor of 0.5 is used with the nonfactored load determined in accordance with ASTM E 1300. This equates to a factor of 2 on the wind loads determined in Section 1609. The resistance for sloped wired glass, patterned glass, or sandblasted glass is to be determined using the nonfactored load from ASTM E 1300. The code fails to specify which type of glazing is appropriate to establish this resistance, but Section 2404.3.1 uses the values for monolithic glass; therefore, it is likely that the intent is to use the nonfactored load for monolithic glass in this section as well.

2404.4 Other designs. For designs outside the scope of this section, an analysis or test data for the specific installation shall be prepared by a registered design professional.

❖ This section recognizes that neither the code nor the referenced standard includes all types of glass and all support conditions. Therefore, if a designer proposes an installation that falls outside the scope of the code and referenced standard, this section requires an analysis to be prepared and submitted to the building official for specific approval.

SECTION 2405
SLOPED GLAZING AND SKYLIGHTS

2405.1 Scope. This section applies to the installation of glass and other transparent, translucent or opaque glazing material installed at a slope more than 15 degrees (0.26 rad) from the vertical plane, including glazing materials in skylights, roofs and sloped walls.

❖ This section covers all applications of glazing and light-transmitting plastics installed on a slope more than 15 degrees (0.26 rad) from vertical. The provisions for determining the combination of wind, snow and dead loads that sloped glazing must resist are found in Section 2404. For additional information on the design of sloped glazing and skylights, refer to *Glass Design for Sloped Glazing*, available from the American Architectural Manufacturers Association (AAMA).

Sloped glazing is defined as any glazing system or material installed on a slope more than 15 degrees (0.26 rad) from vertical. Where glass is installed within 15 degrees (0.26 rad) of vertical, the gravity force (dead load) of the glass normal to its surface is very small and does not contribute materially to the total load. From an engineering standpoint, glass within 15 degrees (0.26 rad) of vertical will perform essentially the same as glass installed vertically.

Sloped glazing, by definition, is not intended to specifically exclude any type of glass or glazing system from the requirements of Chapter 24. Opaque glass, such as spandrel glass, as well as transparent and translucent glass, both opaque and light-transmitting plastics, decorative glass and glazing and any other glazing-type material contained in the chapter are subject to the design and structural requirements of this chapter, Chapter 16 and related chapters elsewhere in code.

Providing occupants with a safe environment is tantamount in the design and construction of buildings and structures. Sound engineering principles should always be employed and building code requirements strictly adhered to when dealing with issues that will affect the life, safety and welfare of people using or occupying buildings or structures. This is especially true when using building materials such as glass, which can cause severe injury to people when it is not used in a safe and proven manner.

2405.2 Allowable glazing materials and limitations. Sloped glazing shall be any of the following materials, subject to the listed limitations.

1. For monolithic glazing systems, the glazing material of the single light or layer shall be laminated glass with a minimum 30-mil (0.76 mm) polyvinyl butyral (or equivalent) interlayer, wired glass, light-transmitting plastic materials meeting the requirements of Section 2607, heat-strengthened glass or fully tempered glass.

2. For multiple-layer glazing systems, each light or layer shall consist of any of the glazing materials specified in Item 1 above.

Annealed glass is permitted to be used as specified within Exceptions 2 and 3 of Section 2405.3.

For additional requirements for plastic skylights, see Section 2610. Glass-block construction shall conform to the requirements of Section 2101.2.5.

❖ All configurations of glazing and plastic are covered in this section. This includes single (monolithic) glass and plastic, insulating glass units consisting of any number of glass panes (but usually two), double-walled plastic skylights and any combination of glass or plastic panels separately framed and inserted into a single opening (such as a prime window and a storm window).

Laminated glass is required to have a minimum 30-mil-thick (0.76 mm) plastic interlayer. Approved plastic materials are those that comply with the provisions of the code for strength, durability, sanitation and fire performance required for installation. Acrylics (e.g., Plexiglas) and polycarbonates (e.g., Lexan) are examples of plastics that are commonly found in sloped glazing installations.

2405.3 Screening. Where used in monolithic glazing systems, heat-strengthened glass and fully tempered glass shall have screens installed below the glazing material. The screens and their fastenings shall: (1) be capable of supporting twice the weight of the glazing; (2) be firmly and substantially fastened to the framing members and (3) be installed within 4 inches (102 mm) of the glass. The screens shall be constructed of a noncombustible material not thinner than No. 12 B&S gage (0.0808 inch) with mesh not larger than 1 inch by 1 inch (25 mm by 25 mm). In a corrosive atmosphere, structurally equivalent noncorrosive screen materials shall be used. Heat-strengthened glass, fully tempered glass and wired glass, when used in multiple-layer glazing systems as the bottom glass layer over the walking surface, shall be equipped with screening that conforms to the requirements for monolithic glazing systems.

Exception: In monolithic and multiple-layer sloped glazing systems, the following applies:

1. Fully tempered glass installed without protective screens where glazed between intervening floors at a slope of 30 degrees (0.52 rad) or less from the vertical plane shall have the highest point of the glass 10 feet (3048 mm) or less above the walking surface.

2. Screens are not required below any glazing material, including annealed glass, where the walking surface below the glazing material is permanently protected from the risk of falling glass or the area below the glazing material is not a walking surface.

3. Any glazing material, including annealed glass, is permitted to be installed without screens in the sloped glazing systems of commercial or detached noncombustible greenhouses used exclusively for growing plants and not open to the public, provided that the height of the greenhouse at the ridge does not exceed 30 feet (9144 mm) above grade.

4. Screens shall not be required within individual dwelling units in Groups R-2, R-3 and R-4 where fully tempered glass is used as single glazing or as both panes in an insulating glass unit, and the following conditions are met:

    4.1. Each pane of the glass is 16 square feet (1.5 m^2) or less in area.

    4.2. The highest point of the glass is 12 feet (3658 mm) or less above any walking surface or other accessible area.

    4.3. The glass thickness is $^3/_{16}$ inch (4.8 mm) or less.

5. Screens shall not be required for laminated glass with a 15-mil (0.38 mm) polyvinyl butyral (or equivalent) interlayer used within individual dwelling units in Groups R-2, R-3 and R-4 within the following limits:

    5.1. Each pane of glass is 16 square feet (1.5 m^2) or less in area.

    5.2. The highest point of the glass is 12 feet (3658 mm) or less above a walking surface or other accessible area.

❖ In sloped applications, certain glass types have a tendency to fall from the opening when broken. As a result, a retaining net or screening is required in specific locations. Regular (annealed) glass may not be used in sloped glazing installations, except as defined in Section 2405.3. Heat-strengthened and fully tempered glazing are two types of glass that may fall from an opening when broken by wind or snow loads. For these glasses, a retaining screen is required on the interior side of the glazing unit. This would apply to single glazing or when the glass is the inboard pane in an insulating glass unit. Other exceptions for fully tempered glass are listed in Section 2405.3.

Wire glass, when used as the inboard pane in an insulating glass unit, has a tendency to crack from thermal stresses. This is due to weakening at the edges where the wire protrudes. Where fire-resistance-rated glazing material is needed, wire glass is mandatory. In this case, because of the increased likelihood of breakage, a retaining screen is also required.

There are both performance and prescriptive re-quirements for the retaining screen. Any screen that meets the performance requirements is acceptable.

Five exceptions to the requirements for a retaining screen are listed. For the purposes of this section, patterned glass that is annealed is to be considered regular (annealed) glass.

Exception 1: Fully tempered glass does not require a protective screen when the glass is sloped 30 degrees (0.52 rad) or less from vertical, and the top edge is less than 10 feet (3048 mm) above a walking surface. The sloped walls in many commercial buildings are within these limitations. The glass has been satisfactory with no known risks. When broken, fully tempered glass will break into small fragmented pieces, which are unlikely to cause harm upon impact [see Figure 2405.3(1)].

Exception 2: In cases where the area below the glass is either permanently protected or inaccessible, falling glass does not present a risk. When these conditions are met, screens are not required below the sloped glazing [see Figure 2405.3(2)].

Exception 3: Glass in greenhouses does not pose a serious risk to the general public. The failure of a sloped glazing system in a private greenhouse is not considered a high risk, as the majority of floor space is taken up by plants and the building is not occupied on a regular basis by a large number of people. Where a greenhouse is not attached to any other structure, has restricted access and meets the specified height limitation, screens are not required below the sloped glazing.

Exception 4: Sloped glazing installations using fully tempered glass are not required to have screens installed below the glass when the glazing units or skylights are installed within individual dwelling units of Group R-2, R-3 and R-4 occupancies, and the glazing complies with the size, thickness and location requirements specified in Items 4.1, 4.2 and 4.3. This exception is applicable only to glazing installations that are located within a dwelling unit; it does not apply to any public or common areas. It is also not applicable to areas in buildings containing two-family and multiple-family dwellings that are common to all occupants (e.g., entry foyers, main corridors, community storage and laundry rooms).

This exception is based on a survey conducted by the glass and glazing industry. The information produced from the survey indicated a 0.081-percent incidence of breakage among fully tempered sun spaces and skylights. More importantly, the survey found no incidences of injury were reported when the glazing was fully tempered glass installed within the limitations specified in the exception.

Exception 5: This exception is permitted because of the protection afforded by the polyvinyl butyral

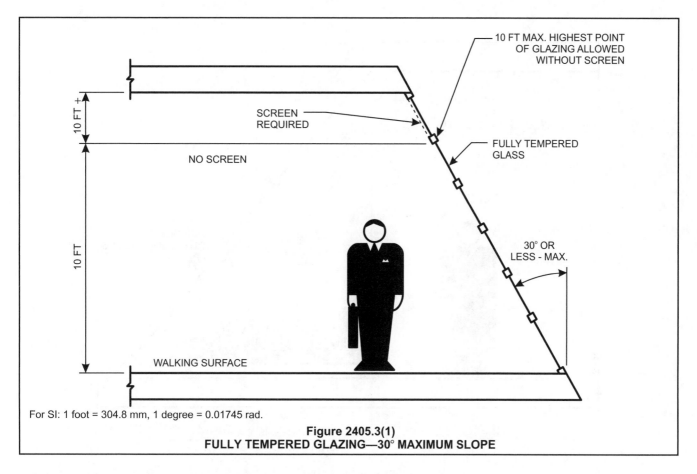

For SI: 1 foot = 304.8 mm, 1 degree = 0.01745 rad.

Figure 2405.3(1)
FULLY TEMPERED GLAZING—30° MAXIMUM SLOPE

interlayer located within the laminated glass panel. The interlayer serves to keep the broken pieces of glass in place so that little or no glass falls when broken by impact. This exception applies only to individual dwelling units within Group R-2, R-3 and R-4 occupancies when sized and located as prescribed in Items 5.1 and 5.2.

2405.4 Framing. In Type I and II construction, sloped glazing and skylight frames shall be constructed of noncombustible materials. In structures where acid fumes deleterious to metal are incidental to the use of the buildings, approved pressure-treated wood or other approved noncorrosive materials are permitted to be used for sash and frames. Framing supporting sloped glazing and skylights shall be designed to resist the tributary roof loads in Chapter 16. Skylights set at an angle of less than 45 degrees (0.79 rad) from the horizontal plane shall be mounted at least 4 inches (102 mm) above the plane of the roof on a curb constructed as required for the frame. Skylights shall not be installed in the plane of the roof where the roof pitch is less than 45 degrees (0.79 rad) from the horizontal.

Exception: Installation of a skylight without a curb shall be permitted on roofs with a minimum slope of 14 degrees (three units vertical in 12 units horizontal) in Group R-3 occupancies. All unit skylights installed in a roof with a pitch flatter than 14 degrees (0.25 rad) shall be mounted at least 4 inches (102 mm) above the plane of the roof on a curb

constructed as required for the frame unless otherwise specified in the manufacturer's installation instructions.

❖ The requirement for noncombustible (metal) framing for glazing systems in Type 1 and 2 construction must be consistent with the requirements for the remainder of the structure. In cases where acidic fumes harmful to metal framing may occur, pressure-treated wood or noncorrosive material may be used.

Regardless of the type of construction, when installed on surfaces having a slope of less than 45 degrees (0.79 rad) from the horizontal, all sloped glazing and skylights are required to be mounted on curbs at least 4 inches (102 mm) high. The raised curb will provide additional protection to the skylight from burnt or burning debris in the event of a fire. Burnt or burning pieces of wood are known technically as "brands." When air borne, they are known as "flying brands." For slopes 45 degrees (0.79 rad) and greater, the protection provided by the curbs is not necessary. In these cases, the glass may be installed flush with the surrounding roof.

The exception for Group R-3 occupancies is based on field experience and general practice for one- and two-family dwellings. For the 3:12 slope typical for Group R-3 occupancies neither flying brands nor water runoff has been a problem.

Figure 2405.3(2)
INACCESSIBLE AREA UNDER SLOPED GLAZING

2405.5 Unit skylights. Unit skylights shall be tested and labeled as complying with AAMA/WDMA/CSA 101/I.S.2/A440. The label shall state the name of the manufacturer, the approved labeling agency, the product designation and the performance grade rating as specified in AAMA/WDMA/CSA 101/I.S.2/A440. If the product manufacturer has chosen to have the performance grade of the skylight rated separately for positive and negative design pressure, then the label shall state both performance grade ratings as specified in AAMA/WDMA/CSA 101/I.S.2/A440 and the skylight shall comply with Section 2405.5.2. If the skylight is not rated separately for positive and negative pressure, then the performance grade rating shown on the label shall be the performance grade rating determined in accordance with AAMA/WDMA/CSA 101/I.S.2/A440 for both positive and negative design pressure and the skylight shall conform to Section 2405.5.1.

❖ This section establishes design provisions for unit skylights (single-panel, factory-assembled units) that are consistent with those provided for sloped glazing in Section 2402, but permit the unit skylights to be designed separately for maximum positive and negative pressure. The most critical load on a skylight is determined by the climate in which it is installed. In a colder climate with heavier snow loads and moderate design wind speeds, the positive load on a skylight from the combined snow and dead loads will be more critical than the negative load from wind uplift. The opposite will be the case in warmer, coastal climates with higher design wind speeds and little or no snow load.

This method of rating skylights is addressed in the referenced standard AAMA 101/I.S.2/A440, *Specification for Windows, Doors and Unit Skylights*. The separate rating system for positive and negative pressure on skylights allows the manufacturer to design and fabricate products that are best suited for the climate in which they will be used. AAMA 101/I.S.2/A440 establishes the performance requirements for skylights based on the desired performance grade rating, which includes minimum requirements for resistance to air leakage, water infiltration and design load pressures. The resulting performance grade rating states the design load pressure used to rate the product, but also includes consideration of these additional performance characteristics. For skylights certified for only one performance grade, the rating is based on the minimum requirements met for both positive and negative design pressure. Skylights certified for two performance grades are rated separately for positive and negative design pressure.

2405.5.1 Unit skylights rated for the same performance grade for both positive and negative design pressure. The design of unit skylights shall be based on the following equation:

$$F_g \leq PG \qquad \textbf{(Equation 24-13)}$$

where:

F_g = Maximum load on the skylight determined from Equations 24-2 through 24-4 in Section 2404.2.

PG = Performance grade rating of the skylight.

❖ For skylights that are not rated separately for positive and negative design pressure, this section requires that the minimum performance grade rating of the skylight must not be exceeded by any of the load combinations given in Section 2404.2.

2405.5.2 Unit skylights rated for separate performance grades for positive and negative design pressure. The design of unit skylights rated for performance grade for both positive and negative design pressures shall be based on the following equations:

$$F_{gi} \leq PG_{Pos} \qquad \textbf{(Equation 24-14)}$$

$$F_{go} \leq PG_{Neg} \qquad \textbf{(Equation 24-15)}$$

where:

PG_{Pos} = Performance grade rating of the skylight under positive design pressure;

PG_{Neg} = Performance grade rating of the skylight under negative design pressure; and

F_{gi} and F_{go} are determined in accordance with the following:

For $W_o \geq D$,

where:

W_o = Outward wind force, psf (kN/m²) as calculated in Section 1609.

D = The dead weight of the glazing, psf (kN/m²) as determined in Section 2404.2 for glass, or by the weight of the plastic, psf (kN/m²) for plastic glazing.

F_{gi} = Maximum load on the skylight determined from Equations 24-3 and 24-4 in Section 2404.2.

F_{go} = Maximum load on the skylight determined from Equation 24-2.

For $W_o < D$,

where:

W_o = Is the outward wind force, psf (kN/m²) as calculated in Section 1609.

D =

The dead weight of the glazing, psf (kN/m²) as determined in Section 2404.2 for glass, or by the weight of the plastic for plastic glazing.

F_{gi} = Maximum load on the skylight determined from Equations 24-2 through 24-4 in Section 2404.2.

F_{go} = 0.

❖ This section establishes requirements for skylights that are rated separately for positive and negative design pressure. The performance grade rating for positive design pressure is not to be exceeded by load combinations that include dead loads, snow and wind acting towards the face of the skylight, and the performance grade rating for negative design pressure is not to be exceeded by the load combination that considers wind acting away from the face of the skylight.

SECTION 2406
SAFETY GLAZING

2406.1 Human impact loads. Individual glazed areas, including glass mirrors, in hazardous locations as defined in Section 2406.3 shall comply with Sections 2406.1.1 through 2406.1.4.

❖ With the increased use of sliding patio doors, glass tub and shower enclosures and similar applications where large pieces of glass are used, cutting and piercing injuries from contact with broken glass have become a safety concern. In recognition of a lack of code provisions controlling these risks, codes included limited requirements that regulated the use of glass. It was recognized that safety glazing materials must be required for glazed areas that would reasonably be exposed to human impact. Some questions arose as to the proper definition of these areas (Section 2406.3 identifies these "hazardous locations"). In response, Section 2406 provides comprehensive regulations for the use of safety glazing materials and installation parameters.

Safety glazing requirements apply to both replacement glass and new construction. Replacement glass includes windows, doors and other assemblies containing glass that are moved, partially or intact, from one location in a building to another location in the same building, or to another building on the same site or a different site.

While glass mirrors are included within the scope of glazing, mirrors with a continuous backing support are not required to be safety glazed according to Exception 7 in Section 2406.3.1.

TABLE 2406.1. See page 24-14.

❖ The classifications in this table are according to the CPSC 16 CFR, Part 1201 regulation. The tests in the federal regulation vary with the category classification. Category I glass is impacted from a drop height of 18 inches (457 mm) and Category II is impacted from a drop height of 48 inches (1219 mm).

TABLE 2406.1
MINIMUM CATEGORY CLASSIFICATION OF GLAZING

EXPOSED SURFACE AREA OF ONE SIDE OF ONE LITE	GLAZING IN STORM OR COMBINATION DOORS (Category class)	GLAZING IN DOORS (Category class)	GLAZED PANELS REGULATED BY ITEM 7 OF SECTION 2406.3 (Category class)	GLAZED PANELS REGULATED BY ITEM 6 OF SECTION 2406.3 (Category class)	DOORS AND ENCLOSURES REGULATED BY ITEM 5 OF SECTION 2406.3 (Category class)	SLIDING GLASS DOORS PATIO TYPE (Category class)
9 square feet or less	I	I	No requirement	I	II	II
More than 9 square feet	II	II	II	II	II	II

For SI: 1 square foot = 0.0929m².

2406.1.1 CPSC 16 CFR 1201. Except as provided in Sections 2406.1.2 through 2406.1.4, all glazing shall pass the test requirements of CPSC 16 CFR 1201, listed in Chapter 35. Glazing shall comply with the CPSC 16 CFR, Part 1201 criteria, for Category I or II as indicated in Table 2406.1.

❖ In 1977, the CPSC, in cooperation with building officials and the glass industry, drafted and approved CPSC 16 CFR, Part 1201. This is the standard recognized in the code. Laminated glass with an 0.030-inch (0.76 mm) interlayer readily meets CPSC test requirements. Sections 2406.1.2 through 2406.1.4 are essentially exceptions to CPSC 16 CFR, Part 1201.

2406.1.2 Plastic glazing. Plastic glazing shall meet the weathering requirements of ANSI Z97.1.

❖ This section has requirements for plastic safety glazing that have been eliminated from CPSC 16 CFR, Part 1201, necessitating the inclusion of this exception for plastic glazing. Requirements for plastic glazing are in Chapter 26.

2406.1.3 Glass block. Glass-block walls shall comply with Section 2101.2.5.

❖ The testing requirements for glass-block walls are described in Chapter 21, which eliminates the CPSC test requirement and references the glass-block wall section of the code. Glass-block walls and panels are also not required to meet the test requirements of CPSC 16 CFR; however, that does not mean there are no safety requirements placed on the installation of glass block. Structural limitations to prevent failure caused by excessive loading, size limitations of panels, location and extent of panels in fire-resistance-rated assemblies and other safety issues are defined elsewhere in the code.

2406.1.4 Louvered windows and jalousies. Louvered windows and jalousies shall comply with Section 2403.5.

❖ Louvered and jalousie windows are discussed in detail in Section 2403.5.

2406.2 Identification of safety glazing. Except as indicated in Section 2406.2.1, each pane of safety glazing installed in hazardous locations shall be identified by a manufacturer's designation specifying who applied the designation, the manufacturer or installer and the safety glazing standard with which it complies, as well as the information specified in Section 2403.1. The designation shall be acid etched, sand blasted, ceramic fired, laser etched, embossed or of a type that once applied, cannot be removed without being destroyed. A label as defined in Section 1702.1 and meeting the requirements of this section shall be permitted in lieu of the manufacturer's designation.

Exceptions:

1. For other than tempered glass, manufacturer's designations are not required, provided the building official approves the use of a certificate, affidavit or other evidence confirming compliance with this code.

2. Tempered spandrel glass is permitted to be identified by the manufacturer with a removable paper designation.

❖ This section requires that each individual piece of safety glazing must be permanently marked with the manufacturer's or installer's designation. The designation is applied directly to the glass by the manufacturer or installer and states that the product or material complies with the specified code standards or requirements as defined in Section 2406.1.

Many manufacturers of safety glazing materials have their products certified by the Safety Glazing Certification Council (SGCC) to meet CPSC 16 CFR, Part 1201. This standard provides the basis for the marking of safety glazing materials.

2406.2.1 Multilight assemblies. Multilight glazed assemblies having individual lights not exceeding 1 square foot (0.09 m²) in exposed areas shall have at least one light in the assembly marked as indicated in Section 2406.2. Other lights in the assembly shall be marked "CPSC 16 CFR 1201."

❖ In the case of glass material, many times the glass is purchased by the fabricator/installer in large sheets from the manufacturer and cut into many pieces to fit various openings. This is particularly true with laminated or plastic glazing material. In this instance, it is impracticable to apply the designation label in such a manner that each cut piece will have a manufacturer's permanent designation. Where this occurs, the code permits multilight glazed assemblies to have only one light marked with the required information.

2406.3 Hazardous locations. The following shall be considered specific hazardous locations requiring safety glazing materials:

1. Glazing in swinging doors except jalousies (see Section 2406.3.1).

2. Glazing in fixed and sliding panels of sliding door assemblies and panels in sliding and bifold closet door assemblies.

3. Glazing in storm doors.

4. Glazing in unframed swinging doors.

5. Glazing in doors and enclosures for hot tubs, whirlpools, saunas, steam rooms, bathtubs and showers. Glazing in any portion of a building wall enclosing these compartments where the bottom exposed edge of the glazing is less than 60 inches (1524 mm) above a standing surface.

6. Glazing in an individual fixed or operable panel adjacent to a door where the nearest exposed edge of the glazing is within a 24-inch (610 mm) arc of either vertical edge of the door in a closed position and where the bottom exposed edge of the glazing is less than 60 inches (1524 mm) above the walking surface.

   Exceptions:

   1. Panels where there is an intervening wall or other permanent barrier between the door and glazing.

   2. Where access through the door is to a closet or storage area 3 feet (914 mm) or less in depth. Glazing in this application shall comply with Section 2406.3, Item 7.

   3. Glazing in walls perpendicular to the plane of the door in a closed position, other than the wall towards which the door swings when opened, in one- and two-family dwellings or within dwelling units in Group R-2.

7. Glazing in an individual fixed or operable panel, other than in those locations described in preceding Items 5 and 6, which meets all of the following conditions:

   7.1. Exposed area of an individual pane greater than 9 square feet (0.84 m^2);

   7.2. Exposed bottom edge less than 18 inches (457 mm) above the floor;

   7.3. Exposed top edge greater than 36 inches (914 mm) above the floor; and

   7.4. One or more walking surface(s) within 36 inches (914 mm) horizontally of the plane of the glazing.

   Exception: Safety glazing for Item 7 is not required for the following installations:

   1. A protective bar $1^1/_2$ inches (38 mm) or more in height, capable of withstanding a horizontal load of 50 pounds plf (730 N/m) without contacting the glass, is installed on the accessible sides of the glazing 34 inches to 38 inches (864 mm to 965 mm) above the floor.

2. The outboard pane in insulating glass units or multiple glazing where the bottom exposed edge of the glass is 25 feet (7620 mm) or more above any grade, roof, walking surface or other horizontal or sloped (within 45 degrees of horizontal) (0.78 rad) surface adjacent to the glass exterior.

8. Glazing in guards and railings, including structural baluster panels and nonstructural in-fill panels, regardless of area or height above a walking surface.

9. Glazing in walls and fences enclosing indoor and outdoor swimming pools, hot tubs and spas where all of the following conditions are present:

   9.1. The bottom edge of the glazing on the pool or spa side is less than 60 inches (1524 mm) above a walking surface on the pool or spa side of the glazing; and

   9.2. The glazing is within 60 inches (1524 mm) horizontally of the water's edge of a swimming pool or spa.

10. Glazing adjacent to stairways, landings and ramps within 36 inches (914 mm) horizontally of a walking surface; when the exposed surface of the glass is less than 60 inches (1524 mm) above the plane of the adjacent walking surface.

11. Glazing adjacent to stairways within 60 inches (1524 mm) horizontally of the bottom tread of a stairway in any direction when the exposed surface of the glass is less than 60 inches (1524 mm) above the nose of the tread.

    Exception: Safety glazing for Item 10 or 11 is not required for the following installations where:

    1. The side of a stairway, landing or ramp which has a guardrail or handrail, including balusters or in-fill panels, complying with the provisions of Sections 1013 and 1607.7; and

    2. The plane of the glass is greater than 18 inches (457 mm) from the railing.

❖ The provisions of this section apply to all occupancies and building types, except as specifically excluded in this chapter. This section identifies 11 locations that are considered by the code to be hazardous and require the use of safety glass for glazed openings or glazed partitioning. Several figures are included to help illustrate how and where safety glazing must be used.

The first four items included in this section identify various types of doors and door assemblies. Four separate types of doors are presented independently to help avoid any confusion as to what types of doors are required to have safety glass. Collectively, Items 1 through 4 can be summarized by saying that any door containing glazing must be glazed with safety glass or other safety glazing material recognized by the code for that intended purpose. Items 1 through 4 are not restricted to egress or exit doors [see Figure 2406.3(1)].

Jalousie assemblies in doors, as described in Section 2403.5, are not required to have safety glazing. There are other limited exceptions discussed in Item 6 that apply to doors and fixed panels.

Because of the increased possibility of injury caused by falling in the areas identified in Section 2406.3, Item 5, the provisions for safety glazing are intentionally stringent. Falling by itself can cause enough injury without adding cutting from broken glass to the list of possible injuries. Glass wall partitions and glass doors that are not part of the primary building wall construction, interior or exterior, used to enclose these areas must be constructed from safety glass or other safety glazing material allowed by the code.

Glazed openings in primary building walls, interior or exterior, located within the enclosed compartment areas identified in Section 2406.3, Item 5, where the bottom edge of the glass is within 60 inches (1524 mm) of the standing or walking surface must be glazed with safety glass or other safety glazing material allowed by the code [see Figure 2406.3(2)]. Glazed openings (i.e., windows) located outside the identified enclosures are not required to be glazed with safety glass, unless where required by other sections in this chapter or elsewhere in the code.

The purpose for including the area in Item 6 as a hazardous location is to provide protection in cases where a person may slip or mistake the glass panel adjacent to a door for a passageway and walk into the glass, or where a person may push against the side-light with one hand for support while opening the door with the other hand. There are reported accidents where a person's hand has slipped from the doorknob, impacted and broken the glass adjacent to the door, thereby causing injury. This item is applicable to glass adjacent to both exterior and interior doors used for passage for all occupancies and types of buildings.

It is not necessary for an entire piece of glass in a glazed wall or opening to be within the 24-inch (610 mm) arc to require it to be safety glass. If any portion of an individual piece of glass is within the arc, that piece of glass must be safety glass [see Figure 2406.3(3)].

There are three exceptions to Item 6 that eliminate the need for safety glazing adjacent or in close proximity to glass doors. These exceptions are included because they eliminate or greatly reduce the possibility of human contact with fixed glazing.

Exception 1: An intervening wall provides a permanent barrier that prevents people from having physical contact with any glass that is beyond the intervening wall, but is still within the 24-inch (610 mm) arc. This barrier eliminates many of the safety issues addressed in Item 6. This exception may be applied to an interior or exterior wall/glazing condition [see Figure 2406.3(4)]. The code does not specify a minimum height requirement for the intervening wall; therefore, where the wall does not extend to the full height of the rooms or spaces where the door/wall is located, the

60-inch (1524 mm) requirement must be applied. If, for example, the top of the wall is 60 inches (1524 mm) above the floor, any individual piece of glass that extends below the top of the wall must be safety glass, or the top of the wall must be extended to provide a barrier for any piece of nonsafety glazing within 60 inches (1524 mm) of the floor. The same must be applied to the 24-inch (610 mm) arc in the plan dimension. Any glazing within the arc must be safety glazing, or the wall must be extended to include glazing not protected by the wall [see Figure 2406.3(5)].

Exception 2: This exception addresses closets and other storage areas that are not considered "walk-ins." The assumption is that at such locations a building occupant would not need to pass through the door opening to access the storage area, thus reducing the risk of impact with any adjacent glazing.

Exception 3: Because the occupants of one- and two-family dwellings are presumed to be familiar with their environment, and walls that are perpendicular to the plane of a door are, by definition, parallel to the direction of travel of an occupant using the door, the risk of impact with glazing is considered to be low. This decreased risk is recognized by Exception 3. Since there is a chance that an opening door could push a person into the wall towards which the door opens, the exception does not apply, and safety glazing would be required at that location.

Individual dwelling units within Group R-2 occupancies are allowed the same exemption; however, it does not extend to the use of glazing in close proximity to doors in public use areas.

The reason for including the type of glazed opening in Item 7 for a hazardous location is to provide protection where the glazed opening could be mistaken for a passageway, or a clear opening that someone might be able to walk through, fall into or otherwise be accidentally forced into. In the case of a child, the opening would not need to be very large to provide the necessary setting to encourage an accident. A glazed opening sized and located within the parameters described in Item 7 and not protected with safety glazing would most certainly present an unsafe condition [see Figure 2406.3(6)].

The requirements in Item 7 are modeled after the criteria established in CPSC 16 CFR, Part 1201, which requires the use of safety glazing where all of the conditions listed in Items 7.1 through 7.4 occur.

In Item 7.1, the 9-square-feet (0.84 m^2) requirement is meant to include an individual piece of glass. It is not intended to limit the number of individual pieces that can be placed next to each other within a single glazed opening. It just requires that every individual piece exceeding the size specified be safety glass.

In Item 7.2, when the bottom edge of a glazed opening is less than 18 inches (457 mm) above a floor or walking surface, it does not present enough of a visual barrier to provide ample warning to prevent someone from mistaking the glazed opening for a passageway or clear opening. People have a natural tendency to look forward while walking rather than looking down; therefore, a more substantial visual barrier is needed at the floor level.

Where Item 7.2 addresses the sill height of glazed openings with no regard to how high the head of the opening may be, it assumes the opening is high enough for an adult to walk through. Item 7.3 addresses the head height of the glazed opening with the assumption that the sill is at the floor level. Item 7.3 is aimed at the same basic set of safety concerns. It is merely directing the focus at a different group of people, that being children. Just as adults can mistake a tall glazed opening for a passageway or clear opening, a small child can do the same for a glazed open-

ing with a low head height. Other examples of potential accidents from impacting glass at this type of opening might include: people in wheelchairs, shopping carts, baby strollers (with or without baby inside), children on tricycles or other small riding toys. The accident potential for this type of glazed opening is high; therefore, safety glass or other approved safety glazing material is demanded accordingly.

The focus of Item 7.4 for requiring safety glass is no different than it has been throughout Item 7, which is preventing unsafe glazed openings that people can mistake for passageways or clear openings. In Items 7.1 through 7.3, it states that where a walking surface is adjacent to the wall where the glazed opening is located, human contact with the glass in the opening is

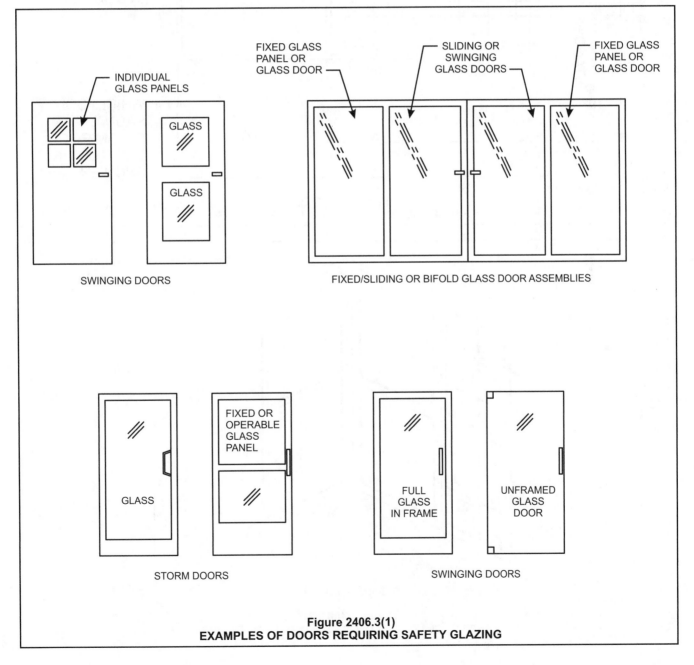

Figure 2406.3(1)
EXAMPLES OF DOORS REQUIRING SAFETY GLAZING

For SI: 1 inch = 25.4 mm.

Figure 2406.3(2)
SAFETY GLAZING IN DOORS AT WET AREAS

PRIMARY BUILDING WALL

SAFETY GLAZING REQ'D WHEN BOTTOM EDGE OF GLAZING IS LESS THAN 60" ABOVE STANDING SURFACE

SAFETY GLAZING REQUIRED IN DOORS

HOT TUB, WHIRLPOOL, SAUNA, STEAM ROOM, BATHTUB OR SHOWER

For SI: 1 inch = 25.4 mm.

① SAFETY GLAZING REQUIRED

60 INCHES

24-INCH RADIUS FROM EDGE OF CLOSED DOOR

LESS THAN 60 INCHES

60 INCHES

Figure 2406.3(3)
SAFETY GLAZING IN CLOSE PROXIMITY TO DOORS

For SI: 1 inch = 25.4 mm.

Figure 2406.3(4)
SAFETY GLAZING AT FULL-HEIGHT INTERVENING WALL

For SI: 1 inch = 25.4 mm.

Figure 2406.3(5)
SAFETY GLAZING AT INTERMEDIATE-HEIGHT INTERVENING WALL

inevitable. Item 7.4 states that if there is not a walking surface within 36 inches (914 mm) horizontally, in conjunction with the other requirements, the risks described throughout Item 7 are minimal, and safety glass is not needed. It is only when walking surfaces are within 36 inches (914 mm) that human contact becomes a safety concern that must be addressed.

Exceptions 1 and 2 to Item 7 are allowed because the concerns of Item 7 are resolved or they do not exist. When these exceptions are met, there is no need for the use of safety glass or other approved safety glazing.

Exception 1: The protective bar described in this exception will provide a visual as well as a physical barrier, both of which will prevent someone from attempting to walk through a glazed opening [see Figure 2406.3(7)].

Exception 2: The outboard pane of insulating glass as described in this exception is well outside the area that any person could possibly come into contact with and certainly not mistake for a passageway or clear opening. It should be noted that this exception applies only to the outboard pane. The criteria for requiring the inboard pane to be safety glass are prescribed elsewhere in this chapter [see Figure 2406.3(8)].

Item 8 states that regardless of any provision for safety glazing or exemption from requiring safety glazing at any location or application contained in this section or elsewhere in the code—where glass or other approved glazing material is used in guards or railings, either wholly or partly, structural or nonstructural, regardless of size, location, type or any other circumstances not specifically mentioned herein—it must be safety glass or other approved safety glazing material. If the type of glazing material is not available as an approved safety glazing material, it may not be used in guards or railings. There are no exceptions to Item 8.

The reason for requiring safety glazing at walls or fencing used for enclosing swimming pools, hot tubs and spas, as stated in Item 9, is the high amount of activity coupled with the wet floor and walking surface conditions that are always present in these types of facilities or spaces. The conditions presented in Item 9 are focused on glazing in openings as viewed from the pool or spa side of the wall, fence or enclosure. Glazed openings in pool or spa walls viewed from the side of the wall away from the pool or spa are discussed elsewhere in this section. As a point of clarity, a wall, fence or enclosure is any wall that encloses a pool or spa area. It can be a wall constructed specifi-

For SI: 1 inch = 25.4 mm, 1 square foot = 0.0929 m².

Figure 2406.3(6)
FIXED OR OPERABLE GLAZED OPENING

cally to enclose an outdoor pool or spa, an existing or new exterior building wall, a residential garage wall, a dressing cabana or a pool equipment shed. It can also be an interior or exterior wall. It applies to any wall that can support a glazed opening.

Item 9.1 has the requirement for safety glass when the sill of the glazed opening is within 60 inches (1524 mm) of a walking surface, which is similar to Item 5 because of the similarity in environmental and usage conditions. In both areas people are exposed to wet floor or walking surfaces in high-activity facilities. Compare the 18-inch (457 mm) bottom edge height in Condition 7.2 of Item 7 to the 60 inches (1524 mm) required here and in Item 5. The 18-inch (457 mm) standard is primarily a visual barrier whereas the 60-inch (1524 mm) standard is a human impact consideration [see Figure 2406.3(9)].

All the things that have been discussed in Item 9 and 9.1 can be applied to Item 9.2 as well. The standard here is for any glazing within 60 inches (1524 mm) horizontally when measured from the edge of the swimming pool or spa to the water.

All of the elements in both of these conditions must be present or safety glazing is not required. Glazing in doors is addressed elsewhere in this and other sections.

Stairways and ramps present users with a greater risk for injury caused by falling than a flat surface. Not only is the risk of falling greater when using a stair, but the injuries are generally more severe. Unlike falling on a flat surface where the floor will, for the most part, break a person's fall, there is nothing to stop someone from continuing to fall until he or she reaches the bottom of the stair. The increased risks inherent in stairways, as well as attempting to be consistent with other chapters in the code that mandate more restrictive requirements when addressing safety issues involving stairs and ramps, account for the more restrictive requirements for glazing in and around stairways and ramps.

According to Item 10, this includes any glazing within 36 inches (914 mm) horizontally of any walking surface when the exposed surface of that glazing is within 60 inches (1524 mm) of the walking surface. The walking surface in question would be part of a stair or ramp itself, including top, bottom and intermediate landings. It does not include adjacent floors or other walking surfaces.

In Item 11, the concern is with any glass that may be located within 60 inches (1524 mm) from the bottom tread in a run of stairs and within 60 inches (1524 mm) vertically of the walking surface of a stair. The code does not distinguish between a bottom tread at the primary floor level or at an intermediate landing. The last

PROTECTIVE BAR MUST NOT CONTACT GLASS WITH LOAD APPLIED

1½-INCH-MINIMUM PROTECTIVE BAR

50 PLF HORIZONTAL DESIGN LOAD

36 INCHES ±2 INCHES

For SI: 1 inch = 25.4 mm, 1 pound per foot = 14.5 N/m.

Figure 2406.3(7)
PROTECTIVE BAR AT GLAZED OPENINGS

tread in a run of steps is the bottom tread. The 60-inch (1524 mm) dimension is from any point on the bottom tread, horizontally in any direction to any surface of any glazing within that range.

There is an exception to hazardous location conditions of Item 10 and 11 that relieves the requirement or safety glazing because they effectively eliminate the risk of a cutting or piercing injury caused by broken glass upon human contact. Both conditions of the exception to Items 10 and 11 must be present to eliminate the need for safety glazing as described in Item 10 and 11.

Guards and railings that comply with the means of egress requirements prescribed in Chapter 10, the structural requirements of Chapter 16 and are located where directed in Item 1 of the exception will effectively serve to prevent someone from crashing into glazed walls or glazed openings in walls located in close vertical proximity to stairs or ramps. Once the guards or railings required in Item 1 are in place, no glazing may be closer than 18 inches (457 mm) to guards or railings when measured horizontally from the guard or railing.

2406.3.1 Exceptions. The following products, materials and uses shall not be considered specific hazardous locations:

1. Openings in doors through which a 3-inch (76 mm) sphere is unable to pass.

2. Decorative glass in Section 2406.3, Item 1, 6 or 7.

3. Glazing materials used as curved glazed panels in revolving doors.

4. Commercial refrigerated cabinet glazed doors.

5. Glass-block panels complying with Section 2101.2.5.

6. Louvered windows and jalousies complying with the requirements of Section 2403.5.

7. Mirrors and other glass panels mounted or hung on a surface that provides a continuous backing support.

❖ There are seven specific glazing applications that are not classified as hazardous locations and are not required to be glazed with safety glass or other approved safety glazing material:

1. Glass view panels 3 inches (76 mm) or less in diameter installed in doors are not likely to create a safety hazard if the glass is broken.

2. Decorative glass and decorative glass assemblies are not required to be safety glass for most applications (see Item 8 of Section 2406.3 regarding guards and railings). Stained glass assemblies in doors or walls are readily visible and easily identified, and are not likely to be mistaken for an open passageway. As a rule, stained glass assemblies

are made from small pieces of glass that would cause little or no damage if broken.

3. The reason for not classifying revolving doors, which are generally made from curved glass panels, as a hazardous location is that the two primary safety concerns are either not present or pose minimal or no unsafe condition in the case of human impact.

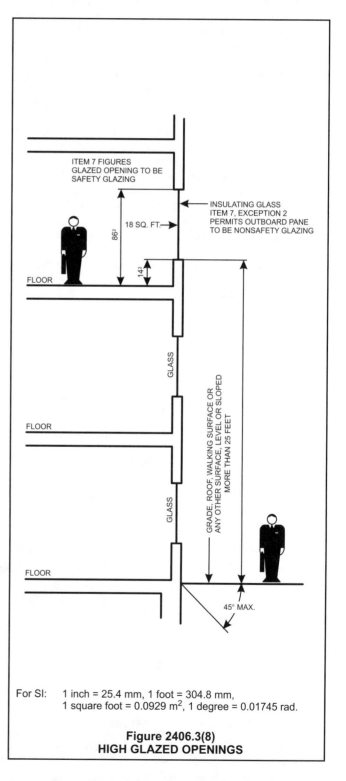

For SI: 1 inch = 25.4 mm, 1 foot = 304.8 mm, 1 square foot = 0.0929 m², 1 degree = 0.01745 rad.

Figure 2406.3(8)
HIGH GLAZED OPENINGS

Several factors must be considered when classifying revolving doors. The first thing to be considered is that they cannot be used as part of the required means of egress. Another thing to consider is that revolving doors are not fixed glass panels that can be mistaken for a passageway or clear opening. In addition, the individual pieces of glass to make the large door panels, by necessity, are required to be very thick. The extra thickness of glass, coupled with the added strength of the curved form and the movable panel, make it highly unlikely that injury could result from glass breakage caused by human impact.

4. This exception is based on CPSC 16 CFR, Part 1201. Technically, glass in doors of refrigerator cabinets should comply with the standard when the door is open. The intent, however, was not to cover doors a person would not ordinarily use for egress. Refrigerated cabinets in food markets would be in this category and are, therefore, exempt. There are also no records of injuries from glass in this application.

5. Glass block, also referred to as "glass unit masonry," is not classified as glass or glazing. It is considered a type of masonry and is subject to comply with Chapter 21 (see Sections 2101.2.5 and 2110).

6. Jalousie and louvered windows are not classified as hazardous locations, and are not required to be glazed with safety glass or other approved safety glazing material. The reasons why these types of glazed openings are exempt are addressed in Section 2403.5.

7. This exception recognizes the potential for cutting or piercing injuries to be minimized when the glass or glazing material is completely supported by a solid, continuous backing material, as is often the case for mounted mirrors or other glass panels. When glazing is fully supported with a solid backing material, it is still subject to breakage by human impact; however, a person's limb or other body part would be protected because the backing material would prevent it from being in the path of falling glass.

2406.4 Fire department access panels. Fire department glass access panels shall be of tempered glass. For insulating glass units, all panes shall be tempered glass.

❖ Because of the need to provide safety for fire-fighting access, this section requires that all glass be safety glazing. When fire-fighting operations include breaking through openings, all glass will obviously be broken, which will subject fire fighters to the hazards of broken glass from all panes.

For SI: 1 inch = 25.4 mm.

Figure 2406.3(9)
GLAZING IN POOL ENCLOSURE

SECTION 2407
GLASS IN HANDRAILS AND GUARDS

2407.1 Materials. Glass used as a handrail assembly or a guard section shall be constructed of either single fully tempered glass, laminated fully tempered glass or laminated heat-strengthened glass. Glazing in railing in-fill panels shall be of an approved safety glazing material that conforms to the provisions of Section 2406.1.1. For all glazing types, the minimum nominal thickness shall be $^1/_4$ inch (6.4 mm). Fully tempered glass and laminated glass shall comply with Category II of CPSC 16 CFR 1201, listed in Chapter 35.

❖ This section provides requirements for glass applications in handrail and guard assemblies.

Glass in handrail and guard applications is often exposed to impact from people, carried objects and other items. Early code requirements were vague regarding the appropriate types of glass for these installations.

Glass used as structural balustrade panels must resist the design loads applied to the handrail component of the guard system with an adequate factor of safety (see Figure 2407.1). On a practical basis, the only glass types that are structurally adequate are single and laminated tempered glass and laminated heat-strengthened glass. Other glass types and all safety plastics would need to be excessively thick to resist the design loads on the railing.

Glass guard infill panels are almost always installed in a hard-setting expanding cement. This is to provide a rigid, secure support for the glass. Large compressive stresses, transferred from the handrail component above, are exerted on the glass at the bottom edge. Only the glass types listed in the previous paragraph will routinely resist these forces without breakage. This is another reason why only single and laminated tempered glass and laminated heat-strengthened glass are allowed.

In-fill panels that do not support the railing have less stringent requirements. All safety glazing materials are allowed. A minimum thickness of $^1/_4$ inch (6.4 mm) is listed to verify reasonable penetration and breakage resistance. All types must meet the requirements of CPSC 16 CFR, Part 1201.

2407.1.1 Loads. The panels and their support system shall be designed to withstand the loads specified in Section 1607.7. A safety factor of four shall be used.

❖ This section requires that the support system for glass guard or handrail assemblies be designed based on a safety factor of 4. Nominally identical panes of glass inherently have a wide variation in strength. The safety factor of 4 is used in the design to minimize the likelihood that breakage will occur below the design loads. It is not intended that an in-place glass guard or handrail system be tested for or capable of withstanding four times the design load.

2407.1.2 Support. Each handrail or guard section shall be supported by a minimum of three glass balusters or shall be otherwise supported to remain in place should one baluster panel

fail. Glass balusters shall not be installed without an attached handrail or guard.

❖ Glass balustrade panels have been used for many decades without a significant history of problems. Glass of any type may break, however, if exposed to severe impact or other unusual or unexpected loads. A guard or handrail system should be designed so that the failure of a single glass panel does not cause the collapse of the assembly. As a result, the code does not allow glass balusters to be used without a guard or handrail, and it requires that guards or handrails be fastened to a minimum of three glass balustrade panels or otherwise supported if one panel fails.

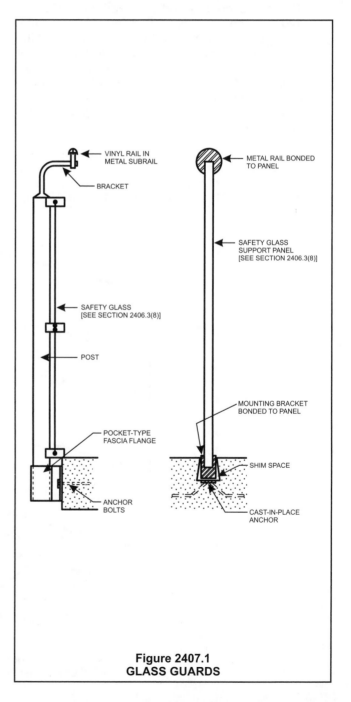

**Figure 2407.1
GLASS GUARDS**

2407.1.3 Parking garages. Glazing materials shall not be installed in handrails or guards in parking garages except for pedestrian areas not exposed to impact from vehicles.

❖ This section is included to remove any ambiguity. The large impact loads from vehicles cannot be resisted by glass balustrades. As a result, they are not allowed under any circumstances where the guard or handrail might reasonably be impacted by vehicles. Glass and plastic in-fill panels are also not allowed because of the high likelihood of failure if impacted by vehicles.

SECTION 2408
GLAZING IN ATHLETIC FACILITIES

2408.1 General. Glazing in athletic facilities and similar uses subject to impact loads, which forms whole or partial wall sections or which is used as a door or part of a door, shall comply with this section.

❖ In most applications of safety glazing materials, human impact into glass is an unexpected and infrequent occurrence. Safety glazing is required to minimize injury associated with the unintentional impact and subsequent breakage of fixed glass or glazing. Glass in walls or doors of enclosures for athletic activity is exposed to frequent and intentional impact. For proper performance, the glass not only must break in a relatively safe manner, but also withstand breakage under normal use. A further requirement is that the glass be sufficiently rigid to provide a resilient playing surface. This section includes the performance requirements for glass in this special application.

The code requirement for safety glazing applies to any athletic facility as well as similar areas where the glass is subject to frequent human impact in the opinion of the building official.

2408.2 Racquetball and squash courts.

❖ The primary focus of this section is the walls and doors of racquetball, handball, squash and volleyball courts. In these applications, the glass is intentionally and frequently impacted in the normal course of its use.

2408.2.1 Testing. Test methods and loads for individual glazed areas in racquetball and squash courts subject to impact loads shall conform to those of CPSC 16 CFR, Part 1201, listed in Chapter 35, with impacts being applied at a height of 59 inches (1499 mm) above the playing surface to an actual or simulated glass wall installation with fixtures, fittings and methods of assembly identical to those used in practice.

Glass walls shall comply with the following conditions:

1. A glass wall in a racquetball or squash court, or similar use subject to impact loads, shall remain intact following a test impact.

2. The deflection of such walls shall not be greater than 1¹/₂ inches (38 mm) at the point of impact for a drop height of 48 inches (1219 mm).

Glass doors shall comply with the following conditions:

1. Glass doors shall remain intact following a test impact at the prescribed height in the center of the door.

2. The relative deflection between the edge of a glass door and the adjacent wall shall not exceed the thickness of the wall plus ¹/₂ inch (12.7 mm) for a drop height of 48 inches (1219 mm).

❖ The acceptance testing for the glass walls is similar in some respects to that included in CPSC 16 CFR, Part 1201. There are some major variations in the impact testing and end-point criteria. There are two major variations in the testing: the glass is impacted at a height of 59 inches (1499 mm) and the actual (or simulated) construction, complete with fixtures and attachments, is tested. The impact level of 59 inches (1499 mm) is the average shoulder height for an adult male as listed in handbooks for architectural design.

The end-point criteria require that the glass assembly not fail upon impact and the maximum deflection not exceed 1¹/₂ inches (38 mm). Although the latter is not related to safety, it is included to provide a reasonable playing surface. Both are unique and are not ordinary requirements for safety glazing materials.

Glass doors in the applications cited have been particularly hazardous in some designs. A player may impact and deflect the access door, while the adjacent wall remains in its original position. Occasionally, a player's fingers have become trapped between the door and the wall when the force on the door is released. This section of the code is directed at reducing this hazard.

Though not stated, the intent is that the impact for the glass panels be at the midpoint of the horizontal dimension.

2408.3 Gymnasiums and basketball courts. Glazing in multipurpose gymnasiums, basketball courts and similar athletic facilities subject to human impact loads shall comply with Category II of CPSC 16 CFR 1201, listed in Chapter 35.

❖ Unlike racquetball and squash courts, the intentional impacting of gymnasium and basketball court enclosures is not anticipated. Nevertheless, due to the nature of their use, these facilities have a higher incidence of unintentional glazing impacts. This section requires any glazing that is subject to human impact loads in gymnasiums, basketball courts and similar athletic facilities to be Category II safety glazing (see commentary, Table 2406.1).

SECTION 2409
GLASS IN ELEVATOR HOISTWAY

2409.1 Glass in elevator enclosures. Glass in elevator enclosures shall be laminated glass conforming to ANSI Z97.1 or 16 CFR Part 1201. Markings as specified in the applicable standard shall be on each separate piece of glass and shall remain visible after installation.

❖ This section provides a level of protection to building occupants from the hazards associated with broken glass as it relates to glass in elevator enclosures. These glazed openings are susceptible to impact from components during elevator operation or servicing. The requirement for laminated glass is consistent with the requirements within the safety code for elevators, ASME A17.1.

Bibliography

The following resource materials are referenced in this chapter or are relevant to the subject matter addressed in this chapter.

AAMA 101/I.S.2/A440-04, *Specifications for Windows, Doors and Unit Skylights*. Schaumberg, IL: American Architectural Manufacturers Association, 2004.

AAMA NAFS 1-02, *Voluntary Specification for Windows, Skylights and Glass Doors*. Schaumburg, IL: American Architectural Manufacturers Association, 2002

ANSI Z97.1-84 (R1994), *Safety Performance Specifications and Methods of Test for Safety Glazing Materials Used in Buildings*. New York: American National Standards Institute, 1984.

ASCE 7-05, *Minimum Design Loads for Buildings and Other Structures*. New York: American Society of Civil Engineers, 2005.

ASTM C 1036-91, *Specification for Flat Glass*. West Conshohocken, PA: ASTM International, 1991.

ASTM C 1048-92, *Heat-Treated Flat Glass: Kind HS, Kind FT Coated and Uncoated Glass*. West Conshohocken, PA: ASTM International, 1992.

ASTM E 1300-04, *Standard Practice for Determining Load Resistance of Glass in Buildings*. West Conshohocken, PA: ASTM International, 2004.

CPSC 16 CFR, Part 1201-00, *Safety Standard for Architectural Glazing*. Washington, D.C.: Consumer Product Safety Commission, 1998.

Glass Design for Sloped Glazing. Des Plaines, IL: American Architectural Manufacturers Association, 1987.

CGSB 101/I.S.2/A440-04, *Specifications for Windows, Doors and Unit Skylights*. Gatineau, Quebec: Canadian General Standards Board, 2004.

WDMA 101/I.S.2/A440-04, *Specifications for Windows, Doors and Unit Skylights.* Des Plaines, IL: Window and Door Manufacturers Association, 2004.

Chapter 25:
Gypsum Board and Plaster

General Comments

Chapter 25 contains the provisions and referenced standards that regulate the design, construction and quality of gypsum board and plaster. These represent the most common interior and exterior finish materials in the building industry, and the provisions dealing with them are contained within this single chapter. This chapter deals primarily with quality-control-related issues with regards to material specifications and installation requirements. This chapter addresses weather ratings, and also provides cross references to the requirements for shear walls constructed of either wood stud or light-gage steel stud framing. In addition, provisions for constructing horizontal gypsum board ceiling diaphragms are included.

Where materials described in this chapter are used or required for fire-resistant construction, the code requires that they also comply with the provisions of Chapter 7.

Section 2501 addresses the scope of this chapter.

Section 2502 provides the meaning of words and phrases that are used frequently throughout this chapter. Additional definitions are listed at the beginning of each chapter and in Chapter 2.

Section 2503 explains the responsibility of the contractor to have work inspected by the building official.

Section 2504 defines the construction and application requirements for gypsum board and plaster for use on walls and ceilings.

Section 2505 allows for the use of lightweight shear walls that are sheathed with gypsum board or lath and plaster.

Section 2506 outlines the quality standards for gypsum board and related materials and accessories.

Section 2507 outlines the quality standards for lath, plaster and related materials and accessories.

Section 2508 includes material and construction standards for the use of gypsum board and gypsum plaster materials with reference to fire-resistance-rated assemblies. It also provides requirements for constructing gypsum board ceiling diaphragms using wood joists.

Section 2509 establishes the requirements for the use of gypsum board in wet areas.

Section 2510 describes the material standards and construction requirements for cement plaster and related accessories.

Section 2511 defines the material standards, construction requirements and limitations for plaster and related accessories when applied to an interior surface.

Section 2512 defines the material standards, construction requirements and limitations for plaster and related accessories when applied to exterior and weather-exposed surfaces.

Section 2513 addresses the use of applied aggregate finishes to interior and exterior plaster.

Gypsum board and plaster products represent a line of building products that are, for the most part, manufactured under the control of industry standards. The building official or inspector only needs to verify that the appropriate product has been used and properly installed for the intended use and location. Several key points of installation have been addressed in this chapter where these practices have been known to create problems or when not addressed by other gypsum or plaster product standards.

Purpose

Chapter 25 provides the minimum design requirements through referenced and industry standards for the use of gypsum board, gypsum plaster and cement plaster materials in construction.

SECTION 2501
GENERAL

2501.1 Scope.

❖ Chapter 25 includes the requirements that govern the materials, design, construction, quality and application for both interior and exterior gypsum board and plaster products. Although these products are most commonly used as interior or exterior finished wall and ceiling covering the surface of buildings, proper design and application are necessary to provide weather resistiveness and required fire protection for both structural and nonstructural building components, as well as structural requirements for gypsum board and plaster products and assemblies.

Provisions for the materials, quality, construction and labeling of all gypsum board and plaster materials used in the construction of wall and ceiling coverings are included, along with their method of fastening and, in the case of plaster, the permitted materials for lath, plaster and aggregate in all buildings and structures. In general, all gypsum board and plaster materials are required to conform to the applicable standards as referenced in this chapter.

2501.1.1 General. Provisions of this chapter shall govern the materials, design, construction and quality of gypsum board, lath, gypsum plaster and cement plaster.

❖ Although plaster has many uses in construction, including ornamental and decorative work, its use in the code is regulated purely as a wall and ceiling covering material. The code regulates the installation of wall and ceiling covering materials, as well as the quality standards for the materials themselves. Chapter 25 regulates gypsum board, lath and plaster.

2501.1.2 Performance. Lathing, plastering and gypsum board construction shall be done in the manner and with the materials specified in this chapter, and when required for fire protection, shall also comply with the provisions of Chapter 7.

❖ The extensive use of gypsum board and plaster throughout the construction industry for many years has led to the development of a comprehensive index of tested and proven performance standards. This chapter is directed at identifying the performance standards required to ensure a quality of construction that will provide a safe, habitable environment for the life of the building or structure. The standards contained and referenced in this chapter and elsewhere in the code establish the minimum performance standards and requirements for all methods and materials contained in this chapter.

Gypsum plaster, portland cement plaster and gypsum board products and materials have inherent fire-resistant qualities that provide outstanding passive fire protection of building components and assemblies. When these materials are used in the construction of fire-resistive assemblies, the details of construction are to comply with the requirements established by the official reports for the fire tests conducted by recognized testing laboratories in accordance with Chapter 7 and the referenced standards listed in Chapter 35.

2501.1.3 Other materials. Other approved wall or ceiling coverings shall be permitted to be installed in accordance with the recommendations of the manufacturer and the conditions of approval.

❖ Gypsum board and plaster are currently the primary building materials being used for wall and ceiling covering, but there are many other materials in use today for covering interior and exterior walls, ceilings and soffits. This chapter makes no attempt to address the use of any other building materials such as wood, concrete, concrete masonry, brick, metal or vinyl, nor is it intended to limit the use of these or any other materials for similar installations. The material standards and construction regulations for other approved building materials are addressed elsewhere in the code.

SECTION 2502
DEFINITIONS

2502.1 Definitions. The following words and terms shall, for the purposes of this chapter and as used elsewhere in this code, have the meanings shown herein.

❖ Definitions of terms can help in the understanding and application of the code requirements. The purpose for including these definitions within this chapter is to provide more convenient access to them without having to refer back to Chapter 2.

For convenience, these terms are also listed in Chapter 2 with a cross reference to this section. Terms that are italicized provide a visual identification throughout the code text that a definition exists for that term.

The use and application of all defined terms, including those defined herein, are set forth in Section 201.

CEMENT PLASTER. A mixture of portland or blended cement, portland cement or blended cement and hydrated lime, masonry cement or plastic cement and aggregate and other approved materials as specified in this code.

❖ Cement plaster (often referred to as "portland cement plaster" or "stucco") is a cementitious-based plaster material with excellent water-resistant properties. It is the only type of plaster that is permitted by the code to be used as an exterior wall covering and as a base coat and finish coat. Cement plaster can also be used as an interior finish wall covering, but is required by the code to be used in all interior wet areas such as toilet rooms, showers, saunas, steam rooms, indoor swimming pools or any other area that will be exposed to excessive amounts of moisture or humidity for prolonged periods of time.

EXTERIOR SURFACES. Weather-exposed surfaces.

❖ Exterior surfaces are the vertical and horizontal surfaces located on the outside of a building or structure and are subject to weather conditions such as wind, changes in temperature, humidity, moisture and dampness. The commentary on "Weather-exposed surfaces" included in this section allows for specific exceptions to areas on the exterior of a building or structure that will not be exposed to direct contact with water, such as rain, sleet and frozen or melting snow.

GYPSUM BOARD. Gypsum wallboard, gypsum sheathing, gypsum base for gypsum veneer plaster, exterior gypsum soffit board, predecorated gypsum board or water-resistant gypsum backing board complying with the standards listed in Tables 2506.2, 2507.2 and Chapter 35.

❖ Gypsum board is the most commonly used material for interior wall covering. Gypsum board is also used for exterior sheathing, plaster lath and ceiling covering. Because it is installed in sheet form, it is less labor intensive and generally considered more cost effective than other wall and ceiling materials, such as plaster.

Gypsum board requires a minimal amount of finishing and will readily accept paint, wallpaper, vinyl fabric, special textured paint and similar surface finish materials.

Gypsum board will be subject to severe failure when placed in direct contact with water or continuous moisture. For this reason, the code does not allow gypsum board to be used in wet areas unless it is provided with a finish material impervious to moisture.

GYPSUM PLASTER. A mixture of calcined gypsum or calcined gypsum and lime and aggregate and other approved materials as specified in this code.

❖ Gypsum plaster is available in several forms and is designed for a variety of applications, including those where fire protection of building components, construction of fire-resistance-rated assemblies within the building or control of sound transmission is needed. Gypsum plaster can be applied using a two- or three-coat method, depending upon the backing or lathing system.

Gypsum plaster provides a hard, smooth finish surface that will readily receive most decorative finishes, such as paint, vinyl wall fabric, wallpaper and textured paint. Various aggregates can be added to the plaster composition to achieve different surface textures, from a smooth troweled to a heavily textured finish.

However, the code does not permit the use of gypsum plaster at exterior locations or at wet areas on the interior of the building.

GYPSUM VENEER PLASTER. Gypsum plaster applied to an approved base in one or more coats normally not exceeding $^1/_4$ inch (6.4 mm) in total thickness.

❖ Gypsum veneer plaster is specifically designed for use as a one-coat plaster veneer finish surface. Gypsum veneer plaster must be applied over a solid base such as gypsum board lath, gypsum base for veneer plaster, concrete or concrete unit masonry. Although gypsum veneer plaster is designed to be applied in one coat, it can be applied in two coats, as long as the maximum thickness does not exceed $^1/_4$ inch (6.4 mm).

Gypsum veneer plaster provides a hard, smooth finish surface, much like three-coat gypsum plaster, that will readily receive a variety of decorative finishes. Like gypsum plaster, the code does not permit the use of gypsum veneer plaster at exterior locations or at wet areas on the interior of the building.

INTERIOR SURFACES. Surfaces other than weather-exposed surfaces.

❖ Interior surfaces are all surfaces, exposed and unexposed to view, on the inside of a building or structure that are not subject to damage or failure of their intended purpose because of the effects of uncontrollable weather conditions, such as rain, sleet, snow and wind. Interior surfaces are generally exposed to mechanically controlled climatic conditions through air conditioning, heating and humidity control. Because of these controls, interior surfaces require far less restrictions than exterior surfaces; therefore, the quantities

and types of materials and finishes permitted for use on interior surfaces are greatly increased.

WEATHER-EXPOSED SURFACES. Surfaces of walls, ceilings, floors, roofs, soffits and similar surfaces exposed to the weather except the following:

1. Ceilings and roof soffits enclosed by walls, fascia, bulkheads or beams that extend a minimum of 12 inches (305 mm) below such ceiling or roof soffits.

2. Walls or portions of walls beneath an unenclosed roof area, where located a horizontal distance from an open exterior opening equal to at least twice the height of the opening.

3. Ceiling and roof soffits located a minimum horizontal distance of 10 feet (3048 mm) from the outer edges of the ceiling or roof soffits.

❖ Outside surfaces of a building or structure are generally exposed to elements of the weather; therefore, it is necessary to take special precautions to protect them from damage. The most common way to protect exterior surfaces from damage caused by rain, sleet, wind, ice and snow or water from nonweather and other sources is by the use of building materials that are designed, manufactured and installed to resist or withstand the extremes produced by weather. There are many building materials other than those identified in this chapter that provide excellent protection from weather damage, and are permitted by the *International Codes*® for use on weather-exposed surfaces. Requirements for other approved materials, such as, but not limited to, brick, concrete, precast concrete panels, concrete unit masonry, wood, vinyl, aluminum, glass and steel, are specified in other chapters of the code.

There are isolated surfaces located on the exterior of a building or structure that are not directly exposed to or in contact with the damaging effects of the weather. These areas are identified in Items 1, 2 and 3 above. Although these surfaces are located on the exterior of the building or structure, they are provided with a permissible amount of protection and are not directly exposed to the weather with regards to moisture. This must be taken into consideration when wall and ceiling covering materials are being selected. It is not required to use waterproof materials at these locations, but materials approved by the code and the product manufacturer for exterior applications must be used.

WIRE BACKING. Horizontal strands of tautened wire attached to surfaces of vertical supports which, when covered with the building paper, provide a backing for cement plaster.

❖ When metal lath attached to exterior wall studs is used as backing for a plaster base coat, additional reinforcing is necessary to stiffen the wall framing system and minimize unwanted movement to the wall itself. Plaster is a fairly rigid material and excessive movement will cause it to crack. Eighteen-gage metal wire attached horizontally to the face of the wall studs and covered with building paper will provide the additional stiffness required to minimize cracking in the finished surface (see Section 2510.5 for additional commentary on the requirements for wire backing).

SECTION 2503
INSPECTION

2503.1 Inspection. Lath and gypsum board shall be inspected in accordance with Section 109.3.5.

❖ This section establishes guidelines and requirements for the inspection of all components of gypsum board and plaster products, assemblies and installations as outlined in the code and as required or requested by the building official.

The building official is allowed to inspect the installation of all plaster lathing system components prior to being concealed from view by the application of a plaster base coat or finished coats. It is essential not only that the lath be the proper type for the plaster system being used, but also that the components of the system are installed properly.

Plaster wall and ceiling assemblies as well as gypsum board wall and ceiling assemblies are quite often, by design, part of a building's passive fire protection system, as they provide not only the fire-resistant construction necessary for the proper separation of defined areas within a structure, but also fire-protected components of the structure. The building official is to be given the opportunity by the permit holder to verify and be assured that the assemblies will serve the purpose for which they are designed. This can only be done by allowing the building official to visually inspect the components of the various systems while they are still accessible to allow for visual and physical inspection.

SECTION 2504
VERTICAL AND HORIZONTAL ASSEMBLIES

2504.1 Scope. The following requirements shall be met where construction involves gypsum board, lath and plaster in vertical and horizontal assemblies.

❖ Section 2504, along with the referenced standards and structural limitations and requirements addressed elsewhere, define the minimum requirements for the structural members supporting the weight of gypsum board and plaster where either or both are used in the construction of walls and ceilings.

Vertical assemblies (walls and partitions) must be designed in accordance with the engineering chapters of the code for proper structural support of gypsum board or plaster wall covering. If the wall or partition

construction and assembly are not designed to resist the loads prescribed in the code, the wall covering materials will perform inadequately. The general design provisions of Chapter 16 and the deflection limitations are particularly important in this regard.

The code requirements for the design and framing of horizontal assemblies (ceilings) are intended to achieve the same results as described for vertical assemblies. Gypsum board and plaster ceilings are generally supported by a suspended metal grid framing system of metal channels or T-bars suspended by steel wires tied to the structure above. The ceiling covering may also be fastened directly to the structural framing member or furring, which is fastened to the structural frame. All ceiling support systems must comply with the structural load requirements of Chapter 16.

2504.1.1 Wood framing. Wood supports for lath or gypsum board, as well as wood stripping or furring, shall not be less than 2 inches (51 mm) nominal thickness in the least dimension.

> **Exception:** The minimum nominal dimension of wood furring strips installed over solid backing shall not be less than 1 inch by 2 inches (25 mm by 51 mm).

❖ The minimum dimensions required by this section for wood supports for gypsum board or lath and plaster are intended to provide adequate support and reduce surface distortion due to warping of wood supports.

Wood furring strips attached to a solid backing may be used to support wall or ceiling covering. The furring must be installed with the wide face against the solid backing. This will provide a nominal 2-inch-wide (51 mm) nailing surface equal to the nailing surface of a standard wood stud, joist or rafter.

2504.1.2 Studless partitions. The minimum thickness of vertically erected studless solid plaster partitions of $^3/_8$-inch (9.5 mm) and $^3/_4$-inch (19.1 mm) rib metal lath or $^1/_2$-inch-thick (12.7 mm) long-length gypsum lath and gypsum board partitions shall be 2 inches (51 mm).

❖ As the name implies, a studless partition is constructed of gypsum board lath or ribbed metal lath with plaster applied to each face, without the stud frame wall that typically provides support for the wall covering [see Figures 2504.1.2(1) and 2504.1.2(2)]. The lath and plaster form the structural elements of the partition that must resist the lateral loads applied to the wall. Studless partitions may only be used as nonload-bearing partitions; however, they must comply with the deflection and general design requirements specified in Chapter 16.

For SI: 1 inch = 25 mm.

Figure 2504.1.2(1)
STUDLESS PLASTER PARTITION WITH GYPSUM BOARD LATH

For SI: 1 inch = 25 mm.

Figure 2504.1.2(2)
STUDLESS PLASTER PARTITION WITH RIBBED METAL LATH

SECTION 2505
SHEAR WALL CONSTRUCTION

2505.1 Resistance to shear (wood framing). Wood-framed shear walls sheathed with gypsum board, lath and plaster shall be designed and constructed in accordance with Section 2306.4 and are permitted to resist wind and seismic loads. Walls resisting seismic loads shall be subject to the limitations in Section 12.2.1 of ASCE 7.

❖ Shear walls designed to resist wind and seismic loads and constructed of wood-framing members and sheathed with gypsum wallboard or lath and plaster are permitted by the code (see Chapter 23 for wood-framing requirements).

- Section 2306.4 defines the structural requirements for wood-framing members, nailing requirements, diaphragm bracing, allowable loads and all other structural requirements and necessary incidentals.
- Section 2306.4.5 includes seismic application limitations for gypsum, plaster and wood-frame shear walls.
- Section 2306.4.5.1 outlines the requirements for the application of gypsum wallboard as a component of a wood-frame shear wall assembly.
- Table 2306.4.5 includes the design guidelines and construction requirements necessary for use of wood-frame shear walls using gypsum board or lath and plaster as the finished wall surface.

The referenced standard for the limitations for basic seismic-force-resisting systems is found in Section 12.2.1 of ASCE 7.

Care must be taken when selecting gypsum board or plaster materials if they are being used as part of an exterior wall that will be exposed to the weather. Gypsum board, gypsum board lath, gypsum plaster and gypsum veneer plaster are not water resistant and are intended for interior applications only. Specific gypsum board materials covered with an appropriate exposed finish material or metal lath with cement plaster base coats and finish coat (stucco) are appropriate for exterior applications exposed to the weather.

2505.2 Resistance to shear (steel framing). Cold-formed steel-framed shear walls sheathed with gypsum board and constructed in accordance with the materials and provisions of Section 2210.5 are permitted to resist wind and seismic loads. Walls resisting seismic loads shall be subject to the limitations in Section 12.2.1 of ASCE 7.

❖ Shear walls designed to resist wind and seismic loads and constructed of light-gage steel stud framing members and sheathed with gypsum board are permitted by the code. Structural design guidelines and construction requirements for these shear walls are provided in Section 2210.5. The referenced standard for the limitations for basic seismic-force-resisting systems is found in Section 12.2.1 of ASCE 7.

SECTION 2506
GYPSUM BOARD MATERIALS

2506.1 General. Gypsum board materials and accessories shall be identified by the manufacturer's designation to indicate compliance with the appropriate standards referenced in this section and stored to protect such materials from the weather.

❖ This section contains the appropriate referenced standards for the proper use and quality of gypsum board materials and accessories.

All gypsum products, including gypsum board materials, are required to be identified with the manufacturer's designation. The manufacturer is required to certify that its product complies with the applicable standard for the gypsum product. This enables the installer and inspector to verify that the product is being used for its intended installation.

Generally, all gypsum products and accessories are delivered to the job site in their original packages, containers or bundles. These products will be identified with the manufacturer's name, trademark and any applicable standard designations. For example, for gypsum board products, the thickness and type are shown on the end-bundling tapes or the board panel. The premixed paste products are identified on their container, and bag products are appropriately identified on the face of the bag. With regard to weather protection, all gypsum products must be kept dry because of the deleterious effect of moisture. If properly stored above ground and fully protected from the weather and exposure to direct sunlight, gypsum products and accessories may be stored outside; however, the storage of gypsum board is not recommended in exterior locations. Exposure to the elements may result in water stain, light discoloration, mildew, loss of structural strength or sagging, all of which could hinder the performance of the product structurally and in fire performance, as well as the ability to be finished in the intended fashion. Gypsum board should be stored flat—never on edge or end.

2506.2 Standards. Gypsum board materials shall conform to the appropriate standards listed in Table 2506.2 and Chapter 35 and, where required for fire protection, shall conform to the provisions of Chapter 7.

❖ Table 2506.2 lists the appropriate ASTM International (ASTM) standards for gypsum board materials and accessories. The most common building material used in Table 2506.2 is gypsum wallboard (also referred to as "gypsum board," "gyp. board" or "sheetrock"). ASTM C 36 provides the design specifications for gypsum board used for walls and ceilings. Gypsum board consists of a noncombustible core, which is primarily gypsum, and is surfaced with paper firmly bonded to the core. The paper surfacing material is less than 0.125-inch (3.18 mm) thick and has a flame spread index less than 25 when tested in accordance with ASTM E 84. This requirement is in accordance with fire-resistant requirements of Chapter 7 for composite materials classified as noncombustible. Gypsum materials that are required to have a fire-resistance rating in accordance with the code must also comply with the applicable provisions of Chapter 7. In addition, Chapter 7 establishes and defines the requirements for the fire-resistance ratings of building assemblies and structural components. Such assemblies are to be constructed in accordance with the test procedures as prescribed in the appropriate referenced standards listed in Chapter 35.

TABLE 2506.2
GYPSUM BOARD MATERIALS AND ACCESSORIES

MATERIAL	STANDARD
Accessories for gypsum board	ASTM C 1047
Adhesives for fastening gypsum wallboard	ASTM C 557
Exterior soffit board	ASTM C 931
Fiber-reinforced gypsum panels	ASTM C 1278
Glass mat gypsum backing panel	ASTM C 1178
Glass mat gypsum substrate	ASTM C 1177
Gypsum backing board and gypsum shaftliner board	ASTM C 442
Gypsum ceiling board	ASTM C 1395
Gypsum sheathing	ASTM C 79
Gypsum wallboard	ASTM C 36
Joint reinforcing tape and compound	ASTM C 474; C 475
Nails for gypsum boards	ASTM C 514, F 547, F 1667
Predecorated gypsum board	ASTM C 960
Steel screws	ASTM C 954; C 1002
Steel studs, load bearing	ASTM C 955
Steel studs, nonload bearing	ASTM C 645
Standard specification for gypsum board	ASTM C 1396
Testing gypsum and gypsum products	ASTM C 22; C 472; C 473
Water-resistant gypsum backing board	ASTM C 630

❖ Table 2506.2 includes a listing of the materials and accessories approved by the code for use in gypsum board construction, along with the appropriate ASTM standard for each. The installation of each of these materials must conform to all the requirements of the referenced standard. It is not the intent of the code, however, by way of this table, to limit or restrict alternative materials or accessories in the use of gypsum board and its associated incidentals as a building material acknowledged and approved by the code. Any and all materials not specifically mentioned, described or referenced, however, are subject to the requirements of the code, as well as the entire family of *International Codes*.

The gypsum industry has established a phase-in program for the use of the new "umbrella" standard ASTM C 1396. Manufacturers have begun to dual-la-

bel products using both the current standard as well as ASTM C 1396. Over time, the individual gypsum product standards will be phased out.

2506.2.1 Other materials. Metal suspension systems for acoustical and lay-in panel ceilings shall conform with ASTM C 635 listed in Chapter 35 and Section 13.5.6 of ASCE 7 for installation in high seismic areas.

❖ Suspended ceiling systems are widely used in building construction of all types, as they provide for a cost effective finished ceiling system that is versatile and adaptable to other building systems, such as heating ventilating and air-conditioning systems (HVAC), fire suppression systems, lighting systems, etc. The standard developed for the manufacture and installation of suspended ceiling systems is ASTM C 635. This standard contains minimum specifications for the metal suspension system, including system types (direct hung, indirect hung and furring bar type), structural classification (light-duty, intermediate-duty and heavy-duty), dimensional tolerances, coatings and physical testing for structural capacities.

For applications in high seismic areas additional installation requirements are applicable. These requirements include installation requirements for Seismic Design Categories C through F and are found in Section 13.5.6 of ASCE 7.

SECTION 2507
LATHING AND PLASTERING

2507.1 General. Lathing and plastering materials and accessories shall be marked by the manufacturer's designation to indicate compliance with the appropriate standards referenced in this section and stored in such a manner to protect them from the weather.

❖ These basic requirements apply to lathing and plastering materials, products and accessories referenced in this chapter. This section requires that all lath and plaster materials be identified with the manufacturer's designation and indicate compliance with the appropriate standards. This enables the installer and the inspector to verify that the product is being used for its intended installation. All accessories must also be marked with a reference to the appropriate standard. The manufacturer is required to certify that its product complies with the applicable standard for lathing and plastering material. Lath and plaster products and accessories are delivered to the job site in their original packages, containers or bundles. These products must be identified with the manufacturer's name, trademark and any applicable standard designations. With regard to weather protection, plaster products must be kept dry because of the deleterious effect of moisture. If properly stored above ground, fully protected from the weather and exposure to direct sunlight and away from condensation and damp surfaces, lath and plaster products and accessories may be stored outside; however, it is not recommended.

2507.2 Standards. Lathing and plastering materials shall conform to the standards listed in Table 2507.2 and Chapter 35 and, where required for fire protection, shall also conform to the provisions of Chapter 7.

❖ The appropriate referenced standards that regulate all lath and plastering materials are contained in this section. It also addresses the proper marking of materials, protection from weather and installation requirements. Additionally, provisions of Chapter 7 are referenced for building assemblies utilizing lath and plastering materials that require a fire-resistance rating; therefore, the assemblies must comply with the provisions of Chapter 7.

TABLE 2507.2
LATH, PLASTERING MATERIALS AND ACCESSORIES

MATERIAL	STANDARD
Accessories for gypsum veneer base	ASTM C 1047
Blended cement	ASTM C 595
Exterior plaster bonding compounds	ASTM C 932
Gypsum base for veneer plasters	ASTM C 588
Gypsum casting and molding plaster	ASTM C 59
Gypsum Keene's cement	ASTM C 61
Gypsum lath	ASTM C 37
Gypsum plaster	ASTM C 28
Gypsum veneer plaster	ASTM C 587
Interior bonding compounds, gypsum	ASTM C 631
Lime plasters	ASTM C 5; C 206
Masonry cement	ASTM C 91
Metal lath	ASTM C 847
Plaster aggregates Sand	ASTM C 35; C 897
Perlite	ASTM C 35
Vermiculite	ASTM C 35
Plastic cement	ASTM C 1328
Portland cement	ASTM C 150
Steel screws	ASTM C 1002; C 954
Steel studs and track	ASTM C 645; C 955
Welded wire lath	ASTM C 933
Woven wire plaster base	ASTM C 1032

❖ Table 2507.2 includes a listing of the materials and accessories approved by the code for use in plaster construction, along with the appropriate ASTM standard for each. The installation of each of these materials must conform to all the requirements of the referenced standard. It is not the intent of the code to limit or restrict alternative materials or accessories in the use of plaster and its associated incidentals as a building material acknowledged and approved by the code. Any

and all materials not specifically mentioned, described or referenced herein are subject to the requirements of the code, as well as the entire family of *International Codes.*

SECTION 2508
GYPSUM CONSTRUCTION

2508.1 General. Gypsum board and gypsum plaster construction shall be of the materials listed in Tables 2506.2 and 2507.2. These materials shall be assembled and installed in compliance with the appropriate standards listed in Tables 2508.1 and 2511.1, and Chapter 35.

❖ This section contains the referenced standards and requirements for the proper installation of gypsum construction.

This section references Table 2506.2 for the appropriate materials required in gypsum board and gypsum plaster construction, as well as the corresponding material standard. Table 2508.1 addresses the appropriate installation standards required for use in gypsum construction. Gypsum board in fire-resistance-rated assemblies is to be installed in accordance with the fastening schedule utilized in the tested assembly. The type and spacing of the fasteners are reflected in the fire test reports of various assemblies, including the summary of fire tests contained in various resource documents that compile summaries of fire test results, such as Underwriters Laboratories' *Fire Resistance Directory,* Warnock Hersey's *Certification Listing* or the Gypsum Association's *Fire Resistance Design Manual D* (GA-600). Refer to Chapter 7 and its commentary for additional information related to fire-resistance-rated assemblies.

TABLE 2508.1
INSTALLATION OF GYPSUM CONSTRUCTION

MATERIAL	STANDARD
Gypsum board	GA-216; ASTM C 840
Gypsum sheathing	ASTM C 1280
Gypsum veneer base	ASTM C 844
Interior lathing and furring	ASTM C 841
Steel framing for gypsum boards	ASTM C 754; C 1007

❖ Table 2508.1 includes the installation requirements from the code for the use of gypsum and gypsum-related construction products, along with the appropriate ASTM standard for each. The installation of each of these materials must conform to all the requirements of the referenced standard. There is no intent to limit or restrict alternative methods of construction acknowledged and approved by the code. Any and all methods not specifically mentioned, described or referenced herein are subject to the requirements of the code, as well as the entire family of *International Codes.*

2508.2 Limitations. Gypsum wallboard or gypsum plaster shall not be used in any exterior surface where such gypsum construction will be exposed directly to the weather. Gypsum wallboard shall not be used where there will be direct exposure to water or continuous high humidity conditions. Gypsum sheathing shall be installed on exterior surfaces in accordance with ASTM C 1280.

❖ Because of the detrimental effect of moisture on gypsum products, gypsum-based materials and construction must be protected from the weather. See Section 2502 for the definition of "Weather-exposed surfaces," which includes some limited conditions where gypsum board is permitted to be used on the outside of a building or structure.

2508.2.1 Weather protection. Gypsum wallboard, gypsum lath or gypsum plaster shall not be installed until weather protection for the installation is provided.

❖ This section relates to the necessary environmental conditions for proper plaster application and hydration. Temperature limitations are needed to provide a uniform heating condition for a minimum of one week prior to the start of plastering, and are the same as the requirements stated in ASTM C 842. This minimizes the possibility of plaster cracking because of movements or stresses caused by thermal changes.

2508.3 Single-ply application. Edges and ends of gypsum board shall occur on the framing members, except those edges and ends that are perpendicular to the framing members. Edges and ends of gypsum board shall be in moderate contact except in concealed spaces where fire-resistance-rated construction, shear resistance or diaphragm action is not required.

❖ The application of gypsum board is specified in this chapter for nonfire-resistant construction or construction where diaphragm (shear wall) action is not required. Chapter 7 and fire test reports will establish the means of fastening and supporting the ends and edges of gypsum board for nonfire-resistant assemblies. In the case of stud walls required for diaphragm action, Chapters 22 and 23 establish structural requirements such as sizes and spacing fasteners, allowable loads and the conditions of end and edge supports.

Gypsum board is the generic name for a family of sheet products consisting of a noncombustible core, primarily composed of gypsum, with a paper surfacing on each face. The installation and finishing of gypsum board materials must comply with the referenced standards in Table 2508.1. GA 216, *Application and Finishing of Gypsum Board*, published by the Gypsum Association, is the most commonly recommended specification for the installation and finishing of gypsum board.

- **Framing:** As with any gypsum board or gypsum sheet product, it is critical to inspect the framing on which the gypsum material is to be attached. Finishing of gypsum board will be made extremely difficult when installed over unlevel, unplumb, out-of-square or structurally inadequate framing. All of these concerns must be satisfied before the installation of any gypsum board product can begin. Gypsum board can be attached to a variety of framing members, including but not necessarily limited to, wood or metal studs; joists and rafters; suspended metal-frame ceiling systems and wood or metal furring on concrete or masonry walls. Framing systems that are not installed square, level and plumb will create stress points where the gypsum board is attached to the framing. Stress points can cause the gypsum board to crack, shear or delaminate. This can become a significant issue when fire-resistance-rated assemblies are compromised.

- **Fastening:** The type of fastening devices and methods for attaching gypsum board to framing are dependent upon the type of framing system being used and the type of assembly being constructed. Screws, nails and construction adhesives are approved types of fasteners. Fasteners are to be selected and installed as required by the specifications referenced in GA 216, except as required elsewhere in the code for fire-resistance-rated assemblies, shear wall construction or where more restrictive requirements are provided by the referenced standards listed in Chapter 35.

- **Finishing:** Gypsum must be finished with materials and accessories designed and approved for the application for which they are being used. Joint tape, joint compound and all accessories are to be approved and installed as recommended by the manufacturer and as required by the referenced standards.

- **Multiple-ply application:** This procedure for gypsum board is generally intended to increase the fire-resistance rating of an assembly, the shear-resisting strength of a shear wall assembly or the sound transmission class for an assembly. The construction and technical requirements for these and similar types of specialized assemblies are covered elsewhere in the *International Codes*. The requirements for gypsum board itself are as defined in this chapter, except where more restrictive requirements are mandated for a specialized assembly.

2508.3.1 Floating angles. Fasteners at the top and bottom plates of vertical assemblies, or the edges and ends of horizontal assemblies perpendicular to supports, and at the wall line are permitted to be omitted except on shear resisting elements or fire-resistance-rated assemblies. Fasteners shall be applied in such a manner as not to fracture the face paper with the fastener head.

❖ Floating angles are the edges of gypsum board that are not directly fastened to a framing member where a change in the plane of the gypsum board occurs. Gypsum board used as part of a shear wall assembly or as a component part of any fire-resistance-rated assembly must have all edges fastened to a framing member. It is given that gypsum board used as part of a vertical assembly (wall) is fastened to the wall studs along the edges and in the field of the gypsum board as defined in GA 216. Adequate support for the weight of the gypsum board itself will be provided, therefore eliminating the need for additional fastening at the top and bottom plates of the wall framing. Fastening is not to be omitted at any inside or outside corner of a wall assembly.

Fasteners may be omitted at the edges of horizontal assemblies (ceiling and soffits) when the edge of the gypsum board is perpendicular to the framing member. As with wall assemblies, it is given that fasteners located along the edges and in the field of the gypsum board in accordance with GA 216 will provide adequate support for the weight of the gypsum board itself. Edges of horizontal gypsum board assemblies that are parallel to framing members must be fastened to the framing member except where the parallel edge is at the intersection of the wall and ceiling. Typically, the gypsum board will be installed on the ceiling before it is installed on the wall. The edge of the gypsum board on the wall will be butted against the underside edge of the ceiling board, providing additional edge support for the ceiling.

This section permits the elimination of unnecessary fastening of gypsum board in specific locations and conditions. It is not the intent of this section to compromise the integrity of the gypsum board support or fastening system. Gypsum board must be fastened to prevent sagging, warping, waviness or any visual defects.

2508.4 Joint treatment. Gypsum board fire-resistance-rated assemblies shall have joints and fasteners treated.

Exception: Joint and fastener treatment need not be provided where any of the following conditions occur:

1. Where the gypsum board is to receive a decorative finish such as wood paneling, battens, acoustical finishes or any similar application that would be equivalent to joint treatment.

2. On single-layer systems where joints occur over wood framing members.

3. Square edge or tongue-and-groove edge gypsum board (V-edge), gypsum backing board or gypsum sheathing.

4. On multilayer systems where the joints of adjacent layers are offset from one to another.

5. Assemblies tested without joint treatment.

❖ This section requires joint and fastener treatment for all gypsum board fire-resistance-rated assemblies. The code also identifies and permits exceptions where the gypsum board is to receive a decorative finish or any similar application that is considered to be equivalent to the joint or fastener treatment.

These exceptions indicate that joint treatment is not required for assemblies tested without it:

1. Where the finish layer of gypsum board is to receive a decorative covering or similar type of applied material that will cover exposed joints, finishing of joints is not required; however, the applied material must provide the same fire-resistance rating of the assembly that it is being applied to.

2. Joint treatment is not required in single-layer applications where joints occur directly over wood framing members. In effect, the additional joint treatment does not materially increase the fire rating of the assembly, and many partitions have been tested and passed the fire test without the added protection of joint and fastener treatment as part of the test design.

3. Gypsum board with factory-designed edges that will prevent the penetration of smoke or fire into a rated assembly does not have to be finished. The factory edges listed will minimize gaps at joints and maintain the integrity of the assembly.

4. Multilayer assemblies with offset joints will eliminate any gaps in the system; therefore, fire smoke and harmful gases cannot penetrate the fire-resistance-rated assembly. This makes finishing of joints unnecessary.

5. Where approved fire tests indicate that joint and fastener treatment is included as a part of the tested assembly, the code requires that joint and fastener treatment be applied in the same manner and extent as the approved test assembly. Where the approved test assembly does not specifically state whether joint or fastener treatment is part of the test assembly, the code requires that joint and fastener treatment is to be applied to any fire-resistance-rated assembly. Assemblies tested without joint treatment need not be finished.

As indicated elsewhere in this section, with regard to gypsum board application, joint and fastener treatment is primarily applied for aesthetic reasons and to provide a satisfactory finish surface for the application of decorative finishes, such as paint, wallpaper, vinyl wall fabric, textured paint or a number of other finishes used for the appearance of a rated or nonrated assembly.

2508.5 Horizontal gypsum board diaphragm ceilings. Gypsum board shall be permitted to be used on wood joists to create a horizontal diaphragm ceiling in accordance with Table 2508.5.

❖ Generally, a gypsum board ceiling does not serve as a load-carrying structural element for other than its own weight. This section provides installation requirements

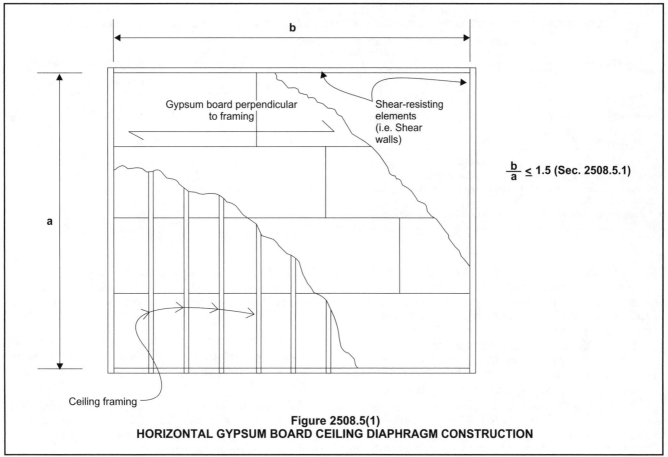

$$\frac{b}{a} \le 1.5 \; (Sec. \; 2508.5.1)$$

Figure 2508.5(1)
HORIZONTAL GYPSUM BOARD CEILING DIAPHRAGM CONSTRUCTION

Figure 2508.5(2)
PARTITION SUPPORTED LATERALLY BY GYPSUM BOARD CEILING DIAPHRAGM

TABLE 2508.5
SHEAR CAPACITY FOR HORIZONTAL WOOD FRAMED GYPSUM BOARD DIAPHRAGM CEILING ASSEMBLIES

MATERIAL	THICKNESS OF MATERIAL (MINIMUM) (inches)	SPACING OF FRAMING MEMBERS (MAXIMUM) (inches)	SHEAR VALUE[a,b] (plf of ceiling)	MIMIMUM FASTENER SIZE
Gypsum board	$^1/_2$	16 o.c.	90	5d cooler or wallboard nail; $1^5/_8$-inch long; 0.086-inch shank; $^{15}/_{64}$-inch head[c]
Gypsum board	$^1/_2$	24 o.c.	70	5d cooler or wallboard nail; $1^5/_8$-inch long; 0.086-inch shank; $^{15}/_{64}$-inch head[c]

For SI: 1 inch = 25.4 mm, 1 pound per foot = 14.59 N/m.

a. Values are not cumulative with other horizontal diaphragm values and are for short-term loading due to wind or seismic loading. Values shall be reduced 25 percent for normal loading.

b. Values shall be reduced 50 percent in Seismic Design Categories D, E and F.

c. $1^1/_4$-inch, No. 6 Type S or W screws are permitted to be substituted for the listed nails.

[refer to Figure 2508.5(1)] that allow the use of a horizontal gypsum board ceiling diaphragm. This permits a wood-frame ceiling to carry limited lateral loads, such as a portion of the component earthquake force on a partition as shown in Figure 2508.5(2).

Table 2508.5 provides the shear capacity that is to be used in designing a horizontal gypsum board ceiling diaphragm. Since the material thickness and fastener size do not vary, the capacity is strictly a function of the framing member spacing. Further qualifications on the tabulated shear capacities that are stated in the footnotes include: ceiling diaphragm capacities are not cumulative with other diaphragm capacities; the capacities given are only for short-term loads due to wind and earthquakes; in structures classified as Seismic Design Category D, E or F the capacities must be reduced by 50 percent.

2508.5.1 Diaphragm proportions. The maximum allowable diaphragm proportions shall be $1^1/_2$:1 between shear resisting elements. Rotation or cantilever conditions shall not be permitted.

❖ Limiting the diaphragm proportions and prohibiting cantilevers and rotation controls the deflection and distortion of the ceiling diaphragm [see Figure 2508.5(1)].

2508.5.2 Installation. Gypsum board used in a horizontal diaphragm ceiling shall be installed perpendicular to ceiling framing members. End joints of adjacent courses of gypsum board shall not occur on the same joist.

❖ The installation requirements of this section are necessary to obtain the shear capacities indicated in Table 2508.5.

2508.5.3 Blocking of perimeter edges. All perimeter edges shall be blocked using a wood member not less than 2-inch by

MINIMUM 2″ NAILING SURFACE

MINIMUM 2 x 6 BLOCKING

½″ GYPSUM BOARD

TOP PLATE

FASTENERS (SEE TABLE 2508.5 AND SECTION 2308.5.5)

WALL STUD

For SI: 1 inch = 25.4 mm.

Figure 2508.5.3
PERIMETER BLOCKING

6-inch (51 mm by 159 mm) nominal dimension. Blocking material shall be installed flat over the top plate of the wall to provide a nailing surface not less than 2 inches (51 mm) in width for the attachment of the gypsum board.

❖ This section states the minimum required edge blocking that is illustrated in Figure 2508.5.3.

2508.5.4 Fasteners. Fasteners used for the attachment of gypsum board to a horizontal diaphragm ceiling shall be as defined in Table 2508.5. Fasteners shall be spaced not more than 7 inches (178 mm) on center (o.c.) at all supports, including perimeter blocking, and not more than $^3/_8$ inch (9.5 mm) from the edges and ends of the gypsum board.

❖ This establishes maximum fastener spacing and minimum edge distances.

2508.5.5 Lateral force restrictions. Gypsum board shall not be used in diaphragm ceilings to resist lateral forces imposed by masonry or concrete construction.

❖ Gypsum board shall not be used in diaphragm ceilings to resist lateral forces imposed by masonry or concrete construction.

This section prohibits the use of a gypsum board ceiling diaphragm in resisting lateral loads due to masonry or concrete.

SECTION 2509
GYPSUM BOARD IN SHOWERS
AND WATER CLOSETS

2509.1 Wet areas. Showers and public toilet walls shall conform to Sections 1210.2 and 1210.3.

❖ Special caution must be taken when gypsum board is used in applications where it will be exposed to moisture, because gypsum is not naturally resistant to water, humidity or other forms of moisture. Failure of gypsum board can be caused by direct or indirect exposure to all forms of moisture. Moisture-resistant gypsum board, which is faced with moisture-resistant paper, must be used in high moisture areas, and additional protection such as paint must be applied to prevent gypsum board from being exposed to moisture. Moisture-resistant gypsum board is known and referred to as "green board," because of the distinctive green moisture-resistant paper that is applied to it, which also makes it readily identifiable as moisture resistant.

Wet areas at interior locations such as those identified in Sections 1210.2 and 1210.3 must be protected from moisture by a nonabsorbent wall or ceiling material. Cement plaster (gypsum plaster is not permitted in wet areas) and tile or other like finish materials will provide moisture protection for the wall surface. An additional waterproofing membrane must be provided where the wall is constructed of wood framing (see related commentary to Section 2511.5).

2509.2 Base for tile. Cement, fiber-cement or glass mat gypsum backers in compliance with ASTM C 1178, C 1288 or C 1325 and installed in accordance with manufacturer recom-

mendations shall be used as a base for wall tile in tub and shower areas and wall and ceiling panels in shower areas. Water-resistant gypsum backing board shall be used as a base for tile in water closet compartment walls when installed in accordance with GA-216 or ASTM C 840 and manufacturer recommendations. Regular gypsum wallboard is permitted under tile or wall panels in other wall and ceiling areas when installed in accordance with GA-216 or ASTM C 840.

❖ In areas that have the potential for exposure to heavy concentrations of water, it is necessary to use cement, fiber-cement or glass mat gypsum as a backing for tile. The code requires that moisture-resistant gypsum board (green board) is to be used as a base for ceramic and other hard tiles for water-closet compartment walls. It can be used in these locations because of its resistance to incidental moisture exposure. All joints in gypsum board that will be covered by tile or wall panels must be taped and finished even though they will be unexposed to view when the installation of the wall-covering material is complete. This is necessary to prevent moisture from migrating through the unfinished joints and causing damage to the gypsum board and the wall-framing system. Corrosion-resistant fasteners must be used in wet areas.

2509.3 Limitations. Water-resistant gypsum backing board shall not be used in the following locations:

1. Over a vapor retarder in shower or bathtub compartments.

2. Where there will be direct exposure to water or in areas subject to continuous high humidity.

3. On ceilings where frame spacing exceeds 12 inches (305 mm) o.c. for $^1/_2$-inch-thick (12.7 mm) water-resistant gypsum backing board and more than 16 inches (406 mm) o.c. for $^5/_8$-inch-thick (15.9 mm) water-resistant gypsum backing board.

❖ Although there are many gypsum board sheet products that are manufactured and approved for use in wet areas or areas exposed to moisture or humidity, there are still some extreme conditions where even water-resistant gypsum board will not provide the level of moisture protection necessary:

1. Gypsum board installed on the walls and ceilings at shower and bathtub areas must be finished to prevent moisture from penetrating the wall or ceiling finish and contacting the gypsum board. The finish applied to the exposed face of the gypsum board, in effect, creates a water-resistant barrier that not only stops water from getting to the gypsum board, but also prevents the release of moisture from within the wall or the gypsum board itself. For this reason, gypsum board must not be installed over the outboard side of any vapor barrier or retarder. This will create a waterproof membrane on both faces of the gypsum board, causing moisture to be trapped in the gypsum board that will ultimately cause it to decompose and fail.

2. Water-resistant gypsum board is not to be used in areas that will be subject to direct exposure to water or continuous exposure to high humidity at locations such as saunas, steam rooms, gang showers or indoor pools. Gypsum board products, including the water-resistant type, are not intended for these extreme conditions and will provide unsatisfactory performance. Nongypsum wall and ceiling materials such as concrete masonry, ceramic tile on cement backer board, cement plaster (stucco) or other materials designed and recommended for high moisture exposure must be used in these locations.

3. The additional weight, due to moisture, that is commonly applied as live loads to ceilings in high moisture or wet areas must be considered when selecting the thickness of the gypsum board and the spacing of the structural support members necessary to prevent sagging of the ceiling. The more restrictive frame spacing and gypsum board thicknesses required in this section are established to allow for the additional weight. Where other structural requirements in the code are less restrictive for the conditions contained in this section, these requirements apply.

SECTION 2510
LATHING AND FURRING FOR
CEMENT PLASTER (STUCCO)

2510.1 General. Exterior and interior cement plaster and lathing shall be done with the appropriate materials listed in Table 2507.2 and Chapter 35.

❖ This section contains the appropriate referenced standards for materials, proper application and protection of interior and exterior lath and furring for the installation of portland cement plaster (stucco).

Table 2507.2 lists the appropriate ASTM standards for portland cement, metal lath and plastering materials. These referenced standards are the minimum material specifications required by the code to meet the requirements of this section, as well as related requirements elsewhere in the code for the proper application and procedures for lath and furring for cement plaster.

2510.2 Weather protection. Materials shall be stored in such a manner as to protect such materials from the weather.

❖ Like gypsum products, portland cement stucco lathing and plastering materials must be stored to prevent the deleterious effect of the weather.

2510.3 Installation. Installation of these materials shall be in compliance with ASTM C 926 and ASTM C 1063.

❖ Table 2511.1.1 refers to ASTM C 1063 as the standard specification for interior and exterior lathing and furring for portland cement-based plastering, as specified in ASTM C 926. Where a fire-resistance rating is re-

quired for plastered assemblies and construction, details of construction must be in accordance with reports of fire tests of assemblies that have met the requirements of the fire-resistance rating imposed. Where a specific degree of sound control is required for plastered assemblies and construction, details of construction must be in accordance with official reports of tests conducted in recognized testing laboratories in accordance with the applicable requirements of ASTM E 90.

Table 2511.1.1 also refers to ASTM C 1063 as the standard specification for the installation of portland cement-based plaster. For exterior (stucco) and interior work, ASTM C 1036 also provides the tables necessary for proportioning various plaster mixes and thicknesses (refer to Section 2512 for additional requirements for cement plaster).

2510.4 Corrosion resistance. Metal lath and lath attachments shall be of corrosion-resistant material.

❖ Because of the amount of water that is present in plaster when it is mixed and applied to a surface, all metal components and accessories, such as metal lath, screws, screeds, grounds, metal runners at edges of the plaster, expansion and control joint molding or any other metal component that will be in contact with the plaster, must be protected from corrosion and rusting from prolonged contact with wet plaster. Corrosion of metal components will cause failure and create voids in the plaster that will allow moisture to get into the wall and cause damage to the structural and nonstructural elements of the building. For these reasons, the code requires metal components be made of corrosion-resistant materials. Table 2507.2 provides a listing of the required ASTM standards for lath and related accessories.

2510.5 Backing. Backing or a lath shall provide sufficient rigidity to permit plaster applications.

❖ Proper application of exterior plaster requires that the base coat is thoroughly pushed through the lath to provide the required mechanical bond. For this reason, it is required by the code that the lath and its backing provide sufficient rigidity to permit the proper application of the plaster. A common method of providing a rigid backing for the lath for exterior plaster over open wood or metal stud framing is to fasten 18-gage wire backing (see Figure 2510.5) to the studs at 6 inches (152 mm) on center horizontally. It is critical to stretch the wire tightly across the face of the studs to ensure the desired rigidity for the backing. The wire backing must be attached to each stud to prevent slippage or movement of the wire.

2510.5.1 Support of lath. Where lath on vertical surfaces extends between rafters or other similar projecting members, solid backing shall be installed to provide support for lath and attachments.

❖ It is important for the lath to be supported continuously along all edges. Continuous edge support is necessary to prevent excessive movement of the finished plaster, which will cause unwanted movement cracking. This is especially critical on exterior vertical surfaces of cement plaster because cracking provides an entry point for water into the building. At locations where the wall plane is interrupted at regular intervals, such as eaves and soffits where rafter ends project beyond the face of the wall, the code requires that the lath be supported by solid backing at all edges. In addition to supporting the edges of the lath, the code requires that all plaster trim, edging, screeds, grounds and other accessories be rigidly supported with solid backing at all edges.

The damage that will be caused to finished plaster on an exterior vertical surface due to unsupported edges is much more obvious than what would be expected on an interior surface; however, failure to interior wall surfaces is of no less concern. For this reason, the code makes no distinction between interior and exterior surfaces. Solid backing is required for lath at all locations, both interior and exterior.

2510.5.2 Use of gypsum backing board.

❖ The code does not permit the use of gypsum board or gypsum lath as a backing for cement plaster, except as noted in Section 2510.5.2.1.

2510.5.2.1 Use of gypsum board as a backing board. Gypsum lath or gypsum wallboard shall not be used as a backing for cement plaster.

> **Exception:** Gypsum lath or gypsum wallboard is permitted, with a water-resistive barrier, as a backing for self-furred metal lath or self-furred wire fabric lath and cement plaster where either of the following conditions occur:
>
> 1. On horizontal supports of ceilings or roof soffits.
>
> 2. On interior walls.

❖ Because of its excellent water-resistant properties, cement plaster is the only type permitted by the code for use as a finished exterior surface exposed to the weather, or at an interior location subject to excessive moisture. Gypsum has none of the water-resistant properties possessed by cement-based plasters or cementitious backer boards, which are designed specifically for use in high moisture areas. Due to the inability of gypsum-based materials to resist moisture, the code does not permit the use of gypsum board or gypsum lath as an interior or exterior backing in direct contact with cement plaster. Gypsum board lath subjected to moisture will decompose and cause failure to the entire plaster system, including the outer or finished coats. Exceptions to this restriction are permitted when gypsum is protected from contact with cement plaster as allowed in specific locations described below.

Gypsum lath and gypsum wallboard are permitted to be used as a backing for cement plaster when a water-resistive barrier, such as Grade D building paper, is installed at the back of the lath or to the face of the gypsum board to prevent the cement plaster from contacting the gypsum backing. The building paper must be securely attached to the face of the gypsum board before installing the self-furring metal lath or self-furring wire fabric to create a water-resistive barrier. Items 1 and 2 include the extent to which exceptions may be applied.

2510.5.2.2 Use of gypsum sheathing backing. Gypsum sheathing is permitted as a backing for metal or wire fabric lath and cement plaster on walls. A water-resistive barrier shall be provided in accordance with Section 2510.6.

Figure 2510.5
THREE-COAT PLASTER ON UNSHEATHED EXTERIOR STUD WALL

❖ Gypsum board sheathing is permitted to be used as a backer for cement plaster when installed with a water-resistive barrier as described in Section 2510.5.2.1 and as required in Section 2510.6.

2510.5.3 Backing not required. Wire backing is not required under expanded metal lath or paperbacked wire fabric lath.

❖ Because metal lath or paper-backed wire fabric lath has reinforcing ribs to provide necessary rigidity, the code requires no additional backing for these materials.

2510.6 Water-resistive barriers. Water-resistive barriers shall be installed as required in Section 1404.2 and, where applied over wood-based sheathing, shall include a water-resistive vapor-permeable barrier with a performance at least equivalent to two layers of Grade D paper.

Exception: Where the water-resistive barrier that is applied over wood-based sheathing has a water resistance equal to or greater than that of 60-minute Grade D paper and is separated from the stucco by an intervening, substantially nonwater-absorbing layer or drainage space.

❖ The code requires that a water-resistive barrier be installed behind exterior plaster for the reasons provided in Section 1404.2. The code also requires that when the barrier is applied over wood-based sheathing such as plywood, the barrier is to be two layers of Grade D building paper. This requirement is based on the observed problems that occur when one layer of typical Type 15-pound (6.8 kg) felt is applied over wood sheathing. The wood sheathing eventually exhibits dry rot due to the penetration of moisture. Cracking is then created in the plaster due to movement of the sheathing caused by alternate expansion and contraction. Field experience has shown that where two layers of building paper are used, the penetration of moisture is considerably decreased, as is the cracking of the plaster due to movement of the sheathing caused by the wet and dry cycles. Grade D building paper is specified because it has the proper water vapor permeability to prevent entrapment of moisture between the paper and the sheathing. The exception to Section 2510.6 allows for the installation of a commonly accepted alternative to the base requirement of two layers of Grade D paper. Where an intervening layer is used, the nonabsorbing characteristics of the layer would need to be substantiated.

2510.7 Preparation of masonry and concrete. Surfaces shall be clean, free from efflorescence, sufficiently damp and rough for proper bond. If the surface is insufficiently rough, approved bonding agents or a portland cement dash bond coat mixed in proportions of not more than two parts volume of sand to one part volume of portland cement or plastic cement shall be applied. The dash bond coat shall be left undisturbed and shall be moist cured not less than 24 hours.

❖ The masonry surface must be properly prepared to receive the cement plaster base coat to ensure proper bond between the masonry base and the plaster. Masonry surfaces must be clean, damp, free of dust and

efflorescence and reasonably flat, without excessive protrusions or indentations. These irregularities will affect the evenness and uniform thickness of the finished plaster. Mortar joints should be struck flush to help minimize any surface irregularities.

Masonry bases for plaster have varying degrees of suction (water absorption rate) but generally provide a good adhesive as well as mechanical bond. Most masonry bases have enough suction so that special bonding agents will not be required to achieve the adequate bond between the plaster and the base. Where denser and harder masonry materials that have very low suction characteristics are used, however, preparation of the surface, similar to that of concrete, may be necessary. Such preparation can include mechanical scoring or the use of chemical bonding agents.

Concrete generally has a denser and tighter finished surface than masonry. Concrete does not have the same suction properties as masonry; therefore, the bond between concrete and plaster must be completely mechanical. Like masonry, the concrete surface must be clean, rough, damp and free of efflorescence and form-releasing agents. If the surface of the concrete is not adequately rough to provide the required mechanical bond, it may be prepared as follows:

- If preparation of the surface is to begin before the concrete is completely set, the surface can be roughened with a metal tool or by using a stiff, heavy-duty wire brush.
- If preparation of the concrete surface must be done after the concrete has hardened, the surface may be bush hammered or roughened with chisels or tools designed for that purpose.
- Another common method, which is specified in this section, is to apply a rich mixture of portland cement, sand and water as a dash coat. This application, in which the material is "dashed" onto the concrete surface with a course brush, may only be used where portland cement plaster is going to be applied to the concrete surface.
- A liquid bonding agent is applied.

SECTION 2511
INTERIOR PLASTER

2511.1 General. Plastering gypsum plaster or cement plaster shall not be less than three coats where applied over metal lath or wire fabric lath and not less than two coats where applied over other bases permitted by this chapter.

Exception: Gypsum veneer plaster and cement plaster specifically designed and approved for one-coat applications.

❖ This section is intended to define the standards and guidelines for the materials and installation of interior plaster. The requirements of this section include gypsum plaster, gypsum veneer plaster and portland cement plaster. This section also addresses plaster applied to nonlath surfaces and the environmental conditions for which interior plaster is stored and installed.

It is the consensus of the industry that multicoat work is necessary for control of plaster thickness and density, particularly where plaster application is done by hand because most of the materials used for plaster compact under hand application as a result of pressure applied to the trowel. Experience has proven that this change in density is more controllable and will be more uniform when the plaster is applied in thin, successive layers. For this reason, the code requires three-coat plastering over metal lath or wire fabric lath and two-coat work applied over other plaster bases [see Figures 2511.1(1) and 2511.1(2)]. Reducing the requirements for plaster bases other than metal or wire lath depends on the rigidity of the plaster base itself. More rigid plaster bases are not as susceptible to variations in thickness and flatness of the surface. It may be considered that the first coat in three-coat work on a flexible base, such as wire lath, is used to stiffen that base to provide the rigidity necessary to attain uniform thickness and surface flatness.

Gypsum veneer plaster as specified in ASTM C 843 is specifically designed to be applied as a one-coat plaster, and is approved by the code for that application. Any one-coat system must be applied over a solid base as approved by the code, the referenced standard or the product manufacturer.

2511.1.1 Installation. Installation of lathing and plaster materials shall conform with Table 2511.1.1 and Section 2507.

❖ This section refers to Section 2508 for the appropriate standards to regulate the proper installation of gypsum lath and related gypsum board materials and accessories. The expected performance of a product or material is dependent on proper installation. Table 2511.1.1 lists the appropriate ASTM standards for plaster construction.

TABLE 2511.1.1
INSTALLATION OF PLASTER CONSTRUCTION

MATERIAL	STANDARD
Gypsum plaster	ASTM C 842
Gypsum veneer plaster	ASTM C 843
Interior lathing and furring (gypsum plaster)	ASTM C 841
Lathing and furring (cement plaster)	ASTM C 1063
Portland cement plaster	ASTM C 926
Steel framing	ASTM C 754; C 1007

❖ Table 2511.1.1 provides the ASTM installation standards required by the code for plaster construction. This table also includes installation standards for related construction materials. Chapter 35 contains the complete title and appropriate issue date for the standard referenced.

2511.2 Limitations. Plaster shall not be applied directly to fiber insulation board. Cement plaster shall not be applied directly to gypsum lath or gypsum plaster except as specified in Sections 2510.5.1 and 2510.5.2.

❖ Fiber insulation board does not have the qualities necessary to provide an acceptable base for the application of plaster. It absorbs excessive moisture from the plaster mix that creates problems of workmanship, and it does not have the structural properties to provide the stability and rigidity required for a proper func-

Figure 2511.1(1)
THREE-COAT PLASTER ON INTERIOR STUD WALL

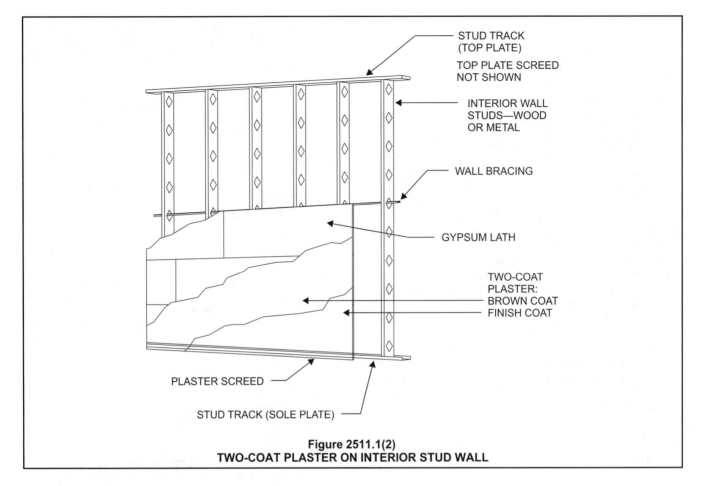

Figure 2511.1(2)
TWO-COAT PLASTER ON INTERIOR STUD WALL

tioning plaster base. In colder and damper climates, fiberboard insulation retains the moisture absorbed from the plaster for a relatively longer period of time than other, more dense bases. This lack of density and excessive retention of moisture will cause the premature failure of the plaster. For these reasons, the code prohibits the use of fiber insulation as a plaster base.

Portland cement plaster will not bond properly to gypsum plaster bases; therefore, the code prohibits the use of portland cement plaster on a gypsum plaster base. The code does, however, permit the use of exterior plaster to be applied on horizontal surfaces, such as ceilings and soffits, over gypsum lath and gypsum board when used as a backing for metal lath.

2511.3 Grounds. Where installed, grounds shall ensure the minimum thickness of plaster as set forth in ASTM C 842 and ASTM C 926. Plaster thickness shall be measured from the face of lath and other bases.

❖ Plaster grounds are utilized to establish and maintain the required thickness of plaster. Grounds are usually wood or metal strips of a known thickness attached to the plaster base. The intent is that plaster grounds are used as a guide for the straightedge in determining the thickness of the plaster. Door or window frames are used as plaster grounds. Dollops of plaster of the desired thickness may also be used as grounds.

2511.4 Interior masonry or concrete. Condition of surfaces shall be as specified in Section 2510.7. Approved specially prepared gypsum plaster designed for application to concrete surfaces or approved acoustical plaster is permitted. The total thickness of base coat plaster applied to concrete ceilings shall be as set forth in ASTM C 842 or ASTM C 926. Should ceiling surfaces require more than the maximum thickness permitted in ASTM C 842 or ASTM C 926, metal lath or wire fabric lath shall be installed on such surfaces before plastering.

❖ Special gypsum plaster and acoustical plaster that is specifically designed and manufactured for direct application to concrete masonry unit walls, concrete walls and ceilings without a gypsum board or metal lath base is permitted by the code. Special care must be taken to properly prepare the masonry or concrete surfaces that are to receive a plaster finish to ensure both proper bond of the plaster and the desired or required levelness of the finished surface. Preparation of masonry and concrete surfaces is discussed in Section 2510.7.

Where a concrete ceiling is to receive a plaster finish, the code requires a total thickness of base coat plaster as required by the ASTM standards referenced in this section. For applications of plaster on concrete ceiling surfaces, a thinner coat is required than for other applications. This is due primarily to the difficulty of achieving an adequate bond between the plaster and the concrete surface. Because of the horizontal application, the weight of the plaster base coat must be kept to a minimum. If it becomes necessary to increase the thickness of the plaster beyond the maximum allowed, due for example to the unevenness of the concrete surface or for some similar reason, metal lath or wire fabric lath must be attached to the concrete ceiling surface to provide anchorage for the weight of the additional plaster.

2511.5 Wet areas. Showers and public toilet walls shall conform to Sections 1210.2 and 1210.3. When wood frame walls and partitions are covered on the interior with cement plaster or tile of similar material and are subject to water splash, the framing shall be protected with an approved moisture barrier.

❖ Wet areas at interior locations such as those identified in Sections 1210.2 and 1210.3 must be protected from moisture by nonabsorbent wall or ceiling material. Cement plaster (gypsum plaster is not permitted in wet areas) and tile or other like materials will provide moisture protection for the wall surface. An additional waterproofing membrane must be provided where the wall is constructed of wood framing.

SECTION 2512
EXTERIOR PLASTER

2512.1 General. Plastering with cement plaster shall not be less than three coats where applied over metal lath or wire fabric lath and not less than two coats where applied over masonry, concrete or gypsum board backing as specified in Section 2510.5. If the plaster surface is to be completely covered by veneer or other facing material, or is completely concealed by another wall, plaster application need be only two coats, provided the total thickness is as set forth in ASTM C 926.

❖ This section is intended to define the standards and guidelines for the type of materials and installation of exterior plaster. The requirements of this section include the limitations of gypsum plaster and gypsum veneer plaster. This section will also address plaster applied to solid backings, alternative application methods, curing and plaster additives. Guidelines and requirements for preventing water infiltration into the building are discussed in this section.

Portland cement plaster (stucco) is the only type approved by the code for the use of exterior plaster finish. Exterior portland cement plasters are required by the code to be applied in no less than three coats when applied over expanded metal or wire fabric lath for the same reasons discussed for interior plaster (see commentary, Section 2511.1.1 and Figure 2512.1). When portland cement plaster is applied over other approved plaster bases, the code permits the application of a two-coat plaster system. The code also allows plaster work that is going to be completely concealed to be applied as a two-coat system, provided the total thickness meets the requirements of ASTM C 926, because the finish coat (second coat) of plaster will be the surface that will receive the exterior finish, such as

Figure 2512.1
THREE-COAT PLASTER ON EXTERIOR STUD WALL

paint, and it is critical for the finish coat to provide a clean, smooth and visually acceptable surface. Where the plaster surface is to be completely concealed, it is not necessary to provide a finish coat.

2512.1.1 On-grade floor slab. On wood framed or steel stud construction with an on-grade concrete floor slab system, exterior plaster shall be applied in such a manner as to cover, but not to extend below, the lath and paper. The application of lath, paper and flashing or drip screeds shall comply with ASTM C 1063.

❖ The code requires that the exterior plaster be installed to completely cover, but not extend below, the lath and paper (moisture barrier membrane) on wood or metal stud exterior wall construction supported by a concrete slab-on-grade floor. This requirement, combined with the requirement for a continuous weep screed as described in Section 2512.1.2, are intended to prevent the entrapment and subsequent channeling of free moisture to the interior of the building.

2512.1.2 Weep screeds. A minimum 0.019-inch (0.48 mm) (No. 26 galvanized sheet gage), corrosion-resistant weep screed with a minimum vertical attachment flange of $3^1/_2$ inches (89 mm) shall be provided at or below the foundation plate line on exterior stud walls in accordance with ASTM C 926. The weep screed shall be placed a minimum of 4 inches (102 mm) above the earth or 2 inches (51 mm) above paved areas and be of a type that will allow trapped water to drain to the exterior of the building. The water-resistive barrier shall lap the attachment flange. The exterior lath shall cover and terminate on the attachment flange of the weep screed.

❖ Water and moisture can penetrate an exterior plaster wall for a variety of reasons and in a number of ways. It is expected that some moisture will penetrate the plaster in an exterior wall; therefore, the design of the wall should include a weep screed, which will provide a way to release the moisture (see Figure 2512.1.2). Once water or moisture penetrates the plaster, it will migrate down the exterior wall face of the water-resistive barrier until it reaches the sill plate or mud sill. At this point, the water will seek a way out of the wall. If the exterior plaster system is not detailed and constructed with provisions to allow the moisture to escape to the exterior, it will find its own way out. This exit will almost certainly be through the interior of the wall and cause leaking, and damage to the interior of the building. For this reason, the code requires a continuous weep screed at the bottom of exterior walls to permit the moisture to escape to the exterior of the building.

2512.2 Plasticity agents. Only approved plasticity agents and approved amounts thereof shall be added to portland cement. When plastic cement or masonry cement is used, no additional lime or plasticizers shall be added. Hydrated lime or the equivalent amount of lime putty used as a plasticizer is permitted to be added to cement plaster or cement and lime plaster in an amount not to exceed that set forth in ASTM C 926.

For SI: 1 inch = 25.4 mm.

Figure 2512.1.2
WEEP SCREED FOR EXTERIOR PLASTER ON STUD WALL AT SLAB ON GRADE

❖ Admixtures such as plasticizers should not be added to portland cement unless approved by the building official. Some admixtures can create harmful effects that more than offset the desired improvement in plasticity. It is preferable that plasticizers be added during the manufacturing of the cement for product uniformity and proper proportions. When plastic cement is used, the code does not allow any further additions of plasticizers because the amount added during the manufacturing process is adequate and is the maximum amount permitted. Hydrated lime and lime putty are time-tested plasticizers used with portland cement plaster, and their use is permitted by the code in the amounts specified in ASTM C 926 as referenced herein.

2512.3 Limitations. Gypsum plaster shall not be used on exterior surfaces.

❖ Gypsum veneer plaster, gypsum plaster or any other similar gypsum-based plaster materials are not permitted for use as an exterior finish material. Gypsum-based plaster is subject to deterioration and ultimate failure when it is exposed to exterior weather conditions such as excessive heat, humidity and moisture, which are natural and unavoidable elements.

2512.4 Cement plaster. Plaster coats shall be protected from freezing for a period of not less than 24 hours after set has occurred. Plaster shall be applied when the ambient temperature is higher than 40°F (4°C), unless provisions are made to keep cement plaster work above 40°F (4°C) during application and 48 hours thereafter.

❖ Portland cement plaster contains a high percentage of water when it is mixed and applied, and before the process of curing is complete. For this reason, it is adversely affected by cold weather and freezing in much the same way that portland cement mortar and portland cement concrete are affected. The application of portland cement plaster to a frozen base or to a base covered with ice or frost will not only weaken the bond of the plaster to its base, but also cause the plaster itself to freeze.

When portland cement plaster is applied in freezing weather or to a frozen base, or is mixed with frozen ingredients, it loses a high proportion of its strength and, therefore, does not comply with the requirements of the code. Plaster materials must be stored to protect them from adverse weather conditions and should only be installed when weather conditions are within the requirements of the code and will remain so for the duration of the curing process.

2512.5 Second-coat application. The second coat shall be brought out to proper thickness, rodded and floated sufficiently rough to provide adequate bond for the finish coat. The second coat shall have no variation greater than $^{1}/_{4}$ inch (6.4 mm) in any direction under a 5-foot (1524 mm) straight edge.

❖ In plaster work, a base coat is any coat beneath the finish coat. This is true whether the plaster is a two-coat or three-coat application. In a three-coat application, the first coat is usually referred to as the "scratch" coat. It is usually applied over a flexible base, such as expanded metal or wire fabric lath, and is intended to stiffen the base and provide a mechanical bond to the base. As the name implies, the first coat is scratched with a scarifying tool to provide a series of horizontal ridges or scratches that are intended to provide mechanical keys for the application of the second coat (or brown coat). The brown coat usually constitutes the major bulk of the plaster and, consequently, materially affects the membrane strength. As a result, proportioning and workability are critical, and the mix should have high plasticity for proper application. The term "brown coat" is utilized by the trade to differentiate the color of the second coat from the finish coat. The finish coat is usually much lighter in color than the second coat.

2512.6 Curing and interval. First and second coats of cement plaster shall be applied and moist cured as set forth in ASTM C 926 and Table 2512.6.

❖ Sufficient time between coats is to be allowed to permit each coat to cure or develop enough rigidity to resist cracking or other physical damage when the next coat is applied. The proper amount of moisture must be provided to the plaster mix to allow for the proper curing time for each coat. The most effective procedure for curing and the time required between each coat will depend on the climate and job conditions. Moist or fog curing will permit and control the continuous hydration of cementitious materials.

TABLE 2512.6
CEMENT PLASTERS[a]

COAT	MINIMUM PERIOD MOIST CURING	MINIMUM INTERVAL BETWEEN COATS
First	48 hours[a]	48 hours[b]
Second	48 hours	7 days[c]
Finish	—	Note c

a. The first two coats shall be as required for the first coats of exterior plaster, except that the moist-curing time period between the first and second coats shall not be less than 24 hours. Moist curing shall not be required where job and weather conditions are favorable to the retention of moisture in the cement plaster for the required time period.

b. Twenty-four-hour minimum interval between coats of interior cement plaster. For alternate method of application, see Section 2512.8.

c. Finish coat plaster is permitted to be applied to interior portland cement base coats after a 48-hour period.

❖ The timing between coats will vary with climate conditions and the type of plaster base being used. Table 2512.6 specifies the minimum curing times between coats; however, temperature and relative humidity will extend or reduce the curing time required between consecutive coats. Cold or wet weather will increase the required curing time, while hot or dry weather will shorten it. Moderate changes in temperature and humidity can be controlled by providing additional heating during cold and damp weather, and by reducing the loss of moisture by prewetting during hot or dry weather.

All the conditions mentioned above must be taken into consideration when selecting the most appropriate method of curing that will provide the best results for the application being considered as well as the job conditions present during the curing process. Any of the three curing methods listed below, or any combination of the three, are permitted by the code:

1. Moist curing is accomplished by applying a fine fog spray of water as often as required to retain moisture and control the rate of curing. Moisture is generally applied two times a day—once in the morning and again in the late afternoon or evening—although job conditions are more of a factor than a predetermined time schedule. Care must be taken to avoid erosion damage to portland cement-based plaster surfaces due to excessive amounts of water running down vertical surfaces. Except in extreme or severe drying conditions, the wetting of the finish coat should be avoided.

2. Plastic film, when taped or weighted down around the perimeter of the plastered work area, will provide a vapor barrier to retain the moisture between the membrane and the plaster. Care must be taken when placing the plastic film around the work area. If the film is placed too soon before the plaster is allowed to stiffen, the film can come into contact with the wet plaster and damage the surface texture. If too much time is allowed before placing the film, an excessive amount of moisture will have escaped causing premature curing, which can cause cracking to the finished plaster.

3. Canvas, cloth or sheet material barriers can be erected to deflect sunlight and wind, both of which will affect the rate of evaporation. If the humidity is very low, this method alone may not provide adequate protection.

2512.7 Application to solid backings. Where applied over gypsum backing as specified in Section 2510.5 or directly to unit masonry surfaces, the second coat is permitted to be applied as soon as the first coat has attained sufficient hardness.

❖ When plaster is applied directly to a solid backing, such as gypsum board lath, concrete or concrete masonry, the minimum time intervals listed in Table 2512.6 are not required between the first (scratch) coat and the second (brown) coat. The brown coat may be applied as soon as the scratch coat has reached sufficient hardness to support the weight of the additional plaster without causing damage to the scratch coat.

2512.8 Alternate method of application. The second coat is permitted to be applied as soon as the first coat has attained sufficiently rigidity to receive the second coat.

❖ While the majority of experts on the application of plaster generally agree that the first two coats of portland cement plaster should be cured as required by Section 2512.8.2, there is an increasingly large number who feel that the brown coat can be applied without requiring moist curing of the scratch coat, but at least allowing the scratch coat to harden to a condition of sufficient rigidity to accept the second coat. This alternative method of application is also considered by its proponents to be stronger than when the scratch coat has been cured as required by the code. This is due to a manufacturer's claim that a superior bond is produced between the two coats of plaster when the second coat is applied to a set, but uncured, coat.

2512.8.1 Admixtures. When using this method of application, calcium aluminate cement up to 15 percent of the weight of the portland cement is permitted to be added to the mix.

❖ In order to develop a quicker set, the code permits the addition of calcium aluminate cement up to 15 percent of the weight of the portland cement. These high-alumina cements are characterized by a faster set time than portland. Because high-alumina cements do not possess the same high strength and qualities as portland cement in plaster mixes, the code limits their proportion of the total mix.

2512.8.2 Curing. Curing of the first coat is permitted to be omitted and the second coat shall be cured as set forth in ASTM C 926 and Table 2512.6.

❖ When applied as specified in ASTM C 926, the complete curing of the first (scratch) coat may be omitted. The second (brown) coat may be applied as soon as the scratch coat achieves sufficient rigidity; however, the second coat must be fully cured as indicated in Table 2512.6 prior to the application of the finished coat (for additional commentary on curing, see Section 2512.6).

2512.9 Finish coats. Cement plaster finish coats shall be applied over base coats that have been in place for the time periods set forth in ASTM C 926. The third or finish coat shall be applied with sufficient material and pressure to bond and to cover the brown coat and shall be of sufficient thickness to conceal the brown coat.

❖ The base coats in plaster work provide the strength for the plaster membrane but generally do not provide a proper texture for a finished surface; therefore, a thin, veneer-like coat of plaster or stucco is applied to the base coats as a finish coat. The finish coat can be applied as a textured, ornamental or decorative finish, or it can be applied as a smooth, flat surface appropriate to receive paint, wallpaper, applied textured finishes or a variety of architectural finishes. Finish plaster coats must be allowed to properly cure before applying any adhered or wet finishes.

SECTION 2513
EXPOSED AGGREGATE PLASTER

2513.1 General. Exposed natural or integrally colored aggregate is permitted to be partially embedded in a natural or colored bedding coat of cement plaster or gypsum plaster, subject to the provisions of this section.

❖ This section is provided in the code to establish the requirements for the use of exposed aggregate plaster, both portland cement plaster and gypsum plaster. Exposed aggregate plaster is approved by the code for both interior and exterior applications, although it is more commonly used with portland cement plaster as an exterior wall finish surface. The plaster systems, accessories, materials, assemblies, installation procedures and installation applications are essentially the same as other plastering systems with the exception of the bedding coat that is prepared to receive the exposed aggregate material.

Exposed aggregate plaster is a plaster finish, which as the name implies, has small chips of exposed aggregate embedded in its finish surface. Exposed aggregate finish is used most commonly on exterior cement plaster walls, although its use is not limited to exterior applications or cement plaster. Exposed aggregate finishes may also be applied to gypsum plaster, but like all other forms of gypsum plaster, it is limited to interior and nonwet areas. The purpose of using an exposed aggregate plaster is to provide a variety of textured and colored surfaces.

2513.2 Aggregate. The aggregate shall be applied manually or mechanically and shall consist of marble chips, pebbles or similar durable, moderately hard (three or more on the Mohs hardness scale), nonreactive materials.

❖ Two of the more common exposed aggregate finishes are a washed pebble finish and a marble chip finish. There are other aggregates available, but the material requirements and application procedures are the same as those contained in this section. The mention of these two aggregates is not meant to exclude other materials from being acceptable.

The code requires that the materials used for the aggregate be nonreactive and moderately hard (three or harder on the Mohs hardness scale for minerals). The mineral that is used as the identifier for the hardness of three is calcite, which in its common form is represented by limestone and marble, to name only two. By requiring at least this hardness, the intent of the code is to approximate ordinary sand aggregate plaster in durability.

The aggregate is applied by rubbing or "dashing" it into or against the bedding coat and then using a trowel or other suitable tool to embed the aggregate firmly into the bedding coat. The aggregate often requires a light tamping to ensure the proper bedding.

2513.3 Bedding coat proportions. The bedding coat for interior or exterior surfaces shall be composed of one-part portland cement, one-part Type S lime and a maximum of three parts of graded white or natural sand by volume. The bedding coat for

interior surfaces shall be composed of 100 pounds (45.4 kg) of neat gypsum plaster and a maximum of 200 pounds (90.8 kg) of graded white sand. A factory-prepared bedding coat for interior or exterior use is permitted. The bedding coat for exterior surfaces shall have a minimum compressive strength of 1,000 pounds per square inch (psi) (6895 kPa).

❖ The plaster mix proportions for the exposed aggregate bedding coat are to be as defined in this section.

For the same reason the code establishes the hardness requirements for the aggregate used in exposed aggregate plaster, it also establishes durability requirements for the bedding coat. The minimum compressive strength of the exterior bedding coat is to be 1,000 pounds per square inch (psi) (6895 kPa).

Type S lime is a special hydrated lime for masonry purposes, and is required in the exterior bedding coat to increase high early plasticity and water retention.

For the increased plasticity required of the bedding coat on an interior application, the code requires the use of a relatively richer gypsum plaster.

2513.4 Application. The bedding coat is permitted to be applied directly over the first (scratch) coat of plaster, provided the ultimate overall thickness is a minimum of $^7/_8$ inch (22 mm), including lath. Over concrete or masonry surfaces, the overall thickness shall be a minimum of $^1/_2$ inch (12.7 mm).

❖ The requirements for the scratch coat used as part of the exposed aggregate plaster are the same as those for other scratch coats described and specified elsewhere in the code. The bedding coat may be used as the second coat, provided the total plaster thicknesses required in this section are maintained through the entire system.

2513.5 Bases. Exposed aggregate plaster is permitted to be applied over concrete, masonry, cement plaster base coats or gypsum plaster base coats installed in accordance with Section 2511 or 2512.

❖ The requirements for the base coat used as part of the exposed aggregate plaster are the same as those for other base coats described and specified elsewhere in the code.

2513.6 Preparation of masonry and concrete. Masonry and concrete surfaces shall be prepared in accordance with the provisions of Section 2510.7.

❖ See the commentary for Section 2510.7 for the preparation of masonry and concrete.

2513.7 Curing of base coats. Cement plaster base coats shall be cured in accordance with ASTM C 926. Cement plaster bedding coats shall retain sufficient moisture for hydration (hardening) for 24 hours minimum or, where necessary, shall be kept damp for 24 hours by light water spraying.

❖ The curing requirements for cement plaster base coat are required to be the same as that shown in Table 2512.6 and as described in the related commentary, except that moist curing is only necessary for 24 hours in lieu of the 48 hours shown in Table 2512.6.

Bibliography

The following resource materials are referenced in this chapter or are relevant to the subject matter addressed in this chapter.

1995 Certification Listing. Ontario, Canada: Warnock Hersey, 1995.

ASTM C 5-03, *Specification for Quicklime for Structural Purposes.* West Conshohocken, PA: ASTM International, 2003.

ASTM C 22/C 22M-00, *Specification for Gypsum.* West Conshohocken, PA: ASTM International, 2000.

ASTM C 28/C 28M-00e01, *Specification for Gypsum Plasters.* West Conshohocken, PA: ASTM International, 2001.

ASTM C 35-95(2001), *Specification for Inorganic Aggregates for Use in Gypsum Plaster.* West Conshohocken, PA: ASTM International, 1995.

ASTM C 36/C 36M-03, *Specification for Gypsum Wallboard.* West Conshohocken, PA: ASTM International, 2003.

ASTM C 37/C 37M-01, *Specification for Gypsum Lath.* West Conshohocken, PA: ASTM International, 2001.

ASTM C 59/C 59M-00, *Specification for Gypsum Casting and Molding Plaster.* West Conshohocken, PA: ASTM International, 2000.

ASTM C 61/C 61M-00, *Specification for Gypsum Keene's Cement.* West Conshohocken, PA: ASTM International, 2000.

ASTM C 79-04a, *Specification for Treated Core and Nontreated Core Gypsum Sheathing Board.* West Conshohocken, PA: ASTM International, 2004.

ASTM C 91-03a, *Specification for Masonry Cement.* West Conshohocken, PA: ASTM International, 2003.

ASTM C 150-04, *Specification for Portland Cement.* West Conshohocken, PA: ASTM International, 2004.

ASTM C 206-03, *Specification for Finishing Hydrated Lime.* West Conshohocken, PA: ASTM International, 2003.

ASTM C 442/C 442M-04, *Specification for Gypsum Backing Board Coreboard and Gypsum Shaftliner Board.* West Conshohocken, PA: ASTM International, 2004.

ASTM C 472-99(2004), *Specification for Standard Test Methods for Physical Testing of Gypsum, Gypsum Plasters and Gypsum Concrete.* West Conshohocken, PA: ASTM International, 2004.

ASTM C 473-03, *Specification for Standard Test Methods for Physical Testing of Gypsum Panel Products.* West Conshohocken, PA: ASTM International, 2003.

ASTM C 474-02, *Test Methods for Joint Treatment Materials for Gypsum Board Construction.* West Conshohocken, PA: ASTM International, 2002.

ASTM C 475-01, *Specification for Joint Compound and Joint Tape for Finishing Gypsum Board.* West Conshohocken, PA: ASTM International, 2001.

ASTM C 514-01, *Specification for Nails for the Application of Gypsum Board.* West Conshohocken, PA: ASTM International, 2001.

ASTM C 557-03, *Specification for Adhesives for Fastening Gypsum Wallboard to Wood Framing.* West Conshohocken, PA: ASTM International, 2003.

ASTM C 587-02, *Specification for Gypsum Veneer Plaster.* West Conshohocken, PA: ASTM International, 2002.

ASTM C 588/C 588M-01, *Specification for Gypsum Base for Veneer Plasters.* West Conshohocken, PA: ASTM International, 2001.

ASTM C 595-03, *Specification for Blended Hydraulic Cements.* West Conshohocken, PA: ASTM International, 2003.

ASTM C 630/C630M-03, *Specification for Water-Resistant Gypsum Backing Board.* West Conshohocken, PA: ASTM International, 2003.

ASTM C 631-00, *Specification for Bonding Compounds for Interior Gypsum Plastering.* West Conshohocken, PA: ASTM International, 2000.

ASTM C 635-00, *Specification for the Manufacture, Performance, and Testing of Metal Suspension Systems for Acoustical Tile and Lay-in Panel Ceilings.* West Conshohocken, PA: ASTM International, 2000.

ASTM C 645-04, *Specification for Nonstructural Steel Framing Members.* West Conshohocken, PA: ASTM International, 2004.

ASTM C 754-04, *Specification for Installation of Steel Framing Members to Receive Screw-Attached Gypsum Panel Products.* West Conshohocken, PA: ASTM International, 2004.

ASTM C 840-04, *Specification for Application and Finishing of Gypsum Board.* West Conshohocken, PA: ASTM International, 2004.

ASTM C 841-03, *Specification for Installation of Interior Lathing and Furring.* West Conshohocken, PA: ASTM International, 2003.

ASTM C 842-99, *Specification for Application of Interior Gypsum Plaster.* West Conshohocken, PA: ASTM International, 1999.

ASTM C 843-99e01, *Specification for Application of Gypsum Veneer Plaster.* West Conshohocken, PA: ASTM International, 2001.

ASTM C 844-99, *Specification for Application of Gypsum Base to Receive Gypsum Veneer Plaster.* West Conshohocken, PA: ASTM International, 1999.

ASTM C 847-00, *Specification for Metal Lath.* West Conshohocken, PA: ASTM International, 2000.

ASTM C 897-00, *Standard Practice for Vapor Attack on Refractories for Furnace Superstructures.* West Conshohocken, PA: ASTM International, 2000.

ASTM C 926-98a, *Specification for Application of Portland Cement-Based Plaster.* West Conshohocken, PA: ASTM International, 1998.

ASTM C 931/C 931M-04, *Specification for Exterior Gypsum Soffit Board.* West Conshohocken, PA: ASTM International, 2004.

ASTM C 932-03, *Specification for Surface-Applied Bonding Agents for Exterior Plastering.* West Conshohocken, PA: ASTM International, 2003.

ASTM C 933-04, *Specification for Welded Wire Lath.* West Conshohocken, PA: ASTM International, 2004.

ASTM C 954-00, *Specification for Steel Drill Screws for the Application of Gypsum Panel Products or Metal Plaster Bases to Steel Studs From 0.033 in. (0.84 mm) to 0.112 in. (2.84 mm) in Thickness.* West Conshohocken, PA: ASTM International, 2000.

ASTM C 955-03, *Specification for Load-Bearing (Transverse and Axial) Steel Studs, Runners (Track), and Bracing or Bridging for Screw Application of Gypsum Panel Products and Metal Plaster Bases.* West Conshohocken, PA: ASTM International, 2003.

ASTM C 960-04, *Specification for Predecorated Gypsum Board.* West Conshohocken, PA: ASTM International, 2004.

ASTM C 1002-01, *Specification for Steel Drill Screws for the Application of Gypsum Panel Products or Metal Plaster Bases.* West Conshohocken, PA: ASTM International, 2001.

ASTM C 1007-04, *Specification for Installation of Load-Bearing (Transverse and Axial) Steel Studs and Related Accessories.* West Conshohocken, PA: ASTM International, 2004.

ASTM C 1032-04, *Specification for Woven Wire Plaster Base.* West Conshohocken, PA: ASTM International, 2004.

ASTM C 1047-99, *Specification for Accessories for Gypsum Wallboard and Gypsum Veneer Base.* West Conshohocken, PA: ASTM International, 1999.

ASTM C 1063-03, *Specification for Installation of Lathing and Furring to Receive Interior and Exterior Portland Cement-Based Plaster.* West Conshohocken, PA: ASTM International, 2003.

ASTM C 1177/C 1177M-04, *Specification for Glass Mat Gypsum Substrate for Use as Sheathing.* West Conshohocken, PA: ASTM International, 2004.

ASTM C 1178/C 1178M-04, *Specification for Glass Mat Water-Resistant Gypsum Backing Panel.* West Conshohocken, PA: ASTM International, 2004.

ASTM C 1280-04, *Specification for Application of Gypsum Sheathing.* West Conshohocken, PA: ASTM International, 2004.

ASTM C 1328-03a, *Specification for Plastic (Stucco) Cement.* West Conshohocken, PA: ASTM International, 2003.

ASTM C 1395/C 1395M-04, *Specification for Gypsum Ceiling Board.* West Conshohocken, PA: ASTM International, 2004.

ASTM C 1396-02, *Standard Specification for Gypsum Board.* West Conshohocken, PA: ASTM International, 2002.

ASTM E 84-04, *Test Method for Surface Burning Characteristics of Building Materials.* West Conshohocken, PA: ASTM International, 2004.

ASTM E 90-04, *Test Method for Laboratory Measurement of Airborne Sound Transmission Loss of Building Partitions and Elements.* West Conshohocken, PA: ASTM International, 2004.

ASTM E 119-00, *Test Methods for Fire Tests of Building Construction and Materials.* West Conshohocken, PA: ASTM International, 2000.

ASTM F 547-01, *Terminology of Nails for Use with Wood and Wood-Base Materials.* West Conshohocken, PA: ASTM International, 2001.

ASTM F 1667-03, *Specification for Driven Fasteners: Nails, Spikes, and Staples.* West Conshohocken, PA: ASTM International, 2003.

GA 216-04, *Application and Finishing of Gypsum Board.* Washington, D.C.: Gypsum Association, 2004.

GA 230-91, *Vapor Retarders Over Water-Resistant Gypsum Backing Board.* Washington, D.C.: Gypsum Association, 1991.

GA 600-03, *Fire Resistance Design Manual.* Washington, D.C.: Gypsum Association, 2003.

Guidelines for Determining the Fireresistance of Building Elements. Country Club Hills, IL: Building Officials and Code Administrators International, Inc., 1993.

UL-96, *Fire Resistance Directory.* Northbrook, IL: Underwriters Laboratories Inc., 1996.

Chapter 26:
Plastic

General Comments

Plastics are a very large group of synthetic materials whose structures are based on the chemistry of carbon. Plastics are also called polymers because they are made of extremely long chains of carbon atoms. Since plastics can be readily molded, extruded, cast into various shapes and forms or drawn into filaments and fibers, they are increasingly being used in the construction of buildings and structures. While progress in polymer technology makes it increasingly difficult to make general statements about these materials, the following properties are characteristic of many plastics:

- Low strength (in comparison to steel and other metals commonly used in construction);
- Low stiffness (modulus of elasticity) as compared to metals (except for reinforced plastics);
- Tendency to creep (increase in length under tensile stress);
- Low hardness (except for formaldehyde plastics);
- Low density (varies, plastics can be expanded or unexpanded);
- Brittleness at low temperatures and loss of strength and hardness at moderately elevated temperatures;
- Flammability (although many plastics do not burn);
- Outstanding electrical characteristics, such as electrical resistance; and
- Degradation of some plastics, due to environmental agents, such as ultraviolet radiation (although most plastics are highly resistant to chemical attack).

When plastics and foam plastics were first proposed to be included in the code, consideration was given to several factors.

The following is the evolution of plastics being introduced into the model codes, ultimately leading to inclusion in the code.

It was recognized that most buildings classified as noncombustible construction contained combustible elements. These elements were not primary structural members but included materials such as: asphalt and pitch (in roof coverings); wood (in flooring systems, as wall and ceiling interior finish and trim and as furring and lath in walls); paper products (as slip sheets and vapor retarders) and cork (as insulation).

There was a sensitivity to the burning characteristics of all plastic materials based on the history of fires involving plastics in warehouses and in building applications

not covered by any of the model codes. It was agreed, therefore, that plastic materials should be required to meet minimum fire performance properties and each application would have appropriate requirements related to characteristics such as rate of burning, surface flammability or smoke development.

This position is somewhat different from the regulation of traditional materials, such as wood, paper and asphalt, which frequently are not required to be separately fire tested when used within an assembly.

Plastic materials to be left exposed on the interior or exterior of buildings would have to meet the existing requirements for other finish materials as well as be limited in area and subject to separation requirements between adjoining areas. The primary concern was that plastic materials would be used in applications in which glass had traditionally been used, such as skylights and light-diffusing ceilings. Since little experience was available at the time, a conservative approach should be taken.

Foam plastics used as thermal insulation would be separated from the interior of buildings by a thermal barrier equivalent to $^1/_2$-inch-thick (12.7 mm) gypsum wallboard. This position was supported by the industry because most of the applications at that time included plaster or wallboard finishes of $^1/_2$ inch (12.7 mm) or greater thickness, and the fire record for such applications was good.

Specific examples of commonly used acceptable applications would be included in the code, such as the use of foam plastics within wood frame and masonry walls.

Finally, the code would include a statement that "diversified tests" would be necessary for approval of any application or material not meeting the prescriptive requirements of the code. The reason for including diversified tests was because it was clear that it would not be possible or even desirable to include all applications specifically in the code. Diversified tests were intended to mean tests related to the end use and usually were full-scale assemblies exposed to fire conditions that were representative of actual fires that might be expected to occur.

Purpose

This chapter establishes minimum requirements for all light-transmitting plastics and foam plastics in all applications regulated by the code. Some plastics exhibit rapid flame spread and heavy smoke density characteristics when exposed to fire. Additionally, exposure to the heat generated by a fire can cause some plastics to deform and undergo a change in their structural stability. The requirements and limitations of this chapter are neccations regulated by the code. Some plastics exhibit rapid flame spread and heavy smoke density characteristics when exposed to fire. Additionally, exposure to the heat generated by a fire can cause some plastics to deform and undergo a change in their structural stability.

The requirements and limitations of this chapter are necessary to control the use of these combustible products such that they do not compromise the safety of building occupants.

SECTION 2601
GENERAL

2601.1 Scope. These provisions shall govern the materials, design, application, construction and installation of foam plastic, foam plastic insulation, plastic veneer, interior plastic finish and trim and light-transmitting plastics. See Chapter 14 for requirements for exterior wall finish and trim.

❖ Chapter 26 governs foam plastics, plastic veneers and light-transmitting plastic materials regardless of their use. Requirements for foam-plastic insulation are found in Section 2603. Requirements for interior finish and trim are found in Section 2604. Requirements for plastic veneer are found in Section 2605. The main requirements for light-transmitting plastics are found in Section 2606. Sections 2607 through 2611 cover a variety of applications of light-transmitting plastics. Essentially, this chapter goes into detail as to how such combustible materials can be used in building construction. This includes wall panels, glazing, roof panels, skylights and interior signs.

SECTION 2602
DEFINITIONS

2602.1 General. The following words and terms shall, for the purposes of this chapter and as used elsewhere in this code, have the meanings shown herein.

❖ Definitions of terms can help in the understanding and application of the code requirements. The purpose for including these definitions within this chapter is to provide more convenient access to them without having to refer back to Chapter 2. For convenience, these terms are also listed in Chapter 2 with a cross reference to this section. The use and application of all defined terms, including those defined herein, are set forth in Section 202.

Plastics include any of the large group of synthetic or natural organic materials that can be shaped, formed and molded at high temperatures and are capable of maintaining their form at low temperatures. Plastics can be divided into two basic classes: thermoplastic and thermosetting. Plastics can also be broken down into Class CC1 or CC2 (see Section 2606). Plastics are combustible materials and some can give off toxic gases when exposed to fire. Plastics can be synthetically designed to exhibit desirable strength, rigidity and flammability characteristics.

FOAM PLASTIC INSULATION. A plastic that is intentionally expanded by the use of a foaming agent to produce a reduced-density plastic containing voids consisting of open or closed cells distributed throughout the plastic for thermal insulating or acoustical purposes and that has a density less than 20 pounds per cubic foot (pcf) (320 kg/m³).

❖ Foam-plastic insulation is plastic that is imbedded with gas.

LIGHT-DIFFUSING SYSTEM. Construction consisting in whole or in part of lenses, panels, grids or baffles made with light-transmitting plastics positioned below independently mounted electrical light sources, skylights or light-transmitting plastic roof panels. Lenses, panels, grids and baffles that are part of an electrical fixture shall not be considered as a light-diffusing system.

❖ Heat-resistant plastics, such as polystyrene and urea, are frequently used for light diffusion.

LIGHT-TRANSMITTING PLASTIC ROOF PANELS. Structural plastic panels other than skylights that are fastened to structural members, or panels or sheathing and that are used as light-transmitting media in the plane of the roof.

❖ Plastic roof panels are primarily used in areas where light transmission rather than visual clarity is the goal (i.e., factory and industrial applications). Even though they are used as light-transmitting media, they must support the loads and forces encountered, just like roof surfaces. Typical applications include factories, warehouses and greenhouses where glass breakage is a practical concern.

LIGHT-TRANSMITTING PLASTIC WALL PANELS. Plastic materials that are fastened to structural members, or to structural panels or sheathing, and that are used as light-transmitting media in exterior walls.

❖ This term is only applicable to plastic wall panels allowing the transmission of light. Even though they are used as light-transmitting media, they must support the loads and forces encountered, just like other wall surfaces. Foam-core insulated panels are regulated by Section 2603 and are not considered plastic wall panels.

PLASTIC, APPROVED. Any thermoplastic, thermosetting or reinforced thermosetting plastic material that conforms to combustibility classifications specified in the section applicable to the application and plastic type.

❖ This term applies to plastic that has properties that conform to the code requirements for its intended use.

PLASTIC GLAZING. Plastic materials that are glazed or set in frame or sash and not held by mechanical fasteners that pass through the glazing material.

❖ In lieu of glass, plastic is permitted to be used as a glazing material. Plastic glazing is required to meet the same load requirements as glass.

REINFORCED PLASTIC, GLASS FIBER. Plastic reinforced with glass fiber having not less than 20 percent of glass fibers by weight.

❖ Most commonly, translucent polyester (plastic) panels are reinforced with glass fibers to achieve a stronger, stiffer and more impact-resistant panel.

THERMOPLASTIC MATERIAL. A plastic material that is capable of being repeatedly softened by increase of temperature and hardened by decrease of temperature.

❖ Thermoplastic materials become plastic and deform easily at higher temperatures. Indeed, many of these materials have limited construction value at high temperatures. Examples of thermoplastics are cellulose acetate, nylon, polycarbonates, polyethylene (three different densities), methyl methacrylate (plexiglass), polypropylene, polybutylene, polystyrene, acrylonitrile-butadiene-styrene (ABS), polytetrafluoroethylene (Teflon®), polyvinyl chloride (PVC), chlorinated polyvinyl chloride (CPVC) and polyethylene terephthalate (polyester). Note there are both thermosetting and thermoplastic polyesters. These materials have high thermal expansion coefficients and, while they may burn slowly in a fire, they decompose rapidly to yield a heavy black smoke and acrid, sometimes very toxic, fumes depending on the specific material and burning conditions.

THERMOSETTING MATERIAL. A plastic material that is capable of being changed into a substantially nonreformable product when cured.

❖ This term is the opposite of a thermoplastic material. Common thermosetting plastic includes epoxy, melamine, urea, polyester and polyurethane.

SECTION 2603
FOAM PLASTIC INSULATION

2603.1 General. The provisions of this section shall govern the requirements and uses of foam plastic insulation in buildings and structures.

❖ Section 2603 contains all of the life safety provisions for foam-plastic insulation products, such as rigid insulation board, spray-foam insulation and pour-in-place insulation. "Foam plastic" is a general term given to insulating products that have been manufactured by intentionally expanding plastic by the use of a foaming agent. Extruded polystyrene (XPS), expanded polystyrene (EPS), polyisocyanurate (PIR or ISO), open-cell polyisocyanurate, phenolic, polyurethane and polypropylene foam are among the many types of foam plastic insulation products subject to the requirements found in this section. In addition to technical requirements listed below, all packages and containers of foam-plastic insulation must bear a label of an approved agency listing the manufacturer's name, the product listing, product identification and information sufficient to determine that the end use will comply with the code requirements.

Compliance with Section 2603 shall be achieved through either a prescriptive method or the performance method. The prescriptive method requires compliance with Sections 2603.4 through 2603.7. The performance method requires compliance with Section 2603.9. Note that both methods require compliance with Section 2603.3, which covers surface-burning characteristics and Section 2603.8, which covers protection against termites.

Figure 2603.1 documents code compliance when foam plastic is used.

2603.2 Labeling and identification. Packages and containers of foam plastic insulation and foam plastic insulation components delivered to the job site shall bear the label of an approved agency showing the manufacturer's name, the product listing, product identification and information sufficient to determine that the end use will comply with the code requirements.

❖ All foam plastics or packages of foam plastics delivered to the construction site must be labeled. Also, labels are required on containers [usually two components in 55-gallon (208 L) drums] of ingredients delivered for the production of foam plastic at the construction site. The label must include identification of the approved and detailed product identification or information describing the performance characteristics of the product. It is intended that labeling printed on board stock or on packaging be acceptable. It is also intended that the label information include the name of the manufacturer or distributor, the type of foam plastic, the performance characteristics required and the name of the approved testing agency.

2603.3 Surface-burning characteristics. Unless otherwise indicated in this section, foam plastic insulation and foam plastic cores of manufactured assemblies shall have a flame spread index of not more than 75 and a smoke-developed index of not more than 450 where tested in the maximum thickness intended for use in accordance with ASTM E 84. Loose fill-type foam plastic insulation shall be tested as board stock for the flame spread index and smoke-developed index.

Exceptions:

1. Smoke-developed index for interior trim as provided for in Section 2604.2.

2. In cold storage buildings, ice plants, food plants, food processing rooms and similar areas, foam plastic insulation where tested in a thickness of 4 inches (102 mm) shall be permitted in a thickness up to 10 inches (254 mm) where the building is equipped throughout with an automatic fire sprinkler system in accordance with Section 903.3.1.1. The approved automatic sprinkler system shall be provided in both the room and that part of the building in which the room is located.

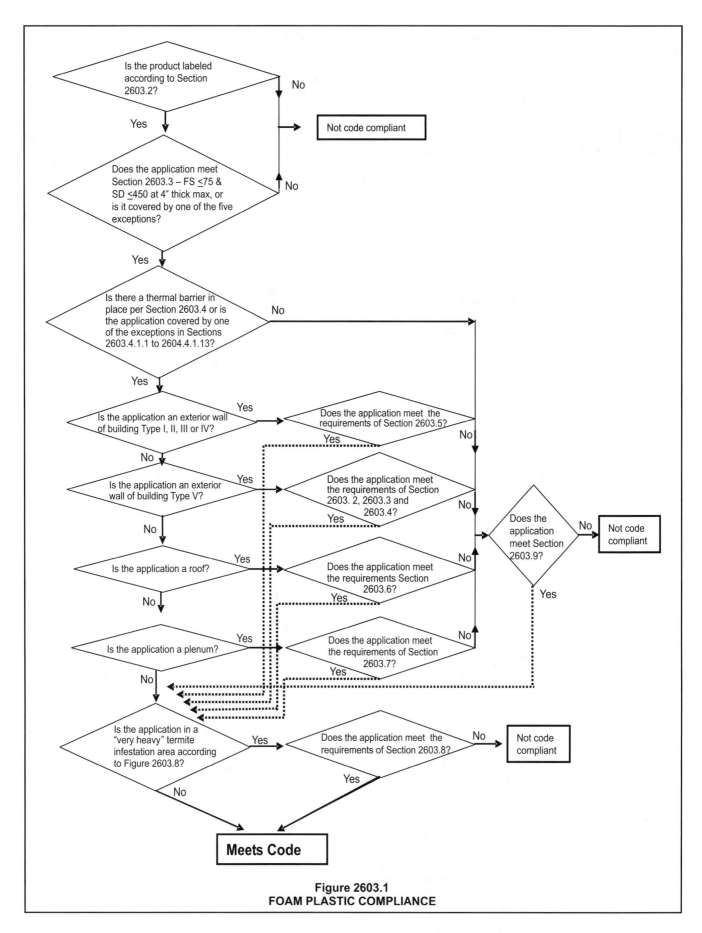

Figure 2603.1
FOAM PLASTIC COMPLIANCE

3. Foam plastic insulation that is a part of a Class A, B or C roof-covering assembly provided the assembly with the foam plastic insulation satisfactorily passes FM 4450 or UL 1256. The smoke-developed index shall not be limited for roof applications.

4. Foam plastic insulation greater than 4 inches (102 mm) in thickness shall have a maximum flame spread index of 75 and a smoke-developed index of 450 where tested at a minimum thickness of 4 inches (102 mm), provided the end use is approved in accordance with Section 2603.9 using the thickness and density intended for use.

5. Flame spread and smoke-developed indexes for foam plastic interior signs in covered mall buildings provided the signs comply with Section 402.15.

❖ Foam plastic insulation or foam plastic cores used as a component in a manufactured assembly are combustible and must be assessed for flame spread index and smoke-developed index. Testing in accordance with ASTM E 84 is required. The foam plastic must be tested in the maximum thickness to be used [up to 4-inch (102 mm) thickness; see Exception 4 for foam thicker than 4 inches (102 mm)] and be limited to a flame spread index of 75 or less and a smoke-developed index of 450 or less. The 4-inch (102 mm) thickness for testing is specifically related to the limitation on testing thickness in ASTM E 84. All foam plastic materials are required to be tested in accordance with ASTM E 84, unless specifically exempted or modified by one of the listed exceptions.

Five exceptions relate to whether the flame spread limitations of 75 and smoke-developed of 450 apply or if the thickness of material is allowed to exceed 4-inches (102 mm) as limited by ASTM E 84:

1. Use as interior trim is not required to meet the smoke-developed index limitation of 450 but must still meet the flame spread index requirement of less than or equal to 75.

2. In fully sprinklered buildings, cold storage construction using foam plastic is allowed to exceed the thickness of 4 inches (102 mm) up to 10 inches (254 mm) even when the material has only been tested at 4 inches (102 mm) with ASTM E 84.

3. When used as part of a Class A, B or C roof covering assembly in which the assembly has successfully passed FM 4450 or UL 1256 testing, foam-plastic insulation is not required to meet the flame spread rating of 75. Also, smoke-developed ratings are not required or limited for any foam-plastic insulation used in roof assemblies. The intent is to recognize that the overall hazards of roofing have already been addressed and additional testing requirements for flame spread would be unnecessary.

4. Foam-plastic insulation thicker than 4 inches (102 mm) is allowed if tested in accordance with Section 2603.9. Such testing would require the full as-installed thickness to be tested. The tests referenced in Section 2603.9 are generally full-scale room corner tests. Such material would also still be required to be tested to ASTM E 84 at a 4-inch (102 mm) thickness and result in a flame spread index no greater than 75 and a smoke-developed index of 450.

5. Use as interior signs in covered mall buildings is required to comply with Section 402.15. Section 402.15 requires compliance with UL 1975, which measures the rate of heat release of burning materials used in the manufacture of such signs. Section 402.15 sets the pass/fail criteria at a maximum heat release rate of 150 kW. Although this test method assesses heat release rate rather than surface burning characteristics, the testing and criteria specified provide material characteristics that address the goal of limiting the spread of fire.

The maximum flame spread value of 75 was chosen on the basis that it is lower than untreated wood (which usually is 100 to 165). The maximum smoke-developed rating of 450 was selected because, at the time, the code permitted interior finish materials that gave off "smoke no more dense than that given off by untreated wood." In selecting the maximum flame spread and smoke values, it was believed that a conservative approach was being taken by requiring an insulation material to meet the same requirements as interior finish, even though the insulation was intended to be covered with an interior finish material. The requirements for surface-burning characteristics of foam plastic apply to foam plastics used as cores of manufactured assemblies. The intent is that, while the finished assemblies are not required to be tested for surface burning characteristics, the foam-plastic core is not exempt from the general requirement; therefore, foam plastic is regulated in factory-manufactured assemblies the same as it is in field-fabricated applications.

2603.4 Thermal barrier. Except as provided for in Sections 2603.4.1 and 2603.9, foam plastic shall be separated from the interior of a building by an approved thermal barrier of 0.5-inch (12.7 mm) gypsum wallboard or equivalent thermal barrier material that will limit the average temperature rise of the unexposed surface to not more than 250°F (120°C) after 15 minutes of fire exposure, complying with the standard time-temperature curve of ASTM E 119. The thermal barrier shall be installed in such a manner that it will remain in place for 15 minutes based on FM 4880, UL 1040, NFPA 286 or UL 1715. Combustible concealed spaces shall comply with Section 717.

❖ The use of an approved thermal barrier to separate foam plastics from the interior of a building is a basic requirement for the use of foam plastics as shown in this section of the code. Sections 2603.4.1 and 2603.9 describe circumstances where the thermal barrier is modified or eliminated for specific uses. The job of a thermal barrier is to limit the temperature rise that the foam plastic will be exposed to. An approved thermal barrier is defined as $^1/_2$-inch (12.7 mm) gypsum wallboard or the equivalent. This section sets forth the test methods and criteria by which alternative thermal barriers are to be qualified to limit the average temperature rise of the unexposed face to 250°F (121°C) for 15 minutes of fire exposure while complying with the time-temperature conditions of ASTM E 119. The thermal barrier must be installed in such a way that it remains in place for 15 minutes when exposed to fire conditions and that combustible concealed spaces comply with Section 717.

Before 1975, experience had shown that foam plastics covered with plaster or $^1/_2$-inch (12.7 mm) gypsum wallboard had performed satisfactorily in building fires. For this reason, $^1/_2$-inch (12.7 mm) gypsum wallboard was included in the code as a minimum requirement. It was recognized that specifying a single material would not be desirable in a performance code; therefore, the words "or equivalent" were added. The thermal barrier test was selected as the appropriate method to determine equivalent performance. Although regular $^1/_2$-inch (12.7 mm) gypsum wallboard happened to have a 15-minute rating in the prescribed thermal barrier test, there was no intent that the 15-minute requirement for the thermal barrier would ensure 15 minutes of escape time in an actual building fire. This is related to the fact that the time-temperature conditions of ASTM E 119 may not reflect actual fire conditions. Gypsum wallboard is still the most commonly used thermal barrier, but equivalent materials are permitted and have been used when shown to be equivalent. Alternative thermal barriers are listed in evaluation reports issued by ICC Evaluation Services (ICC ES).

2603.4.1 Thermal barrier not required. The thermal barrier specified in Section 2603.4 is not required under the conditions set forth in Sections 2603.4.1.1 through 2603.4.1.13.

❖ A thermal barrier is not required for applications described in Sections 2603.4.1.1 through 2603.4.1.13. The prescriptive installation methods described in these sections must be strictly followed in order to serve as an alternative to the thermal barrier. Furthermore, installations of foam plastics in accordance with these sections do not have to be tested via the alternative approval testing requirements of Section 2603.9.

2603.4.1.1 Masonry or concrete construction. A thermal barrier is not required for foam plastic installed in a masonry or concrete wall, floor or roof system where the foam plastic insulation is covered on each face by a minimum of 1 inch (25 mm) thickness of masonry or concrete.

❖ No thermal barrier is required when 1 inch (25 mm) or more of masonry or concrete is placed between the foam plastic and the interior of the building. The intent is to accept 1 inch (25 mm) of masonry or concrete as equal to (or better than) $^1/_2$-inch (12.7 mm) gypsum wallboard. This condition can arise when foam plastics are installed either within a wall or on the exterior side

of a masonry wall. Some common examples are when foam plastics are installed:

- In the cavity of a hollow masonry wall;
- As the core of a concrete-faced panel;
- On the exterior face of a masonry wall and covered with an exterior finish;
- Within the cores of hollow masonry units; or
- Encapsulated within a minimum of 1-inch (25 mm) concrete or masonry wall, floor or roof system, such as in insulated tilt-up or pour-in-place concrete panels.

Note that the exterior surface would be required to comply with Section 2603.5.

2603.4.1.2 Cooler and freezer walls. Foam plastic installed in a maximum thickness of 10 inches (254 mm) in cooler and freezer walls shall:

1. Have a flame spread index of 25 or less and a smoke-developed index of not more than 450, where tested in a minimum 4 inch (102 mm) thickness.

2. Have flash ignition and self-ignition temperatures of not less than 600°F and 800°F (316°C and 427°C), respectively.

3. Have a covering of not less than 0.032-inch (0.8 mm) aluminum or corrosion-resistant steel having a base metal thickness not less than 0.0160 inch (0.4 mm) at any point.

4. Be protected by an automatic sprinkler system. Where the cooler or freezer is within a building, both the cooler or freezer and that part of the building in which it is located shall be sprinklered.

❖ A thermal barrier is not required in cooler and freezer walls insulated with 10 inches (254 mm) or less of foam plastic when all four criteria in Section 2603.4.1.2 are met.

2603.4.1.3 Walk-in coolers. In nonsprinklered buildings, foam plastic having a thickness that does not exceed 4 inches (102 mm) and a maximum flame spread index of 75 is permitted in walk-in coolers or freezer units where the aggregate floor area does not exceed 400 square feet (37 m^2) and the foam plastic is covered by a metal facing not less than 0.032-inch-thick (0.81 mm) aluminum or corrosion-resistant steel having a minimum base metal thickness of 0.016 inch (0.41 mm). A thickness of up to 10 inches (254 mm) is permitted where protected by a thermal barrier.

❖ A thermal barrier is not required for foam plastics up to 4 inches (102 mm) thick when the minimum thickness metal facings are used in coolers or freezers no greater than 400 square feet (37 m^2) in floor area. The required metal facing is intended to act as a barrier against ignition of the foam plastic. It is not intended to serve as a thermal barrier. This section recognizes the relatively small size of walk-in coolers and the accompanying lower hazard. Foam plastic up to 10 inches thick (254 mm) is permitted when protected by a thermal barrier.

2603.4.1.4 Exterior walls—one-story buildings. For one-story buildings, foam plastic having a flame spread index of 25 or less, and a smoke-developed index of not more than 450, shall be permitted without thermal barriers in or on exterior walls in a thickness not more than 4 inches (102 mm) where the foam plastic is covered by a thickness of not less than 0.032-inch-thick (0.81 mm) aluminum or corrosion-resistant steel having a base metal thickness of 0.0160 inch (0.41 mm) and the building is equipped throughout with an automatic sprinkler system in accordance with Section 903.3.1.1.

❖ Foam plastics may be used without a thermal barrier in one-story buildings provided that:

- The material is not more than 4 inches (102 mm) thick;
- It has a flame spread index not greater than 25;
- It has a smoke-developed index of not more than 450;
- It is covered with specified metal facings; and
- The building is equipped throughout with an automatic sprinkler system.

This provision is intended to permit the use of metal-faced panels, primarily in storage buildings other than cold-storage construction. Cold-storage construction was also covered in this section (see also Sections 2603.4.1.2 and 2603.4.1.3, and Exception 2 of Section 2603.3). The limitations are based on the results of large-scale tests performed by independent laboratories and sponsored by the plastics industry.

2603.4.1.5 Roofing. Foam plastic insulation under a roof assembly or roof covering that is installed in accordance with the code and the manufacturer's instructions shall be separated from the interior of the building by wood structural panel sheathing not less than 0.47 inch (11.9 mm) in thickness bonded with exterior glue, with edges supported by blocking, tongue-and-groove joints or other approved type of edge support, or an equivalent material. A thermal barrier is not required for foam plastic insulation that is a part of a Class A, B or C roof-covering assembly, provided the assembly with the foam plastic insulation satisfactorily passes FM 4450 or UL 1256.

❖ No thermal barrier is required when foam plastic is used in a roof assembly over wood structural panel sheathing given that the wood product meets all of the following:

- At least $^{15}/_{32}$ inch (12 mm) thick;
- Manufactured utilizing exterior-grade glue; and
- Properly installed to provide adequate edge support.

Equivalent materials to the wood structural panel sheathing described above are specifically allowed by this section. In any situation, the material must be compatible with the type of construction.

Also, no thermal barrier is required when foam plastic is used with a Class A, B or C roof assembly (as prescribed in Section 1505) when the assembly complies with FM 4450 or UL 1256. The intent is to recognize that roof assemblies tested to the criteria of the referenced standards have adequately demonstrated a resistance to fire from the underside of the roof deck; therefore, no additional testing is necessary, nor is an additional thermal barrier needed. It is important to note that this section applies to a specific roof assembly (i.e., specific roof deck, specific foam insulation, fasteners and specific roof membrane). Passing the prescribed tests with one specific roof assembly does not cover all combinations of roof systems where foam-plastic insulation is used. Users must confirm that the specific roof assembly has achieved a Class A, B or C rating. This coordinates with Exception 4 to Section 2603.3 for surface-burning characteristics.

2603.4.1.6 Attics and crawl spaces. Within an attic or crawl space where entry is made only for service of utilities, foam plastic insulation shall be protected against ignition by 1.5-inch-thick (38 mm) mineral fiber insulation; 0.25-inch-thick (6.4 mm) wood structural panel, particleboard or hardboard; 0.375-inch (9.5 mm) gypsum wallboard, corrosion-resistant steel having a base metal thickness of 0.016 inch (0.4 mm) or other approved material installed in such a manner that the foam plastic insulation is not exposed. The protective covering shall be consistent with the requirements for the type of construction.

❖ In an attic or crawl space where entry is made only for service of utilities, and when foam plastics are used, an ignition barrier may be used in place of a thermal barrier to cover the foam plastic. Multiple materials are listed, which can be used as the ignition barrier:

- 1.5-inch-thick (38 mm) mineral fiber insulation;
- 0.25-inch-thick (6.4 mm) wood structural panels;
- 0.375-inch (9.5 mm) particleboard;
- 0.25-inch (6.4 mm) hardboard;
- 0.375-inch (9.5 mm) gypsum board; or
- Corrosion-resistant steel having a base metal thickness of 0.016 inch (0.406 mm).

Note that the ignition barrier must be consistent with the type of construction, which means that combustible coverings are permitted only where combustible materials are otherwise allowed.

The foam-plastic material covered with the ignition barrier can be on the floor, wall or the ceiling of the attic or crawl space. The phrase "where entry is made only for service of utilities" applies to attics or crawl spaces that only contain mechanical equipment, electrical wiring, fans, plumbing, gas or electric hot water heaters, gas or electric furnaces, etc. Such provisions would not be allowed for such spaces used for storage or general occupancy. The reduced provision (from a thermal barrier to an ignition barrier) provides a protective cover that was the sole purpose to prevent the direct impingement of flame on the foam-plastic insulation (see Figure 2603.4.1.6). Note that Section 2603.4 would still require a thermal barrier between the interior space above (crawl space) or below (attic) in a structure and the foam plastic.

If the foam-plastic insulation has passed large-scale testing in the thickness and density intended for use, in accordance with Section 2603.9, no thermal barrier or ignition barrier is required over the foam-plastic insulation in an attic or crawl space and this section of the code does not apply. It is important to note that the actual configuration must be tested. For example, foam insulation applied to the ceiling of the attic or crawl space must be tested with the foam applied to the ceiling in a room corner test or in an assembly that reflects end use. The same restrictions would apply to those insulations applied to the walls, floors or combination of surfaces.

2603.4.1.7 Doors not required to have a fire protection rating. Where pivoted or side-hinged doors are permitted without a fire protection rating, foam plastic insulation, having a flame spread index of 75 or less and a smoke-developed index of not more than 450, shall be permitted as a core material where the door facing is of metal having a minimum thickness of 0.032-inch (0.8 mm) aluminum or steel having a base metal thickness of not less than 0.016 inch (0.4 mm) at any point.

❖ A thermal barrier is not required when foam plastic is used as a core material for doors, provided that a specified metal facing is used and the door is not required to have a fire-resistance rating. The intent is that doors required to have a fire protection rating must be tested with the foam-plastic cores in place in order to qualify. This section recognizes the limited area of doors compared to the walls in which they are located and regulates foam plastics similar to other materials used in doors.

2603.4.1.8 Exterior doors in buildings of Group R-2 or R-3. In occupancies classified as Group R-2 or R-3, foam-filled exterior entrance doors to individual dwelling units that do not require a fire-resistance rating shall be faced with wood or other approved materials.

❖ Wood or other approved facings on foam-filled exterior entrance doors of Group R-2 and R-3 buildings are permitted when the doors are not required to have a fire-resistance rating. Such doors do not need to comply with Section 2603.4.1.7.

2603.4.1.9 Garage doors. Where garage doors are permitted without a fire-resistance rating and foam plastic is used as a core material, the door facing shall be metal having a minimum thickness of 0.032-inch (0.8 mm) aluminum or 0.010-inch (0.25 mm) steel or the facing shall be minimum 0.125-inch-thick (3.2 mm) wood. Garage doors having facings other than those described above shall be tested in accordance with, and meet the acceptance criteria of, DASMA 107.

Exception: Garage doors using foam plastic insulation complying with Section 2603.3 in detached and attached garages associated with one- and two-family dwellings need not be provided with a thermal barrier.

❖ The prescriptive requirements of this section regulate garage doors with foam-plastic cores used in nonfire-resistance-rated applications. The requirements in Section 2603.4.1.9, which governs overhead sectional, coiling or vertical lift-type garage doors, are slightly different from the requirements of Sections 2603.4.1.7 and 2603.4.1.8, which govern means of egress doors, such as side-swinging doors. These garage doors are not limited to such doors used in conjunction with a private or public garage occupancy but

For SI: 1 inch = 25.4 mm.

Figure 2603.4.1.6
FOAM PLASTIC, ATTIC AND CRAWL SPACES

refer to doors for vehicles in any occupancies.

The minimum door-facing material and thickness requirements for garage doors with foam-plastic cores stated in this section were determined based on the results of a fire testing program sponsored by the National Association of Garage Door Manufacturers (NAGDM). This program exposed a variety of commercially manufactured garage doors to a room corner fire test. The results of the testing indicated that garage doors having facings in the minimum thicknesses stated did not spread fire to the edge of the specimen and did not cause flashover in the test room. Other garage door constructions using foam plastics that do not meet minimum code requirements can be brought into compliance by meeting the requirements of testing the complete garage door assembly to ANSI/DASMA 107, *Room Fire Test Standard for Garage Doors Using Foam Plastic Insulation*. This standard includes a description of the test method, performance data to be obtained and the acceptance criteria to use in evaluating the performance data.

This section exempts exterior garage doors used in garages attached to one- and two-family dwellings and detached garages (Group R-3) from the thermal barrier requirements of Section 2603.4. This exemption is similar to that of Section 2603.4.1.8 for exterior egress doors in Group R-2 and R-3 occupancies.

As noted within the section such doors are not intended for installation in fire-rated exterior walls. There are other products, such as rolling steel fire doors, manufactured for this purpose.

2603.4.1.10 Siding backer board. Foam plastic insulation of not more than 2,000 British thermal units per square feet (Btu/sq. ft.) (22.7 MJ/m²) as determined by NFPA 259 shall be permitted as a siding backer board with a maximum thickness of 0.5 inch (12.7 mm), provided it is separated from the interior of the building by not less than 2 inches (51 mm) of mineral fiber insulation or equivalent or where applied as insulation with re-siding over existing wall construction.

❖ Foam plastic is frequently used in residing applications to provide a leveling surface for new siding, while also bringing additional insulation value to the wall assembly. Also available in the marketplace are products that combine exterior siding material and foam plastic. If these siding/residing products are used on the exterior of a wall and the requirements of Sections 2603.3 and 2603.4 are met, this section of the code does not apply. If a thermal barrier is not used on the interior of the building, then there are limitations placed on the products and their use. In addition to the flame spread limitations of Section 2603.3, other properties of the foam plastic or foam plastic portion of the product include maximum thickness of 0.5 inch (12.7mm) and potential heat of less than 2,000 Btu when tested using NFPA 259. Limitations in foam and siding/foam combination product use includes separation from the interior of the building by 2 inches (51 mm) of mineral fiber insulation, installation over an existing wall finish as part of residing or when the foam plastic insulation is

tested in accordance with Section 2603.9. The removal of the thermal barrier requirement in this section is reasonable considering the separation provided by the existing construction and the limitation of the potential heat of the foam plastic imposed by the code. Also, the code requires fire-resistance ratings from the exterior side of walls only when the fire separation distance is 5 feet (1524 mm) or less (see Section 704.5).

2603.4.1.11 Interior trim. Foam plastic used as interior trim in accordance with Section 2604 shall be permitted without a thermal barrier.

❖ This section specifies that a thermal barrier is not needed for interior trim that meets the limitations of Section 2604 and related subsections, which place limitations on foam-plastic density, thickness, wall and ceiling area coverage and flame spread.

2603.4.1.12 Interior signs. Foam plastic used for interior signs in covered mall buildings in accordance with Section 402.15 shall be permitted without a thermal barrier. Foam plastic signs that are not affixed to interior building surfaces shall comply with Chapter 8 of the *International Fire Code*.

❖ A thermal barrier is not required for interior signs that meet the limits stated in Section 402.14, since the hazard from the interior signs is appropriately addressed by test standard UL 1975 (see commentary, Section 402.14). If the foam-plastic sign is not affixed to an interior surface of the building, then the requirements of Chapter 8 of the *International Fire Code*® (IFC®) must be met. More specifically, Section 808.2 of the IFC requires compliance with UL 1975 when such signs exceed 10 percent of the wall or ceiling area, whichever is less.

2603.4.1.13 Type V construction. Foam plastic spray applied to a sill plate and header of Type V construction is subject to all of the following:

1. The maximum thickness of the foam plastic shall be $3^1/_4$ inches (82.6 mm).

2. The density of the foam plastic shall be in the range of 1.5 to 2.0 pcf (24 to 32 kg/m³).

3. The foam plastic shall have a flame spread index of 25 or less and an accompanying smoke-developed index of 450 or less when tested in accordance with ASTM E 84.

❖ A thermal barrier is required when foam plastic is spray applied to the sill plate and joist header in Type V construction when all of the conditions listed in Section 2603.4.1.13 are met. Because foam-plastic insulation in this application is left exposed, the four conditions listed limit the hazards of the spray-applied foam plastic used in this application:

- Thickness greater than or equal to $3^1/_4$ inches (82.6 mm);

- Density is between 1.5 to 2.0 pcf (24 to 32 kg/m³);

- Flame spread index is less than or equal to 25; and

• Smoke-developed index is less than or equal to 450.

This particular allowance was based upon testing in a room corner fire test that compared the performance of an all-wood floor system to that of a wood floor system with foam plastic sprayed on sill plates and headers (see Figure 2603.4.1.13).

2603.5 Exterior walls of buildings of any height. Exterior walls of buildings of Type I, II, III or IV construction of any height shall comply with Sections 2603.5.1 through 2603.5.7. Exterior walls of cold storage buildings required to be constructed of noncombustible materials, where the building is more than one story in height, shall also comply with the provisions of Sections 2603.5.1 through 2603.5.7. Exterior walls of buildings of Type V construction shall comply with Sections 2603.2, 2603.3 and 2603.4.

❖ All foam plastics used on exterior walls of all types of buildings, except wood frame, are to be installed in accordance with Sections 2603.5.1 through 2603.5.7. Installations on one-story, noncombustible walls of cold storage buildings are required to also comply with the provisions of Sections 2603.5.1 through 2603.5.7. Foam plastics used on exterior walls of Type V buildings are required to comply with Sections 2603.2, 2603.3 and 2603.4. The intent is to regulate the use of an insulating envelope over the exterior of a structure

when the envelope provides no structural support other than the transfer of wind loads. It is recognized that some envelopes will be constructed in place by installing a rigid foam plastic and covering it with an exterior finish while others will be installed as prefabricated panels complete with exterior finish.

2603.5.1 Fire-resistance-rated walls. Where the wall is required to have a fire-resistance rating, data based on tests conducted in accordance with ASTM E 119 shall be provided to substantiate that the fire-resistance rating is maintained.

❖ Foam plastics are permitted in walls that are required to have fire-resistance ratings. Such assemblies must be fire tested with the foam plastic (at the intended maximum thickness and density) in place.

2603.5.2 Thermal barrier. Any foam plastic insulation shall be separated from the building interior by a thermal barrier meeting the provisions of Section 2603.4, unless special approval is obtained on the basis of Section 2603.9.

Exception: One-story buildings complying with Section 2603.4.1.4.

❖ A thermal barrier is required, unless alternative approval is obtained in accordance with Section 2603.9. The intent is to make it clear that the thermal barrier requirement of Section 2603.4 is applicable. The focus of this section is upon external exposure. The excep-

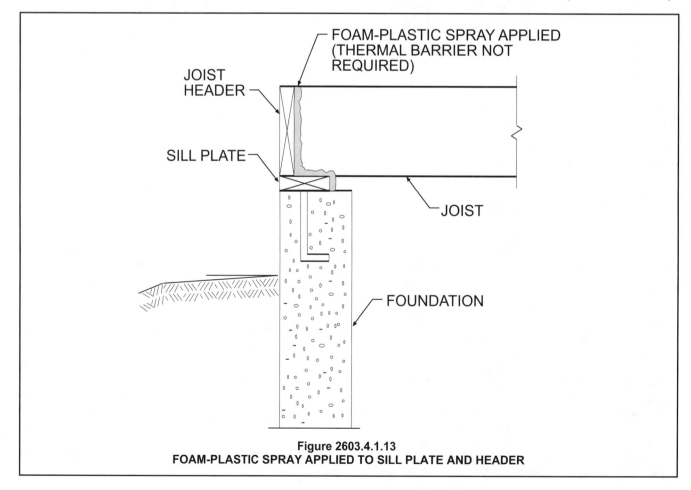

Figure 2603.4.1.13
FOAM-PLASTIC SPRAY APPLIED TO SILL PLATE AND HEADER

tion provides for the reduced requirements of Section 2603.4.1.4 for one-story buildings.

2603.5.3 Potential heat. The potential heat of foam plastic insulation in any portion of the wall or panel shall not exceed the potential heat expressed in Btu per square feet (mJ/m²) of the foam plastic insulation contained in the wall assembly tested in accordance with Section 2603.5.5. The potential heat of the foam plastic insulation shall be determined by tests conducted in accordance with NFPA 259 and the results shall be expressed in Btu per square feet (mJ/m²).

> **Exception:** One-story buildings complying with Section 2603.4.1.4.

❖ This section limits the combustible content of exterior walls based on the potential heat of the foam-plastic insulation. Potential heat is essentially the amount of energy potential a particular material contains in terms of its ability to burn. Generally, the higher the potential heat, the higher the fire hazard. The potential heat must not exceed that as found in the tested wall assembly as required in Section 2603.5.5. It is important to note that NFPA 268 addresses flammability not fire resistances as addressed by Chapter 7.

2603.5.4 Flame spread and smoke-developed indexes. Foam plastic insulation, exterior coatings and facings shall be tested separately in the thickness intended for use, but not to exceed 4 inches (102 mm), and shall each have a flame spread index of 25 or less and a smoke-developed index of 450 or less as determined in accordance with ASTM E 84.

> **Exception:** Prefabricated or factory-manufactured panels having minimum 0.020-inch (0.51 mm) aluminum facings and a total thickness of 0.25 inch (6.4 mm) or less are permitted to be tested as an assembly where the foam plastic core is not exposed in the course of construction.

❖ Each component, including the foam-plastic core, must be tested and have a maximum flame spread of 25 and a maximum smoke-developed index of 450 when tested in accordance with ASTM E 84. The material should be tested with the same thickness for use, except that ASTM E 84 is limited to a 4-inch (102 mm) thickness for testing. Full-scale testing can address the actual thickness and is required by Section 2603.5.5. Materials should not be restricted by the testing limitations of ASTM E 84 when they perform well at a higher thickness in the full scale test. The exception applies to prefabricated panels that can be installed without exposing the foam plastic.

2603.5.5 Test standard. The wall assembly shall be tested in accordance with and comply with the acceptance criteria of NFPA 285.

> **Exception:** One-story buildings complying with Section 2603.4.1.4.

❖ This section is the only provision dealing with propagation of fire due to exposure from an exterior source. Other combustibles are not allowed as components of noncombustible building exterior walls. The focus is flammability versus fire resistance.

2603.5.6 Label required. The edge or face of each piece of foam plastic insulation shall bear the label of an approved agency. The label shall contain the manufacturer's or distributor's identification, model number, serial number or definitive information describing the product or materials' performance characteristics and approved agency's identification.

❖ Each piece of foam plastic must be labeled. This requirement is somewhat different than the general marking requirement of Section 2603.2, which only requires a manufacturer's identifying mark on either the product or the packaging. The reason for this is that properties other than fire performance of the foam plastic, such as bond strength to the exterior finish, are of significance in this type of application. Additionally, the intent is to provide a means for specific product identification not only during construction but also after construction is complete by the removal of a sample product from the finished structure should a question arise.

2603.5.7 Ignition. Exterior walls shall not exhibit sustained flaming where tested in accordance with NFPA 268. Where a material is intended to be installed in more than one thickness, tests of the minimum and maximum thickness intended for use shall be performed.

> **Exception:** Assemblies protected on the outside with one of the following:
>
> 1. A thermal barrier complying with Section 2603.4.
>
> 2. A minimum 1 inch (25 mm) thickness of concrete or masonry.
>
> 3. Glass-fiber-reinforced concrete panels of a minimum thickness of 0.375 inch (9.5 mm).
>
> 4. Metal-faced panels having minimum 0.019-inch-thick (0.48 mm) aluminum or 0.016-inch-thick (0.41 mm) corrosion-resistant steel outer facings.
>
> 5. A minimum 0.875 inch (22.2 mm) thickness of stucco complying with Section 2510.

❖ The foam plastic is not to support continued flaming when tested in accordance with NFPA 268, which looks at ignitability when exposed to a radiant heat source. Where the foam plastic is to be used in more than one thickness, the test is to be performed on minimum and maximum thickness. If the assembly is protected on the outside with one of the specified materials, the test is not required.

2603.6 Roofing. Foam plastic insulation meeting the requirements of Sections 2603.2, 2603.3 and 2603.4 shall be permitted as part of a roof-covering assembly, provided the assembly with the foam plastic insulation is a Class A, B or C roofing assembly where tested in accordance with ASTM E 108 or UL 790.

❖ Foam-plastic insulation that meets the labeling and identification, surface-burning characteristics and thermal barrier criteria are permitted as part of a Class A, B or C roof-covering assembly that has been tested in accordance with ASTM E 108 or UL 790. Both

ASTM E 108 and UL 790 cover exterior spread of flame.

2603.7 Plenums. Foam plastic insulation shall not be used as interior wall or ceiling finish in plenums except as permitted in Section 2604 or when protected by a thermal barrier in accordance with Section 2603.4.

❖ This section clarifies how foam plastics are to be applied in plenums as opposed to their use as duct and pipe insulation. Foam plastic found in a plenum must meet one of the following options:

1. Must have a thermal barrier;

2. Pass an appropriate large scale test (Section 2603.9); or

3. Meet the requirements of Section 2604 for trim and finish (i.e., is dense enough, meets area and thickness limitations and complies with flame spread limitations).

2603.8 Protection against termites. In areas where the probability of termite infestation is very heavy in accordance with Figure 2603.8, extruded and expanded polystyrene, polyisocyanurate and other foam plastics shall not be installed on the exterior face or under interior or exterior foundation walls or slab foundations located below grade. The clearance between foam plastics installed above grade and exposed earth shall be at least 6 inches (152 mm).

Exceptions:

1. Buildings where the structural members of walls, floors, ceilings and roofs are entirely of noncombustible materials or preservatively treated wood.

2. An approved method of protecting the foam plastic and structure from subterranean termite damage is provided.

3. On the interior side of basement walls.

❖ This section of the code addresses the use of foam plastics in areas of "very heavy" termite infestation. Termites have been reported to cause damage to foam insulation in very heavy termite infestation areas. These probabilities are referenced in Figure 2603.8. Foam plastic used in the "very heavy" infestation area is prohibited on the exterior face of below-grade foundations walls or slab foundations, under exterior or interior foundation walls or slab foundations below grade or where located within 6 inches (152 mm) of exposed earth, unless the requirement of one of the three specific exceptions is met. Foam plastics are permitted on the interior side of basement walls, where the structural members of the building are either noncombustible or pressure preservatively treated wood, or where the foam plastic is adequately protected from subterranean termite damage.

2603.9 Special approval. Foam plastic shall not be required to comply with the requirements of Sections 2603.4 through 2603.7 where specifically approved based on large-scale tests such as, but not limited to, NFPA 286 (with the acceptance cri-

teria of Section 803.2), FM 4880, UL 1040 or UL 1715. Such testing shall be related to the actual end-use configuration and be performed on the finished manufactured foam plastic assembly in the maximum thickness intended for use. Foam plastics that are used as interior finish on the basis of special tests shall also conform to the flame spread requirements of Chapter 8. Assemblies tested shall include seams, joints and other typical details used in the installation of the assembly and shall be tested in the manner intended for use.

❖ Foam plastic does not have to comply with the installation and use requirements of Sections 2603.4 through 2603.7 when specific approval is obtained in accordance with this section. This section lists examples of specific large-scale tests, such as: FM 4880, UL 1040, NFPA 286 or UL 1715. Also, other large-scale fire tests related to actual end-use configuration can be used. The intent is to require testing based on the proposed end-use configuration of the foam-plastic assembly with a fire exposure that is appropriate in size and location for the proposed application. These tests are to be performed on full-scale assemblies. The tested assemblies must include typical seams, joints and other details that will occur in the finished installation. The foam plastic must be tested in the maximum thickness and density intended for use. Thorough testing provides an accurate depiction of the in-place fire performance of assemblies and systems using foam plastics.

There are two ways to show code compliance under Section 2603.9. One method is to provide the actual test report that contains a description of the assembly and test results showing that the foam plastic, in the end use application, has passed the test. The second method is to obtain, from the ICC ES, an evaluation report that covers the end-use application. Materials that are being used as interior finish would still need to additionally pass the flame spread requirements of Chapter 8. It should be noted that NFPA 286 is one way of demonstrating compliance with the interior finish requirements of Chapter 8 and with this section.

SECTION 2604
INTERIOR FINISH AND TRIM

2604.1 General. Plastic materials installed as interior finish or trim shall comply with Chapter 8. Foam plastics shall only be installed as interior finish where approved in accordance with the special provisions of Section 2603.9. Foam plastics that are used as interior finish shall also meet the flame spread index requirements for interior finish in accordance with Chapter 8. Foam plastics installed as interior trim shall comply with Section 2604.2.

❖ Plastic materials used as interior finish that are neither light transmitting nor foam are required to comply with Chapter 8. Foam plastic used as interior finish is to be in accordance with the special approval provisions of Section 2603.9 and meet the flame spread index requirements of Chapter 8 for interior finish. Foam plastics installed as interior trim are required to comply with

Section 2604.2. It should be noted that the IFC addresses foam-plastic trim in existing buildings in the same manner.

[F] 2604.2 Interior trim. Foam plastic used as interior trim shall comply with Sections 2604.2.1 through 2604.2.4.

❖ This section identifies the applicable subsections that apply to the interior trim requirements.

[F] 2604.2.1 Density. The minimum density of the interior trim shall be 20 pcf (320 kg/m³).

❖ In order to qualify as interior trim, the density of materials must be at least 20 pounds per cubic foot (pcf) (320 kg/m³). The intent was to separate those materials used for trim from those intended for use as insulation. The 20-pcf (320 kg/m³) value was selected because it was approximately midway between the densities commonly used for insulations and for trim when this provision was added to the code. Most foam-plastic insulation was, and still is, in the range of 1 to 2¹/₂ pcf (16.02 to 40 kg/m³), with very few materials over 5 pcf (81 kg/m³). Foam-plastic trim was manufactured in a density range of 30 to 45 pcf (481 to 721 kg/m³) when the density provision was originally incorporated in the code.

[F] 2604.2.2 Thickness. The maximum thickness of the interior trim shall be 0.5 inch (12.7 mm) and the maximum width shall be 8 inches (204 mm).

❖ Even though other trim materials are not limited in dimension, the maximum thickness and width of foam-plastic trim is limited to ¹/₂ inch (12.7 mm) and 8 inches (203 mm), respectively. These dimensions were selected because they were typical of the maxi-

mums being produced at the time this provision was included in the code.

[F] 2604.2.3 Area limitation. The interior trim shall not constitute more than 10 percent of the aggregate wall and ceiling area of any room or space.

❖ Trim cannot constitute more than 10 percent of the aggregate area of the walls and ceiling of a room. This limitation is simply a restatement of the general requirement for all combustible trim, which appears in Section 806.5.

[F] 2604.2.4 Flame spread. The flame spread index shall not exceed 75 where tested in accordance with ASTM E 84. The smoke-developed index shall not be limited.

❖ The flame spread must not be greater than 75 when tested in accordance with ASTM E 84. The value of 75 was selected to be consistent with the requirement for foam-plastic insulation, even though other materials used as trim are permitted to have flame spread indexes of up to 200 in many locations. The smoke-developed index is not regulated.

SECTION 2605
PLASTIC VENEER

2605.1 Interior use. Where used within a building, plastic veneer shall comply with the interior finish requirements of Chapter 8.

❖ Plastic veneer may be used inside a building if it meets the finish requirements of Chapter 8.

2605.2 Exterior use. Exterior plastic veneer shall be permitted to be installed on the exterior walls of buildings of any type of

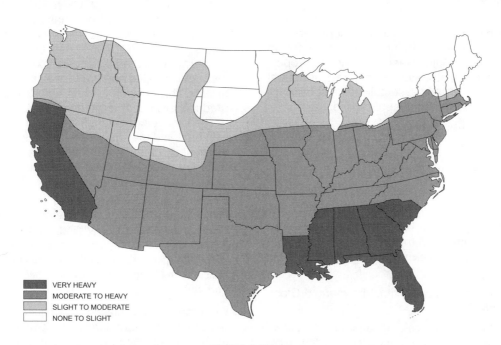

VERY HEAVY
MODERATE TO HEAVY
SLIGHT TO MODERATE
NONE TO SLIGHT

FIGURE 2603.8
TERMITE INFESTATION PROBABILITY MAP

construction in accordance with all of the following requirements:

1. Plastic veneer shall comply with Section 2606.4.

2. Plastic veneer shall not be attached to any exterior wall to a height greater than 50 feet (15 240 mm) above grade.

3. Sections of plastic veneer shall not exceed 300 square feet (27.9 m²) in area and shall be separated by a minimum of 4 feet (1219 mm) vertically.

> **Exception:** The area and separation requirements and the smoke-density limitation are not applicable to plastic veneer applied to buildings constructed of Type VB construction, provided the walls are not required to have a fire-resistance rating.

❖ Exterior plastic veneer must be an approved Class CC1 or CC2 plastic material in accordance with the specifications in Section 2606.4 and meet all three of the requirements in this section for height, area and separation. The classifications within Section 2606.4 are combustibility classifications. The requirements for rigid vinyl siding are listed in Section 1404.9 and ASTM D 3679

SECTION 2606
LIGHT-TRANSMITTING PLASTICS

2606.1 General. The provisions of this section and Sections 2607 through 2611 shall govern the quality and methods of application of light-transmitting plastics for use as light-transmitting materials in buildings and structures. Foam plastics shall comply with Section 2603. Light-transmitting plastic materials that meet the other code requirements for walls and roofs shall be permitted to be used in accordance with the other applicable chapters of the code.

❖ Section 2606 contains the provisions for light-transmitting plastics. It specifies both the requirements for an approved light-transmitting plastic and the fire properties of materials. It also requires adequate structural properties of materials and assemblies and specific installation methods for various applications.

2606.2 Approval for use. Sufficient technical data shall be submitted to substantiate the proposed use of any light-transmitting material, as approved by the building official and subject to the requirements of this section.

❖ Before the material can be used, sufficient data is to be submitted to the building official for approval.

2606.3 Identification. Each unit or package of light-transmitting plastic shall be identified with a mark or decal satisfactory to the building official, which includes identification as to the material classification.

❖ The material or its packaging must be marked Class CC1 or CC2 plastic.

2606.4 Specifications. Light-transmitting plastics, including thermoplastic, thermosetting or reinforced thermosetting plastic material, shall have a self-ignition temperature of 650°F (343°C) or greater where tested in accordance with ASTM D

1929; a smoke-developed index not greater than 450 where tested in the manner intended for use in accordance with ASTM E 84, or not greater than 75 where tested in the thickness intended for use in accordance with ASTM D 2843 and shall conform to one of the following combustibility classifications:

Class CC1: Plastic materials that have a burning extent of 1 inch (25 mm) or less where tested at a nominal thickness of 0.060 inch (1.5 mm), or in the thickness intended for use, in accordance with ASTM D 635,

Class CC2: Plastic materials that have a burning rate of 2.5 inches per minute (1.06 mm/s) or less where tested at a nominal thickness of 0.060 inch (1.5 mm), or in the thickness intended for use, in accordance with ASTM D 635.

❖ An approved material may be either thermoplastic, thermosetting or reinforced thermosetting. Approved plastics require a minimum self-ignition temperature of 650°F (343°C) when tested in accordance with ASTM D 1929. This temperature excludes the use of easily ignited plastic materials, such as cellulose nitrate, which ignites at a temperature of about 300°F (149°C). For comparative purposes, untreated wood and paper ignite at about 500°F (260°C).

In order for a plastic to be approved, its smoke-density rating is limited and must be evaluated by one of two test methods prescribed (smoke-developed index limited to 450 by ASTM E 84 or 75 by ASTM D 2843). For comparative purposes, these smoke-developed values are the maximum that can be expected from some wood products tested under similar conditions; however, there is no intent to indicate a correlation between the results from the two test methods.

Approved light-transmitting plastics are to be tested in accordance with ASTM D 635 and be of combustibility Class CC1 or CC2. Class CC1 plastic materials burn at a slower rate than Class CC2 plastics. Class CC1 generally consists of polycarbonate materials whereas Class CC2 consists of acrylics.

Thermoplastic materials are those plastics that will melt when heated and are formed into their final shape utilizing this property. Most light-transmitting plastic products are flat or formed sheets.

Thermosetting materials are those plastics that cure with heat and may generate heat during curing but will not melt by heating after curing. They are formed into their final product shape before or during curing, except for possible machining.

2606.5 Structural requirements. Light-transmitting plastic materials in their assembly shall be of adequate strength and durability to withstand the loads indicated in Chapter 16. Technical data shall be submitted to establish stresses, maximum unsupported spans and such other information for the various thicknesses and forms used as deemed necessary by the building official.

❖ Light-transmitting plastic applications must meet the necessary structural load requirements. For example, glazing assemblies must be able to withstand the imposed wind loads, and skylight assemblies must be able to support expected snow loads. It is the intent of this section that durability includes weatherability for those applications subjected to outdoor exposure. Technical information to support the structural integrity is to be submitted to the building official as he or she deems necessary.

2606.6 Fastening. Fastening shall be adequate to withstand the loads in Chapter 16. Proper allowance shall be made for expansion and contraction of light-transmitting plastic materials in accordance with accepted data on the coefficient of expansion of the material and other material in conjunction with which it is employed.

❖ The design loads for the assembly, members and fasteners are specified in Chapter 16. Adequate provision must be made for movement caused by the differential expansion expected from the different types of materials connected. This is especially important with plastics since they usually have coefficients of expansion that are greater than the materials to which they are connected.

2606.7 Light-diffusing systems. Unless the building is equipped throughout with an automatic sprinkler system in accordance with Section 903.3.1.1, light-diffusing systems shall not be installed in the following occupancies and locations:

1. Group A with an occupant load of 1,000 or more.

2. Theaters with a stage and proscenium opening and an occupant load of 700 or more.

3. Group I-2.

4. Group I-3.

5. Exit stairways and exit passageways.

❖ Light-diffusing systems are not permitted in the four occupancies above, nor in exits in any group, unless the building is protected by an automatic sprinkler system in accordance with NFPA 13. The intent is to limit the use of combustible materials in occupancies without sprinkler protection. These occupancies are generally recognized as being potentially more hazardous for the building occupants in the event of fire. The specific restriction for nonsprinklered buildings of Groups I-2 and I-3 is based on the physical restraint or incapacitation of the occupants.

2606.7.1 Support. Light-transmitting plastic diffusers shall be supported directly or indirectly from ceiling or roof construction by use of noncombustible hangers. Hangers shall be at least No. 12 steel-wire gage (0.106 inch) galvanized wire or equivalent.

❖ This section specifies that all hangers supporting the plastic diffusers must be of steel wire at least 0.106 inch (2.7 mm) in diameter or equivalent. It is important that noncombustible supports be used because the original test data were generated on these types of assemblies.

2606.7.2 Installation. Light-transmitting plastic diffusers shall comply with Chapter 8 unless the light-transmitting plas-

tic diffusers will fall from the mountings before igniting, at an ambient temperature of at least 200°F (111°C) below the ignition temperature of the panels. The panels shall remain in place at an ambient room temperature of 175°F (79°C) for a period of not less than 15 minutes.

❖ Diffusers must also comply with Chapter 8 unless the plastic panels will fall from their mountings at an ambient temperature of at least 200°F (93°C) below their ignition temperature. The intent is to regulate panels that remain in place in the same as interior finish and to eliminate surface flame spread requirements for materials that will not burn while remaining in place.

The code requires that all panels must remain in place for at least 15 minutes at an ambient room temperature of 175°F (79°C). The intent of this requirement is to avoid a panic situation that could arise from premature dropping of ceiling panels.

2606.7.3 Size limitations. Individual panels or units shall not exceed 10 feet (3048 mm) in length nor 30 square feet (2.79 m²) in area.

❖ This section states the maximum length [10 feet (3048 mm)] and area [30 square feet (2.8 m²)] of any single panel. The intent is to limit the size of panels expected to fall from place if the ambient room temperature at the ceiling were to reach 175°F (79°C), in order to avoid possible obstructions and injury caused by the fallen panels.

2606.7.4 Fire suppression system. In buildings that are equipped throughout with an automatic sprinkler system in accordance with Section 903.3.1.1, plastic light-diffusing systems shall be protected both above and below unless the sprinkler system has been specifically approved for installation only above the light-diffusing system. Areas of light-diffusing systems that are protected in accordance with this section shall not be limited.

❖ When buildings are required to have approved automatic sprinkler systems throughout, sprinkler protection must be provided both above and below the diffusing system. A sprinkler system is permitted to be installed above the diffusing system only when the assembly is designed for this use and specifically approved by the building official. The intent is that when specific approval is granted, it is based on appropriate test results related to end use. When sprinkler protection is provided, the area of the light-diffusing system is not subject to the area limitations of Section 2606.7.3 since the likelihood of such panels dropping is decreased.

2606.7.5 Electrical luminaires. Light-transmitting plastic panels and light-diffuser panels that are installed in approved electrical luminaires shall comply with the requirements of Chapter 8 unless the light-transmitting plastic panels conform to the requirements of Section 2606.7.2. The area of approved light-transmitting plastic materials that are used in required exits or corridors shall not exceed 30 percent of the aggregate area of the ceiling in which such panels are installed, unless the building is equipped throughout with an automatic sprinkler system in accordance with Section 903.3.1.1.

❖ Plastic light-transmitting and diffuser panels in electrical luminaires are required to comply with Chapter 8 or the installation requirements of Section 2606.7.2. Additionally, the aggregate ceiling area of such panels is limited to 30 percent in required exits and corridors, unless the building is equipped throughout with an automatic sprinkler system in accordance with NFPA 13. The intent of this section is to regulate luminaires the same as other ceiling finish panels. It also restricts the use of combustible materials in required exits and corridors while giving recognition to the benefit of automatic sprinkler systems.

2606.8 Partitions. Light-transmitting plastics used in or as partitions shall comply with the requirements of Chapters 6 and 8.

❖ When light-transmitting plastics are used in or as partitions, they must comply with the requirements of Chapters 6 and 8. The intent is to regulate light-transmitting plastics with the same requirements as other materials used within a building.

2606.9 Bathroom accessories. Light-transmitting plastics shall be permitted as glazing in shower stalls, shower doors, bathtub enclosures and similar accessory units. Safety glazing shall be provided in accordance with Chapter 24.

❖ Approved plastics are permitted as glazing in a variety of uses in bathrooms. When an approved Class CC1 or CC2 plastic material is used for glazing shower stalls, shower doors, bathtub enclosures or other bathroom accessories, it must meet the human impact requirements of ANSI Z97.1 under Section 2406.1.3

2606.10 Awnings, patio covers and similar structures. Awnings constructed of light-transmitting plastics shall be constructed in accordance with the provisions specified in Section 3105 and Chapter 32 for projections. Patio covers constructed of light-transmitting plastics shall comply with Section 2606. Light-transmitting plastics used in canopies at motor fuel-dispensing facilities shall comply with Section 2606, except as modified by Section 406.5.2.

❖ Light-transmitting plastics used as awnings and similar structures must meet the general performance requirements of other appropriate sections of the code in addition to the requirements found in Chapter 26.

2606.11 Greenhouses. Light-transmitting plastics shall be permitted in lieu of plain glass in greenhouses.

❖ Greenhouses may be constructed of any type of approved plastic material that meets the requirements of Section 2606.

2606.12 Solar collectors. Light-transmitting plastic covers on solar collectors having noncombustible sides and bottoms shall be permitted on buildings not over three stories in height or 9,000 square feet (836.1 m²) in total floor area, provided the light-transmitting plastic cover does not exceed 33.33 percent

of the roof area for CC1 materials or 25 percent of the roof area for CC2 materials.

Exception: Light-transmitting plastic covers having a thickness of 0.010 inch (0.3 mm) or less or shall be permitted to be of any plastic material provided the area of the solar collectors does not exceed 33.33 percent of the roof area.

❖ Approved light-transmitting plastics can be used for solar collector covers that meet the requirements of Section 2606.12. Basically size and height of the building is limited and depending on the combustibility two different maximums are provided. Any plastic material may be used as solar collector covers provided that the cover thickness does not exceed 0.010 inches (0.254 mm) and the collector does not exceed 33.33 percent of the roof area.

Essentially the fire hazard only exits on the roof which poses minimal threats initially to occupants.

SECTION 2607
LIGHT-TRANSMITTING PLASTIC WALL PANELS

2607.1 General. Light-transmitting plastics shall not be used as wall panels in exterior walls in occupancies in Groups A-1, A-2, H, I-2 and I-3. In other groups, light-transmitting plastics shall be permitted to be used as wall panels in exterior walls, provided that the walls are not required to have a fire-resistance rating and the installation conforms to the requirements of this section. Such panels shall be erected and anchored on a foundation, waterproofed or otherwise protected from moisture absorption and sealed with a coat of mastic or other approved waterproof coating. Light-transmitting plastic wall panels shall also comply with Section 2606.

❖ Section 2607 establishes requirements for the use of plastic panels as exterior walls. Plastic wall panels are permitted to be used in most occupancies except those with dense occupant loads or where the occupants are confined or otherwise impaired. The size and quantity of plastic wall panels are also regulated by this section. These panels must also comply with the applicable provisions of Section 2606, which governs all light-transmitting plastics.

Plastic wall panels are permitted to be used in occupancies classified as Groups A-3, A-4, A-5, B, E, F-1, F-2, I-1, M, R-1, R-2, R-3, R-4, S-1, S-2 and U. They may not be used in occupancies where the generated products of combustion would significantly impair the egress of the impaired, confined or crowded occupants.

As with any exterior covering, plastic wall panels must be adequately fastened in a permanent manner. To reduce the chance of water infiltration into the plastic wall panels, they must be erected and anchored on a base that is protected from water absorption. Additionally, the plastic wall panel itself must be sealed with a waterproof coating. These types of combustible wall panels are not permitted where the exterior walls are required to be fire-resistance rated.

2607.2 Installation. Exterior wall panels installed as provided for herein shall not alter the type of construction classification of the building.

❖ Plastic is a combustible material. The use of light-transmitting plastic wall panels is permitted, however, without downgrading the type of construction. This exception is also clarified in Section 603.1, Item 12, for Type I and II construction; therefore, the use of light-transmitting plastic wall panels is permitted in all types of construction. These types of combustible wall panels, however, are not permitted where the exterior walls are required to be fire-resistance rated (see Section 2607.1).

2607.3 Height limitation. Light-transmitting plastics shall not be installed more than 75 feet (22 860 mm) above grade plane, except as allowed by Section 2607.5.

❖ Plastic wall panels are not permitted to be installed greater than 75 feet (22 860 mm) above grade plane in an unsprinklered building. The 75-foot (22 860 mm) threshold is the highest elevation that can be reached by hose lines on the ground. In structures sprinklered throughout, however, Section 2607.5 permits the maximum height to be unlimited. Be aware that Table 503 also places height limitations on structures. In some cases, the limitations of Table 503 may be more restrictive than those of Sections 2607.3 and 2607.5. At all times, the most restrictive requirement prevails.

2607.4 Area limitation and separation. The maximum area of a single wall panel and minimum vertical and horizontal separation requirements for exterior light-transmitting plastic wall panels shall be as provided for in Table 2607.4. The maximum percentage of wall area of any story in light-transmitting plastic wall panels shall not exceed that indicated in Table 2607.4 or the percentage of unprotected openings permitted by Section 704.8, whichever is smaller.

Exceptions:

1. In structures provided with approved flame barriers extending 30 inches (760 mm) beyond the exterior wall in the plane of the floor, a vertical separation is not required at the floor except that provided by the vertical thickness of the flame barrier projection.

2. Veneers of approved weather-resistant light-transmitting plastics used as exterior siding in buildings of Type V construction in compliance with Section 1406.

3. The area of light-transmitting plastic wall panels in exterior walls of greenhouses shall be exempt from the area limitations of Table 2607.4 but shall be limited as required for unprotected openings in accordance with Section 704.8.

❖ Refer to Table 2607.4 for area limitations of single wall panels and the minimum separation requirements between adjacent panels. The maximum percentage of wall area faced with plastic panels is limited by Table 2607.4, as well as by Section 704.8. At all times, the most restrictive requirement prevails.

Vertical separation of panels (see Table 2607.4) is not required where continuous architectural projections create at least a 30-inch (762 mm) barrier (called a flame barrier) between adjacent panels. Under most conditions, this flame barrier will delay or prevent vertical fire spread from one story to the next. This horizontal barrier is seen as protection equivalent to the vertical separation typically required by Table 2607.4 (see Figure 2607.4).

Veneers of approved exterior light-transmitting plastic siding installed in compliance with Section 1406 on Type V construction are exempt from the area limitations and separation requirements.

Greenhouse walls can have an unlimited area of approved light-transmitting wall panels, unless limited by the unprotected opening requirements of Section 704.8.

TABLE 2607.4. See below.

❖ This table establishes size and separation limitations for exterior plastic wall panels. The requirements of this table are not applicable to plastic exterior siding

TABLE 2607.4
AREA LIMITATION AND SEPARATION REQUIREMENTS FOR
LIGHT-TRANSMITTING PLASTIC WALL PANELS[a]

FIRE SEPARATION DISTANCE (feet)	CLASS OF PLASTIC	MAXIMUM PERCENTAGE AREA OF EXTERIOR WALL IN PLASTIC WALL PANELS	MAXIMUM SINGLE AREA OF PLASTIC WALL PANELS (square feet)	MINIMUM SEPARATION OF PLASTIC WALL PANELS (feet)	
				Vertical	Horizontal
Less than 6	—	Not Permitted	Not Permitted	—	—
6 or more but less than 11	CC1	10	50	8	4
	CC2	Not Permitted	Not Permitted	—	—
11 or more but less than or equal to 30	CC1	25	90	6	4
	CC2	15	70	8	4
Over 30	CC1	50	Not Limited	3[b]	0
	CC2	50	100	6[b]	3

For SI: 1 foot = 304.8 mm, 1 square foot = 0.0929 m^2.

a. For combinations of plastic glazing and plastic wall panel areas permitted, see Section 2607.6.

b. For reductions in vertical separation allowed, see Section 2607.4.

meeting the exception to Section 2607.4. The requirements are determined by the type of plastic material (Class CC1 or CC2) and the fire separation distance. Plastic wall panels are only permitted where a minimum fire separation distance of 6 feet (1829 mm) is provided. The vertical separation requirements can be modified by Section 2607.4 (see Note b of Table 2607.4). The maximum area limitations of a single panel and the maximum percentage area of an exterior wall can be doubled when the structure is completely sprinklered in accordance with NFPA 13 (see Section 2607.5).

2607.5 Automatic sprinkler system. Where the building is equipped throughout with an automatic sprinkler system in accordance with Section 903.3.1.1, the maximum percentage area of exterior wall in any story in light-transmitting plastic wall panels and the maximum square footage of a single area given in Table 2607.4 shall be increased 100 percent, but the area of light-transmitting plastic wall panels shall not exceed 50 percent of the wall area in any story, or the area permitted by Section 704.8 for unprotected openings, whichever is smaller. These installations shall be exempt from height limitations.

❖ The extra protection provided by a complete NFPA 13 automatic fire sprinkler system enables the limitations placed on light-transmitting plastic wall panels to be reduced. In order for these modifications to apply, the structure must be sprinklered throughout. When this protection is achieved, the height to which plastic wall panels can be installed is not limited. Furthermore, the single panel size and exterior wall area limitations can be doubled. At no time, however, can the area of wall panels per story exceed 50 percent of the wall area of that story. The percentage limitation of Section 704.8 for unprotected openings is also applicable.

2607.6 Combinations of glazing and wall panels. Combinations of light-transmitting plastic glazing and light-transmitting plastic wall panels shall be subject to the area, height and percentage limitations and the separation requirements applicable to the class of light-transmitting plastic as prescribed for light-transmitting plastic wall panel installations.

❖ Plastic glazing and plastic wall panels display equivalent fire characteristics; therefore, where plastic glazing (regulated by Section 2608) and plastic wall panels are both used, the aggregate area of plastics must be in compliance with the limitations and requirements of Table 2607.4.

SECTION 2608
LIGHT-TRANSMITTING PLASTIC GLAZING

2608.1 Buildings of Type VB construction. Openings in the exterior walls of buildings of Type VB construction, where not

LIGHT-TRANSMITTING
PLASTIC WALL PANEL

30" MINIMUM

LIGHT-TRANSMITTING
PLASTIC WALL PANEL

NO VERTICAL SEPARATION IS REQUIRED
BETWEEN ADJACENT PANELS

For SI: 1 inch = 25.4 mm.

Figure 2607.4
FLAME BARRIER

required to be protected by Section 704, shall be permitted to be glazed or equipped with light-transmitting plastic. Light-transmitting plastic glazing shall also comply with Section 2606.

❖ This section restricts the use of plastics to openings that are not required to have a fire protection rating. The requirement for walls, and their openings, to be protected is dependent on their fire separation distance (see Section 704).

2608.2 Buildings of other types of construction. Openings in the exterior walls of buildings of types of construction other than Type VB, where not required to be protected by Section 704, shall be permitted to be glazed or equipped with light-transmitting plastic in accordance with Section 2606 and all of the following:

1. The aggregate area of light-transmitting plastic glazing shall not exceed 25 percent of the area of any wall face of the story in which it is installed. The area of a single pane of glazing installed above the first story above grade plane shall not exceed 16 square feet (1.5 m²) and the vertical dimension of a single pane shall not exceed 4 feet (1219 mm).

   Exception: Where an automatic sprinkler system is provided throughout in accordance with Section 903.3.1.1, the area of allowable glazing shall be increased to a maximum of 50 percent of the wall face of the story in which it is installed with no limit on the maximum dimension or area of a single pane of glazing.

2. Approved flame barriers extending 30 inches (762 mm) beyond the exterior wall in the plane of the floor, or vertical panels not less than 4 feet (1219 mm) in height, shall be installed between glazed units located in adjacent stories.

   Exception: Buildings equipped throughout with an automatic sprinkler system in accordance with Section 903.3.1.1.

3. Light-transmitting plastics shall not be installed more than 75 feet (22 860 mm) above grade level.

   Exception: Buildings equipped throughout with an automatic sprinkler system in accordance with Section 903.3.1.1.

❖ In addition to restricting the use of plastic glazing to exterior wall openings that are not required to have a fire protection rating, this section specifies the maximum area within a wall that openings can occupy. An opening that has plastic as the glazing material is essentially treated the same as an unprotected glass opening.

To limit the spread of fire, the allowable area of plastic glazing is limited to the lesser of two conditions. The first condition restricts the area of plastic glazing to 25 percent of the wall area of the story. The second condition restricts the area of plastics to that of the unprotected openings permitted in Section 704.

The exception doubles the amount of wall area that is allowed to be glazed with plastic when an approved automatic sprinkler system is installed. The allowable areas are increased to 50 percent of the wall area of the story with no limit on the maximum dimension or area of a single pane of glazing.

Consistent with other provisions to limit fire spread, this section places restrictions on the size and vertical separation of plastic glazing. The 4-foot (1219 mm) vertical separation is widely accepted as adequate for spandrels and other intervening construction between floors.

The 75-foot (22 860 mm) height limitation is generally accepted as the maximum height attainable by fire-fighting apparatus. If sprinklers are provided, the need for such height restrictions is lessened.

SECTION 2609
LIGHT-TRANSMITTING PLASTIC ROOF PANELS

2609.1 General. Light-transmitting plastic roof panels shall comply with this section and Section 2606. Light-transmitting plastic roof panels shall not be installed in Groups H, I-2 and I-3. In all other groups, light-transmitting plastic roof panels shall comply with any one of the following conditions:

1. The building is equipped throughout with an automatic sprinkler system in accordance with Section 903.3.1.1.

2. The roof construction is not required to have a fire-resistance rating by Table 601.

3. The roof panels meet the requirements for roof coverings in accordance with Chapter 15.

❖ This section establishes general requirements for light-transmitting plastic roof panels. These requirements are similar in nature to the other light-transmitting plastic provisions of this chapter. The requirements are intended to control the use of combustible construction material (i.e., light-transmitting plastics). These provisions specifically detail where roof panels may be used and any accompanying conditions that are required. Occupancies in Group H, I-2 or I-3 may not use light-transmitting plastic roof panels. Buildings or structures containing all other groups may use the roof panels, but at least one of the three conditions listed must be present. The first condition requires the structure under consideration to be equipped with an automatic sprinkler system installed in accordance with NFPA 13. The second condition allows a structure that is not sprinklered to contain plastic roof panels if Table 601 of the code does not require any fire-resistance ratings for the roof construction, including beams, trusses, framing, arches and roof decks (based on the type of construction and the distance between the floor and lowest roof member). The final condition that allows plastic roof panels in certain structures is that panels meet the requirements for roof coverings in Chapter 15.

2609.2 Separation. Individual roof panels shall be separated from each other by a distance of not less than 4 feet (1219 mm) measured in a horizontal plane.

Exceptions:

1. The separation between roof panels is not required in a building equipped throughout with an automatic sprinkler system in accordance with Section 903.3.1.1.

2. The separation between roof panels is not required in low-hazard occupancy buildings complying with the conditions of Section 2609.4, Exception 2 or 3.

❖ The separation requirements for plastic roof panels are established herein. Individual light-transmitting plastic roof panels are limited in area by the provisions of Section 2609.4. This section provides the separation distances necessary between the individual roof panels to isolate the fire hazard associated with this combustible material.

The 4-foot (1219 mm) minimum separation distance required between each individual roof panel must be measured in a horizontal plane (see Figure 2609.2).

There are two exceptions in the code pertaining to the separation requirements. Any structure that is equipped with an automatic sprinkler system installed in accordance with NFPA 13, or low-hazard occupancy

buildings complying with Exception 2 or 3 of Section 2609.4, are exempt from the separation requirements.

2609.3 Location. Where exterior wall openings are required to be protected by Section 704.8, a roof panel shall not be installed within 6 feet (1829 mm) of such exterior wall.

❖ This section provides requirements as to where light-transmitting plastic roof panels may be located in relation to the exterior walls of buildings or structures. Again, limitations are included as to where these panels of combustible construction can be located in relation to adjacent fire hazards. These provisions apply only when they are determined to be applicable from the requirements for fire-resistance-rated exterior walls and their exterior opening protectives found in Section 704.8. When certain openings in the exterior walls are required to be protected, these location requirements must be followed.

When an exterior wall contains required exterior opening protectives, the code text requires plastic roof panels to be installed at least 6 feet (1829 mm) from the wall. There is no exception to this requirement for structures equipped throughout with an approved automatic sprinkler system (see Figure 2609.3).

2609.4 Area limitations. Roof panels shall be limited in area and the aggregate area of panels shall be limited by a percent-

For SI: 1 inch = 25.4 mm, 1 foot = 304.8 mm.

Figure 2609.2
PLASTIC ROOF PANEL SEPARATION FROM OTHER PANELS

age of the floor area of the room or space sheltered in accordance with Table 2609.4.

Exceptions:

1. The area limitations of Table 2609.4 shall be permitted to be increased by 100 percent in buildings equipped throughout with an automatic sprinkler system in accordance with Section 903.3.1.1.

2. Low-hazard occupancy buildings, such as swimming pool shelters, shall be exempt from the area limitations of Table 2609.4, provided that the buildings do not exceed 5,000 square feet (465 m²) in area and have a minimum fire separation distance of 10 feet (3048 mm).

3. Greenhouses that are occupied for growing plants on a production or research basis, without public access, shall be exempt from the area limitations of Table 2609.4 provided they have a minimum fire separation distance of 4 feet (1220 mm).

4. Roof coverings over terraces and patios in occupancies in Group R-3 shall be exempt from the area limitations of Table 2609.4 and shall be permitted with light-transmitting plastics.

❖ This section provides certain area limitations for plastic roof panels. Panels of combustible plastic construction with its high fire-hazard properties must be limited in terms of the area of the structure. These requirements not only list the area limitations for individual panels but also limit the aggregate or sum total of all roof panels located in the structure's roof. The aggregate area limitations are based on a floor area percentage of the room or space enclosed by the roof.

The area limitations, which are based on the class of the plastic roof panels, are detailed in Table 2609.4.

There are four exceptions to these area limitations. When certain additional protection is provided or the panels are located in structures of limited hazard, the area limitations can be altered. In some cases, these exceptions increase the limitations or, in other cases, exempt certain structures from the limitations. Exception 1 increases the area limitations by 100 percent for those structures equipped throughout with an automatic sprinkler system installed in accordance with NFPA 13. Exception 2 allows certain low-hazard buildings to be exempt from the roof panel limitations. These buildings must not exceed 5,000 square feet (465 m²) in area and they must have at least 10 feet (3048 mm) of fire separation distance in order for the exception to apply. Exception 3 is for greenhouses that have at least a 4-foot (1219 mm) fire separation distance. Exception 4 allows occupancies in Group R-3 that have roofed-over terraces and patios to be exempt from the area limitations for roof panels. Such roof panels must be constructed with approved plastic meeting the requirements of Section 2606.

For SI: 1 foot = 304.8 mm.

Figure 2609.3
PLASTIC ROOF PANEL SEPARATION FROM EXTERIOR WALL

TABLE 2609.4
AREA LIMITATIONS FOR LIGHT-TRANSMITTING
PLASTIC ROOF PANELS

CLASS OF PLASTIC	MAXIMUM AREA OF INDIVIDUAL ROOF PANELS (square feet)	MAXIMUM AGGREGATE AREA OF ROOF PANELS (percent of floor area)
CC1	300	30
CC2	100	25

For SI: 1 square foot = 0.0929 m².

❖ The maximum area limitations for both individual panels and aggregate areas are based on the class of plastic. These limitations are needed to restrict the use of light-transmitting plastic in roofs and roof coverings of buildings and structures. Section 2609.4 provides text that describes when these limitations are to be used. Table 2609.4 contains three different headings and only two lines of entry. The first heading is labeled "Class of plastic" and is to be used in conjunction with Section 2606.4 for determining whether the plastic used in roof panels is either Class CC1 (e.g., polycarbonate) or CC2 (e.g., acrylic). Once the class of plastic of the roof panels in question has been determined, the table is to be used to determine the area limitations for such panels.

The second heading in the table is identified as "Maximum area of individual roof panels (square feet)." These values provide the maximum area limitations for each individual roof panel based on the class of plastic used to manufacture the panel. These limitations are given in units of square feet.

The last heading, labeled "Maximum aggregate area of roof panels (percent of floor area)," provides the aggregate amount of area that can be occupied by plastic panels located in a roof. This limitation is based on a percentage of the floor area of the room or space that is sheltered by the roof in question. The floor area in units of square feet is multiplied by the appropriate percentage for the class of plastic used. This product provides the maximum area for all plastic panels used in the roof.

SECTION 2610
LIGHT-TRANSMITTING PLASTIC
SKYLIGHT GLAZING

2610.1 Light-transmitting plastic glazing of skylight assemblies. Skylight assemblies glazed with light-transmitting plastic shall conform to the provisions of this section and Section 2606. Unit skylights glazed with light-transmitting plastic shall also comply with Section 2405.5.

Exception: Skylights in which the light-transmitting plastic conforms to the required roof-covering class in accordance with Section 1505.

❖ This section covers horizontal and sloped applications of light-transmitting plastic skylights. This section contains mounting, size and spacing limitations for these combustible assemblies. Plastic skylights sloped more than 15 degrees (0.26 rad) from vertical are also re-

quired to comply with Section 2405.

Acrylic and polycarbonate plastics are commonly used glazing materials in skylight assemblies. The use of combustible skylights is permitted subject to the requirements of Section 2606. The reference to Section 2405.5 is intended to highlight the specific structural considerations for unit skylights, regardless of construction materials.

The exception eliminates any added requirements for plastic skylights that meet the roofing classification for the surrounding roof covering as described in Chapter 15.

2610.2 Mounting. The light-transmitting plastic shall be mounted above the plane of the roof on a curb constructed in accordance with the requirements for the type of construction classification, but at least 4 inches (102 mm) above the plane of the roof. Edges of light-transmitting plastic skylights or domes shall be protected by metal or other approved noncombustible material, or the light-transmitting plastic dome or skylight shall be shown to be able to resist ignition where exposed at the edge to a flame from a Class B brand as described in ASTM E 108 or UL 790.

Exceptions:

1. Curbs shall not be required for skylights used on roofs having a minimum slope of three units vertical in 12 units horizontal (25-percent slope) in occupancies in Group R-3 and on buildings with a nonclassified roof covering.

2. The metal or noncombustible edge material is not required where nonclassified roof coverings are permitted.

❖ The mounting requirements for plastic skylights are similar to those established in Section 2405.4 for glass skylights. The 4-inch (102 mm) curb requirement protects the skylight from flying brands, provides a wash for water running down the slope and provides a noticeable barrier to caution a person on the roof from stepping on the skylight; however, the skylight must be designed for the structural requirements of Section 2606.5. Since the edge of the plastic is more susceptible to ignition than the surface, noncombustible edge protection is required to reduce the chances of edge ignition. This requirement is only applicable when the plastic is of a type and design that does not resist the flying brand exposure as described in ASTM E 108 or UL 790.

Exception 1 is based on the general field experience with plastic skylights used in one- and two-family dwellings with slopes equal to or greater than three units vertical in 12 units horizontal (3:12). Additionally, skylights in unclassified roofs are not required to be mounted on curbs. Such a skylight need not provide fire performance beyond that required for the surrounding roof covering.

The basis for Exception 2 is similar to that for Exception 1. There is no need to require edge protection for the plastic when the adjacent area is an unclassified roof permitted by Section 1505.5.

2610.3 Slope. Flat or corrugated light-transmitting plastic skylights shall slope at least four units vertical in 12 units horizontal (4:12). Dome-shaped skylights shall rise above the mounting flange a minimum distance equal to 10 percent of the maximum span of the dome but not less than 3 inches (76 mm).

Exception: Skylights that pass the Class B Burning Brand Test specified in ASTM E 108 or UL 790.

❖ As are most of the restrictions for plastic skylights, the minimum slope requirements are based on limiting the risks involved with combustible roof components. Both the slope for flat and corrugated plastic skylights and the rise for domed skylights are selected to shed flying brands. Consistent with other sections, the slope requirements are not applicable when the skylight complies with the specified requirements of ASTM E 108 or UL 790.

2610.4 Maximum area of skylights. Each skylight shall have a maximum area within the curb of 100 square feet (9.3 m²).

Exception: The area limitation shall not apply where the building is equipped throughout with an automatic sprinkler system in accordance with Section 903.3.1.1 or the building is equipped with smoke and heat vents in accordance with Section 910.

❖ The area limitation for unsprinklered buildings minimizes the fire risks for plastic skylights. The 100-square-foot (9 m²) limitation does not apply to buildings provided throughout with an NFPA 13 automatic sprinkler system or equipped with smoke and heat vents in accordance with Section 910. This exception is largely academic, since contemporary plastic skylights do not normally exceed this area limitation.

2610.5 Aggregate area of skylights. The aggregate area of skylights shall not exceed 33¹/₃ percent of the floor area of the room or space sheltered by the roof in which such skylights are installed where Class CC1 materials are utilized, and 25 percent where Class CC2 materials are utilized.

Exception: The aggregate area limitations of light-transmitting plastic skylights shall be increased 100 percent beyond the limitations set forth in this section where the building is equipped throughout with an automatic sprinkler system in accordance with Section 903.3.1.1 or the building is equipped with smoke and heat vents in accordance with Section 910.

❖ This section includes limitations directed to minimize the risks of fire spread. The 33¹/₃- and 25-percent limitations based on floor area for Class CC1 (polycarbonate) and CC2 (acrylic) materials, respectively, are judged to inhibit sufficiently the spread of fire. An exception allows the percentage to be increased to 66 ²/₃ percent (Class CC1) and 50 percent (Class CC2) when an approved automatic sprinkler system is installed throughout the building or is equipped with smoke and heat vents in accordance with Section 910. To comply with the intent of this pro-

vision, the skylights should be distributed over the entire area of the roof and not concentrated in one area.

2610.6 Separation. Skylights shall be separated from each other by a distance of not less than 4 feet (1219 mm) measured in a horizontal plane.

Exceptions:

1. Buildings equipped throughout with an automatic sprinkler system in accordance with Section 903.3.1.1.
2. In Group R-3, multiple skylights located above the same room or space with a combined area not exceeding the limits set forth in Section 2610.4.

❖ As with the other items cited, the minimum 4-foot (1219 mm) spacing minimizes the risk of fire spread for buildings without automatic sprinklers (also see Figure 2609.2). The separation is not necessary when complete fire suppression is provided in accordance with NFPA 13 or in Group R-3, as applicable in Section 2610.4, where the skylights are located above the same room or space with a combined area not to exceed the limits of Section 2610.4.

2610.7 Location. Where exterior wall openings are required to be protected in accordance with Section 704, a skylight shall not be installed within 6 feet (1829 mm) of such exterior wall.

❖ The minimum 6-foot (1829 mm) spacing to a fire-resistance-rated wall is to minimize the risk of fire spread (also see Section 2609.3).

2610.8 Combinations of roof panels and skylights. Combinations of light-transmitting plastic roof panels and skylights shall be subject to the area and percentage limitations and separation requirements applicable to roof panel installations.

❖ This is a precautionary statement to clarify that skylights and roof panels must be considered together when establishing size and separation limitations. When combinations of plastic skylights and roof panels occur, the area limitations of Section 2609.4 govern. As with wall panels and glazing, roof panels and skylights represent a similar combustible aspect of construction and should be addressed together.

SECTION 2611
LIGHT-TRANSMITTING PLASTIC INTERIOR SIGNS

2611.1 General. Light-transmitting plastic interior wall signs shall be limited as specified in Sections 2611.2 through 2611.4. Light-transmitting plastic interior wall signs in covered mall buildings shall comply with Section 402.15. Light-transmitting plastic interior signs shall also comply with Section 2606.

❖ This section contains the scoping statement that requires all light-transmitting plastic materials used for interior wall signage applications to comply with the requirements of Sections 2611.2, 2611.3 and 2611.4. These requirements include the maximum allowable aggregate area of signs, the maximum individual area of signs and encasement requirements. The only exception to these requirements is light-transmitting plastics that are used for interior wall signage in covered mall buildings, since they are regulated by Section 402.14.

These provisions help clarify the requirements for plastic signs in all occupancies other than covered mall buildings and exterior applications. One of the most common uses of light-transmitting plastic in interior signage applications is as the facing of internally illuminated or nonilluminated wall signs (sometimes referred to as "box" or "can" signs). Typical interior applications include tenant signage, building directories, promotional or advertising sign boxes and directional signage (other than exit signage).

2611.2 Aggregate area. The sign shall not exceed 20 percent of the wall area.

❖ The provisions of this section limit the maximum amount of wall area that may be covered by all plastic signs located on that wall, whereas Section 2611.3 governs the maximum permitted size of any individual sign. This limitation on the amount of wall area that can be covered by plastic signs is intended to minimize the potential for the spread of fire along a wall surface through involvement of the plastic signage.

2611.3 Maximum area. The sign shall not exceed 24 square feet (2.23 m²).

❖ When applying the provisions of Sections 2611.2 and 2611.3, the most restrictive of each section must be applied. For example, if a single sign with the maximum area of 24 square feet (2 m²) exceeds 20 percent of the wall area on which it is located, the area of the sign must be reduced to a size where the 20-percent limitation is not exceeded. Figures 2611.3(1) and 2611.3(2) give two examples of these provisions.

2611.4 Encasement. Edges and backs of the sign shall be fully encased in metal.

❖ The provisions of this section are intended to minimize the amount of surface that could be exposed to direct impingement by fire, and thus reduce the potential for ignition of the plastic.

Bibliography

The following resource materials are referenced in this chapter or are relevant to the subject matter addressed in this chapter.

ANSI Z97.1-84 (R1994), *Safety Performance Specifications and Methods of Test for Safety Glazing Materials Used in Buildings.* New York: American National Standards Institute, 1994.

ASTM D 635-03 *Standard Test Method for Rate of Burning and/or Extent and Time of Burning of Plastics in a Horizontal Position.* West Conshohocken, PA: ASTM International, 2000.

AREA OF SIGN #1 = 24 SQ. FT. (MAX. AREA PER SECTION 2611.3)
AGGREGATE AREA = 24 SQ. FT.

AREA OF WALL = 110 SQ. FT.
20% OF 110 SQ. FT. = 22 SQ. FT. (MAX. AREA PER SECTION 2611.2)
24 SQ. FT. ≤ 22 SQ. FT. ∴ AGGREGATE AREA OF SIGNS EXCEEDS
AREA PERMITTED BY SECTION 2611.2

For SI: 1 inch = 25.4 mm, 1 foot = 304.8 mm,
           1 square foot = 0.0929 m².

Figure 2611.3(1)
MAXIMUM AREA OF LIGHT-TRANSMITTING PLASTIC SIGNS—EXAMPLE 1

ASTM D 1929-96 (2001)E01, *Standard Test Method for Ignition Properties of Plastics.* West Conshohocken, PA: ASTM International, 2001.

ASTM D 2843-99 (2004), *Standard Test Method for Density of Smoke from the Burning or Decomposition of Plastics.* West Conshohocken, PA: ASTM International, 2004.

ASTM D 3679-04, *Specification for Rigid Poly [Vinyl Chloride (PVC) Siding].* West Conshohocken, PA: ASTM International, 2004.

ASTM E 84-04, *Standard Test Method for Surface-Burning Characteristics of Building Materials.* West Conshohocken, PA: ASTM International, 2004.

ASTM E 108-04, *Standard Test Method for Fire Tests of Roof Coverings.* West Conshohocken, PA: ASTM International, 2004.

ASTM E 119-00, *Standard Test Methods for Fire Tests of Building Construction and Materials.* West Conshohocken, PA: ASTM International, 2000.

CPSC 16 CFR, Part 1201-77, *Safety Standard for Architectural Glazing.* Washington, D.C.: Consumer Product Safety Commission, 1977.

DASMA 107-98(03), *Room Fire Test Standard for Garage Doors Using Foam Plastic Insulation.* Cleveland, OH: Door and Access Systems Manufactures Association, 2003.

FM 4450 (1989), *Approval Standard for Class 1 Insulated Steel Deck Roofs*—with Supplement (July 1992). Norwood, MA: Factory Mutual Standards Laboratories Department, 1989.

FM 4880 (2001), *Approval Standard for Class 1: a) Insulated Wall or Wall and Roof/Ceiling Panels, b) Plastic Interior Finish Materials, c) Plastic Exterior Building Panels, d) Wall/Ceiling Coating Systems and e) Interior or Exterior Finish Systems.* Norwood, MA: Factory Mutual Standards Laboratories Department, 1994.

NFPA 13-99, *Installation of Sprinkler Systems.* Quincy, MA: National Fire Protection Association, 2001.

NFPA 259-04, *Test Method for Potential Heat of Building Materials.* Quincy, MA: National Fire Protection Association, 2004.

NFPA 265-02, *Method of Fire Tests for Evaluating Room Fire Growth Contribution of Textile Wall Coverings on Full Height Panels and Walls.* Quincy, MA: National Fire Protection Association, 2002.

NFPA 268-01, *Standard Test Method for Determining Ignitability of Exterior Wall Assemblies Using a Radiant Heat Energy Source.* Quincy, MA: National Fire Protection Association, 2001.

AREA OF SIGN #1 = 24 SQ. FT. (MAX. AREA PER SECTION 2611.3)
AREA OF SIGN #2 = 4 SQ. FT.
AREA OF SIGN #3 = 18 SQ. FT.
AGGREGATE AREA = 46 SQ. FT.

AREA OF WALL = 242 SQ. FT.
20% OF 242 SQ. FT. = 48.4 SQ. FT. (MAX. AREA PER SECTION 2611.2)
46 SQ. FT. ≤ 48.4 SQ. FT. ∴ AGGREGATE AREA OF SIGNS COMPLIES
WITH SECTION 2611.2

For SI: 1 inch = 25.4 mm, 1 foot = 304.8 mm,
           1 square foot = 0.0929 m².

Figure 2611.3(2)
MAXIMUM AREA OF LIGHT-TRANSMITTING PLASTIC SIGNS—EXAMPLE 2

NFPA 285-98, *Standard Method of Test for the Evaluation of Fammability Characteristics of Exterior Nonload-bearing Wall Assemblies Containing Combustible Components.* Quincy, MA: National Fire Protection Association, 1998.

NFPA 286-00, *Standard Method of Fire Test for Evaluating Contribution of Wall and Ceiling Interior Finish to Room Fire Growth.* Quincy, MA: National Fire Protection Association, 2000.

UL 790-98, *Tests of Fire Resistance of Roof Covering Materials.* Northbrook, IL: Underwriters Laboratories Inc., 1998.

UL 1040-96, *Fire Test of Insulated Wall Construction—with Revisions through April 2001.* Northbrook, IL: Underwriters Laboratories Inc., 2001.

UL 1256-02, *Standard for Safety Fire Test of Roof Deck Constructions.* Northbrook, IL: Underwriters Laboratories Inc., 2002.

UL 1715-97, *Fire Test of Interior Finish Material.* Northbrook, IL: Underwriters Laboratories Inc., 1997.

UL 1975-96, *Fire Test of Foamed Plastics Used for Decorative Purposes.* Northbrook, IL: Underwriters Laboratories Inc., 1996.

Chapter 27:
Electrical

General Comments

This chapter references the *International Code Council Electrical Code Administrative Provisions* and contains specific provisions for emergency and standby power. The *International Code Council Electrical Code Administrative Provisions* regulates the design, construction, installation, operation, maintenance and use of electrical systems and equipment. The *International Code Council Electrical Code Administrative Provisions* references NFPA 70 in addition to providing administrative and technical provisions. Except for the scope section, this chapter is devoted to establishing where emergency and standby power systems are required.

Purpose

Since electrical systems and components are an integral part of almost all structures, it is necessary for the code to address them to protect life, limb, health and property. In addition to general lighting and power needs, structures depend on electricity for operation of many of the life safety systems required by the code, including fire alarm, smoke control, exhaust, fire suppression, communication and fire command systems. Electricity is also depended upon for elements of egress systems, including elevators, exit signage, egress path illumination, power doors, stair pressurization and smokeproof enclosures.

SECTION 2701
GENERAL

2701.1 Scope. This chapter governs the electrical components, equipment and systems used in buildings and structures covered by this code. Electrical components, equipment and systems shall be designed and constructed in accordance with the provisions of the International Code Council Electrical Code Administrative Provisions.

❖ Chapter 27 regulates the electrical portions of buildings and structures that fall under the scope of the code. Regulation is accomplished by reference to the *International Code Council Electrical Code Administrative Provisions* that, in turn, references NFPA 70.

[F] SECTION 2702
EMERGENCY AND STANDBY POWER SYSTEMS

[F] 2702.1 Installation. Emergency and standby power systems required by this code or the *International Fire Code* shall be installed in accordance with this code, NFPA 110 and 111.

❖ This section is consistent and coordinated with Section 1202.6 of the *International Code Council Electrical Code Administrative Provisions*. Emergency power systems are intended to provide electrical power for life safety systems, such as egress illumination, emergency communications, fire pumps, high-rise building elevators and processes involving the handling and use of hazardous materials. In other words, emergency power is required where the loss of normal power would endanger occupants. Such systems are covered in Article 700 of NFPA 70 and one of their key features is the required response time of 10 seconds

or less. The time between loss of normal power and the provision of emergency power must be kept very short to prevent putting occupants at risk. This is especially important during an emergency event such as a building fire, but is important at all times to prevent occupant panic, which could happen if a crowded building is suddenly plunged into darkness.

Standby power systems are covered in Article 701 of NFPA 70 and are intended to provide electrical power for loads not as critical as those requiring emergency power. Standby power loads include smoke control systems; certain elevators; certain hazardous material operations; smokeproof enclosure systems; illumination; heating, ventilating and air-conditioning (HVAC) systems; refrigeration and sewage pumps. Standby power systems must provide power within 60 seconds of failure of primary power.

Sources of power for emergency power systems (NFPA 70, Section 701-11) include storage batteries, generators, uninterruptible power supplies and separate services. Sources of power for standby systems include those allowed for emergency systems plus a source that is taken from a point of connection ahead of the normal service disconnecting means.

NFPA 110 addresses the performance criteria and "nuts and bolts" of emergency and standby power systems and classifies them into types, classes and levels relative to maximum response time, minimum required operation time and life safety importance factor, respectively. NFPA 111 addresses stored emergency power supply systems and is similar in coverage to NFPA 110. Stored energy systems typically rely on batteries that store chemical energy.

[F] 2702.1.1 Stationary generators. Stationary emergency and standby power generators required by this code shall be listed in accordance with UL 2200.

❖ UL 2200 was developed at the request of building officials to provide a standard to evaluate the safety and reliability of stationery engine generators. UL 2200 establishes a basis for this evaluation.

[F] 2702.2 Where required. Emergency and standby power systems shall be provided where required by Sections 2702.2.1 through 2702.2.20.

❖ Fires can cause the loss of utility power as a result of either damage to equipment and wiring or fire-fighter action to shut off power to eliminate sources of ignition and danger to personnel.

Sections 2702.2.1 through 2702.2.20 dictate where emergency and standby power systems are required based upon the nature of the electrical loads. Note that emergency and standby power systems have different characteristics, and subsequent sections will require one or the other (see commentary, Section 2702.1).

[F] 2702.2.1 Group A occupancies. Emergency power shall be provided for voice communication systems in Group A occupancies in accordance with Section 907.2.1.2.

❖ Emergency voice/alarm systems are designed to allow the fast, efficient and orderly egress of large numbers of people and are, therefore, considered a life safety system worthy of being provided with emergency power (see commentary, Section 907.2.1.2).

[F] 2702.2.2 Smoke control systems. Standby power shall be provided for smoke control systems in accordance with Section 909.11.

❖ Smoke control systems are intended to maintain a tenable environment in a building to allow the occupants ample time to evacuate or relocate to protected areas. As such, smoke control systems are life safety systems and must be dependable (see commentary, Section 909.11).

[F] 2702.2.3 Exit signs. Emergency power shall be provided for exit signs in accordance with Section 1011.5.3.

❖ Emergency power is warranted for exit signage illumination since guiding occupants to the exits is certainly a life safety function (see commentary, Section 1011.5.3).

[F] 2702.2.4 Means of egress illumination. Emergency power shall be provided for means of egress illumination in accordance with Section 1006.3.

❖ The path of travel to all exits must be illuminated to guide occupants and allow for safe egress; therefore, emergency power for illumination is warranted (see commentary, Section 1006.3).

[F] 2702.2.5 Accessible means of egress elevators. Standby power shall be provided for elevators that are part of an accessible means of egress in accordance with Section 1007.4.

❖ Elevators can be a component of an accessible means of egress in accordance with Section 1007.4 and

must, therefore, be dependable at all times. Without backup power, an elevator could be a dead end for someone with physical disabilities who is trying to egress a building (see commentary, Section 1007.4).

[F] 2702.2.6 Accessible means of egress platform lifts. Standby power in accordance with this section or ASME A18.1 shall be provided for platform lifts that are part of an accessible means of egress in accordance with Section 1007.5.

❖ Platform lifts are allowed to be used as part of an accessible means of egress only when they are allowed as part of an accessible route, in accordance with Section 1109.7, Items 1 through 9. When this is the case, the platform lift could be a dead end for someone with physical disabilities if power to the platform lift has been lost; therefore, standby power is required, in accordance with Section 2702 or ASME A18.1.

[F] 2702.2.7 Horizontal sliding doors. Standby power shall be provided for horizontal sliding doors in accordance with Section 1008.1.3.3.

❖ Power-operated doors could be an obstruction to egress if the primary power supply fails; therefore, standby power is required to maintain door operation (see commentary, Section 1008.1.3.3).

[F] 2702.2.8 Semiconductor fabrication facilities. Emergency power shall be provided for semiconductor fabrication facilities in accordance with Section 415.8.10.

❖ Where hazardous materials are utilized, such as in hazardous production materials (HPM) facilities, many systems are depended upon to protect the occupants from exposure to hazardous materials, including exhaust/ventilation systems, gas cabinet exhaust systems, gas detection systems, alarm systems and suppression systems. Loss of power would endanger the occupants; thus, emergency power is essential for these occupancies (see commentary, Sections 415.8.10 and 415.8.10.1).

[F] 2702.2.9 Membrane structures. Standby power shall be provided for auxiliary inflation systems in accordance with Section 3102.8.2. Emergency power shall be provided for exit signs in temporary tents and membrane structures in accordance with the *International Fire Code*.

❖ Air-supported and air-inflated structures would collapse on the occupants if the inflation systems fail. Section 3102.8.1.1 requires redundant inflation equipment that would be as worthless as the primary system in the event of power failure; therefore, standby power is required. Emergency power is required for all exit signs in temporary tents and membrane structures (see commentary, Section 3102.8.2).

[F] 2702.2.10 Hazardous materials. Emergency or standby power shall be provided in occupancies with hazardous materials in accordance with Section 414.5.4.

❖ Where hazardous materials and processes are housed, occupant safety could be dependent upon one or more ventilation, treatment, temperature control, alarm and detection system. Thus, emergency or standby power is required, depending upon the load (see commentary, Section 414.5.4).

[F] 2702.2.11 Highly toxic and toxic materials. Emergency power shall be provided for occupancies with highly toxic or toxic materials in accordance with the *International Fire Code*

❖ See the commentary to Section 2702.2.9 and Chapter 27 of the *International Fire Code*® (IFC®) for additional information.

[F] 2702.2.12 Organic peroxides. Standby power shall be provided for occupancies with silane gas in accordance with the *International Fire Code*.

❖ See the commentary to Section 2702.2.9 and Chapter 39 of the IFC for additional information.

[F] 2702.2.13 Pyrophoric materials. Emergency power shall be provided for occupancies with silane gas in accordance with the *International Fire Code*.

❖ See the commentary to Section 2702.2.9 and Chapter 41 of the IFC for additional information.

[F] 2702.2.14 Covered mall buildings. Standby power shall be provided for voice/alarm communication systems in covered mall buildings in accordance with Section 402.13.

❖ See the commentary to Sections 907.2.20 and 402.13 for additional information.

[F] 2702.2.15 High-rise buildings. Emergency and standby power shall be provided in high-rise buildings in accordance with Sections 403.10 and 403.11.

❖ Occupants of high-rise buildings are at greater risk due to longer egress travel times, lack of fire-fighter access and the danger of vertical spread of fire and upward smoke migration. In accordance with this chapter and Sections 403.10 and 403.11, some loads in a high-rise building will require standby power and some will require emergency power (see commentary, Sections 403.10 and 403.11).

[F] 2702.2.16 Underground buildings. Emergency and standby power shall be provided in underground buildings in accordance with Sections 405.9 and 405.10.

❖ In the event of power failure, occupants could be underground without light, ventilation and the numerous required life safety systems. These structures are analogous to inverted high-rise buildings. Section 405.9 requires standby power for specified loads while Section 405.10 requires emergency power for specified loads (see commentary, Sections 405.9 and 405.10).

[F] 2702.2.17 Group I-3 occupancies. Emergency power shall be provided for doors in Group I-3 occupancies in accordance with Section 408.4.2.

❖ In an emergency, occupants in detention and correctional facilities are at the mercy of door-locking mechanisms and those who control such locks; thus, emergency power is warranted (see commentary, Section 408.4.2).

[F] 2702.2.18 Airport traffic control towers. Standby power shall be provided in airport traffic control towers in accordance with Section 412.1.5.

❖ This requirement is similar in intent to that for standby power in high-rise buildings, but due to the limited number of occupants and other protective features required by Section 412, only standby power is required (see commentary, Section 412.1.5).

[F] 2702.2.19 Elevators. Standby power for elevators shall be provided as set forth in Section 3003.1.

❖ Section 3003.1 provides the "how to" related text for elevator standby power systems, whereas other sections, including Sections 2702.2.5 and 2702.2.15, dictate where such power must be provided (see commentary, Section 3003.1).

[F] 2702.2.20 Smokeproof enclosures. Standby power shall be provided for smokeproof enclosures as required by Section 909.20.

❖ To protect egress elements from the invasion of smoke, mechanical ventilation/pressurization systems must be dependable and thus provided with standby power (see commentary, Section 909.20.6.2).

[F] 2702.3 Maintenance. Emergency and standby power systems shall be maintained and tested in accordance with the *International Fire Code*.

❖ Emergency and standby power systems must be maintained, serviced and tested to ensure dependability (see commentary, Section 604 of the IFC).

Bibliography

The following resource materials are referenced in this chapter or are relevant to the subject matter addressed in this chapter.

International Code Council Electrical Code Administrative Provisions. Falls Church, VA: International Code Council, 2006.

ICC IFC-2006, *International Fire Code.* Falls Church, VA: International Code Council, 2006.

NFPA 70, *National Electrical Code.* Quincy, MA: National Fire Protection Association, 1999.

NFPA 110, *Emergency and Standby Power Systems.* Quincy, MA: National Fire Protection Association, 1999.

NFPA 111, *Stored Electrical Energy Emergency and Standby Power Systems.* Quincy, MA: National Fire Protection Association, 1996.

UL 2200-98, *Stationery Engine Generator Assemblies (Revisions through July 2004).* Northbrook, IL: Underwriters Laboratories, Inc., 2004.

Chapter 28:
Mechanical Systems

General Comments

Chapter 28 provides for the approval, installation, construction, inspection, operation and maintenance of all mechanical appliances, equipment and systems by direct reference to the *International Mechanical Code*® (IMC®) and the *International Fuel Gas Code*® (IFGC®). All mechanical equipment and appliances, other than gas-fired equipment and appliances, must comply with the provisions of the IMC. Fuel-gas piping, gas-fired appliances and gas-fired appliance venting must comply with the IFGC.

Purpose

The purpose of Chapter 28 is to verify that mechanical equipment, appliances and mechanical systems are regulated with respect to design, construction, installation, inspection and maintenance. This chapter references the IMC and IFGC. This eliminates the need to duplicate the text of those codes and utilizes the performance and specification criteria of those codes as the basis for regulation of the construction, inspection and maintenance of all mechanical and fuel gas equipment, appliances and systems.

While this chapter only consists of references to other chapters and codes, it is necessary since almost all newly constructed buildings and structures contain mechanical equipment, appliances and systems. Without this chapter, a necessary link between the code and the IMC and IFGC would not exist.

SECTION 2801
GENERAL

2801.1 Scope. Mechanical appliances, equipment and systems shall be constructed, installed and maintained in accordance with the *International Mechanical Code* and the *International Fuel Gas Code*. Masonry chimneys, fireplaces and barbecues shall comply with the *International Mechanical Code* and Chapter 21 of this code.

❖ The requirements for all mechanical systems are governed by this chapter. This chapter does not contain any requirements for mechanical equipment, appliances and systems; instead it references the IMC and IFGC. These two codes along with nine others, including this code, make up the family of the 2006 editions of the *International Codes*®.

In the process of producing the code, including actual mechanical equipment, appliances and system provisions were considered unnecessary since such provisions would duplicate and possibly conflict with those of the IMC or IFGC. Since most newly constructed buildings contain mechanical equipment, appliances and systems, this chapter is necessary to reference the IMC and IFGC and tie them together with the code.

The IMC and IFGC work together to regulate all mechanical systems. Therefore, whether designing the mechanical system for a building or enforcing the provisions of the code that deal with mechanical systems, both the IMC and IFGC will be required.

The IMC and IFGC are complete codes in themselves, and set forth necessary administrative and enforcement requirements, as well as the requirements for installation, construction, inspection and maintenance of mechanical equipment, appliances and systems. The IMC regulates all aspects of mechanical systems, including: air distribution systems and ductwork; boilers and hydronic systems; exhaust systems; fuel-oil piping systems; combustion air requirements; chimneys and vents; ventilation air requirements; refrigeration systems; solar-powered systems and specific appliances and equipment. The IFGC regulates all aspects of fuel-gas piping systems, fuel-gas utilization equipment and related accessories, including: combustion air requirements for gas appliances; chimneys and vents for gas appliances; fuel-gas piping and specific gas appliances and gas equipment.

A key component of the IMC and IFGC is Chapter 3 of both codes, which requires all mechanical equipment and appliances to be listed and labeled to the standard or standards relevant to that equipment or appliance (see Figure 2801.1).

Another important element of the IMC and IFGC is a set of unique definitions found in Chapter 2 of both codes. These definitions supersede any others with respect to mechanical and gas systems. Especially noteworthy is the definition of "Noncombustible material" in the IMC, which does not include composite materials as provided for in Section 703.4.2 of the code. In the context of the IMC, such composite materials are considered to be combustible.

Figure 2801.1
EXAMPLE LABELS

Bibliography

The following resource material is referenced in this chapter or is relevant to the subject matter addressed in this chapter.

IFGC-2006, *International Fuel Gas Code*. Falls Church, VA: International Code Council, 2006.

IMC-2006, *International Mechanical Code*. Falls Church, VA: International Code Council, 2006.

Chapter 29:
Plumbing Systems

General Comments

Determining the number of fixtures that are necessary for a given occupancy has always been a complicated issue for plumbing codes.

There have been various methods used to evaluate the need for plumbing fixtures.

For office buildings, studies have indicated that design guidelines based on occupancy times, arrival rates, duration and patterns of fixture use provide the number of required plumbing fixtures for a desired level of service. The Building Technology Research Laboratory at the Stevens Institute of Technology conducted a study based on the "queuing theory" for determining the number of plumbing fixtures for a desired level of service. The theory is based on waiting times during peak periods of use, fixture utilization and the probability of finding a vacant fixture. This type of method provides answers for designing service systems. For residential-type buildings and health care facilities, plumbing fixtures are based on the minimum need, resulting in at least one water closet and one lavatory for each dwelling unit, guestroom or hospital room (see Note c to Table 2902.1)

Studies completed by the U.S. military have been used for dormitories and prisons, which determine the number of fixtures based on a simultaneous need in a regimented society. This assumes that everyone rises at approximately the same time and has a limited amount of time to shower, use the water closet and lavatory.

The National Restaurant Association conducted a study based on the difference in use for a restaurant and a nightclub. It should be noted that the study did not take into account today's fast-food style restaurant, nor did it allow for restaurants located along heavily traveled routes, such as highway rest stops.

Fixture requirements for factory and industrial uses are based on the same requirements as for storage facilities. This method establishes a more realistic minimum requirement for factory occupancies. The reasonableness in the number of plumbing fixtures was established through a limited study of factory projects in Henrico County, Virginia.

The fixture needs for the remaining occupancies were determined based on empirical data, experience and tradition. There are no exact definitive studies providing values that are supportable; the values have been periodically modified based on general observations of how occupants behave.

Purpose

The purpose of Chapter 29 is to provide a building with the necessary number and type of plumbing fixtures to adequately serve the anticipated number of occupants. The fixtures must be properly installed to be both accessible and usable by the individuals occupying the building. The quality and design of every fixture must be in accordance with the *International Plumbing Code*® (IPC®).

SECTION 2901
GENERAL

[P] 2901.1 Scope. The provisions of this chapter and the *International Plumbing Code* shall govern the erection, installation, alteration, repairs, relocation, replacement, addition to, use or maintenance of plumbing equipment and systems. Plumbing systems and equipment shall be constructed, installed and maintained in accordance with the *International Plumbing Code*. Private sewage disposal systems shall conform to the *International Private Sewage Disposal Code*.

❖ This section contains the scoping requirements for Chapter 29. Compliance with this chapter will result in a building or structure having adequate plumbing fixtures for the sanitary, hygienic, cleaning, washing and food preparation needs of the occupants. It is important to note that the "P" in brackets before each section number indicates that the entire content of that particular section is maintained by the IPC Development Committee. The entire content of Section 2902 can be found in Section 403 of the IPC. These provisions have been located in the code for the convenience of the architect, who is often responsible for the layout and number of plumbing fixtures when creating the floor plan of a building.

SECTION 2902
MINIMUM PLUMBING FACILITIES

[P] 2902.1 Minimum number of fixtures. Plumbing fixtures shall be provided for the type of occupancy and in the minimum number shown in Table 2902.1 Types of occupancies not shown in Table 2902.1 shall be considered individually by the building official. The number of occupants shall be determined by this code. Occupancy classification shall be determined in accordance with Chapter 3.

❖ Table 2902.1 establishes the minimum number of plumbing fixtures required for each building. Note that this is the same as Table 403.1 in the IPC. The occupant load used for calculating the number of fixtures required is the same occupant load used for determining means of egress. Methods for calculating occupant loads are found in Section 1004. By using the means of egress occupant loads, there is consistency in occupant load calculations for the application of the number of plumbing fixtures and the means of egress provisions. The means of egress occupant loads of a building do not always reflect typical day-to-day occupant loads; however, the table takes this into account by modifying the values for determining the number of fixtures.

TABLE 2902.1. See below.

❖ Table 2902.1 provides simple ratios for determining the minimum number of plumbing fixtures. To aid in the use of the table, the type of building category has been listed by occupancy classification along with a brief description of that occupancy. The occupancy identification parallels the classification listed in Chapter 3.

The user simply locates the occupancy in Table 2902.1 and applies the ratio or ratios to determine the number of water closets, lavatories, bathtubs, showers, drinking fountains and sinks needed for the building.

Section 419.2 of the IPC permits urinals to be substituted for a maximum of 67 percent of the required number of water closets in assembly and educational occupancies and a maximum of 50 percent in all other occupancies. For example, if five water closets are re-

[P] TABLE 2902.1
MINIMUM NUMBER OF REQUIRED PLUMBING FIXTURES[a]
(See Sections 2902.2 and 2902.3)

No.	CLASSIFICATION	OCCUPANCY	DESCRIPTION	WATER CLOSETS (URINALS SEE SECTION 419.2 OF THE INTERNATIONAL PLUMBING CODE) MALE	FEMALE	LAVATORIES MALE	FEMALE	BATHTUBS/ SHOWERS	DRINKING FOUNTAINS (SEE SECTION 410.1 OF THE INTERNATIONAL PLUMBING CODE)	OTHER
1	Assembly (see Sections 2902.2, 2902.4 and 2902.4.1)	A-1[d]	Theaters and other buildings for the performing arts and motion pictures	1 per 125	1 per 65	1 per 200		—	1 per 500	1 service sink
		A-2[d]	Nightclubs, bars, taverns, dance halls and buildings for similar purposes	1 per 40	1 per 40	1 per 75		—	1 per 500	1 service sink
			Restaurants, banquet halls and food courts	1 per 75	1 per 75	1 per 200		—	1 per 500	1 service sink
		A-3[d]	Auditoriums without permanent seating, art galleries, exhibition halls, museums, lecture halls, libraries, arcades and gymnasiums	1 per 125	1 per 65	1 per 200		—	1 per 500	1 service sink
			Passenger terminals and transportation facilities	1 per 500	1 per 500	1 per 750		—	1 per 1,000	1 service sink
			Places of worship and other religious services	1 per 150	1 per 75	1 per 200		—	1 per 1,000	1 service sink
		A-4	Coliseums, arenas, skating rinks, pools and tennis courts for indoor sporting events and activities	1 per 75 for the first 1,500 and 1 per 120 for the remainder exceeding 1,500	1 per 40 for the first 1,500 and 1 per 60 for the remainder exceeding 1,500	1 per 200	1 per 150	—	1 per 1,000	1 service sink
		A-5	Stadiums, amusement parks, bleachers and grandstands for outdoor sporting events and activities	1 per 75 for the first 1,500 and 1 per 120 for the remainder exceeding 1,500	1 per 40 for the first 1,500 and 1 per 60 for the remainder exceeding 1,500	1 per 200	1 per 150	—	1 per 1,000	1 service sink

(continued)

quired in an educational occupancy men's room, the design professional may choose to install two water closets and three urinals, or three water closets and two urinals. If four urinals are installed, however, two water closets are also required. Experience has shown that water closet shortages result with a 67-percent substitution allowance in occupancies other than Groups E and A; thus, a 50-percent maximum allowance was introduced for those occupancies.

In all occupancy classifications, except Group I-2, only one service sink is required for the entire building. The one service sink per building must be available from all portions of the building. For example, a service sink installed in an office building should not be located in an area that is not accessible from all tenant spaces.

Determining Number of Fixtures

Once the occupancy of the building is selected, determining the minimum required number of fixtures is simple and straightforward. The common question asked in applying the table is whether the occupant load is divided regarding the number of each sex and the table is then applied, or whether the table is applied to the entire occupant load and the required number of fixtures is then divided between the sexes. Except for assembly occupancies required to be pro-

[P] TABLE 2902.1—continued
MINIMUM NUMBER OF REQUIRED PLUMBING FIXTURES[a]

No.	CLASSIFICATION	OCCUPANCY	DESCRIPTION	WATER CLOSETS (URINALS SEE SECTION 419.2 OF THE *INTERNATIONAL PLUMBING CODE*)		LAVATORIES		BATHTUBS OR SHOWERS	DRINKING FOUNTAINS (SEE SECTION 410.1 OF THE *INTERNATIONAL PLUMBING CODE*)	OTHER
				MALE	FEMALE	MALE	FEMALE			
2	Business (see Sections 2902.2, 2902.4 and 2902.4.1)	B	Buildings for the transaction of business, professional services, other services involving merchandise, office buildings, banks, light industrial and similar uses	1 per 25 for the first 50 and 1 per 50 for the remainder exceeding 50		1 per 40 for the first 80 and 1 per 80 for the remainder exceeding 80		—	1 per 100	1 service sink
3	Educational	E	Educational facilities	1 per 50		1 per 50		—	1 per 100	1 service sink
4	Factory and industrial	F-1 and F-2	Structures in which occupants are engaged in work fabricating, assembly or processing of products or materials	1 per 100		1 per 100		See Section 411 of the *International Plumbing Code*	1 per 400	1 service sink
5	Institutional	I-1	Residential care	1 per 10		1 per 10		1 per 8	1 per 100	1 service sink
		I-2	Hospitals, ambulatory nursing home patients[b]	1 per per room[c]		1 per per room[c]		1 per 15	1 per 100	1 service sink
			Employees, other than residential care[b]	1 per 25		1 per 35		—	1 per 100	—
			Visitors, other than residential care	1 per 75		1 per 100		—	1 per 500	—
		I-3	Prisons[b]	1 per cell		1 per cell		1 per 15	1 per 100	1 service sink
		I-3	Reformatories, detention centers and correctional centers[b]	1 per 15		1 per 15		1 per 15	1 per 100	1 service sink
		I-4	Adult day care and child care	1 per 15		1 per 15		—	1 per 100	1 service sink
6	Mercantile (see Section 2902.2, 2902.4, 2902.4.1 and 2902.4.2)	M	Retail stores, service stations, shops, salesrooms, markets and shopping centers	1 per 500		1 per 750		—	1 per 1,000	1 service sink

(continued)

vided with different fixture ratios for females and males, it does not matter which way it is calculated as both methods yield the same answer.

The division between male and female water closets is an attempt to resolve what is known as "potty parity" in assembly facilities. The table reflects a more equitable distribution in the number of plumbing fixtures provided at large places of assembly. The increase in the number of water closets required in all assembly occupancies is based on the following criteria:

1. For a variety of social and physical reasons, women generally take a longer period of time to

use the facilities, thereby decreasing turnaround time;

2. Women outnumber men in the general population and this profile is represented in large places of assembly; and

3. Women sometimes require more frequent use of plumbing facilities, thereby creating longer waiting lines.

In the case of assembly occupancies, the total design occupant load obtained from the code is first divided into a 50:50 ratio as required by Section 2902.3, unless statistical data approved by the building official

[P] TABLE 2902.1—continued
MINIMUM NUMBER OF REQUIRED PLUMBING FIXTURES[a]

No.	CLASSIFICATION	OCCUPANCY	DESCRIPTION	WATER CLOSETS (URINALS SEE SECTION 419.2 OF THE *INTERNATIONAL PLUMBING CODE*)		LAVATORIES		BATHTUBS/ SHOWERS	DRINKING FOUNTAINS (SEE SECTION 410.1 OF THE *INTERNATIONAL PLUMBING CODE*)	OTHER
				MALE	FEMALE	MALE	FEMALE			
7	Residential	R-1	Hotels, motels, boarding houses (transient)	1 per sleeping unit		1 per sleeping unit		1 per sleeping unit	—	1 service sink
		R-2	Dormitories, fraternities, sororities and boarding house (not transient)	1 per 10		1 per 10		1 per 8	1 per 100	1 service sink
		R-2	Apartment house	1 per dwelling unit		1 per dwelling unit		1 per dwelling unit	—	1 kitchen sink per dwelling unit; 1 automatic clothes washer connection per 20 dwelling units
		R-3	One- and two-family dwellings	1 per dwelling unit		1 per dwelling unit		1 per dwelling unit	—	1 kitchen sink per dwelling unit; 1 automatic clothes washer connection per 20 dwelling units
		R-4	Residential care/assisted living facilities	1 per 10		1 per 10		1 per 8	1 per 100	1 service sink
8	Storage (see Sections 2902.2, 2902.4 and 2902.4.1)	S-1 S-2	Structures for the storage of goods, warehouses, storehouses and freight depots, low and moderate hazard	1 per 100		1 per 100		See Section 411 of the *International Plumbing Code*	1 per 1,000	1 service sink

a. The fixtures shown are based on one fixture being the minimum required for the number of persons indicated or any fraction of the number of persons indicated. The number of occupants shall be determined by this code.

b. Toilet facilities for employees shall be separate from facilities for inmates or patients.

c. A single-occupant toilet room with one water closet and one lavatory serving not more than two adjacent patient rooms shall be permitted where such room is provided with direct access from each patient room and with provisions for privacy.

d. The occupant load for seasonal outdoor seating and entertainment areas shall be included when determining the minimum number of facilities required.

indicate a different distribution of the sexes. The total number of water closets required for males is then calculated by multiplying the male occupant load by the corresponding water closet ratio found in Table 2902.1. The total number of female water closets is determined using the female occupant load and corresponding ratio. Such an application will result in a larger number of water closets required for females (even though the sex distribution is still considered as 50:50). The requirements for male and female water closets are the same in nightclub and restaurant occupancies because a larger number of fixtures is already required for these types of establishments.

When applying ratios listed in the table, any fraction of the quantity of people requires an additional fixture. For example, when the table requires one fixture per 25 people and the building occupant load is 76, a total of four fixtures would be required.

The total number of required fixtures may result in an odd number. Because the fixtures must be equally distributed between the sexes, an additional fixture is required. For example, if an office building requires a total of three water closets, there must be two in the women's room and two in the men's room. When calculations yield a fractional number of required fixtures, the fraction must be rounded up to a whole number. For example, since 4.5 water closets is an impossibility, it is rounded up to 5 water closets.

In a mixed-use building there are two approaches that can be used to determine the required number of plumbing fixtures. Example 1 illustrates the first approach where the number of fixtures is determined for each occupancy separately and any fractional number of fixtures is rounded up to the next whole number for each occupancy. The total number of fixtures re-quired for the mixed-use building is determined by adding up the whole number of fixtures for each occupancy. Example 2 takes a more liberal approach by determining the fractional number of fixtures for each occupancy, adding up the fractional numbers of fixtures and then rounding up the total number of fixtures to the next whole number.

Example 1: A mixed-use building with a Group B (business) occupant load of 538, Group A-3 (libraries, halls, museums) occupant load of 115 and Group S-1 (storage) occupant load of 82.

The categories to apply are Group B, Group A-3 and Group S-1.

Refer to the table for Group B (business):

Number of water closets:

538 occupants

$50 \times \frac{1}{25} = 2$

$(538 - 50) \times \frac{1}{50} = 488 \times \frac{1}{50} = 9.76$

$9.76 + 2 = 11.76 = 12$ water closets

Divide between the sexes:

12 water closets x 50%

= 6 water closets/women

= 6 water closets/men

Number of lavatories:

538 occupants:

$80 \times \frac{1}{40} = 2$

$(538 - 80) \times \frac{1}{80} = 458 \times \frac{1}{80} = 5.7$

$5.7 + 2 = 7.7 = 8$ lavatories

		EXAMPLE 1 TOTAL NUMBER OF REQUIRED FIXTURES (Mixed-Use Building)									
OCCUPANCY	OCCUPANCY LOAD	WATER CLOSETS				LAVATORIES			DRINKING FOUNTAINS	SERVICE SINK
		RATIO	MEN	RATIO	WOMEN	RATIO	MEN	WOMEN		
Business	538	1 per 25 for the first 50 and 1 per 50 for the remainder exceeding 50	6	1 per 25 for the first 50 and 1 per 50 for the remainder exceeding 50	6	1 per 40 for the first 80 and 1 per 80 for the remainder exceeding 80	4	4	6	1
Libraries, Halls, Museums, etc.	115	1/125	1	1/65	1	1/200	1	1	1	1
Storage	82	1/100	1	1/100	1	1/100	1	1	1	1
Total Required			8		8		6	6	8	1

Note that except for hospitals and nursing homes, one service sink is all that the code intends to require in a building where all potential users of the fixture have access to it.

Divide between the sexes:

8 lavatories x 50%

= 4 lavatories/women

= 4 lavatories/men

Number of drinking fountains:

538 x $^1/_{100}$ = 5.38 = 6 drinking fountains

Refer to the table for Group A-3 libraries, halls and museums:

Number of water closets:

115 occupants x 50% = 57.5

57.5 men x $^1/_{125}$ = .46 = 1 water closet (male)

57.5 women x $^1/_{65}$ = .88 = 1 water closet (female)

Number of lavatories:

115 x $^1/_{200}$ = .58 lavatories = 1

Divide between the sexes:

1 lavatory x 50%

= 0.5 lavatory/women = 1

= 0.5 lavatory/men = 1

Number of drinking fountains:

115 x $^1/_{500}$ = .23 drinking fountains = 1

Refer to the table for Group S storage:

Number of water closets:

82 x $^1/_{100}$ = .82 water closet = 1

Divide between the sexes:

1 water closet x 50%

= 0.5 water closet/women = 1

= 0.5 water closet/men = 1

Number of lavatories:

82 x $^1/_{100}$ = .82 lavatory = 1

Divide between the sexes:

1 lavatory x 50%

= 0.5 water closet/women = 1

= 0.5 water closet/men = 1

Number of drinking fountains:

82 x $^1/_{1,000}$ = .08 drinking fountains = 1

Example 2: A mixed-use building Group B (business) with an occupant load of 940 and a Group A-2 (restaurants, banquet halls and food courts) with an occupant load of 360.

Refer to the table for Group B (business):

Divide between the sexes:

940 x 50% = 470 occupants

Use 470 occupants to determine the number of fixtures for males and females:

50 occupants x $^1/_{25}$ = 2 water closets

(470 - 50) x $^1/_{50}$ =

**EXAMPLE 2
TOTAL NUMBER OF REQUIRED FIXTURES
(Mixed-Use Building)**

OCCUPANCY	OCCUPANT LOAD	WATER CLOSETS				LAVATORIES			DRINKING FOUNTAINS	SERVICE SINK
		RATIO	MEN	RATIO	WOMEN	RATIO	MEN	WOMEN		
Business	940	1 per 25 for the first 50 and 1 per 50 for the remainder exceeding 50	10.4	1 per 25 for the first 50 and 1 per 50 for the remainder exceeding 50	10.4	1 per 40 for the first 80 and 1 per 80 for the remainder exceeding 80	6.88	6.88	9.4	1
Restaurants, Banquet Halls and Food Courts	360	1/75	2.4	1/75	2.4	1/200	.9	.9	.72	1
Subtotal		12.8		12.8		7.78		7.78	10.12	1
Total Required		13		13		8		8	11	1

In Example 2, if the fractional number of fixtures for each occupancy were rounded up before adding them together, a total of 14 water closets would be required for both males and females

$420 \times \frac{1}{50} = 8.4$ water closets

$2 + 8.4 = 10.4$ water closets each for males and females

Number of lavatories:

$80 \times \frac{1}{40} = 2$

$(470 - 80) \times \frac{1}{80} =$

$390 \times \frac{1}{80} = 4.88$ lavatories

$2 + 4.88 = 6.88$ lavatories each for males and females

Number of drinking fountains:

$940 \times \frac{1}{100} = 9.4$ drinking fountains

Refer to the table for Group A-2 restaurants, banquet halls and food courts:

Number of occupants for each sex:

$360 \times 50\% = 180$ occupants

Use 180 occupants to determine the number of fixtures for males and females:

$180 \times \frac{1}{75} = 2.4$ water closets each for males and females

Number of lavatories:

$180 \times \frac{1}{200} = .9$ lavatories each for males and females

Number of drinking fountains:

$360 \times \frac{1}{500} = .72$ drinking fountains

Example 3: A stadium has an occupant load of 5,200 seats. What is the minimum number of fixtures required for each sex?

Refer to the table for Group A-5 (stadiums):

Divide occupant load between the sexes:

$5,200$ occupants $\times 50\%$

$= 2,600$/female occupants

$= 2,600$/male occupants

Number of water closets:

$1500 \times \frac{1}{40} = 37.5$

$(2600-1500) \times \frac{1}{60}$

$1100 \times \frac{1}{60} = 18.3$

$37.5 + 18.3 = 55.8 = 56$ water closets (female)

$1500 \times \frac{1}{75} = 20$

$(2600 - 1500) \times \frac{1}{120} = 1100 \times \frac{1}{120} = 9.16$

$20 + 9.16 = 29.16 = 30$ water closets (male)

Number of lavatories:

Divide between the sexes:

$5,200$ occupants $\times 50\%$

$= 2,600$ men

$= 2,600$ women

$2,600 \times \frac{1}{150} = 17.3 = 18$ lavatories (female)

$2,600 \times \frac{1}{200} = 13$ lavatories (Male)

Number of drinking fountains:

$5,200 \times \frac{1}{1,000} = 5.2 = 6$ drinking fountains

Number of service sinks $= 1$

EXAMPLE 3 TOTAL NUMBER OF REQUIRED FIXTURES		
FIXTURE	WOMEN'S	MEN'S
Water closet	56	30
Lavatory	18	13
Drinking fountain	6	
Service sink	1	

The design professional is permitted to substitute urinals for up to 20 water closets (67 percent of 30 water closets). In this example, the designer chooses to install 20 urinals in the men's room. The resulting number of fixtures would change accordingly.

EXAMPLE 3 TOTAL NUMBER OF REQUIRED FIXTURES		
FIXTURE	WOMEN'S	MEN'S
Water closet	56	10
Urinal	—	20
Lavatory	18	13
Drinking fountain	6	
Service sink	1	

In buildings with large areas or multiple stories, the distribution of plumbing fixtures must satisfy the requirements of the occupants. For example, a 20-story high-rise office building cannot have all of the required plumbing fixtures located on the 10th floor. The path of travel to the plumbing fixtures must neither exceed a travel distance of 500 feet (152 400 mm) nor require travel beyond the adjacent story above or below. This travel distance limitation, both total and vertical distance, can result in fixtures not being located on every floor of a multistory building. When determining the number of fixtures required in a multistory building, the entire building occupant load is used. Fixtures are not required based on the occupant load of each floor. The distribution of fixtures, however, must satisfy the demand for each floor. The percentage of the number of fixtures is to be consistent with the percentage of the occupant load they serve.

In determining the minimum required number of plumbing fixtures in a child care facility or elementary school, a design professional might provide child-size fixtures based on the building's use in an attempt to accommodate children; however, this is subject to the approval of the building official. These facilities must be located so that they can be accessed by the public. It is not the intent of the code to prevent toilet facilities from being located in individual classrooms, but these facilities should be in excess of the fixtures required by Table 2902.1.

[P] 2902.1.1 Unisex toilet and bath fixtures. Fixtures located within unisex toilet bathing rooms complying with Section 404 of the *International Plumbing Code* are permitted to be included in determining the minimum required number of fixtures for assembly and mercantile occupancies.

❖ A unisex toilet room and/or bathing room is a special facility as required by Chapter 11 and is intended for use by individuals who need assistance from a family member or a caregiver. This section permits fixtures located in unisex toilet and bathing rooms to count towards the minimum required number for assembly and mercantile occupancies. Such fixtures can reduce the number required in either male or female toilet rooms, but not both (see Sections 1109.2.1 through 1109.2.1.4).

[P] 2902.2 Separate facilities. Where plumbing fixtures are required, separate facilities shall be provided for each sex.

Exceptions:

1. Separate facilities shall not be required for dwelling units and sleeping units.

2. Separate facilities shall not be required in structures or tenant spaces with a total occupant load, including both employees and customers, of 15 or less.

3. Separate facilities shall not be required in mercantile occupancies in which the maximum occupant load is 50 or less.

❖ Separate facilities shall not be required in mercantile occupancies in which the maximum occupant load is 50 or less.

Separate facilities must be provided for each sex, meaning a separate women's room and men's room. If the required total number of fixtures based on Table 2902.1 is one water closet and one lavatory, two of each fixture must be provided. One group of fixtures must be installed in the women's room and one in the men's room.

Single-occupant toilet rooms for use by both sexes are permitted for:

1. Private residential uses, such as within dwelling units and hotel guestrooms;

2. Small establishments in which the occupant load, including customers and employees, is 15 or fewer; and

3. Mercantile occupancies in which the maximum occupant load is 50 or fewer.

Where fixtures are located in a mercantile occupancy, the number of employees is determined by the building owner. In other facilities, such as an office building, the number of employees is determined by the occupant load requirements in the code.

[P] 2902.3 Number of occupants of each sex. The required water closets, lavatories and showers or bathtubs shall be distributed equally between the sexes based on the percentage of each sex anticipated in the occupant load. The occupant load shall be composed of 50 percent of each sex, unless statistical data approved by the building official indicate a different distribution of the sexes.

❖ This section requires an equal number of plumbing fixtures for each sex, except as otherwise required by the ratios in Table 2902.1 for assembly occupancies. Although this section attempts to provide equality between the sexes for other occupancies, it only regulates the minimum required number of fixtures. A design professional may provide additional plumbing fixtures to one of the sexes, such as a greater number of urinals in the men's room.

Another exception to providing an equal division of required fixtures between the sexes is where a building owner or designer documents that the use of the building involves an unequal distribution of the sexes. An example of this type of building use is a women's health club. It would be inappropriate to provide an equal number of men's plumbing fixtures in this building.

[P] 2902.4 Required public toilet facilities. Customers, patrons and visitors shall be provided with public toilet facilities in structures and tenant spaces intended for public utilization. The accessible route to public facilities shall not pass through kitchens, storage rooms, closets or similar spaces. Employees shall be provided with toilet facilities in all occupancies. Employee toilet facilities shall be either separate or combined employee and public toilet facilities.

❖ Public facilities are required in all spaces to be used by the public, including restaurants, nightclubs, places of public assembly and business occupancies open to the public. It is inappropriate to locate them in a storage area, behind the kitchen or in other areas not open and available to the public, primarily for the safety of the occupants that are from the general public. This provision avoids placing occupants from the general public in an unfamiliar area of the building where they could be confused about the path of egress in an emergency. The code requires employees to be provided with toilet facilities in all occupancies and this can be satisfied by separate employee facilities or shared employee/public facilities.

[P] 2902.4.1 Location of toilet facilities in occupancies other than covered malls. In occupancies other than covered malls, the required public and employee toilet facilities shall be located not more than one story above or below the space required to be provided with toilet facilities and the path of travel to such facilities shall not exceed a distance of 500 feet (152 m).

Exception: The location and maximum travel distances to required employee facilities in factory and industrial occupancies are permitted to exceed that required by this section, provided that the location and maximum travel distance are approved.

❖ In order to prevent health problems for the building occupants related to difficulty in accessing the sanitary facilities in a timely fashion, the code clearly provides two conditions that must be met: (1) the required toilet facilities must be located within a travel distance of 500 feet (152 400 mm); and (2) the building occupant must not be required to travel beyond the next adjacent story above or below his or her location. An exception provides for the unique nature of factory and industrial occupancies where the designer is better equipped to determine the location of employee toilet facilities.

[P] 2902.4.2 Location of toilet facilities in covered malls. In covered mall buildings, the required public and employee toilet facilities shall be located not more than one story above or below the space required to be provided with toilet facilities, and the path of travel to such facilities shall not exceed a distance of 300 feet (91 440 mm). In covered mall buildings, the required facilities shall be based on total square footage, and facilities shall be installed in each individual store or in a central toilet area located in accordance with this section. The maximum travel distance to the central toilet facilities in covered mall buildings shall be measured from the main entrance of any store or tenant space. In covered mall buildings, where employees' toilet facilities are not provided in the individual store, the maximum travel distance shall be measured from the employees' work area of the store or tenant space.

❖ In covered malls, the path of travel to required toilet facilities must not exceed a distance of 300 feet (91 440 mm) and is measured from the main entrance of any store or tenant space. As for other occupancies, the facilities must be located no further than the story above or below the occupant's location. This lower travel distance addresses the fact that covered malls are frequently very congested and occupants are unfamiliar with their surroundings. The minimum number of required plumbing facilities is based on total square footage of the mall, including tenant spaces, and must be installed in each individual store or in a central toilet area. This section does not prohibit the installation of separate toilet facilities in individual tenant spaces; however, such facilities are not deductible from the total common facilities required if the facilities in the tenant space are intended for employees only.

[P] 2902.4.3 Pay facilities. Where pay facilities are installed, such facilities shall be in excess of the required minimum facilities. Required facilities shall be free of charge.

❖ Pay facilities have been included in some public areas to prevent vagrants from loitering in the public bathrooms or vandalizing the facilities. This section does not prevent pay facilities from being installed, but they must be in excess of those required in Table 2902.1.

[P] 2902.5 Signage. A legible sign designating the sex shall be provided in a readily visible location near the entrance to each toilet facility. Signs for accessible toilet facilities shall comply with ICC A117.1.

❖ Public facilities must be designated by a legible sign for each sex that is located near the entrance to such facilities. The code is silent on the specifics of the type of designation; therefore, the type of designation is subject to the approval of the building official.

Bibliography

The following resource materials are referenced in this chapter or are relevant to the subject matter addressed in this chapter.

ASPE-95, *A Critique of Queueing Theory Approaches to Plumbing Design.* Westlake, CA: American Society of Plumbing Engineers, 1995.

ICC A117.1-2003, *Accessible and Usable Buildings and Facilities.* Falls Church, VA: International Code Council, 2003.

IPC-2006, *International Plumbing Code.* Falls Church, VA: International Code Council, 2006.

Chapter 30:
Elevators and Conveying Systems

General Comments

Chapter 30 contains the provisions that regulate vertical and horizontal transportation and material-handling systems installed in buildings. The installation must comply with the requirements in this chapter and the standards referenced herein.

Section 3001.1 contains the scope of the chapter.

Section 3001.2 identifies the standards to which elevators and conveying systems must comply.

Section 3001.3 contains requirements for accessible elevators.

Sections 3002.1 through 3002.7 contain the requirements for elevator hoistway enclosures.

Sections 3003.1 through 3003.2 include requirements for emergency operation of elevators.

Sections 3004.1 through 3004.5 identify requirements for hoistway venting.

Sections 3005.1 through 3005.4 address requirements for conveying systems and personnel and material hoists.

Sections 3006.1 through 3006.6 address machine room requirements.

Purpose

The purpose of this chapter is to regulate the installation, testing, inspection, maintenance, alteration and repair of vertical and horizontal transportation and material-handling systems installed in buildings. Compliance with the requirements in this chapter provides for life safety and promotes public welfare. The chapter is also intended to be used as a minimum safety standard by architects, engineers, insurance companies, manufacturers and contractors and as standard safety practice for owners and managers of buildings where equipment covered by this chapter is installed and used.

SECTION 3001
GENERAL

3001.1 Scope. This chapter governs the design, construction, installation, alteration and repair of elevators and conveying systems and their components.

❖ This section indicates that the requirements in this chapter are applicable to the design, construction, installation, alteration and repair of any elevator, conveying system or component thereof. In addition to requirements for design, construction and installation, requirements for alteration, repair, testing and inspections of elevators and conveying systems are located in the standards referenced in this chapter, and are, therefore, not included in the code.

3001.2 Referenced standards. Except as otherwise provided for in this code, the design, construction, installation, alteration, repair and maintenance of elevators and conveying systems and their components shall conform to ASME A17.1, ASME A90.1, ASME B20.1, ALI ALCTV, and ASCE 24 for construction in flood hazard areas established in Section 1612.3.

❖ The enforceability of a standard is established in this section, and applies wherever the provisions of this chapter do not otherwise indicate a requirement. Elevators installed in buildings and structures located in designated flood hazard areas may be subject to additional flood-resistant design and construction requirements. The application of these additional requirements depends on whether the elevator or components of the elevator equipment are to be located below the design flood elevation. Those not familiar with the subtle differences that make a standard applicable to a specific equipment design should review both the standard's scope and definitions. In general, the applications of the referenced standards are as follows:

- ASME A17.1 applies to transportation equipment that is permanently installed to transport people and freight.

- ASME A90.1 applies specifically to belt-type manlifts.

- ASME B20.1 applies to equipment that is permanently installed to transport freight. People are prohibited from riding this equipment.

- ALI ALCTV applies specifically to automotive lifts, such as those used in automated parking structures.

- ASCE 24 applies to the construction of elevators and conveying systems located in flood hazard areas, as established by application of Section 1612.3 of the code. Section 1612.3 requires flood hazard areas to be established by the governing body through the adoption of a flood hazard map, which locates the flood hazard areas within a given jurisdiction. Chapter 8 of ASCE 24-98 and FEMA FIA-TB #4, *Elevator Installation for Buildings Located in Special Flood Hazard Areas* provides extensive guidance on the proper method of installing elevators in buildings and structures located in designated flood hazard areas. In addition to raising elevator service equipment above the design flood elevation, a concern is determining that the emergency operation requirements of Section 3003 do not require the elevator cab to descend automatically to a level below the design flood elevation during conditions of flooding. The safety of any occupants may be jeopardized and the elevator cab may be damaged if it descends below the design flood elevation during flooding.

Referenced standards have a long history of providing minimum safety requirements for the equipment covered by their scope. As an example, ASME A17.1 was first published in 1921 and 14 subsequent editions have been published.

Interpretations of the referenced standards can usually be obtained from their publisher. The committee responsible for developing and maintaining the standard will respond to all interpretation requests. The ASME A17.1, A90.1 and B20.1 standards committees have also published interpretations and include copies of all recent interpretations to purchasers of these documents. The publisher of ASME A17.1 also has available a handbook that explains and augments the requirements. While every effort has been made to eliminate conflicts between the code and the referenced standards, such conflicts may occasionally occur. Where differences occur, the requirements of the code take precedence (see Section 102.4).

3001.3 Accessibility. Passenger elevators required to be accessible by Chapter 11 shall conform to ICC A117.1.

❖ Section 1109.6 requires elevators that form part of an accessible route to be accessible as required by ICC A117.1. ICC A117.1 contains requirements for elevator location and access; operation and leveling; door operation; door size; door protective and reopening devices; door delay (passenger service time) from hall and car calls; car inside dimensions, controls and position indicators and signals; telephone or intercom systems; floor coverings; minimum illumination; hall buttons; hall lanterns; door jamb markings and clearance between sills.

This standard provides acceptable and appropriate service for all users and makes multistory buildings accessible and usable by people with physical disabilities.

3001.4 Change in use. A change in use of an elevator from freight to passenger, passenger to freight, or from one freight class to another freight class shall comply with Part XII of ASME A17.1.

❖ This section requires enforcement of Part XII of ASME A17.1 whenever the elevator changes in the use or

freight class. The application of these requirements results in an elevator that will operate safely and comply with requirements that are unique to and necessary for the new use or freight class.

SECTION 3002
HOISTWAY ENCLOSURES

3002.1 Hoistway enclosure protection. Elevator, dumbwaiter and other hoistway enclosures shall be shaft enclosures complying with Section 707.

❖ To address the migration of smoke and fire between floor levels in a building, this section requires the elevator hoistway to be protected (fire-resistance rated). Reference is made to Section 707 of the code for the required fire resistance and construction of the hoistway enclosure. Section 707 contains fire-resistance rating requirements, as well as construction requirements for shafts, which is typically the method used to satisfy the requirements for a hoistway enclosure. It should be noted that this section of the code does not require hoistway enclosures for all elevators dumbwaiters, etc. The elevator could, in fact be included in a floor opening that is not required to be protected with a shaft. For example, an elevator located totally within an atrium would not need its own enclosure because the floor openings are protected by virtue of the atrium requirements (smoke control, fire suppression system, etc.).

3002.1.1 Opening protectives. Openings in hoistway enclosures shall be protected as required in Chapter 7.

> **Exception:** The elevator car doors and the associated hoistway enclosure doors at the floor level designated for recall in accordance with Section 3003.2 shall be permitted to remain open during Phase I Emergency Recall Operation.

❖ To address the migration of smoke and fire between floor levels in a building, this section requires elevator hoistway openings to be protected (fire-resistance rated). Reference is made to Chapter 7 for protection of openings in hoistway enclosures. Chapter 7 contains minimum hourly rating requirements for openings based on the fire-resistance rating required for the enclosure. For example, if a hoistway enclosure was required to be 2-hour fire-resistance rated, a $1^1/_2$-hour fire door assembly would be required in accordance with Table 715.3.

3002.1.2 Hardware. Hardware on opening protectives shall be of an approved type installed as tested, except that approved interlocks, mechanical locks and electric contacts, door and gate electric contacts and door-operating mechanisms shall be exempt from the fire test requirements.

❖ In order to verify that hardware used with elevator doors will not affect their fire-resistance rating, door hardware not specifically exempt by this section is required to be part of the door assembly during the fire test. The devices that are exempt from the fire test have minimum fire and physical requirements contained in ASME A17.1. Passenger elevator door hardware consists of a header, track hangers, pendant bolts, floor sill with guides, sill support plates, sill brackets, retaining angles and closer assemblies. Each component may bear a separate label, or the assembly may have one label to cover all components. The label should clearly identify the assembly or component to which it applies. Typically, labels are viewed from inside the hoistway.

3002.2 Number of elevator cars in a hoistway. Where four or more elevator cars serve all or the same portion of a building, the elevators shall be located in at least two separate hoistways. Not more than four elevator cars shall be located in any single hoistway enclosure.

❖ The maximum number of elevator cars in a single hoistway is four. This limits the potential of a single hoistway fire incident causing the removal of all elevators from service, particularly in larger structures, such as high-rise buildings.

3002.3 Emergency signs. An approved pictorial sign of a standardized design shall be posted adjacent to each elevator call station on all floors instructing occupants to use the exit stairways and not to use the elevators in case of fire. The sign shall read: IN FIRE EMERGENCY, DO NOT USE ELEVATOR. USE EXIT STAIRS. The emergency sign shall not be required for elevators that are part of an accessible means of egress complying with Section 1007.4.

❖ Elevators are unsafe during a fire because:

- Persons may push a corridor button and have to wait for an elevator that may never respond. Valuable time in which to leave the building safely is lost.

- Elevators cannot start until the car and hoistway doors are closed. A panic could lead to overcrowding of an elevator and blockage of the doors, thus preventing closing.

- Power failure during a fire can happen at any time and thus lead to entrapment.

- Elevators respond to car and corridor calls. One of these calls may be at the fire floor.

Fatal delivery of the elevator to the fire floor can be caused by:

- An elevator passenger pressing the car button for the fire floor.

- One or both of the corridor call buttons being pushed on the fire floor.

- Heat melting or deforming the corridor push button or its wiring at the fire floor.

- Normal functioning of the elevator, such as high-call or low-call reversal, occurring at the fire floor.

In a fire emergency, an occupant will normally try to exit the building by the route he or she entered (elevators). The pictorial sign will indicate to the occupants not to use the elevators, but rather to use the stairs to exit the building.

The exception to this requirement recognizes that elevators can be an integral part of the means of egress required for persons with physical disabilities. In such cases, the sign required by this section would not be appropriate since it would be communicating a contradictory message. Section 1007.4 dictates specific protection methods to permit the use of elevators as an element of an accessible means of egress. See Figure 3002.3 for an example of an emergency pictorial sign.

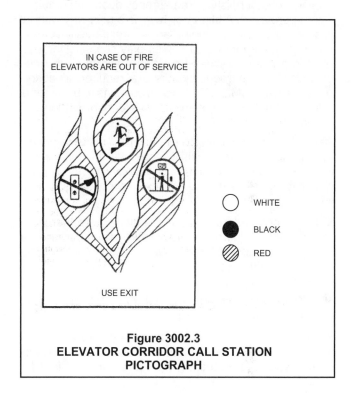

Figure 3002.3
ELEVATOR CORRIDOR CALL STATION PICTOGRAPH

3002.4 Elevator car to accommodate ambulance stretcher. Where elevators are provided in buildings four or more stories above grade plane or four or more stories below grade plane, at least one elevator shall be provided for fire department emergency access to all floors. The elevator car shall be of such a size and arrangement to accommodate a 24-inch by 84-inch (610 mm by 2250 mm) ambulance stretcher in the horizontal, open position and shall be identified by the international symbol for emergency medical services (star of life). The symbol shall not be less than 3 inches (76 mm) high and shall be placed inside on both sides of the hoistway door frame.

❖ To aid fire-fighting rescue efforts in taller buildings (over three stories above grade), one of the provided elevators must be sized to accommodate a typical ambulance stretcher and emergency personnel. Additionally, to aid the emergency personnel in identifying which elevator is sized to accommodate a stretcher, the international symbol (star of life) is required to be located on each side of the hoistway door frame. The size of the ambulance stretcher is based upon the typical size of a stretcher commonly used by fire departments in the U.S. It is necessary to have the patient in a horizontal position on the elevator to avoid further injury to the patient and to facilitate many of the medical tasks being performed by the paramedics while on the elevator.

3002.5 Emergency doors. Where an elevator is installed in a single blind hoistway or on the outside of a building, there shall be installed in the blind portion of the hoistway or blank face of the building, an emergency door in accordance with ASME A17.1.

❖ To allow emergency personnel reasonable access to a blind elevator hoistway, emergency doors are required to be provided. A blind hoistway is a hoistway without openings. For example, an express elevator that opens at the lobby and serves only floors 10 through 20 is a blind hoistway from floors two through nine. Therefore, emergency doors are required in accordance with ASME A17.1, which contains requirements for opening sizes, opening access, door operation and door location.

3002.6 Prohibited doors. Doors, other than hoistway doors and the elevator car door, shall be prohibited at the point of access to an elevator car unless such doors are readily openable from the car side without a key, tool, special knowledge or effort.

❖ This section addresses the situation where there is a door installed in front of the hoistway door, perhaps for security purposes. If that door is locked, a building occupant could become trapped in the area between the set of doors.

3002.7 Common enclosure with stairway. Elevators shall not be in a common shaft enclosure with a stairway.

❖ To avoid the potential of a fire hazard spreading from the elevator or elevator hoistway and affecting an exit stair, separate protective enclosures are required for each.

3002.8 Glass in elevator enclosures. Glass in elevator enclosures shall comply with Section 2409.1.

❖ The purpose of this section is simply to remind the code user that Section 2409.1 contains provisions regarding glass in elevator enclosures. Please see the commentary for Section 2409.1. It should be noted that other provisions of the code may also be applicable to the glass, such as Section 2406, when human impact loads are possible.

[F] SECTION 3003
EMERGENCY OPERATIONS

[F] 3003.1 Standby power. In buildings and structures where standby power is required or furnished to operate an elevator, the operation shall be in accordance with Sections 3003.1.1 through 3003.1.4.

❖ Elevators are the most common and convenient means of providing access in multistory buildings. As such, elevators represent a prime candidate for acces-

sible means of egress from such buildings, especially in light of the difficulties involved in carrying a person in a wheelchair up or down a stairway. For this reason, Chapter 10 requires elevators, which are used to provide an accessible means of egress, to be provided with standby power in accordance with this section. The primary consideration for elevators as an accessible means of egress is that the elevator most likely will be available and protected during a fire event. This section addresses the standby power and operation requirements for such an elevator. The emergency power requirement establishes a higher degree of reliability that the elevator will be available and usable by reducing the likelihood of power loss caused by fire or other conditions.

[F] 3003.1.1 Manual transfer. Standby power shall be manually transferable to all elevators in each bank.

❖ To allow emergency personnel to provide standby power to any elevator(s) at any time, this section requires a manual means of transferring standby power to all elevators in each bank. This manual transfer system could assist emergency personnel in evacuating building occupants.

[F] 3003.1.2 One elevator. Where only one elevator is installed, the elevator shall automatically transfer to standby power within 60 seconds after failure of normal power.

❖ In the case where a building is equipped only with one elevator, a standby power system must automatically restore full operation of the elevator within 60 seconds after failure of the primary power supply.

[F] 3003.1.3 Two or more elevators. Where two or more elevators are controlled by a common operating system, all elevators shall automatically transfer to standby power within 60 seconds after failure of normal power where the standby power source is of sufficient capacity to operate all elevators at the same time. Where the standby power source is not of sufficient capacity to operate all elevators at the same time, all elevators shall transfer to standby power in sequence, return to the designated landing and disconnect from the standby power source. After all elevators have been returned to the designated level, at least one elevator shall remain operable from the standby power source.

❖ In the case where a building is equipped with two or more elevators, this section provides allowable methods of standby power operation. When the provided standby power system is sized to accommodate all of the elevators, then all are required to automatically switch to standby power 60 seconds after failure of the primary power supply. There are provisions for a standby power system to be sized smaller and provide standby power to the elevators in sequence, rather than all at once. However, after all elevators have been returned to their designated landings by standby power, at least one must remain operable through standby power.

[F] 3003.1.4 Venting. Where standby power is connected to elevators, the machine room ventilation or air conditioning shall be connected to the standby power source.

❖ To allow elevator equipment in the machine room to operate normally, the venting system serving the machine room must be connected to the standby power system that serves the elevators.

[F] 3003.2 Fire-fighters' emergency operation. Elevators shall be provided with Phase I emergency recall operation and Phase II emergency in-car operation in accordance with ASME A17.1.

❖ A fire-fighter service is required for all elevators. There are two objectives of fire-fighter service required by ASME A17.1:

- Phase I emergency recall operation is the operation of an elevator wherein it is automatically or manually recalled to a specific landing and removed from normal service because of activation of the fire-fighter service; and

- Phase II emergency in-car operation is the operation of an elevator by fire fighters where such operation is under their control.

Elevators need to have recall capability to reduce the possibility of trapping passengers during a fire. Elevators need to have Phase II emergency in-car operation capability to aid fire fighters in the evacuation of persons with physical disabilities. An elevator needs to provide a usable means of egress for the disabled in a fire emergency.

SECTION 3004
HOISTWAY VENTING

3004.1 Vents required. Hoistways of elevators and dumbwaiters penetrating more than three stories shall be provided with a means for venting smoke and hot gases to the outer air in case of fire.

Exceptions:

1. In occupancies of other than Groups R-1, R-2, I-1, I-2 and similar occupancies with overnight sleeping quarters, venting of hoistways is not required where the building is equipped throughout with an approved automatic sprinkler system installed in accordance with Section 903.3.1.1 or 903.3.1.2.

2. Sidewalk elevator hoistways are not required to be vented.

❖ Ventilation of hoistways is required to prevent the accumulation and spread of hot smoke and gases from a fire to the upper stories of a building. The majority of deaths in fires are a result of smoke. For example, in the fire at the MGM Grand Hotel in Las Vegas, 70 of the 84 deaths occurred on the upper floors where smoke concentration was the greatest—even though the fire was at the first floor. This could be attributed to stack effect, which is a phenomenon that exists in high-rise buildings. Stack effect describes the movement of air inside and outside a building. During a fire, the presence of stack effect generally results in the movement of smoke and combustion products from lower levels to upper levels through shafts in the building. Stack effect can be reversed in an air-conditioned building when the outside temperature is relatively high (see Figure 3004.1).

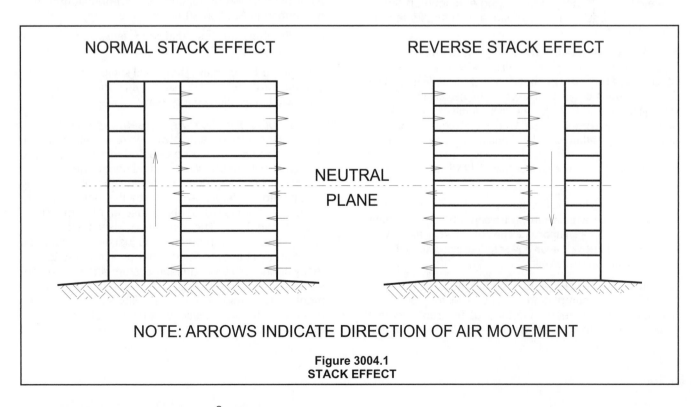

NORMAL STACK EFFECT   REVERSE STACK EFFECT

NEUTRAL PLANE

NOTE: ARROWS INDICATE DIRECTION OF AIR MOVEMENT

Figure 3004.1
STACK EFFECT

Under the conditions listed in Exception 1, sprinklers are permitted as an alternative to the vents in all buildings, except those of the groups noted. The top of sidewalk elevator hoistways is open to the outside; thus, no venting is required as indicated in Exception 2.

3004.2 Location of vents. Vents shall be located at the top of the hoistway and shall open either directly to the outer air or through noncombustible ducts to the outer air. Noncombustible ducts shall be permitted to pass through the elevator machine room, provided that portions of the ducts located outside the hoistway or machine room are enclosed by construction having not less than the fire protection rating required for the hoistway. Holes in the machine room floors for the passage of ropes, cables or other moving elevator equipment shall be limited so as not to provide greater than 2 inches (51 mm) of clearance on all sides.

❖ The vent location is intended to aid in the exhaust of smoke to the exterior of the building and to limit migration of smoke into the machine room. At one time, hoistways were vented into the machine room, and a grate in the machine room floor vented to the exterior of the building. Allowing smoke and heat into the machine room may cause equipment malfunction. Therefore, vents must be located in the hoistway directly below the ceiling, and if ventilation through the machine room is advantageous by design, it must be provided through the use of noncombustible ductwork. Further, when noncombustible ductwork is utilized for the ventilation system, any portion of the ductwork that is outside of the hoistway or the machine room must be enclosed by fire-resistance-rated construction equivalent to that required for the hoistway. The clearance around elevator ropes and other moving equipment is limited in size in order to limit the passage of smoke into the machine room.

3004.3 Area of vents. Except as provided for in Section 3004.3.1, the area of the vents shall be not less than $3^1/_2$ percent of the area of the hoistway nor less than 3 square feet (0.28 m²) for each elevator car, and not less than $3^1/_2$ percent nor less than 0.5 square feet (0.047 m²) for each dumbwaiter car in the hoistway, whichever is greater. Of the total required vent area, not less than one-third shall be permanently open. Closed portions of the required vent area shall consist of openings glazed with annealed glass not greater than 0.125 inch (3.2 mm) in thickness.

Exception: The total required vent area shall not be required to be permanently open where all the vent openings automatically open upon detection of smoke in the elevator lobbies or hoistway, upon power failure and upon activation of a manual override control.

❖ Vent sizes are calculated such that there will not be a significant accumulation of smoke at the top of the hoistway. In general, one-third of the required vent area needs to be permanently open to ensure that

smoke can escape if the openable portions of the vents would not open for some reason. The exception allows that the vents may all be closed, provided that they open automatically upon activation of any of the elevator lobby smoke detectors. This will provide a higher degree of energy conservation while maintaining the ability to exhaust smoke.

3004.3.1 Reduced vent area. Where mechanical ventilation conforming to the *International Mechanical Code* is provided, a reduction in the required vent area is allowed provided that all of the following conditions are met:

1. The occupancy is not in Group R-1, R-2, I-1 or I-2 or of a similar occupancy with overnight sleeping quarters.

2. The vents required by Section 3004.2 do not have outside exposure.

3. The hoistway does not extend to the top of the building.

4. The hoistway and machine room exhaust fan is automatically reactivated by thermostatic means.

5. Equivalent venting of the hoistway is accomplished.

❖ If a mechanical means of ventilation is provided to the hoistway, then a reduction in the vent area is permitted. This option is not applicable to buildings with sleeping quarters, because mechanical systems are subject to malfunctioning, which could allow smoke to infiltrate areas in which occupants are sleeping. The vent arrangement is to be such that the vents will not be exposed to the elements, and thus not subject to malfunctioning due to weather exposure. Usually, this can be accomplished through the use of louvers or hoods. Engineering calculations should be provided as substantiation of venting that is equivalent to that prescribed in Section 3004.3.

3004.4 Plumbing and mechanical systems. Plumbing and mechanical systems shall not be located in an elevator shaft.

Exception: Floor drains, sumps and sump pumps shall be permitted at the base of the shaft provided they are indirectly connected to the plumbing system.

❖ If a pipe or duct that conveys gases, vapors or liquids, and passes through a hoistway were to fail, operation of the equipment could be affected in such a manner as to make it unsafe. Water could cause the malfunctioning of hoistway door interlocks, thereby allowing an elevator to leave a floor with the hoistway door open. A wet brake would also cause an extremely hazardous condition. The exception allows for a floor drain, sump or sump pump to be installed at the base of the shaft. However, the discharge is required to indirectly connect to the plumbing system and not be connected directly to any drain. The intent of this requirement is to eliminate the possibility of sewer gases entering into the hoistway, which could cause an explosion hazard.

SECTION 3005
CONVEYING SYSTEMS

3005.1 General. Escalators, moving walks, conveyors, personnel hoists and material hoists shall comply with the provisions of this section.

❖ Conveying systems include escalators, moving walks, conveyors and personnel and material hoists. Section 3005 contains requirements specific to these types of conveying systems.

3005.2 Escalators and moving walks. Escalators and moving walks shall be constructed of approved noncombustible and fire-retardant materials. This requirement shall not apply to electrical equipment, wiring, wheels, handrails and the use of $^1/_{28}$-inch (0.9 mm) wood veneers on balustrades backed up with noncombustible materials.

❖ The intent of this requirement is to limit the fire fuel loading. The presence of large amounts of combustible material would create a large fuel load, and result in an increased migration of fire and smoke from floor to floor, or from space to space.

3005.2.1 Enclosure. Escalator floor openings shall be enclosed with shaft enclosures complying with Section 707.

❖ Escalator floor openings are required to be enclosed in shaft enclosures unless they are constructed in accordance with Section 707.2, Exception 2, for the enclosure requirements of escalators. Exception 2 gives two methods of protection that would allow the floor opening without a shaft enclosure.

The first method is to provide draft curtains and sprinklers. The area of the floor opening is also limited to twice the area of the escalator. This is necessary since the protection method is not adequate for every size opening, especially when the floor area within the perimeter of the opening is used for stock storage, displays, etc. (the atrium provisions of Section 404 are more appropriate for those use conditions). The use of this method is acceptable for connecting an unlimited number of stories in buildings and occupancies of Groups B and M. In buildings other than Groups B and M, the alternative can only be utilized to connect four stories. This provision acknowledges the current uses of this concept but also limits its use in buildings with higher fuel load and more hazardous occupancy characteristics, such as sleeping occupancies (Groups I and R).

The sprinkler draft curtain method is detailed in NFPA 13. The method consists of surrounding the escalator opening, in an otherwise fully sprinklered building, with an 18-inch-deep (457 mm) draft curtain located on the underside of the floor to which the escalator or moving walk ascends. This serves to delay the heat, smoke and combustion gases (developed in the early stages of a fire) on that floor from entering into the escalator or moving walk well. A row of closely spaced automatic sprinklers located outside of the draftstop also surrounds the opening. When activated by heat, the sprinklers provide a water curtain. A typical installation is shown in Figure 3005.2.1. In combination with the sprinkler system in the building, this system delays fire spread to allow time for evacuation.

The second method is to provide power-operated automatic fire shutters at all floor openings. The rolling shutter completely closes off the opening between floors in the event of a fire. When the shutter is actuated, power to the escalator or moving walk must be automatically disconnected. To provide protection

For SI: 1 inch = 25.4 mm, 1 foot = 304.8 mm.

Figure 3005.2.1
DRAFT AND SPRINKLER WATER CURTAIN

against entrapment of an occupant, the shutter is required to have a sensitive leading edge such that it stops closing when it comes in contact with an obstacle. Once the obstacle is removed, the shutter continues to close.

3005.2.2 Escalators. Where provided in below-grade transportation stations, escalators shall have a clear width of 32 inches (815 mm) minimum.

Exception: The clear width is not required in existing facilities undergoing alterations.

❖ A minimum 32-inch (864 mm) width for escalators servicing below-grade transportation systems, such as subways, is based on minimum accessibility requirements.

3005.3 Conveyors. Conveyors and conveying systems shall comply with ASME B20.1.

❖ ASME B20.1 applies to the design, construction, installation, maintenance, inspection and operation of conveyors and conveying systems in relation to hazards. The conveyors may be of bulk material, package or unit-handling types where the installation is designed for permanent, temporary or portable operation. This standard must apply, along with the requirements noted below, to all conveyor installations.

This standard specifically excludes any conveyor designed, installed or used primarily for the movement of human beings. It does, however, apply to certain conveying devices that have incorporated platforms or control rooms specifically designed for authorized operating personnel within their supporting structure.

This standard does not apply to equipment such as industrial trucks, tractors and trailers, tiering machines (except pallet load tierers), cranes, hoists, power shovels, power scoops, bucket drag lines, trenchers, elevators, manlifts, moving walks, escalators, highway or rail vehicles, cableways, tramways or pneumatic conveyors.

3005.3.1 Enclosure. Conveyors and related equipment connecting successive floors or levels shall be enclosed with shaft enclosures complying with Section 707.

❖ Protection against the migration of smoke and fire must be maintained whenever a conveyor penetrates a floor/ceiling assembly, and thus connects floor levels. This section references Section 707 for the requirements for shaft enclosures.

3005.3.2 Conveyor safeties. Power-operated conveyors, belts and other material-moving devices shall be equipped with automatic limit switches which will shut off the power in an emergency and automatically stop all operation of the device.

❖ Conveyor controls must be arranged such that in an emergency, the conveyor will stop all operation. All safety devices, including wiring of electrical safety devices, are required to be arranged to operate in a "fail-safe" manner; that is, if power failure or failure of the device itself occurs, a hazardous condition must not result.

3005.4 Personnel and material hoists. Personnel and material hoists shall be designed utilizing an approved method that accounts for the conditions imposed during the intended operation of the hoist device. The design shall include, but is not limited to, anticipated loads, structural stability, impact, vibration, stresses and seismic restraint. The design shall account for the construction, installation, operation and inspection of the hoist tower, car, machinery and control equipment, guide members and hoisting mechanism. Additionally, the design of personnel hoists shall include provisions for field testing and maintenance which will demonstrate that the hoist device functions in accordance with the design. Field tests shall be conducted upon the completion of an installation or following a major alteration of a personnel hoist.

❖ In lieu of a referenced standard, personnel and material hoists are to be designed and installed using a method that includes structural stability, anticipated loads, impact vibration, stresses and seismic restraint. The building official should examine construction, installation, operation of the hoist tower, etc., for approval of the design.

Personnel hoist design must also include field tests upon completion of installation or major alteration and maintenance, which demonstrate that hoist device functions are in accordance with the design.

SECTION 3006
MACHINE ROOMS

3006.1 Access. An approved means of access shall be provided to elevator machine rooms and overhead machinery spaces.

❖ A means of access to the elevator machine room is required by this section and must be specifically approved. Things to consider in providing for this access are the maintenance and replacement of the equipment therein, and the necessary inspections to be performed.

3006.2 Venting. Elevator machine rooms that contain solid-state equipment for elevator operation shall be provided with an independent ventilation or air-conditioning system to protect against the overheating of the electrical equipment. The system shall be capable of maintaining temperatures within the range established for the elevator equipment.

❖ Elevator machine rooms may contain equipment that is temperature sensitive. If this equipment is not kept within its safe-temperature range, the possibility exists for a malfunction, which could trap an occupant or occupants within the elevator. Additionally, if this malfunction occurred during an emergency condition, it could impede emergency personnel in their efforts to perform fire-fighting or rescue duties.

3006.3 Pressurization. The elevator machine room serving a pressurized elevator hoistway shall be pressurized upon activation of a heat or smoke detector located in the elevator machine room.

❖ To prevent the migration of smoke and heat into a machine room from a pressurized hoistway enclosure, the machine room is required to be pressurized as well. However, the machine room only needs to be pressurized when heat or smoke is present. The pressurization is required to occur upon activation of a smoke or heat detector located in the machine room.

3006.4 Machine rooms and machinery spaces. Elevator machine rooms and machinery spaces shall be enclosed with fire barriers complying with Section 706 or horizontal assemblies complying with Section 711 having a fire-resistance rating not less than the required rating of the hoistway enclosure served by the machinery. Openings shall be protected with assemblies having a fire-protection rating not less than that required for the hoistway enclosure doors.

❖ This section acknowledges that there will inherently be openings between the hoistway and the machine room. This protection is required to be both vertical and horizontal; therefore, to maintain the fire-resistance rating required for the hoistway, the machine room must be protected with fire barrier and horizontal assemblies, both with opening protectives.

3006.5 Shunt trip. Where elevator hoistways or elevator machine rooms containing elevator control equipment are protected with automatic sprinklers, a means installed in accordance with NFPA 72, Section 3-9.4, Elevator Shutdown, shall be provided to disconnect automatically the main line power supply to the affected elevator prior to the application of water. This means shall not be self-resetting. The activation of sprinklers outside the hoistway or machine room shall not disconnect the main line power supply.

❖ This section acknowledges the hazards of a sprinkler system, mainly electrical malfunction, to an operating elevator and elevator equipment contained in the machine room. Therefore, the main power supply line to the affected elevator(s) is required to automatically disconnect, and not be reset, before the suppression system is activated.

3006.6 Plumbing systems. Plumbing systems shall not be located in elevator equipment rooms.

❖ If a pipe or duct that conveys gases, vapors or liquids and passes through a machine room or machinery space were to fail, the operation of the equipment could be affected in such a manner as to make it unsafe. Water could cause the malfunctioning of hoistway door interlocks, thereby allowing an elevator to leave a floor with the hoistway door open.

Bibliography

The following resource materials are referenced in this chapter or are relevant to the subject matter addressed in this chapter.

ALI ALCTV-98, *Standard for Automobile Lifts—Safety Requirements for the Construction, Testing and Validation.* Indialantic, FL: Automotive Lift Institute, 1998.

ASME A17.1-04, *Safety Code for Elevators and Escalators—with A17.1a-2004 Addenda and A17.1S Sup-*

plement. New York: American Society of Mechanical Engineers, 2004.

ASME A90.1-04, *Safety Standard for Belt Manlifts.* New York: American Society of Mechanical Engineers, 2003.

ASME B20.1 2003, *Safety Standard for Conveyors and Related Equipment.* New York: American Society of Mechanical Engineers, 2003.

FEMA FIA-TB #4, *Elevator Installation for Buildings Located in Special Flood Hazard Areas.* Washington, D.C.: Federal Emergency Management Agency.

ICC A117.1-2003, *Accessible and Usable Buildings and Facilities.* Falls Church, VA: International Code Council, 2003.

NFIP Technical Bulletin Series. Washington, DC: National Flood Insurance Program. [Online]. Available: *http://www.fema.gov/MIT/techbul.htm*

NFPA 13-02, *Installation of Sprinkler Systems.* Quincy, MA: National Fire Protection Association, 2002.

NFPA 72-02, *National Fire Alarm Code.* Quincy, MA: National Fire Protection Association, 2002.

Chapter 31:
Special Construction

General Comments

Chapter 31 contains provisions that govern the construction and protection of structures with unique characteristics not addressed by other sections of the code and that require special consideration. This chapter includes requirements for outdoor signs; membrane structures; temporary structures; pedestrian walkways; radio and television towers; radio and television antennas; canopies and awnings; marquees and swimming pool enclosures.

Section 3101 addresses the scope of the chapter.

Section 3102 regulates membrane structures.

Section 3103 regulates structures that are erected for a period of less than 180 days.

Section 3104 regulates pedestrian walkways between buildings.

Section 3105 regulates the construction of awnings and canopies.

Section 3106 regulates the construction of marquees.

Section 3107 addresses the design, construction and maintenance of projecting signs.

Section 3108 regulates the construction of towers and antennas that are used for radio and television transmission.

Section 3109 regulates swimming pool enclosures.

The provisions of Chapter 31 are to be considered as additional requirements to those contained elsewhere in the code. These provisions may alter the general requirements found in other locations, but are only applicable to the specific element regulated therein. For example, the provisions for temporary construction regulate how the structure is to be built, but do not address means of egress. As such, the provisions of Chapter 10 are applicable.

The regulations contained in this chapter for outdoor signs were prompted by improper design, erection, placement, location and maintenance. Improperly designed and installed signs can pose a hazard to people and property. Improperly wired electrical signs that have not been maintained can cause fires.

The criteria in this chapter are straightforward and, in turn, lend themselves to straightforward enforcement.

Purpose

The purpose of Chapter 31 is to locate in one place the provisions that regulate construction or protection requirements for structures having unique characteristics, such as structures that are unusually tall and not used for human occupancy, structures occupied for short periods of time or recreational structures (swimming pools).

This chapter also provides the owner, designer and installer of outdoor signs with the minimum safe practices presented in performance language whenever possible, thus allowing the widest possible application.

SECTION 3101
GENERAL

3101.1 Scope. The provisions of this chapter shall govern special building construction including membrane structures, temporary structures, pedestrian walkways and tunnels, awnings and canopies, marquees, signs, and towers and antennas.

❖ This section makes it clear that the requirements of this chapter are to be used in conjunction with the general requirements of the code. When a specific provision is contained in Chapter 31, it will take precedence over the general requirement.

SECTION 3102
MEMBRANE STRUCTURES

3102.1 General. The provisions of this section shall apply to air-supported, air-inflated, membrane-covered cable and membrane-covered frame structures, collectively known as membrane structures, erected for a period of 180 days or longer. Those erected for a shorter period of time shall comply with the *International Fire Code*. Membrane structures covering water storage facilities, water clarifiers, water treatment plants, sewage treatment plants, greenhouses and similar facilities not used for human occupancy, are required to meet only the requirements of Sections 3102.3.1 and 3102.7.

❖ This section applies to membrane structures that are erected for a period of 180 days or longer. Some membrane structures are of the inflated type, where the air pressure is within a series of tubes, which then form the structure and keep it upright. In other structures, the interior atmosphere is pressurized and cables may be used to restrain the fabric membrane. Other types include cable-supported membrane structures in which the roof element is supported by cables and air pressure may or may not be provided. These are commonly used in sports stadiums. Membrane structures may also be framework-type structures in which the membrane is stretched over the frame to form the enclosing walls and the roof. Finally, membrane structures may consist of a membrane tensioned to provide weather protection without enclosing walls.

The history of regulating membrane structures dates back to 1944, when efforts were first made to control circus tent flammability after a circus fire in Hartford, Connecticut claimed 168 lives. Early regulations required that the tent fabric be flame resistant, that adequate exiting from both the grandstands and the tent be provided and that sufficient clear space around the tent be provided to control potential ignition sources and to permit rapid emergency exiting.

Large permanent membrane structures began to see significant use in the United States in the early 1960s. Prior to that time, the only membrane structures in common usage were air-inflated and air-supported covers for tennis courts, swimming pools, etc. Because of fire prevention code requirements at that time, such structures were often restricted to temporary use, which was defined as less than 90 days.

The reference to 180 days in Section 3102.1 is intended to establish membrane structures complying with this section as being outside the scope and provisions for temporary structures.

This section includes provisions for construction types, height and area limitations and structural support. Otherwise, the other applicable sections of the code apply.

So that a membrane structure will continue to provide its intended function, it must undergo continuous maintenance. Fire safety of such structures and the premises is regulated by the *International Fire Code®* (IFC®).

This section applies to all membrane structures, regardless of the supporting mechanisms or structure. Membrane structures erected for a period of less than 180 days are required to comply with the provisions of the IFC. Membrane structures that are not intended for human occupancy need only comply with the provisions of Section 3102.3.1 or 3102.7.

3102.2 Definitions. The following words and terms shall, for the purposes of this section and as used elsewhere in this code, have the meanings shown herein:

❖ Definitions of terms that are associated with the content of this section are contained herein. These definitions can help in the understanding and application of the code requirements. It is important to emphasize that these terms are not exclusively related to this section, but are applicable everywhere the term is used in the code. The purpose for including these definitions within this section is to provide more convenient access to them without having to refer back to Chapter 2.

For convenience, these terms are also listed in Chapter 2 with a cross reference to this section.

The use and application of all defined terms, including those defined herein, are set forth in Section 201.

AIR-INFLATED STRUCTURE. A building where the shape of the structure is maintained by air pressurization of cells or tubes to form a barrel vault over the usable area. Occupants of such a structure do not occupy the pressurized area used to support the structure.

❖ This type of membrane structure is characterized by multiple layers of membrane arranged such that small pockets or cells are formed. These pockets or cells are pressurized with air and form the structural capability of the membrane structure. Note that the occupants of the structure are not subjected to the pressurized areas, as they are in air-supported membrane structures.

AIR-SUPPORTED STRUCTURE. A building wherein the shape of the structure is attained by air pressure and occupants of the structure are within the elevated pressure area. Air-supported structures are of two basic types:

❖ An air-supported structure identifies those membrane structures that are completely pressurized for the purposes of supporting the membrane covering. Most "domed" sports arenas use air pressure within the structure to support the membrane covering. The membrane covering can consist of one layer or multiple layers; thus, air-supported structures are classified as either "single skin" or "double skin."

Double skin. Similar to a single skin, but with an attached liner that is separated from the outer skin and provides an airspace which serves for insulation, acoustic, aesthetic or similar purposes.

❖ A double-skin, air-supported structure contains multiple layers of membrane sheathing. The membranes are usually separated by enough distance to allow for pressurized air or other materials to be inserted between the plies. The pressurized air or other materials usually serve to increase the insulating and acoustical properties.

Single skin. Where there is only the single outer skin and the air pressure is directly against that skin.

❖ A single-skin, air-supported structure consists of just one membrane covering that is directly supported by the interior pressurized air. No other membranes are provided for insulating or acoustical purposes. If the membrane covering consists of several laminated plies, such an arrangement is still considered a single-skin, air-supported structure.

CABLE-RESTRAINED, AIR-SUPPORTED STRUCTURE. A structure in which the uplift is resisted by cables or webbings which are anchored to either foundations or dead men. Reinforcing cable or webbing is attached by various methods to the membrane or is an integral part of the membrane. This is not a cable-supported structure.

❖ This definition establishes a variation of the air-supported membrane structure. A single-skin or double-skin membrane is still pressurized by interior air, but does not have the strength or tearing resistance to support the air pressure load. A "fishnet" system of cable wires is placed over the exterior surface of the membrane or is integrally woven into the membrane material. The "fishnet" system is then tied to exterior walls of the building or other structural supports to resist the pressurized air loads.

MEMBRANE-COVERED CABLE STRUCTURE. A nonpressurized structure in which a mast and cable system provides support and tension to the membrane weather barrier and the membrane imparts stability to the structure.

❖ This definition identifies a structure in which the membrane fabric is draped over a cable framework. These structures do not involve any interior air pressurization. The cable framework system is supported by a mast or other tower system. The cables are usually pretensioned to provide structural support for the rest of the exterior structure.

MEMBRANE-COVERED FRAME STRUCTURE. A nonpressurized building wherein the structure is composed of a rigid framework to support a tensioned membrane which provides the weather barrier.

❖ This type of membrane structure is the same as the membrane-covered cable structure, except that a trussed framework constructed of structural shapes provides the support instead of pretensioned cables. Again, no air pressurization is required or provided.

NONCOMBUSTIBLE MEMBRANE STRUCTURE. A membrane structure in which the membrane and all component parts of the structure are noncombustible.

❖ Any membrane structure that is constructed of all noncombustible materials is classified by this definition. To qualify as noncombustible, the materials, including the membrane itself, must be tested in accordance with ASTM E 136 and must satisfy the criteria in Section 703.4.

3102.3 Type of construction. Noncombustible membrane structures shall be classified as Type IIB construction. Noncombustible frame or cable-supported structures covered by an approved membrane in accordance with Section 3102.3.1 shall be classified as Type IIB construction. Heavy timber frame-supported structures covered by an approved membrane in accordance with Section 3102.3.1 shall be classified as Type IV construction. Other membrane structures shall be classified as Type V construction.

Exception: Plastic less than 30 feet (9144 mm) above any floor used in greenhouses, where occupancy by the general public is not authorized, and for aquaculture pond covers is not required to meet the fire propagation performance criteria of NFPA 701.

❖ Because of the limited fire-resistance capabilities of the membrane, membrane structures are usually classified as Type IIB, IV or V. These are defined as follows:

Type IIB—Membrane structures in which the membrane is noncombustible as defined by Section 703.4; or membrane-covered structures in which the frame or cable is noncombustible and the membrane is either noncombustible or flame resistant.

Type IV—Membrane-covered structures in which the frame is made from heavy timber and the membrane is noncombustible or flame resistant.

Type V—Membrane structures in which the membrane is flame resistant and combustible; or membrane-covered structures in which the frame or cable is combustible and the membrane is noncombustible or flame resistant.

In accordance with Section 602.2, a membrane-covered structure is to be considered as Type I or II construction, provided that the frame or cable complies with the provisions for Type I or II construction and the membrane structure is noncombustible, used exclusively as a roof and located more than 20 feet (6096 mm) above any floor, balcony or gallery floor level. This is similar to the provisions in Table 601 that do not require fire-resistance-rated roof construction under similar conditions.

The exception indicates that in recent years it has been common practice to build greenhouses with higher ridge lines. Since this increase in height, there has been no record of related property loss.

3102.3.1 Membrane and interior liner material. Membranes and interior liners shall be either noncombustible as set forth in Section 703.4 or meet the fire propagation performance criteria of NFPA 701 and the manufacturer's test protocol.

Exception: Plastic less than 20 mil (0.5 mm) in thickness used in greenhouses, where occupancy by the general public is not authorized, and for aquaculture pond covers is not required to meet the fire propagation performance criteria of NFPA 701.

❖ The membrane is a thin, flexible, impervious material capable of being supported by an air pressure of $1^1/_2$ inches (38.1 mm) of water column. Membranes may be noncombustible. Except for limited applications of plastic greenhouses and pond covers, combustible membranes are required to be flame resistant.

Noncombustible membranes and liner materials must meet the criteria of Section 703.4, which references ASTM E 136.

Materials that are not noncombustible are required to be flame resistant. Flame resistance is determined by testing the membrane in accordance with NFPA 701.

Although typically not applicable to membrane structures, it should be noted that recent tests have indicated a problem with combining two fabrics that individually passed the test procedure, but in combination would not be acceptable; therefore, if multiple layers are used, testing should demonstrate acceptable performance for the material combination to be utilized.

Both greenhouses that are not open to the public and aquaculture pond covers may be of plastic that is less than 20 mil (500 mm) in thickness and need not be flame resistant. This provision for greenhouses is based on the limited number of occupants.

3102.4 Allowable floor areas. The area of a membrane structure shall not exceed the limitations set forth in Table 503, except as provided in Section 506.

❖ The allowable area per floor of a membrane structure is determined using Sections 503 and 506, based on the construction types established in Section 3102.3.

3102.5 Maximum height. Membrane structures shall not exceed one story nor shall such structures exceed the height limitations in feet set forth in Table 503.

Exception: Noncombustible membrane structures serving as roofs only.

❖ Membrane structures are most commonly used to enclose large, open areas without intermediate structural supports, such as columns. For this reason, as well as the engineering difficulties associated with multistory buildings, membrane structures are limited to one story and the maximum height permitted in Table 503. The height is not limited, however, if a noncombustible membrane is used as a roof membrane only.

3102.6 Mixed construction. Membrane structures shall be permitted to be utilized as specified in this section as a portion of buildings of other types of construction. Height and area limits shall be as specified for the type of construction and occupancy of the building.

❖ The supporting structure of the membrane must be equal to or better than that of the building to which it is attached. Where air-supported or air-inflated structures are attached as an element of a building, their area limitations are based on the building to which they are attached. For example, an air-supported membrane building for a large assembly without a working stage would be limited to 24,000 square feet (2230 m^2). When used as the roof of a Type AV building, such as a football stadium, the area is limited in accordance with the construction classification of the building it covers.

3102.6.1 Noncombustible membrane. A noncombustible membrane shall be permitted for use as the roof or as a skylight of any building or atrium of a building of any type of construction provided it is at least 20 feet (6096 mm) above any floor, balcony or gallery.

❖ See Figure 3102.6.1 for an example of the application of this section.

3102.6.1.1 Membrane. A membrane meeting the fire propagation performance criteria of NFPA 701 shall be permitted to be used as the roof or as a skylight on buildings of Type IIB, III, IV and V construction, provided it is at least 20 feet (6096 mm) above any floor, balcony or gallery.

❖ See Figure 3102.6.1 for an example of the application of this section.

3102.7 Engineering design. The structure shall be designed and constructed to sustain dead loads; loads due to tension or

NONCOMBUSTIBLE MEMBRANE PERMITTED ON ANY TYPE OF CONSTRUCTION

FLAME-RESISTANT MEMBRANE PERMITTED ON TYPES IIB, III, IV and V

For SI: 1 foot = 304.8 mm.

Figure 3102.6.1
NONCOMBUSTIBLE AND FLAME-RESISTANT MEMBRANE CONSTRUCTION

inflation; live loads including wind, snow or flood and seismic loads and in accordance with Chapter 16.

❖ The structure must be designed and constructed to sustain all dead loads, loads caused by tensioning or inflation and live loads, including wind, snow, flood and seismic. Chapter 16 provides criteria for design loads, except those associated with tensioning or inflation.

3102.8 Inflation systems. Air-supported and air-inflated structures shall be provided with primary and auxiliary inflation systems to meet the minimum requirements of Sections 3102.8.1 through 3102.8.3.

❖ While the primary inflation system is essential to the integrity of air-supported and air-inflated buildings, a secondary or auxiliary inflation system is required for all such membrane structures to prevent deflation and to allow the occupants to egress. A back-up structural support system is also required for air-supported and air-inflated structures with an occupant load of more than 50 or that cover a swimming pool because of life safety considerations.

3102.8.1 Equipment requirements. This inflation system shall consist of one or more blowers and shall include provisions for automatic control to maintain the required inflation pressures. The system shall be so designed as to prevent overpressurization of the system.

❖ The structural integrity of air-supported and air-inflated structures depends on a reliable inflation system.

3102.8.1.1 Auxiliary inflation system. In addition to the primary inflation system, in buildings exceeding 1,500 square feet (140 m²) in area, an auxiliary inflation system shall be provided with sufficient capacity to maintain the inflation of the structure in case of primary system failure. The auxiliary inflation system shall operate automatically when there is a loss of internal pressure and when the primary blower system becomes inoperative.

❖ As a safety precaution, such structures that exceed 1,500 square feet (139 m²) must also be provided with an auxiliary inflation system. The auxiliary system is required to operate automatically upon loss of internal pressure or if the primary system becomes inoperative. In accordance with Section 3102.8.2, when an auxiliary system is required, standby power is also to be provided for at least 4 hours.

3102.8.1.2 Blower equipment. Blower equipment shall meet the following requirements:

1. Blowers shall be powered by continuous-rated motors at the maximum power required for any flow condition as required by the structural design.

2. Blowers shall be provided with inlet screens, belt guards and other protective devices as required by the building official to provide protection from injury.

3. Blowers shall be housed within a weather-protecting structure.

4. Blowers shall be equipped with backdraft check dampers to minimize air loss when inoperative.

5. Blower inlets shall be located to provide protection from air contamination. The location of inlets shall be approved.

❖ Safety features of the blower system for the purposes of reliability, prevention of overpressurization and personal safety include:

• Blowers with motors rated for continuous operation at maximum power;

• Inlet screens, belt guards and other protective devices;

• Weather-protected housing for blowers;

• Backdraft check dampers to minimize air loss; and

• Inlets located to prevent air contamination.

The *International Mechanical Code*® (IMC®) requires makeup air inlets to be at least 10 feet (3048 mm) from any potential source of contamination.

3102.8.2 Standby power. Wherever an auxiliary inflation system is required, an approved standby power-generating system shall be provided. The system shall be equipped with a suitable means for automatically starting the generator set upon failure of the normal electrical service and for automatic transfer and operation of all of the required electrical functions at full power within 60 seconds of such service failure. Standby power shall be capable of operating independently for a minimum of 4 hours.

❖ Air-supported and air-inflated structures exceeding 1,500 square feet (139 m²) are to be provided with standby power for the auxiliary inflation system. The standby power system must comply with the provisions of Section 2702 and be capable of operating for a minimum of 4 hours.

3102.8.3 Support provisions. A system capable of supporting the membrane in the event of deflation shall be provided for in air-supported and air-inflated structures having an occupant load of 50 or more or where covering a swimming pool regardless of occupant load. The support system shall be capable of maintaining membrane structures used as a roof for Type I construction not less than 20 feet (6096 mm) above floor or seating areas. The support system shall be capable of maintaining other membranes at least 7 feet (2134 mm) above the floor, seating area or surface of the water.

❖ To allow safe occupant egress and access for fire fighting, air-supported and air-inflated structures that have an occupant load of more than 50 or cover a swimming pool are required to have a standby support system. Regardless of the occupant load, swimming pools are included because of the additional time necessary to egress when in a pool and because of the hazard that would exist if a membrane collapsed on a pool in which a person was swimming.

The support system is to be capable of maintaining all membrane structures at least 7 feet (2134 mm) above the floor, seating area or surface water, whichever is highest, to permit upright egress travel by the occupants, as well as to aid occupants who may be unable to egress. When the membrane structure is used as a roof on a Type I or II building, the support system must be capable of maintaining the membrane structure at least 20 feet (6096 mm) above the floor or seating area so that the contribution of the roof structure to the fire is minimized.

SECTION 3103
TEMPORARY STRUCTURES

3103.1 General. The provisions of this section shall apply to structures erected for a period of less than 180 days. Tents and other membrane structures erected for a period of less than 180 days shall comply with the *International Fire Code*. Those erected for a longer period of time shall comply with applicable sections of this code.

❖ The purpose of Section 3103 is to provide public safety requirements in relation to fire, storm, collapse and crowd behavior in temporary structures. Temporary use is defined as a period of less than 180 days. Section 1024 is applicable to the means of egress and the use of bleachers, grandstands and folding or telescopic seating within the temporary structure. Temporary structures may also include temporary wiring and other fire hazards, which are different from those expected in permanent structures. These items are to be identified in the construction documents.

Tents have been a major fire protection concern since the disastrous circus fire in Hartford, Connecticut on July 6, 1944. This fire, which demonstrated the need for flame-retardant materials, resulted in 168 fatalities.

Any structure that is erected for a period of 180 days or more must comply with all other applicable sections of the code. It should be noted that "temporary" refers to the structure being erected, not the use of the structure; therefore, if a structure is erected for 365 days a year, but is used on a seasonal basis for less than 180 days, this section does not apply. Such structures are considered permanent.

3103.1.1 Permit required. Temporary structures that cover an area in excess of 120 square feet (11.16 m²), including connecting areas or spaces with a common means of egress or entrance which are used or intended to be used for the gathering together of 10 or more persons, shall not be erected, operated or maintained for any purpose without obtaining a permit from the building official.

❖ Permits are required for temporary structures that are greater than 120 square feet (11 m²) and are used for the gathering of 10 or more people. A permit from the building official is not required when the temporary structure is used for recreational camping or by less than 10 persons.

3103.2 Construction documents. A permit application and construction documents shall be submitted for each installation of a temporary structure. The construction documents shall include a site plan indicating the location of the temporary structure and information delineating the means of egress and the occupant load.

❖ Before the installation of any temporary structure, a permit application and construction documents are required to be submitted.

Construction documents, as defined in Chapter 2, include written, graphic and pictorial documents. These documents are to clearly provide pertinent information about the project, including the location, seating capacity, construction and all mechanical and electrical equipment, as a minimum.

3103.3 Location. Temporary structures shall be located in accordance with the requirements of Table 602 based on the fire-resistance rating of the exterior walls for the proposed type of construction.

❖ With respect to exposure hazards, temporary structures are no different from permanent structures; therefore, the fire-resistance rating of the temporary structure's exterior walls will depend upon the occupancy and the fire separation distance.

3103.4 Means of egress. Temporary structures shall conform to the means of egress requirements of Chapter 10 and shall have a maximum exit access travel distance of 100 feet (30 480 mm).

❖ The means of egress in temporary structures is to comply with the requirements of Chapter 10, except the maximum exit access travel distance is limited to 100 feet (30 480 mm). The reduced travel distance applies to all temporary uses and is based on concerns associated with the construction weaknesses, which may accelerate fire spread or rapidly decrease the stability of the structure during a fire.

SECTION 3104
PEDESTRIAN WALKWAYS AND TUNNELS

3104.1 General. This section shall apply to connections between buildings such as pedestrian walkways or tunnels, located at, above or below grade level, that are used as a means of travel by persons. The pedestrian walkway shall not contribute to the building area or the number of stories or height of connected buildings.

❖ Pedestrian walkways are normally seen as bridges connecting adjacent buildings. This section addresses the practical application of pedestrian walkways and tunnels. This section deals with walkways between buildings, including grade-level connections between buildings or corridors around the perimeter of one building. The area of the walkway itself is not to be considered a part of any building it serves, nor is it considered to add to the number of stories or height of any building it serves, provided that the walkway and its construction comply with this section.

3104.2 Separate structures. Connected buildings shall be considered to be separate structures.

Exceptions:

1. Buildings on the same lot in accordance with Section 503.1.2.

2. For purposes of calculating the number of Type B units required by Chapter 11, structurally connected buildings and buildings with multiple wings shall be considered one structure.

❖ Exception 1 to this section states that each building to be served by a pedestrian walkway is to be considered as a separate building except where, in accordance with Section 503.1.3, the owner chooses to have separate buildings on the same lot considered as one building. Exception 1 also applies to a connected building that may be of unlimited area as addressed in Section 507. Exception 2 establishes that, for purposes of evaluation of a building complex for the required number of Type B dwelling units, buildings connected by pedestrian walkways must be evaluated as a single building. This provision was installed to coordinate with the *Fair Housing Accessibility Guidelines* (FHAG). The requirement is based upon recommendations in the *Accessibility Analysis of Model Codes, International Building Code Analysis, Report for Public Comment,* dated October 20, 1999 by the U.S. Department of Housing and Urban Development (HUD). Based upon International Code Council® (ICC®) compliance with these recommendations, HUD has determined that compliance with the 2006 edition of the code is "safe harbor" for responsible parties in the design of a building for accessibility (see commentary, Section 1101).

3104.3 Construction. The pedestrian walkway shall be of noncombustible construction.

Exceptions:

1. Combustible construction shall be permitted where connected buildings are of combustible construction.

2. Fire-retardant-treated wood, in accordance with Table 601, Note c, shall be permitted for the roof construction of the pedestrian walkway where connected buildings are a minimum of Type I or II construction.

❖ The walkway must be of noncombustible construction, unless all of the buildings it connects are of combustible construction.

It is intended that the construction materials of the walkway be consistent with the type of construction classification of the buildings connected by the walkway. Also note that this section does not require structural members of the walkway to have a fire-resistance rating, even if the connected buildings are of a protected construction type. Exception 2 simply allows

fire-retardant-treated wood where it is already allowed for Type I or II construction, because this section does not depend upon Table 601 for determination of the construction requirements for pedestrian walkways.

3104.4 Contents. Only materials and decorations approved by the building official shall be located in the pedestrian walkway.

❖ This restriction on the use of the floor area is similar to the restriction on the use of the floor of an atrium, except there is no relaxation of requirements if an automatic sprinkler system is installed. The reference to low-hazard uses implies that the space created by the walkway is not to be used for any purpose that would incorporate significant amounts of combustible materials. In general, the same parameters that would apply to exit access corridors and atriums can be applied here.

3104.5 Fire barriers between pedestrian walkways and buildings. Walkways shall be separated from the interior of the building by fire-barrier walls with a fire-resistance rating of not less than 2 hours. This protection shall extend vertically from a point 10 feet (3048 mm) above the walkway roof surface or the connected building roof line, whichever is lower, down to a point 10 feet (3048 mm) below the walkway and horizontally 10 feet (3048 mm) from each side of the pedestrian walkway. Openings within the l0-foot (3048 mm) horizontal extension of the protected walls beyond the walkway shall be equipped with devices providing a $^3/_4$-hour fire protection rating in accordance with Section 715.

Exception: The walls separating the pedestrian walkway from a connected building are not required to have a fire-resistance rating by this section where any of the following conditions exist:

1. The distance between the connected buildings is more than 10 feet (3048 mm), the pedestrian walkway and connected buildings, except for open parking garages, are equipped throughout with an automatic sprinkler system in accordance with NFPA 13 and the wall is constructed of a tempered, wired or laminated glass wall and doors subject to the following:

   1.1. The glass shall be protected by an automatic sprinkler system in accordance with NFPA 13 and the sprinkler system shall completely wet the entire surface of interior sides of the glass wall when actuated.

   1.2. The glass shall be in a gasketed frame and installed in such a manner that the framing system will deflect without breaking (loading) the glass before the sprinkler operates.

   1.3. Obstructions shall not be installed between the sprinkler heads and the glass.

2. The distance between the connected buildings is more than 10 feet (3048 mm) and both sidewalls of the pedestrian walkway are at least 50 percent open with the open area uniformly distributed to prevent the accumulation of smoke and toxic gases.

3. Buildings are on the same lot in accordance with Section 503.1.2.

4. Where exterior walls of connected buildings are required by Section 704 to have a fire-resistance rating greater than 2 hours, the walkway shall be equipped throughout with an automatic sprinkler system installed in accordance with NFPA 13.

The previous exceptions shall apply to pedestrian walkways having a maximum height above grade of three stories or 40 feet (12 192 mm), or five stories or 55 feet (16 764 mm) where sprinklered.

❖ It is this requirement for fire-resistance-rated separation that allows buildings and walkways to be considered as separate buildings. The extent of this fire-resistance-rated separation is shown in Figure 3104.5. If the wall of the connected building would not normally extend 10 feet (3048 mm) above the walkway roof, there is no requirement in this section to extend it.

The portion of the exterior wall that separates the walkway from the building is required to have a fire-resistance rating unless one of the exceptions of this section is met.

This fire-resistance rating of the wall separating the walkway from a connected building is not required when one of the following is met:

1. If the distance between the connected buildings is more than 10 feet (3048 mm) and the walkway and all connected buildings are fully sprinklered or are open parking garages, then any glass in the wall between the walkway and a connected building must be protected as described, even though the wall need not be rated.

2. The distance between the connected buildings is more than 10 feet (3048 mm) and both sides of the walls of the pedestrian walkway are open for at least 50 percent of each wall area, with the open area uniformly distributed to prevent the accumulation of smoke and toxic gases. This concept is synonymous with that used for open parking structures where free movement of smoke is achieved through the openness of the structure.

3. The buildings are designed under the provisions of Section 503.1.3 for two buildings on one lot. This exception is based on the concept that the exterior walls are considered to be interior walls for fire-resistance purposes.

4. If the walkway is fully sprinklered and if more than a 2-hour fire-resistance rating is required by Section 704, this exception can be taken and no fire-resistance rating would be required at the end wall of the walkway.

These four exceptions apply to the walkway rather than the buildings that the walkway connects.

¾-HR. OPENING PROTECTIVE

1½-HR. OPENING PROTECTIVE IN ACCORDANCE WITH SECTION 706.7 (UNLESS AN EXCEPTION IS MET)

2-HR. FIRE SEPARATION WALL
* REMAINING WALLS AND WINDOWS PROTECTED IN ACCORDANCE WITH SECTION 704

For SI: 1 foot = 304.8 mm.

Figure 3104.5
PEDESTRIAN WALKWAY—BUILDING INTERFACE

3104.6 Public way. Pedestrian walkways over a public way shall also comply with Chapter 32.

❖ The acceptance of a pedestrian walkway over a public street by the use of a bridge or any other means is required to comply with Chapter 32.

3104.7 Egress. Access shall be provided at all times to a pedestrian walkway that serves as a required exit.

❖ Pedestrian walkways may be used as exits. If they are, their doors must be equipped with hardware that will allow them to function as exits at all times.

3104.8 Width. The unobstructed width of pedestrian walkways shall not be less than 36 inches (914 mm). The total width shall not exceed 30 feet (9144 mm).

❖ A maximum width has been set to prevent entire floors from being extended to adjacent buildings, and thus from being called walkways. This section is intended to encompass bridge-type structures, not full-floor extensions.

The minimum width is also subject to other code requirements. For example, if the walkway provides a means of egress component, it would be sized in accordance with Section 1003.2.3.

3104.9 Exit access travel. The length of exit access travel shall not exceed 200 feet (60 960 mm).

Exceptions:

1. Exit access travel distance on a pedestrian walkway equipped throughout with an automatic sprinkler system in accordance with NFPA 13 shall not exceed 250 feet (76 200 mm).

2. Exit access travel distance on a pedestrian walkway constructed with both sides at least 50 percent open shall not exceed 300 feet (91 440 mm).

3. Exit access travel distance on a pedestrian walkway constructed with both sides at least 50 percent open, and equipped throughout with an automatic sprinkler system in accordance with NFPA 13, shall not exceed 400 feet (122 m).

❖ The maximum length of exit access travel distance within the pedestrian walkway has been set as a means of controlling the size. A 200-foot (60 960 mm) maximum length is consistent with the maximum length of exit access travel distance allowed in unsprinklered buildings for most occupancy classifications (see Table 1004.2.4). This section gives three means by which the exit access travel distance within the pedestrian walkway can be increased. Exception 1 allows the exit access travel distance to be increased to 250 feet (76 200 mm) if the walkway is equipped with an automatic sprinkler system. This increase in travel distance correlates with that allowed in Table 1004.2.4. Exception 2 allows the exit access travel distance to be increased to 300 feet (91 440 mm) if both sides of the walkway have openings to the exterior for at least 50 percent of the wall area, with the open area uniformly distributed to prevent the accumulation of smoke and toxic gases. This increase in travel distance correlates to that allowed in Table 1004.2.4 for an open parking structure (Group S-2). Exception 3 expands on Exception 2 by including the use of an automatic sprinkler system. The exit access travel distance is then increased to 400 feet (122 m²). This increase in travel distance correlates to the allowances in Table 1004.2.4.

3104.10 Tunneled walkway. Separation between the tunneled walkway and the building to which it is connected shall not be less than 2-hour fire-resistant construction and openings therein shall be protected in accordance with Table 715.4.

❖ Tunnels present a greater hazard than other walkways. The concern is greater than for enclosed walkways because a tunnel's underground location makes it less accessible to fire-fighting efforts and it relies on a mechanical ventilation system. To prevent the spread of fire, 2-hour fire-resistant construction with 1¹/₂-hour opening protectives are required at the tunnel and building interface (see Figure 3104.10).

Figure 3104.10
TUNNELED WALKWAY

SECTION 3105
AWNINGS AND CANOPIES

3105.1 General. Awnings or canopies shall comply with the requirements of this section and other applicable sections of this code.

❖ This section addresses the variety of special structures that are defined as canopies and awnings. The intent of this section is that if a canopy or awning is used, it should be soundly designed so as not to present a hazard to its users or the public.

3105.2 Definition. The following term shall, for the purposes of this section and as used elsewhere in this code, have the meaning shown herein.

RETRACTABLE AWNING. A retractable awning is a cover with a frame that retracts against a building or other structure to which it is entirely supported.

❖ The key to this definition is that it provides for an element of building construction that is increasing in popularity, especially in residential construction. The need was to properly identify that this section of the code is applicable to these awnings.

3105.3 Design and construction. Awnings and canopies shall be designed and constructed to withstand wind or other lateral loads and live loads as required by Chapter 16 with due allowance for shape, open construction and similar features that relieve the pressures or loads. Structural members shall be protected to prevent deterioration. Awnings shall have frames of noncombustible material, fire-retardant-treated wood, wood of Type IV size, or 1-hour construction with combustible or noncombustible covers and shall be either fixed, retractable, folding or collapsible.

❖ Any of the structures addressed in Section 3105 are subject to wind and live load considerations under the provisions of Chapter 16. When Chapter 16 is applied, care should be taken to consider the shape of the awning or canopy in question with regard to snow buildup and wind effects.

Structural members of an awning or canopy should be of materials that will resist rust or decay. Chapter 23 should be referenced for wood members that are required to comply with the decay-resistance provision. Steel or aluminum structural members are to be protected to resist rust.

3105.4 Canopy materials. Canopies shall be constructed of a rigid framework with an approved covering that meets the fire propagation performance criteria of NFPA 701 or has a flame spread index not greater than 25 when tested in accordance with ASTM E 84.

❖ Canopies, as regulated by this section, are awnings that have ground support. This section specifically states the types of materials that are to be used and how a canopy is to be constructed. It limits the frame construction to metal and the coverings to only those that are deemed flame resistant when tested in accordance with NFPA 701, or those having a flame spread of 25 or less when tested in accordance with ASTM E 84. The purpose of these limitations is to limit the possible spread of fire on the exterior of the building due to fire from the interior of the building or fire from an exterior source on the street or public way.

SECTION 3106
MARQUEES

3106.1 General. Marquees shall comply with this section and other applicable sections of this code.

❖ The intent of this section is that if a marquee is used, it should be soundly designed so as not to present a hazard to its users or the public.

3106.2 Thickness. The maximum height or thickness of a marquee measured vertically from its lowest to its highest point shall not exceed 3 feet (914 mm) where the marquee projects more than two-thirds of the distance from the property line to the curb line, and shall not exceed 9 feet (2743 mm) where the marquee is less than two-thirds of the distance from the property line to the curb line.

❖ The restrictions placed on the size, projection and clearances (see Figure 3106.2) for marquees are intended to:

1. Prevent interference with the free movement of pedestrians.

2. Prevent interference with trucks and other tall vehicles using the public street.

3. Prevent interference with fire-fighting operations at a building.

4. Prevent interference with utilities.

3106.3 Roof construction. Where the roof or any part thereof is a skylight, the skylight shall comply with the requirements of Chapter 24. Every roof and skylight of a marquee shall be sloped to downspouts that shall conduct any drainage from the marquee in such a manner so as not to spill over the sidewalk.

❖ The purpose of this section is to require proper drainage to prevent buildup on the roof due to a marquee or skylight.

3106.4 Location prohibited. Every marquee shall be so located as not to interfere with the operation of any exterior standpipe, and such that the marquee does not obstruct the clear passage of stairways or exit discharge from the building or the installation or maintenance of street lighting.

❖ A marquee is to be placed so as to not prohibit any type of fire protection, means of egress or maintenance work to the building or streetlights.

3106.5 Construction. A marquee shall be supported entirely from the building and constructed of noncombustible materials. Marquees shall be designed as required in Chapter 16. Structural members shall be protected to prevent deterioration.

❖ Damage to both life and property can occur from improperly constructed marquees. Proper design of a marquee involves a structural analysis by qualified persons. Wind and seismic loads must be considered in the design process, as outlined in Chapter 16. Additionally, all material must be noncombustible.

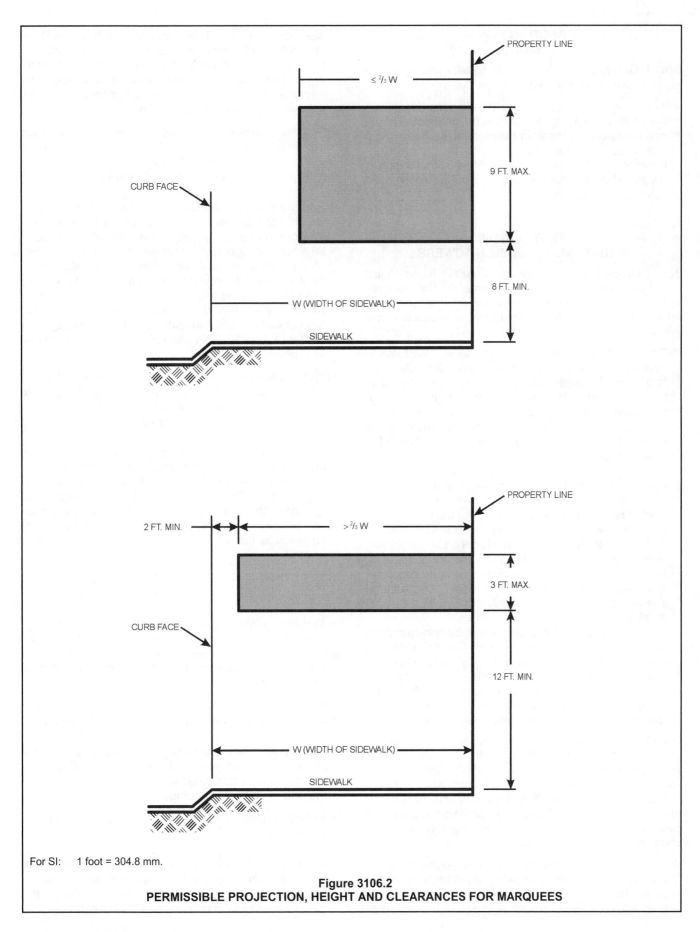

For SI: 1 foot = 304.8 mm.

Figure 3106.2
PERMISSIBLE PROJECTION, HEIGHT AND CLEARANCES FOR MARQUEES

SECTION 3107
SIGNS

3107.1 General. Signs shall be designed, constructed and maintained in accordance with this code.

❖ The design of outdoor signs is to be in accordance with the code. If the design is inadequate, no amount of expert construction, repair or maintenance will result in a safe sign. As can be determined from the code, the enforcement of this section is confined to that which is displayed in any manner out of doors for recognized advertising purposes.

SECTION 3108
RADIO AND TELEVISION TOWERS

3108.1 General. Subject to the provisions of Chapter 16 and the requirements of Chapter 15 governing the fire-resistance ratings of buildings for the support of roof structures, radio and television towers shall be designed and constructed as herein provided.

❖ Radio and television towers are special structures in that they are not normally occupied and are open structures containing little, if any, fuel load. As such, this section provides specific provisions for radio and television towers, is designed to maintain the structural integrity of the tower and to minimize its potential to present a hazard (such as serving as a potential fire source because of inadequate grounding).

Since the towers are not normally occupied and do not typically contain a fuel load, the structure is exempt from other provisions of the code, except for the wind load provisions of Section 3108.4 and the combustibility restrictions of the supporting structures as required by Section 1509 (see Section 3108.1). The height and area limitations of Chapter 5 do not apply; instead, Section 3108.3 provides minimum construction criteria depending on the height of the tower. The structural integrity of the tower is basically governed by Section 3108.4.

Radio and television towers are to be constructed in accordance with the provisions of Section 3108, the wind load provisions of Section 3108.4 and the fire-resistance ratings of the supporting structure, as governed by Section 1509. The provisions of Section 3108.4 minimize the potential for the collapse of a tower as caused by forces associated with wind loads.

3108.2 Location and access. Towers shall be located and equipped with step bolts and ladders so as to provide ready access for inspection purposes. Guy wires or other accessories shall not cross or encroach upon any street or other public space, or over above-ground electric utility lines, or encroach upon any privately owned property without written consent of the owner of the encroached-upon property, space or above-ground electric utility lines.

❖ Access must be provided to the tower so that its condition can be surveyed. Guy wires are to be arranged so as not to represent a hazard to the public, and therefore, not cross or encroach a street, public space or

electric power line. This is because of both the damage that could occur if a guy wire were suddenly released and the need to prohibit the wire from becoming an obstruction. Guy wires are not to encroach on any other property without previous written consent by the owner of the affected property.

3108.3 Construction. Towers shall be constructed of approved corrosion-resistant noncombustible material. The minimum type of construction of isolated radio towers not more than 100 feet (30 480 mm) in height shall be Type IIB.

❖ Radio and television towers are required to be constructed of material that can withstand exposure to the environment and, therefore, be corrosion resistant. Whereas Section 1509 addresses the minimum allowable construction of towers that are roof structures, Section 3108.3 provides minimum construction requirements for all towers. Most radio and television towers are constructed of noncombustible materials. Towers that are isolated and are neither roof structures nor more than 100 feet (30 480 mm) in height may be of Type IIB construction.

3108.4 Loads. Towers shall be designed to resist wind loads in accordance with TIA/EIA-222. Consideration shall be given to conditions involving wind load on ice-covered sections in localities subject to sustained freezing temperatures.

❖ To maintain structural integrity, radio and television towers must be designed to resist wind loads in accordance with TIA/EIA 222, which provides minimum load requirements for the design of buildings and other structures that are subject to code requirements. In areas subject to freezing conditions, consideration must be given to the effects of wind loads on ice-covered towers.

3108.4.1 Dead load. Towers shall be designed for the dead load plus the ice load in regions where ice formation occurs.

❖ To minimize the potential for collapse, a radio and television tower is to be designed to withstand the dead load of the structural elements of the tower, as well as the loads associated with a buildup of ice in areas subject to ice conditions.

3108.4.2 Wind load. Adequate foundations and anchorage shall be provided to resist two times the calculated wind load.

❖ To include a factor of safety, the foundation and anchorage must be designed and constructed to withstand twice the calculated wind uplift so as to minimize the potential for overturning.

3108.5 Grounding. Towers shall be permanently and effectively grounded.

❖ To minimize the potential for a fire resulting from lightning, radio and television towers are to be adequately grounded so that the electric charge is carried to the ground without affecting adjacent structures.

SECTION 3109
SWIMMING POOL ENCLOSURES AND
SAFETY DEVICES

3109.1 General. Swimming pools shall comply with the requirements of this section and other applicable sections of this code.

❖ The provisions of this section apply to both public and private swimming pools. They are intended to provide for the use of swimming pools as well as to limit or delay unauthorized access to swimming pools by small children, particularly those five years old and younger. All but small pools not equipped with water-recirculating systems or those that involve structural materials must comply with the provisions of this section. Requirements are included for pool enclosures and barriers.

Small pools not equipped with water-recirculating systems or those that do not involve structural materials need not comply with the provisions of this section.

3109.2 Definition. The following word and term shall, for the purposes of this section and as used elsewhere in this code, have the meaning shown herein.

❖ The definition of a term that is associated with the content of this section is contained herein. This definition can help in the understanding and application of the code requirements. It is important to emphasize that this term is not exclusively related to this section but is applicable everywhere the term is used throughout the code. The purpose for including this definition within this chapter is to provide more convenient access to it without having to refer back to Chapter 2.

SWIMMING POOLS. Any structure intended for swimming, recreational bathing or wading that contains water over 24 inches (610 mm) deep. This includes in-ground, above-ground and on-ground pools; hot tubs; spas and fixed-in-place wading pools.

❖ The small above-ground pools frequently found in residential yards are exempt, provided that they do not exceed 2 feet (610 mm) in depth.

3109.3 Public swimming pools. Public swimming pools shall be completely enclosed by a fence at least 4 feet (1290 mm) in height or a screen enclosure. Openings in the fence shall not permit the passage of a 4-inch-diameter (102 mm) sphere. The fence or screen enclosure shall be equipped with self-closing and self-latching gates.

❖ Whether the pool is privately owned and operated is irrelevant to whether or not it is a public swimming pool. As long as the pool is not associated with an occupancy in Group R-3, it is considered a public swimming pool. Examples of public swimming pools include a pool in an apartment complex located in a common use area, a pool at a health club and a municipal swimming pool.

A public pool requires a barrier no lower than 4 feet (1219 mm). This barrier/wall limits or delays unauthorized access to the pool by small children, particularly those five years old or younger. It is important to note that this barrier is to enclose the swimming pool completely. Any construction or natural element that does not surround the pool will allow access at some point; hence, any gate, door or other access component must be self-closing and self-latching.

3109.4 Residential swimming pools. Residential swimming pools shall comply with Sections 3109.4.1 through 3109.4.3.

Exception: A swimming pool with a power safety cover or a spa with a safety cover complying with ASTM F 1346.

❖ Although the occupants and guests of a Group R-3 residence have access to the swimming pool regardless of its location, it is important to distinguish what elements separate the pool from the occupants. Pools that are totally within the structure are included in this section. Of critical concern is the easy access afforded to children by an indoor pool. For this reason, Sections 3109.4.1 through 3109.4.3 must be closely followed.

3109.4.1 Barrier height and clearances. The top of the barrier shall be at least 48 inches (1219 mm) above grade measured on the side of the barrier that faces away from the swimming pool. The maximum vertical clearance between grade and the bottom of the barrier shall be 2 inches (51 mm) measured on the side of the barrier that faces away from the swimming pool. Where the top of the pool structure is above grade, the barrier is authorized to be at ground level or mounted on top of the pool structure, and the maximum vertical clearance between the top of the pool structure and the bottom of the barrier shall be 4 inches (102 mm).

❖ The barrier height requirement of 48 inches (1219 mm) above the ground is based on reports that have documented the ability of children under the age of five to climb over barriers that are less than 48 inches (1219 mm) in height. The basis for the 4-inch (102 mm) criterion for an opening between the barrier and the top of pool frame is the same as for guardrail construction (see Figure 3109.4.1).

3109.4.1.1 Openings. Openings in the barrier shall not allow passage of a 4-inch-diameter (102 mm) sphere.

❖ The basis for the 4-inch (102 mm) criterion is the same as for guardrail construction (see Figure 3109.4.1.1). It is based on studies of the body measurements of children 13 to 18 months old.

3109.4.1.2 Solid barrier surfaces. Solid barriers which do not have openings shall not contain indentations or protrusions except for normal construction tolerances and tooled masonry joints.

❖ This provision is intended to reduce the potential for gaining a foothold and climbing the barrier.

3109.4.1.3 Closely spaced horizontal members. Where the barrier is composed of horizontal and vertical members and the distance between the tops of the horizontal members is less than 45 inches (1143 mm), the horizontal members shall be located on the swimming pool side of the fence. Spacing between vertical members shall not exceed 1.75 inches (44 mm) in width. Where there are decorative cutouts within vertical members, spacing within the cutouts shall not exceed 1.75 inches (44 mm) in width.

❖ This more stringent 1¾-inch (44 mm) provision for spacing between vertical members applies when the spacing between horizontal members is less than 45 inches (1143 mm). It acknowledges the potential for a child to gain both a handhold and a foothold on closely spaced horizontal members, and is intended to reduce that potential by limiting the space between the vertical members on the same barrier. If the horizontal members are spaced less than 45 inches (1143 mm), they must also be located on the swimming pool side of the fence (see Figure 3109.4.1.3) so that they are not available to be used to climb the barriers.

3109.4.1.4 Widely spaced horizontal members. Where the barrier is composed of horizontal and vertical members and the distance between the tops of the horizontal members is 45 inches (1143 mm) or more, spacing between vertical members shall not exceed 4 inches (102 mm). Where there are decorative cutouts within vertical members, spacing within the cutouts shall not exceed 1.75 inches (44 mm) in width.

❖ This requirement is the counterpart to Section 3109.4.1.3 in that it permits the opening in the barrier to be 4 inches (102 mm), provided the vertical spacing of the horizontal members equals or exceeds 45 inches (1143 mm) (see Figure 3104.10). It limits openings in the barrier to a 4-inch (102 mm) diameter. The spacing of horizontal members 45 inches (1143 mm) apart precludes them from being used by small children to climb the barrier.

3109.4.1.5 Chain link dimensions. Maximum mesh size for chain link fences shall be a 2.25 inch square (57 mm square)

unless the fence is provided with slats fastened at the top or the bottom which reduce the openings to no more than 1.75 inches (44 mm).

❖ The 2¼-inch (57 mm) dimension is intended to reduce the potential to gain a foothold (see Figure 3109.4.1.5). The opening is permitted to be increased above this 2¼-inch (57 mm) dimension if decorative slats are used that reduce the size of the opening to 1¾ inches (44 mm), since the potential to obtain a foothold or handhold is reduced when such slats are installed.

3109.4.1.6 Diagonal members. Where the barrier is composed of diagonal members, the maximum opening formed by the diagonal members shall be no more than 1.75 inches (44 mm).

❖ A slightly bigger opening is permitted for barriers composed of diagonal members other than chain-link fences on the basis that such barriers would be more difficult to gain a foothold and handhold on than a chain-link fence. The 1¾-inch (44 mm) dimension is consistent with Sections 3109.4.1.3, 3109.4.1.4 and 3109.4.1.5.

3109.4.1.7 Gates. Access gates shall comply with the requirements of Sections 3109.4.1.1 through 3109.4.1.6 and shall be equipped to accommodate a locking device. Pedestrian access gates shall open outward away from the pool and shall be self-closing and have a self-latching device. Gates other than pedestrian access gates shall have a self-latching device. Release mechanisms shall be in accordance with Sections 1008.1.8 and 1109.13. Where the release mechanism of the self-latching device is located less than 54 inches (1372 mm) from the bottom of the gate, the release mechanism shall be

For SI: 1 inch = 25.4 mm.

Figure 3109.4.1
OPENING LIMITATIONS

For SI: 1 inch = 25.4 mm.

Figure 3109.4.1.1
BARRIER OPENINGS

located on the pool side of the gate at least 3 inches (76 mm) below the top of the gate, and the gate and barrier shall have no opening greater than 0.5 inch (12.7 mm) within 18 inches (457 mm) of the release mechanism.

❖ A gate represents the same potential hazard relative to climbing as do other portions of the barrier and, therefore, must be constructed in accordance with Sections 3109.4.1.1 through 3109.4.1.6. Additionally, since the gate represents a potential breach of the barrier due to the ability to be opened, the code provides prescriptive details for its construction and operation. A self-closing pedestrian gate is required to open away from the pool because if the latch fails to operate, a child pushing on the gate will not gain immediate access to the pool. Pushing on the gate may also engage the latch. Large, nonpedestrian gates are not required to be self-closing due to the prohibitive cost coupled with the fact that these gates are typically operated by persons other than small children. The 54-inch (1372 mm) latch height requirement is intended to limit the potential for small children to reach and activate it. If located less than 54 inches (1372 mm), the code's prescriptive location requirements are intended to preclude the latch from being activated by small children who are not on the pool side of the gate.

For SI: 1 inch = 25.4 mm.

Figure 3109.4.1.5
CHAIN-LINK FENCE MESH FOR PRIVATE SWIMMING POOLS

For SI: 1 inch = 25.4 mm.

Figure 3109.4.1.3
PRIVATE SWIMMING POOL BARRIER CONSTRUCTION

3109.4.1.8 Dwelling wall as a barrier. Where a wall of a dwelling serves as part of the barrier, one of the following shall apply:

1. Doors with direct access to the pool through that wall shall be equipped with an alarm that produces an audible warning when the door and/or its screen, if present, are opened. The alarm shall be listed in accordance with UL 2017. The audible alarm shall activate within 7 seconds and sound continuously for a minimum of 30 seconds after the door and/or its screen, if present, are opened and be capable of being heard throughout the house during normal household activities. The alarm shall automatically reset under all conditions. The alarm shall be equipped with a manual means, such as touchpad or switch, to temporarily deactivate the alarm for a single opening. Such deactivation shall last for not more than 15 seconds. In dwellings not required to be Accessible, Type A or Type B units, the deactivation switch shall be located 54 inches (1372 mm) or more above the threshold of the door. In dwellings required to be Accessible, Type A or Type B units, the deactivation switch(es) shall be located at 54 inches (1372 mm) maximum and 48 inches minimum above the threshold of the door.

2. The pool shall be equipped with a power safety cover that complies with ASTM F 1346.

3. Other means of protection, such as self-closing doors with self-latching devices, which are approved by the administrative authority, shall be accepted so long as the degree of protection afforded is not less than the protection afforded by Section 3109.4.1.8, Item 1 or 2.

❖ Many residential settings with backyard pools utilize the dwelling as a portion of the barrier required around the pool, (i.e., the fence bounding the property terminates at the dwelling). This limits access to the pool by unsupervised children around the perimeter of the fence but there is still a potential for children to access the pool from within the dwelling. Indeed, almost half the children involved in drowning or near-drowning accidents gained access to the pool from the dwelling.

The provisions of this section are intended to restrict such access by small children and are applicable to all doors in walls that form a portion of the barrier required around private swimming pools, both outdoor and indoor.

Protection of such door openings to pool areas can be achieved in any one of the methods described in Items 1 through 3. The alarm is configured to allow adults accessing the house to open the door, enter the house and deactivate the system to prevent a "false alarm." The touchpad permitted to deactivate the system is required to be mounted 54 inches (1372 mm) above the floor, which is presumed to be beyond the reach of small children.

An audible alarm specified in Item 1 on the doors leading to the pool area is intended to provide a warning to a supervising adult that the pool area has been entered.

Item 2 permits doors to pool areas to be protected by devices that render the door self-closing and self-latching. The release mechanism for the latching device is required to be located a minimum of 54 inches (1372 mm) above the floor, which is anticipated to be beyond the reach of small children. In addition to self-closing and self-latching devices, doors protected by the method specified in Item 2 are required to open away from the pool area. This is so that if the door failed to latch, a child outside the pool pushing against the door would cause it to close and not swing to an open position.

Item 3 does not require protection of the door itself; but limits access to the pool by means of a power safety cover. The performance criteria specified when this option is selected is to ensure that the power safety cover is an adequate and reliable barrier to the pool.

3109.4.1.9 Pool structure as barrier. Where an aboveground pool structure is used as a barrier or where the barrier is mounted on top of the pool structure, and the means of access is a ladder or steps, then the ladder or steps either shall be capable of being secured, locked or removed to prevent access, or the ladder or steps shall be surrounded by a barrier which meets the requirements of Sections 3109.4.1.1 through 3109.4.1.8. When the ladder or steps are secured, locked or removed, any opening created shall not allow the passage of a 4-inch-diameter (102 mm) sphere.

❖ The code permits the wall of the pool itself to serve as the barrier to the pool, provided that the wall extends at least 48 inches (1219 mm) above the finished ground level around the perimeter of the pool.

3109.4.2 Indoor swimming pools. Walls surrounding indoor swimming pools shall not be required to comply with Section 3109.4.1.8.

❖ Indoor pools are required to have the alarms, etc., required by Section 3108.4.1.8. The pool in an indoor environment is generally under closer scrutiny; therefore, it is less apt to be an area where small children would be allowed to stray into without being noticed.

3109.4.3 Prohibited locations. Barriers shall be located so as to prohibit permanent structures, equipment or similar objects from being used to climb the barriers.

❖ This section is especially important when the wall of the pool itself is used as the barrier (see Section 3109.4.1.9). Pumps and other equipment located adjacent to the pool wall (barrier) present a hazard in that they may provide a means by which small children could climb over the pool wall and access the pool area.

3109.5 Entrapment avoidance. Suction outlets shall be designed to produce circulation throughout the pool or spa. Single-outlet systems, such as automatic vacuum cleaner systems, or other such multiple suction outlets whether isolated by valves or otherwise shall be protected against user entrapment.

❖ Vacuum devices for suction inlet systems in pool water circulation are a safety hazard. Body or hair entrapment can cause drowning and evisceration; therefore, it is important to provide protection against possible entrapment at the pool entrances to suction inlets, as well as vacuum relief for the vacuum system.

3109.5.1 Suction fittings. All pool and spa suction outlets shall be provided with a cover that conforms to ASME A112.19.8M, a 12-inch by 12-inch (305 mm by 305 mm) drain grate or larger, or an approved channel drain system.

Exception: Surface skimmers.

❖ Section 3109.5 is focused on entrapment avoidance. The first line of defense against entrapment of a body is a cover or grate for the opening. ANSI/ASME A112.19.8M requires cover material for pool and spa suction outlets to be tested for structural integrity and for entrapment/entanglement potential. It also requires the cover to be marked with the maximum flow rate for which the cover has been tested. Exceeding the maximum flow rate will increase the potential for a child or small adult being entrapped due to the increased suction. The code also allows 12-inch by 12-inch (305 mm by 305 mm) drain grates or approved channel drain systems as alternative protection methods. Both of these will provide larger surface areas to maintain the desired flow and will minimize the entrapment hazard because it will be difficult to seal off the entire 1-foot-square area.

3109.5.2 Atmospheric vacuum relief system required. All pool and spa single- or multiple-outlet circulation systems shall be equipped with an atmospheric vacuum relief should grate covers located therein become missing or broken. Such vacuum relief systems shall include at least one approved or engineered method of the type specified herein, as follows:

1. Safety vacuum release systems conforming to ASME A112.19.17; or

2. Approved gravity drainage system.

❖ Safety vacuum relief systems (SRVS) are required for all pool and spa circulating systems to serve as a backup safety system in case the suction outlet cover or grate is removed or becomes broken. If the outlet becomes sealed off (perhaps by the action of a person inadvertently sitting on it), the SRVS will activate and eliminate the high vacuum forces at the drain, avoiding body entrapment. An SRVS is not required if the pool or spa has a gravity drain system instead of a pumped circulation system, because a gravity drainage system cannot create the vacuum forces capable of entrapping a body.

3109.5.3 Dual drain separation. Single- or multiple-pump circulation systems shall be provided with a minimum of two suction outlets of the approved type. A minimum horizontal or vertical distance of 3 feet (914 mm) shall separate such outlets. These suction outlets shall be piped so that water is drawn through them simultaneously through a vacuum-relief-protected line to the pump or pumps.

❖ The principle for installing dual or multiple drain systems is to prevent a single drain opening from being the sole inlet to the suction side of the pump. The installation of additional drains effectively divides the suction between the drains, provided the interconnecting piping configuration produces hydraulic balance, thus reducing the amount of suction on any drain and reducing the risk of body entrapment.

3109.5.4 Pool cleaner fittings. Where provided, vacuum or pressure cleaner fitting(s) shall be located in an accessible position(s) at least 6 inches (152 mm) and not greater than 12 inches (305 mm) below the minimum operational water level or as an attachment to the skimmer(s).

❖ Pool cleaner fittings are not required for all pools and spas, but where they are provided, they need to be located below the normal water surface but not more than 12 inches (305 mm) below the surface. This location provides ease of access to the fittings for cleaning and prevents them from contributing to an entrapment situation at the bottom of the pool.

Bibliography

The following resource materials are referenced in this chapter or are relevant to the subject matter addressed in this chapter.

24 CFR, *Fair Housing Accessibility Guidelines* (FHAG). Washington, D.C.: Department of Housing and Urban Development, 1991.

ASTM E 84-04, *Test Method for Surface Burning Characteristics of Building Materials*. West Conshohocken, PA: ASTM International, 2004.

ASTM E 136-99e01, *Test Method for Behavior of Materials in a Vertical Tube Furnace at 750°C*. West Conshohocken, PA: ASTM International, 2001.

IFC-2006, *International Fire Code*. Falls Church, VA: International Code Council, 2006.

IMC-2006, *International Mechanical Code*. Falls Church, VA: International Code Council, 2006.

NFPA 701-99, *Methods of Fire Tests for Flame-Resistant Textiles and Films*. Quincy, MA: National Fire Protection Association, 1999.

NFPA 204 M-98, *Guide for Smoke and Heat Venting*. Quincy, MA: National Fire Protection Association, 1998.

NSPI 1-91 *Standard for Public Swimming Pools*. Alexandria, VA: National Spa and Pool Institute, 1991.

NSPI 4-92 *Standard for Above-Ground/Onground Residential Swimming Pools*. Alexandria, VA: National Spa and Pool Institute, 1992.

NSPI 5-94, *Standards for Residential Swimming Pools*. Alexandria, VA: National Spa and Pool Institute, 1994.

Present, Paula. *Child Drowning Study, A Report on the Epidemiology of Drownings in Residential Pools to Children Under Age 5*. Washington D.C.: Consumer Product Safety Commission, September 1987.

TIA/EIA 222-F-96, *Structural Standards for Steel Antenna Towers and Antenna Supporting Structures.* Arlington, VA: Electronics Industries Association, 1996.

Chapter 32:
Encroachments Into the Public Right-Of-Way

General Comments

Chapter 32 contains provisions that regulate a variety of structure encroachments and projections into the public right-of-way, such as a sidewalk, parking lot or street.

Section 3202 addresses encroachments on lot lines and street lot lines below grade, above grade, below 8 feet (2438 mm) in height and 8 feet (2438 mm) or more above grade.

Chapter 32 establishes the basis for the determination and application of many other code requirements, such as the minimum type of construction. Incorrect determination of the minimum type of construction will result in misapplication of the code.

The requirements of this chapter affect the decisions to be made when applying the rest of the code. Encroachments and projections have to be taken into consideration in addition to means of egress, fire resistance, fire protection or even structural and material considerations.

Purpose

The purpose of Chapter 32 is to regulate projections and encroachments of structures into the public right-of-way.

SECTION 3201
GENERAL

3201.1 Scope. The provisions of this chapter shall govern the encroachment of structures into the public right-of-way.

❖ These provisions apply only to encroachments on the public right of way, not on interior lot lines. The extent and location of the street lot line (public right-of-way) is determined by zoning or planning statutes.

3201.2 Measurement. The projection of any structure or portion thereof shall be the distance measured horizontally from the lot line to the outermost point of the projection.

❖ This section defines how the extent of encroachment is determined. The extent of encroachment of any structure or appendage is to be measured from the lot line to the outermost (furthest) point of the projection.

3201.3 Other laws. The provisions of this chapter shall not be construed to permit the violation of other laws or ordinances regulating the use and occupancy of public property.

❖ The projections permitted in this chapter are also subject to other limitations identified in applicable laws and ordinances. Other applicable laws, for example, are those established by the authorities in regulating right-of-way for highways, streets or public areas adjacent to the structure in question.

The street lot line occurs on each side of a structure that faces or fronts on a street. Many property descriptions have the centerline of the public way as the property line. In these cases, the lot line is defined not by the centerline of the public way, but by the line that identifies where the public way exists. The public right-of-way or other setback lines are usually established by zoning laws. For example, local zoning typically requires that residential uses have setbacks ranging from 25 to 60 feet (7620 to 18 288 mm), whereas business uses may be permitted to build up to the street lot line. While the walls of the structure may not project beyond this street lot line, certain projections are permitted as stated in this chapter.

The building lines for other parts of the structure are also usually established by local zoning ordinances in the form of required side and rear yards. This chapter and most zoning ordinances limit the projections into side setback areas. This chapter does not permit projections beyond interior lot lines (i.e., lot lines that are not street or public way lot lines). Once established, setback lines form the reference point for application of this chapter, especially when the street right-of-way line and setback are close or coincide.

Figure 3201.3 shows the building and street lot lines as established for a multiple-family occupancy.

3201.4 Drainage. Drainage water collected from a roof, awning, canopy or marquee, and condensate from mechanical equipment shall not flow over a public walking surface.

❖ A great number of street projections are canopies or mansard roofs. They have the potential to create a hazardous condition due to improper drainage on the walkways they cover. It is very important that all street projections have proper drainage systems.

For SI: 1 inch = 25.4 mm, 1 foot = 304.8 mm.

Figure 3201.3
LIMITATIONS ON STREET PROJECTIONS

SECTION 3202
ENCROACHMENTS

3202.1 Encroachments below grade. Encroachments below grade shall comply with Sections 3202.1.1 through 3202.1.3.

❖ Allowable projections of below-grade structure components are described in Sections 3202.1.1 through 3202.1.3.

3202.1.1 Structural support. A part of a building erected below grade that is necessary for structural support of the building or structure shall not project beyond the lot lines, except that the footings of street walls or their supports which are located at least 8 feet (2438 mm) below grade shall not project more than 12 inches (305 mm) beyond the street lot line.

❖ One of the permissible street lot line encroachments (footings) is defined in this section. Where a structure is built immediately adjacent to the street lot line, the footing is permitted to project across that line (see Figure 3202.1.1). Projection of footings across side (interior) lot lines and onto adjacent property is not permitted.

3202.1.2 Vaults and other enclosed spaces. The construction and utilization of vaults and other enclosed space below grade shall be subject to the terms and conditions of the authority or legislative body having jurisdiction.

❖ Below-grade vaults and other enclosed spaces that open at the sidewalk level are usually covered with solid structural caps (e.g., concrete slabs) and are permitted to project beyond the building line or street lot line. However, since these structures occur in the public right-of-way, they are subject to local zoning laws and approval of the authorities having jurisdiction.

For SI: 1 inch = 25.4 mm, 1 foot = 304.8 mm.

Figure 3202.1.1
ENCROACHMENT BELOW GRADE

3202.1.3 Areaways. Areaways shall be protected by grates, guards or other approved means.

❖ Areaways, which provide natural light or ventilation to below-grade spaces, are required to be protected by grates, guards or other approved means of protection (see Figure 3202.1.3).

3202.2 Encroachments above grade and below 8 feet in height. Encroachments into the public right-of-way above grade and below 8 feet (2438 mm) in height shall be prohibited except as provided for in Sections 3202.2.1 through 3202.2.3. Doors and windows shall not open or project into the public right-of-way.

❖ This section prohibits all projections into the public right-of-way that are above grade and below 8 feet (2438 mm), except those specifically allowed by Sections 3202.2.1 through 3202.2.3 and 3202.2.4 (i.e., steps, architectural features such as columns or pilasters, awnings, temporary vestibules and storm enclosures). The maximum permitted encroachments and minimum clearances, as well as additional protection requirements, for the specified allowed components are detailed in their respective sections.

3202.2.1 Steps. Steps shall not project more than 12 inches (305 mm) and shall be guarded by approved devices not less than 3 feet (914 mm) high, or shall be located between columns or pilasters.

❖ Steps are permitted to project a maximum of 12 inches (305 mm) into the public right-of-way by this section.

However, since steps may pose a tripping hazard, an approved device (guard), ornamental column or pilaster must be provided at a minimum height of 3 feet (914 mm) to minimize this hazard. Figure 3202.2.1 illustrates compliance with this provision.

3202.2.2 Architectural features. Columns or pilasters, including bases and moldings shall not project more than 12 inches (305 mm). Belt courses, lintels, sills, architraves, pediments and similar architectural features shall not project more than 4 inches (102 mm).

❖ This section addresses the variety of architectural appurtenances that are part of a structure's exterior wall or roof. Consistent with the projection limitations given for steps in Section 3202.2.1, the maximum encroachment into the public way is 12 inches (305 mm) for columns and pilasters (see Figure 3202.2.1). Other noted architectural features are limited to maximum encroachments of 4 inches (102 mm).

3202.2.3 Awnings. The vertical clearance from the public right-of-way to the lowest part of any awning, including valances, shall be 7 feet (2134 mm) minimum.

❖ Awnings are permitted to project into the public right-of-way. However, the clearance to any part of an awning, including its valance, is required to be a minimum of 7 feet (2134 mm) above grade (see Figure 3202.2.3).

Figure 3202.1.3
AREAWAYS

SIDE ELEVATION

For SI: 1 inch = 25.4 mm, 1 foot = 304.8 mm.

Figure 3202.2.1
STEPS

ELEVATION

For SI: 1 inch = 25.4 mm, 1 foot = 304.8 mm.

Figure 3202.2.3
AWNINGS

3202.3 Encroachments 8 feet or more above grade. Encroachments 8 feet (2438 mm) or more above grade shall comply with Sections 3202.3.1 through 3202.3.4.

❖ This section prohibits all projections into the public right-of-way that are 8 feet (2438 mm) or more above grade, except those specifically allowed by Sections 3202.3.1 through 3202.3.4. The maximum permitted encroachments and minimum clearances, as well as additional approval requirements, for the specified allowed components are detailed in their respective sections.

3202.3.1 Awnings, canopies, marquees and signs. Awnings, canopies, marquees and signs shall be constructed so as to support applicable loads as specified in Chapter 16. Awnings, canopies, marquees and signs with less than 15 feet (4572 mm) clearance above the sidewalk shall not extend into or occupy more than two-thirds the width of the sidewalk measured from the building. Stanchions or columns that support awnings, canopies, marquees and signs shall be located not less than 2 feet (610 mm) in from the curb line.

❖ This section contains three distinct requirements. First, awnings, canopies, marquees and signs must be designed in accordance with the provisions of Chapter 16, with regard to the live loads that may be imposed on such structures.

Second, the allowable projection of awnings, cano-pies, marquees and signs with less than 15 feet (4572 mm) of clearance above the sidewalk is limited to two-thirds of the width of the sidewalk measured from the building.

Finally, the location of vertical structural components used for supporting awnings, canopies, marquees and signs (stanchions or columns) is also regulated. A minimum of 2 feet (610 mm) of clearance between the curb line and the vertical structural support for awnings, canopies, marquees and signs is required (see Figure 3202.3.1).

3202.3.2 Windows, balconies, architectural features and mechanical equipment. Where the vertical clearance above grade to projecting windows, balconies, architectural features or mechanical equipment is more than 8 feet (2438 mm), 1 inch (25 mm) of encroachment is permitted for each additional 1 inch (25 mm) of clearance above 8 feet (2438 mm), but the maximum encroachment shall be 4 feet (1219 mm).

❖ Projections into the public right-of-way for appurtenances such as windows, balconies, architectural features or mechanical equipment more than 8 feet (2438 mm) above grade are allowed by this section. Such projections are limited to 1 inch (25 mm) of encroachment for each additional inch of clearance above the minimum 8-foot (2438 mm) clearance required. However, the maximum allowable projection is limited to 4 feet (1219 mm) (see Figure 3202.3.2).

For SI: 1 foot = 304.8 mm.

Figure 3202.3.1
AWNINGS, CANOPIES, MARQUEES AND SIGNS 8 FEET OR MORE ABOVE GRADE

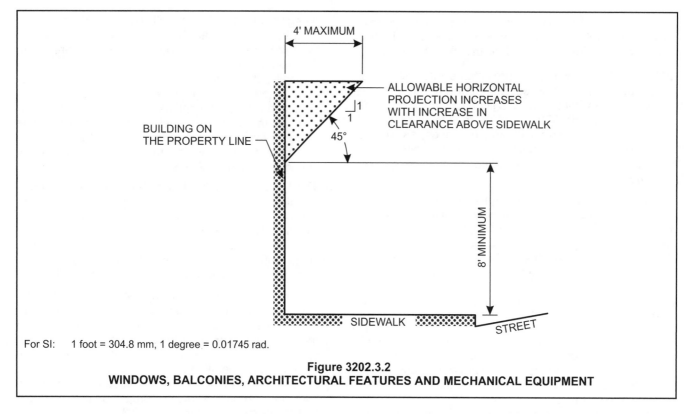

For SI: 1 foot = 304.8 mm, 1 degree = 0.01745 rad.

Figure 3202.3.2
WINDOWS, BALCONIES, ARCHITECTURAL FEATURES AND MECHANICAL EQUIPMENT

3202.3.3 Encroachments 15 feet or more above grade. Encroachments 15 feet (4572 mm) or more above grade shall not be limited.

❖ Because encroachments into the public right-of-way that are 15 feet (4572 mm) or more above grade do not interfere with or impede pedestrian or vehicular traffic, they are not regulated by the provisions of this chapter.

3202.3.4 Pedestrian walkways. The installation of a pedestrian walkway over a public right-of-way shall be subject to the approval of local authority having jurisdiction. The vertical clearance from the public right-of-way to the lowest part of a pedestrian walkway shall be 15 feet (4572 mm) minimum.

❖ As in Section 3202.3.3, projections of pedestrian walkways that have a minimum clearance of 15 feet (4572 mm) above grade do not interfere or impede pedestrian or vehicular traffic and thus are not prohibited by this section. The installation of pedestrian walkways over a public right-of-way is, however, required to be approved by the local authority having jurisdiction.

3202.4 Temporary encroachments. Where allowed by the local authority having jurisdiction, vestibules and storm enclosures shall not be erected for a period of time exceeding 7 months in any one year and shall not encroach more than 3 feet (914 mm) nor more than one-fourth of the width of the sidewalk beyond the street lot line. Temporary entrance awnings shall be erected with a minimum clearance of 7 feet (2134 mm) to the lowest portion of the hood or awning where supported on removable steel or other approved noncombustible support.

❖ This section addresses a special situation in which temporary projections across the building line or street lot line are erected for climatic reasons (i.e., during the winter months some building owners may erect vestibule enclosures or canopies over entrances). The purpose of this section is to allow these energy-saving procedures, yet place limitations on such projections.

Specifically, enclosures temporarily appended to an entranceway are permitted to project not more than 3 feet (914 mm) beyond the building line or street lot line, or up to one-fourth the width of the sidewalk. Whichever dimension results in the lesser projection defines the maximum projection.

Additionally, when temporary awnings are erected, clearance between the sidewalk and the lowest part of the awning must be at least 7 feet (2134 mm), which is consistent with the requirements for permanent awnings in Section 3202.2.3. It is further required that the frame for the hood or awning is to be supported on removable steel or other approved noncombustible material. This last provision of this section is necessitated by unforeseen changes in street lot lines, for example, that would result in the encroachment becoming an obstruction in the public right-of-way.

Chapter 33:
Safeguards During Construction

General Comments

The building construction process involves a number of known and unanticipated hazards. Chapter 33 establishes specific regulations in order to minimize the risk to the public and adjacent property. Some construction failures have resulted during the initial stages of grading, excavation and demolition. During these early stages, poorly designed and installed sheeting and shoring have resulted in ditch and embankment cave-ins. Also, inadequate underpinning of adjoining existing structures or careless removal of existing structures has produced construction failures.

The most critical period in building construction related to the physical safety of those on the job site is during the ongoing process when all building components have not yet been completed. Compounding this incomplete state is the use of dangerous construction methods, materials and equipment.

The importance of reasonable precautions is evidenced by the federal Occupational Safety and Health Act and related state and local regulations.

Purpose

The purpose of Chapter 33 is to cite safety requirements during construction or demolition of buildings and structures. These requirements are intended to protect the public from injury and adjoining property from damage.

SECTION 3301
GENERAL

3301.1 Scope. The provisions of this chapter shall govern safety during construction and the protection of adjacent public and private properties.

❖ The fundamental rationale behind this section is to establish that reasonable safety precautions are to be provided during the construction or demolition process to protect the public from injury and adjoining property from damage.

3301.2 Storage and placement. Construction equipment and materials shall be stored and placed so as not to endanger the public, the workers or adjoining property for the duration of the construction project.

❖ This section requires that construction materials and equipment must be located and protected pursuant to the governing provisions of this chapter so that the public and adjoining property are safeguarded at all times during construction or demolition processes.

SECTION 3302
CONSTRUCTION SAFEGUARDS

3302.1 Remodeling and additions. Required exits, existing structural elements, fire protection devices and sanitary safeguards shall be maintained at all times during remodeling, alterations, repairs or additions to any building or structure.

Exceptions:

1. When such required elements or devices are being remodeled, altered or repaired, adequate substitute provisions shall be made.

2. When the existing building is not occupied.

❖ Demolition and construction operations must not create a hazard for the occupants of a building during an alteration or addition. As such, the existing fire protection, means of egress elements and safety systems must remain in place and be functional. Two options, however, are provided in this section that permit the fire protection, means of egress elements and safety systems to be changed, modified or removed. The first option provides an alternative method that must be approved by the building official. The second option is simply to have an unoccupied building, which means that residents of a dwelling unit or employees of a business are not to be present during the construction or demolition process.

3302.2 Manner of removal. Waste materials shall be removed in a manner which prevents injury or damage to persons, adjoining properties and public rights-of-way.

❖ Safe and sanitary procedures for the removal of building construction and demolition waste must be provided. The method of waste removal must be controlled such that debris will not pose a hazard, eyesore or nuisance to the public and neighboring properties. Examples of acceptable practices include: evidence that a professional disposal service will haul away the debris; limiting areas used for storing and handling demolished materials; enclosing storage areas so that only authorized personnel can gain access; establishing routes that waste removal vehicles are permitted to use; covering or tarping debris to prevent flying objects; providing fully enclosed chutes to control falling objects and dust and scheduling waste removal when adjoining property and the public will be least exposed to unusual and possibly dangerous situations caused by fumes, noise, dust and unfamiliar events involved during a demolition process. Note that the accepted practice of waste removal is subject to the approval of the building official.

SECTION 3303
DEMOLITION

3303.1 Construction documents. Construction documents and a schedule for demolition must be submitted when required by the building official. Where such information is required, no work shall be done until such construction documents or schedule, or both, are approved.

❖ Some processes, methods and materials used during demolition can cause damage to adjoining property and increase the risk of injury to the general public. As such, this section indicates that the building official can request: verification that structural stability of existing buildings on the same lot or adjacent properties is maintained and not compromised; to know the materials and equipment necessary to perform the demolition involved; the timetable in which certain demolition activities will be performed; details to establish if the general public and adjoining property are exposed to any unfamiliar or greater hazard than normal; special inspection reports made by qualified firms, agencies or individuals; qualifications of the laborers involved in the operation, handling and removal of demolition equipment and materials; verification that any federal, state or local statute is followed, etc. Note that the building official must grant approval for any additional information requested before demolition can be started.

3303.2 Pedestrian protection. The work of demolishing any building shall not be commenced until pedestrian protection is in place as required by this chapter.

❖ Demolition must not be started until all of the necessary precautions are taken to protect the general public as indicated throughout this chapter.

3303.3 Means of egress. A party wall balcony or horizontal exit shall not be destroyed unless and until a substitute means of egress has been provided and approved.

❖ The exits provided through a party wall must be maintained operational and usable during demolition; however, based upon the extent of work, an alternative means of egress through a party wall is only permitted when approved by the authority having jurisdiction.

3303.4 Vacant lot. Where a structure has been demolished or removed, the vacant lot shall be filled and maintained to the existing grade or in accordance with the ordinances of the jurisdiction having authority.

❖ A demolition site must be restored if additional building construction or demolition operations are not scheduled to take place. The site must be filled and graded to the level of the surrounding site or maintained in ac-

cordance with local or state statutes, which may set forth other or additional grading requirements, such as provisions for elevations, drainage and flood control.

The time frame imposed or stipulated to abate an imminent hazard created by a vacant lot is subject to the building official. In addition, footings, foundations, basement walls and floors must be removed when the building official determines that a threat is posed to human life or the public welfare or if any portion of the foundation system prevents proper grading from being done.

3303.5 Water accumulation. Provision shall be made to prevent the accumulation of water or damage to any foundations on the premises or the adjoining property.

❖ A vacant lot must be graded in accordance with Section 3303.4 in such a way that water is prevented from ponding and causing damage to structures on the premises or adjacent properties, particularly foundation systems and building elements in contact with the ground. Footings, foundations, basement walls or floors must be removed if this drainage is prevented.

3303.6 Utility connections. Service utility connections shall be discontinued and capped in accordance with the approved rules and the requirements of the authority having jurisdiction.

❖ The procedures for disconnecting and abandoning utility connections in a safe and satisfactory manner must be in accordance with local, state and federal statutes. Utility providers, such as electrical, telephone, water, gas or sewer, often recommend or require notification and have certain guidelines to follow.

A related aspect is the precaution to investigate for any on-site underground obstructions, such as utility service lines and buried oil and gasoline or septic tanks. If the temporary use of some existing service, such as electricity or water, is requested during the demolition project, then temporary permits must be obtained with the appropriate stipulations, including any rerouting or protection requirements from the appropriate authority.

SECTION 3304
SITE WORK

3304.1 Excavation and fill. Excavation and fill for buildings and structures shall be constructed or protected so as not to endanger life or property. Stumps and roots shall be removed from the soil to a depth of at least 12 inches (305 mm) below the surface of the ground in the area to be occupied by the building. Wood forms which have been used in placing concrete, if within the ground or between foundation sills and the ground, shall be removed before a building is occupied or used for any purpose. Before completion, loose or casual wood shall be removed from direct contact with the ground under the building.

❖ A site that is to be excavated or filled must be kept out of contact with dead and decaying foliage. As such, the base of trees, roots and limbs that protrude above the ground as well as 12 inches (305 mm) into the ground must be taken out or separated from the earth where the building or structure will be erected. Additionally, any wood forms used for the concrete must be removed along with any wood product near the building that may decay or rot.

3304.1.1 Slope limits. Slopes for permanent fill shall not be steeper than one unit vertical in two units horizontal (50-percent slope). Cut slopes for permanent excavations shall not be steeper than one unit vertical in two units horizontal (50-percent slope). Deviation from the foregoing limitations for cut slopes shall be permitted only upon the presentation of a soil investigation report acceptable to the building official.

❖ The final grading around a foundation system must slope away from the building or structure, such that the ratio does not exceed 2:1. A trench or hole dug for a foundation system is required to have a stable embankment. As such, the embankment must not have a slope that exceeds a 2:1 ratio; however, due to the existing soil conditions or particular location, it may be impractical to achieve the required slope. This section, therefore, indicates that an engineering analysis can be submitted with recommendations for alternative methodology and safeguards, subject to the approval of the building official.

3304.1.2 Surcharge. No fill or other surcharge loads shall be placed adjacent to any building or structure unless such building or structure is capable of withstanding the additional loads caused by the fill or surcharge. Existing footings or foundations which can be affected by any excavation shall be underpinned adequately or otherwise protected against settlement and shall be protected against later movement.

❖ Gradient stability must be maintained for existing buildings or structures during the excavation, construction or demolition processes. As such, earth, construction materials, debris and even construction equipment must not impose a load that will cause the ground to give way and collapse or damage the foundation system of adjacent buildings or structures; therefore, components such as engineered bracing must be provided for protection if adjacent buildings or structures are to be exposed to loads that exceed those of the existing design.

3304.1.3 Footings on adjacent slopes. For footings on adjacent slopes, see Chapter 18.

❖ Any footings located on adjacent slopes must be handled pursuant to Chapter 18.

3304.1.4 Fill supporting foundations. Fill to be used to support the foundations of any building or structure shall comply with Section 1803.5. Special inspections of compacted fill shall be in accordance with Section 1704.7.

❖ Fill that will be installed to support a building or structure must be compacted and verified in accordance with Section 1803.4 and special inspection regulations stated in Section 1704.7.

SECTION 3305
SANITARY

3305.1 Facilities required. Sanitary facilities shall be provided during construction, remodeling or demolition activities in accordance with the *International Plumbing Code*.

❖ Construction employees must have plumbing facilities available during the construction or demolition process of a building. The facilities must conform to the requirements set forth in the *International Plumbing Code®* (IPC®).

SECTION 3306
PROTECTION OF PEDESTRIANS

3306.1 Protection required. Pedestrians shall be protected during construction, remodeling and demolition activities as required by this chapter and Table 3306.1. Signs shall be provided to direct pedestrian traffic.

❖ Safeguards are required to be in place during construction or demolition operations in accordance with Table 3306.1 and this chapter. In addition, since construction operations alter the familiar setting and path of travel, it is necessary to provide some form of visible directional sign to lead the public toward safety and away from potential hazards.

This table establishes the type of protection required based upon the location to overhead hazards and distance in relation to ground hazards. Once the type of protection is determined from the table, the applicable code section, such as Section 3306.4 for construction railings, Section 3306.5 for barriers and Section 3306.7 for covered walkways, must be provided as regulated.

3306.2 Walkways. A walkway shall be provided for pedestrian travel in front of every construction and demolition site unless the authority having jurisdiction authorizes the sidewalk to be fenced or closed. Walkways shall be of sufficient width to accommodate the pedestrian traffic, but in no case shall they be less than 4 feet (1219 mm) in width. Walkways shall be provided with a durable walking surface. Walkways shall be accessible in accordance with Chapter 11 and shall be designed to support all imposed loads and in no case shall the design live load be less than 150 pounds per square foot (psf) (7.2 kN/m²).

❖ Construction operations must not narrow or impede the normal flow of pedestrian traffic along a walkway by the placement of a fence or other enclosure. Note that the authority having jurisdiction must approve any construction operations that will cause the narrowing or impedance of a walkway, such as the sidewalk. If a walkway is narrowed or enclosed, the construction of another or wider walkway is required for all pedestrians. The walkway must be able to handle the normal anticipated flow of pedestrian traffic and must not be less than the minimum 4-foot (1219 mm) width. In addition, the walkway must be a stable surface that is capable of supporting all imposed loads. The minimum design load of the walkway must not be less than 150 pounds per square feet (psf) (7182 Pa).

TABLE 3306.1
PROTECTION OF PEDESTRIANS

HEIGHT OF CONSTRUCTION	DISTANCE FROM CONSTRUCTION TO LOT LINE	TYPE OF PROTECTION REQUIRED
8 feet or less	Less than 5 feet	Construction railings
	5 feet or more	None
More than 8 feet	Less than 5 feet	Barrier and covered walkway
	5 feet or more, but not more than one-fourth the height of construction	Barrier and covered walkway
	5 feet or more, but between one-fourth and one-half the height of construction	Barrier
	5 feet or more, but exceeding one-half the height of construction	None

For SI: 1 foot = 304.8 mm.

3306.3 Directional barricades. Pedestrian traffic shall be protected by a directional barricade where the walkway extends into the street. The directional barricade shall be of sufficient size and construction to direct vehicular traffic away from the pedestrian path.

❖ Similar to the local, federal and state guidelines that govern provisions to protect employees while doing road work, this section establishes that a barrier must be erected. The barrier must be capable of redirecting and regulating the flow of vehicular traffic where constructed or projected into a street. An example of how to regulate traffic is to provide the necessary warnings, notice of caution and instructions to the operators of motor vehicles with signs and colors that are common to the local, federal and state guidelines. The intent of this visual barrier is to keep pedestrians out of vehicular traffic and prevent vehicles from encroaching on or using the pedestrian walkway as a roadway.

3306.4 Construction railings. Construction railings shall be at least 42 inches (1067 mm) in height and shall be sufficient to direct pedestrians around construction areas.

❖ A barrier consisting of a horizontal rail and supports must be constructed with a minimum height of 42 inches (1067 mm). In addition, this barrier must be capable of controlling the flow of pedestrian traffic by routing travel away from the hazards associated with a construction area.

3306.5 Barriers. Barriers shall be a minimum of 8 feet (2438 mm) in height and shall be placed on the side of the walkway nearest the construction. Barriers shall extend the entire length of the construction site. Openings in such barriers shall be protected by doors which are normally kept closed.

❖ When a barrier is required by Table 3306.1, it must be constructed to impede, separate and obstruct passage of pedestrians onto a construction site. The structure must be a minimum 8 feet (2438 mm) in height and located continuously along the walkway on the side where construction activities are being done. Openings, such as doors or gates, are permitted as long as the only time they are open is during use by authorized personnel. As such, when the doors or gates are not being used, they must remain closed.

3306.6 Barrier design. Barriers shall be designed to resist loads required in Chapter 16 unless constructed as follows:

1. Barriers shall be provided with 2-inch by 4-inch (51 mm by 102 mm) top and bottom plates.

2. The barrier material shall be a minimum of $^3/_4$-inch (19.1 mm) boards or $^1/_4$-inch (6.4 mm) wood structural use panels.

3. Wood structural use panels shall be bonded with an adhesive identical to that for exterior wood structural use panels.

4. Wood structural use panels $^1/_4$ inch (6.4 mm) or $^5/_{16}$ inch (23.8 mm) in thickness shall have studs spaced not more than 2 feet (610 mm) on center (o.c.).

5. Wood structural use panels $^3/_8$ inch (9.5 mm) or $^1/_2$ inch (12.7 mm) in thickness shall have studs spaced not more than 4 feet (1219 mm) o.c., provided a 2-inch by 4-inch (51 mm by 102 mm) stiffener is placed horizontally at midheight where the stud spacing exceeds 2 feet (610 mm) o.c.

6. Wood structural use panels $^5/_8$ inch (15.9 mm) or thicker shall not span over 8 feet (2438 mm).

❖ This section establishes two methods that can be used to design barriers. The first method establishes that the barrier can be designed using the same materials and assembly instructions as indicated in Items 1 through 6. The second method provides for a barrier that will resist the loads imposed on it in order to be part of a designed assembly. The loads, which need to be resisted, are the same as those established in Chapter 16.

3306.7 Covered walkways. Covered walkways shall have a minimum clear height of 8 feet (2438 mm) as measured from the floor surface to the canopy overhead. Adequate lighting shall be provided at all times. Covered walkways shall be designed to support all imposed loads. In no case shall the design live load be less than 150 psf (7.2 kN/m²) for the entire structure.

Exception: Roofs and supporting structures of covered walkways for new, light-frame construction not exceeding two stories in height are permitted to be designed for a live load of 75 psf (3.6kN/m²) or the loads imposed on them, whichever is greater. In lieu of such designs, the roof and supporting structure of a covered walkway are permitted to be constructed as follows:

1. Footings shall be continuous 2-inch by 6-inch (51 mm by 152 mm) members.

2. Posts not less than 4 inches by 6 inches (102 mm by 152 mm) shall be provided on both sides of the roof and spaced not more than 12 feet (3658 mm) o.c.

3. Stringers not less than 4 inches by 12 inches (102 mm by 305 mm) shall be placed on edge upon the posts.

4. Joists resting on the stringers shall be at least 2 inches by 8 inches (51 mm by 203 mm) and shall be spaced not more than 2 feet (610 mm) o.c.

5. The deck shall be planks at least 2 inches (51 mm) thick or wood structural panels with an exterior exposure durability classification at least $^{23}/_{32}$ inch (18.3 mm) thick nailed to the joists.

6. Each post shall be knee braced to joists and stringers by 2-inch by 4-inch (51 mm by 102 mm) minimum members 4 feet (1219 mm) long.

7. A 2-inch by 4-inch (51 mm by 102 mm) minimum curb shall be set on edge along the outside edge of the deck.

❖ Walkways that are provided with a covering that extends out over the pedestrian path of travel, pursuant to Table 3306.1, must be at least 8 feet (2438 mm) in height. The measurement must be taken from the top of the walking surface vertically upward to the underside of the roof covering. The walkway must maintain a satisfactory level of light to the space that will be equal to or better than that provided prior to the installation of the protective covering. The covered walkway must also be structurally capable of resisting the design loads established in Chapter 16, but not less than 150 psf (7182 Pa) for the live load.

Note that this section provides two alternative design options. One option permits the covered walkway to resist the design loads established in Chapter 16, but not less than the minimum 75 psf (3591 Pa) live load if nearby construction does not exceed two stories in height and the construction materials are lightweight, such as light-frame construction. The second option is to construct the covered walkway in accordance with the seven criteria set forth in this section.

3306.8 Repair, maintenance and removal. Pedestrian protection required by this chapter shall be maintained in place and kept in good order for the entire length of time pedestrians may be endangered. The owner or the owner's agent, upon the completion of the construction activity, shall immediately remove walkways, debris and other obstructions and leave such public property in as good a condition as it was before such work was commenced.

❖ Any safeguards required by this chapter must be kept in good functional condition for the entire duration of the construction or demolition activity so that the public will not be placed in harm's way. Once a building or structure is occupiable and the site is properly graded,

the protection must be removed. As such, public property that was affected by the construction activity must be restored to or left in the condition that existed prior to the work.

3306.9 Adjacent to excavations. Every excavation on a site located 5 feet (1524 mm) or less from the street lot line shall be enclosed with a barrier not less than 6 feet (1829 mm) high. Where located more than 5 feet (1524 mm) from the street lot line, a barrier shall be erected when required by the building official. Barriers shall be of adequate strength to resist wind pressure as specified in Chapter 16.

❖ This section establishes that whenever excavation is to take place less than 5 feet (1524 mm) from the edge of a roadway, a barrier must be erected with a minimum height of 6 feet (1829 mm). If the excavation is located greater than 5 feet (1524 mm) from the edge of the roadway, the building official must evaluate the level of hazard for the public and the necessary precautions to take. As such, the building official is the authority to order the construction of a structural barrier. Any barrier erected must maintain other provisions of the code such that it is capable of handling and resisting design wind loads denoted in Chapter 16.

SECTION 3307
PROTECTION OF ADJOINING PROPERTY

3307.1 Protection required. Adjoining public and private property shall be protected from damage during construction, remodeling and demolition work. Protection must be provided for footings, foundations, party walls, chimneys, skylights and roofs. Provisions shall be made to control water runoff and erosion during construction or demolition activities. The person making or causing an excavation to be made shall provide written notice to the owners of adjoining buildings advising them that the excavation is to be made and that the adjoining buildings should be protected. Said notification shall be delivered not less than 10 days prior to the scheduled starting date of the excavation.

❖ This section emphasizes the need to protect all existing public and private property bordering the proposed construction or demolition operations. The term "property" only alludes to existing buildings. As such, any building element or system must be provided with a safeguard that will limit the damage that could be caused from the processes involved to the equipment and materials used. Additionally, soil erosion and land disbursement control resulting from the construction or demolition operations must be provided to prevent spillage and spread of disturbed soil debris. The site must be graded in accordance with Sections 3303.4 and 3303.5 for demolition and must be maintained in a similar manner while there is construction taking place. The owner or owner's agent has the responsibility to provide a written notice 10 days in advance for any demolition or construction activities that may warrant bordering lots to be protected from damage.

SECTION 3308
TEMPORARY USE OF STREETS, ALLEYS AND PUBLIC PROPERTY

3308.1 Storage and handling of materials. The temporary use of streets or public property for the storage or handling of materials or of equipment required for construction or demolition, and the protection provided to the public shall comply with the provisions of the authority having jurisdiction and this chapter.

❖ In addition to jurisdictional regulations, this chapter establishes procedures for the storage of construction materials and equipment to protect access to public safety equipment, utilities and public transportation facilities.

3308.1.1 Obstructions. Construction materials and equipment shall not be placed or stored so as to obstruct access to fire hydrants, standpipes, fire or police alarm boxes, catch basins or manholes, nor shall such material or equipment be located within 20 feet (6096 mm) of a street intersection, or placed so as to obstruct normal observations of traffic signals or to hinder the use of public transit loading platforms.

❖ This section indicates that precautions for material and equipment storage and placement must be provided so as not to block or obstruct access to fire hydrants; standpipes (including fire department siamese connections for sprinklers and standpipes); fire or police alarm boxes; utility boxes and meters; catch basins or manholes or any other vital facility whose function contributes to the health, safety and welfare of the public. Also, storage must not be placed within 20 feet (6096 mm) of a street intersection if it obstructs the normal observation of traffic signals or hinders the use of any mass-transit loading platforms, such as sidewalk bus stops, taxi waiting areas, etc.

3308.2 Utility fixtures. Building materials, fences, sheds or any obstruction of any kind shall not be placed so as to obstruct free approach to any fire hydrant, fire department connection, utility pole, manhole, fire alarm box or catch basin, or so as to interfere with the passage of water in the gutter. Protection against damage shall be provided to such utility fixtures during the progress of the work, but sight of them shall not be obstructed.

❖ Utility fixtures, such as those used by electrical, telephone, water, gas or sewer companies, as well as fire protection devices must be hidden from view and not be blocked from access and use. When construction operations are near a utility fixture, precautions must be provided so as not to cause damage. The utility companies and the authority having jurisdiction, including the fire department, may have specific requirements and guidelines to follow to limit the possibility of damage and obstruction.

SECTION 3309
FIRE EXTINGUISHERS

[F] 3309.1 Where required. All structures under construction, alteration or demolition shall be provided with not less than one

approved portable fire extinguisher in accordance with Section 906 and sized for not less than ordinary hazard as follows:

1. At each stairway on all floor levels where combustible materials have accumulated.

2. In every storage and construction shed.

3. Additional portable fire extinguishers shall be provided where special hazards exist, such as the storage and use of flammable and combustible liquids.

❖ A means to provide fire protection during the construction and demolition process is required. As such, this section indicates provisions for portable fire extinguishers. In addition to provisions of this section, the regulations of Section 906 and the *International Fire Code*® (IFC®) must be maintained.

3309.2 Fire hazards. The provisions of this code and the *International Fire Code* shall be strictly observed to safeguard against all fire hazards attendant upon construction operations.

❖ Methods, procedures and construction materials each contribute, in some way, to creating a fire hazard; therefore, in addition to the provisions of the code, the IFC must be used to regulate the proper means of safety that must be provided for a building that is being demolished, altered or constructed.

SECTION 3310
EXITS

3310.1 Stairways required. Where a building has been constructed to a height greater than 50 feet (15 240 mm) or four stories, or where an existing building exceeding 50 feet (15 240 mm) in height is altered, at least one temporary lighted stairway shall be provided unless one or more of the permanent stairways are erected as the construction progresses.

❖ Temporary stairways must be constructed in accordance with Section 1009, including riser, tread, guard and handrail requirements.

3310.2 Maintenance of exits. Required means of egress shall be maintained at all times during construction, demolition, remodeling or alterations and additions to any building.

Exception: Approved temporary means of egress systems and facilities.

❖ Access to all existing required exits is required to remain unobstructed and usable by any occupants within the existing building while construction or demolition work is being done. Note that depending upon the extent of work, an exit may become obstructed; therefore, an alternative exit must be provided if the blocked exit is a required exit.

[F] SECTION 3311
STANDPIPES

3311.1 Where required. Buildings four stories or more in height shall be provided with not less than one standpipe for use during construction. Such standpipes shall be installed where the progress of construction is not more than 40 feet (12 192 mm) in height above the lowest level of fire department access. Such standpipe shall be provided with fire department hose connections at accessible locations adjacent to usable stairs. Such standpipes shall be extended as construction progresses to within one floor of the highest point of construction having secured decking or flooring.

❖ The scope of this section is to provide fire safety procedures during the construction operation in accordance with the code and the IFC. Standpipes that are required by Section 905 to be a permanent part of the building must be installed and remain functional as construction or demolition progresses. Functional standpipes are required so that fire-fighting capability is available at all times within a reasonable proximity to all potential fire locations. Standpipes must be in place when a building or structure under construction exceeds 40 feet (12 192 mm) in height and thereafter as it progresses to its completed height. During demolition or construction, the standpipe must be operational no lower than one floor below the highest point of construction.

3311.2 Buildings being demolished. Where a building is being demolished and a standpipe exists within such a building, such standpipe shall be maintained in an operable condition so as to be available for use by the fire department. Such standpipe shall be demolished with the building but shall not be demolished more than one floor below the floor being demolished.

❖ Standpipes in buildings under demolition must remain in service during the demolition process so that fire-fighting capability is maintained. Similar to buildings under construction, the standpipe is to remain in operation up to one floor level below the floor being demolished. The purpose of this section is to establish the requirement for fire safety procedures during the demolition process in accordance with the code and the IFC.

3311.3 Detailed requirements. Standpipes shall be installed in accordance with the provisions of Chapter 9.

Exception: Standpipes shall be either temporary or permanent in nature, and with or without a water supply, provided that such standpipes conform to the requirements of Section 905 as to capacity, outlets and materials.

❖ During the construction or demolition process, standpipes complying with Section 3311.1 must be provided in a temporary or permanent location so that fire fighters will have a sufficient means to supply water to their connection. These standpipes must either be connected to a permanent water supply source or have the capability to be connected to one in accordance with Section 3311.4. All standpipes must comply with Section 905 and must be installed and provided in a building pursuant to the regulations of Chapter 9.

3311.4 Water supply. Water supply for fire protection, either temporary or permanent, shall be made available as soon as combustible material accumulates.

❖ This section is to provide a fire safety provision in accordance with the code and the IFC. As such, when a standpipe system is present during the construction process pursuant to Section 3311.1, a water source must be readily available to provide fire-fighting capabilities at all times.

SECTION 3312
AUTOMATIC SPRINKLER SYSTEM

[F] 3312.1 Completion before occupancy. In buildings where an automatic sprinkler system is required by this code, it shall be unlawful to occupy any portion of a building or structure until the automatic sprinkler system installation has been tested and approved, except as provided in Section 110.3.

❖ A certificate of occupancy must not be issued by the building official if a required automatic sprinkler system is not approved; however, a temporary occupancy may be granted at the discretion of the building official.

[F] 3312.2 Operation of valves. Operation of sprinkler control valves shall be permitted only by properly authorized personnel and shall be accompanied by notification of duly designated parties. When the sprinkler protection is being regularly turned off and on to facilitate connection of newly completed segments, the sprinkler control valves shall be checked at the end of each work period to ascertain that protection is in service.

❖ The scope of this section is to provide fire safety procedures during construction operations in accordance with the code and the IFC. A sprinkler system must remain operable unless work is being done on it. In such a case, the water must only be shut off by authorized personnel coupled with the notification of the proper authorities so that a form of check and balance is achieved to make certain the water is turned back on.

Bibliography

The following resource materials are referenced in this chapter or are relevant to the subject matter addressed in this chapter.

DOL 29 CFR; Part 1910-74, *Occupational Safety and Health Standards.* Washington, D.C.: U.S. Department of Labor-Occupational Safety and Health Administration, 1974.

IFC-2006, *International Fire Code.* Falls Church, VA: International Code Council, 2006.

IPC-2006, *International Plumbing Code.* Falls Church, VA: International Code Council, 2006.

Chapter 34
Existing Structures

General Comments

Chapter 34 contains provisions for the alteration, repair, addition and change of occupancy of existing buildings and structures. Chapter 34 includes or refers to the code requirements for existing buildings and structures, exclusive of the administrative provisions presented in Chapter 1 (Section 102.6 regulates unaltered existing buildings). The format of this chapter readily identifies the alternative methods of code compliance for existing buildings and structures.

Section 3401 establishes the provisions for maintenance, repairs and compliance with other technical codes.

Section 3402 provides definitions of terms that are primarily associated with Chapter 34.

Section 3403 states general requirements and also includes provisions and referenced sections that regulate existing structural loads, nonstructural elements and alteration or replacement of existing stairways.

Section 3404 contains the requirements for fire escapes.

Section 3405 contains provisions for the installation of new glass and glass replacement.

Section 3406 describes the conditions that constitute a change of occupancy.

Section 3407 addresses the requirements for historic buildings and structures.

Section 3408 addresses the requirements for structures that are moved into or within a jurisdiction.

Section 3409 contains provisions for accessible routes, as well as accessible elements within existing buildings and structures.

Section 3410 contains provisions for an alternative method of evaluation of additions, alterations or changes of occupancy in existing buildings and structures based on a numerical scoring system involving various safety issues and the degree of code compliance for each issue. This section contains the provisions to be included in the evaluation and the conditions on which that evaluation is based.

There are 18 safety parameters of every existing building that must be evaluated, including: building height; building area; compartmentation; tenant and dwelling unit separations; corridor walls; vertical openings; heating, ventilating and air-conditioning (HVAC) systems; automatic fire detection; fire alarm systems; smoke control; means of egress capacity and number; dead-end pockets and corridors; maximum travel distance to an exit; elevator controls; means of egress emergency lighting; mixed occupancies; sprinklers and incidental use areas.

The procedure for evaluating the 18 safety parameters in a qualitative and quantitative matrix has evolved from several different sources. A particularly strong proponent for this methodology has been the National Institute of Building Sciences (NIBS). NIBS has published eight rehabilitation guidelines for the U.S. Department of Housing and Urban Development (HUD).

Section 3410 also contains the necessary database information, the form for the evaluation process and the mandatory safety scores by which each occupancy must comply. The procedure for determining whether a building meets the pass/fail criteria and which mandatory score is needed for mixed occupancies is also included in this section.

Purpose

A large number of existing buildings and structures do not comply with the current building code requirements for new construction. Although many of these buildings are potentially salvageable, rehabilitation is often cost prohibitive because they may not be able to comply with all the requirements for new construction. At the same time, it is necessary to regulate construction in existing buildings that undergo additions, alterations, renovations, extensive repairs or change of occupancy. Such activity represents an opportunity to ensure that new construction complies with the current building codes and that existing conditions are maintained, at a minimum, to their current level of compliance or are improved as required. To accomplish this objective, and to make the rehabilitation process easier, this chapter allows for a controlled departure from full compliance with the technical codes, without compromising the minimum standards for fire prevention and life safety features of the rehabilitated building.

SECTION 3401
GENERAL

3401.1 Scope. The provisions of this chapter shall control the alteration, repair, addition and change of occupancy of existing structures.

> **Exception:** Existing bleachers, grandstands and folding and telescopic seating shall comply with ICC 300-02.

❖ This section states the scope of this chapter and references alternative methods of code compliance for the alteration, repair, addition and change of occupancy of existing structures. This section also defines the responsibilities for maintenance, repairs, compliance with other codes and periodic testing. These existing buildings or structure features known as bleachers, grandstands and folding and telescopic seating must comply with referenced standard ICC 300-02.

3401.2 Maintenance. Buildings and structures, and parts thereof, shall be maintained in a safe and sanitary condition. Devices or safeguards which are required by this code shall be maintained in conformance with the code edition under which installed. The owner or the owner's designated agent shall be responsible for the maintenance of buildings and structures. To determine compliance with this subsection, the building official shall have the authority to require a building or structure to be reinspected. The requirements of this chapter shall not provide the basis for removal or abrogation of fire protection and safety systems and devices in existing structures.

❖ This section establishes the owner's responsibility to keep the building maintained in accordance with the code and the other referenced *International Codes®*.

The building official has the authority to rule on the performance of maintenance work when public health or safety is affected. The building official also has the authority to require a building to be maintained in compliance with the health and safety provisions required by the code.

Fire protection and safety systems in existing structures are to remain in place and be maintained. The removal of a previously required existing safety system is not permitted, even if the structure would meet the minimum requirements of this chapter without that system. For example, an existing sprinkler system in part of a building must remain functional, even if the system was not required to meet the compliance alternative provisions of Section 3410.

3401.3 Compliance with other codes. Alterations, repairs, additions and changes of occupancy to existing structures shall comply with the provisions for alterations, repairs, additions and changes of occupancy in the *International Fire Code*, *International Fuel Gas Code*, *International Mechanical Code*, *International Plumbing Code*, *International Property Maintenance Code*, *International Private Sewage Disposal Code*, *International Residential Code* and ICC *Electrical Code*.

❖ This section clarifies the relationship between this chapter of the code and the *International Fire Code®* (IFC®), *International Fuel Gas Code®* (IFGC®), *International Mechanical Code®* (IMC®), *International Plumb-*

ing Code® (IPC®) *International Property Maintenance Code®* (IPMC®), *International Private Sewage Disposal Code®* (IPSDC®), *International Code Council Electrical Code Administrative Provisions* and the *International Residential Code®* (IRC®). When alterations and repairs are made to existing mechanical and plumbing systems, the provisions of the International Codes for alterations and repairs must be followed. Those codes indicate the extent to which existing systems must comply with the stated requirements. Where portions of existing building systems, such as plumbing, mechanical and electrical systems, are not being altered or repaired, those systems may continue to exist without being upgraded as long as they are not hazardous or unsafe to the building occupants.

SECTION 3402
DEFINITIONS

3402.1 Definitions. The following term shall, for the purposes of this chapter and as used elsewhere in the code, have the following meaning:

❖ Definitions of terms can be helpful in the understanding and application of the code requirements. The purpose for including this definition within this chapter is to provide more convenient access to it without having to refer back to Chapter 2. For convenience, this term is also in Chapter 2 with a cross reference to this section.

PRIMARY FUNCTION. A primary function is a major activity for which the facility is intended. Areas that contain a primary function include, but are not limited to, the customer service lobby of a bank, the dining area of a cafeteria, the meeting rooms in a conference center, as well as offices and other work areas in which the activities of the public accommodation or other private entity using the facility are carried out. Mechanical rooms, boiler rooms, supply storage rooms, employee lounges or locker rooms, janitorial closets, entrances, corridors and restrooms are not areas containing a primary function.

❖ Primary function areas contain the major activities for the building or space. Determination of what constitutes a primary function space can be somewhat subjective. There can be multiple areas containing a primary function in a single building. Primary function areas are not limited to public use areas. For example, both a bank lobby and the bank's employee areas, such as the teller areas and walk-in safe, are primary function areas. Areas that are not primary function spaces are support and circulation spaces. Determination of the primary function areas for a building will also determine when the route to that area and associated toilet rooms and drinking fountains must be evaluated for accessibility. If these items are not accessible, additional alterations may be necessary (see also commentary, Section 3409.6).

TECHNICALLY INFEASIBLE. An alteration of a building or a facility that has little likelihood of being accomplished because the existing structural conditions require the removal or alteration of a load-bearing member that is an essential part of the structural frame, or because other existing physical or site constraints prohibit modification or addition of elements, spaces or features which are in full and strict compliance with the minimum requirements for new construction and which are necessary to provide accessibility.

❖ This term is defined in order to provide a basis for the application of accessibility provisions to existing buildings (see Section 3409). Bringing any given existing site or building that is altered into full compliance with all accessibility requirements applicable to new construction may require extraordinary effort because of existing physical characteristics. The code utilizes the concept of technical infeasibility to provide a basis for exceptions from strict compliance with the provisions for new construction in an existing building.

SECTION 3403
ADDITIONS, ALTERATIONS OR REPAIRS

3403.1 Existing buildings or structures. Additions or alterations to any building or structure shall comply with the requirements of the code for new construction. Additions or alterations shall not be made to an existing building or structure that will cause the existing building or structure to be in violation of any provisions of this code. An existing building plus additions shall comply with the height and area provisions of Chapter 5. Portions of the structure not altered and not affected by the alteration are not required to comply with the code requirements for a new structure.

❖ The purpose of this section is to establish the guidelines for additions or alterations to existing buildings and structures.

An addition is an increase in the area to an existing building. When a new building is erected immediately adjacent to an existing building, and they are separated by a fire wall, it is considered a separate building, not an addition to the existing structure. The new building must be designed to comply with the technical provisions of Chapters 1 through 33; not with the provisions of this chapter. The existing building must be evaluated considering the elimination of the adjacent open space now occupied by the new building.

An existing structure that is of a type of construction that does not comply with the height and area limitations of Table 503 may not be added to unless the type of construction is upgraded, or the allowable building height and area are increased through the addition of a sprinkler system throughout both the existing building and the addition.

A building with a proposed addition is to be evaluated based on the type of construction of the existing building or the addition, whichever is the lower type. When reviewing for compliance with Table 503, see Section 602.1.1.

Figure 3403.1 shows an example of this situation. If a sprinkler system is not planned for both the existing building and the addition, the proposed addition does not meet the requirements of Table 503. The area limitation in Table 503 for the existing construction Type VB, Group F-1 is 8,500 square feet (789.7 m²) and the total proposed area is 11,000 square feet (1022 m²). The area evaluation is based on Type VB construction,

For SI: 1 inch = 25.4 mm, 1 foot = 304.8 mm,
1 square foot = 0.0929 m².

Figure 3403.1
EXISTING BUILDING WITH ADDITION

even though the proposed addition is Type VA construction, because the allowable area in Table 503 for Type VB construction is less than the allowable area for Type VA construction.

One way to satisfy the requirements of Table 503 is to upgrade the type of construction of the existing building to Type VA, which has an allowable area of 14,000 square feet (130 m²). Another solution is to install an automatic sprinkler system that complies with NFPA 13 throughout the building. The revised allowable area of the building would be:

8,500 square feet + (3 × 8,500 square feet)
= 34,000 square feet (3,158.7 m²)

in accordance with the sprinkler area increase, which is more than the proposed building area of 11,000 square feet (1022 m²).

Alterations include renovations, which implies that something is changed in the structure. For example, the removal, rearrangement or replacement of partition walls in an office building is an alteration because, in part, of possible impact on the means of egress, fire resistance or other life safety features of the building. Conversely, the replacement of damaged trim pieces on a door frame is considered a repair, not an alteration.

Alterations are to conform to the requirements for a new structure. For example, consider the corridor of an office building that is to be extended 18 feet (5486 mm), as shown in Figure 3403.2. The existing office building has no sprinkler system (and none is proposed), so the corridor extension walls and doors are to be fire-resistance rated in accordance with Section 1017. This is applicable even if, for some reason, the walls and doors of the existing corridor system are not fire-resistance-rated construction.

This section also indicates that unaltered portions of a structure are not required to comply with code provisions for new construction. All requirements applicable to existing, unaltered areas are contained in the code (see Sections 102.6 and 115) or the IPMC and the IFC.

Assume that the existing corridor walls in the example office building shown in Figure 3403.2 are not fire-resistance-rated construction.

Should this condition be deemed "unsafe" in accordance with Sections 102.6 and 115? The code gives the building official the authority to make these judgements based on the specific conditions unique to each existing building's characteristics. The IFC and the IPMC do not require fire-resistance-rated construction of existing corridor walls, nor do they require existing doors to be fire rated; however, they do require the doors to be self-closing. The corridor doors, therefore, must be brought into compliance with this requirement. The lack of a fire-resistance rating on an existing corridor is not, in itself, an example of an unsafe condition as defined in Section 115.1.

3403.1.1 Flood hazard areas. For buildings and structures in flood hazard areas established in Section 1612.3, any additions, alterations or repairs that constitute substantial improvement of the existing structure, as defined in Section 1612.2, shall comply with the flood design requirements for new construction, and all aspects of the existing structure shall be brought into compliance with the requirements for new construction for flood design.

❖ Reduction in exposure to flood hazards, including exposure of older buildings, is one of the purposes for regulating floodplain development. Existing buildings or structures located in flood hazard areas are to be brought into compliance with the flood-resistance provisions of Section 1612 when the value of improvements or the repair of damage exceeds a certain value.

Substantial improvement means any reconstruction, rehabilitation, addition or other improvement of a structure in which the cost equals or exceeds 50 percent of the market value of the structure before the start of construction for the improvement. This term includes structures that have incurred "substantial damage" regardless of the amount or value of actual repair work performed. Substantial damage means damage of any origin sustained by a structure whereby the cost of restoring the structure to its before-damaged condition would equal or exceed 50 percent of the market value of the structure before the damage occurred. The terms, defined in Section 1612.2, do not, include either:

1. Any project for improvement of a building or structure to correct existing violations of state or local health, sanitary or safety code specifications that have been previously identified by the building official and that are the minimum necessary to assure safe living conditions; or

2. Any alteration of a historic structure, provided that the alteration will not preclude the structure's continued designation as an historic structure.

Section 108.3 requires the applicant to provide an estimate of permit value, which is to include the total value of work, including materials and labor. If the proposed work will be performed on existing buildings or structures in flood hazard areas, a determination should be made as to whether the proposed work constitutes substantial improvement. For damaged buildings, the value of the proposed work is the value of all work necessary to restore the building to its predamage condition, even if the applicant is proposing less work. For any work on existing buildings and structures, the value of work includes the value of the property owner's labor and the value of donated labor and materials.

To make a determination about whether a proposed repair, reconstruction, rehabilitation, addition or improvement of a building or structure will constitute a substantial improvement or correction of substantial damage, the cost of the proposed work is to be compared to the market value of the building or structure before the work is started. In order to determine market value, the building official may require the applicant to provide such information. For additional guidance, refer to FEMA 213, *Answers to Questions about Substantially Damaged Buildings,* and FEMA 311, *Guidance on Estimating Substantial Damage Using the NFIP Residential Substantial Damage Estimator.*

3403.2 Structural. Additions or alterations to an existing structure shall not increase the force in any structural element by more than 5 percent, unless the increased forces on the element are still in compliance with the code for new structures, nor shall the strength of any structural element be decreased to less than that required by this code for new structures. Where repairs are made to structural elements of an existing building, and uncovered structural elements are found to be unsound or otherwise structurally deficient, such elements shall be made to conform to the requirements for new structures.

❖ Wherever an addition to an existing building is made, the affected members of the original structure must be assessed to determine their ability to resist the increased forces. Because an addition typically adds new loads to the existing building, the architect or engineer responsible for the design of the project must analyze the original construction to determine whether any of the structural components, including foundations, require reinforcement. Where an existing building was initially designed to support the loads of future additions, the building official should seek verification to establish that the proposed addition will result in the affected building complying with the current code requirements. All new construction is also required to comply with current code requirements.

Structural element forces may be increased by no more than 5 percent, but only if the structural element stresses do not exceed those given by the material standards referenced by the code at the time of construction. This restriction does not apply if the structural element meets the current material standards and is permitted by the code as if it were new.

When remodeling or repairing an existing building, structural elements, including any uncovered structural elements, that are found to be deficient must be reinforced or replaced. If the structural integrity has been affected, repairs are to be made in compliance with current code provisions.

3403.2.1 Existing live load. Where an existing structure heretofore is altered or repaired, the minimum design loads for the structure shall be the loads applicable at the time of erection, provided that public safety is not endangered thereby.

❖ Where an existing building is undergoing alterations or repairs requiring changes to the structural loading, the applicable minimum design loads are those required by the code at the time of the original construction. Should the original design loads (typically live loads) and stresses be substantially different from the current requirements, or other pertinent design criteria are added to the code that would lead the building official to believe that the public would be endangered, he or she has the authority to order investigation and correction of the deficiencies based on the load requirements and other applicable provisions of the current code.

3403.2.2 Live load reduction. If the approved live load is less than required by Section 1607, the areas designed for the reduced live load shall be posted in with the approved load. Placards shall be of an approved design.

For SI: 1 inch = 25.4 mm, 1 foot = 304.8 mm.

Figure 3403.2
PROPOSED CORRIDOR EXTENSION

❖ It is not uncommon for the design live load requirements to change from the time a building is originally designed to when it undergoes a renovation or change of occupancy. The live loads used in the original design may have been adequate for the building's initial use and may have been in compliance with all the code requirements that were in effect at that time. Many years and many code changes can change the status of the structural design parameters and code requirements, which does not mean the existing structural system is inadequate and cannot be used when the building is renovated or has a change of occupancy. It just means that when the live loads that were used for the design of the existing building are lower than the those required by current standards, the design live loads used for the original design must be posted.

3403.2.3 Seismic. Additions, alterations or modification or change of occupancy of existing buildings shall be in accordance with this section for the purposes of seismic considerations.

❖ An existing building or structure that is added on to, is altered or renovated or further modified or undergoes a change of use and occupancy must be carefully evaluated for its ability to withstand earthquake loading. The seismic resistance of an existing building or structure cannot be lessened or pose any undue increase in the fire and life safety hazards of the building or structure. Separate provisions are included for additions to existing buildings and alterations.

3403.2.3.1 Additions to existing buildings. An addition that is structurally independent from an existing structure shall be designed and constructed with the seismic requirements for new structures. An addition that is not structurally independent from an existing structure shall be designed and constructed such that the entire structure conforms to the seismic-force-resistance requirements for new structures unless the following conditions are satisfied:

1. The addition conforms with the requirements for new structures,

2. The addition does not increase the seismic forces in any structural element of the existing structure by more than 10 percent cumulative since the original construction, unless the element has the capacity to resist the increased forces determined in accordance with ASCE 7, and

3. Additions do not decrease the seismic resistance of any structural element of the existing structure by more than 10 percent cumulative since the original construction, unless the element has the capacity to resist the forces determined in accordance with ASCE 7. If the building's seismic base shear capacity has been increased since the original construction, the percent change in base shear may be calculated relative to the increased value.

❖ This section of the code gives more specific guidance with respect to the impact of an addition on the seismic resistance of an existing structure. The requirements of this section do not affect an existing structure where the addition is structurally independent. Only the independent addition is required to comply with the seismic requirements for new structures. Additions that are not structurally independent from the existing structure may increase the lateral load effect on the existing structure or the existing structure could impose additional lateral loads on the addition.

Because the addition and the existing structure resist earthquake effects as a single structure, designing the entire structure to resist the total lateral seismic forces is necessary. A small addition that is not structurally independent is permitted without alteration to the seismic-force-resisting system of the existing structure, provided the three conditions cited in the section are met. Condition 1 is reasonable since the addition represents new construction, and requiring compliance of the addition itself is not a hardship.

Condition 2 allows a 10-percent increase of the seismic forces in any structural element. It also allows the addition to increase the element's seismic forces by more than 10 percent as determined by the applicable provisions of referenced standard ASCE 7, provided that the existing structural element has the capacity to resist such increased forces.

Condition 3 permits a maximum cumulative decrease of 10 percent in the seismic resistance of any structural element in existing structures. It also allows the addition to decrease the element's seismic resistance by more that 10 percent, provided that the reduced capacity of the existing structural element is sufficient to resist the seismic forces prescribed for new structure as determined by the applicable provisions of referenced standard ASCE 7.

3403.2.3.2 Alterations. Alterations are permitted to be made to any structure without requiring the structure to comply with Section 1613, provided the alterations conform to the requirements for a new structure. Alterations that increase the seismic force in any existing structural element by more than 10 percent cumulative since the original construction or decrease the design strength of any existing structural element to resist seismic forces by more than 10 percent cumulative since the original construction shall not be permitted unless the entire seismic-force-resisting system is determined to conform to ASCE 7 for a new structure. If the building's seismic base shear capacity has been increased since the original construction, the percent change in base shear may be calculated relative to the increased value.

Exception: Alterations to existing structural elements or additions of new structural elements that are not required by ASCE 7 and are initiated for the purpose of increasing the strength or stiffness of the seismic-force-resisting system of an existing structure need not be designed for forces conforming to ASCE 7, provided that an engineering analysis is submitted indicating the following:

1. The design strength of existing structural elements required to resist seismic forces is not reduced.

2. The seismic force to required existing structural elements is not increased beyond their design strength.

3. New structural elements are detailed and connected to the existing structural elements as required by Chapter 16.

4. New or relocated nonstructural elements are detailed and connected to existing or new structural elements as required by Chapter 16.

5. The alterations do not create a structural irregularity as defined in ASCE 7 or make an existing structural irregularity more severe.

6. The alterations do not result in the creation of an unsafe condition.

❖ The term "alteration" is defined in Section 202 as "any construction or renovation to an existing structure other than repair or addition." This section of the code gives more specific guidance with respect to the impact of an alteration on the seismic resistance of the structure. Alterations that increase the seismic load on existing structural elements by more an 10 percent or decrease the seismic resistance of existing structural elements by more an 5 percent require the entire seismic-force-resisting system to be in compliance with the seismic load provisions applicable to a new structure. Alterations that affect existing structural elements to a lesser extent are permitted without requiring the existing structure to comply with the provisions for new structures, as long as the alteration itself complies.

The exception addresses the issue of upgrading existing structures voluntarily for improved seismic performance. It does not apply to situations where other code sections trigger full compliance with the code. Otherwise, it allows an owner to initiate an improvement to the seismic-force-resisting system to the extent that it is viable to do so and provided the required engineering analysis is furnished.

3403.3 Nonstructural. Nonstructural alterations or repairs to an existing building or structure are permitted to be made of the same materials of which the building or structure is constructed, provided that they do not adversely affect any structural member or the fire-resistance rating of any part of the building or structure.

❖ Where an existing building is altered or repaired, this section permits nonstructural materials that are removed to be replaced with the same type of materials that were used in the existing building. This provision is limited to the replacement use of nonstructural materials only, and only when the code does not specifically require otherwise. This section permits an existing wood glazing frame to be replaced with a new wood frame, but Section 3405.1 requires the replacement glazing to comply with the provisions for new construction.

3403.4 Stairways. An alteration or the replacement of an existing stairway in an existing structure shall not be required to comply with the requirements of a new stairway as outlined in Section 1009 where the existing space and construction will not allow a reduction in pitch or slope.

❖ An existing stair that is rebuilt in place is not required to comply with the provisions of Section 1009.3 for tread and riser dimensions. The rebuilt stair is permitted to have the same tread and risers as the existing stair. The term "rebuilt" in this section is intended to mean extensively repaired or removed and reconstructed in the same location.

SECTION 3404
FIRE ESCAPES

3404.1 Where permitted. Fire escapes shall be permitted only as provided for in Sections 3404.1.1 through 3404.1.4.

❖ Sections 3404.1.1 through 3404.5 address the current requirements for the use of exterior fire escapes. Their use and features as defined in this section should not be confused with the requirements for exterior exit stairways as defined in Section 1023.

The use of exterior fire escapes as a means of egress was popular in building designs of the past. They were used for economy of construction and to "save" usable space in the buildings they served. Being located outside the building, they were also considered safer than the unenclosed interior exit stairways of the time. Over the years, exterior fire escapes lost their appeal for many reasons:

- Fire escapes were never an integral part of the building's design; they were a necessary appendage hung from or attached to the building wall.
- Fire escapes were most commonly constructed of cast-iron or steel, both of which required a high degree of maintenance to protect the corrodible materials from the effects of weather.
- Because fire escapes are "open structures," they are subject to icing during the winter months. This makes them dangerous to use, and in extreme cases they can become completely unusable.
- People with a fear of heights can find exterior fire escapes difficult and sometimes impossible to use.
- Fire escapes were an unpleasant sight and an unwanted element in the architectural design.

The use of exterior fire escapes is all but obsolete except for existing buildings with a clear deficiency in the means of egress that cannot be reasonably rectified in other ways.

This section indicates that fire escapes are permitted only on existing buildings. New fire escapes may be installed on existing buildings only where exterior stairs cannot be used. Section 3404.2 addresses the location of fire escapes, and Section 3404.3 addresses the construction of fire escapes.

3404.1.1 New buildings. Fire escapes shall not constitute any part of the required means of egress in new buildings.

❖ Because of the inherent hazards and lack of dependability, open exterior fire escapes are not permitted as part of the required means of egress in new construction.

3404.1.2 Existing fire escapes. Existing fire escapes shall be continued to be accepted as a component in the means of egress in existing buildings only.

❖ Exterior fire escapes in existing buildings are acceptable as a component of the required means of egress because, in most cases, it would be physically impractical and economically prohibitive to retrofit older buildings with interior or exterior exit stairways that comply with all the code requirements for new construction.

3404.1.3 New fire escapes. New fire escapes for existing buildings shall be permitted only where exterior stairs cannot be utilized due to lot lines limiting stair size or due to the sidewalks, alleys or roads at grade level. New fire escapes shall not incorporate ladders or access by windows.

❖ Continuing corrosion of existing exterior fire escapes, as well as other factors that may affect the safety of the structures, often make replacement of these fire escapes necessary. The building official may determine that the means of egress in an existing building is unsafe, requiring some corrective action, which may be the installation of a new exterior fire escape. This section permits new exterior fire escapes for existing buildings where exterior stairs conforming to the requirements of Section 1023 cannot be used because of either lot lines that limit stair size or encroachments on sidewalks, alleys or roads at grade level.

Because it was the accepted practice of the time, access to exterior fire escapes was quite often through windows located in corridors or individual offices. This practice is not permitted when replacement exterior fire escapes are used. The building must be altered to provide proper exit access to the fire escape.

The code does not permit the use of ladders as a component in the means of egress; therefore, ladders cannot be used as a component part of an exterior fire escape, nor can a ladder be used to access an exterior fire escape.

3404.1.4 Limitations. Fire escapes shall comply with this section and shall not constitute more than 50 percent of the required number of exits nor more than 50 percent of the required exit capacity.

❖ For reasons of overall life safety, exterior fire escapes may be used as a component of the means of egress only when the total number used does not exceed one-half the required number of exits and more than one-half the total required exit capacity.

3404.2 Location. Where located on the front of the building and where projecting beyond the building line, the lowest landing shall not be less than 7 feet (2134 mm) or more than 12 feet (3658 mm) above grade, and shall be equipped with a counterbalanced stairway to the street. In alleyways and thoroughfares less than 30 feet (9144 mm) wide, the clearance under the lowest landing shall not be less than 12 feet (3658 mm).

❖ In the past, exterior fire escapes were allowed to project beyond the property line and extend over sidewalks, alleyways and roads. Where such conditions are still accepted by the code (see Sections 3404.1.2 and 3404.1.3), the clearance to the lowest landing from grade level must be at least 7 feet (2134 mm), and not more than 12 feet (3658 mm) for exterior fire escapes located in front of the building. In alleyways and thoroughfares less than 30 feet (9144 mm) wide, the clearance under the lowest landing must be at least 12 feet (3658 mm). To facilitate these clearance requirements, exterior fire escapes are usually equipped with a counterbalanced stairway that retracts to a horizontal position when the fire escape is not in use.

3404.3 Construction. The fire escape shall be designed to support a live load of 100 pounds per square foot (4788 Pa) and shall be constructed of steel or other approved noncombustible materials. Fire escapes constructed of wood not less than nominal 2 inches (51 mm) thick are permitted on buildings of Type 5 construction. Walkways and railings located over or supported by combustible roofs in buildings of Type 3 and 4 construction are permitted to be of wood not less than nominal 2 inches (51 mm) thick.

❖ Traditionally and typically, exterior fire escapes are constructed of cast iron or steel, although the code does permit exterior fire escapes to be constructed of other noncombustible materials. The use of nominal 2-inch (51 mm) wood is permitted in Type V construction. Nominal 2-inch (51 mm) wood may also be used in Type III and IV construction when the exterior fire escape is supported by wood construction permitted by Table 601.

3404.4 Dimensions. Stairs shall be at least 22 inches (559 mm) wide with risers not more than, and treads not less than, 8 inches (203 mm) and landings at the foot of stairs not less than 40 inches (1016 mm) wide by 36 inches (914 mm) long, located not more than 8 inches (203 mm) below the door.

❖ The normal dimensions for interior and exterior exit stairways (see Section 1009) used in the means of egress do not apply to exterior fire escapes. The dimensional standard for exterior fire escapes has been in place for many years, thus many exterior fire escapes in use were designed to comply with this standard. Many of these fire escapes are installed in places where it would be difficult, and often impossible, to retrofit a new fire escape if a new set of dimensional standards were employed. For this reason, it is judged appropriate to maintain the current standard. The minimum tread, riser, stair width and landing size dimensions for fire escape construction remain unchanged from the past.

3404.5 Opening protectives. Doors and windows along the fire escape shall be protected with $^3/_4$-hour opening protectives.

❖ Safe exit access and safe exits are required for occupants using an exterior fire escape as a means of egress. Door and window openings in exterior walls adjacent to and along the path of travel of the exterior fire escape must be protected from the interior of the building with not less than a $^3/_4$-hour protective. This is similar to the requirements for protection of openings on exterior stairs as described in Sections 1020.1.4 and 1023.6.

Where more restrictive opening protectives are required by other sections elsewhere in the code, the more restrictive protection must be used. The $^3/_4$-hour requirement for the exterior fire escape is the minimum when no other considerations are present (see Tables 601, 602, 704.8 and 715.5).

SECTION 3405
GLASS REPLACEMENT

3405.1 Conformance. The installation or replacement of glass shall be as required for new installations.

❖ The technical provisions for glass and glazing are detailed in Chapter 24. All technical requirements of the code for glass and glazing that apply to new construction also apply to all additions or alterations to existing buildings. There are no exceptions to this requirement. When glazing in an existing building is replaced or relocated within the same building, it must comply with current standards. Section 2401.2 contains specific requirements for replacement glazing.

SECTION 3406
CHANGE OF OCCUPANCY

3406.1 Conformance. No change shall be made in the use or occupancy of any building that would place the building in a different division of the same group of occupancy or in a different group of occupancies, unless such building is made to comply with the requirements of this code for such division or group of occupancy. Subject to the approval of the building official, the use or occupancy of existing buildings shall be permitted to be changed and the building is allowed to be occupied for purposes in other groups without conforming to all the requirements of this code for those groups, provided the new or proposed use is less hazardous, based on life and fire risk, than the existing use.

❖ A change in occupancy in an existing structure may change the level of inherent hazards that the code was initially intended to address.

Regardless of whether the change is to an occupancy considered to be more or less hazardous, this section applies the provisions of the code for new construction to an existing structure having a new occupancy. This is done so that the applicable code requirements adequately address the specific hazards of the new occupancy. For example, a change from an existing mercantile occupancy to a business occupancy renders all Group B provisions applicable to all portions of the structure where the occupancy has changed.

This section is one of the most frequently used provisions in the code for application to existing struc-

tures, since the occupancy in a building or structure is subject to change during the life of the building.

3406.2 Certificate of occupancy. A certificate of occupancy shall be issued where it has been determined that the requirements for the new occupancy classification have been met.

❖ An existing building that has been classified into a new occupancy group must receive a certificate of occupancy before tenancy. The code requirements for one occupancy group are not always the same as those for the new occupancy group. The new occupancy must be inspected to verify that all the applicable code requirements have been meet.

3406.3 Stairways. Existing stairways in an existing structure shall not be required to comply with the requirements of a new stairway as outlined in Section 1009 where the existing space and construction will not allow a reduction in pitch or slope.

❖ A stairway in an existing building does not have to be modified to comply with the dimensional provisions in Section 1009.3 for new stair construction. This provision is for the design of a particular stair and is aimed at addressing that stair only. This exception is not intended to eliminate any of the other technical provisions that might apply to an occupancy group, such as means of egress, accessibility, exit capacity, etc.

For example, when a building is altered or there is a change of tenant and the building is subject to a change in occupancy, the existing stair is permitted to remain as it was prior to the alteration or change of tenant. The means of egress capacity, however, must comply with the requirements for the new occupancy. The capacity may not be lessened to match the existing stair if it is not sufficient to provide the capacity required by the new occupancy. An additional stair or another approved means of egress must be provided to comply with the means of egress requirements for the new occupancy as indicated in Chapter 10.

3406.4 Change of occupancy. When a change of occupancy results in a structure being reclassified to a higher occupancy category, the structure shall conform to the seismic requirements for a new structure.

Exceptions:

1. Specific seismic detailing requirements of this code or ASCE 7 for a new structure shall not be required to be met where it can be shown that the level of performance and seismic safety is equivalent to that of a new structure. Such analysis shall consider the regularity, overstrength, redundancy and ductility of the structure within the context of the existing and retrofit (if any) detailing provided.

2. When a change of use results in a structure being reclassified from Occupancy Category I or II to Occupancy Category III and the structure is located in a seismic map area where $S_{DS} < 0.33$, compliance with the seismic requirements of this code and ASCE 7 are not required.

❖ This new provision of the code goes hand-in-hand with Section 3403.2.3. An existing building or structure that undergoes a change of use and occupancy must be carefully evaluated for its seismic resistance. The occupancy categories listed in Table 1604.5 must be used to determine whether an existing building or structure that now contains a new use and occupancy has been reclassified to a higher occupancy category. Obviously, an existing building or structure that previously contained ordinary uses and occupancies must be able to meet current code requirements relative to seismic loading if it will be used as an "essential facility."

Without Exception 1, it would be impossible in many instances to make an existing structure comply with the seismic requirements for a new structure. Exception 1 provides guidance for the building official and designer of areas that need to be investigated when compliance with the seismic requirements set forth in the code cannot be accomplished.

Exception 2 has it origins in ASCE 7. The purpose is to permit a change from Occupancy Category I or II to III for structures subjected to low earthquake accelerations without meeting the requirements for a new structure.

SECTION 3407
HISTORIC BUILDINGS

3407.1 Historic buildings. The provisions of this code relating to the construction, repair, alteration, addition, restoration and movement of structures, and change of occupancy shall not be mandatory for historic buildings where such buildings are judged by the building official to not constitute a distinct life safety hazard.

❖ This section provides an exception from code requirements when the building in question has historic value. The most important criterion for application of this section is that the building must be essentially accredited as being of historic significance by a qualified party or agency. Usually, this is done by a state or local authority after careful review of the historical value of the building. Most, if not all, states have such authorities, as do many local jurisdictions. The agencies with such authority can be located at the state or local government level or through the local chapter of the American Institute of Architects (AIA). Other considerations include the structural condition of the building (i.e., is the building structurally sound), its proposed use, its impact on life safety and how the intent of the code, if not the letter, will be achieved.

3407.2 Flood hazard areas. Within flood hazard areas established in accordance with Section 1612.3, where the work proposed constitutes substantial improvement as defined in Section 1612.2, the building shall be brought into conformance with Section 1612.

Exception: Historic buildings that are:

1. Listed or preliminarily determined to be eligible for listing in the National Register of Historic Places; or

2. Determined by the Secretary of the U.S. Department of Interior as contributing to the historical significance of a registered historic district or a district preliminarily determined to qualify as an historic district; or

3. Designated as historic under a state or local historic preservation program that is approved by the Department of Interior.

❖ This section is simply a reminder to the user of the code that when substantial improvements (as defined in Section 1612.2) are proposed to historic structures (as defined in Chapter 2), there could be some ramifications regarding a flood hazard that need to be considered, notwithstanding the exception stated in Section 3407.1. In other words, when there are substantial improvements to a historic building in a flood hazard area, the building official is not given the same discretion, as given in Section 3407.1, to waive code requirements for the alteration or restoration of any and all historic buildings. That discretion is only allowed for certain historic buildings that are listed in the exception to Section 3407.2. For additional discussion and guidance, see the commentary for Section G105.3.

SECTION 3408
MOVED STRUCTURES

3408.1 Conformance. Structures moved into or within the jurisdiction shall comply with the provisions of this code for new structures.

❖ Moved structures generally are required to comply with the provisions applicable to new construction. The moved structure may comply with the alternative provisions of Section 3410 instead of the code requirements for new structures, which may be particularly useful if the moved structure is older than the effective date of the adoption of building codes within the jurisdiction. The fire separation distance of the moved structure must comply with requirements for new structures, even if the compliance alternative provisions of Section 3410 are used to meet the code requirements.

SECTION 3409
ACCESSIBILITY FOR EXISTING BUILDINGS

3409.1 Scope. The provisions of Sections 3409.1 through 3409.9 apply to maintenance, change of occupancy, additions and alterations to existing buildings, including those identified as historic buildings.

Exception: Type B dwelling or sleeping units required by Section 1107 are not required to be provided in existing buildings and facilities.

❖ The purpose of Section 3409 is to establish minimum criteria for accessibility when dealing with existing buildings and facilities. The history and efforts involved are similar to that discussed in Chapter 11 (see commentary, Chapter 11, "General Comments"). When a facility or element is altered, it must meet new code requirements. For example, if a door and frame are removed and replaced, the door must meet the requirements for width, height, maneuvering clearances and hardware. If just the doorknob is being removed, it must be replaced with lever hardware.

If the area undergoing alteration does not contain a primary function (see Sections 3409.6 and 3409.7), there are no additional requirements; however, if the area contains a primary function, there are additional criteria that may require work not in the original scope to achieve accessibility. This additional criteria is to provide an accessible route to the altered area, including any toilets and drinking fountains that serve it. Requirements for an accessible route might specify that the door previously discussed be removed and replaced because it did not have adequate width or maneuvering clearances.

The principle behind this approach to upgrading existing buildings is that they will become more accessible over time. A valid time to work towards that goal is when a structure is being altered. Special considerations are offered because of the difficulty involved in dealing with existing facilities that may not have been built with accessibility for physically disabled persons in mind. For example, when a historically registered home is being made into a museum, if changing the front door to allow for wheelchair access would alter the historical significance, alternatives are offered in Section 3409.9. Another example is the alternatives offered in Section 3409.6 if it is technically infeasible to provide full accessibility in an existing building. Please note that the term "technically infeasible" refers to either movement of a major structural element or other physical constraints. For example, a ramp to provide entrance or exit from a particular door may not be possible because of property lines or setback constraints.

Appendix E addresses accessibility in existing buildings for items that are not typically enforceable through the normal building code enforcement approach.

Section 3409.1 addresses accessibility in existing buildings that are being renovated or altered. The exception for Type B dwelling and sleeping units in existing buildings being altered or undergoing a change of occupancy is for consistency with the Fair Housing Amendments Act (FHAA). Additions that contained dwelling and sleeping units required to be Type B in new construction would be required to meet Type B dwelling unit requirements. Accessible and Type A dwelling and sleeping units are required in existing institutional and residential buildings undergoing additions or alterations (see Section 3409.8.7).

3409.2 Maintenance of facilities. A building, facility or element that is constructed or altered to be accessible shall be maintained accessible during occupancy.

❖ Continued compliance with the accessibility requirements of the code is dependent on maintenance of

such facilities throughout the life of the building. For example, drinking fountains that are required to be accessible are of little value if they malfunction through deterioration or failure of any of the working parts. In other cases, inoperable elevators, locked accessible doors and obstructed accessible routes must be maintained such that they are readily usable by individuals with disabilities.

3409.3 Extent of application. An alteration of an existing element, space or area of a building or facility shall not impose a requirement for greater accessibility than that which would be required for new construction.

Alterations shall not reduce or have the effect of reducing accessibility of a building, portion of a building or facility.

❖ Alterations to existing buildings or structures should neither increase the need for greater accessibility, nor make the building or structure any less safe or usable. The scoping requirements for alterations are provided in Section 3409.8, but in no case should those alterations require any greater degree of accessibility than if the building or structure was brand new.

Exception 3 has been added to the code to address the specific circumstances when an existing Type A dwelling unit is undergoing alterations and renovations. If this Type A dwelling unit is "individually owned" such as a condominium, then it only needs to meet the code's requirements as a Type B dwelling unit. The provisions for Type B dwelling units are provided in Chapter 11 of the code and referenced standard ICC A117.1.

3409.4 Change of occupancy. Existing buildings, or portions thereof, that undergo a change of group or occupancy shall have all of the following accessible features:

1. At least one accessible building entrance.

2. At least one accessible route from an accessible building entrance to primary function areas.

3. Signage complying with Section 1110.

4. Accessible parking, where parking is being provided.

5. At least one accessible passenger loading zone, when loading zones are provided.

6. At least one accessible route connecting accessible parking and accessible passenger loading zones to an accessible entrance.

Where it is technically infeasible to comply with the new construction standards for any of these requirements for a change of group or occupancy, the above items shall conform to the requirements to the maximum extent technically feasible. Change of group or occupancy that incorporates any alterations or additions shall comply with this section and Sections 3409.5, 3409.6, 3409.7 and 3409.8.

❖ This section establishes that when an existing building or a portion thereof undergoes a change of occupancy, full compliance with accessibility requirements is expected and reasonable for the portion undergoing the change of occupancy except where technical infeasibility can be demonstrated. If full compliance is technically infeasible, the element must be made accessible to the fullest extent that is feasible. This is consistent with the general approach that has always been taken relative to other matters regulated by the code. Changes of occupancy have always been viewed as a reasonable opportunity to secure a higher level of code compliance than is necessarily deemed acceptable for alterations.

Six items are required in a facility undergoing a change of occupancy. The intent is to provide the bare minimum to get people from a point of arrival into the building and to the area of primary function for the new use. This is not based on any specific provisions of the *Americans with Disabilities Act Accessibility Guidelines* (ADAAG), but parallels the intent of the requirements for removal of barriers. If the area undergoing a change of occupancy contains a primary function, the accessible route provisions in Section 3409.7 are also applicable (see commentary, Section 3409.7).

3409.5 Additions. Provisions for new construction shall apply to additions. An addition that affects the accessibility to, or contains an area of, a primary function shall comply with the requirements in Section 3409.7.

❖ Additions must comply with new construction. An addition, however, is also an alteration to an existing building; therefore, accessible route provisions for existing buildings are applicable (see commentary, Section 3409.7). For example, a new dining area is added in a restaurant. All accessible elements within the parameter of the addition must be constructed accessible. If the route to the addition or the bathrooms is in the existing building, it must be evaluated to see if it needs to be altered.

3409.6 Alterations. A building, facility or element that is altered shall comply with the applicable provisions in Chapter 11 and ICC A117.1, unless technically infeasible. Where compliance with this section is technically infeasible, the alteration shall provide access to the maximum extent technically feasible.

Exceptions:

1. The altered element or space is not required to be on an accessible route, unless required by Section 3409.7.

2. Accessible means of egress required by Chapter 10 are not required to be provided in existing buildings and facilities.

3. The alteration to Type A individually owned dwelling units within a Group R-2 occupancy shall meet the provision for a Type B dwelling unit and shall comply with the applicable provisions in Chapter 11 and ICC/ANSI A117.1.

❖ The code approaches application of accessibility provisions to a facility that is altered by broadly requiring full conformance to new construction, meaning full accessibility is expected. Exceptions are then provided to indicate the conditions under which less than full ac-

cessibility is permitted.

The circumstance under which full compliance with accessibility provisions is not required is when it is deemed to be technically infeasible (see the commentary on the definition of "Technically infeasible" in Section 3402.1). This is considered reasonable since, if not provided for, plans for alterations may be otherwise abandoned by the building owner. The opportunity to upgrade and increase the current level of accessibility in an existing building would then be lost. This concern is also embodied in the requirement that an altered element or space is expected to be made accessible to the extent to which it is technically feasible to do so. In this manner, the code accomplishes the greatest degree of accessibility while recognizing the justifiable difficulties that may be involved in providing full accessibility.

The availability and usability of accessible elements are critically dependent on the presence of an accessible route leading to the accessible elements. The requirement for an accessible route represents one of the potential difficulties in an existing building that does not currently have adequate accessible routes. For example, in a multistory building, the accessible route to upper floors will most often be provided in the form of an elevator. It may be technically infeasible to provide an elevator in an existing building that does not currently contain an elevator. For this reason, Exception 1 indicates that an altered element or space is not required to be on an accessible route, except to the extent required in Section 3409.7.

Exception 2 indicates that accessible means of egress are not required as a result of undertaking alterations to existing buildings. Strict compliance with Section 1007 is often technically infeasible. Note that this is not an exception for accessible entrance requirements.

Exception 3 has been added to the code to address the specific circumstances when an existing Type R dwelling unit is undergoing alterations and innovations. If this Type A dwelling unit is "individually owned" such as a condominium, then it only needs to meet the code's requirements as a Type B dwelling unit. The provisions for Type B dwelling units are provided in Chapter 11 of the code and referenced standard ICC A117.1.

3409.7 Alterations affecting an area containing a primary function. Where an alteration affects the accessibility to, or contains an area of primary function, the route to the primary function area shall be accessible. The accessible route to the primary function area shall include toilet facilities or drinking fountains serving the area of primary function.

Exceptions:

1. The costs of providing the accessible route are not required to exceed 20 percent of the costs of the alterations affecting the area of primary function.

2. This provision does not apply to alterations limited solely to windows, hardware, operating controls, electrical outlets and signs.

3. This provision does not apply to alterations limited solely to mechanical systems, electrical systems, installation or alteration of fire protection systems and abatement of hazardous materials.

4. This provision does not apply to alterations undertaken for the primary purpose of increasing the accessibility of an existing building, facility or element.

❖ An area containing a primary function is one in which a major activity for which the building or facility is intended is carried out. For example, the lobby of a hotel in which the registration and check-out desk is located would be a primary function area. Other examples would be the dining area of a restaurant, the meeting rooms or exhibition halls in a conference center, virtually all office and work areas in a business building and retail display areas in a mercantile occupancy. The key concept is that a primary function area is one that contains a major activity of the facility. Areas that contain activities not related to the main purpose of the facility would not be considered a primary function area. For example, a mechanical equipment room, storage closet, toilet facilities, corridors, lounges and locker rooms would not be considered primary function areas. With this background, it is clear that areas containing a primary function are clearly more critical in terms of the purpose for which people enter and use the facility; therefore, this section reflects that when such areas are altered or added, it is important to require that an accessible route to the primary function area be provided. When an accessible route to a primary function area is required by this section, an accessible route to such facilities, including any restrooms and drinking fountains serving the primary function area, must also be made accessible, even though such facilities and areas may not by themselves be considered primary function areas.

There are conditions under which it may not be reasonable to enforce strictly this requirement for an accessible route to an altered or added primary function area. Exception 1 approaches this by utilizing the cost of the alterations or addition as a basis for determining if providing an accessible route is reasonable. The requirement for a complete accessible route does not apply when the cost of providing it exceeds 20 percent of the cost of the alterations or addition to the primary function area. These costs are intended to be based on the actual costs of the planned alterations or addition to the primary function area before consideration of the cost of providing an accessible route. For example, if the planned alterations will cost $100,000, not including the cost of an accessible route to a primary function area, this exception would apply if the additional cost of providing the accessible route would exceed $20,000.

It is not the intent to exempt all requirements for accessibility when the total cost for providing the accessible route exceeds the 20-percent threshold. Improvements to the accessible route are required to

the extent that costs do not exceed 20 percent of the cost to the planned alteration or addition. It is not required that the full 20 percent be spent. If the accessible route (including accessible bathrooms) is already provided, no additional expenditure is required. Note that there is not a priority list given for where money should be spent on improving the accessible route. The logical progression is access to the site, accessible exterior routes to accessible entrances, access throughout the facility, access to services within the facility, toilet and bathing rooms and, finally, drinking fountains. Evaluation on how and where the money available should best be spent must be made on a case-by-case basis. For example, if an accessible route is not available to an upper level, and the cost of an elevator is more than 20 percent of the cost of the renovation, then other alternatives could be investigated, such as a platform lift or limited access elevator, or adding the elevator pit and shaft at this time with elevator equipment added later. If all such items are in excess of the 20-percent limit, perhaps the money available could be spent towards making the toilet rooms accessible. The idea is that existing buildings would become fully accessible over time.

Exceptions 2 and 3 identify certain alterations that are not intended to trigger the requirement for providing an accessible route to a primary function area. Alterations limited to such elements as windows, hardware, operating controls, electrical outlets, signage, mechanical, electrical and fire protection systems, including alterations for the purpose of abating a hazardous materials circumstance, do not affect the usability of a primary function area in the same manner as alterations that affect the floor plan or the configuration, location or size of rooms or spaces. It is therefore considered unreasonable to require the installation of an accessible route when the scope of alterations is limited to that reflected in these exceptions.

Note that the costs for these items are not "backed out" of the total cost for the alteration before applying Exception 1. Exceptions 2 and 3 are alterations limited to the specific items referenced.

Exception 4 is intended to avoid penalizing a building owner who is undertaking alterations or additions for the purpose of increasing accessibility. It is appropriate to encourage owners to make such alterations without requiring them to do more work simply because they chose to increase the accessibility of the space. This could otherwise have the opposite effect of discouraging such alterations to avoid the expense of undertaking more work and expense than was originally planned. For example, federal law [Americans with Disabilities Act (ADA)] requires that owners of existing buildings remove certain existing barriers to accessibility. Removal of such barriers may require a permit from the building official. It would be unreasonable to have such activity trigger the mandatory requirement for further alterations to accomplish accessibility beyond the original planned work. In principle, the code takes the view that some extent of greater accessibility is positive progress and should be encouraged, not penalized.

3409.8 Scoping for alterations. The provisions of Sections 3409.8.1 through 3409.8.12 shall apply to alterations to existing buildings and facilities.

❖ The specific provisions of this section are intended to reflect conditions under which less than full accessibility, as would be required in new construction, is permitted in altered areas. As previously discussed, Section 3409.6 requires altered areas to comply with the full range of accessibility-related provisions of the code for new construction. This section reflects a reasonable set of conditions under which a different level of accessibility can be provided. Sections 3409.8.1 through 3409.8.12 are part of the code's coordination effort with referenced standard ICC A117.1 and the recommendations for the ADAAG Review Federal Advisory Committee.

3409.8.1 Entrances. Accessible entrances shall be provided in accordance with Section 1105.

> **Exception:** Where an alteration includes alterations to an entrance, and the building or facility has an accessible entrance, the altered entrance is not required to be accessible, unless required by Section 3409.7. Signs complying with Section 1110 shall be provided.

❖ This provision is contained here to point to the accessibility provisions of Chapter 11 for entrances (see commentary, Section 1105.).

3409.8.2 Elevators. Altered elements of existing elevators shall comply with ASME A17.1 and ICC A117.1. Such elements shall also be altered in elevators programmed to respond to the same hall call control as the altered elevator.

❖ Requirements for new construction state that all elevators on an accessible route must be fully accessible in accordance with reference standard ICC A117.1. If a passenger elevator is altered, the altered element must be accessible in accordance with the requirements for existing elevators in Section 407.5 of referenced standard ICC A117.1. If the altered elevator is part of a bank of elevators, the same element must be made accessible in every elevator that is part of that bank. The purpose of this requirement is to have consistency among elevators in a bank so that disabled people are not required to wait for a specific elevator when the general population can take the first available elevator.

3409.8.3 Platform lifts. Platform (wheelchair) lifts complying with ICC A117.1 and installed in accordance with ASME A18.1 shall be permitted as a component of an accessible route.

❖ This section provides for the use of platform (wheelchair) lifts in existing buildings. In order to create an accessible route where there are changes in floor levels, the provisions for new construction would most often require the installation of an elevator or ramp. Platform lifts are allowed in new construction for limited conditions (see Section 1109.7). If the space in an existing building precludes the installation of an elevator or ramp, a platform lift may be the only practical solution. Given the choice between no accessibility or accessibility by a platform lift, accessibility is preferable. Note that in accordance with Section 1007.5, platform lifts are also permitted for an accessible means of egress; however, accessible means of egress are not required in existing buildings undergoing alterations in accordance with Section 3409.6, Exception 2.

3409.8.4 Stairs and escalators in existing buildings. In alterations where an escalator or stair is added where none existed previously, an accessible route shall be provided in accordance with Sections 1104.4 and 1104.5.

❖ If a stair or escalator is added as part of an alteration in a location where one did not previously exist, the alteration must also include an accessible route between the same two levels. If an accessible route is already available between the two levels, or if the stair or escalator is replacing an existing stair or escalator, this requirement is not applicable. In conjunction with Section 3409.3, if the requirement for the accessible route would be in excess of what is required for new construction, such as an accessible route to an area that was exempted by Section 1103.2, 1104.4, 1107 or 1108, this requirement is not applicable. The intent is that if a route is provided between accessible levels for a nondisabled person to use, it is reasonable to also expect an accessible route.

3409.8.5 Ramps. Where steeper slopes than allowed by Section 1010.2 are necessitated by space limitations, the slope of ramps in or providing access to existing buildings or facilities shall comply with Table 3409.8.5.

❖ This section recognizes the circumstances where because of existing site or configuration constraints, a ramp with a slope of one unit vertical in 12 units horizontal (1:12) may not be feasible. A steeper slope is allowed where the elevation change does not exceed 6 inches (152 mm). The remainder of ramp requirements, such as width, landings, etc., are set forth in Section 1010.

TABLE 3409.8.5
RAMPS

SLOPE	MAXIMUM RISE
Steeper than 1:10 but not steeper than 1:8	3 inches
Steeper than 1:12 but not steeper than 1:10	6 inches

For SI: 1 inch = 25.4 mm.

❖ In existing buildings, ramps that rise 3 inches (76 mm) or less may have a slope as steep as one unit vertical in eight units horizontal (1:8). In existing buildings, ramps that rise 6 inches (152 mm) or less may have a slope as steep as one unit vertical in 10 units horizontal (1:10). If it is possible to provide a lesser slope, it is desirable to do so. These steeper slopes should only be utilized when the one unit vertical in 12 units horizontal (1:12) slope is not possible.

3409.8.6 Performance areas. Where it is technically infeasible to alter performance areas to be on an accessible route, at least one of each type of performance area shall be made accessible.

❖ This section recognizes that, because of the existing arrangement and location of performing areas (e.g., stages, platforms, orchestra pits, etc.), it may be infeasible to alter all performing areas to be on an accessible route. In such cases, it is reasonable to require that a minimum of one of each type of performing area be made accessible.

3409.8.7 Dwelling or sleeping units. Where I-1, I-2 , I-3, R-1, R-2 or R-4 dwelling or sleeping units are being altered or added, the requirements of Section 1107 for Accessible or Type A units and Section 907 for accessible alarms apply only to the quantity of spaces being altered or added.

❖ This section sets forth the rate for providing Accessible and Type A dwelling or sleeping units in Groups I-1, I-2, I-3, R-1, R-2 and R-4 when such facilities are altered. Assuming that Accessible or Type A units are not already provided, the number of Accessible or Type A units to be incorporated into each alteration is based on the number being altered. For example, if a nursing home was being altered a portion at a time, 50 percent of the units being altered each time would be required to be wheelchair accessible. It is not the intent that all units being altered are required to be Accessible until 50 percent of the units in the entire facility are accessible. The total number of Accessible units in the facility is not required to exceed that required for new construction, as indicated in Section 3409.3. It is unreasonable to require a greater level of accessibility in an existing building than is required in new construction.

   This section also references visible and audible alarm requirements in Section 907. If the alarm system is part of the alteration, the alarms must comply with Section 907. Sleeping accommodations in Groups I-1 and R-1 are required to have visible alarms in accordance with Table 907.9.1.3. Section 907.9.1.4 also contains requirements for alarms within Group R-2 units. If the alarm system is not part of the alteration, it is not the intent of this section to require the fire alarm system to be upgraded (see Section 3403.1).

3409.8.8 Jury boxes and witness stands. In alterations, accessible wheelchair spaces are not required to be located within the defined area of raised jury boxes or witness stands and shall be permitted to be located outside these spaces where the ramp or lift access restricts or projects into the means of egress.

❖ This exception for jury boxes and witness stands is consistent with Section 232 of ADAAG. The intent is that if ramp access to a jury box or witness stand would have the ramp limiting or blocking the means of egress for the general population in the space, alternative locations for potential jurors or witnesses are viable.

3409.8.9 Toilet rooms. Where it is technically infeasible to alter existing toilet and bathing facilities to be accessible, an accessible unisex toilet or bathing facility is permitted. The unisex facility shall be located on the same floor and in the same area as the existing facilities.

❖ This section deals with circumstances in which it is technically infeasible to alter existing toilet facilities to be accessible. In new construction, both the men's and women's facilities would be required to be accessible. An alternative solution when it is technically infeasible to alter the existing toilet rooms would be the creation of a single unisex toilet or bathing room containing accessible facilities. If this alternative is selected, the room must be located on the same floor and in the same area as the existing toilet or bathroom. This is the best alternative to fully complying, separate men's and women's facilities. One might argue that it is technically infeasible to accomplish either of these alternatives, since the alternative to altering existing facilities involves creation of an additional toilet or bathroom that was not otherwise contemplated. This would not be a persuasive argument since there is likely to be space available somewhere in the facility to commit for use as a toilet or bathroom. In any case, the intent is that some form of accessible toilet room or bathing facility is necessary and must be provided. Signage must be provided at the inaccessible toilet rooms in accordance with Sections 1110.1 and 1110.2 to notify disabled persons when a facility is not accessible and direct them to the nearest accessible facilities. It should be noted that this alternative is not offered as a choice between making the existing separate-sex toilet rooms accessible or providing a unisex accessible toilet room. The existing separate-sex toilet rooms must be altered when it is technically feasible. Consideration of a unisex facility is only available when alteration of the existing toilet rooms is technically infeasible (see the definition for "Technically infeasible" in Section 3402.1).

3409.8.10 Dressing, fitting and locker rooms. Where it is technically infeasible to provide accessible dressing, fitting or locker rooms at the same location as similar types of rooms, one accessible room on the same level shall be provided. Where separate-sex facilities are provided, accessible rooms for each sex shall be provided. Separate-sex facilities are not required where only unisex rooms are provided.

❖ This section takes a similar approach for dressing rooms as provided for in Section 3409.8.9 for toilet rooms and bathing facilities. If it is technically infeasible to alter existing dressing rooms to be accessible, then space elsewhere on the level must be committed to providing no less than one accessible dressing room. In this case, if the existing dressing rooms provide separate rooms for each sex, then no less than one accessible dressing room for each sex must be provided.

3409.8.11 Check-out aisles. Where check-out aisles are altered, at least one of each check-out aisle serving each function shall be made accessible until the number of accessible check-out aisles complies with Section 1109.12.2.

❖ This section deals with accessible check-out aisles. As check-out aisles are altered, at least one serving each function is required to be made accessible until the total number of accessible check-out aisles for the function meets Section 1109.12.2. In facilities with 5,000 square feet (465 m²) or less of selling space, only one check-out aisle is required to be made accessible (see Section 1109.12.2).

3409.8.12 Thresholds. The maximum height of thresholds at doorways shall be ³/₄ inch (19.1 mm). Such thresholds shall have beveled edges on each side.

❖ Thresholds at doorways may be ³/₄ inch (19.1 mm) maximum in existing buildings. In new construction, a typical threshold is ¹/₂ inch (12.7 mm) maximum in accordance with Section 1008.1.6. This section recognizes that such things as differences in floor materials may create changes in elevation greater than that allowed in new construction. Edges of thresholds must be beveled to allow passage of a wheelchair.

3409.9 Historic buildings. These provisions shall apply to buildings and facilities designated as historic structures that undergo alterations or a change of occupancy, unless technically infeasible. Where compliance with the requirements for accessible routes, entrances or toilet facilities would threaten or destroy the historic significance of the building or facility, as determined by the authority having jurisdiction, the alternative requirements of Sections 3409.9.1 through 3409.9.4 for that element shall be permitted.

❖ For this section to be applicable, the building must be registered as historic. Historic buildings are treated much the same as provided for in Sections 3409.4 and 3409.6, in that a historic building that is altered or has undergone a change of occupancy is expected to comply with accessibility requirements, unless technical infeasibility can be demonstrated; however, this section also goes on to acknowledge that the historic character of a building may be adversely affected by strict compliance with accessibility provisions. For example, compliance with door width requirements may necessitate the removal of an existing set of doors that is critical to the historic character of the building. This section is intended to exempt such conditions in order to maintain the historic character of the building. Because limited extent of accessibility is desired in all facilities, Sections 3409.9.1 through 3409.9.4 allow for alternatives.

In an effort to coordinate with ADAAG requirements, items that cannot be enforced through the typical code enforcement process, but are associated with historical buildings, have been included in Appendix E.

3409.9.1 Site arrival points. At least one accessible route from a site arrival point to an accessible entrance shall be provided.

❖ Full compliance would require an accessible route from all site arrival points. If this requirement would adversely affect the historical significance of the building, the alternative available is to provide an accessible route from one site arrival point to an accessible entrance.

3409.9.2 Multilevel buildings and facilities. An accessible route from an accessible entrance to public spaces on the level of the accessible entrance shall be provided.

❖ Full compliance might require an accessible route to levels above or below, as well as throughout the entrance level. If this requirement would adversely affect the historical significance of the building, the alternative is to provide an accessible route from the accessible entrance to all spaces open to the public on the entrance level. If elevators are provided, but are not accessible, signage in accordance with Section 1110.2 is required.

3409.9.3 Entrances. At least one main entrance shall be accessible.

Exceptions:

1. If a main entrance cannot be made accessible, an accessible nonpublic entrance that is unlocked while the building is occupied shall be provided; or

2. If a main entrance cannot be made accessible, a locked accessible entrance with a notification system or remote monitoring shall be provided.

Signs complying with Section 1110 shall be provided at the primary entrance and the accessible entrance.

❖ Full compliance would require 60 percent of the entrances to be accessible. If this requirement would adversely affect the historical significance of the building, only one main entrance is required to be made accessible. If a main entrance cannot be made accessible, then an employee or service entrance may serve as the accessible entrance, provided that it remains unlocked when the building is open. Alternatively, a locked entrance, where monitoring or a notification system is available, could be provided. Signage must be provided at inaccessible entrances in accordance with Sections 1110.1 and 1110.2.

3409.9.4 Toilet and bathing facilities. Where toilet rooms are provided, at least one accessible toilet room complying with Section 1109.2.1 shall be provided.

❖ Full compliance would require an accessible toilet/bathing facility at each location where toilet/bathing facilities are provided. If altering the existing facilities to be accessible would adversely affect the historical significance of the building, only one accessible toilet/bathing facility is required. Signage must be provided at inaccessible toilet rooms in accordance with Section 1110.2.

SECTION 3410
COMPLIANCE ALTERNATIVES

3410.1 Compliance. The provisions of this section are intended to maintain or increase the current degree of public safety, health and general welfare in existing buildings while permitting repair, alteration, addition and change of occupancy without requiring full compliance with Chapters 2 through 33, or Sections 3401.3, and 3403 through 3407, except where compliance with other provisions of this code is specifically required in this section.

❖ Neither the designer nor the building official can physically inspect and evaluate every aspect of an existing building or structure, since many of its features may be concealed within the construction. It is therefore necessary to emphasize those items that can be evaluated. There are 18 critically important elements that can be quantified and evaluated to determine the level of safety for an existing building.

This type of analysis provides the designer and the building official with a rational basis for establishing the safety of an existing building or structure without having physical access to every part of the building, documentation of the original design or the construction history of a building.

3410.2 Applicability. Structures existing prior to [DATE TO BE INSERTED BY THE JURISDICTION. NOTE: IT IS RECOMMENDED THAT THIS DATE COINCIDE WITH THE EFFECTIVE DATE OF BUILDING CODES WITHIN THE JURISDICTION], in which there is work involving additions, alterations or changes of occupancy shall be made to conform to the requirements of this section or the provisions of Sections 3403 through 3407. The provisions in Sections 3410.2.1 through 3410.2.5 shall apply to existing occupancies that will continue to be, or are proposed to be, in Groups A, B, E, F, M, R, S and U. These provisions shall not apply to buildings with occupancies in Group H or I.

❖ The adopting jurisdiction is to insert the desired date as indicated in this section. Section 3410 applies only to structures existing prior to the established date. The date that construction was first regulated through a comprehensive building code in the jurisdiction is recommended because buildings predating any building regulation are often not equipped with the types of systems and features that modern codes require. Newer buildings are more likely to be in compliance with current codes. These older buildings are assumed to face more difficulty in achieving a minimum level of life safety and are more likely to need the greater flexibility provisions of Section 3410.

The occupancies that qualify for the provisions of Section 3410 are listed in Section 3410.2.

Section 3410 does not apply to Groups H (high hazard) and I (institutional).

3410.2.1 Change in occupancy. Where an existing building is changed to a new occupancy classification and this section is applicable, the provisions of this section for the new occupancy shall be used to determine compliance with this code.

❖ When a building undergoes a change of occupancy classification and Section 3410 is applied, the evalua-

tion method in Section 3410 must be applied to the new occupancy for determining whether the existing building meets the compliance alternative in the code.

3410.2.2 Partial change in occupancy. Where a portion of the building is changed to a new occupancy classification, and that portion is separated from the remainder of the building with fire barriers or horizontal assemblies having a fire-resistance rating as required by Table 508.3.3 for the separate occupancies, or with approved compliance alternatives, the portion changed shall be made to conform to the provisions of this section.

Where a portion of the building is changed to a new occupancy classification, and that portion is not separated from the remainder of the building with fire barriers or horizontal assemblies having a fire-resistance rating as required by Table 508.3.3 for the separate occupancies, or with approved compliance alternatives, the provisions of this section which apply to each occupancy shall apply to the entire building. Where there are conflicting provisions, those requirements which secure the greater public safety shall apply to the entire building or structure.

❖ Where a portion of the building is changed to a new occupancy classification and is designed in accordance with Section 3410, the following options may be employed, dependent upon the fire separation of the portion of the building from the remainder of the existing building.

For a full fire separation, the new occupancy portion must be evaluated with the existing or proposed building design to be in full compliance with the provisions of Section 3410. The remainder of the existing building must also be evaluated in accordance with Section 3410. The mandatory safety scores for the new occupancy portion of the building and the existing occupancy are obtained from those listed in Table 3410.8 and are incorporated in the building's final evaluation score (see Table 3410.9).

For a nonfire-resistance-rated assembly of a new occupancy portion of an existing building, the provisions of Section 3410 for each occupancy must apply to the entire building. The requirements offering the greater public safety are applied to the entire building. Any proposed or existing building attributes and modifications must be reviewed in light of these requirements. See the examples described in Sections 3410.6.16 and 3410.9.1 for mixed occupancies.

3410.2.3 Additions. Additions to existing buildings shall comply with the requirements of this code for new construction. The combined height and area of the existing building and the new addition shall not exceed the height and area allowed by Chapter 5. Where a fire wall that complies with Section 705 is provided between the addition and the existing building, the addition shall be considered a separate building.

❖ The requirements in this section are applied in the same way as those in Section 3410.1. This section is included in Section 3410 so that provisions for additions are included for both optional methods of code compliance indicated in Section 3401.1. For a discussion of these options, see the commentary to Section 3401.1.

The evaluation method in Section 3410 may be used for an existing building that has been modified by an addition, provided that it meets the applicability requirements of Section 3410.2; however, the addition must comply with all the requirements of the code for new construction.

3410.2.4 Alterations and repairs. An existing building or portion thereof, which does not comply with the requirements of this code for new construction, shall not be altered or repaired in such a manner that results in the building being less safe or sanitary than such building is currently. If, in the alteration or repair, the current level of safety or sanitation is to be reduced, the portion altered or repaired shall conform to the requirements of Chapters 2 through 12 and Chapters 14 through 33.

❖ An existing building that is altered or repaired may be designed and evaluated in accordance with Section 3410, provided it meets the applicability provisions of Section 3410.2.

When an existing building is altered or repaired, materials or methods consistent with the original construction must be used. The alteration or repair must not cause the building to be less safe or sanitary than when it was originally constructed. Should the alteration or repair cause a reduction in safety or sanitation, such reductions must be corrected to meet the requirements of Chapters 2 through 12 and Chapters 14 through 33.

3410.2.4.1 Flood hazard areas. For existing buildings located in flood hazard areas established in Section 1612.3, if the alterations and repairs constitute substantial improvement of the existing building, the existing building shall be brought into compliance with the requirements for new construction for flood design.

❖ See the commentary to Section 3403.1.1.

3410.2.5 Accessibility requirements. All portions of the buildings proposed for change of occupancy shall conform to the accessibility provisions of Chapter 11.

❖ Any building or part of a building that has a change of occupancy must meet the requirements for accessibility in Chapter 11; however, Section 3409 is an acceptable alternative since Section 3409 applies to all existing buildings that have alterations, repairs, additions or a change of occupancy.

3410.3 Acceptance. For repairs, alterations, additions and changes of occupancy to existing buildings that are evaluated in accordance with this section, compliance with this section shall be accepted by the building official.

❖ When an owner or designer of an existing building decides to apply Section 3410 and complies with all the provisions of the section, including the applicability requirements in Section 3410.2, the building official must accept for review the proposed work or change of occupancy. This mandatory acceptance is tempered by one exception, as stated in Section 3410.3.1.

3410.3.1 Hazards. Where the building official determines that an unsafe condition exists, as provided for in Section 115, such unsafe condition shall be abated in accordance with Section 115.

❖ When the building official finds an unsafe condition in the building that is not being corrected by the proposed work, he or she must order the abatement or correction of the unsafe condition or hazard just as would be ordered in an existing building that is not being renovated, as stipulated by Section 115. This section sets forth a required comprehensive performance objective of abating any condition that is unsafe, insanitary, illegal or of improper occupancy, egress deficient, fire hazardous, poorly maintained or otherwise dangerous to human life or the public welfare. Guidelines for abatement are provided in the code and in Chapter 1 of both the IPMC (see Sections 108 and 110) and the IFC.

3410.3.2 Compliance with other codes. Buildings that are evaluated in accordance with this section shall comply with the *International Fire Code* and *International Property Maintenance Code*.

❖ This section requires an existing building that is subjected to the evaluation scoring process of Section 3410.6 to also comply with the IFC and IPMC. Those codes provide minimum requirements for health and safety that all existing buildings are expected to meet, regardless of whether there are any changes being made to the building or occupancy. Regardless of an existing building's final safety scores, the requirements of these referenced codes must be followed so occupants are safeguarded from hazards. This provision is similar to the requirements of Section 3401.2.

3410.4 Investigation and evaluation. For proposed work covered by this section, the building owner shall cause the existing building to be investigated and evaluated in accordance with the provisions of this section.

❖ This section and the subsequent sections address what must be done by the owner or designer who elects to employ Section 3410 for a proposed rehabilitation program and the corresponding action by the building official to assess the program objectively for approval or denial. The following actions must be taken:

• Fully investigate the building using both on-site inspections and research of all available building construction documents;

• Evaluate the building for conformance to Sections 3410.5 through 3410.9.1;

• Analyze the structure; and

• Submit to the building official documented results of the investigation and evaluation, plus any proposed compliance alternatives.

The building official must then determine whether the existing building, proposed work or change in occupancy complies with Section 3409.

Thus, the election and implementation of Section

3410 by the designer or owner is part of a code compliance option that may be used for existing buildings that the building official has the responsibility to review and assess. A proper and satisfactorily prepared submission by the owner or designer offering qualitative and quantitative data requires review by the building official for approval. If the submission is rejected, the building official must specifically cite any deficiencies and violations.

The owner is to have the existing building and any proposed work investigated and evaluated for compliance with the 18 parameters specified in Sections 3410.6.1 through 3410.6.18.

3410.4.1 Structural analysis. The owner shall have a structural analysis of the existing building made to determine adequacy of structural systems for the proposed alteration, addition or change of occupancy. The existing building shall be capable of supporting the minimum load requirements of Chapter 16.

❖ The owner is required to perform a complete structural analysis of the building to ensure that it can support the required loads. This requires that all interior loads meet the minimum load requirements of Chapter 16. Existing and altered buildings must be shown to support the expected loading. Any existing exterior member not affected by either interior loading or additional exterior loading need not be evaluated, provided that the structural member has, over a period of time, proven its ability to withstand the forces that normally create stress. Loads imposed on existing structural members by alterations, additions or a change of occupancy must be shown to sustain the requirements of Chapter 16, as stated in Sections 3403.2 through 3403.2.3.2. This structural analysis provides the owner and the building official with reasonable assurance that the building is structurally safe.

3410.4.2 Submittal. The results of the investigation and evaluation as required in Section 3410.4, along with proposed compliance alternatives, shall be submitted to the building official.

❖ The results of the investigation, including structural analysis and evaluation, must be submitted to the building official. If alternative methods, materials or equivalency concepts are proposed, these must also be submitted to the building official for review and approval.

3410.4.3 Determination of compliance. The building official shall determine whether the existing building, with the proposed addition, alteration or change of occupancy, complies with the provisions of this section in accordance with the evaluation process in Sections 3410.5 through 3410.9.

❖ When the results of the investigation and evaluation are submitted to the building official, he or she must determine whether the proposed work conforms to the provisions of Section 3410 and whether the evaluation was performed in accordance with Sections 3410.5 through 3410.9.

3410.5 Evaluation. The evaluation shall be comprised of three categories: fire safety, means of egress and general safety, as defined in Sections 3410.5.1 through 3410.5.3.

❖ This section and the subsequent sections address three general areas of safety to be evaluated: fire safety (FS), means of egress (ME) and general safety (GS). Section 3410.6 and the subsequent sections address 18 safety parameters that reflect on those areas. Each of the 18 safety parameters indicated in Sections 3410.6.1 through 3410.6.18 must be carefully reviewed and assigned a numerical value that signifies the degree of safety influence on the three overall general safety categories.

The 18 safety parameters that are given assigned values are:

3410.6.1	Building height
3410.6.2	Building area
3410.6.3	Compartmentation
3410.6.4	Tenant and dwelling unit separations
3410.6.5	Corridor walls
3410.6.6	Vertical openings
3410.6.7	HVAC systems
3410.6.8	Automatic fire detection
3410.6.9	Fire alarm systems
3410.6.10	Smoke control
3410.6.11	Means of egress capacity and number
3410.6.12	Dead ends
3410.6.13	Maximum exit access travel distance
3410.6.14	Elevator control
3410.6.15	Means of egress emergency lighting
3410.6.16	Mixed occupancies
3410.6.17	Automatic sprinklers
3410.6.18	Incidental use

These 18 safety parameters have been determined to be the most critical factors related to the minimum degree of life safety and property protection needed in an existing building.

When mixed occupancies in an existing building are not separated by fire-resistance-rated assemblies or fire walls meeting the most restrictive fire rating of the different occupancies, the entire evaluation must be based on the occupancy with the most restrictive requirements. The evaluation process considers the score for the various occupancies and applies the lowest score to the entire building.

When the mixed occupancies are properly separated in compliance with Section 508.3.3, they are to be evaluated separately and the score for each occupancy will apply to each portion based on its use. Both height and area formulas are computed for each occupancy. See the commentary to Sections 3410.2.2 and 3410.6.16 for a further discussion of application for mixed occupancies.

If there are four different occupancies in the building, the values must be computed for each of the four occupancies. Each occupancy is required to meet only its own mandatory building score.

When mixed occupancies are separated by fire walls complying with the requirements of Section 705, separate buildings are created and must be evaluated separately for all 18 safety parameters. The assigning of numerical values to each of the 18 safety parameters establishes a measurable quantity of what each of the parameters contributes to the overall safety of the building. Some evaluated parameters have a negative influence; others have a positive one. In total, the parameters may or may not result in an acceptable building score. The evaluation will determine whether the existing building has enough positive factors to overcome the negative parameters, or will indicate the negative factors that must be upgraded by alternative modifications.

The evaluation is divided into three categories or areas, which are defined in Sections 3410.5.1 through 3410.5.3.

3410.5.1 Fire safety. Included within the fire safety category are the structural fire resistance, automatic fire detection, fire alarm and fire suppression system features of the facility.

❖ A partial list of the items used to evaluate fire safety in a building is given in this section.

3410.5.2 Means of egress. Included within the means of egress category are the configuration, characteristics and support features for means of egress in the facility.

❖ The means of egress features that are evaluated by Section 3410 fall into the general areas of configuration, characteristics and support features. The specific features include travel distance, dead ends, emergency lighting and exit capacity and number.

3410.5.3 General safety. Included within the general safety category are the fire safety parameters and the means of egress parameters.

❖ This category includes every item that is used in either the fire safety or means of egress evaluation.

3410.6 Evaluation process. The evaluation process specified herein shall be followed in its entirety to evaluate existing buildings. Table 3410.7 shall be utilized for tabulating the results of the evaluation. References to other sections of this code indicate that compliance with those sections is required in order to gain credit in the evaluation herein outlined. In applying this section to a building with mixed occupancies, where the separation between the mixed occupancies does not qualify for any category indicated in Section 3410.6.16, the score for each occupancy shall be determined and the lower score determined for each section of the evaluation process shall apply to the entire building.

Where the separation between the mixed occupancies qualifies for any category indicated in Section 3410.6.16, the score for each occupancy shall apply to each portion of the building based on the occupancy of the space.

❖ This section is the key to understanding the entire evaluation process. The first sentence of this section clearly states that every one of the 18 safety parameters indicated in Sections 3410.6.1 through 3410.6.18 must be evaluated and nothing may be omitted. After the 18 safety parameters have been evaluated and assigned a numerical value, the values are entered in Ta-

ble 3410.7. The values must be tabulated to obtain the building score for each of the three general categories.

Some of the 18 safety parameters listed require mandatory compliance with other sections of the code and establish the foundation for determining a proper evaluation of those parameters, regardless of whether the result is positive or negative. This involves a coordination of mandatory basic requirements with the respective existing building conditions to arrive at the numerical evaluations prescribed in Sections 3410.6.1 through 3410.6.18. Section 3410.6.16 also addresses how mixed occupancies are handled in the evaluation. To apply this section, it is necessary to understand Section 508. See the commentary to Sections 508, 3410.2.2, 3410.6.2.1 and 3410.6.16 for a further discussion of the application of the evaluation procedure for mixed occupancies.

3410.6.1 Building height. The value for building height shall be the lesser value determined by the formula in Section 3410.6.1.1. Chapter 5 shall be used to determine the allowable height of the building, including allowable increases due to automatic sprinklers as provided for in Section 504.2. Subtract the actual building height from the allowable and divide by 12 $^1/_2$ feet. Enter the height value and its sign (positive or negative) in Table 3410.7 under Safety Parameter 3410.6.1, Building Height, for fire safety, means of egress and general safety. The maximum score for a building shall be 10.

❖ As a starting point for the actual evaluation process, this section and Section 3410.6.1.1 define in detail how to perform the building height evaluation. The exact values are to be computed and compared against the mandatory safety scores. The values are not to be rounded. For calculation purposes, two decimal places would be appropriate. It is not necessary to build in any inaccuracy that occurs through the rounding process.

The maximum score for a building is now 10. Regardless of the building's allowable height as compared to its actual height, the intent of the code is to limit all buildings to a maximum score of 10.

3410.6.1.1 Height formula. The following formulas shall be used in computing the building height value.

$$\text{Height value, feet} = \frac{(AH) - (EBH)}{12.5} \times CF$$

$$\text{Height value, stories} = (AS - EBS) \times CF$$

(Equation 34-1)

where:

AH = Allowable height in feet from Table 503.

EBH = Existing building height in feet.

AS = Allowable height in stories from Table 503.

EBS = Existing building height in stories.

CF = 1 if $(AH) - (EBH)$ is positive.

CF = Construction-type factor shown in Table 3410.6.6(2) if $(AH) - (EBH)$ is negative.

Note: Where mixed occupancies are separated and individually evaluated as indicated in Section 3410.6, the values AH, AS, EBH and EBS shall be based on the height of the fire area of the occupancy being evaluated.

❖ Two height formulas for calculating the score to be entered in Table 3410.7 are given. The formulas use the allowable height in both feet and stories from Table 503 and the height of the existing building. The denominator in the formula, 12.5, represents an average story height in feet.

The actual story height and the overall existing building height are to be directly compared to Table 503. Table 503 serves as a datum level that allows for establishing a numerical height value for the existing building by comparing its actual height and its type of construction as represented by a construction factor (CF).

If the existing building actual height is less than or equal to the allowable height of Table 503, then the CF value is 1 (no negative factor is assigned). If the actual height exceeds the Table 503 allowable height, the building is not in compliance with Table 503 and represents a safety deficiency. A deficiency rating is assigned that is dependent on the type of construction. The construction factors to establish this negative value are given in Table 3410.6.6(2).

When a building is not in compliance with Table 503, it is considered less safe than a building that does comply with Table 503. As a result, deficiency points are assessed. Additional safeguards must be provided to compensate for this condition. This is the primary reason that a different construction factor must be used for buildings not in compliance. The construction factor is the equivalent deficiency that must be overcome by providing additional protection in other areas that are to be evaluated.

Example 1:

A six-story building of Type IB construction is 60 feet (18 288 mm) tall. The building is not sprinklered. It contains a Group B business occupancy.

The allowable height from Table 503 is 11 stories, 160 feet (48,800 mm).

AH = 160 feet

AS = 11 stories

EBH = 60 feet

EBS = 6 stories

CF = 1.5 from Table 3410.6.6 (2)

Height value in feet $= \dfrac{160 - 60}{12.5} \times CF$

$= 8 \times 1.5$

$= 12$

Height value in stories $= (11 - 6) \times CF$

$= 5 \times 1.5$

$= 7.5$

The building height value will be 5, because it is the lesser of the two height values.

Example 2:

A seven-story building of Type IIIB construction is 80 feet (24 400 mm) tall. The building is sprinklered. It contains a Group M mercantile occupancy.

The allowable height from Table 503 is four stories, 55 feet (16 775 mm).

AH = 55 + 20 (for sprinklers) = 75 feet

AS = 4 + 1 (for sprinklers) = 5 stories

EBS = 7 stories

EBH = 80 feet

CF = 3.5 from Table 3410.6.6(2)

Height value in feet $= \dfrac{75 - 80}{12.5} \times CF$

$= -0.4 \times 3.5$

$= -1.4$

Height value in stories $= (5 - 7) \times CF$

$= -2 \times 3.5$

$= -7$

The building height value will be -7, because it is the lesser of the two height values.

The above values determine the entries for Table 3410.7 under Safety Parameter 3410.6.1 Building Height.

In Example 1, the height value of 7.5 is entered into the columns for fire safety (FS), means of egress (ME) and general safety (GS).

In Example 2, the height value of -7 is entered into the columns for fire safety (FS), means of egress (ME) and general safety (GS).

The assessed height value parameter is only one of the 18 safety parameters that need to be evaluated. Example 1 shows a positive contribution; however, this may not be enough to result in a sufficient overall building score. Similarly, the negative value in Example 2 may not in itself result in an overall insufficient building score.

3410.6.2 Building area. The value for building area shall be determined by the formula in Section 3410.6.2.2. Section 503 and the formula in Section 3410.6.2.1 shall be used to determine the allowable area of the building. This shall include any allowable increases due to open perimeter and automatic sprinklers as provided for in Section 506. Subtract the actual building area from the allowable area and divide by 1,200 square feet (112 m²). Enter the area value and its sign (positive or negative) in Table 3410.7 under Safety Parameter 3410.6.2, Building Area, for fire safety, means of egress and general safety. In determining the area value, the maximum permitted positive value for area is 50 percent of the fire safety score as listed in Table 3410.8, Mandatory Safety Scores.

❖ In this section, the code user is shown how to calculate the building score for the building area.

This section also requires the maximum score of a building in this category. The maximum value is 50 percent of just the fire safety score listed in Table 3410.8 (see commentary, Section 3410.6.2.2). This maximum score is applicable and equal for the three building score categories—fire safety, means of egress and general safety. The positive score is limited to prevent this one parameter from providing enough points to unjustifiably overcome too many deficiencies in other parameters.

3410.6.2.1 Allowable area formula. The following formula shall be used in computing allowable area:

$$AA = \frac{(SP + OP + 100) \times (\text{area, Table } 503)}{100}$$

(Equation 34-2)

where:

AA = Allowable area.

SP = Percent increase for sprinklers (Section 506.3).

OP = Percent increase for open perimeter (Section 506.2).

❖ The formula used to calculate the allowable area of the building is a direct correlation of the area increases and reductions allowed for new buildings in Section 506. The use of these increases and reductions is addressed in Chapter 5.

3410.6.2.2 Area formula. The following formula shall be used in computing the area value. Determine the area value for each occupancy fire area on a floor-by-floor basis. For each occupancy, choose the minimum area value of the set of values obtained for the particular occupancy.

$$\text{Area value } i = \frac{\text{Allowable area}_i}{1,200 \text{ square feet}} \left[1 - \left(\frac{\text{Actual area}_i}{\text{Allowable area}_i} + \ldots + \frac{\text{Actual area}_n}{\text{Allowable area}_n} \right) \right]$$

(Equation 34-3)

where:

i = Value for an individual separated occupancy on a floor.

n = Number of separated occupancies on a floor.

❖ The area formula provides a numerical value for the actual building area to be entered in Table 3410.7 under Safety Parameter 3410.6.2, Building Area. If the area of the existing building is less than the allowable

area, it is considered to be safer. Credit for area is given. If the area of the existing building is larger than the allowable area, it represents a condition that is judged to be less safe. The building receives a negative score. This can be overcome by providing additional safeguards.

To use the formula for existing buildings that contain mixed occupancies, an area value must be determined for each story of a building. If a building contains just one occupancy or just one occupancy on a particular floor separated from other floors, the formula simply reduces to the allowable area minus the actual area divided by the constant 1,200. The formula is also applicable to a story containing several separated occupancies. In such a situation, the formula requires each actual area to be divided by its respective allowable area. The resulting fractions are added together, subtracted from the constant of 1 and multiplied by the ratio of the allowable area of the particular use divided by the constant 1,200. This method of determining area values for several occupancies that are separated on a story is directly comparable to the unity formula in Sections 506.9.1 and 508.3.3.2.

Example 1: Figure 3410.6.2.2(1) illustrates a Group B building of Type IIA construction. The building is five stories in height, unsprinklered.

The overall building area is 200 feet (60 960 mm) by 200 feet (60 960 mm).

$$\text{Area value} = \frac{56,250}{1,200} \left[1 - \left(\frac{40,000}{56,250} \right) \right]$$

$$= 46.88 \, [1 - 0.71]$$

$$= 13.54$$

Allowable area = 56,250 square feet
Actual area = 40,000 square feet

In Table 3410.7, enter the following values under Safety Parameter 3410.6.2, Building Area:

Fire safety (FS) = 12

Means of egress (ME) = 12

General safety (GS) = 12

These values are equal to 50 percent of the value in Table 3410.8 for the mandatory fire safety score (MFS) of 24 for a Group B occupancy. This is the maximum value credit permitted for the building area parameter of Section 3410.6.2, even though the area value has been computed as 29.2 in Section 3410.6.2.2. The purpose for limiting the positive score is to prevent this one parameter from providing enough points to unjustifiably overcome too many deficiencies in other parameters.

The positive values entered in Table 3410.7 under Safety Parameter 3410.6.2, Building Area (FS = 12, ME = 12 and GS = 12), represents the evaluation for only one of the 18 safety parameters that must be assessed to arriving at the overall building score evaluation of Table 3410.7. As previously described for the building height parameter values under line 3410.6.1 of Table 3410.7, a single parameter of the total 18 safety parameters to be assessed does not, in itself, determine the ultimate acceptable building score.

Example 2: Figure 3410.6.2.2(2) illustrates a separated mixed occupancy building (Groups B, M and S-1) of Type IIIB construction. The building is two stories in height, fully sprinklered.

The overall building area is 100 feet (30 480 mm) by 200 feet (60 960 mm).

Actual area for Group M on first story = 20,000 square feet.

Actual area for Group M on second story = 4,400 square feet.

Actual area for Group B on second story = 10,000 square feet.

Actual area for Group S-1 on second story = 5,600 square feet.

SP = 200% (from Section 506.3)

OP = 5.6% (from Section 506.2)

Tabular areas from Table 503:

Group B = 19,000 square feet

Group M = 12,500 square feet

Group S-1 = 17,500 square feet

Allowable Areas:

$$AA = \frac{(200 + 5.6 + 100) \times \text{Tabular Area}}{100}$$

AA for Group B = 58,064 square feet

AA for Group M = 38,200 square feet

AA for Group S-1 = 53,480 square feet

Area Values:

For Group M, 1st story

$$\text{Area value} = \frac{38,200}{1,200}\left[1 - \left(\frac{20,000}{38,200}\right)\right]$$

$$= 31.83[1 - 0.52]$$

$$= 15.17$$

For Group B, 2nd story

$$\text{Area value} =$$
$$\frac{58,064}{1,200}\left[1 - \left(\frac{10,000}{58,064} + \frac{4,400}{38,200} + \frac{5,600}{53,480}\right)\right]$$

$$= 48.39[1 - 0.395]$$

ACTUAL AREA = 200' × 200' = 40,000 SQ.FT.

SP = 0 (FROM SECTION 506.3)

OP = $[\frac{600}{800} - .25]\frac{30}{30}$ = 50% (FROM SECTION 506.2)

ALLOWABLE AREA IN TABLE 503 = 37,500 SQ.FT.

$$AA = \frac{(0 + 50 + 100) \times 37,500}{100} = \frac{150 \times 37,500}{100} = 56,250 \text{ SQ.FT.}$$

Figure 3410.6.2.2(1)
ALLOWABLE AREA—SINGLE OCCUPANCY BUILDING

Area value = 29.41

For Group M, 2nd story

Area value = $\dfrac{38,200}{1,200}[1 - 0.39]$

Area value = 19.42

For Group S-1, 2nd story

Area value = $\dfrac{53,480}{1,200}[1 - 0.39]$

Area value = 27.19

Since these mixed occupancies are being separated so that one of the categories indicated in Section 3410.6.16 is applicable, a separate score must be computed for each occupancy.

In this example, the area value for the Group B occupancy is calculated to be 29.41, but the maximum area value permitted is 50 percent of the mandatory fire safety score listed in Table 3410.8. For a Group B

occupancy, the mandatory fire safety score is 24; therefore, the maximum positive value that can be entered into Table 3410.7 is 12.

When an occupancy is located in more than one story, a separate area value must be calculated for each story. For input into Table 3410.7, the area value to be used is the lesser of all the individual area values for that occupancy group, but not greater than 50 percent of the mandatory fire safety score. In Figure 3410.6.2.2(2), there is an area classified as a Group M occupancy on both the first and second stories. The area value for the Group M occupancy on the first story is 15.17, and the area value for the second story is 19.42. Fifty percent of the mandatory fire safety score for Group M is 9.5. The lowest value for all the Group M occupant areas in the building is 9.5, which is 50 percent of the mandatory fire safety score.

The area value for the Group S-1 occupancy portion of the building is 27.19. This value, just as the values for the other occupancies in this example, exceeds the 50 percent maximum permitted by the mandatory

Figure 3410.6.2.2(2)
ALLOWABLE AREA-MIXED OCCUPANCY BUILDING

fire safety score. The maximum for Group S-1 occupancy is 7.5, 50 percent of 15.

3410.6.3 Compartmentation. Evaluate the compartments created by fire barriers or horizontal assemblies which comply with Sections 3410.6.3.1 and 3410.6.3.2 and which are exclusive of the wall elements considered under Sections 3410.6.4 and 3410.6.5. Conforming compartments shall be figured as the net area and do not include shafts, chases, stairways, walls or columns. Using Table 3410.6.3, determine the appropriate compartmentation value (*CV*) and enter that value into Table 3410.7 under Safety Parameter 3410.6.3, Compartmentation, for fire safety, means of egress and general safety.

❖ This section establishes and evaluates the compartments contained within an existing building by the effectiveness of the enclosing fire separation assemblies of both walls and floor/ceiling assemblies. Larger compartments are considered to be a greater safety risk than smaller compartments because the larger areas can become involved in a single fire incident affecting a greater portion of the building at one time.

Fire barriers and horizontal assemblies must comply with Sections 3410.6.3.1 and 3410.6.3.2. Fire barriers and horizontal assemblies are exclusive of the other separations or enclosures relating to Sections 3410.6.4 and 3410.6.5.

The evaluation of the compartments contained within an existing building is a linear function allowing interpolation between the various categories (see Table 3410.6.3). This approach allows the compartmentation value to increase or decrease consistent with the actual changes in compartment sizes. Such an adjustment removes the previously built-in bias against smaller-sized buildings. Higher compartmentation values are assigned to buildings with smaller compartments.

TABLE 3410.6.3. See below.

❖ See Commentary to Section 3410.6.3

3410.6.3.1 Wall construction. A wall used to create separate compartments shall be a fire barrier conforming to Section 706

with a fire-resistance rating of not less than 2 hours. Where the building is not divided into more than one compartment, the compartment size shall be taken as the total floor area on all floors. Where there is more than one compartment within a story, each compartmented area on such story shall be provided with a horizontal exit conforming to Section 1022. The fire door serving as the horizontal exit between compartments shall be so installed, fitted and gasketed that such fire door will provide a substantial barrier to the passage of smoke.

❖ This section states that the walls determining the boundary of the compartment need to have fire barrier ratings of no less than 2 hours. These assemblies must be constructed in accordance with Section 706. For an existing building, this may need to be evaluated by both analyzing available plans and on-site investigations with a professional engineering determination of the required 2-hour fire-resistance rating. If the fire-resistance rating is less than 2 hours or cannot be reasonably determined, such walls should not be considered as compartmenting a fire area. The entire floor of an existing building must then be considered the compartment. If 2-hour-rated floor/ceiling assemblies are not present, the compartment size becomes the total area on all floors.

Openings and continuity requirements of Section 706 must be followed to maintain the integrity of the fire-resistance-rated wall assemblies and, thus, the compartments. Horizontal exits and their fire doors must comply with Section 1022.

The evaluation of an existing door condition requires an investigation of available engineering data and on-site inspections. Any uncertainty as to the performance of such doors may result in the openings being considered unprotected, expanding the compartment area or requiring an upgraded modification of the door to meet the current requirements of Section 1022.

3410.6.3.2 Floor/ceiling construction. A floor/ceiling assembly used to create compartments shall conform to Section 711 and shall have a fire-resistance rating of not less than 2 hours.

TABLE 3410.6.3
COMPARTMENTATION VALUES

	CATEGORIES[a]				
OCCUPANCY	a Compartment size equal to or greater than 15,000 square feet	b Compartment size of 10,000 square feet	c Compartment size of 7,500 square feet	d Compartment size of 5,000 square feet	e Compartment size of 2,500 square feet
A-1, A-3	0	6	10	14	18
A-2	0	4	10	14	18
A-4, B, E, S-2	0	5	10	15	20
F, M, R, S-1	0	4	10	16	22

For SI: 1 square foot = 0.093 m².
a. For areas between categories, the compartmentation value shall be obtained by linear interpolation.

❖ The building features that provide the horizontal boundaries of the compartment need to provide effective fire-resistant integrity between floors. The existing floor/ceiling assemblies need to be rated for 2 hours and need to be tight against exterior walls. Penetrations in the floor/ceiling assemblies must be protected in accordance with Section 711 to maintain their fire-resistant integrity. The floor/ceiling assemblies must conform to all of the requirements of Section 711 to create the level of compartmentation required for this evaluation parameter.

3410.6.4 Tenant and dwelling unit separations. Evaluate the fire-resistance rating of floors and walls separating tenants, including dwelling units, and not evaluated under Sections 3410.6.3 and 3410.6.5. Under the categories and occupancies in Table 3410.6.4, determine the appropriate value and enter that value in Table 3410.7 under Safety Parameter 3410.6.4, Tenant and Dwelling Unit Separation, for fire safety, means of egress and general safety.

❖ This parameter is used to evaluate partitions in an existing building other than those used for the creation of compartments in Section 3410.6.3 or the enclosure of corridors in Section 3410.6.5. This section examines the level of separation between tenant spaces and dwelling units. The listed categories specifically reference Sections 706, 708 and 711 not only for fire-resistance ratings but also for continuity and opening purposes. Further credit is provided for existing buildings that have a 2-hour-rated separation between adjacent tenant spaces or dwelling units, which exceeds what is required for new construction.

TABLE 3410.6.4
SEPARATION VALUES

OCCUPANCY	CATEGORIES				
	a	b	c	d	e
A-1	0	0	0	0	1
A-2	-5	-3	0	1	3
R	-4	-2	0	2	4
A-3, A-4, B, E, F, M, S-1	-4	-3	0	2	4
S-2	-5	-2	0	2	4

❖ Table 3410.6.4 provides values for tenant space and dwelling unit separations that resist the spread of flames and smoke. The rationale for considering a nonfire-resistance-rated or incomplete separation as a safety liability is its unassumed role of impeding flame and smoke spread. Separations that compartmentalize the areas with significantly rated fire separation walls and floor/ceiling assemblies are considered the most effective. Buildings containing Group R occupancies have separation values based on the fact that dwelling unit separations are more critical than tenant separations in other occupancies.

This chapter places greater emphasis on compartmentation than on open floor areas when compared to the rest of the code (see commentary, Sections 3410 and 3410.6.4).

3410.6.4.1 Categories. The categories for tenant and dwelling unit separations are:

1. Category a — No fire partitions; incomplete fire partitions; no doors; doors not self-closing or automatic closing.

2. Category b — Fire partitions or floor assembly less than 1-hour fire-resistance rating or not constructed in accordance with Sections 708 or 711, respectively.

3. Category c — Fire partitions with 1 hour or greater fire-resistance rating constructed in accordance with Section 708 and floor assemblies with 1-hour but less than 2-hour fire-resistance rating constructed in accordance with Section 711, or with only one tenant within the fire area.

4. Category d — Fire barriers with 1-hour but less than 2-hour fire-resistance rating constructed in accordance with Section 706 and floor assemblies with 2-hour or greater fire-resistance rating constructed in accordance with Section 711.

5. Category e — Fire barriers and floor assemblies with 2-hour or greater fire-resistance rating and constructed in accordance with Sections 706 and 711, respectively.

❖ Tenant space and dwelling unit separations are categorized by the partitions being evaluated. The values of each category are listed in Table 3410.6.4 by occupancy classifications. Typical illustrations of the types of partitions are shown in Figures 3410.6.4.1(1) and 3410.6.4.1(2). The listed categories provide a graduated level of separation when compared to new construction requirements.

Category a addresses the situation where there is no separation or there are gaps in the separation provided between tenant spaces or dwelling units.

Category b accounts for those tenant spaces or dwelling units that are separated from one another with less than a 1-hour rating.

Category c represents what a newly constructed building would have for tenant space and dwelling unit separation. An existing building meeting this level of compliance gains no benefit or penalty and, therefore, the separation value is zero.

Category d is given additional credit because the existing building space or dwelling unit has tenant separation that is rated higher than what is required by the code for new construction. Additionally, the walls are required to meet the fire barrier requirements of Section 706.

Category e is an added column that provides increased credit for existing buildings that have walls and floor/ceiling assemblies with fire-resistance ratings exceeding what is required for new buildings. This increased level of compartmentation in an existing building is credited accordingly with high separation values.

3410.6.5 Corridor walls. Evaluate the fire-resistance rating and degree of completeness of walls which create corridors serving the floor, and constructed in accordance with Section 1017. This evaluation shall not include the wall elements considered under Sections 3410.6.3 and 3410.6.4. Under the categories and groups in Table 3410.6.5, determine the appropriate value and enter that value into Table 3410.7 under Safety Parameter 3410.6.5, Corridor Walls, for fire safety, means of egress and general safety.

❖ Corridor walls are evaluated as fire partitions possessing an adequate fire-resistance rating and completeness to restrict fire and smoke migration into the corridor. Prescribed requirements are stated in Sections 708.4 and 1017. Existing corridor walls require investigation and analysis to determine equivalency to code requirements. Figures 3410.6.5(1) and 3410.6.5(2) illustrate various corridor wall values. Corridor walls contrast with the compartmentation and tenant/dwelling unit separations of Sections 3410.6.3 and 3410.6.4 by requiring an appropriate fire-resistance rating and continuity (see commentary, Sections 708.4, 3410.6.3 and 3410.6.4).

The corridor wall evaluations do not include partitions required to establish a compartment (see Section 3410.6.3) or tenant and dwelling unit separations (see Section 3410.6.4).

TABLE 3410.6.5
CORRIDOR WALL VALUES

OCCUPANCY	CATEGORIES			
	a	b	c[a]	d[a]
A-1	-10	-4	0	2
A-2	-30	-12	0	2
A-3, F, M, R, S-1	-7	-3	0	2
A-4, B, E, S-2	-5	-2	0	5

a. Corridors not providing at least one-half the travel distance for all occupants on a floor shall use Category b.

FLOOR/CEILING SYSTEM FIRE-RESISTANCE RATED IN ACCORDANCE WITH TABLE 601

FLOOR DECK

PARTITION EXTENDS TO UNDERSIDE OF FLOOR DECK

CEILING MEMBRANE

1- HOUR FIRE-RESISTANCE-RATED ASSEMBLY

• APARTMENTS, OCCUPANCY GROUP R-2
• FLOOR SYSTEM COMPLIES WITH TABLE 601
• SUSPENDED CEILING MEMBRANE IS PART OF RATED FLOOR/CEILING ASSEMBLY
• FLOOR-TO-DECK PARTITIONS

• CLASSIFICATION = CATEGORY c
• SCORE FROM TABLE 3410.6.4
• ENTER 0 IN TABLE 3410.7

Figure 3410.6.4.1(1)
DWELLING UNIT SEPARATION

❖ Table 3410.6.5 assigns values to the different uses based on the relative degree of fire resistance and smoke resistance of corridor walls. Since corridors are enclosed (confined) spaces subject to the rapid buildup of smoke and heat, a degree of protection is necessary to minimize this hazard to the occupants egressing the building. The table reflects this emphasis through substantial negative scores in buildings without properly enclosed corridors. Both Section 3410.6.5 and this table place an emphasis on corridor walls, which differs from that expressed elsewhere in the code. For example, Chapter 10 does not require corridors to be provided except in Group I-2 occupancies. The emphasis here is that when corridors are already present, they must provide basic protection. The table is an assessment of the relative risk represented by the corridor.

Note a to the table further controls the application of Categories c and d to existing buildings with certain means of egress arrangements. Although an existing building may have corridors with significant fire-resistance ratings and protected openings, little credit can be granted when the building occupants are protected

for only a short period of time. Very short corridors in large floor plans or corridors located in just one tenant space of a floor plan provide a benefit to only a portion of the occupant load for a small portion of the overall exit access travel and, thus, do not accrue any positive points. Unless rated corridors are available for all occupants of that particular floor or they provide a protected path of travel for at least one-half of the occupants' overall travel length, negative corridor wall values are assigned. In existing buildings with very short or limited-use corridors, the code user is directed to use Category b, regardless of the corridors' fire-resistance ratings and opening protectives.

3410.6.5.1 Categories. The categories for corridor walls are:

1. Category a — No fire partitions; incomplete fire partitions; no doors; or doors not self-closing.

2. Category b — Less than 1-hour fire-resistance rating or not constructed in accordance with Section 708.4.

3. Category c — 1-hour to less than 2-hour fire-resistance rating, with doors conforming to Section 715 or without corridors as permitted by Section 1017.

- FOOD COURT RESTAURANTS, OCCUPANCY GROUP A-3
- SOME PARTITIONS EXTEND TO CEILING OF FIRE-RESISTANCE-RATED FLOOR/CEILING ASSEMBLY
- NO CLOSERS ON DOORS BETWEEN ADJACENT TENANT SPACES
- SOME 6' PRIVACY PARTITIONS (PARTIAL PARTITIONS)

- CLASSIFICATION = CATEGORY a
- SCORE FROM TABLE 3410.6.4
- ENTER -4 IN TABLE 3410.7

For SI: 1 foot = 304.8 mm.

Figure 3410.6.4.1(2)
TENANT SEPARATION

- OCCUPANCY GROUP B
- NO CLOSERS ON DOORS
- FLOOR SYSTEM COMPLIES WITH TABLE 601
- SUSPENDED CEILING MEMBRANE IS NONFIRE-RESISTANCE RATED
- CORRIDOR WALLS ARE 1-HR. FIRE-RESISTANCE RATED

NOTE: THESE CONDITIONS PRODUCE A CATEGORY a
BECAUSE OF BOTH INCOMPLETE FIRE PARTITIONS
AND NO SELF-CLOSING DOORS

- CLASSIFICATION = CATEGORY a
- SCORE FROM TABLE 3410.6.5 = -5
- ENTER -5 IN TABLE 3410.7

Figure 3410.6.5(1)
CORRIDOR WALL VALUES—CATEGORY a

- OCCUPANCY GROUP B
- CLOSERS ON DOORS
- FIRE-RESISTANCE-RATED FLOOR/CEILING ASSEMBLY
- CORRIDOR WALLS ARE NONFIRE-RESISTANCE RATED
- PARTITIONS EXTEND TO UNDERSIDE OF CEILING MEMBRANE

NOTE: THESE CONDITIONS PRODUCE A CATEGORY b
AS THE PARTITIONS HAVE LESS THAN A 1 HR. RATING

- CLASSIFICATION = CATEGORY b
- SCORE FROM TABLE 3410.6.5 = -2
- ENTER -2 IN TABLE 3410.7

Figure 3410.6.5(2)
CORRIDOR WALL VALUES—CATEGORY b

4. Category d — 2-hour or greater fire-resistance rating, with doors conforming to Section 715.

❖ Corridor walls are categorized by the partitions being evaluated. The values of each category are listed in Table 3410.6.5 by occupancy classification.

3410.6.6 Vertical openings. Evaluate the fire-resistance rating of exit enclosures, hoistways, escalator openings and other shaft enclosures within the building, and openings between two or more floors. Table 3410.6.6(1) contains the appropriate protection values. Multiply that value by the construction-type factor found in Table 3410.6.6(2). Enter the vertical opening value and its sign (positive or negative) in Table 3410.7 under Safety Parameter 3410.6.6, Vertical Openings, for fire safety, means of egress and general safety. If the structure is a one-story building, enter a value of 2. Unenclosed vertical openings that conform to the requirements of Section 707 shall not be considered in the evaluation of vertical openings.

❖ Vertical openings are used to evaluate the fire-resistance ratings of openings between floors of a building and between shaft enclosures, such as stairs, elevator hoistways and escalator openings. This section also gives the formula for determining the score to be entered in Table 3410.7. This section does not apply to any building element that conforms to the requirements of Section 707.2, which requires all floor openings connecting two or more stories to be protected by a shaft enclosure complying with Section 707; however, Section 707.2 includes a number of exceptions to these requirements, such as Section 404 on atriums; a floor opening connecting no more than two floors that is not a means of egress and an escalator opening complying with Section 707.2 (also see commentary, Section 707).

3410.6.6.1 Vertical opening formula. The following formula shall be used in computing vertical opening value.

$$VO = PV \times CF \qquad \textbf{(Equation 34-4)}$$

VO = Vertical opening value

PV = Protection value [Table 3410.6.6(1)]

CF = Construction type factor [Table 3410.6.6(2)]

❖ See Commentary to Section 3410.6.6

TABLE 3410.6.6(1)
VERTICAL OPENING PROTECTION VALUE

PROTECTION	VALUE
None (unprotected opening)	-2 times number floors connected
Less than 1 hour	-1 times number floors connected
1 to less than 2 hours	1
2 hours or more	2

❖ Table 3410.6.6(1) assigns relative protection values based on the fire protection ratings of the vertical openings in a building. The lower the fire protection, the greater the hazard to the rest of the building. The table reflects the varying levels of impact to an existing building based on the number of stories that are inter-

connected by unprotected openings. The greater the number of floors that are interconnected by unprotected openings, the greater the number of negative points assessed in the existing building's evaluation scores. The closer the building comes to meeting code compliance relative to rated shafts, the larger the positive protection value becomes. This table encourages and provides incentives to the existing building renovator to bring noncomplying situations into compliance with new construction requirements.

TABLE 3410.6.6(2)
CONSTRUCTION-TYPE FACTOR

	TYPE OF CONSTRUCTION								
	IA	IB	IIA	IIB	IIIA	IIIB	IV	VA	VB
FACTOR	1.2	1.5	2.2	3.5	2.5	3.5	2.3	3.3	7

❖ Relative values for each type of construction are assigned in Table 3410.6.6(2). These represent the relative degree of fire hazard of each type of construction when compared to other types of construction. Similar factors were considered in the original development of Table 503 for height and area limitations. When one building has two different opening circumstances that individually result in different values, the lower value must be used.

Example 1:

Assume a building of Type IIB construction, three stories in height. The building has a 2-hour fire-resistance-rated exhaust shaft with 1¹/₂-hour fire dampers (complying with UL 555).

$VO = 2 \times 3.5 = 7.0$

Enter 7.0 in Table 3410.7 under Safety Parameter 3410.6.6, Vertical Openings.

Example 2:

Assume a building of Type IA construction, 12 stories in height. The building has a ¹/₂-hour return-air shaft with rated dampers.

$VO = (-1 \times 12) \times 1.2 = -14.4$

Enter -14.4 in Table 3410.7 under Safety Parameter 3410.6.6, Vertical Openings.

3410.6.7 HVAC systems. Evaluate the ability of the HVAC system to resist the movement of smoke and fire beyond the point of origin. Under the categories in Section 3410.6.7.1, determine the appropriate value and enter that value into Table 3410.7 under Safety Parameter 3410.6.7, HVAC Systems, for fire safety, means of egress and general safety.

❖ The provisions of Chapter 34 are intended to evaluate the HVAC system's potential for resisting the movement and spread of fire, smoke or products of combustion. This section does not address HVAC systems that are used exclusively for smoke control in the building. The systems evaluated in this section are those that use either supply air, return air or exhaust air. For example, a typical building might have supply air ducts or shafts; return air ducts or shafts; toilet exhaust ducts or shafts and kitchen exhaust ducts or shafts, all of which are considered HVAC systems. All systems in the building are evaluated and the lowest score obtained by any of the systems is the score that must be assigned to the entire building.

These provisions include two other safety aspects that are also applicable for new construction: plenums and air movement in egress elements, such as exit access corridors and exit stairways. These factors can significantly affect the safety of the occupants of the existing building. In some cases, these safety aspects can be more important than just the number of stories connected by an HVAC system.

3410.6.7.1 Categories. The categories for HVAC systems are:

1. Category a — Plenums not in accordance with Section 602 of the *International Mechanical Code*. -10 points.

2. Category b — Air movement in egress elements not in accordance with Section 1017.4. -5 points.

3. Category c — Both categories a and b are applicable. -15 points.

4. Category d — Compliance of the HVAC system with Section 1017.4 and Section 602 of the *International Mechanical Code*. 0 points.

5. Category e — Systems serving one story; or a central boiler/chiller system without ductwork connecting two or more stories. 5 points.

❖ The five categories that must be used in the evaluation process are defined in this section, along with their applicable values. The values are to be entered in Table 3410.7 under Safety Parameter 3410.6.7, HVAC Systems. These values are not occupancy sensitive, since the spread of fire is dependent on the HVAC system present in the existing building, not on its occupancy classification.

Category a requires a value of -10 for existing buildings that contain plenums not in compliance with the requirements of Section 602 of the IMC. Locations of plenums within the existing buildings, the materials they are built of relative to the building's type of construction and the materials exposed to plenum air must all be evaluated.

Category b corresponds to existing buildings that have exit access corridors or exit stairways that are used to supply, return or exhaust air or for other ventilation purposes. That type of layout puts the existing building's occupants at greater risk and must, therefore, be penalized with a value of -5. An existing building complying with one of the exceptions of Section 1017.4 is not considered as Category b.

Category c is applicable when an existing building has both noncomplying plenums and corridors/stairways used for air movement. A value of -15 must be assigned to such a building because the movement of smoke and fire throughout would pose an even greater hazard to the occupants.

Category d represents the base value of zero. A newly constructed building is required to meet all the provisions of Section 1017.4 of the code and Section 602 of the IMC. An existing building that complies with these provisions is meeting this same minimum compliance level and, therefore, is neither penalized nor given benefit for that compliance.

Category e specifies a value of 5 points for HVAC systems that serve only one story of a building. The hazards of a fire spreading laterally through a story of a building via the HVAC system are minimal; therefore, the code assigns a positive value. A boiler/chiller system also does not lend itself to fire spread as long as there is no air movement in ducts.

Example 1:
Assume a Group R-2 apartment building, six stories in height. Each apartment has its own HVAC equipment located within the dwelling unit. Bathrooms are ventilated by fans connected to a central exhaust shaft; kitchen exhaust hoods connect to a central exhaust shaft serving all floors. Corridors have a direct outside supply-air system on each floor with no exhaust. The HVAC systems are classified as Category b because the corridors are being used as the make-up air source for the bathroom and kitchen exhausts. A score of -5 should be entered in Table 3410.7 under Safety Parameter 3410.6.7, HVAC Systems.

Example 2:
Assume a Group S-2 low-hazard storage occupancy is located in a one-story building. The building is a pre-engineered steel Type IIB construction, and has a suspended gypsum board ceiling. Gypsum board has also been attached to the underside of the ceiling joists to serve as the upper membrane of a return-air plenum.

The plenum also contains plastic fire sprinkler piping for the automatic fire suppression system that is installed throughout the building.

The plenum is classified as noncombustible in accordance with Section 602 of the IMC. As long as the plastic fire sprinkler piping meets the optical density and flame spread limits, the plenum is in compliance with the code; therefore, the HVAC system will be classified as Category d. A value of zero should be entered in Table 3410.7 under Safety Parameter 3410.6.7, HVAC Systems.

3410.6.8 Automatic fire detection. Evaluate the smoke detection capability based on the location and operation of automatic fire detectors in accordance with Section 907 and the *International Mechanical Code*. Under the categories and occupancies in Table 3410.6.8, determine the appropriate value and enter that value into Table 3410.7 under Safety Parameter 3410.6.8, Automatic Fire Detection, for fire safety, means of egress and general safety.

❖ This section considers the use of smoke detectors in a building. To receive credit for the smoke detectors, they must be connected to audible alarms and installed in accordance with Section 907 of the code and Section 606 of the IMC.

TABLE 3410.6.8. See page 34-32.

❖ Table 3410.6.8 assigns values for each occupancy. The use of detectors increases the safety in a building by providing early warning to occupants. The lack of detectors and the associated early warning are considered less than optimum for safety in occupancies with high population densities and high combustible loads. Large deficiency points are accrued if adequate detection systems are not provided.

Example:

Assume a four-story Group R-1 motel with corridors and an elevator lobby on each story. Smoke detectors are installed throughout the corridors, closets, rooms and elevator lobbies. Only single-station detectors are installed in the guestrooms. There are smoke detectors in the HVAC return-air system and a fire alarm system is provided. An analysis of these building characteristics results in the selection of Category d. To be classified in Category e, guestrooms must have detectors connected to the building's emergency electrical system and annunciated by each room at a constantly attended location, such as the front desk. Additionally, the fire alarm system must be capable of being manually activated by the front desk when a smoke detector operates.

Category d value from Table 3410.6.8 = 2.

Enter 2 in Table 3410.7 under Safety Parameter 3410.6.8, Automatic Fire Detection.

TABLE 3410.6.8
AUTOMATIC FIRE DETECTION VALUES

OCCUPANCY	CATEGORIES				
	a	b	c	d	e
A-1, A-3, F, M, R, S-1	-10	-5	0	2	6
A-2	-25	-5	0	5	9
A-4, B, E, S-2	-4	-2	0	4	8

3410.6.8.1 Categories. The categories for automatic fire detection are:

1. Category a — None.

2. Category b — Existing smoke detectors in HVAC systems and maintained in accordance with the *International Fire Code*.

3. Category c — Smoke detectors in HVAC systems. The detectors are installed in accordance with the requirements for new buildings in the *International Mechanical Code*.

4. Category d — Smoke detectors throughout all floor areas other than individual sleeping units, tenant spaces and dwelling units.

5. Category e — Smoke detectors installed throughout the fire area.

❖ The categories are based on the location and completeness of the smoke detectors. The categories represent a graduation in the levels of smoke detection

that ranges from no detectors in Category a to full detection throughout all fire area spaces in Category e.

Category a is a facility that has no automatic smoke detection system.

Category b assumes that there are some smoke detectors in an existing building's HVAC system, but not to the extent required for new construction. If the limited HVAC system detectors are maintained in accordance with the IFC, the detection system qualifies as Category b.

Category c addresses existing buildings that have upgraded HVAC systems with duct detectors, as specified by the IMC. The detection values for Category c are zero, since this is the level of protection required for all new construction.

Category d is the classification for existing buildings that have full detection coverage throughout the public common spaces and secondary rooms and areas.

Category e has smoke detectors throughout fire areas in compliance with the provisions for new construction.

3410.6.9 Fire alarm systems. Evaluate the capability of the fire alarm system in accordance with Section 907. Under the categories and occupancies in Table 3410.6.9, determine the appropriate value and enter that value into Table 3410.7 under Safety Parameter 3410.6.9, Fire Alarm, for fire safety, means of egress and general safety.

❖ This section evaluates the capabilities of building fire alarm systems that are separate from the automatic fire detection system evaluated in Section 3410.6.8. A fire alarm system that is manually operated or activated by smoke detectors or sprinkler water-flow devices alerts the occupants to a fire situation. The fire alarm system will notify the occupants with visible or audible alarms so they may begin their egress and discharge from the building. Buildings of assembly, business, educational or residential occupancies can have large numbers of occupants in rooms with concentrated seating or people who are sleeping.

TABLE 3410.6.9
FIRE ALARM SYSTEM VALUES

OCCUPANCY	CATEGORIES			
	a	b[a]	c	d
A-1, A-2, A-3, A-4, B, E, R	-10	-5	0	5
F, M, S	0	5	10	15

a. For buildings equipped throughout with an automatic sprinkler system, add 2 points for activation by a sprinkler water flow device.

❖ Table 3410.6.9 gives values for each occupancy and type of fire alarm system provided. It reflects the idea that the presence of an alarm system in a building usually creates a safer condition for the occupants when compared to a building without a communication system.

Deficiency points are assigned to high population densities, such as Groups A-1, A-2, A-3, A-4 and E, which fit into Category a.

Example:
Assume a one-story assembly building has a complete manual fire alarm system, a voice/alarm system, a public address system and a fire command station that does not contain status indicators and controls for the air-handling system. The fire command station does not have emergency power or lighting system controls. It also does not have a fire department communication panel. As a result, the building is classified in Category c.

Category c value from Table 3410.6.9 = 0.

Enter 0 in Table 3410.7 under Safety Parameter 3410.6.9, Fire Alarm System.

3410.6.9.1 Categories. The categories for fire alarm systems are:

1. Category a — None.

2. Category b — Fire alarm system with manual fire alarm boxes in accordance with Section 907.3 and alarm notification appliances in accordance with Section 907.9.

3. Category c — Fire alarm system in accordance with Section 907.

4. Category d — Category c plus a required emergency voice/alarm communications system and a fire command center that conforms to Section 403.8 and contains the emergency voice/alarm communications system controls, fire department communication system controls and any other controls specified in Section 911 where those systems are provided.

❖ These categories are defined by the fire alarm system that is provided within the existing building.

Category a means that there is no fire alarm system in the building or it does not conform to all the requirements of Section 907. This includes a system that does not have a secondary power supply in accordance with Section 907 or one where the zones of a floor exceed 22,500 square feet (1858.1 m²) as noted in Section 907.8.

Category b applies when the manual fire alarm boxes comply with Section 907.3 and NFPA 72. The alarm notification appliances, specifically the audible alarms, are in accordance with Section 907.9.2. Location, height and color of the manual fire alarm boxes must be specifically evaluated for compliance, along with the sound levels of the audible alarms when compared to the normal sound levels within the existing building.

Category c requires that the fire alarm system must comply with Section 907 and NFPA 72. This category indicates that a complete fire alarm system is present in the existing building and that the system complies with all the requirements for new construction.

Category d is applicable for a building that is provided with a fire command station for fire department operations that conforms to Section 403.8. Section 403.8 may be used when evaluating any building that complies with its performance requirements. This sec-

tion specifies the fire command operations required, including:

- Voice/alarm communication system panels;
- Fire department communication panels;
- Fire detection and alarm system annunciator panels;
- An annunciator for visually indicating floor locations of elevators and whether they are operational;
- Status indicators and controls for air-handling systems;
- Controls for unlocking all stairway doors simultaneously;
- Sprinkler valve and water-flow detector display panels;
- Emergency and standby power; and
- Status indicators and a telephone for fire department use with controlled access to the public telephone system.

3410.6.10 Smoke control. Evaluate the ability of a natural or mechanical venting, exhaust or pressurization system to control the movement of smoke from a fire. Under the categories and occupancies in Table 3410.6.10, determine the appropriate value and enter that value into Table 3410.7 under Safety Parameter 3410.6.10, Smoke Control, for means of egress and general safety.

❖ This section is used to evaluate characteristics that could limit smoke migration in the building, including operable windows, mechanical exhaust systems or pressurized stair or smokeproof enclosures.

TABLE 3410.6.10
SMOKE CONTROL VALUES

OCCUPANCY	CATEGORIES					
	a	b	c	d	e	f
A-1, A-2, A-3	0	1	2	3	6	6
A-4, E	0	0	0	1	3	5
B, M, R	0	2[a]	3[a]	3[a]	3[a]	4[a]
F, S	0	2[a]	2[a]	3[a]	3[a]	3[a]

a. This value shall be 0 if compliance with Category d or e in Section 3410.6.8.1 has not been obtained.

❖ Table 3410.6.10 assigns values for each occupancy and category of smoke control in the building. The table indicates the relative risk to the building occupants if adequate smoke control methods are not provided in the building.

In occupancies having a higher population density, zero points are indicated if no smoke control or limited smoke control is provided. The table also indicates that smoke control is extremely important in the exit stairs of a building. Note a of the table assigns zero credit points for Groups B, M, R, F and S unless the building complies with Category d or e in Section 3410.6.8.1, which are the automatic fire detection system requirements.

Example 1:

Assume a one-story Group A-3 church sanctuary. It has no operable windows and no smoke control system.

Category a from Table 3410.6.10 = 0.

Enter 0 in Table 3410.7 under Safety Parameter 3410.6.10, Smoke Control, for only means of egress (ME) and general safety (GS) parameters. No entry is to be made under the fire safety (FS) parameter.

Example 2:

Assume a three-story Group E high school. It has operable windows throughout the building. The stairways are interior without windows or a pressurization system.

Category b from Table 3410.6.10 = 0.

Enter 0 in Table 3410.7 under Safety Parameter 3410.6.10, Smoke Control, for means of egress (ME) and general safety (GS) parameters.

Example 3:

Assume a two-story Group A-2 nightclub with a smoke control system that meets the requirements for Category e.

Category e from Table 3410.6.10 = 6.

Enter 6 in Table 3410.7 under Safety Parameter 3410.6.10, Smoke Control, for means of egress and general safety parameters.

Example 4:

Assume a six-story Group B office building with three stairways: one is a smokeproof enclosure conforming to Section 1020.1.7, one is pressurized in accordance with Section 909.20.5; and one has operable exterior windows. The building has smoke detectors throughout all floor spaces and fire areas.

Category f from Table 3410.6.10 = 4.

Enter 4 in Table 3410.7 under Safety Parameter 3410.6.10, Smoke Control, for means of egress (ME) and general safety (GS).

If detectors are omitted from some of the offices of the building, then a score of 0 must be entered in Table 3410.7. This is determined from Note a in Table 3410.6.10.

3410.6.10.1 Categories. The categories for smoke control are:

1. Category a — None.

2. Category b — The building is equipped throughout with an automatic sprinkler system. Openings are provided in exterior walls at the rate of 20 square feet (1.86 m²) per 50 linear feet (15 240 mm) of exterior wall in each story and distributed around the building perimeter at intervals not exceeding 50 feet (15 240 mm). Such openings shall be readily openable from the inside without a key or separate tool and shall be provided with ready access thereto. In lieu of operable openings, clearly and permanently marked tempered glass panels shall be used.

3. Category c — One enclosed exit stairway, with ready access thereto, from each occupied floor of the building. The stairway has operable exterior windows and the building has openings in accordance with Category b.

4. Category d — One smokeproof enclosure and the building has openings in accordance with Category b.

5. Category e — The building is equipped throughout with an automatic sprinkler system. Each fire area is provided with a mechanical air-handling system designed to accomplish smoke containment. Return and exhaust air shall be moved directly to the outside without recirculation to other fire areas of the building under fire conditions. The system shall exhaust not less than six air changes per hour from the fire area. Supply air by mechanical means to the fire area is not required. Containment of smoke shall be considered as confining smoke to the fire area involved without migration to other fire areas. Any other tested and approved design which will adequately accomplish smoke containment is permitted.

6. Category f — Each stairway shall be one of the following: a smokeproof enclosure in accordance with Section 1020.1.7; pressurized in accordance with Section 909.20.5; or shall have operable exterior windows.

❖ The six categories to be evaluated are compared to the occupancies to determine the score to be entered in Table 3410.7 under Safety Parameter 3410.6.10, Smoke Control.

Category a means there is no method of controlling smoke in the building.

Category b means the existing building is sprinklered and there are exterior windows that can be readily opened without the use of keys or tools. This category recognizes the fire safety benefits of the code's high-rise provisions, including automatic sprinkler systems. Operable panels or windows in the exterior walls must be provided at the rate of 20 square feet per 50 lineal feet (1.85 m² per 15 240 mm) of exterior wall in each story. The openings must be distributed around the perimeter of the building at intervals no more than 50 feet (15 240 mm) between windows. A story with very high ceilings may have a strip of windows located with latches or controls that are not easily reachable. These openings would not meet Category b standards, even if they are the correct sizes.

Category c requires at least one enclosed exit stairway to have operable windows that open to the exterior. Additionally, the building must have operable windows complying with the size and spacing requirements of Category b.

Category d requires a minimum of one smokeproof enclosure and operable windows in the building. These operable windows are the same as those addressed in Category b. By definition, a smokeproof enclosure refers to an enclosed interior exit stairway that conforms to Section 1009.3 as designated in Section 1020.1.7.

Section 1020.1.7 required stairways to be protected from smoke by one of two methods: a smokeproof enclosure or stair pressurization design (see Section 909.20.5).

Additional requirements that must be considered when evaluating a smokeproof enclosure are: access (see Section 909.20.1), construction (see Section 909.20.2), ventilating equipment (see Section 909.20.6) and standby power (see Section 909.20.6.2).

Category e recognizes the use of a mechanical smoke control system designed in accordance with the provisions of the code. Each fire area within the building must have a mechanical smoke control system. Return and exhaust air from the system must be discharged directly to the exterior to achieve the necessary level of protection. A specific air change requirement is provided, independent of the fire area's volume or size.

Category f recognizes the merits of smokeproof enclosures, pressurized stairs and stairs with operable exterior windows. To be classified in this category, all stairs in the building must comply with the requirements of any one, or a combination of, the three types of stairs. It is possible for a single building to have a smokeproof enclosure, a pressurized stairway and a stairway with operable exterior windows.

Both the categories and the occupancy of the building are used in determining the score that will be entered in Table 3410.7. This score is determined from Table 3410.6.10. Note a in Table 3410.6.10 can have a significant impact on the scores. The note states that, even if the building has some level of smoke control, it is to receive no credit if it does not have an automatic fire detection system complying with Category d or e in Section 3410.6.8.1.

3410.6.11 Means of egress capacity and number. Evaluate the means of egress capacity and the number of exits available to the building occupants. In applying this section, the means of egress are required to conform to Sections 1003 through 1015 and 1017 through 1024 (except that the minimum width required by this section shall be determined solely by the width for the required capacity in accordance with Table 1005.1). The number of exits credited is the number that are available to each occupant of the area being evaluated. Existing fire escapes shall be accepted as a component in the means of egress when conforming to Section 3404. Under the categories and occupancies in Table 3410.6.11, determine the appropriate value and enter that value into Table 3410.7 under Safety Parameter 3410.6.11, Means of Egress Capacity, for means of egress and general safety.

❖ This section addresses the exit capacity and number of existing exits available to the building occupants. Before a building can be evaluated in this category, and Section 3410 in general, it must comply with the sections listed.

Exits are required to conform to the following sections before evaluation:

• Sections 1014, 1015 and 1017. A mandatory requirement is not placed on the length of travel in accordance with Section 1016, because travel length does not directly impact exit capacity. Overall exit safety is influenced by travel length, however, and is evaluated in Section 3410.6.13.

• Section 1004 establishes the minimum number of occupants the exit facilities must accommodate.

• Section 1005 defines the capacity of the means of egress by identifying a minimum width of egress component per occupant. This is used to calculate the total capacity of the means of egress component.

• Section 1019.1 establishes the minimum number of exits for the existing building occupant load.

• Section 1024 requires exit stairs to discharge directly to the exterior or through recognized exit elements at the level of discharge.

• Section 3404.1.2 permits the continued use of existing fire escapes in existing buildings only.

Evaluation of the means of egress capacity involves all egress components, including exit access, exits and exit discharge. This evaluation is correlated to the requirements for the means of egress for new construction.

TABLE 3410.6.11
MEANS OF EGRESS VALUES

OCCUPANCY	CATEGORIES				
	a[a]	b	c	d	e
A-1, A-2, A-3, A-4, E	-10	0	2	8	10
M	-3	0	1	2	4
B, F, S	-1	0	0	0	0
R	-3	0	0	0	0

a. The values indicated are for buildings six stories or less in height. For buildings over six stories in height, add an additional -10 points.

❖ Table 3410.6.11 assigns values for each occupancy and category from Section 3410.6.11.1. The table gives credit for providing additional exits and exit capacities beyond the minimum. These positive values indicate a greater degree of safety than is required under current code provisions. This additional evaluation contributes to the overall assessment of the building's safety and may offset other safety deficiencies [see Figures 3410.6.11(1) and 3410.6.11(2)]. Note a of the table provides a significant penalty for high-rise buildings that use fire escapes as part of the means of egress. In this case, if the building is seven stories or greater in height and uses fire escapes, -10 points must be added to the values already listed for Category a buildings.

3410.6.11.1 Categories. The categories for means of egress capacity and number of exits are:

1. Category a — Compliance with the minimum required means of egress capacity or number of exits is achieved through the use of a fire escape in accordance with Section 3404.

2. Category b — Capacity of the means of egress complies with Section 1004 and the number of exits complies with the minimum number required by Section 1019.

3. Category c — Capacity of the means of egress is equal to or exceeds 125 percent of the required means of egress

capacity, the means of egress complies with the minimum required width dimensions specified in the code and the number of exits complies with the minimum number required by Section 1019.

4. Category d — The number of exits provided exceeds the number of exits required by Section 1019. Exits shall be located a distance apart from each other equal to not less than that specified in Section 1015.2.

5. Category e — The area being evaluated meets both Categories c and d.

❖ Five categories must be evaluated. These categories and the occupancy of the building will determine the score entered in Table 3410.7.

Category a is appropriate for buildings that comply with either the means of egress capacity or the number of exits, including the use of fire escapes. Although the code allows fire escapes to be used as an egress element in existing buildings, this is the least desirable option in any type of building. The code requires a building using fire escapes to use Category a and its corresponding negative points.

Category b is used for buildings that meet the minimum requirements of Sections 1004 for capacity of means of egress and 1019 for minimum number of exits and continuity.

Category c is used for buildings that exceed the requirements for new construction. This category provides small positive points for existing buildings that meet all of the following requirements:

- The capacity of all the means of egress components is greater than or equal to 125 percent of the required capacity.
- All of the means of egress components comply with the minimum required widths [32 inch (813 mm) clear for doors, corridors that are 44 inches (1118 mm) wide and stairways].
- The minimum number of exits is provided based on the number of occupants on each floor level.

By providing oversized egress capacity and minimum egress width requirements, an existing building has added safety, which is rewarded with a small number of positive points.

Category d simply relates to buildings where more exits are provided than required by Section 1019, which contributes a positive factor to the building's exit capacity.

Before credit can be given to an existing building with a greater number of exits, the exits must be evaluated for their remoteness and independence from each other. All of the exits must be located at least one-half the length of the diagonal from each other before Category d can be used. If the building is equipped throughout with an automatic sprinkler system in accordance with NFPA 13 or 13R, the exits can be located one-third the length of the diagonal from each other.

Category e provides additional credit for existing buildings that have the characteristics of both Categories c and d. Such a building, which has oversized ca-

400'-0"

BUILDING AREA = 80,000 SQ.FT.
OCCUPANT LOAD = 800

44" CLEAR WIDTH EXIT DOORS

200'-0"

• OCCUPANCY GROUP F
• COMPLETE SPRINKLER SYSTEM
• EGRESS WIDTH PER SECTION 1005.1 AND TABLE 1005.1
• THE EXIT CAPACITY OF THE THREE 44"
  CLEAR WIDTH EXIT DOORS = $\frac{3 \times 44"}{0.15}$ = 880 OCCUPANTS

• CLASSIFICATION = CATEGORY b. THE
  EXIT CAPACITY OF 880 EXCEEDS
  THE ALLOWABLE CAPACITY OF 800, COMPLYING
  WITH SECTION 1005
• SCORE = 0 FROM TABLE 3410.6.11
• ENTER 0 IN TABLE 3410.7

For SI: 1 foot = 304.8 mm,
        1 square foot = .0929 m².

Figure 3410.6.11(1)
MEANS OF EGRESS VALUES—CATEGORY b

pacities in its egress elements, minimum required clear widths for the egress components and additional exits that are all remotely located from one another, is considered to have the highest degree of safety; however, this is true for only some of the applicable occupancies. In Group B, F, S and R occupancies, any additional capacity or increased number of exits is not a significant factor. No positive points are awarded for this life safety issue for these occupancies.

3410.6.12 Dead ends. In spaces required to be served by more than one means of egress, evaluate the length of the exit access travel path in which the building occupants are confined to a single path of travel. Under the categories and occupancies in Table 3410.6.12, determine the appropriate value and enter that value into Table 3410.7 under Safety Parameter 3410.6.12, Dead Ends, for means of egress and general safety.

❖ This section is used to evaluate dead-end exit access conditions within the building. This section uses the terminology "confined to a single path of travel." Another way to illustrate the meaning of this is "a single direction of travel to reach an exit." A typical corridor is a single path in which a building occupant can travel in two directions to reach exits. A dead end may be a single path, but the key feature of a dead end is that only one direction is available to reach an exit. When building occupants have only one direction of travel available, a potentially hazardous condition is created because they may become trapped if the direction of travel is blocked by fire or smoke.

This section addresses only dead ends that are a component of exit access travel, which is "that portion of a means of egress that leads to an entrance to an exit." Although a room containing only one door may restrict exiting from the room to only one direction, it is not considered a dead-end condition as described in Section 1017.3. This section notes that the dead-end path of travel is limited to 20 feet (6096 mm) for exit access passageways and corridors that serve more than one exit; therefore, the single-door exit of a room enters an "exit access passageway or corridor," which is subject to a dead-end limitation of 20 feet (6096 mm). The travel distance within the room to the exit access is limited by Section 1014.3 and Table 1015.1. The total travel distance from the remote point within a room to the exit must include the room travel length plus exit access passageway or corridor length (see commentary, Section 3410.6.13). Section 3410.6.12 addresses the condition of dead-end passageways and corridors (see commentary, Section 1017.3).

TABLE 3410.6.12
DEAD-END VALUES

	CATEGORIES[a]		
OCCUPANCY	a	b	c
A-1, A-3, A-4, B, E, F, M, R, S	-2	0	2
A-2, E	-2	0	2

a. For dead-end distances between categories, the dead-end value shall be obtained by linear interpolation.

- OCCUPANCY GROUP A-3
- BUILDING FULLY SPRINKLERED
- NUMBER OF OCCUPANTS = 400
- TABLE 1019.1 REQUIRES TWO EXITS, FOUR EXTERIOR DOORS (EXITS) ARE PROVIDED
- THE 1/3 DIAGONAL = $\frac{224'}{3}$ = 75'
- EACH DOOR IS SPACED 112' APART
- CLASSIFICATION = CATEGORY d FROM TABLE 3410.6.11
- ENTER 8 IN TABLE 3410.7

For SI: 1 foot = 304.8 mm,
1 square foot = .0929 m².

Figure 3410.6.11(2)
MEANS OF EGRESS VALUES—CATEGORY d

❖ The table reflects the relative degree of hazard associated with dead-end passageways and corridors. This is shown by the deficiency points that apply where a dead-end exceeds 35 or 70 feet (10 668 or 21 336 mm) in an occupancy with a relatively high occupant load, such as an assembly occupancy. Note a allows for the interpolation for actual dead-end lengths between the distances specified in the categories. For example, if a building contains a corridor that has a dead-end length of 10 feet (3048 mm) at each end beyond the exits, the value of 1 is used; this is the midpoint between Categories b and c.

3410.6.12.1 Categories. The categories for dead ends are:

1. Category a — Dead end of 35 feet (10 670 mm) in nonsprinklered buildings or 70 feet (21 340 mm) in sprinklered buildings.

2. Category b — Dead end of 20 feet (6096 mm); or 50 feet (15 240 mm) in Group B in accordance with Section 1017.3 exception 2.

3. Category c — No dead ends; or ratio of length to width (l/w) is less than 2.5:1.

❖ This section defines the categories for dead-end exit access conditions.

Category a allows dead-end conditions of up to 35 feet (10 668 mm) for existing buildings that are not fully sprinklered, and up to 70 feet (21 336 mm) for existing buildings that are fully sprinklered. These distances correspond to the dead-end lengths listed in the IFC. Since these lengths far exceed the allowable lengths permitted for new construction, negative values are associated with this category.

Category b is the classification for buildings that comply with the requirements for new construction. This zero-based category provides no extra credit for complying with new construction requirements.

Category c represents conditions that exceed the requirements for new construction. If there are no dead-end corridors or the "corridor" is more of a "space," Category c can be used. In Category c the length-to-width ratio of 2.5 to define a space/corridor is the same as Exception 2 in Section 1017.3. Such a space allows the building user a more circular route and full view of the space; therefore, this does not constitute a dead-end situation.

These categories and occupancies of the building are used in Table 3410.6.12 to determine the score to be entered in Table 3410.7 under Safety Parameter 3410.6.12, Dead Ends [see Figures 3410.6.12.1(1), 3410.6.12.1(2) and 3410.6.12.1(3)].

3410.6.13 Maximum exit access travel distance. Evaluate the length of exit access travel to an approved exit. Determine the appropriate points in accordance with the following equation and enter that value into Table 3410.7 under Safety Parameter 3410.6.13, Maximum Exit Access Travel Distance, for means of egress and general safety. The maximum allowable exit access travel distance shall be determined in accordance with Section 1016.1.

$$\text{Points} = 20 \times \frac{\begin{array}{c}\text{Maximum allowable}\\\text{travel distance}\end{array} - \begin{array}{c}\text{Maximum actual}\\\text{travel distance}\end{array}}{\text{Max. allowable travel distance}}$$

❖ The length of exit access travel distance is evaluated in comparison to the travel distance allowed by Table 1016.1 for a particular occupancy. The exit access travel distance is measured from the most remote point in the building to the nearest exit. Total exit access travel length of Table 1016.1 includes travel distance within a room or space plus exit access corridor travel to the exit. Some modifications to the Table 1016.1 requirements are listed for certain occupancies and buildings as listed in the table's notes.

Figure 3410.6.12.1(1)
TYPICAL DEAD-END CORRIDOR

To determine the value to be assigned for maximum travel distances to an exit, the specified equation must be used. This equation allows a graduated scale to be used to evaluate compliance of an existing situation with new construction travel distance requirements. With the equation, a more definite evaluation can occur with a broader range of scores. Any existing building having overall travel distances less than those specified in Table 1016.1 will achieve a positive credit.

Existing buildings with travel distances greater than those allowed for new construction will be assigned negative points based on the extent the travel distance exceeds the allowable limit.

For example, an existing Group B business building has a travel distance of 150 feet (45 720 mm). The distance is measured from the most remote corner of a partitioned office, down a corridor and to an exit door. Table 1016.1 permits an unsprinklered Group B build-

- OCCUPANCY GROUP B WITH MOVABLE PARTITIONS AND FURNITURE 6'-6" HIGH
- DEAD-END PASSAGEWAY IS 20'-0"
- CLASSIFICATION = CATEGORY b
- VALUE = 0 FROM TABLE 3410.6.12
- ENTER 0 IN TABLE 3410.7

For SI: 1 foot = 304.8 mm.

Figure 3410.6.12.1
DEAD-END VALUES—CATEGORY B

- OCCUPANCY GROUP R
- UNSPRINKLERED BUILDING
- DEAD-END CORRIDOR LENGTH IS 50'-0"
- CLASSIFICATION = CATEGORY a
- VALUE = -2 FROM TABLE 3410.6.12
- ENTER -2 IN TABLE 3410.7

For SI: 1 foot = 304.8 mm.

Figure 3410.6.12.1(3)
DEAD-END VALUES—CATEGORY a

- COMBUSTIBLE STORAGE, OCCUPANCY GROUP S-1
  WITH AUTOMATIC FIRE SUPPRESSION SYSTEM
- MOST REMOTE LENGTH OF EXIT
  ACCESS TRAVEL (SECTION 1016.1)
- TRAVEL DISTANCE IN BUILDING = 350'-0"
- TRAVEL DISTANCE LIMIT FROM TABLE 1016.1 = 250'-0"
- POINTS = $20 \times \dfrac{250 - 350}{250} = -8$

For SI: 1 foot = 304.8 mm.

Figure 3410.6.13(1)
EXIT ACCESS TRAVEL DISTANCE VALUES—OCCUPANCY GROUP S-1

- OCCUPANCY GROUP B
  WITHOUT FIRE SUPPRESSION SYSTEM
- MOST REMOTE TRAVEL LENGTH = 200'-0"
- TRAVEL DISTANCE IN BUILDING = 200'-0"
- TRAVEL DISTANCE LIMIT FROM TABLE 1016.1 = 200'-0"

- POINTS = $20 \times \dfrac{200 - 200}{200} = 0$

For SI: 1 foot = 304.8 mm.

Figure 3410.6.13(2)
EXIT ACCESS TRAVEL DISTANCE VALUES—OCCUPANCY GROUP B

- OCCUPANCY GROUP R
  WITH FIRE SUPPRESSION SYSTEM
- MOST REMOTE TRAVEL LENGTH = 100'-0"
- TRAVEL DISTANCE IN BUILDING = 100'-0"
- TRAVEL DISTANCE LIMIT FROM TABLE 1015.1 = 250'-0"

- POINTS = $20 \times \dfrac{250 - 100}{250} = +12$

For SI: 1 foot = 304.8 mm.

Figure 3410.6.13(3)
EXIT ACCESS TRAVEL DISTANCE VALUES—OCCUPANCY GROUP R

ing to have a travel distance of 200 feet (60 960 mm). The equation yields the following:

$$\text{Points} = 20 \times \left(\frac{200\,\text{feet} - 150\,\text{feet}}{200\,\text{feet}} \right)$$

$$= 5.0$$

See Figures 3410.6.13(1), 3410.6.13(2) and 3410.6.13(3) for other examples.

3410.6.14 Elevator control. Evaluate the passenger elevator equipment and controls that are available to the fire department to reach all occupied floors. Elevator recall controls shall be provided in accordance with the *International Fire Code.* Under the categories and occupancies in Table 3410.6.14, determine the appropriate value and enter that value into Table 3410.7 under Safety Parameter 3410.6.14, Elevator Control, for fire safety, means of egress and general safety. The values shall be zero for a single-story building.

❖ The availability of elevators in the building and their incapability in an emergency are evaluated. This section addresses four different categories of elevators in existing buildings. This section does not address the requirements of Chapter 30. Access must be provided to all occupied floors by passenger elevators for the purposes of this evaluation. Freight elevators cannot be considered, since they may be in locations not readily accessible for fire department use. Elevator recall controls must comply with the IFC for either a Phase I or II category. One-story buildings, including those with mezzanines served by an elevator, are awarded zero value.

TABLE 3410.6.14. See next column.

❖ The table assigns values based on the types of controls the elevators have and the distance the elevators must travel to reach the floors they serve. The 25-foot (7620 mm) threshold of elevator travel is based on the

TABLE 3410.6.14
ELEVATOR CONTROL VALUES

ELEVATOR TRAVEL	CATEGORIES			
	a	b	c	d
Less than 25 feet of travel above or below the primary level of elevator access for emergency fire-fighting or rescue personnel	-2	0	0	+2
Travel of 25 feet or more above or below the primary level of elevator access for emergency fire-fighting or rescue personnel	-4	NP	0	+4

For SI: 1 foot = 304.8 mm.

IFC requirements and reference standard ASME A17.1. Elevator travel distance is based on the level where the elevator is accessed by the fire department. Usually, this is the ground floor or grade-level floor and corresponds to the level where fire command stations or fire alarm system annunciator panels are located. Any elevator that travels 25 feet (7620 mm) or more must always be provided with Phase I and II recall capabilities.

Example:

Assume a Group B business occupancy, three stories in height. The elevator lobbies are equipped with smoke detectors that recall the elevator to the main floor or to an alternate floor if the main floor detector is activated. The elevator travels more than 25 feet (7620 mm) above the main floor.

This building must be placed in Category c. Because the elevator travels more than 25 feet (7620 mm) above the main floor, it must be brought into compliance with the IFC, which requires Phase I or II recall. When the elevators are brought into compliance with the IFC, they are automatically placed in Category c. The value from Table 3410.6.14 is 0; therefore, 0 must be entered in Table 3410.7 under Safety Parameter 3410.6.14, Elevator Control.

3410.6.14.1 Categories. The categories for elevator controls are:

1. Category a — No elevator.

2. Category b — Any elevator without Phase I and II recall.

3. Category c — All elevators with Phase I and II recall as required by the *International Fire Code*.

4. Category d — All meet Category c; or Category b where permitted to be without recall; and at least one elevator that complies with new construction requirements serves all occupied floors.

❖ The categories that a building may be placed into range from buildings with no elevators to buildings with elevators complying with new construction requirements and total elevator recall capability.

Category a is for buildings with no elevators. Without an elevator, fire department personnel are required to use the stairs for rescuing people and accessing fire floors. Category a is assigned a negative value.

Category b is for buildings in which elevators are present but have no recall controls. Without recall or fire department control, the elevators are allowed vertical travel distances of less than 25 feet (7620 mm). This approach is consistent with the IFC, which requires controls for elevators reaching 25 feet (7620 mm) or more.

Category c is used for buildings with elevators that have automatic recall when required by the IFC and, specifically, Section 607.1 of that code.

Category d is used for buildings when the elevator controls comply with either Category b or c and at least one of the elevators in the existing building serves all occupied floor levels. The controls must also comply with all the requirements for new construction. This category recognizes the benefits of having all elevators with Phase I and II controls, as well as having a newly constructed elevator with all the features to facilitate fire-fighting and rescue operations.

Based on the controls provided, or the lack of elevators, the appropriate category is determined. This category and the distance the elevator travels are then used along with Table 3410.6.14 to determine the value score the building will receive for this item. The value is then entered in Table 3410.7.

3410.6.15 Means of egress emergency lighting. Evaluate the presence of and reliability of means of egress emergency lighting. Under the categories and occupancies in Table 3410.6.15, determine the appropriate value and enter that value into Table 3410.7 under Safety Parameter 3410.6.15, Means of Egress Emergency Lighting, for means of egress and general safety.

❖ Lighting throughout the entire means of egress is evaluated in this section. Illumination of the means of egress is essential in a building during normal occu-

pancy. During a fire condition, illumination becomes even more important, since visibility will more than likely be reduced by the buildup of smoke. Failure to provide adequate lighting in a fire condition has led to numerous deaths because occupants become disoriented and are unable to follow the means of egress to safety. It is appropriate to include egress lighting as one of the major building components that must be evaluated. The importance of this lighting is shown by the minimal values assigned by Table 3410.6.15 when egress lighting is not provided.

TABLE 3410.6.15
MEANS OF EGRESS EMERGENCY LIGHTING VALUES

NUMBER OF EXITS REQUIRED BY SECTION 1019	CATEGORIES		
	a	b	c
Two or more exits	NP	0	4
Minimum of one exit	0	1	1

❖ This table is used to assess the relative risk to the occupants of the building when egress lighting is not adequate. The table reinforces the idea that when a building is required to have two or more exits, lighting has more significance. The lighting values are less critical in smaller buildings where only one exit is required because the hazard of a substantial occupant load locating and using the means of egress is minimal.

Example:
A Group A church building with a number of classrooms that will accommodate up to 60 people has at least 1 footcandle (11 lux) of illumination at the floor level throughout the entire means of egress. The means of egress lighting in the sanctuary, corridors and stairs is provided with battery back-up power that is sized to provide lighting for 1 hour. The classroom lighting is wired to the main switch panel in the building and has no emergency power source.

The lighting must be placed in Category a because the classrooms are part of the means of egress (exit access) and the lack of emergency power does not comply with Section 1006.3. The classroom will accommodate up to 60 occupants; therefore, in accordance with Section 1015.1 and Table 1015.1, the rooms are required to have two means of egress. The lighting must be provided with an emergency power source in accordance with Section 1006.3. The value determined from Table 3410.6.15 is "NP." Emergency power must be added to the lighting in the classrooms. The resulting Category b has a value of 0. Enter 0 in Table 3410.7.

3410.6.15.1 Categories. The categories for means of egress emergency lighting are:

1. Category a — Means of egress lighting and exit signs not provided with emergency power in accordance with Section 2702.

2. Category b — Means of egress lighting and exit signs provided with emergency power in accordance with Section 2702.

3. Category c — Emergency power provided to means of egress lighting and exit signs which provides protection in the event of power failure to the site or building.

❖ There are three categories of egress lighting. These categories are consistent with the requirements established by the IFC .

Category a is used when there is no egress lighting in any occupied room or space within the building or when the level of illumination is less than 1 footcandle (11 lux) at the floor level at any location in the means of egress, other than the exempted locations indicated in Section 1006 (see commentary, Section 1006). If means of egress lighting and exit signs are provided but are not connected to emergency power, Category a is also applicable. Without emergency power providing a 90-minute duration after primary power loss, the means of egress lighting and exit signs would not provide as reliable a level of safety to the building occupants.

Category b includes buildings where the means of egress lighting and exit signs comply with the requirements of Sections 1006 and 1011, respectively, including emergency power requirements. Section 1006.3 includes requirements for emergency power to maintain continuous illumination of the means of egress during a power failure. This category represents the zero-based criteria, which is the level required for new construction.

Category c is used when emergency power is provided for means of egress lighting and exit signs in excess of the minimum requirements for new construction. If the emergency power provides full protection to the site or building during power failure, Category c is applicable. Campus-type complexes or buildings that require extra security may have a power plant available to provide complete back-up power for indefinite periods of time. This will qualify for the positive points of this category.

These categories, along with the number of required exits, are used to determine the appropriate value from Table 3410.6.15. The value determined from Table 3410.6.15 is then entered in Table 3410.7.

3410.6.16 Mixed occupancies. Where a building has two or more occupancies that are not in the same occupancy classification, the separation between the mixed occupancies shall be evaluated in accordance with this section. Where there is no separation between the mixed occupancies or the separation between mixed occupancies does not qualify for any of the categories indicated in Section 3410.6.16.1, the building shall be evaluated as indicated in Section 3410.6 and the value for mixed occupancies shall be zero. Under the categories and occupancies in Table 3410.6.16, determine the appropriate value and enter that value into Table 3410.7 under Safety Parameter 3410.6.16, Mixed Occupancies, for fire safety and general safety. For buildings without mixed occupancies, the value shall be zero.

❖ These provisions are used to evaluate mixed occupancies within an existing building and whether the

method of separating mixed occupancies conforms to the requirements of Section 508.3.3. If Section 508.3.3 is not completely understood, see the associated commentary before proceeding with the evaluation of the existing building in this section (see also commentary, Section 3410.6).

This section is applicable only to separated mixed occupancies. If a building is a single occupancy, the applicable value for this section is zero. If an existing mixed occupancy building has no fire-resistance-rated separation between the different uses, or the separation is fire-resistance rated for less than 1 hour, the applicable value is also zero. The building must also be evaluated in accordance with Section 3410.6. The zero-based category for this section is equivalent to full compliance with Section 508.3.3, which is the requirement for new construction.

TABLE 3410.6.16
MIXED OCCUPANCY VALUES[a]

OCCUPANCY	CATEGORIES		
	a	b	c
A-1, A-2, R	-10	0	10
A-3, A-4, B, E, F, M, S	-5	0	5

a. For fire-resistance ratings between categories, the value shall be obtained by linear interpolation.

❖ This table addresses the relative risk of a building in or close to compliance with the provisions for separated mixed occupancies. When mixed occupancies are not separated from each other, the risk from hazards is greater in high-density occupancies, such as Groups A-1 and A-2. This risk is also greater in residential occupancies, because occupants may be sleeping and not fully alert. For this reason, inadequate separation is given greater negative values. In buildings with lower occupant loads, and where the occupants are alert, the risks are relatively lower.

Note a permits linear interpolation of the corresponding values in each category. For example, an unsprinklered, Group B/Group F-2 mixed occupancy building is separated with a $1^{1}/_{2}$-hour fire-resistance-rated assembly. From Table 508.3.3, the required fire-resistance rating of the separation between a Group B and a Group F-2 occupancy for Category b is 2 hours. Since the rating that is provided is only $1^{1}/_{2}$ hours, an interpolation halfway between Category a and Category b can be used. The resulting points are -2.5.

Example 1:
Assume a three-story building. The first floor is a Group M mercantile occupancy and the upper two stories are Group R residential occupancies. The building is Type VB construction, 9,000 square feet (836.1 m²) per floor. It is fully sprinklered and only 25 percent of the perimeter is accessible. Although Type VB construction is not required to be protected, the Group M occupancy on the first floor is separated from the Group R occupancy on the second floor with a 2-hour fire-resistance-rated floor/ceiling assembly. In

this case, Table 508.3.3 requires only 1-hour separation because the building is equipped throughout with an automatic sprinkler system in accordance with Section 903.3.1.1. The provided 2-hour rating is twice that required; therefore, this building qualifies for Category c. A value of 10 points is assigned to the residential portion of the building, and a value of 5 points is assigned to the mercantile portion (see Section 3410.6 for commentary about mixed uses).

Example 2:
Assume a four-story building of Type IIB construction, fully sprinklered in accordance with NFPA 13. It has 15,000 square feet (1393.5 m²) per floor with 25-percent open perimeter. The building is to be a Group R-1 hotel, except for a portion of the first floor that will be used as a Group A-2 nightclub, occupying 1,000 square feet (92.9 m²). Although the building is unprotected Type IIB construction, the floor/ceiling assemblies of the first and second floors are 3 hours and the fire barrier assembly at the first floor between the Group R-1 occupancy and the Group A-2 occupancy is also 3 hours. Considering these building characteristics, the following determinations are made:

Because of the presence of an automatic sprinkler system, Table 508.3.3 requires a 1-hour separation. Although this building is required to have 1-hour separation, it has a 3-hour separation, which is more than double the minimum. This building is classified as Category c. A value of 10 points is awarded to both occupancies.

3410.6.16.1 Categories. The categories for mixed occupancies are:

1. Category a — Minimum 1-hour fire barriers between occupancies.

2. Category b — Fire barriers between occupancies in accordance with Section 508.3.3

3. Category c — Fire barriers between occupancies having a fire-resistance rating of not less than twice that required by Section 508.3.3

❖ This section addresses three different conditions:
Category a is used when the separation of mixed occupancies does not conform to the fire-resistance ratings specified in Section 508.3.3. If the ratings between mixed occupancies are less than specified in Section 508.3.3, but at least a 1-hour rating is provided, Category a is applicable.

Category b is used when the separation of mixed occupancies conforms to Section 508.3.3 for fire-resistance ratings.

Category c is used when the fire-resistance ratings that separate occupancies in an existing building are no less than twice the requirements in Section 508.3.3. Category c gives bonus points or increased credits to existing buildings that have higher fire-resistance ratings.

Section 3410 does not require compliance with Section 508.3.3. Section 3410.6.16 evaluates whether the existing building's compliance with Section 508.3.3 provides relative safety within the building. If the building does comply with Section 508.3.3, Section 3410.6.16 acknowledges that there is no basis for assigning negative points, nor is there any basis for assigning positive points for safety; therefore, a neutral zero value is assigned. Whatever the value, it is entered in Table 3410.7 under Safety Parameter 3410.6.16, Mixed Occupancies.

3410.6.17 Automatic sprinklers. Evaluate the ability to suppress a fire based on the installation of an automatic sprinkler system in accordance with Section 903.3.1.1. "Required sprinklers" shall be based on the requirements of this code. Under the categories and occupancies in Table 3410.6.17, determine the appropriate value and enter that value into Table 3410.7 under Safety Parameter 3410.6.17, Automatic Sprinklers, for fire safety, means of egress divided by 2 and general safety.

❖ These provisions are used to determine the amount of credit that can be applied to the evaluation for the installation of an automatic sprinkler system in an existing building.

Sprinkler safety parameters appear to provide a double credit for automatic sprinkler systems. Basic credits are provided for: building height, Section 3410.6.1; building area, Section 3410.6.2 and maximum travel distance to an exit, Section 3410.6.13. Throughout the code, additional credits are offered for the use of an automatic sprinkler system.

The evaluation of sprinklers in an existing building is based on whether an automatic sprinkler system is both required and installed. The criteria used to determine when an automatic sprinkler system is required are tied to the same requirements for new construction in Section 903. The thresholds listed in Sections 903.2 through 903.2.13 must be used to evaluate whether those characteristics and occupancies are present and whether a sprinkler system is needed.

The exceptions to Section 903; the requirements for stories and basements without openings of Section 903.2.10.1 and the suppression requirements for covered mall buildings, high-rise buildings, public garages and unlimited area buildings must also be evaluated. The determination of whether sprinklers are required in a building or a portion of a building must be done to correctly determine the category applicable to the existing building.

These guidelines for sprinklers allow for a more equitable evaluation of the contribution of that feature to overall building safety. This factor encourages the installation of automatic sprinkler systems in existing buildings by providing substantial negative and positive points.

The values for sprinklers are given in Table 3410.6.17. The appropriate credit values from Table 3410.6.17 are entered in Table 3410.7 under Safety Parameter 3410.6.17, Automatic Sprinklers, for both fire safety (FS) and general safety (GS), but only one-half the value is entered under the means of egress (ME). The one-half credit for egress is allowed because some credits for sprinklers are incorporated into the parameters for means of egress capacity (see Section 3410.6.11), dead-ends (see Section 3410.6.12) and maximum travel distance to an exit (see Section 3410.6.13).

TABLE 3410.6.17
SPRINKLER SYSTEM VALUES

OCCUPANCY	CATEGORIES					
	a	b	c	d	e	f
A-1, A-3, F, M, R, S-1	-6	-3	0	2	4	6
A-2	-4	-2	0	1	2	4
A-4, B, E, S-2	-12	-6	0	3	6	12

❖ This table lists the credit values for the respective categories of Section 3410.6.17.1, based on the occupancy being evaluated in the existing building. The assembly occupancies containing large combustible fuel loads with large occupant loads are included with those occupancies containing large fuel loads. These occupancies represent buildings where experience has shown that adequate sprinkler systems save lives and where an on-site sprinkler system is necessary to supplement the local fire department capabilities.

Group A-2 buildings are contained in a separate line in the table for determining sprinkler system values. This occupancy with its densely packed high occupant loads, ill-defined seating and aisle arrangements and facilities services must be separately evaluated to adequately match its hazards with sprinkler system requirements. Sports arenas and schools, which have high occupant/low-fuel loads, are combined with other low-fuel load occupancies in the last line item in the table. Category c is the zero-based category. The categories to the left of this column contain negative values that define buildings and occupancies that are required to be, but are not, sprinklered or are provided with inadequate sprinkler systems. The three categories to the right of this column contain positive values that represent buildings and occupancies with sprinkler systems that are deemed adequate or comply with current standards and requirements.

3410.6.17.1 Categories. The categories for automatic sprinkler system protection are:

1. Category a — Sprinklers are required throughout; sprinkler protection is not provided or the sprinkler system design is not adequate for the hazard protected in accordance with Section 903.

2. Category b — Sprinklers are required in a portion of the building; sprinkler protection is not provided or the sprinkler system design is not adequate for the hazard protected in accordance with Section 903.

3. Category c — Sprinklers are not required; none are provided.

4. Category d — Sprinklers are required in a portion of the building; sprinklers are provided in such portion; the system is one which complied with the code at the time of

installation and is maintained and supervised in accordance with Section 903.

5. Category e — Sprinklers are required throughout; sprinklers are provided throughout in accordance with Chapter 9.

6. Category f — Sprinklers are not required throughout; sprinklers are provided throughout in accordance with Chapter 9.

❖ Six categories are defined in this section for evaluating the automatic sprinkler system in an existing building. These categories address all aspects, from existing buildings that are unsprinklered but are required to be sprinklered if new construction, to existing buildings that are sprinklered and are not required to be sprinklered if new construction. These categories reduce the impact of providing sprinkler protection required by Chapter 9 to be sprinklered in new construction. This greater range of points increases the flexibility in the use of this evaluation method.

Category a includes buildings or occupancies that exceed the requirements of Section 903 and are required to be sprinklered throughout the building. Category a buildings are provided with either no sprinkler system, or one that is inadequate and does not provide the required level of protection. To evaluate the existing sprinkler system, a trained fire protection engineer should evaluate the system's design against the applicable referenced standards. This is not the same type of consideration needed in Category d, where the existing system actually meets all of the requirements of any earlier edition of the applicable referenced standards. This category is considered the lowest acceptable level of compliance and is therefore associated with the largest negative values in Table 3410.6.17.

Category b is applicable when only a portion of the existing building is required by the provisions of Section 903 to be sprinklered. For example, a multistory building may have two of its stories qualify as windowless stories that require sprinklers, or a mercantile building may have one fire area exceeding 12,000 square feet (1114.8 m²) and require sprinklers. In both cases the buildings are without an automatic sprinkler system, or the sprinkler systems that are in place are inadequate and do not comply with the technical provisions of the code. Since only portions of the buildings are required to be sprinklered, the negative sprinkler system values for this category in Table 3410.6.17 are exactly half of the Category a values.

Category c is the zero-based category for this safety parameter. The existing building is not required by Section 903 to be sprinklered and no sprinkler system is provided. Such buildings are neither penalized nor rewarded by Table 3410.6.17.

Category d is similar to Category b because only a portion of the existing building is required by Section 903 to be sprinklered. In this case, there is a sprinkler system in place in that portion of the building and the system was designed at the time of its installation to comply with the requirements of an earlier edition of the applicable standards in Section 903. An example of this is a sprinkler system designed to comply with the hydraulic design criteria of the 1989 edition of NFPA 13 for an ordinary Group 2 hazard. As long as this existing sprinkler system is still properly maintained and supervised, the building qualifies for the small positive values listed in Table 3410.6.17.

Category e are existing buildings that are required by the provisions of Chapter 9 to be sprinklered throughout, and are protected throughout by a properly designed, installed and supervised sprinkler system. Although this category more closely represents the requirements for new construction, moderate positive values are awarded by Table 3410.6.17. The evaluation rewards existing buildings that are sprinklered with higher points.

Category f is the highest level of protection and is rewarded with maximum positive points in the accompanying table. Existing buildings, which are not required by Section 903 to be sprinklered but are voluntarily provided with a fully designed, installed and supervised system, provide an added level of protection that justifies substantial bonus points that may ultimately determine whether an existing building meets its mandatory safety scores.

3410.6.18 Incidental use. Evaluate the protection of incidental use areas in accordance with Section 508.2. Do not include those where this code requires suppression throughout the building including covered mall buildings, high-rise buildings, public garages and unlimited area buildings. Assign the lowest score from Table 3410.6.18 for the building or fire area being evaluated. If there are no specific occupancy areas in the building or fire area being evaluated, the value shall be zero.

❖ This section includes an evaluation system for the separation/protection requirements indicated for incidental use areas in an existing building. This evaluation is based on Table 508.2. If the building designer chooses to separate/protect the small rooms or areas in an existing building in accordance with this table, the building is classified according to its main use. Because some existing buildings may have been designed in this fashion, an evaluation procedure has been added to account for the level of protection provided. The lowest score must be assigned to the building or fire area for the specific occupancy areas. For example, an existing Group B building has six separate storage rooms. Each room is more than 100 square feet (9.3 m²) in area. If five of the storage rooms are protected with limited area sprinkler systems and separated with construction capable of resisting the passage of smoke, and the sixth room is not protected in the same manner, the lowest score for all the storage rooms is determined by the single unprotected storage room. If the building designer chooses to treat the specific occupancy areas as a separate occupancy, the provisions of Section 3410.9.1 are applica-

ble and a zero is inserted in Table 3410.7 under Safety Parameter 3410.6.18, Incidental Use Area.

TABLE 3410.6.18
INCIDENTAL USE AREA VALUES[a]

PROTECTION REQUIRED BY TABLE 508.2	PROTECTION PROVIDED						
	None	1 Hour	AFSS	AFSS with SP	1 Hour and AFSS	2 Hours	2 Hours and AFSS
2 Hours and AFSS	-4	-3	-2	-2	-1	-2	0
2 Hours, or 1 Hour and AFSS	-3	-2	-1	-1	0	0	0
1 Hour and AFSS	-3	-2	-1	-1	0	-1	0
1 Hour	-1	0	-1	0	0	0	0
1 Hour, or AFSS with SP	-1	0	-1	0	0	0	0
AFSS with SP	-1	-1	-1	0	0	-1	0
1 Hour or AFSS	-1	0	0	0	0	0	0

a. AFSS = Automatic fire suppression system; SP = Smoke partitions (See Section 508.2.2).
NOTE: For Table 3410.7, see next page 23-47.

❖ This table provides a matrix of characteristics arranged in a format of rows and columns for determining values for incidental use areas. The values are inserted into the building score sheet. The left-hand column of the table is arranged for the separation/protection specified by Table 508.2 for a particular occupancy room or area. Table 508.2 must first be consulted to determine the level of separation/protection required by the code for new construction. This same entry is then found in the left-hand column of Table 3410.6.18. The top row represents the actual level of protection that is provided in the existing building. The corresponding occupancy area value is read and then inserted into Table 3410.7 under Safety Parameter 3410.6.18, Incidental Use. Values of zero are assigned to all arrangements that represent compliance with the requirements for new construction. Negative values are assigned based on the degree of noncompliance with the requirements for new construction.

3410.7 Building score. After determining the appropriate data from Section 3410.6, enter those data in Table 3410.7 and total the building score.

❖ This section is the tally sheet for all of the 18 safety parameters evaluated in Sections 3410.6.1 through 3410.6.18, which determine the building's overall safety profile for fire safety (FS), means of egress (ME) and general safety (GS).

This section directs the data and values of the 18 safety parameters of Sections 3410.6.1 through 3410.6.18 to be entered in Table 3410.7 for totaling the building scores.

TABLE 3410.7
SUMMARY SHEET — BUILDING CODE

Existing occupancy _____   Proposed occupancy_____

Year building was constructed _____   Number of stories _____ Height in feet ____

Type of construction _____   Area per floor _____

Percentage of open perimeter ____%   Percentage of height reduction ____%

Completely suppressed: Yes _____ No _____   Corridor wall rating _____

Compartmentation: Yes _____ No _____   Required door closers:_____Yes _____ No

Fire-resistance rating of vertical opening enclosures _____

Type of HVAC system _____, serving number of floors _____

Automatic fire detection: Yes _____ No _____,   Type and location _____

Fire alarm system: Yes _____ No _____,   Type_____

Smoke control: Yes _____ No _____,   Type_____

Adequate exit routes: Yes _____ No _____   Dead ends: _____ Yes _____ No

Maximum exit access travel distance _____   Elevator controls: _____Yes _____No

Means of egress emergency lighting: _____Yes _____No   Mixed occupancies:_____Yes_____ No

SAFETY PARAMETERS	FIRE SAFETY (FS)	MEANS OF EGRESS (ME)	GENERAL SAFETY (GS)
3410.6.1 Building Height 3410.6.2 Building Area 3410.6.3 Compartmentation			
3410.6.4 Tenant and Dwelling Unit Separations 3410.6.5 Corridor Walls 3410.6.6 Vertical Openings			
3410.6.7 HVAC Systems 3410.6.8 Automatic Fire Detection 3410.6.9 Fire Alarm System			
3410.6.10 Smoke control 3410.6.11 Means of Egress 3410.6.12 Dead ends	* * * * * * * * * * * *		
3410.6.13 Maximum Exit Access Travel Distance 3410.6.14 Elevator Control 3410.6.15 Means of Egress Emergency Lighting	* * * * * * * *		
3410.6.16 Mixed Occupancies 3410.6.17 Automatic Sprinklers 3410.6.18 Incidental Use		* * * * ÷ 2 =	
Building score — total value			

* * * *No applicable value to be inserted.

TABLE 3410.7. See page 34-47.

❖ Table 3410.7 is the summary sheet containing all of the relative attributes of the building. The summary sheet also contains a complete listing of the 18 safety parameters that have been evaluated. The upper portion of the summary sheet serves as a guide to the user to catalog and highlight existing building elements that relate to the 18 safety parameters and to the evaluation. The lower portion of the summary sheet is used to record the results of the 18 safety parameters that have been evaluated. These are added to produce the building score total values for fire safety (FS), means of egress (ME) and general safety (GS).

3410.8 Safety scores. The values in Table 3410.8 are the required mandatory safety scores for the evaluation process listed in Section 3410.6.

❖ This section lists the minimum scores for fire safety (FS), means of egress (ME) and general safety (GS) that must be obtained from the evaluation of the 18 safety parameters to be acceptable as a building meeting the code's objectives for public safety and health. This section summarizes the mandatory values of Table 3410.8 that must be met from the evaluation program.

TABLE 3410.8
MANDATORY SAFETY SCORES[a]

OCCUPANCY	FIRE SAFETY (MFS)	MEANS OF EGRESS (MME)	GENERAL SAFETY (MGS)
A-1	16	27	27
A-2	19	30	30
A-3	18	29	29
A-4, E	23	34	34
B	24	34	34
F	20	30	30
M	19	36	36
R	17	34	34
S-1	15	25	25
S-2	23	33	33

a. MFS = Mandatory Fire Safety;
   MME = Mandatory Means of Egress;
   MGS = Mandatory General Safety.

❖ The table lists the minimum mandatory safety scores for the evaluation of various occupancies for the three major mandatory safety scores of fire safety (MFS), means of egress (MME) and general safety (MGS). The mandatory safety values are based on the scores considered to be in compliance with the code for new construction. This is the zero-based concept. The scores have been determined as representing one level of compliance higher than the code's minimum requirements for new construction. The mandatory safety scores are consistent with the idea of establishing an equivalent level of safety, even though the exist-

ing building is evaluated only for the 18 safety parameters.

3410.9 Evaluation of building safety. The mandatory safety score in Table 3410.8 shall be subtracted from the building score in Table 3410.7 for each category. Where the final score for any category equals zero or more, the building is in compliance with the requirements of this section for that category. Where the final score for any category is less than zero, the building is not in compliance with the requirements of this section.

❖ The sections and table that follow are the final steps in the evaluation process. This section also discusses how mixed occupancies must to be treated during the final step of the evaluation process.

This section compares the three building scores from Table 3410.7 to the three mandatory safety scores from Table 3410.8. If the values in all three categories from Table 3410.7 exceed the corresponding mandatory safety scores in Table 3410.8, the building passes and is in compliance with the code. If the score in any one category is less than the mandatory safety score, the building is deemed to have failed and additional measures must be taken to bring the scores to a point that will at least equal the mandatory safety scores.

Table 3410.9
EVALUATION FORMULAS[a]

Formula	Table 3410.7	Table 3410.8	Score	Pass	Fail
FS-MFS $\geq$ 0	23 (FS) -	19 (MFS) =	4	X	___
ME-MME $\geq$ 0	41 (ME) -	36 (MME) =	5	X	___
GS-MGS $\geq$ 0	36 (GS) -	36 (MGS) =	0	X	___

Note a.
FS = Fire Safety   MFS = Mandatory Fire Safety
ME = Means of Egress   MME = Mandatory Means of Egress
GS = General Safety   MGS = Mandatory General Safety

**Figure 3410.9
EXAMPLE OF EVALUATION FORMULAS**

TABLE 3410.9. See page 34-49.

❖ Table 3410.9 shows in simple equations whether a building passes the evaluation by subtracting the mandatory safety score from Table 3410.7. This is done for each of the three general categories of evaluation: fire safety, means of egress and general safety. If the difference for each category is zero or greater, the existing building passes and is considered to comply with the code objectives for public safety.

Example:
A Group M mercantile occupancy receives evaluations for the 18 safety parameters that result in total building scores in Table 3410.7 as follows:

Fire safety (FS) = 23

Means of egress (ME) = 41

General safety (GS) = 36

TABLE 3410.9
EVALUATION FORMULAS[a]

FORMULA	T.3410.7			T.3410.8	SCORE	PASS	FAIL
FS-MFS ≥ 0	_____	(FS)	–	_____ (MFS) =	_____	_____	_____
ME-MME ≥ 0	_____	(ME)	–	_____ (MME) =	_____	_____	_____
GS-MGS ≥ 0	_____	(GS)	–	_____ (MGS) =	_____	_____	_____

a. FS = Fire Safety   MFS = Mandatory Fire Safety
    ME = Means of Egress   MME = Mandatory Means of Egress
    GS = General Safety   MGS = Mandatory General Safety

These scores are compared to the mandatory safety scores for a Group M occupancy from Table 3410.8 (see Figure 3410.9).

Mandatory fire safety (MFS) = 19

Mandatory means of egress (MME) = 36

Mandatory general safety (MGS) = 36

The mandatory score is subtracted from the building score.

Conclusion: The building is acceptable; all of the final safety scores are zero or greater.

3410.9.1 Mixed occupancies. For mixed occupancies, the following provisions shall apply:

1. Where the separation between mixed occupancies does not qualify for any category indicated in Section 3410.6.16, the mandatory safety scores for the occupancy with the lowest general safety score in Table 3410.8 shall be utilized (see Section 3410.6.)

2. Where the separation between mixed occupancies qualifies for any category indicated in Section 3410.6.16, the mandatory safety scores for each occupancy shall be placed against the evaluation scores for the appropriate occupancy.

❖ This section explains how to determine whether a mixed occupancy building passes or fails the process of evaluation. It restates the information in Sections 3410.6 and 3410.6.16. The mixed occupancy evaluation is based on the requirements of Sections 508.3.2 and 508.3.3. The two procedures described are:

Procedure 1:
For nonseparated occupancies in accordance with Section 508.3.2, or for occupancies that are separated with a fire-resistance rating of less than 1 hour, mandatory safety scores for the occupancy with the lowest general safety score of Table 3410.8 apply.

Procedure 2:
For separated occupancies in accordance with Section 508.3.3, or one of the categories listed in Section 3410.6.16, mandatory safety scores for each occupancy are compared to the evaluation scores for the appropriate occupancy. The total building score of Table 3410.7 is computed for each appropriate occupancy.

This section does not include a category or condition for a third option for mixed occupancies, which is the separation of multiple occupancies within a single building or structure with one or more fire walls. Likewise, this option is not addressed in Section 508. A fire wall creates separate and independent buildings; therefore, a separate evaluation must be done for each area within a building that is separated by fire walls.

Example:
A 40-foot (12 192 mm), three-story building has an open perimeter of 25 percent. It is Type IIB unprotected construction, unsprinklered. The building has 9,000 square feet (836.1 m²) per floor. A Group M mercantile is on the first floor with Group R residential occupancies on the second and third floors. Although the building is unprotected Type IIB construction, there is a 2-hour fire-resistance-rated floor/ceiling assembly separating the first and second floors. An analysis of this building will result in the following: "In accordance with Table 508.3.3, the mixed occupancies are separated by the required fire-rated assembly of 2 hours."

This building qualifies for Category b in Section 3410.6.16.1. A separate summary sheet for tabulating the building score in Table 3410.7 has to be completed for each of the two occupancies. The values of the building scores for fire safety, means of egress and general safety for the Group M mercantile occupancy and the Group R residential occupancy must be calculated. The mercantile building scores are compared to the mandatory safety scores of 19, 36 and 36, and the residential building scores are compared to the mandatory safety scores of 17, 34 and 34. If any one of the three building scores from either of the two occupancies does not equal or exceed the applicable safety score, the entire building fails the evaluation. For the building to pass, the safety parameters that are less than the mandatory safety score must be upgraded until a passing score is achieved.

Bibliography

The following resource materials are referenced in this chapter or are relevant to the subject matter addressed in this chapter.

36 CFR Parts 1190 and 1191 Draft, *The Americans with Disabilities Act Accessibility Guidelines* (ADAAG), Washington, DC: Architectural and Transportation Barriers Compliance Board, April 2, 2002.

42 USC 3601-88, *Fair Housing Amendments Act* (FHAA). Washington, D.C.: United States Code, 1988.

"*Accessibility and Egress for People with Physical Disabilities.*" Falls Church, VA: CABO Board for the Coordination of the Model Codes Report, October 5, 1993.

ASME A17.1-00, *Safety Code for Elevators and Escalators.* New York: American Society of Mechanical Engineers, 2000.

ASME A17.3-96, *Safety Code for Existing Elevators and Escalators.* New York: American Society of Mechanical Engineers, 1996.

ASME 18.1-99, *Safety Standard for Platform Lifts and Stairway Chairlifts–with Addenda A18.1a-2001,* New York: American Society of Mechanical Engineers, 1999.

DOJ 28 CFR, Part 36-91, *Americans with Disabilities Act* (ADA). Washington, D.C.: Department of Justice, 1991.

DOJ 28 CFR, Part 36-91(Appendix A), *ADA Accessibility Guidelines for Buildings and Facilities* (ADAAG). Washington, D.C.: Department of Justice, 1991.

FED-STD-795-88, *Uniform Federal Accessibility Standards.* Washington, D.C.: General Services Administration; Department of Defense; Department of Housing and Urban Development; U.S. Postal Service, 1988.

"Final Report, Recommendations for a New ADAAG." Washington D.C: ADAAG Review Federal Advisory Committee, September 30, 1996.

ICC A117.1-05, *Accessible and Usable Buildings and Facilities.* Falls Church, VA: International Code Council, 2003.

IMC-06, *International Mechanical Code.* Falls Church, VA: International Code Council, 2006.

IPMC-06, *International Property Maintenance Code.* Falls Church, VA: International Code Council, 2006.

NFPA 13-02, *Installation of Sprinkler Systems.* Quincy, MA: National Fire Protection Association, 2002.

NFPA 13D-02, *Installation of Sprinkler Systems in One- and Two-Family Dwellings and Manufactured Homes.* Quincy, MA: National Fire Protection Association, 2002.

NFPA 13R-02, *Installation of Sprinkler Systems in Residential Occupancies Up to and Including Four Stories in Height.* Quincy, MA: National Fire Protection Association, 2002.

NFPA 72-02, *National Fire Alarm Code.* Quincy, MA: National Fire Protection Association, 2002.

NFPA 101-03, *Alternative Approaches to Life Safety.* Quincy, MA: National Fire Protection Association, 2003.

Rehabilitation Guidelines/1980. Washington, DC: National Institute of Building Science for the U.S. Department of Housing and Urban Development, 1980.

1. Guideline for Setting and Adopting Standards for Building Rehabilitation;
2. Guideline for Approval of Building Rehabilitation;
3. Statutory Guideline for Building Rehabilitation;
4. Guideline for Managing Official Liability Associated with Building Rehabilitation;
5. Egress Guideline for Residential Rehabilitation;
6. Electrical Guideline for Residential Rehabilitation;
7. Plumbing DWV Guideline for Residential Rehabilitation; and
8. Guideline on Fire Ratings of Archaic Materials and Assemblies.

UL 555-99, *Fire Dampers.* Northbrook, IL: Underwriters Laboratories Inc., 1999.

Chapter 35:
Referenced Standards

General Comments

Chapter 35 contains a comprehensive list of all standards that are referenced in the code. It is organized in a manner that makes it easy to locate specific document references.

This chapter lists the standards that are referenced in various sections of this document. The standards are listed herein by the promulgating agency of the standard, the standard identification, the date and title and the section or sections of this document that reference the standard. The application of the referenced standards shall be as specified in Section 102.4.

It is important to understand that not every document related to building design and construction is qualified to be a "referenced standard." The International Code Council® (ICC®) has adopted a criterion that standards referenced in and intended for adoption into the *International Codes®* must meet in order to qualify as a referenced standard. The policy is summarized as follows:

- Code references: The scope and application of the standard must be clearly identified in the code text.
- Standard content: The standard must be written in mandatory language and appropriate for the subject covered. The standard shall not have the effect of requiring proprietary materials or prescribing a proprietary testing agency.
- Standard promulgation: The standard must be readily available and developed and maintained in a consensus process, such as ASTM or ANSI.

It should be noted that the ICC Code Development Procedures, of which the standards policy is a part, are updated periodically. A copy of the latest version can be obtained from the ICC offices.

Once a standard is incorporated into the code through the code development process, it becomes an enforceable part of the code. When the code is adopted by a jurisdiction, the standard also is part of that jurisdiction's adopted code. It is for this reason that the criteria were developed. Compliance with this policy provides that documents incorporated into the code are, among others, developed through the use of a consensus process, written in mandatory language and do not mandate the use of proprietary materials or agencies. The requirement for a standard to be developed through a consensus process is vital, as it means that the standard will be representative of the most current body of available knowledge on the subject as determined by a broad spectrum of interested or affected parties without dominance by any single interest group. A true consensus process has many attributes, including but not limited to:

- An open process that has formal (published) procedures that allow for the consideration of all viewpoints;
- A definitive review period that allows for the standard to be updated or revised;
- A process of notification to all interested parties; and
- An appeals process.

Many available documents related to design, installation and construction, though useful, are not "standards" and are not appropriate for reference in the code. Often, these documents are developed or written with the intention of being used for regulatory purposes and are unsuitable for use as a regulation due to extensive use of recommendations, advisory comments and nonmandatory terms. Typical examples of such documents include installation instructions, guidelines and practices.

The objective of ICC's standards policy is to provide regulations that are clear, concise and enforceable; thus the requirement for standards to be written in mandatory language. This requirement is not intended to mean that a standard cannot contain informational or explanatory material that will aid the user of the standard in its application. When it is the desire of the standard's promulgating agency for such material to be included, however, the information must appear in a nonmandatory location, such as an annex or appendix, and be clearly identified as not being part of the standard.

Overall, standards referenced by the code must be authoritative, relevant, up to date and, most important, reasonable and enforceable. Standards that comply with ICC's standards policy fulfill these expectations.

Purpose

As a performance-oriented code, the code contains numerous references to documents that are used to regulate materials and methods of construction. The references to these documents within the code text consist of the promulgating agency's acronym and its publication designation (e.g., ASME A17.1) and a further indication that the document being referenced is the one that is listed in Chapter 35. Chapter 35 contains all of the information that is necessary to identify the specific referenced document. Included is the following information on a document's promulgating agency (see Figure 35):

- The promulgating agency (i.e., the agency's title);
- The promulgating agency's acronym; and
- The promulgating agency's address.

For example, a reference to an ASME standard within the code indicates that the document is promulgated by the American Society of Mechanical Engineers (ASME), which is located in New York City. This chapter lists the standards agencies alphabetically for ease of identification.

This chapter also includes the following information on the referenced document itself (see Figure 35):

• The document's publication designation;

• The document's edition year;

• The document's title;

• Any addenda or revisions to the document that are applicable; and

• Every section of the code in which the document is referenced.

For example, a reference to ASME A17.1 indicates that this document can be found in Chapter 35 under the heading ASME. The specific standards designation is A17.1. For convenience, these designations are listed in alphanumeric order. This chapter identifies that ASME A17.1 is titled *Safety Code for Elevators and Escalators,* the applicable edition (i.e., its year of publication) is 2004 and it is referenced in numerous sections of the code.

This chapter will also indicate when a document has been discontinued or replaced by its promulgating agency. When a document is replaced by a different one, a note will appear to tell the user the designation and title of the new document.

The key aspect of the manner in which standards are referenced by the code is that a specific edition of a specific standard is clearly identified. In this manner, the requirements necessary for compliance can be readily determined. The basis for code compliance is, therefore, established and available on an equal basis to the building official, contractor, designer and owner.

This chapter lists the standards that are referenced in various sections of this document. The standards are listed herein by the promulgating agency of the standard, the standard identification, the effective date and title and the section or sections of this document that reference the standard. The application of the referenced standards shall be as specified in Section 102.4.

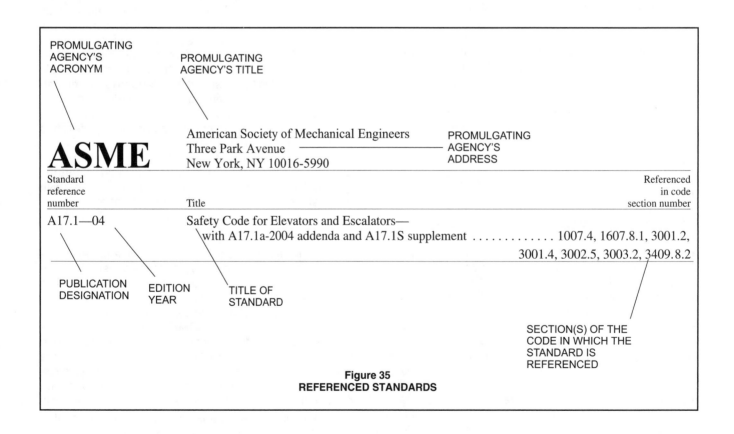

Figure 35
REFERENCED STANDARDS

This chapter lists the standards that are referenced in various sections of this document. The standards are listed herein by the promulgating agency of the standard, the standard identification, the effective date and title, and the section or sections of this document that reference the standard. The application of the referenced standards shall be as specified in Section 102.4.

AA

Aluminum Association
900 - 19th Street N.W., Suite 300
Washington, DC 20006

Standard reference number	Title	Referenced in code section number
ADM 1—00	Aluminum Design Manual: Part 1-A Aluminum Structures, Allowable Stress Design; and Part 1-B— Aluminum Structures, Load and Resistance Factor Design of Buildings and Similar Type Structures	1604.3.5, 2002.1
ASM 35—00	Aluminum Sheet Metal Work in Building Construction (Fourth Edition)	2002.1

AAMA

American Architectural Manufacturers Association
1827 Waldon Office Square, Suite 550
Schaumburg, IL 60173

Standard reference number	Title	Referenced in code section number
1402—86	Standard Specifications for Aluminum Siding, Soffit and Fascia	1404.5.1
101/I.S.2/A440—05	Specifications for Windows, Doors and Unit Skylights	1714.5.1, 2405.5

ACI

American Concrete Institute
P.O. Box 9094
Farmington Hills, MI 48333-9094

Standard reference number	Title	Referenced in code section number
216.1—97	Standard Method for Determining Fire Resistance of Concrete and Masonry Construction Assemblies	Table 720.1(2), 721.1
318—05	Building Code Requirements for Structural Concrete . . . . 1604.3.2, Table 1704.3, 1704.4.1, Table 1704.4, 1708.3, 1805.4.2.6, 1805.9, 1808.2.23.1.1, 1808.2.23.2,1808.2.23.2.1, 1809.2.2.2.2, 1809.2.3.2, 1809.2.3.2.2, 1810.1.2.2, 1812.8,1901.2, 1901.3, 1901.4, 1902I, 1903.1, 1904.1, 1904.2.1, 1904.2.2, 1904.2.3, 1904.3, 1904.4, 1905.1.1, 1905.2, 1905.3, 1905.4, 1905.5, 1905.6.2, 1905.6.3, 1905.6.4, 1905.6.5, 1905.7, 1905.8, 1905.9, 1905.10, 1905.11, 1905.12, 1905.13, 1906.1, 1906.2, 1906.3, 1906.4, 1907.1, 1907.2, 1907.3, 1907.4, 1907.5, 1907.6, 1907.7.1, 1907.7.2, 1907.7.3, 1907.7.4, 1907.7.5, 1907.8, 1907.9, 1907.10, 1907.11, 1907.12, 1907.13, 1908, 1908.1, 1908.1.1, 1908.1.2, 1908.1.3, 1908.1.4, 1908.1.5, 1908.1.6, 1908.1.7, 1908.1.8, 1908.1.9, 1908.1.10, 1908.1.11, 1908.1.12, 1908.1.13, 1908.1.14, 1908.1.15, 1908.1.16, 1909.1, 1909.3, 1909.4, 1909.5, 1909.6, 1912.1, 2108.3, 2205.3
530—05	Building Code Requirements for Masonry Structures . . . . 1405.5, 1405.5.2, 1405.9, 1604.3.4, 1704.5, 1704.5.1, Table 1704.5.1, 1704.5.2, 1704.5.3, Table 1704.5.3, 1708.1.1, 1708.1.2, 1708.1.3, 1708.1.4, 1805.5.2, 1812.7, 2101.2.2, 2101.2.3, 2101.2.4, 2101.2.5, 2101.2.6, 2103.1.3.6, 2106.1, 2106.1.1, 2106.1.1.1, 2106.1.1.2, 2106.1.1.3, 2106.3, 2106.4, 2106.5, 2106.6, 2107.1, 2107.2, 2107.3, 2107.4, 2107.5, 2107.6, 2107.7, 2107.8, 2108.1, 2108.2, 2108.3, 2108.4, 2109.1, 2109.2.3.1, 2109.7.3
530.1—05	Specifications for Masonry Structures . . . . 1405.5.1, Table 1704.5.1, Table 1704.5.3, 2103.13.7, 2104.1, 2104.1.1, 2104.3, 2104.4

AF&PA

American Forest & Paper Association
1111 19th St, NW Suite 800
Washington, DC 20036

Standard reference number	Title	Referenced in code section number
WCD No. 4—89	Wood Construction Data—Plank and Beam Framing for Residential Buildings	2306.1.2
WFCM—01	Wood Frame Construction Manual for One- and Two-family Dwellings	2301.2, 2308.1, 2308.2.1
T.R. No. 7—87	Technical Report—Basic Requirements for Permanent Wood Foundation System	1805.4.6, 1807.2, 2304.9.5

AF&PA—continued

NDS—05	National Design Specification (NDS) for Wood Construction with 2005 Supplement. . . . . .721.6.3.2, 1715.1.1, 1715.1.4, 1805.4.5, 1809.1.4, 2302.1, 2304.12, 2306.1, 2306.2.1, 2306.3.1, Table 2306.3.1, Table 2306.3.2, Table 2306.4.1, Table 2306.4.4, 2306.3.4, 2306.3.5, 2306.4.1, 2307.1, 2307.1.1	
AF&PA—93	Span Tables for Joists and Rafters. 2306.1.1, 2308.8, 2308.10.2, 2308.10.3	
SDPWS—05	AF&PA Supplement Special Design Provisions for Wind and Seismic. 2305.1	

AHA

American Hardwood Association
1210 West N.W. Highway
Palatine, IL 60067

Standard reference number	Title	Referenced in code section number
A135.4—2004	Basic Hardboard. 1404.3.1, 2303.1.6
A135.5—2004	Prefinished Hardboard Paneling. 2303.1.6, 2304.6.2
A135.6—1998	Hardboard Siding 1404.3.2, 2303.1.6

AISC

American Institute of Steel Construction
One East Wacker Drive, Suite 3100
Chicago, IL 60601-2001

Standard reference number	Title	Referenced in code section number
341—05	Seismic Provisions for Structural Steel Buildings, including Supplement No. 1 dated 2006. . . . . .1613.6.2, 1707.2, 1708.4, 2205.2.1, 2205.2.2, 2205.3, 2205.3.1
360—05	Specification for Structural Steel Buildings. . . . . .1604.3.3, Table 1704.3, 2203.2, 2205.1, 2205.3

AISI

American Iron and Steel Institute
1140 Connecticut Avenue
Suite 705
Washington, DC 20036

Standard reference number	Title	Referenced in code section number
NAS—01	North American Specification for the Design of Cold-formed Steel Structural Members, including 2004 Supplement 1604.3.3, 2209.1, 2210.1
General—04	Standard for Cold-formed Steel Framing—General Provisions. 1604.3.3, 2210.1
Header—04	Standard for Cold-formed Steel Framing—Header Design 2210.2
Lateral—04	Standard for Cold-formed Steel Framing—Lateral Design 2210.5
PM—01	Standard for Cold-formed Steel Framing—Prescriptive Method for One- and Two-family Dwellings, including 2004 Supplement. 2210.6
Truss—04	Standard for Cold-formed Steel Framing—Truss Design 1604.3.3, 2210.3
WSD—04	Standard for Cold-formed Steel Framing—Wall Stud Design 2210.4

AITC

American Institute of Timber Construction
Suite 140
7012 S. Revere Parkway
Englewood, CO 80112

Standard reference number	Title	Referenced in code section number
AITC Technical Note 7—96	Calculation of Fire Resistance of Glued Laminated Timbers. 721.6.3.3
AITC 104—03	Typical Construction Details. 2306.1
AITC 110—01	Standard Appearance Grades for Structural Glued Laminated Timber 2306.1
AITC 113—01	Standard for Dimensions of Structural Glued Laminated Timber 2306.1
AITC 117—04	Standard Specifications for Structural Glued Laminated Timber of Softwood Species. 2306.1
AITC 119—96	Standard Specifications for Structural Glued Laminated Timber of Hardwood Species 2306.1

AITC 200—04   Manufacturing Quality Control Systems Manual for Structural Glued Laminated Timber 2306.1

ANSI/AITC A 190.1—02   Structural Glued Laminated Timber 2303.1.3, 2306.1

ALI

Automotive Lift Institute
P.O. Box 85
Courtland, NY 13045

Standard reference number	Title	Referenced in code section number
ALI ALCTV—98	Standard for Automotive Lifts—Safety Requirements for Construction, Testing and Validation (ANSI) 3001.2	

ANSI

American National Standards Institute
25 West 43rd Street, Fourth Floor
New York, NY 10036

Standard reference number	Title	Referenced in code section number
A13.1—96 (Reaffirmed 2002)	Scheme for the Identification of Piping Systems 415.8.6.4	
A108.1A—99	Installation of Ceramic Tile in the Wet-set Method, with Portland Cement Mortar 2103.10	
A108.1B—99	Installation of Ceramic Tile, quarry Tile on a Cured Portland Cement Mortar Setting Bed with Dry-set or Latex-portland Mortar 2103.10	
A108.4—99	Installation of Ceramic Tile with Organic Adhesives or Water-cleanable Tile-setting Epoxy Adhesive 2103.10.6	
A108.5—99	Installation of Ceramic Tile with Dry-set Portland Cement Mortar or Latex-portland Cement Mortar 2103.9.3, 2103.10.1, 2103.10.2	
A108.6—99	Installation of Ceramic Tile with Chemical-resistant, Water Cleanable Tile-setting and -grouting Epoxy 2103.10.3	
A108.8—99	Installation of Ceramic Tile with Chemical-resistant Furan Resin Mortar and Grout 2103.10.4	
A108.9—99	Installation of Ceramic Tile with Modified Epoxy Emulsion Mortar/Grout. 2103.10.5	
A108.10—99	Installation of Grout in Tilework 2103.10.7	
A118.1—99	American National Standard Specifications for Dry-set Portland Cement Mortar. 2103.10.1	
A118.3—99	American National Standard Specifications for Chemical-resistant, Water-cleanable Tile-setting and -grouting Epoxy and Water Cleanable Tile-setting Epoxy Adhesive 2103.10.3	
A118.4—99	American National Standard Specifications for Latex-portland Cement Mortar 2103.10.2	
A118.5—99	American National Standard Specifications for Chemical Resistant Furan Mortar and Grouts for Tile Installation. 2103.10.4	
A118.6—99	American National Standard Specifications for Cement Grouts for Tile Installation. 2103.10.7	
A118.8—99	American National Standard Specifications for Modified Epoxy Emulsion Mortar/Grout 2103.10.5	
A136.1—99	American National Standard Specifications for Organic Adhesives for Installation of Ceramic Tile 2103.10.6	
A137.1—88	American National Standard Specifications for Ceramic Tile 2103.5	
A208.1—99	Particleboard 2303.1.7, 2303.1.7.1	
Z 97.1—84 (R1994)	Safety Glazing Materials Used in Buildings—Safety Performance Specifications and Methods of Test (Reaffirmed 1994) 2406.1.2, 2409.1	

APA

APA - Engineered Wood Association
P.O. Box 11700
Tacoma, WA 98411-0700

Standard reference number	Title	Referenced in code section number
APA PDS—04	Panel Design Specification 2306.1, 2306.3.1, 2306.4.1	
APA PDS Supplement 1—90	Design and Fabrication of Plywood Curved Panels (revised 1995) 2306.1	
APA PDS Supplement 2—92	Design and Fabrication of Plywood-lumber beams (revised 1998) 2306.1	
APA PDS Supplement 3—90	Design and Fabrication of Plywood Stressed-skin Panels (revised 1996) 2306.1	
APA PDS Supplement 4—90	Design and Fabrication of Plywood Sandwich Panels (revised 1993) 2306.1	
APA PDS Supplement 5—95	Design and Fabrication of All-plywood Beams (revised 1995) 2306.1	

APA—continued

EWS R540—96	Builders Tips: Proper Storage and Handling of Glulam Beams	.2306.1
EWS S475—01	Glued Laminated Beam Design Tables	.2306.1
EWS S560—03	Field Notching and Drilling of Glued Laminated Timber Beams	.2306.1
EWS T300—02	Glulam Connection Details	.2306.1
EWS X440—00	Product Guide—Glulam	.2306.1
EWS X450—01	Glulam in Residential Construction —Western Edition	.2306.1

ASAE

American Society of Agricultural Engineers
2950 Niles Road
St. Joseph, MI 49085-9659

Standard reference number	Title	Referenced in code section number
EP 484.2 (1998)	Diaphragm Design of Metal-clad, Post-frame Rectangular Buildings	.2306.1
EP 486.1 (2000)	Shallow-post Foundation Design	.2306.1
EP 559 (1997)	Design Requirements and Bending Properties for Mechanically Laminated Columns	.2306.1

ASCE/SEI

American Society of Civil Engineers
Structural Engineering Institute
1801 Alexander Bell Drive
Reston, VA 20191-4400

Standard reference number	Title	Referenced in code section number
3—91	Structural Design of Composite Slabs	.1604.3.3, 2209.2
5—05	Building Code Requirements for Masonry Structures	1405.5, 1405.5.2, 1405.9, 1604.3.4, 1704.5, 1704.5.1, Table 1704.5.1, 1704.5.2, 1704.5.3, Table 1704.5.3, 1708.1.1, 1708.1.2, 1708.1.3, 1708.1.4, 1805.5.2, 1812.7, 2101.2.2, 2101.2.3, 2101.2.4, 2101.2.5, 2101.2.6, 2103.1.3.6, 2106.1, 2106.1.1, 2106.1.1.1, 2106.1.1.2, 2106.1.1.3, 2106.3, 2106.4, 2106.5, 2106.6, 2107.1, 2107.2, 2107.3, 2107.4, 2107.5, 2107.6, 2107.7, 2170.8, 2108.1, 2108.2, 2108.3, 2108.4, 2109.1, 2109.2.3.1, 2109.7.3
6—05	Specifications for Masonry Structures	1405.5.1, Table 1704.5.1, Table 1704.5.3, 1805.5.2.2, 2103.13.7, 2104.1, 2104.1.1, 2104.3, 2104.4
7—05	Minimum Design Loads for Buildings and Other Structures including Supplement No. 1 and excluding Chapter 14 and Appendix 11A	1602.1, 1604.3, 1604.10, 1605.1, 1605.2.2, 1605.3.1.2, 1605.3.2, 1605.4, 1607.11.1, 1608.1, 1608.2, 1609.1.1, 1609.1.2, 1609.3, 1609.5.1, 1609.5.3, 1611.2, 1612.2, 1613.1, 1613.2, Table 1613.5.3(1), Table 1613.5.3(2), 1613.5.6, 1613.5.6.1, 1613.5.6.2, 1613.6, 1613.6.1, 1613.6.2, 1801.2.1, 1802.2.7, 2205.2.1, 2205.3, 2205.3.1, 2208.1, 2305.1.5, 2305.2.5, 2305.3.1, 2306.4.5, Table 2306.4.5, Table 2308.10.1
8—02	Standard Specification for the Design of Cold-formed Stainless Steel Structural Members	1604.3.3, 2209.1
19—96	Structural Applications of Steel Cables for Buildings	.2207.1, 2207.2
24—05	Flood Resistant Design and Construction	1203.3.2, 1612.4, 1612.5, 3001.2
29—05	Standard Calculation Methods for Structural Fire Protection	.721.1
32—01	Design and Construction of Frost Protected Shallow Foundations	.1805.2.1

ASME

American Society of Mechanical Engineers
Three Park Avenue
New York, NY 10016-5990

Standard reference number	Title	Referenced in code section number
A17.1—04	Safety Code for Elevators and Escalators with A17.1a-2005 addenda and A17.1S Supplement 2005	1007.4, 1607.8.1, 3001.2, 3001.4, 3002.5, 3003.2, 3409.8.2
A18.1—03	Safety Standard for Platform Lifts and Stairway Chairlifts	1109.7, 2702.2.6, 3409.8.3
A90.1—03	Safety Standard for Belt Manlifts	.3001.2
A112.18.19.8M—1987	Suction Fittings for Use in Swimming Pools, Wading Pools, Spas, Hot Tubs and Whirlpool Bathing Appliances	.3109.5.1
A112.19.17—2002	Manufactured Safety Vacuum Release Systems (SVRS) for Residential and Commercial Swimming Pool, Spa, Hot Tub and Wading Pool	.3109.5.2

ASME—continued

B16.18—2001	Cast Copper Alloy Solder Joint Pressure Fittings.	909.13.1
B16.22—2001	Wrought Copper and Copper Alloy Solder Joint Pressure Fittings.	909.13.1
B20.1—2003	Safety Standard for Conveyors and Related Equipment	3001.2, 3005.3
B31.3—2002	Process Piping.	415.8.6.1

ASTM

ASTM International
100 Barr Harbor Drive
West Conshohocken, PA 19428-2959

Standard reference number	Title	Referenced in code section number
A 6/A 6M— 04a	Specification for General Requirements for Rolled Steel, Structural Steel Bars, Plates, Shapes, and Sheet Piling	Table 1704.3
A 36/A 36M—04	Specification for Carbon Structural Steel.	1809.3.1, 2103.13.5
A 82—02	Specification for Steel Wire, Plain, for Concrete Reinforcement	2103.13.5, 2103.13.6
A 123/A 123M—02	Specification for Zinc (Hot-dip Galvanized) Coating on Iron and Steel Products	2103.13.7.2
A 153—03	Specification for Zinc Coating (Hot-dip) on Iron and Steel Hardware.	2103.13.7.2, 2304.9.5
A 185—02	Specification for Steel Welded Wire Reinforcement, Plain for Concrete	2103.13.4, 2103.13.5
A 240—04	Standard Specification for Chromium and Chromium-nickel Stainless Steel Plate, Sheet and Strip for Pressure Vessels and for General Applications	Table 1507.4.3(1), 2103.13.5
A 252—98 (2002)	Specification for Welded and Seamless Steel Pipe Piles.	1809.3.1, 1810.6.1, 1810.8.2
A 283/A 283M—03	Specification for Low and Intermediate Tensile Strength Carbon Steel Plates.	1809.3.1, 1810.6.1
A 307—03	Specification for Carbon Steel Bolts and Studs, 60,000 psi Tensile Strength	1911.1, 2103.13.5
A 416/A 416M—02	Specification for Steel Strand, Uncoated Seven-wire for Prestressed Concrete.	1809.2.3.1, 2103.13.6
A 421/A 421M—02	Specification for Uncoated Stress-relieved Steel Wire for Prestressed Concrete	2103.13.6
A 435/A 435M— 90 (2001)	Specification for Straight-beam Ultrasonic Examination of Steel Plates	1708.4
A 463/A 463M—02a	Specification for Steel Sheet, Aluminum-coated, by the Hot Dip Process	Table 1507.4.3(2)
A 480/A480M—02	Specification for General Requirements for Flat-rolled Stainless and Heat-resisting Steel Plate, Sheet, and Strip	2103.13.5
A 496—02	Specification for Steel Wire, Deformed for Concrete Reinforcement.	2103.13.3
A 497—01	Specification for Steel Welded Reinforcement Deformed, for Concrete.	2103.13.4
A 510—03	Specification for General Requirements for Wire Rods and Coarse Round Wire, Carbon Steel	2103.13.6
A 568/A 568M—03	Specification for Steel, Sheet, Carbon, and High-strength, Low-alloy, Hot-rolled and Cold-rolled, General Requirements for.	Table 1704.3
A 572/A 572M—04	Specification for High-strength Low-alloy Columbium-vanadium Structural Steel.	1809.3.1
A 588/A 588M—04	Specification for High-strength Low-alloy Structural Steel with 50 ksi (345 Mpa) Minimum Yield Point to 4 inches (100 mm) Thick.	1809.3.1
A 615/A 615M—04a	Specification for Deformed and Plain Billet-steel Bars for Concrete Reinforcement	1708.3, 1810.8.2, 1908.1.5, 2103.13.1, 2103.13.6
A 641/A 641M—03	Specification for Zinc-coated (Galvanized) Carbon Steel Wire	2103.13.7.1
A 653/A 653M—04a	Specification for Steel Sheet, Zinc-coated Galvanized or Zinc-iron Alloy-coated Galvannealed by the Hot-dip Process	Table 1507.4.3(1), Table 1507.4.3(2), 2103.13.7.1
A 690—00a	Standard Specification for High Strength Low-alloy Steel H-piles and Sheet Piling for Use in Marine Environments	1809.3.1
A 706/A 706M—04a	Specification for Low-alloy Steel Deformed and Plain Bars for Concrete Reinforcement	1704.4.1, Table 1704.3, 1908.1.5, 2103.13.1, 2103.13.6, 2108.3
A 722/A 722M— 98 (2003)	Specification for Uncoated High-strength Steel Bar for Prestressing Concrete	1810.8.2, 2103.13.6, 2106.1.1.3.1
A 755/A 755M—04	Specification for Steel Sheet, Metallic-coated by the Hot-dip Process and Prepainted by the Coil-coating Process for Exterior Exposed Building Products.	Table 1507.4.3(1), Table 1507.4.3(2)
A 767/A 767M—00b	Specification for Zinc-coated (Galvanized) Steel Bars for Concrete Reinforcement	2103.13.1
A 775/A 775M—04	Specification for Epoxy-coated Steel Reinforcing Bars	2103.13.1
A 792/A 792M—03	Specification for Steel Sheet, 55% Aluminum-zinc Alloy-coated by the Hot-dip Process.	Table 1507.4.3(1), Table 1507.4.3(2)
A 875/A 875M—02a	Standard Specification for Steel Sheet Zinc-5 percent, Aluminum Alloy-coated by the Hot-dip Process.	Table 1507.4.3(2)
A 884—02	Specification for Epoxy-coated Steel Wire and Welded Wire Fabric for Reinforcement	2103.13.7.3
A 898/A 898M— 91 (2001)	Specification for Straight Beam Ultrasonic Examination of Rolled Steel Shapes.	1708.4
A 899—91 (2002)	Specification for Steel Wire Epoxy-coated.	2103.13.7.3

ASTM—continued

A 913/A 913M—04	Specification for High-strength Low-alloy Steel Shapes of Structural Quality, Produced by Quenching and Self-tempering Process (QST)	1809.3.1
A924—04	Standard Specification for General Requirements for Steel Sheet, Metallic-coated by the Hot-dip Process	Table 1507.4.3(1)
A 951—02	Specification for Masonry Joint Reinforcement	2103.13.2
A 992—04a	Standard Specification for Structural Shapes	1809.3.1
A 996/A 996M—04	Specification for Rail-steel and Axle-steel Deformed Bars for Concrete Reinforcement	2103.13.1, 2103.13.6
A1008/A1008M—04b	Specification for Steel, Sheet, Cold-rolled, Carbon, Structural, High-strength Low-alloy and High-strength Low-alloy with Improved Formability	2103.13.5
B 42—02e01	Specification for Seamless Copper Pipe, Standard Sizes	909.13.1
B 43—04	Specification for Seamless Red Brass Pipe, Standard Sizes	909.13.1
B 68—02	Specification for Seamless Copper Tube, Bright Annealed (Metric)	909.13.1
B 88—03	Specification for Seamless Copper Water Tube	909.13.1
B 101—02	Specification for Lead-coated Copper Sheet and Strip for Building Construction	Table 1404.5.3 Table 1507.4.3(1), Table 1507.2.9.2
B 209—04	Specification for Aluminum and Aluminum Alloy Steel and Plate	Table 1507.4.3(1)
B 251—02e01	Specification for General Requirements for Wrought Seamless Copper and Copper-alloy Tube	909.13.1
B 280—03	Specification for Seamless Copper Tube for Air Conditioning and Refrigeration Field Service	909.13.1
B 370—03	Specification for Cold-rolled Copper Sheet and Strip for Building Construction	1404.5.2, Table 1507.2.9.2, Table 1507.4.3(1)
B 695—00	Standard Specification for Coatings of Zinc Mechanically Deposited on Iron and Steel	2304.9.5
C 5—03	Specification for Quicklime for Structural Purposes	Table 2507.2
C 22/C 22M—00	Specification for Gypsum	Table 2506.2
C 27—98 (2002)	Specification for Standard Classification of Fireclay and High-alumina Refractory Brick	2111.5, 2111.8
C 28/C 28M—00e01	Specification for Gypsum Plasters	Table 2507.2
C 31/31M—98	Practice for Making and Curing Concrete Test Specimens in the Field	Table 1704.4
C 33—03	Specification for Concrete Aggregates	721.3.1.4, 721.4.1.1.3
C 34—03	Specification for Structural Clay Load-bearing Wall Tile	2103.2
C 35—95 (2001)	Specification for Inorganic Aggregates for Use in Gypsum Plaster	Table 2507.2
C 36/C 36M—03	Specification for Gypsum Wallboard	Figure 721.5.1(2), Figure 721.5.1(3), Table 2506.2
C 37/C 37M—01	Specification for Gypsum Lath	Table 2507.2
C 55—03	Specification for Concrete Brick	Table 721.3.2, 2103.1, 2105.2.2.1.2
C 56—96 (2001)	Specification for Structural Clay Non-load Bearing Tile	2103.2
C 59/C 59M—00	Specification for Gypsum Casting and Molding Plaster	Table 2507.2
C 61/C 61M—00	Specification for Gypsum Keene's Cement	Table 2507.2
C 62—04	Specification for Building Brick (Solid Masonry Units Made from Clay or Shale)	2103.2, 2105.2.2.1.1
C 67—03ae01	Test Methods of Sampling and Testing Brick and Structural Clay Tile	721.4.1.1.1, 2104.5, 2105.2.2.1.1, 2109.8.1.1
C 73—99a	Specification for Calcium Silicate Face Brick (Sand-lime Brick)	Table 721.3.2, 2103.1
C 79—04a	Specification for Treated Core and Nontreated Core Gypsum Sheathing Board	Table 2506.2
C 90—03	Specification for Loadbearing Concrete Masonry Units	Table 721.3.2, 1805.5.2.2, 2103.1, 2105.2.2.1.2
C 91—03a	Specification for Masonry Cement	Table 2103.8(1), Table 2507.2
C 94/C 94M—04	Specification for Ready-mixed Concrete	109.3.1
C109/C109M—02	Standard Test Method for Compressive Strength of Hydraulic Cement Mortars (Using 2-in. or [50-mm] Cube Specimens)	2103.11.1
C 126—99	Specification for Ceramic Glazed Structural Clay Facing Tile, Facing Brick, and Solid Masonry Units	2103.2
C 140—03	Test Method Sampling and Testing Concrete Masonry Units and Related Units	721.3.1.2, 2105.2.2.1.2
C 150—04	Specification for Portland Cement	Table 2103.8(1), Table 2507.2
C 172—04	Practice for Sampling Freshly Mixed Concrete	Table 1704.4
C 199—84 (2000)	Test Method for Pier Test for Refractory Mortars	2111.5, 2111.8, 2113.12
C 206—03	Specification for Finishing Hydrated Lime	Table 2507.2
C 207—04	Specification for Hydrated Lime for Masonry Purposes	Table 2103.8(1)
C 208—95 (2001)	Specification for Cellulosic Fiber Insulating Board	Table 1508.2, 2303.1.5
C 212—00	Specification for Structural Clay Facing Tile	2103.2
C 216—04a	Specification for Facing Brick (Solid Masonry Units Made from Clay or Shale)	1805.5.2.2, 2103.2, 2105.2.2.1.1
C 270—04	Specification for Mortar for Unit Masonry	2103.8, Table 2103.8(2)
C 315—02	Specification for Clay Flue Linings	2113.11.1, Table 2113.16(1)
C 317/C 317M—00	Specification for Gypsum Concrete	1914.1
C 330—04	Specification for Lightweight Aggregates for Structural Concrete	721.1.1

ASTM—continued

C 331—04	Specification for Lightweight Aggregates for Concrete Masonry Units	721.3.1.4, 721.4.1.1.3
C 406—00	Specification for Roofing Slate	1507.7.4
C 442/C 442M—04	Specification for Gypsum Backing Board and Coreboard and Gypsum Shaftliner Board	Table 2506.2
C 472— 99 (2004)	Specification for Standard Test Methods for Physical Testing of Gypsum, Gypsum Plasters and Gypsum Concrete	Table 2506.2
C 473—03	Test Method for Physical Testing of Gypsum Panel Products	Table 2506.2
C 474—02	Test Methods for Joint Treatment Materials for Gypsum Board Construction	Table 2506.2
C 475—01	Specification for Joint Compound and Joint Tape for Finishing Gypsum Wallboard	Table 2506.2
C 476—02	Specification for Grout for Masonry	2103.12, 2105.2.2.1.1, 2105.2.2.1.2, 2105.2.2.1.3
C 503—03	Specification for Marble Dimension Stone (Exterior)	2103.4
C 514—01	Specification for Nails for the Application of Gypsum Board	Table 720.1(2), Table 720.1(3), Table 2306.4.5, Table 2506.2
C 516—02	Specifications for Vermiculite Loose Fill Thermal Insulation	721.3.1.4, 721.4.1.1.3
C 547—03	Specification for Mineral Fiber Pipe Insulation	Table 720.1(2), Table 720.1(3)
C 549—02	Specification for Perlite Loose Fill Insulation	721.3.1.4, 721.4.1.1.3
C 552—03	Standard Specification for Cellular Glass Thermal Insulation	Table 1508.2
C 557—03	Specification for Adhesives for Fastening Gypsum Wallboard to Wood Framing	Table 2506.2
C 568—03	Specification for Limestone Dimension Stone	2103.4
C 578—04	Standard Specification for Rigid, Cellular Polystyrene Thermal Insulation	Table 1508.2
C 587—02	Specification for Gypsum Veneer Plaster	Table 2507.2
C 588/C 588M—01	Specification for Gypsum Base for Veneer Plasters	Table 2507.2
C 595—03	Specification for Blended Hydraulic Cements	Table 2103.8(1), Table 2507.2
C 615—03	Specification for Granite Dimension Stone	2103.4
C 616—03	Specification for Quartz Dimension Stone	2103.4
C 629—03	Specification for Slate Dimension Stone	2103.4
C 630/C 630M—03	Specification for Water-resistant Gypsum Backing Board	Table 2506.2
C 631—95a (2000)	Specification for Bonding Compounds for Interior Gypsum Plastering	Table 2507.2
C 635—00	Specification for the Manufacturer, Performance and Testing of Metal Suspension Systems for Acoustical Tile and Lay-in Panel Ceilings	803.9.1.1, 2506.2.1
C 636—04	Practice for Installation of Metal Ceiling Suspension Systems for Acoustical Tile and Lay-in Panels	803.9.1.1
C 645—04	Specification for Nonstructural Steel Framing Members	Table 2506.2, Table 2507.2
C 652—04a	Specification for Hollow Brick (Hollow Masonry Units Made from Clay or Shale)	1805.5.2.2, 2103.2, 2105.2.2.1.1
C 728—97[E1]	Standard Specification for Perlite Thermal Insulation Board	Table 1508.2
C 744—99	Specification for Prefaced Concrete and Calcium Silicate Masonry Units	2103.1
C 754—04	Specification for Installation of Steel Framing Members to Receive Screw-attached Gypsum Panel Products	Table 2508.1, Table 2511.1
C 836—03	Specification for High-solids Content, Cold Liquid-applied Elastomeric Waterproofing Membrane for Use with Separate Wearing Course	1507.15.2
C 840—04	Specification for Application and Finishing of Gypsum Board	Table 2508.1, 2509.2
C 841—03	Specification for Installation of Interior Lathing and Furring	Table 2508.1, Table 2511.1
C 842—99	Specification for Application of Interior Gypsum Plaster	Table 2511.1, 2511.3, 2511.4
C 843—99e01	Specification for Application of Gypsum Veneer Plaster	Table 2511.1
C 844—99	Specification for Application of Gypsum Base to Receive Gypsum Veneer Plaster	Table 2508.1
C 847—00	Specification for Metal Lath	Table 2507.2
C 887—79a (2001)	Specification for Packaged, Dry Combined Materials for Surface Bonding Mortar	807.2.2, 2103.9
C 897—00	Specification for Aggregate for Job-mixed Portland Cement-based Plasters	Table 2507.2
C 926—98a	Specification for Application of Portland Cement Based-plaster	2109.8.4.6, 2510.3, Table 2511.1, 2511.3, 2511.4, 2512.1, 2512.1.2 2512.2, 2512.6, 2512.8.2, 2512.9, 2513.7
C 931/C 931M—04	Specification for Exterior Gypsum Soffit Board	Table 2506.2
C 932—03	Specification for Surface-applied Bonding Agents for Exterior Plastering	Table 2507.2
C 933—04	Specification for Welded Wire Lath	Table 2507.2
C 946—91 (2001)	Specification for Practice for Construction of Dry-stacked, Surface-bonded Walls	2103.9, 2109.2.3.2
C 954—00	Specification for Steel Drill Screws for the Application of Gypsum Panel Products or Metal Plaster Bases to Steel Studs from 0.033 inch (0.84 mm) to 0.112 inch (2.84 mm) in Thickness	Table 2506.2, Table 2507.2
C 955—03	Standard Specification for Load-bearing Transverse and Axial Steel Studs, Runners Tracks, and Bracing or Bridging, for Screw Application of Gypsum Panel Products and Metal Plaster Bases	Table 2506.2, Table 2507.2
C 956—04	Specification for Installation of Cast-in-place Reinforced Gypsum Concrete	1914.1

ASTM—continued

C 957—04	Specification for High-solids Content, Cold Liquid-applied Elastomeric Waterproofing Membrane with Integral Wearing Surface	1507.15.2
C 960—04	Specification for Predecorated Gypsum Board	Table 2506.2
C1002—01	Specification for Steel Self-piercing Tapping Screws for the Application of Gypsum Panel Products or Metal Plaster Bases to Wood Studs or Steel Studs	Table 2506.2, Table 2507.2
C1006—84 (2001)	Test Method for Splitting Tensile Strength of Masonry Units	2103.11.1
C1007—04	Specification for Installation of Load Bearing (Transverse and Axial) Steel Studs and Related Accessories	Table 2508.1, Table 2511.1
C1019—03	Test Method of Sampling and Testing Grout	2105.2.2.1.1, 2105.2.2.1.2, 2105.2.2.1.3
C1029—02	Specification for Spray-applied Rigid Cellular Polyurethane Thermal Insulation	1507.14.2
C1032—04	Specification for Woven Wire Plaster Base	Table 2507.2
C1047—99	Specification for Accessories for Gypsum Wallboard and Gypsum Veneer Base	Table 2506.2, Table 2507.2
C1063—03	Specification for Installation of Lathing and Furring to Receive Interior and Exterior Portland Cement Based Plaster	2109.8.4.6, 2510.3, Table 2511.1, 2512.1.1
C 1072—00a	Standard Text Method for Measurement of Masonry Flexural Bond Strength	2103.11.1
C1088—02	Specification for Thin Veneer Brick Units Made from Clay or Shale	2103.2
C1167—03	Specification for Clay Roof Tiles	1507.3.4
C1177/C1177M—04	Specification for Glass Mat Gypsum Substrate for Use as Sheathing	Table 2506.2
C1178/C1178M—04	Specification for Glass Mat Water-resistant Gypsum Backing Panel	Table 2506.2
C1186—02	Specification for Flat Nonasbestos Fiber Cement Sheets	1404.10
C1261—04	Specification for Firebox Brick for Residential Fireplaces	2111.5, 2111.8
C1278/C 1278M—03	Specification for Fiber-reinforced Gypsum Panels	Table 2506.2
C1280—04	Specification for Application of Gypsum Sheathing	Table 2508.1, 2508.2
C1283—03e01	Practice for Installing Clay Flue Liners	2113.12
C1288—01	Standard Specification for Discrete Nonasbestos Fiber-cement Interior Substrate Sheets	2509.2
C1289—03	Standard Specification for Faced Rigid Cellular Polyisocyanurate Thermal Insulation Board	Table 1508.2
C1314—03b	Test Method for Compressive Strength of Masonry Prisms	2105.2.2.2.2, 2105.3.1, 2105.3.2
C1325—04	Standard Specification for Nonasbestos Fiber-mat Reinforced Cement Interior Substrate Sheets	2509.02
C1328—03a	Specification for Plastic (Stucco Cement)	Table 2507.2
C1329—04	Specification for Mortar Cement	Table 2103.8(1)
C1386—98	Specification for Precast Autoclaved Aerated Concrete (PAAC) Wall Construction Units	2102.1, 2103.3, 2105.2.2.1.3
C1395/1395M—04	Specification for Gypsum Ceiling Board	Table 2506.2
C1396—02	Standard Specifications for Gypsum Wallboard	Table 2506.2
C1405—00a	Standard Specification for Glazed Brick (Single Fired, Solid Brick Units)	2103.2
C1492—03	Standard Specification for Concrete Roof Tile	1507.3.5
D 25—99E01	Specification for Round Timber Piles	1809.1.1, 2303.1.11
D 41—94 (2000) e01	Specification for Asphalt Primer Used in Roofing, Dampproofing and Waterproofing	Table 1507.10.2
D 43—94 (2000)	Coal Tar Primer Used in Roofing, Dampproofing and Waterproofing	Table 1507.10.2
D 56—02a	Test Method for Flash Point By Tag Closed Tester	307.2
D 86—04b	Test Method for Distillation of Petroleum Products at Atmospheric Pressure	307.2
D 93—02a	Test Method for Flash Point By Pensky-Martens Closed Cup Tester	307.2
D 225—04	Specification for Asphalt Shingles (Organic Felt) Surfaced with Mineral Granules	1507.2.5
D 226—97a	Specification for Asphalt-saturated Organic Felt Used in Roofing and Waterproofing	1404.2, 1507.2.3, 1507.3.3, 1507.6.3, 1507.7.3, Table 1507.8, 1507.8.3, 1507.9.3, 1507.9.4, Table 1507.10.2
D 227—03	Specification for Coal-tar-saturated Organic Felt Used in Roofing and Waterproofing	Table 1507.10.2
D 312—00	Specification for Asphalt Used in Roofing	Table 1507.10.2
D 422—63 (2002)	Test Method for Particle-size Analysis of Soils	1802.3.2
D 450—96 (2000)e01	Specification for Coal-tar Pitch Used in Roofing, Dampproofing and Waterproofing	Table 1507.10.2
D 635—03	Test Method for Rate of Burning and/or Extent and Time of Burning of Self-Supporting Plastics in a Horizontal Position	2606.4
D1143—81 (1994) E01	Test Method for Piles Under Static Axial Compressive Load	1808.2.8.3
D1227—00	Specification for Emulsified Asphalt Used as a Protective Coating for Roofing	Table 1507.10.2, 1507.15.2
D1557—02	Test Method for Laboratory Compaction Characteristics of Soil Using Modified Effort [56,000 ft-lb/ft^3 (2,700 KN m/m^3)]	1803.5
D1586—99	Specification for Penetration Test and Split-barrel Sampling of Soils	1613.5.5
D1761—88 (2000) e01	Test Method for Mechanical Fasteners in Wood	1715.1.1, 1715.1.2, 1715.1.3
D1863—03	Specification for Mineral Aggregate Used on Built-up Roofs	Table 1507.10.2

ASTM—continued

D1929—96 (2001)e01	Test Method for Determining Ignition Properties of Plastics	402.15.4, 406.5.2, 1407.11.2.1, 2606.4
D1970—01	Specification for Self-adhering Polymer Modified Bituminous Sheet Materials Used as Steep Roof Underlayment for Ice Dam Protection	1507.2.4, 1507.2.9.2, 1507.3.9, 1507.5.6, 1507.8.7, 1507.9.8
D2166—00	Test Method for Unconfined Compressive Strength of Cohesive Soil	1613.5.5
D2178—97a	Specification for Asphalt Glass Felt Used in Roofing and Waterproofing	Table 1507.10.2
D2216—98	Test Method for Laboratory Determination of Water (Moisture) Content of Soil and Rock by Mass	1613.5.5
D2487—00	Practice for Classification of Soils for Engineering Purposes (Unified Soil Classification System)	Table 1610.1, 1802.3.1
D2626—04	Specification for Asphalt Saturated and Coated Organic Felt Base Sheet Used in Roofing	1507.3.3, Table 1507.10.2
D2822—91 (97)e01	Specification for Asphalt Roof Cement	Table 1507.10.2
D2823—90 (97) el	Specification for Asphalt Roof Coatings	Table 1507.10.2
D2843—99 (2004)	Test for Density of Smoke from the Burning or Decomposition of Plastics	2606.4
D2850—03a	Test Method for Unconsolidated, Undrained Triaxial Compression Test on Cohesive Soils	1613.5.5
D2898—94 (1999)	Test Methods for Accelerated Weathering of Fire-Retardant-Treated Wood for Fire Testing	1505.1, 2303.2.1, 2303.2.3
D3019—e01(Supp)	Specification for Lap Cement Used with Asphalt Roll Roofing, Nonfibered, Asbestos Fibered and Nonasbestos Fibered	Table 1507.10.2
D3161—03b	Test Method for a Wind Resistance of Asphalt Shingles (Fan Induced Method)	1504.1.1, 1507.2.5
D3200—74 (2000)	Standard Specification and Test Method for Establishing Recommended Design Stresses for Round Timber Construction Poles	2303.1.11
D3201—94 (2003)	Test Method for Hygroscopic Properties of Fire-retardant-treated Wood and Wood-based Products	2303.2.4
D3278—96e01	Test Methods for Flash Point of Liquids by Small Scale Closed-cup Apparatus	307.2
D3462—04	Specification for Asphalt Shingles Made from Glass Felt and Surfaced with Mineral Granules	1507.2.5
D3468—99	Specification for Liquid-applied Neoprene and Chlorosulfonated Polyethylene Used in Roofing and Waterproofing	1507.15.2
D3679—04	Specification for Rigid Poly [Vinyl Chloride (PVC) Siding]	1404.9, 1405.13
D3689—90 (1995)	Method for Testing Individual Piles Under Static Axial Tensile Load	1808.2.8.5
D3737—03	Practice for Establishing Allowable Properties for Structural Glued Laminated Timber (Glulam)	2303.1.3
D3746—85 (2002)	Test Method for Impact Resistance of Bituminous Roofing Systems	1504.7
D3747—79 (2000e01)	Specification for Emulsified Asphalt Adhesive for Adhering Roof Insulation	Table 1507.10.2
D3909—97b	Specification for Asphalt Roll Roofing (Glass Felt) Surfaced with Mineral Granules	1507.2.9.2, 1507.6.4, Table 1507.10.2
D 3957—03	Standard Practices for Establishing Stress Grades for Structural Members Used in Log Buildings	2303.1.10
D4022—94 (2000)e01	Specification for Coal Tar Roof Cement, Asbestos Containing	Table 1507.10.2
D4272—03	Test Method for Total Energy Impact of Plastic Films by Dart Drop	1504.7
D4318—00	Test Methods for Liquid Limit, Plastic Limit and Plasticity Index of Soils	1613.5.5, 1802.3.2
D4434—04	Specification for Poly (Vinyl Chloride) Sheet Roofing	1507.13.2
D4479—00	Specification for Asphalt Roof Coatings—Asbestos-free	Table 1507.10.2
D4586—00	Specification for Asphalt Roof Cement, Asbestos-free	Table 1507.10.2
D4601—98	Specification for Asphalt-coated Glass Fiber Base Sheet Used in Roofing	Table 1507.10.2
D4637—04	Specification for EPDM Sheet Used in Single-ply Roof Membrane	1507.12.2
D4829—03	Test Method for Expansion Index of Soils	1802.3.2
D4869—04	Specification for Asphalt-saturated (Organic Felt) Underlayment Used in Steep Slope Roofing	1507.2.3, 1507.5.3, 1507.6.3, 1507.7.3, 1507.8.3, 1507.9.3
D4897—01	Specification for Asphalt-coated Glass Fiber Venting Base Sheet Used in Roofing	Table 1507.10.2
D4945—00	Test Method for High-strain Dynamic Testing of Piles	1808.2.8.3
D4990—97a	Specification for Coal Tar Glass Felt Used in Roofing and Waterproofing	Table 1507.10.2
D5019—96e01	Specification for Reinforced Nonvulcanized Polymeric Sheet Used in Roofing Membrane	1507.12.2
D5055—04	Specification for Establishing and Monitoring Structural Capacities of Prefabricated Wood I-Joists	2303.1.2
D5456—03	Specification for Evaluation of Structural Composite Lumber Products	2303.1.9
D5516—03	Test Method of Evaluating the Flexural Properties of Fire-retardant-treated Softwood Plywood Exposed to the Elevated Temperatures	2303.2.2.1
D5643—94 (2000)e01	Specification for Coal Tar Roof Cement, Asbestos-free	Table 1507.10.2
D5664—02	Test Methods for Evaluating the Effects of Fire-retardant Treatment and Elevated Temperatures on Strength Properties of Fire-retardant-treated Lumber	2303.2.2.2
D5665—99a	Specification for Thermoplastic Fabrics Used in Cold-applied Roofing and Waterproofing	Table 1507.10.2
D5726—98	Specification for Thermoplastic Fabrics Used in Hot-applied Roofing and Waterproofing	Table 1507.10.2
D6083—97a	Specification for Liquid Applied Acrylic Coating Used in Roofing	Table 1507.10.2, 1507.15.2
D6162—00A	Specification for Styrene Butadiene Styrene (SBS) Modified Bituminous Sheet Materials Using a Combination of Polyester and Glass Fiber Reinforcements	1507.11.2

ASTM—continued

D6163—00 E01	Specification for Styrene-Butadiene-Styrene (SBS) Modified Bituminous Sheet Materials Using Glass Fiber Reinforcements	1507.11.2
D6164—00	Specification for Styrene-Butadiene-Styrene (SBS) Modified Bituminous Sheet Metal Materials Using Polyester Reinforcements	1507.11.2
D6222—02	Specification for Atactic Polypropylene (APP) Modified Bituminous Sheet Materials Using Polyester Reinforcements	1507.11.2
D6223—02	Specification for Atactic Polypropylene (APP) Modified Bituminous Sheet Materials Using a Combination of Polyester and Glass Fiber Reinforcements	1507.11.2
D6298—00	Specification for Fiberglass Reinforced Styrene-Butadiene-Styrene (SBS) Modified Bituminous Sheets with a Factory Applied Metal Surface	1507.11.2
D6305—02e01	Practice for Calculating Bending Strength Design Adjustment Factors for Fire-retardant-treated Plywood Roof Sheathing	2303.2.2.1
D6380—01[E1]	Standard Specification for Asphalt Roll Roofing (Organic) Felt	1507.2.9.2, 1507.3.3, 1507.6.4
D6381—03	Standard Test Method for Measurement of Asphalt Shingle Mechanical Uplift Resistance	1504.2.1, Table 1504.2.1, 1609.5.2
D6694—01	Standard Specification for Liquid-applied Silicone Coating Used in Spray Polyurethane Foam Roofing	1507.15.2
D6754—02	Standard Specification for Ketone Ethylene Ester Based Sheet Roofing	1507.13.2
D6757—02	Standard Specification for Inorganic Underlayment for Use with Steep Slope Roofing Products	1507.2.3
D6841—03	Standard Practice for Calculating Design Value Treatment Adjustment Factors for Fire-retardant-treated Lumber	2303.2.2.2
D6878—03	Standard Specification for Thermoplastic Polyolefin Based Sheet Roofing	1507.13.2
E 72—02	Standard Test Methods of Conducting Strength Tests of Panels for Building Construction	2103.11.1
E 84—04	Test Methods for Surface Burning Characteristics of Building Materials	402.10, 402.15.4, 406.5.2, 410.3.5.3, 703.4.2, 719.1, 719.4, 802.1, 803.1, 803.5, 803.6.1, 806.5, 1407.9, 1407.10.1, 2303.2, 2603.3, 2603.4.1.13, 2603.5.4, 2604.2.4, 2606.4, 31054
E 90—04	Test Method for Laboratory Measurement of Airborne Sound Transmission Loss of Building Partitions and Elements	1207.2
E 96—00e01	Test Method for Water Vapor Transmission of Materials	1203.2
E 108—04	Test Methods for Fire Tests of Roof Coverings	1505.1, 2603.6, 2610.2, 2610.3
E 119—00	Test Methods for Fire Tests of Building Construction and Materials	410.3.5.2, 703.2, 703.2.1, 703.2.3, 703.3, 704.7, 704.9, 706.2.1, 706.7, 711.3.2, 712.3.1, 713.1, 713.4, 714.7, 715.2, 716.5.2, 716.5.3, 716.6.1, 716.6.2, Table 720.1(1), 1407.10.2, 2103.2, 2603.4, 2603.4.1.13, 2603.5.1
E 136—99e01	Test Method for Behavior of Materials in a Vertical Tube Furnace at 750°C	703.4.1
E 328—02	Methods for Stress Relaxation for Materials and Structures	2103.13.6
E 330—02	Test Method for Structural Performance of Exterior Windows, Curtain Walls and Doors by Uniform Static Air Pressure Difference	1714.5.2
E 331—00	Test Method for Water Penetration of Exterior Windows, Skylights, Doors and Curtain Walls by Uniform Static Air Pressure Difference	1403.2
E 492—04	Test Method for Laboratory Measurement of Impact Sound Transmission Through Floor-ceiling Assemblies Using the Tapping Machine	1207.3
E 518—03	Standard Test Methods for Flexural Bond Strength of Masonry	2101.3.11.1
E 519—02	Standard Test Method for Diagonal Tension (Shear) in Masonry Assemblages	2103.11.1
E 605—00	Test Method for Thickness and Density of Sprayed Fire-resistive Material (SFRM) Applied to Structural Members	1704.10.3, 1704.10.3.1, 1704.10.3.2, 1704.10.4
E 681—04	Test Methods for Concentration Limits of Flammability of Chemical Vapors and Gases	307.2
E 736—00	Test Method for Cohesion/Adhesion of Sprayed Fire-resistive Materials Applied to Structural Members	1704.10.5
E 814—02	Test Method of Fire Tests of Through-penetration Firestops	702.1, 712.3.1.2, 712.4.1.1, 712.4.1.1.2
E 970—00	Test Method for Critical Radiant Flux of Exposed Attic Floor Insulation Using a Radiant Heat Energy Source	719.3.1
E1300—04e01	Practice for Determining Load Resistance of Glass in Buildings	2404.1, 2404.2, 2404.3.1, 2404.3.2, 2404.3.3, 2404.3.4, 2404.3.5
E1592—01	Test Method for Structural Performance of Sheet Metal Roof and Siding Systems by Uniform Static Air Pressure Difference	1504.3.2
E1602—03	Guide for Construction of Solid Fuel-burning Masonry Heaters	2112.2
E1886—04	Test Method for Performance of Exterior Windows, Curtain Walls, Doors and Storm Shutters Impacted by Missiles and exposed to Cyclic Pressure Differentials	1609.1.2
E1966—00	Test Method for Fire-resistant Joint Systems	702.1, 7.13.3
E1996—04	Specification for Performance of Exterior Windows, Glazed Curtain Walls, Doors and Storm Shutters Impacted by Windborne Debris in Hurricanes	1609.1.2, 1609.1.2.1

ASTM—continued

E 2307—04	Standard Test Method for Determining Fire Resistance of Perimeter Fire Barrier Systems Using Intermediate-scale, Multistory Test Apparatus. 713.4	
F 547—01	Terminology of Nails for Use with Wood and Wood-based Materials. . . . . .Table 2506.2	
F1346—91 (2003)	Performance Specification for Safety Covers and Labeling Requirements for All Covers for Swimming Pools, Spas and Hot Tubs 3109.4, 3109.4.1.8	
F1667—03	Specification for Driven Fasteners: Nails, Spikes and Staples . . . . .Table 720.1(2), Table 720.1(3), 1507.2.6, 2303.6, Table 2506.2	
F2006—00	Standard/Safety Specification for Window Fall Prevention Devices for Nonemergency Escape (Egress) and Rescue (Ingress) Windows 1405.12.2	
F2090—01a	Specification for Window Fall Prevention Devices with Emergency Escape (Egress) Release Mechanisms 1405.12.2	
G 152—04	Practice for Operating Open Flame Carbon Arc Light Apparatus for Exposure of Nonmetallic Materials 1504.6	
G 154—00A	Practice for Operating Fluorescent Light Apparatus for UV Exposure of Nonmetallic Materials 1504.6	
G 155—04	Practice for Operating Xenon Arc Light Apparatus for Exposure of Nonmetallic Materials 1504.6	

AWCI

The Association of the Wall and Ceiling Industries International
803 West Broad Street, Suite 600
Falls Church, VA 22046

Standard reference number	Title	Referenced in code section number
12—B—98	Technical Manual 12-B Standard Practice for the Testing and Inspection of Field Applied This—Film Intumescent Fire-resistive Materials; an Annotated Guide, First Edition 1704.11	

AWPA

American Wood-Preservers' Association
P.O. Box 5690
Grandbury, TX 76049

Standard reference number	Title	Referenced in code section number
C1—00	All Timber Products—Preservative Treatment by Pressure Processes. 1505.6	
M4—02	Standard for the Care of Preservative-treated Wood Products 1809.1.2, 2303.1.8	
U1—04	USE CATEGORY SYSTEM: User Specification for Treated Wood Except Section 6, Commodity Specification H . . . . .1403.5, 1505.6, Table 1507.9.5, 1805.4.5, 1805.4.6, 1805.7.1, 1809.1.2, 2303.1.8, 2304.11.2, 2304.11.4, 2304.11.6, 2304.11.7	

AWS

American Welding Society
550 N.W. LeJeune Road
Miami, FL 33126

Standard reference number	Title	Referenced in code section number
D1.1—04	Structural Welding Code—Steel . . . . .Table 1704.3, 1704.3.1, 1708.4	
D1.3—98	Structural Welding Code—Sheet Steel . . . . .Table 1704.3	
D1.4—98	Structural Welding Code—Reinforcing Steel. . . . . .Table 1704.3, Table 1704.4	

BHMA

Builders Hardware Manufacturers' Association
355 Lexington Avenue, 17th Floor
New York, NY 10017-6603

Standard reference number	Title	Referenced in code section number
A 156.10—99	American National Standard for Power Operated Pedestrian Doors 1008.1.3.2	
A 156.19—02	Standard for Power Assist and Low Energy Operated Doors. 1008.1.3.2	

CGSB

Canadian General Standards Board
222 Queens Street
14th Floor, Suite 1402
Ottawa, Ontario, Canada KIA 1G6

Standard reference number	Title	Referenced in code section number
37-GP-52M (1984)	Roofing and Waterproofing Membrane, Sheet Applied, Elastomeric 1504.7, 1507.12.2	
CAN/CGSB 37.54—95	Polyvinyl Chloride Roofing and Waterproofing Membrane 1507.13.2	
37-GP-56M (1980)	Membrane, Modified, Bituminous, Prefabricated and Reinforced for Roofing—with December 1985 Amendment..1507.11.2	

CSA

Canadian Standards Association
5060 Spectrum Way, Suite 100
Mississauga, Ontario, L4W 5N6 Canada

Standard reference number	Title	Referenced in code section number
101/I.S.2/A440—05	Specifications for Windows, Doors and Unit Skylights 1714.5.1, 2405.5	

CPSC

Consumer Product Safety Commission
4330 East West Highway
Bethesada, MD 20814-4408

Standard reference number	Title	Referenced in code section number
16 CFR Part 1201(1977)	Safety Standard for Architectural Glazing Material 2406.1.1, 2406.2.1, 2407.1, 2408.2.1, 2408.3	
16 CFR Part 1209 (1979)	Interim Safety Standard for Cellulose Insulation 719.6	
16 CFR Part 1404 (1979)	Cellulose Insulation...719.6	
16 CFR Part 1500 (1991)	Hazardous Substances and Articles; Administration and Enforcement Regulations...................307.2	
16 CFR Part 1500.44 (2001)	Method for Determining Extremely Flammable and Flammable Solids 307.2	
16 CFR Part 1507 (2001)	Fireworks Devices..307.2	
16 CFR Part 1630 (2000)	Standard for the Surface Flammability of Carpets and Rugs....................................804.4.1	

CSSB

Cedar Shake and Shingle Bureau
P.O. Box 1178
Sumas, WA 98295-1178

Standard reference number	Title	Referenced in code section number
CSSB—97	Grading and Packing Rules for Western Red Cedar Shakes and Western Red Shingles of the Cedar Shake and Shingle Bureau.....................................Table 1507.8.4, Table 1507.9.5	

DASMA

Door and Access Systems Manufacturers
Association International
1300 Summer Avenue
Cleveland, OH 44115-2851

Standard reference number	Title	Referenced in code section number
107—98 (03)	Room Fire Test Standard for Garage Doors Using Foam Plastic Insulation 2603.4.1.9	

DOC

U.S. Department of Commerce
National Institute of Standards and Technology
100 Bureau Drive Stop 3460
Gaithersburg, MD 20899

Standard reference number	Title	Referenced in code section number
PS-1—95	Construction and Industrial Plywood ....2303.1.4, 2304.6.2, Table 2304.7(4), Table 2304.7(5), 2306.3.1, Table 2306.3.1, Table 2306.3.2	
PS-2—92	Performance Standard for Wood-based Structural-use Panels ....2303.1.4, 2304.6.2, Table 2304.7(5), Table 2306.3.1, 2306.3.1, Table 2306.3.2	
PS 20—99	American Softwood Lumber Standard 1809.1.1, 2302.1, 2303.1.1	

DOL

U.S. Department of Labor
c/o Superintendent of Documents
U.S. Government Printing Office
Washington, DC 20402-9325

Standard reference number	Title	Referenced in code section number
29 CFR Part 1910.1000 (1974)	Air Contaminants 902.1	

DOTn

U.S. Department of Transportation
c/o Superintendent of Documents
U.S. Government Printing Office
Washington, DC 20402-9325

Standard reference number	Title	Referenced in code section number
49 CFR—1998	Specification of Transportation of Explosive and Other Dangerous Articles, UN 0335,UN 0336 Shipping Containers 307.2	
49 CFR Parts 173.137 (1990)	Shippers—General Requirements for Shipments and Packaging—Class 8—Assignment of Packing Group 307.2	

FEMA

Federal Emergency Management Agency
Federal Center Plaza
500 C Street S.W.
Washington, DC 20472

Standard reference number	Title	Referenced in code section number
FIA-TB11—01	Crawlspace Construction for Buildings Located in Special Flood Hazard Areas 1807.1.2.1	

FM

Factory Mutual
Standards Laboratories Department
1151 Boston-Providence Turnpike
Norwood, MA 02062

Standard reference number	Title	Referenced in code section number
4450 (1989)	Approval Standard for Class 1 Insulated Steel Deck Roofs—with Supplements through July 1992..............1504.3.1, 1508.1, 2603.3, 2603.4.1.5	
4470 (1992)	Approval Standard for Class 1 Roof Covers..1504.3.1, 1504.7	
4880 (2001)	American National Standard for Evaluating Insulated Wall or Wall and Roof/Ceiling Assemblies, Plastic Interior Finish Materials, Plastic Exterior Building Panels, Wall/Ceiling Coating Systems, Interior and Exterior Finish Systems................................2603.4, 2603.9	

GA

Gypsum Association
810 First Street N.E. #510
Washington, DC 20002-4268

Standard reference number	Title	Referenced in code section number
GA 216—04	Application and Finishing of Gypsum Board.	Table 2508.1, 2509.2
GA 600—03	Fire-Resistance Design Manual, 17th Edition.	Table 720.1(1), Table 720.1(2), Table 720.1(3)

HPVA

Hardwood Plywood Veneer Association
1825 Michael Faraday Drive
Reston, VA 20190-5350

Standard reference number	Title	Referenced in code section number
HP-1—2000	The American National Standard for Hardwood and Decorative Plywood	2303.3, 2304.6.2

ICC

International Code Council, Inc.
500 New Jersey Ave, NW
6th Floor
Washington, D.C. 20001

Standard reference number	Title	Referenced in code section number
ICC/ANSI A117.1—03	Accessible and Usable Buildings and Facilities	406.2.2, 907.9.1.4,1007.6.5, 1010.1, 1010.6.5, 1010.9, 1011.3, 1101.2, 1102.1, 1103.2.14, 1106.7, 1107.2, 1108.2.2, 1109.1, 1109.2, 1109.2.1.1, 1109.2.2, 1109.3, 1109.4, 1109.8, 3001.3, 3409.6, 3409.8.2, 3409.8.3
ICC 300—02	ICC Standard on Bleachers, Folding and Telescopic Seating and Grandstands.	1025.1.1, 3401.1
ICC EC—06	International Code Council Electrical Code Administrative Provisions.	101.4.1, 107.3, 414.5.4, 415.8.2.8.1, 904.3.1, 907.5, 909.11, 909.12.1, 909.16.3, 1205.4.1, 1405.10.4, 2701.1, 2701.1, 3401.3
IECC—06	International Energy Conservation Code®.	101.4.7, 1203.3.2, 1301.1.1, 1403.2
IFC—06	International Fire Code®.	101.4.6, 102.6, 201.3, 307.1, 307.1.1, 307.2, Table 307.1(1), 404.2, 406.5.1, 406.6.1, 410.3.6, 411.1, 412.4.1, 413.1, 414.1.1, 414.1.2, 414.1.2.1, 414.2, 414.2.5, Table 414.2.5(1), Table 414.2.5(2), 414.3, 414.5, 414.5.1, Table 414.5.1, 414.5.2, 414.5.4, 414.5.5, 414.6, 415.1, 415.2, 415.3, 415.3.1, Table 415.3.1, Table 415.3.2, 415.6, 415.6.1, 415.6.1.4, 415.6.2, 415.6.2.3, 415.6.2.5, 415.6.2.7, 415.6.2.8, 415.6.2.9, 415.6.3, 415.6.3.3.3, 415.6.3.5, 415.6.4, 415.7, 415.8.1, 415.8.2.7, 415.8.5.1, 415.8.7.2, 415.8.9.3, 415.8.10.1, 416.1, 420.1, 420.7, 704.8.2, 706.1, 901.2, 901.3, 901.5, 901.6.2, 903.2.6.1, 903.2.11, Table 903.2.13, 903.5, 904.2.1, 905.1, 905.3.6, 906.1, 907.2.5, 907.2.12.2, 907.2.14, 907.2.16, 907.14, 907.19, 909.20, 910.2.2, Table 910.3, 1001.3, 1203.4.2, 1203.5, 2702.1, 2702.2.9, 2702.2.11, 2702.2.12, 2702.2.13, 2702.3, 3102.1, 3103.1, 3309.2, 3401.3, 3410.3.2, 3410.6.8.1, 3410.6.14, 3410.6.14.1
IFGC—06	International Fuel Gas Code®	101.4.2, 201.3, Table 307.1(1), 415.6.3, 2113.11.1.2, 2113.15, 2801.1, 3401.3
IMC—06	International Mechanical Code®	101.4.3, 201.3, 307.1, Table 307.1(1), 406.4.2, 406.6.3, 406.6.5, 409.3, 412.4.6, 414.1.2, 414.3, 415.6.1.4, 415.6.2, 415.6.2.8, 415.6.3, 415.6.4, 415.8.11.1, 416.3, 420.5, 603.1, 707.2, 716.2.2, 716.5.4, 716.6.1, 716.6.2, 716.6.3, 717.5, 719.1,719.7, 903.2.12.1, 904.2.1, 904.11, 908.6, 909.1, 909.10.2, 1015.5, 1017.4.1, 1203.1, 1203.2.1, 1203.4.2, 1203.4.2.1, 1203.5, 1209.3, 2304.5, 2801.1, 3004.3.1, 3401.3, 3410.6.7.1, 3410.6.8, 3410.6.8.1
IPC—06	International Plumbing Code®	101.4.4, 201.3, 415.6.4, 717.5, 903.3.5, 912.5, 1206.3.3, 1503.4, 1807.4.3, 2901.1, Table 2902.1, 2902.1.1, 3305.1, 3401.3
IPMC—06	International Property Maintenance Code®	101.4.5, 102.6, 103.3, 3401.3, 3410.3.2.
IPSDC—06	International Private Sewage Disposal Code®	101.4.4, 2901.1, 3401.3
IRC—06	International Residential Code®	101.2, 308.2, 308.3, 308.5, 310.1, 2308.1, 3401.3
IWUIC—06	International Wildland-Urban Interface Code™	Table 1505.1
SBCCI SSTD 10—99	Standard for Hurricane Resistant Residential Construction	1609.1.1, 1609.1.1.1, 2308.2.1
SBCCI SSTD 11—97	Test Standard for Determining Wind Resistance of Concrete or Clay Roof Tiles	1715.2.1, 1715.2.2

ISO

International Standards Organization
ISO Central Secretariat1, rue de Varembee, Case postale 56
CH-1211 Geneva 20, Switzerland

Standard reference number	Title	Referenced in code section number
ISO 8115—86	Cotton Bales–Dimensions and Density . . . . .Table 415.8.2.1.1	

NAAMM

National Association of Architectural
Metal Manufacturers
8 South Michigan Ave
Chicago, IL 60603

Standard reference number	Title	Referenced in code section number
FP 1001—97	Guide Specifications for Design of Metal Flag Poles 1609.1.1	

NCMA

National Concrete Masonry Association
2302 Horse Pen Road
Herndon, VA 22071-3499

Standard reference number	Title	Referenced in code section number
TEK 5-84 (1996)	Details for Concrete Masonry Fire Walls . . . . .Table 720.1(2)	

NFPA

National Fire Protection Association
1 Batterymarch Park
Quincy, MA 02269-9101

Standard reference number	Title	Referenced in code section number
11—02	Low Expansion Foam 904.7	
12—00	Carbon Dioxide Extinguishing Systems 904.8, 904.11	
12A—04	Halon 1301 Fire Extinguishing Systems 904.9	
13—02	Installation of Sprinkler Systems 707.2, 903.3.1.1, 903.3.2, 903.3.5.1.1, 903.3.5.2, 904.11, 905.3.4, 907.8, 3104.5, 3104.9	
13D—02	Installation of Sprinkler Systems in One- and Two-family Dwellings and Manufactured Homes 903.3.1.3, 903.3.5.1.1	
13R—02	Installation of Sprinkler Systems in Residential Occupancies Up to and Including Four Stories in Height 903.3.1.2, 903.3.5.1.1, 903.3.5.1.2, 903.4	
14—03	Installation of Standpipe and Hose System 905.2, 905.3.4, 905.4.2, 905.8	
16—03	Installation Foam-water Sprinkler and Foam-water Spray Systems 904.7, 904.11	
17—02	Dry Chemical Extinguishing Systems 904.6, 904.11	
17A—02	Wet Chemical Extinguishing Systems 904.5, 904.11	
30—03	Flammable and Combustible Liquids Code 415.3	
31—01	Installation of Oil-burning Equipment 2113.15	
32—00	Dry Cleaning Plants 415.6.4	
40—01	Storage and Handling of Cellulose Nitrate Film. 409.1	
61—99	Prevention of Fires and Dust Explosions in Agricultural and Food Product Facilities 415.6.1	
72—02	National Fire Alarm Code 901.6, 903.4.1, 904.3.5, 907.2, 907.2.1.1, 907.2.10, 907.2.10.4, 907.2.11.2, 907.2.11.3, 907.2.12.2.3, 907.2.12.3, 907.4, 907.5, 907.9.2,907.10, 907.14,907.16, 907.17, 911.1, 3006.5	
80—99	Fire Doors and Fire Windows 508.2.2.1, 715.4, 715.4.5, 715.4.6.1, 715.4.7.2, 715.5, 1008.1.3.3	
85—04	Boiler and Combustion System Hazards Code . . . . .415.6.1 (Note: NFPA 8503 has been incorporated into NFPA 85)	
92B—05	Smoke Management Systems in Malls, Atria and Large Spaces 909.8	
101—03	Life Safety Code. 1025.6.2	
105—03	Standard for the Installation of Smoke Door Assemblies 405.4.2, 715.4.3.1, 909.20.4.1	
110—02	Emergency and Standby Power Systems 2702.1	
111—01	Stored Electrical Energy Emergency and Standby Power Systems 2702.1	

NFPA—continued

PCI

Precast Prestressed Concrete Institute
175 W. Jackson Boulevard, Suite 1859
Chicago, IL 60604-9773

PTI

Post-Tensioning Institute
1717 W. Northern Avenue, Suite 114
Phoenix, AZ 85021

RMI

Rack Manufacturers Institute
8720 Red Oak Boulevard, Suite 201
Charlotte, NC 28217

Standard reference number	Title	Referenced in code section number
RMI (2002)	Specification for Design, Testing and Utilization of Industrial Steel Storage Racks	2208.1

SJI

Steel Joist Institute
3127 10th Avenue, North
Myrtle Beach, SC 29577-6760

Standard reference number	Title	Referenced in code section number
JG-1.1—05	Standard Specification for Joist Girders	1604.3.3, 2206.1
K-1.1—05	Standard Specification for Open Web Steel Joists, K-Series	1604.3.3, 2206.1
LH/DLH-1.1—05	Standard Specification for Longspan Steel Joists, LH Series and Deep Longspan Steel Joists, DLH Series	1604.3.3, 2206.1

SPRI

Single-Ply Roofing Institute
77 Rumford Ave.
Suite 3-B
Waltham, MA 02453

Standard reference number	Title	Referenced in code section number
ES-1—03	Wind Design Standard for Edge Systems Used with Low Slope Roofing Systems	1504.5
RP-4—02	Wind Design Guide for Ballasted Single-ply Roofing Systems	1504.4

TIA

Telecommunications Industry Association
2500 Wilson Boulevard
Arlington, VA 22201-3834

Standard reference number	Title	Referenced in code section number
TIA/EIA-222-F—96	Structural Standards for Steel Antenna Towers and Antenna Supporting Structures	1609.1.1, 3108.4

TMS

The Masonry Society
3970 Broadway, Unit 201-D
Boulder, CO 80304-1135

Standard reference number	Title	Referenced in code section number
0216—97	Standard Method for Determining Fire Resistance of Concrete and Masonry Construction Assemblies	Table 720.1(2), 721.1
402—05	Building Code Requirements for Masonry Structures	1405.5, 1405.5.2, 1405.9, 1604.3.4, 1704.5, 1704.5.1, Table 1704.5.1, 1704.5.2, 1704.5.3, Table 1703.4.5.3, 1708.1.1, 1708.1.2, 1708.1.3, 1708.1.4, 1805.5.2, 1812.7, 2101.2.2, 2101.2.3, 2101.2.4, 2101.2.5, 2101.2.6, 2103.1.3.6, 2106.1, 2106.1.1, 2106.1.1.1, 2106.1.1.2, 2106.1.1.3, 2106.3, 2106.4, 2106.5, 2106.6, 2107.1, 2107.2, 2107.3, 2107.4, 2107.5, 2107.6, 2107.7, 2107.8, 2108.1, 2108.2, 2108.3, 2108.4, 2109.1, 2109.2.3.1, 2109.7.3
602—05	Specification for Masonry Structures	1405.5.1, Table 1704.5.1, Table 1704.5.3, 2103.13.7, 2104.1, 2104.1.1, 2104.3, 2104.4

TPI

Truss Plate Institute
583 D'Onofrio Drive, Suite 200
Madison, WI 53719

Standard reference number	Title	Referenced in code section number
TPI 1—2002	National Design Standards for Metal-plate-connected Wood Truss Construction	2303.4.2, 2306.1

UL

Underwriters Laboratories, Inc.
333 Pfingsten Road
Northbrook, IL 60062-2096

Standard reference number	Title	Referenced in code section number
10A—98	Tin Clad Fire Doors—with Revisions through March 2003	715.4
10B—97	Fire Tests of Door Assemblies—with Revisions through October 2001	715.4.2
10C—98	Positive Pressure Fire Tests of Door Assemblies—with Revisions through November 2001	715.4.1, 715.4.3
14B—98	Sliding Hardware for Standard Horizontally Mounted Tin Clad Fire Doors— with Revisions through July 2000	715.4
14C—99	Swinging Hardware for Standard Tin Clad Fire Doors Mounted Singly and in Pairs	715.4
94—96	Test for Flammability of Plastic Materials for Parts in Devices and Appliances	402.11.1
103—01	Factory-built Chimneys, for Residential Type and Building Heating Appliances	717.2.5
127—96	Factory-built Fireplaces—with Revisions through November1999	717.2.5, 2111.11
199E—04	Outline of Investigation for Fire Testing of Sprinklers and Water Spray Nozzles for Protection of Deep Fat Fryers	904.11.4.1
217—97	Single and Multiple Station Smoke Alarms—with revisions through January 2004	907.2.10
268—96	Smoke Detectors for Fire Protective Signaling Systems—with Revisions through January 1999	407.6, 907.2.6.2
300—96	Fire Testing of Fire Extinguishing Systems for Protection of Restaurant Cooking Areas — with Revisions through December 1998	904.11
555—99	Fire Dampers—with Revisions through January 2002	716.3
555C—96	Ceiling Dampers	716.3, 716.6.2
555S—99	Smoke Dampers—with Revisions through January 2002	716.3, 716.3.1.1
580—94	Test for Uplift Resistance of Roof Assemblies—with Revisions through February 1998	1504.3.1, 1504.3.2
641—95	Type L Low-temperature Venting Systems—with Revisions through April 1999	2113.11.1.4
710B—2004	Recirculating Systems	904.11
790—98	Tests for Fire Resistance of Roof Covering Materials—with Revisions through July 1998	1505.1, 2603.6, 2610.2, 2610.3
793—97	Standard for Automatically Operated Roof Vents for Smoke and Heat	910.3.1
864—03	Standard For Control Units and Accessories for Fire Alarm Systems— with Revisions through October 2003	909.12
1040—96	Fire Test of Insulated Wall Construction—with Revisions through April 2001	1407.10.3, 2603.4, 2603.9
1256—02	Fire Test of Roof Deck Construction	1508.1, 2603.3, 2603.4.1.5
1479—94	Fire Tests of Through-penetration Fire stops— with Revisions through August 2000	712.3.1.2, 712.4.1.1.2, 712.5
1482—98	Solid-fuel Type Room Heater—with Revisions through January 2000	2112.2, 2112.5
1715—97	Fire Test of Interior Finish Material—with Revisions through October 2002	1407.10.2, 1407.10.3, 2603.4, 2603.9
1777—04	Chimney Liners—with Revisions through July 1998	2113.11.1, 2113.19
1784—95	Air Leakage Tests of Door Assemblies	707.14.1, 710.5.2, 715.4.3.1, 715.4.5.1, 715.4.5.3
1897—98	Uplift Tests for Roof Covering Systems—with Revisions through November 2002	1504.3.1
1975—96	Fire Test of Foamed Plastics Used for Decorative Purposes	402.10, 402.11.1, 402.15.5
2017—2000	Standard for General-purpose Signaling Devices and Systems—with Revisions through June 2004	3109.4.1.8
2079—98	Tests for Fire Resistance of Building Joint Systems	702.1, 713.3, 713.6
2200—98	Stationary Engine Generator Assemblies. (Revisions through July 2004)	2702.1.1
2390—04	Test Method for Measuring the Wind Uplift Coefficients for Asphalt Shingles	1504.2.1, 1609.5.2

ULC

Underwriters Laboratories of Canada
7 Crouse Road
Scarborough, Ontario, Canada M1R3A9

Standard reference number	Title	Referenced in code section number
CAN/ULC S102.2— 1988	Standard Method of Test for Surface Burning Characteristics of Flooring, Floor Coverings and Miscellaneous Materials and Assemblies with 2000 Revisions	719.4

USC

United States Code
c/o Superintendent of Documents
U.S. Government Printing Office
Washington, DC 20402-9325

Standard reference number	Title	Referenced in code section number
18 USC Part 1, Ch.40	Importation, Manufacture, Distribution and Storage of Explosive Materials.	307.2

WDMA

Window and Door Manufacturers Association
1400 East Touhy Avenue #470
Des Plaines, IL 60018

Standard reference number	Title	Referenced in code section number
101/I.S.2/A440—05	Specifications for Windows, Doors and Unit Skylights	1714.5.1, 2405.5

WRI

Wire Reinforcement Institute, Inc.
203 Loudon Street, S.W.
2nd Floor, Suite 203C
Leesburg, VA 22075

Standard reference number	Title	Referenced in code section number
WRI/CRSI—81	Design of Slab-on-ground Foundations—with 1996 Update.	1805.8.2

Appendix A:
Employee Qualifications

The provisions contained in this appendix are not mandatory unless specifically referenced in the adopting ordinance.

General Comments

When adopted (see Section 101.2.1), this appendix provides jurisdictions with training, experience and certification requirements for those employees of the department of building safety responsible for enforcing the provisions of the building code, including the codes referenced in Section 101.4. The authorization to create these positions is found in Section 103 of the code. Note that jurisdictions may also be mandated by applicable state laws to employ only persons licensed by the state to perform certain duties, and may or may not be able to impose additional requirements.

Purpose

The purpose of this appendix is to provide optional criteria for qualifications for employees who enforce the code. A jurisdiction that wants to make this appendix a mandatory part of the code needs to specifically list this appendix in its adoption ordinance (see page v of the code for a sample ordinance for adoption).

SECTION A101
BUILDING OFFICIAL QUALIFICATIONS

A101.1 Building official. The building official shall have at least 10 years' experience or equivalent as an architect, engineer, inspector, contractor or superintendent of construction, or any combination of these, five years of which shall have been supervisory experience. The building official should be certified as a building official through a recognized certification program. The building official shall be appointed or hired by the applicable governing authority.

❖ As the executive official in charge of administering the code enforcement program of the jurisdiction, the building official is required to possess significant experience in the construction industry, including at least five years as a supervisor. Although not mandated, it is recommended that the person hold certification as a building official, such as the Certified Building Official® (C.B.O.®) issued by the International Code Council® (ICC®). People with this certification have passed examinations that measure knowledge of legal and management principles, as well as code technology. This position may have a different title, as described in the commentary to Section 103.1.

A101.2 Chief inspector. The building official can designate supervisors to administer the provisions of the *International Building, Mechanical* and *Plumbing Codes, International Fuel Gas Code* and ICC *Electrical Code.* Each supervisor shall have at least 10 years' experience or equivalent as an architect, engineer, inspector, contractor or superintendent of construction, or any combination of these, five years of which shall have been in a supervisory capacity. They shall be certified through a recognized certification program for the appropriate trade.

❖ People who supervise inspectors and plans examiners that review construction for conformance to the building and referenced codes are expected to have considerable experience in their area of expertise, since they are responsible for overseeing the day-to-day operation of their divisions within the department. Large departments may have a separate division for each of the codes, while those of lesser size may group two or more together. Because of the responsibility placed on these persons, it is required that they have demonstrated knowledge of the codes that they administer by being certified in the appropriate category. The ICC provides certification programs that cover the range of categories involved.

A101.3 Inspector and plans examiner. The building official shall appoint or hire such number of officers, inspectors, assistants and other employees as shall be authorized by the jurisdiction. A person shall not be appointed or hired as inspector of construction or plans examiner who has not had at least 5 years' experience as a contractor, engineer, architect, or as a superintendent, foreman or competent mechanic in charge of construction. The inspector or plans examiner shall be certified through a recognized certification program for the appropriate trade.

❖ Departments typically have inspectors and plans examiners who specialize in codes that regulate the various aspects of building construction. Additionally, there will be employees who handle clerical and various administrative duties. The building official will hire the number of employees authorized by the jurisdiction's elected officials. Newly hired inspectors and plans examiners are not necessarily expected to have had previous experience in code enforcement, but should be familiar with the construction industry, with at least five years experience in one or more of the fields listed. They also must possess certification in the job category in which they are employed. ICC provides certification programs that cover the range of categories involved.

A101.4 Termination of employment. Employees in the position of building official, chief inspector or inspector shall not be removed from office except for cause after full opportunity has been given to be heard on specific charges before such applicable governing authority.

❖ This section establishes that once employed, the building official and other technical employees of the department cannot be removed, except for cause, subject to a due process review of the specific charges.

SECTION A102
REFERENCED STANDARDS

IBC-06	International Building Code	A101.2
IMC-06	International Mechanical Code	A101.2
IPC-06	International Plumbing Code	A101.2
IFGC-06	International Fuel Gas Code	A101.2
ICC EC-06	ICC Electrical Code	A101.2

Appendix B:
Board Of Appeals

The provisions contained in this appendix are not mandatory unless specifically referenced in the adopting ordinance.

General Comments

When adopted (see Section 101.2.1), this appendix provides jurisdictions with detailed appeals board member qualifications and administrative procedures to supplement the basic requirements found in Section 112.

Purpose

The purpose of this appendix is to provide optional criteria for administrative procedures of the board of appeals and board member qualifications. A jurisdiction that wants to make this appendix a mandatory part of the code needs to specifically list this appendix in its adoption ordinance (see page v of the code for sample ordinance for adoption).

SECTION B101
GENERAL

B101.1 Application. The application for appeal shall be filed on a form obtained from the building official within 20 days after the notice was served.

❖ A party aggrieved by a decision of the building official and desiring to appeal that decision must file a written request for a hearing in a timely manner, in this case 20 days from the date of the decision. The building official's decision also needs to be in writing, as specified in Sections 105.3.1 for a permit application, 109.6 for an inspection and 113.2 for a violation, or in a letter written regarding a request for an approval, depending on the action being appealed. The request must be on the jurisdiction's standard appeals form, which requires the appellant to provide the information needed by the board members to make a decision. This is often supplemented by whatever additional information the appellant wants to provide to support his or her case.

B101.2 Membership of board. The board of appeals shall consist of persons appointed by the chief appointing authority as follows:

1. One for five years; one for four years; one for three years; one for two years; and one for one year.

2. Thereafter, each new member shall serve for five years or until a successor has been appointed.

The building official shall be an ex officio member of said board but shall have no vote on any matter before the board.

❖ The board of appeals is to consist of five members appointed by the "chief appointing authority," typically the mayor or city manager. One member must be ap-

pointed for five years, one for four, one for three, one for two and one for one year. This method of appointment allows for a smooth transition of board of appeals members, providing continuity of action over the years. The building official also sits as a member of the board, but cannot vote on matters regarding his or her own decisions.

B101.2.1 Alternate members. The chief appointing authority shall appoint two alternate members who shall be called by the board chairperson to hear appeals during the absence or disqualification of a member. Alternate members shall possess the qualifications required for board membership and shall be appointed for five years, or until a successor has been appointed.

❖ This section authorizes the chief appointing authority to appoint two alternate members who must be available if the principal members of board are absent or disqualified. Alternate members must possess the same qualifications as the principal members and are appointed for a term of five years, or until such time that a successor is appointed.

B101.2.2 Qualifications. The board of appeals shall consist of five individuals, one from each of the following professions or disciplines:

1. Registered design professional with architectural experience or a builder or superintendent of building construction with at least ten years' experience, five of which shall have been in responsible charge of work.

2. Registered design professional with structural engineering experience

3. Registered design professional with mechanical and plumbing engineering experience or a mechanical con-

tractor with at least ten years' experience, five of which shall have been in responsible charge of work.

4. Registered design professional with electrical engineering experience or an electrical contractor with at least ten years' experience, five of which shall have been in responsible charge of work.

5. Registered design professional with fire protection engineering experience or a fire protection contractor with at least ten years' experience, five of which shall have been in responsible charge of work.

❖ The board of appeals consists of five persons with the qualifications and experience indicated in this section. One must be a registered design professional (see Item 2) with structural experience. The others must be registered design professionals, construction superintendents or contractors with experience in the various areas of building construction. This is necessary because the board will be hearing appeals regarding the codes referenced in Section 101.4 as well as the building code. These requirements are important in that technical people rule on technical matters. The board of appeals is not the place for policy or political deliberations. It is intended that these matters be decided purely on their technical merits, with due regard for state-of-the-art construction technology.

B101.2.3 Rules and procedures. The board is authorized to establish policies and procedures necessary to carry out its duties.

❖ The board can set forth the detailed procedures by which it operates, supplementing those required by Section B101.3.2.

B101.2.4 Chairperson. The board shall annually select one of its members to serve as chairperson.

❖ It is customary to determine a chairperson annually so that a regular opportunity is available to evaluate and either reappoint the current chairperson or appoint a new one.

B101.2.5 Disqualification of member. A member shall not hear an appeal in which that member has a personal, professional or financial interest.

❖ All members must disqualify themselves regarding any appeal in which they have a personal, professional or financial interest.

B101.2.6 Secretary. The chief administrative officer shall designate a qualified clerk to serve as secretary to the board. The secretary shall file a detailed record of all proceedings in the office of the chief administrative officer.

❖ The chief administrative officer is to designate a qualified clerk, typically an employee in the department, to serve as secretary to the board. The secretary is required to keep detailed records of the proceed-

ings. These may be needed for any future review of the board's decision.

B101.2.7 Compensation of members. Compensation of members shall be determined by law.

❖ Members of the board of appeals are not required to be compensated unless required by state or local law.

B101.3 Notice of meeting. The board shall meet upon notice from the chairperson, within 10 days of the filing of an appeal or at stated periodic meetings.

❖ In order that an appellant's request be heard in a timely manner, the board must meet within 10 days of the filing of an appeal. In large jurisdictions, where there are likely to be more appeals, the board will often set a regular schedule of meeting dates, such as monthly, and the 10-day rule would not apply.

B101.3.1 Open hearing. All hearings before the board shall be open to the public. The appellant, the appellant's representative, the building official and any person whose interests are affected shall be given an opportunity to be heard.

❖ All hearings before the board must be open to the public. The appellant, the appellant's representative, the building official and any person whose interests are affected must be heard.

B101.3.2 Procedure. The board shall adopt and make available to the public through the secretary procedures under which a hearing will be conducted. The procedures shall not require compliance with strict rules of evidence, but shall mandate that only relevant information be received.

❖ The board is required to establish and make available to the public written procedures detailing how hearings are to be conducted. Additionally, this section provides that although strict rules of evidence are not applicable, the information presented must be deemed relevant so the hearings are not unnecessarily long.

B101.3.3 Postponed hearing. When five members are not present to hear an appeal, either the appellant or the appellant's representative shall have the right to request a postponement of the hearing.

❖ When all five members of the board are not present, either the appellant or the appellant's representative may request that the hearing be postponed. Note that this is not a requirement, and that the appellant can elect to have the case heard before the smaller board.

B101.4 Board decision. The board shall modify or reverse the decision of the building official by a concurring vote of two-thirds of its members.

❖ A concurring vote of two-thirds of the board is needed to modify or reverse the decision of the building official. This means that at least four members must vote to modify or reverse the building official, even if less than the full board is present in accordance with Section B101.3.3.

B101.4.1 Resolution. The decision of the board shall be by resolution. Certified copies shall be furnished to the appellant and to the building official.

❖ The building official is to be bound by the action of the board of appeals, unless it is the opinion of the building official that the board of appeals has acted improperly. In such cases, relief through the court having jurisdiction may be sought by the jurisdiction's counsel.

B101.4.2 Administration. The building official shall take immediate action in accordance with the decision of the board.

❖ To avoid any undue hindrance in the progress of construction, the building official is required to act without delay based on the board's decision. This action may be to enforce the decision or to seek judicial relief if the board's action can be demonstrated to be inappropriate.

Appendix C:
Group U - Agricultural Buildings

The provisions contained in this appendix are not mandatory unless specifically referenced in the adopting ordinance.

General Comments

The provisions of this appendix are supplemental to the remainder of the code's requirements for agricultural buildings and are only applicable when specifically adopted, as stated in Section 101.2.1. This appendix contains provisions that may alter requirements found elsewhere in the code; however, the general requirements of the code still apply unless modified within this appendix. For example, the height and area limitations established in Chapter 5 apply to all Group U buildings and structures unless this appendix is specifically adopted by the jurisdiction. In such a case, the allowable height and area limitations in Table C102.1 would supersede those in Table 503.

As with all of the appendices to the code, the provisions of this appendix are not applicable unless the appendix is explicitly included in an adopting ordinance by the jurisdiction. Section 101.2.1 establishes the relationship of all appendices (see the commentary to that section for further discussion).

Section 312.1 of the code already contains a use and occupancy classification for buildings and structures considered as utility or miscellaneous in nature. Included in the Group U classification are agricultural buildings, barns, grain silos, greenhouses, livestock shelters, sheds and stables. Complete and detailed requirements are contained in the code for Group U buildings and structures; however, in most cases these provisions are more restrictive than a Group S-1 occupancy containing moderate-hazard contents that are likely to burn with moderate rapidity. This appendix provides alternative provisions that can be adopted by a jurisdiction to realistically match safeguards with the actual hazards associated with buildings and structures that are minimally occupied by persons and pose a limited fire risk.

Purpose

This appendix contains the requirements for protecting agricultural buildings. The provisions in this appendix reflect those occupancies and uses that deserve special consideration based on their limited occupant load. The code's foundation is the equivalent risk theory. Type of construction, fire-resistance ratings and means of egress requirements are balanced with the risks associated with agricultural buildings. This appendix contains general requirements, allowable heights and areas, mixed use provisions and exit requirements. The purpose of these provisions is to modify or supersede the basic code requirements for agricultural buildings. This approach allows the jurisdiction an alternative method that is more liberal and to address just the requirements for agricultural buildings that are included in this appendix.

SECTION C101
GENERAL

C101.1 Scope. The provisions of this appendix shall apply exclusively to agricultural buildings. Such buildings shall be classified as Group U and shall include the following uses:

1. Livestock shelters or buildings, including shade structures and milking barns.

2. Poultry buildings or shelters.

3. Barns.

4. Storage of equipment and machinery used exclusively in agriculture.

5. Horticultural structures, including detached production greenhouses and crop protection shelters.

6. Sheds.

7. Grain silos.

8. Stables.

❖ This section contains the scope and applicability requirements for agricultural buildings. A specific list is provided to identify which agricultural buildings are intended to be controlled by the alternative requirements of the appendix. This list expands and clarifies agricultural buildings as stated in Section 312.1.

This section of the code contains an explicit list of those types of agricultural buildings that are eligible for the alternative requirements of this appendix. Section 312.1 contains a similar list of utility and miscellaneous uses classified as Group U. Livestock shelters, barns, greenhouses, sheds, grain silos and stables are contained in both lists. Additionally, the term "livestock shelters" has been expanded to include weather protection structures and those used as milking barns. All of these buildings and structures are united in that they are used exclusively for agricultural purposes. As such, the occupant load for a Group U building is minimal. Although some of these structures may contain a considerable fuel load associated with their contents and uses, they would pose little fire hazards to persons or other occupiable buildings. The intent of the code is to match minimal requirements with these buildings based on the minimal fire and life hazards they pose.

SECTION C102
ALLOWABLE HEIGHT AND AREA

C102.1 General. Buildings classified as Group U Agricultural shall not exceed the area or height limits specified in Table C102.1.

❖ This section provides general building height and area limitations for Group U occupancies, just like Chapter 5 of the code. The height limitations, in terms of number of stories, and the area limitations from Table 503 for Group U buildings are in all cases the same or significantly less than those established for a Group S-1 structure. Those types of buildings are characterized by significant fuel loads that usually mandate the presence of an on-site automatic fire sprinkler system to abate the hazards to acceptable levels. A Group U building typically contains a lesser fire hazard and therefore, the allowable heights and areas should be greater than what is permitted for Group S-1 occupancies.

Revised entries for Table 503 are provided in this section for Group U buildings. Additionally, there are two modifications provided that supplement the requirements of Sections 507.1 and 507.3 for unlimited area buildings.

The provisions for governing the heights and areas of agricultural buildings are established in this section and are based on the Group U occupancy classification and the building's type of construction. Table

C102.1 is referenced here as the primary determinant for the minimum type of construction required for Group U buildings.

TABLE C102.1. See below.

❖ This table is configured similar to Table 503. The nine types of construction are listed across the top of the table. Included are separate lines for the allowable area, allowable height in terms of the number of stories and allowable height in terms of feet. The values listed for height in terms of feet are the same values listed in Table 503. Only the allowable areas and heights in terms of number of stories have been increased. Although not specifically stated here, the rest of the provisions of Chapter 5 are still applicable for determining any modifications (see Sections 504 and 506).

C102.2 One-story unlimited area. The area of a one-story Group U agricultural building shall not be limited if the building is surrounded and adjoined by public ways or yards not less than 60 feet (18 288 mm) in width.

❖ Just as additional allowances are made for unlimited area buildings in Section 507, there are similar provisions for agricultural buildings. A single-story building without a basement level containing agricultural uses represents a low fire hazard due to the limited number of occupants that may be present. If such a building is further isolated with sufficient clear open space around it, the building area is not limited. No other structures may be located within the 60 feet (18 288 mm) of clear open space required around the building. The open space must be located on the same lot as the agricultural building to allow access for fire-fighting operations. The type of construction is not restricted with this option.

C102.3 Two-story unlimited area. The area of a two-story Group U agricultural building shall not be limited if the building is surrounded and adjoined by public ways or yards not less than 60 feet (18 288 mm) in width and is provided with an approved automatic sprinkler system throughout in accordance with Section 903.3.1.1.

❖ This provision of the appendix, in effect, expands the allowance of Section 507.3. A two-story unlimited area

TABLE C102.1
BASIC ALLOWABLE AREA FOR A GROUP U, ONE STORY IN HEIGHT AND MAXIMUM HEIGHT OF SUCH OCCUPANCY

I		II		III and IV		V	
A	B	A	B	III A and IV	III B	A	B
ALLOWABLE AREA (square feet)[a]							
Unlimited	60,000	27,100	18,000	27,100	18,000	21,100	12,000
MAXIMUM HEIGHT IN STORIES							
Unlimited	12	4	2	4	2	3	2
MAXIMUM HEIGHT IN FEET							
Unlimited	160	65	55	65	55	50	40

For SI: 1 square foot = 0.0929 m².
a. See Section C102 for unlimited area under certain conditions.

agricultural building is permitted with the same open space requirement of Section C102.2 when it is equipped throughout with an automatic NFPA 13 fire sprinkler system. An agricultural building represents a lesser hazard than a Group B, F, M or S building. Therefore, a two-story fully suppressed agricultural building can also be of unlimited area as long as it is isolated. Once again, the type of construction is not restricted in these buildings.

SECTION C103
MIXED OCCUPANCIES

C103.1 Mixed occupancies. Mixed occupancies shall be protected in accordance with Section 508.

❖ The appendix includes a reference for any mixed occupancies that may exist within the agricultural building; for example, a large milking barn with integral offices and lab areas used for the transaction of farm business and testing of milk products. Since these areas are occupiable spaces, they are not classified as part of the Group U portions. Most likely these areas are classified as Group B. As such, the building must be protected in accordance with Chapter 3. Another example would be a barn with sleeping accommodations for migrant farm workers. The sleeping rooms would be considered as guestrooms and would need to be classified as Group R.

Any of the options listed in Chapter 3 are available to the building designer to address the mixed uses and occupancies that may be present within an agricultural building. This would include accessory use areas in Section 302.2, nonseparated uses in Section 302.3.1 or separated uses in Section 302.3.2.

SECTION C104
EXITS

C104.1 Exit facilities. Exits shall be provided in accordance with Chapters 10 and 11.

Exceptions:

1. The maximum travel distance from any point in the building to an approved exit shall not exceed 300 feet (91 440 mm).

2. One exit is required for each 15,000 square feet (1393.5 m²) of area or fraction thereof.

❖ The last section of this appendix provides a reminder to the permit applicant that the means of egress requirements of the code are still applicable to agricultural buildings. Although these buildings have a minimal number of occupants, provisions must still be made for their egress. This section includes two modifications to the Chapter 10 requirements that acknowledge the low occupant load and the limited hazards the occupants are exposed to.

The means of egress for agricultural buildings must comply with the applicable provisions of Chapter 10. For example, Table 1004.1.2 specifies an occupant load for agricultural buildings based on a rate of one person for every 300 square feet (28 m²) of gross floor area. Typically, Group U buildings are exempt from the requirements of Chapter 11 for accessibility to buildings by physically disabled persons. Section 1103.2.5 only mandates that access is required to paved work areas and those areas within a building or structure that are open to the general public.

The two exceptions modify or reinforce certain requirements of Chapter 10. Table 1015.1 limits the exit access travel distance for Group U buildings to 300 feet (91 440 mm). Exception 1 restates this 300-foot (91 440 mm) limit.

Exception 2 refines Table 1018.2 in determining the number of exits required for agricultural buildings. An exit is required for every 15,000 square feet (1,393 m²) of floor area. This works out to the same 50 occupants specified in Table 1018.2; however, the single-story height limitation and 75-foot (22 860 mm) travel distance are deleted with Exception 2.

Appendix D:
Fire Districts

The provisions contained in this appendix are not mandatory unless specifically referenced in the adopting ordinance.

General Comments

The provisions contained in this appendix are not mandatory unless specifically referenced in the adopting ordinance, as stated in Section 101.2.1.

Purpose

Fire districts, in one form or another, have been in the model codes for years. Over time, they have been revised, and in some cases deleted, as the codes have evolved to address the conflagration hazards that the fire district concept was intended to abate. However, there are jurisdictions that continue to utilize the provisions for fire districts. It is for this reason, in order to transition from the previously adopted model code to the *International Building Code*® (IBC®), that the provisions have been incorporated into the appendix.

SECTION D101
GENERAL

D101.1 Scope. The fire district shall include such territory or portion as outlined in an ordinance or law entitled "An Ordinance (Resolution) Creating and Establishing a Fire District." Wherever, in such ordinance creating and establishing a fire district, reference is made to the fire district, it shall be construed to mean the fire district designated and referred to in this appendix.

❖ The establishment of fire districts is among the oldest recognized means of controlling the threat of urban conflagrations. (The term "conflagration" is used to describe a fire that involves three or more buildings and spreads across a natural or constructed barrier, such as a stream, street or other open space.) In fact, the use of fire district ordinances by local governments could be considered an early form of land use control similar to zoning as it is now practiced. Establishment of fire districts became popular at the beginning of the 20th century as a means of preventing the destruction of entire cities by fire. Massive urban conflagrations struck Chicago (1871), Baltimore (1872 and 1904) and San Francisco (1906), destroying all or most of the urbanized areas as they then existed just as these cities were experiencing perhaps their greatest development. Fire districts usually include areas, designated by ordinance or resolution by the local governing body, which are considered susceptible to mass fires involving multiple buildings or groups of buildings. While it may be true that fire district ordinances and building code adoption have significantly reduced the incidence of conflagrations in densely built or populated urbanized areas, continuing development of suburban and rural communities in areas where nearby wildlands are preserved or protected has spurred the increased incidence and severity of fires in urban/wildland interface and intermix areas, which often lie outside designated fire districts. This appendix is currently not designed to address this hazard. This appendix establishes regulatory guidelines for controlling construction, use and occupancy of buildings and conduct of dangerous operations or processes in urbanized areas that may be especially prone to the danger of fires spreading from one building to another. This appendix may be particularly useful in managing fire danger in areas built prior to the adoption and enforcement of a building code but, as stated, may not be suitable for managing the threat of urban/wildland interface and intermix area fire hazards. See the *International Wildland-Urban Interface Code™* (IWUC™) for guidance on managing fire hazards in urban/wildland interface and intermix areas.

D101.1.1 Mapping. The fire district complying with the provisions of Section D101.1 shall be shown on a map that shall be available to the public.

❖ The requirement to prepare a map facilitates communication to the public of the additional controls found in these guidelines. In effect, establishing a fire district is a form of land use regulation, and many communities include fire district controls in their zoning ordinance or establish fire districts through overlay zoning. Zoning maps are frequently prepared and used to communicate land use controls to the public and to illustrate the relationships between land uses, infrastructure and other features of the community.

D101.2 Establishment of area. For the purpose of this code, the fire district shall include that territory or area as described in Sections D101.2.1 through D101.2.3.

❖ The following sections describe the conditions typical of fire districts established by ordinances that reference this appendix. Note that these sections are not customarily intended to include an entire jurisdiction. In fact, this is contrary to the fundamental purpose of a fire district—protection from catastrophic fires that could destroy an entire town. By dividing all jurisdictions into at least two separate fire districts, a city presumably reduces its risk to destruction of just a portion of the city or town. In newer communities developed largely after the adoption of local land use regulations and building codes, fire districts provide little added protection against conflagrations.

If a community is so compact and densely developed that designation of separate areas is not possible or practical, an entire jurisdiction could be designated a fire district as a means of controlling hazardous occupancies and regulating combustible construction.

When adopted for use in areas developed prior to the adoption of a building code, fire districts may serve as an effective means of abating the hazard posed when no single building constitutes a public nuisance, but taken in aggregate, the hazard of groups of similarly noncode-conforming buildings is considered unreasonable. By preventing new construction and new occupancies from contributing to the nonconformity of existing hazards and isolating these nonconforming districts from new development, a city may gain a measure of security from a mass fire.

D101.2.1 Adjoining blocks. Two or more adjoining blocks, exclusive of intervening streets, where at least 50 percent of the ground area is built upon and more than 50 percent of the built-on area is devoted to hotels and motels of Group R-1;

Group B occupancies; theaters, nightclubs, restaurants of Group A-1 and A-2 occupancies; garages, express and freight depots, warehouses and storage buildings used for the storage of finished products (not located with and forming a part of a manufactured or industrial plant); or Group S occupancy. Where the average height of a building is two and one-half stories or more, a block should be considered if the ground area built upon is at least 40 percent.

❖ The fire district should include areas densely developed with light and moderate fire hazard occupancies that have either significant fuel loads or significant transient occupant loads (see Figure D101.2.1). Nightclubs; theaters; restaurants; hotels and motels; certain types of Group B occupancies and warehouses not associated with industrial operations have been the sites of some of the nation's largest conflagrations. Intentionally set fires and delayed detection or notification of fire-fighting forces have contributed significantly to these losses. Occupancies and occupants typical of those listed in this section may be particularly prone to both of these factors.

Although Group F and M occupancies and Group S warehouses containing raw or waste materials are not listed, they are typically found in these areas as well. One- and two-family dwellings; multiple-family housing; schools or day care centers and institutional occupancies are not included in the description because they are typically found in less densely developed areas.

Local officials should recognize that redevelopment of industrial and warehouse buildings in existing fire districts for new residential and mercantile uses is increasingly common (see Section D103.2). These activities typically do not pose the same degree of mass fire danger in such buildings, despite the greater frequency of fires in residential occupancies. This may be due to reduced fuel loading and increased occupant activity, which aids in early detection and notification of a fire.

MORE THAN 50% OF ALL BLOCKS IS BUILT UPON

⧄ - BUILT UPON GROUND AREA   ☐ - FIRE DISTRICT

Figure D101.2.1
EXAMPLE OF FIRE DISTRICTS

D101.2.2 Buffer zone. Where four contiguous blocks or more comprise a fire district, there shall be a buffer zone of 200 feet (60 960 mm) around the perimeter of such district. Streets, rights-of-way and other open spaces not subject to building construction can be included in the 200-foot (60 960 mm) buffer zone.

❖ The buffer areas described by this section serve as fire or (preferably) fuel breaks by prohibiting the construction of buildings or structures that could become involved, allowing the fire to spread (see Figure D101.2.2). The green space or buffer zones may include the width of streets, streams or other dedicated rights-of-way that ensure segregation between adjacent fire district zones.

D101.2.3 Developed blocks. Where blocks adjacent to the fire district have developed to the extent that at least 25 percent of the ground area is built upon and 40 percent or more of the built-on area is devoted to the occupancies specified in Section D101.2.1, they can be considered for inclusion in the fire district, and can form all or a portion of the 200-foot (60 960 mm) buffer zone required in Section D101.2.2.

❖ Adjacent blocks may be added to the fire district as development of buffer spaces progresses (see Figure D101.2.3). However, care should be taken to ensure that the densities of buffer spaces do not increase to such an extent as to provide an avenue of fire spread between adjacent fire districts. If building densities remain between 25 and 50 percent of the land area, danger of fire spread between buildings in separate fire districts remains relatively low, as long as building separations required by Section D102.2 for the construction type are provided and maintained and combustible building features, such as roofing, sheathing, trim, canopies and signs, are similarly controlled.

Figure D101.2.2
BUFFER ZONES

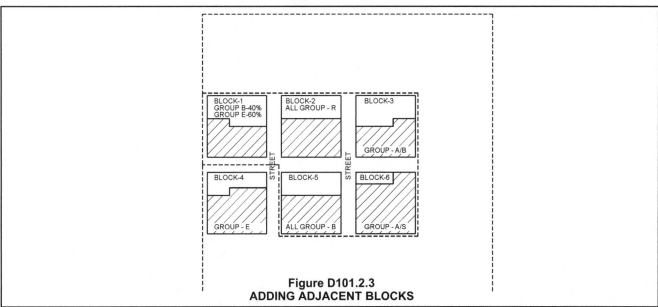

Figure D101.2.3
ADDING ADJACENT BLOCKS

SECTION D102
BUILDING RESTRICTIONS

D102.1 Types of construction permitted. Within the fire district every building hereafter erected shall be either Type I, II, III or IV, except as permitted in Section D104.

❖ The construction of Type V buildings is prohibited within a fire district. All other construction types are permitted, including unprotected Type IIB and IIIB structures with the further restrictions of Section D102.2 below.

D102.2 Other specific requirements.

D102.2.1 Exterior walls. Exterior walls of buildings located in the fire district shall comply with the requirements in Table 601 except as required in Section D102.2.6.

❖ Fire-resistant separations between buildings, either by distance or barriers, as described by Table 601, provide protection against fire spread by radiant and convective heat transfer and, to a limited extent, from small flying brands.

D102.2.2 Group H prohibited. Group H occupancies shall be prohibited from location within the fire district.

❖ Group H-1, H-2, H-3 and H-5 hazardous occupancies are prohibited within the fire district because of the fire hazard posed by these occupancies themselves. Group H-4 occupancies are prohibited due to the inherent danger of release and contamination posed by fire should occupancies containing highly toxic, toxic, radioactive or other health-hazard materials be exposed.

D102.2.3 Construction type. Every building shall be constructed as required based on the type of construction indicated in Chapter 6.

❖ This section reinforces the need for new buildings to be constructed as required based on height and area and type of construction considerations in the body of the code.

D102.2.4 Roof covering. Roof covering in the fire district shall conform to the requirements of Class A or B roof coverings as defined in Section 1505.

❖ Class C and wood-shake shingle roofs are prohibited within the fire district due to the danger of fire spread and involvement through production of or exposure to burning brands.

D102.2.5 Structural fire rating. Walls, floors, roofs and their supporting structural members shall be a minimum of 1-hour fire-resistance-rated construction.

Exceptions:

1. Buildings of Type IV construction.

2. Buildings equipped throughout with an automatic sprinkler system in accordance with Section 903.3.1.1.

3. Automobile parking structures.

4. Buildings surrounded on all sides by a permanently open space of not less than 30 feet (9144 mm).

5. Partitions complying with Section 603.1(8).

❖ Unprotected Type IIB and IIIB construction is permitted within the fire district only for fully sprinklered buildings, automobile parking garages and buildings with 30 feet (9144 mm) of open space on all sides. All other buildings, except those of Type IV construction, must be of fire-resistance-rated construction.

D102.2.6 Exterior walls. Exterior load-bearing walls of Type II buildings shall have a fire-resistance rating of 2 hours or more where such walls are located within 30 feet (9144 mm) of a common property line or an assumed property line. Exterior nonload-bearing walls of Type II buildings located within 30 feet (9144 mm) of a common property line or an assumed property line shall have fireresistance ratings as required by Table 601, but not less than 1 hour. Exterior walls located more than 30 feet (9144 mm) from a common property line or an assumed property line shall comply with Table 601.

Exception: In the case of one-story buildings that are 2,000 square feet (186 m²) or less in area, exterior walls located more than 15 feet (4572 mm) from a common property line or an assumed property line need only comply with Table 601.

❖ Although Type II buildings may be capable of resisting ignition because their components are generally noncombustible, their structural integrity could be threatened or contents ignited by radiant heat exposure from a fire in an adjacent building. Moreover, since Section 603.1, Item 1, permits the use of fire-retardant-treated wood for nonbearing partitions and roof systems of Type II buildings, these structures should not be assumed to be of noncombustible construction.

Small accessory buildings [less than 2,000 square feet (186 m²)] are exempt from the exterior wall fireresistance requirements of this section, provided they comply with Table 601 and are located at least 15 feet (4572 mm) from an adjacent or assumed property line. Bear in mind that, except as provided in Section D105, such structures may not be of Type V construction. Such buildings present little danger of contributing to a mass fire, provided they comply with the requirements of the code.

D102.2.7 Architectural trim. Architectural trim on buildings located in the fire district shall be constructed of approved noncombustible materials or fire-retardant-treated wood.

❖ The ignition of architectural trim on the exterior of buildings from radiant heat transfer may pilot the ignition of other building components or contents of the building, providing an avenue for further fire spread. Many types of architectural trim may also catch flying brands from a fire, sparking the ignition of the exposed building.

D102.2.8 Permanent canopies. Permanent canopies are permitted to extend over adjacent open spaces provided:

1. The canopy and its supports shall be of noncombustible material, fire-retardant-treated wood, Type IV construction or of 1-hour fire-resistance-rated construction.

Exception: Any textile covering for the canopy shall meet the fire propagation performance criteria of NFPA 701 after both accelerated water leaching and accelerating weathering.

2. Any canopy covering, other than textiles, shall have a flame spread index not greater than 25 when tested in accordance with ASTM E 84 in the form intended for use.

3. The canopy shall have at least one long side open.

4. The maximum horizontal width of the canopy shall not exceed 15 feet (4572 mm).

5. The fire resistance of exterior walls shall not be reduced.

❖ Although combustible exterior canopies may become involved in fire from radiant heat transfer, their relatively low mass and method of support and attachment to the building should minimize the risk that they will pilot the ignition of exterior walls or building contents. The requirements of this section restrict certain features of exterior canopies to ensure that their construction contributes little to fire growth or spread in the fire district.

D102.2.9 Roof structures. Structures, except aerial supports 12 feet (3658 mm) high or less, flagpoles, water tanks and cooling towers, placed above the roof of any building within the fire district shall be of noncombustible material and shall be supported by construction of noncombustible material.

❖ Vertical structures with little combustible mass, or which will be cooled by their contents, are permitted, provided their failure in a fire does not pose a significant secondary hazard. Of course, rooftop water tanks supplying fire protection systems should not be of combustible construction because their failure may undermine the effectiveness of required fire protection systems.

D102.2.10 Plastic signs. The use of plastics complying with Section 2611 for signs is permitted provided the structure of the sign in which the plastic is mounted or installed is noncombustible.

❖ Free-standing and projecting signs are unlikely to contribute to fire spread between buildings, provided their structural supports are noncombustible and structural collapse of the sign supports will not directly expose a building or its contents. Although rooftop, marquee and wall-mounted signs may present a more direct hazard to buildings, Sections 2611.2 and 2611.3 limit the area of any plastic components and require all plastic materials to meet specified combustibility criteria (see Section 2606.4).

D102.2.11 Plastic veneer. Exterior plastic veneer is not permitted in the fire district.

❖ All plastic veneers, including vinyl siding, are prohibited in the fire district. Most plastics deform under high radiant heat conditions, some melt and drip and all will burn. Plastics represent a wide range of fire hazard conditions, generally all of which are unfavorable. This prohibition is consistent with restrictions on

other combustible exterior elements, such as roof and architectural trim. Although this appendix does not explicitly prohibit other combustible exterior veneers, Section D105.1, Item 11, excludes wood veneer from these provisions, and the use of such combustible cladding materials remains severely restricted.

SECTION D103
CHANGES TO BUILDINGS

D103.1 Existing buildings within the fire district. An existing building shall not hereafter be increased in height or area unless it is of a type of construction permitted for new buildings within the fire district or is altered to comply with the requirements for such type of construction. Nor shall any existing building be hereafter extended on any side, nor square footage or floors added within the existing building unless such modifications are of a type of construction permitted for new buildings within the fire district.

❖ Bearing in mind that the principal purpose of Section D103 is to control the fire hazard posed by existing nonconforming construction by isolating areas where such buildings are located, it would be imprudent to allow any increase in the hazard within the fire district once it is established. Therefore, any new construction must comply with the requirements of the code, including limits on construction, area and height and fire protection (see Section D102.2.3).

D103.2 Other alterations. Nothing in Section D103.1 shall prohibit other alterations within the fire district provided there is no change of occupancy that is otherwise prohibited and the fire hazard is not increased by such alteration.

❖ All construction and any alteration of existing buildings must comply with the provisions of the code (see Section 101.2). Assuming all new work complies with the relevant requirements of the code and all applicable referenced standards, the hazard of any building in the fire district should not be increased.

Change of occupancy from one group to another is allowed within the fire district, provided that the new occupancy is permitted by the code for the type of construction, fire protection, means of egress and other requirements and is not a hazardous (Group H) occupancy (see Section D102.2.2).

D103.3 Moving buildings. Buildings shall not hereafter be moved into the fire district or to another lot in the fire district unless the building is of a type of construction permitted in the fire district.

❖ Relocated buildings, like new construction, must be of a type of construction permitted within the fire district. The type of construction will be determined by the area, height and occupancy of the building after its relocation. However, certain structural components and systems may require higher fire-resistance ratings than specified in Table 601, in accordance with Section D102.2.5.

SECTION D104
BUILDINGS LOCATED PARTIALLY
IN THE FIRE DISTRICT

D104.1 General. Any building located partially in the fire district shall be of a type of construction required for the fire district, unless the major portion of such building lies outside of the fire district and no part is more than 10 feet (3048 mm) inside the boundaries of the fire district.

❖ Buildings that lie principally outside the fire district and project no more than 10 feet (3048 mm) within a fire district boundary are exempt from the fire district requirements and should be considered part of the buffer zone described in Sections D101.2.2 and D101.2.3.

In Figure D104(A), the building does not need to meet the provisions of Appendix D, since the majority of the building is outside and no more than 10 feet (3048 mm) of the building lies within the boundaries of the fire district.

In Figure D104(B), the building shall comply with the provisions of Appendix D, since more than 10 feet (3048 mm) of the building is within the boundaries of the fire district.

SECTION D105
EXCEPTIONS TO RESTRICTIONS
IN FIRE DISTRICT

D105.1 General. The preceding provisions of this appendix shall not apply in the following instances:

1. Temporary buildings used in connection with duly authorized construction.

2. A private garage used exclusively as such, not more than one story in height, nor more than 650 square feet (60 m²) in area, located on the same lot with a dwelling.

3. Fences not over 8 feet (2438 mm) high.

4. Coal tipples, material bins and trestles of Type IV construction.

5. Water tanks and cooling towers conforming to Sections 1509.3 and 1509.4.

6. Greenhouses less than 15 feet (4572 mm) high.

7. Porches on dwellings not over one story in height, and not over 10 feet (3048 mm) wide from the face of the building, provided such porch does not come within 5 feet (1524 mm) of any property line.

8. Sheds open on a long side not over 15 feet (4572 mm) high and 500 square feet (46 m²) in area.

9. One- and two-family dwellings where of a type of construction not permitted in the fire district can be extended 25 percent of the floor area existing at the time of inclusion in the fire district by any type of construction permitted by this code.

10. Wood decks less than 600 square feet (56 m²) where constructed of 2-inch (51 mm) nominal wood, pressure treated for exterior use.

11. Wood veneers on exterior walls conforming to Section 1405.4.

12. Exterior plastic veneer complying with Section 2605.2 where installed on exterior walls required to have a fire-resistance rating not less than 1 hour, provided the exterior plastic veneer does not exhibit sustained flaming as defined in NFPA 268.

❖ With the exception of Items 11 and 12, excluded structures are generally accessory to other buildings and pose little hazard of contributing to fire spreading from one building to another or between buildings in adjacent fire districts. Excluding these structures from the provisions of this appendix does not exempt any new

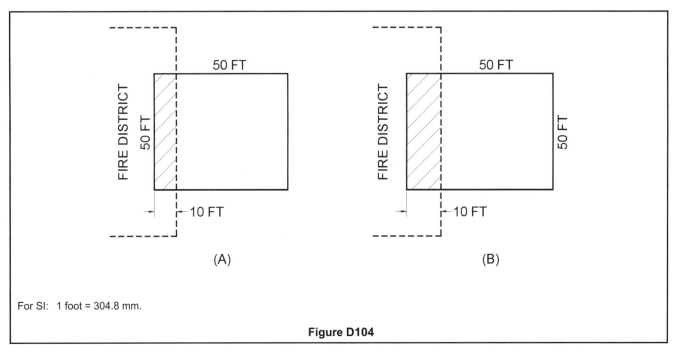

For SI: 1 foot = 304.8 mm.

Figure D104

structures of the types or uses listed from complying with the requirements of the code.

Water tanks for fire protection located within the fire district may not be governed by the requirements of this appendix.

Wood veneer applied to the exterior of structures must comply with the restrictions of Section 1405.4. The referenced provisions restrict the use of wood veneer to installation on noncombustible exterior walls of the fire resistance required by Table 601, and not exceeding two stories in height.

SECTION D106
REFERENCED STANDARDS

ASTM E 84-04	Test Method for Surface Burning Characteristics of Building Materials	D102.2.8
NFPA 268-01	Test Method for Determining Ignitability of Exterior Wall Assemblies Using a Radiant Heat Energy Source	D105.1
NFPA 701-99	Methods of Fire Tests for Flame-Propagation of Textiles and Films	D102.2.8

Appendix E:
Supplementary Accessibility Requirements

The provisions contained in this appendix are not mandatory unless specifically referenced in the adopting ordinance.

General Comments

As stated in Section 101.2.1, the provisions of the appendices do not apply unless specifically adopted.

Chapter 11 sets forth requirements for accessibility to buildings and their associated sites and facilities for people with physical disabilities. This appendix was added to address accessibility in construction for items that were not typically enforceable through the traditional code enforcement process.

Section E101 contains the broad scope statement of the appendix and identifies the baseline criteria for accessibility as being in compliance with this appendix and ICC/ANSI A117.1-2003, *Accessible and Usable Buildings and Facilities* (ICC A117.1). ICC A117.1 is the consensus national standard that sets forth the details, dimensions and construction specifications for accessibility.

Section E102 contains definitions of terms that are associated with accessibility that are not found in Chapter 11.

Section E103 contains additional requirements for interior accessible routes. An accessible route is a key component of the built environment that provides a person with a disability access to spaces, elements, facilities and buildings.

Section E104 contains various accessibility requirements that are unique to specific occupancies and are applicable in addition to other general requirements of this chapter. Specific provisions unique to transient lodging and Group I-3 facilities are included.

Section E105 contains various requirements that are applicable to special equipment provided for public use, including requirements for portable toilet and bathing rooms; laundry equipment; vending machines; automatic teller machines (ATMs) and fare machines and two-way communication systems.

Section E106 contains special criteria for public telephones, including TTY requirements.

Section E107 sets forth requirements for signage identifying permanent designations, such as room numbers in a hospital, directional and informational signs, and signage specifically related to the transportation facilities identified in Sections E108, E109 and E110.

Section E108 lists criteria specific to bus stops and terminals.

Section E109 lists criteria specific to fixed transportation facilities and stations, such as train stations.

Section E110 lists criteria specific to airports.

Section E111 enumerates the standards referenced in this appendix.

Purpose

The Architectural and Transportation Barriers Compliance Board (Access Board) has revised and updated its accessibility guidelines for buildings and facilities covered by the Americans with Disabilities Act of 1990 (ADA) and the Architectural Barriers Act of 1968 (ABA). The final ADA/ABA Guidelines, published by the Access Board in September 2004, will serve as the basis for the minimum standards when adopted by other federal agencies responsible for issuing enforceable standards. The plan is to eventually use this new document in place of the Uniform Federal Accessibility Standard (UFAS) and the Americans with Disabilities Act Accessibility Guidelines (ADAAG). Refer to the Access Board website (www.access-board.gov) for the current status of this adoption process.

Appendix E includes scoping requirements in the new ADA/ABA Guidelines that are not in Chapter 11 not otherwise mentioned or mainstreamed throughout the code. Items in the appendix deal with topics not typically addressed in building codes. For example, items in Sections E102 through E107 are not part of the "built" environment, as they can be added, modified or removed without a building permit and without notifying the building official. Sections E108 through E110 deal specifically with transportation facilities.

This appendix is being included in the code for the convenience of building owners and designers who are required to comply with new ADA/ABA Guidelines and for the benefit of jurisdictions that may wish to go beyond the traditional boundary of the code and formally adopt these additional ADA/ABA Guidelines requirements. The requirements within the code indicate what is required and enforceable by the building official. The appendix includes items that are outside the scope of code enforcement, but must be complied with in addition to the code requirements to satisfy ADA/ABA Guidelines.

SECTION E101
GENERAL

E101.1 Scope. The provisions of this appendix shall control the supplementary requirements for the design and construction of facilities for accessibility to physically disabled persons.

❖ This section establishes the scope of the appendix as providing for design and construction of facilities for accessibility to disabled persons other than items covered under Chapter 11. Items in the appendix are not enforceable under the typical code review process because these items can be altered or removed without notification to the building official. Items in this appendix are included in new ADAAG.

E101.2 Design. Technical requirements for items herein shall comply with this code and ICC A117.1.

❖ This section establishes the primary and fundamental relationship of ICC A117.1 to the code. The code text is intended to "scope" or provide thresholds for application of required accessibility features. The referenced standard contains technical provisions indicating how compliance with the code is achieved. In short, this appendix specifies what, when and how many accessible features are required; the referenced standard ICC A117.1 indicates how to make these features accessible. Compliance with both the scoping and technical standard is necessary for accessibility.

SECTION E102
DEFINITIONS

E102.1 General. The following words and terms shall, for the purposes of this appendix, have the meanings shown herein.

❖ This section contains definitions of terms that are associated with the subject matter of this appendix.
   Definitions of terms can help in the understanding and application of the code requirements. The purpose for including these definitions within this appendix is because appendices are not considered part of the code requirements unless specifically referenced in the adopting ordinance.

CLOSED-CIRCUIT TELEPHONE. A telephone with a dedicated line such as a house phone, courtesy phone or phone that must be used to gain entrance to a facility.

❖ Telephones described in this definition are utilized in hotels, apartments, airports, hospitals and similar facilities. Access to these systems is important to provide communication throughout a facility.

MAILBOXES. Receptacles for the receipt of documents, packages or other deliverable matter. Mailboxes include, but are not limited to, post office boxes and receptacles provided by commercial mail-receiving agencies, apartment houses and schools.

❖ The definition for "Mailboxes" clarifies that the scope of this provision is meant to address all types of facilities where a person could receive correspondence, packages or written notifications. The description includes post office boxes, mailboxes in the lobby of an apartment building or mailboxes in the lobby of a college dorm.

TRANSIENT LODGING. A building, facility or portion thereof, excluding inpatient medical care facilities and long-term care facilities, that contains one or more dwelling units or sleeping units. Examples of transient lodging include, but are not limited to, resorts, group homes, hotels, motels, dormitories, homeless shelters, halfway houses and social service lodging.

❖ Transient lodging facilities, such as hotels, typically provide short-term stays; however, this definition also includes facilities that provide longer term stays, such as dormitories. The 30-day or less stay typically associated with transient lodging is not applicable in this definition. The key is whether sleeping or dwelling units are provided. The exclusions are for inpatient and long-term care facilities, such as hospitals and nursing homes. Additionally, this would not be applicable to permanent lodging facilities with dwelling units, such as apartments, condominiums or townhouses.

SECTION E103
ACCESSIBLE ROUTE

E103.1 Raised platforms. In banquet rooms or spaces where a head table or speaker's lectern is located on a raised platform, an accessible route shall be provided to the platform.

❖ An accessible route is required to a temporary raised platform utilized for a speaker or head table. An accessible route would not be required to a raised platform used for other purposes, such as a band. Access to permanently raised platforms or stages is addressed in Chapter 11.

SECTION E104
SPECIAL OCCUPANCIES

E104.1 General. Transient lodging facilities shall be provided with accessible features in accordance with Sections E104.2 and E104.3. Group I-3 occupancies shall be provided with accessible features in accordance with Sections E104.3 and E104.4.

❖ In the following sections there are bed and communication feature requirements for transient lodging facilities. Also included are communication and visitation booth requirements for jails (Group I-3).

E104.2 Accessible beds. In rooms or spaces having more than 25 beds, five percent of the beds shall have a clear floor space complying with ICC A117.1.

❖ If dormitory-type transient lodging facilities have 25 or more beds in a room, 5 percent of the beds must have adequate clearance to be accessible by a person using a wheelchair. If separate-sex facilities are provided, accessible beds must be provided for both sexes. Clear floor space requirements are found in Section 305 of ICC A117.1.

E104.2.1 Sleeping areas. A clear floor space complying with ICC A117.1 shall be provided on both sides of the accessible bed. The clear floor space shall be positioned for parallel approach to the side of the bed.

Exception: This requirement shall not apply where a single clear floor space complying with ICC A117.1 positioned for parallel approach is provided between two beds.

❖ To be considered accessible, the bed must be approachable from either side in order to accommodate wheelchair users who may need to transfer from a particular side due to their mobility impairment. The alternative applies when a bed is provided on both sides of a single-approach location. Clear floor space requirements are found in Section 305 of ICC A117.1.

Some people require a type of lift system to get them into bed. Some hotels have their beds on platforms. Beds on platforms do not allow for the supports for these types of lifts to move under the beds. While not a requirement, it may be advisable to have some accessible beds on frames to allow for the use of these lifts.

E104.3 Communication features. Communication features complying with ICC A117.1 shall be provided in accordance with Sections E104.3.1 through E104.3.4.

❖ Examples of communication features are telephones, doorbells and peepholes. In units required to have accessible communication features, the designer should consider compatibility with different types of adaptive equipment used by people with hearing impairments. Telephone interface jacks should be compatible with both digital and analog signal use. If an audio headphone jack is provided on a speaker phone, a cutoff switch can be included in the jack so that insertion of the jack cuts off the speaker. If a telephone-like handset is used, the external speakers can be turned off when the handset is removed from the cradle. For headset or external amplification system compatibility, a standard subminiature jack installed in the telephone will provide the most flexibility. Peepholes in doors should be installed at heights for both persons who are standing and persons using a wheelchair.

E104.3.1 Transient lodging. In transient lodging facilities, sleeping units with accessible communication features shall be provided in accordance with Table E104.3.1. Units required to comply with Table E104.3.1 shall be dispersed among the various classes of units.

❖ In transient lodging facilities, an alternative notification system for persons with hearing impairments must be offered in the number of dwelling and sleeping units specified in Table E104.3.1 (see the definition for "Dwelling units" and "Sleeping units" in Chapter 2). The number of dwelling or sleeping units that are required to provide accessible communication features depends on the number of units provided. The number of accessible communication features is based on the number of units, not the number of beds. The rooms

with accessible communication features are not required to be the same rooms that contain accessible beds. Requirements for communication features within dwelling and sleeping units are found in Section 1005 of ICC A117.1.

E104.3.2 Group I-3. In Group I-3 occupancies at least 2 percent, but no fewer than one of the total number of general holding cells and general housing cells equipped with audible emergency alarm systems and permanently installed telephones within the cell, shall comply with Section E104.3.4.

❖ Jails are required to have a fire alarm system in accordance with Section 907.6.2. When inmates or detainees are allowed independent means of egress, a minimum of 2 percent of the cells must have visible Section 702.1 of as well as audible alarm notification. The code and Section 702.1of ICC A117.1 both reference NFPA 72 for installation requirements. If cells also have permanently installed telephones, 2 percent of the telephones shall have a visible notification device as well as volume control and a place to plug in a TTY. The visual notification device for the phone cannot be connected to the visual alarms for the fire alarm system.

TABLE E104.3.1. See page E-4.

❖ This table specifies the number of dwelling and sleeping units in transient lodging facilities that are required to provide accessible communication features. In accordance with the definition of "Sleeping unit," bedrooms in dwelling units are not considered sleeping units.

E104.3.3 Dwelling units and sleeping units. Where dwelling units and sleeping units are altered or added, the requirements of Section E104.3 shall apply only to the units being altered or added until the number of units with accessible communication features complies with the minimum number required for new construction.

❖ When transient lodging dwelling and sleeping units are altered or added, only these units are counted to determine the number of units that will require accessible communication features. Once the entire building meets the number required for new construction, no additional accessible communication features are required. This is consistent with the intent of Sections 3409.5.1 and 3409.7.7.

E104.3.4 Notification devices. Visual notification devices shall be provided to alert room occupants of incoming telephone calls and a door knock or bell. Notification devices shall not be connected to visual alarm signal appliances. Permanently installed telephones shall have volume controls and an electrical outlet complying with ICC A117.1 located within 48 inches (1219 mm) of the telephone to facilitate the use of a TTY.

❖ A type of visual notification, such as a flashing light, must be provided in some sleeping accommodations. The purpose of the visual notification is to alert a person with a hearing impairment to incoming phone calls or someone knocking at the door. The notification cannot be connected to any type of visual alarm system. If phones are provided within the dwelling or sleeping unit, they must be equipped with a volume control and be located within 48 inches (1219 mm) of an electrical outlet to accommodate a TTY. Specific requirements for telephones are provided in Section 704 of ICC A117.1.

E104.4 Partitions. Solid partitions or security glazing that separates visitors from detainees in Group I-3 occupancies shall provide a method to facilitate voice communication. Such methods are permitted to include, but are not limited to, grilles, slats, talk-through baffles, intercoms or telephone handset devices. The method of communication shall be accessible to individuals who use wheelchairs and individuals who have difficulty bending or stooping. Hand-operable communication devices, if provided, shall comply with Section E106.3.

❖ Section 1109.11 requires 5 percent of cubicles or counters in jail (Group I-3) visiting areas to be accessible in accordance with the built-in counter/work surface requirements. This section deals with the visiting situation where solid partitions of security glazing separates visitors from detainees. To facilitate communication, the system may include some type of handset, headphones or microphone. The system must consider in its design that either user may use a wheelchair or have difficulty bending or stooping. Any type of telephone or headphone system must have a volume control in accordance with Section 106.3.

SECTION E105
OTHER FEATURES AND FACILITIES

E105.1 Portable toilets and bathing rooms. Where multiple single-user portable toilet or bathing units are clustered at a single location, at least 5 percent, but not less than one toilet unit or bathing unit at each cluster, shall comply with ICC A117.1. Signs containing the International Symbol of Accessibility and complying with ICC A117.1 shall identify accessible portable toilets and bathing units.

Exception: Portable toilet units provided for use exclusively by construction personnel on a construction site.

❖ When portable toilets are provided for special events, at least 5 percent must be accessible. A "cluster" is a group of toilets adjacent or in close proximity to one another. This is consistent with the requirements for sinks in Section 1109.3. Appropriate signage is also required. If portable toilets are provided for personnel on a construction site, they are not required to be accessible. This is consistent with Section 1103.2.6. Technical requirements for bathroom and signage are found in Sections 602 and 703 of ICC A117.1, respectively.

E105.2 Laundry equipment. Where provided in spaces required to be accessible, washing machines and clothes dryers shall comply with this section.

❖ When laundry facilities are provided that are open to the general public, or are associated with Accessible, Type A or B units, the washers and dryers must be accessible in accordance with Sections E105.3.1 and E105.3.2. This is consistent with Section 1107.3. The requirements are based on the washers or dryers provided in each space, not the total within the building. For example, a dorm with a single washer and dryer on each floor of a building should provide an accessible washer and dryer in each location.

E105.2.1 Washing machines. Where three or fewer washing machines are provided, at least one shall comply with ICC A117.1. Where more than three washing machines are provided, at least two shall comply with ICC A117.1.

❖ When three or fewer washing machines are provided in a laundry room, at least one must meet the requirements for accessibility in Section 611 in ICC A117.1 for

TABLE E104.3.1
DWELLING OR SLEEPING UNITS WITH ACCESSIBLE COMMUNICATION FEATURES

TOTAL NUMBER OF DWELLING OR SLEEPING UNITS PROVIDED	MINIMUM REQUIRED NUMBER OF DWELLING OR SLEEPING UNITS WITH ACCESSIBLE COMMUNICATION FEATURES
1	1
2 to 25	2
26 to 50	4
51 to 75	7
76 to 100	9
101 to 150	12
151 to 200	14
201 to 300	17
301 to 400	20
401 to 500	22
501 to 1,000	5% of total
1,001 and over	50 plus 3 for each 100 over 1,000

clear floor space, operable parts and height. In larger facilities, at least two washers must be accessible. Requirements include criteria for clear floor space to access the unit; operable parts that meet height, graspability and force requirements and the location of doors to access the interior of the appliance.

E105.2.2 Clothes dryers. Where three or fewer clothes dryers are provided, at least one shall comply with ICC A117.1. Where more than three clothes dryers are provided, at least two shall comply with ICC A117.1.

❖ The same criteria for washers is expected for dryers (see commentary, Section E105.2.1).

E105.3 Depositories, vending machines, change machines and similar equipment. Where provided, at least one of each type of depository, vending machine, change machine and similar equipment shall comply with ICC A117.1.

   Exception: Drive-up-only depositories are not required to comply with this section.

❖ All types of self-service machinery must be accessible. Examples include the book return at a library, a change machine at a laundry mat or a soda machine. A drive-up-only depository, such as a drive-up night deposit at a bank, is not required to be accessible. To be accessible, the equipment must meet the provisions for clear floor space, height and have operable parts within reach ranges in accordance with Section 309 of ICC A117.1.

E105.4 Mailboxes. Where mailboxes are provided in an interior location, at least 5 percent, but not less than one, of each type shall comply with ICC A117.1. In residential and institutional facilities, where mailboxes are provided for each dwelling unit or sleeping unit, mailboxes complying with ICC A117.1 shall be provided for each unit required to be an Accessible unit.

❖ Whenever mailboxes are provided, such as in an office building lobby, at least 5 percent are required to be accessible. In residential and institutional facilities, if mailboxes are provided for each unit, the mailboxes for the Accessible units must have adequate clear floor space and meet operable parts provisions. Examples of such types of facilities would be dorms, hotels, assisted living facilities or nursing homes. The number of Accessible units depends on the type of facility (see Section 1107 for requirements). To be accessible, the mailbox must meet the provisions for clear floor space, height and have operable parts within reach ranges in accordance with Section 309 of ICC A117.1.

E105.5 Automatic teller machines and fare machines. Where automatic teller machines or self-service fare vending, collection or adjustment machines are provided, at least one machine of each type at each location where such machines are provided shall be accessible. Where bins are provided for envelopes, wastepaper or other purposes, at least one of each type shall be accessible.

❖ When such machines are provided, such as at banks and in bus or train stations, at least one of each type of machine must be accessible. If a site provides both interior and exterior ATMs, each such installation is considered a separate location. Accessible ATMs, including those with speech and those that are within reach of people who use wheelchairs, must provide all the functions provided to customers at that location at all times. For example, it is unacceptable for the accessible ATM only to provide cash withdrawals while inaccessible ATMs in that area also sell theater tickets. Specific criteria is provided in Section 707 of ICC A117.1.

E105.6 Two-way communication systems. Where two-way communication systems are provided to gain admittance to a building or facility or to restricted areas within a building or facility, the system shall comply with ICC A117.1.

❖ Apartments or condominiums with secured entrances often have some type of two-way communication system at the entry so the residents can screen persons entering the building. Specific criteria for systems connected with dwelling or sleeping units are provided in Sections 1005.6 and 1005.7 of ICC A117.1. Similar provisions would be applicable to two-way communication systems provided at other locations, such as office buildings, courthouses and other facilities where access to the building or restricted spaces are dependent on two-way communication systems.

SECTION E106
TELEPHONES

E106.1 General. Where coin-operated public pay telephones, coinless public pay telephones, public closed-circuit telephones, courtesy phones or other types of public telephones are provided, accessible public telephones shall be provided in accordance with Sections E106.2 through E106.5 for each type of public telephone provided. For purposes of this section, a bank of telephones shall be considered two or more adjacent telephones.

❖ Many types of public phones are provided, such as pay telephones, courtesy phones throughout airports, house phones in hotels, entrance phones in apartment complexes, etc. Wherever such phones are provided, at least one of each type must comply with Sections E106.2 through E106.5. A bank of telephones is defined as two or more telephones in one location.

E106.2 Wheelchair-accessible telephones. Where public telephones are provided, wheelchair-accessible telephones complying with ICC A117.1 shall be provided in accordance with Table E106.2.

   Exception: Drive-up-only public telephones are not required to be accessible.

❖ For each type of public telephone provided, the number of phones specified in Table E106.2 must be accessible. Specific criteria are provided in Section 704.2 of ICC A117.1.

TABLE E106.2. See page E-6.

❖ If single telephones are provided in several locations or one bank of telephones is provided on a floor level or over a site, then at least one of each type must be accessible. The accessible telephone can be accessed using a parallel approach or a forward approach. If two or more banks are provided on a floor level or over a site, at least one telephone at each location must be accessible.

E106.3 Volume controls. All public telephones provided shall have volume control complying with ICC A117.1.

❖ Every public telephone must have volume controls and be identified with appropriate signage. Amplifiers on pay phones are located in the base of the handset or are built into the telephone. Most are operated by pressing a button or key. If the microphone in the handset is not being used, a mute button that temporarily turns off the microphone can also reduce the amount of background noise that the person can hear in the earpiece. Consider compatibility issues when matching an amplified handset with a phone or phone system. Amplified handsets that can be switched with pay telephone handsets are available. Specific criteria is found in Section 704.3 of ICC A117.1.

E106.4 TTYs. TTYs complying with ICC A117.1 shall be provided in accordance with Sections E106.4.1 through E106.4.9.

❖ TTYs are defined in ICC A117.1 as "machinery or equipment that employs interactive, graphic communications through the transmission of coded signals across the standard telephone network." The term "TTY" also refers to devices known as text telephones and TDDs. Text telephones enable a person with a hearing impairment to utilize a telephone to send and receive text messages. Consideration should be given to phone systems that can accommodate both digital and analog transmissions for compatibility with digital and analog TTYs. Sections E106.4.1 through E106.4.4 indicate when TTYs are required within a bank of telephones, on a floor, within a building or on a site. More than one section could apply. Sections E106.4.5 through E106.4.8 are special requirements to specific types of sites. Section E106.4.9 deals with appropriate signage.

E106.4.1 Bank requirement. Where four or more public pay telephones are provided at a bank of telephones, at least one public TTY shall be provided at that bank.

Exception: TTYs are not required at banks of telephones located within 200 feet (60 960 mm) of, and on the same floor as, a bank containing a public TTY.

❖ A TTY is required wherever four or more telephones are located together. If another bank is located both on

the same floor and within 200 feet (60 960 mm) of the TTY, that bank is not required to contain a TTY.

E106.4.2 Floor requirement. Where four or more public pay telephones are provided on a floor of a privately owned building, at least one public TTY shall be provided on that floor. Where at least one public pay telephone is provided on a floor of a publicly owned building, at least one public TTY shall be provided on that floor.

❖ This section is similar to Section E106.4.1; however, the number of phones on the floor is counted, instead of within a bank. The term "publicly owned building" is intended to apply to facilities such as government buildings, courthouses, libraries, schools, police stations, fire stations, park district buildings, etc.

E106.4.3 Building requirement. Where four or more public pay telephones are provided in a privately owned building, at least one public TTY shall be provided in the building. Where at least one public pay telephone is provided in a publicly owned building, at least one public TTY shall be provided in the building.

❖ This section is similar to Sections E106.4.1 and E106.4.2; however, the number of phones in a building is counted, instead of within a bank or on a floor. In some situations, one TTY can satisfy more than one requirement. For example, the TTY required for a bank of phones could also satisfy the requirement for the TTY required for the building. The term "publicly owned building" is intended to apply to facilities such as government buildings, courthouses, libraries, schools, police stations, fire stations, park district buildings, etc.

E106.4.4 Site requirement. Where four or more public pay telephones are provided on a site, at least one public TTY shall be provided on the site.

❖ This section is similar to Sections E106.4.1, E106.4.2 and E106.4.3; however, the number of phones on a site is counted. In some situations, one TTY can satisfy more than one requirement. For example, the TTY required for a bank of phones could also satisfy the requirement for the TTY required for the building. However, the requirement for a TTY on an exterior site cannot be met by installing a TTY inside the building due to the phones in the building typically not being available 24 hours a day.

E106.4.5 Rest stops, emergency road stops, and service plazas. Where a public pay telephone is provided at a public rest stop, emergency road stop or service plaza, at least one public TTY shall be provided.

TABLE E106.2
WHEELCHAIR-ACCESSIBLE TELEPHONES

NUMBER OF TELEPHONES PROVIDED ON A FLOOR, LEVEL OR EXTERIOR SITE	MINIMUM REQUIRED NUMBER OF WHEELCHAIR-ACCESSIBLE TELEPHONES
1 or more single unit	1 per floor, level and exterior site
1 bank	1 per floor, level and exterior site
2 or more banks	1 per bank

❖ A TTY is required at these types of facilities regardless of the number of phones provided. The possibility of emergencies or a need for assistance at these locations increases the need for someone to be able to make a telephone call.

E106.4.6 Hospitals. Where a public pay telephone is provided in or adjacent to a hospital emergency room, hospital recovery room or hospital waiting room, at least one public TTY shall be provided at each such location.

❖ In hospitals, when phones are located for public use in an emergency room, recovery room or waiting room, a TTY is required at each location. The possibility of emergencies or a need for assistance at these locations increases the need for someone to be able to make a telephone call.

E106.4.7 Transportation facilities. Transportation facilities shall be provided with TTYs in accordance with Sections E109.2.5 and E110.2 in addition to the TTYs required by Sections E106.4.1 through E106.4.4.

❖ Transportation facilities, such as bus stations, train stations and airports, have specific requirements as stated in Sections E109.2.5 and E110.2. Additionally, they must comply with the requirements based on the number of telephones in a bank, on a floor, within a building or on a site.

E106.4.8 Detention and correctional facilities. In detention and correctional facilities, where a public pay telephone is provided in a secured area used only by detainees or inmates and security personnel, then at least one TTY shall be provided in at least one secured area.

❖ Specific criteria is provided for the detainee side of detention and correctional facilities. Phones available on the public side should comply with Sections E106.4.1 through E106.4.4.

E106.4.9 Signs. Public TTYs shall be identified by the International Symbol of TTY complying with ICC A117.1. Directional signs indicating the location of the nearest public TTY shall be provided at banks of public pay telephones not containing a public TTY. Additionally, where signs provide direction to public pay telephones, they shall also provide direction to public TTYs. Such signs shall comply with ICC A117.1 and shall include the International Symbol of TTY.

❖ Signage must be provided indicating the location of TTYs. Technical requirements for signage can be found in Section 703.6 of ICC A117.1.

E106.5 Shelves for portable TTYs. Where a bank of telephones in the interior of a building consists of three or more public pay telephones, at least one public pay telephone at the bank shall be provided with a shelf and an electrical outlet in accordance with ICC A117.1.

Exceptions:

1. In secured areas of detention and correctional facilities, if shelves and outlets are prohibited for purposes of security or safety shelves and outlets for TTYs are not required to be provided.

2. The shelf and electrical outlet shall not be required at a bank of telephones with a TTY.

❖ If a bank of telephones has three or more public telephones, a shelf and electrical outlet must be provided to allow a person with a hearing impairment to utilize his or her own portable TTY. In a bank of four or more phones a public TTY is required (see Section E106.4.1); therefore, a shelf and outlet for a portable TTY are not required within the same bank. Due to security concerns, there may be situations in jails where the shelf and outlet for a portable TTY is not required. Technical requirements for the shelf can be found in Section 704.6 of ICC A117.1.

SECTION E107
SIGNAGE

E107.1 Signs. Required accessible portable toilets and bathing facilities shall be identified by the International Symbol of Accessibility.

❖ Accessible portable toilets must be identified (see Section E105.2).

E107.2 Designations. Interior and exterior signs identifying permanent rooms and spaces shall be tactile. Where pictograms are provided as designations of interior rooms and spaces, the pictograms shall have tactile text descriptors. Signs required to provide tactile characters and pictograms shall comply with ICC A117.1.

Exceptions:

1. Exterior signs that are not located at the door to the space they serve are not required to comply.

2. Building directories, menus, seat and row designations in assembly areas, occupant names, building addresses and company names and logos are not required to comply.

3. Signs in parking facilities are not required to comply.

4. Temporary (seven days or less) signs are not required to comply.

5. In detention and correctional facilities, signs not located in public areas are not required to comply.

❖ This section applies to signs that provide designations, labels or names for interior rooms or spaces where the sign is not likely to change over time. Examples include room numbers in hospitals or hotels, restrooms, room names in conference centers, etc. Where permanent signage is provided, tactile signage consisting of both raised letters and Braille must also be provided. When pictograms are utilized to identify a space, such as the women's restroom, they must also have tactile signage. Pictograms that provide information about a space, such as "no smoking" or the International Symbol of Accessibility, are not required to have text descriptors. For signage requirements, reference Section 703 of ICC A117.1. Exceptions are provided for types of signs where it may not be effective or practical to provide tactile signage. Examples would be the remote sign at the street indicating the name or occupant of the building, the menu above the take-out counter, temporary signage or building directories.

E107.3 Directional and informational signs. Signs that provide direction to, or information about, permanent interior spaces of the site and facilities shall contain visual characters complying with ICC A117.1.

> **Exception:** Building directories, personnel names, company or occupant names and logos, menus and temporary (seven days or less) signs are not required to comply with ICC A117.1.

❖ Directional information must meet requirements for font style, location, finish and contrast in order for signage to be readable by persons with limited vision. The characters should contrast as much as possible with their background. Additional factors affecting the readability of signage include lighting, surface glare and text uniformity. When making color choices, remember that a percentage of the population is color blind; therefore, some color combinations may not have adequate contrast for these individuals (for example, red and blue). Color blindness is typically a genetic condition, and it is much more common in men than in women. Approximately one in 12 men has at least some color perception problems. Less common are acquired deficiencies that stem from injury, disease or the aging process. Also, although not called "color blindness," when people age, their corneas typically turn yellowish, severely hampering their ability to see violet and blue colors. The exception is for signage that may be changeable, such as building directories, personnel names, etc. For requirements for this signage, reference Section 703.2 of ICC A117.1.

E107.4 Other signs. Signage indicating special accessibility provisions shall be provided as follows:

1. At bus stops and terminals, signage must be provided in accordance with Section E108.4.

2. At fixed facilities and stations, signage must be provided in accordance with Sections E109.2.2 through E109.2.2.3.

3. At airports, terminal information systems must be provided in accordance with Section E110.3.

❖ This section is a reference to additional signage requirements found in subsequent sections. This was done in an effort to include all signage requirements in one section.

SECTION E108
BUS STOPS

E108.1 General. Bus stops shall comply with Sections E108.2 through E108.5.

❖ Bus stops and terminals provide access to public transportation and are considered an essential ser-

vice; therefore, access in these facilities is of special concern. Technical requirements for transportation facilities are found in Section 805 of ICC A117.1.

E108.2 Bus boarding and alighting areas. Bus boarding and alighting areas shall comply with Sections E108.2.1 through E108.2.4.

❖ "Kneeling" buses (those with a lift) deploy a ramp to allow access by persons with physical disabilities. When new bus stop pads are constructed to allow use of these buses, the requirements in Sections E108.2.1 through E108.2.4 are applicable. At bus stops where a shelter is provided, the bus stop pad can be located either within or outside the shelter.

E108.2.1 Surface. Bus boarding and alighting areas shall have a firm, stable surface.

❖ To allow proper deployment of the ramp and approach by a person in a wheelchair, a firm, stable surface must be used for the pad. Typically, this will be some form of concrete or asphalt.

E108.2.2 Dimensions. Bus boarding and alighting areas shall have a clear length of 96 inches (2440 mm) minimum, measured perpendicular to the curb or vehicle roadway edge, and a clear width of 60 inches (1525 mm) minimum, measured parallel to the vehicle roadway.

❖ To allow maneuvering for both the ramp and wheelchair user, a bus stop pad should be a minimum of 96 inches (2438 mm) deep and 60 inches (1524 mm) wide.

E108.2.3 Connection. Bus boarding and alighting areas shall be connected to streets, sidewalks or pedestrian paths by an accessible route complying with Section 1104.

❖ All bus stop pads must have an accessible route leading from public access points, such as the sidewalk, to the bus stop pad. A pad allowing for kneeling buses is not beneficial if the pad itself is not accessible.

E108.2.4 Slope. Parallel to the roadway, the slope of the bus boarding and alighting area shall be the same as the roadway, to the maximum extent practicable. For water drainage, a maximum slope of 1:48 perpendicular to the roadway is allowed.

❖ The bus stop pad should be as flat as possible to allow a person in a wheelchair to wait at this location comfortably. A maximum slope of one unit vertical in 48 units horizontal (1:48) is allowed for drainage of rainwater to the curb and gutter drainage system provided along the edge of the road.

E108.3 Bus shelters. Where provided, new or replaced bus shelters shall provide a minimum clear floor or ground space complying with ICC A117.1, Section 305, entirely within the shelter. Such shelters shall be connected by an accessible route to the boarding area required by Section E108.2.

❖ A wheelchair parking space [30 inches by 48 inches (762 mm by 1219 mm)] must be provided within the bus shelter. If the shelter is small enough to meet the alcove requirements in Section 305.7 of ICC A117.1, additional space is required for maneuvering. In addition, an accessible route is required from the bus shelter to the bus stop (see Section E108.2).

E108.4 Signs. New bus route identification signs shall have finish and contrast complying with ICC A117.1. Additionally, to the maximum extent practicable, new bus route identification signs shall provide visual characters complying with ICC A117.1.

> **Exception:** Bus schedules, timetables and maps that are posted at the bus stop or bus bay are not required to meet this requirement.

❖ The bus routes must be identified with visual characters that have light/dark contrast and a nonglare surface so they are easily readable. As much as practicable, the signage must also meet the requirements in Section 703.2 of ICC A117.1, for character form and spacing, line spacing and mounting height. There is an exception for schedules, timetables and maps due to the amount of information provided.

E108.5 Bus stop siting. Bus stop sites shall be chosen such that, to the maximum extent practicable, the areas where lifts or ramps are to be deployed comply with Sections E108.2 and E108.3.

❖ Whenever possible, bus stop sites should be chosen based on efforts to comply with the previous requirements for bus stop pads and bus shelters. Site constraints, such as a steeply inclined street, may not always make this practical.

SECTION E109
TRANSPORTATION FACILITIES AND STATIONS

E109.1 General. Fixed transportation facilities and stations shall comply with the applicable provisions of Section E109.2.

❖ Fixed transportation facilities and stations, such as train or light rail stations, provide access to public transportation and are considered an essential service; therefore, access in these facilities is of special concern. Technical requirements for transportation facilities can be found in Section 805 of ICC A117.1.

E109.2 New construction. New stations in rapid rail, light rail, commuter rail, intercity rail, high speed rail and other fixed guideway systems shall comply with Sections E109.2.1 through E109.2.8.

❖ All new construction involved with fixed guideway systems, including stations and terminals, is required to comply with this section. Criteria for existing buildings is listed in Section E109.3.

E109.2.1 Station entrances. Where different entrances to a station serve different transportation fixed routes or groups of fixed routes, at least one entrance serving each group or route shall comply with Section 1104 and ICC A117.1.

❖ A fixed guideway system, such as a commuter train system, may have separate platforms or entrances for a group of routes. When this is the case, at least one accessible route leading from public access points (e.g., public sidewalk, public parking, etc.) to the tran-

sit platform utilized for each group must be provided. When transit systems are underground, an accessible route must provide access to each subway station. As much as possible, the accessible entrance should be the same as the main entrance to the facility.

E109.2.2 Signs. Signage in fixed transportation facilities and stations shall comply with Sections E109.2.2.1 through E109.2.2.3.

❖ Specific requirements are listed for tactile, identification and informational signs.

E109.2.2.1 Tactile signs. Where signs are provided at entrances to stations identifying the station or the entrance, or both, at least one sign at each entrance shall be tactile. A minimum of one tactile sign identifying the specific station shall be provided on each platform or boarding area. Such signs shall be placed in uniform locations at entrances and on platforms or boarding areas within the transit system to the maximum extent practicable. Tactile signs shall comply with ICC A117.1.

> **Exceptions:**
> 1. Where the station has no defined entrance but signs are provided, the tactile signs shall be placed in a central location.
> 2. Signs are not required to be tactile where audible signs are remotely transmitted to hand-held receivers, or are user or proximity actuated.

❖ When identification signs are provided at a facility entrance, a tactile sign for the visually impaired must also be provided. A tactile sign must also be provided at each platform or boarding area. The location of tactile signage should be consistent throughout the rail system. Tactile signage must include both raised letters and Braille and meet the requirements of Section 703 of ICC A117.1. Tactile signage is not required when information is provided by alternative audible means. The purpose of Exception 2 is to allow for emerging technologies such as an audible sign systems using infrared transmitters and receivers. These new systems may provide greater accessibility in the transit environment than traditional Braille and raised letter signs. The transmitters are placed on or next to print signs and transmit their information to an infrared receiver that is held by a person. By scanning an area, the person will hear the sign. This means that signs can be placed well out of reach of Braille readers, even on parapet walls and on walls beyond barriers. Additionally, such signs can be used to provide wayfinding information that cannot be efficiently conveyed on Braille signs.

E109.2.2.2 Identification signs. Stations covered by this section shall have identification signs containing visual characters complying with ICC A117.1. Signs shall be clearly visible and within the sightlines of a standing or sitting passenger from within the train on both sides when not obstructed by another train.

❖ When identification signs are provided so that persons on the trains can identify the station, the signs must meet the requirements of Section 703.2 of ICC A117.1

for visual characters including font style, finish and contrast.

E109.2.2.3 Informational signs. Lists of stations, routes and destinations served by the station which are located on boarding areas, platforms or mezzanines shall provide visual characters complying with ICC A117.1 Signs covered by this provision shall, to the maximum extent practicable, be placed in uniform locations within the transit system.

❖ Signs indicating stations, routes and destinations served by the train system accessed from that platform must comply with Section 703.2 of ICC A117.1 for visual characters including font style, finish and contrast. The location of such signage should be consistent throughout the system. Route maps are not required to comply with the informational sign requirements.

E109.2.3 Fare machines. Self-service fare vending, collection and adjustment machines shall comply with ICC A117.1, Section 707. Where self-service fare vending, collection or adjustment machines are provided for the use of the general public, at least one accessible machine of each type provided shall be provided at each accessible point of entry and exit.

❖ Self-service fare machines must be accessible in accordance with Section 707 of ICC A117.1. The requirements are the same as for ATM machines. If self-service machines are provided, at least one accessible machine must be provided along the accessible route between the entrance and the platform.

E109.2.4 Rail-to-platform height. Station platforms shall be positioned to coordinate with vehicles in accordance with the applicable provisions of 36 CFR, Part 1192. Low-level platforms shall be 8 inches (250 mm) minimum above top of rail.

Exception: Where vehicles are boarded from sidewalks or street level, low-level platforms shall be permitted to be less than 8 inches (250 mm).

❖ The height and position of a platform must be coordinated with the floor of the vehicles it serves to minimize the vertical and horizontal gaps, in accordance with the *ADA Accessibility Guidelines for Transportation Vehicles* (36 CFR Part 1192). The vehicle guidelines, categorized by bus, van, light rail, rapid rail, commuter rail and intercity rail are available at www.access-board.gov. The preferred alignment is a high platform, level with the vehicle floor. In some cases, the vehicle guidelines permit use of a low platform in conjunction with a lift or ramp. Most such low platforms must have a minimum height of 8 inches (203 mm) above the top of the rail. Some vehicles are designed to be boarded from a street or the sidewalk along the street and the exception permits such boarding areas to be less than 8 inches (203 mm) high.

E109.2.5 TTYs. Where a public pay telephone is provided in a transit facility (as defined by the Department of Transportation) at least one public TTY complying with ICC A117.1, Section 704.4, shall be provided in the station. In addition, where one or more public pay telephones serve a particular entrance

to a transportation facility, at least one TTY telephone complying with ICC A117.1, Section 704.4, shall be provided to serve that entrance.

❖ Due to the essential nature of the facility, at least one public telephone in a transit facility is required to have a permanent TTY in accordance with Section 704.4 of ICC A117.1. When a bank of four or more is provided at an entrance, at least one is required to have a permanent TTY in accordance with Section 704.4 of ICC A117.1 (see also commentary, Section E106.4.7).

E109.2.6 Track crossings. Where a circulation path serving boarding platforms crosses tracks, an accessible route complying with ICC A117.1 shall be provided.

Exception: Openings for wheel flanges shall be permitted to be $2^{1}/_{2}$ inches (64 mm) maximum.

❖ If people utilizing the train station must cross tracks to access the platform, it is necessary to provide an accessible route that takes into account the changes in elevation and gaps associated with such a route. Front wheels of wheelchairs are small, and if a person in a wheelchair got stuck while crossing the tracks, the consequences could be severe. If it is not possible to meet the criteria specified, an alternative accessible route must be provided. This requirement is not a requirement for track crossings remote from the boarding platform.

E109.2.7 Public address systems. Where public address systems convey audible information to the public, the same or equivalent information shall be provided in a visual format.

❖ If a public address system is utilized within a facility to provide information to the passengers, there must be an alternative means for providing the same information to persons with a hearing impairment.

E109.2.8 Clocks. Where clocks are provided for use by the general public, the clock face shall be uncluttered so that its elements are clearly visible. Hands, numerals and digits shall contrast with the background either light-on-dark or dark-on-light. Where clocks are mounted overhead, numerals and digits shall comply with ICC A117.1, Section 703.2.

❖ Clocks provided in transportation facilities must be readily distinguishable. When clocks are provided overhead, the numerals must comply with Section 703.2 of ICC A117.1 for visual characters.

SECTION E110
AIRPORTS

E110.1 New construction. New construction of airports shall comply with Sections E110.2 through E110.4.

❖ Airports provide access to public transportation and are considered an essential service; therefore, access in these facilities is of special concern. Technical requirements for transportation facilities can be found in Section 805 of ICC A117.1.

E110.2 TTYs. Where public pay telephones are provided, at least one TTY shall be provided in compliance with ICC A117.1, Section 704.4. Additionally, if four or more public pay telephones are located in a main terminal outside the security areas, a concourse within the security areas or a baggage claim area in a terminal, at least one public TTY complying with ICC A117.1, Section 704.4, shall also be provided in each such location.

❖ Due to the essential nature of the facility, at least one public telephone in an airport facility is required to have a permanent TTY in accordance with Section 704.4 of ICC A117.1. Given the restricted areas in an airport, when a bank of four or more phones is provided at either side of security checkpoints or in a baggage claim area, at least one phone in each bank is required to have a permanent TTY in accordance with Section 704.4 of ICC A117.1 (see also commentary, Section E106.4.7).

E110.3 Terminal information systems. Where terminal information systems convey audible information to the public, the same or equivalent information shall be provided in a visual format.

❖ If a public address system is utilized within a facility to provide information to passengers, there must be an alternative means for providing the same information to persons with a hearing impairment.

E110.4 Clocks. Where clocks are provided for use by the general public, the clock face shall be uncluttered so that its elements are clearly visible. Hands, numerals and digits shall contrast with their background either light-on-dark or dark-on-light. Where clocks are mounted overhead, numerals and digits shall comply with ICC A117.1, Section 703.2.

❖ Clocks provided in transportation facilities must be readily distinguishable. When clocks are provided overhead, the numerals must comply with Section 703.2 of ICC A117.1 for visual characters.

SECTION E111
REFERENCED STANDARDS

DOJ 36 CFR Part 1192	Americans with Disabilities Act (ADA) Accessibility Guidelines for Transportation Vehicles (ADAAG). Washington, D.C.: Department of Justice, 1991	E109.2.4
ICC/ANSI A117.1-03	Accessible and Usable Buildings and Facilities	E101.2, E104.2, E104.2.1, E104.3, E104.3.4, E105.1, E105.2.1, E105.2.2, E105.3, E105.4 E105.6, E106.2 E106.3, E106.4, E106.4.9, E106.5, E107.2, E107.3, E108.3, E108.4, E109.2.1, E109.2.2.1, E109.2.2.2, E109.2.2.3, E109.2.3

❖ This section lists the standards that are referenced in Appendix E. They are listed here and not in Chapter 35 of the code, since appendices are not considered part of the code unless specifically referenced in the adopting ordinance.

Bibliography

The following resource materials are referenced in this chapter or are relevant to the subject matter addressed in this chapter.

"Accessibility and Egress for People with Physical Disabilities." CABO Board for the Coordination of the Model Codes Report, October 5, 1993.

Architectural and Transportation Barriers Compliance Board, 36 CFR Parts 1190 and 1191 Final Rule, *The Americans with Disabilities Act (ADA) Accessibility Guidelines); Architectural Barriers Act (ABA) Accessibility Guidelines.* Washington, D.C., July 23, 2004.

DOJ 28 CFR, Part 36-91, *Americans with Disabilities Act (ADA).* Washington, D.C.: Department of Justice, 1991.

DOJ 28 CFR, Part 36-91 (Appendix A), *ADA Accessibility Guidelines for Buildings and Facilities* (ADAAG). Washington, D.C.: Department of Justice, 1991.

DOJ 36 CFR Part 1192/DOT 49 CFR Part 38, *ADA Accessibility Guidelines for Transportation Vehicles* (ADAAG). Washington, D.C.: Department of Justice, 1991.

DOT 49 CFR Part 37, *Transportation Services for Individuals with Disabilities* (ADA). Washington, D.C.: Department of Transportation, 1999.

FED-STD-795-88, *Uniform Federal Accessibility Standards.* Washington, D.C.: General Services Administration; Department of Defense; Department of Housing and Urban Development; U.S. Postal Service, 1988.

ICC A117.1-03, *Accessible and Usable Buildings and Facilities.* Falls Church, VA: International Code Council, 2003.

Appendix F:
Rodentproofing

The provisions contained in this appendix are not mandatory unless specifically referenced in the adopting ordinance.

General Comments

The requirements of this appendix apply only where it has been specifically adopted by the governing jurisdiction, as stated in Section 101.2.1.

The provisions of this appendix are minimum mechanical methods to prevent the entry of rodents. If a food supply is available, a rodent problem cannot be eliminated solely by these methods. However, removal of the food supply and general cleanliness and maintenance of the premises will have a significant impact.

Purpose

The purpose of this appendix is to provide regulations for rodentproofing for health purposes.

SECTION F101
GENERAL

F101.1 General. Buildings or structures and the walls enclosing habitable or occupiable rooms and spaces in which persons live, sleep or work, or in which feed, food or foodstuffs are stored, prepared, processed, served or sold, shall be constructed in accordance with the provisions of this section.

❖ In an effort to protect the public health and welfare and to prevent the spread of disease, this section requires all rooms and spaces used for living, sleeping or working and all spaces used for the storage, preparation or serving of food to be protected from rodents in the manner described in the code. Note that a building used for the storage of materials other than foodstuffs is not required to comply with the requirements of this section.

F101.2 Foundation wall ventilation openings. Foundation wall ventilator openings shall be covered for their height and width with perforated sheet metal plates no less than 0.070 inch (1.8 mm) thick, expanded sheet metal plates not less than 0.047 inch (1.2 mm) thick, cast iron grills or grating, extruded aluminum load-bearing vents or with hardware cloth of 0.035 inch (0.89 mm) wire or heavier. The openings therein shall not exceed $^1/_4$ inch (6.4 mm).

❖ The purpose of this section is to protect ventilation openings in the foundation walls from the entry of rodents. The thickness of the metal shields and the maximum openings that are specified in this section are necessary for the shields to be effective.

F101.3 Foundation and exterior wall sealing. Annular spaces around pipes, electric cables, conduits, or other openings in the walls shall be protected against the passage of rodents by closing such openings with cement mortar, concrete masonry or noncorrosive metal.

❖ During construction, the openings in the foundation and exterior walls that are made for pipes, electrical cables, conduits and the like are significantly larger than the component that passes through the wall. These openings around the component are required to be closed for rodentproofing.

F101.4 Doors. Doors on which metal protection has been applied shall be hinged so as to be free swinging. When closed, the maximum clearance between any door, door jambs and sills shall not be greater than $^3/_8$ inch (9.5 mm).

❖ This section addresses wood doors located at ground or basement floor levels. Solid sheet metal protection is needed to stop rodents from gnawing their way through the wood into areas where food is located. The limitation on the clearance between the door and the adjacent door jamb and sill is to deter rodent entry.

F101.5 Windows and other openings. Windows and other openings for the purpose of light or ventilation located in exterior walls within 2 feet (610 mm) above the existing ground level immediately below such opening shall be covered for their entire height and width, including frame, with hardware cloth of at least 0.035 inch (0.89 mm) wire or heavier.

❖ The requirement for wire screens or "hardware cloth" for rodent protection applies only to openings 2 feet (610 mm) or less from the ground. This is a minimum requirement and will not keep all rodents out of the building.

F101.5.1 Rodent-accessible openings. Windows and other openings for the purpose of light and ventilation in the exterior walls not covered in this chapter, accessible to rodents by way of exposed pipes, wires, conduits and other appurtenances, shall be covered with wire cloth of at least 0.035 inch (0.89 mm) wire. In lieu of wire cloth covering, said pipes, wires, conduits and other appurtenances shall be blocked from rodent usage by installing solid sheet metal guards 0.024 inch (0.61 mm) thick or heavier. Guards shall be fitted around pipes, wires, conduits or other appurtenances. In addition, they shall be fastened securely to and shall extend perpendicularly from the exterior wall for a minimum distance of 12 inches (305 mm) beyond and on either side of pipes, wires, conduits or appurtenances.

❖ This section addresses screen requirements for windows or other openings that are higher than 2 feet (610 mm) from grade and are made rodent accessible by pipes, wires, conduits and the like. The window or other opening is required to be screened or rodent guards are to be installed on the pipes, wires or conduits so that the window or other opening is not accessible to rodents.

F101.6 Pier and wood construction.

❖ The requirements of Sections F101.6.1 and F101.6.2 apply only if the building does not have a continuous foundation wall of either treated wood, concrete or coated metal. An example of such a building is a post and beam wood-frame building with a slab on grade. Also note that these requirements apply if the building contains habitable or occupiable rooms in which persons live, sleep or work, or in which foodstuffs are stored, prepared, processed, served or sold. Thus, a storage or utility building used for storage of materials other than foodstuffs is not required to meet this section, even though it does not have a continuous foundation wall.

F101.6.1 Sill less than 12 inches above ground. Buildings not provided with a continuous foundation shall be provided with protection against rodents at grade by providing either an apron in accordance with Section F101.6.1.1 or a floor slab in accordance with Section 101.6.1.2.

❖ See the commentary to Section F101.6 for an explanation of when the requirements of this section apply. This section cites two alternative methods that meet the requirements: an apron or a concrete grade floor. Either of these methods provides effective rodent protection.

F101.6.1.1 Apron. Where an apron is provided, the apron shall not be less than 8 inches (203 mm) above, nor less than 24 inches (610 mm) below, grade. The apron shall not terminate below the lower edge of the siding material. The apron shall be constructed of an approved nondecayable, water-resistant rodentproofing material of required strength and shall be installed around the entire perimeter of the building. Where constructed of masonry or concrete materials, the apron shall not be less than 4 inches (102 mm) in thickness.

❖ The apron is usually a concrete, masonry or treated wood wall. Each material meets the requirement of this section. The extension into the ground is to prevent rodents from burrowing underground into the building. Note that a concrete floor slab is not required where the apron is used for rodent protection in accordance with Section F101.6.2.

F101.6.1.2 Grade floors. Where continuous concrete grade floor slabs are provided, open spaces shall not be left between the slab and walls, and openings in the slab shall be protected.

❖ This is the second option for rodent protection in Section 101.6.1. When a grade floor that meets the requirements of this section is installed, the apron described in Section F101.6.1.2 is not required.

The grade floor is required to have any openings protected from rodents. For example, if the floor slab has an opening where drain pipes go through it, the area around the pipe is to be sealed with material that will keep rodents out of the building.

F101.6.2 Sill at or above 12 inches above ground. Buildings not provided with a continuous foundation and which have sills 12 or more inches (305 mm) above the ground level shall be provide with protection against rodents at grade in accordance with any of the following:

1. Section F101.6.1.1 or F101.6.1.2;

2. By installing solid sheet metal collars at least 0.024 inch (0.6 mm) thick at the top of each pier or pile and around each pipe, cable, conduit, wire or other item which provides a continuous pathway from the ground to the floor; or

3. By encasing the pipes, cables, conduits or wires in an enclosure constructed in accordance with Section F101.6.1.1.

❖ See the commentary to Section F101.6 for an explanation of when the requirements of this section apply. Where the sill height is 12 inches (305 mm) or more above the ground, this section provides two additional acceptable methods for rodent protection versus methods for instances where the sill height is less than 12 inches (305 mm).

Methods 2 and 3 described in this section provide the rodent barrier directly around the pipes, cables, conduits or wires that would otherwise be a continuous pathway from the ground to the raised floor.

Appendix G:
Flood-Resistant Construction

The provisions contained in this appendix are not mandatory unless specifically referenced in the adopting ordinance.

General Comments

The provisions contained in this appendix are not mandatory unless specifically referenced in the adopting ordinance, as stated in Section 101.2.1. This appendix is intended to fulfill the flood-plain management and administrative requirements of the National Flood Insurance Program (NFIP) that are not included in the code. Communities that adopt the code and this appendix without modification, will meet the minimum requirements of NFIP as set forth in Title 44 of the Code of Federal Regulations (CFR). This appendix includes administrative requirements of NFIP and requirements concerning modifications to watercourses, permits for flood hazard area development, conditions for the issuance of variances from flood-plain management requirements and site improvements, subdivision planning and installation of manufactured homes, recreation vehicles and tanks in flood hazard areas. It is important to note that many states and communities regulate flood-plain development to higher standards than the minimum requirements of the code and this appendix. Prior to adopting the code or this appendix, communities are advised to consult their state NFIP coordinator or Federal Emergency Management Agency (FEMA) regional office to determine additional actions that may be necessary to provide for continued participation in NFIP.

Purpose

The purpose of this appendix is to provide optional criteria for flood-resistant construction. A jurisdiction that wants to make this appendix mandatory needs to include it in its adoption ordinance. See page v of the code for a sample ordinance for adoption.

SECTION G101
ADMINISTRATION

G101.1 Purpose. The purpose of this appendix is to promote the public health, safety and general welfare and to minimize public and private losses due to flood conditions in specific flood hazard areas through the establishment of comprehensive regulations for management of flood hazard areas designed to:

1. Prevent unnecessary disruption of commerce, access and public service during times of flooding;

2. Manage the alteration of natural flood plains, stream channels and shorelines;

3. Manage filling, grading, dredging and other development which may increase flood damage or erosion potential;

4. Prevent or regulate the construction of flood barriers which will divert floodwaters or which can increase flood hazards; and

5. Contribute to improved construction techniques in the flood plain.

❖ This section sets forth the purposes of this appendix. Communities that administer flood-plain management regulations such as in this appendix achieve multiple objectives beyond reducing physical damage to buildings.

G101.2 Objectives. The objectives of this appendix are to protect human life, minimize the expenditure of public money for flood control projects, minimize the need for rescue and relief efforts associated with flooding, minimize prolonged business interruption, minimize damage to public facilities and utilities, help maintain a stable tax base by providing for the sound use and development of flood-prone areas, contribute to improved construction techniques in the flood plain and ensure that potential owners and occupants are notified that property is within flood hazard areas.

❖ This appendix, combined with the provisions in the code, achieve several objectives intended to reduce the impacts of flooding. Management of flood hazard areas in ways that reduce exposure to damage also protects health and safety. Additionally, the provisions of this appendix promote sustainable development in communities subject to flooding. Flooding has both short-term and long-term impacts. Some impacts are obvious, such as damaged homes and businesses. Others are less apparent, including increases in flood levels due to changes in the flood plain; scour and erosion; impaired public and private water and sewage systems and reductions in the natural and beneficial functions of flood plains, including wetland areas.

G101.3 Scope. The provisions of this appendix shall apply to all proposed development in a flood hazard area established in Section 1612 of this code, including certain building work exempt from permit under Section 105.2.

❖ In communities that participate in NFIP, its minimum requirements must be adopted and applied to all development in flood hazard areas. The definition of "Development," as stated in Section G201.2, is inclusive. This broad definition captures all activities that take place in flood hazard areas that are subject to the requirements of the code or this appendix. Because NFIP requires regulation of all development within special flood hazard areas that could have an impact on flooding, a code that applies only to buildings or structures does not fulfill the minimum requirements for participation in the program. Work that is exempt from the requirement to obtain a permit, listed in Section 105.2, is subject to the requirements of this appendix in order to satisfy the NFIP definition of "Development," described in the commentary for Sections G102.1 and G201.2.

G101.4 Violations. Any violation of a provision of this appendix, or failure to comply with a permit or variance issued pursuant to this appendix or any requirement of this appendix, shall be handled in accordance with Section 113.

❖ Violations of the provisions of this appendix are to be corrected or adjudicated under the provisions of Section 113. Communities participating in NFIP enter into an agreement with FEMA: in exchange for the availability of federally backed flood insurance and many forms of federal disaster assistance, communities agree to adopt and effectively enforce the flood-plain management and administrative provisions of NFIP. A significant part of community responsibility takes the form of enforcing flood-plain management and other flood damage reduction ordinances and laws. The procedures set forth in Section 113 are to be followed if activities take place in a flood hazard area without a permit, or if a permit applicant fails to comply with a condition of a permit or variance. If reasonable enforcement actions do not result in compliance, the community is encouraged to contact its state NFIP coordinator or appropriate FEMA regional office for further guidance.

SECTION G102
APPLICABILITY

G102.1 General. This appendix, in conjunction with the *International Building Code*, provides minimum requirements for

development located in flood hazard areas, including the subdivision of land; installation of utilities; placement and replacement of manufactured homes; new construction and repair, reconstruction, rehabilitation or additions to new construction; substantial improvement of existing buildings and structures, including restoration after damage; and certain building work exempt from permit under Section 105.2.

❖ The definition of "Development" in Section G201.2 encompasses a wide array of activities that are to be regulated if they are proposed to take place within a designated flood hazard area. Some activities have the potential to increase flooding on other properties. Other activities, if planned in a way that recognizes flood hazards, have the potential to allow reasonable use of land without placing people and site improvements at significant risk.

G102.2 Establishment of flood hazard areas. Flood hazard areas are established in Section 1612.3 of the *International Building Code*, adopted by the governing body on [INSERT DATE].

❖ The flood hazard areas established in Section 1612.3 are subject to the additional provisions of this appendix. The flood hazard map to be adopted is, at a minimum, the flood insurance rate map (FIRM), supported by the flood insurance study prepared by FEMA. A community participating in NFIP may specify another map, provided it designates flood hazard areas that are the same or more extensive and regulatory flood elevations that are the same or higher than those shown on the FIRM. Regardless of which map is specified, flood elevations are the design flood elevations. From time to time, FEMA revises and republishes FIRMs. A formal review process is followed, including community and public comment. When new FIRMs are published, communities are required to use them. The flood hazard areas shown on maps prepared by FEMA are determined using the base flood, which is defined as having a 1-percent chance (one chance in 100) of occurring in any given year. The maps are not intended to show the worst-case flood, or the "flood of record," which usually refers to the most severe flood in the history of the community. Although a 1-percent chance seems fairly remote, larger floods occur regularly throughout the United States. Application of the provisions of this appendix is not intended to prevent or eliminate all future flood damage. These provisions are intended to represent a reasonable balance of the knowledge and awareness of flood hazards, methods to guide development to less hazard-prone locations, methods of design and construction intended to resist flood damage and each community's and landowner's reasonable expectations to use the land. Some coastal communities have FIRMs that show areas that are designated as units of the Coastal Barrier Resource System (CBRS) established by the Coastal Barrier Resource Act (CoBRA) of 1982 and subsequent amendments. Within these designated areas, NFIP is prohibited from offering flood insurance on new or substantially improved buildings. The community is responsible for application of the flood-resistant provisions of the code and this appendix in all designated flood hazard areas, whether or not flood insurance is made available.

SECTION G103
POWERS AND DUTIES

G103.1 Permit applications. The building official shall review all permit applications to determine whether proposed development sites will be reasonably safe from flooding. If a proposed development site is in a flood hazard area, all site development activities (including grading, filling, utility installation and drainage modification), all new construction and substantial improvements (including the placement of prefabricated buildings and manufactured homes) and certain building work exempt from permit under Section 105.2 shall be designed and constructed with methods, practices and materials that minimize flood damage and that are in accordance with this code and ASCE 24.

❖ The building official is empowered to examine all permit applications to determine if the proposed activities will take place in designated flood hazard areas on the community's flood hazard map. Such activities are to be designed and constructed in accordance with both the code and this appendix, as applicable. For additional guidance for development activities that involve the placement of fill in a flood hazard area, refer to FEMA FIA-TB #10, *Ensuring that Structures Built on Fill In or Near Special Flood Hazard Areas are Reasonably Safe from Flooding.*

G103.2 Other permits. It shall be the responsibility of the building official to assure that approval of a proposed development shall not be given until proof that necessary permits have been granted by federal or state agencies having jurisdiction over such development.

❖ Other federal, state and local regulatory authorities may have jurisdiction over activities within designated flood hazard areas. Examples at the federal level include permitting under Section 404 of the Clean Water Act of 1977 and Section 10 of the Rivers and Harbors Act of 1899 and consultation or permitting under the Endangered Species Act of 1973. State and regional agencies may also regulate activities in flood hazard areas, including activities that impact wetlands, forestry resources, dunes, the coastal zone, subaquatic vegetation, threatened and endangered species, navigation and waterways. The intent is to provide coordination among various levels of government and to avoid added costs to applicants in situations where multiple permits are required. Building officials may satisfy this requirement by either withholding the permit until other permits have been obtained, or by issu-

ing the permit contingent on the applicant obtaining other specified permits.

G103.3 Determination of design flood elevations. If design flood elevations are not specified, the building official is authorized to require the applicant to:

1. Obtain, review and reasonably utilize data available from a federal, state or other source, or

2. Determine the design flood elevation in accordance with accepted hydrologic and hydraulic engineering techniques. Such analyses shall be performed and sealed by a registered design professional. Studies, analyses and computations shall be submitted in sufficient detail to allow review and approval by the building official. The accuracy of data submitted for such determination shall be the responsibility of the applicant.

❖ Many FIRMs show flood hazard areas without specifying the base flood elevations (BFE). These areas, often referred to as "unnumbered A or V Zones," are subject to the flood-plain management requirements of the code and the appendix, although the minimum height to which buildings and structures are to be elevated has not been determined by FEMA. An important step in regulating these flood hazard areas is determination of the design flood elevation (DFE). As defined in the code, at a minimum the DFE is the BFE shown on a community's flood insurance rate map. In some instances, flood elevation information may have been developed by sources other than FEMA, including other federal or state agencies. The building official is to obtain the information or require the permit applicant to do so. Some communities develop flood hazard information and provide it to applicants so that all development in an area is conducted based on the same level of risk. If flood elevation information is not available, the building official may require the applicant to develop the DFE in accordance with accepted engineering practices. Local officials unfamiliar with establishing DFEs in unnumbered A and V Zones are encouraged to contact the state NFIP coordinator or appropriate FEMA regional office. For additional guidance, refer to FEMA 265, *Managing Floodplain Development in Approximate Zone A Areas: A Guide for Obtaining and Developing Base (100-Year) Flood Elevations.*

G103.4 Activities in riverine flood hazard areas. In riverine flood hazard areas where design flood elevations are specified but floodways have not been designated, the building official shall not permit any new construction, substantial improvement or other development, including fill, unless the applicant demonstrates that the cumulative effect of the proposed development, when combined with all other existing and anticipated flood hazard area encroachment, will not increase the design flood elevation more than 1 foot (305 mm) at any point within the community.

❖ Although FEMA has provided floodways along many rivers and streams shown on a community's FIRM, many other riverine flood hazard areas have BFEs but not designated floodways. In these areas the potential effects that flood-plain activities may have on flood elevations have not been evaluated. If FEMA has not designated a regulatory floodway on a community's FIRM, the community is responsible for regulating development so as not to increase flood elevations by more than 1 foot (305 mm) at any point in the community. In effect, this means a community must either prepare a hydraulic analysis for proposed activities or require permit applicants to do so. Several states have more restrictive requirements, which can be determined by contacting the state NFIP coordinator.

G103.5 Floodway encroachment. Prior to issuing a permit for any floodway encroachment, including fill, new construction, substantial improvements and other development or land-disturbing activity, the building official shall require submission of a certification, along with supporting technical data, that demonstrates that such development will not cause any increase of the level of the base flood.

❖ The floodway is that part of the riverine flood plain that must be reserved in order to convey the base flood without cumulatively increasing the water surface elevation more than a designated height specified in the flood insurance study provided by FEMA. Generally, this designated height is 1 foot (305 mm), although some states and communities have a more restrictive requirement. Development in the floodway could obstruct flood flows and increase flood levels, causing additional damage. Usually, the floodway is where the water will be deepest and move the fastest. As required in Section G103.5, communities are to prohibit any floodway encroachment, including fill, new construction and substantial improvement, if such activities will cause any increase in flood levels. Permit applications are to include a certification supported with engineering analyses to demonstrate the anticipated impacts. To be acceptable, the certification must demonstrate that the proposed development will not impact BFEs, floodway elevations or floodway widths and is to be signed and sealed by a registered professional engineer. The building official is to review the certification and determine that it meets accepted engineering practices. If necessary to make this determination, the building official is advised to seek technical assistance from the community's engineer, the state NFIP coordinator or the FEMA regional office. In limited situations, a community may decide to permit development in the floodway that causes an increase in flood elevations. This may be appropriate for dams or other water resource structures and for bridges and culverts where the most cost-effective alternative results in an increase in flood elevations. Section 103.5 allows for these activities, provided the community or applicant requests and obtains a conditional letter of map revision (CLOMR) and a floodway revision from FEMA. The proposed development may not increase flooding of existing buildings and all impacted property owners must be notified. A permit for the development may be issued after the CLOMR is approved by FEMA

and after the community adopts the increased flood elevations. When the project is completed, the community is required to submit as-built certifications to FEMA so that a final map revision can be issued. For guidance on preparing and certifying revisions to NFIP map products, see FEMA Form 81-89 Series, *Revisions to National Flood Insurance Rate Maps: Application/Certification Forms and Instructions for Conditional Letters of Map Revision, Letters of Map Revision and Physical Map Revisions.* Although each floodway proposal is to be reviewed carefully and some states have specific and more restrictive regulations that apply within floodways, there are some activities and uses that may be permitted without extensive engineering analysis. The most important factor is that the fill and grading do not change the shape of the land. Possible uses include:

- Agricultural uses not involving buildings or structures;
- At-grade uses such as parking, loading areas and small airport landing strips;
- Passive recreational uses, such as hiking, biking, horse trails, wildlife and nature preserves and hunting and fishing areas;
- Active recreational uses (where improvements are anchored to prevent flotation), such as picnic and playground areas, ball fields, boat launch ramps, swimming areas, target shooting ranges and similar uses; and
- Uses incidental to residential buildings, such as lawns, gardens, parking areas, playgrounds and tot lots.

G103.5.1 Floodway revisions. A floodway encroachment that increases the level of the base flood is authorized if the applicant has applied for a conditional Flood Insurance Rate Map (FIRM) revision and has received the approval of the Federal Emergency Management Agency (FEMA).

❖ Under certain circumstances, a permit may be issued for a proposed floodway encroachment that, based on analysis and supporting technical data, will cause an increase in the level of the base flood. The community is to require that the analysis be submitted to FEMA in accordance with the guidance described in Section 103.5 (FEMA Form 81-89). The permit may be issued if FEMA approves the proposal by issuance of a conditional revision to the flood hazard area map. Prior to issuance of a final FIRM revision, FEMA will require that the applicant submit post-construction documentation to demonstrate that the work was conducted as proposed.

G103.6 Watercourse alteration. Prior to issuing a permit for any alteration or relocation of any watercourse, the building official shall require the applicant to provide notification of the proposal to the appropriate authorities of all affected adjacent government jurisdictions, as well as appropriate state agencies. A copy of the notification shall be maintained in the permit records and submitted to FEMA.

❖ Relocation or alteration of the channel of a river or stream can significantly and adversely impact the flood plain. Because such impact may extend a significant distance from a proposed relocation or alteration, NFIP requires notification of adjacent communities, appropriate state agencies and FEMA.

G103.6.1 Engineering analysis. The building official shall require submission of an engineering analysis which demonstrates that the flood-carrying capacity of the altered or relocated portion of the watercourse will not be decreased. Such watercourses shall be maintained in a manner which preserves the channel's flood-carrying capacity.

❖ Prior to issuing a permit that allows modification of a waterway, the applicant is to submit a hydraulic analysis demonstrating that the ability of the altered channel and flood plain to carry flood discharges is not diminished. An important condition of a permit, if issued, is that the applicant or other designated entity is to maintain the altered channel in order to preserve its flood-carrying capacity and to avoid increasing future flood risks.

G103.7 Alterations in coastal areas. Prior to issuing a permit for any alteration of sand dunes and mangrove stands in flood hazard areas subject to high velocity wave action, the building official shall require submission of an engineering analysis which demonstrates that the proposed alteration will not increase the potential for flood damage.

❖ As part of the review of proposed site work in coastal areas, a community is to examine proposals for possible impacts to sand dunes or mangrove stands, which mitigate the effects of coastal flooding. If a proposal is to alter a sand dune or mangrove stand, an engineering analysis of the effect on flood damages is required as part of the documentation. If engineering analysis indicates that the potential for flood damage will be increased due to the proposed alteration, the building official is to require the permit applicant to make necessary modifications to the proposed site work to avoid such increases.

G103.8 Records. The building official shall maintain a permanent record of all permits issued in flood hazard areas, including copies of inspection reports and certifications required in Section 1612.

❖ The building official's responsibility to maintain department records is established in Section 104.7. This section extends that responsibility to require that records for permits in flood hazard areas be maintained permanently. Specifically identified for retention are the inspection reports and certifications required to be submitted in Section 1612. As a condition for participation in the NFIP, a community agrees that FEMA or its designee may examine records relating to administration of the community's flood management regulations.

SECTION G104
PERMITS

G104.1 Required. Any person, owner or authorized agent who intends to conduct any development in a flood hazard area shall first make application to the building official and shall obtain the required permit.

❖ The definition of "Development" in Section G201 encompasses a wide array of human activities that may be proposed within designated flood hazard areas. Communities are to require that all such activities are authorized by permits issued in compliance with the applicable flood-plain management requirements of the code and this appendix.

G104.2 Application for permit. The applicant shall file an application in writing on a form furnished by the building official. Such application shall:

1. Identify and describe the development to be covered by the permit.

2. Describe the land on which the proposed development is to be conducted by legal description, street address or similar description that will readily identify and definitely locate the site.

3. Include a site plan showing the delineation of flood hazard areas, floodway boundaries, flood zones, design flood elevations, ground elevations, proposed fill and excavation and drainage patterns and facilities.

4. Indicate the use and occupancy for which the proposed development is intended.

5. Be accompanied by construction documents, grading and filling plans and other information deemed appropriate by the building official.

6. State the valuation of the proposed work.

7. Be signed by the applicant or the applicant's authorized agent.

❖ This section details the information to be shown on or included in an application for a permit to conduct activities within a flood hazard area. The site plan is to show sufficient detail and information about the designated flood hazard area, including floodway, to allow for a complete review of the proposed activities. The horizontal boundary of the designated flood hazard area shown on the community's flood hazard map is an approximate boundary based on topographic information that was available when the map was prepared by FEMA. The determination of the actual boundary of the designated flood hazard is to be based on the best topographic information that is available at the time of the application. All land area below the DFE is subject to regulation even if it is not shown on the community's flood hazard map as being within the designated flood hazard area. Along with necessary flood hazard information, the storm drainage conveyance requirements are to be shown on the site plan.

G104.3 Validity of permit. The issuance of a permit under this appendix shall not be construed to be a permit for, or approval of, any violation of this appendix or any other ordinance of the jurisdiction. The issuance of a permit based on submitted documents and information shall not prevent the building official from requiring the correction of errors. The building official is authorized to prevent occupancy or use of a structure or site which is in violation of this appendix or other ordinances of this jurisdiction.

❖ This section clarifies that it is the owner's responsibility to comply with the code. Permits are approved based on a review of the materials submitted and a determination that the proposed activities are in accordance with the provisions of the code. If errors or deviations are made during construction, or it is discovered that the information and documentation submitted to obtain the permit were in error, the code official has the authority to require correction and compliance.

G104.4 Expiration. A permit shall become invalid if the proposed development is not commenced within 180 days after its issuance, or if the work authorized is suspended or abandoned for a period of 180 days after the work commences. Extensions shall be requested in writing and justifiable cause demonstrated. The building official is authorized to grant, in writing, one or more extensions of time, for periods not more than 180 days each.

❖ Permits issued for activities in a flood hazard area are to expire after 180 days if work either has not commenced during that period of time or is abandoned for any 180-day period after it commences. The building official may grant extensions to the permit for up to 180 days at a time. Section 1612.2 defines "Start of construction" as "the date of permit issuance for new construction and substantial improvements to existing structures, provided the actual start of construction, repair, reconstruction, rehabilitation, addition placement or other improvement is within 180 days after the date of issuance. The actual start of construction means the first placement of permanent construction of a building (including a manufactured home) on a site, such as the pouring of a slab or footings, installation of pilings or construction of columns."

Building officials may not grant permit extensions in situations where construction has not begun during the 180-day period if FEMA has issued a revised FIRM that shows increases in flood hazard areas or DFEs at the location of the permitted activity. A new permit is required if the revised FIRM changes the flood-risk information at the site of the permit. The new application is to be reviewed based on the revised flood-risk information.

G104.5 Suspension or revocation. The building official is authorized to suspend or revoke a permit issued under this appendix wherever the permit is issued in error or on the basis of incorrect, inaccurate or incomplete information, or in violation of any ordinance or code of this jurisdiction.

❖ A permit issued in error or based on incorrect, inaccurate or incomplete information is to be suspended or revoked. This provision does not preclude issuance of a new or revised permit that is based on accurate information. Violation of a permit or permit condition may also prompt suspension or revocation.

SECTION G105
VARIANCES

G105.1 General. The board of appeals established pursuant to Section 112 shall hear and decide requests for variances. The board of appeals shall base its determination on technical justifications, and has the right to attach such conditions to variances as it deems necessary to further the purposes and objectives of this appendix and Section 1612.

❖ Section 112 empowers the board of appeals to hear and decide requests for variances. Variances from the flood plain management requirements of the code and this appendix may place people and property at significant risk. Therefore, communities are cautioned to carefully evaluate the impacts of issuing a variance from the flood-resistant construction provisions of the code and this appendix, particularly the requirements to elevate or floodproof buildings to the DFE. The impacts to be evaluated include impacts on the site, the permit applicant and other parties that may be affected, such as adjacent property owners and the community. Flood-plain development that is not undertaken in accordance with the flood-resistant construction provisions of the code and this appendix will be exposed to increased flood damages. As a consequence, flood insurance premium rates will be significantly higher. Variance decisions made by the board of appeals are to be based solely on technical justifications outlined in this section, not on the personal circumstances of an owner or applicant.

Applicants sometimes request variances to the minimum elevation requirements for the lowest floor of buildings in flood hazard areas. Such requests may be based on the need to improve access for the disabled and elderly. Generally, variances of this nature are not to be granted since these are personal circumstances that will change as the property changes ownership. Not only would persons of limited mobility be at risk from flooding, but the building would continue to be exposed to flood damage long after the personal need for a variance changes. More appropriate alternatives are to be considered to serve the needs of disabled or elderly persons, such as varying setbacks to allow construction on less flood-prone portions of sites or installing personal elevators.

G105.2 Records. The building official shall maintain a permanent record of all variance actions, including justification for their issuance.

❖ The permanent records of the community that document official actions on permit applications and variance requests concerning flood-plain management requirements are to include the justifications on which variance determinations are based. These records will be examined periodically by FEMA or its designee in an effort to determine if the community is effectively administering its flood management provisions consistent with the minimum requirements of NFIP.

G105.3 Historic structures. A variance is authorized to be issued for the repair or rehabilitation of a historic structure upon a determination that the proposed repair or rehabilitation will not preclude the structure's continued designation as a historic structure, and the variance is the minimum necessary to preserve the historic character and design of the structure.

Exception: Within flood hazard areas, historic structures that are not:

1. Listed or preliminarily determined to be eligible for listing in the National Register of Historic Places; or

2. Determined by the Secretary of the U.S. Department of Interior as contributing to the historical significance of a registered historic district or a district preliminarily determined to qualify as an historic district; or

3. Designated as historic under a state or local historic preservation program that is approved by the Department of Interior.

❖ This provision recognizes the importance and value of historic buildings and structures. If an owner proposes repair or restoration of a historic building or structure that would otherwise be found to be a substantial improvement, the board of appeals may grant a variance to perform the work without requiring full compliance with the flood-resistant provisions of the code and this appendix. This is not to say that work towards flood-resistant provisions should not be performed. It is intended that the historic building should be brought into compliance as much as possible without adversely affecting the historic significance. To qualify for this treatment, the building or structure must be individually listed or eligible to be listed as a historic structure, or be certified as contributing to the historic significance of a historic district. Merely being located in a historic district is insufficient justification for a variance.

A variance may be issued, provided the proposed work does not change the historic designation of the building or structure. The board of appeals or the building official may require that the permit applicant obtain a written review and determination to that effect from the appropriate state or local organization responsible for the historic designation.

All variances are to be the minimum necessary to afford relief [see Section G105.7(4)]. There are a number of actions that may be taken to reduce future flood damage to substantially damaged or substantially improved historic buildings or structures, while maintaining their historic and cultural value. For example, if elevation of a historic building or structure would negatively affect its character or cause a loss of its his-

toric designation, a variance from the elevation requirement may be appropriate. However, it may still be possible to require that building utility systems be elevated or otherwise protected from flood damage to the maximum extent possible. A condition of a variance may require that building materials, including interior finishes, be resistant to flood damage (refer to FEMA FIA-TB #2, *Flood Resistant Materials Requirements for Buildings Located in Special Flood Hazard Areas*). Another condition may require that building contents be elevated to the DFE, or the maximum extent possible. For further information, contact the state NFIP coordinator, state historic preservation officer or appropriate FEMA regional office.

G105.4 Functionally dependent facilities. A variance is authorized to be issued for the construction or substantial improvement of a functionally dependent facility provided the criteria in Section 1612.1 are met and the variance is the minimum necessary to allow the construction or substantial improvement, and that all due consideration has been given to methods and materials that minimize flood damages during the design flood and create no additional threats to public safety.

❖ Functionally dependent facilities include those that must be located in close proximity to water to fulfill their intended purpose, such as shipbuilding, ship or boat repair, as well as docks and port facilities for loading or unloading cargo or passengers. Some facilities that often are located close to water are specifically excluded from the definition, including long-term storage and manufacturing, sales or service facilities that are associated with water activities. Functionally dependent facilities represent a special case recognized by NFIP as warranting treatment under the variance provisions. Under some circumstances it may impossible for a facility to perform its function unless one or more of the flood-resistant construction provisions of the code or this appendix are varied. For example, temporary storage facilities may need to be at dock level or a building in a designated flood hazard area subject to high-velocity wave action may need to be elevated on armored fill rather than on piles or columns.

At a minimum, functionally dependent facilities are to be wet floodproofed. This type of flood protection reduces flood damage while allowing floodwaters to enter the building or structure. The keys to successful wet floodproofing are to design the building or structure, to the maximum extent possible, to resist flood and wind loads anticipated during conditions of the design flood; to use flood damage-resistant materials and to elevate or use damage-resistant building utility systems as required by the flood-resistant construction provisions of the code and this appendix. For further guidance, refer to FEMA FIA-TB #1, *Openings in Foundation Walls for Buildings Located in Special Flood Hazard Areas*; FEMA FIA-TB #2, *Flood Resistant Material Requirements for Buildings Located in Special Flood Hazard Areas* and FEMA FIA-TB #7, *Wet Floodproofing Requirements for Structures Located in Special Flood Hazard Areas*.

Within functionally dependent facilities, only those functions that must be carried out below the DFE are to be included within the scope of the variance. For example, a boat repair facility may have areas that are used for a variety of purposes. The actual boat repair activities may need to be conducted at an elevation below the DFE. Other areas that are not functionally dependent, such as employee locker rooms, bathrooms and offices, must either be elevated or dry floodproofed (A Zones only) to or above the DFE as required by the code.

G105.5 Restrictions. The board of appeals shall not issue a variance for any proposed development in a floodway if any increase in flood levels would result during the base flood discharge.

❖ The board of appeals may not issue variances to the requirements in Section G103.5, which addresses activities in floodways that may increase flood elevations. The commentary to Section G103.5 describes submission of engineering analyses and approval by FEMA prior to approving activities that increase flood elevations in designated floodways.

G105.6 Considerations. In reviewing applications for variances, the board of appeals shall consider all technical evaluations, all relevant factors, all other portions of this appendix and the following:

1. The danger that materials and debris may be swept onto other lands resulting in further injury or damage;

2. The danger to life and property due to flooding or erosion damage;

3. The susceptibility of the proposed development, including contents, to flood damage and the effect of such damage on current and future owners;

4. The importance of the services provided by the proposed development to the community;

5. The availability of alternate locations for the proposed development that are not subject to flooding or erosion;

6. The compatibility of the proposed development with existing and anticipated development;

7. The relationship of the proposed development to the comprehensive plan and flood plain management program for that area;

8. The safety of access to the property in times of flood for ordinary and emergency vehicles;

9. The expected heights, velocity, duration, rate of rise and debris and sediment transport of the floodwaters and the effects of wave action, if applicable, expected at the site; and

10. The costs of providing governmental services during and after flood conditions including maintenance and repair of public utilities and facilities such as sewer, gas, electrical and water systems, streets and bridges.

❖ This section provides factors to be taken into account by the board of appeals when it reviews requests for variances from the flood-resistant construction and flood-plain management provisions of the code and this appendix. Section G105.2 requires that documentation be maintained as part of the permanent record to indicate how the board of appeals addressed each consideration.

G105.7 Conditions for issuance. Variances shall only be issued by the board of appeals upon:

1. A technical showing of good and sufficient cause that the unique characteristics of the size, configuration or topography of the site renders the elevation standards inappropriate;

2. A determination that failure to grant the variance would result in exceptional hardship by rendering the lot undevelopable;

3. A determination that the granting of a variance will not result in increased flood heights, additional threats to public safety, extraordinary public expense, nor create nuisances, cause fraud on or victimization of the public or conflict with existing local laws or ordinances;

4. A determination that the variance is the minimum necessary, considering the flood hazard, to afford relief; and

5. Notification to the applicant in writing over the signature of the building official that the issuance of a variance to construct a structure below the base flood level will result in increased premium rates for flood insurance up to amounts as high as $25 for $100 of insurance coverage, and that such construction below the base flood level increases risks to life and property.

❖ Each of the listed conditions for issuance of variances is to be addressed by the board of appeals, especially the requirement that it determine that failure to grant the variance would result in exceptional hardship by rendering the lot undevelopable. By itself, this determination may be insufficient to result in exceptional hardship if other conditions for issuance of a variance cannot be met. The determination of hardship is to be based on the unique characteristics of the site and not the personal circumstances of the applicant.

In guidance materials, FEMA cautions that economic hardship alone is not to be considered an exceptional hardship. Building officials and boards of appeals are cautioned that granting a variance does not affect how the building will be rated for the purposes of NFIP flood insurance. Even if circumstances justify granting a variance to build a floor that is below the design flood elevation, the rate used to calculate the cost of a flood insurance policy will be based on the risk to the building. Flood insurance, required by certain mortgage lenders, may be extremely expensive, costing as much as $25 for each $100 of insurance coverage. Although the applicant may not be required to purchase flood insurance, the requirement may be imposed on subsequent owners. The building official is to provide the applicant a written notice to this effect, along with the other conditions listed in this section.

SECTION G201
DEFINITIONS

G201.1 General. The following words and terms shall, for the purposes of this appendix, have the meanings shown herein. Refer to Chapter 2 for general definitions.

❖ Definitions of terms can help in the understanding and application of the code requirements. Terms specific to this appendix are defined herein. The use and application of the general definitions of the code are found in Section 202.

G201.2 Definitions.

DEVELOPMENT. Any man-made change to improved or unimproved real estate, including but not limited to, buildings or other structures, temporary or permanent storage of materials, mining, dredging, filling, grading, paving, excavations, operations and other land disturbing activities.

❖ Any activity within a flood hazard area has the potential to damage and thereby alter the characteristics of the area. Some alterations may increase flood frequency or flood levels that in turn may adversely impact adjacent property owners. This term is broadly defined so that communities examine and regulate all activities for potential flood hazard area impacts. Land development activities that are not buildings and structures are also subject to the provisions of this appendix.

FUNCTIONALLY DEPENDENT FACILITY. A facility which cannot be used for its intended purpose unless it is located or carried out in close proximity to water, such as a docking or port facility necessary for the loading or unloading of cargo or passengers, shipbuilding or ship repair. The term does not include long-term storage, manufacture, sales or service facilities.

❖ By the very nature of their use, certain buildings and structures would not be able to fulfill their intended functions if located away from bodies of water. In addition, their functions may be hampered if they are elevated in full compliance with the code. Very specific facilities are defined as functionally dependent facilities and other facilities that often are located near bodies of water are specifically excluded from the definition.

MANUFACTURED HOME. A structure that is transportable in one or more sections, built on a permanent chassis, designed for use with or without a permanent foundation when attached to the required utilities, and constructed to the Federal Mobile Home Construction and Safety Standards and rules and regulations promulgated by the U.S. Department of Housing and Urban Development. The term also includes mobile homes, park trailers, travel trailers and similar transportable structures that are placed on a site for 180 consecutive days or longer.

❖ Manufactured homes are defined separately from buildings and approvals for them are administered differently by some local jurisdictions. In flood hazard areas, manufactured homes that are not installed according to the provisions of the code will be more vulnerable to flood damage.

MANUFACTURED HOME PARK OR SUBDIVISION. A parcel (or contiguous parcels) of land divided into two or more manufactured home lots for rent or sale.

❖ Typically, manufactured home parks and subdivisions are developed by a single owner and, in most cases, the utilities, pads and foundations are installed before the lots are rented or sold. These development activities are included in this appendix (see Sections G301.1 and G301.2) so that manufactured homes that are placed or replaced will comply with the requirements for flood hazard areas.

RECREATIONAL VEHICLE. A vehicle that is built on a single chassis, 400 square feet (37.16 m²) or less when measured at the largest horizontal projection, designed to be self-propelled or permanently towable by a light-duty truck, and designed primarily not for use as a permanent dwelling but as temporary living quarters for recreational, camping, travel or seasonal use. A recreational vehicle is ready for highway use if it is on its wheels or jacking system, is attached to the site only by quick disconnect-type utilities and security devices and has no permanently attached additions.

❖ Recreational vehicles are intended to be used temporarily and, therefore, typically are not placed on permanent foundations that are designed and constructed to resist flood damage. Due to the nature of their placement and susceptibility to movement, this appendix (see Section G601.1) provides that they not be placed in flood hazard areas subject to high-velocity wave action (V Zones) and floodways.

VARIANCE. A grant of relief from the requirements of this section which permits construction in a manner otherwise prohibited by this section where specific enforcement would result in unnecessary hardship.

❖ Within flood hazard areas, a variance, especially to build below the design flood elevation, exposes people and property to higher risks and increased potential for flood damage. For the purpose of handling requests for variances to the flood-resistant design and construction provisions of the code, specific restrictions, considerations and conditions are set forth in this appendix (see Section G105). Note that repair or rehabilitation of historic structures, as well as construction or substantial improvements of functionally dependent facilities, may be handled by variance.

VIOLATION. A development that is not fully compliant with this appendix or Section 1612, as applicable.

❖ Work that is approved by issuance of a permit is to be conducted in accordance with the code and conditions of the permit. If the actual work is not in compliance, then it is in violation. Violations are to be handled in accordance with the procedures set forth in the code (see Section 113) and this appendix (see Sections G101.4, G104.3 and G104.5).

SECTION G301
SUBDIVISIONS

G301.1 General. Any subdivision proposal, including proposals for manufactured home parks and subdivisions, or other proposed new development in a flood hazard area shall be reviewed to assure that:

1. All such proposals are consistent with the need to minimize flood damage;

2. All public utilities and facilities, such as sewer, gas, electric and water systems are located and constructed to minimize or eliminate flood damage; and

3. Adequate drainage is provided to reduce exposure to flood hazards.

❖ When land is subdivided, the opportunity arises to recognize flood hazard areas and design the lot layout to reduce future flood damage. It is during this process that many communities elect to guide development to less hazardous locations. At a minimum, public utilities are to be designed and located to avoid or minimize impairment and damage. For additional guidance, refer to APA PAS #473.

G301.2 Subdivision requirements. The following requirements shall apply in the case of any proposed subdivision, including proposals for manufactured home parks and subdivisions, any portion of which lies within a flood hazard area:

1. The flood hazard area, including floodways and areas subject to high velocity wave action, as appropriate, shall be delineated on tentative and final subdivision plats;

2. Design flood elevations shall be shown on tentative and final subdivision plats;

3. Residential building lots shall be provided with adequate buildable area outside the floodway; and

4. The design criteria for utilities and facilities set forth in this appendix and appropriate *International Codes* shall be met.

❖ To facilitate the review of subdivision proposals with respect to flood hazards, tentative and final subdivision plats are to include adequate information. Where the land to be subdivided includes areas that are not within the designated flood hazard area, lots may be laid out to avoid or minimize flood-plain impacts. Residential lots are to be laid out so that building footprints and fills or other obstructions are outside the floodway.

SECTION G401
SITE IMPROVEMENT

❖ The definition of "Development" in Section G201 includes site work necessary to place buildings or structures, as well as site improvements that do not involve buildings or structures. When buildings and structures cannot be located outside of the designated flood hazard area, they are to be elevated so that the lowest floor is at or above the DFE. The most common ways to elevate buildings or structures are on fill, on solid foundation walls surrounding crawl spaces or on posts, pilings or columns. Compacted fill, placed in accordance with the code, may be placed to raise a building pad above the DFE. Note that fill is not to be used to elevate buildings in flood hazard areas subject to high-velocity wave action (V Zones).

Before allowing fill in a designated flood hazard area, the building official is to determine that the fill will not increase flooding or cause drainage problems on neighboring properties. In areas with known drainage and flooding problems, the building official has the authority to require submission of an engineering analysis. Under Section G103.4, this analysis is required when a floodway has not been delineated and, under Sections G103.5 and G401.1, if fill will encroach into a designated floodway. Some states and communities require encroachment analyses for all large fill proposals. In areas that are already developed, increased flood levels resulting from flood hazard area activities may be considered threats to public safety and are to be mitigated to the maximum extent possible by the permit applicant.

Permit applicants may propose placing fill in designated flood hazard areas with the intent of later constructing buildings with excavated basements. The definition of "Basement" in Section 1612.2 is "the portion of a building having its floor subgrade (below ground level) on all sides." When excavated into fill, basements may be subject to damage, especially in designated flood hazard areas where waters remain high for more than a few hours. Fill materials can become saturated and provide inadequate support or water pressure can collapse below-grade walls. Basements below residential buildings are not to be constructed below the DFE, even if excavated into fill that is placed above the DFE. Basements beneath nonresidential buildings are similarly not to be built unless they are designed, constructed and certified to be dry floodproofed in accordance with flood-resistant construction provisions of the code and ASCE 24.

If elevated on individual fill pads in designated flood hazard areas, buildings will be surrounded by water during general conditions of flooding. To address emergency management concerns about evacuation and public safety, it may be appropriate to consult with local emergency personnel responsible for evacuations. This could be especially important when reviewing applications for large subdivisions or critical facilities such as hospitals and care facilities. Many states and communities have provisions that require uninterrupted access to all buildings during design flood conditions. These requirements may be administered by agencies responsible for permitting the construction of public rights-of-way.

G401.1 Development in floodways. Development or land disturbing activity shall not be authorized in the floodway unless it has been demonstrated through hydrologic and hydraulic analyses performed in accordance with standard engineering practice that the proposed encroachment will not result in any increase in the level of the base flood.

❖ See the commentary to Section G103.5 for more information.

G401.2 Flood hazard areas subject to high velocity wave action.

1. Development or land disturbing activity shall only be authorized landward of the reach of mean high tide.

2. The use of fill for structural support of buildings is prohibited.

❖ All new construction within flood hazard areas subject to high-velocity wave action (commonly called V Zones) is to be located landward of the reach of mean high tide. The use of fill to provide structural support is prohibited because areas subject to wave action are more likely to experience flood-related erosion, which may cause failure of foundations on fill. For additional guidance, refer to FEMA FIA-TB #5, *Free of Obstruction Requirements for Buildings Located in Coastal High Hazard Areas* and FEMA 55, *Coastal Construction Manual: Principles and Practices of Planning, Siting, Designing, Constructing, and Maintaining Residential Buildings in Coastal Areas.*

G401.3 Sewer facilities. All new or replaced sanitary sewer facilities, private sewage treatment plants (including all pumping stations and collector systems) and on-site waste disposal systems shall be designed in accordance with Chapter 8, ASCE 24, to minimize or eliminate infiltration of floodwaters into the facilities and discharge from the facilities into floodwaters, or impairment of the facilities and systems.

❖ New and replacement sewer facilities are to be designed in accordance with Chapter 7 of ASCE 24. Adhering to these requirements will minimize damage to such facilities, as well as the likelihood of contamination from such facilities.

G401.4 Water facilities. All new replacement water facilities shall be designed in accordance with the provisions of Chapter 8, ASCE 24, to minimize or eliminate infiltration of floodwaters into the systems.

❖ New and replacement water supply facilities are to be designed in accordance with Chapter 7 of ASCE 24. Adhering to these requirements will minimize damage to water supply facilities and infiltration of contamination, which may be carried by floodwaters.

G401.5 Storm drainage. Storm drainage shall be designed to convey the flow of surface waters to minimize or eliminate damage to persons or property.

❖ In addition to designing a site to minimize damage from flood conditions, storm runoff is to be addressed to minimize or eliminate related damage. Storm drainage conveyance patterns and facilities are to be shown on the site plan described in Section G104.2.

G401.6 Streets and sidewalks. Streets and sidewalks shall be designed to minimize potential for increasing or aggravating flood levels.

❖ The layout of streets and sidewalks may impact both storm runoff and flood flows. The design and construction of streets and roadways, including bridges and culverts, are to reduce or eliminate damage from both storm runoff and flooding. In addition to removing flood storage from designated flood hazard areas, the placement of fill associated with the construction of roads and sidewalks has the potential to alter the flow of both storm runoff and floodwater. Bridges and culverts are to be sized to minimize their effects on flooding. Refer to Section G401.1 if such activities are proposed within floodways.

SECTION G501
MANUFACTURED HOMES

❖ This section requires that all new and substantially improved manufactured homes are to be elevated such that their lowest floors are at or above the DFE. Replacement of an existing manufactured home is considered "new construction" and must comply with the provisions of this section. In effect, manufactured homes are to meet the same flood-resistant construction requirements as other buildings and structures. In part, this provision exceeds the minimum requirements of NFIP, which allows, in some limited situations in existing manufactured home parks and subdivisions, replacement manufactured homes to be located below the DFE. Manufactured homes are to be anchored to permanent foundations to resist flotation, collapse and lateral movement. Foundations are to be reinforced and elevated so that the floor of the manufactured home is at or above the DFE. Foundations are to meet all applicable requirements to resist wind, seismic and flood loads prescribed in the code.

Manufactured homes may be placed either on prepared sites in manufactured home parks or subdivisions or on individually owned parcels of land. Manufactured home parks have a single owner, and the pads, foundations and utility hookups are rented to the manufactured home owners. As in a typical subdivision, a manufactured home subdivision normally is developed by a single developer and the lots are sold to individuals for use as manufactured home sites.

For additional guidance on installation, foundations, anchoring and ties for manufactured homes in flood hazard areas, refer to FEMA 85, *Manufactured Homes in Flood Hazard Area.*

G501.1 Elevation. ll new and replacement manufactured homes to be placed or substantially improved in a flood hazard area shall be elevated such that the lowest floor of the manufactured home is elevated to or above the design flood elevation.

❖ The elevation requirements set forth in the Section 1612 and ASCE 24 apply to manufactured homes, thereby treating this type of construction the same as other buildings and structures. The certification of the lowest floor elevation, as required in Section 1612.5, also applies to manufactured homes that are placed in designated flood hazard areas.

G501.2 Foundations. All new and replacement manufactured homes, including substantial improvement of existing manufactured homes, shall be placed on a permanent, reinforced foundation that is designed in accordance with Section 1612.

❖ Flood loads experienced in most designated flood hazard areas include hydrostatic and hydrodynamic components. Impacts from floating debris and waves can impart additional loads. Additionally, in flood hazard areas subject to high-velocity wave action (V Zones), flood and wind loads acting simultaneously, as required in Section 1603, are to be addressed in foundation design. Permanent, reinforced foundations for manufactured homes are necessary to resist anticipated loads. A permanent foundation system for a manufactured home includes the following:

- A below-grade footing capable of providing resistance against overturning;
- Footing depth below the frost line;
- Reinforced piers, driven pilings, embedded posts or poured concrete or reinforced block foundation walls; and
- Anchors and connections (see Section G501.3).

G501.3 Anchoring. All new and replacement manufactured homes to be placed or substantially improved in a flood hazard area shall be installed using methods and practices which minimize flood damage. Manufactured homes shall be securely anchored to an adequately anchored foundation system to resist flotation, collapse and lateral movement. Methods of anchoring are authorized to include, but are not limited to, use of over-the-top or frame ties to ground anchors. This requirement is in addition to applicable state and local anchoring requirements for resisting wind forces.

❖ The manufactured home anchoring requirement has two elements:

1. Manufactured homes are to be anchored to a permanent, reinforced foundation to limit movement and resist uplift and overturning due to flood, wind and seismic forces; and

2. Manufactured homes are to have ties firmly attached to ground anchors to transfer loads to the ground. Manufacturer specifications for ground anchor depths typically do not include flood loads. Steel strapping, cable, chain or other approved material may be used for ties, which are to be fastened to ground anchors and drawn tight with turnbuckles or other adjustable tensioning devices. Manufacturer specifications for ties typically do not include flood loads. Ties may be over-the-top straps or frame ties, as appropriate to the manufactured home.

SECTION G601
RECREATIONAL VEHICLES

G601.1 Placement prohibited. The placement of recreational vehicles shall not be authorized in flood hazard areas subject to high velocity wave action and in floodways.

❖ By their very nature, recreational vehicles are not normally elevated or permanently attached to a founda-

tion and, therefore, may be very susceptible to flooding. This provision limits the permanent placement of recreational vehicles in designated flood hazard areas where velocities may be high and flood loads may be significant, such as floodways and flood hazard areas subject to high-velocity wave action.

G601.2 Temporary placement. Recreational vehicles in flood hazard areas shall be fully licensed and ready for highway use, and shall be placed on a site for less than 180 consecutive days.

❖ Unless otherwise prohibited under Section G601.1, recreational vehicles may be temporarily placed in designated flood hazard areas if they are fully licensed, remain highway ready and can be moved when flooding threatens. "Highway ready" means that a recreational vehicle is on its wheels or internal jacking system, is connected to the site only by quick disconnect-type utilities and security devices and has no permanently attached additions.

G601.3 Permanent placement. Recreational vehicles that are not fully licensed and ready for highway use, or that are to be placed on a site for more than 180 consecutive days, shall meet the requirements of Section G501 for manufactured homes.

❖ Recreational vehicles that do not meet the requirements of Section G601.2 are considered permanent buildings and are to be installed in accordance with the requirements for manufactured homes in Section G501.

SECTION G701
TANKS

❖ This section addresses requirements for tanks located in flood hazard areas. Physical damage as well as environment contamination and health risks result when above-ground and underground tanks are not adequately protected and restrained during conditions of flooding. Many hazardous materials, such as gas or oil, are lighter than water and, if released due to damage or unprotected vents and fill opening, contribute to contamination and risks to public health and safety. For additional guidance, reference FEMA 348 and ASCE 24.

G701.1 Underground tanks. Underground tanks in flood hazard areas shall be anchored to prevent flotation, collapse or lateral movement resulting from hydrostatic loads, including the effects of buoyancy, during conditions of the design flood.

❖ Most tanks will not be entirely filled at all times, and thus will contain at least some air that contributes to buoyancy. When the surrounding ground becomes saturated during conditions of flooding, underground tanks may be dislodged or shifted, causing damage and loss of product. Installation designs and methods are to account for buoyancy.

G701.2 Above-ground tanks. Above-ground tanks in flood hazard areas shall be elevated to or above the design flood elevation or shall be anchored or otherwise designed and constructed to prevent flotation, collapse or lateral movement resulting from hydrodynamic and hydrostatic loads, including the effects of buoyancy, during conditions of the design flood.

❖ Above-ground tanks are to be raised above the DFE, which may be accomplished by installation on platforms or other foundations that are designed to resist flood loads. Above-ground tanks that are not elevated are subject to flood loads imposed by moving water and buoyancy and are to be adequately anchored to resist flood damage. Tanks that are exposed to floodwaters may be impacted by floating debris or become dislodged, resulting in loss of product and contributing to contamination and becoming floating debris themselves.

G701.3 Tank inlets and vents. In flood hazard areas, tank inlets, fill openings, outlets and vents shall be:

1. At or above the design flood elevation or fitted with covers designed to prevent the inflow of floodwater or outflow of the contents of the tanks during conditions of the design flood.

2. Anchored to prevent lateral movement resulting from hydrodynamic and hydrostatic loads, including the effects of buoyancy, during conditions of the design flood.

❖ Tank inlets and vents are to be protected from the entry of floodwater either by raising them above the DEF or by protective covers or devices. In addition, inlets and vents, and the piping that serve them, are to be anchored and protected from flood damage.

SECTION G801
OTHER BUILDING WORK

❖ Certain activities that are specified in Section 105.2 are exempt from the requirement to obtain a permit, although they must not be conducted in a manner that violates the code. As explained in Section G101.3, this appendix addresses flood-resistant construction provisions that, in conjunction with provisions of the code that relate to flood-resistant design and construction, can be used to satisfy the requirements of the NFIP. Under the NFIP, all development is to meet the requirements for flood-resistant construction. In order to ensure that activities that are exempt from a permit meet specific flood-resistant provisions, those activities are specifically listed in this appendix.

G801.1 Detached accessory structures. Detached accessory structures shall be anchored to prevent flotation, collapse or lateral movement resulting from hydrostatic loads, including the effects of bouyancy, during conditions of the design flood. Fully enclosed accessory structures shall have flood openings to allow for the automatic entry and exit of flood waters.

❖ Detached accessory structures are not required to be elevated, provided they are constructed or installed to minimize damage to the structures themselves, and to avoid flotation that may adversely impact other structures or block the flow of floodwaters and drainage.

G801.2 Fences. Fences in floodways that may block the passage of floodwaters, such as stockade fences and wire mesh fences, shall meet the requirement of Section G103.5.

❖ Fences that are solid or that may trap floating leaves and debris alter the flow of water and may increase flooding on adjacent properties. Such fences in floodways are to be avoided or, if unavoidable, their impact on flooding and flood heights is to be examined as floodway encroachments

G801.3 Oil derricks. Oil derricks located in flood hazard areas shall be designed in conformance with the flood loads in Sections 1603.1.6 and 1612.

❖ As permanent structures, if located in flood hazard areas, oil derricks are to be designed to resist the flood loads. If located in floodways, analyses are required to determine the impact on flood levels. Attendant equipment and utilities are to meet the requirements in ASCE 24, a referenced standard in Section 1612 of the code.

G801.4 Retaining walls, sidewalks and driveways. Retaining walls, sidewalks and driveways shall meet the requirements of Section 1803.4.

❖ Retaining walls, sidewalks and driveways may involve grading and fill. When constructed in flood hazard areas they may be subject to saturation or the erosive action of moving water and waves. If placed in floodways, they may contribute to increased flood levels and flooding of adjacent properties. Section 1803.4 outlines requirements, including engineering analyses, which are applicable depending on the nature of the flood hazard area.

G801.5 Prefabricated swimming pools. Prefabricated swimming pools in floodways shall meet the requirements of Section G103.5.

❖ Prefabricated swimming pools are designed for ready placement in a variety of situations. However, if placed in a floodway, they would constitute an encroachment that could increase flood levels and flood damage on adjacent properties. The definition for "Swimming pool" is found in Section 3109.2.

SECTION G901
REFERENCED STANDARDS

ASCE 24–05	Flood Resistance Design and Construction	G103.1, G401.3, G401.4
HUD 24 CFR Part 3280 (1994)	Manufactured Home Construction and Safety Standards	G201
IBC-06	International Building Code	G102.2

Bibliography

The following resource materials are referenced in this appendix or are relevant to the subject matter addressed in this appendix. See the commentary to Chapter 16 for information on obtaining additional related publications from FEMA.

APA PAS #473, *Subdivision Design in Flood Hazard Areas.* Washington, D.C.: American Planning Association, 1997.

FEMA 44 CFR Parts 59-73, *National Flood Insurance Program* (NFIP). Washington, D.C.: Federal Emergency Management Agency.

FEMA 55, *Coastal Construction Manual: Principles and Practices of Planning, Siting, Designing, Constructing, and Maintaining Residential Buildings in Coastal Areas.* Washington, D.C.: Federal Emergency Management Agency, 2000.

FEMA 85, *Manufactured Homes in Flood Hazard Areas.* Washington, D.C.: Federal Emergency Management Agency, 1985.

FEMA 265, *Managing Floodplain Development in Approximate Zone A Areas: A Guide for Obtaining and Developing Base (100-Year) Flood Elevations.* Washington, D.C.: Federal Emergency Management Agency, 1995.

FEMA 348, *Protecting Building Utilities from Flood Damage: Principles and Practices for the Design and Construction of Flood Resistant Building Utility Systems.* Washington, D.C.: Federal Emergency Management Agency, 1999.

FEMA FIA-TB #1, *Openings in Foundation Walls for Buildings Located in Special Flood Hazard Areas. Washington, D.C.:* Federal Emergency Management Agency, 1993.

FEMA FIA-TB #2, *Flood Resistant Material Requirements for Buildings Located in Special Flood Hazard Areas.* Washington, D.C.: Federal Emergency Management Agency, 1993.

FEMA FIA-TB #3, *Non-Residential Floodproofing Requirements and Certification for Buildings Located in Special Flood Hazard Areas.* Washington, D.C.: Federal Emergency Management Agency, 1993.

FEMA FIA-TB #4, *Elevator Installation for Buildings Located in Special Flood Hazard Areas.* Washington, D.C.: Federal Emergency Management Agency, 1993.

FEMA FIA-TB #5, *Free of Obstruction Requirements for Buildings Located in Coastal High Hazard Areas.* Washington, D.C.: Federal Emergency Management Agency, 1993.

FEMA FIA-TB #6, *Below Grade Parking Requirements for Buildings Located in Special Flood Hazard Areas.* Washington, D.C.: Federal Emergency Management Agency, 1993.

FEMA FIA-TB #7, *Wet Floodproofing Requirements for Structures Located in Special Flood Hazard Areas.* Washington, D.C.: Federal Emergency Management Agency, 1993.

FEMA FIA-TB #8, *Corrosion Protection for Metal Connectors In Coastal Areas for Structures Located in Special Flood Hazard Areas.* Washington, D.C.: Federal Emergency Management Agency,1996.

FEMA FIA-TB #9, *Design and Construction Guidance for Breakaway Walls Below Elevated Coastal Buildings .* Washington, D.C.: Federal Emergency Management Agency,1999.

FEMA FIA-TB #10, *Ensuring that Structures Built on Fill In or Near Special Flood Hazard Areas are Reasonably Safe from Flooding.* Washington, D.C.: Federal Emergency Management Agency, 2001.

FEMA FIA-TB #11, *Crawlspace Construction for Buildings Located in Special Flood Hazard Areas.* Washington, D.C.: Federal Emergency Management Agency, 2001.

FEMA Form 81-89 Series, *Revisions to National Flood Insurance Rate Maps: Application/Certification Forms and Instructions for Conditional Letters of Map Revision, Letters of Map Revision and Physical Map Revisions.* Washington, D.C.: Federal Emergency Management Agency, 2002.

FEMA (various dates). *NFIP Technical Bulletin Series.* Washington, D.C.: National Flood Insurance Program. [Online]. Available: http://www.fema.gov/MIT/techbul.htm

Appendix H:
Signs

The provisions contained in this appendix are not mandatory unless specifically referenced in the adopting ordinance.

General Comments

The provisions contained in this appendix are not mandatory unless specifically referenced in the adopting ordinance, as stated in Section 101.2.1.

The regulations contained in this appendix were prompted by improper design, erection, placement, location and maintenance of outdoor signs. Improperly designed and installed signs pose a hazard to people and property. Improperly wired signs or improperly maintained electrical signs can cause fires. The criteria in this chapter are straightforward that, in turn, facilitates straightforward enforcement. Section H102 contains definitions for nine distinct types of signs.

Purpose

The purpose of this appendix is to locate in one place the provisions that regulate construction or protection requirements for outdoor signs. This appendix also provides the owner, designer, installer and maintainer of outdoor signs with minimum safe practices presented in performance language whenever possible, thus allowing the widest possible application.

SECTION H101
GENERAL

H101.1 General. A sign shall not be erected in a manner that would confuse or obstruct the view of or interfere with exit signs required by Chapter 10 or with official traffic signs, signals or devices. Signs and sign support structures, together with their supports, braces, guys and anchors, shall be kept in repair and in proper state of preservation. The display surfaces of signs shall be kept neatly painted or posted at all times.

❖ The provisions of this appendix are confined to that which is displayed in any manner out of doors for recognized advertising purposes. An orderly procedure is to be followed in erecting and maintaining outdoor signs. Outdoor signs must not block the view of signs regulated in Section 1011 that identify the location and path of travel to exits or street signs regulated by the jurisdiction. Signs must be properly maintained in accordance with this section, including corrosion prevention on all parts of the sign and its supports. Corrosion of a structural member results in loss of cross section, which could render that member incapable of handling the load of the sign or loads (snow, wind, etc.) on the sign. The building official has the authority to order the removal of a sign that is not maintained properly, as it can represent a hazard to the public and, therefore, can be determined to be unsafe (see Section 115).

H101.2 Signs exempt from permits. The following signs are exempt from the requirements to obtain a permit before erection:

1. Painted nonilluminated signs.

2. Temporary signs announcing the sale or rent of property.

3. Signs erected by transportation authorities.

4. Projecting signs not exceeding 2.5 square feet (0.23 m²).

5. The changing of moveable parts of an approved sign that is designed for such changes, or the repainting or repositioning of display matter shall not be deemed an alteration.

❖ The objective of this section is to describe when a permit is not required to regulate the erection, construction, alteration and maintenance of a new sign. Note that this section applies to new signs only, except for existing signs that are to be altered, and that the owner of an exempt sign is responsible for erecting and maintaining it in a safe manner (see Section H104.1 for a related issue).

A permit is not required for the erection of wall signs (commonly of limited area) without electrical service that, when not maintained, can be painted over or the surface repaired to its original state.

A sale or rental sign is temporary in nature, pertains to the premises on which it is placed and its installation is not likely to represent a hazard; for example, it is substantially fixed in place and without electrical service.

Regulation of street signs in the code would be redundant, since signs erected by the jurisdiction inevitably undergo one or more levels of regulatory review before installation. Directional (informational) signs in conjunction with transit lines, such as subways, trains, buses, ferries and airports, are exempt from permit requirements since it is generally not necessary for a jurisdiction to take out a permit for an activity it is performing.

Projecting signs that do not exceed an area of 2.5 square feet (0.23 m²) do not require a permit, as the size limitation allows access for maintenance. The sign will likely be constructed so as not to create a hazard, for example, lack of headroom for pedestrians, protrusions into a walking surface and electrical supply disrepair.

When an existing sign is altered (i.e., enlarged, relocated, etc.), the alterations must comply with the requirements for new signs. A permit is required for such alterations. The exceptions to the requirements for alterations apply to a sign that has been designed and previously authorized (through the permit process) to have changeable or movable parts. Repainting or reposting of display matter (lettering) is not considered an alteration. If a sign is enlarged or relocated without a permit and, in turn, without a field inspection, it could add excessive loads to a roof, interfere with the exhaust or intake ventilation facilities from a building or create a potential hazard to the public.

SECTION H102
DEFINITIONS

H102.1 General. Unless otherwise expressly stated, the following words and terms shall, for the purposes of this appendix, have the meanings shown herein. Refer to Chapter 2 of the *International Building Code* for general definitions.

❖ Definitions of terms that are associated with the content of this section are contained herein. These definitions can help in the understanding and application of the code requirements. The use and application of all defined terms are set forth in Section 202 of the code.

COMBINATION SIGN. A sign incorporating any combination of the features of pole, projecting and roof signs.

❖ The purpose of this definition is to identify which combination of features constitutes a combination sign. These signs are characterized by being simultaneously supported partly by one or more poles, partly by the wall, partly by the roof of a structure or any combination thereof.

DISPLAY SIGN. The area made available by the sign structure for the purpose of displaying the advertising message.

❖ A display sign constitutes the surface of a sign where the alphabetic or pictographic components of a message are arranged for display.

ELECTRIC SIGN. A sign containing electrical wiring, but not including signs illuminated by an exterior light source.

❖ This definition states that an electrical sign is any sign activated or illuminated by means of electrical energy. Electric signs are characterized by the use of artificial light projecting through, but not reflecting off, its surface(s).

GROUND SIGN. A billboard or similar type of sign which is supported by one or more uprights, poles or braces in or upon the ground other than a combination sign or pole sign, as defined by this code.

❖ This definition identifies those signs that are not part of other buildings or structures. The structural supports of these signs are such that the wind loading is directly transferred into its own foundation system rather than into other buildings or structures. Section H109.1 contains the specific material and height requirements of ground signs.

POLE SIGN. A sign wholly supported by a sign structure in the ground.

❖ Pole signs are those where the structural supports are constructed so that loads are directly transferred to the ground.

PORTABLE DISPLAY SURFACE. A display surface temporarily fixed to a standardized advertising structure which is regularly moved from structure to structure at periodic intervals.

❖ This definition identifies the surfaces available for displaying commercial or noncommercial advertising messages that are not permanently attached to a structure and are periodically placed at different locations by means of manual or remote input.

PROJECTING SIGN. A sign other than a wall sign, which projects from and is supported by a wall of a building or structure.

❖ A projecting sign is the opposite of a wall sign in that it is projecting from the wall that provides its support. Section H112 contains requirements for projecting signs to regulate their materials, maximum projections, clearances and structural capabilities.

ROOF SIGN. A sign erected upon or above a roof or parapet of a building or structure.

❖ This definition identifies that any exterior sign supported on the roof of a building or structure is considered a roof sign. Such a sign directly impacts the design of the roof because of increased wind and snow drifting loads. Section H110.1 contains explicit material and clearance requirements along with height limitations for the erection of roof signs.

SIGN. Any letter, figure, character, mark, plane, point, marquee sign, design, poster, pictorial, picture, stroke, stripe, line, trademark, reading matter or illuminated service, which shall be constructed, placed, attached, painted, erected, fastened or manufactured in any manner whatsoever, so that the same shall be used for the attraction of the public to any place, subject, person, firm, corporation, public performance, article, machine or merchandise, whatsoever, which is displayed in any manner outdoors. Every sign shall be classified and conform to the requirements of that classification as set forth in this chapter.

❖ This definition provides the basis for determining which building components are considered as signs, and thus must comply with the provisions of this appendix. For the purposes of the code, signs are always considered as exterior elements, since the code addresses the hazards associated with exterior exposure. Interior signs must meet the specific code requirements for the materials involved, along with the code provisions for interior finishes and trim (see Chapter 8). The code further subdivides signs into different groups, with specific requirements for each group.

SIGN STRUCTURE. Any structure which supports or is capable of supporting a sign as defined in this code. A sign structure is permitted to be a single pole and is not required to be an integral part of the building.

❖ Structures supporting signs, which may be affixed to the ground and include one or more columns, poles or braces placed in or upon the ground, are not required to be part of the building structural system.

WALL SIGN. Any sign attached to or erected against the wall of a building or structure, with the exposed face of the sign in a plane parallel to the plane of said wall.

❖ This definition details those signs that are part of the wall surfaces of a building or structure, or which are separate signs and are then fastened to the wall surface. Wall signs are separately fastened signs of limited extension from the wall surface; otherwise, they become projecting signs. The code allows certain-sized wall signs to be exempt from permit requirements (see Section H101.2). Section H111.1 specifies the material requirements and extension limitations of wall signs.

SECTION H103
LOCATION

H103.1 Location restrictions. Signs shall not be erected, constructed or maintained so as to obstruct any fire escape or any window or door or opening used as a means of egress or so as to prevent free passage from one part of a roof to any other part thereof. A sign shall not be attached in any form, shape or manner to a fire escape, nor be placed in such manner as to interfere with any opening required for ventilation.

❖ The intent of this section is to prohibit the installation of signs at locations that may interfere with either the evacuation process during an emergency situation or free movement throughout the roof of a building. Fire escapes and openings that are part of the means of egress or required for ventilation are to be free from obstructions that may affect their intended use.

SECTION H104
IDENTIFICATION

H104.1 Identification. Every outdoor advertising display sign hereafter erected, constructed or maintained, for which a permit is required shall be plainly marked with the name of the person, firm or corporation erecting and maintaining such sign and shall have affixed on the front thereof the permit number issued

for said sign or other method of identification approved by the building official.

❖ All signs must bear the name of the owner to identify the party responsible for installation and maintenance. The construction documents on file must show where such identification is located on each sign. If the sign is erected improperly or becomes damaged, the field inspector will be able to identify the responsible person to whom any notification is to be directed.

SECTION H105
DESIGN AND CONSTRUCTION

H105.1 General requirements. Signs shall be designed and constructed to comply with the provisions of this code for use of materials, loads and stresses.

❖ This section addresses the general requirements pertaining to the construction, design loads and working stresses of signs. The design and construction of signs must be in accordance with all of the pertinent requirements of the code. In many instances, structural calculations and the resulting design must be submitted showing the type of materials used, their strength capabilities and the loads and resultant stresses placed on the sign and its supports (see Chapter 16).

H105.2 Permits, drawings and specifications. Where a permit is required, as provided in Chapter 1, construction documents shall be required. These documents shall show the dimensions, material and required details of construction, including loads, stresses and anchors.

❖ The intent of this section is to require construction documents when a permit is required for the erection of a sign. In order to grant a permit to the owner, the sign is to be delineated on construction documents, such as plans and specifications, in sufficient detail to allow the determination of code compliance. Section H104.1 affects this section in that it explains that the owner of a sign (new or altered) is not relieved from obtaining the necessary permit, even if he or she holds a building or other type of permit at the time the sign is under construction. The sign permit is usually a separate permit and may also require other related permits (e.g., electrical).

H105.3 Wind load. Signs shall be designed and constructed to withstand wind pressure as provided for in Chapter 16.

❖ A reference is made to Section 1609, which relates to wind loads. Although a reference is not made to snow loads (see Section 1608), consideration should also be given thereto, especially for signs with a significant horizontal surface that will have snow accumulation, such as marquee signs.

H105.4 Seismic load. Signs designed to withstand wind pressures shall be considered capable of withstanding earthquake loads, except as provided for in Chapter 16.

❖ The designer and plan reviewer are to consider the provisions set forth in Section 1622 before the sign and its construction can be considered in compliance with the code.

H105.5 Working stresses. In outdoor advertising display signs, the allowable working stresses shall conform to the requirements of Chapter 16. The working stresses of wire rope and its fastenings shall not exceed 25 percent of the ultimate strength of the rope or fasteners.

Exceptions:

1. The allowable working stresses for steel and wood shall be in accordance with the provisions of Chapters 22 and 23.

2. The working strength of chains, cables, guys or steel rods shall not exceed one-fifth of the ultimate strength of such chains, cables, guys or steel.

❖ Consideration is to be given to the imposed loads and resultant working stresses in the design of signs. The allowable working stresses are to be evaluated in accordance with the provisions of Chapter 16. Structural members are to be of the appropriate size to support the loads applied to the sign (snow, wind, etc.) or the load of the sign itself. All members and connections are to be sufficient to carry all design loads of Chapter 16 without exceeding the specified design values set forth in the referenced chapter.

An exception to this requirement applies only to steel and wood allowable working stresses, which are to comply with the provisions of Chapters 22 and 23. A limitation is placed on wire rope and fastening utilized as structural members supporting signs, where the working stresses are to be below 25 percent of the ultimate strength of the member. An exception to this limitation applies to chains, cables, guys or steel rods where the working stresses of such members may not exceed one-fifth of the ultimate strength of the members.

H105.6 Attachment. Signs attached to masonry, concrete or steel shall be safely and securely fastened by means of metal anchors, bolts or approved expansion screws of sufficient size and anchorage to safely support the loads applied.

❖ In accordance with this section, metal connectors are the prescriptive requirement for the support and transfer of loads when signs are bearing on walls, floors and roofs of masonry, concrete or steel construction. Metal connectors provide points of anchorage at intersecting components and are also used as bonding elements. Section 2103.11 contains the standards and material requirements for joint reinforcement, wire fabric, wire accessories, metal anchors, ties and accessories and the required corrosion protection of these components.

SECTION H106
ELECTRICAL

H106.1 Illumination. A sign shall not be illuminated by other than electrical means, and electrical devices and wiring shall be installed in accordance with the requirements of the ICC *Electrical Code.* Any open spark or flame shall not be used for display purposes unless specifically approved.

❖ The reference to the *International Code Council Electrical Code Administrative Provisions* when dealing with electrical aspects is common throughout the *International Codes®* (see commentary, Chapter 27). Improperly installed lighting and wiring are a potential fire or electrocution hazard. Other illumination, utilizing such means as flame or sparks, must be approved by the building official so that such an arrangement does not induce a fire hazard for which commensurate safety precautions are not considered.

H106.1.1 Internally illuminated signs. Except as provided for in Sections 402.14 and 2611, where internally illuminated signs have facings of wood or approved plastic, the area of such facing section shall be not more than 120 square feet (11.16 m²) and the wiring for electric lighting shall be entirely enclosed in the sign cabinet with a clearance of not less than 2 inches (51 mm) from the facing material. The dimensional limitation of 120 square feet (11.16 m²) shall not apply to sign facing sections made from flame-resistant-coated fabric (ordinarily known as "flexible sign face plastic") that weighs less than 20 ounces per square yard (678 g/m²) and that, when tested in accordance with NFPA 701, meets the fire propagation performance requirements of both Test 1 and Test 2 or that when tested in accordance with an approved test method, exhibits an average burn time of 2 seconds or less and a burning extent of 5.9 inches (150 mm) or less for 10 specimens.

❖ This section applies to outdoor signs that are internally illuminated. The exception is a reminder that the area limitations of this section are not applicable to internally illuminated signs located within a building. Wood and approved plastic are limited because of the need to control the amount of combustibles incorporated into the sign. Once the amount of combustibles is limited, maintenance of a 2-inch (51 mm) clearance is still necessary between this material and electrical wiring. Square-foot limitations are not imposed on flame-resistant controlled fabrics that weigh less than 20 ounces per square yard (678 g/m²), provided that the material meets the specific criteria listed. This is because of the low fuel content of such fabrics and the resulting lower potential for fire propagation. One standard that is referenced, NFPA 701, deals with not only average flame time and average length of char, but also outdoor conditions and weathering.

H106.2 Electrical service. Signs that require electrical service shall comply with the ICC *Electrical Code.*

❖ This section is concerned with sign wiring being installed in a safe and protected manner so as not to create a potential tripping or electrical shock hazard. Electrical safety is as important as structural safety when dealing with illuminated signs. This section emphasizes the need for control by the building official so that the referenced *International Code Council Electrical Code Administrative Provisions* is utilized properly.

SECTION H107
COMBUSTIBLE MATERIALS

H107.1 Use of combustibles. Wood, approved plastic or plastic veneer panels as provided for in Chapter 26, or other materials of combustible characteristics similar to wood, used for moldings, cappings, nailing blocks, letters and latticing, shall comply with Section H109.1, and shall not be used for other ornamental features of signs, unless approved.

❖ The use of combustibles for ornamentation and sign facings is addressed in Sections H107.1 and H107.1.2. The control of combustible signs and portions of signs is necessary to limit the fuel load that the combustible features add to a structure. The use of the term "approved" means approved by the building official in accordance with the provisions of Chapter 26 and other applicable code sections. The reference to Section H109.1 controls the height at which combustibles may be used (see Section H109.1 for details).

H107.1.1 Plastic materials. Notwithstanding any other provisions of this code, plastic materials which burn at a rate no faster than 2.5 inches per minute (64 mm/s) when tested in accordance with ASTM D 635 shall be deemed approved plastics and can be used as the display surface material and for the letters, decorations and facings on signs and outdoor display structures.

❖ These provisions specify those plastic materials, mostly consisting of acrylics, that burn at a faster rate when tested in accordance with ASTM D 635 are to be considered approved plastics for the purpose of installation as display surface materials and alphabetic characters, ornaments and coverings on signs. ASTM D 635 measures the burning rate of a small sample held horizontally and exposed to a Bunsen burner flame to compare the relative linear rate of burning, extent and time of burning or both of plastics in the horizontal position.

H107.1.2 Electric sign faces. Individual plastic facings of electric signs shall not exceed 200 square feet (18.6 m²) in area.

❖ Consistent with other provisions of the code to limit the spread of fire, this section places restrictions on the size of plastic facings utilized in signs activated by electrical means. This limitation is intended to control the amount of combustible materials that potentially soften or melt when subject to heat.

H107.1.3 Area limitation. If the area of a display surface exceeds 200 square feet (18.6 m²), the area occupied or covered by approved plastics shall be limited to 200 square feet (18.6 m²) plus 50 percent of the difference between 200 square feet (18.6 m²) and the area of display surface. The area of plastic on a display surface shall not in any case exceed 1,100 square feet (102 m²).

❖ This section provides requirements for how to determine the permissible amount of combustible materials when the area of a display surface exceeds the 200-square-foot (18.6 m²) threshold. As an additional safeguard against exterior fire spread, the aggregate area of combustible materials covering a display surface should not exceed the sum of 200 square feet (18.6 m²) and 50 percent of the difference between the limitation and the area of the display surface. However, at no time can the aggregate area of plastic in display surfaces occupy more than 1,100 square feet (102 m²).

H107.1.4 Plastic appurtenances. Letters and decorations mounted on an approved plastic facing or display surface can be made of approved plastics.

❖ Under the provisions of this section, combustible lettering and ornaments are permitted to be mounted on display surfaces constructed of approved plastic materials.

SECTION H108
ANIMATED DEVICES

H108.1 Fail-safe device. Signs that contain moving sections or ornaments shall have fail-safe provisions to prevent the section or ornament from releasing and falling or shifting its center of gravity more than 15 inches (381 mm). The fail-safe device shall be in addition to the mechanism and the mechanism's housing which operate the movable section or ornament. The fail-safe device shall be capable of supporting the full dead weight of the section or ornament when the moving mechanism releases.

❖ Animated signs may be operated for many hours at a time, perhaps even continuously. Thus, it is imperative that if the operating mechanism malfunctions, the animated part of the sign will not release or fall, creating a hazard or damaging property.

SECTION H109
GROUND SIGNS

H109.1 Height restrictions. The structural frame of ground signs shall not be erected of combustible materials to a height of more than 35 feet (10668 mm) above the ground. Ground signs constructed entirely of noncombustible material shall not be erected to a height of greater than 100 feet (30 480 mm) above the ground. Greater heights are permitted where approved and located so as not to create a hazard or danger to the public.

❖ Combustible materials exceeding 35 feet (10 668 mm) in height are prohibited in a sign's structural frame. This correlates with the height limitation for structures of Type V construction in Table 503. The primary concern is the sign and its structure's potential involvement in a fire conflagration. The height of a noncombustible sign structure is limited to 100 feet (30 480 mm) so as to limit reasonably the area exposed to a sign or the sign's structural failure. The building official is permitted to approve signs that exceed 100 feet (30 480 mm) in height, provided that a public hazard is not created. An example of this is an area remote from other buildings and occupied areas, where structural failure would not jeopardize the public.

H109.2 Required clearance. The bottom coping of every ground sign shall be not less than 3 feet (914 mm) above the ground or street level, which space can be filled with platform decorative trim or light wooden construction.

❖ This section regulates the bottom clearance of ground signs. This clear space of 3 feet (914 mm) is permitted to be decorated with nonstructural or decorative trim or light wooden construction, provided this arrangement will not create a hazard.

H109.3 Wood anchors and supports. Where wood anchors or supports are embedded in the soil, the wood shall be pressure treated with an approved preservative.

❖ The intent of this section is to require wood support members in direct contact with the earth to be preservative treated. This provision is consistent with the requirements of Section 2304.11.4. Typically, wood species that are untreated are susceptible to decay. However, when preservative treated with chemicals in accordance with standardized procedures, wood becomes less susceptible to failure due to decay and rot.

SECTION H110
ROOF SIGNS

H110.1 General. Roof signs shall be constructed entirely of metal or other approved noncombustible material except as provided for in Sections H106.1.1 and H107.1. Provisions shall be made for electric grounding of metallic parts. Where combustible materials are permitted in letters or other ornamental features, wiring and tubing shall be kept free and insulated therefrom. Roof signs shall be so constructed as to leave a clear space of not less than 6 feet (1829 mm) between the roof level and the lowest part of the sign and shall have at least 5 feet (1524 mm) clearance between the vertical supports thereof. No portion of any roof sign structure shall project beyond an exterior wall.

Exception: Signs on flat roofs with every part of the roof accessible.

❖ Section H110.1 controls the materials, bottom clearances and arrangements of roof signs (refer to definition of "Sign" in the Section H102.1 for special sign descriptions). Noncombustible materials are required for roof signs. This arrangement will not add to the fuel load present on the building (see Sections H106.1.1 and H107.1 for exceptions that deal with sign facings, electrical wiring and electrical clearances). Grounding of metal signs for lightning control and preventing electrical shock and fire is mandatory. A 6-foot (1829 mm) clear space is to be maintained between the bottom of the sign and the roof. This arrangement will enable wind to travel under the sign, thereby minimizing the collection of snow and debris. It will also facilitate roof repair. An exception to these requirements applies to signs constructed on flat roofs where every portion of the roof is available for access.

H110.2 Bearing plates. The bearing plates of roof signs shall distribute the load directly to or upon masonry walls, steel roof girders, columns or beams. The building shall be designed to avoid overstress of these members.

❖ The provisions of this section require bearing plates of roof signs to transfer the loads directly to masonry or steel structural members. The intent is to minimize the possibility of imposing additional loads on bearing plates not prepared to carry them, which may affect the structural stability of the roof sign. The structural design is to consider the load-carrying capability of the bearing plates to avoid distress of these elements.

H110.3 Height of solid signs. A roof sign having a solid surface shall not exceed, at any point, a height of 24 feet (7315 mm) measured from the roof surface.

❖ This section specifies a height limitation on roof signs of solid construction. The 24-foot (7315 mm) threshold provides a conservative criteria where solid signs are to remain stable against the forces of wind.

H110.4 Height of open signs. Open roof signs in which the uniform open area is not less than 40 percent of total gross area shall not exceed a height of 75 feet (22 860 mm) on buildings of Type 1 or Type 2 construction. On buildings of other construction types, the height shall not exceed 40 feet (12 192 mm). Such signs shall be thoroughly secured to the building upon which they are installed, erected or constructed by iron, metal anchors, bolts, supports, chains, stranded cables, steel rods or braces and they shall be maintained in good condition.

❖ An open sign is the opposite of a "closed sign" (see Section H110.5) in that the membranes and edges are substantially open, thus allowing wind loads to pass through without any significant structural effects. The code allows increased heights for open signs located on the roofs of buildings based on the type of construction. On buildings of combustible construction, the height is limited to 40 feet (12.2 m), whereas on buildings of noncombustible construction, the height is permitted to be 75 feet (22.9 m). Consistent with other provisions of the code, the use of noncombustible materials will not add to the fuel load present in the building. Open signs are to be properly anchored to their supporting structure and structural supports are to be kept in proper state of preservation and free from rust to avoid failure of the members.

H110.5 Height of closed signs. A closed roof sign shall not be erected to a height greater than 50 feet (15 240 mm) above the roof of buildings of Type 1 or Type 2 construction, nor more than 35 feet (10 668 mm) above the roof of buildings of Type 3, 4 or 5 construction.

❖ As in the previous section, there are different height limitations for closed signs located on the roofs of buildings. A closed sign usually consists of membranes containing the lettering or design fastened to the structural framework with its edges also enclosed. The 50-foot (15 240 mm) threshold identifies the point at which wind loading on the sign becomes significant and a fire could go undetected for some time.

SECTION H111
WALL SIGNS

H111.1 Materials. Wall signs which have an area exceeding 40 square feet (3.72 m²) shall be constructed of metal or other approved noncombustible material, except for nailing rails and as provided for in Sections H106.1.1 and H107.1.

❖ This section regulates the materials used and the arrangement of wall signs (see the definition of "Wall sign"). Signs in excess of 40 square feet (3.72 m²) must be constructed of noncombustible materials. Nailing rails and the exceptions regarding sign facings, electrical wiring and clearances to electrical wires, as stipulated in Section H107.1, are permitted. Signs of less than 40 square feet (3.72 m²) may be of combustible construction. To date, experience has shown that fire exposure from a combustible sign of this size is minimal.

H111.2 Exterior wall mounting details. Wall signs attached to exterior walls of solid masonry, concrete or stone shall be safely and securely attached by means of metal anchors, bolts or expansion screws of not less than $^3/_8$ inch (9.5 mm) diameter and shall be embedded at least 5 inches (127 mm). Wood blocks shall not be used for anchorage, except in the case of wall signs attached to buildings with walls of wood. A wall sign shall not be supported by anchorages secured to an unbraced parapet wall.

❖ This section indicates the size, type and location of masonry anchors intended to be used for the attachment of signs to exterior walls of masonry construction. Note that the anchorage required is a minimum $^3/_8$-inch-diameter (9.5 mm) metal anchor, bolt or expansion screw embedded a minimum of 5 inches (127 mm) in solid masonry, concrete or stone construction. The use of wood blocking for attachment of wall signs is prohibited except in buildings where the exterior walls are constructed of wood. The anchors of wall signs are not permitted to be supported by unbraced parapet walls that are not providing adequate structural support for load transfer.

H111.3 Extension. Wall signs shall not extend above the top of the wall, nor beyond the ends of the wall to which the signs are attached unless such signs conform to the requirements for roof signs, projecting signs or ground signs.

❖ Signs cannot extend above the top of the wall (access to the roof must be maintained for repair and fire fighting) or beyond the ends of the wall. The signs must be kept on the property, unless exceptions for other types of signs apply.

SECTION H112
PROJECTING SIGNS

H112.1 General. Projecting signs shall be constructed entirely of metal or other noncombustible material and securely attached to a building or structure by metal supports such as bolts, anchors, supports, chains, guys or steel rods. Staples or nails shall not be used to secure any projecting sign to any building or structure. The dead load of projecting signs not par-

allel to the building or structure and the load due to wind pressure shall be supported with chains, guys or steel rods having net cross-sectional dimension of not less than $^3/_8$ inch (9.5 mm) diameter. Such supports shall be erected or maintained at an angle of at least 45 percent (0.78 rad) with the horizontal to resist the dead load and at angle of 45 percent (0.78 rad) or more with the face of the sign to resist the specified wind pressure. If such projecting sign exceeds 30 square feet (2.8 m²) in one facial area, there shall be provided at least two such supports on each side not more than 8 feet (2438 mm) apart to resist the wind pressure.

❖ This section establishes explicit provisions for the construction and anchorage of projecting signs (see the definition of "Projecting sign"). Projecting signs are required to be constructed of metal or approved noncombustible materials and are to be anchored to a building or structure only by metal bolts, anchors, supports, chains, guys or steel rods. Metal chains, guys or steel bolts with a minimum $^3/_8$-inch (9.5 mm) diameter are specifically prescribed to support the dead loads and wind pressures imposed on projecting signs perpendicular to the building or structure. These anchorage requirements are based on common construction techniques and are considered to provide a minimum resistance to pressures caused by wind loads. To resist the dead loads, these structural supports are to be located at a minimum angle of 45 degrees (0.78 rad). However, to resist wind pressures, these supports are to be at an angle of more than 45 degrees (0.78 rad) with the face of the sign. The use of staples or nails is prohibited for the attachment of projecting signs to buildings or structures. These limitations recognize that nails, when subjected to wind forces, may not provide adequate connection strength and structural resistance. A minimum of two supports is to be provided with a space between them of at least 8 feet (2438 mm) when the facial area of projecting signs exceeds 30 square feet (2.8 m²). The intent is to provide a minimum number of structural elements so that adequate load transfer is provided.

H112.2 Attachment of supports. Supports shall be secured to a bolt or expansion screw that will develop the strength of the supporting chains, guys or steel rods, with a minimum $^5/_8$-inch (15.9 mm) bolt or lag screw, by an expansion shield. Turn buckles shall be placed in chains, guys or steel rods supporting projecting signs.

❖ This section stipulates that structural supports of projecting signs are to be secured to a bolt or lag screw with a diameter of at least $^5/_8$ inch (15.9 mm). The bolt or expansion screw is to develop the strength of the supporting members. Structural supports are to be provided with turnbuckles to adjust the tension of the members.

H112.3 Wall mounting details. Chains, cables, guys or steel rods used to support the live or dead load of projecting signs are permitted to be fastened to solid masonry walls with expansion bolts or by machine screws in iron supports, but such supports shall not be attached to an unbraced parapet wall. Where the supports must be fastened to walls made of wood, the supporting anchor bolts must go through the wall and be plated or fastened on the inside in a secure manner.

❖ This section specifies the use of expansion bolts or machine screws in iron supports to fasten tension members required to support the live or dead loads of projecting signs at all locations of solid masonry walls, except on unbraced parapets. Where walls constructed of wood are provided for fastening the supports, anchor bolts are required to penetrate the wall and be securely plated or fastened on the inside face of the wall.

H112.4 Height limitation. A projecting sign shall not be erected on the wall of any building so as to project above the roof or cornice wall or above the roof level where there is no cornice wall; except that a sign erected at a right angle to the building, the horizontal width of which sign is perpendicular to such a wall and does not exceed 18 inches (457 mm), is permitted to be erected to a height not exceeding 2 feet (610 mm) above the roof or cornice wall or above the roof level where there is no cornice wall. A sign attached to a corner of a building and parallel to the vertical line of such corner shall be deemed to be erected at a right angle to the building wall.

❖ Consistent with other provisions of the code, projecting signs cannot extend beyond the ends of the wall or above the roof or cornice wall of a building. An exception to this limitation is projecting signs constructed perpendicular to the building, with a horizontal width of 18 inches (457 mm) and a height not exceeding 2 feet (610 mm) above the top of the roof or cornice wall or roof level without a cornice wall. In accordance with this section, signs attached to a corner of a building and parallel to the corner's vertical line are considered projecting signs constructed at a right angle to the building wall. Thus, their extension beyond the end of the wall is permitted within the limitations prescribed in this section.

H112.5 Additional loads. Projecting sign structures which will be used to support an individual on a ladder or other servicing device, whether or not specifically designed for the servicing device, shall be capable of supporting the anticipated additional load, but not less than a 100-pound (445 N) concentrated horizontal load and a 300-pound (1334 N) concentrated vertical load applied at the point of assumed or most eccentric loading. The building component to which the projecting sign is attached shall also be designed to support the additional loads.

❖ The intent of this section is to require additional structural capability in projecting signs that may be subjected to loads during maintenance procedures as shown in Figure H112.5. The face of the sign and the supports, therefore, must be capable of withstanding not only an additional concentrated horizontal load anywhere on the face of the sign, but also a 300-pound (1334 N) concentrated load acting vertically down at the point most eccentric from the sign fasteners. The building or building component to which the sign is attached must be capable of withstanding the resultant load.

SECTION H113
MARQUEE SIGNS

H113.1 Materials. Marquee signs shall be constructed entirely of metal or other approved noncombustible material except as provided for in Sections H106.1.1 and H107.1.

❖ A marquee sign is always supported by the building or structure and projects from the building into surrounding spaces. Marquee signs are usually located on canopies or other extensions of the building or structure to obtain a greater visual effect. The code has specific provisions for marquee signs concerning allowable materials. Marquee signs must be of metal or ap-

For SI: 1 pound = 4.4 N.

Figure H112.5
ADDITIONAL SIGN LOADS

proved noncombustible materials. Sections H113.2 through H113.4 regulate the attachment, dimensions and height limitations of a marquee sign. Section H107.1 describes exceptions regarding sign facings, electrical wiring and clearances to electrical wires.

H113.2 Attachment. Marquee signs shall be attached to approved marquees that are constructed in accordance with Section 3106.

❖ Marquee signs are attached to a structure called a "marquee." The marquee is to be designed and constructed in accordance with Section H113.1. A marquee sign should not be confused with a projecting sign (see Section H112.1).

H113.3 Dimensions. Marquee signs, whether on the front or side, shall not project beyond the perimeter of the marquee.

❖ Marquees accommodate a specially designed marquee sign arrangement. In other words, the sign fits into the marquee. A projecting sign cannot be placed on a marquee because the sign would project beyond the perimeter of the marquee.

H113.4 Height limitation. Marquee signs shall not extend more than 6 feet (1829 mm) above, nor 1 foot (305 mm) below such marquee, but under no circumstances shall the sign or signs have a vertical dimension greater than 8 feet (2438 mm).

❖ This section requires marquee signs to project less than 6 feet (1829 mm) above and not more than 1 foot (305 mm) below the marquee structure. The total vertical dimension of the marquee sign is limited to 8 feet (2438 mm). The intent of this section is to place a limitation on structures with a higher fuel load due to decorations incorporated on the sign that are subject to outdoor conditions and weathering.

SECTION H114
PORTABLE SIGNS

H114.1 General. Portable signs shall conform to requirements for ground, roof, projecting, flat and temporary signs where such signs are used in a similar capacity. The requirements of this section shall not be construed to require portable signs to have connections to surfaces, tie-downs or foundations where provisions are made by temporary means or configuration of the structure to provide stability for the expected duration of the installation.

❖ Portable signs must conform to the general regulations as set forth in this appendix for all signs, except that temporary support or stability methods may be utilized when approved by the building official. Portable signs must also conform to the requirements of this section commensurate to the manner in which the sign is used. This section establishes criteria for those types of signs that are neither mounted on buildings nor provided with a permanent foundation support. A typical example of a portable sign is a trailer-mounted illuminated sign. Such signs are used temporarily at locations and do not require any permanent support to resist wind overturning loads because of their limited

size. The code contains electrical supply requirements for portable signs.

TABLE 4-A
SIZE, THICKNESS AND TYPE OF GLASS PANELS IN SIGNS

MAXIMUM SIZE OF EXPOSED PANEL		MINIMUM THICKNESS OF GLASS (inches)	TYPE OF GLASS
Any dimension (inches)	Area (square inches)		
30	500	$^1/_8$	Plain, plate or wired
45	700	$^3/_{16}$	Plain, plate or wired
144	3,600	$^1/_4$	Plain, plate or wired
> 144	> 3,600	$^1/_4$	Wired glass

For SI: 1 inch = 25.4 mm, 1 square inch = 645.16 mm².

❖ Glazing materials installed in all types of signs are required to conform to the limitations of Table 4-A. This table details the maximum allowable size, thickness limitations and type of glass panels to be used in signs. Glass subject to unusual loading conditions may require special engineering analysis to determine its size.

TABLE 4-B
THICKNESS OF PROJECTION SIGN

PROJECTION (feet)	MAXIMUM THICKNESS (feet)
5	2
4	2.5
3	3
2	3.5
1	4

For SI: 1 foot = 304.8 mm.

❖ The requirements of Table 4-B are applicable only for projecting signs constructed in accordance with Section H112.

SECTION H115
REFERENCED STANDARDS

ASTM D 635-03	Test Method for Rate of Burning and/or Extent and Time of Burning of Self-Supporting Plastics in a Horizontal Position	H107.1.1
ICC EC-06	ICC Electrical Code	H106.1, H106.2
NFPA 701-99	Methods of Fire Test for Flame Propagation of Textiles and Films	H106.1.1

Appendix I:
Patio Covers

The provisions contained in this appendix are not mandatory unless specifically referenced in the adopting ordinance.

General Comments

These provisions are optional. They apply only if specifically adopted by the jurisdiction as stated in Section 101.2.1. The requirements are intended to apply only to patio covers associated with residential dwelling units.

Purpose

These provisions for patio covers are intended to simplify the requirements for residential patio installations.

SECTION I101
GENERAL

I101.1 General. Patio covers shall be permitted to be detached from or attached to dwelling units. Patio covers shall be used only for recreational, outdoor living purposes and not as carports, garages, storage rooms or habitable rooms. Openings shall be permitted to be enclosed with insect screening, approved translucent or transparent plastic not more that 0.125 inch (3.2 mm) in thickness, glass conforming to the provisions of Chapter 24 or any combination of the foregoing.

❖ The provisions of this appendix are intended to be used in conjunction with Groups R-1, R-2, R-3 and U with attached or detached patio covers. Patio covers must be used for recreational or outdoor living purposes only and not for the uses generally ascribed to Group U and R occupancies. If used for storage or as a garage, an entire new fuel load is introduced, thereby qualifying the patio cover as a private garage governed by Sections 312 and 406.

SECTION I102
DEFINITIONS

I102.1 General. The following word and term shall, for the purposes of this appendix, have the meaning shown herein.

❖ Definitions of terms that are associated with the content of this appendix are contained herein. This definition can help in the understanding and application of the code requirements. It is important to emphasize that this term is not exclusively related to this appendix, but is applicable everywhere the term is used throughout the code. The purpose for including this definition within this appendix is to provide more convenient access to it rather than referring back to Chapter 2.

PATIO COVERS. One story structures not exceeding 12 feet (3657 mm) in height. Enclosure walls shall be permitted to be

of any configuration, provided the open or glazed area of the longer wall and one additional wall is equal to at least 65 percent of the area below a minimum of 6 feet 8 inches (2032 mm) of each wall, measured from the floor.

❖ A patio cover is defined in order to make a distinction between carports, garages and covered awnings. The code does not consider a patio a hazardous condition with a high fuel load; therefore, special construction or consideration is not necessary. The maximum height a patio cover is allowed without being considered a separate structure is 12 feet (3658 mm) (see Figure I102.1).

SECTION I103
EXTERIOR OPENINGS

I103.1 Light, ventilation and emergency egress. Exterior openings required for light and ventilation shall be permitted to open into a patio structure. However, the patio structure shall be unenclosed if such openings are serving as emergency egress or rescue openings from sleeping rooms. Where such exterior openings serve as an exit from the dwelling unit, the patio structure, unless unenclosed, shall be provided with exits conforming to the provision of Chapter 10.

❖ A patio is allowed to be placed over an opening that is to be used as a means of natural light or ventilation. Once the patio covers an opening used for egress by the resident, the patio must provide an appropriate means of egress conforming with Chapter 10.

SECTION I104
STRUCTURAL PROVISIONS

I104.1 Design loads. Patio covers shall be designed and constructed to sustain, within the stress limits of this code, all dead loads plus a minimum vertical live load of 10 pounds per square foot (0.48 kN/m²) except that snow loads shall be used where such snow loads exceed this minimum. Such patio covers shall be designed to resist the minimum wind and seismic loads set forth in this code.

❖ The minimum value for the live load of the patio is to be either 10 pounds per square foot (psf) (0.48 kN/m²) or the roof snow load, whichever is greater. Additionally, where required, the patio must be constructed to be stabilized against the forces of wind and seismic loading conditions.

I104.2 Footings. In areas with a frost depth of zero, a patio cover shall be permitted to be supported on a concrete slab on grade without footings, provided the slab conforms to the provisions of Chapter 19 of this code, is not less than $3^1/_2$ inches (89 mm) thick and further provided that the columns do not support loads in excess of 750 pounds (3.36 kN) per column.

❖ This allowance, which permits foundations to be constructed without footings, is due to the minor value of these structures and the lack of any substantive safety issue involved. This exception addresses the exemption of footings in regard to frost penetration; it does not address allowable load-bearing capacity of the soil.

For SI: 1 inch = 25.4 mm, 1 foot = 304.8 mm.

Figure I102.1
PATIO COVERS

Appendix J:
Grading

The provisions contained in this appendix are not mandatory unless specifically referenced in the adopting ordinance.

General Comments

This appendix contains grading provisions that address soil-related hazards such as slope failure, landslides and erosion. Jurisdictions that include mountainous or hilly terrain often encounter grading work in connection with large commercial developments or the development of residential subdivisions that can pose grading problems that are not addressed by any other section of the code. Note that only excavation, grading and fill in connection with foundation construction is regulated by Section 1803.

In western states the *Uniform Building Code*® (UBC) appendix chapter on excavating and grading has found very widespread usage in many areas where topography is an issue. While the need for this appendix is based on the extensive use of that UBC appendix, which was originally developed in the 1960s, suggestions from numerous individuals and groups involved in the code development process have been incorporated into this grading appendix to make it compatible for use with the code.

Purpose

This appendix provides a set of grading requirements that can be utilized in lieu of (or in addition to) developing local regulations. Where grading is an important consideration, a comprehensive code is available simply by adopting this appendix. It is intended to provide consistent and uniform code requirements anywhere grading is considered an issue.

SECTION J101
GENERAL

J101.1 Scope. The provisions of this chapter apply to grading, excavation and earthwork construction, including fills and embankments. Where conflicts occur between the technical requirements of this chapter and the soils report, the soils report shall govern.

❖ This section states the scope of the chapter and also states that the soils report, which is required by Section J104.3, would supercede the requirements of this chapter when a conflict exists. This gives the professional responsible for the report much latitude to provide recommendations that are appropriate for each individual site.

J101.2 Flood hazard areas. The provisions of this chapter shall not apply to grading, excavation and earthwork construction, including fills and embankments, in floodways within flood hazard areas established in Section 1612.3 or in flood hazard areas where design flood elevations are specified but floodways have not been designated, unless it has been demonstrated through hydrologic and hydraulic analyses performed in accordance with standard engineering practice that the proposed work will not result in any increase in the level of the base flood.

❖ As the definition in Section 1612.2 indicates, a floodway is that portion of a flood hazard area that is reserved for the discharge of the design flood event. The National Flood Insurance Program (NFIP) requires that the impact of development or encroachment on the floodway be considered. If the authority having jurisdiction has not established a floodway, then the earthwork construction, grading or excavation is prohibited in the flood hazard area, unless appropriate analysis has been done. The intent of this section is to permit grading in a floodway only if is demonstrated that this activity will not adversely affect surrounding areas by increasing the base flood elevation.

SECTION J102
DEFINITIONS

J102.1 Definitions. For the purposes of this appendix chapter, the terms, phrases and words listed in this section and their derivatives shall have the indicated meanings.

❖ This section defines terms used in this appendix for uniformity of meaning to facilitate administration of these grading provisions.

BENCH. A relatively level step excavated into earth material on which fill is to be placed.

❖ This term is used to identify the steps that are cut into an existing slope prior to placing fill over that slope (see Figure J107.3).

COMPACTION. The densification of a fill by mechanical means.

❖ This definition aides the application of Section J107.5, which is applicable to all fill material.

CUT. See Excavation.

❖ Earthwork is typically referred to as "cut" and "fill." The term "cut" is synonymous with excavation.

DOWN DRAIN. A device for collecting water from a swale or ditch located on or above a slope, and safely delivering it to an approved drainage facility

❖ The inclusion of a definition for "Down drain" helps to clarify the term and make the requirements of Section J109.2 enforceable.

EROSION. The wearing away of the ground surface as a result of the movement of wind, water or ice.

❖ This definition describes what is a naturally occurring process of wearing away the ground surface. The provisions of this chapter regulate grading with the intent to minimize the susceptibility to erosion of man-made slopes primarily due to surface water.

EXCAVATION. The removal of earth material by artificial means, also referred to as a cut.

❖ Excavations are regulated by Section J106.

FILL. Deposition of earth materials by artificial means.

❖ Earthwork is typically referred to as "cut" and "fill." Fill is regulated by Section J107.

GRADE. The vertical location of the ground surface.

❖ The elevation of the existing ground surface is usually established by topographic surveys. This must be done in order to methodically evaluate a site to determine the need for regrading. The process of developing an appropriate grading plan leads to the required finished grade that must be achieved either by excavating where the existing grade is above the desired finished grade or by filling where the existing grade is below the desire finished grade.

GRADE, EXISTING. The grade prior to grading.

❖ See the definition of "Grade."

GRADE, FINISHED. The grade of the site at the conclusion of all grading efforts.

❖ See the definition of "Grade."

GRADING. An excavation or fill or combination thereof.

❖ This definition clarifies that the term "grading" refers to all earthwork.

KEY. A compacted fill placed in a trench excavated in earth material beneath the toe of a slope.

❖ This term refers to the level area that creates a notch at the bottom of a slope requiring benching in accordance with Section J107.3 (see Figure J107.3).

All slope references in the chapter have been modified to show the horizontal/vertical relationship.

SLOPE. An inclined surface, the inclination of which is expressed as a ratio of horizontal distance to vertical distance.

❖ This definition establishes that slopes in this chapter are expressed in the form of the horizontal-to-vertical distance ratio.

TERRACE. A relatively level step constructed in the face of a graded slope for drainage and maintenance purposes.

❖ This definition helps to clarify the term "terrace" and makes the requirements of Section J109.2 enforceable.

SECTION J103
PERMITS REQUIRED

J103.1 Permits required. Except as exempted in Section J103.2, no grading shall be performed without first having obtained a permit therefor from the building official. A grading permit does not include the construction of retaining walls or other structures.

❖ A grading permit does not include the construction of retaining walls or other structures, since that is covered by Section 1803.

J103.2 Exemptions. A grading permit shall not be required for the following:

1. Grading in an isolated, self-contained area, provided there is no danger to the public, and that such grading will not adversely affect adjoining properties.

2. Excavation for construction of a structure permitted under this code.

3. Cemetery graves.

4. Refuse disposal sites controlled by other regulations.

5. Excavations for wells, or trenches for utilities.

6. Mining, quarrying, excavating, processing or stockpiling rock, sand, gravel, aggregate or clay controlled by other regulations, provided such operations do not affect the lateral support of, or significantly increase stresses in, soil on adjoining properties.

7. Exploratory excavations performed under the direction of a registered design professional.

Exemption from the permit requirements of this appendix shall not be deemed to grant authorization for any work to be done in any manner in violation of the provisions of this code or any other laws or ordinances of this jurisdiction.

❖ The exemptions listed are for grading work that is relatively minor in nature (e.g., graves) or grading work that is normally regulated by other laws (e.g., utility trenching, mining, etc.).

SECTION J104
PERMIT APPLICATION AND SUBMITTALS

J104.1 Submittal requirements. In addition to the provisions of Section 105.3, the applicant shall state the estimated quantities of excavation and fill.

❖ Section 105.3 lists the information that is required for a building permit application, which also applies for a grading permit. Additionally, the quantities of earthwork, commonly referred to as "cut" and "fill," must be furnished.

J104.2 Site plan requirements. In addition to the provisions of Section 106, a grading plan shall show the existing grade and finished grade in contour intervals of sufficient clarity to indicate the nature and extent of the work and show in detail that it complies with the requirements of this code. The plans shall show the existing grade on adjoining properties in sufficient detail to identify how grade changes will conform to the requirements of this code.

❖ It is necessary for the permit applicant to furnish all applicable construction documents that are required in Section 106. Additionally, grading plans must provide sufficient grading information so that compliance with these requirements is verifiable.

J104.3 Soils report. A soils report prepared by registered design professionals shall be provided which shall identify the nature and distribution of existing soils; conclusions and recommendations for grading procedures; soil design criteria for any structures or embankments required to accomplish the proposed grading; and, where necessary, slope stability studies, and recommendations and conclusions regarding site geology.

Exception: A soils report is not required where the building official determines that the nature of the work applied for is such that a report is not necessary.

❖ This section requires the applicant to furnish a soils report unless the building official determines that it is not necessary because the proposed grading is of relatively minor consequence. The intent is that the appropriate professional provides a report that assesses the hazards of a particular site and includes recommendations for accomplishing the proposed grading. Note that Section J101.1 states that the soils report would supercede the requirements of this chapter when a conflict exists, giving great weight to the professional's recommendations.

J104.4 Liquefaction study. For sites with mapped maximum considered earthquake spectral response accelerations at short periods (S_s) greater than 0.5g as determined by Section 1613, a study of the liquefaction potential of the site shall be provided, and the recommendations incorporated in the plans.

Exception: A liquefaction study is not required where the building official determines from established local data that the liquefaction potential is low.

❖ This section intends to provide a guideline for the building official to determine if a liquefaction study is necessary. The exception provides the latitude to waive this requirement where it is evidenced that liq-

uefaction is a low risk.

The need for a liquefaction study is triggered by the mapped spectral acceleration at short periods (see Section 1615.1) rather than seismic design category (see Section 1616.3), which is based on the nature of an occupancy in addition to the mapped spectral accelerations modified for the site soil classification. While seismic design category is an entirely appropriate criteria for such a requirement in connection with a building foundation under Section 1802, the mapped spectral acceleration is more direct and appropriate for the application of this grading appendix since occupancy is most likely not applicable. While the mapped acceleration is used as the trigger for a study, it is ultimately the soil profile that determines the susceptibility to liquefaction.

SECTION J105
INSPECTIONS

J105.1 General. Inspections shall be governed by Section 109 of this code.

❖ A general reference is made to the inspection requirements in Section 109 of the code. The type and frequency of inspections specific to grading are not enumerated and are, therefore, left to the judgement of the building official.

J105.2 Special inspections. The special inspection requirements of Section 1704.7 shall apply to work performed under a grading permit where required by the building official.

❖ The special inspection requirements of Section 1704.7 that are referred to in this section concern site preparation and fill placement.

SECTION J106
EXCAVATIONS

J106.1 Maximum slope. The slope of cut surfaces shall be no steeper than is safe for the intended use, and shall be no steeper than 2 horizontal to 1 vertical (50 percent) unless the applicant furnishes a soils report justifying a steeper slope.

Exceptions:

1. A cut surface may be at a slope of 1.5 horizontal to 1 vertical (67 percent) provided that all the following are met:

   1.1. It is not intended to support structures or surcharges.

   1.2. It is adequately protected against erosion.

   1.3. It is no more than 8 feet (2438 mm) in height.

   1.4. It is approved by the building official.

2. A cut surface in bedrock shall be permitted to be at a slope of 1 horizontal to 1 vertical (100 percent).

❖ The code establishes a maximum two units horizontal to one unit vertical (50 percent) slope even though many professionals experienced in hillside urban development are likely to consider a one and one-half units horizontal to one unit vertical (67 percent) slope to be adequate for stability and erosion control. The two units horizontal to one unit vertical (50 percent) slope gradient is required to provide an extra margin of safety and easier maintenance capability, due to the fact that slope failures of cuts (as well as fills) with one and one-half units horizontal to one unit vertical (67 percent) slopes have been observed in the past. The cause of these failures was frequently traced to a lack of geotechnical inspection during grading or of proper maintenance after they were constructed.

The building official may allow steeper cut slopes if they are justified by the soils report. Exception 1 allows the steeper cut surface slope (discussed above) of one and one-half units horizontal to one unit vertical (67 percent) where the listed conditions are met, indicating a lower slope failure hazard. A steeper cut surface slope in bedrock is permitted because it is inherently more stable.

SECTION J107
FILLS

J107.1 General. Unless otherwise recommended in the soils report, fills shall conform to provisions of this section.

❖ These provisions govern all fills. As is indicated in Section J101.1, the soils report takes precedence in the event of conflicts with this chapter.

J107.2 Surface preparation. The ground surface shall be prepared to receive fill by removing vegetation, topsoil and other unsuitable materials, and scarifying the ground to provide a bond with the fill material.

❖ Fill must only be placed onto sound bedrock or other competent material to minimize subsidence or settlement; therefore, the removal of all unsuitable materials is necessary.

J107.3 Benching. Where existing grade is at a slope steeper than 5 horizontal to 1 vertical (20 percent) and the depth of the fill exceeds 5 feet (1524 mm) benching shall be provided in accordance with Figure J107.3. A key shall be provided which is at least 10 feet (3048 mm) in width and 2 feet (610 mm) in depth.

❖ Benching requirements are illustrated in Figure J107.3. The fill over an existing slope is potentially unstable unless constructed properly. Benching creates a series of (temporary) level steps on which the fill can be reliably placed and compacted. It is required for fills over 5 feet (1524 mm) deep where the existing slope is steeper than five units horizontal to one unit vertical (20 percent). The toe-of-fill bench is referred to as a key and the code specifies a minimum depth and width. The area of ground surface left at the planned toe-of-fill slope should be graded to drain away from the fill (see Section J108.3).

This figure illustrates the intention of the benching requirements. It shows the key that is located at the toe-of-slope and the series of level benches that must

TOP OF FILL

FILL SLOPE

5 FT. (1524 mm) OR GREATER

NATURAL SLOPE

REMOVE UNSUITABLE MATERIAL

2 FT. (610 mm) MINIMUM

KEY

1

≤5

For SI: 1 foot = 304.8 mm.

FIGURE J107.3
BENCHING DETAILS

be established in order to accomplish the proposed earthwork.

J107.4 Fill material. Fill material shall not include organic, frozen or other deleterious materials. No rock or similar irreducible material greater than 12 inches (305 mm) in any dimension shall be included in fills.

❖ Organic materials should not be allowed in fills because they decompose over time, causing settlement or creating planes of weakness. Rocks of limited size are permitted in fills. Where doing so, they should be placed to ensure that the fill material can be placed in the voids around the rocks and compacted as required.

J107.5 Compaction. All fill material shall be compacted to 90 percent of maximum density as determined by ASTM D 1557, Modified Proctor, in lifts not exceeding 12 inches (305 mm) in depth.

❖ This section specifies a minimum soil density to be achieved uniformly throughout the fill as well as the test to be used. This sets minimum acceptable levels of fill quality to achieve performance that is consistent with the purpose of this chapter.

J107.6 Maximum slope. The slope of fill surfaces shall be no steeper than is safe for the intended use. Fill slopes steeper than 2 horizontal to 1 vertical (50 percent) shall be justified by soils reports or engineering data.

❖ The maximum gradient of two units horizontal to one unit vertical (50 percent) for fill slopes minimizes the risk of slope failures (see commentary, Section J106.1). Compaction of fill can be accomplished more

readily on slopes of two units horizontal to one unit vertical (50 percent) or less. Also, such slopes are more accessible, increasing the likelihood they will be maintained. Greater slopes are only permitted if substantiated by the soils report or other engineering data.

SECTION J108
SETBACKS

J108.1 General. Cut and fill slopes shall be set back from the property lines in accordance with this section. Setback dimensions shall be measured perpendicular to the property line and shall be as shown in Figure J108.1, unless substantiating data is submitted justifying reduced setbacks.

❖ This section establishes minimum setbacks from property lines between the property being graded and adjacent properties. It is intended to apply at the exterior boundary of a tract and should not be applied, for instance, at the interior lot lines of a proposed development.

The figure specifies the setback requirements at the top and bottom of graded slopes.

J108.2 Top of slope. The setback at the top of a cut slope shall not be less than that shown in Figure J108.1, or than is required to accommodate any required interceptor drains, whichever is greater.

❖ The intention of the top-of-slope setback is to protect the adjacent property. The setback allows for some soil erosion without losing lateral support of the adjacent property, thus protecting structures on that property. It also provides some space for an interceptor

For SI: 1 foot = 304.8 mm.

FIGURE J108.1
DRAINAGE DIMENSIONS

drain if needed (see Section J109.3). The top-of-slope setback requirements are illustrated in Figure J108.1.

J108.3 Slope protection. Where required to protect adjacent properties at the toe of a slope from adverse effects of the grading, additional protection, approved by the building official, shall be included. Such protection may include but shall not be limited to:

1. Setbacks greater than those required by Figure J108.1.

2. Provisions for retaining walls or similar construction.

3. Erosion protection of the fill slopes.

4. Provision for the control of surface waters.

❖ The minimum toe-of-fill setback from the adjacent property line is specified in Figure J108.1. Additional clearance must be provided as necessary, for example, to allow for the construction of retaining walls. For example, it may also be necessary to provide an area to collect slope runoff along the toe of slope and to construct a drain if needed.

SECTION J109
DRAINAGE AND TERRACING

J109.1 General. Unless otherwise recommended by a registered design professional, drainage facilities and terracing shall be provided in accordance with the requirements of this section.

Exception: Drainage facilities and terracing need not be provided where the ground slope is not steeper than 3 horizontal to 1 vertical (33 percent).

❖ This section establishes minimum requirements for drainage that are largely based on past experience. They are minimum standards and the design professional should use them as such, evaluating each site based on its own merits. Consideration also should be given to long-term performance in addition to the likelihood of inadequate maintenance of slopes in residential developments. Controlling drainage onto and off the proposed site is one of the most important considerations in grading design, since even a well-compacted fill or graded cut slope can be compromised by inadequate drainage.

J109.2 Terraces. Terraces at least 6 feet (1829 mm) in width shall be established at not more than 30-foot (9144 mm) vertical intervals on all cut or fill slopes to control surface drainage and debris. Suitable access shall be provided to allow for cleaning and maintenance.

Where more than two terraces are required, one terrace, located at approximately mid-height, shall be at least 12 feet (3658 mm) in width.

Swales or ditches shall be provided on terraces. They shall have a minimum gradient of 20 horizontal to 1 vertical (5 percent) and shall be paved with concrete not less than 3 inches (76 mm) in thickness, or with other materials suitable to the application. They shall have a minimum depth of 12 inches (305 mm) and a minimum width of 5 feet (1524 mm).

A single run of swale or ditch shall not collect runoff from a tributary area exceeding 13,500 square feet (1256 m²) (projected) without discharging into a down drain.

❖ Terraces are required where a graded slope has a height over 30 feet (9144 mm). They provide level surfaces along the graded slope to allow access for slope maintenance and accommodate drainage.

The terrace ditch gradient of 5 percent helps make the drain more self-cleansing and recognizes a typical lack of maintenance, since most of the hillside dwellers will not climb up or down the slope to keep a drain clear of debris. Swales serving larger drainage areas must discharge into a down drain, which is defined in Section J102.1.

J109.3 Interceptor drains. Interceptor drains shall be installed along the top of cut slopes receiving drainage from a tributary width greater than 40 feet (12 192 mm), measured horizontally. They shall have a minimum depth of 1 foot (305 mm) and a minimum width of 3 feet (915 mm). The slope shall be approved by the building official, but shall not be less than 50 horizontal to 1 vertical (2 percent). The drain shall be paved with concrete not less than 3 inches (76 mm) in thickness, or by other materials suitable to the application. Discharge from the drain shall be accomplished in a manner to prevent erosion and shall be approved by the building official.

❖ Interceptor drains are important in protecting the face of cut slopes from excessive erosion and are, therefore, required under certain conditions. The discharge of an interceptor drain must not contribute to erosion and requires the building official's approval.

J109.4 Drainage across property lines. Drainage across property lines shall not exceed that which existed prior to grading. Excess or concentrated drainage shall be contained on site or directed to an approved drainage facility. Erosion of the ground in the area of discharge shall be prevented by installation of nonerosive down drains or other devices.

❖ This section protects adjacent properties from exposure to added runoff as a result of the grading as well as any concentrated runoff created by the grading. It requires collection of any excess or concentrated drainage and allows it to be held on site or discharged by approved methods.

SECTION J110
EROSION CONTROL

J110.1 General. The faces of cut and fill slopes shall be prepared and maintained to control erosion. This control shall be permitted to consist of effective planting.

Exception: Erosion control measures need not be provided on cut slopes not subject to erosion due to the erosion-resistant character of the materials.

Erosion control for the slopes shall be installed as soon as practicable and prior to calling for final inspection.

❖ Adequate provision should be made to prevent surface water from damaging the face of a graded slope.

Effective planting is a lasting method of minimizing erosion. Other measures may include berms, interceptor drains and terrace drains.

J110.2 Other devices. Where necessary, check dams, cribbing, riprap or other devices or methods shall be employed to control erosion and provide safety.

❖ This requires use of the alternative means of providing erosion control where necessary.

SECTION J111
REFERENCED STANDARDS

ASTM D 1557-e01   Test Method for Laboratory   J107.6
Compaction Characteristics
of Soil Using Modified Effort
[56,000 ft-lb/ft^3
(2,700kN-m/m^3)].

❖ This section lists the standards referenced by this appendix chapter. Since appendices are not considered part of the code unless specifically referred to in the adopting ordinance, the standards that are referenced in this chapter are listed here rather than in Chapter 35.

Bibliography

The following resource materials are referenced in this appendix or are relevant to the subject matter addressed in this appendix.

ASTM D 1557-00, *Test Method for Laboratory Compaction Characteristics of Soil Using Modified Effort (56,000ft-lb/ ft).* West Conshohocken, PA: ASTM International, 2000.

Scullin, C. Michael. *Excavation and Grading Code Administration, Inspection and Enforcement.* Englewood Cliffs, NJ: Prentice-Hall, Inc., 1983.

UBC-97, *Uniform Building Code.* Whittier, CA: International Conference of Building Officials, 1997.

Appendix K
ICC Electrical Code

General Comments

In September 2005, at the 2005 Code Development Final Action Hearings in Detroit, the International Code Council® (ICC®) membership voted to include the entire *International Code Council Electrical Code Administrative Provisions* in the appendices of the code. This appendix is simply a replica of the 2006 ICC *Electrical Code—Administrative Provisions*. For a discussion of the relationship of the code to the *International Code Council Electrical Code Administrative Provisions*, please see the commentary on Chapter 27.

Purpose

The purpose for Appendix K is to give communities that are in the process of adopting the *International Codes®* the opportunity to simply adopt this appendix rather than adopt an entirely separate code that is produced under a separate cover.

TABLE OF CONTENTS

CHAPTER K1

SCOPE

SECTION K101
GENERAL

K101.1 Title. These regulations shall be known as the *Electrical Code–Administrative Provisions* of [NAME OF JURISDICTION] and shall be cited as such and will be referred to herein as "this code."

K101.2 Purpose. The purpose of this code is to provide minimum standards to safeguard life or limb, health, property and public welfare by regulating and controlling the design, construction, installation, quality of materials, location, operation, and maintenance or use of electrical systems and equipment.

K101.3 Scope. This code shall regulate the design, construction, installation, alteration, repairs, relocation, replacement, addition to, use or maintenance of electrical systems and equipment.

SECTION K102
APPLICABILITY

K102.1 General. The provisions of this code shall apply to all matters affecting or relating to structures and premises, as set forth in Section K101.

K102.1.1 Existing installations. Except as otherwise provided for in this chapter, a provision in this code shall not require the removal, alteration or abandonment of, nor prevent the continued utilization and maintenance of, existing electrical systems and equipment lawfully in existence at the time of the adoption of this code.

K102.1.2 Maintenance. Electrical systems, equipment, materials and appurtenances, both existing and new, and parts thereof shall be maintained in proper operating condition in accordance with the original design and in a safe, hazard-free condition. Devices or safeguards that are required by this code shall be maintained in compliance with the code edition under which installed. The owner or the owner's designated agent shall be responsible for the maintenance of the electrical systems and equipment. To determine compliance with this provision, the code official shall have the authority to require that the electrical systems and equipment be reinspected.

K102.1.3 Additions, alterations and repairs. Additions, alterations, renovations and repairs to electrical systems and equipment shall conform to that required for new electrical systems and equipment without requiring that the existing electrical systems or equipment comply with all of the requirements of this code. Additions, alterations and repairs shall not cause existing electrical systems or equipment to become unsafe, hazardous or overloaded.

Minor additions, alterations, renovations and repairs to existing electrical systems and equipment shall meet the provisions for new construction, except where such work is performed in the same manner and arrangement as was in the existing system, is not hazardous and is approved.

K102.1.4 Change in occupancy. It shall be unlawful to make a change in the occupancy of any structure that will subject the structure to any special provision of this code applicable to the new occupancy without approval. The code official shall certify that such structure meets the intent of the provisions of law governing building construction for the proposed new occupancy and that such change of occupancy does not result in any hazard to public health, safety or welfare.

K102.1.5 Moved buildings. Electrical systems and equipment that are a part of buildings or structures moved into or within the jurisdiction shall comply with the provisions of this code for new installations.

K102.2 Differences. Where, in any specific case, different sections of this code specify different materials, methods of construction or other requirements, the most restrictive shall govern. Where there is a conflict between a general requirement and a specific requirement, the specific requirement shall be applicable.

K102.3 Other laws. The provisions of this code shall not be deemed to nullify any provisions of local, state or federal law.

K102.4 Validity. In the event any part or provision of this code is held to be illegal or void, this shall not have the effect of making void or illegal any of the other parts or provisions thereof, which are determined to be legal; and it shall be presumed that this code would have been adopted without such illegal or invalid parts or provisions.

K102.4.1 Segregation of invalid provisions. Any invalid part of this code shall be segregated from the remainder of this code by the court holding such part invalid, and the remainder shall remain effective.

K102.5 Application of references. References to chapter or section numbers, or to provisions not specifically identified by number, shall be construed to refer to such chapters, sections or provisions of this code.

K102.6 Referenced codes and standards. The codes and standards referenced in this code shall be those that are listed in Chapter K13 and such codes and standards shall be considered part of the requirements of this code to the prescribed extent of each such reference. Where differences occur between provisions of this code and referenced codes or standards, the provisions of this code shall apply.

Exception: Where enforcement of a code provision would violate the conditions of the listing of the equipment or appliance, the conditions of the listing and manufacturer's instructions shall apply.

K102.7 Appendices. Provisions in the appendices shall not apply unless specifically referenced in the adopting ordinance.

K102.8 Subjects not regulated by this code. Where no applicable standards or requirements are set forth in this code, or are contained within other laws, codes, regulations, ordinances or bylaws adopted by the jurisdiction, compliance with applicable standards of nationally recognized standards as are approved shall be deemed as prima facie evidence of compliance with the intent of this code. Nothing herein shall derogate from the authority of the code official to determine compliance with codes or standards for those activities or installations within the code official's jurisdiction or responsibility.

DEFINITIONS

SECTION K201
GENERAL

K201.1 Scope. Unless otherwise expressly stated, the following words and terms shall, for the purposes of this code, have the meanings indicated in this chapter.

K201.2 Interchangeability. Words used in the present tense include the future; words in the masculine gender include the feminine and neuter; the singular number includes the plural and the plural, the singular.

K201.3 Terms defined in other codes. Where terms are not defined in this code and are defined in the *International Building Code, International Energy Conservation Code, International Fire Code, International Fuel Gas Code, International Mechanical Code, International Plumbing Code, International Private Sewage Disposal Code, International Property Maintenance Code, International Residential Code, International Zoning Code* or NFPA 70, such terms shall have meanings ascribed to them as in those codes.

K201.4 Terms not defined. Where terms are not defined through the methods authorized by this section, such terms shall have ordinarily accepted meanings such as the context implies.

SECTION K202
GENERAL DEFINITIONS

APPROVED. Approved by the code official or other authority having jurisdiction.

APPROVED AGENCY. An established and recognized agency regularly engaged in conducting tests or furnishing inspection services, where the agency has been approved by the code official.

CODE OFFICIAL. The officer or other designated authority charged with the administration and enforcement of this code, or a duly authorized representative.

LISTED AND LISTING. Equipment, appliances or materials included in a list published by a nationally recognized testing laboratory, inspection agency or other organization concerned with product evaluation that maintains periodic inspection of the production of listed equipment, appliances or materials, and whose listing states either that the equipment, appliances or materials meet nationally recognized standards, or has been tested and found suitable for use in a specified manner. Not all testing laboratories, inspection agencies and other organizations concerned with product evaluation use the same means for identifying listed equipment, appliances or materials. Some do not recognize equipment, appliances or materials as listed unless they are also labeled. The authority having jurisdiction shall utilize the system employed by the listing organization to identify a listed product.

OCCUPANCY. The purpose for which a building, or part thereof, is utilized or occupied.

CHAPTER K3

ORGANIZATION AND ENFORCEMENT

SECTION K301
DEPARTMENT OF ELECTRICAL INSPECTION

K301.1 Creation of enforcement agency. The department of electrical inspection is hereby created and the official in charge thereof shall be known as the code official. The function of the department shall be to assist the code official in the administration and enforcement of the provisions of this code.

K301.2 Appointment. The code official shall be appointed by the chief appointing authority of the jurisdiction.

K301.3 Deputies. In accordance with the prescribed procedures of this jurisdiction and with the concurrence of the appointing authority, the code official shall have the authority to appoint a deputy code official, the related technical officers, inspectors, plans examiners and other employees. Such employees shall have powers as delegated by the code official.

SECTION K302
DUTIES AND POWERS OF THE
CODE OFFICIAL

K302.1 General. The code official is hereby authorized and directed to enforce the provisions of this code. The code official shall have the authority to render interpretations of this code, and to adopt policies, procedures, rules and regulations in order to clarify the application of its provisions. Such interpretations, policies, procedures, rules and regulations shall be in compliance with the intent and purpose of this code. Such policies and procedures shall not have the effect of waiving requirements specifically provided for in this code.

K302.2 Rule-making authority. The code official shall have authority as necessary in the interest of public health, safety and general welfare, to adopt and promulgate rules and regulations and to designate requirements applicable because of local climatic or other conditions. Such rules shall not have the effect of waiving requirements specifically provided for in this code, or of violating accepted engineering methods involving public safety.

K302.3 Applications and permits. The code official is authorized to receive applications, review construction documents and issue permits for the installation of electrical systems and equipment, inspect the premises for which such permits have been issued, and enforce compliance with the provisions of this code.

K302.4 Notices and orders. The code official is authorized to issue all necessary notices or orders in accordance with Chapter K10 as are required to effect compliance with this code.

K302.5 Inspections. The code official shall make all of the inspections necessary to determine compliance with the provisions of this code in accordance with Chapter K7.

K302.6 Identification. The code official shall carry proper identification as required by Section K702.4.1.

K302.7 Right of entry. The code official is authorized to enter the structure or premises at reasonable times to inspect or perform the duties imposed by this code in accordance with Section K702.4.

K302.8 Department records. The code official shall keep official records of applications received, permits and certificates issued, fees collected, reports of inspections, notices and orders issued, and as required by this code, such records shall be retained in the official records for the period required for retention of public records.

K302.8.1 Approvals and modifications. A record of approvals and modifications granted shall be maintained by the code official and shall be available for public inspection during business hours in accordance with applicable laws.

K302.8.2 Inspections. The code official shall keep a record of each inspection made, including notices and orders issued, showing the findings and disposition of each.

K302.8.3 Alternative methods or materials. The application for modification, alternative methods or materials, and the final decision of the code official shall be in writing and shall be officially recorded in the permanent records of the code official.

K302.9 Liability. The code official, officer or employee charged with the enforcement of this code, while acting for the jurisdiction in good faith and without malice in the discharge of the duties required by this code or other pertinent law or ordinance, shall not thereby be rendered liable personally, and is hereby relieved from all personal liability for any damage accruing to persons or property as a result of any act or by reason of an act or omission in the discharge of official duties. Any suit instituted against any officer or employee because of an act performed by that officer or employee in the lawful discharge of duties and under the provisions of this code shall be defended by the legal representative of the jurisdiction until the final termination of the proceedings.

The code official or any subordinate shall not be liable for costs in any action, suit or proceeding that is instituted in pursuance of the provisions of this code; and any official, officer or employee, acting in good faith and without malice, shall be free from liability for acts performed under any of its provisions or by reason of any act or omission in the performance of official duties in connection therewith.

SECTION K303
CERTIFICATE OF OCCUPANCY

K303.1 Use and occupancy. No building or structure shall be used or occupied until a certificate of occupancy has been provided in accordance with the *International Building Code*.

CHAPTER K4
PERMITS AND FEES

SECTION K401
GENERAL

K401.1 Permits required. Permits required by this code shall be obtained from the code official. Permit fees, if any, shall be paid prior to issuance of the permit. Issued permits shall be kept on the premises designated therein at all times and shall be readily available for inspection by the code official.

K401.2 Types of permits. An owner, authorized agent or contractor who desires to construct, enlarge, alter, repair, move, demolish or change the occupancy of a building or structure, or to erect, install, enlarge, alter, repair, remove, convert or replace electrical systems or equipment, the installation of which is regulated by this code, or to cause such work to be done, shall first make application to the code official and obtain the required permit for the work.

> **Exception:** Where repair or replacement of electrical systems or equipment must be performed in an emergency situation, the permit application shall be submitted within the next working business day of the department of electrical inspection.

K401.3 Work exempt from permit. The following work shall be exempt from the requirement for a permit:

1. Listed cord and plug connected temporary decorative lighting.

2. Reinstallation of attachment plug receptacles, but not the outlets therefor.

3. Repair or replacement of branch circuit overcurrent devices of the required capacity in the same location.

4. Temporary wiring for experimental purposes in suitable experimental laboratories.

5. Electrical wiring, devices, appliances, apparatus or equipment operating at less than 25 volts and not capable of supplying more than 50 watts of energy.

Exemption from the permit requirements of this code shall not be deemed to grant authorization for work to be done in violation of the provisions of this code or other laws or ordinances of this jurisdiction.

SECTION K402
APPLICATION

K402.1 General. The code official is authorized to receive applications for and issue permits as required by this code.

K402.2 Application. Application for a permit required by this code shall be made to the code official in such form and detail as prescribed by the code official. Applications for permits shall be accompanied by such plans as prescribed by the code official.

K402.3 Action on application. The code official shall examine or cause to be examined applications for permits and

amendments thereto within a reasonable time after filing. If the application or the construction documents do not conform to the requirements of pertinent laws, the code official shall reject such application in writing, stating the reasons therefor. If the code official is satisfied that the proposed work conforms to the requirements of this code and laws and ordinances applicable thereto, the code official shall issue a permit therefor as soon as practicable.

K402.4 Inspection required. Before a permit is issued, the code official is authorized to inspect and approve the systems, equipment, buildings, devices, premises, and spaces or areas to be used.

K402.5 Time limitation of application. An application for a permit for any proposed work or operation shall be deemed to have been abandoned 180 days after the date of filing, unless such application has been pursued in good faith or a permit has been issued; except that the code official is authorized to grant one or more extensions of time for additional periods not exceeding 90 days each. The extension shall be requested in writing and justifiable cause demonstrated.

SECTION K403
CONDITIONS

K403.1 Conditions of a permit. A permit shall constitute permission to conduct work as set forth in this code in accordance with the provisions of this code. Such permission shall not be construed as authority to violate, cancel or set aside any of the provisions of this code or other applicable regulations or laws of the jurisdiction.

K403.2 Expiration. Every permit issued shall become invalid unless the work on the site authorized by such permit is commenced within 180 days after its issuance, or if the work authorized on the site by such permit is suspended or abandoned for a period of 180 days after the time the work is commenced. The code official is authorized to grant, in writing, one or more extensions of time, for periods not more than 180 days each. The extension shall be requested in writing and justifiable cause demonstrated.

K403.3 Extensions. The code official is authorized to grant, in writing, one or more extensions of the time period of a permit for periods of not more than 90 days each. Such extensions shall be requested by the permit holder in writing and justifiable cause demonstrated.

K403.4 Posting the permit. Issued permits shall be kept on the premises designated therein at all times and shall be readily available for inspection by the code official.

K403.5 Validity. The issuance or granting of a permit shall not be construed to be a permit for, or an approval of, any violation of any of the provisions of this code or of any other ordinance of the jurisdiction. Permits presuming to give authority to violate or cancel the provisions of this code or other ordinances of the jurisdiction shall not be valid. The issuance of a permit based on construction documents and other data shall not prevent the code official from requiring the correction of errors in the construction documents and other data. The code official is also authorized to prevent occupancy or use of a structure where in violation of this code or of any other ordinances of this jurisdiction.

K403.6 Information on the permit. The code official shall issue all permits required by this code on an approved form furnished for that purpose. The permit shall contain a general description of the operation or occupancy and its location and any other information required by the code official. Issued permits shall bear the signature of the code official.

K403.7 Suspension or revocation. The code official is authorized to suspend or revoke a permit issued under the provisions of this code wherever the permit is issued in error, on the basis of incorrect, inaccurate or incomplete information; in violation of any ordinance, regulation or any of the provisions of this code; or if any one of the following conditions exist:

1. The permit is used for a location or establishment other than that for which it was issued.

2. The permit is used for a condition or activity other than that listed in the permit.

3. Conditions and limitations set forth in the permit have been violated.

4. There have been any false statements or misrepresentations as to the material fact in the application for permit or plans submitted or a condition of the permit.

5. The permit is used by a different person or firm than the name for which it was issued.

6. The permittee failed, refused or neglected to comply with orders or notices duly served in accordance with the provisions of this code within the time provided therein.

7. The permit was issued in error or in violation of an ordinance, regulation or this code.

SECTION K404
FEES

K404.1 Payment of fees. A permit shall not be valid until the fees prescribed by law have been paid. Nor shall an amendment to a permit be released until the additional fee, if any, has been paid.

K404.2 Schedule of permit fees. A fee for each permit shall be paid as required, in accordance with the schedule as established by the applicable governing authority. The fees for electrical work shall be as indicated in the following schedule.

[JURISDICTION TO INSERT APPROPRIATE SCHEDULE]

K404.3 Work commencing before permit issuance. Any person who commences any work before obtaining the necessary permits shall be subject to an additional fee established by the code official, which shall be in addition to the required permit fees.

K404.4 Related fees. The payment of the fee for the construction, alteration, removal or demolition for work done in connection with, or concurrently with, the work authorized by a permit shall not relieve the applicant or holder of the permit from the payment of other fees that are prescribed by law.

K404.5 Refunds. The code official is authorized to establish a refund policy.

CHAPTER K5

CONSTRUCTION DOCUMENTS

SECTION K501
GENERAL

K501.1 Submittal documents. Construction documents, special inspection and structural observation programs, and other data shall be submitted in one or more sets with each application for a permit. The construction documents shall be prepared by a registered design professional where required by the statutes of the jurisdiction in which the project is to be constructed. Where special conditions exist, the code official is authorized to require additional construction documents to be prepared by a registered design professional.

> **Exception:** The code official is authorized to waive the submission of construction documents and other data not required to be prepared by a registered design professional if it is found that the nature of the work applied for is such that reviewing of construction documents is not necessary to determine compliance with this code.

K501.2 Information on construction documents. Construction documents shall be drawn to scale upon suitable material. Electronic media documents are permitted to be submitted where approved by the code official. Construction documents shall be of sufficient clarity to indicate the location, nature and extent of the work proposed and show in detail that it will conform to the provisions of this code and relevant laws, ordinances, rules and regulations, as determined by the code official.

K501.2.1 Penetrations. Construction documents shall indicate where penetrations will be made for electrical systems and shall indicate the materials and methods for maintaining required structural safety, fire-resistance rating and fireblocking.

K501.2.2 Load calculations. Where an addition or alteration is made to an existing electrical system, an electrical load calculation shall be prepared to determine if the existing electrical service has the capacity to serve the added load.

K501.3 Site plan. The construction documents submitted with the application for permit shall be accompanied by a site plan showing to scale the size and location of new construction and existing structures on the site, distances from lot lines, the established street grades and the proposed finished grades; and it shall be drawn in accordance with an accurate boundary line survey. In the case of demolition, the site plan shall show construction to be demolished and the location and size of existing structures and construction that are to remain on the site or plot. The code official is permitted to waive or modify the requirement for a site plan where the application for permit is for alteration or repair or where otherwise warranted.

SECTION K502
EXAMINATION OF DOCUMENTS

K502.1 General. The code official shall examine or cause to be examined the accompanying construction documents and shall ascertain by such examinations whether the construction indicated and described is in accordance with the requirements of this code and other pertinent laws or ordinances.

K502.2 Approval of construction documents. When the code official issues a permit, the construction documents shall be approved, in writing or by stamp, as "Reviewed for Code Compliance." One set of construction documents so reviewed shall be retained by the code official. The other set shall be returned to the applicant, shall be kept at the site of work and shall be open to inspection by the code official or the authorized representative.

K502.2.1 Previous approvals. This code shall not require changes in the construction documents, construction or installation of electrical systems or equipment for which a lawful permit has been heretofore issued or otherwise lawfully authorized, and the construction of which has been pursued in good faith within 180 days after the effective date of this code and has not been abandoned.

K502.2.2 Phased approval. The code official is authorized to issue a permit for the installation of part of an electrical system before the construction documents for the electrical system have been submitted, provided that adequate information and detailed statements have been filed complying with pertinent requirements of this code. The holder of such permit shall proceed at the holder's own risk with the building operation and without assurance that a permit for the entire system will be granted.

SECTION K503
DESIGN PROFESSIONAL IN
RESPONSIBLE CHARGE

K503.1 General. Where it is required that documents be prepared by a registered design professional, the code official shall require the owner to engage and designate on the permit application a registered design professional who shall act as the registered design professional in responsible charge. If the circumstances require, the owner shall be permitted to designate a substitute registered design professional in responsible charge who shall perform the duties required of the original registered design professional in responsible charge. The code official shall be notified in writing by the owner if the registered design professional in responsible charge is changed or is unable to perform the duties.

The registered design professional in responsible charge shall be responsible for reviewing and coordinating submittal documents prepared by others, including phased and deferred submittal items, for compatibility with the design of the system.

SECTION K504
HANDLING SUBMITTALS

K504.1 Deferred submittals. For the purposes of this section, deferred submittals are defined as those portions of the design that are not submitted at the time of the application and that are to be submitted to the code official within a specified period.

Deferral of any submittal items shall have the prior approval of the code official. The registered design professional in responsible charge shall list the deferred submittals on the construction documents for review by the code official.

Submittal documents for deferred submittal items shall be submitted to the registered design professional in responsible charge, who shall review them and forward them to the code official with a notation indicating that the deferred submittal documents have been reviewed and that they have been found to be in general compliance with the design of the system. The deferred submittal items shall not be installed until their design and submittal documents have been approved by the code official.

K504.2 Amended construction documents. Work shall be installed in accordance with the reviewed construction documents, and any changes made during construction which are not in compliance with the approved construction documents shall be resubmitted for approval as an amended set of construction documents.

K504.3 Retention of construction documents. One set of approved construction documents shall be retained by the code official for a period of not less than 180 days from date of completion of the permitted work, or as required by state or local laws.

CHAPTER K6

APPROVAL

SECTION K601
GENERAL

K601.1 Approved materials and equipment. All materials, equipment and devices approved by the code official shall be constructed and installed in accordance with such approval.

K601.1.1 Technical assistance. To determine the acceptability of technologies, processes, products, facilities, materials and uses attending the design, operation or use of a building or premises subject to the inspection of the department, the code official is authorized to require the owner or the person in possession or control of the building or premises to provide, without charge to the jurisdiction, a technical opinion and report. The opinion and report shall be prepared by a qualified engineer, specialist, laboratory or organization acceptable to the code official and shall analyze the properties of the design, operation or use of the building or premises and the facilities and appurtenances situated thereon, to recommend necessary changes. The code official is authorized to require design submittals to be prepared by and bear the stamp of, a registered design professional.

K601.2 Modifications. Wherever there are practical difficulties involved in carrying out the provisions of this code, the code official shall have the authority to grant modifications for individual cases, provided the code official shall first find that special individual reason makes the strict letter of this code impractical and that the modification is in compliance with the intent and purpose of this code, and that such modification does not lessen health, life and fire-safety requirements. The details of action granting modifications shall be recorded and entered in the files of the department of electrical inspection.

K601.3 Alternative materials, methods, equipment and appliances. The provisions of this code are not intended to prevent the installation of any material or to prohibit any method of construction not specifically prescribed by this code, provided that any such alternative has been approved. An alternative material or method of construction shall be approved where the code official finds that the proposed design is satisfactory and complies with the intent of the provisions of this code, and that the material, method or work offered is, for the purpose intended, at least the equivalent of that prescribed in this code in quality, strength, effectiveness, fire resistance, durability and safety.

K601.4 Material, equipment and appliance reuse. Materials, equipment, appliances and devices shall not be reused unless such elements have been reconditioned, tested and placed in good and proper working condition and approved.

SECTION K602
TESTING

K602.1 Required testing. Wherever there is insufficient evidence of compliance with the provisions of this code, or evidence that a material or method does not conform to the requirements of this code, or in order to substantiate claims for alternative materials or methods, the code official shall have the authority to require tests as evidence of compliance to be made at no expense to the jurisdiction.

K602.2 Test methods. Test methods shall be as specified in this code or by other recognized test standards. In the absence of recognized and accepted test methods, the code official shall approve the testing procedures.

K602.3 Testing agency. All tests shall be performed by an approved agency.

K602.4 Test reports. Reports of tests shall be retained by the code official for the period required for retention of public records.

SECTION K603
ALTERNATIVE ENGINEERED DESIGN

K603.1 General. The design, documentation, inspection, testing and approval of an alternative engineered design electrical system shall comply with this section.

K603.2 Design criteria. An alternative engineered design shall conform to the intent of the provisions of this code and shall provide an equivalent level of quality, strength, effectiveness, fire resistance, durability and safety. Materials, equipment or components shall be designed and installed in accordance with the manufacturer's installation instructions.

K603.3 Submittal. The registered design professional shall indicate on the permit application that the electrical system is an alternative engineered design. The permit and permanent permit records shall indicate that an alternative engineered design was part of the approved installation.

K603.4 Technical data. The registered design professional shall submit sufficient technical data to substantiate the proposed alternative engineered design and to prove that the performance meets the intent of this code.

K603.5 Construction documents. The registered design professional shall submit to the code official two complete sets of signed and sealed construction documents for the alternative engineered design. The construction documents shall include floor plans and a diagram of the work.

K603.6 Design approval. Where the code official determines that the alternative engineered design conforms to the intent of this code, the electrical system shall be approved. If the alternative engineered design is not approved, the code official shall notify the registered design professional in writing, stating the reasons therefor.

K603.7 Inspection and testing. The alternative engineered design shall be tested and inspected in accordance with the requirements of this code.

CHAPTER K7

INSPECTIONS AND TESTING

SECTION K701
GENERAL

K701.1 General. The code official is authorized to conduct inspections that are deemed necessary to determine the extent of compliance with the provisions of this code and to approve reports of inspection by approved agencies or individuals. All reports of such inspections shall be prepared and submitted in writing for review and approval. Inspection reports shall be certified by a responsible officer of such approved agency or by the responsible individual. The code official is authorized to engage such expert opinion as deemed necessary to report upon unusual, detailed or complex technical issues subject to the approval of the governing body.

SECTION K702
REQUIRED INSPECTIONS

K702.1 General. The code official, upon notification, shall make the inspections set forth in this section.

K702.1.1 Evaluation and follow-up inspection services. Prior to the approval of a prefabricated construction assembly having concealed electrical work and the issuance of an electrical permit, the code official shall require the submittal of an evaluation report on each prefabricated construction assembly, indicating the complete details of the electrical system, including a description of the system and its components, the basis upon which the system is being evaluated, test results and similar information, and other data as necessary for the code official to determine conformance to this code.

K702.1.1.1 Evaluation service. The code official shall designate the evaluation service of an approved agency as the evaluation agency, and review such agency's evaluation report for adequacy and conformance to this code.

K702.1.1.2 Follow-up inspection. Except where ready access is provided to electrical systems, service equipment and accessories for complete inspection at the site without disassembly or dismantling, the code official shall conduct the in-plant inspections as frequently as necessary to ensure conformance to the approved evaluation report or shall designate an independent, approved inspection agency to conduct such inspections. The inspection agency shall furnish the code official with the follow-up inspection manual and a report of inspections upon request, and the electrical system shall have an identifying label permanently affixed to the system indicating that factory inspections have been performed.

K702.1.1.3 Test and inspection records. Required test and inspection records shall be available to the code official at all times during the fabrication of the electrical system and the erection of the building; or such records as the code official designates shall be filed.

K702.1.2 Concealed work. Work shall remain accessible and exposed for inspection purposes until approved. It shall be the duty of the permit applicant to cause the work to remain accessible and exposed for inspection purposes. Wherever any installation subject to inspection prior to use is covered or concealed without having first been inspected, the code official shall have the authority to require that such work be exposed for inspection. Neither the code official nor the jurisdiction shall be liable for expense entailed in the removal or replacement of any material required to allow inspection.

K702.1.3 Underground. Underground inspection shall be made after trenches or ditches are excavated and bedded, piping and conductors installed, and before backfill is put in place. Where excavated soil contains rocks, broken concrete, frozen chunks and other rubble that would damage or break the raceway, cable or conductors, or where corrosive action will occur, protection shall be provided in the form of granular or selected material, approved running boards, sleeves or other means.

K702.1.4 Rough-in. Rough-in inspection shall be made after the roof, framing, fireblocking and bracing are in place and all wiring and other components to be concealed are complete, and prior to the installation of wall or ceiling membranes.

K702.1.5 Other inspections. In addition to the inspections specified above, the code official is authorized to make or require other inspections of any construction work to ascertain compliance with the provisions of this code and other laws, which are enforced by the department of electrical inspection.

K702.1.6 Final inspection. The final inspection shall be made after all work required by the permit is completed.

K702.1.7 Inspection record card. Work requiring a permit shall not be commenced until the permit holder or an agent of the permit holder shall have posted or otherwise made available an inspection record card such as to allow the code official to make conveniently the required entries thereon regarding inspection of the work. This card shall be maintained by the permit holder until final approval has been granted by the code official.

K702.1.8 Approval required. Work shall not be performed beyond the point indicated in each successive inspection and test without first obtaining the approval of the code official. The code official, upon notification, shall make the requested inspections and tests and shall either indicate the portion of the construction that is satisfactory as completed, or shall notify the permit holder or an agent of the permit holder wherein the same fails to comply with this code. Any portions that do not comply shall be corrected and such portion shall not be covered or concealed until authorized by the code official.

K702.2 Validity. Approval as a result of an inspection shall not be construed to be an approval of a violation of the provisions of this code or of other ordinances of the jurisdiction. Inspections presuming to give authority to violate or cancel the provisions of this code or of other ordinances of the jurisdiction shall not be valid.

K702.3 Preliminary inspection. Before issuing a permit, the code official is authorized to examine or cause to be examined buildings, structures and sites for which an application has been filed. The code official shall be notified when the installation is ready for inspection and is authorized to conduct the inspection within a reasonable period of time.

K702.4 Entry. The code official is authorized to enter and examine any building, structure, marine vessel, vehicle or premises in accordance with Section K702.4.3 for the purpose of enforcing this code.

K702.4.1 Identification. The code official shall carry proper identification issued by the governing authority where inspecting structures, premises or facilities in the performance of duties under this code and shall be identified by proper credentials issued by this governing authority.

K702.4.2 Impersonation prohibited. A person shall not impersonate the code official through the use of a uniform, identification card, badge or any other means.

K702.4.3 Right of entry. Where it is necessary to make an inspection to enforce the provisions of this code, or where the code official has reasonable cause to believe that there exists in a structure or upon any premises a condition that is contrary to or in violation of this code, which makes the structure or premises unsafe, dangerous or hazardous, the code official is authorized to enter the structure or premises at reasonable times to inspect or to perform the duties imposed by this code, provided that if such structure or premises be occupied, that credentials be presented to the occupant and entry requested. If such structure or premises is unoccupied, the code official is authorized to first make a reasonable effort to locate the owner or other person having charge or control of the structure or premises and request entry. If entry is refused, the code official shall have recourse to the remedies provided by law to secure entry.

K702.5 Inspection agencies. The code official is authorized to accept reports of approved inspection agencies, provided such agencies satisfy the requirements as to qualifications and reliability.

K702.6 Inspection requests. It shall be the duty of the person doing the work authorized by a permit to notify the code official that such work is ready for inspection. It shall be the duty of the person requesting any inspections required by this code to provide access to and means for inspection of such work.

K702.7 Assistance from other agencies. The assistance and cooperation of police, building, fire and health department officials and all other officials shall be available as required in the performance of duties.

K702.8 Contractors' responsibilities. It shall be the responsibility of every contractor who enters into contracts for the installation or repair of electrical systems for which a permit is required to comply with adopted state and local rules and regulations concerning licensing.

SECTION K703
TESTING

K703.1 General. Electrical work shall be tested as required in this code. Tests shall be performed by the permit holder and observed by the code official.

K703.2 Apparatus, material and labor for tests. Apparatus, material and labor required for testing an electrical system or part thereof shall be furnished by the permit holder.

K703.3 Reinspection and testing. Where any work or installation does not pass an initial test or inspection, the necessary corrections shall be made so as to achieve compliance with this code. The work or installation shall then be resubmitted to the code official for inspection and testing.

CHAPTER K8

SERVICE UTILITIES

SECTION K801
GENERAL

K801.1 Connection of service utilities. No person shall make connections from a utility, source of energy, fuel or power to any building or system that is regulated by this code for which a permit is required, until released by the code official.

K801.2 Temporary connection. The code official shall have the authority to authorize the temporary connection of the building or system to the utility source of energy, fuel or power.

K801.3 Authority to disconnect service utilities. The code official shall have the authority to authorize disconnection of utility services or energy sources to the building, structure or system regulated by this code in case of an emergency where it is necessary to eliminate an immediate hazard to life or property. The code official shall notify the serving utility and, wherever possible, the owner and occupant of the building, structure or service system of the decision to disconnect prior to taking such action. If not notified prior to disconnecting, the owner or occupant of the building, structure or service system shall be notified in writing as soon as practical thereafter.

K801.3.1 Connection after order to disconnect. A person shall not make utility service or energy source connections to systems regulated by this code, which have been disconnected or ordered to be disconnected by the code official, or the use of which has been ordered to be discontinued by the code official until the code official authorizes the reconnection and use of such systems.

CHAPTER K9

UNSAFE SYSTEMS AND EQUIPMENT

SECTION K901
CONDITIONS

K901.1 Unsafe electrical systems. An electrical system that is unsafe, constitutes a fire or health hazard, or is otherwise dangerous to human life, as regulated by this code, is hereby declared as an unsafe electrical system. Use of an electrical system regulated by this code constituting a hazard to health, safety or welfare by reason of inadequate maintenance, dilapidation, fire hazard, disaster, damage or abandonment is hereby declared an unsafe use. Such unsafe equipment and appliances are hereby declared to be a public nuisance and shall be abated by repair, rehabilitation, demolition or removal.

K901.2 Authority to condemn electrical systems. Wherever the code official determines that any electrical system, or portion thereof, regulated by this code has become hazardous to life, health or property, the code official shall order in writing that such electrical systems either be removed or restored to a safe condition. A time limit for compliance with such order shall be specified in the written notice. A person shall not use or maintain a defective electrical system or equipment after receiving such notice.

Where such electrical system is to be disconnected, written notice as prescribed in this code shall be given. In cases of immediate danger to life or property, such disconnection shall be made immediately without such notice.

K901.3 Dangerous conditions. Wherever the code official shall find in any structure or upon any premises dangerous or hazardous conditions or materials, the code official is authorized to order such dangerous conditions or materials to be removed or remedied in accordance with the provisions of this code.

K901.4 Record. The code official shall cause a report to be filed on an unsafe condition. The report shall state the occupancy of the structure and the nature of the unsafe condition.

K901.5 Notice. If an unsafe condition is found, the code official shall serve on the owner, agent or person in control of the structure, a written notice that describes the condition deemed unsafe and specifies the required repairs or improvements to be made to abate the unsafe condition, or that requires the unsafe condition to be removed within a stipulated time. Such notice shall require the person thus notified to declare immediately to the code official acceptance or rejection of the terms of the order.

K901.6 Method of service. Such notice shall be deemed properly served if a copy thereof is: (a) delivered to the owner personally; or (b) sent by certified or registered mail addressed to the owner at the last known address with the return receipt requested. If the certified or registered letter is returned showing that the letter was not delivered, a copy thereof shall be posted in a conspicuous place in or about the structure affected by such notice. Service of such notice in the foregoing manner upon the owner's agent or upon the person responsible for the structure shall constitute service of notice upon the owner.

CHAPTER K10

VIOLATIONS

SECTION K1001
UNLAWFUL ACTS

K1001.1 General. It shall be unlawful for any person, firm or corporation to erect, construct, alter, extend, repair, move, remove, demolish or occupy any system or equipment regulated by this code, or cause same to be done, in conflict with or in violation of any of the provisions of this code.

SECTION K1002
NOTICE OF VIOLATION

K1002.1 Issuance. Where the code official finds any building, premises, vehicle, system or equipment that is in violation of this code, the code official is authorized to issue corrective orders.

K1002.2 Notice. Wherever the code official determines violations of this code or observes an apparent or actual violation of a provision of this code or other codes or ordinances under the code official's jurisdiction, the code official is authorized to prepare a written notice of violation describing the conditions deemed unsafe and, where compliance is not immediate, specifying a time for reinspection. Such order shall direct the discontinuance of the illegal action or condition and the abatement of the violation.

K1002.3 Service. Any order or notice issued pursuant to this code shall be served upon the owner, operator, occupant or other person responsible for the condition or violation, either by personal service, mail or by delivering the same to, and leaving it with, some person of responsibility upon the premises. For unattended or abandoned locations, a copy of such order or notice shall be posted on the premises in a conspicuous place at or near the entrance to such premises, and the order or notice shall be mailed by certified mail with return receipt requested or a certificate of mailing, to the last known address of the owner, occupant or both.

K1002.4 Compliance with orders and notices. Orders and notices issued or served as provided by this code shall be complied with by the owner, operator, occupant or other person responsible for the condition or violation to which the order or notice pertains.

K1002.5 Failure to correct violations. If the notice of violation is not complied with, the code official is authorized to request the legal counsel of the jurisdiction to institute the appropriate legal proceedings to restrain, correct or abate such violation or to require removal or termination of the unlawful occupancy of the structure in violation of the provisions of this code or of any order or direction made pursuant thereto.

K1002.6 Failure to comply. Failure to comply with an abatement notice or other corrective notice issued by the code official shall result in each day that such violation continues being regarded as a new and separate offense.

K1002.7 Unauthorized tampering. Signs, tags or seals posted or affixed by the code official shall not be mutilated, destroyed or tampered with or removed without authorization from the code official.

SECTION K1003
PENALTIES

K1003.1 Penalties. Any person who fails to comply with the provisions of this code or who fails to carry out an order made pursuant of this code or violates any condition attached to a permit, approval or certificate shall be subject to the penalties established by this jurisdiction.

K1003.2 Abatement of violation. The imposition of the penalties herein described shall not prevent the legal officer of the jurisdiction from instituting appropriate action to prevent unlawful construction or to restrain, correct or abate a violation; or to prevent illegal occupancy of a structure or premises; or to stop an illegal act, conduct of business or occupancy of a structure on or about any premises.

SECTION K1004
STOP WORK ORDER

K1004.1 Issuance. Upon notice from the code official that any electrical work is being done contrary to the provisions of this code or in a dangerous or unsafe manner, such work shall immediately cease. Such notice shall be in writing and shall be given to the owner of the property, or to the owner's agent, or to the person doing the work. The notice shall state the conditions under which work is authorized to resume.

K1004.2 Emergencies. Where an emergency exists, the code official shall not be required to give a written notice prior to stopping the work.

K1004.3 Unlawful continuance. Any person who shall continue any work in or about the structure after having been served with a stop work order, except such work as that person is directed to perform to remove a violation or unsafe condition, shall be subject to penalties as prescribed by law.

CHAPTER K11

MEANS OF APPEAL

SECTION K1101
GENERAL

K1101.1 Board of appeals established. In order to hear and decide appeals of orders, decisions or determinations made by the code official relative to the application and interpretation of this code, there shall be and is hereby created a board of appeals. The board of appeals shall be appointed by the governing body and shall hold office at its pleasure. The board shall adopt rules of procedure for conducting its business, and shall render all decisions and findings in writing to the appellant with a duplicate copy to the code official.

K1101.2 Limitations on authority. An application for appeal shall be based on a claim that the true intent of this code or the rules legally adopted thereunder have been incorrectly interpreted, the provisions of this code do not fully apply, or an at least equivalent method of protection or safety is proposed. The board shall have no authority to waive the requirements of this code.

SECTION K1102
MEMBERSHIP

K1102.1 Membership of board. The board of appeals shall consist of five members appointed by the chief appointing authority as follows: one for five years; one for four years; one for three years; one for two years and one for one year. Thereafter, each new member shall serve for five years or until a successor has been appointed.

K1102.2 Qualifications. The board of appeals shall consist of five individuals, one from each of the following professions or disciplines.

1. Registered design professional who is a registered architect; or a builder or superintendent of building construction with at least 10 years' experience, five of which shall have been in responsible charge of work.

2. Registered design professional with structural engineering or architectural experience.

3. Registered design professional with mechanical, plumbing or fuel-gas engineering experience; or a mechanical, plumbing or fuel-gas contractor with at least 10 years' experience, five of which shall have been in responsible charge of work.

4. Registered design professional with electrical engineering experience; or an electrical contractor with at least 10 years' experience, five of which shall have been in responsible charge of work.

5. Registered design professional with fire protection engineering experience; or a fire protection contractor with at least 10 years' experience, five of which shall have been in responsible charge of work.

6. The code official shall be an ex officio member of said board, but shall have no vote on any matter before the board.

K1102.3 Alternate members. The chief appointing authority shall appoint two alternate members who shall be called on by the board chairman to hear appeals during the absence or disqualification of a member. Alternate members shall possess the qualifications required for board membership and shall be appointed for five years, or until a successor has been appointed.

K1102.4 Chairman. The board shall annually select one of its members to serve as chairman.

K1102.5 Disqualification of members. A member shall not hear an appeal in which that member has a personal, professional or financial interest.

K1102.6 Secretary. The chief appointing authority shall designate a qualified clerk to serve as secretary to the board. The secretary shall file a detailed record of all proceedings in the office of the chief appointing authority.

K1102.7 Compensation of members. Compensation of members shall be determined by law.

SECTION K1103
PROCEDURES

K1103.1 Application for appeal. A person shall have the right to appeal a decision of the code official to the board of appeals. An application for appeal shall be based on a claim that the true intent of this code or the rules legally adopted thereunder, have been incorrectly interpreted, the provisions of this code do not fully apply, or an equally good or better form of construction is proposed. The application shall be filed on a form obtained from the code official within 20 days after the notice was served.

K1103.2 Notice of meeting. The board shall meet upon notice from the chairman, within 10 days of the filing of an appeal, or at stated periodic meetings.

K1103.3 Open hearing. All hearings before the board shall be open to the public. The appellant, the appellant's representative, the code official and any person whose interests are affected shall be given an opportunity to be heard.

K1103.4 Rules of procedure. The board shall adopt and make available to the public through the secretary rules of procedure under which a hearing will be conducted. The procedures shall not require compliance with strict rules of evidence, but shall mandate that only relevant information be received.

K1103.5 Postponed hearing. Where five members are not present to hear an appeal, either the appellant or the appellant's representative shall have the right to request a postponement of the hearing.

K1103.6 Decisions. The board shall modify or reverse the decision of the code official by a concurring vote of three members.

> **K1103.6.1 Resolution.** The decision of the board shall be by resolution. Certified copies shall be furnished to the appellant and to the code official.

> **K1103.6.2 Administration.** The code official shall take immediate action in accordance with the decision of the board.

K1103.7 Court review. Any person, whether or not a previous party of the appeal, shall have the right to apply to the appropriate court for a writ of certiorari to correct errors of law. Application for review shall be made in the manner and time required by law following the filing of the decision in the office of the chief appointing authority.

CHAPTER K12

ELECTRICAL PROVISIONS

SECTION K1201
GENERAL

K1201.1 Scope. This chapter governs the design and construction of electrical systems and equipment.

K1201.1.1 Adoption. Electrical systems and equipment shall be designed and constructed in accordance with the *International Residential Code* or NFPA 70 as applicable, except as otherwise provided in this code.

[F] K1201.2 Abatement of electrical hazards. All identified electrical hazards shall be abated. All identified hazardous electrical conditions in permanent wiring shall be brought to the attention of the code official responsible for enforcement of this code. Electrical wiring, devices, appliances and other equipment which is modified or damaged and constitutes an electrical shock or fire hazard shall not be used.

[F] K1201.3 Appliance and fixture listing. Electrical appliances and fixtures shall be tested and listed in published reports of inspected electrical equipment by an approved agency and installed in accordance with all instructions included as part of such listing.

SECTION K1202
PROVISIONS

K1202.1 General. The provisions of this section shall apply to the design, construction, installation, use and maintenance of electrical systems and equipment. Where differences occur between provisions of this code and referenced codes or standards, the provisions of this code shall apply.

K1202.2 Nonmetallic-sheathed cable. The use of Type NM, NMC and NMS (nonmetallic sheathed) cable wiring methods shall not be limited based on height, number of stories or construction type of the building or structure.

K1202.3 Cutting, notching and boring. The cutting, notching and boring of wood and steel framing members, structural members and engineered wood products shall be in accordance with the *International Building Code.*

K1202.4 Penetrations. Penetrations of walls, floors, ceilings and assemblies required to have a fire-resistance rating, shall be protected in accordance with the *International Building Code.* Where cables, conductors and raceways penetrate fireblocking or draftstopping, such penetrations shall be protected by filling the annular space with an approved fireblocking material.

K1202.5 Smoke detector circuits. Smoke detectors required by the *International Building Code* and installed within dwelling units shall not be connected as the only load on a branch circuit. Such detectors shall be supplied by branch circuits having lighting loads consisting of lighting outlets in habitable spaces.

[M] K1202.6 Appliance access. Where appliances requiring access are installed in attics or underfloor spaces, a luminaire controlled by a switch located at the required passageway opening to such space and a receptacle outlet shall be provided at or near the appliance location.

[FG] K1202.7 Prohibited grounding electrode. Fuel gas piping shall not be used as a grounding electrode.

[F] K1202.8 Emergency and standby power. Emergency and standby power systems required by the *International Building Code* or *International Fire Code* shall be installed in accordance with the *International Building Code,* the *International Fire Code,* NFPA 110, NFPA 111 and this code.

[F] K1202.9 Smoke control systems. Smoke control systems required by the *International Building Code* or *International Fire Code* shall be supplied with two sources of power. Primary power shall be the normal building power systems. Secondary power shall be from an approved standby source complying with this code. The standby power source and its transfer switches shall be in a separate room from the normal power transformers and switch gear, and shall be enclosed in a room constructed of not less than 1-hour fire-resistance-rated fire barriers, ventilated directly to and from the exterior. Power distribution from the two sources shall be by independent routes. Transfer to full standby power shall be automatic and within 60 seconds of failure of the primary power.

[F] K1202.9.1 Power sources and power surges. Elements of the smoke management system relying on volatile memories or the like shall be supplied with uninterruptable power sources of sufficient duration to span 15-minute primary power interruption. Elements of the smoke management system susceptible to power surges shall be suitably protected by conditioners, suppressors or other approved means.

[F] K1202.9.2 Wiring. In addition to meeting the requirements of this code, all signal and control wiring for smoke control systems, regardless of voltage, shall be fully enclosed within continuous raceways.

[M] K1202.10 Wiring in plenums. Combustible electrical or electronic wiring methods and materials, optical fiber cable, and optical fiber raceway exposed within plenums regulated by Section 602 of the *International Mechanical Code* shall have a peak optical density not greater than 0.50, an average optical density not greater than 0.15, and a flame spread not greater than 5 feet (1524 mm) when tested in accordance with NFPA 262. Only type OFNP (plenum-rated nonconducive optical fiber cable) shall be installed in plenum-rated optical fiber raceways. Wiring, cable and raceways addressed in this section shall be listed and labeled as plenum rated and shall be installed in accordance with this code.

[M] K1202.10.1 Combustible electrical equipment. Combustible electrical equipment exposed within plenums regulated by Section 602 of the *International Mechanical Code* shall have a peak rate of heat release not greater than 100 kilowatts (kW), a peak optical density not greater than

0.50, and an average optical density not greater than 0.15 when tested in accordance with UL 2043. Combustible electrical equipment shall be listed and labeled.

[M] K1202.11 Engine and gas turbine-powered equipment and appliances. Permanently installed equipment and appliances powered by internal combustion engines and turbines shall be installed in accordance with the manufacturer's installation instructions, the *International Mechanical Code, International Fuel Gas Code* and NFPA 37.

[F] K1202.12 Stationary fuel cell power systems. Stationary fuel cell power systems having a power output not exceeding 10 MW shall be tested in accordance with ANSI CSA America FC1 and shall be installed in accordance with the manufacturer's installation instructions and NFPA 853.

[M] K1202.13 Boiler control requirements. The power supply to the electrical control system for boilers shall be from a two-wire branch circuit that has a grounded conductor or from an isolation transformer with a two-wire secondary. Where an isolation transformer is provided, one conductor of the secondary winding shall be grounded. Control voltage shall not exceed 150 volts nominal, line to line. Control and limit devices shall interrupt the ungrounded side of the circuit. A means of manually disconnecting the control circuit shall be provided, and controls shall be arranged so that when deenergized, the burner shall be inoperative. Such disconnecting means shall be capable of being locked in the off position and shall be provided with ready access.

[F] K1202.14 Equipment and door labeling. Doors into electrical control panel rooms shall be marked with a plainly visible and legible sign stating ELECTRICAL ROOM or similar approved wording. The disconnecting means for each service, feeder or branch circuit originating on a switchboard or panelboard shall be legibly and durably marked to indicate its purpose unless such purpose is clearly evident.

[F] K1202.15 Smoke alarm power source. In new construction, required smoke alarms shall receive their primary power from the building wiring where such wiring is served from a commercial source and shall be equipped with a battery backup. Smoke alarms shall emit a signal when the batteries are low. Wiring shall be permanent and without a disconnecting switch other than as required for overcurrent protection.

> **Exception:** Smoke alarms are not required to be equipped with battery backup in Group R-1 where they are connected to an emergency electrical system.

[F] K1202.16 Smoke alarm interconnection. Where more than one smoke alarm is required to be installed within an individual dwelling unit or sleeping unit in Group R-2, R-3 or R-4, or within an individual sleeping unit in Group R-1, the smoke alarms shall be interconnected in such a manner that the activation of one alarm will activate all of the alarms in the individual unit. The alarm shall be clearly audible in all bedrooms over background noise levels with all intervening doors closed.

SECTION K1203
EXISTING ELECTRICAL FACILITIES

[PM] K1203.1 Existing buildings. This section shall apply to buildings and structures that are within the scope of the *International Property Maintenance Code*. Every occupied building shall be provided with an electrical system in compliance with the requirements of Sections K1203.1.1 through K1203.1.5.

[PM] K1203.1.1 Service. The size and usage of appliances and equipment shall serve as a basis for determining the need for additional facilities in accordance with this code. Dwelling units shall be served by a three-wire, 120/240 volt, single-phase electrical service having a rating of not less than 60 amperes.

[PM] K1203.1.2 Electrical system hazards. Where it is found that the electrical system in a structure constitutes a hazard to the occupants or the structure by reason of inadequate service, improper fusing, insufficient receptacle and lighting outlets, improper wiring or installation, deterioration or damage, or for similar reasons, the code official shall require the defects to be corrected to eliminate the hazard.

[PM] K1203.1.3 Installation. All electrical equipment, wiring and appliances shall be properly installed and maintained in a safe and approved manner.

[PM] K1203.1.4 Receptacles. Every habitable space in a dwelling shall be provided with at least two separate and remote receptacle outlets. Every laundry area shall be provided with at least one grounding-type receptacle outlet or a receptacle outlet with ground fault circuit interrupter protection. Every bathroom shall contain at least one receptacle outlet. Any new bathroom receptacle outlet shall have ground fault circuit interrupter protection.

[PM] K1203.1.5 Luminaires. Every public hall, interior stairway, toilet room, kitchen, bathroom, laundry room, boiler room and furnace room shall be provided with at least one electric luminaire.

CHAPTER K13

REFERENCED STANDARDS

This chapter lists the standards that are referenced in various sections of this document. The standards are listed herein by the promulgating agency of the standard, the standard identification, the effective date and title, and the section or sections of this document that reference the standard. The application of the referenced standards shall be as specified in Section K102.6.

ICC

International Code Council
Suite 600
5203 Leesburg Pike
Falls Church, VA 22041

Standard reference number	Title	Referenced in code section number
IBC—06	International Building Code®	K201.3, K303.1, K1202.3, K1202.4, K1202.5, K1202.5, K1202.8, K1202.9
IECC—06	International Energy Conservation Code®	K201.3
IFC—06	International Fire Code®	K201.3, K1202.8, K1202.9
IFGC—06	International Fuel Gas Code®	K201.3, K1202.11
IMC—06	International Mechanical Code®	K201.3, K1202.10, K1202.10.1, K1202.11
IPC—06	International Plumbing Code®	K201.3
IPMC—06	International Property Maintenance Code®	K201.3, K1203.1
IPSDC—06	International Private Sewage Disposal Code®	K201.3
IRC—06	International Residential Code®	K201.3, K1201.1.1
IZC—06	International Zoning Code®	K201.3

NFPA

National Fire Protection Association
1 Batterymarch Park
Quincy, MA 02269

Standard reference number	Title	Referenced in code section number
37—02	Installation and Use of Stationary Combustion Engines and Gas Turbines	K1202.11
70—05	National Electrical Code	K201.3, K1201.1.1
110—99	Emergency and Standby Power Systems	K1202.8
111—01	Stored Electrical Energy Emergency and Standby Power Systems	K1202.8
262—99	Standard Method of Test for Flame Travel and Smoke of Wires and Cables for Use in Air-Handling Spaces	K1202.10
853—03	Installation of Stationary Fuel Cell Power Systems	K1202.12

UL

Underwriters Laboratories
333 Pfingsten Road
Northbrook, IL 60062–2096

Standard reference number	Title	Referenced in code section number
2043—96	Fire Test for Heat and Visible Smoke Release for Discrete Products and their Accessories Installed in Air-Handling Spaces—with Revisions through February 1998	K1202.10.1

INDEX

Note: This is taken from the index for the 2006 International Building Code. Volume II of the IBC Commentary only includes Chapters 16 through 35.